MULTIVARIABLE MATHEMATICS

MULTIVARIABLE MATHEMATICS

FOURTH EDITION

Richard E. Williamson
Dartmouth College

Hale F. Trotter
Princeton University

PEARSON EDUCATION, INC.
Upper Saddle River, New Jersey 07458

Library of Congress Cataloging-in-Publication Data

Williamson, Richard E.
    Multivariable mathematics / Richard E. Williamson, Hale F. Trotter–4th ed.
        p.   cm.
    Includes index.
    ISBN 0-13-067276-9
    1. Algebras, Linear. 2. Differential Equations. 3. Calculus.   I. Trotter, Hale F. II. Title.

QA184.W54 2004
512′.15–dc21   2003049839

Acquisitions Editor: *George Lobell*
Editor in Chief: *Sally Yagan*
Vice-President/Director of Production and Manufacturing: *David W. Riccardi*
Executive Managing Editor: *Kathleen Schiaparelli*
Senior Managing Editor: *Linda Mihatov Behrens*
Assistant Managing Editor: *Bayani Mendoza de Leon*
Production Editor/Interior Designer: *Jeanne Audino*
Manufacturing Buyer: *Michael Bell*
Marketing Manager: *Halee Dinsey*
Marketing Assistant: *Rachel Beckman*
Director of Creative Services: *Paul Belfanti*
Art Editor: *Tom Benfatti*
Art Director: *Jayne Conte*
Cover Designer: *Bruce Kenselaar*
Editorial Assistant: *Jennifer Brady*
Art Studio & Composition: *Laserwords Private Limited*

Cover Image: *Provided by Richard Williamson. It is a trajectory of the three-species system described in Exercise 34 in Chapter 12, Section 4.*

© 2004, 1996, 1979, 1974 by Pearson Education, Inc.
Pearson Education, Inc.
Upper Saddle River, New Jersey 07458

Printed in the United States of America

10 9 8 7 6 5 4 3 2 1

ISBN 0-13-067276-9

Pearson Education LTD., *London*
Pearson Education Australia PTY, Limited, *Sydney*
Pearson Education Singapore, Pte. Ltd
Pearson Education North Asia Ltd., *Hong Kong*
Pearson Education Canada, Ltd., *Toronto*
Pearson Educación de Mexico, S.A. de C.V.
Pearson Education–Japan, *Tokyo*
Pearson Education Malaysia, Pte. Ltd

CONTENTS

■ CHAPTER 3   Vector Spaces and Linearity
                (Optional Chapter)   102

Appendix: Finding Indefinite Integrals   735

Answers to Odd-Numbered Exercises   741

Index   833

PREFACE

This book covers material that is often studied after a first course in one-variable calculus, namely the algebra and geometry of vectors and matrices, multivariable and vector calculus, and differential equations, including systems. The branches of these three areas are strongly intertwined and we've designed our treatment to display the connections effectively. Our aim has been to teach basic problem solving, both pure and applied, in a framework that is mathematically coherent, while allowing for selective emphasis on traditional rigor.

While the sequence of topics follows rather traditional lines of mathematical classification, the actual route taken may vary widely from course to course. An underlying theme is the encouragement of geometric thinking in two and three dimensions, extended to arbitrary dimension when it's useful to do so. Thus, most of Chapters 1 and 2 on vectors and matrices is prerequisite for the rest of the book, but otherwise there is considerable flexibility for course scheduling. Chapter 3 on linear algebra, with an introduction to general vector spaces and linear transformations, is included for those who want to cover this material at some point, but none of it is prerequisite for later chapters. In particular the material on differentiability in Chapter 5 is organized so that the motivation for the definition depends on gradient vectors rather than linear transformations.

For this edition the exposition has been completely rewritten in many places and, in addition to Chapter 3, a number of topics that are optional additions to a basic course have been added, as follows:

Additional emphasis on scientific applications in Section 1B of Chapter 2.
Subsection on vector integrals in Chapter 4, Section 1.
Subsections on quadric surfaces in Chapter 4.
Subsection on flow lines in Chapter 6, Section 1.
Subsection on use of the chain rule in coordinate changes.
Expanded treatment of the second-derivative criterion for extrema.
Section 5 on centroids and moments in Chapter 7.
Section 6 on application of improper integrals in Chapter 7.
Section 4, Chapter 8 relating flow lines, divergence, and curl.
Subsection on finding potentials in Chapter 9, Section 2.
Additional subsection on flows in Chapter 12, Section 1.
More efficient computation of exponential matrices in Chapter 13, Section 2.

Chapter on infinite series, with applications to differential equations.
Sections in Chapter 4 on computer plotting of curves and surfaces.
Subsection on the steepest ascent method.
Subsection on the midpoint and Simpson rules for multiple integrals.
Subsection on Newton's method for vector functions.
Java applets for the graphical and numerical work.

The figures are a salient feature of the text, including those in the answer section and the one on the cover, which represents a trajectory of a Lotka-Volterra system modified for three species and discussed in Exercise 34 in Chapter 12, Section 4.

The impetus for this edition came from George Lobell, whose knowledge of the field and continued support has helped us a great deal. Allan Gunter read the entire text, making insightful suggestions and working all the exercises; his collaboration was invaluable. Jeanne Audino's experienced and good humored oversight of the production has made working out final details a pleasure rather than a chore. Corrections can be sent to *rewn@dartmouth.edu* or *hft@math.princeton.edu*.

Richard E. Williamson
Hale F. Trotter

SYLLABUS SUGGESTIONS

The vertical listings by chapter and section are by no means exhaustive but display variety in emphasis to give some feeling for the book's flexibility. Unlisted sections or subsections selected from the table of contents can, of course, be included at an instructor's discretion.

Chapter Number	Basic Calculus	Vector Calculus	Differential Equations	Emphasis or Linear Algeb
1	1–6	1–6	1–6	1–6
2	1–5	1–5	1–5	1–5
3				1–8
4	1 A–D, 2–4	1 A–D, 2–4	1 A–F, 2–4	1 A, B, 2
5	1–3	1–3		
6	1, 2, 4, 5	1, 2, 4, 5		1, 2
7	1 A, B, 2 A–D, 4	1 A, B, 2 A–D, 4		
8		1		
9		1, 3, 4		
10	1–3		1–3	1–3
11	1–3		1–3	
12	1 A–C, 2, 3	1 A–D	1 A–C, 2, 3	1, 2
13	1		1–3	1–3
14	1, 2		1, 2, 5–7, 9	

CHAPTER 1

VECTORS

Originally vectors were conceived of as geometric objects with magnitude and direction, suitable for representing physical quantities such as displacements, velocities, or forces. Later on, introduction of a more general algebraic concept of vector unified and simplified various topics in pure and applied mathematics. This first chapter introduces vectors in algebraic terms but is chiefly concerned with their geometric interpretation. The ideas are fundamental for the rest of the book, because the possibility of visualizing multivariable problems geometrically is one of the major advantages of using vectors.

SECTION 1   COORDINATE VECTORS

We use the symbol $\mathbb{R}$ to denote the set of all real numbers, $\mathbb{R}^2$ for the set of ordered pairs (x_1, x_2) of real numbers, $\mathbb{R}^3$ for the set of ordered triples (x_1, x_2, x_3), and in general $\mathbb{R}^n$ for the set of n-tuples $(x_1, x_2, \ldots, x_n)$. To save writing subscripts, we often write general pairs and triples simply as (x, y) and (x, y, z). We'll refer to pairs, triples, and n-tuples as **vectors** and denote them by boldface letters **x**, **y**, **z**, etc., while ordinary lightface letters stand for the single numbers that we sometimes refer to as **scalars** when we want to distinguish them from vectors. The term *scalar* is common in physics, to distinguish numerical quantities like mass or temperature that we measure on "scales" of numbers from vector quantities such as velocity or force that have both magnitude and direction. Though complex numbers don't have such a direct physical interpretation, we can also use them as scalars, and we have good reason to do that in Chapters 3, 11, 12, and 13. Exercise 24 on page 43 tells a little more about the origin of these terms.

For a scalar r and a vector $\mathbf{x} = (x_1, x_2, \ldots, x_n)$, we define the **scalar multiple** $r\mathbf{x}$ to be the vector that we get by multiplying each entry x_k by r, so

$$r\mathbf{x} = (rx_1, rx_2, \ldots, rx_n).$$

EXAMPLE 1   If we take $\mathbf{x} = (1, 2)$ in $\mathbb{R}^2$ and $r = 3$, then

$$r\mathbf{x} = 3(1, 2) = (3, 6).$$

Similarly, with $\mathbf{x} = (1, 2, -3)$ in $\mathbb{R}^3$ and $r = -2$,

$$r\mathbf{x} = -2(1, 2, -3) = (-2, -4, 6).$$

For two vectors $\mathbf{x} = (x_1, x_2, \ldots, x_n)$ and $\mathbf{y} = (y_1, y_2, \ldots, y_n)$ in $\mathbb{R}^n$, we define the **sum** to be the vector

$$\mathbf{x} + \mathbf{y} = (x_1 + y_1, x_2 + y_2, \ldots, x_n + y_n),$$

in which each entry $x_k + y_k$ is the sum of the corresponding entries x_k and y_k. Note that the sum is defined only for vectors with the same number of entries. For example, the sum of $(1, 2)$ and $(3, 4, 5)$ is undefined because corresponding entries can't be matched up.

EXAMPLE 2   With $\mathbf{x} = (1, 2)$ and $\mathbf{y} = (2, -3)$ both in $\mathbb{R}^2$,

$$\mathbf{x} + \mathbf{y} = (1, 2) + (2, -3) = (3, -1).$$

In $\mathbb{R}^3$, with $\mathbf{x} = (0, 2, 4)$ and $\mathbf{y} = (-1, -2, 2)$, we have

$$\mathbf{x} + \mathbf{y} = (0, 2, 4) + (-1, -2, 2) = (-1, 0, 6).$$

The two basic operations on vectors, scalar multiplication and vector addition, are extensions of the operations of addition and multiplication of single real numbers, and in combination they help us express multiple operations concisely.

EXAMPLE 3   If $\mathbf{x} = (2, -1, 0)$ and $\mathbf{y} = (0, -1, -2)$, then $2\mathbf{x} + \mathbf{y}$ expresses the result of three multiplications and three additions:

$$2\mathbf{x} + \mathbf{y} = 2(2, -1, 0) + (0, -1, -2)$$
$$= (4, -2, 0) + (0, -1, -2) = (4, -3, -2).$$

Similarly, $3\mathbf{x} - 2\mathbf{y}$ involves six multiplications and three additions:

$$3\mathbf{x} - 2\mathbf{y} = 3(2, -1, 0) - 2(0, -1, -2)$$
$$= (6, -3, 0) + (0, 2, 4) = (6, -1, 4).$$

We customarily write $-\mathbf{x}$ for the scalar multiple $(-1)\mathbf{x}$, and $\mathbf{x} - \mathbf{y}$ as an abbreviation for $\mathbf{x} + (-\mathbf{y})$, and use $\mathbf{0}$ to denote an n-tuple whose entries are all zero. Then for an arbitrary vector $\mathbf{x}$, $\mathbf{x} - \mathbf{x} = \mathbf{0}$, as in

$$(1, 2, 3) - (1, 2, 3) = (1, 2, 3) + (-1, -2, -3) = (0, 0, 0) = \mathbf{0}.$$

The notation $\mathbf{0}$ is ambiguous since $\mathbf{0}$ may stand for $(0, 0)$ in one formula and for $(0, 0, 0)$ in another. The ambiguity disappears in context since only one interpretation will make sense. For instance, if $\mathbf{z} = (-2, 0, 3)$, then in the formula $\mathbf{z} + \mathbf{0}$, the $\mathbf{0}$ must stand for $(0, 0, 0)$, since addition is defined only for n-tuples with the same number of entries.

Formulas 1 to 9 below are valid for arbitrary $\mathbf{x}$, $\mathbf{y}$, and $\mathbf{z}$ in $\mathbb{R}^n$ and arbitrary real numbers r, s. They state rules for our new operations of addition and scalar multiplication very closely analogous to the familiar distributive, commutative, and associative laws for ordinary addition and multiplication of numbers.

1. $r\mathbf{x} + s\mathbf{x} = (r + s)\mathbf{x}$
2. $r\mathbf{x} + r\mathbf{y} = r(\mathbf{x} + \mathbf{y})$

3. $r(s\mathbf{x}) = (rs)\mathbf{x}$

4. $\mathbf{x} + \mathbf{y} = \mathbf{y} + \mathbf{x}$

5. $(\mathbf{x} + \mathbf{y}) + \mathbf{z} = \mathbf{x} + (\mathbf{y} + \mathbf{z})$

6. $\mathbf{x} + \mathbf{0} = \mathbf{x}$

7. $\mathbf{x} + (-\mathbf{x}) = \mathbf{0}$

8. $1\mathbf{x} = \mathbf{x}$

9. $0\mathbf{x} = \mathbf{0}$

Note that the 0 on the left side of Formula 9 is the real number zero, while the $\mathbf{0}$ on the right side is the zero vector in $\mathbb{R}^n$ for some n.

These formulas are straightforward consequences of the definitions of the vector operations and of the laws of arithmetic. For illustration, we give a formal proof of the second one.

Proof of Formula 2.   Let $\mathbf{x} = (x_1, x_2, \ldots, x_n)$ and $\mathbf{y} = (y_1, y_2, \ldots, y_n)$, and let r be a real number. Then

$$r\mathbf{x} = (rx_1, rx_2, \ldots, rx_n), \qquad \text{[definition of scalar multiplication]}$$

$$r\mathbf{y} = (ry_1, ry_2, \ldots, ry_n), \qquad \text{[definition of scalar multiplication]}$$

so

$$r\mathbf{x} + r\mathbf{y} = (rx_1 + ry_1, rx_2 + ry_2, \ldots, rx_n + ry_n). \qquad \text{[definition of addition]}$$

On the other hand,

$$\mathbf{x} + \mathbf{y} = (x_1 + y_1, x_2 + y_2, \ldots, x_n + y_n), \qquad \text{[definition of addition]}$$

so

$$r(\mathbf{x} + \mathbf{y}) = \big(r(x_1 + y_1), r(x_2 + y_2), \ldots, r(x_n + y_n)\big). \qquad \text{[definition of scalar multiplication]}$$

By the distributive law for numbers, $r(x_1 + y_1) = rx_1 + ry_1, r(x_2 + y_2) = rx_2 + ry_2$, etc., so the n-tuples $r\mathbf{x} + r\mathbf{y}$ and $r(\mathbf{x} + \mathbf{y})$ are the same.   ∎

A set with operations of addition and multiplication by scalars defined in such a way that the Formulas 1 through 9 hold is called a **vector space**, and its elements are called **vectors**. We use this more general point of view in Chapter 3, but elsewhere in the book the term *vector* may be taken to refer to an element of $\mathbb{R}^n$.

As with numbers and other algebraic expressions such as polynomials, the commutative and associative properties of vector addition stated in Formulas 4 and 5 imply that we can reorder and regroup the terms in a sum without changing the value of the sum. Consequently we can simply write a sum such as $\mathbf{x}_1 + \cdots + \mathbf{x}_n$ without putting in parentheses to show how the terms are grouped, because the grouping doesn't affect the value. Also, the distributive laws stated for two-term sums in 1 and 2 hold for sums of more than two terms.

A sum of scalar multiples $a_1\mathbf{x}_1 + \cdots + a_k\mathbf{x}_k$ is called a **linear combination** of the vectors $\mathbf{x}_1, \ldots, \mathbf{x}_k$. Formulas 1 through 9 justify manipulating and simplifying linear combinations in much the same way as other algebraic expressions, as illustrated in the following example.

EXAMPLE 4   For vectors $\mathbf{u}$, $\mathbf{v}$, and $\mathbf{w}$, the expression

$$3(2\mathbf{u} + \mathbf{v} + \mathbf{w}) - (\mathbf{u} + 2\mathbf{v} + 3\mathbf{w})$$

simplifies to $(6 - 1)\mathbf{u} + (3 - 2)\mathbf{v} + (3 - 3)\mathbf{w} = 5\mathbf{u} + \mathbf{v}$.

We take this kind of manipulation for granted from now on, but here is an outline of a formal justification on the basis of Formulas 1 through 9. By Formula 2 (extended to 3-term sums),

$$3(2\mathbf{u} + \mathbf{v} + \mathbf{w}) + (-1)(\mathbf{u} + 2\mathbf{v} + 3\mathbf{w})$$
$$= 3(2\mathbf{u}) + 3\mathbf{v} + 3\mathbf{w} + (-1)\mathbf{u} + (-1)2\mathbf{v} + (-1)3\mathbf{w}.$$

By Formula 3, this is equal to $6\mathbf{u} + 3\mathbf{v} + 3\mathbf{w} + (-1)\mathbf{u} + (-2)\mathbf{v} + (-3)\mathbf{w}$. Rearranging and regrouping using Formulas 4 and 5 gives

$$(6\mathbf{u} + (-1)\mathbf{u}) + (3\mathbf{v} + (-2)\mathbf{v}) + (3\mathbf{w} + (-3)\mathbf{w}).$$

Applying Formula 1 gives $(6 - 1)\mathbf{u} + (3 - 2)\mathbf{v} + (3 - 3)\mathbf{w} = 5\mathbf{u} + 1\mathbf{v} + 0\mathbf{w}$, which simplifies by Formulas 6, 8, and 9 to $5\mathbf{u} + \mathbf{v}$.

The special vectors

$$\mathbf{e}_1 = (1, 0), \qquad \mathbf{e}_2 = (0, 1) \qquad\qquad \text{in } \mathbb{R}^2,$$
$$\mathbf{e}_1 = (1, 0, 0), \quad \mathbf{e}_2 = (0, 1, 0), \quad \mathbf{e}_3 = (0, 0, 1) \qquad \text{in } \mathbb{R}^3,$$

and, in general,

$$\mathbf{e}_1 = (1, 0, \ldots, 0), \mathbf{e}_2 = (0, 1, \ldots, 0), \ldots, \mathbf{e}_n = (0, 0, \ldots, 1) \text{ in } \mathbb{R}^n$$

have the property that if $\mathbf{x} = (x_1, \ldots, x_n)$ is an arbitrary vector, then

$$\mathbf{x} = x_1\mathbf{e}_1 + \cdots + x_n\mathbf{e}_n.$$

Note that these are the only coefficients that express $\mathbf{x}$ as a linear combination of $\mathbf{e}_1, \ldots, \mathbf{e}_n$, for if

$$x_1\mathbf{e}_1 + \cdots + x_n\mathbf{e}_n = y_1\mathbf{e}_1 + \cdots + y_n\mathbf{e}_n,$$

then $x_k = y_k$ for $k = 1, \ldots, n$. In other words, every vector in $\mathbb{R}^n$ appears in just one way as a linear combination of the vectors $\mathbf{e}_1, \ldots, \mathbf{e}_n$, and the coefficients in the linear combination are simply the entries in the vector. Because of these properties, the set $\{\mathbf{e}_1, \ldots, \mathbf{e}_n\}$ is called the **standard basis** for $\mathbb{R}^n$. The numbers $x_1, \ldots, x_n$ are called the **coordinates** of $\mathbf{x}$ with respect to the standard basis, and the vectors $x_1\mathbf{e}_1, \ldots, x_n\mathbf{e}_n$ are called the **components** of $\mathbf{x}$ with respect to this basis.

In $\mathbb{R}^3$ we often write the standard basis vectors as $\mathbf{i}$, $\mathbf{j}$, and $\mathbf{k}$ instead of $\mathbf{e}_1$, $\mathbf{e}_2$, and $\mathbf{e}_3$, and in $\mathbb{R}^2$ we may use $\mathbf{i}$ and $\mathbf{j}$ instead of $\mathbf{e}_1$ and $\mathbf{e}_2$. The **ijk**-notation appears most often in geometric and physical applications. The notation $\mathbf{e}_k$ by itself is ambiguous because it could in principle refer to a vector in $\mathbb{R}^n$ for any $n \geq k$. It will always be clear from the context how many entries are meant for a vector $\mathbf{e}_k$.

EXAMPLE 5   Given a vector $\mathbf{x} = (x_1, x_2)$ in $\mathbb{R}^2$,

$$(x_1, x_2) = x_1(1, 0) + x_2(0, 1)$$
$$= x_1\mathbf{e}_1 + x_2\mathbf{e}_2 = x_1\mathbf{i} + x_2\mathbf{j}.$$

In particular, $(2, -3) = 2\mathbf{e}_1 - 3\mathbf{e}_2 = 2\mathbf{i} - 3\mathbf{j}$.

In $\mathbb{R}^3$, we have

$$(x_1, x_2, x_3) = x_1(1, 0, 0) + x_2(0, 1, 0) + x_3(0, 0, 1)$$
$$= x_1\mathbf{e}_1 + x_2\mathbf{e}_2 + x_3\mathbf{e}_3$$
$$= x_1\mathbf{i} + x_2\mathbf{j} + x_3\mathbf{k}.$$

In particular, $(1, 2, -7) = \mathbf{e}_1 + 2\mathbf{e}_2 - 7\mathbf{e}_3 = \mathbf{i} + 2\mathbf{j} - 7\mathbf{k}$.

EXAMPLE 6   The equation

$$(2, 3) = 2\mathbf{e}_1 + 3\mathbf{e}_2$$

shows $(2, 3)$ written as a linear combination of $\mathbf{e}_1$ and $\mathbf{e}_2$ in $\mathbb{R}^2$. The vector $(2, 3)$ is also a linear combination of $(1, 1)$ and $(1, -1)$ as follows:

$$(2, 3) = \tfrac{5}{2}(1, 1) - \tfrac{1}{2}(1, -1).$$

In $\mathbb{R}^3$, the equation

$$(2, 3, 4) = 2\mathbf{e}_1 + 3\mathbf{e}_2 + 4\mathbf{e}_3$$
$$= 4(1, 1, 1) - 1(1, 1, 0) - 1(1, 0, 0)$$

shows the vector $(2, 3, 4)$ represented as a linear combination of the vectors $(1, 1, 1)$, $(1, 1, 0)$, and $(1, 0, 0)$.

EXAMPLE 7   To express $(1, 3)$ as a linear combination of $(1, 1)$ and $(3, 4)$, we look for numbers x and y such that

$$x(1, 1) + y(3, 4) = (1, 3), \quad \text{or} \quad (x + 3y, x + 4y) = (1, 3).$$

We need to solve the pair of equations

$$x + 3y = 1$$
$$x + 4y = 3$$

for x and y. Subtracting the first equation from the second gives $y = 2$. Then setting $y = 2$ in the first equation gives $x = -5$. So $(1, 3)$ equals the linear combination

$$(1, 3) = -5(1, 1) + 2(3, 4).$$

EXAMPLE 8   To express $(1, 3, 8)$ as a linear combination of $(1, 1, 1)$ and $(3, 4, 5)$ we look for numbers x and y such that

$$(1, 3, 8) = x(1, 1, 1) + y(3, 4, 5)$$
$$= (x + 3y, x + 4y, x + 5y).$$

Now solve
$$x + 3y = 1$$
$$x + 4y = 3$$
$$x + 5y = 8$$

for x and y. The first two equations are the same as in the previous example, and the calculations there showed that $x = -5$ and $y = 2$ are the only values that satisfy both equations. Substituting these values in the third equation we find $-5 + 5(2) = 5 \neq 8$, so there are no values for x and y that satisfy all three equations. We conclude that $(1, 3, 8)$ is not a linear combination of $(1, 1, 1)$ and $(3, 4, 5)$.

The last two examples show that answering questions about linear combinations may require solving systems of first-degree equations. Chapter 2 describes routines for solving such equations with many variables. Equations that come up in examples and exercises in this chapter will be simple enough to be solved by common-sense methods as in these examples.

In this book we emphasize applications to geometry and physics in two and three dimensions, and most of our examples in this chapter involve vectors in $\mathbb{R}^2$ or $\mathbb{R}^3$. However, we'll be applying the concepts and methods illustrated here to vectors in $\mathbb{R}^n$ for arbitrary values of n later on. Here is a nongeometric example.

EXAMPLE 9   A model for an economy might use a vector $\mathbf{p}$ to represent annual production, with entries p_i giving the year's production for each of n commodities considered in the model. Thus $\mathbf{p}$ would be a vector in $\mathbb{R}^n$, where n might be as large as several hundred in an elaborate model. Similarly, vectors $\mathbf{c}$ and $\mathbf{b}$ might represent annual consumption and the amount in inventory at the beginning of the year for each commodity. Then the amounts in inventory at the end of the year would be given by the vector $\mathbf{b} + \mathbf{p} - \mathbf{c}$.

Calculations with vectors in $\mathbb{R}^n$ are impractical to do by hand when n is large, but computers do them efficiently. Interactive programs such as MATLAB, Maple, and Mathematica, and programming languages such as Basic, C, Pascal, and others, provide for computations with vectors, often called arrays in the programming context. The entries $x_1, x_2, \ldots$ in a vector $\mathbf{x}$ usually appear as $x(1), x(2), \ldots$ or $x[1], x[2], \ldots$, depending on the program language. Many languages allow direct specification of scalar multiplication and addition using mathematical notation; in others, such as C and Pascal, subroutines are used to carry out these operations.

EXERCISES

1. Let $\mathbf{x} = (-3, 4)$ and $\mathbf{y} = (2, 2)$. Compute
- **(a)** $\mathbf{x} + \mathbf{y}$
- **(b)** $2\mathbf{x} + 3\mathbf{y}$
- **(c)** $-\mathbf{x} + \mathbf{y} - (1, 4)$

2. Let $\mathbf{u} = (2, -1)$ and $\mathbf{v} = (-3, 1)$. Compute
- **(a)** $\mathbf{u} - 2\mathbf{v}$
- **(b)** $3\mathbf{u} + 2\mathbf{v}$
- **(c)** $4\mathbf{u} + \mathbf{v} - (-1, 3)$

3. Let $\mathbf{x} = (3, -1, 0)$, $\mathbf{y} = (0, 1, 5)$, and $\mathbf{z} = (2, 5, -1)$. Compute
- **(a)** $3\mathbf{x}$
- **(b)** $4\mathbf{x} - 2\mathbf{y} + 3\mathbf{z}$
- **(c)** $-\mathbf{y} + (1, 2, 1)$

4. Let $\mathbf{u} = (1, 2, 3)$, $\mathbf{v} = (0, -1, 0)$, $\mathbf{w} = (-3, 0, 2)$. Compute
- **(a)** $2\mathbf{u} + (0, -3, 1) + \mathbf{w}$
- **(b)** $2\mathbf{u} + 2\mathbf{v} - 3\mathbf{w}$
- **(c)** $3\mathbf{u} + 2\mathbf{v} + \mathbf{w}$

In Exercises 5 to 8, write the given vector as a linear combination of $\mathbf{i}$ and $\mathbf{j}$ in $\mathbb{R}^2$, or of $\mathbf{i}$, $\mathbf{j}$, and $\mathbf{k}$ in $\mathbb{R}^3$.

5. $2(1, 2) - 3(-1, 4)$ **6.** $(1, 0, 1) + 3(2, 3, -1)$

7. $(1, 4) - (2c, d)$ **8.** $(x, y, z) + (z, y, x)$

9. Find numbers a and b such that $a\mathbf{x} + b\mathbf{y} = (9, -1, 10)$, where $\mathbf{x} = (3, -1, 0)$ and $\mathbf{y} = (0, 1, 5)$. Is there more than one solution?

10. Find numbers a and b such that $a\mathbf{u} + b\mathbf{v} = (9, -3, 6)$, where $\mathbf{u} = (3, -1, 2)$ and $\mathbf{v} = (-6, 2, -4)$. Is there more than one solution?

11. Show that no choice of numbers a and b can make $a\mathbf{x} + b\mathbf{y} = (3, 0, 0)$, where $\mathbf{x}$ and $\mathbf{y}$ are as in Exercise 9. For what values of c is $a\mathbf{x} + b\mathbf{y} = (3, 0, c)$ possible?

12. Show that no choice of numbers a and b can make $a\mathbf{u} + b\mathbf{v} = (6, 2, 4)$, where $\mathbf{u}$ and $\mathbf{v}$ are as in Exercise 10. For what values of c, if any, is $a\mathbf{u} + b\mathbf{v} = (6, c, 4)$ possible? What about $a\mathbf{u} + b\mathbf{v} = (6, 2, c)$?

13. Let $\mathbf{x} = \mathbf{i} + \mathbf{j}$, $\mathbf{y} = 2\mathbf{i} + \mathbf{j} + \mathbf{k}$, and $\mathbf{z} = -2\mathbf{i} + \mathbf{j} + 2\mathbf{k}$. Calculate
- **(a)** $-\mathbf{x} + 2\mathbf{y} - \mathbf{z}$
- **(b)** $6\mathbf{x} - 2\mathbf{y} + \mathbf{z}$
- **(c)** $-4\mathbf{x} + 3\mathbf{y} + \mathbf{z}$

14. Let $\mathbf{u} = \mathbf{i} - \mathbf{j} + 3\mathbf{k}$, $\mathbf{v} = 2\mathbf{j} + \mathbf{k}$, and $\mathbf{w} = -2\mathbf{i} + 2\mathbf{j} - \mathbf{k}$. Calculate
- **(a)** $-\frac{2}{5}\mathbf{u} + \frac{1}{2}\mathbf{v} - \frac{7}{10}\mathbf{w}$
- **(b)** $-\frac{1}{5}\mathbf{u} + \frac{1}{2}\mathbf{v} - \frac{1}{10}\mathbf{w}$
- **(c)** $\frac{2}{5}\mathbf{u} + \frac{1}{5}\mathbf{w}$

15. Write out a proof for Formula 3 on page 3, giving precise justification for each step.

16. Write out a proof for Formula 4 on page 3, giving precise justification for each step.

In Exercises 17 to 20, simplify the given expression, showing which of the Formulas 1 through 9 for vector algebra justify each step.

17. $2(3\mathbf{x} - 2\mathbf{y} + \mathbf{z}) - 4\mathbf{x}$ **18.** $\mathbf{x} + (\mathbf{x} + (\mathbf{x} + \mathbf{y}))$

19. $\frac{1}{2}(\mathbf{x} + \mathbf{y}) - \mathbf{y}$ **20.** $(2\mathbf{x} + 3\mathbf{y}) + (3\mathbf{x} - \mathbf{z})$

In Exercises 21 to 24, represent the first vector given as a linear combination of the remaining vectors, either by inspection or by solving a system of equations.

21. $(-2, 3)$; $\mathbf{e}_1, \mathbf{e}_2$

22. $(2, 0, -5)$; $\mathbf{e}_1, \mathbf{e}_2, \mathbf{e}_3$

23. $(2, -7)$; $(1, 1), (1, -1)$

24. $(2, 3, 4)$; $(1, 1, 1), (1, 2, 1), (-1, 1, 2)$

25. Let $\mathbf{x} = 2\mathbf{i}$, $\mathbf{y} = \mathbf{i} - 3\mathbf{j}$, and $\mathbf{z} = 3\mathbf{i} + 2\mathbf{j} - 2\mathbf{k}$
- **(a)** Express $\mathbf{i}$ in terms of $\mathbf{x}$.
- **(b)** Express $\mathbf{j}$ as a linear combination of $\mathbf{x}$ and $\mathbf{y}$.
- **(c)** Express $\mathbf{k}$ as a linear combination of $\mathbf{x}$, $\mathbf{y}$, and $\mathbf{z}$. [The answer to part (a) helps in (b); the answers to (a) and (b) help in (c).]

26. Let $\mathbf{u} = 2\mathbf{i} - 3\mathbf{j} + \mathbf{k}$, $\mathbf{v} = \mathbf{j} + 2\mathbf{k}$, and $\mathbf{w} = \mathbf{j} - \mathbf{k}$.
- **(a)** Express $\mathbf{j}$ and $\mathbf{k}$ as linear combinations of $\mathbf{v}$ and $\mathbf{w}$.

(b) Express **i** as a linear combination of **u**, **v**, and **w**. [First use the answer to part (a) to express **u** in terms of **i**, **v**, and **w**.]

27. Let $\mathbf{x} = (5, 500, 10)$ represent the amount of ink, paper, and binding material needed to produce a single copy of some book and let $\mathbf{y} = (4, 800, 90)$ be the same vector for some other book. What does $100\mathbf{x} + 50\mathbf{y}$ represent?

28. A small factory produces products of four different kinds. The vectors $\mathbf{w} = (50, 75, 100, 190)$ and $\mathbf{r} = (100, 150, 200, 300)$ give the wholesale and retail prices in dollars for a single unit of each kind. The vector $\mathbf{p} = (25, 25, 15, 10)$ gives the number of units of each product produced in a day.

(a) What vector gives the retailer's profit per unit for the four products?

(b) If the wholesale price vector is doubled, what happens to the retailer's profit vector?

(c) What happens to the retailer's profit vector if the retail prices each increase by 10% and the wholesale prices stay unchanged?

(d) What vector gives the total production for a five-day week?

29. A computer monitors the temperatures recorded by sensors at 50 sites in a building. Suppose that $x_k(t)$ is the temperature at the kth site at time t as measured on a 24-hour clock. Then the vector $\mathbf{x}(t) = (x_1(t), \ldots, x_{50}(t))$ represents the profile of temperatures in the entire building at time t. Write an expression in terms of $\mathbf{x}(t)$ for the vector that represents the average temperatures for the day at the 50 sites, using the readings at $t = 2, 8, 14$, and 20.

*30. Give a formal proof that the extension of Formula 2 on page 3 to sums of m terms follows from the given Formulas 1 to 9, that is, prove that for given vectors $\mathbf{x}_1, \ldots, \mathbf{x}_m$ and scalar r,

$$r(\mathbf{x}_1 + \cdots + \mathbf{x}_m) = r\mathbf{x}_1 + \cdots + r\mathbf{x}_m.$$

Use mathematical induction: starting with $m = 2$, assume the formula true with m terms and prove that it's true with $m + 1$ terms.

SECTION 2   GEOMETRIC VECTORS

2A   Points and Vectors

We can find geometric representations of $\mathbb{R} = \mathbb{R}^1$ as a line, of $\mathbb{R}^2$ as a plane, and of $\mathbb{R}^3$ as three-dimensional space by using coordinates. To represent elements of $\mathbb{R}$ as points of a line, we specify a point on the line to be identified with the number 0 and called the origin, a unit of distance, and a direction on the line to be called positive, with the opposite direction called negative. A positive number x corresponds to the point that is x units of distance in the positive direction from the origin, and a negative number x corresponds to the point that is $|x|$ units from the origin in the negative direction. The number line is usually pictured as horizontal with the positive direction to the right. With this standard convention, we obtain the familiar Figure 1.1 in which the arrow indicates the positive direction.

In the plane, we pick an origin, a unit of distance and, a pair of perpendicular lines called **axes** through the origin, with a positive direction on each axis. Given a vector in $\mathbb{R}^2$, that is, a pair $\mathbf{x} = (x_1, x_2)$ of numbers, the procedure described in the preceding paragraph determines a point $(x_1, 0)$ on the first axis corresponding to the number x_1 and a point $(0, x_2)$ on the second axis corresponding to the number x_2. The **projection** of a point $\mathbf{p}$ on a line L is defined as the foot of the perpendicular from $\mathbf{p}$ to L, unless $\mathbf{p}$ happens to be a point of L, in which case the projection of $\mathbf{p}$ on L is $\mathbf{p}$ itself. Thus the vector $\mathbf{x} = (x_1, x_2)$ corresponds to the point in the plane whose projection onto the first axis is $(x_1, 0)$ and whose projection on the second axis is $(0, x_2)$.

FIGURE 1.1
Oriented line.

The conventional choice is to take the first axis horizontal with the positive direction to the right, and the second axis vertical with the positive direction upward. This leads to the usual picture shown in Figure 1.2(a), where the pairs $(-3, 2)$ and $(2, 1)$ appear along with their projections on the axes.

FIGURE 1.2

(a) Positions. (b) Directed
arrows.

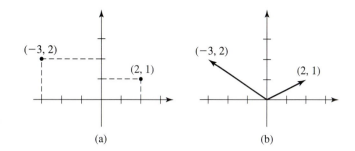

(a)   (b)

Here is an alternative geometric interpretation for elements of $\mathbb{R}$, $\mathbb{R}^2$, and $\mathbb{R}^3$. What we do is draw an arrow from the origin to a point x_1 in $\mathbb{R}$, or to (x_1, x_2) in $\mathbb{R}^2$, or to (x_1, x_2, x_3) in $\mathbb{R}^3$. In $\mathbb{R}$ a positive number x corresponds to an arrow pointing to the right as in Figure 1.1, while if x is negative it corresponds to an arrow pointing to the left.

Similarly, a pair $\mathbf{x} = (x, y)$ in $\mathbb{R}^2$ corresponds to an arrow from the origin to the point with coordinates (x, y), as in Figure 1.2(b), which shows arrows corresponding to $(-3, 2)$ and $(2, 1)$. You can get from the tail at $(0, 0)$ to the tip of the arrow for (x, y) by going to x on the horizontal axis and then to (x, y) vertically.

When referring to the point or vector that represents an element $\mathbf{x} = (x, y)$ of $\mathbb{R}^2$, we usually say simply "the point $\mathbf{x}$" or "the vector $\mathbf{x}$" instead of the more cumbersome "the point with coordinates $\mathbf{x}$" or "the vector with coordinates $\mathbf{x}$."

Everything we have said about points and vectors in the plane extends to three dimensions. Choose an origin, three perpendicular axes through it, and a positive direction on each. If $\mathbf{x} = (x_1, x_2, x_3)$ is a triple in $\mathbb{R}^3$, the point $\mathbf{x}$ with coordinates (x_1, x_2, x_3) is then the point whose projections on the three axes correspond to the numbers x_1, x_2, and x_3, and the geometric vector with coordinates (x_1, x_2, x_3) is the arrow from the origin to this point.

It's customary to take the first two axes in a horizontal plane and the third axis vertical, labeled as in Figure 1.3, with the positive directions as shown. Some people interchange the x_1 and x_2 axes, but the convention illustrated in Figure 1.3 is more commonly used and is the one we'll follow. Points and geometric vectors are closely related; every point has associated with it a **position vector**, defined as the vector represented by the arrow from the origin to the point.

Direct geometric visualization is possible only in dimensions up to three, so many of our examples will be set in $\mathbb{R}^2$ or $\mathbb{R}^3$. With the exception of the cross product described in Section 6, the algebraic equivalents of geometric concepts that we

FIGURE 1.3

(a) Position. (b) Directed arrow.

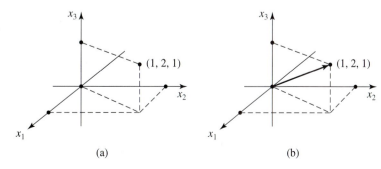

(a)   (b)

FIGURE 1.4

(a) d in $\mathbb{R}^2$. (b) d in $\mathbb{R}^3$.

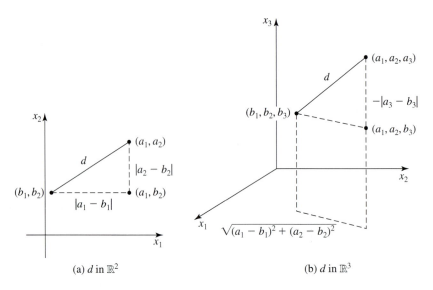

(a) d in $\mathbb{R}^2$ (b) d in $\mathbb{R}^3$

introduce are valid in $\mathbb{R}^n$ for all values of n, allowing us to extend the geometric concepts to higher dimensions in ways that are essential for applying geometric reasoning to problems with many variables. In particular, Sections 3 and 4 will extend and justify the intuitive geometric ideas of line and perpendicularity that we used to introduce axes and projections onto axes.

2B   Distance and Length

We define the **distance** between the points $\mathbf{x} = (x_1, \ldots, x_n)$ and $\mathbf{y} = (y_1, \ldots, y_n)$ in $\mathbb{R}^n$ to be

$$d = \sqrt{(x_1 - y_1)^2 + \cdots + (x_n - y_n)^2}.$$

This definition agrees with the usual notion of distance in $\mathbb{R}^1$, $\mathbb{R}^2$, and $\mathbb{R}^3$. When $n = 1$, $\sqrt{(x_1 - y_1)^2} = |x_1 - y_1|$, the *absolute value* of $x_1 - y_1$, which is the natural distance between points x and y in $\mathbb{R}$. Application of the theorem of Pythagoras to the right triangles in Figures 1.4(a) and (b) shows that the formula defines in $\mathbb{R}^2$ and $\mathbb{R}^3$ the usual geometric distance. This agreement motivates the definition. We'll see in Section 4 that distance defined here has other basic properties of distance that we would want in $\mathbb{R}^n$.

EXAMPLE 1

In $\mathbb{R}$, the distance between 2 and -5 is $|2 - (-5)| = 7$. In $\mathbb{R}^2$, the distance between $(2, 4)$ and $(-3, 1)$ is

$$\sqrt{(2 - (-3))^2 + (4 - 1)^2} = \sqrt{5^2 + 3^2} = \sqrt{34}.$$

In $\mathbb{R}^3$, the distance between $(1, -1, 2)$ and $(-1, 1, 1,)$ is

$$\sqrt{(1 - (-1))^2 + (-1 - 1)^2 + (2 - 1)^2} = \sqrt{2^2 + 2^2 + 1^2} = 3.$$

The **length** of a vector $\mathbf{x} = (x_1, \ldots, x_n)$ is denoted by $|\mathbf{x}|$ and defined to be

$$|\mathbf{x}| = \sqrt{x_1^2 + \cdots + x_n^2}.$$

The length is equal to the distance from the origin to the point $\mathbf{x}$ and is the geometric length of an arrow representing the vector $\mathbf{x}$. The distance between points $\mathbf{x}$ and $\mathbf{y}$ is

$$|\mathbf{x} - \mathbf{y}| \quad \text{or} \quad |\mathbf{y} - \mathbf{x}|,$$

since these two numbers are both equal to

$$\sqrt{(x_1 - y_1)^2 + \cdots + (x_n - y_n)^2}.$$

EXAMPLE 2

(a) The length of $(1, -2, 1)$ in $\mathbb{R}^3$ is

$$|(1, -2, 1)| = \sqrt{1^2 + (-2)^2 + 1^2} = \sqrt{6}.$$

(b) In $\mathbb{R}^4$, the length of $(1, 2, 5, -1)$ is

$$|(1, 2, 5, -1)| = \sqrt{1^2 + 2^2 + 5^2 + (-1)^2} = \sqrt{31}.$$

(c) The distance in $\mathbb{R}^2$ between $(1, 2)$ and $(3, 4)$ is

$$|(1, 2) - (3, 4)| = |(-2, -2)|$$
$$= \sqrt{(-2)^2 + (-2)^2} = \sqrt{8}.$$

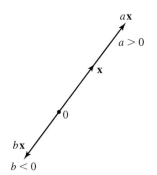

FIGURE 1.5
$a\mathbf{x}, a > 0; b\mathbf{x}, b < 0.$

2C   Scalar Multiplication

We can picture scalar multiplication in terms of the arrows representing vectors. Figure 1.5 illustrates scalar multiplication by both a positive number a and a negative number b. If $a > 0$, the arrow representing $a\mathbf{x}$ has the same direction as the arrow for $\mathbf{x}$ and is a times as long. If $b < 0$, then $b\mathbf{x}$ points in the parallel but opposite direction to $\mathbf{x}$ and is $|b|$ times as long. One reason for drawing the arrows $a\mathbf{x}$ and $b\mathbf{x}$ this way is that for a vector $\mathbf{x}$ in $\mathbb{R}^n$, $|\mathbf{x}|$, the length of $\mathbf{x}$, and $|r\mathbf{x}|$, the length of $r\,\mathbf{x}$, are related by

2.1
$$|r\mathbf{x}| = |r||\mathbf{x}| \ ;$$

that is, the length of $r\mathbf{x}$ is $|r|$ times the length of $\mathbf{x}$. To prove Equation 2.1, we only have to observe that

$$|r\mathbf{x}| = \sqrt{(rx_1)^2 + \cdots + (rx_n)^2}$$
$$= \sqrt{r^2}\sqrt{x_1^2 + \cdots + x_n^2} = |r||\mathbf{x}|.$$

EXAMPLE 3

Multiplying the position vector of a point by 2 moves the point directly away from the origin in the direction of the position vector to double its distance from the origin.

FIGURE 1.6
Scalar multiples.

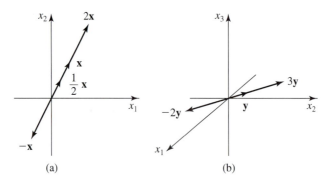

Multiplying the position vector of a point by $\frac{1}{2}$ moves the point halfway directly toward the origin. For instance, let $\mathbf{x} = (1, 2)$. Then $\frac{1}{2}\mathbf{x} = (\frac{1}{2}, 1)$, $2\mathbf{x} = (2, 4)$, and $-\mathbf{x} = (-1, -2)$, as shown in Figure 1.6(a). Figure 1.6(b) shows $\mathbf{y} = (1, 2, 1)$, $3\mathbf{y} = (3, 6, 3)$, and $-2\mathbf{y} = (-2, -4, -2)$.

FIGURE 1.7
Midpoints.

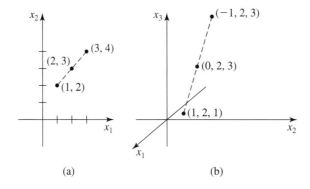

The point midway between two points will be useful in Section 2D for arriving at a geometric interpretation for the sum $\mathbf{x} + \mathbf{y}$. We define the **midpoint m** between $\mathbf{x}$ and $\mathbf{y}$ by $\mathbf{m} = \frac{1}{2}(\mathbf{x} + \mathbf{y}) = \frac{1}{2}\mathbf{x} + \frac{1}{2}\mathbf{y}$, motivated by observing that the coordinates of $\mathbf{m}$ will then be the averages of the coordinates of $\mathbf{x}$ and $\mathbf{y}$. Furthermore, the distances from $\mathbf{x}$ to $\mathbf{m}$ and $\mathbf{y}$ to $\mathbf{m}$ are

$$|\mathbf{m} - \mathbf{x}| = \left|\tfrac{1}{2}(\mathbf{x} + \mathbf{y}) - \mathbf{x}\right| = \left|\tfrac{1}{2}\mathbf{y} - \tfrac{1}{2}\mathbf{x}\right| = \tfrac{1}{2}|\mathbf{y} - \mathbf{x}|,$$

$$|\mathbf{m} - \mathbf{y}| = \left|\tfrac{1}{2}(\mathbf{x} + \mathbf{y}) - \mathbf{y}\right| = \left|\tfrac{1}{2}\mathbf{x} - \tfrac{1}{2}\mathbf{y}\right| = \tfrac{1}{2}|\mathbf{x} - \mathbf{y}|,$$

each of which is half the distance between $\mathbf{x}$ and $\mathbf{y}$. Thus the sum of the distances from $\mathbf{x}$ to $\mathbf{m}$ and from $\mathbf{m}$ to $\mathbf{y}$ is equal to the distance from $\mathbf{x}$ to $\mathbf{y}$, so $\mathbf{x}$, $\mathbf{m}$, and $\mathbf{y}$ are collinear instead of being the vertices of a triangle. This justifies calling $\mathbf{m}$ the midpoint. See Figure 1.7.

EXAMPLE 4   The midpoint $\mathbf{m}$ between $(1, 2)$ and $(3, 4)$ in $\mathbb{R}^2$ is

$$\mathbf{m} = \tfrac{1}{2}((1, 2) + (3, 4)) = \tfrac{1}{2}(4, 6) = (2, 3).$$

FIGURE 1.8
Parallelogram law.

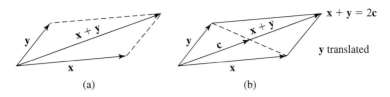

(a)   (b)

The midpoint **m** between $(1, 2, 1)$ and $(-1, 2, 5)$ in $\mathbb{R}^3$ is

$$\mathbf{m} = \tfrac{1}{2}((1, 2, 1) + (-1, 2, 5)) = \tfrac{1}{2}(0, 4, 6) = (0, 2, 3).$$

2D   Vector Addition

Figure 1.8(a) shows $\mathbf{x} + \mathbf{y}$ as the arrow from the origin to the opposite corner of the parallelogram having as two adjacent sides the arrows representing $\mathbf{x}$ and $\mathbf{y}$; this geometric rule for adding vectors is called the **parallelogram rule** of addition. To see why the parallelogram rule is valid, consider the arrow from **0** to the midpoint $\mathbf{m} = \tfrac{1}{2}(\mathbf{x} + \mathbf{y})$ of $\mathbf{x}$ and $\mathbf{y}$. The three arrows with tails at the origin and tips at $\mathbf{x}$, $\mathbf{y}$, and $\mathbf{m}$ all lie in the plane determined by the three points **0**, $\mathbf{x}$, and $\mathbf{y}$. But $\mathbf{x} + \mathbf{y}$ is the arrow twice as long as $\mathbf{m} = \tfrac{1}{2}(\mathbf{x} + \mathbf{y})$ and pointing in the same direction, so the arrow for $\mathbf{x} + \mathbf{y}$ also lies in the same plane as $\mathbf{x}$ and $\mathbf{y}$. Furthermore, opposite sides of the figure with corners at **0**, $\mathbf{x}$, $\mathbf{y}$, and $\mathbf{x} + \mathbf{y}$ have equal lengths because

$$|(\mathbf{x} + \mathbf{y}) - \mathbf{y}| = |\mathbf{x}| \quad \text{and} \quad |\mathbf{x} + \mathbf{y} - \mathbf{x}| = |\mathbf{y}|.$$

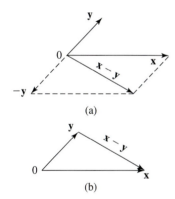

(a)

(b)

FIGURE 1.9
Differences.

Hence the four-sided figure is a parallelogram.

Another way to look at the parallelogram rule appears in Figure 1.9(b). Starting at the tip of the arrow for $\mathbf{x}$, draw an arrow parallel to $\mathbf{y}$ and of the same length as $\mathbf{y}$; in other words, we **translate** $\mathbf{y}$, keeping it parallel to itself, to the tip of $\mathbf{x}$. Then $\mathbf{x} + \mathbf{y}$ is the arrow from the origin to the tip of this translation of $\mathbf{y}$. This form of the rule applies literally to a pair of vectors, whereas the parallelogram rule as previously stated doesn't strictly speaking apply if the arrows representing $\mathbf{x}$ and $\mathbf{y}$ lie on the same line.

The difference $\mathbf{x} - \mathbf{y}$ of vectors $\mathbf{x}$ and $\mathbf{y}$ has an interesting geometric interpretation. By definition, $\mathbf{x} - \mathbf{y}$ is $\mathbf{x} + (-\mathbf{y})$, and the parallelogram law provides an arrow for it, as shown in Figure 1.9(a). But it's simpler to think of $\mathbf{x} - \mathbf{y}$ as the vector $\mathbf{v}$ such that $\mathbf{y} + \mathbf{v} = \mathbf{x}$. The tip-to-tail rule implies that if arrows for $\mathbf{x}$ and $\mathbf{y}$ start at the same point, forming two sides of a triangle, then the third side of the triangle gives an arrow for $\mathbf{x} - \mathbf{y}$ as in Figure 1.9(b). Note the direction carefully; the arrow for $\mathbf{x} - \mathbf{y}$ goes from the tip of $\mathbf{y}$ to the tip of $\mathbf{x}$.

EXAMPLE 5 Let $\mathbf{x} = (-1, 2)$ and $\mathbf{y} = (3, 1)$. Figure 1.10(a) shows $\mathbf{x} + \mathbf{y}$, $\mathbf{x} - \mathbf{y}$, and $\mathbf{x} + 2\mathbf{y}$. Figure 1.10(b) shows $\mathbf{x} + \mathbf{y}$ and $\mathbf{x} - \mathbf{y}$ when $\mathbf{x} = (1, 2, 4)$ and $\mathbf{y} = (1, -1, 1)$.

EXAMPLE 6 Choose coordinate axes so that **i** points east, **j** points north, and **k** points up. Find a vector of length 1 that points in the direction of the sun

(a) when the sun is due south and at angle $\alpha = 60°$ above the horizon.

(b) when the sun is southwest and at angle $\beta = 30°$ above the horizon.

FIGURE 1.10

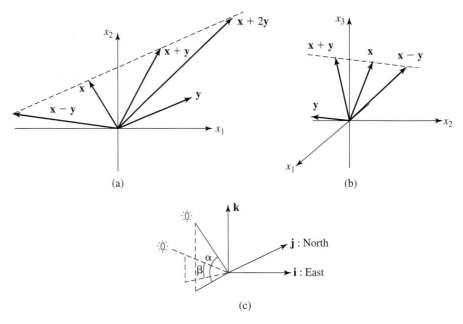

(a)

(b)

(c)

In each case consider the required vector as an arrow starting at the origin that's the hypotenuse of a right triangle whose other sides are vertical and horizontal, as in Figure 1.10(c).

In case (a), the vertical side has length $\sin 60° = \sqrt{3}/2$ and the horizontal side has length $\cos 60° = 1/2$. The required vector is the sum of the horizontal side considered as an arrow pointing south and the vertical side pointing up. Thus the required vector is $-\frac{1}{2}\mathbf{j} + \frac{\sqrt{3}}{2}\mathbf{k}$.

In case (b), the vertical component has length $\sin 30° = 1/2$ and is $(1/2)\mathbf{k}$. The horizontal component has length $\cos 30° = \sqrt{3}/2$ and points southwest, which is the direction of $-\mathbf{i} - \mathbf{j}$. Since $|-\mathbf{i} - \mathbf{j}| = \sqrt{2}$, we have to multiply $-\mathbf{i} - \mathbf{j}$ by $(\sqrt{3}/2)/\sqrt{2} = \sqrt{6}/4$ to get a vector of length $\sqrt{3}/2$, so the required vector is the sum $-\frac{\sqrt{6}}{4}\mathbf{i} - \frac{\sqrt{6}}{4}\mathbf{j} + \frac{1}{2}\mathbf{k}$ of the two components.

EXAMPLE 7

To illustrate how vector calculations can prove geometric theorems, we show that the diagonals of a parallelogram intersect at their midpoints. We choose one vertex of the parallelogram to be the origin $\mathbf{0}$ and let the adjacent vertices have position vectors $\mathbf{a}$ and $\mathbf{b}$. The fourth vertex then has position vector $\mathbf{a} + \mathbf{b}$. The midpoint between $\mathbf{a}$ and $\mathbf{b}$ has position vector $\frac{1}{2}\mathbf{a} + \frac{1}{2}\mathbf{b}$, while the midpoint between $\mathbf{0}$ and $\mathbf{a} + \mathbf{b}$ has position vector $\frac{1}{2}\mathbf{0} + \frac{1}{2}(\mathbf{a} + \mathbf{b})$. Since these vectors are equal, the two midpoints are the same.

2E   Points, Arrows, and Vectors

Numerical calculations involving vectors typically leave us with an n-tuple in $\mathbb{R}^n$, where n is 2, 3, or some unspecified value n. Geometric interpretation of the result is not always pertinent, but when it is we have to decide what interpretation is appropriate. If the result is meant to describe the location of a particle, then we use the point interpretation. If the result is meant to describe a force with magnitude and direction, then we represent it by an arrow with the tail translated to the point where

FIGURE 1.11

$\mathbf{z}(t)$ and $\mathbf{z}(s)$ with $s = \frac{1}{2}, t = \frac{3}{5}$.

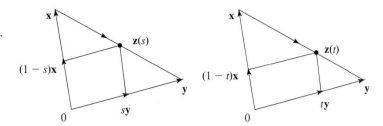

the force is applied. The choice should always be made to create the picture, mental or visible, that conveys the underlying idea in the most useful way. Experience is the best guide, and examples point the way.

Arrows. We've discussed arrows, their tails, and their tips in an intuitive way without a formal definition. We'll clarify the mathematical connections among these geometric ideas as follows. For vectors $\mathbf{x}$ and $\mathbf{y}$ in $\mathbb{R}^n$, the **arrow** or **directed line segment** from $\mathbf{x}$ to $\mathbf{y}$ consists of all points of $\mathbb{R}^n$ of the form $\mathbf{z}(t) = (1 - t)\mathbf{x} + t\mathbf{y}$, where $0 \leq t \leq 1$. We call $\mathbf{x}$ the **tail** of the arrow and $\mathbf{y}$ the **tip**. To see that the definition of $\mathbf{z}(t)$ is appropriate, we need to show that $\mathbf{z}(t)$ covers all the points from tail to tip as t increases from 0 to 1. To prove this consider the distance relations

$$|\big((1 - t)\mathbf{x} + t\mathbf{y}\big) - \mathbf{x}| = t|\mathbf{y} - \mathbf{x}|, \quad \text{and} \quad |\big((1 - t)\mathbf{x} + t\mathbf{y}\big) - \mathbf{y}| = (1 - t)|\mathbf{x} - \mathbf{y}|,$$

generalizing the case $t = \frac{1}{2}$ that justifies the midpoint formula in Section 2C. The equations tell us that a typical point $(1-t)\mathbf{x}+t\mathbf{y}$ of the segment is at distance $t|\mathbf{y}-\mathbf{x}|$ along the segment from $\mathbf{x}$ to $\mathbf{y}$ and is distance $(1 - t)|\mathbf{x} - \mathbf{y}|$ along the segment from $\mathbf{y}$ to $\mathbf{x}$. All of these points are collinear with $\mathbf{x}$ and $\mathbf{y}$, because the two distances add up to the distance between $\mathbf{x}$ and $\mathbf{y}$. As t runs from 0 to 1, $\mathbf{z}(t)$ covers exactly the points from $\mathbf{x}$ to $\mathbf{y}$ in that order as shown in Figure 1.11, so the segment inherits the order of the points in the interval $0 \leq t \leq 1$ in the real number line. To reverse direction we just interchange $\mathbf{x}$ and $\mathbf{y}$ in the formula for $\mathbf{z}(t)$.

Equivalence. Two arrows, one from $\mathbf{x}$ to $\mathbf{y}$ and a second from $\mathbf{z}$ to $\mathbf{w}$, are called **equivalent** if the second is a **translation** of the other, which means that there is a vector $\mathbf{v}$ such that $\mathbf{z} = \mathbf{x} + \mathbf{v}$ and $\mathbf{w} = \mathbf{y} + \mathbf{v}$. If this is so, then $(1 - t)\mathbf{z} + t\mathbf{w} = (1-t)\mathbf{x}+t\mathbf{y}+\mathbf{v}$ so corresponding points on the two arrows are all translated by the same vector $\mathbf{v}$. Each arrow is equivalent to infinitely many translations of itself, one for each vector $\mathbf{v}$. An arrow and a few of its translates appear in Figure 1.12.

Every arrow from $\mathbf{x}$ to $\mathbf{y}$ is equivalent to one with its tail at 0; just take $\mathbf{v} = -\mathbf{x}$ to get the arrow from $\mathbf{x} - \mathbf{x} = \mathbf{0}$ to $\mathbf{y} - \mathbf{x}$; in other words, subtract the tail from the tip. Among all the arrows equivalent to a given one, we have singled out the one with its tail at 0 as the equivalent **position vector**, because it's natural to identify this special one with the point at its tip, and hence with its coordinate vector in $\mathbb{R}^n$. The identification of arrows with elements of $\mathbb{R}^n$ via position vectors allows us to do numerical computations involving arrows, even when it's better to think of their tails as somewhere other than at $\mathbf{0}$; just translate each arrow to the location of the equivalent position vector. Then identify each position vector with the unique element of $\mathbb{R}^n$ that corresponds to it, and perform the desired operations in $\mathbb{R}^n$. Finally, use the reverse identification to find a position vector or some other equivalent arrow, if that conveys more in a picture.

FIGURE 1.12
Translates of (a) (3, 1) and
(b) (1, −1, 1).

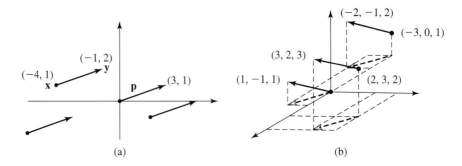

(a)   (b)

EXAMPLE 8   The arrows in a plane shown in Figure 1.12(a) are mutually equivalent and the unique position vector associated with each one of them is $\mathbf{p} = (3, 1)$. The arrow from $\mathbf{x}$ to $\mathbf{y}$ is related to $\mathbf{p}$ by the translation vector $\mathbf{v} = (-4, 1)$.

Drawing arrows in 3-dimensional perspective often requires more care. One strategy is to locate the projections of the tail and tip of an arrow in the horizontal plane and then mark off the corresponding vertical coordinates along a lightly traced line. Then join the tail to the tip.

EXAMPLE 9   Figure 1.12(b) shows the position vector $(1, -1, 1)$ translated once so that its tail is at $(2, 3, 2)$ and again so that its tail is at $(-3, 0, 1)$.

EXERCISES

For each pair of elements in $\mathbb{R}^2$ or $\mathbb{R}^3$ given in Exercises 1 to 4, sketch coordinate axes and mark the points for which they are the coordinates. Draw the line segment connecting the points and calculate its length. Also calculate the coordinates of the midpoint of the segment, and mark it in the drawing.

1. $(1, 1), (-2, 2)$ **2.** $(-1, 4), (2, -1)$

3. $(1, 1, 1), (1, -1, 1)$ **4.** $(1, 0, 0), (1, 2, 1)$

For each vector in $\mathbb{R}^2$ or $\mathbb{R}^3$ given in Exercises 5 to 8, draw arrows starting at the origin representing $\mathbf{x}$, $\frac{1}{2}\mathbf{x}$, and $-2\mathbf{x}$. Find $|\mathbf{x}|$, $|\frac{1}{2}\mathbf{x}|$, and $|-2\mathbf{x}|$.

5. $(1, 1)$ **6.** $(-1, 2)$ **7.** $(1, 2, 2)$ **8.** $(-1, 1, 1)$

For each pair of vectors in $\mathbb{R}^2$ or $\mathbb{R}^3$ given in Exercises 9 to 12, find the vectors $\mathbf{x} + \mathbf{y}$, $\mathbf{x} - \mathbf{y}$, and $\mathbf{x} + 2\mathbf{y}$. Sketch coordinate axes and draw arrows representing the vectors you have found.

9. $\mathbf{x} = (1, 1), \mathbf{y} = (1, -1)$

10. $\mathbf{x} = (2, 4), \mathbf{y} = (-1, -2)$

11. $\mathbf{x} = (1, 1, 1), \mathbf{y} = (1, 1, -1)$

12. $\mathbf{x} = (0, 0, -1), \mathbf{y} = (0, 1, 1)$

For each pair of vectors given in Exercises 13 to 16, draw coordinate axes and arrows representing $\mathbf{x}$ and $\mathbf{y}$. Then use geometric constructions to draw arrows for $2\mathbf{x} + \mathbf{y}$ and $\mathbf{x} - \mathbf{y}$.

13. $\mathbf{x} = (-2, 1), \mathbf{y} = (1, 2)$

14. $\mathbf{x} = (1, 0), \mathbf{y} = (-2, -1)$

15. $\mathbf{x} = (1, 0, 1), \mathbf{y} = (2, 1, -1)$

16. $\mathbf{x} = (0, 1, -1), \mathbf{y} = (1, 1, 2)$

In Exercises 17 to 20, draw the arrows with the given tails and tips. Then draw the equivalent position vector.

17. Tip $(1, 2)$, tail $(2, 1)$

18. Tip $(-1, 1)$, tail $(2, 2)$

19. Tip $(1, 0, 0)$, tail $(0, 1, 1)$

20. Tip $(1, 2, 0)$, tail $(-1, 0, 1)$

In Exercises 21 to 24, draw the arrow that represents the position vector identified with the given element $\mathbf{u}$ of $\mathbb{R}^2$ or $\mathbb{R}^3$. Then draw the arrow equivalent to $\mathbf{u}$ with tail at $\mathbf{v}$.

21. $\mathbf{u} = (1, 2); \mathbf{v} = (2, 1)$

22. $\mathbf{u} = (-1, 1); \mathbf{v} = (-2, 0)$

23. $\mathbf{u} = (1, 1, 1); \mathbf{v} = (2, 1, 0)$

24. $\mathbf{u} = (1, 0, 1); \mathbf{v} = (1, 0, -1)$

25. Verify that $|((1 - t)\mathbf{x} + t\mathbf{y}) - \mathbf{x}| = t|\mathbf{y} - \mathbf{x}|$ and $|((1 - t)\mathbf{x} + t\mathbf{y}) - \mathbf{y}| = (1 - t)|\mathbf{y} - \mathbf{x}|$ for $\mathbf{x}$, $\mathbf{y}$ in $\mathbb{R}^n$ and $0 \leq t \leq 1$. Why is the condition $0 \leq t \leq 1$ needed?

26. Suppose that the tail $\mathbf{z}$ and tip $\mathbf{w}$ of one arrow are related to the tail and tip $\mathbf{x}$ and $\mathbf{y}$ of another arrow by $\mathbf{z} = \mathbf{x} + \mathbf{v}$ and $\mathbf{w} = \mathbf{y} + \mathbf{v}$. Check that the same relation holds between elements $(1 - t)\mathbf{z} + t\mathbf{w}$ and $(1 - t)\mathbf{x} + t\mathbf{y}$ of the two arrows.

27. Draw the square that has the origin and the point $(2, 2)$ as one pair of diagonally opposite vertices. Mark arrowheads on the four sides so that following the arrows takes you around the square in the counterclockwise direction. Write down the vectors represented by the four arrows and calculate their sum.

28. Do the same as in Exercise 27 for the three sides of the triangle with vertices at $(-2, 0)$, $(3, 1)$, and $(0, 4)$.

29. Can you state a general condition on a set of arrows that implies that the vectors they represent add up to $\mathbf{0}$? (The condition should apply to the previous two problems.) Does the condition work in $\mathbb{R}^3$ as well as in the plane?

30. (a) Draw a triangle with one vertex at the origin and the other two at $\mathbf{a} = 5\mathbf{i}$ and $\mathbf{b} = 2\mathbf{i} + 4\mathbf{j}$. Find the midpoints of the sides joining the origin to $\mathbf{a}$ and $\mathbf{b}$, and the vector represented by the arrow from one midpoint to the other. Show that it is $\frac{1}{2}$ times the vector represented by an arrow along the third side of the triangle.

(b) Do a vector calculation to show that the conclusion of part (a) holds for an arbitrary triangle, that is, for arbitrary choices of $\mathbf{a}$ and $\mathbf{b}$.

31. It is a theorem of geometry that a point $\mathbf{c}$ is between the points $\mathbf{a}$ and $\mathbf{b}$ if and only if the distance from $\mathbf{a}$ to $\mathbf{b}$ is the sum of the distances from $\mathbf{a}$ to $\mathbf{c}$ and from $\mathbf{c}$ to $\mathbf{b}$.

In Section 2E we saw that $\mathbf{z}(t) = t\mathbf{x} + (1 - t)\mathbf{y}$ covers the points between $\mathbf{x}$ and $\mathbf{y}$ as t ranges over the interval $0 < t < 1$.

(a) Show that $\mathbf{c}$ is between $\mathbf{a}$ and $\mathbf{b}$ only if $0 \leq t \leq 1$.

(b) For what values of t is $\mathbf{a}$ between $\mathbf{b}$ and $\mathbf{c}$?

(c) For what values of t is $\mathbf{b}$ between $\mathbf{a}$ and $\mathbf{c}$?

In Exercises 32 and 33, a surveyor is standing at the origin of a coordinate system that has its axes oriented so that $\mathbf{i}$ points east, $\mathbf{j}$ points north, and $\mathbf{k}$ points up. (You'll need a calculator or trigonometric tables, and answers should be given to three significant figures.)

32. A corner of the base of a building is 200 feet away from the surveyor, in a direction $40°$ north of east, and the building is 150 feet high. What are the coordinates of the point at the top of the building directly above the corner?

33. The base of a flagpole is in a direction $20°$ west of north from the surveyor and the top is $20°$ up from the horizon. Find the unit vector pointing at the top of the flagpole.

Exercises 34 to 38 involve locating places on the surface of the earth in terms of degrees of latitude north or south of the equator and degrees of longitude east or west of a reference meridian that passes near London, England. Take a coordinate system with origin at the center of the earth, with $\mathbf{i}$ pointing toward the location with latitude $0°$, longitude $0°$ (which is in the Atlantic Ocean), and with $\mathbf{k}$ pointing toward the north pole. In doing these problems, assume that the surface of the earth is a sphere of radius 4000 miles (which is a fairly good approximation).

34. What are the latitude and longitude of the location pointed toward by $\mathbf{j}$?

35. Los Angeles has latitude $34°$ north and longitude $118°$ west (approximately). Find its position vector.

36. Do the same for New York, which has latitude $41°$ north and longitude $74°$ west (approximately).

37. Find the distance between New York and Los Angeles, measured in a straight line (which would pass below the earth's surface).

38. Find the straight-line distance between New York and Paris, France (which has approximate latitude and longitude $48°$ north and $2°$ east).

SECTION 3   LINES AND PLANES

In the previous section, we took the ideas of line and plane for granted and used them informally to justify the geometric interpretation of vector operations. Now we take a more formal point of view and define lines and planes in terms of vectors. The informal geometry we've been using will motivate the definitions.

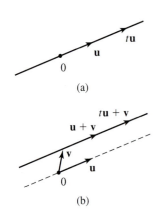

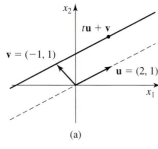

FIGURE 1.13

Arrows on a line.

3A   Lines

In both the plane and $\mathbb{R}^3$ there are two natural ways of specifying a line, either as passing through a particular point in a particular direction or as passing through two particular points. We begin with the first of these.

A nonzero vector specifies a direction. Motivated by the earlier discussion of scalar multiplication, we define two nonzero vectors **u** and **v** to have the **same direction** if $\mathbf{u} = t\mathbf{v}$ for some positive number t and the **opposite direction** if $\mathbf{u} = t\mathbf{v}$ for some negative number t. We say that **u** and **v** are **parallel** if they have either the same or the opposite direction.

Let **u** be a nonzero vector. We define the **line through the origin in the direction of u** to consist of all points whose position vectors are multiples of **u**, as illustrated in Figure 1.13(a). Translating all the points of such a line by a fixed vector **v** gives a parallel line through the point whose position vector is **v** as in Figure 1.13(b). Hence we make the following definition.

A **line** is a set consisting of all the points $t\mathbf{u}+\mathbf{v}$, where **u** and **v** are fixed position vectors, while $\mathbf{u} \neq 0$ and t ranges over all real numbers. Each value of t corresponds to a point of the line. A variable t used in this way is called a **parameter**, and the expression $t\mathbf{u} + \mathbf{v}$ is called the **parametric representation** of the line in vector form.

Although motivated by geometric intuition in $\mathbb{R}^2$ and $\mathbb{R}^3$, this definition of a line applies in $\mathbb{R}^n$, as do the definitions of same and opposite direction and parallelism.

EXAMPLE 1

Given the vectors $\mathbf{u} = (2, 1)$ and $\mathbf{v} = (-1, 1)$ in $\mathbb{R}^2$, suppose that we want to sketch the line of points $\mathbf{x} = t\mathbf{u} + \mathbf{v}$. We think of the direction of the line as determined by the vector **u** and so sketch the line $t\,\mathbf{u}$ through the origin. Then draw the line parallel to $t\,\mathbf{u}$ that passes through the tip of the arrow representing **v**; this gives the line of points $\mathbf{x} = t\mathbf{u} + \mathbf{v}$, shown in Figure 1.14(a). Alternatively, plot two points on the line, for example, those obtained by setting $t = 0$, which gives $\mathbf{x} = \mathbf{v} = (-1, 1)$, and $t = 1$, which gives $\mathbf{x} = \mathbf{u} + \mathbf{v} = (2, 1) + (-1, 1) = (1, 2)$. Then draw the line through these two points as shown in Figure 1.14(b).

If we let (x, y) be the coordinates of **x**, the vector equation $\mathbf{x} = t\mathbf{u} + \mathbf{v}$, or $(x, y) = t(2, 1) + (-1, 1)$, is equivalent to the pair of numerical equations

$$x = 2t - 1$$
$$y = t + 1,$$

which are scalar parametric equations for the line. We'll usually use the more concise vector representation, but both forms are equally valid.

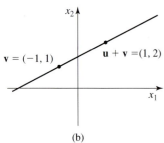

FIGURE 1.14

Points on a line.

To find a parametric representation for the line through two distinct points with position vectors **a** and **b**, recall that the vector from **a** to **b** is $\mathbf{b} - \mathbf{a}$. Thus $\mathbf{b} - \mathbf{a}$ gives the direction of the line, and **a** is a point on it, so

$$\mathbf{x} = t(\mathbf{b} - \mathbf{a}) + \mathbf{a}$$

is a parametric representation of the line. A more symmetrical rewriting of this expression is

3.1 $$\mathbf{x} = (1-t)\mathbf{a} + t\mathbf{b}.$$

When $t = 0$, $\mathbf{x} = \mathbf{a}$ and when $t = 1$, $\mathbf{x} = \mathbf{b}$. For t between 0 and 1 we saw in Section 2E that $\mathbf{x}$ is on the line segment between $\mathbf{a}$ and $\mathbf{b}$. Thus if $t = \frac{1}{2}$, $\mathbf{x}$ is the midpoint.

EXAMPLE 2

To find a parametric representation for the line in $\mathbb{R}^3$ through $(-1, 1, 0)$ and $(2, 2, 1)$, we find $(2, 2, 1) - (-1, 1, 0) = (3, 1, 1)$ as the vector from the first point to the second and obtain

$$\mathbf{x} = t(3, 1, 1) + (-1, 1, 0).$$

To find out whether a point such as $(1, -2, 2)$ lies on this line, we check to see whether there is a value of t such that

$$(1, -2, 2) = t(3, 1, 1) + (-1, 1, 0)$$

or, subtracting $(-1, 1, 0)$ from each side,

$$t(3, 1, 1) = (2, -3, 2).$$

The second coordinates match only if $t = -3$, and the third coordinates match only if $t = 2$, so the equation has no solution and the point is not on the line.

If we ask instead about the point $(-10, -2, -3)$, we check the equation

$$(-10, -2, -3) = t(3, 1, 1) + (-1, 1, 0)$$

or

$$t(3, 1, 1) = (-9, -3, -3).$$

This equation has the solution $t = -3$, so the point is on the line.

Two lines are defined to be **parallel** if they have representations $t\mathbf{u}_1 + \mathbf{v}_1$, $t\mathbf{u}_2 + \mathbf{v}_2$ in which the direction vectors $\mathbf{u}_1$ and $\mathbf{u}_2$ are parallel, that is, if $\mathbf{u}_2$ is a scalar multiple of $\mathbf{u}_1$. In plane geometry, it's usual to define lines to be parallel if they do not intersect, but that definition doesn't work in higher dimensions. For an example in $\mathbb{R}^3$, consider the x-axis and a line parallel to the y-axis, but one unit above it. The lines don't meet, but they aren't parallel either. (See Exercise 24.)

EXAMPLE 3

Let $\mathbf{a} = (1, 1, 0)$ and $\mathbf{b} = (2, -1, 0)$, and let L be the line through them. The direction of L is $\mathbf{b} - \mathbf{a} = (2, -1, 0) - (1, 1, 0) = (1, -2, 0)$. So L has the representation $t(1, -2, 0) + (1, 1, 0)$. The line through the origin parallel to L has the representation $t(1, -2, 0)$ and the line through $(1, 2, 3)$ parallel to L has the representation $t(1, -2, 0) + (1, 2, 3)$.

The line through $(-1, 2, 3)$ and $(1, -2, 3)$ is parallel to L, because it has $(-1, 2, 3) - (1, -2, 3) = (-2, 4, 0)$ for a direction vector, and the vector $(-2, 4, 0) = 2(1, -2, 0)$ is a scalar multiple of $(1, -2, 0)$. The line through $(-1, 2, 3)$ and the origin is not parallel to L, because it has $(-1, 2, 3)$ for a direction vector and $(-1, 2, 3)$ is not a scalar multiple of $(1, -2, 0)$.

A given line has many different representations. In the expression $t\mathbf{u}+\mathbf{v}$, the vector $\mathbf{u}$ giving the direction is replaceable by a nonzero multiple $r\mathbf{u}$, and $\mathbf{v}$ is replaceable by the position vector of another point on the line without changing the line being represented. A way to check whether two representations give the same line is to take two distinct points in one representation and see whether they are also given by the other, as in the following example.

| EXAMPLE 4 | Consider the parametric representations

$$\mathbf{x} = t(1, 2) + (1, 1),$$

$$\mathbf{x} = t(2, 4) + (0, -1).$$

Setting $t = 0$ and $t = 1$ in the second representation gives the points $(0, -1)$ and $(2, 3)$. These points are both given by the first representation because

$$t(1, 2) + (1, 1) = (2, 3)$$

when $t = 1$, and

$$t(1, 2) + (1, 1) = (0, -1)$$

when $t = -1$. Hence the two representations give the same line.

Instead of viewing $t\mathbf{u}+\mathbf{v}$ just as a way to describe a line as a set of points, we can think of it as a function $\mathbf{x}(t) = t\mathbf{u} + \mathbf{v}$ associating particular points with particular values of t. If we let $\mathbf{x}(t)$ be the position of an object at time t, the function describes motion of the object along the straight line parameterized by the function $\mathbf{x}(t)$ with parameter t. Different representations of the same line correspond to motions along the same line, but with different velocities $\mathbf{u}$ and different starting positions $\mathbf{v}$.

| EXAMPLE 5 | A motorboat starts out from a dock on a lake at 10 miles per hour in the direction given by $4\mathbf{i}+3\mathbf{j}$. (We take coordinates with origin at the dock and $\mathbf{i}$ pointing east and $\mathbf{j}$ pointing north, with the unit of distance 1 mile.) At the same time a rowboat starts north at 4 miles per hour from a point 2 miles east of the dock. The boats move along the lines with parametric representations $t_1(4\mathbf{i} + 3\mathbf{j})$ and $2\mathbf{i} + t_2\mathbf{j}$. The lines intersect where $t_1(4\mathbf{i} + 3\mathbf{j}) = 2\mathbf{i} + t_2\mathbf{j}$, giving $4t_1 = 2$ and $3t_1 = t_2$. Then $t_1 = \frac{1}{2}, t_2 = \frac{3}{2}$, and the point of intersection is $\frac{1}{2}(4\mathbf{i} + 3\mathbf{j}) = 2\mathbf{i} + \frac{3}{2}\mathbf{j}$.

The representations we have used tell us the paths followed by the boats but take no account of how fast they go. The vector $4\mathbf{i}+3\mathbf{j}$ has length 5; if we double it, and let $\mathbf{m}(t) = t(8\mathbf{i}+6\mathbf{j})$, we have a function describing the motion of a boat that goes 10 miles in 1 hour and is at the origin (the dock) when $t = 0$. Similarly $\mathbf{r}(t) = 2\mathbf{i}+4t\mathbf{j}$ describes the motion of the rowboat. The motorboat reaches the point $2\mathbf{i} + \frac{3}{2}\mathbf{j}$ when $t = \frac{1}{4}$, and the rowboat reaches it when $t = \frac{3}{8}$, which is $\frac{1}{8}$ hour $= 7\frac{1}{2}$ minutes later, so the boats do not run into each other.

| EXAMPLE 6 | Recall from Section 2E that the points on the segment directed from $\mathbf{x}$ to $\mathbf{y}$ were represented in the form $\mathbf{z}(t) = (1 - t)\mathbf{x} + t\mathbf{y}$ with the parameter t increasing over the interval $0 \leq t \leq 1$. Rewriting $\mathbf{z}(t)$ shows that $\mathbf{z}(t) = t(\mathbf{y} - \mathbf{x}) + \mathbf{x}$, so letting t

run over all real numbers gives the points on the line through $\mathbf{x}$ in the direction of $\mathbf{y} - \mathbf{x}$, assuming $\mathbf{y} - \mathbf{x} \neq 0$. When $t > 1$ the equations

$$\left|\big((1-t)\mathbf{x} + t\mathbf{y}\big) - \mathbf{x}\right| = t|\mathbf{y} - \mathbf{x}| \quad \text{and} \quad \left|\big((1-t)\mathbf{x} + t\mathbf{y}\big) - \mathbf{y}\right| = (t-1)|\mathbf{x} - \mathbf{y}|$$

show that $|\mathbf{z}(t) - \mathbf{x}| = |\mathbf{x} - \mathbf{y}| + |\mathbf{y} - \mathbf{z}(t)|$, so $\mathbf{y}$ is between $\mathbf{x}$ and $\mathbf{z}(t)$. Thus the values $t > 1$ give the points on the line that lie beyond $\mathbf{y}$ in the direction of $\mathbf{y} - \mathbf{x}$. A similar argument with $t < 0$ shows that $\mathbf{x}$ is between $\mathbf{y}$ and $\mathbf{z}(t)$, so $\mathbf{z}(t)$ also gives the points on the line that lie beyond $\mathbf{x}$ in the opposite direction.

3B   Planes

If we pick two nonzero vectors and picture them as arrows from the origin, then unless they lie along the same line, there is a unique plane that contains both arrows. We define a **plane through the origin** to be the set of linear combinations

$$\mathbf{x} = t_1\mathbf{u}_1 + t_2\mathbf{u}_2,$$

where neither $\mathbf{u}_1$ nor $\mathbf{u}_2$ is a scalar multiple of the other. We say that $\mathbf{u}_1$ and $\mathbf{u}_2$ are **linearly independent** if neither one is a scalar multiple of the other. Geometrically this means that the vectors aren't parallel as defined on page 18 and that neither one is the zero vector.

A part of such a plane appears in Figure 1.15(a). In Figure 1.15(b) the plane through the origin has been translated by adding a fixed vector $\mathbf{v}$ to every point of the plane through the origin. Formally, we define a **plane** to be a set of points whose position vectors have the form $t_1\mathbf{u}_1 + t_2\mathbf{u}_2 + \mathbf{v}$, where $\mathbf{u}_1, \mathbf{u}_2$, and $\mathbf{v}$ are fixed vectors, $\mathbf{u}_1$ and $\mathbf{u}_2$ are linearly independent, and t_1 and t_2 range over all real numbers. The variables t_1 and t_2 are **parameters** and $t_1\mathbf{u}_1 + t_2\mathbf{u}_2 + \mathbf{v}$ is called a **parametric representation** of a plane in vector form.

This definition of plane is motivated by the way we picture planes in $\mathbb{R}^3$, but like our definition of line, it applies in $\mathbb{R}^n$. A characteristic property of geometric planes is that they are *flat*, that is, if $\mathbf{p}$ and $\mathbf{q}$ are two distinct points in a plane, then the entire line through $\mathbf{p}$ and $\mathbf{q}$ lies in the plane. To see that planes as we have defined them have this property, let $\mathbf{p} = p_1\mathbf{u}_1 + p_2\mathbf{u}_2 + \mathbf{v}$ and $\mathbf{q} = q_1\mathbf{u}_1 + q_2\mathbf{u}_2 + \mathbf{v}$ be two points in the plane that has parametric representation $t_1\mathbf{u}_1 + t_2\mathbf{u}_2 + \mathbf{v}$. By Formula 3.1, every point $\mathbf{x}$ on the line through $\mathbf{p}$ and $\mathbf{q}$ has a position vector of the

FIGURE 1.15

Generating planes.

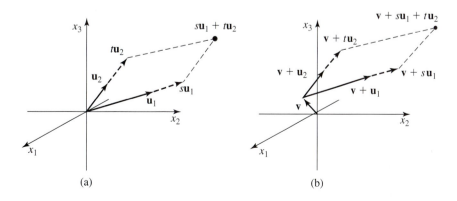

(a)   (b)

FIGURE 1.16
Parallel planes.

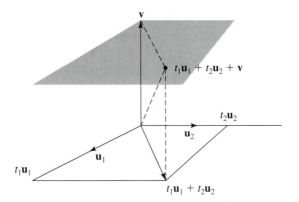

form $(1-t)\mathbf{p} + t\mathbf{q}$ for some number t. Substituting for $\mathbf{p}$ and $\mathbf{q}$ in this formula and rearranging the terms gives

$$\mathbf{x} = (1-t)(p_1\mathbf{u}_1 + p_2\mathbf{u}_2 + \mathbf{v}) + t(q_1\mathbf{u}_1 + q_2\mathbf{u}_2 + \mathbf{v})$$
$$= \big((1-t)p_1 + tq_1\big)\mathbf{u}_1 + \big((1-t)p_2 + q_2\big)\mathbf{u}_2 + \mathbf{v}.$$

This shows that $\mathbf{x}$ is in the plane because it has the form $t_1\mathbf{u}_1 + t_2\mathbf{u}_2 + \mathbf{v}$ with $t_1 = (1-t)p_1 + tq_1$ and $t_2 = (1-t)p_2 + q_2$.

EXAMPLE 7   Let $\mathbf{u}_1 = (1,0,0)$, $\mathbf{u}_2 = (0,1,0)$, and $\mathbf{v} = (0,0,2)$. The vectors $t_1\mathbf{u}_1 + t_2\mathbf{u}_2$ are just the vectors $(t_1, t_2, 0)$ making up the xy-coordinate plane. The vectors

$$t_1\mathbf{u}_1 + t_2\mathbf{u}_2 + \mathbf{v} = (t_1, t_2, 2)$$

give a parallel plane two units above the coordinate plane. These planes appear in Figure 1.16.

Here's how to find a representation for the plane through three given points. For three points $\mathbf{x}_1, \mathbf{x}_2$, and $\mathbf{x}_3$ to determine a unique plane passing through them, the three points must not lie on a line. If the three points do not lie on a line, the vectors $\mathbf{x}_3 - \mathbf{x}_1$ and $\mathbf{x}_2 - \mathbf{x}_1$ are linearly independent; otherwise $\mathbf{x}_3 - \mathbf{x}_1 = t(\mathbf{x}_2 - \mathbf{x}_1)$, for some value of t. Then $\mathbf{x}_3 = t(\mathbf{x}_2 - \mathbf{x}_1) + \mathbf{x}_1$, and $\mathbf{x}_3$ would lie on the line through $\mathbf{x}_1$ and $\mathbf{x}_2$.

Let $\mathbf{x}_1, \mathbf{x}_2$, and $\mathbf{x}_3$ be three points that don't lie on a line. We just observed that the vectors $\mathbf{u}_1 = \mathbf{x}_3 - \mathbf{x}_1$ and $\mathbf{u}_2 = \mathbf{x}_2 - \mathbf{x}_1$ are linearly independent. Then

$$\mathbf{x} = t_1\mathbf{u}_1 + t_2\mathbf{u}_2 + \mathbf{x}_1$$

is the parametric representation of a plane; this is the plane containing the three given points because $\mathbf{x} = \mathbf{x}_1$ when $t_1 = t_2 = 0$, $\mathbf{x} = \mathbf{x}_2$ when $t_1 = 0$ and $t_2 = 1$, and $\mathbf{x} = \mathbf{x}_3$ when $t_1 = 1$ and $t_2 = 0$.

EXAMPLE 8   Let $\mathbf{x}_1 = (1,2,1)$, $\mathbf{x}_2 = (-1,2,0)$, and $\mathbf{x}_3 = (1,1,2)$. Let

$$\mathbf{u}_1 = \mathbf{x}_3 - \mathbf{x}_1 = (0,-1,1) \text{ and } \mathbf{u}_2 = \mathbf{x}_2 - \mathbf{x}_1 = (-2,0,-1).$$

FIGURE 1.17
(a) Arrows. (b) Points.

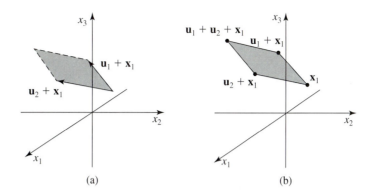

(a)   (b)

Then the parametric representation for the plane through the three points is

$$\mathbf{x} = t_1(0, -1, 1) + t_2(-2, 0, -1) + (1, 2, 1).$$

To picture the plane relative to rectangular coordinate axes in $\mathbb{R}^3$, we can either draw the translated vectors $\mathbf{u}_1 + \mathbf{x}_1$ and $\mathbf{u}_2 + \mathbf{x}_1$ as in Figure 1.17(a) or plot three or four noncollinear points on the plane as in Figure 1.17(b).

EXERCISES

In Exercises 1 to 4, represent the described line in the parametric form $\mathbf{x} = t\mathbf{u} + \mathbf{v}$ and sketch the line.

1. The line in $\mathbb{R}^2$ parallel to the vector $(2, 1)$ and passing through the point $(-1, 2)$

2. The line in $\mathbb{R}^2$ through the points $(1, 0)$ and $(0, 1)$

3. The line in $\mathbb{R}^3$ passing through the points $(2, 2, 3)$ and $(1, 2, 2)$

4. The line in $\mathbb{R}^3$ parallel to the vector $(1, 1, 2)$ and passing through the point $(2, 0, 1)$

5. Let $\mathbf{a} = (-1, 1)$, $\mathbf{b} = (0, 1)$, $\mathbf{c} = (2, 1)$, and $\mathbf{d} = (-3, 2)$.
   (a) Sketch the lines $t\mathbf{a} + \mathbf{b}$ and $s\mathbf{c} + \mathbf{d}$.
   (b) Find the point $\mathbf{p}$ where the lines in part (a) intersect by finding values of s and t for which

   $$\mathbf{p} = t\mathbf{a} + \mathbf{b} = s\mathbf{c} + \mathbf{d}.$$

   (c) Change $\mathbf{c}$ to $(2, -2)$ and show that then the lines $t\mathbf{a} + \mathbf{b}$ and $s\mathbf{c} + \mathbf{d}$ do not intersect.

6. Let $\mathbf{a} = (-3, 0, 1)$, $\mathbf{b} = (0, 1, 2)$, $\mathbf{c} = (2, -1, 1)$, and $\mathbf{d} = (1, 2, 0)$.
   (a) Sketch the lines $t\mathbf{a} + \mathbf{b}$ and $s\mathbf{c} + \mathbf{d}$.
   (b) Find the point of intersection of the lines in part (a), or show that they do not intersect.
   (c) What is the answer to part (b) if $\mathbf{d}$ is changed to $(-1, 0, 4)$?

For the pairs of lines given in Exercises 7 to 10, find out whether the two lines are the same, and if they aren't, whether they are parallel.

7. $t(1, 2) + (2, 1)$ and $t(-2, -4) + (3, 3)$

8. $t(2, -1) + (2, 1)$ and $t(-1, 2) + (2, -1)$

9. $t(2, 3, -1) + (-1, -1, 1)$ and $t(-4, -6, 2) + (1, 1, -1)$

10. $t(4, 2, 2) + (2, 0, 1)$ and $t(2, 1, 1) + (2, 2, 1)$

Which of the pairs of vectors given in Exercises 11 to 14 are linearly independent?

11. $(1, 2), (2, 1)$ 12. $(2, -1), (-2, 1)$

13. $(3, 1, 3), (1, 3, 1)$ 14. $(-1, 3, 1), (2, -6, -2)$

In Exercises 15 to 18, find a representation for the given plane in the parametric form $t_1\mathbf{u}_1 + t_2\mathbf{u}_2 + \mathbf{v}$, and sketch the plane.

15. The plane parallel to the vectors $(1, 1, 0)$ and $(0, 1, 1)$ and passing through the origin

16. The plane parallel to the vectors $\mathbf{e}_1$ and $\mathbf{e}_2$ in $\mathbb{R}^3$ and passing through the point $(0, 0, 2)$

17. The plane passing through the three points $(1, 0, 0)$, $(0, 1, 0)$, and $(0, 0, 1)$

18. The plane passing through the three points $(1, 1, 0)$, $(-3, 0, 2)$, and $(2, 4, 7)$

19. A plane P in $\mathbb{R}^3$ contains the line $t(1, -1, 2) + (1, 2, 1)$ and the point $(3, 0, 1)$.

  (a) Find a vector between two points of P that is linearly independent of $(1, -1, 2)$.

  (b) Find a parametric representation for P.

20. Let the vertices of a triangle be the points **a**, **b**, and **c** and let **p**, **q**, and **r** be the midpoints of the sides opposite **a**, **b**, and **c**, respectively. A line joining a vertex to the midpoint of the opposite side is called a **median** of the triangle. Show that the point $\mathbf{p} = \frac{1}{3}\mathbf{a} + \frac{2}{3}\mathbf{p}$ is on the median that joins **a** and **p**. Express **p** in terms of **a**, **b**, and **c**. Show also that **p** is on the other two medians of the triangle.

21. Show that if **a**, **b**, **c**, and **d** are the position vectors of the vertices of a quadrilateral, not necessarily lying in a plane, then the midpoints of the four sides are vertices of a parallelogram and do lie in a plane.

22. Show that two lines are parallel if and only if for two distinct points $\mathbf{v}_1$ and $\mathbf{v}_2$ on the first line, and two distinct points $\mathbf{w}_1$ and $\mathbf{w}_2$ on the second line, the difference $\mathbf{w}_1 - \mathbf{w}_2$ is a multiple of $\mathbf{v}_1 - \mathbf{v}_2$.

23. **(a)** Show that if $t\mathbf{u} = \mathbf{0}$, for **u** a vector in $\mathbb{R}^n$, then either $t = 0$ or $\mathbf{u} = \mathbf{0}$. Can you derive the same result using only the laws 1 through 9 on page 3?

  (b) Show that if $\mathbf{u}_1$ is a scalar multiple of $\mathbf{u}_2$, and is not zero, then $\mathbf{u}_2$ is a scalar multiple of $\mathbf{u}_1$.

  (c) Show that $t\mathbf{u}_1 + \mathbf{v}_1$ and $t\mathbf{u}_2 + \mathbf{v}_2$ represent the same line if and only if both $\mathbf{u}_2$ and $\mathbf{v}_2 - \mathbf{v}_1$ are scalar multiples of $\mathbf{u}_1$.

24. Let $\mathbf{u} = (u_1, u_2)$, $\mathbf{v} = (v_1, v_2)$ be nonzero vectors in $\mathbb{R}^2$.

  (a) Show that **u** and **v** are parallel if and only if $u_1 v_2 = u_2 v_1$.

  (b) Let **a** and **b** be vectors in $\mathbb{R}^2$, so that $t\mathbf{u} + \mathbf{a}$ and $s\mathbf{v} + \mathbf{b}$ are two lines in the plane. Show that if **u** and **v** are not parallel, then there are values of s and t such that $t\mathbf{u} + \mathbf{a} = s\mathbf{v} + \mathbf{b}$, so the lines intersect. (Write out the vector equation $t\mathbf{u} + \mathbf{a} = s\mathbf{v} + \mathbf{b}$ as a pair of scalar equations and show that they can always be solved if $u_1 v_2 \neq u_2 v_1$.)

25. A subset S of $\mathbb{R}^n$ is **convex** if whenever it contains **a** and **b** it also contains the line segment joining them, that is, all points $t\mathbf{a} + (1 - t)\mathbf{b}$ with $0 \leq t \leq 1$. Show that if S and T are convex subsets of $\mathbb{R}^n$, then the set $S + T$ of all sums $\mathbf{x} + \mathbf{y}$, with **x** in S and **y** in T is also convex.

SECTION 4   DOT PRODUCTS

The **dot product** of two vectors $\mathbf{x} = (x_1, \ldots, x_n)$ and $\mathbf{y} = (y_1, \ldots, y_n)$ in $\mathbb{R}^n$ is defined to be the number given by the formula

4.1
$$\mathbf{x} \cdot \mathbf{y} = x_1 y_1 + \cdots + x_n y_n.$$

This simple formula has a geometric interpretation that allows us solve problems involving lengths and angles. We discuss other applications in this section, and still others appear in later chapters.

We often use the following properties of dot products in both theoretical calculations and applications.

4.2 Theorem.   Positivity: $\mathbf{x} \cdot \mathbf{x} > 0$, except that $\mathbf{0} \cdot \mathbf{0} = 0$

Symmetry: $\mathbf{x} \cdot \mathbf{y} = \mathbf{y} \cdot \mathbf{x}$

Additivity: $(\mathbf{x} + \mathbf{y}) \cdot \mathbf{z} = \mathbf{x} \cdot \mathbf{z} + \mathbf{y} \cdot \mathbf{z}$

Homogeneity: $(r\mathbf{x}) \cdot \mathbf{y} = r(\mathbf{x} \cdot \mathbf{y})$

Proof.   To prove positivity, note that in the sum $\mathbf{x} \cdot \mathbf{x} = x_1^2 + \cdots + x_n^2$, each term $x_i^2 \geq 0$. If $\mathbf{x} = \mathbf{0}$ then all the terms x_k^2 are 0 so the sum is 0. Otherwise, at least one term is greater than 0 and the sum is greater than 0.

Symmetry holds because in the sums for $\mathbf{x} \cdot \mathbf{y} = x_1 y_1 + \cdots + x_n y_n$ and $\mathbf{y} \cdot \mathbf{x} = y_1 x_1 + \cdots + y_n x_n$, corresponding terms $x_i y_i$ and $y_i x_i$ are equal by the commutative law for ordinary multiplication, and therefore the sums are equal.

Proofs for additivity and homogeneity also follow directly from the definition of dot product and the laws of ordinary arithmetic and are left as Exercise 11.   ∎

Because of the symmetry of the dot product, it follows immediately that additivity and homogeneity hold for the second vector also; that is,

$$\mathbf{x}{\cdot}(\mathbf{y} + \mathbf{z}) = \mathbf{x} \cdot \mathbf{y} + \mathbf{x} \cdot \mathbf{z} \quad \text{and} \quad \mathbf{x} \cdot (r\mathbf{y}) = r(\mathbf{x} \cdot \mathbf{y}).$$

4A   Lengths and Angles

The dot product of a vector $\mathbf{x}$ with itself is

$$\mathbf{x} \cdot \mathbf{x} = x_1^2 + \cdots + x_n^2 \,.$$

Referring to our earlier definition of the length of $\mathbf{x}$ in Section 2B as

$$|\mathbf{x}| = \sqrt{x_1^2 + \cdots + x_n^2},$$

we see that the length of $\mathbf{x}$ equals the square root of the dot product of $\mathbf{x}$ with itself:

4.3 $$|\mathbf{x}| = \sqrt{\mathbf{x} \cdot \mathbf{x}}.$$

FIGURE 1.18

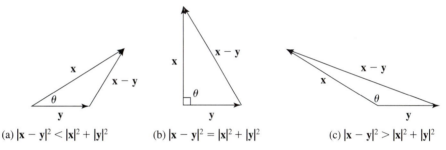

(a) $|\mathbf{x} - \mathbf{y}|^2 < |\mathbf{x}|^2 + |\mathbf{y}|^2$ (b) $|\mathbf{x} - \mathbf{y}|^2 = |\mathbf{x}|^2 + |\mathbf{y}|^2$ (c) $|\mathbf{x} - \mathbf{y}|^2 > |\mathbf{x}|^2 + |\mathbf{y}|^2$

We get a relation between dot products and angles by recalling from Section 2 that if two sides of a triangle represent vectors $\mathbf{x}$ and $\mathbf{y}$, then the third side represents the vector $\mathbf{x} - \mathbf{y}$ from the tip of $\mathbf{y}$ to the tip of $\mathbf{x}$ as illustrated in Figure 1.18. The key to the relation is the familiar trigonometry formula called the law of cosines that relates the length of one side of a triangle to the lengths of the other sides and the angle between them. Applied to the triangles in the figure it gives the following:

4.4   Law of Cosines.

$$|\mathbf{x} - \mathbf{y}|^2 = |\mathbf{x}|^2 + |\mathbf{y}|^2 - 2|\mathbf{x}||\mathbf{y}| \cos \theta \,,$$

where θ is the angle opposite the side $\mathbf{x} - \mathbf{y}$.

We now use the law of cosines to derive the following fundamental geometric property of the dot product.

4.5   Theorem.   The dot product of two nonzero vectors is equal to the product of their lengths times the cosine of the angle between them. In other words, if θ is the angle between vectors $\mathbf{x}$ and $\mathbf{y}$, then

$$\mathbf{x} \cdot \mathbf{y} = |\mathbf{x}||\mathbf{y}| \cos \theta.$$

Proof.   Using the additivity and homogeneity of the dot product, we have

$$|\mathbf{x} - \mathbf{y}|^2 = (\mathbf{x} - \mathbf{y}) \cdot (\mathbf{x} - \mathbf{y})$$
$$= \mathbf{x} \cdot (\mathbf{x} - \mathbf{y}) - \mathbf{y} \cdot (\mathbf{x} - \mathbf{y})$$
$$= \mathbf{x} \cdot \mathbf{x} - \mathbf{x} \cdot \mathbf{y} - \mathbf{y} \cdot \mathbf{x} + \mathbf{y} \cdot \mathbf{y}$$
$$= |\mathbf{x}|^2 + |\mathbf{y}|^2 - 2(\mathbf{x} \cdot \mathbf{y}),$$

where the last step uses the symmetry property $\mathbf{y} \cdot \mathbf{x} = \mathbf{x} \cdot \mathbf{y}$. Comparing this with the value for $|\mathbf{x} - \mathbf{y}|^2$ given by the law of cosines shows that $\mathbf{x} \cdot \mathbf{y} = |\mathbf{x}||\mathbf{y}| \cos \theta$. ∎

As illustrated in Figure 1.18, the formulas for $|\mathbf{x} - \mathbf{y}|^2$ in the proof of Theorem 4.5 show that when $\mathbf{x} \cdot \mathbf{y} > 0$ and $\cos \theta > 0$ then $|\mathbf{x} - \mathbf{y}|^2 < |\mathbf{x}|^2 + |\mathbf{y}|^2$; similarly, when $\mathbf{x} \cdot \mathbf{y} < 0$ and $\cos \theta < 0$, then $|\mathbf{x} - \mathbf{y}|^2 > |\mathbf{x}|^2 + |\mathbf{y}|^2$. The special case of perpendicular vectors $\mathbf{x}$ and $\mathbf{y}$ shown in Figure 1.18(b), when $\cos \theta$ and $\mathbf{x} \cdot \mathbf{y}$ are both 0, is particularly simple and important and we emphasize it as follows:

4.6   Theorem of Pythagoras.   If $\mathbf{x} \cdot \mathbf{y} = 0$, then $|\mathbf{x} - \mathbf{y}|^2 = |\mathbf{x}|^2 + |\mathbf{y}|^2$.

Note that if we replace $\mathbf{y}$ by $-\mathbf{y}$ throughout, the conclusion of Theorem 4.6 becomes the more symmetrical statement $|\mathbf{x} + \mathbf{y}|^2 = |\mathbf{x}|^2 + |\mathbf{y}|^2$. We say that two vectors are **orthogonal** to each other if their dot product is 0. Thus two vectors are orthogonal if they have perpendicular directions or if one of them is $\mathbf{0}$.

It follows from Theorem 4.5 that $\cos \theta = (\mathbf{x} \cdot \mathbf{y})/(|\mathbf{x}||\mathbf{y}|)$ and hence that the angle θ between $\mathbf{x}$ and $\mathbf{y}$ is

4.7
$$\theta = \arccos \frac{\mathbf{x} \cdot \mathbf{y}}{|\mathbf{x}||\mathbf{y}|}.$$

If either $\mathbf{x}$ or $\mathbf{y}$ is $\mathbf{0}$, the right side of Equation 4.7 is undefined, which is appropriate because the zero vector has no direction and so can't form an angle with another vector.

EXAMPLE 1   To find the angle θ between $\mathbf{x} = (1, 3)$ and $\mathbf{y} = (-1, 1)$ in $\mathbb{R}^2$ we compute

$$\mathbf{x} \cdot \mathbf{x} = 1^2 + 3^2 = 10, \quad \mathbf{y} \cdot \mathbf{y} = (-1)^2 + 1^2 = 2, \quad \text{and } \mathbf{x} \cdot \mathbf{y} = -1 + 3 = 2.$$

Then

$$|\mathbf{x}| = \sqrt{10}, \quad |\mathbf{y}| = \sqrt{2}, \quad \text{and} \quad \cos \theta = \frac{2}{\sqrt{20}} = \frac{1}{\sqrt{5}}.$$

Using a calculator we find $\theta \approx \arccos .447 \approx 1.1$ radians, or about $63°$.

To find the angle θ between $\mathbf{x} = (1, 2, 0, -2)$ and $\mathbf{y} = (0, -6, 3, 2)$ in $\mathbb{R}^4$ we compute

$$\mathbf{x} \cdot \mathbf{x} = 1^2 + 2^2 + 0^2 + (-2)^2 = 1 + 4 + 0 + 4 = 9,$$
$$\mathbf{y} \cdot \mathbf{y} = 0^2 + (-6)^2 + 3^2 + 2^2 = 2 = 0 + 36 + 9 + 4 = 49, \quad \text{and}$$
$$\mathbf{x} \cdot \mathbf{y} = 0 - 12 + 0 - 4 = -16.$$

Then $|\mathbf{x}| = 3$, $|\mathbf{y}| = 7$, and $\cos\theta = -16/21$, giving $\theta \approx \arccos(-.762) \approx 2.44$ radians, or about $139.6°$.

FIGURE 1.19

Perpendicular vectors.

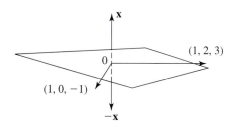

EXAMPLE 2

Problem: Find a vector $\mathbf{x}$ of length 2 perpendicular to $\mathbf{a} = (1, 2, 3)$ and $\mathbf{b} = (1, 0, -1)$. Note that if $\mathbf{x}$ is a solution, so is $-\mathbf{x}$, as in Figure 1.19. The perpendicularity conditions $\mathbf{x} \cdot \mathbf{a} = \mathbf{x} \cdot \mathbf{b} = 0$ give the equations

$$x_1 + 2x_2 + 3x_3 = 0,$$
$$x_1 \qquad\quad - \; x_3 = 0.$$

The second equation gives $x_1 = x_3$, and substituting in the first equation gives $4x_1 + 2x_2 = 0$ or $x_2 = -2x_1$. Taking $x_1 = 1$ gives $\mathbf{x} = (1, -2, 1)$ as one solution, and a scalar multiple of $\mathbf{x}$ is also perpendicular to $\mathbf{a}$ and $\mathbf{b}$. The length of $\mathbf{x}$ is $\sqrt{6}$, so to get a vector of length 2 we must multiply by $\pm 2/\sqrt{6} = \pm\sqrt{2/3}$, giving $\mathbf{x} = \pm\left(\sqrt{\frac{2}{3}}, -\sqrt{\frac{8}{3}}, \sqrt{\frac{2}{3}}\right)$.

EXAMPLE 3

To illustrate how to use the dot product to prove geometric theorems, we'll show that in a parallelogram with all four sides the same length the diagonals are perpendicular. Figure 1.20 shows such a parallelogram with diagonals parallel to the vectors $\mathbf{x} + \mathbf{y}$ and $\mathbf{x} - \mathbf{y}$. To show that the diagonals are perpendicular we need to show that $(\mathbf{x} + \mathbf{y}) \cdot (\mathbf{x} - \mathbf{y}) = 0$. Using the additivity property of the dot product twice allows us to multiply out, getting

$$(\mathbf{x} + \mathbf{y}) \cdot (\mathbf{x} - \mathbf{y}) = \mathbf{x} \cdot (\mathbf{x} - \mathbf{y}) + \mathbf{y} \cdot (\mathbf{x} - \mathbf{y})$$
$$= \mathbf{x} \cdot \mathbf{x} + \mathbf{x} \cdot (-\mathbf{y}) + \mathbf{y} \cdot \mathbf{x} + \mathbf{y} \cdot (-\mathbf{y}).$$

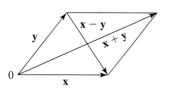

By homogeneity, $\mathbf{x} \cdot (-\mathbf{y}) = -(\mathbf{x} \cdot \mathbf{y})$ and $\mathbf{y} \cdot (-\mathbf{y}) = -(\mathbf{y} \cdot \mathbf{y})$. Since the lengths of $\mathbf{x}$ and $\mathbf{y}$ are equal, $\mathbf{x} \cdot \mathbf{x} = |\mathbf{x}|^2 = |\mathbf{y}|^2 = \mathbf{y} \cdot \mathbf{y}$, so the right side is zero, as we wanted to show.

FIGURE 1.20

Perpendicular diagonals.

4B   Properties of $\mathbf{x} \cdot \mathbf{y}$ and $|\mathbf{x}|$

The following inequality expresses a fundamental property of the dot product that comes up in many areas of mathematics.

4.8   Cauchy–Schwarz Inequality.

$$|\mathbf{x} \cdot \mathbf{y}| \le |\mathbf{x}||\mathbf{y}|$$

Proof.   Theorem 4.5 gives $\mathbf{x} \cdot \mathbf{y} = |\mathbf{x}||\mathbf{y}| \cos \theta$ for some angle θ. Taking absolute values, and noting that the absolute value of a product of real numbers equals the product of their absolute values, we have

$$|\mathbf{x} \cdot \mathbf{y}| = |\mathbf{x}||\mathbf{y}|| \cos \theta| \leq |\mathbf{x}||\mathbf{y}|$$

because $| \cos \theta| \leq 1$ for all θ.

For a purely algebraic proof that treats vectors only as n-tuples of numbers and makes no use of their geometric interpretation, see Exercise 29.   ∎

4.9   Theorem.   The length function defined on $\mathbb{R}^n$ by $|\mathbf{x}| = \sqrt{\mathbf{x} \cdot \mathbf{x}}$ has the properties:

$$\textbf{Positivity:} \quad |\mathbf{x}| > 0 \text{ except that } |\mathbf{0}| = 0$$

$$\textbf{Homogeneity:} \quad |r\mathbf{x}| = |r||\mathbf{x}|$$

$$\textbf{Triangle Inequality:} \quad |\mathbf{x} + \mathbf{y}| \leq |\mathbf{x}| + |\mathbf{y}|$$

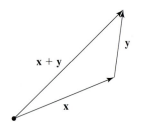

Proof.   Positivity is an immediate consequence of the positivity property of dot products, for since $\mathbf{x} \cdot \mathbf{x} > 0$ unless $\mathbf{x} = \mathbf{0}$, the same is true of $|\mathbf{x}| = \sqrt{\mathbf{x} \cdot \mathbf{x}}$. We leave the proof of homogeneity to the reader in Exercise 12.

Geometrically, the triangle inequality is equivalent to the theorem that a side of a triangle cannot be longer that the sum of the lengths of the other two sides, which is why it's called the triangle inequality. See Figure 1.21.

For an algebraic proof, we start with the equation

$$|\mathbf{x} + \mathbf{y}|^2 = (\mathbf{x} + \mathbf{y}) \cdot (\mathbf{x} + \mathbf{y}) = |\mathbf{x}|^2 + |\mathbf{y}|^2 + 2\mathbf{x} \cdot \mathbf{y}.$$

FIGURE 1.21
Triangle inequality.

From the Cauchy–Schwarz inequality, $\mathbf{x} \cdot \mathbf{y} \leq |\mathbf{x}||\mathbf{y}|$, so

$$|\mathbf{x} + \mathbf{y}|^2 \leq |\mathbf{x}|^2 + |\mathbf{y}|^2 + 2|\mathbf{x}||\mathbf{y}| = (|\mathbf{x}| + |\mathbf{y}|)^2.$$

Taking square roots, we get

$$|\mathbf{x} + \mathbf{y}| \leq |\mathbf{x}| + |\mathbf{y}|.$$ ∎

4C   Unit Vectors and Projections

A vector is called a **unit vector** if it has length 1. Any nonzero vector may be used to specify a direction, but it's often convenient to use a unit vector for the purpose. If $\mathbf{v} \neq \mathbf{0}$ then $\mathbf{n} = \mathbf{v}/|\mathbf{v}|$ is the unique unit vector with the same direction as $\mathbf{v}$ and is called the **normalization** of $\mathbf{v}$.

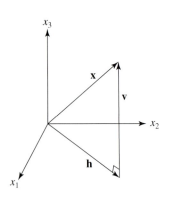

We often need to decompose a force vector into the sum of two vectors, one having a given direction and the other perpendicular to it. For example, in $\mathbb{R}^3$ every vector $\mathbf{x}$ is the sum of a vertical component $\mathbf{v}$ parallel to the z-axis and a horizontal component $\mathbf{h}$ in the xy-plane perpendicular to the z-axis, as in Figure 1.22. Other examples and applications are given in this section and in the next section. The following theorem tells how to find such a decomposition.

FIGURE 1.22
$\mathbf{x} = \mathbf{h} + \mathbf{v}$.

FIGURE 1.23
$\mathbf{x} = \mathbf{p} + \mathbf{q}$.

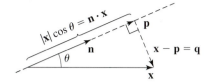

4.10   Theorem.   Let $\mathbf{x}$ and $\mathbf{n}$ be vectors in $\mathbb{R}^n$, with $|\mathbf{n}| = 1$. There are unique vectors $\mathbf{p}$ and $\mathbf{q}$ such that $\mathbf{x} = \mathbf{p} + \mathbf{q}$, with $\mathbf{p}$ parallel to $\mathbf{n}$ and $\mathbf{q}$ perpendicular to $\mathbf{n}$. The vector $\mathbf{p}$ equals $(\mathbf{x} \cdot \mathbf{n})\mathbf{n}$, with length $|\mathbf{x} \cdot \mathbf{n}|$, and $\mathbf{q} = \mathbf{x} - \mathbf{p}$, with length $|\mathbf{q}| = \sqrt{|\mathbf{x}|^2 - |\mathbf{p}|^2}$.

Proof.   Since $\mathbf{p}$ is parallel to $\mathbf{n}$ there is a scalar r such that $\mathbf{p} = r\mathbf{n}$. For $\mathbf{q} = \mathbf{x} - \mathbf{p}$ to be perpendicular to $\mathbf{n}$, we need $(\mathbf{x} - \mathbf{p}) \cdot \mathbf{n} = 0$. But since $\mathbf{n} \cdot \mathbf{n} = 1$,

$$(\mathbf{x} - \mathbf{p}) \cdot \mathbf{n} = (\mathbf{x} - r\mathbf{n}) \cdot \mathbf{n} = \mathbf{x} \cdot \mathbf{n} - r(\mathbf{n} \cdot \mathbf{n}) = \mathbf{x} \cdot \mathbf{n} - r = 0,$$

so $(\mathbf{x} - \mathbf{p}) \cdot \mathbf{n} = 0$ if and only if $r = \mathbf{n} \cdot \mathbf{x}$. By Pythagoras $|\mathbf{q}|^2 = |\mathbf{x}|^2 - |\mathbf{p}|^2$.   ∎

The vector $\mathbf{p}$ in the statement of Theorem 4.10 is the **component** of $\mathbf{x}$ in the direction of $\mathbf{n}$, or the **projection** of $\mathbf{x}$ on $\mathbf{n}$. The vector $\mathbf{q} = \mathbf{x} - (\mathbf{n} \cdot \mathbf{x})\mathbf{n}$ is the **component** of $\mathbf{x}$ **perpendicular** to $\mathbf{n}$. The scalar $\mathbf{x} \cdot \mathbf{n}$ is the **coordinate** of $\mathbf{x}$ in the direction of $\mathbf{n}$.

Figure 1.23 illustrates this decomposition for a vector $\mathbf{x}$ making an angle θ with $\mathbf{n}$. The vector $\mathbf{p}$ is the component of $\mathbf{x}$ in the direction of $\mathbf{n}$. We see geometrically that its length is $|\mathbf{x}||\cos\theta|$ and that it has the same direction as $\mathbf{n}$ when $\mathbf{n} \cdot \mathbf{x} > 0$ and $0 \leq \theta < \pi/2$ and the opposite direction to $\mathbf{n}$ when $\mathbf{n} \cdot \mathbf{x} < 0$ and $\pi/2 < \theta \leq \pi$. Since $\mathbf{n}$ is a unit vector, this agrees with the formula $\mathbf{p} = (\mathbf{n} \cdot \mathbf{x})\mathbf{n}$ and the geometric interpretation of the dot product in Equation 4.5.

EXAMPLE 4

The standard basis vectors $\mathbf{e}_1, \mathbf{e}_2, \ldots, \mathbf{e}_n$ for $\mathbb{R}^n$ are unit vectors and are orthogonal to each other. In particular, in $\mathbb{R}^3$ we have $\mathbf{i} \cdot \mathbf{i} = \mathbf{j} \cdot \mathbf{j} = \mathbf{k} \cdot \mathbf{k} = 1$ and $\mathbf{i} \cdot \mathbf{j} = \mathbf{j} \cdot \mathbf{k} = \mathbf{k} \cdot \mathbf{i} = 0$. If $\mathbf{x} = (x_1, x_2, \ldots, x_n) = x_1\mathbf{e}_1 + \cdots + x_n\mathbf{e}_n$ is an arbitrary vector in $\mathbb{R}^n$ then its coordinate in the direction of $\mathbf{e}_i$ is $\mathbf{x} \cdot \mathbf{e}_i = x_i$. In $\mathbb{R}^3$, if $\mathbf{v} = x\mathbf{i} + y\mathbf{j} + z\mathbf{k}$ then $x = \mathbf{v} \cdot \mathbf{i}$, $y = \mathbf{v} \cdot \mathbf{j}$, and $z = \mathbf{v} \cdot \mathbf{k}$.

A mechanical force has magnitude and direction and so is a vector with its arrow's tail usually drawn at the point of application of the force. It's often useful to express a force as a sum of perpendicular components because they act independently of each other, as in the next example.

EXAMPLE 5

Here we analyze the effect of gravity on a 1-pound brick held in place by friction on a roof that slopes so that $\mathbf{n} = \frac{2}{7}\mathbf{i} + \frac{3}{7}\mathbf{j} + \frac{6}{7}\mathbf{k}$ is the unit vector perpendicular to the roof, as in Figure 1.24. Gravity exerts a downward vertical force of 1 pound, which as a vector is $-\mathbf{k}$. The brick doesn't move because the roof exerts an opposing force $\mathbf{F} = \mathbf{k}$ on it. The component of $\mathbf{F}$ in the direction of $\mathbf{n}$ is

$$\mathbf{p} = (\mathbf{F} \cdot \mathbf{n})\mathbf{n} = \frac{6}{7}\mathbf{n} = \frac{12}{49}\mathbf{i} + \frac{18}{49}\mathbf{j} + \frac{36}{49}\mathbf{k}$$

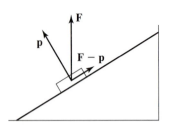

FIGURE 1.24
Brick on a roof.

and is the force with which the roof presses directly against the brick. The other force component $\mathbf{F} - \mathbf{p} = -\frac{12}{49}\mathbf{i} - \frac{18}{49}\mathbf{j} + \frac{13}{49}\mathbf{k}$ is perpendicular to $\mathbf{n}$ and thus parallel to the roof, and is the frictional force that keeps the brick from sliding.

The vector $\mathbf{v} = 2\mathbf{i} + 3\mathbf{j} + 6\mathbf{k}$ has the same direction as $\mathbf{n}$ and could have been given instead of $\mathbf{n}$ to specify the way the roof slopes. In that case we would have to start by calculating $|\mathbf{v}| = 7$ and finding the unit vector $\mathbf{n} = \mathbf{v}/|\mathbf{v}|$.

The work done by a force in moving an object a certain distance in a straight line is the product of the distance and the magnitude of the force, provided the motion is in the direction of the force. More generally, the work done is the distance times the coordinate of the force in the direction of the motion. Suppose a force $\mathbf{F}$ moves an object through a displacement $\mathbf{d}$ so the distance moved is $d = |\mathbf{d}|$. If $\mathbf{n}$ is the unit vector in the direction of $\mathbf{d}$, then $\mathbf{d} = d\mathbf{n}$ and the coordinate of $\mathbf{F}$ in the direction of $\mathbf{d}$ is $\mathbf{F} \cdot \mathbf{n}$. The work done is $(\mathbf{F} \cdot \mathbf{n})d = (\mathbf{F} \cdot d\mathbf{n}) = \mathbf{F} \cdot \mathbf{d}$.

EXAMPLE 6   The bottom of a snow-covered slope is at the origin, and the top at $(100, 20)$, with units measured in feet. Pulling a child's sled up the slope takes a force given by the vector $(8, 3)$, in units of pounds. The work done is $(100, 20) \cdot (8, 3) = 800 + 60 = 860$ foot-pounds.

We use the dot product to analyze the flow of fluid, or heat, or radiant energy described by a vector-valued function $\mathbf{v}(x,y,z)$ that gives the direction and magnitude of the flow, that is, the **flow velocity**, at each point (x, y, z). We consider here only the simplest kind of flow, in which $\mathbf{v}$ is a constant vector so that the flow is uniform along parallel lines.

For a given flow there will be a rate of flow per time unit through a surface in its path, which is called the **flux** through the surface. This is illustrated in Figure 1.25 showing a horizontally placed parallelogram P, and a flow vector $\mathbf{v}$ down and from the right. The vector $\mathbf{n}$ is a unit vector perpendicular to P. The shaded region indicates the volume flowing through P in one time unit, and is a solid B bounded by six parallelograms with horizontal top and bottom, with its other four edges of length $|\mathbf{v}|$ and parallel to $\mathbf{v}$. The volume $V(B)$ is the area $A(P)$ of its top times the vertical height h of B. Since $h = \mathbf{n} \cdot \mathbf{v}$, the coordinate of $\mathbf{v}$ in the direction of $\mathbf{n}$, the flux through the top parallelogram is $A(P)\mathbf{n} \cdot \mathbf{v}$. Note that the sign of the flux through P would change if we reversed the direction of $\mathbf{n}$.

Similar considerations apply to other planar figures R perpendicular to $\mathbf{n}$ and of area $A(R)$. If a flow velocity in $\mathbb{R}^3$ is the constant vector $\mathbf{v}$ at every point, the flow

FIGURE 1.25
Flow's flux equals $V(B)$.

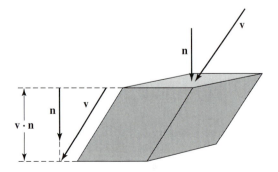

through R is $A(R)\mathbf{n} \cdot \mathbf{v}$. In Chapter 9, Section 3C we'll define the flux of a variable flow through a curved surface S that we approximate near each point of S by a tangent parallelogram.

EXAMPLE 7   We'll compute the flow of air through a window of area 10 square feet that is perpendicular to the vector $\mathbf{w} = 3\mathbf{i} - 4\mathbf{j}$ when the wind velocity (in feet per second) is $20\mathbf{i}$. Since $|\mathbf{w}| = 5$, we find $\mathbf{n} = \frac{1}{5}\mathbf{w}$ as the unit vector perpendicular to the plane of the window, so $A\mathbf{n} = 10\mathbf{n} = 2\mathbf{w} = 6\mathbf{i} - 8\mathbf{j}$. The flux through the window is then $20\mathbf{i} \cdot (6\mathbf{i} - 8\mathbf{j}) = 120$ cubic feet per second.

EXAMPLE 8   This example uses the dot product in a nongeometric context. Suppose a manufacturer produces four different models of widgets. We can write information about the models as vectors in $\mathbb{R}^4$, with each entry of a vector corresponding to one of the four models. Suppose unit production costs are given by a vector $\mathbf{c} = (2, 4, 5, 7)$, meaning that it costs 2 dollars to produce each model 1 widget, 4 dollars for each model 2 widget, and so on. Similarly let the unit wholesale prices be $\mathbf{w} = (3, 6, 7, 10)$ and retail prices be $\mathbf{r} = (5, 9, 11, 18)$. If $\mathbf{p} = (100, 30, 10, 5)$ gives the number of each model produced in a day and $\mathbf{s} = (80, 40, 8, 3)$ the number sold at retail, then the day's total manufacturing cost is $\mathbf{p} \cdot \mathbf{c} = (100)(2) + (30)(4) + (10)(5) + (5)(7) = 200 + 120 + 50 + 35 = 405$ dollars, and the retailer's gross income (before expenses) is $\mathbf{s} \cdot (\mathbf{r} - \mathbf{w}) = (80, 40, 8, 3) \cdot (2, 3, 4, 8) = 160 + 120 + 32 + 24 = 436$ dollars for the day.

EXERCISES

In Exercises 1 to 4, compute $\mathbf{x} \cdot \mathbf{y}$ for the given vectors.

1. $\mathbf{x} = (1, 3), \mathbf{y} = (-2, 4)$

2. $\mathbf{x} = (\sqrt{2}, \sqrt{3}), \mathbf{y} = (\sqrt{3}, \sqrt{2})$

3. $\mathbf{x} = (-1, -1, 2), \mathbf{y} = (1, 6, 1)$

4. $\mathbf{x} = (1, 2, 1, 3), \mathbf{y} = (0, 1, 2, 1)$

In Exercises 5 to 8, for the given vectors $\mathbf{u}$ and $\mathbf{v}$ find (a) $\mathbf{u} \cdot \mathbf{v}$, (b) $|\mathbf{u}|$ and $|\mathbf{v}|$, and (c) the angle between $\mathbf{u}$ and $\mathbf{v}$.

5. $\mathbf{u} = (1, 1), \mathbf{v} = (1, 0)$

6. $\mathbf{u} = (\sqrt{3}, 1), \mathbf{v} = (1, \sqrt{3})$

7. $\mathbf{u} = (2, 1, 2), \mathbf{v} = (1, 2, 2)$

8. $\mathbf{u} = (3, 1, 1), \mathbf{v} = (4, 1, 0)$

In Exercises 9 and 10, use the information and approximate coordinate system of Exercises 34 to 38 on page 17.

9. Find the angle between the position vectors of New York and Los Angeles, and find the approximate airline distance between the two cities.

10. Do the same as in Exercise 9, for New York and Paris.

11. Prove that the dot product has the additivity and homogeneity properties in Theorem 4.2.

12. Prove the homogeneity property of length listed in Theorem 4.9.

In Exercises 13 to 16, find (a) the coordinate and (b) the component of the vector $\mathbf{x}$ in the direction of the vector $\mathbf{v}$, and also the component of $\mathbf{x}$ perpendicular to $\mathbf{v}$.

13. $\mathbf{x} = (1, -1, 2), \mathbf{v} = (1/\sqrt{3}, 1/\sqrt{3}, 1/\sqrt{3})$

14. $\mathbf{x} = (1, -1, 2), \mathbf{v} = (2/7, -3/7, 6/7)$

15. $\mathbf{x} = (2, -3, 1), \mathbf{v} = (1, 3, -2)$

16. $\mathbf{x} = (-4, 0, -1), \mathbf{v} = (0, -3, 2)$

In Exercises 17 and 18, for the triangle with the given vertices A, B, C, find the lengths of its sides, and determine which of its angles are acute, obtuse, or right angles.

17. $A = (2, -3, 6), B = (1, 3, -2), C = (1, 7, 1)$

18. $A = (1, 2, 4), B = (-2, -1, 2), C = (4, 2, -3)$

19. (a) Show that for a vector $\mathbf{x}$ in $\mathbb{R}^3$,

$$\mathbf{x} = (\mathbf{x} \cdot \mathbf{e}_1)\mathbf{e}_1 + (\mathbf{x} \cdot \mathbf{e}_2)\mathbf{e}_2 + (\mathbf{x} \cdot \mathbf{e}_3)\mathbf{e}_3.$$

(b) If the vector $\mathbf{x}$ in part (a) is a unit vector, that is, a vector $\mathbf{u}$ of length 1, show that $\mathbf{u} \cdot \mathbf{e}_i = \cos\alpha_i$, where α_i is the angle between $\mathbf{u}$ and $\mathbf{e}_i$. The coordinates $\cos\alpha_i$ are called the **direction cosines** of $\mathbf{u}$ relative to the standard basis vectors $\mathbf{e}_i$. If $\mathbf{x}$ is a nonzero vector, the direction cosines of $\mathbf{x}$ are defined to be the direction cosines of the unit vector $\mathbf{x}/|\mathbf{x}|$.

(c) Find the direction cosines of $(1, 2, 1)$.

20. Find the direction cosines of $(6, -3, -2)$

21. Show that the standard basis vectors satisfy $\mathbf{e}_i \cdot \mathbf{e}_j = 0$ if $i \neq j$, and $\mathbf{e}_i \cdot \mathbf{e}_j = 1$ if $i = j$.

22. Show that if $\mathbf{x} \neq 0$, then the vector $(1/|\mathbf{x}|)\mathbf{x}$ has length 1.

23. A solar energy collector with area 15 square meters is mounted so that its panels are perpendicular to the vector $4\mathbf{i}+3\mathbf{k}$. At what rate does solar energy fall on the collector if the vector $\mathbf{i} + \mathbf{j} + 3\mathbf{k}$ gives the direction of the sun and the rate of energy falling on a surface perpendicular to the sun's rays is 80 watts per square meter? (Use the method of Example 7, treating radiation as a flow of energy.)

24. At what rate does solar energy fall on the collector in Exercise 23 later in the day when the direction of the sun is $-3\mathbf{j}+\mathbf{k}$?

25. A wind blowing from the northwest exerts a force of 15 pounds on a bicycle rider who follows a road that goes 400 feet west and then 500 feet in a direction $30°$ north of west. In a coordinate system in which $\mathbf{i}$ points east and $\mathbf{j}$ points north, find a vector representing the force of the wind and vectors representing the two parts of the road. Calculate the work that the rider does against the wind in cycling along each part. If there were a road running straight from the starting point to the finish, would taking it make a difference in the total work done?

26. Suppose that a factory produces each day four different items in amounts represented by the production vector $\mathbf{p} = (25, 25, 15, 10)$ and that these items are sold according to the wholesale dollar price vector $\mathbf{w} = (100, 150, 200, 300)$. What is the total revenue for the factory from selling all of each day's production?

27. Derive the inequality

$$|\mathbf{x}| - |\mathbf{y}| \leq |\mathbf{x} - \mathbf{y}|$$

from the triangle inequality, and then show that

$$\big||\mathbf{x}| - |\mathbf{y}|\big| \leq |\mathbf{x} - \mathbf{y}|.$$

This inequality is sometimes called the **reversed triangle inequality**.

28. Show that the sum of the squares of the lengths of the four sides of a parallelogram is equal to the sum of the squares of the lengths of the diagonals. [*Hint*: Write $|\mathbf{x} \pm \mathbf{y}|^2 = (\mathbf{x} \pm \mathbf{y}) \cdot (\mathbf{x} \pm \mathbf{y})$ and multiply out.]

29. Here is one way of proving the Cauchy-Schwarz inequality in $\mathbb{R}^n$ without appealing to geometry or trigonometry. Recall that if $b^2 - 4ac > 0$ and $a \neq 0$, then the quadratic equation $at^2 + bt + c = 0$ has two distinct real roots, and there are some values of t that make the expression $at^2 + bt + c$ negative. We suppose two vectors $\mathbf{x}$ and $\mathbf{y}$ are given in $\mathbb{R}^n$, and we want to show that $|(\mathbf{x} \cdot \mathbf{y})| \leq |\mathbf{x}||\mathbf{y}|$.

(a) Show that if either $\mathbf{x}$ or $\mathbf{y}$ is 0, then the inequality is true because both sides of it are 0. From now on we may assume that neither $\mathbf{x}$ nor $\mathbf{y}$ is 0.

(b) Using the properties 4.2 of the dot product, show that

$$|t\mathbf{x} + \mathbf{y}|^2 = (t\mathbf{x} + \mathbf{y}) \cdot (t\mathbf{x} + \mathbf{y})$$
$$= |\mathbf{x}|^2 t^2 + 2(\mathbf{x} \cdot \mathbf{y})t + |\mathbf{y}|^2$$
$$\geq 0 \quad \text{for all values of } t.$$

(c) Use the remark at the beginning of the problem to conclude that

$$4(\mathbf{x} \cdot \mathbf{y})^2 - 4|\mathbf{x}|^2|\mathbf{y}|^2 \leq 0.$$

(d) Derive the Cauchy–Schwarz inequality from (c).

Once the inequality is established, we know that for nonzero vectors in $\mathbb{R}^n$ the ratio $(\mathbf{x} \cdot \mathbf{y})/(|\mathbf{x}||\mathbf{y}|)$ is always between -1 and 1 and therefore equal to $\cos\theta$ for a unique angle θ, with $0 \leq \theta \leq \pi$. Since the angle between two vectors is measured in the plane containing the vectors, we use Equation 4.7 to define angle in $\mathbb{R}^n$.

30. If equality holds in the Cauchy–Schwarz inequality, that is, if $|(\mathbf{x} \cdot \mathbf{y})| = |\mathbf{x}||\mathbf{y}|$, then one of the vectors is a scalar multiple of the other. Prove this

(a) for vectors in $\mathbb{R}^2$ and $\mathbb{R}^3$ using Equation 4.5

(b) in general, using ideas from the proof outlined in Exercise 29

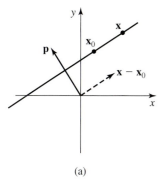

(a)

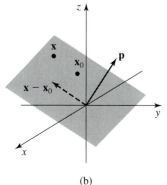

(b)

FIGURE 1.26
Line and plane.

SECTION 5   EUCLIDEAN GEOMETRY

A basic fact of analytic geometry is that the points in the plane whose coordinates (x, y) satisfy an equation of the form $ax + by = c$ lie on a straight line and that every straight line has such an equation. There is a similar correspondence between planes in $\mathbb{R}^3$ and equations of the form $ax + by + cz = d$. With the help of the dot product we'll find a geometric interpretation for the coefficients in these equations and extend the idea to higher dimensions.

5A   Equations for Lines and Planes

Suppose that $\mathbf{x}_0$ is a fixed point on a line in $\mathbb{R}^2$ or a plane in $\mathbb{R}^3$, and $\mathbf{p}$ is a vector perpendicular to the line or plane. Then a point $\mathbf{x}$ is on the line or plane if and only if the vectors $\mathbf{p}$ and $\mathbf{x} - \mathbf{x}_0$ are perpendicular. In other words, for every point $\mathbf{x}$ on the line or plane, we must have

5.1 $$\mathbf{p} \cdot (\mathbf{x} - \mathbf{x}_0) = 0.$$

Figure 1.26 shows the relation between these vectors in $\mathbb{R}^2$ and $\mathbb{R}^3$. In $\mathbb{R}^2$ we can write $\mathbf{p} = (a, b)$, $\mathbf{x} = (x, y)$, and $\mathbf{x}_0 = (x_0, y_0)$. Then Equation 5.1 becomes

$$(a, b) \cdot (x - x_0, y - y_0) = 0 \quad \text{or} \quad a(x - x_0) + b(y - y_0) = 0.$$

Letting $ax_0 + by_0 = d$ gives the standard equation for a line in $\mathbb{R}^2$, in which (a, b) is still a vector perpendicular to the line:

$$(a, b) \cdot (x, y) = (a, b) \cdot (x_0. y_0) \quad \text{or} \quad ax + by = d.$$

EXAMPLE 1 Find an equation of the line in $\mathbb{R}^2$ that is perpendicular to $\mathbf{p} = (1, 2)$ and passes through $(3, 4)$. The answer is

$$(1, 2) \cdot (x - 3, y - 4) = 0 \quad \text{or} \quad (x - 3) + 2(y - 4) = 0,$$

which simplifies to $x + 2y = 11$.

For a plane in $\mathbb{R}^3$, we let $\mathbf{p} = (a, b, c)$, $\mathbf{x} = (x, y, z)$, and $\mathbf{x}_0 = (x_0, y_0, z_0)$ in Equation 5.1 to get

$$(a, b, c) \cdot (x - x_0, y - y_0, z - z_0) = 0$$

or

$$a(x - x_0) + b(y - y_0) + c(z - z_0) = 0.$$

Letting $ax_0 + by_0 + cz_0 = d$ gives

$$ax + by + cz = d$$

for an equation of a plane perpendicular to (a, b, c).

FIGURE 1.27
Plane perpendicular to $(1, 1, 1)$.

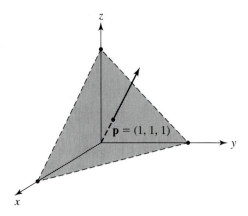

EXAMPLE 2

In $\mathbb{R}^3$ let $\mathbf{p} = (1, 1, 1)$ and $\mathbf{x}_0 = (1, 0, 0)$. The plane perpendicular to $\mathbf{p}$ and passing through $\mathbf{x}_0$ has equation

$$(1, 1, 1) \cdot (x - 1, y, z) = 0 \quad \text{or} \quad (x - 1) + y + z = 0$$

or

$$x + y + z = 1.$$

To get an idea of how the plane lies in $\mathbb{R}^3$, we can pick three points on the plane, for example $(1, 0, 0)$, $(0, 1, 0)$, and $(0, 0, 1)$, and sketch the triangle formed by them as in Figure 1.27. We can find points on a plane by picking values for two coordinates and solving the plane's equation for the third. In this example, we found the plane's intercepts, the points where the plane intersects the axes. Intercepts are easy to spot since they have two coordinates equal to zero.

The angle between two planes is the angle between vectors perpendicular to the planes, but as the next example shows, some care is needed in specifying the angle and choosing which way the vectors point.

EXAMPLE 3

The sides of a shallow trough lie in the planes with equations $-y + 5z = 0$ and $y + 5z = 0$, meeting on the x-axis as in Figure 1.28(a), which is a cross-sectional view showing the y- and z-axes, with the x-axis coming straight out from the page. What is the angle between the sides?

The vectors $\mathbf{p} = -\mathbf{j} + 5\mathbf{k}$ and $\mathbf{q} = \mathbf{j} + 5\mathbf{k}$ are normal to the sides. For the angle θ between $\mathbf{p}$ and $\mathbf{q}$ we have $\cos\theta = (\mathbf{p} \cdot \mathbf{q})/|\mathbf{p}||\mathbf{q}|$. Since $\mathbf{p} \cdot \mathbf{q} = 24$ and

FIGURE 1.28
Trough.

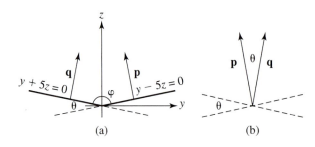

$|\mathbf{p}| = |\mathbf{q}| = \sqrt{26}$, $\cos\theta = 24/26 \approx 0.923$ and (from tables or a calculator) $\theta \approx 22.6°$. This looks reasonable for the angle between $\mathbf{p}$ and $\mathbf{q}$, as shown in Figure 1.28(b), but not for the angle between the sides of the trough, labeled φ in Figure 1.28(a). Instead, θ is the angle between one side of the trough and the extension of the plane containing the other, shown as a dotted line in the figure, and $\varphi = 180° - \theta \approx 157.4°$. Note that this is the angle between $\mathbf{p}$ and $-\mathbf{q}$, which could have been chosen as normals instead of $\mathbf{p}$ and $\mathbf{q}$. In the abstract, any one of $\pm\mathbf{p}$ and $\pm\mathbf{q}$ could be chosen as normals to the planes, and θ or φ could be taken as the angle between the planes; the appropriate choice in a concrete problem is best made with the help of a sketch.

FIGURE 1.29

Distance to line and plane.

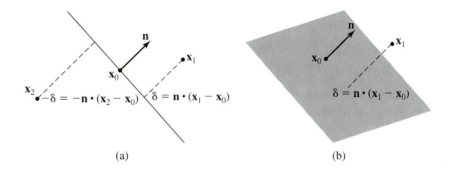

(a)   (b)

5B   Distance to a Line in $\mathbb{R}^2$ or a Plane in $\mathbb{R}^3$

The representation of a line or plane by an equation $\mathbf{p} \cdot (\mathbf{x} - \mathbf{x}_0) = 0$ is not unique, because another point on the line or plane can replace $\mathbf{x}_0$, and a nonzero scalar multiple of $\mathbf{p}$ can replace $\mathbf{p}$ without changing the set of points $\mathbf{x}$ that satisfy the equation. We can use some of this freedom to advantage by requiring the vector $\mathbf{p}$ to have length 1. We call the result a **normalized equation**, which then becomes $\mathbf{n} \cdot (\mathbf{x} - \mathbf{x}_0) = 0$, where $\mathbf{n} = \mathbf{p}/|\mathbf{p}|$. Alternatively we can write $\mathbf{n} \cdot \mathbf{x} - c = 0$ or $\mathbf{n} \cdot \mathbf{x} = c$ where $c = \mathbf{n} \cdot \mathbf{x}_0$. Using a normalized equation we can find the distance, measured perpendicularly, from a point to a line or plane. Figure 1.29 shows some examples of how this works for both lines and planes.

5.2   Theorem.   Let $\mathbf{n} \cdot (\mathbf{x} - \mathbf{x}_0) = 0$, or $\mathbf{n} \cdot \mathbf{x} - c = 0$, be a normalized equation of a line or plane P and let $\mathbf{x}_1$ be a point in the same space. Then the distance from $\mathbf{x}_1$ to P is

$$\delta = \mathbf{n} \cdot (\mathbf{x}_1 - \mathbf{x}_0) = \mathbf{n} \cdot \mathbf{x}_1 - c$$

if $\mathbf{x}_1$ is on the side of P where the tip of $\mathbf{n}$ is and is $-\delta$ if $\mathbf{x}_1$ is on the other side of P.

Proof.   Since $|\mathbf{n}| = 1$, we have

$$\delta = \mathbf{n} \cdot (\mathbf{x}_1 - \mathbf{x}_0) = |\mathbf{n}||\mathbf{x}_1 - \mathbf{x}_0| \cos\theta = |\mathbf{x}_1 - \mathbf{x}_0| \cos\theta,$$

where θ is the angle between $\mathbf{n}$ and the vector $\mathbf{x}_1 - \mathbf{x}_0$. If $0 \le \theta < \pi/2$, then δ is the length of the projection of $\mathbf{x}_1 - \mathbf{x}_0$ on a line parallel to $\mathbf{n}$; thus δ is the perpendicular distance from $\mathbf{x}_0$ to P. If $\pi/2 < \theta \le \pi$, we have $\cos\theta < 0$, so δ is negative and the actual length is $-\delta$. If $\theta = \pm\pi/2$ then $\mathbf{x}_0$ lies on P and $\delta = -\delta = 0$.   ∎

EXAMPLE 4

The line $(3, 4) \cdot (x, y) = 1$, or $3x + 4y = 1$, in $\mathbb{R}^2$ is perpendicular to $\mathbf{p} = (3, 4)$ of length 5, so the normalized equation is $\frac{3}{5}x + \frac{4}{5}y - \frac{1}{5} = 0$. The distance from $x_1 = (2, 3)$ to the line is then $\delta = \frac{6}{5} + \frac{12}{5} - \frac{1}{5} = \frac{17}{5}$. On the other hand, if $\mathbf{x}_1 = (1, -1)$ we find $\delta = \frac{3}{5} - \frac{4}{5} - \frac{1}{5} = -\frac{2}{5}$; this tells us that $(1, -1)$ is on the other side of the line from $(2, 3)$ at a distance of $\frac{2}{5}$.

EXAMPLE 5

The plane $x + y - 2z = -1$ is perpendicular to $\mathbf{p} = (1, 1, -2)$ of length $\sqrt{6}$, so the normalized equation is

$$\frac{1}{\sqrt{6}}x + \frac{1}{\sqrt{6}}y - \frac{2}{\sqrt{6}}z + \frac{1}{\sqrt{6}} = 0.$$

The distance from the origin $\mathbf{x}_1 = (0, 0, 0)$ to the plane is then $1/\sqrt{6}$. Note that if we put $\mathbf{x}_1 = (1, 1, 1)$ instead, we get the same result, so this point is on the same side of the plane as the origin and lies at the same distance from the plane.

EXERCISES

In Exercises 1 and 2, find an equation for the line described in $\mathbb{R}^2$, and sketch it.

1. Perpendicular to $\mathbf{e}_2$ in $\mathbb{R}^2$ and passing through $(2, 3)$

2. Perpendicular to $(2, -3)$ and passing through $(1, 1)$

In Exercises 3 and 4, find an equation for the plane described, and sketch it.

3. Perpendicular to $(1, 2, 4)$ and passing through $(-1, 0, 0)$

4. Perpendicular to $\mathbf{e}_2 - \mathbf{e}_3$ in $\mathbb{R}^3$ and passing through $(0, 1, 0)$

In Exercises 5 to 8, describe the point or set of points that the plane in $\mathbb{R}^3$ with equation $2x + 3y - z = 2$ has in common with the line having the given parameterization.

5. $(x, y, z) = t(1, -1, 4) + (1, 0, -2)$

6. $(x, y, z) = t(-1, 0, 1) + (-1, 1, -1)$

7. $(x, y, z) = t(-1, 1, 1) + (-1, 1, -1)$

8. $(x, y, z) = t(1, 1, 5) + (0, 0, -2)$

9. Find the point of intersection in $\mathbb{R}^3$ of the plane perpendicular to $(1, -1, 2)$ and passing through $(0, -1, 0)$, and the line parallel to $(1, 0, 1)$ and passing through $(1, 1, 1)$.

10. Find the point of intersection in $\mathbb{R}^3$ of the plane perpendicular to $(0, 2, -1)$ and passing through $(0, -1, 0)$, and the line passing through the points $(1, 2, 1)$ and $(-1, 3, 1)$.

In Exercises 11 and 12, find the cosine of the angle between the given lines in $\mathbb{R}^2$.

11. The lines with equations $x - 2y = 1$ and $2x + y = 3$

12. The line through $(0, 0)$ and $(1, 2)$ and the line through $(1, 2)$ and $(2, 3)$

13. Find the cosine of the angle between the planes $2x + y + z = 1$ and $x - y - z = -1$ in $\mathbb{R}^3$. [*Hint*: Look at vectors perpendicular to the planes.]

14. Find the cosine of the angle between the plane $2x + y + z = 1$ and a line parallel to $(1, 2, 1)$.

In Exercises 15 to 18, sketch the plane in $\mathbb{R}^3$ with the given equation.

15. $x + y - z = 1$

16. $2x - y + 3z = 0$

17. $y + 2z = 1$

18. $x - z = -1$

19. Find an equation for the plane parallel to the plane $3x - 2y + 5z = 2$ and passing through $(2, 1, 1)$.

20. Find a unit vector $\mathbf{n}$ perpendicular to the plane passing through $\mathbf{a} = (1, 0, 1)$, $\mathbf{b} = (2, 1, 0)$, and $\mathbf{c} = (1, 1, 1)$. [*Hint*: $\mathbf{n} \cdot (\mathbf{b} - \mathbf{a}) = 0$ and $\mathbf{n} \cdot (\mathbf{c} - \mathbf{a}) = 0$.]

21. Find an equation for the plane through
    (a) $(0, 0, 0)$, $(0, 1, 2)$, $(-1, 0, 1)$
    (b) $(1, -1, 1)$, $(1, 1, 1)$, $(1, 2, 1)$

22. Find an equation for the plane through the point $(2, 3, 5)$ and perpendicular to $(-1, -4, 1)$.

For each of the points and planes listed in Exercises 23 to 28, find the distance from the point to the plane or line. Is the point on the same side of the plane or line as the origin, or on the opposite side? Is the point above the plane or line, or below it (where "up" is the direction of the positive y-axis in $\mathbb{R}^2$ and the direction of the positive z-axis in $\mathbb{R}^3$)?

23. point $(2, -1)$; line $2x + y = 2$

24. point $(-1, 2)$; line $2x + y = 2$

25. point $(1, 0, -1)$; plane $(1, 1, 1) \cdot \mathbf{x} = 1$

26. point $(1, 1, 1)$; plane $(1, 1, 1) \cdot \mathbf{x} = 3$

27. point $(1, 0, -1)$; plane $x + 2y + 3z = 1$

28. point $(-2, 1, 0)$; plane $x - y + z = 2$

29. Let $ax + by + cz = d$ be the normalized equation of a plane in $\mathbb{R}^3$, so $a^2 + b^2 + c^2 = 1$; what is the distance from the plane to the origin?

SECTION 6   THE CROSS PRODUCT

The cross product is a construction defined only for vectors in $\mathbb{R}^3$ that's useful in a variety of geometric problems and also has applications in physics, especially in the study of electromagnetism. Our first use of it is as a convenient way to find a vector perpendicular to two given vectors, and Exercise 23 shows how the following formula arises naturally in trying to solve this problem.

We begin with the formal definition. The **cross product** of two vectors $\mathbf{u} = (u_1, u_2, u_3)$ and $\mathbf{v} = (v_1, v_2, v_3)$ in $\mathbb{R}^3$ is defined to be

6.1
$$\mathbf{u} \times \mathbf{v} = (u_2 v_3 - u_3 v_2, \; u_3 v_1 - u_1 v_3, \; u_1 v_2 - u_2 v_1).$$

As a help in remembering the formula, note that the pattern of subscripts in each component comes from the one before it by the substitutions $1 \to 2 \to 3 \to 1$. Formal computation with 2-by-2 and 3-by-3 determinants, treated more generally in Chapter 2, Section 5, makes the cross product easier to remember in the form

$$\mathbf{u} \times \mathbf{v} = \begin{vmatrix} \mathbf{i} & \mathbf{j} & \mathbf{k} \\ u_1 & u_2 & u_3 \\ v_1 & v_2 & v_3 \end{vmatrix} = \begin{vmatrix} u_2 & u_3 \\ v_2 & v_3 \end{vmatrix} \mathbf{i} - \begin{vmatrix} u_1 & u_3 \\ v_1 & v_3 \end{vmatrix} \mathbf{j} + \begin{vmatrix} u_1 & u_2 \\ v_1 & v_2 \end{vmatrix} \mathbf{k}.$$

EXAMPLE 1   The cross product of $\mathbf{u} = (1, -3, 2)$ and $\mathbf{v} = (2, 4, -5)$ is

$$\begin{vmatrix} \mathbf{i} & \mathbf{j} & \mathbf{k} \\ 1 & -3 & 2 \\ 2 & 4 & -5 \end{vmatrix} = \begin{vmatrix} -3 & 2 \\ 4 & -5 \end{vmatrix} \mathbf{i} - \begin{vmatrix} 1 & 2 \\ 2 & -5 \end{vmatrix} \mathbf{j} + \begin{vmatrix} 1 & -3 \\ 2 & 4 \end{vmatrix} \mathbf{k}$$

$$= (15 - 8)\mathbf{i} - (-5 - 4)\mathbf{j} + (4 + 6)\mathbf{k} = 7\mathbf{i} + 9\mathbf{j} + 10\mathbf{k}.$$

Working directly from the definition of cross product, the computation is less transparent:

$$(1, -3, 2) \times (2, 4, -5) = \big((-3)(-5) - (2)(4), \; (2)(2) - (1)(-5), \; (1)(4) - (-3)(2)\big)$$

$$= (15 - 8, \; 4 + 5, \; 4 + 6) = (7, 9, 10).$$

An important property of the cross product of $\mathbf{u}$ and $\mathbf{v}$ is that it's perpendicular to both $\mathbf{u}$ and $\mathbf{v}$ as shown in Figure 1.30(a). In other words,

6.2
$$\mathbf{u} \cdot (\mathbf{u} \times \mathbf{v}) = 0, \quad \text{and} \quad \mathbf{v} \cdot (\mathbf{u} \times \mathbf{v}) = 0.$$

FIGURE 1.30
Axis orientation.

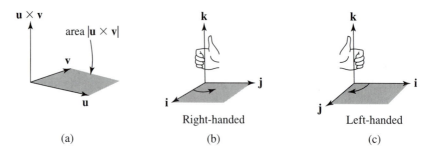

To check the first formula we write

$$\mathbf{u} \cdot (\mathbf{u} \times \mathbf{v}) = (u_1, u_2, u_3) \cdot (u_2v_3 - u_3v_2, \ u_3v_1 - u_1v_3, \ u_1v_2 - u_2v_1)$$

$$= u_1u_2v_3 - u_1u_3v_2 + u_2u_3v_1 - u_2u_1v_3 + u_3u_1v_2 - u_3u_2v_1.$$

Now observe that each term matches another with the opposite sign so that the sum is zero. A similar calculation shows that $\mathbf{v} \cdot (\mathbf{u} \times \mathbf{v}) = 0$.

Note that interchanging $\mathbf{u}$ and $\mathbf{v}$ exchanges the two terms in each coordinate entry of $\mathbf{u} \times \mathbf{v}$ and so has the effect of changing the sign of the entry. Hence the cross product is not commutative. Instead we have

6.3 $$\mathbf{v} \times \mathbf{u} = -\mathbf{u} \times \mathbf{v}.$$

EXAMPLE 2   We can illustrate Equation 6.2 by calculating

$$\mathbf{i} \times \mathbf{j} = (1, 0, 0) \times (0, 1, 0)$$

$$= \big((0)(0) - (0)(1), \ (0)(0) - (1)(0), \ (1)(1) - (0)(0)\big)$$

$$= 0\mathbf{i} + 0\mathbf{j} + 1\mathbf{k} = \mathbf{k}$$

and noting that $\mathbf{k}$ is perpendicular to both $\mathbf{i}$ and $\mathbf{j}$. Similar calculations give $\mathbf{j} \times \mathbf{k} = \mathbf{i}$ and $\mathbf{k} \times \mathbf{i} = \mathbf{j}$. From Equation 6.3 we then have $\mathbf{j} \times \mathbf{i} = -\mathbf{k}, \mathbf{k} \times \mathbf{j} = -\mathbf{i}$, and $\mathbf{i} \times \mathbf{k} = -\mathbf{j}$. Recall that we've already adopted the right-handed orientation for labelling axes shown in Figure 1.30(b). The algebra is the same regardless of what orientation we choose, but the picture would look different if we had chosen the left-handed orientation.

EXAMPLE 3   Let us find an equation for the plane that has parametric representation $\mathbf{x} = s\mathbf{u} + t\mathbf{v} + \mathbf{w}$ with $\mathbf{u} = (3, -1, 2), \mathbf{v} = (2, 5, -2)$ and $\mathbf{w} = (0, 0, 4)$. We need a vector $\mathbf{p}$ perpendicular to the plane. Since $\mathbf{u}$ and $\mathbf{v}$ are parallel to the plane, and $\mathbf{u} \times \mathbf{v}$ is perpendicular to both of them, we can take $\mathbf{p} = \mathbf{u} \times \mathbf{v} = ((-1)(-2) - (2)(5), (2)(2) - (3)(-2), (3)(5) - (-1)(2)) = (-8, 10, 17)$ to obtain a vector perpendicular to the plane. Then $\mathbf{p} \cdot \mathbf{x} = \mathbf{p} \cdot \mathbf{w}$, or $-8x + 10y + 17z = 68$, is the required equation.

EXAMPLE 4   Unless they are parallel or coincide, two planes in $\mathbb{R}^3$ intersect in a line. Here is how to find a parametric representation for the line of intersection. As an example, we take the planes with equations $2x - y + z = 10$ and $-3x + 2y - z = -7$, rewriting these equations as $\mathbf{p} \cdot \mathbf{x} = 10$ and $\mathbf{q} \cdot \mathbf{x} = -7$, where $\mathbf{p} = (2, -1, 1)$ and

$\mathbf{q} = (-3, 2, -1)$ are perpendicular to the first and second plane, respectively. The line of intersection lies in both planes and is therefore perpendicular to both $\mathbf{p}$ and $\mathbf{q}$, so we calculate

$$\mathbf{v} = \mathbf{p} \times \mathbf{q}$$
$$= ((-1)(-1) - (1)(2), (1)(-3) - (2)(-1), (2)(2) - (-1)(-3))$$
$$= (-1, -1, 1)$$

to get a vector with the direction of the line. We still need to find a point on the line. Usually it is simplest to find the point where a line meets one of the coordinate planes $x = 0$, $y = 0$, or $z = 0$. For instance, taking $x = 0$ in the equations for the planes gives $-y + z = 10$ and $2y - z = -7$, which we solve to give $y = 3$ and $z = 13$. The point $(0, 3, 13)$ is therefore on both planes and so on the line of intersection. We now know the direction of the line of intersection and a point on it, and have $t(-1, -1, 1) + (0, 3, 13)$ as a parametric representation for the line.

EXAMPLE 5   To find the point of intersection of the line $t(-1, 2, 4) + (0, -2, 3)$ with the plane through $(4, 2, 3)$, $(-2, 0, 1)$, and $(1, 3, -1)$, we first find an equation for the plane. The vectors $\mathbf{p} = (4, 2, 3) - (-2, 0, 1) = (6, 2, 2)$ and $\mathbf{q} = (4, 2, 3) - (1, 3, -1) = (3, -1, 4)$ are parallel to the plane, so $\mathbf{v} = \mathbf{p} \times \mathbf{q} = (10, -18, -12)$ is perpendicular to the plane. An equation for the plane is $\mathbf{v} \cdot \mathbf{x} = \mathbf{v} \cdot (4, 2, 3) = (10, -18, -13) \cdot (4, 2, 3) = -32$. For $\mathbf{x}$ on the line we have

$$\mathbf{v} \cdot \mathbf{x} = (10, -18, -12) \cdot (t(-1, 2, 4) + (0, -2, 3)) = -94t + 0,$$

and the point is also on the plane when $-94t = -32$, that is, when $t = \frac{16}{47}$. The point of intersection is then $\frac{16}{47}(-1, 2, 4) + (0, -2, 3) = (-\frac{16}{47}, -\frac{62}{47}, \frac{205}{47})$.

EXAMPLE 6   Suppose we want to find an equation for the plane parallel to the two vectors $\mathbf{x}_1 = (1, 2, -3)$ and $\mathbf{x}_2 = (2, 0, 1)$, and containing the point $\mathbf{x}_0 = (1, 1, 1)$. A vector $\mathbf{p}$ perpendicular to $\mathbf{x}_1$ and $\mathbf{x}_2$ is $\mathbf{x}_1 \times \mathbf{x}_2$:

$$\mathbf{p} = ((2)(1) - (-3)(0), (-3)(2) - (1)(1), (1)(0) - (2)(2))$$
$$= (2, -7, -4).$$

Writing $\mathbf{x} = (x, y, z)$, we require that

$$\mathbf{p} \cdot (\mathbf{x} - \mathbf{x}_0) = 0, \quad \text{that is,} \quad (2, -7, -4) \cdot (x - 1, y - 1, z - 1) = 0.$$

According to the definition of the dot product, this last equation is

$$2(x - 1) - 7(y - 1) - 4(z - 1) = 0, \quad \text{or} \quad 2x - 7y - 4z + 9 = 0.$$

The following formulas are sometimes useful in calculating with expressions involving cross products.

6.4   Let $\mathbf{u}$ and $\mathbf{v}$ be vectors in $\mathbb{R}^3$ and let r be a scalar. Then

$$\text{\textbf{Anticommutativity}:} \quad \mathbf{v} \times \mathbf{u} = -\mathbf{u} \times \mathbf{v}$$

$$\text{\textbf{Additivity}:} \quad \mathbf{u} \times (\mathbf{v} + \mathbf{w}) = \mathbf{u} \times \mathbf{v} + \mathbf{u} \times \mathbf{w}$$

$$(\mathbf{u} + \mathbf{v}) \times \mathbf{w} = \mathbf{u} \times \mathbf{w} + \mathbf{v} \times \mathbf{w}$$

$$\text{\textbf{Homogeneity}:} \quad r(\mathbf{u} \times \mathbf{v}) = (r\mathbf{u}) \times \mathbf{v} = \mathbf{u} \times (r\mathbf{v})$$

Proof.   Anticommutativity has already been justified as Equation 6.3. Proofs of the other properties also follow directly from definition 6.1 and are left as exercises. Note also that $(\mathbf{u} \times \mathbf{v}) \times \mathbf{w}$ is usually not equal to $\mathbf{u} \times (\mathbf{v} \times \mathbf{w})$, so the cross product is not associative. See Exercises 25 to 28.   ∎

We have already seen that $\mathbf{u} \times \mathbf{v}$ is perpendicular to both $\mathbf{u}$ and $\mathbf{v}$. Two more properties are needed to fully characterize the cross product geometrically. The first of these gives the geometric meaning of the length of the cross product, expressed in the formula

6.5 $$|\mathbf{u} \times \mathbf{v}| = A(P),$$

where $A(P)$ is the area of the parallelogram P that has arrows $\mathbf{u}$ and $\mathbf{v}$ as adjacent edges as shown in Figure 1.30(a). In other words, the length of the cross product of two vectors is equal to the area of the parallelogram having the vectors as adjacent edges. This property makes the cross product useful in computing areas and volumes in $\mathbb{R}^3$, and plays a role in defining the area $A(S)$ of a smooth surface in Chapter 9, Section 3B. A proof of Equation 6.5 in straightforward steps is outlined in Exercise 15. The area of the triangle with $\mathbf{u}$ and $\mathbf{v}$ as adjacent edges is half that of the parallelogram and is therefore equal to $\frac{1}{2}|\mathbf{u} \times \mathbf{v}|$.

EXAMPLE 7   We saw in the previous example that the cross product of the vectors $\mathbf{x}_1 = (1, 2, -3)$ and $\mathbf{x}_2 = (2, 0, 1)$ is the vector $(2, -7, -4)$. Hence the area of the parallelogram P with edges $\mathbf{x}_1$ and $\mathbf{x}_2$ is

$$A(P) = |(2, -7, -4)| = \sqrt{2^2 + (-7)^2 + (-4)^2} = \sqrt{69} \approx 8.3,$$

while the triangle with these edges, that is, the triangle with vertices at the origin, $(2, 0, 1)$, and $(2, -7, -4)$, has area $\frac{1}{2}\sqrt{69}$, chosen to be consistent with our choice of right-hand coordinate axes orientation.

Perpendicularity to $\mathbf{u}$ and $\mathbf{v}$ and having length given by Equation 6.5 isn't enough to characterize $\mathbf{u} \times \mathbf{v}$ completely, because $-\mathbf{u} \times \mathbf{v}$ has the same properties. The choice between the two possible directions of $\mathbf{u} \times \mathbf{v}$ relative to the pair $(\mathbf{u}, \mathbf{v})$ is settled by the following rule:

6.6   Right-Hand Rule for the Cross Product.   The vector $\mathbf{u} \times \mathbf{v}$ points in the direction of the thumb when the fingers of the right hand curl from $\mathbf{u}$ to $\mathbf{v}$.

This rule is illustrated in Figure 1.31(b). If you look at the plane containing $\mathbf{u}$ and $\mathbf{v}$ from the side away from which $\mathbf{u} \times \mathbf{v}$ points, then it takes a counterclockwise rotation of less than $180°$ to rotate $\mathbf{u}$ to point in the same direction as $\mathbf{v}$.

FIGURE 1.31

Orientation of **u**, **v**, **w**.

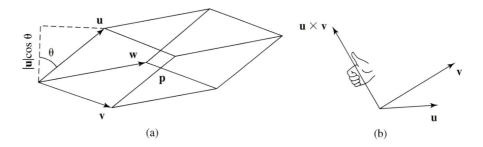

(a)   (b)

EXAMPLE 8   We have already computed that for the unit coordinate vectors, $\mathbf{i} \times \mathbf{j} = \mathbf{k}$. If you hold your right hand with the thumb pointing up (in the direction of **k**), then its fingers naturally curl in the counterclockwise direction taking **i** to **j** as in Figure 1.30(b). Equivalently, if you look down at the xy-plane from above, it takes a positive (counterclockwise) rotation to carry **i** to **j**.

The choice of the right-hand rule for the cross product instead of a left-hand rule is an arbitrary convention, but we have already made that choice in drawing the coordinate axes as we did in Figure 1.4 in Section 1. If we had used the left-hand rule for orienting the coordinate axes, then the cross product would obey the left-hand rule shown in Figure 1.30(c).

The cross product is also linked to volumes. Three vectors **u**, **v**, **w** in $\mathbb{R}^3$ that don't lie in a plane determine a solid region B with **u**, **v**, **w** for adjacent edges; B is called a **parallelepiped** because it's bounded by three pairs of congruent parallelograms, illustrated in Figure 1.31(a). Each edge of B is parallel to one of the vectors **u**, **v**, **w**, and B looks like a lop-sided box. The **scalar triple product** of **u**, **v**, and **w** in that order, is defined to be $\mathbf{u} \cdot (\mathbf{v} \times \mathbf{w})$. We'll show that this number equals either plus or minus the volume $V(B)$ of B, where B is the parallelepiped determined by **u**, **v**, and **w**. The precise statement is:

6.7   Theorem.   Let B be the parallelepiped with the vectors $\mathbf{u}, \mathbf{v}, \mathbf{w}$ as adjacent edges. Then $\mathbf{u} \cdot (\mathbf{v} \times \mathbf{w}) = V(B)$ if the three vectors obey the right-hand rule and otherwise $\mathbf{u} \cdot (\mathbf{v} \times \mathbf{w}) = -V(B)$.

Proof.   To verify Theorem 6.7, note that $\mathbf{u} \cdot (\mathbf{v} \times \mathbf{w}) = |\mathbf{v} \times \mathbf{w}||\mathbf{u}| \cos \theta$, where θ is the angle between **u** and $\mathbf{v} \times \mathbf{w}$ as in Figure 1.31(a). We take the base of B to be the parallelogram P with area $A(P) = |\mathbf{v} \times \mathbf{w}|$ determined by **v** and **w**. The figure shows that the vertical height of B is $|\mathbf{u}|| \cos \theta|$. Hence the volume of B equals the triple product if $\pi/2 \le \theta \le \pi$ and equals minus the triple product if $0 \le \theta \le \pi/2$. The sign of the scalar triple product is positive when $0 < \theta < \frac{\pi}{2}$, that is, when **u** and $\mathbf{v} \times \mathbf{w}$ are on the same side of the plane containing **v** and **w**. Since **v**, **w**, $\mathbf{v} \times \mathbf{w}$ is a right-handed system, this happens just when **u**, **v**, **w** is also a right-handed system. When **u**, **v**, **w** is a left-handed system, **u** and $\mathbf{v} \times \mathbf{w}$ are on opposite sides of the plane and $\frac{\pi}{2} < \theta \le \pi$, so $\cos \theta < 0$ and the triple product is negative.   ∎

EXAMPLE 9   Let $\mathbf{u} = (1, 1, 1), \mathbf{v} = (1, 2, -3), \mathbf{w} = (2, 1, 1)$. The triple product of these three vectors is $(1, 1, 1) \cdot ((1, 2, -3) \times (2, 1, 1)) = (1, 1, 1) \cdot (5, -7, -3) = -5$, so the volume of the parallelepiped B determined by the vectors is $V(B) = 5$. Since the sign of the triple product is negative, **u**, **v**, and **w** form a left-handed system.

EXAMPLE 10	In the discussion before Example 7 of Section 4 we defined the flux of a flow with constant velocity vector $\mathbf{v}$. There we found the flux across a parallelogram P with unit normal vector $\mathbf{n}$ to be $\Phi = A(P)\mathbf{n} \cdot \mathbf{v}$, where $A(P)$ is the area of P. We now see that this flux is a scalar triple product: $\Phi = \mathbf{v} \cdot (\mathbf{p} \times \mathbf{q})$, where the vectors $\mathbf{p}$ and $\mathbf{q}$ represent adjacent edges of P. The reason is that $A(P) = |\mathbf{p} \times \mathbf{q}|$ and $\mathbf{n} = (\mathbf{p} \times \mathbf{q})/|\mathbf{p} \times \mathbf{q}|$, so Φ is the scalar triple product of $\mathbf{v}$, $\mathbf{p}$, and $\mathbf{q}$:

$$\Phi = A(P)\mathbf{n} \cdot \mathbf{v} = |\mathbf{p} \times \mathbf{q}| \frac{\mathbf{p} \times \mathbf{q}}{|\mathbf{p} \times \mathbf{q}|} \cdot \mathbf{v} = \mathbf{v} \cdot (\mathbf{p} \times \mathbf{q}).$$

We use this formula to generalize flux to nonconstant flows across curved surfaces in Chapter 9, Section 3C.

If you choose three vectors $\mathbf{u}$, $\mathbf{v}$, $\mathbf{w}$ in $\mathbb{R}^3$ that form a right-handed system and test $\mathbf{v}$, $\mathbf{w}$, $\mathbf{u}$ and $\mathbf{w}$, $\mathbf{u}$, $\mathbf{v}$ by the right-hand rule, you'll find that they are also right-handed systems. Consequently,

6.8 $$\mathbf{u} \cdot (\mathbf{v} \times \mathbf{w}) \;=\; \mathbf{v} \cdot (\mathbf{w} \times \mathbf{u}) \;=\; \mathbf{w} \cdot (\mathbf{u} \times \mathbf{v}).$$

In other words, permuting $\mathbf{u}, \mathbf{v}, \mathbf{w}$ cyclically so $\mathbf{u} \to \mathbf{v} \to \mathbf{w} \to \mathbf{u}$ doesn't change their scalar triple product. But interchanging any two of the vectors changes the sign of the triple product because that changes the sign of the cross product in the triple product. Exercise 17 asks you to check Equation 6.8 algebraically, though we'll see also in Chapter 2, Section 5 that this follows from a basic property of determinants.

EXERCISES

In Exercises 1 to 4, find the cross product of $\mathbf{u}$ and $\mathbf{v}$, and sketch the three vectors $\mathbf{u}$, $\mathbf{v}$, and $\mathbf{u} \times \mathbf{v}$

1. $\mathbf{u} = \mathbf{e}_2, \mathbf{v} = \mathbf{e}_1$

2. $\mathbf{u} = \mathbf{j}, \mathbf{v} = -\mathbf{k}$

3. $\mathbf{u} = (0, 1, 2), \mathbf{v} = (-1, 0, 1)$

4. $\mathbf{u} = (\sqrt{2}, 1, \sqrt{3}), \mathbf{v} = (\sqrt{2}, \sqrt{2}, 1)$

In Exercises 5 and 6, find the area of the parallelogram with $\mathbf{u}$ and $\mathbf{v}$ as adjacent edges.

5. $\mathbf{u} = \mathbf{i}, \mathbf{v} = \mathbf{j} + \mathbf{k}$

6. $\mathbf{u} = (-1, 0, 0), \mathbf{v} = (0, -1, 0)$

In Exercises 7 and 8, find the area of the triangle with $\mathbf{u}$ and $\mathbf{v}$ as adjacent edges.

7. $\mathbf{u} = (3, -1, 2), \mathbf{v} = (-1, 0, 1)$

8. $\mathbf{u} = (1, 1, 1), \mathbf{v} = (1, 2, 1)$

In Exercises 9 and 10, use the cross product to find an equation of the form $ax + by + cz = d$ for the plane parallel to both $\mathbf{u}$ and $\mathbf{v}$ that contains the point $(-1, -1, 1)$.

9. $\mathbf{u} = (1, 2, 4), \mathbf{v} = (4, 2, 1)$

10. $\mathbf{u} = (-3, 0, 1), \mathbf{v} = (0, 1, -4)$

11. Verify that (a) $\mathbf{i} \times \mathbf{j} = \mathbf{k}$, (b) $\mathbf{j} \times \mathbf{k} = \mathbf{i}$, (c) $\mathbf{k} \times \mathbf{i} = \mathbf{j}$, and (d) $\mathbf{u} \times \mathbf{u} = 0$ for all $\mathbf{u}$ in $\mathbb{R}^3$.

12. Verify properties 6.4(b) and 6.4(c) for arbitrary vectors $\mathbf{u}, \mathbf{v}, \mathbf{w}$ and scalar r.

13. The triangle with vertices $(-2, 1, 0), (2, 3, 0), (2, -1, 0)$ forms the base of an irregular pyramid with apex at $(0, 0, 2)$.

(a) Make a sketch of the pyramid.

(b) Find a vector perpendicular to each of the three sides.

(c) Find the cosine of the angle between each of the three sides and the base.

14. Verify using coordinates the second of Equations 6.2 of the text: $\mathbf{v} \cdot (\mathbf{u} \times \mathbf{v}) = 0$.

*15. (a) Verify by direct coordinate computation that

$$|\mathbf{u} \times \mathbf{v}|^2 = |\mathbf{u}|^2|\mathbf{v}|^2 - (\mathbf{u} \cdot \mathbf{v})^2.$$

(b) Use the result of part (a) to show that

$$|\mathbf{u} \times \mathbf{v}| = |\mathbf{u}||\mathbf{v}|\sin\theta,$$

where θ is the angle between $\mathbf{u}$ and $\mathbf{v}$ that satisfies $0 \le \theta \le \pi$.

(c) Show that $|\mathbf{u}||\mathbf{v}|\sin\theta$ is the area $A(P)$ of the parallelogram with edges $\mathbf{u}$ and $\mathbf{v}$, and hence by part (b) that the length of $\mathbf{u} \times \mathbf{v}$ is equal to $A(P)$, as stated in Equation 6.5 of the text.

16. Compute the volume of the parallelepiped with adjacent edges $(2, 1, 3)$, $(-1, -2, 4)$, $(3, 3, 2)$.

17. Verify Equation 6.8 algebraically by writing out the product in terms of coordinates for three vectors $\mathbf{u},\mathbf{v},\mathbf{w}$.

18. Explain geometrically why Equation 6.8 is consistent with Theorem 6.7.

In Exercises 19 and 20, find the area of the parallelogram in $\mathbb{R}^3$ with the given edges and make a sketch of it.

19. $(1, 1, 0)$, $(0, 1, 2)$ **20.** $(0, 1, 2)$, $(-3, 5, -1)$

21. Find the volume of the parallelepiped B in $\mathbb{R}^3$ with edges $(1, 1, 0)$, $(0, 1, 2)$, and $(-3, 5, -1)$. Make a sketch of B.

22. If $\mathbf{u} = (2, 1, 3)$, $\mathbf{v} = (0, 2, 1)$, and $\mathbf{w} = (1, 1, 1)$, compute
(a) $\mathbf{u} \times \mathbf{v}$ **(b)** $\mathbf{u} \cdot (\mathbf{v} \times \mathbf{w})$
(c) $(\mathbf{u} \times \mathbf{v}) \cdot \mathbf{w}$ **(d)** $(\mathbf{u} \times \mathbf{v}) \times \mathbf{w}$
(e) $\mathbf{u} \times (\mathbf{v} \times \mathbf{w})$ **(f)** $(\mathbf{u} \cdot \mathbf{v})\mathbf{w}$

23. This exercise shows how the formula for the cross product comes up naturally if you try directly to find a vector perpendicular to two given vectors. Let $\mathbf{u} = (u_1, u_2, u_3)$ and $\mathbf{v} = (v_1, v_2, v_3)$ be nonparallel vectors in $\mathbb{R}^3$. To find a vector $\mathbf{x} = (x, y, z)$ perpendicular to $\mathbf{u}$ and $\mathbf{v}$, we want $\mathbf{u} \cdot \mathbf{x} = \mathbf{v} \cdot \mathbf{x} = 0$, that is,

$$u_1 x + u_2 y + u_3 z = 0$$

$$v_1 x + v_2 y + v_3 z = 0.$$

(a) Multiply the first equation by v_1 and the second by u_1 and subtract to get

$$(u_2 v_1 - u_1 v_2)y = (u_1 v_3 - u_3 v_1)z.$$

(b) Do a calculation similar the one in part (a) to get

$$(u_1 v_2 - u_2 v_1)x = (u_2 v_3 - u_3 v_2)z.$$

(c) Use (a) and (b) to express x and y in terms of z, and choose a value for z that avoids fractions in the result. Compare the triple (x, y, z) that you obtain with $\mathbf{u} \times \mathbf{v}$.

Sir William Rowan Hamilton introduced the terms *scalar* and *vector* in 1846 in connection with his invention of **quaternions**, which he defined to be expressions of the form $q = a + b\mathbf{i} + c\mathbf{j} + d\mathbf{k}$, where $a, b, c,$ and d are real numbers. He called a the scalar part of q, representing a single magnitude on the scale of real numbers, and called $b\mathbf{i} + c\mathbf{j} + d\mathbf{k}$ the vector part, representing a line segment with both magnitude and direction in $\mathbb{R}^n$. The word *vector* came from astronomy; in the eighteenth century the line from the sun to a planet was called the planet's radius vector. Hamilton defined the algebraic operations on quaternions so they became an extension of those for complex numbers, adding them as linear combinations of the basis $\{1, \mathbf{i}, \mathbf{j}, \mathbf{k}\}$ and defining the product of two quaternions by multiplying out assuming the distributive law and then simplifying using the rules

$$\mathbf{i}^2 = \mathbf{j}^2 = \mathbf{k}^2 = -1, \quad \mathbf{ij} = \mathbf{k} = -\mathbf{ij},$$

$$\mathbf{jk} = \mathbf{i} = -\mathbf{kj}, \quad \mathbf{ki} = \mathbf{j} = -\mathbf{ik}$$

for multiplying $\mathbf{i}$, $\mathbf{j}$, and $\mathbf{k}$. Note that quaternion multiplication is not commutative.

24. Show that if $q_1 = s_1 + \mathbf{v}_1$ and $q_2 = s_2 + \mathbf{v}_2$ are two quaternions with scalar parts s_1, s_2 and vector parts $\mathbf{v}_1, \mathbf{v}_2$, then their product is the quaternion

$$q_1 q_2 = (s_1 s_2 - \mathbf{v}_1 \cdot \mathbf{v}_2) + (s_1 \mathbf{v}_2 + s_2 \mathbf{v}_1 + \mathbf{v}_1 \times \mathbf{v}_2).$$

Thus the product of two quaternions $\mathbf{v}_1,\mathbf{v}_2$ with zero scalar parts yields scalar part $-\mathbf{v}_1 \cdot \mathbf{v}_2$ and vector part $\mathbf{v}_1 \times \mathbf{v}_2$. Quaternions fell from favor among physical scientists after Josiah Willard Gibbs later introduced the more convenient dot and cross products.

The remaining exercises for this section deal with associativity of multiplication, which fails in general for the cross product but holds for quaternion products.

25. The associative law for the cross product holds only for some choices of vectors; verify this by comparing the following products.
(a) $(\mathbf{i} \times \mathbf{i}) \times \mathbf{j}$ and $\mathbf{i} \times (\mathbf{i} \times \mathbf{j})$
(b) $(\mathbf{i} \times \mathbf{j}) \times \mathbf{i}$ and $\mathbf{i} \times (\mathbf{j} \times \mathbf{i})$
(c) $(\mathbf{i} \times \mathbf{j}) \times \mathbf{k}$ and $\mathbf{i} \times (\mathbf{j} \times \mathbf{k})$
(d) $(\mathbf{i} \times \mathbf{i}) \times \mathbf{i}$ and $\mathbf{i} \times (\mathbf{i} \times \mathbf{i})$

*26. Let $\mathbf{u}$, $\mathbf{v}$, and $\mathbf{w}$ be arbitrary vectors in $\mathbb{R}^3$.
(a) Using geometric properties of the cross product, show that there are scalars a and b such that $\mathbf{u} \times (\mathbf{v} \times \mathbf{w}) = a\mathbf{v} + b\mathbf{w}$.

(b) By taking the dot product of both sides of the equation in part(a) with **u**, show that $a(\mathbf{u} \cdot \mathbf{v}) = -b(\mathbf{u} \cdot \mathbf{w})$.

(c) Verify using coordinates that
$$\mathbf{u} \times (\mathbf{v} \times \mathbf{w}) = (\mathbf{u} \cdot \mathbf{w})\mathbf{v} - (\mathbf{u} \cdot \mathbf{v})\mathbf{w}.$$

*27. Show that the cross product is associative for three vectors **u**, **v**, **w** in that order, that is, $\mathbf{u} \times (\mathbf{v} \times \mathbf{w}) = (\mathbf{u} \times \mathbf{v}) \times \mathbf{w}$, if and only if **v** and $\mathbf{u} \times \mathbf{w}$ are linearly dependent. [*Hint*: Use part (c) of Exercise 26 and other properties of the cross product to find a relation between $\mathbf{u} \times (\mathbf{v} \times \mathbf{w}) - (\mathbf{u} \times \mathbf{v}) \times \mathbf{w}$ and $\mathbf{v} \times (\mathbf{u} \times \mathbf{w})$.]

*28. Use the results of Exercise 24, part (c) of Exercise 26, and Equation 6.8 to show that quaternion multiplication is associative, so $q_1(q_2 q_3) = (q_1 q_2)q_3$ for three quaternions q_1, q_2, q_3.

Chapter 1 REVIEW

Exercises 1 to 4 refer to vectors $\mathbf{a} = \mathbf{i} + \mathbf{j} + \mathbf{k}$, $\mathbf{b} = \mathbf{i} + 2\mathbf{j} + 2\mathbf{k}$, and $\mathbf{c} = 2\mathbf{i} + 3\mathbf{j} + 6\mathbf{k}$.

1. Which is longer, 4**a** or **c**?

2. In the triangle with vertices **a**, **b**, and **c**, which side is longest?

3. Express $\mathbf{b} - \mathbf{a}$ and $\mathbf{c} - \mathbf{b} - \mathbf{a}$ in terms of **i**, **j**, and **k**.

4. Express **k**, **j**, and **i** in terms of **a**, **b**, and **c**.

In Exercises 5 to 8, express the first vector given as a linear combination of the other vectors, or show that it's impossible to do so.

5. $(2, -1, 3, 2)$; $\mathbf{e}_1, \mathbf{e}_2, \mathbf{e}_3, \mathbf{e}_4$

6. $(1, 2)$; $(2, 3)$, $(4, 6)$, $(-6, -9)$

7. $(4, 1, -2)$; $(1, 2, 3)$, $(6, 5, 4)$

8. $(3, 1, -2)$; $(1, 2, 3)$, $(6, 5, 4)$, $(5, 3, 1)$

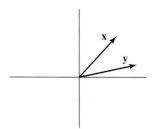

FIGURE 1.32

9. Copy Figure 1.32 and then draw arrows representing (a) $\mathbf{x} + 2\mathbf{y}$, (b) $\mathbf{x} - \mathbf{y}$, and (c) $2\mathbf{x} - 3\mathbf{y}$. Label the arrows with (a), (b), and (c) to show which is which.

10. Let points **a** and **b** be position vectors of diagonally opposite vertices of a parallelogram and let **c** be the position vector of a third vertex. Express the position vector of the fourth vertex in terms of **a**, **b**, and **c**.

11. Find representations $s\mathbf{u} + \mathbf{a}$ and $t\mathbf{v} + \mathbf{b}$ for the line through $(1, 2)$ and $(2, 1)$, and the line through $(4, 5)$ and $(-1, -2)$, and find where the lines intersect.

12. Show that for a pair of points **a** and **b**, the points $\mathbf{p} = \frac{1}{3}\mathbf{a} + \frac{2}{3}\mathbf{b}$ and $\mathbf{q} = \frac{2}{3}\mathbf{a} + \frac{1}{3}\mathbf{b}$ are on the line segment joining **a** and **b**, and divide it into three equal parts.

13. Find a vector function of t that describes a particle moving along the line through $(1, 2, 3)$ and $(-5, 3, 4)$ at unit speed.

14. Find a vector function of t that describes a particle moving in a straight line at constant speed that passes through $(2, -3)$ when $t = 1$ and through $(5, 4)$ when $t = 3$. What is the speed of the motion?

15. Suppose two boats start out at positions $\mathbf{u}_1$ and $\mathbf{u}_2$ when $t = 0$ and maintain constant velocities $\mathbf{v}_1$ and $\mathbf{v}_2$. What functions $\mathbf{p}_1(t)$ and $\mathbf{p}_2(t)$ give their positions at time t? Let $\mathbf{d}(t)$ be the vector displacement from the first boat to the second as a function of t.

(a) Show that if the boats are on a collision course, then the direction of $\mathbf{d}(t)$ doesn't change with time.

(b) Suppose the direction of $\mathbf{d}(t)$ doesn't change with time. When will collision occur if it does? Under what circumstances will it not occur?

16. Consider airplanes moving in three dimensions instead of boats moving on a two-dimensional water surface as in Exercise 15. Does this make a difference in the answers to (a) or (b)?

17. Here are descriptions of four lines K, L, M, and N in the plane:

K: the line through the points $(3, 4)$ and $(-2, 3)$

L: the line with parametric representation $t(\mathbf{i} + 2\mathbf{j}) + \mathbf{i} - 3\mathbf{j}$

M: the line through the point $(8, 5)$ parallel to the vector $5\mathbf{i} + \mathbf{j}$

N: the line through the origin and the point $(2, 4)$

Which of the lines are the same? Which of the lines are parallel?

18. Here are descriptions of four lines K, L, M, and N in $\mathbb{R}^3$:

K: the line through the points $(1, 2, 3)$ and $(4, -5, 6)$.

L: the line with parametric representation $t(3\mathbf{i} - 7\mathbf{j} + 3\mathbf{k}) - 2\mathbf{i} + 9\mathbf{j}$.

M: the line through the point $(0, 1, 2)$ parallel to the vector $3\mathbf{i} + \mathbf{k}$.

N: the line through the origin and the point $(1, 2, 3)$.

Which of the lines are the same? Which of the lines are parallel?

In Exercises 19 to 22, find a parametric representation for the given plane.

19. the plane containing $(3, 0, 0)$, $(0, 2, 0)$, and $(0, 0, 5)$.

20. the plane parallel to the one in Exercise 19 and containing $(0, -1, 3)$.

21. the plane parallel to the x- and y-axes and containing $(1, 2, 3)$.

22. the plane containing the origin, $(1, 2, 3)$, and $(-2, -3, 1)$.

For each set of three points given in Exercises 23 to 26, determine whether the points lie on a line. If they do, find a parametric representation of the line. If they do not, find a parametric representation of the plane in which they lie.

23. $(3, 1, 2)$, $(0, 0, 0)$, $(-6, -2, -4)$

24. $(1, 1, 0)$, $(0, 1, 1)$, $(1, 0, 1)$

25. $(1, 2, 3)$, $(3, 2, 1)$, $(4, 2, 2)$

26. $(1, 2, -3)$, $(10, 5, -9)$, $(-5, 0, 1)$

27. Let $\mathbf{a} = \mathbf{i} + \mathbf{j} + \mathbf{k}$, $\mathbf{b} = \mathbf{i} + 2\mathbf{j} + 2\mathbf{k}$, $\mathbf{c} = 2\mathbf{i} + 3\mathbf{j} + 6\mathbf{k}$, as in Exercises 1 to 4.
 (a) Which is larger, the angle between $\mathbf{a}$ and $\mathbf{b}$ or the angle between $\mathbf{a}$ and $\mathbf{c}$?
 (b) Find a nonzero vector perpendicular to both $\mathbf{a}$ and $\mathbf{b}$.
 (c) What is the area of the triangle whose vertices are the origin and the points $\mathbf{a}$ and $\mathbf{b}$?
 (d) What is the area of the triangle whose vertices are the points $\mathbf{a}$, $\mathbf{b}$, and $\mathbf{c}$?

28. Express $\mathbf{i} + 3\mathbf{j} - 2\mathbf{k}$ as the sum of
 (a) a vector parallel to $2\mathbf{i} - 6\mathbf{j} - 3\mathbf{k}$ and a vector perpendicular to it

(b) a vector in the plane $x - y + 4z = 0$ and a vector perpendicular to it

29. A force $\mathbf{F} = \mathbf{i} + 3\mathbf{j} - 2\mathbf{k}$ drags an object from $(1, 2, 3)$ to $(4, 5, 0)$. Find the work done.

30. If the wind velocity is $10\mathbf{i} + 20\mathbf{j}$ (in feet per second),
 (a) What is the speed of the wind?
 (b) How much air blows through a triangular opening with vertices at $(-2, 2, 0)$, $(3, -4, 0)$, and $(0, 0, 5)$ in one second? (Coordinates are in feet.)

31. Let L be the line $(2, 3) \cdot \mathbf{x} = 6$.
 (a) Find the distance of the origin from L.
 (b) Find a parametric representation for L.
 (c) Find all vectors of length 4 that are perpendicular to L.

32. Let P be the plane $(1, 1, 1) \cdot \mathbf{x} = 3$.
 (a) Find the distance of the origin from P.
 (b) Find a parametric representation for P.
 (c) Find all vectors of length 4 that are perpendicular to P.

33. Let L be the line $t(2, 1) + (-2, 0)$.
 (a) Find a parametric representation for the line K that is perpendicular to L and passes through the origin.
 (b) Find a unit vector $\mathbf{n}$ and number c such that L has the equation $\mathbf{n} \cdot \mathbf{x} = c$.
 (c) What is the distance from L to $(3, 5)$? Is $(3, 5)$ on the same side of L as the origin?

34. Let P be the plane $s(3, 2, 1) + t(2, 1, 1) + (-2, 0, 1)$.
 (a) Find a parametric representation for the line L that is perpendicular to P and passes through the origin.
 (b) Find a unit vector $\mathbf{n}$ and number c such that P has the equation $\mathbf{n} \cdot \mathbf{x} = c$.
 (c) What is the distance from P to $(3, 5, -2)$? Is $(3, 5, -2)$ on the same side of P as the origin?

35. Find a parametric representation for the line of intersection of the planes with equations $x + y + z = 3$ and $2x - y - z = 5$, and find the point where the line intersects the plane with equation $x - y = 2$.

CHAPTER 2

EQUATIONS AND MATRICES

Many specific problems about vectors and their applications require solving systems of first-degree equations. This chapter is mainly about how to solve such systems, but also features applications and an optional look in Section 2D at geometric interpretation of the solutions using the vector geometry of Chapter 1. In practice, the matrix methods introduced in Section 2 of this chapter often provide the most efficient way to do the necessary computations.

SECTION 1   SYSTEMS OF LINEAR EQUATIONS

In this section we look at examples of systems of linear equations and find their solutions in a systematic way that foreshadows the matrix method to be introduced in Section 2. We start with a **linear system** of a type that occurs repeatedly in the rest of the book:

$$a_{11}x + a_{12}y + a_{13}z = b_1$$
$$a_{21}x + a_{22}y + a_{23}z = b_2,$$
$$a_{31}x + a_{32}y + a_{33}z = b_3,$$

where the a_{ij} and b_j are given numbers. All linear systems of equations have this form, though the number of equations and the number of variables may differ. We want to find all the triples of numbers (x, y, z) that satisfy all three equations. Stated geometrically, we want to find all points or vectors in $\mathbb{R}^3$ that satisfy the system. Recall that each of these equations represents a plane in $\mathbb{R}^3$, so we're looking for the points (x, y, z) that are on all three planes. Geometrically we know that three distinct planes intersect either in a single point, or in a line, or else have no points in common. We'll see how all three cases occur algebraically, and show how to represent all solutions parametrically.

1A   Elimination Method

Two systems are called **equivalent** if they have the same set of solutions. For example, a system representing three planes intersecting in a single line will be equivalent to the system we get by discarding one of the three planes. Our procedure will be to alter a system in a sequence of steps to arrive at an equivalent system for which the solutions are obvious. The following example illustrates the process.

EXAMPLE 1   To find all x, y, z simultaneously satisfying the equations

$$3x + 12y + 9z = 3$$
$$2x + 5y + 4z = 4$$
$$-x + 3y + 2z = -5,$$

multiply the first equation by $\frac{1}{3}$, which makes the coefficient of x equal to 1:

$$\begin{aligned}
x + 4y + 3z &= 1 \\
2x + 5y + 4z &= 4 \\
-x + 3y + 2z &= -5.
\end{aligned}$$

Add (-2) times the first equation to the second, and replace the second equation by the result; this eliminates x from the second equation by making its coefficient equal to 0:

$$\begin{aligned}
x + 4y + 3z &= 1 \\
-3y - 2z &= 2 \\
-x + 3y + 2z &= -5.
\end{aligned}$$

Add the first equation to the third, and replace the third equation by the result:

$$\begin{aligned}
x + 4y + 3z &= 1 \\
-3y - 2z &= 2 \\
7y + 5z &= -4.
\end{aligned}$$

Next multiply the second equation by $-\frac{1}{3}$:

$$\begin{aligned}
x + 4y + 3z &= 1 \\
y + \tfrac{2}{3}z &= -\tfrac{2}{3} \\
7y + 5z &= -4.
\end{aligned}$$

Add -4 times the second equation to the first, and -7 times the second equation to the third to get

$$\begin{aligned}
x + \tfrac{1}{3}z &= \tfrac{11}{3} \\
y + \tfrac{2}{3}z &= -\tfrac{2}{3} \\
\tfrac{1}{3}z &= \tfrac{2}{3}.
\end{aligned}$$

Multiply the third equation by 3 to get

$$\begin{aligned}
x + \tfrac{1}{3}z &= \tfrac{11}{3} \\
y + \tfrac{2}{3}z &= -\tfrac{2}{3} \\
z &= 2.
\end{aligned}$$

Add $(-\frac{1}{3})$ times the third equation to the first and $(-\frac{2}{3})$ times the third equation to the second to get

$$\begin{aligned}
x &= 3 \\
y &= -2 \\
z &= 2.
\end{aligned}$$

Hence the system has one solution, the point $(x, y, z) = (3, -2, 2)$ in $\mathbb{R}^3$.

We can verify by substitution into the initial system of equations above that we've found a solution, but this verification doesn't rule out the possibility that the original equations might still have other solutions. We'll dispose of this idea once and for all by singling out the two operations we use to solve linear systems and then proving that the solution set of a system remains unchanged after applying them to the system.

1.1 Definition   An **elementary multiplication** multiplies both sides of a single equation by a nonzero scalar r. Since $r \neq 0$ the operation is reversed by the corresponding **inverse** elementary multiplication by r^{-1}. An **elementary modification** of a single equation adds a scalar multiple by r of another equation to the equation to be modified. The operation is reversed by the corresponding **inverse** elementary modification that uses the scalar $-r$.

In the previous example we used only elementary operations, and the following theorem guarantees that the one solution we found is the only solution.

1.2   Theorem.   Applying elementary operations to a system of linear equations does not change the solution set of the system.

Proof.   If a set of numbers satisfies an equation then the equation is still satisfied by the same set of numbers after multiplication by a scalar. Similarly, if the same set of numbers satisfies two equations it also satisfies the sum of the two equations. Thus every solution of a system also satisfies the equations resulting from a sequence of the two elementary operations.

Conversely, starting from a system that has been modified by elementary operations, we can apply the inverse of each of the operations in the reverse order and get back the original system. Thus every solution of the modified system also satisfies the original system.   ∎

EXAMPLE 2   Consider just the first two equations in the previous example:

$$3x + 12y + 9z = 3$$
$$2x + 5y + 4z = 4.$$

These equations represent two planes, so they may intersect in a line. When we solved these two equations along with the third equation in Example 1, it so happened that we avoided until the end adding a multiple of the third equation to either of the others, so we can go through the first steps of that example simply ignoring the third equation. Three steps from the end we arrive at

$$x + \tfrac{1}{3}z = \tfrac{11}{3}$$
$$y + \tfrac{2}{3}z = -\tfrac{2}{3}.$$

There are infinitely many triples (x, y, z) that satisfy the equations, because for an arbitrary value of z the last two equations determine corresponding x and y values

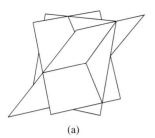

(a)

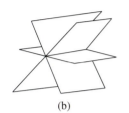

(b)

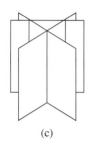

(c)

FIGURE 2.1

such that (x, y, z) is a solution. We introduce a parameter t and write

$$z = t$$

to denote an arbitrary value of z. Then we can write the equations in final form as

$$x = -\tfrac{1}{3}t + \tfrac{11}{3}$$
$$y = -\tfrac{2}{3}t - \tfrac{2}{3}$$
$$z = t.$$

In vector form the equations become

$$(x, y, z) = t\left(-\tfrac{1}{3}, -\tfrac{2}{3}, 1\right) + \left(\tfrac{11}{3}, -\tfrac{2}{3}, 0\right).$$

This is a parametric representation for a line, just what we thought we might get from the intersection of two planes.

We remarked that the intersection determined by three scalar equations in three unknowns could consist of a line; just start with the equations of three different planes all of which contain the line. The various possibilities for three distinct equations are in Figure 2.1: three planes intersecting in a point, three planes intersecting in a line, and three planes with no point in common. The third case, in which there are no solutions, comes from what is called an **inconsistent system**. In addition to distinct planes that intersect in three parallel lines as in Figure 2.1(c), planes such that some pair, or all three, are parallel also illustrate inconsistent systems.

EXAMPLE 3   For an example of an inconsistent system, consider the equations

$$\begin{aligned} x + y - 2z &= 1 \\ -3x + 2y + z &= 0 \\ -x + 4y - 3z &= 1. \end{aligned}$$

To try to solve this system, we add 3 times the first equation to the second and 1 times the first to the third. The result is

$$\begin{aligned} x + y - 2z &= 1 \\ 5y - 5z &= 3 \\ 5y - 5z &= 2. \end{aligned}$$

There are no values of y and z that can satisfy the last two equations simultaneously, so we conclude that the system has no solutions, that is, that the system is inconsistent. Geometrically, we see that the given system turns out to be equivalent to one in which the last two equations represent distinct parallel planes, so there is no point of intersection.

Example 2 illustrates a general principle that we'll prove in Section 2B: When a system of linear equations has more than one solution it always has infinitely many solutions, solutions that we can find by assigning arbitrary values to some of the unknowns, and then determining values for the other unknowns in terms of these arbitrary values. As in Example 2, this always happens for a consistent system with more unknowns than equations. The simplest case is that of one equation in several unknowns, for example,

$$2x - 3y + z = 1.$$

If we let $x = s$ and $y = t$, then

$$z = -2s + 3t + 1,$$

so all solutions are of the form

$$(x, y, z) = (s, t, -2s + 3t + 1)$$
$$= s(1, 0, -2) + t(0, 1, 3) + (0, 0, 1),$$

a parametric representation for a plane.

We'll see in Section 2C that **homogeneous systems**, in which all the constant terms are zero, are the key to describing multiple solutions. Such systems always have a **zero solution**, in which all the unknowns equal zero. The question then is whether there are other solutions. If a system is not only homogeneous but has more unknowns than equations, we'll see in Section 2B that there are always infinitely many nonzero solutions.

EXAMPLE 4

In the homogeneous system

$$x + y - z = 0$$
$$x - y + z = 0,$$

set $z = t$. Then the system becomes

$$x + y = t$$
$$x - y = -t.$$

Subtracting the first equation from the second gives

$$x + y = t$$
$$ - 2y = -2t.$$

Hence $y = t$ from the second equation, so $x = 0$ from the first. The solutions are

$$(x, y, z) = (0, t, t)$$
$$= t(0, 1, 1),$$

which exhibits the infinitely many solutions as a parametric representation of the points on a line in $\mathbb{R}^3$.

EXAMPLE 5

Many questions about lines and planes come down to solving systems of linear equations. For example, two lines in $\mathbb{R}^3$ with parametric representations $s\mathbf{u}_1 + \mathbf{v}_1$ and $t\mathbf{u}_2 + \mathbf{v}_2$ intersect if and only if there are values of s and t such that the vector equation $s\mathbf{u}_1 + \mathbf{v}_1 = t\mathbf{u}_2 + \mathbf{v}_2$ holds, and this amounts to a system of linear equations for s and t. If we take $\mathbf{u}_1 = (3, 2, 1)$, $\mathbf{v}_1 = (-1, 0, 1)$, $\mathbf{u}_2 = (0, 2, 1)$, and $\mathbf{v}_2 = (-4, 2, 2)$, then the vector equation is equivalent to

$$
\begin{aligned}
3s - 1 &= \quad -4 \\
2s &= 2t + 2 \\
s + 1 &= \quad t + 2.
\end{aligned}
$$

The first equation gives $s = -1$, and then the second gives $t = -2$. These values also satisfy the third equation, so the lines do intersect. The point of intersection is obtained by putting $s = -1$ in $s\mathbf{u}_1 + \mathbf{v}_1$ (or $t = -2$ in $t\mathbf{u}_2 + \mathbf{v}_2$) and is $-(3, 2, 1) + (-1, 0, 1) = (-4, -2, 0)$. If we change $\mathbf{v}_2$ to $(-4, 2, 0)$, the third equation is changed to $s + 1 = t$ and is not satisfied by the values of s and t that satisfy the first two equations, so the lines don't intersect.

EXAMPLE 6   Sometimes we know the degree of a polynomial $f(x)$ but don't know the coefficients. If we know values of $f(x)$ for enough values of x, it may be possible to find the coefficients by solving a system of equations. For example, when a particle moves in the plane under the influence of a constant force that is parallel to the y-axis, its path is a parabola with an equation of the form $y = f(x) = ax^2 + bx + c$. If the points $(x, y) = (0, 1)$, $(2, 4)$, and $(3, 3)$ are known to be on the path then we can find a, b, and c by solving the equations

$$
0a + 0b + c = 1
$$

$$
4a + 2b + c = 4
$$

$$
9a + 3b + c = 3.
$$

Exercise 11 asks you to finish the calculation to find the coefficients of $f(x)$.

EXERCISES

Some of the systems of equations in Exercises 1 to 6 have one solution, some have more, and some have no solutions. If there are solutions, find all of them and interpret them as an intersection of lines or planes. In the cases where there is no solution, give a geometric explanation.

1. $\begin{aligned} x + y &= 1 \\ x - y &= 2 \end{aligned}$

2. $\begin{aligned} 2x - y &= 2 \\ -2x + y &= 2 \end{aligned}$

3. $\begin{aligned} x + y + z &= 0 \\ x - y \phantom{{}+z} &= 0 \\ y + z &= 0 \end{aligned}$

4. $\begin{aligned} x + y + z &= 0 \\ x - y \phantom{{}+z} &= 0 \\ 2x + \phantom{y+{}} z &= 0 \end{aligned}$

5. $\begin{aligned} x + y + z &= 0 \\ x + y - z &= 1 \\ x + y + 2z &= 2 \end{aligned}$

6. $\begin{aligned} x - 2y &= 1 \\ 2x + y &= -1 \\ x - 7y &= 4 \end{aligned}$

In Exercises 7 to 10, find a point of intersection of the two given lines, or else show that they do not intersect.

7. $\mathbf{x} = t(1, -1, 2) + (1, 1, 1)$, $\mathbf{x} = s(3, 2, 1) + (-2, -6, 5)$

8. $\mathbf{x} = t(1, 1, 2) + (0, 1, 1)$, $\mathbf{x} = s(-2, 1, 1) + (2, 1, 2)$

9. $\mathbf{x} = t(1, 2) + (2, 1)$, $\mathbf{x} = s(1, 3) + (3, -1)$

10. $\mathbf{x} = t(-1, 1, 1) + (1, 0, 2)$, $\mathbf{x} = s(2, 0, 2) + (1, 2, 2)$

11. **(a)**   Finish the calculation in Example 6 in the text, by finding the values of a, b, c, and $f(x)$.
    (b)   Is there a value y such that if the path of the object in Example 6 passed through $(0, 1)$, $(2, 4)$, and $(3, y)$, then the value of a would be 0? Is the path a parabola in this case?

12. Suppose $f(x) = c_1 e^x + c_2 e^{-x} + c_3$, where c_1, c_2, and c_3 are constants. How should these constants be chosen so that $f(0) = 1$, $f'(0) = 1$, and $f''(0) = 2$?

13. Suppose that a mixture of sand and cinders contains 10 cubic yards and weighs 34 tons. If the sand weighs 4 tons per cubic yard and the cinders weigh 1 ton per cubic yard, how much of each does the mixture contain?

14. Suppose that various mixtures are made of substances S_1, S_2, and S_3 having densities 2, 3, and 1 respectively, measured in grams per cubic centimeter. Suppose also that the price of each substance in cents per cubic centimeter is 4, 3, and 1, respectively. Is it possible to make a mixture weighing 10 grams, with a volume of 20 cubic centimeters and costing 1 dollar?

Recall from Chapter 1, Section 1 that a vector **b** is a *linear combination* of $a_1, \ldots, a_n$ if there are scalars $x_1, \ldots, x_n$ such that $b = x_1 a_1 + \cdots + x_n a_n$. In Exercises 15 to 18, find coefficient x's to express **b** as a linear combination of the **a**'s, or show that no such coefficients exist. First convert the vector equation into a system of linear equations for the x_i.

15. $a_1 = \begin{pmatrix} 1 \\ 2 \end{pmatrix}$, $a_2 = \begin{pmatrix} 2 \\ 2 \end{pmatrix}$; $b = \begin{pmatrix} -1 \\ 1 \end{pmatrix}$.

16. $a_1 = \begin{pmatrix} 1 \\ 2 \end{pmatrix}$, $a_2 = \begin{pmatrix} 2 \\ 1 \end{pmatrix}$; $b = \begin{pmatrix} 0 \\ 1 \end{pmatrix}$.

17. $a_1 = \begin{pmatrix} 2 \\ 1 \\ 1 \end{pmatrix}$, $a_2 = \begin{pmatrix} 2 \\ 1 \\ 2 \end{pmatrix}$, $a_3 = \begin{pmatrix} 1 \\ 2 \\ 2 \end{pmatrix}$; $b = \begin{pmatrix} 1 \\ 0 \\ 0 \end{pmatrix}$.

18. $a_1 = \begin{pmatrix} 0 \\ 1 \\ 1 \end{pmatrix}$, $a_2 = \begin{pmatrix} 1 \\ 1 \\ 0 \end{pmatrix}$, $a_3 = \begin{pmatrix} 1 \\ 2 \\ 1 \end{pmatrix}$; $b = \begin{pmatrix} 1 \\ 2 \\ 3 \end{pmatrix}$.

1B   Applications

Here are some problems in various areas of applied mathematics that require solving systems of linear equations.

EXAMPLE 7   An assembly of electrical conductors is called a **network** if each pair J_i, J_j of junctions in the assembly is contained in a closed loop, or **circuit** where each segment represents a connecting wire with a given electrical resistance. When such a network is connected to a power source, currents flow in the segments, and a voltage is measurable at each junction. **Ohm's law** states that the current flow in a wire is proportional to the difference in voltage at its two ends, the constant of proportionality being the reciprocal of the wire's resistance:

$$c_{ij} = \frac{(v_i - v_j)}{r_{ij}}, \tag{1}$$

where c_{ij} is the current flowing from junction J_i to junction J_j, r_{ij} is the resistance of the connection between junctions J_i and J_j, and v_i and v_j are the values of the voltages at junctions J_i and J_j. The standard units of measurement are amperes for current, ohms for resistance, and volts for voltage. A negative value for the current from J_i to J_j indicates a current flowing from J_j to J_i.

Figure 2.2(a) shows a network with four junctions and five segments, with the resistance of each segment indicated beside it. Suppose external power source terminals connected at junctions 1 to 4 maintain values $v_1 = 12$ and $v_4 = 0$. Since junction J_2 has no external connection, the current flowing in must balance the current flowing out, so that if signs are taken into account, the sum of the currents out of junction J_2 must be zero. Using Equation (1), we get the equation

$$\tfrac{1}{2}(v_2 - v_1) + (v_2 - v_3) + \tfrac{1}{6}(v_2 - v_4) = 0.$$

Rewriting in the form

$$\left(\tfrac{1}{2} + 1 + \tfrac{1}{6}\right) v_2 = \tfrac{1}{2}v_1 + v_3 + \tfrac{1}{6}v_4,$$

FIGURE 2.2
Electric networks.

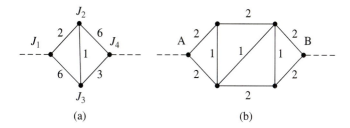

(a)   (b)

we see that v_2 is a weighted average of v_1, v_3, and v_4, with coefficients that are the reciprocals of the resistances in the lines joining J_2 to the others. A similar equation holds at each junction that doesn't have an external connection. Thus at J_3

$$\left(\tfrac{1}{6} + 1 + \tfrac{1}{3}\right) v_3 = \tfrac{1}{6}v_1 + v_2 + \tfrac{1}{3}v_4.$$

Since we assumed external voltages $v_1 = 12$ and $v_4 = 0$, we reorder terms in the previous two equations, and get

$$\tfrac{5}{3}v_2 - v_3 = 6$$
$$-v_2 + \tfrac{3}{2}v_3 = 2.$$

Solving this system gives $v_2 = \tfrac{22}{3} \approx 7.33$ and $v_3 = \tfrac{59}{6} \approx 6.22$. Once we know the voltages, we can find the currents from (1). Thus the current from J_1 to J_2 is

$$c_{12} = (v_1 - v_2)/r_{12} \approx \tfrac{1}{2}(12 - 7.33) \approx 2.34.$$

Similarly the current from junction J_1 to junction J_3 is

$$c_{13} = (v_1 - v_3)/r_{13} \approx \tfrac{1}{6}(12 - 6.22) \approx 0.96.$$

The total current from junction J_1 into the rest of the network is then about $2.3 + 0.96 = 3.3$ amperes, which is the current flowing into junction J_1 from the outside source.

EXAMPLE 8

We may regard vectors in $\mathbb{R}^2$ or $\mathbb{R}^3$ as representing forces acting at some point which for convenience we take to be the origin. The direction of the arrow is the direction in which the force acts, and the length of the arrow is the magnitude of the force. Our fundamental physical assumption here is that if more than one force acts at a point then the resulting force acting at the point is represented by the sum **R** of the separate force vectors acting there. In Figure 2.3 we have two different pictures, and the resultant arrow **R** appears only in Figure 2.3(a). For example, suppose that the force vectors in Figure 2.3(a) lie in a plane, which we take to be $\mathbb{R}^2$ with the origin at the point of action. If we have

$$\mathbf{F}_1 = (-1, 3), \ \mathbf{F}_2 = (4, 3), \ \mathbf{F}_3 = (-2, -4), \tag{2}$$

then by definition

$$\mathbf{R} = \mathbf{F}_1 + \mathbf{F}_2 + \mathbf{F}_3 = (-1, 3) + (4, 3) + (-2, -4) = (1, 2).$$

FIGURE 2.3

Force vectors.

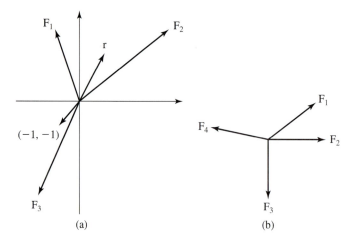

(a)   (b)

Suppose we are given only the directions of the three force vectors and are asked to find corresponding forces that will produce a given resultant, say, $\mathbf{R} = (-1, -1)$. In other words, suppose we want to find nonnegative numbers c_1, c_2, c_3 such that

$$c_1\mathbf{F}_1 + c_2\mathbf{F}_2 + c_3\mathbf{F}_3 = (-1, -1). \tag{3}$$

(Having some $c_i < 0$ would reverse the direction of the corresponding force.) The vector equation is equivalent to the system of equations we get by substituting for the $\mathbf{F}_i$ the given vectors (1) that determine the force directions. We want

$$c_1(-1, 3) + c_2(4, 3) + c_3(-2, -4) = (-1, -1), \quad \text{or} \quad \begin{matrix} -c_1 + 4c_2 - 2c_3 = -1 \\ 3c_l + 3c_2 - 4c_3 = -1. \end{matrix}$$

Since we have two equations and three unknowns, we would expect in general to be able to specify one of the c_i and then solve for the others. However recall that the c_i are to be nonnegative. In particular, a glance at Figure 2.3(a) shows that we could not get a resultant equal to $(-1, -1)$ unless c_3 is positive. Hence we try $c_3 = 1$. This choice leads to the pair of equations

$$\begin{matrix} -c_1 + 4c_2 = 1 \\ c_1 + c_2 = 1. \end{matrix}$$

These equations have the unique solution $c_1 = \frac{3}{5}$, $c_2 = \frac{2}{5}$. Thus the triple $(c_1, c_2, c_3) = (\frac{3}{5}, \frac{2}{5}.1)$ is one possible solution, and the three force vectors are

$$c_1\mathbf{F}_1 = \left(-\frac{3}{5}, \frac{9}{5}\right), \quad c_2\mathbf{F}_2 = \left(\frac{8}{5}, \frac{6}{5}\right), \quad c_3\mathbf{F}_3 = (-2, -4),$$

with magnitudes $|c_1\mathbf{F}_1| = \frac{3}{5}\sqrt{10}$, $|c_2\mathbf{F}_2| = 2$, $|c_3\mathbf{F}_3| = 2\sqrt{5}$.

We could equally well have asked for an assignment of force magnitudes that would put the system in equilibrium, so that the resultant $\mathbf{R}$ is the zero vector. We would then have replaced the vector $(-1, -1)$ on the right side of Equation (3) by $(0, 0)$ and solved the new system in a similar way.

EXAMPLE 9

The analysis in this example is useful for distinguishing completely random behavior of a rat in a maze from purposeful or conditioned behavior. We define for this example a **random walk** to be a process such that a rat proceeds between fixed positions along sequences of paths, each path having a given probability of being used. Specifically we assume about random behavior that the probability of leaving some position along a particular path is the same for all paths heading away from that position. Some sample mazes appear in Figure 2.4.

A probability is a number p in the interval $0 \leq p \leq 1$, and the probability or likelihood of a particular event is equal to the sum of the probabilities of the various distinct ways that event can occur. Thus in Figure 2.4(a), since we assume that all paths heading away from a_5 are equally likely, it follows that the probability of leaving a_5 along each of the two possible paths is $\frac{1}{2}$. Similarly, each of the three paths from a_1 has probability $\frac{1}{3}$, as do the three paths from a_4. Note also that the probability of going from a_1 to a_2 is less than the probability of going from a_2 to a_1. We also assume for this example that the probability of two successive events is equal to the product of their respective probabilities. Thus going from a_2 to a_1 to a_4 has the probability $\frac{1}{2} \cdot \frac{1}{3} = \frac{1}{6}$.

We can now ask a question such as the following: What is the probability p_k of starting at a_k and arriving at the specified position a_5 without first going to a_4? We see that starting at a_1 we can go to a_5 directly with probability $\frac{1}{3}$, or we can go to a_2 with probability $\frac{1}{3}$ and then go to a_5 with probability p_2, the probability of going from a_2 to a_5 without going to a_4. Thus

$$p_1 = \tfrac{1}{3} + \tfrac{1}{3} p_2.$$

Similarly, because going to a_4 does not occur in the events we are watching,

$$p_2 = \tfrac{1}{2} p_1 + \tfrac{1}{2} p_3$$
$$p_3 = \tfrac{1}{2} p_2.$$

We rewrite the previous three equations as

$$p_1 - \tfrac{1}{3} p_2 = \tfrac{1}{3}$$
$$-\tfrac{1}{2} p_l + p_2 - \tfrac{1}{2} p_3 = 0$$
$$-\tfrac{1}{2} p_2 + p_3 = 0$$

and solve them by routine methods. We get $p_1 = \frac{3}{7}$, $p_2 = \frac{2}{7}$, $p_3 = \frac{1}{7}$. It appears that the closer we start to a_5 the more likely we are to get to a_5 without going to a_4, but the exact probabilities depend on the entire maze.

FIGURE 2.4

Rat mazes.

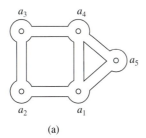

(a)

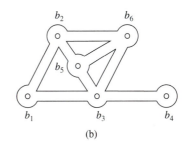

(b)

EXAMPLE 10

In a network of interconnected water pipes the junctions are pipe joints. It's usual in a pipe network to assign each pipe a positive flow direction with an arrow as in Figure 2.5. With this understanding a positive number r_k will be a flow rate in the direction assigned to the kth pipe, while a negative number $-r_k$ will be a flow of equal rate in the opposite direction. We'll separate the flow rates into internal rates r_k and rates t_k from or to external sources or drains. Specifying an external rate $t_k = 0$ at a joint closes off the external pipe there. We also assume that the inflow at a joint equals the outflow. Thus at the upper left corner in Figure 2.5(a) we find $t_1 = r_1 + r_2$, while at the lower left we find $r_3 = r_2 + t_3$, or $-r_2 + r_3 = t_3$. Checking each external joint of the network of Figure 2.5(a), we find the entire set of equations relating the rates t_k to the rates r_k:

$$\begin{aligned}
r_1 + r_2 & & & = t_1 \\
-r_1 & - r_4 + r_5 & & = t_2 \\
& - r_2 + r_3 & & = t_3 \\
& - r_3 + r_4 & + r_6 & = t_4 \\
& & r_5 + r_6 & = t_5.
\end{aligned} \qquad (4)$$

From these equations we conclude that specifying the flows r_k in the internal pipes completely determines the flows t_k at the external joints.

Turning the problem around, we can ask to what extent specifying the flows t_k at the external joints will specify the flows r_k in the pipes. In particular, we can try specifying that the exterior flow t_k at each joint should be zero. This leads to the system of five equations in six unknowns:

$$\begin{aligned}
r_1 + r_2 & & & = 0 \\
-r_1 & - r_4 + r_5 & & = 0 \\
& - r_2 + r_3 & & = 0 \\
& - r_3 + r_4 & + r_6 & = 0 \\
& & r_5 + r_6 & = 0.
\end{aligned} \qquad (5)$$

We can let $r_6 = a$ be an arbitrary number, so we get $r_5 = -a$ from the last equation. Similarly, let $r_4 = b$ be an arbitrary number. Noting from the first and third equations that $r_3 = r_2 = -r_1$, the remaining two equations for r_1 and r_4 both reduce to $r_1 = -a - b$, so the solution vector is

$$\mathbf{r}_0 = (-a - b, \ a + b, \ a + b, \ b, \ -a, \ a).$$

FIGURE 2.5

Water pipes.

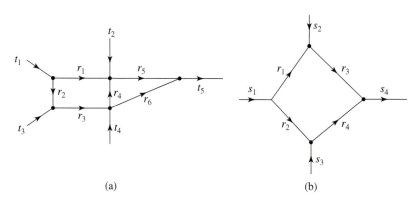

(a)   (b)

It follows that there are infinitely many pipe flows, depending on the parameters a and b, that will produce external flows $t_k = 0$ for $k = 1, 2, 3, 4, 5$. We'll see in Section 2C that every solution $\mathbf{r}$ of a system such as Equation (4) is the sum $\mathbf{r} = \mathbf{r}_p + \mathbf{r}_0$ of one particular solution $\mathbf{r}_p$ and one of the solutions in the 2-parameter family $\mathbf{r}_0$.

EXAMPLE 11 The derivation of Simpson's rule for approximate integration is based on the requirement that it should give exact results when applied to quadratic polynomials. The rule gives an approximation to the integral of a function over an interval $a-h \leq x \leq a+h$ in terms of the values of the function at the points $a - h$, a, and $a + h$. The general form of the approximation is

$$\int_{a-h}^{a+h} f(x)\,dx \approx Af(a - h) + Bf(a) + Cf(a + h),$$

where A, B, and C are constants. If the formula is to be correct for all polynomials of degree less than or equal to 2, it must in particular be correct for the polynomials $f_0(x) = 1$, $f_1(x) = x$, and $f_2(x) = x^2$. Each of these requirements leads to an equation for A, B, and C. For instance, with $f_0(x) = 1$ we have

$$\int_{a-h}^{a+h} f_0(x)\,dx = 2h \quad \text{and} \quad f_0(a - h) = f_0(a) = f_0(a + h) = 1,$$

so we require $A + B + C = 2h$. Similarly, we obtain two more equations using $f_1(x) = x$ and $f_2(x) = x^2$, and have

$$A + B + C = 2h$$

$$(a - h)A + aB + (a + h)C = 2ah,$$

$$(a^2 - 2ah + h^2)A + a^2B + (a^2 + 2ah + h^2)C = 2a^2h + \tfrac{2}{3}h^3.$$

It's straightforward to check that $A = C = \tfrac{1}{3}h$, $B = \tfrac{4}{3}h$ satisfy these three equations, and Exercise 17 asks you to verify that these are the only solutions.

   Thus we have a rule that is correct for the particular polynomials f_0, f_1, and f_2. Its correctness for an arbitrary quadratic polynomial $f(x) = px^2 + qx + r$ follows without additional computation from the following general observation. Let $E(f)$ be the error committed when the rule is applied to a general continuous function $f(x)$, so

$$E(f) = \int_{a-h}^{a+h} f(x)\,dx - \tfrac{1}{3}hf(a - h) - \tfrac{4}{3}hf(a) - \tfrac{1}{3}hf(a + h).$$

Note that $E(f_0) = E(f_1) = E(f_2) = 0$ holds by the way we chose A, B and C. But then elementary properties of the integral and the form of the approximation as a linear combination of values of $f(x)$ show that

$$E(pf_2 + qf_1 + rf_0) = pE(f_2) + qE(f_1) + rE(f_0) = 0.$$

We can use the same method to derive a variety of formulas in the field of numerical analysis, as in Exercise 18.

EXERCISES

1. Figure 2.2(b) shows an electrical network with the resistance in ohms of each edge marked on it. Suppose an external power supply maintains junction A at 10 volts and junction B at 4 volts. Following the procedure of Example 7 in the text, set up equations for the voltages at the other junctions and solve them. From the results, calculate the current flowing into the network at junction A.

The edges and vertices of a 3-dimensional cube form a network with 8 junctions and 12 edges. Suppose that each edge is a wire of resistance 1 ohm and that there are just two external connections, which maintain a voltage of 1 at one of the vertices and 0 at another. In Exercises 2 to 4, find the values of the voltages at the other vertices and the current flowing in the external connections under the stated conditions.

2. The two vertices with external connections are at opposite corners of the cube.

3. The vertices with external connections are at the two ends of an edge of the cube.

4. The vertices with external connections are at opposite corners of a face of the cube.

5. If forces in $\mathbb{R}^2$ act at the origin parallel to $(2, 1)$, $(2, 2)$, and $(-3, -1)$, find magnitudes we can assign to the forces so their sum will be zero.

In Exercises 6 and 7, suppose that three forces acting at the origin in $\mathbb{R}^3$ have directions parallel to $(1, 0, 0)$, $(1, 1, 0)$, and $(1, 1, 1)$.

6. Find examples of magnitudes for forces acting parallel to these directions so the resultant force vector will be $(-1, 2, 4)$.

7. Can an arbitrary force vector $\mathbf{F} = (a, b, c)$ be the resultant of forces acting in the actual directions specified in the preamble? Explain.

In Exercises 8 to 10, suppose that a random walk traverses the paths shown in Figure 2.4(a).

8. What is the probability p_1 that a walk starting at a_1 goes to a_4 without passing through a_5?

9. What is the probability p_2 that a walk starting at a_2 goes to a_4 without passing through a_5?

10. What is the probability p_3 that a walk starting at a_3 goes to a_4 without passing through a_5?

In Exercise Exercises 11 to 13, assume a rat traces a random walk on the paths shown in Figure 2.4(b). Let p_k be the probability of going from b_k to b_6 without going through b_5.

11. Find p_k for $k = 1, 2, 3, 4$.

12. Modify Figure 2.4(b) in the text so b_4 and the path from it to b_3 are eliminated. Then compute the resulting new values for p_k for $k = 1, 2, 3$.

13. Modify Figure 2.4(b) in the text by introducing a new path from b_4 to b_6. Then compute the resulting new values for p_k for $k = 1, 2, 3, 4$.

14. (a) Suppose the vector $\mathbf{t} = (t_1, t_2, t_3, t_4, t_5)$ in Equations 4 of the text is specified to be $\mathbf{t} = (-1, 0, 1, 2, 1)$. Find a vector $\mathbf{r}$ that determines consistent internal flow rates.

    (b) Solve Equations 5 to verify that the vector $\mathbf{r} = (-a - b, a + b, a + b, b, -a, a)$, with arbitrary a and b, describes all solutions that are consistent with external flow $\mathbf{t} = 0$ in Figure 2.5(a).

15. Let the external flow vector in Figure 2.5(b) be $\mathbf{s} = (1, 1, 2, 4)$. Show that there is more than one consistent internal flow vector $\mathbf{r}$, and find all of them in terms of an arbitrarily assigned value.

16. If the external flow vector in Figure 2.5(b) is $\mathbf{s} = (1, 0, 1, 1)$, show that there is no consistent internal flow vector.

17. Carry out the solution of the equations for A, B, C given in Example 11 of the text. [*Suggestion:* Start by subtracting a times the first equation from the second and a^2 times the first from the third.]

18. Use the method of Example 11, to find constants A, B, C, D such that

$$\int_a^{a+3h} f(x)\,dx = Af(a) + Bf(a + h) + Cf(a + 2h)$$
$$+ Df(a + 3h)$$

is exact whenever $f(x)$ is $1, x, x^2$ and x^3, and so as in Example 5 is also exact for a polynomial of degree at most 3.

SECTION 2   MATRIX METHODS

In Section 1 we used elementary operations to solve systems of linear equations in an ad hoc way that's hard to adapt to large systems. In Sections 2A and 2B we introduce matrix equations and an effective solution routine. In Sections 2C and 2D the emphasis is on geometric ideas. The matrix operations appear again in Sections 4 and 5 for computing inverse matrices and determinants, as well as later on in Chapters 6 and 13.

2A   Matrix Equations and Elementary Operations

A **matrix** is simply a rectangular array of numbers. Here are some examples:

$$\begin{pmatrix} 0 & 5 \\ -1 & \sqrt{2} \\ 0 & 4 \end{pmatrix}, \quad \begin{pmatrix} 1 & 0.7 & 3 \\ 0.9 & 0 & 2.8 \end{pmatrix}, \quad \begin{pmatrix} 1 & 0 \\ 0 & -1 \end{pmatrix}, (2,5,0), \begin{pmatrix} \sqrt{2} \\ 4 \\ \sqrt{3} \\ 1 \end{pmatrix}.$$

The horizontal lines of numbers in a matrix are called its **rows**, and the vertical lines are called its **columns**.

The numbers of rows and columns in a matrix determine its **dimensions**, and for consistency the number of rows is always designated before the number of columns. The five examples just given have dimensions 3-by-2, 2-by-3, 2-by-2, 1-by-3, and 4-by-1. A matrix is **square** if it has the same number of rows as columns, so it has dimensions n-by-n for some n. The 1-by-n matrices are called **n-dimensional row vectors**, and n-by-1 matrices are called **n-dimensional column vectors**, so we may regard the rows or columns of an m-by-n matrix as vectors in $\mathbb{R}^n$ or $\mathbb{R}^m$, respectively, as in Definition 2.1.

EXAMPLE 1   Matrices occur naturally for representing systems of linear equations. In the system

$$\begin{aligned} 2x + 3y - 4z &= 1 \\ x - y + 2z &= -1, \end{aligned}$$

we temporarily regard the letters x, y, and z just as placemarkers, so we can describe the two sides of the equations using the two matrices

$$\begin{pmatrix} 2 & 3 & -4 \\ 1 & -1 & 2 \end{pmatrix} \quad \text{and} \quad \begin{pmatrix} 1 \\ -1 \end{pmatrix}.$$

The 2-by-3 matrix is the **coefficient matrix** of the system, and the 2-by-1 matrix is the **right side**.

When writing the coefficient matrix of a system it's important to have the variables lined up in the same order in all the equations and to use the coefficient 0 in the matrix to indicate the absence of a variable. For example, to put the system

$$\begin{aligned} 2x + y &= 4 \\ z - y &= 2 \end{aligned}$$

in matrix form, it is a good idea to make clear the place that each coefficient has in the system by first rewriting it as

$$\begin{aligned} 2x + y \quad &= 4 \\ - y + z &= 2. \end{aligned}$$

The coefficient matrix and right side are then

$$\begin{pmatrix} 2 & 1 & 0 \\ 0 & -1 & 1 \end{pmatrix} \quad \text{and} \quad \begin{pmatrix} 4 \\ 2 \end{pmatrix}.$$

We can use dot products to relate a system's matrix and variables algebraically.

2.1 Definition   The **product** $A\mathbf{x}$ of an m-by-n matrix A and an n-dimensional column vector $\mathbf{x}$ is the m-dimensional column vector whose entries are the successive dot products of the rows of A with the vector $\mathbf{x}$. The product $A\mathbf{x}$ isn't defined if the number of columns in A doesn't equal the number of entries in $\mathbf{x}$.

EXAMPLE 2   If $A = \begin{pmatrix} 4 & 3 \\ -1 & 2 \end{pmatrix}$ and $\mathbf{x} = \begin{pmatrix} x_1 \\ x_2 \end{pmatrix}$,

then $A\mathbf{x} = \begin{pmatrix} 4 & 3 \\ -1 & 2 \end{pmatrix} \begin{pmatrix} x_1 \\ x_2 \end{pmatrix} = \begin{pmatrix} 4x_1 + 3x_2 \\ -x_1 + 2x_2 \end{pmatrix}.$

If $B = \begin{pmatrix} a & b & c \\ d & e & f \end{pmatrix}$ and $y = \begin{pmatrix} 1 \\ 2 \\ 1 \end{pmatrix}$,

then $By = \begin{pmatrix} a & b & c \\ d & e & f \end{pmatrix} \begin{pmatrix} 1 \\ 2 \\ 1 \end{pmatrix} = \begin{pmatrix} a + 2b + c \\ d + 2e + f \end{pmatrix}.$

If $C = \begin{pmatrix} 1 & 0 \\ 2 & -1 \\ 3 & 2 \end{pmatrix}$ and $\mathbf{z} = \begin{pmatrix} 1 \\ 2 \end{pmatrix}$,

then $C\mathbf{z} = \begin{pmatrix} 1 & 0 \\ 2 & -1 \\ 3 & 2 \end{pmatrix} \begin{pmatrix} 1 \\ 2 \end{pmatrix} = \begin{pmatrix} 1 \\ 0 \\ 7 \end{pmatrix}.$

Using products of matrices and column vectors, we write systems of linear equations in the vector form $A\mathbf{x} = \mathbf{b}$, where A is an m-by-n matrix, $\mathbf{x}$ is an n-dimensional column vector with unknown entries, and $\mathbf{b}$ is an m-dimensional column vector.

EXAMPLE 3   Each system on the left is equivalent to the matrix equation on the right. Each variable corresponds to a column of the matrix.

$$\begin{aligned} 4x + 3y &= 1 \\ -x + 2y &= 2 \end{aligned} \qquad \begin{pmatrix} 4 & 3 \\ -1 & 2 \end{pmatrix} \begin{pmatrix} x \\ y \end{pmatrix} = \begin{pmatrix} 1 \\ 2 \end{pmatrix}$$

$$2x + y + 2z = -1$$
$$x + 2y + z = 0$$

$$\begin{pmatrix} 2 & 1 & 2 \\ 1 & 2 & 1 \end{pmatrix} \begin{pmatrix} x \\ y \\ z \end{pmatrix} = \begin{pmatrix} -1 \\ 0 \end{pmatrix}$$

$$x_1 + x_2 = 1$$
$$x_1 - x_2 = 0$$
$$x_1 + 2x_2 = 1$$

$$\begin{pmatrix} 1 & 1 \\ 1 & -1 \\ 1 & 2 \end{pmatrix} \begin{pmatrix} x_1 \\ x_2 \end{pmatrix} = \begin{pmatrix} 1 \\ 0 \\ 1 \end{pmatrix}$$

The operations we used in Section 1 to solve systems of linear equations were **elementary multiplication** of an equation by a nonzero scalar and **elementary modification** that adds a scalar multiple of another equation. The resulting system was **equivalent** to the system we started with in that both systems had precisely the same solutions. Here we apply these same operations to systems $A\mathbf{x} = \mathbf{b}$ described in terms of matrices A and vectors $\mathbf{b}$, noting their effect on the corresponding scalar equations in a system. In particular, the operations have no effect on the vector $\mathbf{x}$, consistent with the invariance of the solution set of the system. Thus we can if we like omit $\mathbf{x}$ and just operate simultaneously on A and $\mathbf{b}$.

EXAMPLE 4

To illustrate how operations on matrices correspond to operations on equations, consider the first system in Example 3. In either form the operations must be applied to *both sides* of the equations. We start with

$$4x + 3y = 1$$
$$-x + 2y = 2$$
with matrix form
$$\begin{pmatrix} 4 & 3 \\ -1 & 2 \end{pmatrix} \begin{pmatrix} x \\ y \end{pmatrix} = \begin{pmatrix} 1 \\ 2 \end{pmatrix}.$$

Multiplying the first rows of the matrices by $\frac{1}{4}$ has the same effect as multiplying the first equation by $\frac{1}{4}$:

$$x + \tfrac{3}{4}y = \tfrac{1}{4}$$
$$-x + 2y = 2$$
with matrix form
$$\begin{pmatrix} 1 & \frac{3}{4} \\ -1 & 2 \end{pmatrix} \begin{pmatrix} x \\ y \end{pmatrix} = \begin{pmatrix} \frac{1}{4} \\ 2 \end{pmatrix}.$$

Adding the first rows of these matrices to their second rows has the same effect as adding the first equation to the second:

$$x + \tfrac{3}{4}y = \tfrac{1}{4}$$
$$\tfrac{11}{4}y = \tfrac{9}{4}$$
with matrix form
$$\begin{pmatrix} 1 & \frac{3}{4} \\ 0 & \frac{11}{4} \end{pmatrix} \begin{pmatrix} x \\ y \end{pmatrix} = \begin{pmatrix} \frac{1}{4} \\ \frac{9}{4} \end{pmatrix}.$$

Multiplying the second rows of the constant matrices by $\frac{4}{11}$ has the effect of multiplying the second equation by $\frac{4}{11}$:

$$x + \tfrac{3}{4}y = \tfrac{1}{4}$$
$$y = \tfrac{9}{11}$$
with matrix form
$$\begin{pmatrix} 1 & \frac{3}{4} \\ 0 & 1 \end{pmatrix} \begin{pmatrix} x \\ y \end{pmatrix} = \begin{pmatrix} \frac{1}{4} \\ \frac{9}{11} \end{pmatrix}.$$

Adding $-\frac{3}{4}$ times the second rows of the constant matrices to the first has the effect of replacing the first equation by $x = -\frac{4}{11}$:

$$x = -\frac{4}{11}$$
$$y = \frac{9}{11}$$

with matrix form

$$\begin{pmatrix} 1 & 0 \\ 0 & 1 \end{pmatrix}\begin{pmatrix} x \\ y \end{pmatrix} = \begin{pmatrix} x \\ y \end{pmatrix} = \begin{pmatrix} -\frac{4}{11} \\ \frac{9}{11} \end{pmatrix}.$$

The unique solution is evident in either scalar equation or matrix form.

You can imagine trying to solve a large system this way, faced with a large number of possible steps to take. We'll describe a fail-safe routine, one that Theorem 2.2 in Section 2B proves will always work. It's helpful to refer to the first nonzero entry in a row of a matrix as the **leading entry** in that row. Here's the routine:

Step 1. Pick a column of A that has leading entry r, by definition nonzero, in some chosen row **r**. Multiply that row by $1/r$ to make the leading entry in **r** equal 1.

Step 2. If a row other than **r** in Step 1 has an entry $c \neq 0$ in the same column as the leading entry 1 in **r**, subtract c times **r** from the other row. Continue this process until all other entries in the column of the leading entry are 0. This column is then said to be **reduced**.

Step 3. Repeat Steps 1 and 2 until every leading entry is 1 and every column that has a leading entry is reduced. The entire matrix is then said to be **reduced**. Note that a row without a leading entry has 0 for every entry.

EXAMPLE 5 | We repeat the calculations of Example 2 of Section 1 to see the solution process in matrix form for a system with infinitely many solutions.

$$\begin{array}{rcl} 3x + 12y + 9z &=& 3 \\ 2x + 5y + 4z &=& 4 \\ -1x + 2y + z &=& -5 \end{array} ; \quad \begin{pmatrix} 3 & 12 & 9 \\ 2 & 5 & 4 \\ -1 & 2 & 0 \end{pmatrix}\begin{pmatrix} x \\ y \\ z \end{pmatrix} = \begin{pmatrix} 3 \\ 4 \\ -5 \end{pmatrix}$$

To change the system so that x appears only in the first equation, we multiply the first equation by $\frac{1}{3}$, adding (-2) times the new first equation to the second, and also adding it to the third. In matrix terms, we apply elementary multiplication by $\frac{1}{3}$ to the first rows of the coefficient matrix and the right side, then apply elementary modifications adding multiples of the first row to the second and third:

$$\begin{array}{rcl} x + 4y + 3z &=& 1 \\ -3y - 2z &=& 2 \\ 6y + 4z &=& -4 \end{array} ; \quad \begin{pmatrix} 1 & 4 & 3 \\ 0 & -3 & -2 \\ 0 & 6 & 4 \end{pmatrix}\begin{pmatrix} x \\ y \\ z \end{pmatrix} = \begin{pmatrix} 1 \\ 2 \\ -4 \end{pmatrix}.$$

The first variable x appears only in the first equation with coefficient 1; correspondingly the first column of the matrix has 1 in the first row and 0 elsewhere.

Similarly, to isolate y in the second equation, multiply the second row of the matrix and of the right side by $-\frac{1}{3}$, and then perform elementary modifications, adding (-4) times the new second row to the first row and adding (-6) times the new second row to the third row, treating the right-side entries similarly:

$$\begin{array}{rcl} x + \frac{1}{3}z &=& \frac{11}{3} \\ y + \frac{2}{3}z &=& -\frac{2}{3} \\ 0 &=& 0 \end{array} ; \quad \begin{pmatrix} 1 & 0 & \frac{1}{3} \\ 0 & 1 & \frac{2}{3} \\ 0 & 0 & 0 \end{pmatrix}\begin{pmatrix} x \\ y \\ z \end{pmatrix} = \begin{pmatrix} \frac{11}{3} \\ -\frac{2}{3} \\ 0 \end{pmatrix}.$$

The second column of the coefficient matrix now has 1 in the second row and 0 in the other rows, corresponding to y appearing only in the second equation.

All possible values of the variables satisfy the third equation, so we can ignore it. Because x appears only in the first equation and y appears only in the second, we get a solution in which $z = t$ has an arbitrary value and $x = -\frac{1}{3}t + \frac{11}{3}$ and $y = -\frac{2}{3}t - \frac{2}{3}$. As in Example 2, the set of solutions has the parametric representation of a line in $\mathbb{R}^3$: $(x, y, z) = t\left(-\frac{1}{3}, -\frac{2}{3}, 1\right) + \left(\frac{11}{3}, -\frac{2}{3}, 0\right)$.

In a reduced matrix R a leading entry is the only nonzero entry in its column, and the variable associated with that column in the system $R\mathbf{x} = \mathbf{c}$ is called a **leading variable**; all other variables in the system are called **nonleading**.

EXAMPLE 6

In the matrix equations with $\mathbf{x} = (x, y, z)$ all the real information is in the 3-by-3 matrices and the constant column vectors; the vector $\mathbf{x}$ and the equal sign just remind us of the context, so in principle could be dropped.

$$\begin{pmatrix} 1 & -2 & -3 \\ \frac{1}{2} & -2 & -\frac{13}{2} \\ -3 & 5 & 4 \end{pmatrix} \begin{pmatrix} x \\ y \\ z \end{pmatrix} = \begin{pmatrix} 2 \\ 7 \\ 0 \end{pmatrix}$$

We reduce the first column by adding $(-\frac{1}{2})$ times the first row to the second row, and 3 times the first row to the third row to get

$$\begin{pmatrix} 1 & -2 & -3 \\ 0 & -1 & -5 \\ 0 & -1 & -5 \end{pmatrix} \begin{pmatrix} x \\ y \\ z \end{pmatrix} = \begin{pmatrix} 2 \\ 6 \\ 6 \end{pmatrix}.$$

To reduce the second column we multiply the second row by (-1) to get

$$\begin{pmatrix} 1 & -2 & -3 \\ 0 & 1 & 5 \\ 0 & -1 & -5 \end{pmatrix} \begin{pmatrix} x \\ y \\ z \end{pmatrix} = \begin{pmatrix} 2 \\ -6 \\ 6 \end{pmatrix},$$

and then add 2 times the second row to the first and 1 times the second row to the third to get

$$\begin{pmatrix} 1 & 0 & 7 \\ 0 & 1 & 5 \\ 0 & 0 & 0 \end{pmatrix} \begin{pmatrix} x \\ y \\ z \end{pmatrix} = \begin{pmatrix} -10 \\ -6 \\ 0 \end{pmatrix}.$$

The leading variables are x and y, while the nonleading variable is z, and we have the two nonzero equations

$$\begin{aligned} x \quad\quad + 7z &= -10 \\ y + 5z &= -6. \end{aligned}$$

Giving z an arbitrary value, we find the unique solution with $z = t$ is $x = -10 - 7t$, $y = -6 - 5t$, $z = t$. The solutions in vector form are the points

$$(x, y, z) = (-10, -6, 0) + t(-7, 5, 1)$$

on a line in $\mathbb{R}^3$ containing $(-10, -6, 0)$ and parallel to $(-7, 5, 1)$. If the 0 on the right side of the last matrix equation had turned out to be 2, the third equation would have been $0 = 2$, so the system would have been inconsistent, with no solutions.

2B   Reduced Matrices

The examples suggest that row operations on the equations in a linear system produce solutions or else tell us that there are none. The choices we made may have looked a bit ad hoc so it may not be clear that the process outlined in Steps 1, 2, and 3 in the previous subsection always works for systems of arbitrary size. We'll now show that they provide a guaranteed routine for displaying a system in a form that makes it easy to read off the solutions. We repeat the definition of the key terms we used to describe the process, that a **leading entry** in a matrix is the first nonzero entry in a row and that a matrix is **reduced** if the following two conditions hold:

(i) Every column containing a leading entry is zero except for the leading entry.
(ii) Every leading entry is 1.

EXAMPLE 7   If

$$A = \begin{pmatrix} 1 & 2 & 0 \\ 0 & 0 & 1 \\ 0 & 0 & 0 \end{pmatrix} \quad \text{and} \quad B = \begin{pmatrix} 0 & 0 & 0 \\ 1 & 1 & 1 \\ 0 & 2 & 0 \end{pmatrix},$$

the matrix A is reduced because the top two rows have leading entry 1 with only zeros elsewhere in the columns containing the leading entries. Note that the zero row has no leading entry. The matrix B is not in reduced form because the conditions (i) and (ii) are both violated; a reduced form for B would have the 1 and 2 in the middle column replaced by 0 and 1, respectively. The reduced form of A gives us the solutions to a matrix equation $A\mathbf{x} = \mathbf{b}$ such as

$$\begin{pmatrix} 1 & 2 & 0 \\ 0 & 0 & 1 \\ 0 & 0 & 0 \end{pmatrix} \begin{pmatrix} x \\ y \\ z \end{pmatrix} = \begin{pmatrix} 1 \\ 2 \\ 0 \end{pmatrix} \quad \text{or in scalar form} \quad \begin{matrix} x + 2y = 1 \\ z = 2. \\ 0 = 0 \end{matrix}$$

Letting $y = t$ in the top scalar equation we find $x = 1-2t$, $y = t$, $z = 2$, representing the line $(x, y, z) = t(-2, 1, 0) + (1, 0, 2)$ in $\mathbb{R}^3$. On the other hand, if we replace the 0-entry on the right side of the last scalar equation by 1, the last row is $0 = 1$ so the system becomes inconsistent, with no solutions.

The Steps 1, 2, 3 listed earlier for applying elementary row operations are the main ideas we need to prove the following theorem.

2.2   Theorem.   Given a matrix A, there is a sequence of elementary operations that converts A to a reduced matrix R, namely a matrix that satisfies conditions (i) and (ii).

Proof.   Suppose the matrix A is not yet reduced. Then there must be some column containing a leading entry such that either (i) or (ii) or both fail to hold. If that column contains the leading entry r for the ith row $\mathbf{r}_i$, multiplying $\mathbf{r}_i$ by r^{-1} will make the leading entry 1. (Since r was a leading entry, it couldn't be zero, though it might be

1 to begin with.) If other entries in the column are nonzero, we can replace them by zero by adding suitable multiples of the ith row to the other rows. Another column that already satisfied (i) and (ii) before these operations must have a zero for its ith entry and therefore is unaltered by the operations. Applying this process to an unreduced matrix A increases the number of columns that satisfy the conditions (i) and (ii). If the resulting matrix is still not reduced, we repeat the process, and we obtain a reduced matrix after at most n steps, where n is the number of columns in A.   ∎

Theorem 2.2 shows that we can always use row reduction to convert a system of linear equations to a system with a reduced coefficient matrix that has the same solutions. If the reduced system has no zero rows, or if any zero rows correspond to zero entries on the right side, then the system is consistent and the solutions are all given by assigning arbitrary values to the nonleading variables.

EXAMPLE 8   The system

$$
\begin{pmatrix} 1 & -2 & 1 & 2 \\ 0 & 0 & 1 & -1 \end{pmatrix} \begin{pmatrix} x \\ y \\ z \\ w \end{pmatrix} = \begin{pmatrix} -1 \\ 1 \end{pmatrix}; \qquad \begin{aligned} x - 2y + z + 2w &= -1 \\ z - w &= 1 \end{aligned}
$$

is not reduced, but we can reduce it by subtracting the second row from the first:

$$
\begin{pmatrix} 1 & -2 & 0 & 3 \\ 0 & 0 & 1 & -1 \end{pmatrix} \begin{pmatrix} x \\ y \\ z \\ w \end{pmatrix} = \begin{pmatrix} -2 \\ 1 \end{pmatrix}; \qquad \begin{aligned} x - 2y \quad + 3w &= -2 \\ z - w &= 1 \end{aligned}.
$$

We can assign arbitrary values to the variables y and w, so for each of the values s and t there is just one solution with $y = s$ and $w = t$, obtained by then putting $x = 2s - 3t - 2$ and $z = t + 1$. The solutions are given as vectors by

$$
\begin{pmatrix} x \\ y \\ z \\ w \end{pmatrix} = \begin{pmatrix} 2s - 3t - 2 \\ s \\ t+1 \\ t \end{pmatrix} = s \begin{pmatrix} 2 \\ 1 \\ 0 \\ 0 \end{pmatrix} + t \begin{pmatrix} -3 \\ 0 \\ 1 \\ 0 \end{pmatrix} + \begin{pmatrix} -2 \\ 0 \\ 1 \\ 0 \end{pmatrix}.
$$

The general form $\mathbf{x} = s\mathbf{u}_1 + t\mathbf{u}_2 + \mathbf{v}$ for solutions of our original system $A\mathbf{x} = \mathbf{b}$ is significant in a fundamental way discussed in Section 2C. Note that if the constant vectors in this linear combination were in $\mathbb{R}^3$ instead of $\mathbb{R}^4$, we could assert that the solutions form a plane containing the point $\mathbf{v}$ in $\mathbb{R}^3$; Section 2D shows how to extend the possibility of this geometric interpretation to solutions of all systems.

2C   Homogeneous Systems

The planar solutions $\mathbf{x} = s\mathbf{u}_1 + t\mathbf{u}_2 + \mathbf{v}$ to the system $A\mathbf{x} = \mathbf{b}$ of Example 8 illustrate an important decomposition for solutions of linear systems. Setting $s = t = 0$ shows that $\mathbf{x} = \mathbf{v}$ is a solution of the original system. But if we let $\mathbf{x} = \mathbf{u}_1$ or $\mathbf{x} = \mathbf{u}_2$ for that example, we find that instead of $A\mathbf{u}_1 = \mathbf{b}$ or $A\mathbf{u}_2 = \mathbf{b}$ we get $A\mathbf{u}_1 = A\mathbf{u}_2 = \mathbf{0}$. Thus $\mathbf{x} = \mathbf{u}_1$ and $\mathbf{x} = \mathbf{u}_2$ are solutions of the **homogeneous** equation $A\mathbf{x} = \mathbf{0}$ that we get when we set $\mathbf{b} = \mathbf{0}$ in $A\mathbf{x} = \mathbf{b}$. To explain what's going on here we start with the following property of matrix-vector products.

2.3   Linearity of Matrix Multiplication.   If the products are defined, then

$$A(s\mathbf{u} + t\mathbf{v}) = sA\mathbf{u} + tA\mathbf{v},$$

which by repeated application implies

$$A(t_1\mathbf{u}_1 + t_2\mathbf{u}_2 + \cdots + t_k\mathbf{u}_k) = t_1 A\mathbf{u}_1 + t_2 A\mathbf{u}_1 + \cdots + t_k A\mathbf{u}_k.$$

Proof.   Let $\mathbf{r}_i$ be the ith row of A, so the ith entries in $A(s\mathbf{u}+t\mathbf{v})$, $A\mathbf{u}$, and $A\mathbf{v}$ are $\mathbf{r}_i \cdot (s\mathbf{u} + t\mathbf{v})$, $\mathbf{r}_i \cdot \mathbf{u}$ and $\mathbf{r}_i \cdot \mathbf{v}$. By additivity and homogeneity of the dot product,

$$\mathbf{r}_i \cdot (s\mathbf{u} + t\mathbf{v}) = \mathbf{r}_i \cdot (s\mathbf{u}) + \mathbf{r}_i \cdot (t\mathbf{v}) = s(\mathbf{r}_i \cdot \mathbf{u}) + t(\mathbf{r}_i \cdot \mathbf{v}).$$

The expression on the right is the ith entry in $sA\mathbf{u} + tA\mathbf{v}$. To get the more general equation, apply the two-term version to $t_1\mathbf{u}_1 + (t_2\mathbf{u}_2 + \cdots + t_k\mathbf{u}_k)$, and then successively split off one more term at a time.   ■

Remark.   The term *linearity* applied to the property of matrix-vector multiplication in Theorem 2.3 stems from the observation that multiplication of the points on a line $t\mathbf{u} + \mathbf{v}$ in $\mathbb{R}^2$ or $\mathbb{R}^3$ by A carries the line into another line, or possibly just a point. The reason is that, by Theorem 2.3, $A(t\mathbf{u} + \mathbf{v}) = tA\mathbf{u} + A\mathbf{v}$. Thus if $A\mathbf{u} \neq \mathbf{0}$ the result of applying A is a line through the point $A\mathbf{v}$ and parallel to the vector $A\mathbf{u}$. If $A\mathbf{u} = \mathbf{0}$ we get only the point $A\mathbf{v}$.

Here is the basic theorem about the structure of solutions of $A\mathbf{x} = \mathbf{b}$.

2.4   Theorem.   Every solution of the matrix equation $A\mathbf{x} = \mathbf{b}$ has the form $\mathbf{x}_h + \mathbf{x}_p$, where $\mathbf{x}_p$ is some particular solution and $\mathbf{x}_h$ is a solution of the homogeneous equation $A\mathbf{x} = \mathbf{0}$.

Proof.   Let $\mathbf{x}_a$ be an arbitrary solution of $A\mathbf{x} = \mathbf{b}$. Then by Theorem 2.3 applied to just two terms we have

$$A(\mathbf{x}_a - \mathbf{x}_p) = A\mathbf{x}_a - A\mathbf{x}_p = \mathbf{b} - \mathbf{b} = \mathbf{0}.$$

Thus $\mathbf{x}_a - \mathbf{x}_p = \mathbf{x}_h$ for some solution $\mathbf{x}_h$ of $A\mathbf{x} = \mathbf{0}$, so $\mathbf{x}_a = \mathbf{x}_h + \mathbf{x}_p$.   ■

EXAMPLE 9

In Example 8 we exhibited the solutions of a system $A\mathbf{x} = \mathbf{b}$ in the form $\mathbf{x} = s\mathbf{u}_1 + t\mathbf{u}_2 + \mathbf{v}$. This illustrates Theorem 2.4 because we get a particular solution $\mathbf{x}_p = \mathbf{v}$ by setting $s = t = 0$, and the other part of the solution, $s\mathbf{u}_1 + t\mathbf{u}_2$, consists of solutions of the homogeneous system. A vector $\mathbf{w}$ such that $\mathbf{x} = \mathbf{w} + \mathbf{v}$ is also a solution must necessarily be a solution of $A\mathbf{x} = \mathbf{0}$ because, by Theorem 2.3,

$$A(\mathbf{w} + \mathbf{v}) = A\mathbf{w} + A\mathbf{v} = A\mathbf{w} + \mathbf{b} = \mathbf{b}.$$

Subtracting $\mathbf{b}$ from both sides of the last equality shows that $A\mathbf{w} = \mathbf{0}$.

The decomposition $\mathbf{x} = \mathbf{x}_h + \mathbf{x}_p$ of solutions in Theorem 2.4 reduces to just $\mathbf{x} = \mathbf{x}_p$ precisely when $\mathbf{x} = \mathbf{0}$ is the only solution of $A\mathbf{x} = \mathbf{0}$. We now examine in general how to tell whether $A\mathbf{x} = 0$ has multiple solutions or only the zero solution.

A zero row in a matrix R represents the scalar equation $0 = 0$ in the system $R\mathbf{x} = \mathbf{0}$, and has no effect on the solutions of the system. We call such equations *trivial* and don't count them as belonging to the system of scalar equations.

2.5   Theorem.   A homogeneous system $A\mathbf{x} = \mathbf{0}$ has infinitely many nonzero solutions if it has more variables than equations. It also has infinitely many nonzero solutions if an equivalent reduced system $R\mathbf{x} = \mathbf{0}$ has more variables than it has nontrivial equations. It has only the zero solution if $R\mathbf{x} = \mathbf{0}$ has at least as many nontrivial equations as variables.

Proof.   Let A have n columns and m rows, with $m < n$. When we convert A to a reduced form R, the zero rows in R will be consistent because the right side is zero also. There can be at most m leading entries, so at most m columns with leading entries. We may specify arbitrary values for each variable that corresponds to a column having no leading entry, then solve in terms of these for the variables that correspond to leading entries. Given the infinitely many arbitrary values for at least one variable, we get infinitely many solutions. The same argument applies to $R\mathbf{x} = \mathbf{0}$ if we don't count trivial equations $0 = 0$. Finally, a reduced system with at least as many nontrivial equations as variables has a leading entry in every column, so has only the zero solution.   ∎

EXAMPLE 10

The system

$$\begin{pmatrix} 1 & -2 & 1 \\ 2 & 1 & -3 \end{pmatrix} \begin{pmatrix} x \\ y \\ z \end{pmatrix} = \begin{pmatrix} 0 \\ 0 \end{pmatrix} \quad \text{or} \quad \begin{array}{r} x - 2y + z = 0 \\ 2x + y - 3z = 0 \end{array}$$

is an example of Theorem 2.5, because it is homogeneous and has more unknowns than equations. We can regard the solutions as the intersection of two planes, and they include the trivial zero solution $(x, y, z) = (0, 0, 0)$. Hence the planes intersect in at least an entire line through the origin. It's straightforward to check that the solutions are of the form $(x, y, z) = t(1, 1, 1)$, where t ranges over all real numbers; to see that these represent *all* solutions just set $z = t$ in the reduced form

$$\begin{pmatrix} 1 & 0 & -1 \\ 0 & 1 & -1 \end{pmatrix} \begin{pmatrix} x \\ y \\ z \end{pmatrix} = \begin{pmatrix} 0 \\ 0 \end{pmatrix} \quad \text{or} \quad \begin{array}{r} x \quad + -z = 0 \\ y + -z = 0. \end{array}$$

Thus with $z = t$ the line of solutions is $(x, y, z) = t(1, 1, 1)$.

EXAMPLE 11

In an example like this one

$$\begin{pmatrix} 1 & 0 & 0 \\ 0 & 1 & 0 \end{pmatrix} \begin{pmatrix} x \\ y \\ z \end{pmatrix} = \begin{pmatrix} 0 \\ 0 \end{pmatrix} \quad \text{or} \quad \begin{array}{r} x \quad + 0z = 0 \\ y + 0z = 0, \end{array}$$

in which z appears only with zero coefficients, we still have to remember it's there and set $z = t$ to get all solutions $(x, y, z) = (0, 0, t)$.

Theorem 2.4 links multiple solutions of a system $A\mathbf{x} = \mathbf{b}$ with multiple solutions of the associated homogeneous system $A\mathbf{x} = \mathbf{0}$. Thus $A\mathbf{x} = \mathbf{b}$ has a unique solution precisely when $\mathbf{x} = \mathbf{0}$ is the only solution to $A\mathbf{x} = \mathbf{0}$, so it's important to be able to tell when the latter happens. Theorem 2.5 gives one way of answering the question, and Sections 4 and 5 present special criteria that apply when A is a square matrix. In the following discussion we assume nothing about the dimensions of A.

If A has columns $\mathbf{u}_j$ with entries u_{ij}, then

2.6
$$A\mathbf{x} = \begin{pmatrix} x_1 u_{11} + & \cdots & + x_n u_{1n} \\ \vdots & \vdots & \vdots \\ x_1 u_{m1} + & \cdots & + x_n u_{mn} \end{pmatrix} = x_1 \mathbf{u}_1 + \cdots + x_n \mathbf{u}_n.$$

From this equation we see that $A\mathbf{x} = 0$ is equivalent to $x_1 \mathbf{u}_1 + \cdots + x_n \mathbf{u}_n = \mathbf{0}$. If the scalars x_k aren't all zero, this relation among the columns of A gives rise to multiple solutions to $A\mathbf{x} = \mathbf{0}$, while the lack of such a relation gives only the solution $\mathbf{x} = \mathbf{0}$. Here is the standard generalization of our earlier definition of linear independence, stated in Chapter 1 for just two vectors $\mathbf{u}_1$ and $\mathbf{u}_2$.

> **2.7 Definition**   Vectors $\mathbf{u}_1, \ldots, \mathbf{u}_n$ are **linearly independent** if the equation $x_1 \mathbf{u}_1 + \cdots + x_n \mathbf{u}_n = \mathbf{0}$ is satisfied only by choosing all $x_k = 0$, or equivalently by Equation 2.6, if the only solution to $A\mathbf{x} = \mathbf{0}$ is $\mathbf{x} = \mathbf{0}$, where A is the matrix with the $\mathbf{u}_k$ for columns. Otherwise the vectors are **linearly dependent**.

Definition 2.7 becomes more intuitive and is often easier to apply, in a form that explicitly contains our original definition for two vectors:

> **2.7′ Definition**   Vectors $\mathbf{u}_1, \ldots, \mathbf{u}_n$ are **linearly independent** if no one of them is a linear combination of the other $n-1$ vectors. Otherwise the vectors are **linearly dependent**.

The following theorem allows us to use whichever of the two definitions is more convenient at a given point.

2.8   Theorem.   Definitions 2.7 and 2.7′ of linear independence are equivalent.

Proof.   The equation $x_1 \mathbf{u}_1 + \cdots + x_n \mathbf{u}_n = \mathbf{0}$ is equivalent to

$$x_k \mathbf{u}_k = -x_1 \mathbf{u}_1 - \cdots - x_{k-1} \mathbf{u}_{k-1} - x_{k+1} \mathbf{u}_{k+1} - \cdots - x_n \mathbf{u}_n.$$

If $\mathbf{u}_k$ is a linear combination of the other $\mathbf{u}$'s, then both equations hold with $x_k = 1 \neq 0$. But if the equations hold with some $x_k \neq 0$, then dividing the second equation by x_k shows that $\mathbf{u}_k$ is a linear combination of the other $\mathbf{u}$'s.   ∎

EXAMPLE 12   Let $A = \begin{pmatrix} 1 & 3 \\ 2 & 2 \\ 3 & 1 \end{pmatrix}$ and $B = \begin{pmatrix} 1 & 3 & 4 \\ 2 & 2 & 4 \\ 3 & 1 & 4 \end{pmatrix}$. The columns of A are linearly independent, because neither is a scalar multiple of the other, so the equation $A\mathbf{x} = \mathbf{0}$ has only the solution $\mathbf{x} = \mathbf{0}$. The columns of B are linearly dependent, because the third is the sum of the first two, so the equation $B\mathbf{x} = \mathbf{0}$ has multiple solutions.

EXERCISES

Note. A linear system may have just one solution, or infinitely many solutions, or else no solutions if it's inconsistent. Solving a system requires finding all solutions or showing that there are none if that's the case.

In Exercises 1 to 4, (a) write the system of equations in matrix form; that is, find a matrix A and a vector $\mathbf{b}$ such that the system is equivalent to the equation $A\mathbf{x} = \mathbf{b}$, and (b) solve the system.

1. $3x - 2y = 1$
$\quad\ x - 3y = 2$

2. $3x + y + z = 1$
$\quad\ x - y - z = 0$

3. $x + y \quad\ = 1$
$\quad\quad y - z = 1$
$\quad x \quad\ + z = 0$

4. $\quad x + 2y = 0$
$\quad\ x - \ \ y = 0$
$\quad -x + \ \ y = 1$

In Exercises 5 to 8, (a) write a system of equations equivalent to the given matrix equation and (b) solve the system.

5. $\begin{pmatrix} 1 & 2 \\ 3 & 1 \end{pmatrix} \begin{pmatrix} x \\ y \end{pmatrix} = \begin{pmatrix} 1 \\ 0 \end{pmatrix}$

6. $\begin{pmatrix} -1 & 2 \\ 0 & 1 \end{pmatrix} \mathbf{x} = \begin{pmatrix} 0 \\ 0 \end{pmatrix}$

7. $\begin{pmatrix} 1 & 0 & 1 \\ 0 & 1 & 0 \\ 1 & 1 & 0 \end{pmatrix} \mathbf{x} = \begin{pmatrix} 0 \\ 1 \\ 0 \end{pmatrix}$

8. $\begin{pmatrix} 2 & 1 \\ 1 & 2 \\ 3 & 3 \end{pmatrix} \mathbf{x} = \begin{pmatrix} 1 \\ 0 \\ 2 \end{pmatrix}$

In Exercises 9 and 10, solve the given system.

9. $\begin{pmatrix} 1 & 1 & 1 \\ -1 & 2 & -4 \\ 1 & 3 & 9 \end{pmatrix} \begin{pmatrix} x \\ y \\ z \end{pmatrix} = \begin{pmatrix} 2 \\ 2 \\ 0 \end{pmatrix}$

10. $\begin{pmatrix} 1 & 1 & 0 \\ -2 & 1 & 1 \\ 3 & -1 & 2 \end{pmatrix} \begin{pmatrix} u \\ v \\ w \end{pmatrix} = \begin{pmatrix} 0 \\ 1 \\ -1 \end{pmatrix}$

In Exercises 11 to 14, (a) find a reduced matrix equivalent to A, (b) solve the system $A\mathbf{x} = 0$, (c) solve the system $A\mathbf{x} = \mathbf{e}_1 + \mathbf{e}_2$.

11. $A = \begin{pmatrix} 1 & -2 & 1 \\ 2 & 1 & -3 \end{pmatrix}$

12. $A = \begin{pmatrix} 0 & 1 & 1 \\ 1 & 1 & 1 \end{pmatrix}$

13. $A = \begin{pmatrix} 1 & 1 & 1 \\ 1 & 1 & 0 \\ 1 & 0 & 0 \end{pmatrix}$

14. $A = \begin{pmatrix} 0 & 0 & 2 \\ 2 & 0 & 0 \\ 0 & 0 & 3 \end{pmatrix}$

15. Express the vectors $\mathbf{i}, \mathbf{j}$ in $\mathbb{R}^2$ as linear combinations of $(1, 2)$ and $(2, 3)$ by solving an appropriate system of equations for the coefficients of combination.

16. Express the vectors $\mathbf{i}, \mathbf{j}, \mathbf{k}$ in $\mathbb{R}^3$ as linear combinations of $(1, 1, 1)$, $(1, 1, 0)$, and $(1, 0, 0)$.

17. Express the vector $(5, 0, 1, 2)$ as a linear combination of $(1, 2, 1, 0)$ and $(2, -1, 0, 1)$.

18. Can an arbitrary vector in $\mathbb{R}^4$ be expressed as a linear combination of the last two vectors in Exercise 17? Explain.

19. Solve the system $\begin{pmatrix} 1 & 2 & 3 & 4 & 5 \\ 0 & 1 & 2 & 3 & 4 \\ 0 & 0 & 1 & 2 & 3 \\ 0 & 0 & 0 & 1 & 2 \\ 0 & 0 & 0 & 0 & 1 \end{pmatrix} \mathbf{x} = \begin{pmatrix} 6 \\ 5 \\ 4 \\ 3 \\ 2 \end{pmatrix}$.

20. Solve the system
$\begin{pmatrix} 1 & 2 & 3 & 4 & 5 \\ 0 & 1 & 2 & 3 & 4 \\ 1 & 0 & -1 & -2 & -3 \\ 0 & 0 & 0 & 1 & 2 \\ 2 & 3 & 4 & 6 & 8 \end{pmatrix} \mathbf{x} = \begin{pmatrix} 6 \\ 5 \\ -4 \\ 3 \\ 10 \end{pmatrix}$.

In Exercises 21 to 24, determine whether or not the vector $\mathbf{v}$ is a linear combination of the other vectors given.

21. $\mathbf{v} = 2\mathbf{i} + 3\mathbf{j}; \mathbf{a} = 2\mathbf{i} - \mathbf{j}, \mathbf{b} = 2\mathbf{i} + \mathbf{j}$

22. $\mathbf{v} = 2\mathbf{i} + 3\mathbf{j} + 4\mathbf{k}; \mathbf{a} = 2\mathbf{i} - \mathbf{j}, \mathbf{b} = \mathbf{i} + \mathbf{j} + \mathbf{k},$
$\quad \mathbf{c} = \mathbf{j} - 2\mathbf{k}$

23. $\mathbf{v} = (-1, 0, -1); \mathbf{a} = (2, -1, 2), \mathbf{b} = (1, 1, -3), \mathbf{c} = (0, 1, 2)$

24. $\mathbf{v} = (3, -1, 0, -1); \mathbf{a} = (2, -1, 3, 2), \mathbf{b} = (-1, 1, 1, -3),$
$\quad \mathbf{c} = (1, 1, 9, -5)$

In Exercises 25 and 26, parametric representations are given for two lines, L_1 and L_2. Show that there is just one line L_3 that intersects both L_1 and L_2 at right angles, and find a parametric representation for it. [*Hint:* If the parameters s and t have values corresponding to the points where L_3 intersects L_1 and L_2, then the vector from one point to the other must be perpendicular to both L_1 and L_2. Show that this leads to two linear equations that s and t must satisfy.]

25. L_1: $s(3, 2, 1) + (-1, 0, 1)$, L_2: $t(0, 2, 1) + (-4, 2, 0)$

26. L_1: $s(-1, 3, 0)$, L_2: $t(4, 0, 1) + (-2, 1, 1)$

*****27.** Show that if L_1 and L_2 are two lines in $\mathbb{R}^3$ that do not intersect and are not parallel, then there is a unique third line L_3 that intersects both of them at right angles, as in

Exercise 25. What if the lines are parallel? What if they intersect?

28. Show that if $\mathbf{v}_1, \dots, \mathbf{v}_k$ are solutions of a homogeneous system $A\mathbf{x} = \mathbf{0}$, then every linear combination of them is also a solution.

29. Let $\mathbf{x} = \mathbf{v}$ be one solution of the system $A\mathbf{x} = \mathbf{b}$. Show that $\mathbf{w}$ is also a solution if and only if $\mathbf{w} - \mathbf{v}$ is a solution of the homogeneous system $A\mathbf{x} = \mathbf{0}$. This version of linearity is sometimes called the **superposition principle**.

30. Show that if $\mathbf{x} = \mathbf{v}$ and $\mathbf{x} = \mathbf{u} + \mathbf{v}$ both satisfy $A\mathbf{x} = \mathbf{b}$, then $A\mathbf{u} = \mathbf{0}$.

31. Show that if $\mathbf{x}_1$ satisfies $A\mathbf{x} = \mathbf{b}_1$ and $\mathbf{x}_2$ satisfies $A\mathbf{x} = \mathbf{b}_2$, then $t_1\mathbf{x}_1 + t_2\mathbf{x}_2$ satisfies $A\mathbf{x} = t_1\mathbf{b}_1 + t_1\mathbf{b}_2$.

2D   Geometry of Solution Sets

In the systems we've solved so far, solution sets have been either empty, a single point, a line, or a plane. We'll show here that the solution set of every system of linear equations has a similar form. Crucial to calling a set of points $\mathbf{x} = s\mathbf{u} + t\mathbf{v}$, a plane is that neither $\mathbf{u}$ nor $\mathbf{v}$ is a scalar multiple of the other, which we called linear independence of $\mathbf{u}$ and $\mathbf{v}$, now seen as a special case of Definition 2.7′ in Section 2C. Linear independence of the set $S = \{\mathbf{u}_1, \mathbf{u}_2, \dots, \mathbf{u}_k\}$ implies that the set of linear combinations $t_1\mathbf{u}_1 + t_2\mathbf{u}_2 + \cdots + t_k\mathbf{u}_k$ can't be collapsed into the set of linear combinations of a smaller subset of S by replacing a vector $\mathbf{u}_i$ by a linear combination of the others. This collapse is what happens when a parametric representation $\mathbf{x} = t_1\mathbf{u}_1 + t_2\mathbf{u}_2 + \mathbf{v}$ of what appears superficially to be a plane collapses to a line $\mathbf{x} = (t_1 + ct_2)\mathbf{u}_1 + \mathbf{v}$ if $\mathbf{u}_2 = c\mathbf{u}_1$, or to a single point $\mathbf{x} = \mathbf{v}$ if $\mathbf{u}_1 = \mathbf{u}_2 = \mathbf{0}$.

> **2.9 Definition**   A subset of $\mathbb{R}^n$ is called a k-**plane** if it consists of all points $\mathbf{x} = t_1\mathbf{u}_1 + \cdots + t_k\mathbf{u}_k + \mathbf{v}$, where $\mathbf{v}$ is a fixed vector, and the $\mathbf{u}_i$ are fixed linearly independent vectors, with the t_i varying over all real numbers.

Comparing Definition 2.9 with the special cases in Chapter 1, we see that an ordinary plane is a 2-plane and a line is a 1-plane. We can even take $k = 0$ and regard a single point as a 0-plane. Every example of a system of linear equations that we have treated has had for its solution set a k-plane for $k = 0$, 1, or 2, or else has had no solutions at all.

EXAMPLE 13   In Example 8, $A\mathbf{x} = \mathbf{b}$ stood for two equations in four variables:

$$\begin{pmatrix} 1 & -2 & 0 & 3 \\ 0 & 0 & 1 & -1 \end{pmatrix}\mathbf{x} = \begin{pmatrix} -2 \\ 1 \end{pmatrix}; \qquad \begin{aligned} x_1 - 2x_2 \quad + 3x_4 &= -2 \\ x_3 - x_4 &= 1 \end{aligned}.$$

For arbitrary scalars s and t there is a unique solution in which the two nonleading variables, x_2 and x_4 have the values $x_2 = s$ and $x_4 = t$, namely,

$$\mathbf{x} = \begin{pmatrix} 2s - 3t - 2 \\ s \\ t+1 \\ t \end{pmatrix} = s\begin{pmatrix} 2 \\ 1 \\ 0 \\ 0 \end{pmatrix} + t\begin{pmatrix} -3 \\ 0 \\ 1 \\ 0 \end{pmatrix} + \begin{pmatrix} -2 \\ 0 \\ 1 \\ 0 \end{pmatrix}.$$

This represents a 2-plane $\mathbf{x} = s\mathbf{u}_1 + t\mathbf{u}_2 + \mathbf{v}$ if we take

$$\mathbf{u}_1 = \begin{pmatrix} 2 \\ 1 \\ 0 \\ 0 \end{pmatrix}, \ \mathbf{u}_2 = \begin{pmatrix} -3 \\ 0 \\ 1 \\ 0 \end{pmatrix}, \ \text{and } \mathbf{v} = \begin{pmatrix} -2 \\ 0 \\ 1 \\ 0 \end{pmatrix}.$$

To check that $\mathbf{u}_1$ and $\mathbf{u}_2$ are linearly independent, note that the second entries in $\mathbf{u}_1$ and $\mathbf{u}_2$ are respectively 0 and 1, while the fourth entries are respectively 1 and 0. Thus neither $\mathbf{u}_1$ nor $\mathbf{u}_2$ can be a scalar multiple of the other, so the set of solutions is a 2-plane in $\mathbb{R}^4$ parallel to $\mathbf{u}_1$ and $\mathbf{u}_2$, and containing the point $\mathbf{v}$. Note that $\mathbf{x}_h = s\mathbf{u}_1 + t\mathbf{u}_2$ solves $A\mathbf{x} = \mathbf{0}$ while $\mathbf{x}_p = \mathbf{v}$ solves $A\mathbf{x} = \mathbf{b}$, illustrating the decomposition of solutions into homogeneous plus particular in Theorem 2.4.

In a reduced matrix with n columns and r nonzero rows, the number of nonleading variables is $k = n - r$, because every nonzero row of a reduced matrix contains just one leading entry and corresponds to just one variable. Thus a system of m linear equations in n variables has a solution set that is an $(n-m)$-plane unless the result of applying row reduction to the coefficient matrix produces one or more zero rows. In particular, the solution set of a single linear equation in $\mathbb{R}^n$ is an $(n-1)$-plane, sometimes called a **hyperplane** in $\mathbb{R}^n$.

EXAMPLE 14

We may write a single linear equation in the form $\mathbf{a} \cdot \mathbf{x} = b$. For an example in $\mathbb{R}^4$, consider $(0, -3, -2, 1) \cdot \mathbf{x} = 3$, with $\mathbf{x} = (x_1, x_2, x_3, x_4)$. Reduction of the single matrix row is just division by -3, giving the equivalent equation

$$(0, 1, \tfrac{2}{3}, -\tfrac{1}{3}) \cdot \mathbf{x} = -1 \quad \text{or} \quad x_2 + \tfrac{2}{3}x_3 - \tfrac{1}{3}x_4 = -1.$$

The only leading variable is x_2, and the nonleading variables are x_1, x_3, and x_4. Proceeding as in the previous example, we set $x_1 = s$, $x_3 = t$, and $x_4 = u$ to find as solution set the hyperplane in $\mathbb{R}^4$ consisting of the points

$$\mathbf{x} = \begin{pmatrix} s \\ -\tfrac{2}{3}t + \tfrac{1}{3}u - 1 \\ t \\ u \end{pmatrix} = s\begin{pmatrix} 1 \\ 0 \\ 0 \\ 0 \end{pmatrix} + t\begin{pmatrix} 0 \\ -\tfrac{2}{3} \\ 1 \\ 0 \end{pmatrix} + u\begin{pmatrix} 0 \\ \tfrac{1}{3} \\ 0 \\ 1 \end{pmatrix} + \begin{pmatrix} 0 \\ -1 \\ 0 \\ 0 \end{pmatrix}.$$

The first three terms in the sum are linearly independent since each one has a 1 in the entry where the other two have only zero, so their linear combination is the general solution $\mathbf{x}_h$ of $\mathbf{a} \cdot \mathbf{x} = 0$. The fourth vector is a solution $\mathbf{x}_p$ of the nonhomogeneous equation, so the solutions form a hyperplane containing $\mathbf{x}_p$ in $\mathbb{R}^n$.

Checking for Independence.   For arbitrary sets of vectors the simple check for independence we used at the end of the previous example is often impossible, but there is a routine check. The method depends on knowing that applying elementary row operations to a matrix A with columns $\mathbf{a}_1, \dots \mathbf{a}_m$ preserves a dependence relation among the columns. For instance if we get B from A by applying row operations, then $\mathbf{a}_1 = \mathbf{a}_2 - 2\mathbf{a}_3$ if and only if $\mathbf{b}_1 = \mathbf{b}_2 - 2\mathbf{b}_3$. The reason is that the row operations (i) multiplication by $r \neq 0$ and (ii) adding a multiple of one row to another, preserve a dependence relation in each affected row. More formally, we have the following.

2.10   Theorem.   If a matrix B is obtained from a matrix A by application of row operations, then a dependence relation among columns of one matrix holds for the corresponding columns of the other.

Proof.   We can express a linear relation $x_1\mathbf{a}_1 + \cdots + x_n\mathbf{a}_n = \mathbf{0}$ among the columns $\mathbf{a}_j$ of A as $A\mathbf{x} = \mathbf{0}$, where $x_j = 0$ if $\mathbf{a}_j$ isn't involved. Since solution vectors $\mathbf{x}$ of $A\mathbf{x} = \mathbf{0}$ are unchanged by row operations on A, a linear relation among the columns of A carries over to the same relation among the corresponding columns of B.   ∎

2.11   Theorem.   Let $\mathbf{u}_1, \ldots, \mathbf{u}_m$ be m vectors in $\mathbb{R}^n$, and let A be the n-by-m matrix with the $\mathbf{u}_j$ as columns. Then $\mathbf{u}_1, \ldots, \mathbf{u}_m$ are linearly independent if and only if every column in a reduced form R of A contains a leading entry. A column of A with no leading entry in the corresponding column of R is a linear combination of the columns of A corresponding to columns that do have leading entries in R.

Proof.   A column $\mathbf{r}_k$ of R with a leading entry can't be a linear combination of other columns containing leading entries, because $\mathbf{r}_k$ has a 1 where the others have zeros. Thus the columns are independent if every column of R contains a leading entry. To write a column $\mathbf{r}_k$ of R with only nonleading entries as a linear combination of the columns with leading entries 1, multiply each such column $\mathbf{r}_j$ by the corresponding nonleading entry in $\mathbf{r}_k$ and add the results to get $\mathbf{r}_k$. The same dependence relation then holds among the corresponding columns of A by Theorem 2.10.   ∎

EXAMPLE 15

To find out whether $\mathbf{a} = (1, 2, 3, -1)$, $\mathbf{b} = (0, 1, -1, 1)$, and $\mathbf{c} = (-1, 0, 2, -1)$ are linearly independent, we form the matrix A that has them as columns, with reduced form R for which we omit the details:

$$A = \begin{pmatrix} 1 & 0 & -1 \\ 2 & 1 & 0 \\ 3 & -1 & 2 \\ -1 & 1 & -1 \end{pmatrix}; \quad R = \begin{pmatrix} 1 & 0 & 0 \\ 0 & 1 & 0 \\ 0 & 0 & 0 \\ 0 & 0 & 1 \end{pmatrix}.$$

Every column of R has a leading entry, so $\mathbf{a}$, $\mathbf{b}$, and $\mathbf{c}$ are linearly independent.

Theorem 2.11 tells us that if some column of a reduced matrix we're checking for dependent columns has no leading entry, that column will be a linear combination of the columns that do have only leading entries, as in the following example.

EXAMPLE 16

To check the vectors $\mathbf{a} = (1, 0, 1, 0)$, $\mathbf{b} = (0, 1, 1, 0)$, $\mathbf{c} = (1, 1, 2, 1)$, and $\mathbf{d} = (-3, -4, -7, -3)$ for independence, we form the matrix A that has them as columns, showing also the system $A\mathbf{x} = 0$ to aid understanding the dependence relations:

$$A = \begin{pmatrix} 1 & 0 & 1 & -3 \\ 0 & 1 & 1 & -4 \\ 1 & 1 & 2 & -7 \\ 0 & 0 & 1 & -3 \end{pmatrix}; \quad \begin{array}{r} x_1 + x_3 - 3x_4 = 0 \\ x_2 + x_3 - 4x_4 = 0 \\ x_1 + x_2 + 2x_3 - 7x_4 = 0 \\ x_3 - 3x_4 = 0 \end{array}.$$

A row reduction left as Exercise 5(b) gives an R and a reduced system $R\mathbf{x} = \mathbf{0}$:

$$R = \begin{pmatrix} 1 & 0 & 0 & 0 \\ 0 & 1 & 0 & -1 \\ 0 & 0 & 0 & 0 \\ 0 & 0 & 1 & -3 \end{pmatrix}; \qquad \begin{aligned} x_1 & & & = 0 \\ x_2 & & - x_4 & = 0 \\ & & 0 & = 0 \\ x_3 & - 3x_4 & & = 0. \end{aligned}$$

Solving the corresponding system $R\mathbf{x} = \mathbf{0}$, we can set the sole nonleading variable x_4 equal to 1, so $x_1 = 0$, $x_2 = 1$, and $x_3 = 3$. Hence $\mathbf{b} + 3\mathbf{c} + \mathbf{d} = \mathbf{0}$. Evidently $\mathbf{d} = -\mathbf{b} - 3\mathbf{c}$, so $\{\mathbf{a}, \mathbf{b}, \mathbf{c}, \mathbf{d}\}$ is a linearly dependent set of vectors, and even the smaller set $\{\mathbf{b}, \mathbf{c}, \mathbf{d}\}$ is dependent. But $\{\mathbf{a}, \mathbf{b}, \mathbf{c}\}$ is an independent set. What about $\{\mathbf{a}, \mathbf{b}, \mathbf{d}\}$? By looking carefully at R, you can answer this question without referring to the system $R\mathbf{x} = \mathbf{0}$.

EXERCISES

1. Solve the equation $w + 3x - 2y + z = 3$ by expressing the solutions parametrically as a 3-plane in $\mathbb{R}^4$.

2. Solve the equation $u + v + w - x - y = 1$ by expressing the solutions parametrically as a 4-plane in $\mathbb{R}^5$.

3. Carry out the row reduction indicated in text Example 15.

4. Find a reduced form R for the matrix

$$\begin{pmatrix} 1 & 0 & -1 & 3 \\ 2 & 1 & 0 & -1 \\ 3 & -1 & 2 & 0 \\ -1 & 1 & -1 & 2 \end{pmatrix},$$

and use R to show that the columns of the matrix are linearly independent.

5. In Example 16 in the text, the four vectors $\mathbf{a} = (1, 0, 1, 0)$, $\mathbf{b} = (0, 1, 1, 0)$, $\mathbf{c} = (1, 1, 2, 1)$, and $\mathbf{d} = (-3, -4, -7, -3)$ were found to be linearly dependent, while the three vectors $\mathbf{a}, \mathbf{b}, \mathbf{c}$ were found to be linearly independent.
   (a) Test the sets of three vectors $\mathbf{a}, \mathbf{b}, \mathbf{d}$; $\mathbf{a}, \mathbf{c}, \mathbf{d}$; and $\mathbf{b}, \mathbf{c}, \mathbf{d}$ for independence.
   (b) Carry out the row reduction we used in Example 16.

6. Let $\mathbf{x} = (1, 3, 9)$, $\mathbf{y} = (0, -4, 6)$, $\mathbf{z} = (1, 7, 3)$, and $\mathbf{w} = (1, 2, 4)$. Find which of the sets $\{\mathbf{x}, \mathbf{y}, \mathbf{z}, \mathbf{w}\}$, $\{\mathbf{x}, \mathbf{y}, \mathbf{w}\}$, and $\{\mathbf{x}, \mathbf{y}, \mathbf{z}\}$ are linearly independent.

7. Show that two nonzero vectors $\mathbf{u}, \mathbf{v}$ in $\mathbb{R}^3$ are linearly independent if and only if they don't both lie on a line through the origin.

8. Show that three nonzero vectors $\mathbf{u}, \mathbf{v}, \mathbf{w}$ in $\mathbb{R}^3$ are linearly independent if and only if each pair determines a plane that doesn't contain the third vector.

9. Let $A = \begin{pmatrix} 1 & 2 & -1 \\ 0 & 1 & 3 \end{pmatrix}$. Show that points $\mathbf{x}$ such that $A\mathbf{x} = \mathbf{0}$ form a line.

10. Show that the system's solutions form a line, or 1-plane, containing $\mathbf{0}$ in $\mathbb{R}^3$:

$$\begin{pmatrix} 1 & 0 & -2 \\ 0 & 1 & -3 \\ -1 & 1 & -1 \end{pmatrix} \begin{pmatrix} x \\ y \\ z \end{pmatrix} = \begin{pmatrix} 1 \\ 1 \\ 0 \end{pmatrix}.$$

11. Show that the system's solutions form a 2-plane containing $\mathbf{0}$ in $\mathbb{R}^4$:

$$\begin{pmatrix} 1 & 0 & 0 & 1 \\ 0 & 0 & 1 & 1 \\ 1 & 0 & 1 & 2 \\ 1 & 0 & 2 & 3 \end{pmatrix} \begin{pmatrix} x \\ y \\ z \\ w \end{pmatrix} = \begin{pmatrix} 0 \\ 0 \\ 0 \\ 0 \end{pmatrix}.$$

A matrix R is in **echelon form** if it's not only reduced but the leading entries shift to the right as you go down the columns and all zero rows are at the bottom. It can be shown that a given matrix A is reducible to a *unique* matrix in echelon form.

12. Show that given a reduced matrix R there is a sequence of row interchanges that will put R in echelon form.

13. Show that if an n-by-n matrix is in echelon form then either it has one or more zero rows, or else its columns are the standard basis vectors $\mathbf{e}_1, \ldots, \mathbf{e}_n$.

SECTION 3   MATRIX ALGEBRA

Operations of addition and multiplication by scalars work for matrices very much as they do for vectors. Similarly, the product of a matrix and a column vector defined in the previous section has a natural extension to a more general product of matrices. This section is about the properties of these extended operations, properties that subsume and organize the arithmetic of multivariable systems of equations, not only linear ones but the calculus of nonlinear systems also. To be specific, matrix algebra plays an important role in establishing the properties of inverse matrices in the next section, in extending the meaning of differentiability for real-valued functions $\mathbb{R}^n \xrightarrow{f} \mathbb{R}$ to vector-valued functions $\mathbb{R}^n \xrightarrow{f} \mathbb{R}^m$ in Chapter 5, Section 4, in extending Newton's method for root-finding to vector-valued functions in Chapter 5, Section 5, and in solving linear systems of differential equations throughout Chapter 13.

If a matrix is named by a capital letter, we denote its entries by the corresponding lowercase letter with a pair of subscripts. The first subscript labels the row where the entry occurs and the second labels the column. Thus the first index increases across the rows and the second increases down the columns in

$$A = \begin{pmatrix} a_{11} & a_{12} \\ a_{21} & a_{22} \\ a_{31} & a_{32} \end{pmatrix} \quad \text{or} \quad B = \begin{pmatrix} b_{11} & b_{12} & b_{13} \\ b_{21} & b_{22} & b_{23} \end{pmatrix}.$$

In general, a_{ij} is called the ijth **entry** of A and stands for the entry in the ith row and the jth column of the matrix. For a row vector or a column vector we usually use only one subscript and write, for example, $\mathbf{a} = (a_1, a_2, \ldots, a_n)$.

EXAMPLE 1   To illustrate the notation more concretely, let

$$P = \begin{pmatrix} 7 & -1 \\ -3 & 2 \end{pmatrix} \quad \text{and} \quad Q = \begin{pmatrix} 11 & 12 & 13 \\ 21 & 22 & 23 \end{pmatrix}.$$

Then in P the entries are $p_{11} = 7$, $p_{12} = -1$, $p_{21} = -3$, $p_{22} = 2$. We can write a formula for the entries in Q, namely $q_{ij} = 10i + j$ for $i = 1, 2$ and $j = 1, 2, 3$.

3A   Sum and Scalar Multiple

The addition and scalar multiplication that were defined in Chapter 1 for vectors in $\mathbb{R}^n$ extend to matrices: If A and B have the same dimensions, then the **sum** $A + B$ is defined to be the matrix C with $c_{ij} = a_{ij} + b_{ij}$.

EXAMPLE 2   Here are two examples of matrix addition.

$$\begin{pmatrix} 1 & 1 \\ 0 & 2 \end{pmatrix} + \begin{pmatrix} -1 & 1 \\ 1 & 2 \end{pmatrix} = \begin{pmatrix} 0 & 2 \\ 1 & 4 \end{pmatrix}$$

$$\begin{pmatrix} 1 & 2 & 1 \\ -1 & 1 & 0 \end{pmatrix} + \begin{pmatrix} -1 & -2 & -1 \\ 1 & -1 & 0 \end{pmatrix} = \begin{pmatrix} 0 & 0 & 0 \\ 0 & 0 & 0 \end{pmatrix}$$

There's no reasonable way to define addition for matrices of different dimensions, so we can't add

$$\begin{pmatrix} 1 & 1 \\ 0 & 2 \end{pmatrix} \quad \text{and} \quad \begin{pmatrix} 1 & 2 & 1 \\ -1 & 1 & 0 \end{pmatrix}.$$

For a matrix A and number r, the **scalar multiple** rA is defined to be the matrix C with entries $c_{ij} = ra_{ij}$.

EXAMPLE 3

Here are two examples of scalar multiplication.

$$-2\begin{pmatrix} 1 & 1 \\ 0 & 2 \end{pmatrix} = \begin{pmatrix} -2 & -2 \\ 0 & -4 \end{pmatrix}$$

$$3\begin{pmatrix} 1 & 2 & 1 \\ -1 & 1 & 0 \end{pmatrix} = \begin{pmatrix} 3 & 6 & 3 \\ -3 & 3 & 0 \end{pmatrix}$$

Using both addition and scalar multiplication, we can write linear combinations of matrices that have the same dimensions. For example, using 2-by-2 matrices,

$$2\begin{pmatrix} 1 & -1 \\ 2 & 3 \end{pmatrix} - 3\begin{pmatrix} 0 & 1 \\ 2 & 1 \end{pmatrix} = \begin{pmatrix} 2 & -2 \\ 4 & 6 \end{pmatrix} + \begin{pmatrix} 0 & -3 \\ -6 & -3 \end{pmatrix}$$

$$= \begin{pmatrix} 2 & -5 \\ -2 & 3 \end{pmatrix}.$$

As with vectors in $\mathbb{R}^n$, we write $-A$ for $(-1)A$ and $A - B$ for $A + (-1)B$. Also, for every m and n there is an m-by-n **zero matrix**, denoted by O, with all entries equal to zero, such that $A + O = A$.

Notational warning. When O is used to denote a zero matrix, we depend on the context to make clear what the dimensions are intended for the matrix. For example, if $A = \begin{pmatrix} 1 & 2 \\ 4 & 3 \end{pmatrix}$, then the O in $A + O$ must stand for $\begin{pmatrix} 0 & 0 \\ 0 & 0 \end{pmatrix}$, because that is the only zero matrix we can add to a 2-by-2 matrix.

The formulas 1 to 9 concerning linear combinations of vectors in $\mathbb{R}^n$ stated in Section 1 of Chapter 1 (page 2) are equally valid for linear combinations of matrices.

3B   Matrix Multiplication

In Section 2A we defined the product $A\mathbf{x}$ of an m-by-n matrix A and n-by-1 column vector $\mathbf{x}$ to be the m-by-1 column vector with entries the dot products of the rows of A with $\mathbf{x}$. Thus writing systems of linear equations we saw that if

$$A = \begin{pmatrix} -1 & 2 \\ 0 & -3 \\ 5 & 1 \end{pmatrix} \text{ and } \mathbf{x} = \begin{pmatrix} x_1 \\ x_2 \end{pmatrix}, \text{ then } A\mathbf{x} = \begin{pmatrix} -1 & 2 \\ 1 & -3 \\ 5 & 1 \end{pmatrix}\begin{pmatrix} x_1 \\ x_2 \end{pmatrix} = \begin{pmatrix} -x_1 + 2x_2 \\ x_1 - 3x_2 \\ 5x_1 + x_2 \end{pmatrix}.$$

Our general definition of the matrix product AB depends on the special case $A\mathbf{x}$.

3.1 Definition   If A is m-by-n and B is n-by-p, the **matrix product** AB is defined to be the m-by-p matrix whose columns are the products of A with the successive columns of B. In other words, if $C = AB$, then c_{ij} is the dot product of the ith row of A and the jth column of B.

EXAMPLE 4   If

$$A = \begin{pmatrix} 4 & 3 \\ -1 & 2 \\ 1 & -1 \end{pmatrix} \quad \text{and} \quad B = \begin{pmatrix} b_{11} & b_{12} \\ b_{21} & b_{22} \end{pmatrix},$$

then

$$AB = \begin{pmatrix} 4 & 3 \\ -1 & 2 \\ 1 & -1 \end{pmatrix} \begin{pmatrix} b_{11} & b_{12} \\ b_{21} & b_{22} \end{pmatrix} = \begin{pmatrix} 4b_{11}+3b_{21} & 4b_{12}+3b_{22} \\ -b_{11}+2b_{21} & -b_{12}+2b_{22} \\ b_{11}-b_{21} & b_{12}-b_{22} \end{pmatrix}.$$

For a numerical example, let $C = \begin{pmatrix} 1 & 2 \\ 4 & 5 \end{pmatrix}$. Then

$$AC = \begin{pmatrix} 4 & 3 \\ -1 & 2 \\ 1 & -1 \end{pmatrix} \begin{pmatrix} 1 & 2 \\ 4 & 5 \end{pmatrix} = \begin{pmatrix} 16 & 23 \\ 7 & 8 \\ -3 & -3 \end{pmatrix}.$$

The entry in the first row and second column of AD is the dot product given by $(4, 3) \cdot (2, 5) = (4)(2) + (3)(5) = 23$; you should check that the other entries are correct as shown.

Matrix multiplication is sometimes called **row-by-column multiplication**, and schematically the process looks like this, putting the dot product of the second row and the fourth column in the corresponding row and column of the product:

$$\begin{pmatrix} * & * & * & * \\ * & * & * & * \\ * & * & * & * \end{pmatrix} \begin{pmatrix} * & * & * & * & * \\ * & * & * & * & * \\ * & * & * & * & * \\ * & * & * & * & * \end{pmatrix} = \begin{pmatrix} * & * & * & * & * \\ * & * & * & * & * \\ * & * & * & * & * \end{pmatrix}.$$

For basic questions about matrix products it's often essential to write the entries in the product of $A = (a_{ij})$ and $B = (b_{ij})$ by using the summation notation. In this notation the ijth entry in AB is

$$\sum_{k=1}^{n} a_{ik}b_{kj},$$

an expression that is read as "the sum for $k = 1$ to n of $a_{ik}b_{kj}$" and means the same as

$$a_{i1}b_{1j} + a_{i2}b_{2j} + \cdots + a_{in}b_{nj}.$$

Note that the summation index k runs over the column index of A (that is, across a row) and over the row index of B (that is, down a column).

It is important to note that matrix multiplication is *not* in general commutative. Thus even if the products AB and BA are defined and have the same dimensions they may not be equal. This is why we need the first two laws, with factors in opposite orders, in Theorem 3.2.

EXAMPLE 5   Let $A = \begin{pmatrix} 1 & 2 \\ 3 & 4 \end{pmatrix}$ and $B = \begin{pmatrix} 1 & 0 \\ 1 & 1 \end{pmatrix}$. Then $AB = \begin{pmatrix} 3 & 2 \\ 7 & 4 \end{pmatrix}$ and $BA = \begin{pmatrix} 1 & 2 \\ 4 & 6 \end{pmatrix}$, so $AB \neq BA$. Exercises 43 to 48 show other ways that matrix operations differ from the analogous scalar operations.

3.2   Theorem.   Let A, B, and C be matrices having the proper dimensions for the sums and products to be defined, and let t be a scalar. The basic properties of matrix products are

    1. $(A + B)C = AC + BC$ (Right distributive law)
    2. $C(A + B) = CA + CB$ (Left distributive law)
    3. $(tA)B = t(AB) = A(tB)$ (Scalar commutativity law)
    4. $A(BC) = (AB)C$ (Associative law)

Proof.   We prove only property 4 since it is the most complicated; proofs of the others are left as exercises. Suppose A is m-by-n. For the products AB to be defined, B must have n rows and so must be n-by-p for some p. Similarly, C must be p-by-q for some q in order for BC to be defined. To prove two matrices equal we have to show that corresponding entries are the same in both. The rjth entry of BC is $\sum_{s=1}^{p} b_{rs} c_{sj}$, so the ijth entry of $A(BC)$ is

$$\sum_{r=1}^{n} a_{ir} \left(\sum_{s=1}^{p} b_{rs} c_{sj} \right) = \sum_{r=1}^{n} \sum_{s=1}^{p} a_{ir} b_{rs} c_{sj}.$$

Similarly, the ijth entry of $(AB)C$ is

$$\sum_{s=1}^{p} \left(\sum_{r=1}^{n} a_{ir} b_{rs} \right) c_{sj} = \sum_{s=1}^{p} \sum_{r=1}^{n} a_{ir} b_{rs} c_{sj}.$$

The sums on the right in these two equations consist of the same terms added in different orders, so corresponding entries in $A(BC)$ and $(AB)C$ are equal.   ∎

Formulas 1 through 4 in Theorem 3.2 state laws that have the same form as some of the laws of ordinary arithmetic, with matrices replaced by scalars. Because of the associative law for matrices it makes sense to write the product ABC instead of $(AB)C$ or $A(BC)$ since the result is independent of the order in which the products are formed, though as we saw in Example 5 not necessarily independent of the order of the factors. In the next section we'll define an inverse operator to multiplication by A, denoted A^{-1} that is defined only for some square matrices.

EXAMPLE 6   To illustrate the associative and distributive laws in the next two examples, let

$$A = \begin{pmatrix} 1 & 2 \\ -1 & 2 \end{pmatrix}, \quad B = \begin{pmatrix} 0 & 2 \\ 2 & 1 \end{pmatrix}, \quad C = \begin{pmatrix} -1 & 0 \\ 3 & 2 \end{pmatrix}, \quad \text{and } D = \begin{pmatrix} -1 & 1 & 2 \\ 1 & 2 & 1 \end{pmatrix}.$$

Then $AB = \begin{pmatrix} 1 & 2 \\ -1 & 2 \end{pmatrix} \begin{pmatrix} 0 & 2 \\ 2 & 1 \end{pmatrix} = \begin{pmatrix} 4 & 4 \\ 4 & 0 \end{pmatrix}$ and $(AB)C = \begin{pmatrix} 4 & 4 \\ 4 & 0 \end{pmatrix} \begin{pmatrix} -1 & 0 \\ 3 & 2 \end{pmatrix} = \begin{pmatrix} 8 & 8 \\ -4 & 0 \end{pmatrix}$, while $BC = \begin{pmatrix} 0 & 2 \\ 2 & 1 \end{pmatrix}\begin{pmatrix} -1 & 0 \\ 3 & 2 \end{pmatrix} = \begin{pmatrix} 6 & 4 \\ 1 & 2 \end{pmatrix}$ and $A(BC) = \begin{pmatrix} 1 & 2 \\ -1 & 2 \end{pmatrix}\begin{pmatrix} 6 & 4 \\ 1 & 2 \end{pmatrix} = \begin{pmatrix} 8 & 8 \\ -4 & 0 \end{pmatrix}.$

Thus $(AB)C = A(BC)$, as the associative law states for A, B, and C in that order.

EXAMPLE 7 To illustrate the first distributive law we calculate $(A + B)D$ as

$$\left(\begin{pmatrix} 1 & 2 \\ -1 & 2 \end{pmatrix} + \begin{pmatrix} 0 & 2 \\ 2 & 1 \end{pmatrix}\right)\begin{pmatrix} -1 & 1 & 2 \\ 1 & 2 & 1 \end{pmatrix} = \begin{pmatrix} 1 & 4 \\ 1 & 3 \end{pmatrix}\begin{pmatrix} -1 & 1 & 2 \\ 1 & 2 & 1 \end{pmatrix}$$

$$= \begin{pmatrix} 3 & 9 & 6 \\ 2 & 7 & 5 \end{pmatrix}.$$

This equals $AD + BD$, since

$$\begin{pmatrix} 1 & 2 \\ -1 & 2 \end{pmatrix}\begin{pmatrix} -1 & 1 & 2 \\ 1 & 2 & 1 \end{pmatrix} + \begin{pmatrix} 0 & 2 \\ 2 & 1 \end{pmatrix}\begin{pmatrix} -1 & 1 & 2 \\ 1 & 2 & 1 \end{pmatrix}$$

$$= \begin{pmatrix} 1 & 5 & 4 \\ 3 & 3 & 0 \end{pmatrix} + \begin{pmatrix} 2 & 4 & 2 \\ -1 & 4 & 5 \end{pmatrix} = \begin{pmatrix} 3 & 9 & 6 \\ 2 & 7 & 5 \end{pmatrix}.$$

Note that the product $D(A + B)$ is not defined.

3C   Identity Matrices

A square matrix of the form

$$I = \begin{pmatrix} 1 & 0 \\ 0 & 1 \end{pmatrix} \quad \text{or} \quad I = \begin{pmatrix} 1 & 0 & 0 \\ 0 & 1 & 0 \\ 0 & 0 & 1 \end{pmatrix} \quad \text{or} \quad I = \begin{pmatrix} 1 & 0 & \cdots & 0 & 0 \\ 0 & 1 & \cdots & 0 & 0 \\ \vdots & \vdots & \ddots & \vdots & \vdots \\ 0 & 0 & \cdots & 0 & 1 \end{pmatrix}$$

that has 1s on its **main diagonal** and 0s elsewhere is called an **identity matrix**. An identity matrix I has the property that both

$$IA = A, \text{ and } BI = B$$

for matrices A and B such that the products are defined. Thus it is an identity element for matrix multiplication somewhat as the number 1 is an identity for multiplication of numbers. There is an n-by-n identity matrix for every value of n, and as with zero matrices, we depend on the context to determine the dimensions of the identity matrix denoted by an occurrence of I in a formula.

EXAMPLE 8 You should check the following matrix products:

$$\begin{pmatrix} 1 & 0 \\ 0 & 1 \end{pmatrix}\begin{pmatrix} 1 & 2 & 3 \\ 4 & 5 & 6 \end{pmatrix} = \begin{pmatrix} 1 & 2 & 3 \\ 4 & 5 & 6 \end{pmatrix}\begin{pmatrix} 1 & 0 & 0 \\ 0 & 1 & 0 \\ 0 & 0 & 1 \end{pmatrix} = \begin{pmatrix} 1 & 2 & 3 \\ 4 & 5 & 6 \end{pmatrix}$$

$$\begin{pmatrix} 1 & 0 & 0 \\ 0 & 1 & 0 \\ 0 & 0 & 1 \end{pmatrix}\begin{pmatrix} 1 & 2 \\ 3 & 4 \\ 5 & 6 \end{pmatrix} = \begin{pmatrix} 1 & 2 \\ 3 & 4 \\ 5 & 6 \end{pmatrix}\begin{pmatrix} 1 & 0 \\ 0 & 1 \end{pmatrix} = \begin{pmatrix} 1 & 2 \\ 3 & 4 \\ 5 & 6 \end{pmatrix}.$$

Notice that while we use the letter I to denote the identity matrix when it occurs on either side of a given matrix A, these identity matrices will have different dimensions if A is not a square matrix.

If $\mathbf{x}$ is an n-dimensional column vector, then $I\mathbf{x} = \mathbf{x}$ looks like this when $n = 3$:

$$\begin{pmatrix} 1 & 0 & 0 \\ 0 & 1 & 0 \\ 0 & 0 & 1 \end{pmatrix} \begin{pmatrix} x_1 \\ x_2 \\ x_3 \end{pmatrix} = \begin{pmatrix} x_1 \\ x_2 \\ x_3 \end{pmatrix}.$$

3D   Matrix Polynomials

We may multiply together any number of matrices of the same dimensions n-by-n to get a square matrix with the same dimensions. In particular, if A is a square matrix we may multiply it by itself repeatedly, and we define

$$A^2 = AA, \quad A^3 = AAA = AA^2, \ \ldots, \quad A^k = AA^{k-1}.$$

Note that if B is not square, for example, 2-by-3, then not even B^2 makes sense.

When A has dimensions n-by-n it is natural to define A^0 to be the n-by-n identity matrix I, since then the rule $A^j A^k = A^{j+k}$ holds for all nonnegative integers j and k. Since the powers of A all have the same dimensions, we can add scalar multiples of them to form polynomials in A such as $A^2 + 3A + 5I$.

EXAMPLE 9   If $A = \begin{pmatrix} 2 & 1 \\ 0 & 3 \end{pmatrix}$, then

$$A^2 = \begin{pmatrix} 2 & 1 \\ 0 & 3 \end{pmatrix} \begin{pmatrix} 2 & 1 \\ 0 & 3 \end{pmatrix} = \begin{pmatrix} 4 & 5 \\ 0 & 9 \end{pmatrix},$$

$$A^3 = \begin{pmatrix} 2 & 1 \\ 0 & 3 \end{pmatrix} \begin{pmatrix} 4 & 5 \\ 0 & 9 \end{pmatrix} = \begin{pmatrix} 8 & 19 \\ 0 & 27 \end{pmatrix}, \text{etc.}$$

If $p(x)$ is the polynomial $x^2 + 4x + 2$, then

$$p(A) = A^2 + 4A + 2I = \begin{pmatrix} 4 & 5 \\ 0 & 9 \end{pmatrix} + 4\begin{pmatrix} 2 & 1 \\ 0 & 3 \end{pmatrix} + 2\begin{pmatrix} 1 & 0 \\ 0 & 1 \end{pmatrix}$$

$$= \begin{pmatrix} 4 & 5 \\ 0 & 9 \end{pmatrix} + \begin{pmatrix} 8 & 4 \\ 0 & 12 \end{pmatrix} + \begin{pmatrix} 2 & 0 \\ 0 & 2 \end{pmatrix} = \begin{pmatrix} 14 & 9 \\ 0 & 23 \end{pmatrix}.$$

The possibility of replacing the variable x in a polynomial $p(x)$ by a square matrix X raises some interesting questions. In Exercise 47 we look at the question of solving for X in a quadratic matrix equation $aX^2 + bX + cI = O$, and in Exercise 50 we look at a way of defining $f(X)$ for some functions other than polynomials, $f(X) = e^X$ in particular, which we'll find useful in Chapter 13.

EXERCISES

In Exercises 1 to 10, compute the given expressions using the following matrices:

$$A = \begin{pmatrix} -1 & 2 \\ 0 & 1 \end{pmatrix}, \quad B = \begin{pmatrix} 1 & 2 \\ 1 & 4 \end{pmatrix},$$

$$C = \begin{pmatrix} 1 & 1 \\ 1 & 2 \end{pmatrix}.$$

1. $3A - B$ **2.** $A + 2B$ **3.** $2A + B + C$

4. $A + 5C$ **5.** AB **6.** BC

7. ABC **8.** $AB - 2B$ **9.** $C + AC$

10. $A + B + C^2$

In Exercises 11 to 20, compute the given expression if it is defined, or else give a reason why it is not defined, using the following matrices.

$$A = \begin{pmatrix} 1 & 3 \\ -4 & 2 \end{pmatrix}, \quad B = \begin{pmatrix} 0 & -2 & 1 \\ -1 & 3 & 0 \end{pmatrix},$$

$$C = \begin{pmatrix} -2 & 0 & 1 \\ 0 & 3 & 0 \\ 2 & 3 & -1 \end{pmatrix}, \quad D = \begin{pmatrix} 2 & -4 \\ 0 & 0 \\ 3 & 3 \end{pmatrix},$$

$$G = \begin{pmatrix} 1 & -1 & 2 \\ 1 & 0 & 3 \end{pmatrix}$$

11. $2B - 3G$ **12.** AB **13.** BA

14. BD **15.** DB **16.** $CD + 3DB$

17. $2AB - 5G$ **18.** $2GC - 4AB$ **19.** CDC

20. DCD

In Exercises 21 to 26, with A, B, C, D the same as in the preceding group of exercises, determine what the dimensions of X and Y would have to be for each of the following equations to be possible. (In some cases there may be no possible dimensions; in other cases there may be more than one possibility.)

21. $AX = B + Y$ **22.** $(D + 2X)YC = O$

23. $AX = YD$ **24.** $CX + DY = O$

25. $AX = YC$ **26.** $AX = CY$

27. Using the matrix C of the preceding group of exercises, compute $C\mathbf{i}$, $C\mathbf{j}$, and $C\mathbf{k}$, taking the basis vectors $\mathbf{i}$, $\mathbf{j}$, and $\mathbf{k}$ as column vectors in $\mathbb{R}^3$.

28. Show that if A is a m-by-n matrix, then $A\mathbf{e}_j$ is the jth column of A, where $\mathbf{e}_1, \ldots, \mathbf{e}_n$ are the standard basis vectors in $\mathbb{R}^n$, considered as column vectors. What are the products $\mathbf{e}_i A$, considering $\mathbf{e}_i$ as a row vector?

In Exercises 29 to 34, compute the given matrix products.

29. $\begin{pmatrix} 1 & 2 & 3 \\ 4 & 5 & 6 \\ 7 & 8 & 9 \end{pmatrix} \begin{pmatrix} 0 \\ 1 \\ 0 \end{pmatrix}$ **30.** $\begin{pmatrix} 0 & 1 & 1 \\ 1 & 0 & 1 \\ 1 & 1 & 0 \end{pmatrix} \begin{pmatrix} 2 \\ 1 \\ 3 \end{pmatrix}$

31. $\begin{pmatrix} 2 & 1 & 4 \end{pmatrix} \begin{pmatrix} 3 \\ 5 \\ 7 \end{pmatrix}$ **32.** $\begin{pmatrix} 3 \\ 5 \\ 7 \end{pmatrix} \begin{pmatrix} 2 & 1 & 4 \end{pmatrix}$

33. $\begin{pmatrix} 2 & 1 \\ 5 & 6 \\ 3 & 4 \end{pmatrix} \begin{pmatrix} 1 & -1 & 1 \\ -1 & 1 & -1 \end{pmatrix}$

34. $\begin{pmatrix} 2 & 0 & 0 \\ 0 & 4 & 0 \\ 0 & 0 & 5 \end{pmatrix} \begin{pmatrix} -1 \\ 1 \\ -1 \end{pmatrix}$

35. Show that for a matrix A and zero matrices of appropriate dimensions,

$$AO = O \quad \text{and} \quad OA = O.$$

If A is m-by-n, for what possible dimensions of zero matrices are AO and OA defined, and what are the dimensions of the products?

36. Prove the left distributive law for matrices, $C(A + B) = CA + CB$.

37. Let $\mathbf{r}$ be a 1-by-n row vector and $\mathbf{c}$ an n-by-1 column vector.

  (a) Show that the matrix product $\mathbf{rc}$ is the same as the dot product $\mathbf{r} \cdot \mathbf{c}$.

  (b) Describe in general the product $\mathbf{cr}$.

38. A company makes m grades of a product in n different factories.

  (a) Let a_{ij} be the number of tons per day of grade i produced in plant j.

  (b) Let d_j be the number of days per month that plant j operates.

  (c) Let p_i be the wholesale price per ton of grade i.

Let $A = (a_{ij})$, $\mathbf{d} = \begin{pmatrix} d_1 \\ \vdots \\ d_n \end{pmatrix}$, and $\mathbf{p} = \begin{pmatrix} p_1 \\ \vdots \\ p_m \end{pmatrix}$, and

let $\mathbf{u}$ be the column vector of all 1s in $\mathbb{R}^n$ and $\mathbf{v}$ be the column vector of all 1s in $\mathbb{R}^m$. Give interpretations for the following expressions.

(a) $A\mathbf{u}$ (b) $(A\mathbf{u}) \cdot \mathbf{v}$ (c) $A\mathbf{d}$

(d) $(A\mathbf{d}) \cdot \mathbf{v}$ (e) $(A\mathbf{d}) \cdot \mathbf{p}$

In Exercises 39 to 42, compute A^2, A^3, and $p(A)$, where $p(x) = 2x^2 - 3x + 3$.

39. $A = \begin{pmatrix} 1 & 1 \\ -2 & 3 \end{pmatrix}$

40. $A = \begin{pmatrix} 0 & 1 & 1 \\ 0 & 0 & 1 \\ 0 & 0 & 0 \end{pmatrix}$

41. $A = \begin{pmatrix} -1 & 0 & 0 \\ 0 & 2 & 0 \\ 0 & 0 & 3 \end{pmatrix}$

42. $A = \begin{pmatrix} 0 & 0 & 3 \\ -1 & 0 & 0 \\ 0 & 5 & 0 \end{pmatrix}$

43. Let $U = \begin{pmatrix} -1 & 2 \\ 2 & -4 \end{pmatrix}$ and $V = \begin{pmatrix} 2 & 6 \\ 1 & 3 \end{pmatrix}$. Compute UV and VU. Are they the same? Is it possible for the product of two matrices to be zero without either factor being zero?

44. (a) Show that if A is an n-by-n matrix and I is the n-by-n identity matrix, then $(A - I)(A + I) = A^2 - I$.

(b) Give examples of 2-by-2 matrices A and B such that $(A - B)(A + B) \neq A^2 - B^2$.

45. (a) Show that for A and I as in Exercise 44, $(A + I)^2 = A^2 + 2A + I$.

(b) Find examples of 2-by-2 matrices A and B such that $(A + B)^2 \neq A^2 + 2AB + B^2$.

46. Find A^2 and A^3 for $A = \begin{pmatrix} 1 & 0 & -1 \\ -1 & 0 & 1 \\ 2 & 1 & -1 \end{pmatrix}$. Zero is the only number whose cube is 0. Thus this exercise illustrates another difference between the arithmetic of numbers and of matrices.

47. The factorization $X^2 - 3X + 2I = (X - I)(X - 2I)$ shows that the matrix equation $X^2 - 3X + 2I = O$ has solutions $X = I$ and $X = 2I$. Show that if X and I are 2-by-2, then $X = \begin{pmatrix} 2 & a \\ 0 & 1 \end{pmatrix}$ is also a solution for every scalar a.

48. We know that if the scalar equation $xy = 0$ holds then at least one of the numbers x and y must equal 0. Thus by

factoring the left side we know that the scalar equation $x^2 - 1 = 0$ has $x = 1$ and $x = -1$ as its only solutions.

(a) Show that if $A = I$ or $A = -I$, then $A^2 - I = O$.

(b) Show that $\begin{pmatrix} a & b \\ c & -a \end{pmatrix}^2 = \begin{pmatrix} 1 & 0 \\ 0 & 1 \end{pmatrix}$ if $a^2 + bc = 1$. Thus the equation $A^2 - I = O$ has infinitely many different solutions in the set of 2-by-2 matrices.

(c) Show that every 2-by-2 matrix A for which $A^2 = I$ is either I, $-I$, or one of the matrices described in part (b).

49. Given numbers a, b, c, d, let $p(x) = x^2 - (a + d)x + (ad - bc)$, and let A be the matrix $\begin{pmatrix} a & b \\ c & d \end{pmatrix}$.

(a) Take $a = 1, b = -3, c = 1, d = -1$, and verify that $p(A) = O$ in this case.

(b) Show that, for arbitrary values of a, b, c, d, we have $p(A) = O$.

***50.** Let A be an n by n matrix. We define an n-by-n matrix e^A by

$$e^A = \lim_{N \to \infty} \sum_{k=0}^{N} \frac{1}{k!} A^k = \sum_{k=0}^{\infty} \frac{1}{k!} A^k.$$

Note that the finite sum from 0 to N is a polynomial $P_N(A)$ and so is itself an n by n matrix in which we compute the limit separately in each matrix entry.

(a) Compute the matrix $P_N(A)$ for $A = \begin{pmatrix} 1 & 1 \\ 0 & 1 \end{pmatrix}$. Then compute the limit matrix e^A by letting $N \to \infty$.

(b) Using the commutativity of scalar-matrix multiplication, redo the computation in part (a) with the 2-by-2 matrix A replaced by tA to find the 2-by-2 matrix e^{tA}.

(c) Since we can compute limits of matrices one entry at a time, we can differentiate one entry at a time. Show that $\dfrac{d}{dt} e^{tA} = A e^{tA}$ for the 2-by-2 matrix of part (a).

SECTION 4   INVERSE MATRICES

This section completes our discussion of the algebra of matrix operations by describing the limited extent to which we can perform an analogue of division by a square matrix.

4A   Invertibility

If A and B are square matrices with the same dimensions such that

$$AB = BA = I,$$

then we say that B is an **inverse** of A, and that A is an **invertible** matrix. As we show in Theorem 4.5, a matrix A can have at most one inverse, so we can speak of *the* inverse of A and denote it by A^{-1}. Thus, if A is invertible,

$$AA^{-1} = A^{-1}A = I.$$

EXAMPLE 1   If $A = \begin{pmatrix} 1 & 2 \\ 3 & 7 \end{pmatrix}$, we can check that $A^{-1} = \begin{pmatrix} 7 & -2 \\ -3 & 1 \end{pmatrix}$:

$$AA^{-1} = \begin{pmatrix} 1 & 2 \\ 3 & 7 \end{pmatrix} \begin{pmatrix} 7 & -2 \\ -3 & 1 \end{pmatrix} = \begin{pmatrix} 1 & 0 \\ 0 & 1 \end{pmatrix}$$

and

$$A^{-1}A = \begin{pmatrix} 7 & -2 \\ -3 & 1 \end{pmatrix} \begin{pmatrix} 1 & 2 \\ 3 & 7 \end{pmatrix} = \begin{pmatrix} 1 & 0 \\ 0 & 1 \end{pmatrix}.$$

A 2-by-2 matrix $\begin{pmatrix} a & b \\ c & d \end{pmatrix}$ is invertible if its **determinant**, $ad - bc$, is not zero. In that case

4.1
$$\begin{pmatrix} a & b \\ c & d \end{pmatrix}^{-1} = \frac{1}{ad - bc} \begin{pmatrix} d & -b \\ -c & a \end{pmatrix}.$$

In Example 1 we used

$$\begin{pmatrix} 1 & 2 \\ 3 & 7 \end{pmatrix}^{-1} = \frac{1}{1} \begin{pmatrix} 7 & -2 \\ -3 & 1 \end{pmatrix} = \begin{pmatrix} 7 & -2 \\ 3 & 1 \end{pmatrix}.$$

Formula 4.1 is worth remembering and is the special 2-by-2 case of Theorem 5.8 in Section 5E. Exercise 32 asks you to verify that the formula is correct.

If A is an n-by-n matrix, then the matrix equation $Ax = b$ is equivalent to a system of n linear equations in n variables. If A happens to be an invertible matrix with inverse A^{-1}, then we can solve the system in matrix form by multiplying both sides on the left by A^{-1} to get $A^{-1}Ax = A^{-1}b$. Since $A^{-1}A = I$, we have $A^{-1}Ax = Ix = x$, so

4.2   Theorem.   If A is invertible, then $Ax = b$ has a unique solution $x = A^{-1}b$. In particular, $x = 0$ is the only solution to $Ax = 0$.

Theorem 4.4 will show that if A is a square matrix and $x = 0$ is the only solution to $Ax = 0$, then A invertible.

EXAMPLE 2   The system

$$\begin{array}{l} x + 2y = 3 \\ 3x + 7y = -4 \end{array} \quad \text{is equivalent to} \quad \begin{pmatrix} 1 & 2 \\ 3 & 7 \end{pmatrix} \begin{pmatrix} x \\ y \end{pmatrix} = \begin{pmatrix} 3 \\ -4 \end{pmatrix}.$$

By Formula 4.1,

$$\begin{pmatrix} 1 & 2 \\ 3 & 7 \end{pmatrix}^{-1} = \begin{pmatrix} 7 & -2 \\ -3 & 1 \end{pmatrix}$$

Multiplying the two sides of the equation by the inverse gives

$$\begin{pmatrix} 7 & -2 \\ -3 & 1 \end{pmatrix} \begin{pmatrix} 1 & 2 \\ 3 & 7 \end{pmatrix} \begin{pmatrix} x \\ y \end{pmatrix} = \begin{pmatrix} 1 & 0 \\ 0 & 1 \end{pmatrix} \begin{pmatrix} x \\ y \end{pmatrix} = \begin{pmatrix} x \\ y \end{pmatrix}$$

on the left and

$$\begin{pmatrix} x \\ y \end{pmatrix} = \begin{pmatrix} 7 & -2 \\ -3 & 1 \end{pmatrix} \begin{pmatrix} 3 \\ -4 \end{pmatrix} = \begin{pmatrix} 29 \\ -13 \end{pmatrix}$$

on the right. Thus $(x, y) = (29, -13)$ is the unique solution.

4B   Computing Inverses

Formula 4.1 gives an easy way to find inverses of 2-by-2 matrices, and Section 5E has formulas for A^{-1} using determinants when A has larger dimensions, but it's often more efficient, particularly for large matrices, to use row operations as in the process described below. To use the method we need to introduce another type of row operation on a matrix A: **row rearrangement**, which changes the order of the rows in the matrix. As with elementary multiplication and elementary modification, applying the same row rearrangement to A and $\mathbf{b}$ leaves the solution set of $A\mathbf{x} = \mathbf{b}$ unchanged; this is so because the solutions of a system are unchanged by writing the scalar equations of the system in different orders. We'll first describe the process and give an example, then prove some theorems that show it always works, either finding the inverse of a matrix A or showing that A has no inverse.

4.3   Matrix Inversion Process.   Given a square matrix A, apply row operations to obtain a reduced matrix R, applying the same operations in the same order to I to obtain a matrix B. Then

If a row of R is a zero row, A is not invertible.

If R has no zero rows, then A is invertible. Apply row rearrangement to convert R to I, and apply the same rearrangement to B. Then B is A^{-1}.

EXAMPLE 3

We illustrate how the process works on the matrix A on the left below, row reducing A and performing the same row operations on I as we go along. We start with

$$A = \begin{pmatrix} 2 & 4 & 8 \\ 1 & 0 & 0 \\ 1 & -3 & -7 \end{pmatrix}, \quad I = \begin{pmatrix} 1 & 0 & 0 \\ 0 & 1 & 0 \\ 0 & 0 & 1 \end{pmatrix}.$$

Add -2 times the second row to the first and -1 times the second to the third:

$$\begin{pmatrix} 0 & 4 & 8 \\ 1 & 0 & 0 \\ 0 & -3 & -7 \end{pmatrix}, \begin{pmatrix} 1 & -2 & 0 \\ 0 & 1 & 0 \\ 0 & -1 & 1 \end{pmatrix}.$$

Multiply the first row by $\frac{1}{4}$ and then add 3 times the first row to the third:

$$\begin{pmatrix} 0 & 1 & 2 \\ 1 & 0 & 0 \\ 0 & 0 & -1 \end{pmatrix}, \begin{pmatrix} \frac{1}{4} & -\frac{1}{2} & 0 \\ 0 & 1 & 0 \\ \frac{3}{4} & -\frac{5}{2} & 1 \end{pmatrix}.$$

Multiply the third row by -1 and then add -2 times the third row to the first:

$$\begin{pmatrix} 0 & 1 & 0 \\ 1 & 0 & 0 \\ 0 & 0 & 1 \end{pmatrix}, \begin{pmatrix} \frac{7}{4} & -\frac{11}{2} & 2 \\ 0 & 1 & 0 \\ -\frac{3}{4} & \frac{5}{2} & -1 \end{pmatrix}.$$

The matrix on the left is reduced. Interchanging the first and second rows gives

$$R = \begin{pmatrix} 1 & 0 & 0 \\ 0 & 1 & 0 \\ 0 & 0 & 1 \end{pmatrix}, \quad B = \begin{pmatrix} 0 & 1 & 0 \\ \frac{7}{4} & -\frac{11}{2} & 2 \\ -\frac{3}{4} & \frac{5}{2} & -1 \end{pmatrix}.$$

You can multiply A on the right and left by B to verify that B is an inverse for A. If R had a zero row, the system $R\mathbf{x} = 0$ would have been one with two equations in three variables having nonzero solutions. Thus A could not have been invertible, by Theorem 4.2.

Since $AA^{-1} = I$, A^{-1} is a solution of the matrix equation $AX = I$, and one way to find A^{-1} is to solve that equation. Let $\mathbf{x}_k$ be the kth column of X, so the kth column of AX is $A\mathbf{x}_k$. The kth column of I is the standard basis vector $\mathbf{e}_k$ in $\mathbb{R}^n$, so solving $AX = I$ amounts to solving all the systems $A\mathbf{x}_1 = \mathbf{e}_1, \ldots, A\mathbf{x}_n = \mathbf{e}_n$ at once. We use this idea in the proof of the following theorem.

4.4   Theorem.   Let A be a square matrix. If $\mathbf{x} = \mathbf{0}$ is the only solution of $A\mathbf{x} = \mathbf{0}$ then the inversion process 4.3 yields a square matrix B such that $AB = I$. If $A\mathbf{x} = \mathbf{0}$ has nonzero solutions, A is not invertible and the inversion process produces a reduced matrix R containing at least one zero row.

Proof.   Let R be the result of applying row reduction operations to A, and B the result of applying the same operations to I, as in the inversion process 4.3.

First suppose that $\mathbf{x} = \mathbf{0}$ is the only solution of $A\mathbf{x} = \mathbf{0}$. Then the same is true of the system $R\mathbf{x} = \mathbf{0}$, and by Theorem 2.5 the system has at least as many nontrivial equations as it has variables. Since R is square, this implies that R has no zero row. Thus every row and column contains a leading entry 1 with 0s in the rest of the column, and the rows of R can be rearranged to produce the identity matrix I. Rearrange the rows of B in the same way, and write $\mathbf{b}_k$ for its kth column. Row operations on a matrix apply simultaneously to all its columns, so we've converted the equations $A\mathbf{x}_k = \mathbf{e}_k$ to an equivalent system $I\mathbf{x}_k = \mathbf{b}_k$. Thus $\mathbf{x}_k = \mathbf{b}_k$ is the kth column of X, and B is a solution of $AX = I$. In other words, $AB = I$.

Otherwise, suppose that the system $A\mathbf{x} = \mathbf{0}$ has nonzero solutions. Then the same is true of the system $R\mathbf{x} = \mathbf{0}$ and, again by Theorem 2.5, the reduced system has more variables than it has nontrivial equations, which implies that R has at least one zero row.   ∎

Theorem 4.4 implies that if A is invertible, then the inversion process 4.3 produces a matrix B such that $AB = I$. To prove that $B = A^{-1}$ we need to show that $BA = I$ as well. Also, many different sequences of row operations will reduce A to I, so it isn't obvious that the process used in Theorem 4.4 always gives the same matrix B as a result. The next theorem settles both these questions.

4.5   Theorem.   If A is an invertible matrix, and B is the matrix produced by the inversion process 4.3 such that $AB = I$, then $BA = I$ as well, so B is inverse to A. Finally $B = A^{-1}$ is the only inverse of A.

Proof.   To show that $BA = I$, note that we got B by row operations on I, so we can get I back again from B by applying the inverses of these operations in the opposite order. Thus if we apply the inversion process to B, we get a matrix C such that $BC = I$. Then $A = AI = A(BC) = (AB)C = IC = C$, so $A = C$ and $BA = BC = I$.

To show the uniqueness of A^{-1}, suppose A is invertible and both B and C satisfy the conditions for being an inverse of A, namely $AB = BA = I$ and $AC = CA = I$. Then $AB = AC$ because they are both equal to I. To show $B = C$ multiply $AB = AC$ by B on the left to get $BAB = BAC$. Since $BA = I$, this gives $IB = IC$ and $B = C$.   ∎

4C   Special Matrices

Invertibility is very easy to determine for some matrices. One such type is the class of **diagonal matrices**, with all entries zero except on the main diagonal:

$$\begin{pmatrix} a_1 & 0 \\ 0 & a_2 \end{pmatrix}, \quad \begin{pmatrix} b_1 & 0 & 0 \\ 0 & b_2 & 0 \\ 0 & 0 & b_3 \end{pmatrix}, \quad \text{etc.}$$

In particular, identity matrices are diagonal matrices in which all the diagonal entries are 1. The notation $\operatorname{diag}(t_1, t_2, \ldots, t_n)$ is convenient for the n-by-n diagonal matrix with entries $t_1, \ldots, t_n$ on the main diagonal. Diagonal matrices are easy to multiply. We have for dimensions n-by-n,

$$\operatorname{diag}(a_1, a_2, \ldots, a_n) \operatorname{diag}(b_1, \ldots, b_n) = \operatorname{diag}(a_1 b_1, \ldots, a_n b_n).$$

It follows that a diagonal matrix is invertible if and only if the main diagonal entries are all nonzero, and in that case

4.6 $(\operatorname{diag}(t_1, \ldots, t_n))^{-1} = \operatorname{diag}(t_1^{-1}, \ldots, t_n^{-1}).$

EXAMPLE 4

$$\begin{pmatrix} 2 & 0 \\ 0 & 3 \end{pmatrix}^{-1} = \begin{pmatrix} \frac{1}{2} & 0 \\ 0 & \frac{1}{3} \end{pmatrix}$$

The **upper triangular** matrices are those of the form

$$\begin{pmatrix} a_{11} & a_{12} \\ 0 & a_{22} \end{pmatrix}, \begin{pmatrix} b_{11} & b_{12} & b_{13} \\ 0 & b_{22} & b_{23} \\ 0 & 0 & b_{33} \end{pmatrix}, \quad \text{etc.,}$$

with all entries below the main diagonal equal to zero. Just as with diagonal matrices, an upper triangular matrix is invertible if and only if the main diagonal entries are all nonzero. The reason is that if there is no 0 on the diagonal of an upper triangular matrix, we can transform it into the identity matrix by elementary operations as in the following example.

| EXAMPLE 5 | With upper triangular matrices, it is simpler to reduce the columns working from right to left instead of from left to right.

$$A = \begin{pmatrix} 2 & 1 & 3 \\ 0 & 1 & 2 \\ 0 & 0 & 4 \end{pmatrix}, \quad \begin{pmatrix} 1 & 0 & 0 \\ 0 & 1 & 0 \\ 0 & 0 & 1 \end{pmatrix}$$

$$\begin{pmatrix} 2 & 1 & 3 \\ 0 & 1 & 2 \\ 0 & 0 & 1 \end{pmatrix}, \quad \begin{pmatrix} 1 & 0 & 0 \\ 0 & 1 & 0 \\ 0 & 0 & \frac{1}{4} \end{pmatrix}$$

$$\begin{pmatrix} 2 & 1 & 0 \\ 0 & 1 & 0 \\ 0 & 0 & 1 \end{pmatrix}, \quad \begin{pmatrix} 1 & 0 & -\frac{3}{4} \\ 0 & 1 & -\frac{1}{2} \\ 0 & 0 & \frac{1}{4} \end{pmatrix}$$

$$\begin{pmatrix} 2 & 0 & 0 \\ 0 & 1 & 0 \\ 0 & 0 & 1 \end{pmatrix}, \quad \begin{pmatrix} 1 & -1 & -\frac{1}{4} \\ 0 & 1 & -\frac{1}{2} \\ 0 & 0 & \frac{1}{4} \end{pmatrix}$$

$$\begin{pmatrix} 1 & 0 & 0 \\ 0 & 1 & 0 \\ 0 & 0 & 1 \end{pmatrix}, \quad \begin{pmatrix} \frac{1}{2} & -\frac{1}{2} & -\frac{1}{8} \\ 0 & 1 & -\frac{1}{2} \\ 0 & 0 & \frac{1}{4} \end{pmatrix} = A^{-1}$$

EXERCISES

In Exercises 1 to 4, either find the inverse of the given matrix or show that it does not have one.

1. $\begin{pmatrix} 1 & 1 \\ 1 & 2 \end{pmatrix}$ **2.** $\begin{pmatrix} 3 & 6 \\ 2 & 4 \end{pmatrix}$

3. $\begin{pmatrix} \frac{1}{2} & \frac{1}{4} \\ \frac{1}{4} & \frac{1}{5} \end{pmatrix}$ **4.** $\begin{pmatrix} -7 & -5 \\ 12 & 9 \end{pmatrix}$

In Exercises 5 and 6, solve the matrix equation $A\mathbf{x} = \mathbf{b}$ by multiplying by A^{-1}.

5. $A = \begin{pmatrix} 2 & -1 \\ 3 & 4 \end{pmatrix}$; $\mathbf{b} = \begin{pmatrix} 1 \\ 1 \end{pmatrix}$

6. $A = \begin{pmatrix} 7 & 2 \\ 1 & 1 \end{pmatrix}$; $\mathbf{b} = \begin{pmatrix} -2 \\ 4 \end{pmatrix}$

In Exercises 7 to 10, use row operations to find the inverse of the given matrix, or to show that its inverse doesn't exist.

7. $\begin{pmatrix} 1 & 0 & 0 \\ 3 & 1 & 5 \\ -2 & 0 & 1 \end{pmatrix}$ 8. $\begin{pmatrix} 1 & 2 & 3 \\ -1 & 1 & 0 \\ 0 & 3 & 3 \end{pmatrix}$

9. $\begin{pmatrix} 2 & 6 & 10 \\ 1 & 1 & 1 \\ 1 & -3 & -7 \end{pmatrix}$ 10. $\begin{pmatrix} 2 & 4 & 8 \\ 1 & 0 & 0 \\ 1 & -3 & -7 \end{pmatrix}$

11. Solve the matrix equation $\begin{pmatrix} 1 & 2 \\ 5 & 6 \end{pmatrix} X = \begin{pmatrix} 0 & -3 & 4 \\ 1 & 2 & 0 \end{pmatrix}$ for X.

12. Show that if $B = A^{-1}$, then $B^{-1} = A$.

13. Show that if A and B are invertible matrices with the same dimensions, then AB is invertible and $(AB)^{-1} = B^{-1}A^{-1}$.

14. Show that if $A_1, \ldots, A_n$ are invertible matrices with the same dimensions, then the n-fold matrix product $A_1 A_2 \cdots A_n$ is invertible and

$$(A_1 A_2 \cdots A_n)^{-1} = A_n^{-1} \cdots A_2^{-1} A_1^{-1}.$$

15. Show that if $A^3 = O$, then $I + A + A^2$ is the inverse of $I - A$.

16. Prove that an upper triangular matrix is invertible if and only if it has no zeros on the main diagonal.

Find the inverses for the following special matrices.

17. $\begin{pmatrix} 1 & 0 & 0 \\ 0 & 2 & 0 \\ 0 & 0 & 1 \end{pmatrix}$ 18. $\begin{pmatrix} 1 & -1 & 1 \\ 0 & -1 & 1 \\ 0 & 0 & 1 \end{pmatrix}$

19. $\begin{pmatrix} 1 & 2 & -1 & 3 \\ 0 & 2 & 0 & 1 \\ 0 & 0 & 1 & 1 \\ 0 & 0 & 0 & 4 \end{pmatrix}$ 20. $\begin{pmatrix} 1 & 0 & 0 & 0 \\ 0 & 2 & 0 & 0 \\ 0 & 0 & 3 & 0 \\ 0 & 0 & 0 & 4 \end{pmatrix}$

The **transpose** A^t of a matrix $A = (a_{ij})$ is defined by $A^t = (a_{ji})$. In Exercises 21 to 24, prove the stated equations.

21. $(A^t)^t = A$ 22. $(AB)^t = B^t A^t$

23. $(A + B)^t = A^t + B^t$ 24. $(A^t)^{-1} = (A^{-1})^t$

25. Theorem 4.4 shows in particular that if A and B are square matrices such that $AB = I$, then $BA = I$ also, so A is invertible and $A^{-1} = B$. Use this result to show that if we know only that $BA = I$ then $A^{-1} = B$ by showing first that $A^t B^t = I$. (This is an alternative to the proof given for Theorem 4.5.)

A square matrix Q is **orthogonal** if its column vectors, or equivalently its row vectors, are mutually perpendicular and all have length 1. In Exercises 26 to 29, show that the given matrix is orthogonal.

26. $\begin{pmatrix} 1/\sqrt{2} & -1/\sqrt{2} \\ 1\sqrt{2} & 1/\sqrt{2} \end{pmatrix}$

27. $\begin{pmatrix} \cos\theta & -\sin\theta \\ \sin\theta & \cos\theta \end{pmatrix}$

28. $\begin{pmatrix} 0 & 0 & 1 \\ 1 & 0 & 0 \\ 0 & 1 & 0 \end{pmatrix}$

29. $\begin{pmatrix} 1/\sqrt{3} & -1/\sqrt{2} & 1/\sqrt{6} \\ 1/\sqrt{3} & 1/\sqrt{2} & 1/\sqrt{6} \\ 1/\sqrt{3} & 0 & -2/\sqrt{6} \end{pmatrix}$

30. Verify that the inverse of each of the orthogonal matrices Q in Exercise 26 and Exercise 28 is the transposed matrix Q^t, as defined just before Exercise 21.

31. (a) Show that every orthogonal matrix Q is invertible, and that $Q^{-1} = Q^t$.
    (b) Use part (a) and Theorem 4.4 to show that the columns of a square matrix are perpendicular and have length 1 if and only if the same is true of the rows.

32. (a) Let $A = \begin{pmatrix} a & b \\ c & d \end{pmatrix}$ be a 2-by-2 matrix with $ad \neq bc$. Prove Formula 4.1 by verifying that if A^{-1} is given by the formula then $AA^{-1} = A^{-1}A = I$. (This proves that A is invertible if its determinant, $ad - bc$, is not zero.)
    (b) Try to find the inverses of the following 2-by-2 matrices using Formula 4.1.

$$\begin{pmatrix} -1 & 1 \\ 2 & 1 \end{pmatrix}, \quad \begin{pmatrix} 0 & 1 \\ 1 & 0 \end{pmatrix}, \quad \begin{pmatrix} 2 & 6 \\ 1 & 3 \end{pmatrix}$$

What goes wrong in the third one?

33. A square matrix A is **symmetric** if $A^t = A$ and **skew symmetric** if $A^t = -A$, where A^t is the transpose of A, defined just before Exercise 21.
    (a) Show that if A is a square matrix, then $A + A^t$ is symmetric and that $A - A^t$ is skew symmetric.
    (b) Using part (a), show that every square matrix is the sum of a symmetric matrix and a skew-symmetric matrix.
    (c) Show that if A is invertible and symmetric, then A^{-1} is symmetric. What if A is invertible and skew symmetric?

34. This exercise shows that we can use matrix products to carry out the elementary row operations on a matrix M

that we used in Section 2 for solving systems and in this section for inverting square matrices.

(a) Let $D_i(r)$ be the matrix that equals the identity matrix I of some given dimensions except that the 1 in the ith diagonal entry is replaced by r. Show that the matrix product $D_i(r)M$ equals the result of multiplying the ith row of M by r.

(b) Let E_{ij} be a matrix with 1 for its ijth entry and 0's elsewhere. For example, we have the 3-by-3 matrices

$$E_{13} = \begin{pmatrix} 0 & 0 & 1 \\ 0 & 0 & 0 \\ 0 & 0 & 0 \end{pmatrix}, \quad \text{and } E_{21} = \begin{pmatrix} 0 & 0 & 0 \\ 1 & 0 & 0 \\ 0 & 0 & 0 \end{pmatrix}.$$

Show that if M is m-by-n while E_{ij} and I are both m-by-m, then $(I + rE_{ij})M$ is the elementary modification of M that we get by adding r times the jth row to the ith row.

(c) Let T_{ij} be a matrix resulting from the interchange of the ith and jth rows of I. Show that interchanging the ith and jth rows, a row rearrangement operation on M, yields $T_{ij}M$ as a result.

(d) A matrix $D_i(r)$ or $(I + rE_{ij})$ or T_{ij} is called an **elementary matrix**. Show that each of the three types of elementary operation is reversible in the sense of Definition 1.1 in Section 1 by verifying that, assuming $i \neq j$, $D_i^{-1}(r) = D_i(1/r)$, $(I + rE_{ij})^{-1} = (I - rE_{ij})$, $T_{ij}^{-1} = T_{ji}$.

*35. Let $p(x) = a_0 + a_1 x + \cdots + a_n x^n$ be a polynomial of degree at most n with real coefficients. An algebra theorem says that if there are more than n distinct values of x such that $p(x) = 0$, then all its coefficients $a_k = 0$. Use this theorem to prove that if $b_0, \ldots, b_n$ are scalars and $x_0, \ldots, x_n$ are distinct scalars, then there is exactly one polynomial of degree at most n such that $p(x_k) = b_k$ for $k = 0, \ldots, n$. [*Hint*: Consider a system of linear equations with $a_0, \ldots, a_n$ as variables, then apply Theorem 4.4 to the associated homogeneous system.]

SECTION 5   DETERMINANTS

Determinants were originally invented as a device for solving systems of linear equations, but they turned out to have both geometric and algebraic significance which make them important in many fields of pure and applied mathematics. Apart from this section, we use determinants in Chapters 3, 7, 11, and 13 in a variety of contexts that will arise naturally there. A determinant is a scalar $\det A$ defined for each square n-by-n matrix A. A common notation for the determinant of a matrix is to replace the parentheses enclosing the matrix by vertical bars. Thus

$$\begin{vmatrix} 1 & 4 & 5 \\ 6 & 7 & -3 \\ -2 & 1 & 0 \end{vmatrix} = \det \begin{pmatrix} 1 & 4 & 5 \\ 6 & 7 & -3 \\ -2 & 1 & 0 \end{pmatrix}, \quad \text{and} \quad \begin{vmatrix} a & b \\ c & d \end{vmatrix} = \det \begin{pmatrix} a & b \\ c & d \end{pmatrix}.$$

5A   Definition

In our definition of determinant we'll define $\det A$ first for 1-by-1 matrices, and then for each n define the determinant of an n-by-n matrix in terms of determinants of $(n-1)$-by-$(n-1)$ submatrices called minors. For a matrix A, the matrix obtained by deleting the ith row and jth column of A is called the ijth **minor** of A and is denoted by A_{ij}. Recall that we use the small letter a_{ij} to denote the ijth entry of a matrix A. Thus the ijth minor A_{ij} corresponds to the entry a_{ij} in a natural way, because the minor is obtained by deleting the row and column containing a_{ij}.

EXAMPLE 1   Let

$$A = \begin{pmatrix} -5 & -6 & 7 \\ 8 & -9 & 0 \\ -3 & 4 & 2 \end{pmatrix}, \quad \text{and} \quad B = \begin{pmatrix} 1 & 2 \\ 3 & 4 \end{pmatrix}.$$

Some examples of entries and corresponding minors of A and B are

$$a_{11} = -5 \qquad A_{11} = \begin{pmatrix} -9 & 0 \\ 4 & 2 \end{pmatrix}$$

$$a_{23} = 0 \qquad A_{23} = \begin{pmatrix} -5 & -6 \\ -3 & 4 \end{pmatrix}$$

$$b_{11} = 1 \qquad B_{11} = (4)$$

$$b_{12} = 2 \qquad B_{12} = (3).$$

For a 1-by-1 matrix, $A = (a)$, we define $\det A = a$. For an n-by-n matrix, $A = (a_{ij})$, $i, j = 1, \ldots, n$, we define

5.1 $$\det A = a_{11} \det A_{11} - a_{12} \det A_{12} + \cdots + (-1)^{n+1} a_{1n} \det A_{1n}.$$

The definition is *inductive* in the sense that the determinant of an n-by-n matrix A is defined in terms of the determinants of the $(n-1)$-by-$(n-1)$ minors A_{ij}. Starting with the simple definition for the 1-by-1 case allows us to go on to 2-by-2, then 3-by-3, and so on. In words, the formula says that $\det A$ is the sum, with alternating signs, of the elements of the first row of A, each multiplied by the determinant of its corresponding minor. For this reason the numbers

$$\det A_{11}, \; -\det A_{12}, \; \ldots, \; (-1)^{n+1} \det A_{1n}$$

are called the *cofactors* of the corresponding elements of the first row of A. In general, the **cofactor** of the entry a_{ij} in A is defined to be $(-1)^{i+j} \det A_{ij}$. Thus in Example 1 the entry $a_{23} = 0$ in the matrix A has cofactor

$$(-1)^{2+3} \det \begin{pmatrix} -5 & -6 \\ -3 & 4 \end{pmatrix} = 38.$$

The factor $(-1)^{i+j}$ associates plus and minus signs with $\det A_{ij}$ according to the pattern

$$\begin{pmatrix} + & - & + & - & \cdots \\ - & + & - & + & \cdots \\ + & - & + & - & \cdots \\ - & + & - & + & \cdots \\ \vdots & \vdots & \vdots & \vdots & \end{pmatrix}.$$

EXAMPLE 2   (a) $\det \begin{pmatrix} 1 & 2 \\ 3 & 4 \end{pmatrix} = (1)(4) - (2)(3) = 4 - 6 = -2$

(b)

$$\det \begin{pmatrix} -5 & -6 & 7 \\ 8 & -9 & 0 \\ -3 & 4 & 2 \end{pmatrix} = -5 \det \begin{pmatrix} -9 & 0 \\ 4 & 2 \end{pmatrix} - (-6) \det \begin{pmatrix} 8 & 0 \\ -3 & 2 \end{pmatrix}$$

$$+ 7 \det \begin{pmatrix} 8 & -9 \\ -3 & 4 \end{pmatrix}$$

$$= (-5)(-18 - 0) + (6)(16 - 0)$$

$$+ 7(32 - 27)$$

$$= 90 + 96 + 35 = 221$$

(c) $\det \begin{pmatrix} a & b \\ c & d \end{pmatrix} = ad - bc$, in agreement with the definition given earlier in connection with Formula 4.1 for inverting a 2-by-2 matrix. Stated in words, $\det \begin{pmatrix} a & b \\ c & d \end{pmatrix}$ is the product of the entries on the main diagonal minus the product of the other two entries. We can often compute 2-by-2 determinants mentally, and consequently find 3-by-3 determinants in one or two lines.

Geometric Interpretation.   Interpreting either the rows, or else the columns, of a square matrix A as vectors leads to a geometric interpretation of $\det A$. Here we'll consider just the 2-by-2 and 3-by-3 cases. Recall from Chapter 1, Section 6 that the cross product of two vectors in the xy-plane of $\mathbb{R}^3$ is

$$(a, b, 0) \times (c, d, 0) = \begin{vmatrix} \mathbf{i} & \mathbf{j} & \mathbf{k} \\ a & b & 0 \\ c & d & 0 \end{vmatrix} = \det \begin{pmatrix} a & b \\ c & d \end{pmatrix} \mathbf{k}.$$

The length of this vector is the absolute value of the 2-by-2 determinant. On the other hand, from Chapter 1 we know that the length of the cross product equals the area $A(P)$ of the parallelogram P with the vectors (a, b) and (c, d) in the xy-plane for adjacent sides. Thus

$$\det \begin{pmatrix} a & b \\ c & d \end{pmatrix} = \pm A(P).$$

Since we can interchange rows and columns in the 2-by-2 matrix without changing the determinant, $A(P)$ is also equal to the area of the parallelogram with the vectors (a, c) and (b, d) for adjacent edges. In either case the sign is "+" if the angle from the first vector to the second is positive, and "−" otherwise.

Instead of the signed area we get in the 2-by-2 case, the determinant of a 3-by-3 matrix is the signed volume $V(P)$ of a solid region P called a parallelepiped, having congruent parallelograms for opposite faces. To see this, recall from Chapter 1, Section 6 that the signed volume of a parallelpiped P with $\mathbf{u}$, $\mathbf{v}$, and $\mathbf{w}$ for adjacent edges is equal to a scalar triple product of the three vectors:

$$\mathbf{u} \cdot (\mathbf{v} \times \mathbf{w}) = (u_1\mathbf{i} + u_2\mathbf{j} + u_3\mathbf{k}) \cdot \begin{vmatrix} \mathbf{i} & \mathbf{j} & \mathbf{k} \\ v_1 & v_2 & v_3 \\ w_1 & w_2 & w_3 \end{vmatrix} = \det \begin{pmatrix} u_1 & u_2 & u_3 \\ v_1 & v_2 & v_3 \\ w_1 & w_2 & w_3 \end{pmatrix}.$$

Hence $\det \begin{pmatrix} u_1 & u_2 & u_3 \\ v_1 & v_2 & v_3 \\ w_1 & w_2 & w_3 \end{pmatrix} = \pm V(P)$, where the sign is "+" if the three column vectors in the given order form a right-hand system and is "−" otherwise.

5B   Row and Column Expansions

It is an important fact, the proof of which we omit, that if in Equation 5.1 the elements and cofactors of the first row are replaced by the elements and cofactors of any other row, or of any column, then the expansion is still valid. Here is the formal statement.

5.2   Theorem.   If A is a square matrix, then

$$\det A = \sum_{j=1}^{n}(-1)^{i+j}a_{ij}\,\det A_{ij} \quad \text{expansion by } i\text{th row}$$

and

$$\det A = \sum_{i=1}^{n}(-1)^{i+j}a_{ij}\,\det A_{ij} \quad \text{expansion by } j\text{th column.}$$

Equation 5.1, which we used to define determinants, appears as a special case of the first equation in the theorem when we let $i = 1$. The alternating pattern of cofactor signs applies to all expansions by row or column, and one way to think of the connection between the two expansion formulas is that the second one, by columns, is a row expansion of $\det A^{t}$, where A^{t} is the matrix you get from A by interchanging rows and columns. Thus if a_{ij} is the entry in the ith row and jth column of A, then the number a_{ji} is the entry in the ith row and jth column of the transpose A^{t}.

The expansion of $\det A^{t}$ by the first row looks just like the expansion of $\det A$ by the first column, except that the minors of $\det A^{t}$ are the transposes of the minors of A. Hence if $\det A^{t} = \det A$ holds for all $(n-1)$-by-$(n-1)$ matrices, it will hold for all n-by-n matrices. It obviously holds for 1-by-1 matrices, and so holds by induction for square matrices of all dimensions.

EXAMPLE 3   An alternative to the determinant computation in part (b) of Example 2 is to use the elements and cofactors of the second row:

$$\det \begin{pmatrix} -5 & -6 & 7 \\ 8 & -9 & 0 \\ -3 & 4 & 2 \end{pmatrix}$$

$$= -(8)\det\begin{pmatrix} -6 & 7 \\ 4 & 2 \end{pmatrix} + (-9)\det\begin{pmatrix} -5 & 7 \\ -3 & 2 \end{pmatrix} - (0)\det\begin{pmatrix} -5 & -6 \\ -3 & 4 \end{pmatrix}$$

$$= -8(-12 - 28) - 9(-10 + 21) = 221.$$

Computing this determinant using the elements and cofactors of the third column, which equals the expansion of the transposed matrix by the third row, we get

$$\det \begin{pmatrix} -5 & -6 & 7 \\ 8 & -9 & 0 \\ -3 & 4 & 2 \end{pmatrix} = \begin{pmatrix} -5 & 8 & -3 \\ -6 & -9 & 4 \\ 7 & 0 & 2 \end{pmatrix}^{t}$$

$$= 7\det\begin{pmatrix} 8 & -9 \\ -3 & 4 \end{pmatrix} - (0)\det\begin{pmatrix} -5 & -6 \\ -3 & 4 \end{pmatrix} + 2\det\begin{pmatrix} -5 & -6 \\ 8 & -9 \end{pmatrix}$$

$$= 7(32 - 27) + 2(45 + 48) = 221.$$

5C   Basic Properties

We can always evaluate a determinant using the definition, but this involves a lot of arithmetic if the dimensions of the matrix are at all large. Some of the theorems we prove will justify other methods of calculation that usually work better than row or column expansions if n is greater than 2 or 3.

5.3   Theorem.   If B is obtained from A by multiplying some row or column by a number r, then $\det B = r \det A$. If A has a zero row or column, then $\det A = 0$.

Proof.   If the ith row of A is multiplied by r, the expansion of B by the ith row is

$$\det B = \sum_{j=1}^{n}(-1)^{i+j}ra_{ij}\, \det A_{ij}$$

$$= r\sum_{j=1}^{n}(-1)^{i+j}a_{ij}\, \det A_{ij} = r\, \det A.$$

A similar argument using a column expansion proves the column version. The inductive expansion by a zero row or column gives zero, so $\det A = 0$ in that case.   ∎

EXAMPLE 4   Let

$$A = \begin{pmatrix} 1 & 2 & 3 \\ -1 & 2 & 4 \\ 0 & 1 & 2 \end{pmatrix}, \quad B = \begin{pmatrix} 1 & 2r & 3 \\ -1 & 2r & 4 \\ 0 & r & 2 \end{pmatrix}.$$

Notice that B is obtained by multiplying the second column of A and r. The theorem says that

$$\det B = \det \begin{pmatrix} 1 & 2r & 3 \\ -1 & 2r & 4 \\ 0 & r & 2 \end{pmatrix} = r\, \det \begin{pmatrix} 1 & 2 & 3 \\ -1 & 2 & 4 \\ 0 & 1 & 2 \end{pmatrix}$$

$$= r\, \det A.$$

5.4   Theorem.   Let A, B, and C be matrices that are identical except in one row or column; and suppose that in the exceptional row or column the entries in C are the sums of the corresponding entries in A and B. Then $\det C = \det A + \det B$.

Proof.   Suppose the special row or column is the jth column. Then expansion using that column gives

$$\det C = \sum_{i=1}^{n}(-1)^{i+j}(a_{ij} + b_{ij})\, \det C_{ij}$$

$$= \sum_{i=1}^{n}(-1)^{i+j}a_{ij}\, \det C_{ij} + \sum_{i=1}^{n}(-1)^{i+j}b_{ij}\, \det C_{ij}.$$

But the minor C_{ij} contains only entries that are the same in both A and B, so $\det A_{ij} = \det B_{ij}$. Hence

$$\det C_{ij} = \det A_{ij} + \det B_{ij}.$$

A similar argument using a row expansion proves the row version.   ■

EXAMPLE 5   Let

$$A = \begin{pmatrix} 1 & 2 & 3 \\ -1 & 2 & 4 \\ 0 & 1 & 2 \end{pmatrix}, \quad B = \begin{pmatrix} 1 & 3 & 3 \\ -1 & 1 & 4 \\ 0 & -2 & 2 \end{pmatrix},$$

$$C = \begin{pmatrix} 1 & 5 & 3 \\ -1 & 3 & 4 \\ 0 & -1 & 2 \end{pmatrix}.$$

The matrices A, B, and C are identical except in the second column, and that column of C is the sum of the corresponding columns of A and B. We have

$$\det A = (1)(4 - 4) - 2(-2 - 0) + 3(-1 - 0)$$
$$= 0 + 4 - 3 = 1,$$
$$\det B = (1)(2 + 8) - 3(-2 - 0) + 3(2 - 0)$$
$$= 10 + 6 + 6 = 22,$$
$$\det C = (1)(6 + 4) - 5(-2 - 0) + 3(1 - 0)$$
$$= 10 + 10 + 3 = 23 = \det A + \det B.$$

Sign Changes.   Another basic property of determinants is that if two rows (or columns) of a matrix A are interchanged then $\det A$ changes sign. This property is sometimes expressed by saying that the determinant is **alternating** as a function of its rows (or of its columns). For 2-by-2 matrices, the property amounts to the observation that

$$\det \begin{pmatrix} a & b \\ c & d \end{pmatrix} = ad - bc = -\det \begin{pmatrix} b & a \\ d & c \end{pmatrix}$$
$$= -\det \begin{pmatrix} c & d \\ a & b \end{pmatrix}.$$

Here is the general statement.

5.5   Theorem.   If B is obtained from A by exchanging two rows or two columns, then $\det B = -\det A$. If A has two rows or columns proportional, then $\det A = 0$.

Proof.   We proceed inductively, assuming that the theorem has been proved for $(n - 1)$-by-$(n - 1)$ matrices, and proving it for n-by-n matrices. We have already observed that the theorem is true for 2-by-2 matrices. Assuming $n \geq 3$, we expand det A using some row different from the two that are to be interchanged, say the kth row. Then

$$\det A = \sum_{j=1}^{n}(-1)^{k+j}a_{kj} \det A_{kj}.$$

But interchanging two rows different from the kth in A interchanges two rows different from the kth in each $(n-1)$-by-$(n-1)$ minor A_{kj}. Since all determinants $\det A_{kj}$ then change sign by the induction assumption, and nothing else changes in the expansion, it follows that $\det A$ changes sign. A similar argument using a column expansion proves the column version. Finally, if A has two rows or columns proportional, we can factor out a proportionality constant r, and get $\det A = r \det B$, where B has two rows or columns the same. Then $\det B = -\det B$ so $\det B = 0$. Hence $\det A = r \det B = 0$. ∎

EXAMPLE 6

The matrix

$$A = \begin{pmatrix} 2 & 1 & 3 \\ -1 & 0 & 5 \\ -6 & -3 & -9 \end{pmatrix}$$

has its first and third rows proportional, with the third being (-3) times the first. We verify, expanding by the second row, that

$$\det = -(-1) \det \begin{pmatrix} 1 & 3 \\ -3 & -9 \end{pmatrix} - 5 \det \begin{pmatrix} 2 & 1 \\ -6 & -3 \end{pmatrix}$$

$$= (-9+9) - 5(-6+6) = 0.$$

5D   Computing Determinants
The following fact is useful in computing the determinants of large matrices.

5.6   Theorem.   Adding a scalar multiple of one row (or column) of A to another leaves $\det A$ unchanged.

Proof.   Let the ith row or column of A be the one affected by adding r times the kth row or column, and denote by C the modified matrix. Then we look at $\det C$ as a function of its rows (or columns) $\mathbf{c}_1, \ldots, \mathbf{c}_i, \ldots, \mathbf{c}_n$, so that $\mathbf{c}_i = \mathbf{a}_i + r\mathbf{a}_k$.

$$\det C = \det(\mathbf{c}_1, \ldots, \mathbf{c}_i, \ldots, \mathbf{c}_n)$$
$$= \det(\mathbf{a}_1, \ldots, \mathbf{a}_i + r\mathbf{a}_k, \ldots, \mathbf{a}_k, \ldots, \mathbf{a}_n)$$
$$= \det(\mathbf{a}_1, \ldots, \mathbf{a}_i, \ldots, \mathbf{a}_n)$$
$$= \det(\mathbf{a}_1, \ldots, r\mathbf{a}_k, \ldots, \mathbf{a}_k, \ldots, \mathbf{a}_n)$$
$$= \det A + \det(\mathbf{a}_1, \ldots, r\mathbf{a}_k, \ldots, \mathbf{a}_k, \ldots, \mathbf{a}_n)$$

The last matrix has two rows or columns proportional, so its determinant is zero. Thus $\det C = \det A$, as was to be shown. ∎

EXAMPLE 7 | Let

$$A = \begin{pmatrix} 1 & 3 & -2 \\ 2 & -4 & 1 \\ 3 & 5 & -2 \end{pmatrix}, \quad C = \begin{pmatrix} 1 & 3 & 0 \\ 2 & -4 & 5 \\ 3 & 5 & 4 \end{pmatrix}.$$

The third column of C is equal to the third column of A plus 2 times the first column. We compute

$$\det C = (1)(-16 - 25) - (3)(8 - 15) + (0)(10 + 12) = -20.$$

It follows that $\det A = -20$ also.

EXAMPLE 8 | Let

$$A = \begin{pmatrix} 2 & 4 & -1 & 0 \\ 3 & 0 & 2 & 3 \\ -1 & 2 & 3 & 1 \\ 0 & 1 & -2 & -1 \end{pmatrix}.$$

By adding 2 times column 3 to column 1, and 4 times column 3 to column 2, we obtain

$$B = \begin{pmatrix} 0 & 0 & -1 & 0 \\ 7 & 8 & 2 & 3 \\ 5 & 14 & 3 & 1 \\ -4 & -7 & -2 & -1 \end{pmatrix},$$

and by Theorem 5.6, $\det A = \det B$. The expansion of $\det B$ has only one nonzero term and we get

$$
\begin{aligned}
\det B \ &= (-1)\det \begin{pmatrix} 7 & 8 & 3 \\ 5 & 14 & 1 \\ -4 & -7 & -1 \end{pmatrix} \\[4pt]
&= -\det \begin{pmatrix} 7 & 1 & 3 \\ 5 & 9 & 1 \\ -4 & -3 & -1 \end{pmatrix} \quad \text{[subtracting column 1 from column 2]} \\[4pt]
&= -\det \begin{pmatrix} 0 & 1 & 0 \\ -58 & 9 & -26 \\ 17 & -3 & 8 \end{pmatrix} \quad \begin{array}{l}\text{[subtract 7 times column 2} \\ \text{from column 1 and 3 times column 2} \\ \text{from column 3].}\end{array}
\end{aligned}
$$

Then $\det B = -[(-1)((-58)(8) - (17)(-26))] = -22.$

Theorems 5.3, 5.5, and 5.6 describe the effect on $\det A$ of the elementary row operations of Section 2A and of the interchange of two rows or columns:

1. An elementary multiplication of a row of A by r gives $r \det A$. (Theorem 5.3).
2. An elementary modification of A leaves $\det A$ unchanged (Theorem 5.6).

3. Interchanging two rows or two columns of A changes the sign of $\det A$ (Theorem 5.5).

Thus in putting a matrix A in reduced form R, all we have to do is to keep track of the interchanges and multiplications. Then $\det A = k \det R$, where k is plus or minus the product of the elementary multipliers, depending on whether the number of interchanges used was even or odd.

5E   Invertible Matrices
To find out whether a large square matrix is invertible, it is usually simplest to try to reduce it to the identity by row operations: By Theorems 4.3 and 4.4, if the reduction is possible the matrix is invertible, and otherwise it is not. But for 2-by-2, 3-by-3, and some special matrices of larger dimensions it may be easier to use the following criterion. This theorem is critical for computing eigenvalues in Chapter 13.

5.7   Theorem.   A square matrix A is invertible if and only if $\det A \neq 0$.

Proof.   Suppose that A has been put in reduced form R by elementary row operations, so that $\det A = k \det R$ for some constant $k \neq 0$. If $\det A \neq 0$, then R cannot have a zero row. A reduced square matrix R with no zero row is equivalent to I, so A is invertible. If $\det A = 0$, then R contains a zero row, and A is not invertible, because the equivalent systems $A\mathbf{x} = \mathbf{0}$ and $R\mathbf{x} = \mathbf{0}$ have nonzero solutions.   ∎

EXAMPLE 9   The matrix A in Example 7 is invertible because $\det A = -20 \neq 0$.

Here we generalize to n-by-n matrices the Formula 4.1 for inverting 2-by-2 matrices. Let $\tilde{a}_{ij} = (-1)^{i+j} \det A_{ij}$, the cofactor of a_{ij} that we use in row or column expansions of $\det A$, where A has entry a_{ij} in its ith row and jth column. Let $\tilde{A}$ be the square matrix with ijth entry $\tilde{a}_{ij}$ and form the **transpose** $\tilde{A}^t$, which is the matrix we get by interchanging rows and columns in the matrix $\tilde{A}$. Thus if the ijth entry in $\tilde{A}$ is $\tilde{a}_{ij}$, the ijth entry in $\tilde{A}^t$ is $\tilde{a}_{ji}$.

5.8   Theorem.   If A is invertible, then $A^{-1} = \dfrac{1}{\det A}\tilde{A}^t$.

Proof.   The key to the proof is the equation

$$\sum_{j=1}^{n}(-1)^{i+j}a_{kj} \det A_{ij} = \begin{cases} \det A & \text{if } k = i, \\ 0 & \text{if } k \neq i. \end{cases} \tag{1}$$

If $k = i$, the left side of Equation 1 is just the expansion of $\det A$ by the elements of the ith row, so we get $\det A$. If $k \neq i$, we can still regard the sum as an expansion by the ith row of the determinant of a matrix in which the ith row is the same as the kth row, thus giving determinant 0 by the last part of Theorem 5.5.

To finish the proof we look at the kith entry in the matrix product $A\tilde{A}^t$:

$$\sum_{j=1}^{n}a_{kj}\tilde{a}_{ji}^t = \sum_{j=1}^{n}a_{kj}\tilde{a}_{ij} = \sum_{j=1}^{n}(-1)^{i+j}a_{kj} \det A_{ij} = \begin{cases} \det A & \text{if } k = i, \\ 0 & \text{if } k \neq i, \end{cases}$$

the last two equalities following from the definition of $\tilde{a}_{ij}$ and Equation (1). Hence $A\tilde{A}^t = (\det A)I$, so dividing by $\det A$ gives $A\big((\det A)^{-1}\tilde{A}^t\big) = I$. Now apply A^{-1} to both sides to get $(\det A)^{-1}\tilde{A}^t = A^{-1}$.   ∎

EXAMPLE 10

To invert the matrix

$$A = \begin{pmatrix} 2 & 3 & 4 \\ 5 & 6 & 7 \\ 8 & 9 & 0 \end{pmatrix} \quad \text{first compute the matrix} \quad \begin{pmatrix} -63 & -56 & -3 \\ -36 & -32 & -6 \\ -3 & -6 & -3 \end{pmatrix}$$

with entries $\det A_{ij}$; thus $\det A_{11} = \begin{vmatrix} 6 & 7 \\ 9 & 0 \end{vmatrix} = -63$, and $\det A_{12} = \begin{vmatrix} 5 & 7 \\ 8 & 0 \end{vmatrix} = -56$. To get the matrix of cofactors insert the factors $(-1)^{i+j}$, changing the sign of every second entry and giving

$$\begin{pmatrix} -63 & 56 & -3 \\ 36 & -32 & 6 \\ -3 & 6 & -3 \end{pmatrix}.$$

Finally, transpose this matrix by reflecting across its main diagonal, then divide by $\det A = 30$, found for example by expanding $\det A$ by the last row. The result is

$$A^{-1} = \frac{1}{30}\begin{pmatrix} -63 & 36 & -3 \\ 56 & -32 & 6 \\ -3 & 6 & -3 \end{pmatrix} = \begin{pmatrix} -\frac{21}{10} & \frac{6}{5} & -\frac{1}{10} \\ \frac{28}{15} & -\frac{16}{15} & \frac{1}{5} \\ -\frac{1}{10} & \frac{1}{5} & -\frac{1}{10} \end{pmatrix}.$$

If a square matrix A is invertible, the system of linear equations that the vector equation $A\mathbf{x} = \mathbf{b}$ represents has the unique solution $\mathbf{x} = A^{-1}\mathbf{b}$. We can combine this solution formula with the formula for A^{-1} in the previous Theorem 5.8 to get $\mathbf{x} = (\det A)^{-1}\tilde{A}^t\mathbf{b}$. This formula leads to the following rule.

5.9   Cramer's Rule.   If $\det A \neq 0$ and $\mathbf{x} = (x_1, \ldots, x_n)$, then the jth coordinate of the solution of $A\mathbf{x} = \mathbf{b}$ is

$$x_j = \frac{\det B^{(j)}}{\det A},$$

where $B^{(j)}$ is A with its jth column replaced by the entries $b_1, \ldots, b_n$ in $\mathbf{b}$.

Proof.   As we noted before, the statement of Cramer's rule $\mathbf{x} = (\det A)^{-1}\tilde{A}^t\mathbf{b}$. The jith entry in the matrix $\tilde{A}^t$ is $\tilde{a}^t_{ji} = \tilde{a}_{ij} = (-1)^{i+j}\det A_{ij}$. Thus

$$x_j = (\det A)^{-1}\sum_{i=1}^{n}\tilde{a}^t_{ji}b_i = (\det A)^{-1}\sum_{i=1}^{n}(-1)^{i+j}\det A_{ij}b_i, \quad j = 1, \ldots, n.$$

But the sum on the right is just the expansion of $B^{(j)}$ by the jth column.   ∎

EXAMPLE 11   We'll solve this system using Cramer's rule:

$$
\begin{aligned}
x_1 & -2x_2 & +4x_3 & = 1 \\
-x_1 & +x_2 & -x_3 & = 2 \\
2x_1 & +3x_2 & -x_3 & = 3.
\end{aligned}
$$

The relevant matrices are

$$
A = \begin{pmatrix} 1 & -2 & 4 \\ 1 & 1 & -1 \\ 2 & 3 & -1 \end{pmatrix}, \quad
B^{(1)} = \begin{pmatrix} 1 & -2 & 4 \\ 2 & 1 & -1 \\ 3 & 3 & -1 \end{pmatrix},
$$

$$
B^{(2)} = \begin{pmatrix} 1 & 1 & 4 \\ 1 & 2 & -1 \\ 2 & 3 & -1 \end{pmatrix}, \quad
B^{(3)} = \begin{pmatrix} 1 & -2 & 1 \\ 1 & 1 & 2 \\ 2 & 3 & 3 \end{pmatrix}.
$$

Expanding the determinants by their first rows gives

$$
\det A = (1)(2) - (-2)(3) + (4)(-5) = 2 + 6 - 20 = -12
$$

$$
\det B^{(1)} = (1)(2) - (-2)(1) + (4)(3) = 2 + 2 + 12 = 16
$$

$$
\det B^{(2)} = (1)(1) - (1)(3) + (4)(-7) = 1 - 3 - 28 = -30
$$

$$
\det B^{(3)} = (1)(-3) - (-2)(-7) + (1)(-5) = -3 - 14 - 5 = -22.
$$

Then $x_1 = -\frac{16}{12} = -\frac{4}{3}$, $x_2 = \frac{30}{12} = \frac{5}{2}$, $x_3 = \frac{22}{12} = \frac{11}{6}$.

EXERCISES

In Exercises 1 and 2, evaluate $\det A$ and also $\det(2A)$.

1. $A = \begin{pmatrix} 1 & -2 & 3 \\ 3 & 1 & 4 \\ 5 & 6 & 7 \end{pmatrix}$

2. $A = \begin{pmatrix} 1 & 0 & 1 & 0 \\ 0 & 3 & 1 & 4 \\ 1 & 1 & 4 & 0 \\ -1 & -1 & 2 & 3 \end{pmatrix}$

3. What is the relation between $\det A$ and $\det(2A)$? (See Exercises 1 and 2.)

4. What is the relation between $\det A$ and $\det(-A)$?

In Exercises 5 and 6, use the method of Example 8 of the text to evaluate the determinants of the given matrices.

5. $\begin{pmatrix} -1 & 0 & 1 & 2 \\ 0 & 1 & 2 & -1 \\ 1 & 2 & -1 & 0 \\ 2 & -1 & 0 & 1 \end{pmatrix}$

6. $\begin{pmatrix} 1 & 1 & 1 & 1 \\ 1 & 2 & 4 & 8 \\ 1 & 3 & 9 & 27 \\ 1 & 4 & 16 & 64 \end{pmatrix}$

The **product rule** for determinants states that if A and B are square matrices with the same dimensions, then $\det(AB) = \det(A)\det(B)$. In Exercises 7 and 8, find AB, BA, and the determinants of A, B, AB, and BA, and verify that the product rule holds for these examples.

7. $A = \begin{pmatrix} 1 & -2 \\ 3 & 1 \end{pmatrix}$, $B = \begin{pmatrix} 0 & 1 \\ 2 & -3 \end{pmatrix}$

8. $A = \begin{pmatrix} 2 & 0 & 0 \\ 0 & 3 & 0 \\ 0 & 0 & 4 \end{pmatrix}$, $B = \begin{pmatrix} -1 & 0 & 1 \\ 2 & -1 & -3 \\ 0 & 3 & 5 \end{pmatrix}$

9. Apply the product rule of the preceding exercises to show that if A is invertible, then $\det A \neq 0$ and $\det(A^{-1}) = (\det A)^{-1}$.

10. Show that if D is the diagonal matrix diag$(d_1, \ldots, d_n)$, in the notation used in Theorem 4.6 in the previous section, then det D is the product of the diagonal elements, $d_1 d_2 \cdots d_n$. In particular, det $I = 1$.

11. Compute the determinant of the matrix
$$\begin{pmatrix} 1 & 2 & 3 & 4 \\ 0 & -1 & 5 & 6 \\ 0 & 0 & 3 & -1 \\ 0 & 0 & 0 & 4 \end{pmatrix}.$$

12. Show that for an upper triangular matrix like the one in Exercise 11, in which every element below the diagonal is 0, the determinant is equal to the product of the diagonal elements.

In Exercises 13 to 20, use Theorem 5.7 to determine which of the following matrices have inverses. For those that are invertible use Theorem 5.8 to find the inverse.

13. $\begin{pmatrix} 1 & 0 & 0 \\ 3 & 1 & 5 \\ -2 & 0 & 1 \end{pmatrix}$
14. $\begin{pmatrix} 1 & 2 & 3 \\ -1 & 1 & 0 \\ 0 & 3 & 3 \end{pmatrix}$

15. $\begin{pmatrix} 2 & 4 & 8 \\ 1 & 0 & 0 \\ 1 & -3 & -7 \end{pmatrix}$
16. $\begin{pmatrix} 1 & 2 & 3 \\ 4 & 5 & 6 \\ 7 & 8 & 9 \end{pmatrix}$

17. $\begin{pmatrix} 1 & 2 & 1 \\ 0 & 0 & 1 \\ 0 & 0 & 3 \end{pmatrix}$
18. $\begin{pmatrix} 1 & -1 & 1 \\ 0 & -1 & 1 \\ 0 & 0 & 1 \end{pmatrix}$

19. $\begin{pmatrix} 1 & 2 & -1 & 3 \\ 0 & 2 & 0 & 1 \\ 0 & 0 & 1 & 1 \\ 0 & 0 & 0 & 4 \end{pmatrix}.$
20. $\begin{pmatrix} 1 & 0 & 1 & 0 \\ 0 & 2 & 0 & 0 \\ 0 & 0 & 3 & 0 \\ 0 & 0 & 0 & 4 \end{pmatrix}$

In Exercises 21 to 24, use Theorem 5.7 to determine for which values of the real variable t the given matrix fails to have an inverse.

21. $\begin{pmatrix} 1-t & 2 & 0 \\ 0 & 2-t & 5 \\ 0 & 0 & 3-t \end{pmatrix}$

22. $\begin{pmatrix} e^t \cos t & -e^t \sin t & 0 \\ e^t \sin t & e^t \cos t & 0 \\ 0 & 0 & e^{3t} \end{pmatrix}$

23. $\begin{pmatrix} 2 & 4 & 8 \\ 1 & 0 & 0 \\ 1 & 2 & t \end{pmatrix}$

24. $\begin{pmatrix} 2-t & 4 & 8 \\ 1 & -t & 0 \\ 1 & 2 & -t \end{pmatrix}$

25. Show that the cross product of vectors $\mathbf{u} = (u_1, u_2, u_3)$ and $\mathbf{v} = (v_1, v_2, v_3)$ has the form of a 3-by-3 determinant:

$$\mathbf{u} \times \mathbf{v} = \det \begin{pmatrix} \mathbf{i} & \mathbf{j} & \mathbf{k} \\ u_1 & u_2 & u_3 \\ v_1 & v_2 & v_3 \end{pmatrix}.$$

26. Show that the **scalar triple product** of the ordered triple of vectors $\mathbf{u} = (u_1, u_2, u_3)$, $\mathbf{v} = (v_1, v_2, v_3)$, $\mathbf{w} = (w_1, w_2, w_3)$ is expressible as a 3-by-3 determinant:

$$\mathbf{u} \cdot (\mathbf{v} \times \mathbf{w}) = \det \begin{pmatrix} u_1 & u_2 & u_3 \\ v_1 & v_2 & v_3 \\ w_1 & w_2 & w_3 \end{pmatrix}.$$

27. Let A be an m-by-m matrix and B an n-by-n matrix. Consider the $(m+n)$-by-$(m+n)$ matrix $\begin{pmatrix} A & O \\ O & B \end{pmatrix}$, which has A in the upper left corner, B in the lower right corner, and zeros elsewhere; show that its determinant is equal to $(\det A)(\det B)$. [*Hint:* Consider the cases $A = I$ and $B = I$. Then use the product rule of Exercise 7.]

28. Explain how Formula 4.1 for inverting 2-by-2 matrices is a special case of Theorem 5.8 for inverting n-by-n matrices.

Chapter 2 REVIEW

In Exercises 1 to 8 let

$$A = \begin{pmatrix} 3 & 0 \\ -2 & 3 \end{pmatrix}, \qquad B = \begin{pmatrix} 4 & 1 & 0 \\ -2 & -1 & 2 \end{pmatrix},$$

$$C = \begin{pmatrix} 1 & -3 & -1 \\ 2 & 1 & 0 \end{pmatrix}, \qquad D = \begin{pmatrix} 2 & 0 & -2 \\ 1 & 0 & 5 \\ 0 & 0 & 3 \end{pmatrix},$$

$$E = \begin{pmatrix} 2 & -1 \\ 0 & -4 \\ -1 & 3 \end{pmatrix}, \qquad F = \begin{pmatrix} 0 & -3 \\ 1 & 2 \\ -2 & 0 \end{pmatrix},$$

and evaluate each of the following expressions, or else explain why it's not defined.

1. $A + B$ 2. AB 3. $B + 2C$

4. $A + BE$ **5.** $A + EB$ **6.** $CD + EA$

7. CD^{-1} **8.** $BF + FB$

In Exercises 9 to 12, let $A, B, C,$ and X be square matrices of the same shape. Solve the given equation for X, stating what conditions you have to assume to produce a unique solution. [For example, $AX + BX = C$ has the solution $(A + B)^{-1}C$, provided that $A + B$ is invertible.]

9. $AX + 2B = CX$ **10.** $XA + 2X = B$

11. $AX = 3A$ **12.** $XA + XB = C$

True or false? In Exercises 13 to 20, if the statement is always true or always false, give a reason. If it is sometimes true and sometimes false, give an example of each possibility. Assume that A and B are square matrices of the same dimensions in all cases.

13. If $AB = 0$, then $BA = 0$.

14. If $AB = I$, then $BA = I$.

15. If $AB = A$, then $BA = A$.

16. If $AB = BA$, then $B = A^{-1}$.

17. If $A = O$, then $A^2 = O$.

18. If $A^2 = O$, then $A = O$.

19. If $A^2 = I$, then $A = I$.

20. If $A^2 = O$, then $(I + A)^{-1} = I - A$.

21. Let $A = \begin{pmatrix} 1 & 0 & 1 \\ 0 & 1 & 1 \\ 0 & 0 & 0 \end{pmatrix}$. Describe the set of solutions for the system of equations $A\mathbf{x} = \mathbf{b}$ when (a) $\mathbf{b} = \mathbf{0}$, (b) $\mathbf{b} = (1, 2, 0)$, (c) $\mathbf{b} = (0, 1, 2)$. In general, for what vectors $\mathbf{b}$ does the system have a solution?

22. Convert the matrix $\begin{pmatrix} 1 & 0 & s \\ 0 & s & 0 \\ 1 & 1 & 2 \end{pmatrix}$ to reduced form by elementary row operations. (The operations allowed at each stage may be different for a few special values of s. Consider different cases if necessary.) For which values of s is the matrix invertible?

23. Write the system of equations

$$\begin{aligned} 2x + y - z &= 0 \\ 3y - z &= b \\ x - y &= 1 \end{aligned}$$

in matrix form, and apply row operations to get an equivalent system with a reduced coefficient matrix. Give one solution for each value of b for which a solution exists.

24. Do the same as in Exercise 23 after changing the third equation to $x + y = 1$.

In Exercises 25 to 28, express the vector $\mathbf{v}$ as a linear combination of the other vectors given, or show that it is impossible to do so.

25. $\mathbf{v} = 2\mathbf{i} + 3\mathbf{j}; \mathbf{a} = 2\mathbf{i} - \mathbf{j}, \mathbf{b} = 2\mathbf{i} + \mathbf{j}$

26. $\mathbf{v} = 2\mathbf{i} + 3\mathbf{j} + 4\mathbf{k}; \mathbf{a} = 2\mathbf{i} - \mathbf{j}, \mathbf{b} = \mathbf{i} + \mathbf{j} + \mathbf{k}, \mathbf{c} = \mathbf{j} - 2\mathbf{k}$

27. $\mathbf{v} = (3, -1, 0, -1); \mathbf{a} = (2, -1, 3, 2), \mathbf{b} = (-1, 1, 1, -3), \mathbf{c} = (1, 1, 9, -5)$

28. $\mathbf{v} = (3, 0, 0, -2); \mathbf{a} = (0, -1, 1, 2), \mathbf{b} = (-1, 1, 1, 0), \mathbf{c} = (1, 1, 0, -2)$

In Exercises 29 to 35, find the inverse of the given matrix or show that the inverse doesn't exist.

29. $\begin{pmatrix} 3 & 5 \\ -2 & 1 \end{pmatrix}$ **30.** $\begin{pmatrix} 6 & -10 \\ 3 & 5 \end{pmatrix}$

31. $\begin{pmatrix} 5 & -2 & 3 \\ 4 & -3 & 2 \\ -3 & 4 & -1 \end{pmatrix}$ **32.** $\begin{pmatrix} 3 & -2 & 0 \\ 1 & 0 & 5 \\ -2 & 3 & 0 \end{pmatrix}$

33. $\begin{pmatrix} -1 & 0 & 3 & 2 \\ -2 & 1 & 6 & 4 \\ 0 & 1 & 0 & 0 \\ 3 & -2 & 1 & 3 \end{pmatrix}$ **34.** $\begin{pmatrix} 1 & 2 & 0 & 0 & 0 \\ 0 & 1 & 2 & 0 & 0 \\ 0 & 0 & 0 & 1 & 2 \\ 0 & 0 & 0 & 0 & 1 \end{pmatrix}$

35. $\begin{pmatrix} 1 & 2 & 0 & 0 & 0 \\ 1 & 1 & 2 & 0 & 0 \\ 0 & 1 & 1 & 2 & 0 \\ 0 & 0 & 1 & 1 & 2 \\ 0 & 0 & 0 & 1 & 1 \end{pmatrix}$

36. Evaluate $p(A)$ when $p(t) = t^2 - 2$ and $A = \text{diag}(a, b, c)$. What is the general rule for evaluating $q(A)$ for a polynomial $q(t)$ and diagonal matrix A?

37. **(a)** Show that if E and F are diagonal matrices of the same dimensions, then $EF = FE$.
   (b) Let $D = \text{diag}(a, b, c)$ and write down DA and AD, where A is the matrix in part (c) of Exercise 8. Describe in words the effect of multiplying a matrix by a diagonal matrix, when the diagonal matrix is on the left, and when it is on the right.
   (c) Assuming $a, b,$ and c are all different, and $D = \text{diag}(a, b, c)$, what can you say about matrices B such that $DB = BD$? What if $a = b = c$? What if $a = b \neq c$?

38. Let $p(x)$ be the polynomial $a + bx + cx^2$. By setting up and solving a system of linear equations, find values for the coefficients a, b, c so that $p(-1) = 1, p(0) = 0, p(1) = 1$. Given numbers r, s, t is it always possible to find values for a, b, c to make $p(-1) = r, p(0) = s, p(1) = t$?

39. Let $f(x) = a \sin x + b \cos x + c$. Find a, b, and c so that $f(0) = 1$, $f'(0) = 2$, and $f''(0) = 3$.

40. Let K be the plane through $(1, 2, 3)$, $(-1, 5, 2)$, and $(2, -6, 10)$, and let L be the plane with equation $2x - 3y + z = 2$.

    (a) Find an equation for K by solving a system of linear equations, without using the cross product.

    (b) Find the intersection of K and L with the plane that has equation $x + 2y + az = 0$. (The result will depend on a.) Are there values of a for which the three planes do not intersect?

In Exercises 41 and 42, find all vectors in $\mathbb{R}^4$ that are perpendicular to the given vectors.

41. $(1, 2, 1, 2)$, $(1, 2, 3, 4)$, and $(1, 0, 0, 1)$

42. $(1, 2, 1, 2)$, $(1, 2, 3, 4)$, and $(0, 0, 1, 1)$

Evaluate the determinants of the matrices given in Exercises 43 to 46.

43. $\begin{pmatrix} 1 & 2 & 0 \\ 0 & 1 & 2 \\ 2 & 0 & 1 \end{pmatrix}$

44. $\begin{pmatrix} 3 & 2 & 0 \\ 0 & 1 & -2 \\ 1 & 0 & 1 \end{pmatrix}$

45. $\begin{pmatrix} 1 & 2 & 0 & 0 \\ 0 & 1 & 2 & 0 \\ -2 & -3 & 0 & 2 \\ 2 & 0 & -1 & 3 \end{pmatrix}$

46. $\begin{pmatrix} 0 & 0 & 2 & 3 \\ 1 & -3 & 0 & 2 \\ 1 & -2 & 1 & 0 \\ -2 & 0 & -2 & 1 \end{pmatrix}$

CHAPTER 3

VECTOR SPACES AND LINEARITY (OPTIONAL CHAPTER)

This chapter extends and generalizes the linear algebra developed in Chapters 1 and 2, but the additional material isn't used in later chapters.

SECTION 1   LINEAR FUNCTIONS ON $\mathbb{R}^n$

In Chapter 2 we used matrix multiplication as a way to describe systems of linear equations, so for instance, the system of equations

$$2x + 3y = 5$$
$$x - y = 3$$

is equivalent to the matrix equation $A\mathbf{x} = \mathbf{b}$ with

$$A = \begin{pmatrix} 2 & 3 \\ 1 & -1 \end{pmatrix}, \quad \mathbf{x} = \begin{pmatrix} x \\ y \end{pmatrix}, \quad \text{and} \quad \mathbf{b} = \begin{pmatrix} 5 \\ 3 \end{pmatrix}.$$

Another way of looking at the matrix product $A\mathbf{x}$ in this example is as a function $\mathbf{y} = A\mathbf{x}$ assigning a value $\mathbf{y} = (z, w)$ in $\mathbb{R}^2$ to a given $\mathbf{x} = (x, y)$ in $\mathbb{R}^2$. For this particular matrix A, the function would be given by

$$\begin{matrix} z = 2x + 3y \\ w = x - y \end{matrix} \quad \text{or} \quad \begin{pmatrix} z \\ w \end{pmatrix} = \begin{pmatrix} 2 & 3 \\ 1 & -1 \end{pmatrix} \begin{pmatrix} x \\ y \end{pmatrix}.$$

1A   Matrix Representation

There's a close connection between systems of linear equations, either in scalar or matrix form as illustrated above, and the natural generalization of the function $y = ax$ from $\mathbb{R}$ to $\mathbb{R}$ to functions from $\mathbb{R}^n$ to $\mathbb{R}^m$. If A is an m-by-n matrix, setting $f(\mathbf{x}) = A\mathbf{x}$ defines a function $\mathbf{y} = f(\mathbf{x})$ from $\mathbb{R}^n$ to $\mathbb{R}^m$. We proved in Theorem 2.3 of Chapter 2, Section 2C that a matrix-vector product $A\mathbf{x}$ has the property we called **linearity**, meaning that for a given linear combination $s\mathbf{x} + t\mathbf{y}$ we have $A(s\mathbf{x} + t\mathbf{y}) = sA\mathbf{x} + tA\mathbf{y}$. In terms of the function $f(\mathbf{x}) = A\mathbf{x}$, this says

1.1 $$f(s\mathbf{x} + t\mathbf{y}) = sf(\mathbf{x}) + tf(\mathbf{y}),$$

and a function f from $\mathbb{R}^n$ to $\mathbb{R}^m$ is called **linear** if it satisfies Equation 1.1. Repeated application of Equation 1.1 shows that a linear function $f(\mathbf{x})$ always satisfies the more general condition

1.2 $\qquad f(t_1\mathbf{x}_1 + t_2\mathbf{x}_2 + \cdots + t_k\mathbf{x}_k) = t_1 f(\mathbf{x}_1) + t_2 f(\mathbf{x}_1) + \cdots + t_k f(\mathbf{x}_k).$

The rest of this section, and indeed this chapter, is about understanding linearity.

EXAMPLE 1 For given $A = (a_{ij})$ the following systems define the same function $f(\mathbf{x}) = A\mathbf{x}$ from $\mathbb{R}^n$ to $\mathbb{R}^m$:

$$
\begin{array}{cccc}
y_1 = a_{11}x_1 + & \cdots & + a_{1n}x_n \\
\vdots & \vdots & \vdots \\
y_m = a_{m1}x_1 + & \cdots & + a_{mn}x_n
\end{array}
\quad \text{or} \quad
\begin{pmatrix} y_1 \\ \vdots \\ y_m \end{pmatrix}
=
\begin{pmatrix} a_{11} & \cdots & a_{1n} \\ \vdots & \vdots & \vdots \\ a_{m1} & \cdots & a_{mn} \end{pmatrix}
\begin{pmatrix} x_1 \\ \vdots \\ x_m \end{pmatrix}.
$$

The next theorem shows that as a consequence of Equation 1.1, all linear functions $\mathbf{y} = f(\mathbf{x})$ from $\mathbb{R}^n$ to $\mathbb{R}^m$ are expressible in the two forms of the previous example, in other words in the form $f(\mathbf{x}) = A\mathbf{x}$.

1.3 Theorem.   If f is a linear function from $\mathbb{R}^n$ to $\mathbb{R}^m$, and A is the m-by-n matrix whose columns are the vectors $f(\mathbf{e}_1), \dots, f(\mathbf{e}_n)$, then $f(\mathbf{x}) = A\mathbf{x}$ for every $\mathbf{x}$ in $\mathbb{R}^n$.

Proof.   The product of a matrix and the standard basis vector $\mathbf{e}_j$ always gives the jth column of the matrix, so the definition of A implies $A\mathbf{e}_j = f(\mathbf{e}_j)$ for $j = 1, \dots, n$. Now consider an arbitrary vector $\mathbf{x} = (x_1, \dots, x_n) = x_1\mathbf{e}_1 + \cdots + x_n\mathbf{e}_n$ in $\mathbb{R}^n$. To check that $f(\mathbf{x}) = A\mathbf{x}$, we use the linearity of f and the linearity of matrix multiplication (Theorem 2.3 in Chapter 2). Since we can write $\mathbf{x} = x_1\mathbf{e}_1 + \cdots + x_n\mathbf{e}_n$,

$$
\begin{aligned}
f(\mathbf{x}) &= f(x_1\mathbf{e}_1 + \cdots + x_n\mathbf{e}_n) \\
&= x_1 f(\mathbf{e}_1) + \cdots + x_n f(\mathbf{e}_n) && \text{[by linearity of } f] \\
&= x_1 A\mathbf{e}_1 + \cdots + x_n A\mathbf{e}_n && \text{[by definition of } A] \\
&= A(x_1\mathbf{e}_1 + \cdots + x_n\mathbf{e}_n) = A\mathbf{x}, && \text{[by linearity of matrix multiplication]}
\end{aligned}
$$

so $f(\mathbf{x}) = A\mathbf{x}$. $\qquad\blacksquare$

EXAMPLE 2 For an example of how the theorem works in a particular case, suppose f is a linear function from $\mathbb{R}^3$ to $\mathbb{R}^2$ with

$$
f(\mathbf{e}_1) = \begin{pmatrix} 2 \\ 1 \end{pmatrix}, \quad f(\mathbf{e}_2) = \begin{pmatrix} 3 \\ -1 \end{pmatrix}, \quad f(\mathbf{e}_3) = \begin{pmatrix} -4 \\ 2 \end{pmatrix}.
$$

We form the matrix $A = \begin{pmatrix} 2 & 3 & -4 \\ 1 & -1 & 2 \end{pmatrix}$ from these column vectors. Then

$$
A\mathbf{e}_1 = \begin{pmatrix} 2 & 3 & -4 \\ 1 & -1 & 2 \end{pmatrix} \begin{pmatrix} 1 \\ 0 \\ 0 \end{pmatrix} = \begin{pmatrix} 2 \\ 1 \end{pmatrix},
$$

$$
A\mathbf{e}_2 = \begin{pmatrix} 2 & 3 & -4 \\ 1 & -1 & 2 \end{pmatrix} \begin{pmatrix} 0 \\ 1 \\ 0 \end{pmatrix} = \begin{pmatrix} 3 \\ -1 \end{pmatrix},
$$

$$Ae_3 = \begin{pmatrix} 2 & 3 & -4 \\ 1 & -1 & 2 \end{pmatrix} \begin{pmatrix} 0 \\ 0 \\ 1 \end{pmatrix} = \begin{pmatrix} -4 \\ 2 \end{pmatrix},$$

and for $\mathbf{x} = (x_1, x_2, x_3) = x_1 \mathbf{e}_1 + x_2 \mathbf{e}_2 + x_3 \mathbf{e}_3$,

$$f(\mathbf{x}) = x_1 f(\mathbf{e}_1) + x_2 f(\mathbf{e}_2) + x_3 f(\mathbf{e}_3) = x_1 \begin{pmatrix} 2 \\ 1 \end{pmatrix} + x_2 \begin{pmatrix} 3 \\ -1 \end{pmatrix} + x_3 \begin{pmatrix} -4 \\ 2 \end{pmatrix}$$

$$= \begin{pmatrix} 2x_1 \\ x_1 \end{pmatrix} + \begin{pmatrix} 3x_2 \\ -x_2 \end{pmatrix} + \begin{pmatrix} -4x_3 \\ 2x_3 \end{pmatrix} = \begin{pmatrix} 2x_1 + 3x_2 - x_3 \\ x_1 - x_2 + 2x_3 \end{pmatrix} = \begin{pmatrix} 2 & 3 & -4 \\ 1 & -1 & 2 \end{pmatrix} \begin{pmatrix} x_1 \\ x_2 \\ x_3 \end{pmatrix}.$$

Note. In the context of real-valued functions of one variable the term "linear function" often means a function of the form $f(x) = ax + b$, because such functions have straight line graphs in the xy-plane. Such functions are linear in the present context of functions with vector variables only if $b = 0$. In this book the term "linear function" will always mean a function satisfying Equation 1.1.

Before looking at more examples of linear functions, we introduce some terminology that's useful for talking about functions in general, whether or not they are linear. We revisit these terms at the beginning of Chapter 4 applied to nonlinear functions.

A function f is defined on a set D called the **domain** of f and takes its values within some set R called the **range** of f. Thus for every $\mathbf{x}$ in D, $f(\mathbf{x})$ is some $\mathbf{y}$ in R. We say that f is a function from D to R, and use the notation

$$f : D \longrightarrow R$$

to indicate that f is a function with domain D and range R.

If E is a subset of the domain of f we write $f(E)$ for the subset of the range of f consisting of the values $f(\mathbf{x})$ taken on by f as $\mathbf{x}$ varies over the set E, and call it the **image** of E under f. The image $f(D)$ of the entire domain of f is called simply the **image** of f.

Variant terminology. Some people use the term "range of f" to mean what we call the "image of f", but for clarity we distinguish between the two terms as we just indicated. For us, the range of a function is the space, such as $\mathbb{R}^n$, within which its values are defined to lie, while the image consists of the values the function actually takes on. The image is contained in the range but need not be all of it.

The following examples illustrate Theorem 1.3 on matrix representation for some typical linear functions $f : \mathbb{R}^n \longrightarrow \mathbb{R}^m$, including some that we've already seen in Chapter 1.

EXAMPLE 3   Defining $f(x)$ to be the scalar multiple $x(1, 2, 1)$ gives a function $f : \mathbb{R}^1 \longrightarrow \mathbb{R}^3$ that is linear because the coordinates of the image of f in $\mathbb{R}^3$ are $y_1 = x$, $y_2 = 2x$, $y_3 = x$, and these have the simplest possible form of the first display in Example 1. The image of f consists of all scalar multiples of $(1, 2, 1)$ and is a line through the origin in $\mathbb{R}^3$ in the direction of the vector $(1, 2, 1)$. If E is the interval $[0, 1]$, then $f(E)$, the image of E under f, is the line segment joining $(0, 0, 0)$ and $(1, 2, 1)$.

EXAMPLE 4   Let $f:\mathbb{R}^2\longrightarrow\mathbb{R}^3$ be the linear function such that $f\begin{pmatrix}1\\0\end{pmatrix}=\begin{pmatrix}1\\2\\4\end{pmatrix}$ and $f\begin{pmatrix}0\\1\end{pmatrix}=\begin{pmatrix}-1\\0\\1\end{pmatrix}$. The matrix of f is $\begin{pmatrix}1&-1\\2&0\\4&1\end{pmatrix}$, and for arbitrary (s,t) in $\mathbb{R}^2$, we have

$$f(s,t)=\begin{pmatrix}1&-1\\2&0\\4&1\end{pmatrix}\begin{pmatrix}s\\t\end{pmatrix}=\begin{pmatrix}s-t\\2s\\4s+t\end{pmatrix}=s\begin{pmatrix}1\\2\\4\end{pmatrix}+t\begin{pmatrix}-1\\0\\1\end{pmatrix}.$$

The image F of f is the plane through the origin in $\mathbb{R}^3$ containing the vectors $(1,2,4)$ and $(-1,0,1)$.

EXAMPLE 5   For a linear function with a different kind of geometric interpretation, consider the function from $\mathbb{R}$ to $\mathbb{R}$ that sends x to $2x$ and has the geometric effect of stretching the real number line by the factor 2. In two dimensions we can define a linear function $f:\mathbb{R}^2\longrightarrow\mathbb{R}^2$ that stretches horizontal distances by a factor of 3 and vertical distances by a factor of 2. To do this we need $f(\mathbf{e}_1)=3\mathbf{e}_1$ and $f(\mathbf{e}_2)=2\mathbf{e}_2$, so f is represented by a diagonal matrix as

$$f\begin{pmatrix}x_1\\x_2\end{pmatrix}=\begin{pmatrix}3&0\\0&2\end{pmatrix}\begin{pmatrix}x_1\\x_2\end{pmatrix}.$$

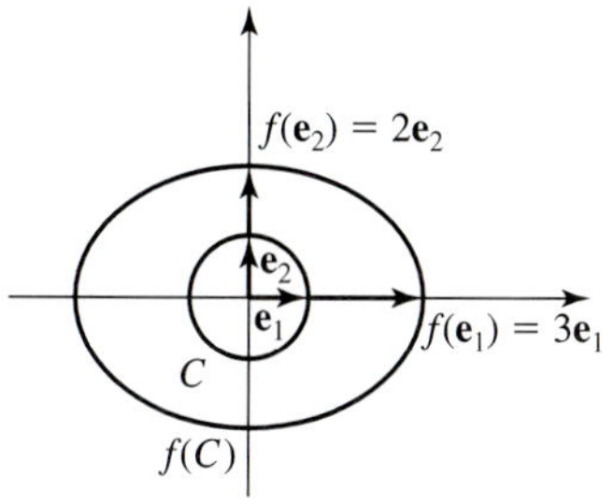

FIGURE 3.1

Unequal expansions.

Figure 3.1 shows the geometric effect of f, where C is the unit circle $x_1^2+x_2^2=1$, and $f(C)$, the image of C under f, is an ellipse. If $\begin{pmatrix}u_1\\u_2\end{pmatrix}$ is the image of $\begin{pmatrix}x_1\\x_2\end{pmatrix}=\begin{pmatrix}3x_1\\2x_2\end{pmatrix}$ under f, then $x_1=\frac{1}{3}u_1$ and $x_2=\frac{1}{2}u_2$. Hence if $\begin{pmatrix}u_1\\u_2\end{pmatrix}$ is in $f(C)$, then $\dfrac{u_1^2}{9}+\dfrac{u_2^2}{4}=1$; this is the equation of the ellipse with semimajor axis 3 and semiminor axis 2 shown in Figure 3.1.

EXAMPLE 6   The projections of one vector on another defined in Chapter 1, Section 4C are geometric examples of linear functions $f:\mathbb{R}^n\longrightarrow\mathbb{R}^n$. Let $\mathbf{n}$ be a unit vector in $\mathbb{R}^n$. Define the function $P_\mathbf{n}:\mathbb{R}^n\longrightarrow\mathbb{R}^n$ by $P_\mathbf{n}(\mathbf{x})=(\mathbf{x}\cdot\mathbf{n})\mathbf{n}$. Then $P_\mathbf{n}$ is linear because using properties of the dot product shows it satisfies Equation 1.1:

$$P_\mathbf{n}(s\mathbf{u}+t\mathbf{v})=((s\mathbf{x}+t\mathbf{y})\cdot\mathbf{n})\mathbf{n}=(s\mathbf{x}\cdot\mathbf{n}+t\mathbf{y}\cdot\mathbf{n})\mathbf{n}$$

$$=s(\mathbf{x}\cdot\mathbf{n})\mathbf{n}+t(\mathbf{y}\cdot\mathbf{n})\mathbf{n}=sP_\mathbf{n}(\mathbf{x})+tP_\mathbf{n}(\mathbf{y}).$$

As we saw in Theorem 4.10 of Chapter 1, $P_\mathbf{n}(\mathbf{x})$ is the projection of $\mathbf{x}$ on the line through the origin in the direction of $\mathbf{n}$, and the image of $P_\mathbf{n}$ is the same line.

If $\mathbf{e}_i$ is one of the standard basis vector in $\mathbb{R}^n$ then $\mathbf{e}_i\cdot\mathbf{e}_i=1$ and $\mathbf{e}_i\cdot\mathbf{e}_j=0$ if $i\neq j$. Thus if $\mathbf{n}=\mathbf{e}_i$, then $P_\mathbf{n}(\mathbf{e}_i)=\mathbf{e}_i$ and $P_\mathbf{n}(\mathbf{e}_j)=\mathbf{0}$ if $i\neq j$. Consequently, the matrix of $P_{\mathbf{e}_i}$ is 0 except for having $\mathbf{e}_i$ in the ith column. For the matrix of $P_\mathbf{n}$ for an arbitrary unit vector $\mathbf{n}$ see Exercises 19 and 20.

Rotations are another class of linear functions from a space to itself. We view a rotation of the plane around the origin as a function $f:\mathbb{R}^2\longrightarrow\mathbb{R}^2$ with $f(\mathbf{x})$ defined as the result of rotating $\mathbf{x}$ through an angle θ, where both $f(\mathbf{x})$ and $\mathbf{x}$ are pictured as arrows with tails at the origin. To see that such a rotation is a linear function, recall the geometric interpretation of vector addition and scalar multiplication in Section 2 of Chapter 1. Figure 3.2(a) shows vectors $\mathbf{u}$, $\mathbf{v}$, and $\mathbf{w} = \mathbf{u} + \mathbf{v}$ and their images under f. By the parallelogram law for addition, the arrow representing $\mathbf{w}$ is the diagonal of the parallelogram with sides $\mathbf{u}$ and $\mathbf{v}$. The rotation carries the entire parallelogram to a congruent one with sides $f(\mathbf{u})$ and $f(\mathbf{v})$ and diagonal $f(\mathbf{w})$, so $f(\mathbf{w}) = f(\mathbf{u}) + f(\mathbf{v})$.

Similarly, Figure 3.2(b), in which $\mathbf{q}$ is a scalar multiple $s\mathbf{p}$ of $\mathbf{p}$, shows that $f(\mathbf{q}) = sf(\mathbf{p})$ because both $\mathbf{q}$ and $\mathbf{p}$ are rotated through the same angle, and their lengths are not changed.

EXAMPLE 7

For a simple example of a rotation, consider turning the plane $90°$ counterclockwise around the origin. This takes $\mathbf{e}_1$ to $\mathbf{e}_2$ and $\mathbf{e}_2$ to $-\mathbf{e}_1$ as shown in Figure 3.3(a). The rotation is then a linear function $f:\mathbb{R}^2\longrightarrow\mathbb{R}^2$ with $f(\mathbf{e}_1) = \mathbf{e}_2$ and $f(\mathbf{e}_2) = -\mathbf{e}_1$, so its matrix has $\mathbf{e}_2$ and $-\mathbf{e}_1$ as columns. For a vector $\mathbf{u} = \begin{pmatrix} x \\ y \end{pmatrix}$,

$$f(\mathbf{u}) = f\begin{pmatrix} x \\ y \end{pmatrix} = \begin{pmatrix} 0 & -1 \\ 1 & 0 \end{pmatrix}\begin{pmatrix} x \\ y \end{pmatrix} = \begin{pmatrix} -y \\ x \end{pmatrix}.$$

The image of a vector $\mathbf{u} = (x, y)$ under a $90°$ rotation should have the same length as $\mathbf{u}$ and be at right angles to it. We can check this algebraically by computing some dot products. We have $|f(\mathbf{u})|^2 = f(\mathbf{u}) \bullet f(\mathbf{u}) = (-y)^2 + x^2 = x^2 + y^2 = |\mathbf{u}|^2$, so $f(\mathbf{u})$ and $\mathbf{u}$ have the same length, and $f(\mathbf{u}) \bullet \mathbf{u} = -yx + xy = 0$ so $f(\mathbf{u})$ and $\mathbf{u}$ are perpendicular to each other.

EXAMPLE 8

Consider rotating the plane counterclockwise around the origin through an arbitrary angle θ. As shown in Figure 3.3(b) this rotation carries $\mathbf{e}_1$ and $\mathbf{e}_2$ to the vectors that form the columns of the matrix

$$R_\theta = \begin{pmatrix} \cos\theta & -\sin\theta \\ \sin\theta & \cos\theta \end{pmatrix}.$$

Computing dot products as in the previous example shows that $R_\theta\mathbf{x}$ has the same length as $\mathbf{x}$, and the cosine of the angle between $R_\theta\mathbf{x}$ and $\mathbf{x}$ is equal to $\cos\theta$. (See Exercises 9 and 10.)

Here is a general statement about the image of a linear function $f(\mathbf{x}) = A\mathbf{x}$.

1.4   Theorem.   Let $f: \mathbb{R}^n \to \mathbb{R}^m$ be a linear function with matrix A. Then the image of f consists of all linear combinations of the columns of A.

Proof.   We can write a vector $\mathbf{x}$ in $\mathbb{R}^n$ as a linear combination

$$\mathbf{x} = x_1\mathbf{e}_1 + \cdots + x_n\mathbf{e}_n,$$

and then use the linearity of f to write

$$f(\mathbf{x}) = x_1 f(\mathbf{e}_1) + \cdots + x_n f(\mathbf{e}_n).$$

Since $\mathbf{x}$ is an arbitrary vector in the domain of f, the last equation expresses an arbitrary vector in the image of f as a linear combination of the vectors $f(\mathbf{e}_1), \dots, f(\mathbf{e}_n)$. By Theorem 1.3, these vectors are just the columns of the matrix of f, so the image of f consists of all linear combinations of the columns of A. ∎

EXAMPLE 9

To illustrate Theorem 1.4, consider a plane P through the origin in $\mathbb{R}^3$. P consists of all linear combinations of two vectors $\mathbf{y}_1$ and $\mathbf{y}_2$ in $\mathbb{R}^3$. You can check that the function $f \colon \mathbb{R}^2 \to \mathbb{R}^3$ defined by

$$f \begin{pmatrix} u \\ v \end{pmatrix} = u\mathbf{y}_1 + v\mathbf{y}_2$$

is linear. Theorem 1.3 implies that the matrix of f has as columns the two vectors

$$f \begin{pmatrix} 1 \\ 0 \end{pmatrix} = \mathbf{y}_1 \quad \text{and} \quad f \begin{pmatrix} 0 \\ 1 \end{pmatrix} = \mathbf{y}_2.$$

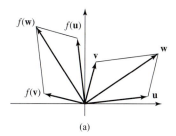

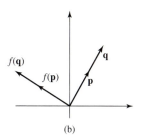

For example, if

$$\mathbf{y}_1 = \begin{pmatrix} 1 \\ 2 \\ 1 \end{pmatrix} \quad \text{and} \quad \mathbf{y}_2 = \begin{pmatrix} -1 \\ 0 \\ 1 \end{pmatrix},$$

then

$$f \begin{pmatrix} u \\ v \end{pmatrix} = \begin{pmatrix} 1 & -1 \\ 2 & 0 \\ 1 & 1 \end{pmatrix} \begin{pmatrix} u \\ v \end{pmatrix} = \begin{pmatrix} u - v \\ 2u \\ u + v \end{pmatrix}.$$

FIGURE 3.2

Thus the image of f in $\mathbb{R}^3$ is the plane P determined by the two column vectors of the 3-by-2 matrix.

More generally, the definition of a k-plane containing the origin in $\mathbb{R}^n$ given in Section 2D of Chapter 2 amounts to saying that a k-plane through the origin is the image of a linear function $f \colon \mathbb{R}^k \longrightarrow \mathbb{R}^n$ given by $f(\mathbf{x}) = A\mathbf{x}$, where A is an n-by-k matrix with k linearly independent columns.

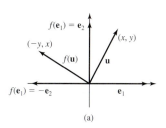

1B   Composition

If f and g are two functions, not necessarily linear, such that the image of f overlaps the domain of g, then the **composition** $g \circ f$ of f and g is defined to be the function obtained by first applying f and then applying g:

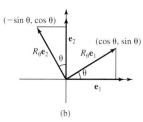

$$(g \circ f)(\mathbf{x}) = g\bigl(f(\mathbf{x})\bigr).$$

The domain of $(g \circ f)$ consists of all $\mathbf{x}$ such that both $f(\mathbf{x})$ and $g\bigl(f(\mathbf{x})\bigr)$ are defined, and is the same as the domain of f when the image of f is contained in the domain of g.

The following theorem states an important connection between composition of linear functions and matrix multiplication that motivates the definition of matrix multiplication.

FIGURE 3.3

1.5 Theorem.   Let $f : \mathbb{R}^n \rightarrow \mathbb{R}^m$ and $g : \mathbb{R}^m \rightarrow \mathbb{R}^p$ be linear functions with matrices A and B, respectively. Then the composition $g \circ f$ is defined, and

$$(g \circ f)(\mathbf{x}) = (BA)\mathbf{x},$$

for all $\mathbf{x}$ in $\mathbb{R}^n$, so $g \circ f$ has matrix BA and is a function from $\mathbb{R}^n$ to $\mathbb{R}^p$. It follows from Theorem 2.3 in Chapter 2, Section 2C that $g \circ f$ is linear.

In Theorem 1.5 the image of f is contained in the domain of g, which is $\mathbb{R}^m$, so the domain of $g \circ f$ is all of $\mathbb{R}^n$. Note also that A is an m-by-n matrix and B is a p-by-m matrix, so the product BA is defined.

Proof.   Suppose that

$$f(\mathbf{x}) = A\mathbf{x} \quad \text{and} \quad g(\mathbf{y}) = B\mathbf{y}$$

are linear functions with matrices A and B. Then

$$(g \circ f)\mathbf{x} = g(f(\mathbf{x})) = g(A\mathbf{x}) = B(A\mathbf{x}) = (BA)\mathbf{x},$$

where at the last step we used the associative law for matrix multiplication. ∎

EXAMPLE 10   Let $f : \mathbb{R}^2 \rightarrow \mathbb{R}^2$ and $g : \mathbb{R}^2 \rightarrow \mathbb{R}^2$ be defined by

$$f(\mathbf{x}) = \begin{pmatrix} 1 & 2 \\ -1 & -4 \end{pmatrix} \mathbf{x} \quad \text{and} \quad g(\mathbf{y}) = \begin{pmatrix} 0 & 1 \\ -2 & -3 \end{pmatrix} \mathbf{y}.$$

Then

$$(g \circ f)(\mathbf{x}) = \begin{pmatrix} 0 & 1 \\ -2 & -3 \end{pmatrix} \begin{pmatrix} 1 & 2 \\ -1 & -4 \end{pmatrix} \mathbf{x} = \begin{pmatrix} -1 & -4 \\ 1 & 8 \end{pmatrix} \mathbf{x}.$$

EXAMPLE 11   Consider the rotation matrix R_θ of Example 8. On geometric grounds we would expect that $(R_\theta)^2 = R_{2\theta}$, because a rotation through angle θ followed by another one is a rotation through angle 2θ. Algebraically, all we have to check is that multiplying the matrix R_θ by itself gives the matrix of $R_{2\theta}$:

$$\begin{pmatrix} \cos\theta & -\sin\theta \\ \sin\theta & \cos\theta \end{pmatrix} \begin{pmatrix} \cos\theta & -\sin\theta \\ \sin\theta & \cos\theta \end{pmatrix} = \begin{pmatrix} \cos^2\theta - \sin^2\theta & -2\sin\theta\cos\theta \\ 2\sin\theta\cos\theta & \cos^2\theta - \sin^2\theta \end{pmatrix}$$

$$= \begin{pmatrix} \cos 2\theta & -\sin 2\theta \\ \sin 2\theta & \cos 2\theta \end{pmatrix}.$$

In the last step we used the trigonometric identities $\sin 2\theta = 2\sin\theta\cos\theta$ and $\cos 2\theta = \cos^2\theta - \sin^2\theta$.

1C   Inverse Functions

A function $f : \mathbb{R}^n \longrightarrow \mathbb{R}^m$ is **one-to-one** if for every $\mathbf{y}$ in the image F of f in $\mathbb{R}^m$ there is a unique $\mathbf{x}$ in $\mathbb{R}^n$ such that $f(\mathbf{x}) = \mathbf{y}$. If f is one-to-one, its **inverse function** $f^{-1} : F \longrightarrow \mathbb{R}^n$ is defined by setting $f^{-1}(\mathbf{y}) = \mathbf{x}$, where $\mathbf{x}$ is the unique $\mathbf{x}$ in $\mathbb{R}^n$ such that $f(\mathbf{x}) = \mathbf{y}$. Thus for all $\mathbf{x}$ in $\mathbb{R}^n$ and all $\mathbf{y}$ in F,

$$f^{-1}(f(\mathbf{x})) = \mathbf{x} \quad \text{and} \quad f(f^{-1}(\mathbf{y})) = \mathbf{y}.$$

If f is one to one, so is f^{-1}, since if $f^{-1}(\mathbf{y}_1) = f^{-1}(\mathbf{y}_2)$, applying f to both sides shows that $\mathbf{y}_1 = \mathbf{y}_2$. Thus there is a unique $\mathbf{y}$ such that f^{-1} has the value $f^{-1}(\mathbf{y})$.

The following theorem describes linear functions that are one to one.

1.6   Theorem.   A linear function $f(\mathbf{x}) = A\mathbf{x}$ is one to one if and only if the columns of A are linearly independent vectors. If in addition A is a square matrix then A is invertible, and $f^{-1}(\mathbf{y}) = A^{-1}\mathbf{y}$ for all $\mathbf{y}$ in $\mathbb{R}^m$.

Proof.   If f is one-to-one, then since $f(\mathbf{0}) = A\mathbf{0} = \mathbf{0}$, the only solution of $A\mathbf{x} = \mathbf{0}$ is $\mathbf{x} = \mathbf{0}$. Conversely, suppose $\mathbf{x} = \mathbf{0}$ is the only solution of $A\mathbf{x} = \mathbf{0}$. Since $A\mathbf{x}_1 = A\mathbf{x}_2$ if and only if $A(\mathbf{x}_1 - \mathbf{x}_2) = \mathbf{0}$, it follows that if $A\mathbf{x}_1 = A\mathbf{x}_2$ then $\mathbf{x}_1 - \mathbf{x}_2 = \mathbf{0}$ and $\mathbf{x}_1 = \mathbf{x}_2$, so f is one-to-one. By the Definition 2.7 of linear independence in Chapter 2, the columns of A are independent if and only if the system $A\mathbf{x} = \mathbf{0}$ has only $\mathbf{x} = \mathbf{0}$ for its solution. Following Inversion Process 4.3 in Chapter 2, a square matrix A is invertible if and only if $A\mathbf{x} = \mathbf{0}$ has only the zero solution. In that case $A\mathbf{x} = \mathbf{y}$ if and only if $\mathbf{x} = A^{-1}\mathbf{y}$, so $f^{-1}(\mathbf{y}) = A^{-1}\mathbf{y}$.   ∎

EXAMPLE 12

For an example of a function given by an invertible matrix A, let $f(\mathbf{x}) = A\mathbf{x} = \begin{pmatrix} 1 & 3 \\ 1 & 4 \end{pmatrix}\mathbf{x}$. Since the columns of A are independent A^{-1} exists, so

$$f^{-1}(\mathbf{x}) = \begin{pmatrix} 1 & 3 \\ 1 & 4 \end{pmatrix}^{-1}\mathbf{x} = \begin{pmatrix} 4 & -3 \\ -1 & 1 \end{pmatrix}\mathbf{x}.$$

In this example the image F of f is all of $\mathbb{R}^2$, since if $\mathbf{y}_0$ is in $\mathbb{R}^2$, then $f(\mathbf{x}_0) = \mathbf{y}_0$, where $\mathbf{x}_0 = A^{-1}\mathbf{y}_0$.

EXAMPLE 13

We can see geometrically that rotating a vector in $\mathbb{R}^2$ through an angle θ and then through the angle $-\theta$ puts it back in its original position, so the functions given by the rotation matrices

$$R_\theta = \begin{pmatrix} \cos\theta & -\sin\theta \\ \sin\theta & \cos\theta \end{pmatrix} \quad \text{and} \quad R_\theta^{-1} = \begin{pmatrix} \cos\theta & \sin\theta \\ -\sin\theta & \cos\theta \end{pmatrix}.$$

are examples of functions that are inverses of each other. As it should, multiplying the matrices in either order gives the identity matrix:

$$\begin{pmatrix} \cos^2\theta + \sin^2\theta & 0 \\ 0 & \sin^2\theta + \cos^2\theta \end{pmatrix} = \begin{pmatrix} 1 & 0 \\ 0 & 1 \end{pmatrix}.$$

EXAMPLE 14

In Example 9 we considered the parametric representation of a plane P containing the origin in $\mathbb{R}^n$ given by

$$f\begin{pmatrix} u \\ v \end{pmatrix} = \begin{pmatrix} 1 & -1 \\ 2 & 0 \\ 1 & 1 \end{pmatrix}\begin{pmatrix} u \\ v \end{pmatrix} \quad \text{or} \quad \begin{aligned} x &= u - v \\ y &= 2u \\ z &= u + v. \end{aligned}$$

The function $f(u, v)$ is one-to-one by Theorem 1.6 because the columns of the 3-by-2 matrix are independent, neither one being a scalar multiple of the other. The inverse function is defined only on P, a plane whose equation in $\mathbb{R}^3$ is $x - y + z = 0$. Given a point on P satisfying $x - y + z = 0$ you can find the coordinates of the corresponding point $f^{-1}(x, y, z) = (u, v)$ by solving $u - v = x$, $u + v = z$ to get $u = \frac{1}{2}(x + z)$, $v = \frac{1}{2}(z - x)$. These formulas for u and v define a function on all of $\mathbb{R}^3$, but we get a one-to-one correspondence that has an inverse only by restricting the domain of $f(x, y, z)$ to P, since u and v are independent of y and (x, y, z) and $(x, 0, z)$ give the same values for u and v for all values of y.

The form of the inverse correspondence in the previous example suggests that f^{-1} is a linear function from P to $\mathbb{R}^2$. That this is so follows from

1.7   Theorem.   If $f : \mathbb{R}^n \longrightarrow \mathbb{R}^m$ is linear and one-to-one then f^{-1} is also linear.

Proof.   Suppose $f(\mathbf{u}) = \mathbf{x}$ and $f(\mathbf{v}) = \mathbf{y}$. Since f is linear, $f(s\mathbf{u} + t\mathbf{v}) = s\mathbf{x} + t\mathbf{y}$ for all scalars s and t. Thus linear combinations of elements $\mathbf{x}$, $\mathbf{y}$ in the image F of f are also in F, and

$$f^{-1}(s\mathbf{x} + t\mathbf{y}) = s\mathbf{u} + t\mathbf{v} = s f^{-1}(\mathbf{x}) + t f^{-1}(\mathbf{y}).$$

This says that $f^{-1} : F \longrightarrow \mathbb{R}^n$ is linear.   ∎

EXERCISES

Exercises 1 to 4 give information about linear functions f. In each case find the matrix A that represents f in the form $f(\mathbf{x}) = A\mathbf{x}$ and determine whether the function is one-to-one.

1. $f\begin{pmatrix} 1 \\ 0 \end{pmatrix} = \begin{pmatrix} 1 \\ 2 \end{pmatrix}$, $f\begin{pmatrix} 0 \\ 1 \end{pmatrix} = \begin{pmatrix} 2 \\ 4 \end{pmatrix}$

2. $f\begin{pmatrix} 1 \\ 0 \end{pmatrix} = \begin{pmatrix} 2 \\ 1 \end{pmatrix}$, $f\begin{pmatrix} 0 \\ 1 \end{pmatrix} = \begin{pmatrix} 1 \\ 1 \end{pmatrix}$

3. $f\begin{pmatrix} 1 \\ 0 \end{pmatrix} = \begin{pmatrix} 1 \\ 2 \\ 3 \end{pmatrix}$, $f\begin{pmatrix} 0 \\ 1 \end{pmatrix} = \begin{pmatrix} 3 \\ 2 \\ 1 \end{pmatrix}$

4. $f\begin{pmatrix} 1 \\ 0 \\ 0 \end{pmatrix} = \begin{pmatrix} -1 \\ 1 \end{pmatrix}$, $f\begin{pmatrix} 0 \\ 1 \\ 0 \end{pmatrix} = \begin{pmatrix} 1 \\ -1 \end{pmatrix}$,

$f\begin{pmatrix} 0 \\ 0 \\ 1 \end{pmatrix} = \begin{pmatrix} 2 \\ 1 \end{pmatrix}$

Exercises 5 to 8 give information about linear functions f. In each case find $f(\mathbf{e}_k)$ for the standard basis vectors $\mathbf{e}_k$ in the domain of f by first expressing each $\mathbf{e}_k$ as a linear combination of the domain vectors $\mathbf{x}$ for which $f(\mathbf{x})$ is given.

5. $f\begin{pmatrix} 1 \\ 1 \end{pmatrix} = \begin{pmatrix} 2 \\ 1 \end{pmatrix}$, $f\begin{pmatrix} -1 \\ 1 \end{pmatrix} = \begin{pmatrix} 1 \\ -1 \end{pmatrix}$

6. $f\begin{pmatrix} 2 \\ 1 \end{pmatrix} = \begin{pmatrix} 1 \\ 1 \end{pmatrix}$, $f\begin{pmatrix} 1 \\ -1 \end{pmatrix} = \begin{pmatrix} -1 \\ 1 \end{pmatrix}$

7. $f\begin{pmatrix} 1 \\ 1 \\ 1 \end{pmatrix} = \begin{pmatrix} 1 \\ 2 \end{pmatrix}$, $f\begin{pmatrix} 1 \\ 1 \\ 0 \end{pmatrix} = \begin{pmatrix} 2 \\ 1 \end{pmatrix}$,

$f\begin{pmatrix} 1 \\ 0 \\ 0 \end{pmatrix} = \begin{pmatrix} 2 \\ 3 \end{pmatrix}$

8. $f\begin{pmatrix} 2 \\ 1 \\ 0 \end{pmatrix} = \begin{pmatrix} 1 \\ 2 \\ 3 \end{pmatrix}$, $f\begin{pmatrix} 1 \\ 2 \\ 1 \end{pmatrix} = \begin{pmatrix} 0 \\ 1 \\ 1 \end{pmatrix}$,

$f\begin{pmatrix} 0 \\ 1 \\ 1 \end{pmatrix} = \begin{pmatrix} 2 \\ 0 \\ 0 \end{pmatrix}$

Exercises 9 and 10 refer to the rotation matrix $R_\theta = \begin{pmatrix} \cos\theta & -\sin\theta \\ \sin\theta & \cos\theta \end{pmatrix}$ of Example 8 in this section.

9. Show that $|R_\theta \mathbf{x}| = |\mathbf{x}|$ for every $\mathbf{x}$ in $\mathbb{R}^2$, so R_θ preserves the lengths of vectors.

10. Show that $\mathbf{x} \cdot R_\theta \mathbf{x} = |\mathbf{x}|^2 \cos\theta$. Assuming the result of Exercise 9, what does this say about the cosine of the angle between $\mathbf{x}$ and $R_\theta \mathbf{x}$?

For each of the pairs of linear functions in Exercises 11 to 14, find the matrix that represents the composition $g \circ f$. Also, say what the domain and range of $g \circ f$ are.

11. $f(\mathbf{x}) = \begin{pmatrix} 2 & -1 \\ 3 & 1 \end{pmatrix}\mathbf{x}, \quad g(\mathbf{y}) = \begin{pmatrix} 1 & 0 \\ 2 & 1 \end{pmatrix}\mathbf{y}$

12. $f(\mathbf{x}) = \begin{pmatrix} 1 & 2 & 0 \\ 2 & 2 & 3 \end{pmatrix}\mathbf{x}, \quad g(\mathbf{y}) = \begin{pmatrix} 2 & 0 \\ 3 & -1 \end{pmatrix}\mathbf{y}$

13. $f(\mathbf{x}) = \begin{pmatrix} 1 & 0 & 2 \\ -1 & 1 & 3 \\ 2 & 1 & 0 \end{pmatrix}\mathbf{x}, \quad g(\mathbf{y}) = \begin{pmatrix} 1 & -1 & 1 \\ -1 & 1 & 1 \\ 1 & 1 & -1 \end{pmatrix}\mathbf{y}$

14. $f(\mathbf{x}) = (2 \ \ 1 \ \ 3)\mathbf{x}, \quad g(\mathbf{y}) = 2\mathbf{y}$

15. (a) Show that the matrix $\begin{pmatrix} 0 & 1 \\ 1 & 0 \end{pmatrix}$ gives a linear function from $\mathbb{R}^2$ to $\mathbb{R}^2$ that corresponds geometrically to reflection in the line through the origin $45°$ counterclockwise from the horizontal.

(b) What matrix corresponds to reflection in the line through the origin $135°$ counterclockwise from the horizontal?

(c) Compute the product of the matrices in parts (a) and (b) and interpret the result geometrically.

16. A counterclockwise rotation in $\mathbb{R}^2$ through an angle α is described by the matrix $R_\alpha = \begin{pmatrix} \cos\alpha & -\sin\alpha \\ \sin\alpha & \cos\alpha \end{pmatrix}$. Let β be another angle, and compute the product $R_\alpha R_\beta$. The composition of a rotation through angle α with one through angle β is a rotation through the angle $\alpha + \beta$. What is the relation between $R_\alpha R_\beta$ and $R_{\alpha+\beta}$?

17. (a) Show that

$$U = \begin{pmatrix} 1 & 0 & 0 \\ 0 & 0 & -1 \\ 0 & 1 & 0 \end{pmatrix} \quad \text{and} \quad V = \begin{pmatrix} 0 & 0 & 1 \\ 0 & 1 & 0 \\ -1 & 0 & 0 \end{pmatrix}$$

represent $90°$ rotations of $\mathbb{R}^3$ about the x_1-axis and x_2-axis, respectively. Find the matrix W that represents a $90°$ rotation about the x_3-axis. Also find U^{-1} and V^{-1}, which represent rotations in the opposite direction.

(b) Compute UVU^{-1} and VUV^{-1} and interpret the results geometrically by checking out what they do to basis vectors.

*18. Let f and g be defined by $f(\mathbf{x}) = A\mathbf{x} + \mathbf{b}$ and $g(\mathbf{x}) = C\mathbf{x} + \mathbf{d}$ for given matrices A and C and vectors $\mathbf{b}$ and $\mathbf{d}$. Find a matrix P and vector $\mathbf{q}$, expressed in terms of A,

$\mathbf{b}$, C, and $\mathbf{d}$, such that $(f \circ g)(\mathbf{x}) = P\mathbf{x} + \mathbf{q}$. When is $f \circ g$ the same function as $g \circ f$?

19. Let $\mathbf{n}$ be the unit vector $(\frac{3}{7}, \frac{6}{7}, \frac{2}{7})$, and let $P_\mathbf{n}: \mathbb{R}^3 \longrightarrow \mathbb{R}^3$ be the associated projection function as in Example 6. Find the matrix of $P_\mathbf{n}$ by finding the image of each of the standard basis vectors under it.

*20. In this exercise, we consider a vector $\mathbf{x}$ in $\mathbb{R}^n$ to be a column vector, that is, an n-by-1 matrix, and we write $\mathbf{x}^t$ for its transpose, that is, the 1-by-n matrix obtained by writing $\mathbf{x}$ as a row vector. With this understanding, if $\mathbf{x}$ and $\mathbf{y}$ are in $\mathbb{R}^n$, the matrix product $\mathbf{x}^t\mathbf{y}$ is just the dot product $\mathbf{x} \cdot \mathbf{y}$, while $\mathbf{x}\mathbf{y}^t$ is an n-by-n matrix.

Show that if $\mathbf{n}$ is a unit vector then the matrix of the projection function $P_\mathbf{n}$ is $\mathbf{n}\mathbf{n}^t$.

In Exercises 21 to 26, find out whether the image of the given function is a line, a plane, or some other subset of its range.

21. $f(x, y) = (2x - y, 6x - 3y)$

22. $f(x, y) = (x - y, x - 3y, x + y)$

23. $f(x, y, z) = (x + y - z, -x - y + z)$

24. $f(x, y, z) = (x + y - z, x - y + z)$

25. $f(x, y, z) = (x - 2y + z, -2x - y - z, -5y + z)$

26. $f(x, y, z) = (y, z, x)$

Given functions $f: \mathbb{R}^n \to \mathbb{R}^m$ and $g: \mathbb{R}^p \to \mathbb{R}^q$, the composition $g \circ f$ doesn't make sense unless the image F of f lies in the domain $\mathbb{R}^n$ of g, in particular, unless $m = p$. For f and g of the types given in Exercises 27 to 30, decide whether $g \circ f$, $f \circ g$, both, or neither, makes sense.

27. $f: \mathbb{R}^1 \longrightarrow \mathbb{R}^2, g: \mathbb{R}^2 \longrightarrow \mathbb{R}^1$

28. $f: \mathbb{R}^2 \longrightarrow \mathbb{R}^3, g: \mathbb{R}^1 \longrightarrow \mathbb{R}^3$

29. $f: \mathbb{R}^2 \longrightarrow \mathbb{R}^2, g: \mathbb{R}^2 \longrightarrow \mathbb{R}^3$

30. $f: \mathbb{R}^3 \longrightarrow \mathbb{R}^1, g: \mathbb{R}^1 \longrightarrow \mathbb{R}^2$

31. Show that if $f: \mathbb{R}^n \to \mathbb{R}$ is a linear function, then there is a vector $\mathbf{a}$ in $\mathbb{R}^n$ such that $f(\mathbf{x}) = \mathbf{a} \cdot \mathbf{x}$ for all $\mathbf{x}$ in $\mathbb{R}^n$.

32. Show that if A is an m-by-n matrix such that $A\mathbf{x} = 0$ for every $\mathbf{x}$ in $\mathbb{R}^n$, then all entries in A are zero.

33. The linear function $f_a: \mathbb{R}^2 \to \mathbb{R}^2$ defined by $f_a(x, y) = (x + ay, y)$, for fixed $a \neq 0$, is an example of a **shear transformation**.

(a) What is the matrix of f_a?

(b) Find the points $f_a(1, 0)$, $f_a(0, 1)$, $f_a(-1, 0)$, and $f_a(0, -1)$, and sketch their relation to the corresponding domain points when $a = 1$.

(c) Show that if $a > 0$, then f_a moves points above the x-axis to the right and points below the x-axis to the left. What happens if $a < 0$?

(d) What points are always left fixed by f_a?

(e) For which lines L in the plane (not necessarily through the origin) is the image $f(L)$ equal to L?

(f) What is the composition of two shear transformations f_a and f_b?

SECTION 2   VECTOR SPACES

In Chapter 1 we defined operations of addition and scalar multiplication on $\mathbb{R}^n$ and noted that they obey the following analogues of familiar laws of arithmetic, where r, s, and 0 are scalars, $\mathbf{x}$ and $\mathbf{y}$ are vectors in $\mathbb{R}^n$, and $\mathbf{0}$ is the zero vector in $\mathbb{R}^n$.

1. $r\mathbf{x} + s\mathbf{x} = (r + s)\mathbf{x}$
2. $r\mathbf{x} + r\mathbf{y} = r(\mathbf{x} + \mathbf{y})$
3. $r(s\mathbf{x}) = (rs)\mathbf{x}$
4. $\mathbf{x} + \mathbf{y} = \mathbf{y} + \mathbf{x}$
5. $(\mathbf{x} + \mathbf{y}) + \mathbf{z} = \mathbf{x} + (\mathbf{y} + \mathbf{z})$
6. $\mathbf{x} + \mathbf{0} = \mathbf{x}$
7. $\mathbf{x} + (-\mathbf{x}) = \mathbf{0}$
8. $1\mathbf{x} = \mathbf{x}$
9. $0\mathbf{x} = \mathbf{0}$

We now take a more general point of view and define a **vector space** $\mathcal{V}$ **over the real numbers** to be a set with operations of addition and scalar multiplication defined so that they behave like the familiar operations on $\mathbb{R}^n$. Specifically, for $\mathbf{x}$ and $\mathbf{y}$ in $\mathcal{V}$ and r in $\mathbb{R}$, the sum $\mathbf{x} + \mathbf{y}$ and the scalar multiple $r\mathbf{x}$ must also be elements of $\mathcal{V}$. In addition, $\mathcal{V}$ has to contain a zero vector, and formulas 1 through 9 above have to hold for all real numbers r, s and all $\mathbf{x}$, $\mathbf{y}$, and $\mathbf{z}$ in $\mathcal{V}$. As we have done all along with vectors in $\mathbb{R}^n$, we write $-\mathbf{x}$ for the scalar multiple $(-1)\mathbf{x}$, $\mathbf{x} - \mathbf{y}$ for $\mathbf{x} + (-\mathbf{y})$, and $\mathbf{0}$ for the zero vector.

We form linear combinations of vectors in a vector space by adding scalar multiples of vectors, and routine calculations such as

$$(2\mathbf{u} - 3\mathbf{v} + \mathbf{w}) + 3(\mathbf{u} + 2\mathbf{v} - \mathbf{w}) = (2+3)\mathbf{u} + (-3+6)\mathbf{v} + (1-3)\mathbf{w} = 5\mathbf{u} + 3\mathbf{v} - 2\mathbf{w}$$

follow from Formulas 1 through 9, just as was illustrated for vectors in $\mathbb{R}^n$ in Example 4 at the beginning of Chapter 1.

2A   Examples of Vector Spaces

EXAMPLE 1

The set of all n-tuples of real numbers, with addition and multiplication by a scalar defined as in Chapter 1, forms the vector space $\mathbb{R}^n$. From the point of view of this chapter, when we showed that the operations defined on $\mathbb{R}^n$ have the properties 1 through 9, we were showing that $\mathbb{R}^n$ is a vector space.

EXAMPLE 2

For fixed m and n the set of all m-by-n matrices forms a vector space $\mathcal{M}_{m,n}$. The vector operations are matrix addition and scalar multiplication of matrices as defined in Chapter 2. For example, in $\mathcal{M}_{2,3}$ we have

$$\begin{pmatrix} 1 & 0 & -1 \\ 2 & 1 & 3 \end{pmatrix} + \begin{pmatrix} 1 & 1 & 1 \\ 2 & 2 & 2 \end{pmatrix} = \begin{pmatrix} 2 & 1 & 0 \\ 4 & 3 & 5 \end{pmatrix}$$

$$\text{and} \quad 2\begin{pmatrix} 1 & 1 & 1 \\ 2 & 2 & 2 \end{pmatrix} = \begin{pmatrix} 2 & 2 & 2 \\ 4 & 4 & 4 \end{pmatrix}.$$

Each m-by-n matrix has mn entries and so corresponds to the element of $\mathbb{R}^{mn}$ we get by lining up the successive rows end to end, so sums and scalar multiples of elements of $\mathcal{M}_{m,n}$ correspond to sums and scalar multiples of elements of $\mathbb{R}^{mn}$. The zero vector is the zero matrix O, and it follows that rules 1 through 9 hold for $\mathcal{M}_{m,n}$ simply because they hold for $\mathbb{R}^{mn}$. We remark that the operation of multiplying one matrix by another plays no part in making $\mathcal{M}_{m,n}$ a vector space.

EXAMPLE 3   Let $\mathcal{V}$ be the set of all infinite sequences of real numbers $\{x_1, x_2, \ldots\}$. If $\mathbf{a} = \{a_1, a_2, \ldots\}$ and $\mathbf{b} = \{b_1, b_2, \ldots\}$ are two elements of $\mathcal{V}$, define $\mathbf{a} + \mathbf{b}$ as the sequence $\{(a_1 + b_1), (a_2 + b_2), \ldots\}$ and $r\mathbf{a}$ as the sequence $\{ra_1, ra_2, \ldots\}$. By a natural analogy with the spaces $\mathbb{R}^n$, we call this space $\mathbb{R}^\infty$, the **space of sequences**.

EXAMPLE 4   Let $\mathcal{P}_n$ be the set of all polynomials of degree at most n, which is the set of all functions p having the form

$$p(x) = a_0 + a_1 x + \ldots + a_n x^n,$$

where $a_0, \ldots, a_n$ are constants. Define addition and scalar multiplication as usual for polynomials, by collecting terms with like powers of x. For example, when $n = 2$ we have

$$(1 + 2x + 3x^2) + (2 - 3x) = 3 - x + 3x^2$$

and

$$3(1 + 2x + 3x^2) = 3 + 6x + 9x^2.$$

$\mathcal{P}_n$ and $\mathbb{R}^{n+1}$ are very much alike as vector spaces, under the correspondence

$$(a_0 + a_1 x + \ldots + a_n x^n) \longleftrightarrow (a_0, a_1, \ldots, a_n).$$

Sums and scalar multiples are preserved by the correspondence, so Formulas 1 through 9 hold for $\mathcal{P}_n$ because they hold for $\mathbb{R}^{n+1}$. The zero element of $\mathcal{P}_n$ is the identically zero polynomial. We'll use $\mathcal{P}$ to denote the space of all polynomials, and $\mathcal{P}$ is also a vector space with addition and scalar multiplication defined as in $\mathcal{P}_n$.

EXAMPLE 5   Let $\mathcal{V}$ be the set of all continuous real-valued functions on $[0, 1]$. For f and g in $\mathcal{V}$ and r in $\mathbb{R}$, define $f + g$ and rf as the functions whose values at a point x in $[0, 1]$ are $f(x) + g(x)$ and $rf(x)$, respectively. We learn in Chapter 5 that sums and constant multiples of continuous functions are continuous, a theorem often assumed in calculus. Thus the defined operations do produce elements of $\mathcal{V}$ as they should.

FIGURE 3.4
(a) Sum. (b) Scalar multiples.

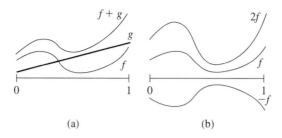

(a)   (b)

Formulas 1 through 9 follow from the same kind of argument used in Chapter 1 to show that they hold for $\mathbb{R}^n$. The **0**-element of $\mathcal{V}$ required by formula 6 is the zero function z defined by $z(x) = 0$ for all x in $[0, 1]$.

The vector space described in Example 5 is commonly denoted by $C[0, 1]$, the space of **continuous functions** defined on the interval $[0, 1]$. More generally, the continuous functions on an interval $[a, b]$ form a vector space called $C[a, b]$. The notation $C(-\infty, \infty)$ denotes the space of continuous functions on the entire real line. Figure 3.4 illustrates the two vector space operations in $C[0, 1]$.

2B   Subspaces

The vector space $\mathbb{R}^2$ fits in a natural way inside $\mathbb{R}^3$ if we identify (x, y) in $\mathbb{R}^2$ with $(x, y, 0)$ in $\mathbb{R}^3$. The following example shows that all planes though the origin are subspaces of $\mathbb{R}^3$.

EXAMPLE 6   Let $\mathcal{V}$ consist of all the vectors in $\mathbb{R}^3$ that lie in a plane $ax + by + cz = 0$. The sum of two vectors in $\mathcal{V}$ is in $\mathcal{V}$ and the same is true for scalar multiples of vectors in $\mathcal{V}$. We see this geometrically from the parallelogram law of addition and the geometric interpretation of scalar multiplication. For this example we can also check algebraically that if

$$ax_1 + by_1 + cz_1 = 0 \quad \text{and} \quad ax_2 + by_2 + cz_2 = 0$$

then

$$a(x_1 + x_2) + b(y_1 + y_2) + c(z_1 + z_2) = 0 \quad \text{and} \quad s(ax_1 + sy_1 + sz_1) = 0$$

for all scalars s. Thus we can think of addition and scalar multiplication as restricted just to $\mathcal{V}$ so $\mathcal{V}$ is a vector space. If $a = b = 0$ and $c = 1$ we get the xy-plane of vectors $(x, y, 0)$ as a special case.

The previous example generalizes as follows. Let $\mathcal{W}$ be a vector space and let $\mathcal{V}$ be a subset of it. We say that $\mathcal{V}$ is **closed under addition** if $\mathbf{x} + \mathbf{y}$ is in $\mathcal{V}$ whenever $\mathbf{x}$ and $\mathbf{y}$ are, and **closed under scalar multiplication** if every scalar multiple $s\mathbf{x}$ is in $\mathcal{V}$ whenever $\mathbf{x}$ is. If $\mathcal{V}$ is closed under both operations, $\mathcal{V}$ is called a **subspace** of $\mathcal{W}$.

2.1   Theorem.   If $\mathcal{V}$ is a nonempty subset of a vector space $\mathcal{W}$, and $\mathcal{V}$ is closed under addition and scalar multiplication as defined on $\mathcal{W}$, then $\mathcal{V}$ is a vector space.

Proof.   To prove that $\mathcal{V}$ is a vector space we have to show that the formulas 1 through 9 hold. First, closure under scalar multiplication implies that if $\mathbf{x}$ is in $\mathcal{V}$ so are $-\mathbf{x} = (-1)\mathbf{x}$ and $\mathbf{0} = 0\mathbf{x}$. Then all the formulas hold for vectors in $\mathcal{V}$ because the vectors are also in $\mathcal{W}$ and the formulas hold because $\mathcal{W}$ is a vector space.   ∎

The next theorem gives an alternative condition for a subset of a vector space to be a subspace.

2.2   Theorem.   If $\mathcal{V}$ is a non-empty subset of a vector space $\mathcal{W}$, then $\mathcal{V}$ is a subspace if and only if every linear combination of elements of $\mathcal{V}$ is also in $\mathcal{V}$.

Proof.   If $\mathcal{V}$ is a subspace and vectors $\mathbf{x}_1, \ldots, \mathbf{x}_n$ belong to it, then repeated application of the closure conditions shows that every linear combination

$$a_1\mathbf{x}_1 + \ldots + a_n\mathbf{x}_n$$

also belongs to $\mathcal{V}$. On the other hand, if linear combinations of elements of $\mathcal{V}$ belong to $\mathcal{V}$, so do sums and scalar multiples of elements of $\mathcal{V}$, because $\mathbf{x}_1 + \mathbf{x}_2$ and $r\mathbf{x}_1$ are special cases of the linear combination $a_1\mathbf{x}_1 + a_2\mathbf{x}_2$ obtained by setting $a_1 = a_2 = 1$ and $a_1 = r, a_2 = 0$, respectively.   ∎

EXAMPLE 7   For a simple example of a subspace, let $\mathcal{V}$ be the subset of $\mathbb{R}^3$ consisting of all vectors $(x, y, 0)$ that have third coordinate zero. Sums and scalar multiples of such vectors also have this property, so $\mathcal{V}$ is a subspace of $\mathbb{R}^3$. We can visualize $\mathcal{V}$ as the horizontal xy-plane in $\mathbb{R}^3$.

EXAMPLE 8   In this example, we'll describe all possible subspaces of $\mathbb{R}^3$.

1. Since $\mathbf{0} + \mathbf{0} = \mathbf{0}$ and every scalar multiple $s\mathbf{0} = \mathbf{0}$, the subset consisting of the zero vector alone satisfies the closure conditions and is a subspace.
2. If $\mathcal{V}$ is a subspace containing some vector $\mathbf{x}_1 \neq \mathbf{0}$ then, since $\mathcal{V}$ is closed under scalar multiplication, it contains all multiples $t\mathbf{x}_1$, where t ranges over the real numbers. In other words, $\mathcal{V}$ contains the line through the origin with parametric representation

$$\mathbf{x} = t\mathbf{x}_1.$$

   It may be that $\mathcal{V}$ contains no other vectors, in which case our subspace $\mathcal{V}$ is identical with the line, shown in Figure 3.5.
3. Otherwise, $\mathcal{V}$ contains a vector $\mathbf{x}_2$ different from all the vectors $t\mathbf{x}_1$. Since $\mathcal{V}$ contains all scalar multiples and sums of vectors in it, $\mathcal{V}$ contains all linear combinations $t\mathbf{x}_1 + u\mathbf{x}_2$, where (t, u) ranges over $\mathbb{R}^2$. In other words, $\mathcal{V}$ contains the plane through the origin with parametric representation

$$\mathbf{x} = t\mathbf{x}_1 + u\mathbf{x}_2.$$

   It may be that $\mathcal{V}$ contains no other vectors, in which case $\mathcal{V}$ is identical with the plane, partly shown in Figure 3.5.
4. Suppose finally that $\mathcal{V}$ in addition contains a vector $\mathbf{x}_3$ different from all the vectors $t\mathbf{x}_1 + u\mathbf{x}_2$. Then because $\mathcal{V}$ is closed under addition and scalar

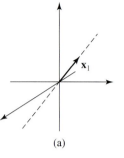

(a)

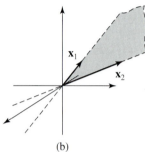

(b)

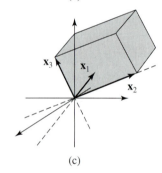

(c)

FIGURE 3.5
Subspaces.

multiplication, $\mathcal{V}$ contains all linear combinations

$$t\mathbf{x}_1 + u\mathbf{x}_2 + v\mathbf{x}_3,$$

where (t, u, v) ranges over $\mathbb{R}^3$. Because $\mathbf{x}_1, \mathbf{x}_2$, and $\mathbf{x}_3$ don't all lie in a plane through the origin, every vector in $\mathbb{R}^3$ is a linear combination of them, something that's apparent geometrically from Figure 3.5(c), where the shaded box shows the vectors with all three coefficient values between 0 and 1. Example 10 in Section 5C shows that we really get all of $\mathbb{R}^3$ this way.

In a vector space $\mathcal{W}$ the zero vector by itself is always a subspace, for the reasons given in item 1 of the preceding example. It is called the **zero subspace**, or sometimes the **trivial subspace**, of $\mathcal{W}$. $\mathcal{W}$ is itself closed under the vector operations, and so is technically a subspace of itself. Subspaces of a vector space $\mathcal{W}$ other than $\mathcal{W}$ itself are called **proper** subspaces of $\mathcal{W}$. We summarize the results of Example 8 as follows, where the proper subspaces of $\mathbb{R}^3$ are the ones listed in (1) to (3).

2.3   Theorem.   The subspaces of $\mathbb{R}^3$ are (1) The zero subspace, (2) the lines through the origin, (3) the planes through the origin, and (4) the space $\mathbb{R}^3$ itself. More generally we'll see in Example 10 of Section 6C, the subspaces of $\mathbb{R}^n$ are the zero subspace, the k-planes through the origin for $k = 1, \ldots, n - 1$, and the space $\mathbb{R}^n$ itself.

The subspaces of $\mathbb{R}^3$ that we found in Example 8 were constructed by forming all linear combinations of certain sets of vectors in $\mathbb{R}^3$. This is a general method of producing subspaces. If S is a subset, not necessarily a subspace, of a vector space $\mathcal{W}$, we define the **span** of S to be the set $\mathcal{V}$ of all linear combinations of vectors from S. When the set S has only finitely many vectors $\mathbf{x}_1, \ldots, \mathbf{x}_n$ in it, the span of S is just all linear combinations

$$\mathbf{x} = a_1\mathbf{x}_1 + \ldots + a_n\mathbf{x}_n.$$

If $\mathcal{V}$ is the span of S, we say that S **spans** $\mathcal{V}$.

EXAMPLE 9

For examples of spans, consider the following:

(a) The span of $\{\mathbf{e}_1, \mathbf{e}_2\}$ in $\mathbb{R}^2$ is all of $\mathbb{R}^2$.
(b) The span of $\{\mathbf{e}_1, \mathbf{e}_2\}$ in $\mathbb{R}^3$ is the xy-plane in $\mathbb{R}^3$.
(c) Let $S = \{\mathbf{x}_1, \mathbf{x}_2\}$ consist of two vectors (arrows) in $\mathbb{R}^3$ that don't lie on a line. The span of S is all linear combinations $t\mathbf{x}_1 + u\mathbf{x}_2$ and is a subspace that is a plane P through the origin. Alternatively, we can say that $\{\mathbf{x}_1, \mathbf{x}_2\}$ spans P.

Here is the formal statement and proof that the span of a subset of a vector space is always a sub*space*.

2.4   Theorem.   Let S be an arbitrary set of vectors in a vector space $\mathcal{W}$ and let $\mathcal{V}$ be the span of S. Then $\mathcal{V}$ is a subspace of $\mathcal{W}$.

Proof.   We need to show that $\mathcal{V}$ is closed under addition and scalar multiplication. Let $\mathbf{x}$ and $\mathbf{y}$ be vectors in the span of S, so each of them is a linear combination of a finite number of vectors in S. Suppose that $\{\mathbf{v}_1, \dots, \mathbf{v}_k\}$ contains all the vectors in S needed in the linear combinations for both $\mathbf{x}$ and $\mathbf{y}$. Then there are scalars a_i and b_i such that

$$\mathbf{x} = a_1\mathbf{v}_1 + \dots + a_k\mathbf{v}_k$$

$$\mathbf{y} = b_1\mathbf{v}_1 + \dots + b_k\mathbf{v}_k.$$

(Some of the a's and b's may be zero if not all of the $\mathbf{v}$'s are needed for both $\mathbf{x}$ and $\mathbf{y}$.) Then $\mathbf{x} + \mathbf{y} = (a_1 + b_1)\mathbf{v}_1 + \dots + (a_k + b_k)\mathbf{v}_k$ and $r\mathbf{x} = ra_1\mathbf{v}_1 + \dots + ra_k\mathbf{v}_k$, so $\mathbf{x} + \mathbf{y}$ and $r\mathbf{x}$ are also linear combinations of vectors in S. Hence $\mathcal{V}$ is closed under addition and scalar multiplication and is therefore a subspace of $\mathcal{W}$. ∎

EXAMPLE 10 In the space $\mathcal{P}$ of polynomials discussed in Example 4, the span of the set $S = \{1, x, x^2, \dots, x^n\}$ is the subspace $\mathcal{P}_n$ of polynomials of degree at most n. If $m \leq n$, then $\mathcal{P}_m$ is a subspace of $\mathcal{P}_n$, and if $m < n$, then $\mathcal{P}_m$ is a proper subspace of $\mathcal{P}_n$. The whole space $\mathcal{P}$ is the span of the infinite set $\{1, x, x^2, \dots\}$.

Here are some examples of subspaces of $C[a, b]$, the space of continuous functions on $[a, b]$ discussed in Example 5.

EXAMPLE 11 We can get a subspace of $C[a, b]$ by taking the span of a set of functions in it. For instance, the span of $\{1, x, x^2\}$, is a subspace of $C[a, b]$ consisting of all polynomials of degree at most 2. Another example is the set of all functions of the form

$$a_0 + a_1 \cos x + b_1 \sin x + a_2 \cos 2x + b_2 \sin 2x + a_3 \cos 3x + b_3 \sin 3x,$$

which is the span of the set $\{1, \cos x, \sin x, \cos 2x, \sin 2x, \cos 3x, \sin 3x\}$.

In the next two examples we have subspaces of $C[0, 1]$ that are not described as the span of a set.

EXAMPLE 12 Let $\mathcal{V}$ consist of all functions in $C[a, b]$ for which $\int_0^1 f(x)\,dx = 0$. Using familiar properties of integration, if f and g are in $\mathcal{V}$, then

$$\int_a^b \big(f(x) + g(x)\big)\,dx = \int_a^b f(x)\,dx + \int_a^b g(x)\,dx = 0 + 0 = 0$$

and
$$\int_a^b sf(x)\,dx = s\int_a^b f(x)\,dx = s0 = 0.$$

Thus $\mathcal{V}$ is closed under addition and scalar multiplication and is a subspace of $C[a, b]$.

EXAMPLE 13 Consider the subset of functions in $C[a, b]$ that have continuous first derivatives; we denote this set by $C^{(1)}[a, b]$. To verify that $C^{(1)}[a, b]$ is a subspace of $C[a, b]$ all we have to do is observe that since

$$\frac{d}{dx}(f + g) = \frac{df}{dx} + \frac{dg}{dx} \quad \text{and} \quad \frac{d(rf)}{dx} = r\frac{df}{dx},$$

sums and scalar multiples of functions with continuous derivatives also have continuous derivatives.

We denote the set of functions whose first k derivatives are continuous by $C^{(k)}[a, b]$. Repeated application of the argument used for $C^{(1)}[a, b]$ shows that it is a subspace of $C[a, b]$. For $l \geq k$, $C^{(l)}[a, b]$ is a subspace of $C^{(k)}[a, b]$. For $l > k$, $C^{(l)}[a, b]$ is a *proper* subspace of $C^{(k)}[a, b]$. A proof is outlined in Exercise 33.

EXERCISES

In each of Exercises 1 to 6, let S be the set of all vectors (x, y, z) in $\mathbb{R}^3$ whose entries satisfy the given conditions. In each case, either show that the subset is a subspace of $\mathbb{R}^3$ by verifying the closure conditions, or show that it is not a subspace by finding some linear combination of elements of S that is not in S.

1. $x + 2y = 0$
2. $x + z = 2$

3. $x + y = 0$ and $z = 0$
4. $x + y = 0$ or $z = 0$

5. $x = y^3$
6. $x + y = 0$ and $x = y^3$

In Exercises 7 to 10, let S be the subset of the vector space of 2-by-2 matrices, $\mathcal{M}_{2,2}$, consisting of the matrices $A = \begin{pmatrix} x & y \\ z & w \end{pmatrix}$ whose entries satisfy the given conditions. Show either that S is a subspace of $\mathcal{M}_{2,2}$ or that it is not.

7. $x = w$
8. $x = -w$

9. $y = z = 1$
10. $\det(A) = xw - yz = 0$

11. (a) Show that the set of vectors (x, y, z) in $\mathbb{R}^3$ such that $x + 2y - z = 0$ is a subspace of $\mathbb{R}^3$.
   (b) By finding a parametric representation for the solutions of $x + 2y - z = 0$, find two vectors that span the subspace in part (a).

12. Let **a** be a fixed nonzero vector in $\mathbb{R}^n$.
   (a) Show that the set S of all vectors **x** such that $\mathbf{a} \cdot \mathbf{x} = 0$ is a subspace of $\mathbb{R}^n$.
   (b) Show that if k is a *nonzero* real number, then the set $\mathcal{A}$ of all vectors **x** such that $\mathbf{a} \cdot \mathbf{x} = k$ is not a subspace.

13. Let S be a subset of $\mathbb{R}^n$, and let $S^\perp$ (pronounced "S perpendicular", or "S perp" for short) be the set of all vectors **p** in $\mathbb{R}^n$ such that $\mathbf{p} \cdot \mathbf{s} = 0$ for all **s** in S. Show that $S^\perp$ is always a subspace of $\mathbb{R}^n$.

14. Show that for a subset S of $\mathbb{R}^n$, the span of S is contained in $(S^\perp)^\perp$. [*Hint:* First show that S is contained in $(S^\perp)^\perp$.]

*15. Let $\mathbf{e}_i$ be the sequence in $\mathbb{R}^\infty$ (Example 3) having 1 in the ith place and 0 elsewhere, so $\mathbf{e}_1 = (1, 0, 0, \dots)$, $\mathbf{e}_2 = (0, 1, 0, \dots)$, etc. Which vectors in $\mathbb{R}^\infty$ are in the span of $\{\mathbf{e}_1, \mathbf{e}_2, \dots, \mathbf{e}_n\}$? Which are in the span of the set of all the $\mathbf{e}_i$? Give an example of a vector in $\mathbb{R}^\infty$ that is *not* in the span of all the $\mathbf{e}_i$.

In Exercises 16 to 19, determine whether the set of all polynomials p in $\mathcal{P}_3$ that satisfy the given conditions is a subspace of $\mathcal{P}_3$.

16. $p(0) = 1$
17. $p(1) = 0$

18. $p(0) = p(1)$

19. $p(1) = p'(2)$, where p' is the derivative of p

20. In the space $\mathcal{P}$ of polynomials, let A be the set of all p such that $p(x) = -p(-x)$, and let B be the set of p such that $p(x) = p(-x)$. Show that A is the span of $\{x, x^3, x^5, \dots\}$, and find a spanning set for B.

In Exercises 21 to 24, determine whether the given subset of $C^{(1)}(-\infty, \infty)$ is also a subspace.

21. All f such that $f'(0)$ exists

22. All f such that $f'(0) = 2$

23. All f such that $f'(0) = f(2)$

24. All f such that $f(x) = f(-x)$ for every value of x

25. Let $C[a, b]$ be the vector space of continuous real-valued functions defined on the interval $[a, b]$. Let $C_0[a, b]$ be the set of functions f in $C[a, b]$ such that $f(a) = f(b) = 0$. Show that $C_0[a, b]$ is a subspace of $C[a, b]$.

In Exercises 26 and 27, show that S and T have the same span in $\mathbb{R}^3$ by showing that the vectors in S are in the span of T and vice versa. [*Hint:* You can do this by solving systems of linear equations.]

26. $S = \{(1, 0, 0), (0, 1, 0)\}$, $T = \{(1, 2, 0), (2, 1, 0)\}$

27. $S = \{(2, 3, 1), (1, 2, 3)\}$, $T = \{(3, 5, 4), (1, 1, -2)\}$

28. (a) Show that the plane P of points $(1, 1, 1) + s(1, 2, 0) + t(-2, 1, 1)$ is not a subspace by finding two vectors in P whose sum is not in P.
   (b) Show that the plane P of points $(-1, 3, 1) + s(1, 2, 0) + t(-2, 1, 1)$ is a subspace.
   (c) What is different about cases (a) and (b)? For which vectors **b** do the points $\mathbf{b} + s(1, 2, 0) + t(-2, 1, 1)$ form a subspace?

29. Show that if S is a subset of a vector space W and V is a subspace of W that contains S, then the span of S is a subset of V. (Another way of stating this is to say that the span of S is the smallest subspace of W that contains S.)

30. Show that the intersection of two subspaces of a vector space V is always a subspace of V.

*31. Exercise 4 shows that the union of two subspaces is not always a subspace. Show that the union of two subspaces is a subspace if and only if one of them is contained in the other.

32. Given two subsets A and B of a vector space, let $A + B$ stand for the set of all vectors that are equal to sums $\mathbf{a} + \mathbf{b}$ with $\mathbf{a}$ in A and $\mathbf{b}$ in B. Show that if A and B are subspaces, then so is $A + B$.

*33. This exercise outlines a proof that for the spaces of functions $C^{(k)}[a, b]$ defined in Example 13 that $C^{(l)}[a, b]$ is a *proper* subspace of $C^{(k)}[a, b]$ when $l > k$.

   (a) Show that $C^{(1)}[a, b]$ is a proper subspace of $C[a, b]$ by giving an example of a function that is continuous on the interval $[a, b]$ but doesn't have a derivative that is continuous on the interval.

   (b) One version of the fundamental theorem of calculus states that if f is continuous on an interval $[a, b]$ then $F(x) = \int_a^x f(t)\, dt$ is a function with derivative $F'(x) = f(x)$. Use this and your example from part (a) to find a function that is in $C^{(1)}[a, b]$ but not in $C^{(2)}[a, b]$.

   (c) Show by induction that for $k = 1, 2, \ldots,$ there is a function in $C^{(k)}[a, b]$ that is not in

$C^{(k+1)}[a, b]$, so $C^{(k+1)}[a, b]$ is a proper subspace of $C^{(k)}[a, b]$.

*34. Let $C^{(\infty)}$ be the vector space of infinitely often differentiable functions of a real variable. Show that $C^{(\infty)}$ is a proper subspace of $C^{(k)}$ for $k = 1, 2, \ldots$.

35. Suppose a linear function $f : \mathbb{R}^3 \to \mathbb{R}$ has $f(\mathbf{e}_1) = 1$, $f(\mathbf{e}_2) = 2$, and $f(\mathbf{e}_3) = 1$. Show that the equation $f(\mathbf{x}) = 1$ has solutions consisting precisely of the points in the plane perpendicular to $(1, 2, 1)$ and passing through $(1, 0, 0)$.

In each of Exercises 36 to 41, say whether the given statement is always true or sometimes false. If the statement is always true, give a reason why; otherwise give an example for which it is false.

36. If S is a subspace of a vector space and $\mathbf{x}$ is in S, then $-\mathbf{x}$ is in S.

37. If S is a subspace of a vector space W and $\mathbf{x}$ is in W but not in S, then the set of all sums $\mathbf{x} + \mathbf{y}$ with $\mathbf{y}$ in S is not a subspace of W.

38. If S is a subspace of $\mathbb{R}^n$ and S contains more than one vector, then S contains a line through the origin.

39. If S_1 and S_2 are two different proper subspaces of a vector space W, then W has a proper subspace that contains both S_1 and S_2.

40. There is no subspace of $\mathbb{R}^n$ such that $|\mathbf{x}| \leq 1$ for all $\mathbf{x}$ in the subspace.

41. No subspace S of $\mathbb{R}^3$ has the property that $\mathbf{x} \cdot (1, 2, 1) = 1$ for all $\mathbf{x}$ in S.

SECTION 3   LINEAR FUNCTIONS

In Section 1 of this chapter we studied functions f from $\mathbb{R}^n$ to $\mathbb{R}^m$ that satisfied Equation 1.1 stating that for a linear combination $s\mathbf{x} + t\mathbf{y}$,

3.1
$$f(s\mathbf{x} + t\mathbf{y}) = sf(\mathbf{x}) + tf(\mathbf{y}).$$

We'll now be looking at linear functions from one vector space to another without assuming that these spaces are the standard coordinate spaces $\mathbb{R}^n$, so we can't always assume a matrix representation $f(\mathbf{x}) = A\mathbf{x}$ for a linear function.

In general, we define a function to be **linear** if its domain and range are vector spaces, and it satisfies Equation 3.1. When checking whether a function f is linear it's sometimes more convenient to check separately that

3.2
   (a) $f(\mathbf{x} + \mathbf{y}) = f(\mathbf{x}) + f(\mathbf{y})$ for all vectors $\mathbf{x}, \mathbf{y}$.
   (b) $f(s\mathbf{x}) = sf(\mathbf{x})$ for all vectors $\mathbf{x}$ and scalars s.

In Section 1 we saw that a linear function $f(\mathbf{x}) = A\mathbf{x}$ with m-by-n matrix A is one-to-one, and so has an inverse, if and only if the columns of A are linearly independent. Matrix representations for linear functions in general are not readily available, but the definition of linear independence in Definition 2.7 in Chapter 2 provides a useful substitute that doesn't depend on properties of $\mathbb{R}^n$.

3.3   Theorem.   A linear function f is one-to-one if and only if $\mathbf{x} = \mathbf{0}$ is the only vector such that $f(\mathbf{x}) = \mathbf{0}$.

Proof.   By linearity, $f(\mathbf{0}) = f(0\mathbf{x}) = 0f(\mathbf{x}) = \mathbf{0}$, and $f(\mathbf{x}_1) = f(\mathbf{x}_2)$ if and only if $f(\mathbf{x}_1 - \mathbf{x}_2) = \mathbf{0}$. If f is one-to-one, then $\mathbf{x} = \mathbf{0}$ is the only vector such that $f(\mathbf{x}) = \mathbf{0}$. If f is not one-to-one, there are vectors such that $f(\mathbf{x}_1) = f(\mathbf{x}_2)$ but $\mathbf{x}_1 \neq \mathbf{x}_2$, and then $\mathbf{x}_1 - \mathbf{x}_2$ is a nonzero solution of $f(\mathbf{x}_1 - \mathbf{x}_2) = \mathbf{0}$.   ∎

3A   Examples of Linear Functions

We now give some examples of linear functions just to illustrate various possibilities that come under the definition. After giving some specific examples, we consider general ways of combining linear functions to get others.

EXAMPLE 1   For our first example, we simply recall Theorems 2.3 in Chapter 2, Section 2C and 1.3 of Section 1 of this chapter. If $f : \mathbb{R}^n \longrightarrow \mathbb{R}^m$ is a linear function, then $f(\mathbf{x}) = A\mathbf{x}$, where A is the m-by-n matrix whose jth column is the vector $f(\mathbf{e}_j)$. This theorem is significant in that it gives us a concrete computational description of *all* possible linear functions from $\mathbb{R}^n$ to $\mathbb{R}^m$.

This direct description of linear functions by matrices only works for functions whose domain and range are standard coordinate spaces $\mathbb{R}^n$, $\mathbb{R}^m$. As we'll see later in Section 5B, many vector spaces are very much like the spaces $\mathbb{R}^n$, and there is a way to associate linear functions on them with matrices, but in other cases such as the next couple of examples this is not possible.

We sometimes use the term **transformation** to refer to a function from one vector space to another, and use the term **operator** to refer to a function from a vector space to itself. These terms help to avoid confusion when we deal with vector spaces such as $C(-\infty, \infty)$ whose elements are themselves functions. For example, differentiation is a **differential operator** from infinitely often differentiable functions such as $f(x) = \sin x$ that operates on $f(x)$ to produce $f'(x) = \cos x$. We use this terminology in several of the examples that follow.

EXAMPLE 2   Let $\mathcal{P}$ be the vector space of all polynomials, as in Example 4 in the previous section. Because the derivative of a polynomial is a polynomial, we can define the differential operator $D : \mathcal{P} \to \mathcal{P}$ by setting $Dp(x) = p'(x)$ for every polynomial $p(x)$ in $\mathcal{P}$. For example, $D(2 + x - x^3) = 1 - 3x^2$. Checking that D is linear is a matter of observing that if $p(x)$ and $q(x)$ are polynomials, and r and s are numbers, then

$$D\big(rp(x) + sq(x)\big) = \big(rp(x) + sq(x)\big)' \text{ and}$$
$$rD\big(p(x)\big) + sD\big(q(x)\big) = rp'(x) + sq'(x),$$

and these are equal by familiar rules of differentiation.

The linear operator D is not one-to-one, because the derivative of every constant polynomial is the identically zero polynomial. The image of D is all of $\mathcal{P}$, because every polynomial is the derivative of some other polynomial.

FIGURE 3.6

Linear actions on $u(x)$.

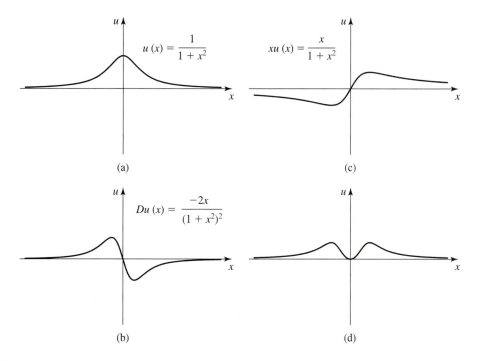

$$u(x) = \frac{1}{1 + x^2}$$

(a)

$$xu(x) = \frac{x}{1 + x^2}$$

(c)

$$Du(x) = \frac{-2x}{(1 + x^2)^2}$$

(b)

(d)

EXAMPLE 3

The discussion of the differential operator D in Example 2 applies somewhat differently if we consider D as a transformation from $C^{(1)}(-\infty, \infty)$, the space of continuously differentiable functions $f(x)$, to $C(-\infty, \infty)$, the continuous functions. The linearity of D follows just as in Example 2. But note that while here $f(x)$ is assumed to have a continuous derivative, $Df(x) = f'(x)$ may only be continuous but not differentiable. As with differentiation of polynomials D is still not one-to-one, for the same reason as before, namely, that the derivative of every constant function is the identically zero function. Figures 3.6(a) and (b) show the graphs of a function $u(x)$ and its derivative $Du(x)$. The linearity of D as an operator on $u(x)$ isn't at all obvious from looking at the pictures.

EXAMPLE 4

Let $C(-\infty, \infty)$ denote the space of continuous real-valued functions $u(x)$ and let $q(x)$ be a fixed function in $C(-\infty, \infty)$. We define an operator $Q : C(-\infty, \infty) \to C(-\infty, \infty)$ by

$$Qu(x) = q(x)u(x).$$

Figure 3.6(c) shows the effect of multiplying by $q(x) = x$ when $u(x) = (1 + x^2)^{-1}$; Figure 3.6(d) shows the effect when $q(x) = x^2(1 + x^2)^{-1}$ instead. Checking that Q is a linear transformation amounts to observing that

$$Q\bigl(ru(x) + sv(x)\bigr) = rQu(x) + sQv(x),$$

or in other words, that

$$q(x)\bigl(ru(x) + su(x)\bigr) = rq(x)u(x) + sq(x)u(x),$$

which follows from the ordinary arithmetic used in combining functions.

Putting the operator Q and the differential operator D in the single equation $Du = Qu$ gives an example of a differential equation:

$$Du = Qu \quad \text{or} \quad u' = xu,$$

where we have chosen $q(x) = x$ for concreteness. It's straightforward to verify that for a constant k the function

$$u(x) = ke^{x^2/2}$$

satisfies the equation. The preceding formula gives all solutions, as we show in Chapter 10, Section 3. In this example the domain of D is the subspace of $C(-\infty, \infty)$ consisting of the vector space of continuously differentiable functions.

EXAMPLE 5   In this example we again use an m-by-n matrix to describe a linear function, but it is *not* the same as Example 1. For one thing, the domain is not $\mathbb{R}^n$ and the range is not $\mathbb{R}^m$. Let $\mathcal{M}_{n,p}$ be the vector space of n-by-p matrices discussed in Example 2 of the previous section. If A is a fixed m-by-n matrix, then for each n-by-p matrix M, the product AM is defined and is an m-by-p matrix. Thus we obtain a function $f_A : \mathcal{M}_{n,p} \to \mathcal{M}_{m,p}$ by defining

$$f_A(M) = AM.$$

The function f_A is linear because of the properties of matrix multiplication given in Theorem 3.2 of Chapter 2. By the right distributive law in that theorem,

$$f_A(M + N) = A(M + N)$$
$$= AM + AN = f_A(M) + f_A(N),$$

and by the scalar commutativity law of the same theorem,

$$f_A(rM) = A(rM) = r(AM) = rf_A(M).$$

Hence f_A is a linear function.

In the preceding examples, formal verification that the transformations were linear was straightforward, and in the future we'll often leave such routine checks to the reader.

A formal proof that a function f is *not* linear involves finding some $\mathbf{x}$ and $\mathbf{y}$ in its domain such that $f(\mathbf{x} + \mathbf{y}) \neq f(\mathbf{x}) + f(\mathbf{y})$, or a scalar r and an $\mathbf{x}$ in the domain of f such that $f(r\mathbf{x}) \neq rf(\mathbf{x})$. For example, $f : \mathbb{R}^1 \longrightarrow \mathbb{R}^1$ defined by $f(x) = x^2$ certainly looks nonlinear. To *prove* that $f(x) = x^2$ is not linear, it's enough to note, for instance, that $f(1 + 1) = f(2) = 4$ while $f(1) + f(1) = 1 + 1 = 2 \neq 4$, or $f((-1)(3)) = f(-3) = 9$ while $-f(3) = -9 \neq 9$. Usually a function that looks nonlinear is nonlinear, but care is sometimes required, as in the next example.

EXAMPLE 6   Define $f : \mathbb{R}^2 \longrightarrow \mathbb{R}^2$ by $f(x, y) = (3x - y, (x+y+2)^2 - (x+y)^2 - 4)$. The function f appears nonlinear at first glance, but a second look shows that $(x+y+2)^2 - (x+y)^2 - 4$ simplifies to $4x + 4y$, so f is linear after all.

The previous example is rather artificial. A more natural situation is to have a family of functions that are nonlinear in general but linear in exceptional cases that may be overlooked. For instance, $ax^2 + bx$ is nonlinear only if $a \neq 0$. The exact domain of a function may also make a difference, as in the following example.

EXAMPLE 7

Let $f: \mathbb{R}^2 \longrightarrow \mathbb{R}^2$ be the function defined by the second-degree formula $f(x, y) = ((x + 1)^2 - (y + 1)^2, 3x^2 + 5xy + 2y^2)$. It certainly looks nonlinear, and we leave it to the reader to check that it is.

Now let $\mathcal{V}$ be the subspace of $\mathbb{R}^2$ consisting of all scalar multiples of $(1, -1)$, and define another function $g: \mathcal{V} \longrightarrow \mathbb{R}^2$ given by the same formula as f, but with domain restricted to $\mathcal{V}$. For (x, y) in $\mathcal{V}$, we have $y = -x$, so for (x, y) in $\mathcal{V}$,

$$g(x, y) = f(x, -x) = ((x + 1)^2 - (-x + 1)^2, 3x^2 - 5x^2 + 2x^2) = (4x, 0),$$

and g is linear on its domain.

3B   Composition and Linear Combination

Composing linear functions by applying first one and then another is a way of producing many other examples of linear functions. The **composition** $g \circ f$ of g and f has been defined by

$$(g \circ f)(\mathbf{x}) = g(f(\mathbf{x}))$$

whenever the right side is defined. According to Theorem 1.5 in Section 1 of this chapter, the composition of linear functions $g: \mathbb{R}^m \longrightarrow \mathbb{R}^p$ and $f: \mathbb{R}^p \longrightarrow \mathbb{R}^n$ that have standard coordinate spaces for domain and range corresponds to matrix multiplication, so that if $f(\mathbf{x}) = A\mathbf{x}$ and $g(\mathbf{y}) = B\mathbf{y}$, then $(g \circ f)(\mathbf{x}) = BA\mathbf{x}$. It then follows from Theorem 2.3 in Chapter 2, Section 2C that $g \circ f$ is linear. The same conclusion holds for linear functions on vector spaces in general.

3.4   Theorem.   If $f: \mathcal{U} \longrightarrow \mathcal{V}$ and $g: \mathcal{V} \longrightarrow \mathcal{W}$ are linear functions, then their composition $g \circ f: \mathcal{U} \longrightarrow \mathcal{W}$, defined by $(g \circ f)(\mathbf{x}) = g(f(\mathbf{x}))$ is also linear.

Proof.   We check the linearity of $g \circ f$ by the following calculation, using first the linearity of f and then the linearity of g.

$$(g \circ f)(r\mathbf{x} + s\mathbf{y}) = g(f(r\mathbf{x} + s\mathbf{y}))$$
$$= g(rf(\mathbf{x}) + sf(\mathbf{y}))$$
$$= rg(f(\mathbf{x})) + sg(f(\mathbf{y})) = r(g \circ f)(\mathbf{x}) + s(g \circ f)(\mathbf{y}) \quad \blacksquare$$

If $f: \mathcal{V} \longrightarrow \mathcal{V}$ is a function whose domain and range are the same space, then $f \circ f$ is defined. We often write f^2 for $f \circ f$. Since f^2 is again a function from $\mathcal{V}$ to $\mathcal{V}$, we can also define $f^3 = f \circ f^2$, and so on. For instance, we write D^2 instead of $D \circ D$ for the second derivative operator so $D^2 f$ means the same as f''.

For functions with the same domain S and the same vector space for their range, **sums** and **scalar multiples** are naturally defined by

$$(f + g)(\mathbf{x}) = f(\mathbf{x}) + g(\mathbf{x}) \quad \text{and} \quad rf(\mathbf{x}) = r(f(\mathbf{x})),$$

whether or not S is a vector space. When the domain of the functions is a vector space, and the functions are linear, we have the following:

3.5   Theorem.   If $f:\mathcal{V} \longrightarrow \mathcal{W}$ and $g:\mathcal{V} \longrightarrow \mathcal{W}$ are linear functions, then the sum $f + g:\mathcal{V} \longrightarrow \mathcal{W}$ and scalar multiple $rf:\mathcal{V} \longrightarrow \mathcal{W}$ are linear also.

The proof amounts to checking linearity by using the definition in much the same way as in the proof of Theorem 3.4, and we leave it as an exercise.

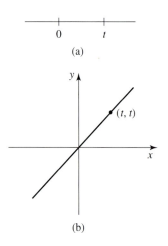

(a)

(b)

FIGURE 3.7
Image line.

| EXAMPLE 8 | We define linear differential operators using both composition and linear combination of transformations. For example, suppose that $p(x)$, $q(x)$, and $r(x)$ are continuous functions, and D is the differentiation operator. Then

$$p(x)D^2 + q(x)D + r(x)$$

acts on twice continuously differentiable functions $u(x)$ as a linear transformation L from $C^{(2)}(-\infty, \infty)$ to $C(-\infty, \infty)$, by

$$L(u) = pu'' + qu' + ru.$$

Typically, we specify a continuous function f and ask for a solution u in $C^{(2)}(-\infty, \infty)$ that satisfies $L(u) = f$.

3C   Inverse Functions

Recall from Section 1 that a function $f:\mathbb{R}^n \longrightarrow \mathbb{R}^m$ has an **inverse function** if there is a function f^{-1} whose domain is the image F of f in $\mathbb{R}^m$, such that

$$f^{-1}(f(\mathbf{x})) = \mathbf{x} \quad \text{and} \quad f(f^{-1}(\mathbf{y})) = \mathbf{y},$$

for every $\mathbf{x}$ in the domain of f and for every $\mathbf{y}$ in the image set F of f. Thus f has an inverse precisely when f is one-to-one; hence we have, by Theorem 3.3:

3.6   Theorem.   If f is linear then f^{-1} exists if and only if $f(\mathbf{x}) = \mathbf{0}$ is satisfied only by $\mathbf{x} = \mathbf{0}$.

The following theorem generalizes Theorem 1.7.

3.7   Theorem.   If $f : \mathcal{V} \to \mathcal{W}$ is linear then its image F is a subspace of $\mathcal{W}$, and if f is one-to-one then f^{-1} is also linear.

Proof.   If $\mathbf{x}$ and $\mathbf{y}$ are in the image F of f, which is the domain of f^{-1}, then there exist $\mathbf{u}$ and $\mathbf{v}$ in $\mathcal{V}$ such that $f(\mathbf{u}) = \mathbf{x}$ and $f(\mathbf{v}) = \mathbf{y}$, so $\mathbf{u} = f^{-1}(\mathbf{x})$ and $\mathbf{v} = f^{-1}(\mathbf{y})$. Because f is linear, if s and t are scalars, $s\mathbf{x} + t\mathbf{y} = f(s\mathbf{u} + t\mathbf{v})$, so $s\mathbf{x} + t\mathbf{y}$ is also in F. This shows that the domain of f^{-1} is a vector subspace of $\mathcal{W}$. The range of f^{-1} is the vector space $\mathcal{V}$. Apply f^{-1} to both sides of $s\mathbf{x} + t\mathbf{y} = f(s\mathbf{u} + t\mathbf{v})$ to get

$$f^{-1}(s\mathbf{x} + \mathbf{y}) = f^{-1}(f(s\mathbf{u} + t\mathbf{v})) = s\mathbf{u} + t\mathbf{v} = sf^{-1}(\mathbf{x}) + tf^{-1}(\mathbf{y}),$$

so f^{-1} is linear.   ∎

Here are some examples of linear functions that have inverses.

EXAMPLE 9   Let $f:\mathbb{R}^2 \longrightarrow \mathbb{R}^2$ be the linear function defined for vectors $\mathbf{x}$ in $\mathbb{R}^2$ by

$$f(\mathbf{x}) = A\mathbf{x}, \quad \text{where} \quad A = \begin{pmatrix} 2 & 3 \\ 1 & 2 \end{pmatrix}.$$

Theorem 4.1 of Chapter 2 tells us that A has an inverse matrix

$$A^{-1} = \begin{pmatrix} 2 & -3 \\ -1 & 2 \end{pmatrix}$$

with the property that $AA^{-1} = A^{-1}A = I$. It follows immediately that f^{-1} exists and $f^{-1}(\mathbf{x}) = A^{-1}\mathbf{x}$.

EXAMPLE 10   Let $\mathbb{R}^2$ be defined for numbers t in $\mathbb{R}$ by $f(t) = (t, t)$. Then f is linear and its image is the line $\mathbb{R}^2$ with equation $x = y$, as shown in Figure 3.7(b). Thus in this case the image of f is a proper subspace $\mathcal{V}$ of $\mathbb{R}^2$, that is one that is not all of $\mathbb{R}^2$. Since the equation $f(t) = (0, 0)$ is satisfied only by $t = 0$, the function f is one-to-one, so has inverse f^{-1}, given by $f^{-1}(t, t) = t$ for all vectors (t, t) in $\mathcal{V}$.

EXAMPLE 11   The function S from $C[0, 1]$ to $C[0, 1]$ defined by $Su(x) = \int_0^x u(t)\,dt$, has an image consisting of the continuously differentiable functions u in $C[0, 1]$ for which $u(0) = 0$. S is one-to-one because Su is identically zero only if u is also. According to the fundamental theorem of calculus, the inverse of S is the differentiation operator D restricted to the functions u such that $u(0) = 0$ and du/dx is continuous.

EXERCISES

In Exercises 1 to 4, a value of n and some information about a linear function $f:\mathbb{R}^n \longrightarrow \mathbb{R}^n$ are given. In each case find the matrix A such that $f(\mathbf{x}) = A\mathbf{x}$ for all $\mathbf{x}$ in $\mathbb{R}^n$.

1. $n = 2$, $f(\mathbf{e}_1) = (1, 2)$, $f(\mathbf{e}_2) = (2, 1)$

2. $n = 3$, $f(\mathbf{e}_1) = (1, 2, 0)$, $f(\mathbf{e}_2) = (-1, 2, 0)$, $f(\mathbf{e}_3) = (0, 0, 1)$

3. $n = 2$, $f(1, 1) = (1, 2)$, $f(2, 1) = (2, 1)$

4. $n = 3$, $f(\mathbf{e}_1) = \mathbf{e}_2$, $f(\mathbf{e}_2) = 2\mathbf{e}_3$, $f(\mathbf{e}_3) = 3\mathbf{e}_1$

Each of Exercises 5 to 8 defines a function from $\mathbb{R}^\infty$ to $\mathbb{R}^\infty$, where $\mathbb{R}^\infty$ is the vector space of sequences (x_k), $k = 1, 2, 3, \ldots$ of Example 3 in Section 2. In each case, show that the function is linear and state whether the function is one-to-one or not. If it is one-to-one then describe its inverse and the domain of the inverse.

5. $f(x_1, x_2, x_3, \ldots) = 2(x_1, x_2, x_3, \ldots)$

6. $g(x_1, x_2, x_3, \ldots) = (x_1, 2x_2, 3x_3, \ldots)$

7. $h(x_1, x_2, x_3, \ldots) = (x_2, x_3, x_4, \ldots)$

8. $p(x_1, x_2, x_3, \ldots) = (0, x_1, x_2, x_3, \ldots)$

In Exercises 9 to 12, determine the effect on a sequence $(x_1, x_2, x_3, \ldots)$ of the given combinations of the functions defined in Exercises 5 to 8.

9. $f \circ g$ and $g \circ f$ **10.** $g \circ h$ and $h \circ g$

11. $g \circ p$ and $p \circ g$ **12.** $h \circ p$ and $p \circ h$

13. In analogy with Example 5 of the text, define for each fixed m-by-n matrix B, the function $g_B : \mathcal{M}_{q,m} \to \mathcal{M}_{q,n}$ by $g_B(M) = MB$. Show that g_B is linear.

14. If A is m-by-p and B is q-by-n, what are the domain and range of $h_{A,B}$ as defined by $h_{A,B}(M) = AMB$ for all M in $\mathcal{M}_{p,q}$? Is $h_{A,B}$ linear?

In Exercises 15 and 16, Let D be the differentiation operator d/dx. For each given function $u(x)$, find $Du(x)$, $xu(x)$, $D(xu(x))$, and $xDu(x)$.

15. $u(x) = 2x^3 - 4x$ **16.** $u(x) = e^{3x}$

17. Let $D = d/dx$ act as a transformation from $C^{(1)}(-\infty, \infty)$ to $C(-\infty, \infty)$.

   (a) If $u(x) = 2x^3$, find $(Dx - xD)u(x)$, where the operator Dx first multiplies by x and then applies D.

   (b) Show that $Dx - xD = I$, where I is the identity operator defined by $Iu = u$ for all u.

(c) Is $D^2 - x^2$ equal to $(D + x)(D - x)$? To find out, apply both operators to a general function $u(x)$ in $C^{(2)}(-\infty, \infty)$ and see if you get the same result.

18. (a) Show that the equation

$$e^x(D + 1)u(x) = De^x u(x)$$

is satisfied by all functions u in $C^{(1)}(-\infty, \infty)$.

(b) Show that the equation

$$(D + 1)u(x) = 0$$

is satisfied by all functions of the form $u(x) = ce^{-x}$, where c is a constant.

(c) Show that the equation $(D + 1)u(x) = 0$ has only the solutions given in part (b).

In Exercises 19 and 20, show that the given function $S: C[0, 1] \longrightarrow C[0, 1]$ is linear.

19. $Su(x) = \int_0^x u(t)\, dt$ **20.** $Su(x) = \int_0^x e^{-t}u(t)\, dt$

In Exercises 21 to 25, $L : \mathcal{V} \to \mathcal{W}$ is a linear function from some specified vector space to another. Use the given information to answer the questions.

21. $L:\mathbb{R}^2 \longrightarrow \mathbb{R}$, $L(\mathbf{u}_0) = 1$, $L(\mathbf{u}_1) = -2$. What is $L(3\mathbf{u}_0 - 4\mathbf{u}_1)$?

22. $L:\mathbb{R}^2 \longrightarrow \mathbb{R}^2$, $L(1, 2) = (2, 3)$, $L(-1, 1) = (1, -1)$. Find a vector $\mathbf{u}$ in $\mathbb{R}^2$ such that $L\mathbf{u} = (3, 7)$.

23. $L:\mathbb{R}^3 \longrightarrow \mathbb{R}^2$, $L(\mathbf{e}_1) = (1, 2)$, $L(\mathbf{e}_2) = (-1, 0)$, $L(\mathbf{e}_3) = (2, 2)$. What is $L(-1, 3, 2)$?

24. $L : C[0, 1] \to C[0, 1]$, $L(1) = 1$, $L(x) = x$, $L(x^2) = x^2 + 2$. What is $L(2x^2 + x - 1)$?

25. $L : C[0, 1] \to C[0, 1]$, $L(1) = x$, $L(x) = x^2$. What is $L(2x + 3)$?

In Exercises 26 to 29, $L : \mathcal{V} \to \mathcal{V}$ is a linear function from a specified vector space $\mathcal{V}$ to itself. Use the given information to answer the questions.

26. $L:\mathbb{R}^2 \longrightarrow \mathbb{R}^2$, $L(x, y) = (x + 2y, 2x + 4y)$. Is there an (x, y) such that $L(x, y) = (-1, 2)$?

27. $L : C^{(1)}[0, 1] \to C[0, 1]$, $Lf = f' - 2f$. Is $f(x) = |x|$ in the domain of L?

28. $L : C^{(2)}[0, 1] \to C[0, 1]$, $Lf(x) = 2f''(x)$. Is there an f in the domain of L such that $Lf(x) = x^2$?

29. $L : C[0, 1] \to C[0, 1]$, $Lf(x) = xf(x)$. Is there an f in the domain of L such that $Lf(x) = x^2 + 1$?

In Exercises 30 to 35, for the given linear function, determine whether it has an inverse function; if it has, describe the inverse by using a matrix, or in some other way. Specify the domain of the inverse function.

30. $f:\mathbb{R}^2 \longrightarrow \mathbb{R}^2$, $f\begin{pmatrix} x \\ y \end{pmatrix} = \begin{pmatrix} 1 & 3 \\ -1 & 2 \end{pmatrix}\begin{pmatrix} x \\ y \end{pmatrix}$

31. $f:\mathbb{R}^2 \longrightarrow \mathbb{R}^2$, $f\begin{pmatrix} x \\ y \end{pmatrix} = \begin{pmatrix} 6 & 3 \\ 4 & 2 \end{pmatrix}\begin{pmatrix} x \\ y \end{pmatrix}$

32. $f:\mathbb{R}^1 \longrightarrow \mathbb{R}^2$, $f(x) = (x, -2x)$

*33. $f:\mathcal{V} \longrightarrow \mathbb{R}^2$, $f\begin{pmatrix} x \\ y \\ z \end{pmatrix} = \begin{pmatrix} 1 & 1 & 3 \\ -1 & 2 & 2 \end{pmatrix}\begin{pmatrix} x \\ y \\ z \end{pmatrix}$, where $\mathcal{V}$ is the subspace of $\mathbb{R}^3$ consisting of all linear combinations of $(1, 1, 1)$ and $(1, 2, 3)$.

*34. $D:\mathcal{V} \longrightarrow C(-\infty, \infty)$, where $Du = u'$, and $\mathcal{V}$ is the subspace of $C^{(1)}(-\infty, \infty)$ consisting of the continuously differentiable functions with $u(0) = 0$

*35. $D:\mathcal{V} \longrightarrow C(-\infty, \infty)$, where $Du = u'$, and $\mathcal{V}$ is the subspace of $C^{(1)}C(-\infty, \infty)$ consisting of the continuously differentiable functions with $u(0) = u(1)$

36. Prove Theorem 3.5: If f and g are linear, so are the sum $f + g$ and scalar multiple rf, for a scalar r.

SECTION 4   IMAGE AND NULL-SPACE

4A   Image

Recall from Section 1 that for a function f and a subset E of the domain of f, the set of values taken on by f as u ranges over E is called the image of E under f and denoted by $f(E)$. The image of the entire domain of f is the image of f.

The images of linear functions are themselves vector spaces, as the following theorem shows.

4.1   Theorem.   Let $f : \mathcal{V} \longrightarrow \mathcal{W}$ be linear and let $\mathcal{U}$ be a subspace of $\mathcal{V}$. Then $f(\mathcal{U})$ is a subspace of $\mathcal{W}$. In particular, $f(\mathcal{V})$, the image of f, is a subspace of $\mathcal{W}$.

Proof.   If $\mathbf{x}$ and $\mathbf{y}$ are in $f(\mathcal{U})$, then there are vectors $\mathbf{u}$ and $\mathbf{v}$ in $\mathcal{U}$ such that $f(\mathbf{u}) = \mathbf{x}$ and $f(\mathbf{v}) = \mathbf{y}$. For scalars r and s, $r\mathbf{u} + s\mathbf{v}$ is also in $\mathcal{U}$ because $\mathcal{U}$ is a subspace, so $f(r\mathbf{u}+s\mathbf{v})$ is in $f(\mathcal{U})$. Because f is linear,

$$f(r\mathbf{u}+s\mathbf{v}) = rf(\mathbf{u}) + sf(\mathbf{v}) = r\mathbf{x} + s\mathbf{y},$$

so $r\mathbf{x}+s\mathbf{y}$ is in $f(\mathcal{U})$. We have shown that $r\mathbf{x}+s\mathbf{y}$ is in $f(\mathcal{U})$ whenever $\mathbf{x}$ and $\mathbf{y}$ are, so $f(\mathcal{U})$ is a subspace of $\mathcal{W}$.   ■

FIGURE 3.8

Image of a square.

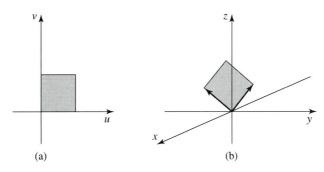

(a)   (b)

EXAMPLE 1

We already know a whole class of linear functions that illustrate Theorem 4.1. If $f:\mathbb{R}^n \longrightarrow \mathbb{R}^m$ is linear then by Theorem 1.4, its image is the span of the columns of the matrix that represents f and is therefore a subspace, by Theorem 2.4. For example, if

$$f\begin{pmatrix} u \\ v \end{pmatrix} = \begin{pmatrix} 1 & 1 \\ 1 & -1 \\ 1 & 1 \end{pmatrix} \begin{pmatrix} u \\ v \end{pmatrix} = u\begin{pmatrix} 1 \\ 1 \\ 1 \end{pmatrix} + v\begin{pmatrix} 1 \\ -1 \\ 1 \end{pmatrix},$$

then the image F of f is the plane spanned by $(1, 1, 1)$ and $(1, -1, 1)$. Figure 3.8(b) shows the images in $\mathbb{R}^3$ of the standard basis vectors $\mathbf{e}_1$ and $\mathbf{e}_2$ in $\mathbb{R}^2$. The shaded parallelogram is the image in $\mathbb{R}^3$ of the square in $\mathbb{R}^2$ with opposite corners at $(0, 0)$ and $(1, 1)$ shown in part (a) of the figure.

EXAMPLE 2

In this example we have to use other reasoning to find the image. If $D = d/dx$ acts in $C^{(1)}(-\infty, \infty)$, the space of continuously differentiable functions, then the image of a single function u in $C^{(1)}(-\infty, \infty)$ is a continuous function v in $C(-\infty, \infty)$. The image of D is all of $C(-\infty, \infty)$, because if v is an arbitrary element of $C(-\infty, \infty)$, then

$$u(x) = \int_0^x v(t)\, dt$$

defines an element of $C^{(1)}(-\infty, \infty)$ such that $Du = v$.

4B   Null-Space

Besides the image, another important subspace associated with a linear function $f:\mathcal{U} \longrightarrow \mathcal{V}$ is the subset of its domain consisting of the vectors $\mathbf{u}$ in $\mathcal{U}$ such that $f(\mathbf{u}) = 0$. As will be shown in Theorem 4.2, it is a subspace of the domain of f. It is called the **null-space** of f.

We look at some examples of null-spaces before giving the formal theorem.

EXAMPLE 3

The formula $f(x, y) = x + 2y$ defines a linear function from $\mathbb{R}^2$ to $\mathbb{R}$. The null-space of f consists of all vectors (x, y) in $\mathbb{R}^2$ such that $x + 2y = 0$. These vectors form a line through the origin of slope $-\frac{1}{2}$, spanned by, for example, the vector $(2, -1)$.

EXAMPLE 4

For each fixed vector $\mathbf{a}$ in $\mathbb{R}^3$, the formula $f(\mathbf{x}) = \mathbf{a} \cdot \mathbf{x}$ defines a linear function $f : \mathbb{R}^3 \longrightarrow \mathbb{R}$. If $\mathbf{a}$ is the zero vector, then the null-space of f is the entire domain $\mathbb{R}^3$. If $\mathbf{a} \neq 0$, then the null-space of f is a plane through the origin in $\mathbb{R}^3$ consisting of all vectors perpendicular to $\mathbf{a}$.

EXAMPLE 5

Finding the null-space of a linear function given by a matrix amounts to solving a homogeneous system of equations. For example, suppose $f : \mathbb{R}^2 \longrightarrow \mathbb{R}^2$ is $f(\mathbf{x}) = A\mathbf{x}$ with 2-by-2 matrix

$$A = \begin{pmatrix} 1 & 4 \\ 3 & 12 \end{pmatrix}.$$

To find the null-space $\mathcal{N}$ of f, we solve the system $A\mathbf{x} = \mathbf{0}$ in the form

$$x + 4y = 0$$
$$3x + 12y = 0,$$

by the row-reduction method of Chapter 2 (or by inspection if you notice that the second equation is just 3 times the first). The solutions are of the form $(x, y) = t(-4, 1)$, where t ranges over all real numbers. In other words, $\mathcal{N}$ is a line through the origin in $\mathbb{R}^2$ with slope $-\frac{1}{4}$.

EXAMPLE 6

In this example, finding the null-space requires knowledge of calculus. If $D = d/dx$ acts on $C^{(1)}(-\infty, \infty)$, the null-space of D consists of all functions with derivative identically zero. It is a theorem of calculus that a function has derivative 0 on an interval if and only if the function is constant on the interval. Hence the null-space of D is the subspace consisting of constant functions. Since sums and multiples of constant functions are constant, they do indeed form a vector subspace of $C^1(-\infty, \infty)$.

For a second example with the same domain and range, if multiplication by x produces a continuous function $xu(x)$ that is identically zero, then u must have been identically zero. It follows that the operation of multiplication by x, acting on $C(-\infty, \infty)$, has its null-space consisting of the zero function in $C(-\infty, \infty)$.

Here is the formal statement of the theorem illustrated in Examples 3 to 6.

4.2   Theorem.   Let $f : \mathcal{V} \longrightarrow \mathcal{W}$ be linear and let $\mathcal{N}$ be the set of all $\mathbf{v}$ in $\mathcal{V}$ such that $f(\mathbf{v}) = \mathbf{0}$. Then $\mathcal{N}$ is a subspace of $\mathcal{V}$.

Proof.   Let $\mathbf{u}$ and $\mathbf{v}$ be vectors in $\mathcal{N}$. Then for a given linear combination $r\mathbf{u} + s\mathbf{v}$ we have

$$f(r\mathbf{u} + s\mathbf{v}) = rf(\mathbf{u}) + sf(\mathbf{v}) = r\mathbf{0} + s\mathbf{0} = \mathbf{0}.$$

Thus $r\mathbf{u} + s\mathbf{v}$ is also in $\mathcal{N}$, so $\mathcal{N}$ is a subspace of $\mathcal{V}$.   ∎

The following theorem is a criterion for a linear function f to be one-to-one in terms of the null-space of f. It is simply a restatement of Theorem 3.3.

4.3   Theorem.   A linear function is one-to-one if and only if its null-space is the zero subspace consisting of the vector $\mathbf{0}$ alone.

EXAMPLE 7

The integration operator defined on $C(-\infty, \infty)$ by

$$v(x) = \int_0^x u(t)\, dt$$

produces a function v in $C(-\infty, \infty)$. Furthermore, if v is identically zero, it follows that its derivative u is also identically zero. That is, the null-space of the integration operator consists of zero alone, so the operator is one-to-one.

4C   Nonhomogeneous Equations

The null-space of a linear function f plays a central role in the description of all solutions of the equation

$$f(\mathbf{x}) = \mathbf{b},$$

where $\mathbf{b}$ is a fixed vector in the image of f. The equation

$$f(\mathbf{x}) = \mathbf{0}$$

is called the **associated homogeneous equation** of $f(\mathbf{x}) = \mathbf{b}$, and the null-space of f is therefore the set of all solutions of the homogeneous equation.

We used the following theorem for solving linear systems of numerical equations in Chapter 2; this more generally applicable version has formally the same proof.

4.4   Theorem.   If $\mathbf{x}_0$ is an arbitrary solution of the linear equation $f(\mathbf{x}) = \mathbf{b}$, then the set S of all solutions consists of all vectors $\mathbf{x}_0 + \mathbf{v}$, where $\mathbf{v}$ ranges over the solutions of the associated homogeneous equation.

Proof.   Suppose that $f(\mathbf{x}_0) = \mathbf{b}$ and also that $f(\mathbf{u}) = \mathbf{b}$. Since f is linear,

$$f(\mathbf{u} - \mathbf{x}_0) = f(\mathbf{u}) - f(\mathbf{x}_0)$$
$$= \mathbf{b} - \mathbf{b} = \mathbf{0}.$$

It follows that $\mathbf{u} - \mathbf{x}_0 = \mathbf{v}$ for some vector $\mathbf{v}$ in the null-space of f. But then $\mathbf{u} = \mathbf{x}_0 + \mathbf{v}$ as we wanted.   ∎

EXAMPLE 8

The linear system

$$x + 4y = 5$$
$$3x + 12y = 15$$

has one solution that we can guess by inspection: $(x, y) = (1, 1)$. In Example 5 we found all the solutions of the associated homogeneous system

$$x + 4y = 0$$
$$3x + 12y = 0$$

to be of the form $(x, y) = t(-4, 1)$. Hence all solutions of the given system are of the form $(x, y) = t(-4, 1) + (1, 1)$.

EXAMPLE 9   One application of Theorem 4.4 is very familiar in elementary calculus. Suppose we want to solve the linear equation

$$Dy = g,$$

where $D = d/dx$ stands for differentiation and g is a continuous function on some interval. If G satisfies $G'(x) = g(x)$, then $y_0(x) = G(x)$ defines one solution. Since the null-space N of D consists of all functions v such that $v'(x)$ is identically zero, N consists of the constant functions $v(x) = C$. Thus every solution of the differential equation has the form

$$y(x) = y_0(x) + v(x)$$
$$= G(x) + C.$$

EXERCISES

In Exercises 1 to 6, a function $f : \mathbb{R}^n \longrightarrow \mathbb{R}^m$ is specified by a formula. In each case state which $\mathbb{R}^m$ is the range of f, and describe the image of f in $\mathbb{R}^m$. Is it a subspace of $\mathbb{R}^m$? Also state whether or not the function is linear, and if it is linear, find its null-space.

1. $f(x, y) = (x, y, x + y)$

2. $f(t) = (t, 2t, 3t)$

3. $f(u, v) = (u, v, 2u + v + 1)$

4. $f(x, y) = (x + y, x - y)$

5. $f(x, y, z) = (x + 4y + 3z, 2x + 5y + 4z)$

6. $f(t) = (t, 0, 1)$

In each of Exercises 7 to 10, describe carefully the image of the given transformation F, state whether the function is linear, and if it is linear, describe its null-space.

7. $F : C(-\infty, \infty) \to C(-\infty, \infty)$, where $F(u)(x) = u(x) + x$.

8. $F : C(-\infty, \infty) \to C(-\infty, \infty)$, where $F(u)(x) = e^{u(x)}$.

9. $F : C(-\infty, \infty) \to C^{(1)}(-\infty, \infty)$, where $F(u)(x) = \int_0^x e^{-t} u(t)\, dt$.

10. $F : C^{(1)}(-\infty, \infty) \to C(-\infty, \infty)$, where $F(u)(x) = u'(x) + u(x)$.

In Exercises 11 to 14, describe the image and the null-space of the function defined by $f(\mathbf{x}) = A\mathbf{x}$ for the given matrix A.

11. $A = \begin{pmatrix} 1 & 1 \\ 0 & 1 \end{pmatrix}$.

12. $A = \begin{pmatrix} 2 & 6 \\ 1 & 3 \end{pmatrix}$.

13. $A = \begin{pmatrix} 2 & 4 & 1 \\ 0 & 1 & 0 \\ 2 & 1 & 1 \end{pmatrix}$.

14. $A = \begin{pmatrix} 0 & 0 & 1 \\ 0 & 1 & 0 \\ 1 & 0 & 0 \end{pmatrix}$.

15. (a) Find all solutions of the homogeneous equation
$$2x - 5y = 0.$$

(b) Verify that a linear combination of two solutions is also a solution.

(c) Find a single solution of the nonhomogeneous equation
$$2x - 5y = 7;$$

then use Theorem 4.4 to represent all solutions of the nonhomogeneous equation.

16. Let $f : \mathbb{R}^n \to \mathbb{R}^m$ be linear.

  (a) If f is not identically zero, show that the image of f contains a line through the origin.

  (b) If $n > m$, show that the null-space of f contains a line through the origin.

17. Show that the null-space of a linear function $f : \mathbb{R}^n \to \mathbb{R}$ is the set of all vectors orthogonal to some fixed vector $\mathbf{x}_0$ in $\mathbb{R}^n$.

18. (a) Find all solutions of the pair of homogeneous equations

$$x + y + z = 0$$
$$x - y + z = 0.$$

  (b) Verify that a linear combination of two solutions is also a solution.

  (c) Verify that $(x, y, z) = (1, 1, -1)$ is a solution of the pair of nonhomogeneous equations

$$x + y + z = 1$$
$$x - y + z = -1.$$

Then use Theorem 4.4 to represent *all* solutions of the pair of nonhomogeneous equations.

19. Consider the homogeneous differential equation $(D - 1)y = 0$.

  (a) Verify that the operator $(D-1) : C^{(1)}(-\infty, \infty) \longrightarrow C(-\infty, \infty)$ is linear.

  (b) Chapter 10, Section 3 shows that all solutions of $(D - 1)y = 0$ have the form $y(x) = ce^x$ for some constant c. Verify that $y(x) = 1 + x$ is a solution of the nonhomogeneous equation $(D - 1)y = -x$, and use Theorem 4.4 to represent all solutions of the nonhomogeneous equation.

20. (a) Verify that for each constant c the function $y(x) = \frac{1}{27}(x + c)^3$ is a solution of the differential equation $Dy - y^{2/3} = 0$.

  (b) Verify that a linear combination of two solutions may not be a solution.

  (c) The conclusion of Theorem 4.4 doesn't hold for $f(y) = Dy - y^{2/3}$. Why doesn't the theorem apply to $f(y)$?

21. Define a function G from $C(-\infty, \infty)$ to $C(-\infty, \infty)$ by

$$Gu(x) = \int_0^x tu(t)\, dt.$$

  (a) Show that G is linear.

  (b) Show that G is one-to-one.

  (c) Describe the image under G of the subspace of $\mathcal{P}$ consisting of all polynomials, $p(x) = a_0 + a_1 x + \ldots + a_n x^n$ of degree $\leq n$.

  (d) Describe the image under G of all of $\mathcal{P}$.

  (e) Describe the inverse of G.

  (f) Find an element of $\mathcal{P}$ that is not in the image of G.

22. If F is a function from a set $\mathcal{A}$ to a set $\mathcal{B}$, and B is a subset of $\mathcal{B}$, then the **inverse image of B under F**, denoted by $F^{-1}(B)$, is defined to be the set of all a in $\mathcal{A}$ for which $F(a)$ is in B. Show that if F is a linear function from one vector space to another, and U is a sub*space* of its range, then $F^{-1}(U)$ is a subspace of the domain of f. What is the connection with Theorem 4.2?

23. A function $u(x)$ in $C(-\infty, \infty)$, the space of continuous functions on $(-\infty, \infty)$ is called **even** if $u(x) = u(-x)$ for all x. It is called **odd** if $u(x) = -u(-x)$ for all x. (For example, $\cos x$ is even and $\sin x$ is odd.) Let R be the operator defined on $C(-\infty, \infty)$ by $(Ru)(x) = u(-x)$. Let I be the identity operator: $(Iu)(x) = u(x)$.

  (a) Show that the graph of Ru is the reflection of the graph of u in the y-axis.

  (b) Show that R and I are linear operators and that $R^2 = I$.

  (c) Let $F_e = \frac{1}{2}(I + R)$. Show that the image of F_e consists of the even functions and that its null-space consists of the odd functions.

  (d) Find the image and null-space of $F_o = I - F_e = \frac{1}{2}(I - R)$.

  (e) Find F_e^2 and F_o^2 in terms of F_e and F_o.

SECTION 5   COORDINATES AND DIMENSION

We say a line has dimension 1 because it takes one coordinate to specify the position of a point on it. It takes two coordinates to specify a point in a plane, so a plane has two dimensions. Similarly, solid space is 3-dimensional. Although our direct spatial experience doesn't go beyond three dimensions, it seems appropriate to say that $\mathbb{R}^n$ has dimension n because it takes an n-tuple of coordinates to specify a point in it. This section has two main goals. One is to show how to introduce coordinates and use them to do computations with vectors and linear functions in vector spaces other than $\mathbb{R}^n$. The other is to show how to define dimension for a vector space for which we lack immediate geometric intuition. The goals are related because both depend on first generalizing the standard basis $\{\mathbf{e}_1, \ldots, \mathbf{e}_n\}$ in $\mathbb{R}^n$.

5A   Bases and Coordinates

Recall that if S is a subset of a vector space $\mathcal{V}$, the **span** of S is defined to consist of all vectors $\mathbf{w}$ that are linear combinations of vectors in S. We showed in Theorem 2.4 of Section 2 that the span of S is always a subspace of $\mathcal{V}$. If the span of S is all of $\mathcal{V}$ we say that S **spans** $\mathcal{V}$ or is a **spanning set** for $\mathcal{V}$.

EXAMPLE 1

Figure 3.9 shows the span of each of three different sets of vectors in $\mathbb{R}^3$. In Figure 3.9(a), the span of $\mathbf{x}_1$ and $\mathbf{x}_2$ is the line containing those two vectors. In Figure 3.9(b), the span of $\mathbf{y}_1$ and $\mathbf{y}_2$ is the plane through the origin containing $\mathbf{y}_1, \mathbf{y}_2$. In Figure 3.9(c), the vector $\mathbf{y}_3 = \mathbf{y}_1 + \mathbf{y}_2$ adds nothing to the span of $\mathbf{y}_1$ and $\mathbf{y}_2$ because $\mathbf{y}_1 + \mathbf{y}_2$ is already in the span. Thus the span of $\mathbf{x}_1$ and $\mathbf{x}_2$ looks one dimensional. The span of $\mathbf{y}_1$ and $\mathbf{y}_2$ looks two dimensional, and since $\mathbf{y}_3$ is a linear combination of $\mathbf{y}_1$ and $\mathbf{y}_2$, the span of $\mathbf{y}_1, \mathbf{y}_2$, and $\mathbf{y}_3$ looks the same. The span of $\mathbf{e}_1, \mathbf{e}_2$, and $\mathbf{e}_3$ looks three dimensional.

Recall from Definitions 2.7 and 2.7$'$ of Chapter 2 that a set of vectors in $\mathbb{R}^n$ is **linearly independent** if no single vector in the set is a linear combination of other vectors in the set, or equivalently, if the only way to express $\mathbf{0}$ as a linear combination of vectors in the set is by taking all the coefficients equal to zero.

The definitions of *spanning set* and *linearly independent* set make sense for general vector spaces, so we can make the following definitions.

5.1   Definition.   A **basis** for a vector space $\mathcal{V}$ is a set of vectors in $\mathcal{V}$ that is linearly independent and spans $\mathcal{V}$.

5.2   Definition.   If $\mathcal{V}$ has a finite basis $\{\mathbf{b}_1, \ldots, \mathbf{b}_n\}$ consisting of n vectors, then $\mathcal{V}$ has **dimension** n, written $\dim(\mathcal{V}) = n$. If $\mathcal{V}$ consists of the zero vector alone we define $\dim(\mathcal{V}) = 0$. If $\mathcal{V}$ isn't spanned by a finite set then $\mathcal{V}$ is **infinite dimensional**.

Note.   It's conceivable that $\mathcal{V}$ might have two bases with unequal numbers of elements. We prove in Section 5C that this can't happen, so if $\mathcal{V}$ has a finite basis, $\dim(\mathcal{V})$ is obtained by counting the vectors in an arbitrary finite basis for $\mathcal{V}$, and it doesn't matter which basis is used.

FIGURE 3.9

Spanning sets.

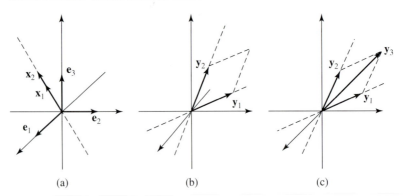

(a)   (b)   (c)

EXAMPLE 2   Starting in Chapter 1 we've referred to the vectors $\mathbf{e}_1, \dots, \mathbf{e}_n$, where $\mathbf{e}_k$ has 1 in entry k and 0 elsewhere, as the **standard basis** for $\mathbb{R}^n$. They span $\mathbb{R}^n$ because a vector $(x_1, \dots, x_n)$ is equal to the linear combination $x_1\mathbf{e}_1 + \cdots + x_n\mathbf{e}_n$. They are a linearly independent set because each $\mathbf{e}_k$ has 1 in position k and so can't be a linear combination of the other $\mathbf{e}_i$ because they all have 0 in that position. Consequently $\{\mathbf{e}_1, \dots, \mathbf{e}_n\}$ is a basis for $\mathbb{R}^n$ according to Definition 5.1, and $\mathbb{R}^n$ officially has dimension n according to Definition 5.2.

EXAMPLE 3   In the vector space $\mathcal{P}_n$ of polynomials of degree at most n, the set S_n consisting of the $n + 1$ polynomials

$$p_0(x) = 1, \ p_1(x) = x, \ \dots, \ p_n(x) = x^n$$

is a basis for $\mathcal{P}_n$. If $p(x) = a_0 + a_1 x + \cdots + a_n x^n$, then $p(x)$ is a linear combination of $p_0(x), p_1(x), \dots, p_n(x)$, so S_n spans $\mathcal{P}_n$. Also, if for all x a linear combination

$$q(x) = a_0 p_0(x) + \cdots + a_k p_k(x) + \cdots + a_n p_n(x) = 0,$$

then all coefficients must be zero, because otherwise $q(x)$ would be a polynomial of degree at most n with more than n roots. Hence $\mathcal{P}_n$ has dimension $n + 1$. The vector space $\mathcal{P}$ of all polynomials is spanned by the infinite set $\{1, x, x^2, \dots\}$, but no finite subset is adequate to span $\mathcal{P}$, so $\mathcal{P}$ is infinite dimensional. The set $\{1, x, x^2, \dots\}$ is linearly independent as required by Definition 5.1, because every finite subset is linearly independent.

The next theorem shows that a basis $\{\mathbf{b}_1, \dots, \mathbf{b}_n\}$ for $\mathcal{V}$ generates a one-to-one correspondence $\mathbf{v} \leftrightarrow (v_1, \dots, v_n)$ between vectors $\mathbf{v}$ in $\mathcal{V}$ and vectors $(v_1, \dots, v_n)$ in $\mathbb{R}^n$ that is linear, that is, it preserves addition and scalar multiplication. Using a different basis will produce a different correspondence of the same kind.

5.3   Theorem.   Let $B = \{\mathbf{b}_1, \dots, \mathbf{b}_n\}$ be a basis for the vector space $\mathcal{V}$. Then for every $\mathbf{v}$ in $\mathcal{V}$, there are unique scalars $v_1, \dots, v_n$ such that $\mathbf{v} = v_1\mathbf{b}_1 + \cdots + v_n\mathbf{b}_n$. The correspondence is linear, that is, If $\mathbf{u} \leftrightarrow (u_1, \dots, u_n)$ and $\mathbf{v} \leftrightarrow (v_1, \dots, v_n)$, then

$$\mathbf{u} + \mathbf{v} \longleftrightarrow (u_1 + v_1, \dots, u_n + v_n) \quad \text{and} \quad r\mathbf{u} \longleftrightarrow (ru_1, \dots, ru_n).$$

Proof.   Since B spans $\mathcal{V}$, every vector $\mathbf{v}$ is some linear combination $v_1\mathbf{b}_1 + \cdots + v_n\mathbf{b}_n$. To prove uniqueness we have to show that if $\mathbf{v}$ is also equal to $w_1\mathbf{b}_1 + \cdots + w_n\mathbf{b}_n$ then the v's are the same as the w's. If both linear combinations are equal to $\mathbf{v}$, then their difference is $\mathbf{0}$, so

$$(v_1\mathbf{b}_1 + \cdots + \mathbf{b}_n) - (w_1\mathbf{b}_1 + \cdots + w_n\mathbf{b}_n)$$
$$= (v_1 - w_1)\mathbf{b}_1 + \cdots + (v_n - w_n)\mathbf{b}_n v = \mathbf{0}.$$

Since B is an independent set, the coefficients $v_k - w_k$ must all be 0, which makes $w_k = v_k$ for $k = 1, \dots, n$. To prove linearity of the correspondence we verify two correspondences:

$$\mathbf{u} + \mathbf{v} = (u_1\mathbf{b}_1 + \cdots + u_n\mathbf{b}_n) + (v_1\mathbf{b}_1 + \cdots + u_n\mathbf{b}_n)$$

$$= \big((u_1 + v_1)\mathbf{b}_1 + \cdots + (u_n + v_n)\mathbf{b}_n\big) \leftrightarrow (u_1 + v_1, \ldots, u_n + v_n),$$

$$r\mathbf{u} = r(u_1\mathbf{b}_1 + \cdots + u_n\mathbf{b}_n) = (ru_1\mathbf{b}_1 + \cdots + ru_n\mathbf{b}_n) \leftrightarrow (ru_1, \ldots, ru_n). \ \blacksquare$$

If $\{\mathbf{b}_1, \ldots, \mathbf{b}_n\}$ is a basis, then the unique n-tuple $(x_1, \ldots, x_n)$ such that

$$\mathbf{x} = x_1\mathbf{b}_1 + \cdots + x_n\mathbf{b}_n$$

is called the n-tuple of **coordinates** of $\mathbf{x}$ relative to the basis. The coordinate n-tuple of $\mathbf{x}$ depends on the order of the basis vectors. The n-tuple $(x_1, \ldots, x_n)$ is a vector in $\mathbb{R}^n$, and is called the **coordinate vector** of $\mathbf{x}$ relative to the ordered basis $\{\mathbf{b}_1, \ldots, \mathbf{b}_n\}$. The previous theorem shows that if $\mathcal{V}$ has a basis with n elements, then we can apply what we know about algebra in $\mathbb{R}^n$ to vectors in $\mathcal{V}$ by using the corresponding n-tuples in $\mathbb{R}^n$. In Section 7 we consider products that serve in a vector space $\mathcal{V}$ the way the dot product does in $\mathbb{R}^n$.

EXAMPLE 4

Some vector spaces have natural bases relative to which coordinates are particularly simple. For example, in $\mathbb{R}^n$ the coordinate vector of the n-tuple $(x_1, \ldots, x_n)$ relative to the standard basis $\{\mathbf{e}_1, \ldots, \mathbf{e}_n\}$ of $\mathbb{R}^n$ is just the n-tuple itself, because $(x_1, \ldots, x_n) = x_1\mathbf{e}_1 + \cdots + x_n\mathbf{e}_n$.

Similarly, in the space $\mathcal{P}_n$ of polynomials of degree at most n, the coordinate vector of a polynomial $p(x) = a_0 + a_1 x + \cdots + a_n x^n$ relative to the basis $\{1, x, \ldots, x^n\}$ of Example 3 is simply the $(n+1)$-tuple of coefficients, $(a_0, a_1, \ldots, a_n)$.

EXAMPLE 5

Giving a basis for a subspace is often a good way to describe the subspace. For example, the functions e^x and e^{-x} span the subspace $\mathcal{S}$ of $C(-\infty, \infty)$ consisting of all functions expressible in the form

$$c_1 e^x + c_2 e^{-x},$$

c_1 and c_2 constant. Neither function is a constant multiple of the other, so they are linearly independent, and $\{e^x, e^{-x}\}$ is a basis for the 2-dimensional space $\mathcal{S}$.

The functions $\sinh x = \frac{1}{2}e^x - \frac{1}{2}e^{-x}$ and $\cosh x = \frac{1}{2}e^x + \frac{1}{2}e^{-x}$ are in $\mathcal{S}$, so relative to the basis $\{e^x, e^{-x}\}$ their respective coordinate vectors are $(\frac{1}{2}, -\frac{1}{2})$ and $(\frac{1}{2}, \frac{1}{2})$. Exercise 13 asks you to show that the pair $\{\cosh x, \sinh x\}$ is also a basis for $\mathcal{S}$.

5B   Linear Functions

Theorem 5.3 shows that by fixing bases in finite-dimensional vector spaces $\mathcal{V}$ and $\mathcal{W}$ we can calculate the results of vector operations in them by working with coordinate vectors in $\mathbb{R}^n$ and $\mathbb{R}^m$. The next theorem is a generalization of Theorem 1.3 in Section 1, and it shows how to use coordinates to represent a linear function $f: \mathcal{V} \longrightarrow \mathcal{W}$ by a matrix. In Sections 6 and 7 we take up some ways of finding bases that lead to particularly simple matrix representations for linear functions we are interested in.

5.4   Theorem.   Let $V = \{\mathbf{v}_1, \ldots, \mathbf{v}_n\}$ be a basis for $\mathcal{V}$ and $W = \{\mathbf{w}_1, \ldots, \mathbf{w}_m\}$ a basis for $\mathcal{W}$. Let f be a linear function with domain $\mathcal{V}$ and range $\mathcal{W}$ such that

$$f(v_1\mathbf{v}_1 + \cdots + v_n\mathbf{v}_n) = w_1\mathbf{w}_1 + \cdots + w_m\mathbf{w}_m.$$

Then the coordinates v_k and w_k are related by

$$\begin{pmatrix} a_{11} & a_{12} & \cdots & a_{1n} \\ \vdots & & & \\ a_{m1} & a_{m2} & \cdots & a_{mn} \end{pmatrix} \begin{pmatrix} v_1 \\ v_2 \\ \vdots \\ v_n \end{pmatrix} = \begin{pmatrix} w_1 \\ \vdots \\ w_m \end{pmatrix},$$

where the jth column of the m-by-n matrix consists of the coordinates of $f(\mathbf{v}_j)$ relative to the basis W as given by $f(\mathbf{v}_j) = a_{1j}\mathbf{w}_1 + \cdots + a_{mj}\mathbf{w}_m$.

Note.   Using standard bases in $V = \mathbb{R}^n$ and $W = \mathbb{R}^m$ we get Theorem 1.3.

Proof.   We can combine f with the linear correspondences between vectors and coordinates given by Theorem 5.3 to obtain a function $F: \mathbb{R}^n \longrightarrow \mathbb{R}^m$ by defining $F(v_1, \ldots v_n)$ to be the coordinate vector $(w_1 \ldots, w_m)$ such that

$$w_1\mathbf{w}_1 + \cdots + w_m\mathbf{w}_m = f(v_1\mathbf{v}_1 + \cdots + v_n\mathbf{v}_n).$$

F is then the composition

$$(v_1, \ldots v_n) \longrightarrow v_1\mathbf{v}_1 + \cdots + v_n\mathbf{v}_n \overset{f}{\longrightarrow} w_1\mathbf{w}_1 + \cdots + w_m\mathbf{w}_m \longrightarrow (w_1 \ldots w_m),$$

and is therefore linear by Theorem 3.4 because it is a composition of linear functions. All that remains to be done is to check that the matrix $A = (a_{ij})$ of F does the right thing to the standard basis vectors $\mathbf{e}_1, \ldots, \mathbf{e}_n$. But the jth column of A is $A\mathbf{e}_j$, and by hypothesis this vector is also the coordinate m-tuple $(w_1, \ldots, w_m)$ that represents $f(\mathbf{v}_j)$ relative to the basis W and so is equal to $F(\mathbf{e}_j)$.   ∎

EXAMPLE 6   In applications of Theorem 5.4 the spaces V and W are very often the same and we use the same basis to represent both domain and image vectors. For example, suppose $V = W = \mathbb{R}^2$ and the single basis is $\{\mathbf{v}_1, \mathbf{v}_2\} = \{(1, 1), (1, 2)\}$. If it's given that

$$f(1, 1) = (2, 3) \quad \text{and} \quad f(1, 2) = (0, -1),$$

then in terms of $\{\mathbf{v}_1, \mathbf{v}_2\}$ these equations say that

$$f(\mathbf{v}_1) = \mathbf{v}_1 + \mathbf{v}_2 \quad \text{and} \quad f(\mathbf{v}_2) = \mathbf{v}_1 - \mathbf{v}_2.$$

Hence relative to $\{\mathbf{v}_1, \mathbf{v}_2\}$-coordinates, f has matrix

$$A = \begin{pmatrix} 1 & 1 \\ 1 & -1 \end{pmatrix}.$$

For example, $f(2\mathbf{v}_1 - 3\mathbf{v}_2)$ has $\{\mathbf{v}_1, \mathbf{v}_2\}$-coordinates

$$\begin{pmatrix} 1 & 1 \\ 1 & -1 \end{pmatrix} \begin{pmatrix} 2 \\ -3 \end{pmatrix} = \begin{pmatrix} -1 \\ 5 \end{pmatrix},$$

so that $f(2\mathbf{v}_1 - 3\mathbf{v}_2) = -\mathbf{v}_1 + 5\mathbf{v}_2$.

The matrix A has inverse

$$A^{-1} = \begin{pmatrix} \frac{1}{2} & \frac{1}{2} \\ \frac{1}{2} & -\frac{1}{2} \end{pmatrix},$$

so f^{-1} exists and has matrix A^{-1} relative to the basis $\{\mathbf{v}_1, \mathbf{v}_2\}$.

We know from Theorem 1.4 that the image of a linear function $f(\mathbf{x}) = A\mathbf{x}$ is spanned by the columns of the matrix A. To get a basis for the image, we need to check for linear dependencies among the column vectors in A. On the other hand, to find a basis for the null-space of f we need to represent the solutions of $A\mathbf{x} = \mathbf{0}$ as a linear combination of independent vectors. We'll do both in the next example. A routine procedure, proved to work in all cases in Theorem 5.10 in the following Section 5C, is to apply elementary row operations to A to get a reduced matrix R for which dependence relations are easier to see. Since solution vectors $\mathbf{x}$ of $A\mathbf{x} = \mathbf{0}$ are unchanged by row operations on A, a linear relation among the columns of A carries over to the same relation among the corresponding columns of R.

EXAMPLE 7 We'll find bases for the null-space and image of the linear function $f:\mathbb{R}^4 \longrightarrow \mathbb{R}^3$ defined by the matrix equation $f(\mathbf{x}) = A\mathbf{x}$, where

$$A\mathbf{x} = \begin{pmatrix} 1 & 3 & 4 & 2 \\ -1 & 2 & 1 & 3 \\ 1 & 0 & 1 & -1 \end{pmatrix}\mathbf{x}.$$

Row reduction of A takes only a few operations, and the resulting equation $R\mathbf{x} = \mathbf{0}$, equivalent to $A\mathbf{x} = \mathbf{0}$, is easier to analyze:

$$R\mathbf{x} = \begin{pmatrix} 1 & 0 & 1 & -1 \\ 0 & 1 & 1 & 1 \\ 0 & 0 & 0 & 0 \end{pmatrix}\mathbf{x} = \mathbf{0}, \quad \text{or} \quad \begin{aligned} x_1 \quad +x_3 -x_4 &= 0, \\ x_2 +x_3 +x_4 &= 0. \end{aligned}$$

The null-space of $f(\mathbf{x}) = A\mathbf{x}$ consists of the vectors $\mathbf{x} = (x_1, x_2, x_3, x_4)$ that satisfy $A\mathbf{x} = \mathbf{0}$, or $R\mathbf{x} = \mathbf{0}$. Setting the nonleading variables $x_3 = s$ and $x_4 = t$, we find leading variables $x_1 = -s - t$ and $x_2 = -s + t$. Thus all solutions are

$$\mathbf{x} = s\begin{pmatrix} -1 \\ -1 \\ 1 \\ 0 \end{pmatrix} + t\begin{pmatrix} -1 \\ 1 \\ 0 \\ 1 \end{pmatrix}.$$

The two vectors span the null-space of f, and they're independent because, based on the first two entries alone, neither is a scalar multiple of the other. Hence the null-space of f is 2-dimensional with these two vectors as a basis.

For the image of f, we know that it's the span of the columns. The first two columns of R are evidently independent, and the third is their sum while the fourth is second minus the first. Hence the first two columns of A suffice to span the image of f. Since these vectors are independent, the image of f is 2-dimensional, with basis $\{(1, -1, 1), (3, 2, 0)\}$ in $\mathbb{R}^3$.

The dimensions of the null-space and image add up, $2 + 2$, to the dimension 4 of the domain of f. We prove in Section 5C that this relation generalizes to all linear functions $f : \mathcal{V} \longrightarrow \mathcal{W}$ if $\mathcal{V}$ has a finite basis.

EXAMPLE 8   The differential operator $D : \mathcal{P}_3 \longrightarrow \mathcal{P}_3$ acts linearly. Using the basis $\{1, x, x^2, x^3\}$ for both the domain and range, the natural coordinates to use are the respective coefficient vectors, (a_0, a_1, a_2, a_3) and (b_0, b_1, b_2, b_3) for polynomials $a_0 + a_1 x + a_2 x^2 + a_3 x^3$ and $b_0 + b_1 x + b_2 x^2 + b_3 x^3$ in $\mathcal{P}_3$. The 4-by-4 matrix operation that carries out differentiation in terms of these coordinates is

$$\begin{pmatrix} 0 & 1 & 0 & 0 \\ 0 & 0 & 2 & 0 \\ 0 & 0 & 0 & 3 \\ 0 & 0 & 0 & 0 \end{pmatrix} \begin{pmatrix} a_0 \\ a_1 \\ a_2 \\ a_3 \end{pmatrix} = \begin{pmatrix} b_0 \\ b_1 \\ b_2 \\ b_3 \end{pmatrix}.$$

Reducing the matrix just replaces the 2 and 3 by 1's, so a_0 is the single nonleading variable, while a_1, a_2, and a_3 are leading variables. To find the null-space we set the $b_k = 0$ and solve, setting $a_0 = t$ and finding $a_1 = 0$, $2a_2 = 0$, and $3a_3 = 0$. Thus the null-space consists of the constant polynomials, with one-element basis $\{1\}$. Since the matrix has just three independent columns, the image is 3-dimensional and consists of the polynomials with coefficients $b_0 = a_1$, $b_1 = 2a_2$, $b_2 = 2a_3$, and $b_3 = 0$. Thus $\{1, x, x^2\}$ is a basis for the image. As in the previous example, we see here that again the sum, $1 + 3$, of the dimensions of the null-space and image is 4, the dimension of the domain.

We have gone through this extensive description of the simple relation $D(a_0 + a_1 x + a_2 x^2 + a_3 x^3) = a_1 + 2a_2 x + 3a_3 x^2$ to illustrate some general principles in an abstract setting that is familiar enough to be thoroughly understood.

EXERCISES

In Exercises 1 to 4, show that the given set of vectors forms a basis for $\mathbb{R}^n$ of the appropriate dimension by showing (a) spanning, and (b) independence.

1. $\{(-1, 1), (1, 1)\}$

2. $\{(1, 2), (1, -2)\}$

3. $\{(1, 0, 0,), (1, 1, 0), (1, 1, 1)\}$

4. $\{(1, 2, 3), (0, 0, 1), (2, 2, 4)\}$

In Exercises 5 to 8, find the dimension of the subspaces of $\mathbb{R}^2$ or $\mathbb{R}^3$ spanned by the given vectors. [*Hint:* If a set of vectors is already independent, it forms a basis for the subspace it spans.]

5. $(-1, 1), (1, -1)$

6. $(1, 2), (1, 3)$

7. $(1, 0, 1), (0, 0, 1), (1, 0, 2)$

8. $(1, 2, 3), (2, 3, 4), (3, 4, 5)$

In Exercises 9 to 12, show that the given subset of $C(-\infty, \infty)$ is linearly independent.

9. $\{e^x, e^{2x}, e^{3x}\}$

10. $\{x, e^x, e^{-x}\}$

11. $\{\cos x, \sin x\}$

12. $\{\cos x, x \cos x, x^2 \cos x\}$

13. Let $\mathcal{S}$ be the subspace of $C(-\infty, \infty)$ with basis $\{e^x, e^{-x}\}$.

    (a) Show that the pair $\{\cosh x, \sinh x\}$ is another basis for $\mathcal{S}$.

    (b) Find the coordinates of e^x and e^{-x} relative to the basis in part (a).

14. Let $\mathcal{S}$ be the subspace of $C(-\infty, \infty)$ with basis $\{e^x, e^{-x}\}$. What are the coordinates of the function $3e^x - 4e^{-x}$ relative to the basis $\{\cosh x, \sinh x\}$?

15. **(a)** Show that the set $\{1, x + 1, (x + 1)^2\}$ is a basis for the space $\mathcal{P}_2$ of polynomials of degree at most 2.

(b) What are the coordinates of the polynomial x^2+x+1 relative to the basis given in part (a)?

In Exercises 16 to 19, let $B_n = \{1, \cos x, \sin x, \cos 2x, \sin 2x, \ldots, \cos nx, \sin nx\}$. Functions in the subspace $\mathcal{T}_n$ of $C(-\infty, \infty)$ spanned by B_n are called **trigonometric polynomials of degree** $\leq n$.

16. We'll show in Theorem 7.4 of Section 7B that B_n is a linearly independent set, so it is actually a basis for $\mathcal{T}_n$. What is the dimension of $\mathcal{T}_n$?

17. Show that $\cos^2 x$ and $\sin^2 x$ are in $\mathcal{T}_2$, and find their coordinates relative to the basis B_2. (Use trigonometric identities.)

*18. Show that for given integers p and q, the product $(\cos px)(\sin qx)$ is in $\mathcal{T}_{p+q}$ and find its coordinates relative to the basis B_{p+q}.

*19. Show that if $f(x)$ is in $\mathcal{T}_p$ and $g(x)$ is in $\mathcal{T}_q$, then $f(x)g(x)$ is in $\mathcal{T}_{p+q}$.

In Exercises 20 to 23, find a basis for the vector space consisting of all linear combinations of each of the following sets of functions.

20. $\{1, x, x-1, x^2+1\}$

21. $\{e^x, e^{-x}, \sinh x, \cosh x\}$

22. $\{\sin^2 x, \cos^2 x, 1\}$

23. $\{\cos x, \sin x, \sin 2x\}$

In Exercises 24 to 27, let the linear function $f \colon \mathbb{R}^n \to \mathbb{R}^m$ be $f(\mathbf{x}) = A\mathbf{x}$. Find a basis, if there is one, for (a) the image of f and (b) the null-space of f.

24. $A = \begin{pmatrix} 2 & 8 \\ 3 & 12 \end{pmatrix}$

25. $A = \begin{pmatrix} 2 & 1 \\ 1 & 2 \end{pmatrix}$

26. $A = \begin{pmatrix} 0 & 0 & 1 \\ 0 & 1 & 0 \\ 1 & 0 & 0 \end{pmatrix}$

27. $A = \begin{pmatrix} 2 & 4 & 2 \\ 0 & 1 & 1 \\ 1 & 3 & 2 \end{pmatrix}$

28. Let $f \colon \mathcal{V} \to \mathcal{W}$ be both linear and one to one.
  (a) Show that if $\mathcal{S}$ is a linearly independent set of vectors in $\mathcal{V}$, then $f(\mathcal{S})$, the image of f in $\mathcal{W}$, is a linearly independent set of vectors in $\mathcal{W}$.
  (b) Show that $\mathcal{V}$ and $f(\mathcal{V})$ have the same dimension.

In Exercises 29 and 30, find the dimension of the subspace of $\mathbb{R}^3$ consisting of all solutions of the given equation or equations.

29. $\begin{array}{l} x + y - z = 0, \\ 2x + y = 0 \end{array}$

30. $x + y - z = 0$

31. Show that if $p_1(x), \ldots, p_k(x)$ are k nonzero polynomials whose degrees are all different, then they form a linearly independent set in the vector space $\mathcal{P}$ of all polynomials.

In Exercises 32 and 33, let $\mathcal{P}_5$ be the space of polynomials of degree at most 5 and let $\mathcal{O}$ be the subspace of $\mathcal{P}_5$ consisting of odd polynomials, that is, polynomials $p(x)$ such that $p(-x) = -p(x)$.

32. Find a basis for $\mathcal{O}$. What is the dimension of $\mathcal{O}$?

33. Is there a polynomial $p(x)$ such that $\{x - x^3, x^3 + x^5, p(x)\}$ is a basis for $\mathcal{O}$?

In Exercises 34 to 37, determine whether $\mathbf{x}$ is in the span of S.

34. $\mathbf{x} = (17, -6, 13)$ and $S = \{(1, -6, 2), (4, 8, 1)\}$

35. $\mathbf{x} = (3, -4, 5, 2)$ and $S = \{(1, -2, 1, 1), (2, 1, -2, 1), (3, 1, 1, 1)\}$

36. $\mathbf{x} = \sin(x + \pi/7)$ and $S = \{\cos x, \sin x\}$

37. $\mathbf{x} = \cos 2x$ and $S = \{1, \cos x, \cos^2 x\}$

38. Let f be the linear function from $\mathbb{R}^2$ to $\mathbb{R}^2$ whose matrix relative to the natural basis is $\begin{pmatrix} 1 & 2 \\ 2 & -1 \end{pmatrix}$. Thus $f(1, 0) = (1, 2)$ and $f(0, 1) = (2, -1)$. Find the matrix of f relative to the basis $\{\mathbf{v}_1, \mathbf{v}_2\} = \{(1, 1), (1, 2)\}$ for $\mathbb{R}^2$.

39. Let g be the linear function from $\mathbb{R}^2$ to $\mathbb{R}^3$ whose matrix relative to the natural basis $\{\mathbf{e}_1, \mathbf{e}_2\}$ for $\mathbb{R}^2$ and natural basis $\{\mathbf{e}_1, \mathbf{e}_2, \mathbf{e}_3\}$ for $\mathbb{R}^3$ is $\begin{pmatrix} 2 & -1 \\ 1 & 2 \\ -2 & 2 \end{pmatrix}$. [Thus $g(1, 0) = (2, 1, -2)$ and $g(0, 1) = (-1, 2, 2)$.] Find the matrix of g relative to the bases $\{\mathbf{v}_1, \mathbf{v}_2\} = \{(1, 1), (1, 2)\}$ for $\mathbb{R}^2$ and $\{\mathbf{w}_1, \mathbf{w}_2, \mathbf{w}_3\} = \{(1, 0, 0), (1, 1, 0), (1, 1, 1)\}$ for $\mathbb{R}^3$.

40. Let $D = d/dx$ be the linear operation of differentiation on the vector space $\mathcal{P}_2$ of polynomials of degree at most 2.
  (a) Find the matrix of D relative to the basis $\{1, x, x^2\}$ for $\mathcal{P}_2$.
  (b) What are the matrices of D^2 and D^3 relative to the same basis for $\mathcal{P}_2$?

41. Let S be the shift operator defined on the vector space $\mathcal{P}_2$ of polynomials of degree at most 2 by $Sp(x) = p(x+1)$. For example, $S(2x + 1) = 2x + 3$.
  (a) Show that S is a linear operator on $\mathcal{P}_2$.
  (b) Show that $S = I + D + \frac{1}{2}D^2$, where $D = d/dx$, and I stands for the identity operator.
  (c) Show that on $\mathcal{P}_n$, the space of polynomials of degree at most n, the shift operator is related to D by $S = I + D + \frac{1}{2!}D^2 + \cdots + \frac{1}{n!}D^n$. [*Hint:* Use Taylor's theorem to expand $p(x + h)$ with $h = 1$.]

5C   Dimension Theorems

A nonzero vector space $\mathcal{V}$ has infinitely many different bases, even when $\mathcal{V} = \mathbb{R}^1$, where each nonzero number is a one-element basis, so it's conceivable that a vector space might have two bases with unequal numbers of vectors. We'll prove here in Theorem 5.6 that if a vector space $\mathcal{V}$ has one finite basis $\{\mathbf{b}_1, \dots, \mathbf{b}_n\}$, then every basis for $\mathcal{V}$ also contains exactly n vectors. This theorem justifies Definition 5.2 that we made in Section 5A, allowing us to say without ambiguity that a nonzero vector space $\mathcal{V}$ has **dimension** n if it has a basis consisting of n vectors.

The next theorem justifies the crucial step we need to prove the Theorem 5.6.

5.5   Theorem.   If $\mathcal{V}$ has a basis with n elements, then a subset of $\mathcal{V}$ with more than n elements is linearly dependent.

Proof.   Let $V = \{\mathbf{v}_1, \dots, \mathbf{v}_n\}$ be a basis for $\mathcal{V}$, and let $\{\mathbf{x}_1, \dots, \mathbf{x}_m\}$ be a subset of $\mathcal{V}$ with $m > n$. We need to show that there are numbers $r_1, \dots, r_m$, not all zero, such that

$$r_1 \mathbf{x}_1 + \cdots + r_m \mathbf{x}_m = 0. \tag{1}$$

Since V is a basis for $\mathcal{V}$, each $\mathbf{x}_k$ is a linear combination

$$\mathbf{x}_k = a_{k1} \mathbf{v}_1 + \cdots + a_{kn} \mathbf{v}_n, \tag{2}$$

with some numbers a_{kj} as coefficients. Substitution of (2) into (1) and interchanging the order of summation gives

$$\sum_{k=1}^{m} r_k \left(\sum_{j=1}^{n} a_{kj} \mathbf{v}_j \right) = \sum_{j=1}^{n} \left(\sum_{k=1}^{m} a_{kj} r_k \right) \mathbf{v}_j = 0.$$

Because the $\mathbf{v}$'s form a basis they are independent, so the coefficients of the $\mathbf{v}$'s are all zero, and the m variables $r_1, \dots, r_m$ satisfy the equations

$$\sum_{k=1}^{m} a_{kj} r_k = 0, \quad j = 1, \dots, n.$$

Since $m > n$ these equations form a homogeneous system with more variables than equations. By Theorem 2.5 of Chapter 2, there are infinitely many nonzero solutions for the r's, which is what we wanted to show.   ∎

We now have a short proof that the dimension of $\mathcal{V}$ is independent of whatever finite set of basis vectors we count to compute $\dim(\mathcal{V})$.

5.6   Theorem.   Let $\mathcal{V}$ be a vector space having a basis with n elements. Then every basis for $\mathcal{V}$ has n elements.

Proof.   Let $\{v_1, \dots, v_n\}$ and $\{u_1, \dots, u_k\}$ be two bases for $\mathcal{V}$; in particular, each set is independent. We can't have $k > n$ because then the $\mathbf{u}$'s would be dependent by the previous theorem, and we can't have $n > k$ because then the **v**'s would be dependent. Hence $n = k$.   ∎

It follows that to find the dimension of a vector space all we have to do is pick a basis and count the number of vectors in it; we always get the same number, no matter what basis we count. Thus we've proved that the definition of dimension in Section 5A assigns a clearly defined number called the **dimension** of $\mathcal{V}$ to every vector space $\mathcal{V}$ that has a finite basis. A vector space with a finite basis is called **finite-dimensional**, and all others are called **infinite-dimensional**.

| EXAMPLE 9 | The vector space $\mathcal{P}$ of all polynomials doesn't have a finite basis. If it did have one with n elements, then the linear independence of the $n+1$ functions $1, x, x^2, \ldots, x^n$ would contradict Theorem 5.5. However $\mathcal{P}$ has a basis consisting of the infinite sequence of independent functions $\{1, x, x^2, \ldots\}$, because every polynomial is a unique linear combination of some finite subset of them.

5.7 Theorem.   If $\mathcal{V}$ is spanned by nonzero vectors $\mathbf{x}_1, \ldots, \mathbf{x}_n$, then some subset of the **x**'s is a basis for $\mathcal{V}$.

Proof.   If $\mathbf{x}_1, \ldots, \mathbf{x}_n$ is an independent set, then that set itself is a basis for $\mathcal{V}$. Otherwise, some relation $r_1\mathbf{x}_1 + \cdots + r_n\mathbf{x}_n = 0$ holds with at least one r, say r_k, different from 0. Dividing by r_k, we get $\mathbf{x}_k = -(r_1/r_k)\mathbf{x}_1 - \cdots - (r_n/r_k)\mathbf{x}_n$. Substituting the right side for $\mathbf{x}_k$ in a linear combination of all the **x**'s, we get a linear combination from which $\mathbf{x}_k$ has been eliminated. It follows that we can delete $\mathbf{x}_k$, and the span of the remaining vectors will still be all of $\mathcal{V}$. If the resulting subset of **x**'s is not independent, we repeat the process until we do arrive at an independent set, which is then a basis for $\mathcal{V}$.   ∎

The previous theorem says we can get a basis from a finite spanning set by deleting some vectors. For a finite-dimensional space the next theorem shows that we can get a basis from a linearly independent set by putting more vectors in the set.

5.8 Theorem.   Let $S = \{\mathbf{x}_1, \ldots, \mathbf{x}_k\}$ be a linearly independent set in a vector space $\mathcal{V}$. If S is not a basis for $\mathcal{V}$, we can include more vectors in S, possibly arriving at a finite basis for $\mathcal{V}$, in which case $\mathcal{V}$ is finite-dimensional. Otherwise we can extend S with an infinite sequence of independent vectors so $\mathcal{V}$ is infinite-dimensional.

Proof.   Suppose $\mathbf{x}_1, \ldots, \mathbf{x}_k$ are linearly independent but don't span all of $\mathcal{V}$. Then there is some vector **y** that is not a linear combination of $\mathbf{x}_1, \ldots, \mathbf{x}_k$. Take $\mathbf{x}_{k+1} = \mathbf{y}$. We claim that the set $\mathbf{x}_1, \ldots, \mathbf{x}_k, \mathbf{x}_{k+1}$ is linearly independent. Suppose instead that $r_1\mathbf{x}_1 + \cdots + r_{k+1}\mathbf{x}_{k+1} = 0$. We must show that all the r's are 0. If r_{k+1} were not 0, we could write $\mathbf{x}_{k+1} = -(r_1/r_{k+1})\mathbf{x}_1 - \cdots - (r_k/r_{k+1})\mathbf{x}_k$, which is impossible because $\mathbf{x}_{k+1}$ is not a linear combination of the other **x**'s. Therefore we have $r_{k+1} = 0$ and $r_1\mathbf{x}_1 + \cdots + r_k\mathbf{x}_k = \mathbf{0}$. Since $\mathbf{x}_1, \ldots, \mathbf{x}_k$ are independent, the last equation implies $r_1 = \cdots = r_k = 0$. Thus if a linearly independent set S doesn't span $\mathcal{V}$, we can add a vector to S so that the resulting set is also independent.

Repeating the process, we may reach a spanning set S in a finite number of steps, so S becomes a basis and $\mathcal{V}$ is finite dimensional. Otherwise we can find an arbitrarily large independent set so $\mathcal{V}$ is infinite dimensional.   ∎

We extend the definition of k-plane in Chapter 2, Section 2D from the spaces $\mathbb{R}^n$ to other vector spaces $\mathcal{V}$ by saying that a k-**plane** is either a k-dimensional proper subspace $\mathcal{S}$ of $\mathcal{V}$, as in the next example, or else a translation of $\mathcal{S}$ by a fixed vector **v** in $\mathcal{V}$.

EXAMPLE 10

Suppose $\dim(\mathcal{V}) = n$ and that $\mathcal{S}$ is a subspace of $\mathcal{V}$. If $\mathcal{S}$ isn't the **0**-subspace, then it contains a vector $\mathbf{x} \neq 0$ that we can extend to a linearly independent set S as in the previous theorem. By Theorem 5.5 applied to $\mathcal{V}$, $\mathcal{S}$ can't be infinite-dimensional, so S is a finite basis, and $\dim(\mathcal{S}) = k$ for some k with $0 < k \leq n$. If $k < n$, then $\mathcal{S}$ is a k-plane containing **0**. Thus every subspace of a finite-dimensional vector space $\mathcal{V}$ is either a k-plane containing **0** or else $\mathcal{V}$ itself. In particular the proper subspaces of $\mathbb{R}^n$ are the k-planes containing **0**.

We summarize much of what we have proved about finite bases and dimension as follows. It is proved by a straightforward application of the previous theorems.

5.9   Theorem.   Let $\dim(\mathcal{V}) = n$. If B is a subset of $\mathcal{V}$ with two of the following properties, then B has the third property and so is a basis for $\mathcal{V}$.

(a) B contains exactly n vectors.
(b) B is a linearly independent set.
(c) B spans $\mathcal{V}$.

Theorem 5.4 in Section 2B shows that introducing bases and coordinates in finite-dimensional vector spaces allows us to study linear functions $f: \mathcal{V} \longrightarrow \mathcal{W}$ by looking at linear functions $f: \mathbb{R}^n \longrightarrow \mathbb{R}^m$ of the form $f(\mathbf{x}) = A\mathbf{x}$. The following theorem gives us an effective description of the null-space and image of such a linear function f as k-planes containing **0** in $\mathcal{V}$ and $\mathcal{W}$; in particular, the dimension of the null-space of f is the number s of columns with nonleading entries in a reduced form R of A, and the dimension of the image is the number r of columns with leading entries. Since every column in R is of one kind or the other, it follows that $s + r = n$, where n is the dimension of the domain of f.

5.10   Theorem.   Let $\mathcal{V}$ and $\mathcal{W}$ be finite-dimensional. A linear function $f: \mathcal{V} \longrightarrow \mathcal{W}$ has null-space of dimension k, with k the number of nonleading variables in a reduced form R of a matrix of f with respect to some bases in $\mathcal{V}$ and $\mathcal{W}$. The image of f in $\mathcal{W}$ is a subspace of dimension r, with r the number of leading variables in R.

Proof.   We'll assume that reducing the matrix of f gives a reduced matrix R with k nonleading variables. The null-space of f consists of the vectors whose coordinates satisfy the system $R\mathbf{x} = \mathbf{0}$. Since the system $R\mathbf{x} = \mathbf{0}$ always has $\mathbf{x} = \mathbf{0}$ as one solution, if there are no nonleading variables then the unique solution to $R\mathbf{x} = \mathbf{0}$ is $\mathbf{x} = \mathbf{0}$, so the null-space is the single point **0**, a 0-plane. If $k > 0$, there is for each $i = 1, \ldots, k$, a unique solution $\mathbf{u}_i$ of $R\mathbf{x} = \mathbf{0}$ in which we choose the ith nonleading variable to be 1 and the other nonleading variables to be 0. These k vectors $\mathbf{u}_i$ are linearly independent, because $\mathbf{u}_i$ has 1 in the position of the ith nonleading variable where the remaining $k-1$ vectors have 0. Hence $\mathbf{u}_i$ can't be a linear combination of the other $\mathbf{u}_j$. The vectors $\mathbf{x} = t_1 \mathbf{u}_1 + \cdots + t_k \mathbf{u}_k$ are the solutions of $R\mathbf{x} = \mathbf{0}$. To see this note first that every such $\mathbf{x}$ is a solution, because by the linearity of matrix-vector multiplication

$$R\mathbf{x} = t_1 R\mathbf{u}_1 + \cdots + t_k R\mathbf{u}_k = \mathbf{0},$$

because $R\mathbf{u}_1 = \ldots = R\mathbf{u}_k = \mathbf{0}$. Second, the nonleading variables have some values v_i or other in every solution $\mathbf{v}_0$, so using $t_i = v_i$ in the formula for $\mathbf{x}$ shows that the solution $\mathbf{v}_0$ has the form $\mathbf{x}$.

Each of the $r = n - k$ leading variables, necessarily the same as the number of nonzero rows in R, corresponds to a column containing a single entry 1, each in a different row and with the other entries 0. These columns are standard basis vectors, spanning the other columns, hence spanning the image of f. Thus the image of f is an r-plane containing $\mathbf{0}$ in $\mathcal{W}$.   ∎

EXERCISES

1. (a) Let E_{ij} be the m-by-n matrix with 1 in the ijth position and zeros elsewhere. Show that the E_{ij} form a basis for the space of all m-by-n matrices. What is the dimension of the space?

(b) What is the dimension of the space of diagonal n-by-n matrices?

2. Which of the following statements is true for every linear function f? Prove your answer.

(a) If $\mathbf{x}_1$ and $\mathbf{x}_2$ are linearly independent, then so are $f(\mathbf{x}_1)$ and $f(\mathbf{x}_2)$.

(b) If $f(\mathbf{x}_1)$ and $f(\mathbf{x}_2)$ are linearly independent, then so are $\mathbf{x}_1$ and $\mathbf{x}_2$.

3. Show that if f is a one-to-one linear function, then the set $\{f(\mathbf{x}_1), \ldots, f(\mathbf{x}_k)\}$ is linearly independent if and only if $\{\mathbf{x}_1, \ldots, \mathbf{x}_k\}$ is linearly independent. What does this imply about the dimensions of the image and domain of f?

4. Let $\mathbf{x}_1 = (1, 2, 3)$, $\mathbf{x}_2 = (-1, 2, 1)$, $\mathbf{x}_3 = (1, 1, 1)$, and $\mathbf{x}_4 = (1, 1, 0)$.

(a) Without doing any computation, give a reason why $\mathbf{x}_1, \mathbf{x}_2, \mathbf{x}_3$, and $\mathbf{x}_4$ form a linearly dependent set.

(b) Express $\mathbf{x}_1$ as a linear combination of $\mathbf{x}_2, \mathbf{x}_3$, and $\mathbf{x}_4$.

5. Prove that two planes that contain $\mathbf{0}$ in $\mathbb{R}^3$ intersect in a line or coincide.

6. Prove Theorem 5.9, that is, prove that if $\mathcal{V}$ has dimension n, then (a) and (b) imply (c), (a) and (c) imply (b), and (b) and (c) imply (a).

7. Prove that if $\dim(\mathcal{W}) = n$ and $\mathcal{V}$ is a subspace of $\mathcal{W}$, then $\dim(\mathcal{V}) \leq n$.

8. Let a vector space $\mathcal{V}$ have dimension n and have an $(n-1)$-dimensional subspace $\mathcal{W}$ with basis $\{\mathbf{x}_1, \ldots, \mathbf{x}_{n-1}\}$. Let $\mathbf{x}_n$ be in $\mathcal{V}$ but not in $\mathcal{W}$. Define $f(a_1\mathbf{x}_1 + \cdots + a_n\mathbf{x}_n) = a_n$.

(a) Show that $f: \mathcal{V} \longrightarrow \mathbb{R}$ is a linear function with null-space precisely $\mathcal{W}$.

(b) Show that if $g: \mathcal{V} \longrightarrow \mathbb{R}$ is another linear function with null-space $\mathcal{W}$, then $g = cf$ for some constant c. [*Hint:* Show that $g(\mathbf{x}) - g(\mathbf{x}_n)f(\mathbf{x}) = 0$ for all $\mathbf{x}$ in $\mathcal{V}$.]

9. (a) Show that if $\mathcal{S}$ is a k-dimensional subspace of $\mathbb{R}^n$, then $\mathcal{S}$ is the null-space of some linear function $f: \mathbb{R}^n \longrightarrow \mathbb{R}^{n-k}$. [*Hint:* Begin by picking a basis for $\mathcal{S}$ and extending it to a basis for $\mathbb{R}^n$.]

(b) Use part (a) to show that every k-dimensional subspace $\mathcal{S}$ of $\mathbb{R}^n$ is the intersection of $n-k$ hyperplanes through the origin, that is, of $(n-1)$-dimensional subspaces of $\mathbb{R}^n$.

10. Assume $\mathcal{V}$ and $\mathcal{W}$ are finite dimensional and that $f: \mathcal{V} \longrightarrow \mathcal{W}$ is linear. Prove that if $\mathcal{N}$ is the null-space of f, then there is a subspace $\mathcal{S}$ of $\mathcal{V}$ such that $\mathcal{S} \cap \mathcal{N} = \mathbf{0}$, and f restricted to $\mathcal{S}$ is one-to-one. [*Hint:* Pick a basis for $\mathcal{S}$ and extend it to be a basis for $\mathcal{V}$.]

11. Assume $\mathcal{V}$ and $\mathcal{W}$ are finite dimensional, and $f: \mathcal{V} \longrightarrow \mathcal{W}$ is linear. Let $\mathcal{N}$ and $f(\mathcal{V})$ be the null-space and image of f. Use Theorem 5.10 to prove the equation $\dim(\mathcal{N}) + \dim(f(\mathcal{V})) = \dim(\mathcal{V})$.

12. Prove that if $\mathcal{V}$ is finite-dimensional and $f: \mathcal{V} \longrightarrow \mathcal{W}$ is linear, then the inverse image of a vector $\mathbf{w}$ in the image of f is a k-plane in $\mathcal{V}$, where k is the dimension of the null-space of f. (For the definition of inverse image, see Exercise 22 in Section 4.)

13. (a) Prove that if $f: \mathcal{V} \longrightarrow \mathcal{W}$ is linear with $\dim(\mathcal{V}) > \dim(\mathcal{W})$, then $\dim(\mathcal{N}) > 0$, where $\mathcal{N}$ is the null-space of f.

(b) Use part (a) to explain why there can't be an m-by-n matrix A and an n-by-m matrix B such that $BA = I$ if $m < n$.

14. Find a 3-by-2 matrix A and a 2-by-3 matrix B such that $BA = I$.

SECTION 6   EIGENVALUES AND EIGENVECTORS

This section is about a way to better understand linear operators $L : \mathcal{V} \to \mathcal{V}$, on a finite-dimensional vector space $\mathcal{V}$. We'll see that finding a basis for $\mathcal{V}$ consisting of vectors $\mathbf{x}$ such that $f(\mathbf{x}) = \lambda\mathbf{x}$, for some scalar λ associated with $\mathbf{x}$, allows us to write the matrix of f in the diagonal form

$$\begin{pmatrix} \lambda_1 & 0 & \cdots & 0 \\ 0 & \lambda_2 & & 0 \\ \vdots & & \ddots & \vdots \\ 0 & 0 & \cdots & \lambda_n \end{pmatrix}.$$

The main advantage is that diagonal matrices display properties of an operator that are easy to read from the matrix. We'll mostly be concerned with linear operators on $\mathbb{R}^n$, represented by square matrices, but some of the ideas apply as well to other linear operators, such as the differential operators $D = d/dx$ and $D^2 = d^2/dx^2$.

6A   Definitions and Examples
For a given linear operator $L : \mathcal{V} \longrightarrow \mathcal{V}$ and a vector $\mathbf{x}$, there is usually no simple relation between $\mathbf{x}$ and $L(\mathbf{x})$. But for special vectors $\mathbf{u}$ it may happen that $L(\mathbf{u})$ is a scalar multiple of $\mathbf{u}$, so that

6.1 $L(\mathbf{u}) = \lambda\mathbf{u}.$

This equation makes no sense unless $\mathbf{u}$ is in both the image and domain of L. If $\mathbf{u} \neq \mathbf{0}$ and satisfies $L(\mathbf{u}) = \lambda\mathbf{u}$, we say that $\mathbf{u}$ is an **eigenvector** of the linear operator L and that λ is the **eigenvalue** of L associated with $\mathbf{u}$, or alternatively that $\mathbf{u}$ is an eigenvector associated with the eigenvalue λ. The terms **characteristic vector $\mathbf{u}$** and **characteristic value** λ are also common. Because L is linear, a nonzero multiple $c\mathbf{u}$ of $\mathbf{u}$ is an eigenvector associated with the same eigenvalue. The vector $\mathbf{0}$ is never called an eigenvector, and the scalar 0 is an eigenvalue of L only if $L(\mathbf{u}) = 0\mathbf{u} = \mathbf{0}$ for some vector $\mathbf{u} \neq \mathbf{0}$.

| EXAMPLE 1 |

Let L be the linear operator on $\mathbb{R}^2$ defined by the matrix

$$\begin{pmatrix} 1 & 1 \\ 4 & 1 \end{pmatrix}.$$

Thus

$$L\begin{pmatrix} x \\ y \end{pmatrix} = \begin{pmatrix} 1 & 1 \\ 4 & 1 \end{pmatrix}\begin{pmatrix} x \\ y \end{pmatrix} = \begin{pmatrix} x+y \\ 4x+y \end{pmatrix}.$$

It's routine to verify that

$$\begin{pmatrix} 1 & 1 \\ 4 & 1 \end{pmatrix}\begin{pmatrix} 1 \\ 2 \end{pmatrix} = 3\begin{pmatrix} 1 \\ 2 \end{pmatrix}$$

and that

$$\begin{pmatrix} 1 & 1 \\ 4 & 1 \end{pmatrix} \begin{pmatrix} 1 \\ -2 \end{pmatrix} = (-1) \begin{pmatrix} 1 \\ -2 \end{pmatrix}.$$

That is, the vector $\mathbf{u} = (1, 2)$ in $\mathbb{R}^2$ is an eigenvector associated with the eigenvalue 3, and the vector $\mathbf{v} = (1, -2)$ is an eigenvector associated with the eigenvalue -1. Likewise, nonzero multiples of these two vectors will be eigenvectors with the same two eigenvalues.

Before discussing how to find eigenvectors, we'll show how they're useful. Suppose that L is a linear operator on a vector space $\mathcal{V}$ and that $\mathbf{u}_1, \ldots, \mathbf{u}_k$ are eigenvectors of L with associated eigenvalues $\lambda_1, \ldots, \lambda_k$. In other words,

$$L(\mathbf{u}_1) = \lambda_1 \mathbf{u}_1, \quad L(\mathbf{u}_2) = \lambda_2 \mathbf{u}_2, \quad \ldots, \quad L(\mathbf{u}_k) = \lambda_k \mathbf{u}_k.$$

If $\mathbf{u} = c_1 \mathbf{u}_1 + \ldots + c_k \mathbf{u}_k$ is a linear combination of the eigenvectors, we have

$$L(\mathbf{u}) = c_1 L(\mathbf{u}_1) + \ldots + c_k L(\mathbf{u}_k)$$
$$= c_1 \lambda_1 \mathbf{u}_1 + \ldots + c_k \lambda_k \mathbf{u}_k.$$

In terms of eigenvectors $\mathbf{u}_j$ of L the effect of L is to multiply each term in the linear combination by the associated eigenvalue:

6.2 If $L(\mathbf{u}_j) = \lambda_j \mathbf{u}_j$ for $j = 1, 2, \ldots, k$, then $L\left(\sum_{j=1}^{k} c_j \mathbf{u}_j \right) = \sum_{j=1}^{k} \lambda_j c_j \mathbf{u}_j$.

These formulas are most useful when there are enough linearly independent eigenvectors to form a basis for the space, since then every vector is a linear combination of the eigenvectors and Equation 6.2 applies to the basis representation of every $\mathbf{x}$. In the general case of an n-dimensional vector space with a basis of n eigenvectors the second equation in terms of matrices and coordinates becomes

$$L(\mathbf{x}) = \begin{pmatrix} \lambda_1 & 0 & \cdots & 0 \\ 0 & \lambda_2 & & 0 \\ \vdots & & \ddots & \vdots \\ 0 & 0 & \cdots & \lambda_n \end{pmatrix} \begin{pmatrix} c_1 \\ c_2 \\ \vdots \\ c_n \end{pmatrix}.$$

EXAMPLE 2   Returning to the linear operator L of Example 1, we express an arbitrary $\mathbf{x}$ in $\mathbb{R}^2$ as a linear combination of the two eigenvectors $\mathbf{u} = (1, 2)$ and $\mathbf{v} = (1, -2)$ with respective eigenvalues 3 and -1. Suppose

$$\mathbf{x} = u \begin{pmatrix} 1 \\ 2 \end{pmatrix} + v \begin{pmatrix} 1 \\ -2 \end{pmatrix}.$$

Since $L(\mathbf{u}) = 3\mathbf{u}$ and $L(\mathbf{v}) = -\mathbf{v}$, the linearity of L implies

$$L(\mathbf{x}) = u L \begin{pmatrix} 1 \\ 2 \end{pmatrix} + v L \begin{pmatrix} 1 \\ -2 \end{pmatrix} = 3u \begin{pmatrix} 1 \\ 2 \end{pmatrix} - v \begin{pmatrix} 1 \\ -2 \end{pmatrix}.$$

FIGURE 3.10

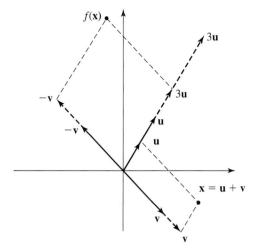

Using u and v as coordinates gives the diagonal matrix representation

$$L(\mathbf{x}) = \begin{pmatrix} 3 & 0 \\ 0 & -1 \end{pmatrix} \begin{pmatrix} u \\ v \end{pmatrix}.$$

Figure 3.10 shows the effect of L on each of the two eigenvectors $\mathbf{u}$ and $\mathbf{v}$, so the image $L(\mathbf{x})$ of a vector $\mathbf{x}$ depends geometrically on the parallelogram law. We express the effect of L by saying that L is a composition of two operators:

1. A stretch by a factor of 3 away from the line through $\mathbf{v}$ along the lines parallel to $\mathbf{u}$ and with eigenvector coordinate matrix $\begin{pmatrix} 3 & 0 \\ 0 & 1 \end{pmatrix}$.

2. A reversal of direction on lines parallel to $\mathbf{v}$, leaving points on the line through $\mathbf{u}$ fixed, with eigenvector coordinate matrix $\begin{pmatrix} 1 & 0 \\ 0 & -1 \end{pmatrix}$. Operators (1) and (2) together produce the same end result in either order:

$$\begin{pmatrix} 3 & 0 \\ 0 & 1 \end{pmatrix} \begin{pmatrix} 1 & 0 \\ 0 & -1 \end{pmatrix} = \begin{pmatrix} 1 & 0 \\ 0 & -1 \end{pmatrix} \begin{pmatrix} 3 & 0 \\ 0 & 1 \end{pmatrix} = \begin{pmatrix} 3 & 0 \\ 0 & -1 \end{pmatrix}.$$

EXAMPLE 3   To compute the eigenvalues and associated eigenvectors for the function L of the previous examples, we proceed as follows. We need to find vectors $\mathbf{u} \neq \mathbf{0}$, and numbers λ such that

$$L(\mathbf{u}) - \lambda\mathbf{u} = \mathbf{0}.$$

In matrix form, this equation is

$$\begin{pmatrix} 1 & 1 \\ 4 & 1 \end{pmatrix} \begin{pmatrix} x \\ y \end{pmatrix} - \lambda \begin{pmatrix} x \\ y \end{pmatrix} = \begin{pmatrix} 0 \\ 0 \end{pmatrix}$$

or

$$\begin{pmatrix} 1 & 1 \\ 4 & 1 \end{pmatrix} \begin{pmatrix} x \\ y \end{pmatrix} - \begin{pmatrix} \lambda & 0 \\ 0 & \lambda \end{pmatrix} \begin{pmatrix} x \\ y \end{pmatrix} = \begin{pmatrix} 0 \\ 0 \end{pmatrix}$$

or

$$\begin{pmatrix} (1-\lambda) & 1 \\ 4 & (1-\lambda) \end{pmatrix} \begin{pmatrix} x \\ y \end{pmatrix} = \begin{pmatrix} 0 \\ 0 \end{pmatrix}. \tag{1}$$

If this 2-by-2 matrix has an inverse, then the only solution of the equation (1) is $x = 0$ and $y = 0$. Hence we must try to find values of λ for which the matrix isn't invertible. By Theorem 5.7 of Chapter 2, Section 5E, this will occur precisely when

$$\det \begin{pmatrix} (1-\lambda) & 1 \\ 4 & (1-\lambda) \end{pmatrix} = (1-\lambda)^2 - 4 = 0.$$

This quadratic equation in λ has roots $\lambda = 3$ and $\lambda = -1$, as we see by inspection, by factoring, or by using the quadratic formula. To find eigenvectors associated with $\lambda = 3$ and $\lambda = -1$, we must find x and y, not both zero, satisfying Equation (1). Thus we consider

$$\lambda = 3: \quad \begin{pmatrix} -2 & 1 \\ 4 & -2 \end{pmatrix} \begin{pmatrix} x \\ y \end{pmatrix} = \begin{pmatrix} 0 \\ 0 \end{pmatrix}$$

and

$$\lambda = -1: \quad \begin{pmatrix} 2 & 1 \\ 4 & 2 \end{pmatrix} \begin{pmatrix} x \\ y \end{pmatrix} = \begin{pmatrix} 0 \\ 0 \end{pmatrix}.$$

Each of these systems reduces to a single equation:

$$\begin{aligned} \lambda &= 3: & -2x + y &= 0 \\ \lambda &= -1: & 2x + y &= 0. \end{aligned}$$

It follows that there are many solutions, but all we need is one nonzero solution for each eigenvalue. We choose for simplicity

$$\lambda = 3: \quad \mathbf{u} = \begin{pmatrix} x \\ y \end{pmatrix} = \begin{pmatrix} 1 \\ 2 \end{pmatrix}; \quad \lambda = -1: \quad \mathbf{v} = \begin{pmatrix} x \\ y \end{pmatrix} = \begin{pmatrix} 1 \\ -2 \end{pmatrix}.$$

In principle, a nonzero numerical multiple of either vector would do as well.

We summarize the method of Example 3 as follows. To find eigenvalues of a linear operator L on $\mathbb{R}^n$ with matrix A, solve the **characteristic equation**

6.3 $$\det(A - \lambda I) = 0,$$

with **characteristic roots** $\lambda_1, \ldots, \lambda_n$. Then for each root $\lambda_1, \ldots, \lambda_n$, try to find nonzero vectors $\mathbf{x}_1, \ldots, \mathbf{x}_n$ that satisfy the matrix equation

6.4 $$(A - \lambda_k I)\mathbf{x}_k = 0.$$

Because Equations 6.3 and 6.4 are expressed in terms of the matrix A of L, we sometimes refer to eigenvectors and eigenvalues of the *matrix* rather than the function L. However the distinction between the matrix A and the operator L is particularly important here, because even if A is a real matrix, some of the roots λ_k may be complex numbers. In that case, the matrix $A - \lambda_k I$ can't be interpreted as an operator on $\mathbb{R}^n$, because it will have complex entries. In Section 6B we'll see that we can still get interesting information by operating instead on the space $\mathbb{C}^n$ of n-tuples of complex numbers.

For differential operators the general definitions and principles of Equations 6.1 and 6.2 remain the same, but the determinants and matrices of Equations 6.3 and 6.4 are replaced by calculus computations as in the next example.

EXAMPLE 4   If r is a constant, then $(d/dx)\, e^{rx} = re^{rx}$. If we consider $u(x) = e^{rx}$ as a vector in the space $C^{(\infty)}$, and let D be the differentiation operator, then

$$Du = ru,$$

so e^{rx} is an eigenvector for D associated with the eigenvalue r. In particular, $De^x = e^x$, $De^{2x} = 2e^{2x}$, and $De^{-3x} = -3e^{-3x}$, so the functions $c_1 e^x, c_2 e^{2x}, c_3 e^{-3x}$ are eigenvectors for the differentiation operator D if the c_i are nonzero constants. The associated eigenvalues are 1, 2, and -3, so

$$D(c_1 e^x + c_2 e^{2x} + c_3 e^{-3x}) = c_1 e^x + 2c_2 e^{2x} - 3c_3 e^{-3x},$$

as we expect from the rules of calculus.

EXAMPLE 5   Here we'll combine the results of Examples 1 to 4 to solve the following system of differential equations for $x(t)$ and $y(t)$:

$$\frac{dx}{dt} = x + y, \qquad \text{or in vector-matrix form} \quad \frac{d\mathbf{x}}{dt} = \begin{pmatrix} 1 & 1 \\ 4 & 1 \end{pmatrix} \mathbf{x},$$
$$\frac{dy}{dt} = 4x + y,$$

where $\mathbf{x}(t) = \big(x(t),\, y(t)\big)$ and $d\mathbf{x}(t)/dt = \big(x'(t),\, y'(t)\big)$.

If we use u and v as coordinates with respect to the two eigenvectors $\mathbf{u}$ and $\mathbf{v}$, then $\mathbf{x} = u\mathbf{u} + v\mathbf{v}$. Example 2 shows that in terms of u and v,

$$L(\mathbf{x}) = \begin{pmatrix} 3 & 0 \\ 0 & -1 \end{pmatrix} \begin{pmatrix} u \\ v \end{pmatrix}.$$

In these same coordinates $\dfrac{d\mathbf{x}}{dt} = \begin{pmatrix} u' \\ v' \end{pmatrix}$, so the original system of differential equations becomes

$$\begin{pmatrix} u' \\ v' \end{pmatrix} = \begin{pmatrix} 3 & 0 \\ 0 & -1 \end{pmatrix} \begin{pmatrix} u \\ v \end{pmatrix} \quad \text{or} \quad \begin{matrix} u' = 3u, \\ v' = -v. \end{matrix}$$

Example 4 shows that we can take $u = c_1 e^{3t}$ and $v = c_2 e^{-t}$. Section 3 of Chapter 10 shows that these are the only possibilities. Switching back to xy-coordinates shows that our solution is

$$\mathbf{x}(t) = c_1 e^{3t} \begin{pmatrix} 1 \\ 2 \end{pmatrix} + c_2 e^{-t} \begin{pmatrix} 1 \\ -2 \end{pmatrix} \quad \text{or} \quad \begin{matrix} x(t) = c_1 e^{3t} + c_2 e^{-t}, \\ y(t) = 2c_1 e^{3t} - 2c_2 e^{-t}. \end{matrix}$$

For this analysis to work, the eigenvectors of L had to be a basis for $\mathbb{R}^2$, but that follows from their linear independence.

EXERCISES

1. The linear operator L from $\mathbb{R}^2$ to $\mathbb{R}^2$ with matrix

$$\begin{pmatrix} 1 & 12 \\ 3 & 1 \end{pmatrix}$$

has eigenvalues 7 and -5. Which of the following vectors is an eigenvector of L? For those that are, what is the associated eigenvalue?

$$\begin{pmatrix} 2 \\ 1 \end{pmatrix}, \quad \begin{pmatrix} -2 \\ 1 \end{pmatrix}, \quad \begin{pmatrix} -4 \\ -2 \end{pmatrix}, \quad \begin{pmatrix} -2 \\ 2 \end{pmatrix}, \quad \begin{pmatrix} 1 \\ 1 \end{pmatrix}$$

In Exercises 2 to 7, find all the eigenvalues of each of the linear operators defined by the following matrices, and for each eigenvalue find an associated eigenvector.

2. $\begin{pmatrix} 1 & 4 \\ 1 & 1 \end{pmatrix}$
3. $\begin{pmatrix} 0 & 4 \\ 1 & 0 \end{pmatrix}$

4. $\begin{pmatrix} 0 & 1 \\ 0 & 2 \end{pmatrix}$
5. $\begin{pmatrix} 2 & 4 \\ 1 & 2 \end{pmatrix}$

6. $\begin{pmatrix} 1 & 0 & 0 \\ 2 & 1 & 0 \\ 0 & 1 & 2 \end{pmatrix}$
7. $\begin{pmatrix} 0 & 0 & 1 \\ 0 & 1 & 0 \\ 0 & 0 & 2 \end{pmatrix}$

8. Show that 0 is an eigenvalue of a linear operator L if and only if L is not one to one.

9. Show that if f is a one-to-one linear operator having λ for an eigenvalue, then f^{-1}, the inverse of f, has $1/\lambda$ for an eigenvalue.

10. Let f be a linear operator having λ for an eigenvalue.
  (a) Show that λ^2 is an eigenvalue associated with $f \circ f$.
  (b) Show that λ^n is an eigenvalue associated with the function we get by composing f with itself n times.

11. Let $C^{(\infty)}(\mathbb{R})$ be the vector space of infinitely often differentiable functions $f(x)$ for x in $\mathbb{R}$. Then the differential operator D^2 acts linearly from $C^{(\infty)}(\mathbb{R})$ to $C^{(\infty)}(\mathbb{R})$. This exercise is about the eigenvectors of the operator D^2.
  (a) For $\lambda > 0$, let $k = \sqrt{\lambda}$ and show that any linear combination of e^{kx} and e^{-kx} is an eigenvector for λ.

(b) For $\lambda < 0$, let $k = \sqrt{-\lambda}$ and show that any linear combination of $\cos kx$ and $\sin kx$ is an eigenvector for λ.

(c) Find two linearly independent functions such that any linear combination of them is an eigenvector for $\lambda = 0$.

(d) We'll show in Chapter 11 that the functions listed in parts (a), (b), and (c) are the only eigenvectors for the operator D^2. Show that the only functions f(x) that satisfy the condition $f(0) = f(\pi) = 0$ and are eigenvectors of D^2 are multiples of the functions $\sin kx$, for k a positive integer. What are the associated eigenvalues? (This question comes up in studying the small vibrations of a string anchored at the points $x = 0$ and $x = \pi$.)

12. (a) Find the eigenvalues and an associated pair of eigenvectors for the linear operator L on $\mathbb{R}^2$ having matrix
$$\begin{pmatrix} 2 & 0 \\ 0 & 3 \end{pmatrix}.$$

  (b) Show that the eigenvectors of the function L in part (a) form a basis for $\mathbb{R}^2$, and use this to give a geometric description of the action of L on $\mathbb{R}^2$, as in Example 2.

  (c) Generalize the results you found for (a) and (b) to a linear operator L from $\mathbb{R}^n$ to $\mathbb{R}^n$ having a diagonal matrix $\operatorname{diag}(a_1, a_2, \ldots, a_n)$.

13. Find the eigenvalues of the operator G on $\mathbb{R}^2$ with matrix $\begin{pmatrix} 1 & 2 \\ 1 & 1 \end{pmatrix}$, show that the associated eigenvectors span $\mathbb{R}^2$, and describe the action of G, as in Example 2.

In Exercises 14 to 17, solve the system of differential equations using eigenvalues and eigenvectors as in Example 5 in the text. The matrices are the same as the ones in Exercises 2 to 5, and you may use the results of those exercises if you have already worked them out.

14. $\dfrac{d\mathbf{x}}{dt} = \begin{pmatrix} 1 & 4 \\ 1 & 1 \end{pmatrix}\mathbf{x}$
15. $\dfrac{d\mathbf{x}}{dt} = \begin{pmatrix} 0 & 4 \\ 1 & 0 \end{pmatrix}\mathbf{x}$

16. $\dfrac{d\mathbf{x}}{dt} = \begin{pmatrix} 0 & 1 \\ 0 & 2 \end{pmatrix}\mathbf{x}$
17. $\dfrac{d\mathbf{x}}{dt} = \begin{pmatrix} 2 & 4 \\ 1 & 2 \end{pmatrix}\mathbf{x}$

6B   Bases of Eigenvectors

In Section 6A we saw that the effect of a linear operator on a linear combination of eigenvectors is particularly simple. Here we'll look at conditions under which a finite-dimensional space $\mathcal{V}$ has a basis of eigenvectors for a given operator $L(\mathbf{x})$ on

$\mathcal{V}$ so we can take advantage of Equation 6.2 and, if $\mathcal{V}$ is finite-dimensional, express L by a diagonal matrix acting on coordinates $u_1, \dots, u_n$ in $\mathbf{x} = u_1\mathbf{u}_1 + \cdots + u_n\mathbf{u}_n$.

6.5 Theorem.   The matrix of an operator L with respect to a basis $\{\mathbf{u}_1, \dots, \mathbf{u}_n\}$ is diagonal if and only if each basis vector $\mathbf{u}_i$ is an eigenvector of L. If the matrix is diagonal, then the ith entry on the diagonal is the eigenvalue associated with $\mathbf{u}_i$.

Proof.   Recall that the matrix of a linear function L with respect to given bases in its domain and range is defined so that its jth column gives the coordinates with respect to the basis in the range of $L(\mathbf{u}_j)$, where $\mathbf{u}_j$ is the jth basis vector of the domain. Here we have an operator L, whose range is the same as its domain, and we can use the basis $\{\mathbf{u}_1, \dots, \mathbf{u}_n\}$ for both domain and range.

If $\mathbf{u}_j$ is an eigenvector, then $L(\mathbf{u}_j) = \lambda_j\mathbf{u}_j$, where λ_j is the associated eigenvalue. To express $\lambda_j\mathbf{u}_j$ as a linear combination $c_1\mathbf{u}_1 + \dots + c_n\mathbf{u}_n$, we take $c_j = \lambda_j$ and all the other c's equal to 0. Thus the coordinate vector of $L(\mathbf{u}_j)$, namely the jth column of the matrix of L, is zero, except for having λ_j in the jth place. The matrix is diagonal if and only if this condition holds for every column. On the other hand, if the entries in the jth column of the matrix are zero except for a value λ_j in the jth place, we have $L(\mathbf{u}_j) = \lambda_j\mathbf{u}_j$ and $\mathbf{u}_j$ is an eigenvector. Thus if the matrix is diagonal, every basis vector $\mathbf{u}_j$ is an eigenvector.   ∎

EXAMPLE 6

In Example 5 of Section 6A we saw that the operator L on $\mathbb{R}^2$ that has the matrix

$$A = \begin{pmatrix} 1 & 1 \\ 4 & 1 \end{pmatrix}$$

with respect to the standard basis has eigenvectors

$$\mathbf{u} = \begin{pmatrix} 1 \\ 2 \end{pmatrix} \qquad \text{with eigenvalue } \lambda_1 = 3$$

and

$$\mathbf{v} = \begin{pmatrix} 1 \\ -2 \end{pmatrix} \qquad \text{with eigenvalue } \lambda_2 = -1.$$

According to Theorem 6.5, the matrix of L with respect to the basis $\{\mathbf{u}_1, \mathbf{u}_2\}$ is the diagonal matrix

$$\begin{pmatrix} 3 & 0 \\ 0 & -1 \end{pmatrix}.$$

Let us check by computing $L(\mathbf{x})$, where $\mathbf{x} = \mathbf{u}_1 + 2\mathbf{u}_2$, by using each matrix. The coordinate vector of $\mathbf{x}$ in the basis $\{\mathbf{u}_1, \mathbf{u}_2\}$ is $\begin{pmatrix} 1 \\ 2 \end{pmatrix}$ and $\begin{pmatrix} 3 & 0 \\ 0 & -1 \end{pmatrix}\begin{pmatrix} 1 \\ 2 \end{pmatrix} = \begin{pmatrix} 3 \\ -2 \end{pmatrix}$ is the coordinate vector of $L(\mathbf{x})$ in the same basis. That means that

$$L(\mathbf{x}) = 3\mathbf{u}_1 - 2\mathbf{u}_2$$

$$= 3\begin{pmatrix} 1 \\ 2 \end{pmatrix} - 2\begin{pmatrix} 1 \\ -2 \end{pmatrix} = \begin{pmatrix} 1 \\ 10 \end{pmatrix}.$$

On the other hand, $\mathbf{x} = \begin{pmatrix} 1 \\ 2 \end{pmatrix} + 2 \begin{pmatrix} 1 \\ -2 \end{pmatrix} = \begin{pmatrix} 3 \\ -2 \end{pmatrix}$ in the standard basis, and

$$L(\mathbf{x}) = \begin{pmatrix} 1 & 1 \\ 4 & 1 \end{pmatrix} \begin{pmatrix} 3 \\ -2 \end{pmatrix} = \begin{pmatrix} 1 \\ 10 \end{pmatrix}$$

by direct calculation.

The next theorem shows that eigenvectors associated with different eigenvalues are linearly independent. We apply it to obtain a condition under which an operator is guaranteed to have a basis of eigenvectors.

6.6 Theorem.   Let $\mathbf{u}_1, \ldots, \mathbf{u}_k$ be eigenvectors of a linear operator L, associated with the respective eigenvalues $\lambda_1, \ldots, \lambda_k$. If $\lambda_1, \ldots, \lambda_k$ are all different from each other, then $\mathbf{u}_1, \ldots, \mathbf{u}_k$ are linearly independent.

Proof.   We proceed by induction on k, the number of eigenvectors in the set. If $k = 1$, then $\mathbf{u}_1$ forms an independent set, since $\mathbf{u}_1 \neq \mathbf{0}$. Now suppose that every set of k eigenvectors of L associated with k different eigenvalues is independent. We'll show that $\mathbf{u}_1, \ldots, \mathbf{u}_{k+1}$ is an independent set. Let constants $c_1, \ldots, c_{k+1}$ be chosen so that

$$c_1 \mathbf{u}_1 + \cdots + c_{k+1} \mathbf{u}_{k+1} = \mathbf{0}.$$

Apply L to both sides of this equation to get

$$c_1 \lambda_1 \mathbf{u}_1 + \cdots + c_{k+1} \lambda_{k+1} \mathbf{u}_{k+1} = \mathbf{0},$$

where we have used $L(\mathbf{u}_j) = \lambda_j \mathbf{u}_j$. Now multiply the previous equation by λ_1, and subtract from this equation to get

$$c_2 (\lambda_2 - \lambda_1) \mathbf{u}_2 + \cdots + c_{k+1} (\lambda_{k+1} - \lambda_1) \mathbf{u}_{k+1} = \mathbf{0}.$$

The k vectors $\mathbf{u}_2, \ldots, \mathbf{u}_{k+1}$ are independent by assumption, so $c_j (\lambda_j - \lambda_1) = 0$ for $j = 2, \ldots, k+1$. Since $\lambda_j - \lambda_1 \neq 0$, we conclude that $c_j = 0$ for $j = 2, \ldots, k+1$. Then the first equation implies that $c_1 \mathbf{u}_1 = \mathbf{0}$, so $c_1 = 0$ also. Hence $\mathbf{u}_1, \ldots, \mathbf{u}_{k+1}$ is an independent set.   ∎

EXAMPLE 7   Theorem 6.6 is not restricted to operators on finite-dimensional spaces. In particular, it applies to the differentiation operator D on the infinite-dimensional space $C^{(1)}$ of continuously differentiable functions. The function e^{rx} is an eigenvector for D associated with the eigenvalue r, because $De^{rx} = re^{rx}$; this then implies that the functions $e^{r_1 x}, \ldots, e^{r_k x}$ are linearly independent, provided that the numbers $r_1, \ldots, r_k$ are all different.

If A is an n-by-n matrix and λ_k is a *real* root of the characteristic equation $\det(A - \lambda I) = 0$, then $A - \lambda_k I$ represents an operator on $\mathbb{R}^n$. Since $\det(A - \lambda_k I) = 0$, the operator $A - \lambda_k I$ is not invertible, and there is some nonzero vector $\mathbf{u}$ such that $(A - \lambda_k I)\mathbf{u} = \mathbf{0}$. Hence

$$A\mathbf{u} = \lambda_k \mathbf{u},$$

and λ_k is an eigenvalue. If we consider a matrix A as representing an operator on the set $\mathbb{C}^n$ of n-tuples of complex numbers, then $A - \lambda_k I$ represents an operator on $\mathbb{C}^n$ for an arbitrary real or *complex* value of λ_k. Thus a characteristic root of A is an eigenvalue of A considered as an operator on $\mathbb{C}^n$. For the reasons just mentioned, the next theorem has a different statement for the real and complex cases.

6.7   Theorem.   If L is a linear operator on $\mathbb{R}^n$, with matrix A, and its characteristic equation $\det(A - \lambda I) = 0$ has n distinct *real* roots, then $\mathbb{R}^n$ has a basis consisting of eigenvectors of L. If L is a linear operator on $\mathbb{C}^n$ and its characteristic equation has n distinct roots, then $\mathbb{C}^n$ has a basis consisting of eigenvectors of L.

Proof.   Let $\mathbf{v}_1, \ldots, \mathbf{v}_n$ be eigenvectors associated with the distinct roots $\lambda_1, \ldots, \lambda_n$. By Theorem 6.6, they are linearly independent. But n linear independent vectors in an n-dimensional space form a basis, by Theorem 5.9.   ∎

EXAMPLE 8   The function L of Example 1 illustrates Theorem 6.7. It is an operator on a 2-dimensional space, and its characteristic equation has the two roots 3 and -1. The eigenvectors $(1, 2)$ and $(1, -2)$ are independent and so form a basis.

EXAMPLE 9   An operator may have a basis of eigenvectors without having a full set of distinct eigenvalues. A simple example is the function from $\mathbb{R}^3$ to $\mathbb{R}^3$ given by the matrix

$$\begin{pmatrix} 2 & 0 & 0 \\ 0 & 2 & 0 \\ 0 & 0 & 3 \end{pmatrix}.$$

The characteristic equation is $(2-\lambda)^2(3-\lambda)$, and the only roots are 2 and 3. We still have a basis of eigenvectors because in this case there are two linearly independent eigenvectors, $\mathbf{e}_1$ and $\mathbf{e}_2$, associated with the eigenvalue 2.

EXAMPLE 10   This example shows one way that a matrix can fail to have enough eigenvectors to form a basis. Consider the linear function from $\mathbb{R}^2$ to $\mathbb{R}^2$ with the matrix

$$A = \begin{pmatrix} 0 & -5 \\ 2 & 2 \end{pmatrix}.$$

To find eigenvalues, we must solve the equation

$$\det(A - \lambda I) = \det\begin{pmatrix} -\lambda & -5 \\ 2 & 2 - \lambda \end{pmatrix} = (-\lambda)(2 - \lambda) + 10$$

$$= \lambda^2 - 2\lambda + 10 = 0.$$

The formula for solving quadratic equations gives the complex roots $1 + 3i$ and $1 - 3i$. Therefore, the linear function from $\mathbb{R}^2$ to $\mathbb{R}^2$ has no real eigenvalues and consequently no eigenvectors in $\mathbb{R}^2$ if we use only real numbers for scalars.

EXAMPLE 11   If we use complex scalars and consider the matrix A of the previous example A as defining a linear function from the complex 2-dimensional coordinate space $\mathbb{C}^2$ into itself, we can use the eigenvalues $1 \pm 3i$. To find the eigenvectors, we proceed just as we have done before with real eigenvalues. For $\lambda = 1 + 3i$, the equation $(A - \lambda I)\mathbf{x} = \mathbf{0}$ becomes

$$\begin{pmatrix} -1 - 3i & -5 \\ 2 & 1 - 3i \end{pmatrix} \begin{pmatrix} x \\ y \end{pmatrix} = 0.$$

Dividing the first row by $-1 - 3i$ gives

$$\begin{pmatrix} 1 & \frac{1}{2} - \frac{3}{2}i \\ 2 & 1 - 3i \end{pmatrix} \begin{pmatrix} x \\ y \end{pmatrix} = 0.$$

Subtracting twice the first row from the second leaves

$$\begin{pmatrix} 1 & \frac{1}{2} - \frac{3}{2}i \\ 0 & 0 \end{pmatrix} \begin{pmatrix} x \\ y \end{pmatrix} = 0,$$

and we see that $\begin{pmatrix} x \\ y \end{pmatrix} = \begin{pmatrix} 1 - 3i \\ -2 \end{pmatrix}$ is one solution. For $\lambda = 1 - 3i$, a similar calculation leads to $\begin{pmatrix} 1 + 3i \\ -2 \end{pmatrix}$ as an associated eigenvector. Thus viewed as an operator on a complex vector space, A does have a basis of eigenvectors—namely, $\begin{pmatrix} 1 - 3i \\ -2 \end{pmatrix}, \begin{pmatrix} 1 + 3i \\ -2 \end{pmatrix}$, associated with the eigenvalues $1 + 3i, 1 - 3i$.

The elementary arithmetic operations of addition, subtraction, multiplication, and division are governed by the same rules in both the real and complex numbers, so the theorems about vector spaces and linear functions in general that we've proved so far are valid whichever set of scalars we assume. But in the two previous examples the choice between real and complex scalars made a genuine difference, the reason being that to find eigenvalues we solve polynomial equations, so the results depend on the number system we allow.

The characteristic equation $\det(A - \lambda I) = 0$ for an n-by-n matrix A is a polynomial equation of degree n. There are two ways in which a polynomial may fail to have n distinct real roots so that $\mathbb{R}^n$ may fail to have a basis of eigenvectors of A. One way is for the polynomial to have complex roots as shown in Example 10. Another is for the polynomial to have one or more multiple roots, and Example 9 showed that in this case a basis of eigenvectors may still exist. The next example shows that a basis of eigenvectors may fail to exist if there multiple roots.

EXAMPLE 12   For the operator on $\mathbb{R}^2$ with matrix

$$A = \begin{pmatrix} -1 & 2 \\ -2 & 3 \end{pmatrix}$$

the characteristic equation $\det(A - \lambda I) = 0$ is

$$(-1 - \lambda)(3 - \lambda) - (-2)(2) = \lambda^2 - 2\lambda + 1$$
$$= (\lambda - 1)^2 = 0.$$

The only root is $\lambda = 1$. To find the associated eigenvectors we solve

$$(A - I)\mathbf{x} = \begin{pmatrix} -2 & 2 \\ -2 & 2 \end{pmatrix} \mathbf{x} = \mathbf{0}.$$

The solution set is the 1-dimensional subspace of $\mathbb{R}^2$ consisting of all multiples of the vector $\begin{pmatrix} 1 \\ 1 \end{pmatrix}$. Two eigenvectors associated with the eigenvalue 1 are linearly dependent, so can't form a basis. The result is exactly the same if we consider A as the matrix of an operator on the complex space $\mathbb{C}^2$; we still fail to have a basis of eigenvectors.

It's possible to make up the deficiency in the previous example by generalizing the definition of eigenvector, but in our application to differential equations in Chapter 13 we avoid the need for this by using exponential matrices e^{tA} defined for arbitrary square matrices A.

6C   Changing Coordinates

Theorem 5.4 of Section 6 showed us how to find the matrix representation for a linear function $f: V \longrightarrow W$ relative to bases $\{\mathbf{v}_1, \ldots, \mathbf{v}_n\}$ in V and $\{\mathbf{w}_1, \ldots, \mathbf{w}_m\}$ in W. Here we assume we have a basis different from the standard basis in $\mathbb{R}^n$ and show how the matrix of a linear operator $F: \mathbb{R}^n \longrightarrow \mathbb{R}^n$ relative to the standard basis is related to the matrix of the same operator relative to the nonstandard basis. We'll then use this result to show how using a basis of eigenvectors may simplify the matrix of an operator.

6.8   Theorem.   Let $\mathbf{x} = (x_1, \ldots, x_n)$ be the standard coordinates of a vector in $\mathbb{R}^n$, and let $\mathbf{y} = (y_1, \ldots, y_n)$ be the coordinates of the same vector relative to a basis $\{\mathbf{u}_1, \ldots, \mathbf{u}_n\}$ for $\mathbb{R}^n$. Then $\mathbf{x} = U\mathbf{y}$, where U is the n-by-n matrix with $\mathbf{u}_j$ for its jth column. If A is the n-by-n matrix of a linear operator $F: \mathbb{R}^n \longrightarrow \mathbb{R}^n$, then the matrix B of F relative to the basis $\{\mathbf{u}_1, \ldots, \mathbf{u}_n\}$ for $\mathbb{R}^n$ is related to A by

$$A = UBU^{-1}, \quad \text{or} \quad B = U^{-1}AU.$$

Proof.   First observe that $\mathbf{x} = y_1\mathbf{u}_1 + \cdots + y_n\mathbf{u}_n = U\mathbf{y}$. Since the columns of U are basis vectors, they're independent, so U^{-1} exists and $\mathbf{y} = U^{-1}\mathbf{x}$. Then $A\mathbf{x} = UB\mathbf{y} = UBU^{-1}\mathbf{x}$ for all $\mathbf{x}$, in particular when $\mathbf{x} = \mathbf{e}_k$. Hence $A = UBU^{-1}$.   ∎

EXAMPLE 13   Suppose we define an operator $F: \mathbb{R}^2 \longrightarrow \mathbb{R}^2$ by $F(\mathbf{x}) = A\mathbf{x} = \begin{pmatrix} 1 & 1 \\ 4 & 1 \end{pmatrix}\begin{pmatrix} x \\ y \end{pmatrix}$.

Using instead the nonstandard basis $\left\{ \begin{pmatrix} 1 \\ 1 \end{pmatrix}, \begin{pmatrix} -1 \\ 1 \end{pmatrix} \right\}$, Theorem 6.8 tells us that the matrix becomes

$$B = U^{-1}AU = \begin{pmatrix} \frac{1}{2} & \frac{1}{2} \\ -\frac{1}{2} & \frac{1}{2} \end{pmatrix}\begin{pmatrix} 1 & 1 \\ 4 & 1 \end{pmatrix}\begin{pmatrix} 1 & -1 \\ 1 & 1 \end{pmatrix} = \begin{pmatrix} -\frac{3}{2} & -\frac{3}{2} \\ \frac{3}{2} & \frac{7}{2} \end{pmatrix}.$$

EXAMPLE 14   If $A = \begin{pmatrix} 1 & 1 \\ 4 & 1 \end{pmatrix}$, then Example 1 shows that $\mathbb{R}^2$ has a basis of eigenvectors

$\left\{ \begin{pmatrix} 1 \\ 2 \end{pmatrix}, \begin{pmatrix} 1 \\ -2 \end{pmatrix} \right\}$, with associated eigenvalues $\lambda_1 = 3, \lambda_2 = -1$. Then $\Lambda = $

$\text{diag}(3, -1)$, and we can check directly that $A = U\Lambda U^{-1}$, where $U = \begin{pmatrix} 1 & 1 \\ 2 & -2 \end{pmatrix}$,

that is $\begin{pmatrix} 1 & 1 \\ 4 & 1 \end{pmatrix} = \begin{pmatrix} 1 & 1 \\ 2 & -2 \end{pmatrix} \begin{pmatrix} 3 & 0 \\ 0 & -1 \end{pmatrix} \begin{pmatrix} \frac{1}{2} & \frac{1}{4} \\ \frac{1}{2} & -\frac{1}{4} \end{pmatrix}$. In the previous example

there's no apparent advantage to using the nonstandard basis, but using the eigenvector basis for the same operator allows us to simplify by operating with a diagonal matrix. This change will be useful in Chapter 13.

EXERCISES

The matrices 1 to 6 are operators on $\mathbb{R}^2$ or $\mathbb{R}^3$, but may also be regarded as operators on $\mathbb{C}^2$ or $\mathbb{C}^3$. Find their eigenvalues and state whether or not Theorem 6.7 guarantees a basis of eigenvectors in $\mathbb{R}^2$ or $\mathbb{R}^3$, in $\mathbb{C}^2$ or $\mathbb{C}^3$. For the matrices for which Theorem 6.7 does not guarantee a basis of eigenvectors, find out whether a basis of eigenvectors exists.

1. $\begin{pmatrix} 0 & 1 \\ 1 & 1 \end{pmatrix}$ 2. $\begin{pmatrix} 1 & 0 \\ 1 & 1 \end{pmatrix}$

3. $\begin{pmatrix} 0 & 1 \\ -1 & 0 \end{pmatrix}$ 4. $\begin{pmatrix} 0 & 1 & 0 \\ 0 & 0 & 1 \\ 1 & 0 & 0 \end{pmatrix}$

5. $\begin{pmatrix} 3 & -2 & -2 \\ -2 & -2 & 1 \\ 2 & 2 & -2 \end{pmatrix}$ 6. $\begin{pmatrix} 1 & 0 & 1 \\ 0 & 1 & 0 \\ 0 & 0 & 1 \end{pmatrix}$

7. A rotation about the origin in $\mathbb{R}^2$ through angle θ has matrix

$$R_\theta = \begin{pmatrix} \cos\theta & -\sin\theta \\ \sin\theta & \cos\theta \end{pmatrix}.$$

    (a)   Show that the only values of θ with $0 \leq \theta < 2\pi$ for which R_θ has real eigenvalues are $\theta = 0$ and $\theta = \pi$. What are the eigenvectors when there are real eigenvalues?

    (b)   Explain the results of (a) in geometric terms.

8. Show that the matrix R_θ of Exercise 7 has complex eigenvalues $\cos\theta \pm i\sin\theta$. Also find associated independent complex eigenvectors.

9. Show that if $A = \begin{pmatrix} 0 & 2 \\ -1 & 3 \end{pmatrix}$, then $A^{10} = \begin{pmatrix} -1022 & 2046 \\ -1023 & 2047 \end{pmatrix}$. [*Hint:* Write $A = U\Lambda U^{-1}$, where Λ is diagonal.]

In each of Exercises 15 to 18, the given n-by-n matrix defines an operator L relative to the natural basis in $\mathbb{R}^n$ that has n independent real eigenvectors. Find a basis for $\mathbb{R}^n$ consisting of eigenvectors of L, and find the diagonal matrix that represents L relative to that basis.

10. $\begin{pmatrix} -1 & 3 \\ -2 & 4 \end{pmatrix}$ 11. $\begin{pmatrix} 4 & -3 \\ 2 & -1 \end{pmatrix}$

12. $\begin{pmatrix} 3 & -1 & 0 \\ -1 & 2 & -1 \\ 0 & -1 & 3 \end{pmatrix}$ 13. $\begin{pmatrix} -1 & 0 & 0 \\ -1 & 0 & 0 \\ -1 & -1 & 1 \end{pmatrix}$

In each of Exercises 19 to 22, find a square matrix U and a diagonal matrix Λ such that the given matrix $A = U\Lambda U^{-1}$. In Exercise 17 you need to use complex numbers.

14. $A = \begin{pmatrix} 3 & -1 \\ 0 & 2 \end{pmatrix}$ 15. $A = \begin{pmatrix} 0 & 0 \\ -4 & 2 \end{pmatrix}$

16. $A = \begin{pmatrix} 1 & 1 & -1 \\ 0 & 2 & -1 \\ 0 & 0 & 1 \end{pmatrix}$ 17. $A = \begin{pmatrix} 0 & 1 \\ -4 & 2 \end{pmatrix}$

18. (a)   Let $\mathbf{u}_1, \ldots, \mathbf{u}_n$ be independent vectors in $\mathbb{R}^n$, and let U be the n-by-n matrix with kth column $\mathbf{u}_k$. Let $\lambda_1, \ldots, \lambda_n$ be real numbers, and let Λ be the diagonal matrix with kth diagonal entry λ_k. Show

that the matrix $A = U\Lambda U^{-1}$ satisfies $A\mathbf{u}_k = \lambda_k \mathbf{u}_k$ for $k = 1, \ldots, n$.

(b) Find a 2-by-2 matrix with eigenvalues $\lambda_1 = 2, \lambda_2 = 3$ and with associated eigenvectors $\mathbf{u}_1 = (1, 2), \mathbf{u}_2 = (2, 5)$.

In Exercises 19 to 22, show that the image of the given subspace of $C^{(1)}(-\infty, \infty)$ under the differentiation operator $D = d/dx$ is contained in the same subspace, so D can be considered as an operator restricted to having the subspace for both domain and range. Find the matrix A of D with respect to the given basis and find the eigenvalues of A. Then find a basis of eigenvectors for the given subspace, using complex coefficients if necessary.

19. The subspace with basis e^x and e^{-x}

20. The subspace with basis 1, x, and x^2

21. The subspace with basis $\sin x$ and $\cos x$

22. The subspace with basis $e^x \cos x$ and $e^x \sin x$

SECTION 7   INNER PRODUCTS

In Chapter 1 we used the dot product in $\mathbb{R}^n$ to introduce length and angle. Here we discuss a more widely applicable generalization called an *inner product*, defined as a function having the same formal properties as the dot product, but defined on a more general class of vector spaces than the spaces $\mathbb{R}^n$. We then define length, angle, and orthogonality in terms of inner products. Though these definitions no longer coincide precisely with the usual geometric concepts, the analogies with intuitive geometry are still useful.

7A   General Properties

Recall that the dot product $\mathbf{x} \cdot \mathbf{y}$ of vectors $\mathbf{x} = (x_1, \ldots, x_n)$ and $\mathbf{y} = (y_1, \ldots, y_n)$ in $\mathbb{R}^n$ is defined by $\mathbf{x} \cdot \mathbf{y} = x_1 y_1 + \cdots + x_n y_n$, and has the properties

Positivity:	$\mathbf{x} \cdot \mathbf{x} > 0$, except that $\mathbf{0} \cdot \mathbf{0} = 0$
Symmetry:	$\mathbf{x} \cdot \mathbf{y} = \mathbf{y} \cdot \mathbf{x}$
Additivity:	$(\mathbf{x} + \mathbf{y}) \cdot \mathbf{z} = \mathbf{x} \cdot \mathbf{z} + \mathbf{y} \cdot \mathbf{z}$
Homogeneity:	$(r\mathbf{x}) \cdot \mathbf{y} = r(\mathbf{x} \cdot \mathbf{y})$

For elements of a vector space we define an **inner product** $\langle \mathbf{x}, \mathbf{y} \rangle$ to be a real-valued function having the same properties:

7.1

Positivity:	$\langle \mathbf{x}, \mathbf{x} \rangle > 0$, except that $\langle \mathbf{0}, \mathbf{0} \rangle = 0$
Symmetry:	$\langle \mathbf{x}, \mathbf{y} \rangle = \langle \mathbf{y}, \mathbf{x} \rangle$
Additivity:	$\langle \mathbf{x} + \mathbf{y}, \mathbf{z} \rangle = \langle \mathbf{x}, \mathbf{z} \rangle + \langle \mathbf{y}, \mathbf{z} \rangle$
Homogeneity:	$\langle r\mathbf{x}, \mathbf{y} \rangle = r \langle \mathbf{x}, \mathbf{y} \rangle$

The dot product on $\mathbb{R}^n$ has Properties 7.1 and so is an example of an inner product. We need a notation different from $\mathbf{x} \cdot \mathbf{y}$ for a general inner product both to make it clear which product we're talking about, and because we sometimes use both together, as in Section 7C.

In Definition 7.1 we assumed homogeneity only in the first entry, but using symmetry twice we get $\langle \mathbf{x}, r\mathbf{y} \rangle = \langle r\mathbf{y}, \mathbf{x} \rangle = r\langle \mathbf{y}, \mathbf{x} \rangle = r\langle \mathbf{x}, \mathbf{y} \rangle$. Hence $\langle \mathbf{x}, r\mathbf{y} \rangle$ also equals $r\langle \mathbf{x}, \mathbf{y} \rangle$. Similarly, we assumed additivity only in the first entry, but we also have it in the second entry also: $\langle \mathbf{x}, \mathbf{y} + \mathbf{z} \rangle = \langle \mathbf{x}, \mathbf{y} \rangle + \langle \mathbf{x}, \mathbf{z} \rangle$ as a consequence of additivity in the first entry and symmetry.

We define the **length**, or **norm**, of a vector by

$$\|\mathbf{x}\| = \sqrt{\langle \mathbf{x}, \mathbf{x} \rangle}.$$

The dot product and length in $\mathbb{R}^n$ satisfy the Cauchy–Schwarz inequality

$$|\mathbf{x} \cdot \mathbf{y}| \leq |\mathbf{x}||\mathbf{y}| \,.$$

To aid geometric intuition we used a proof of the inequality in Chapter 1, Section 4 that depended on the law of cosines. Here we give a different proof that uses only the Properties 7.1 and so works for other inner products as well. As a result the law of cosines holds for general inner products as well. (See Exercise 12.)

7.2   Cauchy–Schwarz Inequality. $|\langle \mathbf{x}, \mathbf{y} \rangle| \leq \|\mathbf{x}\|\|\mathbf{y}\|.$

Proof.   If either vector is **0**, the inequality is satisfied because both sides are 0, so for the rest of the proof we assume that neither vector is **0**. Suppose first that **x** and **y** are unit vectors, that is, that $\|\mathbf{x}\| = \|\mathbf{y}\| = 1$. Then

$$0 \leq \|\mathbf{x} - \mathbf{y}\|^2 = \langle \mathbf{x} - \mathbf{y}, \mathbf{x} - \mathbf{y} \rangle$$
$$= \|\mathbf{x}\|^2 - 2\langle \mathbf{x}, \mathbf{y} \rangle + \|\mathbf{y}\|^2 = 2 - 2\langle \mathbf{x}, \mathbf{y} \rangle,$$

or $0 \leq 2 - 2\langle \mathbf{x}, \mathbf{y} \rangle$, so $\langle \mathbf{x}, \mathbf{y} \rangle \leq 1$. For nonzero **x** and **y**, $\mathbf{x}/\|\mathbf{x}\|$ and $\mathbf{y}/\|\mathbf{y}\|$ are unit vectors, so $\left\langle \dfrac{\mathbf{x}}{\|\mathbf{x}\|}, \dfrac{\mathbf{y}}{\|\mathbf{y}\|} \right\rangle \leq 1$ and multiplying by $\|\mathbf{x}\|\|\mathbf{y}\|$ and using the homogeneity property gives $\langle \mathbf{x}, \mathbf{y} \rangle \leq \|\mathbf{x}\|\|\mathbf{y}\|$. Now replace **x** by $-\mathbf{x}$ to get $-\langle \mathbf{x}, \mathbf{y} \rangle \leq \|\mathbf{x}\|\|\mathbf{y}\|$. The last two inequalities imply the inequality with the absolute value in it.   ∎

When **x** and **y** are not zero we can write the Cauchy–Schwarz inequality as $\dfrac{|\langle \mathbf{x}, \mathbf{y} \rangle|}{\|\mathbf{x}\|\|\mathbf{y}\|} \leq 1$, so there is an angle θ between 0 and π such that $\cos\theta = \dfrac{\langle \mathbf{x}, \mathbf{y} \rangle}{\|\mathbf{x}\|\|\mathbf{y}\|}$. We now define the angle between **x** and **y** to be this angle θ, and the definition will always make sense when the vectors are nonzero. As we did for the dot product in Chapter 1, we say that **x** and **y** are **orthogonal** if $\langle \mathbf{x}, \mathbf{y} \rangle = 0$, allowing for the possibility that **x** or **y** is zero.

Here are the characteristic properties of length as defined by $\|\mathbf{x}\|$.

	Positivity :	$\|\mathbf{x}\| > 0,$ except that $\|\mathbf{0}\| = 0$
7.3	**Homogeneity** :	$\|r\mathbf{x}\| = |r|\|\mathbf{x}\|, r$ real
	Triangle Inequality :	$\|\mathbf{x} + \mathbf{y}\| \leq \|\mathbf{x}\| + \|\mathbf{y}\|$

The proofs are the same as the ones given for the properties 4.9 in Chapter 1, Section 4, with $\langle \mathbf{x}, \mathbf{y} \rangle$ replacing $\mathbf{x} \cdot \mathbf{y}$ and $\|\mathbf{x}\|$ replacing $|\mathbf{x}|$, and we won't repeat them here.

EXAMPLE 1   We've already seen that the dot product $\mathbf{x} \cdot \mathbf{y}$ in $\mathbb{R}^n$ is an example of an inner product. In particular, if $\mathbf{x} = (x_1, x_2)$ and $\mathbf{y} = (y_1, y_2)$,

$$\mathbf{x} \cdot \mathbf{y} = x_1 y_1 + x_2 y_2$$

is an inner product in $\mathbb{R}^2$. If we define

$$\langle \mathbf{x}, \mathbf{y} \rangle = x_1 y_1 + 2x_2 y_2$$

FIGURE 3.11

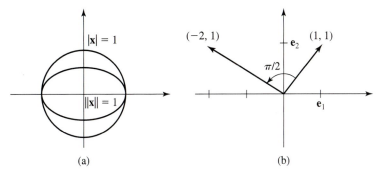

(a)   (b)

instead, we get another inner product for **x** and **y** in $\mathbb{R}^2$. To see this, all we have to do is check that the relations given under 7.1 are satisfied, which we leave as an exercise. Relative to this inner product, length is defined by

$$\|\mathbf{x}\| = \sqrt{x_1^2 + 2x_2^2}.$$

Thus the "unit circle" defined by $\|\mathbf{x}\| = 1$ consists of all points (x_1, x_2) in $\mathbb{R}^2$ that satisfy $x_1^2 + 2x_2^2 = 1$; this ellipse appears in Figure 3.11(a) along with the Euclidean unit circle $|\mathbf{x}| = 1$, or $x_1^2 + x_2^2 = 1$. This definition of length adds more weight to the second coordinate, so to get a really circular unit circle, $|x_2|$ has to be smaller.

We use the inner product $\langle \mathbf{x}, \mathbf{y} \rangle$ to define the angle θ between two vectors as above by the formula

$$\cos \theta = \frac{\langle \mathbf{x}, \mathbf{y} \rangle}{\|\mathbf{x}\| \|\mathbf{y}\|}.$$

Then **x** and **y** are orthogonal or perpendicular when $\cos \theta = \langle \mathbf{x}, \mathbf{y} \rangle = x_1 y_1 + 2x_2 y_2 = 0$. Thus the vectors $\mathbf{e}_1 = (1, 0)$ and $\mathbf{e}_2 = (0, 1)$ are still orthogonal, just as they were relative to the dot product, but the vectors $(1, 1)$ and $(-1, 1)$ are not. Also the vectors $(1, 1)$ and $(-2, 1)$, shown in Figure 3.11(b), are orthogonal relative to this inner product, but not relative to the dot product.

EXAMPLE 2 Let $C[-\pi, \pi]$ stand for the vector space of continuous real-valued functions $f(x)$ defined for $-\pi \le x \le \pi$. The space $C[-\pi, \pi]$ is infinite-dimensional because it contains the infinite linearly independent set $\{1, x, x^2, \dots\}$. We define a useful inner product on $C[-\pi, \pi]$ by the integral

$$\langle f, g \rangle = \int_{-\pi}^{\pi} f(x)g(x)\, dx.$$

We have $\langle f, g \rangle = \langle g, f \rangle$ simply because $f(x)g(x) = g(x)f(x)$ for $-\pi \le x \le \pi$. The other properties of the inner product depend on properties of definite integrals; the verification is left as Exercise 11. The importance of this example depends partly on the formulas

$$\langle \cos kx, \cos lx \rangle = \int_{-\pi}^{\pi} \cos kx \cos lx\, dx = 0, \quad k \ne l,$$

$$\langle \sin kx, \sin lx \rangle = \int_{-\pi}^{\pi} \sin kx \sin lx\, dx = 0, \quad k \ne l,$$

$$\langle \cos kx, \sin lx \rangle = \int_{-\pi}^{\pi} \cos kx \sin lx\, dx = 0,$$

where k and l are integers. These formulas follow in a straightforward way using trigonometric identities; their significance here is that in terms of the inner product $\langle f, g \rangle$, and the orthogonality relation $\langle f, g \rangle = 0$, they assert that certain trigonometric functions are orthogonal. We're not claiming that the graphs of $\cos kx$ and $\sin lx$ intersect at right angles, but rather that their ordinary product has average value zero over the interval $-\pi \le x \le \pi$. If $k = l$ direct computation shows that $\langle \cos kx, \cos kx \rangle$ and $\langle \sin kx, \sin kx \rangle$ both equal π for $k \ge 1$. (If $k = 0$, $\cos kx = 1$ and $\sin kx = 0$ so the integrals are 2π and 0 instead.) We'll return to this example in the next section.

EXERCISES

In Exercises 1 to 4, determine whether the given formula defines an inner product on $\mathbb{R}^2$. Verify your answer by showing either that the Properties 7.1 on page 155 are satisfied or that at least one of them fails. Here $\mathbf{x} = (x_1, x_2)$ and $\mathbf{y} = (y_1, y_2)$.

1. $\langle \mathbf{x}, \mathbf{y} \rangle = x_1 y_1 + 2 x_2 y_2$

2. $\langle \mathbf{x}, \mathbf{y} \rangle = x_1 y_1 - x_2 y_2$

3. $\langle \mathbf{x}, \mathbf{y} \rangle = x_1 y_1 + x_1 y_2 + x_2 y_1 + 2 x_2 y_2$

4. $\langle \mathbf{x}, \mathbf{y} \rangle = x_1 y_1$

5. Sketch the "unit circle" determined by $\|\mathbf{x}\| = 1$, if the norm is determined by the inner product $\langle \mathbf{x}, \mathbf{y} \rangle = 3 x_1 y_1 + 2 x_2 y_2$, where $\mathbf{x} = (x_1, x_2)$, $\mathbf{y} = (y_1, y_2)$.

6. (a) Let $\mathcal{V}$ be a finite-dimensional vector space with basis $\{\mathbf{v}_1, \ldots, \mathbf{v}_n\}$. Let

$$\mathbf{x} = x_1 \mathbf{v}_1 + \cdots + x_n \mathbf{v}_n$$

$$\mathbf{y} = y_1 \mathbf{v}_1 + \cdots + y_n \mathbf{v}_n$$

be representations of $\mathbf{x}$ and $\mathbf{y}$ in the given basis. Show that

$$\langle \mathbf{x}, \mathbf{y} \rangle = (x_1, \ldots, x_n) \bullet (y_1, \ldots, y_n)$$

defines an inner product on $\mathcal{V}$.

(b) Show that with the inner product as defined in part (a), the basis elements $\mathbf{v}_1, \ldots, \mathbf{v}_n$ satisfy $\langle \mathbf{v}_i, \mathbf{v}_j \rangle = 0$ if $i \neq j$ and $\langle \mathbf{v}_i, \mathbf{v}_j \rangle = 1$.

7. Verify the orthogonality relations in text Example 2 by using trigonometric identities.

8. Suppose an inner product defined on $\mathbb{R}^2$ has the values $\langle (-1, 2), (-1, 2) \rangle = 2$, $\langle (2, -5), (2, -5) \rangle = 3$, and $\langle (-1, 2), (2, -5) \rangle = 5$. Calculate the lengths $\|\mathbf{e}_1\|$ and $\|\mathbf{e}_2\|$ of the standard basis vectors for the length function associated with the given inner product. [*Hint*: Express $\mathbf{e}_1$ and $\mathbf{e}_2$ in terms of the vectors $(-1, 2)$ and $(2, -5)$, and use homogeneity and additivity of the inner product.]

9. Show that there is no inner product on $\mathbb{R}^3$ such that $\langle \mathbf{e}_1, \mathbf{e}_1 \rangle = \langle \mathbf{e}_2, \mathbf{e}_2 \rangle = 1$, $\langle \mathbf{e}_3, \mathbf{e}_3 \rangle = 5$, $\langle \mathbf{e}_1, \mathbf{e}_2 \rangle = 0$, and $\langle \mathbf{e}_1, \mathbf{e}_3 \rangle = \langle \mathbf{e}_2, \mathbf{e}_3 \rangle = 2$. [*Hint*: Show that if homogeneity and additivity hold, then $\langle (2, 2, -1), (2, 2, -1) \rangle$ is negative, so positivity fails.]

10. Let $\mathcal{V}$ be a 2-dimensional vector space with an inner product and a basis $\{\mathbf{u}, \mathbf{v}\}$, and let $\langle \mathbf{u}, \mathbf{u} \rangle = a$, $\langle \mathbf{u}, \mathbf{v} \rangle = b$, and $\langle \mathbf{v}, \mathbf{v} \rangle = c$.

(a) Let $\mathbf{x} = p\mathbf{u} + q\mathbf{v}$ and $\mathbf{y} = r\mathbf{u} + s\mathbf{v}$ be vectors in $\mathcal{V}$. Use additivity and homogeneity of the inner product to show that

$$\langle \mathbf{x}, \mathbf{y} \rangle = (p \quad q) \begin{pmatrix} a & b \\ b & c \end{pmatrix} \begin{pmatrix} r \\ s \end{pmatrix}.$$

(b) Show that $a > 0$ and $c > 0$, and that the Cauchy–Schwarz inequality implies that $b^2 < ac$.

(c) Show that if a, b, and c satisfy the conditions of part(b) and $\langle \mathbf{x}, \mathbf{y} \rangle$ is defined by the formula in part(a) then $\langle \mathbf{x}, \mathbf{y} \rangle$ satisfies the conditions for being an inner product. [*Hint*: To show positivity, write out $\langle \mathbf{x}, \mathbf{x} \rangle$ in terms of a, b, c, p, and q and use the technique of completing the square.]

11. (a) Verify that the formula $\int_{-\pi}^{\pi} f(x) g(x)\,dx$ defines an inner product on the space $C[-\pi, \pi]$ of real-valued continuous functions on $[-\pi, \pi]$. To show that $\langle f, f \rangle > 0$ unless $f = 0$ you may assume that if $f(x) \ge 0$ is continuous but not identically zero on an interval $a \le x \le b$, then $\int_a^b h(x)\,dx > 0$.]

(b) Write out explicitly the meaning of the Cauchy–Schwarz inequality for the inner product in part (a).

12. Prove the law of cosines for general inner products by expanding $\|\mathbf{x} - \mathbf{y}\|$ and using the definition of the cosine of the angle θ between two vectors.

13. Prove that the Pythagorean relation $\|\mathbf{x} - \mathbf{y}\|^2 = \|\mathbf{x}\|^2 + \|\mathbf{y}\|^2$ holds if and only if $\mathbf{x}$ and $\mathbf{y}$ are orthogonal.

14. Prove that if $\|\mathbf{x}\| = \sqrt{\langle \mathbf{x}, \mathbf{x} \rangle}$ is the norm defined by an inner product $\langle \mathbf{x}, \mathbf{y} \rangle$, then $\langle \mathbf{x}, \mathbf{y} \rangle = \frac{1}{4}(\|\mathbf{x} + \mathbf{y}\|^2 - \|\mathbf{x} - \mathbf{y}\|^2)$.

7B   Orthogonal Bases

The standard basis vectors $E = \{\mathbf{e}_1, \dots, \mathbf{e}_n\}$ in $\mathbb{R}^n$ form an orthogonal set since $\mathbf{e}_j \cdot \mathbf{e}_k = 0$ if $j \neq k$. If in addition all the vectors in an orthogonal set have length 1, the set is said to be **orthonormal**. Thus the set E is orthonormal since $|\mathbf{e}_k| = \mathbf{e}_k \cdot \mathbf{e}_k = 1$ for $k = 1, \dots, n$. The more restrictive orthonormality is a useful property for a basis $\{\mathbf{u}_1, \dots, \mathbf{u}_n\}$ to have, because we'll see we can then compute the coordinates in a basis representation $\mathbf{x} = u_1\mathbf{u}_1 + \cdots + u_n\mathbf{u}_n$ directly as inner products $\langle \mathbf{x}, \mathbf{u}_k \rangle$ without solving systems of linear equations and also compute inner products $\langle \mathbf{x}, \mathbf{y} \rangle$ as dot products of coordinate vectors. We'll also see how to construct an orthonormal basis starting from an arbitrary given basis in a finite-dimensional space with an inner product.

EXAMPLE 3   The standard basis $\{\mathbf{e}_1, \dots, \mathbf{e}_n\}$ is an orthonormal basis for $\mathbb{R}^n$ relative to the usual dot product in $\mathbb{R}^n$ because $\mathbf{e}_i \cdot \mathbf{e}_j = 0$ if $i \neq j$, and $\mathbf{e}_i \cdot \mathbf{e}_i = 1$.

A finite-dimensional vector space $\mathcal{V}$ has dimension n if $\mathcal{V}$ has a basis $\{\mathbf{v}_1, \dots, \mathbf{v}_n\}$ with n elements, so according to Theorem 5.9 every linearly independent set of n vectors in $\mathcal{V}$ is a basis for $\mathcal{V}$. The following theorem shows further that if $\dim(\mathcal{V}) = n$ and $\mathcal{V}$ contains an orthonormal set $\{\mathbf{u}_1, \dots, \mathbf{u}_n\}$ of n vectors then this set is automatically linearly independent and so is a basis for $\mathcal{V}$. In addition we compute the coordinates $(u_1, \dots, u_n)$ of a vector $\mathbf{x}$ relative to this basis from $u_j = \langle \mathbf{x}, \mathbf{u}_j \rangle$.

7.4   Theorem.   Suppose $\dim(\mathcal{V}) = n$ and that $B = \{\mathbf{u}_1, \dots, \mathbf{u}_n\}$ is an orthonormal set in $\mathcal{V}$. Then the set B is a basis for $\mathcal{V}$, and if $\mathbf{x} = u_1\mathbf{u}_1 + \cdots + u_n\mathbf{u}_n$, then $u_j = \langle \mathbf{x}, \mathbf{u}_j \rangle$ for $j = 1, \dots, n$.

Proof.   To show independence of the $\mathbf{u}_j$ suppose $u_1\mathbf{u}_1 + \cdots + u_j\mathbf{u}_j + \cdots + u_n\mathbf{u}_n = \mathbf{0}$. Using distributivity and homogeneity, the inner product of both sides with $\mathbf{u}_j$ is

$$u_1\langle \mathbf{u}_1, \mathbf{u}_j \rangle + \cdots + u_j\langle \mathbf{u}_j, \mathbf{u}_j \rangle + \cdots + u_n\langle \mathbf{u}_n, \mathbf{u}_j \rangle = u_1\langle \mathbf{u}_1, \mathbf{u}_j \rangle = u_j = 0,$$

since $\langle \mathbf{u}_j, \mathbf{u}_j \rangle = 1$ and $\langle \mathbf{u}_j, \mathbf{u}_k \rangle = 0$ if $k \neq j$. Hence $u_j = 0$ for $j = 1, \dots, n$, so the u_k are independent and form a basis for $\mathcal{V}$ by Theorem 5.9. If $\mathbf{x} = u_1\mathbf{u}_1 + \cdots + u_n\mathbf{u}_n$ a similar computation assuming orthonormality shows that for $j = 1, \dots, n$

$$\langle \mathbf{x}, \mathbf{u}_j \rangle = u_1\langle \mathbf{u}_1, \mathbf{u}_j \rangle + \cdots + u_j\langle \mathbf{u}_j, \mathbf{u}_j \rangle + \cdots + u_n\langle \mathbf{u}_n, \mathbf{u}_j \rangle = u_1\langle \mathbf{u}_1, \mathbf{u}_j \rangle = u_j.$$

Hence $u_j = \langle \mathbf{x}, \mathbf{u}_j \rangle$.   ■

In the next two examples we use the dot product in $\mathbb{R}^2$ and $\mathbb{R}^3$.

EXAMPLE 4   As we remarked in Example 3, if we use the dot product the vectors $\mathbf{e}_1 = (1, 0)$ and $\mathbf{e}_2 = (0, 1)$ form an orthonormal basis for $\mathbb{R}^2$. The pair $\mathbf{u}_1 = (\frac{1}{\sqrt{2}}, \frac{1}{\sqrt{2}})$ and $\mathbf{u}_2 = (\frac{-1}{\sqrt{2}}, \frac{1}{\sqrt{2}})$, shown in Figure 3.12(a) is also orthonormal. To compute coordinates relative to the basis $\{\mathbf{u}_1, \mathbf{u}_2\}$ in $\mathbb{R}^2$, we use the second part of Theorem 7.4. For example, the coordinates of the vector $(-2, 3)$ are

$$u_1 = (-2, 3) \cdot \left(\tfrac{1}{\sqrt{2}}, \tfrac{1}{\sqrt{2}} \right) = \tfrac{1}{\sqrt{2}}$$

$$u_2 = (-2, 3) \cdot \left(\tfrac{-1}{\sqrt{2}}, \tfrac{1}{\sqrt{2}} \right) = \tfrac{5}{\sqrt{2}}.$$

Hence $(-2, 3) = \tfrac{1}{\sqrt{2}}\mathbf{u}_1 + \tfrac{5}{\sqrt{2}}\mathbf{u}_2$.

FIGURE 3.12

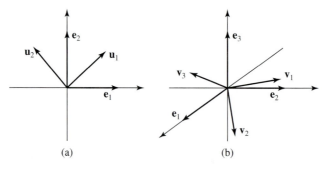

(a)   (b)

EXAMPLE 5

Figure 3.12(b) shows the three orthogonal unit vectors

$$\mathbf{u}_1 = \left(\tfrac{2}{7}, \tfrac{6}{7}, \tfrac{3}{7}\right), \mathbf{u}_2 = \left(\tfrac{6}{7}, -\tfrac{3}{7}, \tfrac{2}{7}\right), \mathbf{u}_3 = \left(\tfrac{3}{7}, \tfrac{2}{7}, -\tfrac{6}{7}\right).$$

Relative to the basis $\{\mathbf{u}_1, \mathbf{u}_2, \mathbf{u}_3\}$ in $\mathbb{R}^3$, the coordinates of $(1, 2, -1)$ are

$$u_1 = (1, 2, -1) \cdot \left(\tfrac{2}{7}, \tfrac{6}{7}, \tfrac{3}{7}\right) = \tfrac{11}{7}$$

$$u_2 = (1, 2, -1) \cdot \left(\tfrac{6}{7}, -\tfrac{3}{7}, \tfrac{2}{7}\right) = -\tfrac{2}{7}$$

$$u_3 = (1, 2, -1) \cdot \left(\tfrac{3}{7}, \tfrac{2}{7}, -\tfrac{6}{7}\right) = \tfrac{13}{7}.$$

Hence $(1, 2, -1) = \tfrac{11}{7}\mathbf{u}_1 - \tfrac{2}{7}\mathbf{u}_2 + \tfrac{13}{7}\mathbf{u}_3$.

EXAMPLE 6

Consider the infinite-dimensional subspace $\mathcal{S}$ of $C[-\pi, \pi]$ spanned by the set of functions

$$\{1, \cos x, \sin x, \cos 2x, \sin 2x, \dots\}.$$

In Example 2 of the previous section, we observed that relative to the inner product

$$\langle f, g \rangle = \int_{-\pi}^{\pi} f(x) g(x)\, dx,$$

the orthogonality relations

$$\langle \cos kx, \cos lx \rangle = 0, \quad k \neq l; k, l = 0, 1, 2, \dots$$

$$\langle \sin kx, \sin lx \rangle = 0, \quad k \neq l; k, l = 1, 2, \dots$$

$$\langle \cos kx, \sin lk \rangle = 0, \quad k = 0, 1, 2, \dots; l = 1, 2, \dots$$

all hold. It follows from Theorem 7.4 that these functions are linearly independent. We could divide each of the functions in the orthogonal set by its length to get an orthonormal set, but we are mainly interested in computing the coefficients in a linear combination of $\cos kx$ and $\sin kx$, so what's usually done is to alter the inner product by a constant positive factor that absorbs the normalization constants We can do this because with one exception all these factors are the same:

$$\langle 1, 1 \rangle = \int_{-\pi}^{\pi} dx = 2\pi,$$

$$\langle \cos kx, \cos kx \rangle = \int_{-\pi}^{\pi} \cos^2 kx \, dx = \pi,$$

$$\langle \sin kx, \sin kx \rangle = \int_{-\pi}^{\pi} \sin^2 kx \, dx = \pi.$$

Because the factor π occurs in each of these numbers, it's customary to alter the definition of the inner product by dividing by π and put

$$\langle f, g \rangle = \frac{1}{\pi} \int_{-\pi}^{\pi} f(x)g(x) \, dx.$$

This doesn't change the orthogonality of the set, but now we have

$$\langle 1, 1 \rangle = 2, \quad \langle \cos kx, \cos kx \rangle = 1, \quad \langle \sin kx, \sin kx \rangle = 1.$$

Defining the **Fourier coefficients** a_k for $k = 0, 1, 2, \ldots$ and b_k for $k = 1, 2, \ldots$ by

$$a_k = \frac{1}{\pi} \int_{-\pi}^{\pi} f(x) \cos kx \, dx,$$

$$b_k = \frac{1}{\pi} \int_{-\pi}^{\pi} f(x) \sin kx \, dx,$$

allows us to write a **trigonometric polynomial of degree** n

$$T_n(x) = \frac{a_0}{2} + a_1 \cos x + b_1 \sin x + \cdots + a_n \cos nx + b_n \sin nx$$

derived from the Fourier coefficients of an arbitrary integrable function $f(x)$ defined for $-\pi \leq x \leq \pi$. Note that $\cos kx = 1$ when $k = 0$, and that the 2 under the a_0 comes from dividing by $\langle 1, 1 \rangle = 2$. We'll see in Theorem 7.5 that T_n is, in a sense made precise there, the best approximation to f by a trigonometric polynomial of degree less than or equal to n.

EXAMPLE 7   Let $f(x) = |x|$ for $-\pi \leq x \leq \pi$. Then

$$a_k = \frac{1}{\pi} \int_{-\pi}^{\pi} |x| \cos kx \, dx, \quad b_k = \frac{1}{\pi} \int_{-\pi}^{\pi} |x| \sin kx \, dx.$$

Now $|x| \sin kx$ has integral zero over $[-\pi, \pi]$, because it's an odd function. Hence $b_k = 0$ for $k = 1, 2, \ldots$. On the other hand, the graph of $|x| \cos kx$ is symmetric about the y-axis, so we can just double the integral over $[0, \pi]$. For $k \neq 0$ we integrate by parts, getting

$$a_k = \frac{2}{\pi} \int_0^{\pi} x \cos kx \, dx$$

$$= \frac{2}{\pi} \left[\frac{x \sin kx}{k} \right]_0^{\pi} - \frac{2}{k\pi} \int_0^{\pi} \sin kx \, dx$$

$$= \left[\frac{2}{k^2 \pi} \cos kx \right]_0^\pi = \frac{2}{k^2 \pi} (\cos k\pi - 1)$$

$$= \frac{2}{k^2 \pi} ((-1)^k - 1) = \begin{cases} 0, & k = 2, 4, 6, \dots, \\ -\dfrac{4}{k^2 \pi}, & k = 1, 3, 5, \dots . \end{cases}$$

When $k = 0$, we have $a_0 = \dfrac{2}{\pi} \displaystyle\int_0^\pi x \, dx = \pi$. To summarize,

$$a_0 = \pi, \quad a_k = \begin{cases} 0, & k = 2, 4, 6, \dots, \\ -\dfrac{4}{k^2 \pi}, & k = 1, 3, 5, \dots, \end{cases}$$
$$b_k = 0, \quad k = 1, 2, 3, \dots .$$

The constant term is $a_0/2$ so the nth Fourier approximation $T_n(x)$ for odd n is

$$T_n(x) = \frac{\pi}{2} - \frac{4}{\pi} \cos x - \frac{4}{\pi} \frac{\cos 3x}{3^2} - \cdots - \frac{4}{\pi} \frac{\cos nx}{n^2}.$$

Approximation of the values $f(x)$ by $T_n(x)$ in the previous example is taken up in Chapter 14, Section 8, but in the present context we consider instead another kind of approximation that's measured by the norm of a vector, be it a function $f(x)$ in $C[-\pi, \pi]$ or a vector $\mathbf{x}$ in $\mathbb{R}^n$. The main theorem is as follows, and it's one of the principle reasons that orthonormal bases are important.

7.5   Theorem.   Let $S = \{\mathbf{u}_1, \dots, \mathbf{u}_n\}$ be an orthonormal set in a vector space $\mathcal{V}$ with inner product $\langle \mathbf{x}, \mathbf{y} \rangle$ and associated norm $\|\mathbf{x}\|$. Then for every $\mathbf{x}$ in $\mathcal{V}$ the distance

$$d_n = \|\mathbf{x} - (u_1 \mathbf{u}_1 - \cdots - u_n \mathbf{u}_n)\|$$

between $\mathbf{x}$ and points in the span of S is minimized by choosing the scalar coefficients to be $u_k = \langle \mathbf{x}, \mathbf{u}_k \rangle$; thus the minimum is $d_n = 0$ if $\mathbf{x}$ is in the span of S. If $\mathbf{x}$ is not in the span of S the minimum d_n is positive and satisfies

$$d_n^2 = \|\mathbf{x}\|^2 - \sum_{k=1}^n \langle \mathbf{x}, \mathbf{u}_k \rangle^2.$$

If S is infinite it is called a **complete** orthonormal set if d_n tends to zero as n tends to infinity.

Proof.   We'll work with d^2, using distributivity and homogeneity of the dot product:

$$\|\mathbf{x} - (u_1 \mathbf{u}_1 - \cdots - u_n \mathbf{u}_n)\|^2 = \langle \mathbf{x} - (u_1 \mathbf{u}_1 - \cdots - u_n \mathbf{u}_n), \mathbf{x} - (u_1 \mathbf{u}_1 - \cdots - u_n \mathbf{u}_n) \rangle$$

$$= \|\mathbf{x}\|^2 - 2 \sum_{k=1}^n c_k \langle \mathbf{x}, \mathbf{u}_k \rangle + \sum_{k=1}^n c_k^2.$$

Adding and subtracting $\sum_{k=1}^{n} \langle \mathbf{x}, \mathbf{u}_k \rangle^2$ to the last expression gives a sum of squares:

$$d^2 = \|\mathbf{x}\|^2 - \sum_{k=1}^{n} \langle \mathbf{x}, \mathbf{u}_k \rangle^2 + \sum_{k=1}^{n} \langle \mathbf{x}, \mathbf{u}_k \rangle^2 - 2 \sum_{k=1}^{n} c_k \langle \mathbf{x}, \mathbf{u}_k \rangle + \sum_{k=1}^{n} c_k^2$$

$$= \|\mathbf{x}\|^2 - \sum_{k=1}^{n} \langle \mathbf{x}, \mathbf{u}_k \rangle^2 + \sum_{k=1}^{n} (\langle \mathbf{x}, \mathbf{u}_k \rangle - c_k)^2.$$

The c_k's occur only in the last sum on the right, which is always nonnegative and takes on its minimum value of 0 just when $c_k = \langle \mathbf{x}, \mathbf{u}_k \rangle$ for every value of k from 1 to n, so these are the values that minimize d^2.   ∎

EXAMPLE 8 In the previous example the inner product and norm on the vector space $C[-\pi, \pi]$ were respectively

$$\langle f, g \rangle = \frac{1}{\pi} \int_{-\pi}^{\pi} f(x) g(x) \, dx \quad \text{and} \quad \|f\| = \left(\frac{1}{\pi} \int_{-\pi}^{\pi} f^2(x) \, dx \right)^{1/2}.$$

For $f(x) = |x|$ we found the nth-degree trigonometric approximation for odd n to be

$$T_n(x) = \frac{\pi}{2} - \frac{4}{\pi} \cos x - \frac{4}{\pi} \frac{\cos 3x}{3^2} - \cdots - \frac{4}{\pi} \frac{\cos nx}{n^2}.$$

Since the function $f(x) = |x|$ isn't differentiable it certainly isn't in the span of the trigonometric system so $\|f - T_n\|$ is always positive. Since $\|f\| = \frac{1}{\pi} \int_{-\pi}^{\pi} x^2 \, dx = \frac{2}{3}\pi^2$,

$$d_n^2 = \|f - T_n\|^2 = \frac{2\pi^2}{3} - \left(\frac{\pi^2}{2} + \frac{16}{\pi^2} \sum_{k=1}^{(n+1)/2} \frac{1}{(2k-1)^4} \right).$$

We're not in a position to prove it here, but the trigonometric system is complete so d_n^2 tends to zero as n tends to infinity.

In Chapter 1, Section 5B we saw how to find the distance between a point $\mathbf{x}$ in $\mathbb{R}^2$ or $\mathbb{R}^3$ and a line or plane when these were given in the respective forms $ax + by = c$ or $ax + by + cz = d$. Using the previous theorem, we can find the distance from a point to a general k-plane if we first represent the k-plane parametrically using an orthonormal set $\{\mathbf{u}_1, \ldots, \mathbf{u}_k\}$.

EXAMPLE 9 To find the distance from the point $(2, 3, 4)$ in $\mathbb{R}^3$ to the line $\mathbf{x} = t(1, 1, 1)$, first rewrite the line as $\mathbf{x} = s\left(\frac{1}{\sqrt{3}}, \frac{1}{\sqrt{3}}, \frac{1}{\sqrt{3}} \right)$. According to Theorem 7.5, the point on the line that produces the minimum distance to $(2, 3, 4)$ is the one for which the single scalar coordinate is $s = (2, 3, 4) \cdot \left(\frac{1}{\sqrt{3}}, \frac{1}{\sqrt{3}}, \frac{1}{\sqrt{3}} \right) = \frac{9}{\sqrt{3}}$. Hence the minimizing point is $(3, 3, 3)$ and the minimum distance is $|(2, 3, 4) - (3, 3, 3)| = |(-1, 0, 1)| = \sqrt{2}$. If the line had been shifted so it didn't contain the origin, for example, $\mathbf{x} = t(1, 1, 1) + (1, 2, 2)$, we would have instead minimized the distance between the point $(2, 3, 4) - (1, 2, 2) = (1, 1, 2)$ and the line $\mathbf{x} = s\left(\frac{1}{\sqrt{3}}, \frac{1}{\sqrt{3}}, \frac{1}{\sqrt{3}}\right)$.

Theorem 7.5 shows that if the vectors in an n-dimensional space are represented using an orthonormal basis, then norms are all computable using Euclidean lengths of coordinate vectors. This is true for inner products also, from which follows the result for norms. See also Exercise 14.

7.6 Theorem. Let $\{\mathbf{u}_1, \ldots, \mathbf{u}_n\}$ be an orthonormal set in a vector space with corresponding inner product $\langle \mathbf{x}, \mathbf{y} \rangle$. If

$$\mathbf{x} = x_1\mathbf{u}_1 + \cdots + x_n\mathbf{u}_n \quad \text{and} \quad \mathbf{y} = y_1\mathbf{u}_1 + \cdots + y_n\mathbf{u}_n \, ,$$

then

$$\langle \mathbf{x}, \mathbf{y} \rangle = (x_1, \ldots, x_n) \bullet (y_1, \ldots, y_n) \quad \text{and} \quad \|\mathbf{x}\| = |(x_1, \ldots, x_n)|.$$

Proof. Using distributivity of the inner product and orthonormality of the $\mathbf{u}_k$, we have

$$\langle \mathbf{x}, \mathbf{y} \rangle = \langle x_1\mathbf{u}_1 + \cdots + x_n\mathbf{u}_n, \ y_1\mathbf{u}_1 + \cdots + y_n\mathbf{u}_n \rangle$$

$$= x_1 y_1 + \cdots + x_k y_k + \cdots + x_n y_n.$$

Setting $\mathbf{y} = \mathbf{u}_k$ and using $\langle \mathbf{u}_k, \mathbf{u}_k \rangle = 1$ and $\langle \mathbf{u}_j, \mathbf{u}_k \rangle = 0$ if $j \neq k$, we get $\langle \mathbf{x}, \mathbf{u}_k \rangle = x_k$. Similarly, $y_k = \langle \mathbf{y}, \mathbf{u}_k \rangle$. ∎

A specific orthonormal basis is often the most natural basis for a vector space, but if all we have to start with is a natural inner product and a basis that isn't orthogonal, then Equations 7.7 and 7.8 below serve as recipes for finding orthonormal bases using what's called the **Gram–Schmidt process**. We start by describing the process in detail. It assumes we have in hand a linearly independent set of vectors $X = \{\mathbf{x}_1, \ldots, \mathbf{x}_n, \ldots\}$, and our aim is to produce from this an orthonormal set $Y = \{\mathbf{y}_1, \ldots, \mathbf{y}_n, \ldots\}$ that spans the same vector subspace that X does. First we pick one of the $\mathbf{x}$'s, say $\mathbf{x}_1$, and normalize $\mathbf{x}_1$ to get

$$\mathbf{u}_1 = \mathbf{x}_1 / \|\mathbf{x}_1\|, \quad \text{so} \quad \|\mathbf{u}_1\| = \|\mathbf{x}_1\| / \|\mathbf{x}_1\| = 1.$$

Next we pick another of the $\mathbf{x}$'s, say $\mathbf{x}_2$, and form its **projection** on $\mathbf{u}_1$, shown in Figure 3.13(a) as a geometric vector defined by $\langle \mathbf{x}_2, \mathbf{u}_1 \rangle \mathbf{u}_1$. The justification for using the term *projection* is that the vector $\mathbf{y}_2$, defined as the difference $\mathbf{y}_2 = \mathbf{x}_2 - \langle \mathbf{x}_2, \mathbf{u}_1 \rangle \mathbf{u}_1$, is orthogonal to $\mathbf{u}_1$:

$$\langle \mathbf{y}_2, \mathbf{u}_1 \rangle = \langle \mathbf{x}_2 - \langle \mathbf{x}_2, \mathbf{u}_1 \rangle \mathbf{u}_1, \mathbf{u}_1 \rangle$$

$$= \langle \mathbf{x}_2, \mathbf{u}_1 \rangle - \langle \mathbf{x}_2, \mathbf{u}_1 \rangle \langle \mathbf{u}_1, \mathbf{u}_1 \rangle = 0.$$

FIGURE 3.13

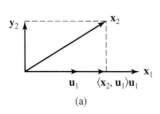

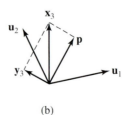

(a)　　　　　　　　　(b)

The vector $\mathbf{y}_2$ can't be zero, because by its definition that would imply $\mathbf{x}_2$ and $\mathbf{u}_1$ to be linearly dependent, which they are not. Thus the vector $\mathbf{u}_2 = \mathbf{y}_2/\|\mathbf{y}_2\|$ has length 1.

Next we take $\mathbf{x}_3$ from our independent set and form its **projection p** on the subspace spanned by $\mathbf{u}_1$ and $\mathbf{u}_2$, defined by

$$\mathbf{p} = \langle\mathbf{x}_3, \mathbf{u}_1\rangle\mathbf{u}_1 + \langle\mathbf{x}_3, \mathbf{u}_2\rangle\mathbf{u}_2,$$

and illustrated in Figure 3.13(b). We define $\mathbf{y}_3$ by subtracting $\mathbf{p}$ from $\mathbf{x}_3$:

$$\mathbf{y}_3 = \mathbf{x}_3 - \langle\mathbf{x}_3, \mathbf{u}_1\rangle\mathbf{u}_1 - \langle\mathbf{x}_3, \mathbf{u}_2\rangle\mathbf{u}_2.$$

We can check that $\mathbf{y}_3$ is orthogonal to both $\mathbf{u}_1$ and $\mathbf{u}_2$ because $\langle\mathbf{u}_1, \mathbf{u}_1\rangle = 1$ and $\langle\mathbf{u}_1, \mathbf{u}_2\rangle = 0$. We get

$$\langle\mathbf{y}_3, \mathbf{u}_1\rangle = \langle\mathbf{x}_3, \mathbf{u}_1\rangle - \langle\mathbf{x}_3, \mathbf{u}_1\rangle\langle\mathbf{u}_1, \mathbf{u}_1\rangle = 0.$$

Also, because $\langle\mathbf{u}_2, \mathbf{u}_2\rangle = 1$,

$$\langle\mathbf{y}_3, \mathbf{u}_2\rangle = \langle\mathbf{x}_3, \mathbf{u}_2\rangle - \langle\mathbf{x}_3, \mathbf{u}_2\rangle\langle\mathbf{u}_2, \mathbf{u}_2\rangle = 0.$$

If $\mathbf{y}_3 = 0$, then $\mathbf{x}_3, \mathbf{u}_1$, and $\mathbf{u}_2$ would be dependent; but this is impossible, because the subspace spanned by $\mathbf{u}_1$ and $\mathbf{u}_2$ is the same as that spanned by $\mathbf{x}_1$ and $\mathbf{x}_2$; therefore, $\mathbf{x}_3, \mathbf{x}_2$, and $\mathbf{x}_1$ would be dependent. Hence $\mathbf{u}_3 = \mathbf{y}_3/\|\mathbf{y}_3\|$ has length 1.

In this way we successively compute $\mathbf{u}_1, \mathbf{u}_2, \ldots, \mathbf{u}_k$. To get $\mathbf{y}_{k+1}$ we set

7.7 $$\mathbf{y}_{k+1} = \mathbf{x}_{k+1} - \langle\mathbf{x}_{k+1}, \mathbf{u}_1\rangle\mathbf{u}_1 - \cdots - \langle\mathbf{x}_{k+1}, \mathbf{u}_k\rangle\mathbf{u}_k.$$

We can verify as before that $\mathbf{y}_{k+1}$ is orthogonal to $\mathbf{u}_1, \mathbf{u}_2, \ldots, \mathbf{u}_k$. To obtain a unit vector, we can normalize $\mathbf{y}_{k+1}$ to $\mathbf{u}_{k+1} = \mathbf{y}_{k+1}/\|\mathbf{y}_{k+1}\|$. In practice, it may be more convenient to compute the $\mathbf{y}$'s from the equivalent formula

7.8 $$\mathbf{y}_{k+1} = \mathbf{x}_{k+1} - \frac{\langle\mathbf{x}_{k+1}, \mathbf{y}_1\rangle}{\|\mathbf{y}_1\|^2}\mathbf{y}_1 - \cdots - \frac{\langle\mathbf{x}_{k+1}, \mathbf{y}_k\rangle}{\|\mathbf{y}_k\|^2}\mathbf{y}_k.$$

EXAMPLE 10   The vectors $\mathbf{x}_1 = (1, -1, 2)$ and $\mathbf{x}_2 = (1, 0, -1)$ span a plane $\mathcal{P}$ in $\mathbb{R}^3$ because they are linearly independent. To find an orthogonal basis for $\mathcal{P}$, we apply the Gram–Schmidt process to the basis $\{\mathbf{x}_1, \mathbf{x}_2\}$ for $\mathcal{P}$. We set $\mathbf{y}_1 = \mathbf{x}_1$ and

$$\mathbf{y}_2 = \mathbf{x}_2 - \frac{(\mathbf{x}_2 \cdot \mathbf{y}_1)}{|\mathbf{y}_1|^2}\mathbf{y}_1$$

$$= (1, 0, -1) - \frac{(-1)}{6}(1, -1, 2)$$

$$= (\tfrac{7}{6}, -\tfrac{1}{6}, -\tfrac{2}{3}).$$

Thus the plane $\mathcal{P}$ was defined as the set of all linear combinations

$$s\mathbf{x}_1 + t\mathbf{x}_2,$$

FIGURE 3.14

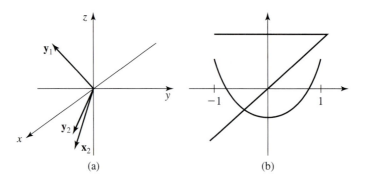

(a)   (b)

but can also be represented as the set of all linear combinations of $\mathbf{y}_1 = \mathbf{x}_1$ and $\mathbf{y}_2$, namely,

$$u\mathbf{y}_1 + v\mathbf{y}_2,$$

where $\{\mathbf{y}_1, \mathbf{y}_2\}$ is the orthogonal pair $\{(1, -1, 2), (\frac{7}{6}, -\frac{1}{6}, -\frac{2}{3})\}$, shown in Figure 3.14(a).

   If we add a third vector $\mathbf{x}_3$ that is linearly independent of $\mathbf{x}_1$ and $\mathbf{x}_2$, we can go on to find $\mathbf{y}_3$ so that $\{\mathbf{y}_1, \mathbf{y}_2, \mathbf{y}_3\}$ is an orthogonal basis for $\mathbb{R}^3$, and $\{\mathbf{y}_1, \mathbf{y}_2\}$ is an orthogonal basis for $\mathcal{P}$. For instance, if we take $\mathbf{x}_3 = \mathbf{e}_1$, $\mathbf{y}_3$ works out to be $(\frac{1}{11}, \frac{3}{11}, \frac{1}{11})$.

EXAMPLE 11

Let $\mathcal{P}_n[-1, 1]$ be the vector space of polynomials $f(x) = a_0 + a_1 x + \cdots + a_n x^n$, but restricted to $-1 \le x \le 1$. We define an inner product

$$\langle f, g \rangle = \int_{-1}^{1} f(x)g(x)\, dx.$$

The argument used in Example 3 of Section 5A to show that the functions $1, x, \ldots, x^n$ form a basis for $\mathcal{P}_n$, works also to show that they form a basis for $\mathcal{P}_n[-1, 1]$ when restricted to $[-1, 1]$. To find an orthogonal basis, let $y_0(x) = 1$. Then let

$$y_1(x) = x - \frac{\langle x, 1 \rangle}{\langle 1, 1 \rangle} 1 = x,$$

because $\langle x, 1 \rangle = 0$. Next let

$$y_2(x) = x^2 - \frac{\langle x^2, 1 \rangle}{\langle 1, 1 \rangle} 1 - \frac{\langle x^2, x \rangle}{\langle x, x \rangle} x = x^2 - \frac{\left(\frac{2}{3}\right)}{2} = x^2 - \frac{1}{3};$$

this is so because $\langle x^2, 1 \rangle = \frac{2}{3}$, $\langle x^2, x \rangle = 0$, and $\langle 1, 1 \rangle = 2$. The graphs of the three polynomials $y_0(x) = 1, y_1(x) = x, y_2(x) = x^2 - (\frac{1}{3})$ are illustrated in Figure 3.14(b). We get the corresponding orthonormal set by dividing successively by $\|1\| = \sqrt{2}$, $\|x\| = \sqrt{\frac{2}{3}}$, and $\|x^2 - \frac{1}{3}\| = \sqrt{\frac{8}{45}}$, to get $\{\sqrt{\frac{1}{2}}, \sqrt{\frac{3}{2}}x, \sqrt{\frac{45}{8}}(x^2 - \frac{1}{3})\}$. Thus we have an orthonormal basis for $\mathcal{P}_3[-1, 1]$. The resulting polynomials are called **normalized Legendre polynomials** .

EXERCISES

1. Find a vector (x, y, z) in $\mathbb{R}^3$ such that the triple of vectors $(1, 1, 1)$, $(-1, \frac{1}{2}, \frac{1}{2})$, (x, y, z) forms an orthogonal basis for $\mathbb{R}^3$. Then normalize this basis by dividing each vector by its length.

2. The vectors $(1, 1, 1)$ and $(1, 2, 1)$ span a plane $\mathcal{P}$ in $\mathbb{R}^3$. Use the Gram–Schmidt process to find an orthogonal basis for $\mathbb{R}^3$ in which the first two vectors form an orthogonal basis for $\mathcal{P}$.

3. Find an orthogonal basis for $\mathbb{R}^4$ in which the first three vectors form a basis for the subspace $\mathcal{S}$ spanned by $(1, 2, 1, 1)$, $(-1, 0, 1, 0)$, and $(0, 1, 0, 2)$.

4. Let $\mathcal{P}_2$ be the three-dimensional space of quadratic polynomials $p(x) = a + bx + cx^2$, restricted so that $0 \le x \le 1$. If $\mathcal{P}_2$ is given the inner product

$$\langle p, q \rangle = \int_0^1 p(x)q(x)\,dx,$$

find an orthonormal basis for $\mathcal{P}_2$. [*Hint:* One basis for $\mathcal{P}_2$ is $\{1, x, x^2\}$.)]

5. Show that applying the Gram–Schmidt process to the three vectors $(3, 0, 0)$, $(1, 1, 0)$, and $(1, 1, 1)$ in order produces an orthogonal basis that normalizes to the standard basis $\{\mathbf{e}_1, \mathbf{e}_2, \mathbf{e}_3\}$.

6. (a) Show that applying the Gram–Schmidt process to an orthonormal set in order gives the orthonormal set back again.

   (b) Let $\{\mathbf{u}_1, \ldots, \mathbf{u}_n\}$ and $\{\mathbf{v}_1, \ldots, \mathbf{v}_n\}$ be two orthonormal sets in a vector space, such that the subspaces spanned by $\{\mathbf{u}_1, \ldots, \mathbf{u}_k\}$ and $\{\mathbf{v}_1, \ldots, \mathbf{v}_k\}$ are the same for $k = 1, 2, \ldots, n$. Show that $\mathbf{u}_k = \pm\mathbf{v}_k$ for $k = 1, 2, \ldots, n$.

7. Show that if $\{\mathbf{u}_1, \ldots, \mathbf{u}_n\}$ and $\{\mathbf{v}_1, \ldots, \mathbf{v}_n\}$ are two orthonormal bases for a vector space, then the matrix M used to change from one set of coordinates to another has columns that form an orthonormal set in $\mathbb{R}^n$. [*Hint:* Express $\mathbf{u}_k$ as a linear combination of $\mathbf{v}_1, \ldots, \mathbf{v}_n$, and use Theorem 7.6.]

8. Find the distance between the point $(2, 3, 4)$ and the plane parametrized by $\mathbf{x} = u(1, 2, 1) + v(1, -1, 1) + (1, 1, 2)$. Note that conveniently $(1, 2, 1) \cdot (1, -1, 1) = 0$.

9. Find the distance between the point $(2, 3, 4)$ and the plane parametrized by $\mathbf{x} = u(1, 1, 1) + v(1, -1, 1)$. Since $(1, 1, 1) \cdot (1, -1, 1) \ne 0$, use the Gram–Schmidt process first.

10. Let $S = \{\mathbf{u}_1, \ldots, \mathbf{u}_n\}$ be an orthonormal set. Prove that the vector $\sum_{k=1}^n u_k\mathbf{u}_k$ is orthogonal to $\mathbf{x} - \sum_{k=1}^n u_k\mathbf{u}_k$ if and only if $u_k = \langle \mathbf{x}, \mathbf{u}_k \rangle$. This is another way to characterize the choice of coefficients in Theorem 7.5, showing that the nearest point to $\mathbf{x}$ in the span of S is the perpendicular projection of $\mathbf{x}$ onto the span of S.

11. For a given inner product $\langle \mathbf{x}, \mathbf{y} \rangle$ on $\mathbb{R}^n$, let A be the n-by-n matrix defined to have entries $a_{ij} = \langle \mathbf{e}_i, \mathbf{e}_j \rangle$ for $i, j = 1 \ldots, n$. Show that for arbitrary $\mathbf{x}$ and $\mathbf{y}$ in $\mathbb{R}^n$, $\langle \mathbf{x}, \mathbf{y} \rangle = \mathbf{x} \cdot A\mathbf{y}$.

12. An n-by-n matrix $A = (a_{ij})$ is **symmetric** if $a_{ij} = a_{ji}$ for $i, j = 1, \ldots, n$. Prove that if A is symmetric then $A\mathbf{x} \cdot \mathbf{y} = \mathbf{x} \cdot A\mathbf{y}$ for all $\mathbf{x}$ and $\mathbf{y}$ in $\mathbb{R}^n$.

13. Show that if A is symmetric and $\langle \mathbf{x}, \mathbf{y} \rangle$ is defined to be $\mathbf{x} \cdot A\mathbf{y}$ for $\mathbf{x}$ and $\mathbf{y}$ in $\mathbb{R}^n$, then $\langle \mathbf{x}, \mathbf{y} \rangle$ has all the properties of an inner product except possibly the positivity property: $\langle \mathbf{x}, \mathbf{x} \rangle > 0$ unless $\mathbf{x} = \mathbf{0}$.

14. A is called a **positive definite** matrix if it is symmetric and $\langle \mathbf{x}, \mathbf{y} \rangle = \mathbf{x} \cdot A\mathbf{y}$ has the positivity property of an inner product. Show that a diagnoal matrix is positive definite if and only if all its diagonal entries are positive.

*15. Show that a symmetric 2-by-2 matrix $A = \begin{pmatrix} a & b \\ b & c \end{pmatrix}$ is positive definite if and only if a and $\det A = ac - b^2$ are both positive.

*16. Even if $\langle \mathbf{x}, \mathbf{y} \rangle$ doesn't have the positivity property, but has the other properties of an inner product, Formula 7.8 in the Gram–Schmidt process still makes sense unless $\langle \mathbf{y}_k, \mathbf{y}_k \rangle = 0$ at some stage. This observation leads to an efficient method for determining whether a symmetric matrix is positive definite or not.

   (a) Suppose that $\langle \mathbf{x}, \mathbf{y} \rangle$ is defined to be $\mathbf{x} \cdot A\mathbf{y}$ for a symmetric matrix A, as in Exercise 13. Show that if applying the formulas of the Gram–Schmidt process to the standard basis vectors $\{\mathbf{e}_1, \ldots, \mathbf{e}_n\}$ leads at some stage to a $\mathbf{y}_k$ with $\langle \mathbf{y}_k, \mathbf{y}_k \rangle \le 0$ then A is not positive definite.

   (b) Show that if $\langle \mathbf{y}_k, \mathbf{y}_k \rangle > 0$ at every stage then A is positive definite.

*17. Show that the result of Exercise 12 remains valid if $\mathbf{x}$ and $\mathbf{y}$ are allowed to have complex entries, and use this to prove that all the eigenvalues of a real symmetric matrix are real numbers. [*Hint:* If $A\mathbf{y} = \lambda\mathbf{y}$ with λ and $\mathbf{y}$ complex, show that $A\bar{\mathbf{y}} = \bar{\lambda}\bar{\mathbf{y}}$ and put $\mathbf{x} = \bar{\mathbf{y}}$ in the equation of Exercise 12.]

7C   Rotation and Reflection

In Section 1 we looked at examples of rotation operators in $\mathbb{R}^2$ that preserve the lengths of vector arrows with tails at the origin as well as the angles between the arrows. We saw that rotation by the angle θ about the origin in $\mathbb{R}^2$ is a linear function with matrix $R_\theta = \begin{pmatrix} \cos\theta & -\sin\theta \\ \sin\theta & \cos\theta \end{pmatrix}$. For the same reasons, rotation about a line through the origin in $\mathbb{R}^3$ is a linear function $R\colon \mathbb{R}^3 \longrightarrow \mathbb{R}^3$ and is therefore given by multiplication by a matrix A whose columns are the images $R(\mathbf{e}_j)$ of the standard basis vectors. Since R preserves lengths and angles, the image vectors $R(\mathbf{e}_j)$ have length 1 and are mutually perpendicular, so like the standard basis vectors, they form an orthonormal basis. Moving $\{\mathbf{e}_1, \mathbf{e}_2, \mathbf{e}_3\}$ continuously to its new position $\{R(\mathbf{e}_1), R(\mathbf{e}_2), R(\mathbf{e}_3)\}$ we see that the image vectors continue to form a right-handed system, so their scalar triple product in that order equals $+1$. Since the scalar triple product also equals $\det(A) = 1$, we'll see in Chapter 7 that a rotation R in $\mathbb{R}^3$ preserves volumes.

EXAMPLE 12   Given what we know about rotations in $\mathbb{R}^2$ it's easy to find the matrix of a rotation in $\mathbb{R}^3$ about one of the coordinate axes. If R is a rotation through angle θ about the x_1-axis, then $R(\mathbf{e}_1) = \mathbf{e}_1$ and R rotates $\mathbf{e}_2$ and $\mathbf{e}_3$ through the angle θ in the x_2x_3-plane, so its matrix is $M = \begin{pmatrix} 1 & 0 & 0 \\ 0 & \cos\theta & -\sin\theta \\ 0 & \sin\theta & \cos\theta \end{pmatrix}$, in which the submatrix in the lower right corner is the same as the matrix R_θ of Example 8 in Section 1.

More generally, if $\{\mathbf{u}_1, \mathbf{u}_2, \mathbf{u}_3\}$ is an orthonormal basis in $\mathbb{R}^3$ and R is a rotation about an axis in the direction of $\mathbf{u}_1$ through the angle θ then the matrix of R relative to the basis $\{\mathbf{u}_1, \mathbf{u}_2, \mathbf{u}_3\}$ is the matrix M of Example 12.

To find the matrix of R relative to the standard basis, we use Theorem 6.8 in Section 6C. If A is the matrix of R relative to the standard basis and M its matrix relative to the basis $\{\mathbf{u}_1, \mathbf{u}_2, \mathbf{u}_3\}$, then $A = UMU^{-1}$, and $U^{-1}AU = M$, where U is the matrix whose columns are $\mathbf{u}_1, \mathbf{u}_2$, and $\mathbf{u}_3$.

EXAMPLE 13   For a numerical example, we'll find the matrix of a rotation R through $60°$ about an axis in the direction of $\mathbf{a} = (1, 1, 1)$. Figure 3.15(a) shows the standard basis vectors and their images under R, with the axis of rotation shown as a dotted line.

The first step is to find an orthonormal basis $\mathbf{u}_1, \mathbf{u}_2$, and $\mathbf{u}_3$ with $\mathbf{u}_1$ in the same direction as $\mathbf{a}$. We could use the Gram–Schmidt process here, but the following method is less work. We'll start by finding an orthogonal basis containing $\mathbf{a}$. For a vector orthogonal to $\mathbf{a}$ we can take an arbitrary $\mathbf{b} = (b_1, b_2, b_3)$ with $\mathbf{a} \cdot \mathbf{b} = b_1 + 2b_2 + 2b_3 = 0$, for instance, $\mathbf{b} = (0, 1, -1)$. Since we are working in $\mathbb{R}^3$, we can get a third vector perpendicular to $\mathbf{a}$ and $\mathbf{b}$ by finding the cross product $\mathbf{c} = \mathbf{a} \times \mathbf{b}$, which works out to be $(-2, 1, 1)$. Now we normalize by dividing each of $\mathbf{a}$, $\mathbf{b}$, and $\mathbf{c}$ by its length to get unit vectors $\mathbf{u}_1 = (1/\sqrt{3}, 1/\sqrt{3}, 1/\sqrt{3})$, $\mathbf{u}_2 = (0, 1/\sqrt{2}, -1/\sqrt{2})$, and $\mathbf{u}_3 = (-2/\sqrt{6}, 1/\sqrt{6}, 1/\sqrt{6})$. Using these vectors as the columns of a matrix gives

$$U = \begin{pmatrix} 1/\sqrt{3} & 0 & -2/\sqrt{6} \\ 1/\sqrt{3} & 1/\sqrt{2} & 1/\sqrt{6} \\ 1/\sqrt{3} & -1/\sqrt{2} & 1/\sqrt{6} \end{pmatrix}.$$ Since $\cos 60° = 1/2$ and $\sin 60° = \sqrt{3}/2$,

FIGURE 3.15

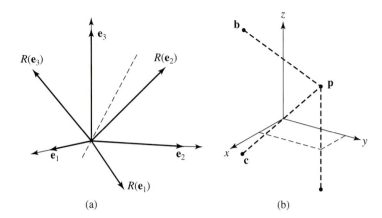

(a)   (b)

$$M = \begin{pmatrix} 1 & 0 & 0 \\ 0 & 1/2 & -\sqrt{3}/2 \\ 0 & \sqrt{3}/2 & 1/2 \end{pmatrix}$$ is the matrix of R relative to the basis $\{\mathbf{u}_1, \mathbf{u}_2, \mathbf{u}_3\}$.

Then the matrix of R relative to the standard basis is $A = UMU^{-1}$ according to the formula given above. We can use orthonormality to find U^{-1} without any computation. It is simply U^t, the **transpose** of U, obtained by flipping U about its main diagonal, so that the rows of U become the columns of U^t and vice versa. In this case, $U^t = \begin{pmatrix} 1/\sqrt{3} & 1/\sqrt{3} & 1/\sqrt{3} \\ 0 & 1/\sqrt{2} & -1/\sqrt{2} \\ -2/\sqrt{6} & 1/\sqrt{6} & 1/\sqrt{6} \end{pmatrix}$. To see that $U^{-1} = U^t$ when U is a square matrix with orthonormal columns, note that row i of U^t (which is column i of U) is just $\mathbf{u}_i$, and column j of U is $\mathbf{u}_j$. The ijth entry in the product U^tU is therefore $\mathbf{u}_i \cdot \mathbf{u}_j$, which is 1 when $i = j$ and 0 when $i \neq j$ because $\{\mathbf{u}_1, \mathbf{u}_2, \mathbf{u}_3\}$ is an orthonormal basis. Thus U^tU is the identity matrix and therefore $U^{-1} = U^t$. Finally,

$$A = UMU^{-1} = UMU^t$$

$$= \begin{pmatrix} 1/\sqrt{3} & 0 & -2/\sqrt{6} \\ 1/\sqrt{3} & 1/\sqrt{2} & 1/\sqrt{6} \\ 1/\sqrt{3} & -1/\sqrt{2} & 1/\sqrt{6} \end{pmatrix} \begin{pmatrix} 1 & 0 & 0 \\ 0 & 1/2 & -\sqrt{3}/2 \\ 0 & \sqrt{3}/2 & 1/2 \end{pmatrix} \begin{pmatrix} 1/\sqrt{3} & 1/\sqrt{3} & 1/\sqrt{3} \\ 0 & 1/\sqrt{2} & -1/\sqrt{2} \\ -2/\sqrt{6} & 1/\sqrt{6} & 1/\sqrt{6} \end{pmatrix}$$

$$= \begin{pmatrix} 1/\sqrt{3} & 0 & -2/\sqrt{6} \\ 1/\sqrt{3} & 1/\sqrt{2} & 1/\sqrt{6} \\ 1/\sqrt{3} & -1/\sqrt{2} & 1/\sqrt{6} \end{pmatrix} \begin{pmatrix} 1/\sqrt{3} & 1/\sqrt{3} & 1/\sqrt{3} \\ 1/\sqrt{2} & 0 & -1/\sqrt{2} \\ -1/\sqrt{6} & 2/\sqrt{6} & -1/\sqrt{6} \end{pmatrix}$$

$$= \begin{pmatrix} 2/3 & -1/3 & 2/3 \\ 2/3 & 2/3 & -1/3 \\ -1/3 & 2/3 & 2/3 \end{pmatrix}$$

is the matrix of R relative to the standard basis. As a partial check on the calculation, you can verify that $A\mathbf{u}_1$ is equal to $\mathbf{u}_1$ as it should be.

EXAMPLE 14

We'll now see how to find the axis of a rotation R if we're given its matrix A. For a numerical example we'll take the matrix $A = \begin{pmatrix} \frac{2}{7} & \frac{6}{7} & \frac{3}{7} \\ \frac{6}{7} & -\frac{3}{7} & \frac{2}{7} \\ \frac{3}{7} & \frac{2}{7} & -\frac{6}{7} \end{pmatrix}$, whose columns are the orthonormal vectors $\mathbf{v}_1, \mathbf{v}_2, \mathbf{v}_3$ of Example 4, and which you can check has determinant 1.

If $\mathbf{a}$ has the same direction as the axis of rotation, then $R(\mathbf{a}) = A\mathbf{a} = \mathbf{a}$, so $\mathbf{a}$ is an eigenvector of A associated with the eigenvalue 1, and

$$(A - I)\mathbf{a} = \begin{pmatrix} -\frac{5}{7} & \frac{6}{7} & \frac{3}{7} \\ \frac{6}{7} & -\frac{10}{7} & \frac{2}{7} \\ \frac{3}{7} & \frac{2}{7} & -\frac{13}{7} \end{pmatrix} \begin{pmatrix} a_1 \\ a_2 \\ a_3 \end{pmatrix} = \mathbf{0}.$$

Row reduction of $(A - I)$ gives a matrix with two rows $\begin{pmatrix} 1 & 0 & -3 \\ 0 & 1 & -2 \end{pmatrix}$ and a third row of zeros, from which we can read off $\mathbf{a} = (3, 2, 1)$ as a solution of $(A - I)\mathbf{a} = \mathbf{0}$, which is therefore an eigenvector of A associated with the eigenvalue 1.

One way to find the angle of rotation of R is to find its matrix relative to an orthonormal basis whose first basis vector has the direction of $\mathbf{a}$, as in Example 13. See Exercise 1 for details.

For a quicker way of finding the angle of a rotation directly from its matrix, see Exercise 2.

EXAMPLE 15

Orthonormal bases make it easy to describe the geometric operation of reflection in a subspace. Let V be a vector space with an inner product and an orthonormal basis $\{\mathbf{u}_1, \ldots, \mathbf{u}_n\}$, and let $\mathcal{U}_k$ be the subspace spanned by the first k basis vectors. Also let $\mathcal{U}_0$ be the zero subspace. The linear function $r: V \longrightarrow V$ defined by

$$f(x_1\mathbf{u}_1 + \cdots + x_k\mathbf{u}_k + x_{k+1}\mathbf{u}_{k+1} + \cdots + \mathbf{u}_n x_n)$$

$$= x_1\mathbf{u}_1 + \cdots + x_k\mathbf{u}_k - x_{k+1}\mathbf{u}_{k+1} - \cdots - \mathbf{u}_n x_n$$

is called **reflection** in the subspace $\mathcal{U}_k$. When n is 2 or 3, and $k = n - 1$ this gives the familiar notion of reflection in a line in $\mathbb{R}^2$ or plane in $\mathbb{R}^3$. For instance, using the standard basis in $\mathbb{R}^3$, $\mathcal{U}_2$ is the xy-plane, and reflection in it sends (x, y, z) to its mirror image $(x, y, -z)$. When $k = 0$ we get reflection through the origin.

Figure 3.15(b) shows a point $\mathbf{p}$, and its reflections in the xy-plane, the z-axis, and the origin. If $\mathbf{p}$ has coordinates (x, y, z) then $\mathbf{a} = (x, y, -z)$, $\mathbf{b} = (-x, -y, z)$, and $\mathbf{c} = (-x, -y, -z)$.

We leave it as Exercise 6 to show that the operation of reflection in a subspace depends only on the subspace, and not on the particular orthonormal basis used in the preceding definition.

EXERCISES

1. This exercise refers to the rotation R with matrix A of Example 14 of the text. Show that the columns of $U = \begin{pmatrix} 3/\sqrt{14} & 0 & 5/\sqrt{70} \\ 2/\sqrt{14} & -1/\sqrt{5} & -6/\sqrt{70} \\ 1/\sqrt{14} & 2/\sqrt{5} & -3/\sqrt{70} \end{pmatrix}$ are an orthonormal set,

and that the axis of the rotation R has the direction of $\mathbf{u}_1$. Find the matrix M of the rotation R relative to the basis $\{\mathbf{u}_1, \mathbf{u}_2, \mathbf{u}_3\}$ using the relation $M = U^{-1}AU = U^tAU$. Compare the result with the matrix M_θ of Example 12 in the text to find the angle of rotation.

2. The **trace** of a square matrix A is defined to be the sum of the entries on its main diagonal, $a_{11} + \cdots + a_{nn}$.
   (a) Show that if A and B are square matrices of the same dimension, then Trace $(AB) = \text{Trace}(BA)$.
   (b) Show that if P and U are square matrices of the same dimension, and U has an inverse, then $\text{Trace}(U^{-1}PU) = \text{Trace}(P)$.
   (c) Show that if A is the matrix of a rotation through an angle θ in $\mathbb{R}^3$, then $\text{Trace}(A) = 1 + 2\cos\theta$. [*Hint*: See Exercise 1 and Example 12 in this section.]
   (d) Use part (c) to find the rotation angle of the matrix A in Example 14 in the text.

3. (a) Find matrices R and S for the rotations through $90°$ about the x- and y-axes in $\mathbb{R}^3$.
   (b) Find the axis of the rotation obtained by first doing the rotation about the x-axis and then the rotation about the y-axis.

4. Do the same as in Exercise 3, but with rotations through the angle θ with $\cos\theta = \frac{3}{5}$ and $\sin\theta = \frac{4}{5}$ instead of $90°$.

5. Find the matrix of a rotation about the axis $(1, 1, 0)$ through the angle θ of Exercise 4.

6. Let $r: V \longrightarrow V$ be reflection in a subspace $\mathcal{U}$ of a space V with an inner product as in Example 15.
   (a) Show that r has the property that $r(\mathbf{x}) = \mathbf{x}$ for $\mathbf{x}$ in $\mathcal{U}$ and $r(\mathbf{x}) = -\mathbf{x}$ if $\mathbf{x}$ is perpendicular to every vector in $\mathcal{U}$.
   (b) Show that every linear function from V to V with the property in part (a) must be identical with r.

The following exercises outline the steps in a proof that every function from $\mathbb{R}^2$ to $\mathbb{R}^2$ or from $\mathbb{R}^3$ to $\mathbb{R}^3$ that preserves lengths and takes the origin to itself is either a rotation, a reflection, or the composition of a rotation and a reflection. We pointed out just before Examples 7 and

8 in Section 7C that such a function is linear and that its matrix relative to an orthonormal basis has columns that form an orthonormal set.

7. Let $M = \begin{pmatrix} a & b \\ c & d \end{pmatrix}$ be a matrix whose columns form an orthonormal set in $\mathbb{R}^2$.
   (a) Show that $a^2 + c^2 = 1$ and either $(c, d) = (-b, a)$ or $(c, d) = (b, -a)$.
   (b) Show that if $(c, d) = (-b, a)$, then M is the matrix of a rotation about the origin.
   (c) Show that if $(c, d) = (b, -a)$, then M is the matrix of a reflection in a line through the origin in the direction of an eigenvector of M.

In Exercises 8 and 9, let $f: \mathbb{R}^3 \longrightarrow \mathbb{R}^3$ be a function that preserves lengths, and let A be its matrix relative to the standard basis.

8. Show that because f preserves lengths, every eigenvalue of f has absolute value 1, so every real eigenvalue is either 1 or -1.

*9. Because the characteristic polynomial of A has degree three, it has at least one real root, so f has at least one eigenvalue $\lambda = \pm 1$. Let M be the matrix of f relative to an orthonormal basis $\{\mathbf{u}_1, \mathbf{u}_2, \mathbf{u}_3\}$ in which $\mathbf{u}_1$ is an eigenvector of U associated with λ. The columns of M are an orthonormal set, because it is the matrix of f relative to an orthonormal basis.
   (a) Show that M has the form $\begin{pmatrix} \lambda & 0 & 0 \\ 0 & a & b \\ 0 & c & d \end{pmatrix}$, with the submatrix $\begin{pmatrix} a & b \\ c & d \end{pmatrix}$ having columns forming an orthonormal set.
   (b) Show that if $\lambda = +1$, then f is either a rotation with axis $\mathbf{u}_1$ or a reflection in a plane containing $\mathbf{u}_1$.
   (c) Show that if $\lambda = -1$, then f is either reflection in a line perpendicular to $\mathbf{u}_1$ or the composition of a rotation with axis $\mathbf{u}_1$ and reflection in the plane perpendicular to $\mathbf{u}_1$.

Chapter 3 REVIEW

In Exercises 1 to 4 determine whether the given set of vectors is independent.

1. $\mathbf{x} = (2, 3), \mathbf{y} = (-1, 2), \mathbf{z} = (1, 0)$

2. $\mathbf{x} = (0, 1, 2), \mathbf{y} = (0, 0, 1), \mathbf{z} = (1, 0, 0)$

3. $\mathbf{x} = (-1, 2, 0, 3), \mathbf{y} = (0, 1, -1, 4), \mathbf{z} = (-1, 3, -1, 1)$

4. $\mathbf{x} = (0, 0, 0), \mathbf{y} = (1, 0, 0), \mathbf{z} = (0, 1, 0)$

In Exercises 5 to 10, find the matrix of a linear function f with the given properties.

5. The null-space of $f: \mathbb{R}^3 \longrightarrow \mathbb{R}^3$ is the plane spanned by $(1, 1, 1)$ and $(1, -1, 1)$, and $f(0, 0, 1) = (0, 0, 1)$.

6. The image of $f:\mathbb{R}^2 \longrightarrow \mathbb{R}^3$ is the plane spanned by $(1, 1, 1)$ and $(1, -1, 1)$.

7. The null-space of $f:\mathbb{R}^3 \longrightarrow \mathbb{R}^2$ is the line $\mathbf{x} = t(1, 1, 1)$, and $f(1, 0, 0) = (1, 1)$ and $f(0, 1, 0,) = (1, 2)$.

8. $f:\mathbb{R}^3 \longrightarrow \mathbb{R}^3$ has $f(1, 0, 0) = (0, 0, 1)$, $f(0, 0, 1) = (-1, 0, 1)$, and $f(\mathbf{x}) = \mathbf{x}$ for all points $\mathbf{x} = (0, t, 0)$.

9. $f:\mathbb{R}^3 \longrightarrow \mathbb{R}^3$ has $f(2, 2, 2) = (1, 0, 1)$, and $f(\mathbf{x}) = \mathbf{x}$ for $\mathbf{x}$ in the plane spanned by $(0, 1, 1)$ and $(1, 1, 0)$.

10. $f:\mathbb{R}^4 \longrightarrow \mathbb{R}^4$ has $f(\mathbf{x}) = \mathbf{x}$ for $\mathbf{x}$ in the subspace spanned by $(1, 1, 1, 1)$ and $(2, 3, 1, 2)$ and $f(\mathbf{x}) = -\mathbf{x}$ for $\mathbf{x}$ perpendicular to the same subspace. [*Hint*: Find an orthogonal basis whose first two vectors span the given subspace.]

In Exercises 11 to 14, $\mathcal{M}_{m,n}$ is the vector space of m-by-n matrices described in Example 2 in Section 2A, and $B_{m,n}$ is the basis for $\mathcal{M}_{m,n}$ consisting of the mn matrices $\{E_{11}, E_{12}, \ldots, E_{mn}\}$, where E_{ij} is the m-by-n matrix with 1 in the ijth position and zeros elsewhere.

11. Let $f:\mathcal{M}_{2,2} \longrightarrow \mathcal{M}_{2,3}$ be defined by $f(M) =$ $M \begin{pmatrix} -1 & 0 & 2 \\ 3 & 1 & 2 \end{pmatrix}$ for M in $\mathcal{M}_{2,2}$. Find the matrix of f relative to the bases $B_{2,2}$ and $B_{2,3}$.

12. Find a basis for the image of the function f of Exercise 11.

13. Let $f:\mathcal{M}_{2,2} \longrightarrow \mathcal{M}_{2,2}$ be defined by $f(M) =$ $\begin{pmatrix} 1 & 1 \\ 1 & -2 \end{pmatrix} M \begin{pmatrix} -1 & 1 \\ 1 & -1 \end{pmatrix}$ for M in $\mathcal{M}_{2,2}$. Find the matrix of f relative to the basis $B_{2,2}$.

14. Find a basis for the null-space of the function f of Exercise 13.

15. (a) Find matrices R and S for the rotations through $90°$ about the x-axis and an angle θ about the y-axis in $\mathbb{R}^3$.

    (b) Find the axis and angle of the rotation with matrix $R^{-1}SR$.

16. Find the matrix of a rotation through $30°$ about the line $t(1, 1, 0)$ in $\mathbb{R}^3$.

In Exercises 17 to 20, find the eigenvalues of the matrix; then find associated eigenvectors and determine whether there is a basis of eigenvectors for the matrix operator.

17. $\begin{pmatrix} 1 & -3 \\ -2 & -4 \end{pmatrix}$ 18. $\begin{pmatrix} -3 & 0 \\ -4 & 1 \end{pmatrix}$

19. $\begin{pmatrix} 1 & -2 & 4 \\ 0 & 2 & 0 \\ 0 & -1 & 3 \end{pmatrix}$ 20. $\begin{pmatrix} 0 & 1 & 1 \\ 0 & 2 & 0 \\ 0 & 1 & 2 \end{pmatrix}$

21. Let $A = \begin{pmatrix} 0 & a & -b \\ -a & 0 & c \\ b & -c & 0 \end{pmatrix}$, with $a, b,$ and c real numbers.

    (a) Find the eigenvalues of A, and show that one of them is always real.

    (b) Find an eigenvector associated with the real eigenvalue.

    *(c) Find three linearly independent eigenvectors of A when $a = 1$, $b = 2$, and $c = -2$. (You'll need to use complex numbers.)

22. (a) Show that the polynomials such that $p(0) = p(1)$ form a subspace of the vector space $\mathcal{P}_3[-1, 1]$ defined in Example 11 of Section 8D, and find a basis for it.

    (b) For the same subspace find a basis that is orthogonal relative to the inner product defined in Example 11 of the text.

23. Let M be an n-by-n matrix with $M^2 = M$. Let $\{\mathbf{v}_1, \ldots, \mathbf{v}_r\}$ be a basis for the null-space of M, and $\{\mathbf{w}_1, \ldots, \mathbf{w}_s\}$ a basis for its image. Show that $\{\mathbf{v}_1, \ldots, \mathbf{v}_r, \mathbf{w}_1, \ldots, \mathbf{w}_s\}$ is a basis for $\mathbb{R}^n$, so that $r + s = n$.

24. (a) Find the 2-by-2 matrix C that converts coordinates in $\mathbb{R}^2$ relative to the standard basis $\{\mathbf{e}_1, \mathbf{e}_2\}$ into coordinates relative to the basis $\{\mathbf{v}_1, \mathbf{v}_2\} = \{(1, 1), (-1, 1)\}$.

    (b) Suppose that $f : \mathbb{R}^2 \longrightarrow \mathbb{R}^2$ has matrix $A = \begin{pmatrix} 5 & -1 \\ -1 & 5 \end{pmatrix}$ relative to the standard basis in $\mathbb{R}^2$. Compute the matrix CAC^{-1}, where C is the matrix found in part (a), and interpret the result as a matrix of the transformation f.

CHAPTER 4

DERIVATIVES

A real-valued function $f(x)$ defined on some interval $a < x < b$ has a **derivative** at a point x in its domain interval, denoted by $f'(x)$, if

$$f'(x) = \lim_{h \to 0} \frac{f(x+h) - f(x)}{h}.$$

Fundamental interpretations such as velocity and slope give the derivative a primary place in applied mathematics and in geometry, and the formulas of one-variable calculus provide techniques for dealing with the functions such as the trigonometric and exponential functions that arise in these areas.

We'll assume that the reader knows the rules and elementary examples of calculus, and that pictures are familiar that show the graph of a function $f(x)$ along with its tangent line at a point x_0 as in Figure 4.1. The purpose of this chapter is to begin extending the definition and interpretations of the derivative from real-valued functions of a real variable, associated with formulas such as $y = f(x)$, to vector-valued functions of a vector variable. The resulting notational change required in this last formula is fairly slight, since we'll now write $\mathbf{y} = f(\mathbf{x})$, but the supply of applications and interpretations will increase considerably. The elementary techniques and examples of one-variable calculus will continue to play an important role throughout the rest of the book. Thus what we're about to do will provide a review of that material.

FIGURE 4.1

Graph with tangent line.

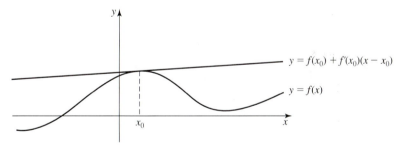

$y = f(x_0) + f'(x_0)(x - x_0)$

$y = f(x)$

When we speak of a function f with **domain** D in $\mathbb{R}^n$ and **range** or **image** in $\mathbb{R}^m$, we mean a function defined for all x in D and taking its values in some subset of $\mathbb{R}^m$. We'll sometimes use the notation

$$f : D \to \mathbb{R}^m$$

for such functions. When we don't need to call attention to the precise domain D of f we'll sometimes use the notation

$$\mathbb{R}^n \xrightarrow{\ f\ } \mathbb{R}^m$$

to describe a function with domain a subset of $\mathbb{R}^n$ and image a subset of $\mathbb{R}^m$.

We often describe a function f from $\mathbb{R}^n$ to $\mathbb{R}^m$ by a collection of m real-valued functions called **coordinate functions**. The coordinate functions will have the same domain as f itself, and if

$$f(\mathbf{x}) = (f_1(\mathbf{x}), \dots, f_k(\mathbf{x}), \dots, f_m(\mathbf{x})),$$

then the real-valued function f_k is called the kth coordinate function of f. For example, if

$$f(x, y) = (x^2 - y^2, 2xy),$$

then $f_1(x, y) = x^2 - y^2$ and $f_2(x, y) = 2xy$ are the coordinate functions of f. The **linear** functions $f(\mathbf{x})$ from $\mathbb{R}^n$ to $\mathbb{R}^m$ are just those functions with domains the whole space $\mathbb{R}^n$ whose m real-valued coordinate functions have the simple form

$$f_i(x_1, \dots, x_n) = a_{i1}x_1 + a_{i2}x_2 + \cdots + a_{in}x_n.$$

In terms of an m-by-n matrix $A = (a_{ij})$, the image points $\mathbf{y}$ of a linear function f are just the matrix-vector product $f(\mathbf{x}) = A\mathbf{x}$ of Definition 2.1 in Chapter 2, where $\mathbf{x}$ is in $\mathbb{R}^n$. For example the function $f : \mathbb{R}^2 \longrightarrow \mathbb{R}^2$ given by

$$\begin{aligned} y_1 &= x_1 + x_2 \\ y_2 &= x_1 - x_2 \end{aligned} \quad \text{has matrix form} \quad \begin{pmatrix} y_1 \\ y_2 \end{pmatrix} = \begin{pmatrix} 1 & 1 \\ 1 & -1 \end{pmatrix} \begin{pmatrix} x_1 \\ x_2 \end{pmatrix}.$$

SECTION 1   FUNCTIONS OF ONE VARIABLE

Here we take up the most straightforward generalization of calculus for real-valued functions of one real variable: vector-valued functions $\mathbf{x} = f(t)$ of a real variable t. An important difference between vector-valued functions and real-valued functions is one of geometric interpretation: for vector functions we usually study the *image* of f, namely the vector values actually taken by f, rather than the graph of an equation $\mathbf{x} = f(t)$. For notation we'll sometimes write vectors as columns instead of rows with comma separations; this practice sometimes results in a more readable display and is often required in the context of matrix multiplication.

1A   Derivatives

If a point moves in space so as to occupy various positions at a progression of times, then its position at time t generates a vector-valued position function f with values $f(t)$. In particular if the position of a point in $\mathbb{R}^3$ at time t is given by

$$f(t) = t\mathbf{x}_1 + \mathbf{x}_0,$$

where $\mathbf{x}_1$ and $\mathbf{x}_0$ are fixed vectors in $\mathbb{R}^3$, then the point is moving on a straight line in $\mathbb{R}^3$ parallel to $\mathbf{x}_1$ and passing through $\mathbf{x}_0$ as in Figure 4.2(a). More generally, a function f taking values in $\mathbb{R}^n$ is typically defined in the form

$$f(t) = \big(f_1(t), \dots, f_n(t)\big),$$

where the coordinate functions $f_1(t), \dots, f_n(t)$ denote the real-valued coordinates of a point in $\mathbb{R}^n$ at times t. This generalization to dimensions higher than 2 or 3 is

FIGURE 4.2

(a) Line, (b) Parabola.

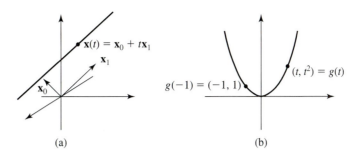

(a)   (b)

not a purely theoretical concept; for instance it's crucial for describing the dynamics of planetary motion in Chapter 12, Section 3.

EXAMPLE 1   If $\mathbf{x}_1 = (x_1, y_1, z_1)$ and $\mathbf{x}_0 = (x_0, y_0, z_0)$ are points in $\mathbb{R}^3$, then the function $\mathbb{R} \xrightarrow{f} \mathbb{R}^3$ with image the points $\mathbf{x}(t)$ given by

$$\mathbf{x}(t) = f(t) = t(x_1, y_1, z_1) + (x_0, y_0, z_0)$$
$$= (tx_1 + x_0, ty_1 + y_0, tz_1 + z_0)$$

gives a parametric representation of a line in $\mathbb{R}^3$ as in Figure 4.2(a).

EXAMPLE 2   The function g from $\mathbb{R}$ to $\mathbb{R}^2$ for which

$$g(t) = (t, t^2)$$

describes a curve in $\mathbb{R}^2$. Because the coordinates $x = t$ and $y = t^2$ satisfy the relation $y = x^2$, the point (t, t^2) always lies on the parabola with equation $y = x^2$, shown in Figure 4.2(b).

We define the **limit** of a vector-valued function f with values in $\mathbb{R}^n$ by using limits of the real-valued coordinate functions f_k of f. Thus if

$$f(t) = \big(f_1(t), \ldots, f_n(t)\big)$$

is defined for an interval $a < t < b$ containing t_0, we write

$$\lim_{t \to t_0} f(t) = \left(\lim_{t \to t_0} f_1(t), \ldots, \lim_{t \to t_0} f_n(t) \right).$$

Similarly a function with values in $\mathbb{R}^n$ is said to be **continuous** if its real-valued coordinate functions are all continuous on their common domain interval. These definitions are treated more generally in Chapter 5, Section 1.

EXAMPLE 3   The function defined by $g(t) = (t, t^2)$ has limit vector $(2, 4)$ at $t = 2$ because

$$\lim_{t \to 2}(t, t^2) = \left(\lim_{t \to 2} t, \lim_{t \to 2} t^2 \right)$$
$$= (2, 4).$$

FIGURE 4.3

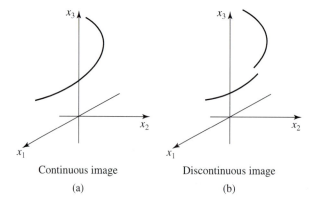

Continuous image   Discontinuous image
      (a)   (b)

The function g is continuous for all real t because the coordinate functions t and t^2 are continuous.

   The intuitive idea behind continuity of a vector-valued function is similar to that for a real-valued function: The values of the function should not change abruptly. Figure 4.3(a) shows the image of a continuous function and Figure 4.3(b) shows the image of a discontinuous function, both from domain in $\mathbb{R}$ to range space $\mathbb{R}^3$. For now we'll consider only continuous functions $\mathbb{R} \xrightarrow{g} \mathbb{R}^n$ with $g(t)$ defined on an open interval $a < t < b$. We'll first define the derivative of g and show how it leads to a definition of tangent line to the image curve of g.

   A function $g(t)$ has a **derivative** $g'(t)$ at a point t in an interval $a < t < b$ if

$$g'(t) = \lim_{h \to 0} \frac{g(t+h) - g(t)}{h},$$

assuming the limit exists. If the limit exists for each t in (a, b), then $g'(t)$ determines a new function $\mathbb{R} \xrightarrow{g'} \mathbb{R}^n$, just as in the case $n = 1$. The derivative is often written dg/dt.

EXAMPLE 4   Let $g(t) = (t^2, t^3)$. Then writing $g(t)$ as a column vector, we have

$$\lim_{h \to 0} \frac{g(t+h) - g(t)}{h} = \lim_{h \to 0} \frac{1}{h} \begin{pmatrix} (t+h)^2 - t^2 \\ (t+h)^3 - t^3 \end{pmatrix}$$

$$= \lim_{h \to 0} \begin{pmatrix} \dfrac{(t+h)^2 - t^2}{h} \\ \dfrac{(t+h)^3 - t^3}{h} \end{pmatrix}.$$

The two entries in this vector have as limits the derivatives of t^2 and t^3, respectively. Hence by the definition of the derivatives,

$$\lim_{h \to 0} \frac{(t+h)^2 - t^2}{h} = 2t \quad \text{and} \quad \lim_{h \to 0} \frac{(t+h)^3 - t^3}{h} = 3t^2.$$

By the definition of vector limit the vector limit $g'(t)$ exists, and $g'(t) = (2t, 3t^2)$.

Example 4 suggests that a function $\mathbb{R} \xrightarrow{g} \mathbb{R}^n$ has a derivative at a point t if and only if each coordinate function of g has a derivative there. This is true, and we have

1.1
$$\text{If } g(t) = \begin{pmatrix} g_1(t) \\ \vdots \\ g_n(t) \end{pmatrix}, \quad \text{then} \quad g'(t) = \begin{pmatrix} g'_1(t) \\ \vdots \\ g'_n(t) \end{pmatrix},$$

where each derivative $g'_k(t)$ is an ordinary derivative of a real-valued function of a real variable t. The resulting vector expression for $g'(t)$ is an immediate consequence of the definition of the limit function in terms of limits of its coordinate functions.

EXAMPLE 5   If $g(t) = \begin{pmatrix} \cos t \\ \sin t \end{pmatrix}$, then $g'(t) = \begin{pmatrix} -\sin t \\ \cos t \end{pmatrix}$. Note that as t varies the curve traced in $\mathbb{R}^2$ by $g(t)$ is a circle of radius 1 centered at the origin. Indeed $|g(t)| = \sqrt{\cos^2 t + \sin^2 t} = 1$, and the geometric definition of $\cos t$ and $\sin t$ is based on the interpretation of t as the counterclockwise angle that the radius at $g(t)$ makes with the positive horizontal axis, as shown in Figure 4.4(a). By a similar argument, $g(-t)$ also traces the same circle, but in the clockwise direction.

EXAMPLE 6   If $h(t) = \begin{pmatrix} t \\ t^2 \\ t^3 \end{pmatrix}$, then $h'(t) = \begin{pmatrix} 1 \\ 2t \\ 3t^2 \end{pmatrix}$. For $0 \le t \le 1$, the points $h(t)$ trace the curve in $\mathbb{R}^3$ shown in Figure 4.4(b). As a guide for sketching this curve, observe that its perpendicular projection into the xy-plane is the parabola, $y = x^2$, and its projection into the xz-plane is the cubic $z = x^3$. The projection $y = z^{2/3}$ into the yz-plane is less familiar; for that interesting curve see Example 9.

Figure 4.4(c) shows that, as h tends to 0, the vector $g(t+h) - g(t)$ has a direction that should tend to what we would like to call the tangent direction to the curve γ at $g(t)$. However, since g is assumed continuous,

$$\lim_{h \to 0} g(t + h) - g(t) = 0,$$

and the zero vector that we get as a limit has no direction. The standard way to overcome this difficulty is to divide by h before letting h tend to zero. Observe that

FIGURE 4.4

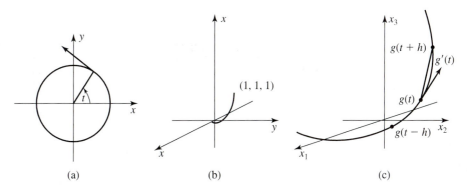

(a)   (b)   (c)

division by h will not change the direction of $g(t + h) - g(t)$ if h is positive; it will reverse it if h is negative. A glance at Figure 4.4(c) shows that this reversal is desirable for our purposes, because we want the tangent vector to point in the direction of increasing t along the curve. (What would happen if we divided by $|h|$ instead?) If the derivative $g'(t)$ exists and is not zero, then $g'(t)$ defines the **standard tangent vector** to γ at $g(t)$. A positive multiple of $g'(t)$ has the same direction as the standard tangent and so is also called a tangent vector. In this context we'll often use the notation $\mathbf{x}(t)$ for position instead of $g(t)$ and also use Newton's **overdot notation** $\dot{\mathbf{x}}(t) = g'(t)$ for the vector derivative, particularly when t is interpreted as a time parameter.

The tangent vector arrow $\dot{\mathbf{x}}(t_0)$ is usually pictured so that its tail is at $\mathbf{x}(t_0)$ as in Figure 4.4(c). The line with direction vector $\dot{\mathbf{x}}(t) = g'(t)$ and passing through $g(t)$ is called the **tangent line** to γ at $\mathbf{x}(t)$. Thus if $\mathbf{x}(t_0)$ is a particular point on a curve, the tangent line at $\mathbf{x}(t_0)$ will have a parametric representation of the form

$$\mathbf{t}(t) = t\dot{\mathbf{x}}(t_0) + \mathbf{x}(t_0).$$

EXAMPLE 7 The circle of Example 5 has points $\mathbf{x}(t) = (\cos t, \sin t)$ and tangent vector $\dot{\mathbf{x}}(t) = (-\sin t, \cos t)$. A typical tangent vector appears in Figure 4.4(a).

Definition If a curve has a parametric representation as the image of a function $g(t)$ such that the derivative $g'(t)$ is (i) continuous, and (ii) never zero, then the curve is called **smooth**.

The condition that $g'(t)$ be nonzero requires the curve to have a well-defined tangent line at every point.

The condition that $g'(t)$ be continuous means that the direction and length of the tangent vector $g'(t)$ change continuously as the point $g(t)$ moves along the curve. Here's an example of a smooth curve that we'll encounter often.

EXAMPLE 8 The image curve defined parametrically by $\mathbf{x}(t) = (\cos t, \sin t, t)$ lies on the cylinder of radius 1 shown in Figure 4.5(a). The image curve is called a **helix**. If we temporarily set the third coordinate function of $\mathbf{x}(t)$ equal to 0, the image is a circle of radius 1 centered at $(0, 0, 0)$, because

$$|(\cos t, \sin t, 0)| = \sqrt{\cos^2 t + \sin^2 t} = 1.$$

But with t again in the third coordinate, the image point rises as t increases, in the direction of the z-axis, as shown in Figure 4.5(a), lying all the while above the points of the circle. A tangent vector is

$$\dot{\mathbf{x}}(t) = (-\sin t, \cos t, 1)$$

and the tangent line to the helix at $\mathbf{x}(0) = (1, 0, 0)$ has the parametric representation

$$\mathbf{t}(t) = t\dot{\mathbf{x}}(0) + \mathbf{x}(0)$$
$$= t(0, 1, 1) + (1, 0, 0).$$

Note that $\dot{\mathbf{x}}(t)$ is a continuous function, and is never zero, so the helix is a smooth curve.

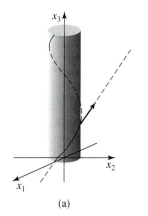

EXAMPLE 9

If a point moves in the plane so that at time t its position is $\mathbf{x}(t) = (t^2, t^3)$, then the tangent vector is $\dot{\mathbf{x}}(t) = (2t, 3t^2)$, with length $|\dot{\mathbf{x}}(t)| = (4t^2 + 9t^4)^{1/2}$. In particular, $\dot{\mathbf{x}}(0) = 0$. The sketch of the path traced by $\mathbf{x}(t)$ is in Figure 4.5(b) for $-1 \le t \le 1$. In making the picture it's helpful to observe that the coordinates of a point on the path satisfy the equation $x = y^{2/3}$; since $x = t^2 \ge 0$ here we also have $y = x^{3/2}$.

The tangent vector shrinks to zero in this example as $\mathbf{x}(t)$ approaches the origin because, with continuously varying $\dot{\mathbf{x}}(t)$, its length becomes instantaneously zero at the abrupt change in the direction of motion shown in Figure 4.5(b). In this way the parametrization describes the geometric situation well. The curve certainly doesn't deserve to be called smooth at $(0, 0)$, and is said to have a **cusp** there.

We list here some useful formulas that hold if two vector-valued functions $\mathbf{x}(t) = f(t)$ and $\mathbf{y}(t) = g(t)$ have vector derivatives on an interval $a < t < b$; we assume $\phi(t)$ and $u(t)$ are real valued and differentiable on the same interval.

1.2
$$\frac{d}{dt}(\mathbf{x} + \mathbf{y}) = \dot{\mathbf{x}} + \dot{\mathbf{y}}, \quad \frac{d}{dt}(c\mathbf{x}) = c\dot{\mathbf{x}}, \quad c \text{ constant}$$

1.3
$$\frac{d}{dt}(\phi\mathbf{x}) = \phi\dot{\mathbf{x}} + \phi'\mathbf{x}$$

1.4
$$\frac{d}{dt}(\mathbf{x} \cdot \mathbf{y}) = \dot{\mathbf{x}} \cdot \mathbf{y} + \mathbf{x} \cdot \dot{\mathbf{y}}$$

1.5 $\dfrac{d}{dt}\big(\mathbf{x}(u)\big) = u'(t)\dot{\mathbf{x}}(u)$,   with u taking values in the domain of $\mathbf{x}(t)$

1.6 $\dfrac{d}{dt}(\mathbf{x} \times \mathbf{y}) = \dot{\mathbf{x}} \times \mathbf{y} + \mathbf{x} \times \dot{\mathbf{y}}$,   if $\mathbf{x}$ and $\mathbf{y}$ take values in $\mathbb{R}^3$

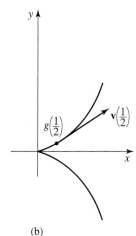

FIGURE 4.5

(a) Helix, (b) Cusp.

The preceding formulas all follow from writing $\mathbf{x} = f(t)$ and $\mathbf{y} = g(t)$ in terms of their coordinate functions and then applying the corresponding differentiation formulas for real-valued functions along with Formula 1.1. For example, the proof of 1.5, a version of the chain rule for differentiation, goes like this:

$$\frac{d}{dt}\mathbf{x}(u) = \big(f_1(u), \dots, f_n(u)\big)'$$
$$= ([f_1(u)]', \dots, [f_n(u)]')$$
$$= (f_1'(u)u', \dots, f_n'(u)u') = u'\dot{\mathbf{x}}(u).$$

1B   Velocity and Speed

One reason for singling out $\dot{\mathbf{x}}(t) = g'(t)$ for special attention as the *standard* tangent vector, rather than some multiple of it, is that we often want to consider the parameter t as a time variable, with $\mathbf{x}(t)$ tracing the path of a point moving in $\mathbb{R}^n$. Under this interpretation, the Euclidean length $|\dot{\mathbf{x}}(t)| = |g'(t)|$ is the natural definition for the

speed of motion along the path γ described by $g(t)$ as t varies. To justify the use of the term *speed*, we observe that, for small h, the number $|g(t+h) - g(t)|/|h|$ is close to the average rate of traversal of γ over a sufficiently short interval from t to $t + h$. In addition, if $g'(t)$ exists, we'll now show that

$$\lim_{h \to 0} \frac{|g(t+h) - g(t)|}{|h|} = |g'(t)|.$$

By the triangle inequality in the reversed form $\big||\mathbf{x}| - |\mathbf{y}|\big| \leq |\mathbf{x} - \mathbf{y}|$, (see p. 32),

$$\left|\frac{|g(t+h) - g(t)|}{|h|} - |g'(t)|\right| \leq \left|\frac{g(t+h) - g(t)}{h} - g'(t)\right|.$$

The right side tends to zero as h tends to zero by the definition of $g'(t)$. Hence the left side tends to zero also. Thus $|g'(t)|$ is a limit of average rates over arbitrarily small time intervals. It's for this reason that the real-valued function v defined by $v(t) = |g'(t)|$ is called the speed of g. It follows that it's natural to call the vector $\mathbf{v}(t) = g'(t)$ the **velocity vector** of the motion at the point $g(t)$. Note that the vector $\mathbf{v}(t)$ is identical to what we called the standard tangent vector to γ at $g(t)$ if $\mathbf{v}(t) \neq 0$. Velocity $\mathbf{v}(t) = \mathbf{0}$ indicates speed zero and no direction at time t.

| EXAMPLE 10 |

Let $\mathbf{x}(t) = (a\cos t, a\sin t, bt)$ with a and b nonzero constants. This is a more general helix than the one in Example 8, where we took $a = b = 1$. Figure 4.6(a) shows the choice $a = 1, b = \frac{1}{2}$ along with $a = -1, b = \frac{1}{2}$ as a dotted curve. The two together outline the general configuration of the double helix portion of the DNA molecule. The velocity at time t is $\dot{\mathbf{x}}(t) = \mathbf{v}(t) = (-a\sin t, a\cos t, b)$. It follows that the velocity vector is always perpendicular to the vector $\mathbf{r}(t) = (a\cos t, a\sin t, 0)$, which points horizontally from the axis of the spiral to $\mathbf{x}(t)$. To see this just check that

$$\mathbf{v}(t) \bullet \mathbf{r}(t) = (-a\sin t, a\cos t, b) \bullet (a\cos t, a\sin t, 0) = 0.$$

The speed at time t is equal to the constant $\sqrt{a^2 + b^2}$, because

$$|\mathbf{v}(t)| = |(-a\sin t, a\cos t, b)|$$

$$= \sqrt{a^2 \sin^2 t + a^2 \cos^2 t + b^2} = \sqrt{a^2 + b^2}.$$

FIGURE 4.6

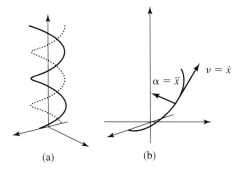

(a)   (b)

1C   Higher-Order Derivatives; Acceleration

If $\mathbb{R} \xrightarrow{g} \mathbb{R}^n$ has a derivative $\mathbb{R} \xrightarrow{g'} \mathbb{R}^n$, then we can ask for the derivative of g', which we denote by g'' or d^2g/dt^2. It may happen that g'' is defined at fewer points than g or g'. We write $g^{(3)}$ or d^3g/dt^3, and so on for higher-order derivatives, though these will occur rarely in what follows. The reason is that derivatives of order $n > 2$ don't have as interesting and intuitively appealing interpretations as the first-order and second-order ones.

EXAMPLE 11

Suppose that $g(t) = (r \cos \omega t, r \sin \omega t, ct)$, where r, c, and ω are positive constants. Then computing first and second derivatives one coordinate at a time gives

$$g'(t) = (-r\omega \sin \omega t, r\omega \cos \omega t, c) \quad \text{and} \quad g''(t) = (-r\omega^2 \cos \omega t, -r\omega^2 \sin \omega t, 0).$$

Suppose $\mathbb{R} \xrightarrow{g} \mathbb{R}^3$ describes a path in $\mathbb{R}^3$ with velocity vector $\mathbf{v}(t) = g'(t)$. If we assume that g' itself has a derivative, we define the **acceleration vector** at $g(t)$ by $\mathbf{a}(t) = g''(t)$. If $\mathbf{x}(t)$ is used to denote the image points of some curve, then along with $\dot{\mathbf{x}}(t)$ for velocity vectors we may denote acceleration vectors by $\ddot{\mathbf{x}}(t)$.

The physical significance of acceleration $\mathbf{a}(t)$ is that if $\mathbf{x}(t)$ describes the motion of a particle of constant mass m, then $\mathbf{F}(t) = m\mathbf{a}(t)$ is by definition the **force vector** acting on the particle. If we denote by $a(t)$ the length of $\mathbf{a}(t)$, then $a(t)$ is called the **magnitude** of the acceleration, and $ma(t)$ is called the **magnitude** of the force acting on the particle. We detect the presence of acceleration that isn't parallel to the velocity at $\mathbf{x}(t)$ by observing a bending of the path of motion away from the straight line through the tangent vector $\dot{\mathbf{x}}(t)$ and toward the direction of $\ddot{\mathbf{x}}(t)$. Look at Figure 4.6(b) and think of the sideways pull that you feel when going around a tight curve at high speed. The previous example illustrates the basic idea; there the acceleration points in a direction perpendicular to the vertical axis of the helix since the third coordinate of $g''(t)$ is always zero. Section 1F provides another illustration, and there is a more general treatment in Chapter 8, Section 3.

EXAMPLE 12

If the vector $\mathbf{x}(t) = (r \cos \omega t, r \sin \omega t, ct)$ gives the position at time t of a particle of mass m in $\mathbb{R}^3$, then the velocity and acceleration vectors, $\mathbf{v}(t) = g'(t)$ and $\mathbf{a}(t) = g''(t)$, are as computed in Example 11, namely

$$\dot{\mathbf{x}}(t) = (-r\omega \sin \omega t, r\omega \cos \omega t, c) \quad \text{and} \quad \ddot{\mathbf{x}}(t) = (-r\omega^2 \cos \omega t, -r\omega^2 \sin \omega t, 0).$$

Typical velocity and acceleration vectors are shown in Figure 4.6(b), located appropriately with tails at $\mathbf{x}(t)$. The lengths of these vectors just *happen* to be constant as functions of time t, depending only on the constants r, w and c as follows:

$$|\mathbf{v}(t)| = \sqrt{(r\omega \sin \omega t)^2 + (r\omega \cos \omega t)^2 + c^2} = \sqrt{r^2\omega^2 + c^2}$$

$$|\mathbf{a}(t)| = \sqrt{(r\omega^2 \cos \omega t)^2 + (r\omega^2 \sin \omega t)^2} = r\omega^2.$$

Note that while the speed and velocity depend on c, the acceleration doesn't.

1D   Arc Length

For a point moving with constant speed v along a curve, the distance covered between time t_0 and time t_1 should turn out to be v times $t_1 - t_0$. More generally, it's natural to obtain the distance along a parametrized curve γ between $t = t_0$ and $t = t_1$ by defining it to be the integral of the speed $v(t)$ with respect to time to get **arc length** $l(\gamma)$:

$$l(\gamma) = \int_{t_0}^{t_1} v(t) \, dt.$$

EXAMPLE 13　If a circle of radius a is parametrized in $\mathbb{R}^2$ by

$$g(t) = (a \cos \omega t, a \sin \omega t), \quad \omega > 0,$$

then

$$v(t) = |g'(t)|$$
$$= |(-a\omega \sin \omega t, a\omega \cos \omega t)|$$
$$= a\omega\sqrt{\sin^2 \omega t + \cos^2 \omega t} = a\omega.$$

Thus the distance covered between times t_0 and t_1 is

$$l = \int_{t_0}^{t_1} a\omega \, dt = a\omega(t_1 - t_0).$$

The constant ω is called the **angular speed** of the motion on the circle.

　　Image curves that appear to be the same may have different parametrizations that yield different arc lengths. The following simple example shows that a little care is needed to avoid producing inconsistent results from different parametrizations.

EXAMPLE 14　Let $g(t) = (\cos t, \sin t)$, for $0 \le t \le 2\pi$ and $h(u) = (\cos u, \sin u)$, for $0 \le u \le 4\pi$. These functions have the same image, namely a circle of radius 1, but h traces the circle twice. Since $|g'(t)| = |h'(u)| = 1$,

$$\int_0^{2\pi} |g'(t)| \, dt = \int_0^{2\pi} dt = 2\pi, \quad \text{while} \quad \int_0^{4\pi} |h'(u)| \, du = \int_0^{4\pi} du = 4\pi.$$

Verification of some formal conditions that guarantee equal arc lengths for different parametrizations is left as Exercise 44.

EXERCISES

Find the derivatives $f'(t)$ and $f''(t)$ for each of the following functions 1 to 6 at the indicated point. Then find a parametric representation for the tangent line at each indicated point.

1. $f(t) = (1 + t^2, 1 + t^3)$, when $t = 2$

2. $f(t) = (t \cos t, t \sin t)$, when $t = \pi/2$

3. $f(t) = \begin{pmatrix} e^t \\ e^{2t} \end{pmatrix}$, when $t = -1$

4. $f(t) = (t + t^2, t^2 + t^3, t^3 + t^4)$, when $t = -1$

5. $f(t) = (\cos t, \cos 2t, \cos 3t, \cos 4t)$ when $t = \pi/2$

6. $f(t) = t\mathbf{i} + t^2\mathbf{j} + t^3\mathbf{k}$ when $t = 1$

Sketch the curves defined parametrically by the following functions 7 to 12.

7. $f(t) = t(1, 2, 0) + (1, 1, 1), -\infty < t < \infty$

8. $f(t) = (t, t^2, t^3), 0 \le t \le 1$

9. $f(t) = (2t, t), -1 \le t \le 1$

10. $h(t) = t\mathbf{i} + t\mathbf{j} + t^2\mathbf{k}, -1 \le t \le 2$

11. $f(t) = (2t, |t|), -1 \le t \le 2$

12. $f(t) = (\cos t, \sin t, t), 0 \le t \le 2\pi$

13. Suppose that temperature at a point (x, y, z) in $\mathbb{R}^3$ is $T(x, y, z) = x^2 + y^2 + z^2$. A particle moves so that at time t its location is given by $(x, y, z) = (t, t^2, t^3)$. Find the temperature at the point occupied by the particle at $t = \frac{1}{2}$. What is the rate of change of the temperature at the particle when $t = \frac{1}{2}$?

14. Show that $d(\mathbf{x} \cdot \mathbf{x})/dt = 2\mathbf{x} \cdot \dot{\mathbf{x}}$ if $\dot{\mathbf{x}}$ exists.

15. Use the result of the previous exercise to show that if a curve is traced with constant speed, then the velocity and acceleration vectors are always perpendicular.

16. A point has position at time t given by $(t, t^2, 1 + t^2)$ for $0 \le t \le 1$. At time $t = 1$ the point leaves this curve and flies off along the tangent line while maintaining the constant speed attained at $t = 1$. Where is the point at $t = 2$?

17. If $g(t) = (e^t, t)$ for all real t, sketch in $\mathbb{R}^2$ the curve described by g together with the tangent vectors $g'(0)$ and $g'(1)$.

18. Let $f(t) = (t, t^2, t^3)$ for $0 \le t \le 1$.
  (a) Sketch the curve described by f in $\mathbb{R}^3$ and the tangent line at $\left(\frac{1}{2}, \frac{1}{4}, \frac{1}{8}\right)$.
  (b) Find $|f'(t)|$.

19. If $f(t) = (t, t^2, t^3)$ for all real t, find all points of the curve described by f at which the tangent vector is parallel to the vector $(4, 4, 3)$. Are there points at which the tangent is perpendicular to $(4, 4, 3)$?

20. Sketch the curve represented by $(x, y) = (t^3, t^5)$, and show that the parametrization fails to assign a tangent vector at the origin. Find a parametrization of the curve that does assign a tangent at the origin. Is the curve smooth?

21. Let $g(t) = (\sin 2t, 2\sin^2 t, 2\cos t)$ and show that the image curve lies on a sphere centered at the origin in $\mathbb{R}^3$. Find the length of the velocity vector $\mathbf{v}(t)$ and show that the projection of this vector into the xy-plane has a constant length.

22. Show that if f is vector valued, differentiable, and never zero for $a < t < b$, then
  (a) $f \cdot \dfrac{df}{dt} = |f|\dfrac{d|f|}{dt}$
  (b) $|f|$ is constant if and only if $f \cdot f' = 0$

Sketch the following four parametrized curves between the given parameter values t_0 and t_1. Then find the speed $v(t)$ and calculate the arc length between t_0 and t_1.

23. $f(t) = (3t, 4t); t_0 = 0, t_1 = 4$

24. $g(t) = (2\cos t, 2\sin t); t_0 = 0, t_1 = \pi/2$

25. $h(t) = (t, 2t^{3/2}); t_0 = 0, t_1 = \frac{5}{3}$

26. $c(t) = (t, \cosh t), 0 \le t \le a$

In 27 and 28 a planet orbits a fixed star in a circular path of radius a with constant angular speed ω. We parametrize the orbital motion by $\mathbf{x}(t) = (a\cos\omega t, a\sin\omega t)$. A moon orbits the planet in a circular path in the plane of the planet's path and of radius $b < a$ with constant angular speed δ relative to the planet. The relative masses of the three bodies are such that we neglect the gravitational attraction between moon and the star.

27. Find a parametric representation for the path of the moon relative to the fixed star assuming that the three bodies are in line at $t = 0$.

28. Find the speed $v(t)$ of the moon at time t. Under what conditions is $v(t)$ constant?

Differentiation Formula 1.5 listed at the end of Section 1A is proved in the text. Using similar ideas, do 29 to 32.

29. Prove Formula 1.2 in the text.

30. Prove Formula 1.3 in the text.

31. Prove Formula 1.4 in the text.

32. Prove Formula 1.6, where $\mathbf{x}$ and $\mathbf{y}$ take values in $\mathbb{R}^3$ and are differentiable on an interval.

33. Show that if $\mathbb{R} \xrightarrow{g} \mathbb{R}^n$ has a derivative and $g'(t) = \mathbf{0}$ for $a < t < b$, then $g(t)$ is a constant vector on that interval. [*Hint:* Consider the coordinate functions one at a time.]

In 34 and 35 let a differentiable function $g(t)$ represent the position in $\mathbb{R}^3$ at time t of a particle of possibly varying mass $m(t)$. The vector function $\mathbf{P}(t) = m(t)\mathbf{v}(t)$ is called the **linear momentum** of the particle. The **force vector** is $\mathbf{F}(t) = \big(m(t)\dot{\mathbf{v}}(t)\big)$. The **angular momentum** about the origin is $\mathbf{L}(t) = g(t) \times \mathbf{P}(t)$, and the **torque** about the origin is $\mathbf{N}(t) = g(t) \times \mathbf{F}(t)$. Apply these ideas to the next two exercises.

34. Show that if $\mathbf{F}$ is identically zero, then $\mathbf{P}$ is constant. This is called the **law of conservation of linear momentum.**

35. Show that $\mathbf{L}'(t) = \mathbf{N}(t)$, and hence that if $\mathbf{N}$ is identically zero, then $\mathbf{L}$ is constant. This is called the **law of conservation of angular momentum.**

36. Show that if a particle has an acceleration vector $\mathbf{a}(t)$ at time t and $v(t) \ne 0$, then $v' = \mathbf{t} \cdot \mathbf{a}$, where $\mathbf{t}$ is the unit vector $(1/v)\mathbf{v}$.

37. Show that if a, b and ω are positive constants, then the parametrization $\mathbf{x}(t) = (a\cos\omega t, b\sin\omega t)$ traces the same ellipse $x^2/a^2 + y^2/b^2 = 1$ in $\mathbb{R}^2$ regardless of the size of ω.

38. Find the velocity and acceleration vectors for $\mathbf{x}(t)$ in Exercise 37. Show that the velocity and acceleration are never zero.

Sketch the following four curves for the indicated time intervals. Then add to your sketch the velocity and acceleration vectors at the designated times.

39. $\mathbf{x}(t) = (t, t, t^2), 0 \le t \le 1; t = 0, \frac{1}{2}, 1$

40. $\mathbf{x}(t) = (2\cos t)\mathbf{i} + (\sin t)\mathbf{j}, 0 \le t \le 2\pi; t = 0, \pi/2, \pi$

41. $\mathbf{x}(t) = (t, t^2, t^3), \ 0 \le t \le 1; t = 0, \frac{1}{2}, 1$

42. $\mathbf{x}(t) = (\cos t)\mathbf{i} + (\sin t)\mathbf{j} + t\mathbf{k}, \ 0 \le t \le 2\pi; t = 0, \pi, 2\pi$

43. The **normal lapse rate** for temperature above the surface of the earth assumes a steady drop in air temperature of 3°F per 1000 feet of increase in elevation. Under this assumption, with ground temperature 32°F, and assuming negligible air resistance, estimate the temperature at time t at the height of a projectile fired straight up with an initial speed of 300 feet per second. What is the minimum temperature attained?

*44. Parametrizations $\mathbf{x} = g(t), a \le t \le b$ and $\mathbf{x} = h(u), \alpha \le u \le \beta$ are called **equivalent** if there is a continuously differentiable ϕ with $\phi' > 0$ from $[\alpha, \beta]$ onto $[a, b]$ such that $g(\phi(u)) = h(u)$. (Note that $\phi' > 0$ implies ϕ strictly increasing.)

(a) Use Equation 1.5 to show that if g and h are equivalent then $|h'(u)| = |g'(\phi(u))||\phi'(u)|$.

(b) Use part (a) to change variable in the arc-length integral for h and show that equivalent parametrizations yield equal arc lengths.

(c) Show that $g(t) = (t, t)$ for $-1 \le t \le 1$ and $h(u) = (-\cos u, -\cos u)$ for $0 \le u \le 5\pi/2$ are not equivalent parametrizations of the line segment from $(-1, -1)$ to $(1, 1)$ in $\mathbb{R}^2$ by showing that they yield different arc lengths. This example shows that the condition that the function $\phi(u)$ be increasing can't be omitted from the definition of equivalence if equal arc length is to be a consequence.

1E   Computer Plotting of Space Curves

Short of making a wire model of a 3-dimensional curve, our best recourse for depicting a curve is a perspective drawing. Standard textbooks have such drawings printed on their pages. A computer screen, like a blackboard or a piece of paper, presents us with a flat 2-dimensional surface on which we depict geometric objects. Software designed to make perspective drawings of objects in 3-dimensional space is widely available. The discussion presented here is independent of any particular software, but it serves to indicate schematically the logical routine for drawing space curves in 3-dimensional perspective. The following algorithm is typical:

```
DEFINE g₁(t)  =  2  sin (t)
DEFINE g₂(t)  =  3  cos (t)
DEFINE g₃(t)  =  .4t
FOR t  =  0 TO 4π STEP .01
PLOT3D (g₁(t), g₂(t), g₃(t))
NEXT t
```

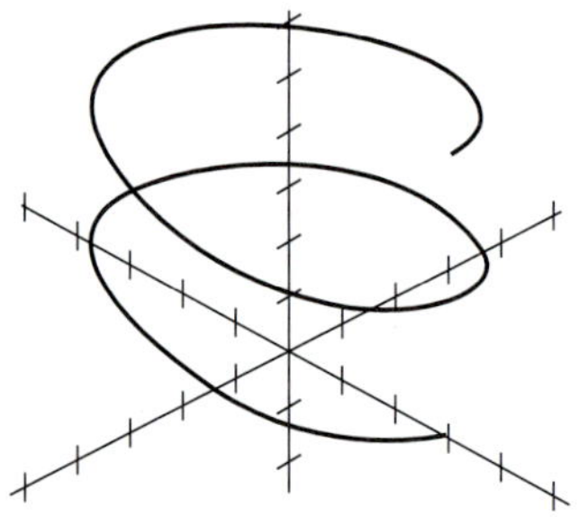

(b)

FIGURE 4.7

This algorithm plots points on an elliptical helix shown in Figure 4.7, and defined by three equations of the form $x = g_1(t), y = g_2(t), z = g_3(t)$ with $a \le t \le b$. In our particular example we get the picture shown, for which $g_1(t) = 2\sin t$, $g_2(t) = 3\cos(t), g_3(t) = 0.4t$, and $a = 0, b = 4\pi$. The viewing direction here is along a line joining the point $(1, 1, 1)$ to the origin. The order of the sine and cosine in the first two coordinate functions makes the helix turn clockwise instead of counterclockwise as it winds up around the vertical axis. Note that the curve winds around an elliptical cylinder rather than a circular one.

The decision to plot a picture by hand or by computer will usually favor the computer if a fairly high degree of accuracy is needed for some reason or if the picture is just too complicated to draw by hand. Otherwise, a quick pencil drawing may convey the necessary information with less fuss. Some of the information you might want to convey is that you understand the basic ideas of graphical representation, and this may best be done, for example on an examination, with a careful pencil drawing. For this reason it's a good idea not to become overly dependent on having computer software do your thinking for you until you've become reasonably adept at doing it for yourself. The assigned exercises will require a mixture of both approaches.

EXERCISES

Plot the following parametrically defined curves 1 to 4.

1. $g(t) = (t \cos t, t \sin t), 0 \le t \le 2\pi$

2. $f(t) = (t, \frac{1}{3}t^3, \frac{1}{2}t^4), 0 \le t \le 1$

3. $g(t) = (\sin 2t, 2 \sin^2 t, 2 \cos t), 0 \le t \le 2\pi$

4. $g(t) = (|t|, 2|t - 1|, 3|t + 1|), -2 \le t \le 2$

5. Prove that the curve in Exercise 3 lies on a sphere centered at the origin.

Plot the image curves 6 to 9 subject to the given conditions.

6. $\mathbf{x} = (\cos t, \sin t, t^2), 0 \le z \le 3$

7. $\mathbf{x} = (2 \cos t, 3 \sin t, e^t), 1 \le z \le 2$

8. $\mathbf{x} = (t, t \cos t, t \sin t), x^2 + y^2 + z^2 \le 1$

9. $\mathbf{x} = (t^2, t^3, t^4), |y| \le 5$

10. Make computer plots of lines $\mathbf{x} = t\mathbf{a} + \mathbf{b}$ in $\mathbb{R}^3$ for $c \le t \le d$ for a variety of choices of the vector and scalar parameters.

1F   Vector Integration

We've seen in Section 1D that if you know the speed $|\dot{\mathbf{x}}(t)|$ of some point moving in space, you integrate speed with respect to t to find distance measured along the path of motion from some chosen point. Finding the actual path of motion requires prior knowledge of more than just the speed; for that we need to know the velocity vector $\mathbf{v}(t) = \dot{\mathbf{x}}(t)$. Since we get from position $\mathbf{x}(t)$ to velocity $\dot{\mathbf{x}}(t)$ by vector differentiation it follows that recovering position from velocity is done by vector integration. Given a vector valued function $f(t)$ with n real-valued coordinate functions $f_1(t), \ldots, f_n(t)$, each integrable over some common interval, the indefinite **vector integral** of f is defined by

$$\int f(t)\, dt + \mathbf{c} = \left(\int f_1(t)\, dt + c_1, \int f_2(t)\, dt + c_2, \ldots, \int f_n(r)\, dt + c_n \right).$$

The vector constant of integration is $\mathbf{c} = (c_1, c_2, \ldots, c_n)$.

EXAMPLE 15   If $f(t) = (1, t^2)$, then

$$\int f(t)\, dt = \left(\int 1\, dt, \int t^2\, dt \right) = \left(t + c_1, \frac{1}{3}t^3 + c_2 \right) = \left(t, \frac{1}{3}t^3 \right) + \mathbf{c}.$$

We interpret he relationship between $f(t)$ and $F(t) = \int f(t)\, dt + \mathbf{c}$ as follows. For whatever choice of $\mathbf{c}$, the tangent vector to the image curve of F at $F(t)$ is the vector $f(t)$, usually pictured with its tail at $F(t)$. Figure 4.8(a) shows two choices for $\mathbf{c}$.

EXAMPLE 16   Suppose $\mathbf{a}$ and $\mathbf{b}$ are constant vectors in $\mathbb{R}^n$ and we want to find the position function $\mathbf{x}(t)$ consistent with velocity $\dot{\mathbf{x}}(t) = t\mathbf{a} + \mathbf{b}$, as well as with the initially specified position $\mathbf{x}(t_0)$. Since $\mathbf{x}(t)$ is an integral of $\dot{\mathbf{x}}(t)$, we have

$$\mathbf{x}(t) = \int \dot{\mathbf{x}}(t)\, dt = \int (t\mathbf{a} + \mathbf{b})\, dt = \tfrac{1}{2}t^2\mathbf{a} + t\mathbf{b} + \mathbf{c}.$$

To determine $\mathbf{c}$, we note that $\mathbf{x}(t_0) = \tfrac{1}{2}t_0^2\mathbf{a} + t_0\mathbf{b} + \mathbf{c}$, so $\mathbf{c} = \mathbf{x}(t_0) - \tfrac{1}{2}t_0^2\mathbf{a} - t_0\mathbf{b}$. Thus

$$\mathbf{x}(t) = \tfrac{1}{2}\left(t^2 - t_0^2\right)\mathbf{a} + (t - t_0)\mathbf{b} + \mathbf{x}(t_0).$$

As an alternative we use a definite integral:

$$\mathbf{x}(t) - \mathbf{x}(t_0) = \int_{t_0}^{t} \dot{\mathbf{x}}(u)\, dt = \int_{t_0}^{t} (u\mathbf{a} + \mathbf{b})\, du = \tfrac{1}{2}u^2\mathbf{a} + u\mathbf{b}\big|_{t_0}^{t}$$

$$= \tfrac{1}{2}\left(t^2 - t_0^2\right)\mathbf{a} + (t - t_0)\mathbf{b}.$$

FIGURE 4.8

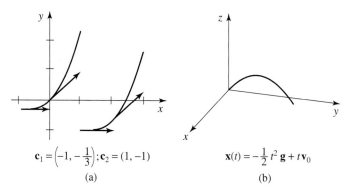

$\mathbf{c}_1 = \left(-1, -\tfrac{1}{3}\right);\ \mathbf{c}_2 = (1, -1)$ $\mathbf{x}(t) = -\tfrac{1}{2}t^2\,\mathbf{g} + t\mathbf{v}_0$

(a)   (b)

EXAMPLE 17   Suppose we fire a projectile from ground level and are willing to ignore the effects of air resistance on the flight of the projectile. (The retarding effect of air resistance is taken into account in Chapter 12, Section 3.) Thus the only acceleration we need to take into account after the initial release of the projectile is that of the vector $-\mathbf{g} = (0, 0, -g)$, where g is the magnitude of gravitational acceleration near our location on earth. Denote by $\mathbf{x} = \mathbf{x}(t)$ the position of the projectile at time t after firing, so that the velocity vector is $\mathbf{v} = \dot{\mathbf{x}}(t)$ and the acceleration vector is $\mathbf{a} = \ddot{\mathbf{x}}(t)$. Equating our two expressions for acceleration gives $\ddot{\mathbf{x}} = -\mathbf{g}$. Writing this equation is the critical step in predicting the projectile's path. To solve the equation we integrate both sides twice with respect to t getting successively

$$\dot{\mathbf{x}} = \mathbf{v}(t) = -t\mathbf{g} + \mathbf{c}_1 \quad \text{and} \quad \mathbf{x} = -\tfrac{1}{2}t^2\mathbf{g} + t\mathbf{c}_1 + \mathbf{c}_2.$$

The vectors $\mathbf{c}_1$ and $\mathbf{c}_2$ are constants of integration determined by imposing initial conditions at time $t = 0$. We place the origin at the firing point, so that $\mathbf{x}(0) = 0$. It follows that $\mathbf{c}_2 = 0$. Denote the initial velocity vector by $\mathbf{v}_0$, so that $\mathbf{x}'(0) = \mathbf{v}(0) = \mathbf{v}_0$. It follows that $\mathbf{c}_1 = \mathbf{v}_0$. Thus the solution to our problem is $\mathbf{x}(t) = -\tfrac{1}{2}t^2\mathbf{g} + t\mathbf{v}_0$. If the initial velocity vector were directed parallel to the unit vector $(\tfrac{1}{2}, \tfrac{2}{3}, \tfrac{2}{3})$ with speed v_0, we would have $\mathbf{v}_0 = \left(\tfrac{1}{3}v_0, \tfrac{2}{3}v_0, \tfrac{2}{3}v_0\right)$. A sketch of the resulting trajectory is in Figure 4.8(b), assuming $g = 32$ and $v_0 = 8$. Note that Figure 4.8(b) does *not* show the relation between time t and position $\mathbf{x}$ on the projectile's trajectory.

EXERCISES

In Exercises 1 to 6, compute the indefinite integrals $F(t) = \int f(t)\, dt + \mathbf{c}$; then determine the constant of integration $\mathbf{c}$ so that the associated condition is satisfied.

1. $f(t) = (t^2 + 1, t^3 - 1); F(1) = (2, 2)$

2. $f(t) = (t, t^2, t^3); F(0) = (1, 2, 1)$

3. $f(t) = (t \cos t, t \sin t); F(0) = (1, 1)$

4. $f(t) = \big(1/(t^2 + 1), t/(t^2 + 1)\big); F(0) = (0, 1)$

5. $f(t) = (1, t^2, -1, t^2); F(1) = (2, 2, 2, 2)$

6. $f(t) = t\mathbf{a} - t^2\mathbf{b}; F(t_0) = \mathbf{x}_0$

In Exercises 7 to 14, given $\dot{\mathbf{x}}(t)$ or $\ddot{\mathbf{x}}(t)$, find the $\mathbf{x}(t)$ that satisfies the initial conditions.

7. $\dot{\mathbf{x}}(t) = (t, -t^2); \mathbf{x}(0) = (2, 1)$

8. $\dot{\mathbf{x}}(t) = t(1, -1); \mathbf{x}(1) = (1, 1)$

9. $\dot{\mathbf{x}}(t) = (\cos t, \sin 2t); \mathbf{x}(\pi/2) = (-1, 1)$

10. $\dot{\mathbf{x}}(t) = (e^t, t); \mathbf{x}(0) = (e, 1)$

11. $\dot{\mathbf{x}}(t) = (t, t, t^2); \mathbf{x}(1) = (1, -1, 1)$

12. $\dot{\mathbf{x}}(t) = t(1, 1, t); \mathbf{x}(0) = (2, 1, 2)$

13. $\ddot{\mathbf{x}}(t) = (t, -t^2); \mathbf{x}(0) = (2, 1), \dot{\mathbf{x}}(0) = (1, 1)$

14. $\ddot{\mathbf{x}}(t) = (t, t^2, e^{-t}); \mathbf{x}(1) = (1, 0, 0), \dot{\mathbf{x}}(1) = (0, 1, 0)$

15. Suppose you want to kick a ball over an h-foot vertical fence a feet away from you in such a way that it just barely gets over the top of the fence and lands on the ground b feet from the fence on the other side. Assuming air resistance neglected, what should be the initial angle of elevation θ of your kick, and what should its initial speed be?

16. A target is suspended over level ground at height h_0, to be released to fall earthward under constant vertical acceleration $-g$. Simultaneously with the release of the target, a gun aimed directly at the suspended target is fired from ground level at a horizontal distance l from the point directly below the target. Assume that the speed of bullet and target are not reduced by air resistance.
    (a) Show that the bullet's trajectory will intersect the vertical path of the target only if $2v_0^2 h_0 \geq g(l^2 + h_0^2)$.
    (b) Show that the bullet will hit the target if the condition in part (a) is met.
    (c) One feature of the conclusion in part (b) is that it happens independently of the size of v_0 as long as it satisfies the condition in part (a). However the distance d_1 that the target has fallen when it is hit does depend on v_0. Find d_1 assuming $d_1 > h_0$.

17. Superman, while standing atop a 200-foot-high building, sees a scoundrel drop a victim out a window 50 feet across the street from his building and 100 feet above the pavement below. Reacting instantly, Superman gives himself a mighty push in just the right direction to plunge under the influence of gravity and effect a dramatic rescue just before the victim hits the pavement. Neglecting air resistance, estimate Superman's initial velocity vector and speed. Also estimate Superman's and the victim's speeds at the time of rescue.

18. Someone wants to kick a ball on level ground so it falls back to earth a feet away.
    (a) Show that we can do this with infinitely many different initial angles of elevation as long as the initial speed v_0 at which the ball is kicked is at least $\sqrt{ag}$.
    (b) Suppose in addition that the ball is to be lobbed over a vertical fence of height h halfway between the initial and terminal points on the ground, Show that barely clearing the fence requires initial angle of elevation $\theta = \arctan(4h/a)$ and initial speed $v_0 = \sqrt{g(a^2 + 16h^2)/(8h)}$.

*19. Suppose you want to stand at distance a from the base of a vertical building wall of height h and then kick a ball in such a way that it lands at distance b back from the edge on the building's flat roof, having just grazed the edge of the roof as it went by. Show that the initial angle of elevation of your kick is $\theta = \arctan\big(h/a + h/(a+b)\big)$ and its initial speed is $v_0 = \sqrt{ga(a + b)/(2h\cos^2\theta)}$. [Hint: Find a parabola containing three crucial points.]

*20. A projectile is fired up from the surface of the earth with initial velocity (u_0, v_0). Under the influence of constant vertical acceleration $-g$ the projectile reaches height $h_{\max}$ and then falls back to earth. Neglecting air resistance, show that the fraction of time during its trajectory that the projectile spends above height h_1 is $|v_1|/v_0$, where (u_1, v_1) is the projectile's velocity vector at height h_1.

21. **Big Bertha** In World War I, Paris was bombarded by guns from the unprecedented distance of 75 miles away, shells taking 186 seconds to complete their trajectories. Estimate the angle of elevation at which the gun was fired and the maximum height of the trajectory, assuming negligible air resistance. During a substantial part of the trajectory the altitude was high enough that air resistance was negligible there.

*22. Fox and Rabbit Suppose that a rabbit runs with constant speed $v > 0$ on a circular path of radius a, and that a fox, also running with constant speed v, pursues the rabbit by starting at the center of the circle, always maintaining a position on the radius from the center to the rabbit. Show that it takes the fox time $t = \pi a/(2v)$ to catch the rabbit and that the fox's path is a semicircle.

23. Suppose that in $\mathbb{R}^3$, constant masses of size $m_1, m_2, \ldots, m_n$ are concentrated at the respective points $\mathbf{x}_1, \mathbf{x}_2, \ldots, \mathbf{x}_n$. The **center of mass** of the system is defined to be the point

$$\mathbf{c} = \frac{m_1\mathbf{x}_1 + \cdots + m_n\mathbf{x}_n}{m_1 + \cdots + m_n}.$$

The **momentum** of the system is defined to be the vector

$$\mathbf{p} = \frac{d}{dt}(m_1\mathbf{x}_1 + \cdots + m_n\mathbf{x}_n) = m_1\frac{d\mathbf{x}_1}{dt} + \cdots + m_n\frac{d\mathbf{x}_n}{dt}.$$

Thus the momentum of the system is the velocity vector of the center of mass multiplied by the sum of the masses. Show that if the momentum of such a system is a constant $\mathbf{p}_0$, then the center of mass either remains fixed or moves with constant speed along a fixed line parallel to $\mathbf{p}_0$.

24. Consider the vector differential equation $\ddot{\mathbf{x}} + a\dot{\mathbf{x}} + b\mathbf{x} = 0$ to be solved for vector functions $\mathbf{x} = g(t)$. We assume a and b are scalar constants.

   (a)   Suppose the scalar equation $r^2 + ar + b = 0$ has roots r_1 and r_2. Show by substitution that $\mathbf{x}(t) = e^{r_1 t}\mathbf{c}_1 + e^{r_2 t}\mathbf{c}_2$ satisfies the vector differential equation for fixed arbitrary choices for the vectors $\mathbf{c}_1$ and $\mathbf{c}_2$ and for all t.

   (b)   If the roots r_1 and r_2 of part (a) happen to be equal, the two terms in $\mathbf{x}(t)$ collapse into a single term with arbitrary coefficient $\mathbf{c}_1 + \mathbf{c}_2$. Show that in that case additional solutions are given by $\mathbf{x}(t) = e^{r_1 t}\mathbf{c}_1 + te^{r_1 t}\mathbf{c}_2$.

25. Let $\mathbb{R} \xrightarrow{f} \mathbb{R}^n$ be a function defined for $a \le t \le b$. If the coordinate functions $f_1, \ldots, f_n$ of f are integrable, we define the integral of f over the interval $[a, b]$ by

$$\int_a^b f(t)\, dt = \left(\int_a^b f_1(t)\, dt, \ldots, \int_a^b f(t)\, dt\right).$$

   (a)   If $f(t) = (\cos t, \sin t)$ for $0 \le t \le \pi/2$, compute $\int_0^{\pi/2} f(t)\, dt$.

   (b)   If $g(t) = (t, t^2, t^3)$ for $0 \le t \le 1$, compute $\int_0^1 g(t)\, dt$.

26. If $\mathbb{R} \xrightarrow{f} \mathbb{R}^n$ and $\mathbb{R} \xrightarrow{g} \mathbb{R}^n$ are both integrable over $[a, b]$, show by using the corresponding properties of integrals of real-valued functions that

$$\int_a^b kf(t)\, dt = k\int_a^b f(t)\, dt, \quad k \text{ a real number,}$$

$$\int_a^b \big(f(t) + g(t)\big)\, dt = \int_a^b f(t)\, dt + \int_a^b g(t)\, dt,$$

where the integrals are defined as in the previous exercise.

27. If $\mathbb{R} \xrightarrow{f} \mathbb{R}^n$ is defined for $a \le t \le b$, and f' is continuous there, prove the following extension of the fundamental theorem of calculus:

$$\int_a^b f'(t)\, dt = f(b) - f(a).$$

28. Suppose $\mathbf{x} = \mathbf{x}(t)$ has two continuous derivatives on an interval, and that $\ddot{\mathbf{x}}(t) = r\dot{\mathbf{x}}(t)$ for some scalar constant $r \ne 0$, so that the acceleration vector is parallel to the velocity vector. The purpose of this exercise is to show that the motion of $\mathbf{x}(t)$ is confined to a line.

   (a)   Verify that the equation $\ddot{\mathbf{x}}(t) = r\dot{\mathbf{x}}(t)$ is equivalent to

$$\frac{d}{dt}\left(e^{-rt}\dot{\mathbf{x}}(t)\right) = 0.$$

   (b)   Show that part (a) implies $\dot{\mathbf{x}}(t) = e^{rt}\mathbf{c}$ for some constant vector $\mathbf{c}$.

   (c)   Show that part (b) implies that $\mathbf{x}(t) = (1/r)e^{rt}\mathbf{c} + \mathbf{d}$ for constant vectors $\mathbf{c}$ and $\mathbf{d}$, and hence that $\mathbf{x}(t)$ stays on a line.

*29. This exercise generalizes the previous one. Suppose $\mathbf{x} = \mathbf{x}(t)$ has two continuous derivatives on an interval $a \le t \le b$ and that $\ddot{\mathbf{x}}(t) = g(t)\dot{\mathbf{x}}(t)$ for some continuous real-valued function $g(t)$. Thus the acceleration vector, if not zero, is parallel to the velocity vector. The purpose of this exercise is to show that the motion of $\mathbf{x}(t)$ is confined to a line.

   (a)   Verify that the equation $\ddot{\mathbf{x}}(t) = g(t)\dot{\mathbf{x}}(t)$ is equivalent to

$$\frac{d}{dt}\left(e^{-h(t)}\dot{\mathbf{x}}(t)\right) = 0, \quad \text{where} \quad h(t) = \int_a^t g(u)\, du.$$

   (b)   Show that part (a) implies $\dot{\mathbf{x}}(t) = e^{h(t)}\mathbf{c}$ for some constant vector $\mathbf{c}$.

   (c)   Show that part (b) implies that $\mathbf{x}(t) = H(t)\mathbf{c} + \mathbf{d}$ for constant vectors $\mathbf{c}$ and $\mathbf{d}$, where $H'(t) = e^{h(t)}$. Hence show that $\mathbf{x}(t)$ stays on a line.

SECTION 2   SEVERAL INDEPENDENT VARIABLES

2A   Graph of a Function

The **graph** of a function f is the set of all ordered pairs $\big(\mathbf{x}, f(\mathbf{x})\big)$, where $\mathbf{x}$ is in the domain of f. The graph of f is then said to be represented **explicitly** by f.

EXAMPLE 1

The graph of the function $f : [0, 1] \to \mathbb{R}$ defined by $f(x) = x^2 - 1$ for $0 \le x \le 1$ is the set of points $(x, x^2 - 1)$, where $0 \le x \le 1$. Thus the graph of f is the part of the parabola shown in Figure 4.9(a).

Apart from real-valued functions of a real variable, the functions we picture most effectively by their graphs are the functions $\mathbb{R}^2 \xrightarrow{f} \mathbb{R}$ with graphs in $\mathbb{R}^3$ consisting of the points

$$(x, y, z) = \big(x, y, f(x, y)\big),$$

where (x, y) is in the domain of f. A typical graph of such a function is in Figure 4.9(b). We often describe the relation between (x, y) and z by writing $z = f(x, y)$.

FIGURE 4.9

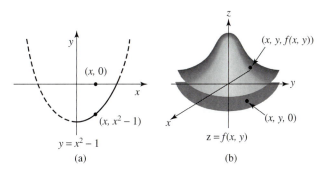

$(x, 0)$

$(x, x^2 - 1)$

$y = x^2 - 1$

(a)

$(x, y, f(x, y))$

$(x, y, 0)$

$z = f(x, y)$

(b)

FIGURE 4.10

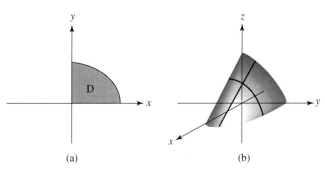

D

(a)

(b)

EXAMPLE 2

Here's how to sketch the part of the graph of

$$f(x, y) = 1 - x - y^2$$

for which $x \ge 0$, $y \ge 0$, and $1 - x - y^2 = z \ge 0$. First observe that the domain D of the function that we are interested in has been restricted to the part of the xy-plane in the first quadrant for which $1 - x - y^2 \ge 0$, or $x \le 1 - y^2$. This domain appears in Figure 4.10(a), and again in Figure 4.10(b) under the graph of f. To sketch the graph of f itself, it helps to notice that cross sections of the graph obtained by holding $y = y_0$ fixed and letting x vary are lines whose projections onto the xz-plane satisfy $z = 1 - x - y_0^2$. Each of these lines joins a point in the yz-plane, where $x = 0$ and $z = 1 - y^2$, to one in the x, y-plane, where $z = 0$ and $1 - x - y^2 = 0$. Such lines are in Figure 4.10(b). We could also include cross sections of the graph of f taken parallel to the yz-plane; such curves are parabolic in shape, with projections onto the yz-plane satisfying $z = 1 - x_0 - y^2$ for values of x_0 between 0 and 1.

EXAMPLE 3

The graph of $f(x, y) = x^2 + y^2$ has the property that f is constant on each circle of a given radius in the xy-plane and centered at the origin. In other words, cross sections of the graph taken with planes parallel to the xy-plane are circles, shown in Figure 4.11(a). All these circles pass through the parabola in the yz-plane with equation $z = y^2$, because $z = f(0, y) = y^2$.

EXAMPLE 4

A function $f : \mathbb{R}^2 \to \mathbb{R}$ is given by

$$f(x, y) = -2x - y + 2.$$

Setting $z = f(x, y)$, we get

$$z = -2x - y + 2 \quad \text{or} \quad 2x + y + z = 2;$$

we see that the graph of f is a plane in $\mathbb{R}^3$. To sketch it, we take cross sections parallel to the yz-plane, which project into that plane as lines with equations $y + z = 2 - 2x_0$. Or we may also take cross sections parallel to the xz-plane. Both are shown in Figure 4.11(b) for $x \geq 0$, $y \geq 0$, $z \geq 0$.

A more direct way to sketch the plane is to locate three points on it by setting, for example, $(x, y) = (0, 0)$, $(1, 0)$, and $(0, 1)$. The corresponding points on the graph are $(x, y, z) = (0, 0, 2)$, $(1, 0, 0)$, and $(0, 1, 1)$, shown as dots in Figure 4.11(b). Joining these dots by lines in this plane gives some idea of the position of the plane. Alternatively, we find the points where the plane intersects the axes by setting two of the coordinates equal to zero and solving for the third; doing this we find $(1, 0, 0)$, $(0, 2, 0)$ and $(0, 0, 2)$.

Note that we my also write the plane's equation as

$$2x + y + (z - 2) = (2, 1, 1) \bullet (x, y, z - 2) = 0.$$

This equation shows that our plane is realized as all points (x, y, z) such that the line joining (x, y, z) to $(0, 0, 2)$ is perpendicular to the vector $(2, 1, 1)$, a normal vector to the plane.

2B   Level Sets

Drawing the graph of a function $\mathbb{R}^3 \xrightarrow{f} \mathbb{R}$ is impossible because the graph is a subset of $\mathbb{R}^4$. Even in $\mathbb{R}^3$ a graph may be too complicated to draw easily. But in both cases we may settle for trying to draw the sets on which f is a constant, thereby getting some picture of the behavior of the function. If f is a real-valued function, and k is a point in the image of f, the **level set** of f at **level** k is the set of all **x** in the domain of f such that $f(\mathbf{x}) = k$. Level sets of f are defined **implicitly** by f, that is, by regarding them as solution sets of $f(\mathbf{x}) = k$, for some k. The

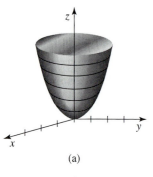

(a)

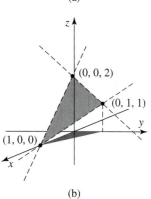

(b)

FIGURE 4.11

FIGURE 4.12

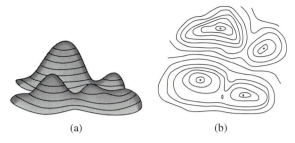

(a)   (b)

implicitly defined level set associated with $f(\mathbf{x}) = k$ is sometimes called the **graph** of the equation $f(\mathbf{x}) = k$, whereas the graph of the function $f(\mathbf{x})$ always refers to the equation $\mathbf{y} = f(\mathbf{x})$.

Topographical maps display terrain elevations by showing level curves at equally spaced levels as in Figure 4.12. Such displays have the advantage over perspective drawings that foreground features don't obscure what lies in back of them. See Figure 4.12(a), which shows the terrain levels, and Figure 4.12(b), which shows the corresponding level curves.

EXAMPLE 5   The function $f(x, y) = x^2 + y^2$ of Example 3 has concentric circles for level sets. At level $k = 1$ we get $f(x, y) = x^2 + y^2 = 1$, which represents a circle of radius 1 about $(0, 0)$ in the xy-plane. In general, at a level $k > 0$ we get a level curve $x^2 + y^2 = k$, which is a circle of radius $\sqrt{k}$. See Figure 4.13(a), where the values of $\sqrt{k}$ are nearly equally spaced. As the surface rises more steeply the level lines will get closer together. If we don't label the level curves with numerical level value k, we can't tell from level curves alone whether the surface is rising or falling as we go out from the center.

EXAMPLE 6   The function $f : \mathbb{R}^3 \to \mathbb{R}$ defined by $f(x, y, z) = x^2 + y^2 + z^2$ has level sets in $\mathbb{R}^3$ consisting of points (x, y, z) that satisfy an equation of the form

$$x^2 + y^2 + z^2 = k$$

for some fixed real number k. If $k > 0$, we get a sphere of radius $\sqrt{k}$ centered at $(0, 0, 0)$, because $x^2 + y^2 + z^2$ is the square of the distance from (x, y, z) to $(0, 0, 0)$. If $k = 0$, the equation is satisfied only by $(0, 0, 0)$. If $k < 0$, the corresponding level set is empty. Some level sets are shown in Figure 4.13(b) as concentric spheres. The *graph* of f is a subset of $\mathbb{R}^4$ and can't be pictured.

FIGURE 4.13

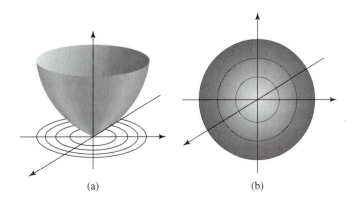

(a)   (b)

EXAMPLE 7   The linear function $g: \mathbb{R}^3 \to \mathbb{R}$ defined by

$$g(x, y, z) = x + y + z$$

has a graph in $\mathbb{R}^4$, so we can't draw it. The level sets of g are the parallel planes with equations

$$x + y + z = k,$$

one for each real number k. Three of the planes are shown in Figure 4.14. Note that each plane is perpendicular to the vector $(1, 1, 1)$, because the equation also takes a form showing $(1, 1, 1)$ and $(x, y, z) - (0, 0, k)$ perpendicular:

$$(1, 1, 1) \cdot (x, y, z - k) = 0.$$

Note that the graph of $f: \mathbb{R}^2 \to \mathbb{R}$ is the set of points (x, y, z) in $\mathbb{R}^3$ such that $z = f(x, y)$, and that this set is the same as the level set at level $k = 0$ of the function $g: \mathbb{R}^3 \to \mathbb{R}$ given by $g(x, y, z) = z - f(x, y)$. Whichever point of view we take, we get the same picture.

FIGURE 4.14

EXERCISES

1. Consider the function $f(x, y) = \sqrt{4 - x^2 - y^2}$.
   (a) Sketch the domain of f, making it as large as possible.
   (b) Sketch the graph of f.
   (c) Sketch the image of f.

2. Consider the function $g(x, y) = \ln(x + y)$.
   (a) Describe the domain of g, making it as large as possible.
   (b) For what values of (x, y) does the graph of g lie above the xy plane?
   (c) Describe the image of g.

Sketch the graphs of the following functions 3 to 8.

3. $f(x, y) = 2 - x^2 - y^2$

4. $h(x, y) = \dfrac{1}{x^2 + y^2}$

5. $g(x, y) = \sin x$

6. $f(x, y) = 0$

7. $f(x, y) = e^{x+y}$

8. $g(x, y) = \left\{ \begin{array}{ll} 1 & \text{if } |x| < |y| \\ 0 & \text{if } |x| \geq |y| \end{array} \right\}$

Sketch the following implicitly defined level sets 9 to 14 in $\mathbb{R}^2$ or $\mathbb{R}^3$.

9. $f(x, y) = x + y = 1$

10. $g(x, y) = x^2 + 2y^2 = 1$

11. $f(x, y) = (x^2 + y^2 + 1)^2 - 4x^2 = 0$

12. $f(x, y, z) = x + y + z = 1$

13. $f(x, y, z) = xyz = 0$

14. $f(x, y, z) = x^2 - y^2 = 2$

In Exercises 15 to 18, sketch the level sets of each of the functions $f: \mathbb{R}^3 \to \mathbb{R}$ for the indicated levels k.

15. $f(x, y, z) = x + y, k = 0, 1, 2$

16. $f(x, y, z) = x^2 + y^2 - z^2, k = 0, 1$

17. $f(x, y, z) = \sqrt{x^2 + y^2 + z^2}, k = 0, 1$

18. $f(x, y, z) = x + y + z, k = 0, 1$

In Exercises 19 to 22, we consider a function $f: \mathbb{R}^3 \to \mathbb{R}^2$ with coordinate functions f_1, f_2 determined by $f(x, y, z) = \big(f_1(x, y, z), f_2(x, y, z) \big)$. For each vector $\mathbf{k} = (k_1, k_2)$ in the image of f, the equation $f(x, y, z) = \mathbf{k}$ determines a level set in $\mathbb{R}^3$ that is the intersection of the level sets determined by the pair of equations

$$f_1(x, y, z) = k_1$$
$$f_2(x, y, z) = k_2.$$

Using this point of view, sketch the following level sets (curves) in $\mathbb{R}^3$.

19. $\begin{pmatrix} x - y \\ y + z \end{pmatrix} = \begin{pmatrix} 0 \\ 0 \end{pmatrix}$

20. $\begin{pmatrix} y + z \\ x - z \end{pmatrix} = \begin{pmatrix} 2 \\ 3 \end{pmatrix}$

21. $\begin{pmatrix} x^2 + y^2 + z^2 \\ x - z \end{pmatrix} = \begin{pmatrix} 1 \\ 0 \end{pmatrix}$

22. $\begin{pmatrix} x^2 + y^2 + z^2 \\ y - z \end{pmatrix} = \begin{pmatrix} 4 \\ 0 \end{pmatrix}$

23. Suppose that the density per unit area of a thin film, referred to in (x, y)-coordinates, is given by the formula $d(x, y) = x^2 + 2y^2 - x + 1$ for $-1 \leq x \leq 1$ and $-1 \leq y \leq 1$. Sketch the set of points at which the film has density $\frac{7}{4}$.

24. Let the density per unit of volume in a cubical box of side length 2 vary directly as the distance from the center and

inversely as $1 + t^2$, where t is time. If the box is described in $\mathbb{R}^3$ by $|x| \leq 1, |y| \leq 1, |z| \leq 1$, and if the density at a corner of the box is 1 when $t = 0$, find a formula for the density at a given point and time. What is the rate of change of the density at a point $\frac{1}{2}$ unit from the center of the box at time $t = 1$?

25. Suppose the region D in $\mathbb{R}^3$ consists of all points (x, y, z) satisfying both $x^2 + y^2 \leq 4$ and $0 \leq z \leq 5$. Suppose the temperature at a point (x, y, z) in D is $T(x, y, z) = x^2 + y^2 - z$.

(a)   Sketch the region D.

(b)   Sketch the set of points in D for which the temperature is -1 degree.

2C   Computer-Generated Graphs

Some graphs of functions $f(x, y)$ are fairly easy to draw by hand. For example, the graph of $z = \sqrt{1 - x^2 - y^2}$ is a hemisphere of radius 1 over the domain $x^2 + y^2 \leq 1$. A few examples that have been done by a computer are shown in Figure 4.15. But whether sketching is done by hand or computer, the technique illustrated here is fundamentally the same in that it consists of drawing curves on the function's graph that are traced by holding one variable fixed and varying the other.

To describe this technique another way, sketching the graph of $z = f(x, y)$ is possible by plotting some carefully chosen curves that lie on the surface, as in Figure 4.15. The simplest curves to draw are often the ones that are images under f of line segments in the domain of f that are parallel to the x and y axes, as in Figure 4.15(b); this approach allows us to use either function values $f(x, y_0)$, with y_0 fixed and x varying, or else $f(x_0, y)$ with x_0 fixed and y varying. Thus a rectangular domain for f such as $0 \leq x \leq 2, \pi/4 \leq y \leq 3\pi/2$ for $f(x, y) = x \cos y$ might be treated using the following routine. This "program" is not intended to run in a particular language, but is presented only as a compact way of indicating the rough structure of such a program.

```
DEFINE f(x, y)  =  x * cos(y)
   [First plot y-varying curves, spaced by x = 1/4.]
FOR x  =  0 TO 2 STEP 0.25
FOR y  =  π/4 TO 3π/2 STEP 0.01
   PLOT3D (x,  y,  f(x, y))
NEXT y
NEXT x
   [Then Plot x-varying curves, spaced by y = π/4.]
FOR y  =  π/4 TO 3π/2 STEP π/4
FOR x  =  0 TO 2 STEP 0.01
   PLOT3D (x,  y,  f(x, y))
NEXT x
NEXT y
```

Following this routine produces the picture shown in Figure 4.16. These drawings are in a style that can in principle be drawn by hand, drawing one curve on the

FIGURE 4.15

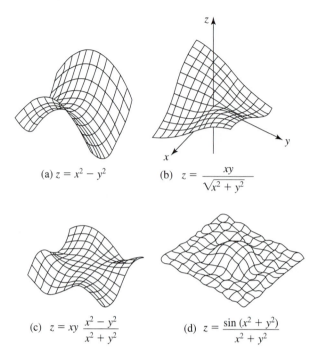

(a) $z = x^2 - y^2$ (b) $z = \dfrac{xy}{\sqrt{x^2 + y^2}}$

(c) $z = xy \dfrac{x^2 - y^2}{x^2 + y^2}$ (d) $z = \dfrac{\sin(x^2 + y^2)}{x^2 + y^2}$

graph at a time, with only one of the variables actually varying. Applications such as Maple, Matlab, and Mathematica make drawings such as this with additional sophistication. The Web site **http://math.dartmouth.edu/~rewn/** also provides some Java programs in the style of the graphical techniques we use here.

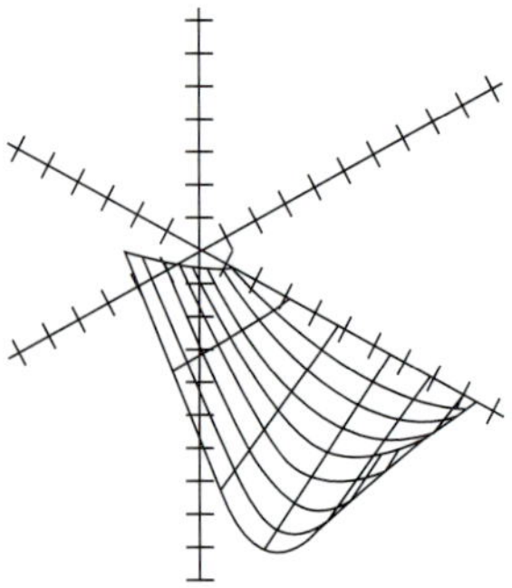

   The Java programs are designed to ignore error-producing values such as square roots of negative numbers or undefined function values that may arise from trying to plot the graph of a function like $f(x, y) = \sqrt{1 - x^2 - y^2}$ over a rectangle that contains the circular disk $x^2 + y^2 \leq 1$. [Here for example, $f(1, 1) = \sqrt{-1}$.] The natural plotting domain of the programs we use is a rectangle with edges parallel to rectangular axes, but we may want to plot only over a domain with some other shape such as a circular or triangular one. We do this easily using the **Heaviside unit step function** defined by

FIGURE 4.16
$z = x \cos y$.

$$H(x) = \begin{cases} 1, & \text{if } x \geq 0, \\ 0, & \text{if } x < 0. \end{cases}$$

EXAMPLE 8   Suppose we want a picture of the graph of the function $f(x, y) = \sin(x^2 + y^2)/(x^2 + y^2)$ with its domain restricted to the part of the first quadrant inside the circle $x^2 + y^2 \leq 9$ of radius 3. Using the Heaviside function we define a new function of two variables $h(x, y)$ by writing $h(x, y) = H(9 - x^2 - y^2)$. It follows that

$$h(x, y) = \begin{cases} 1, & \text{if } x^2 + y^2 \leq 9, \\ 0, & \text{if } x^2 + y^2 > 9. \end{cases}$$

Thus $h(x, y)$ takes the value 1 inside and on the circle of radius 1 centered at the origin and the value 0 outside the circle. Then the product $h(x, y) f(x, y)$ will take

FIGURE 4.17

$z = \sin(x^2 + y^2)/(x^2 + y^2)$, $0 \le x \le 3$, $0 \le y \le 3$.

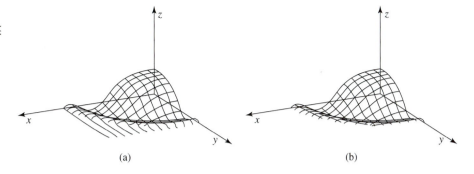

(a)   (b)

on the value 0 outside the circle $x^2 + y^2 = 9$ and will be equal to $f(x, y)$ inside and on the circle. If we sketch the graph of $z = h(x, y)f(x, y)$ over the square $0 \le x \le 3$, $0 \le x \le 3$ we get a picture like Figure 4.17(a). By suppressing the zero values we get Figure 4.17(b).

In the previous example $f(x, y)$ has not been defined at $(x, y) = (0, 0)$ but we've successfully avoided the issue. One way to do this is to define $f(0, 0) = 1$, which incidentally will make f continuous at $(0, 0)$. Another way is to incorporate a feature in the plotting program that allows it to ignore points in the domain that would normally produce an error message. This is what has been done with the Java program GPLOT available at the Web site referred to previously.

EXERCISES

1. Sketch the graph of $f(x, y) = x^2 - y^2$, for $|x| \le 2$, $|y| \le 2$.

2. Sketch the graph of $f(x, y) = x^2 - y^2$ for $0 \le x \le 2$, $0 \le y \le 2$.

3. Sketch the plane $z = 1 - x - y$ for $0 \le x \le 2, 0 \le y \le 2$.

4. Sketch the plane $x + 2y + z = 2$ for $1 \le x \le 2$, $1 \le y \le 2$.

5. Sketch the graph of $f(x, y) = x^2 + y^3$ for $|x| \le 3$, $|y| \le 3$.

6. Sketch the graph of $f(x, y) = x^2 + y^2$ for $|x| \le 1$, $|y| \le 1$.

7. Sketch the graph of $f(x, y) = x + y$ for $0 \le x \le 1$, $0 \le y \le 2$.

8. Sketch the graph of $f(x, y) = y^2 - x^3$, $0 \le x \le 2$, $0 \le y \le 1$.

9. Sketch the graph of $f(x, y) = \cos x \sin y$, $0 \le x \le 2\pi$, $0 \le y \le 2\pi$.

10. Sketch the graph of $f(x, y) = \exp(-x - 2y)$, $0 \le x \le 2$, $0 \le y \le 2$.

11. Let $f(x, y) = xy(x^3 + y^3)/(x^2 + y^2)$. Sketch its graph for $-1 \le x \le 1, -1 \le y \le 1$. What is the difficulty at $(x, y) = (0, 0)$?

Using modifications of the Heaviside function $H(x)$, form a product of functions that we'll refer to as $P(x, y)$,

assuming the value 1 on each region as described in 12 to 17 and assuming the value zero elsewhere in $\mathbb{R}^2$.

12. $P(x, y)$ takes the value 1 where $x \ge 0$, $y \ge 0$, and $y \ge x$.

13. $P(x, y)$ takes the value 1 where $1 \ge x \ge 0$, $1 \ge y \ge 0$, and $y \ge x$.

14. $P(x, y)$ takes the value 1 where $x^2 + y^2 \le 1$.

15. $P(x, y)$ takes the value 1 where $x^2 + y^2 \le 1$ and $x \ge y$.

16. $P(x, y)$ takes the value 1 on the triangular region in $\mathbb{R}^2$ with vertices $(0, 0)$, $(0, 1)$ and $(1, 0)$.

17. $P(x, y)$ takes the value 1 on the square in $\mathbb{R}^2$ with vertices $(0, 0)$, $(0, 1)$, $(1, 0)$ and $(1, 1)$.

18. Sketch the graph of $f(x, y) = (x - y)^3$ for values of (x, y) simultaneously satisfying $0 \le x \le 2$, $0 \le y \le 2$, and $y \le x$.

19. Sketch the graph of $f(x, y) = y^2 - x^2$ for values of (x, y) simultaneously satisfying $0 \le x \le 2$, $0 \le y \le 2$, and $x \le y$.

20. Sketch the graph of $f(x, y) = \cos(x^2 + y^2)/(1 + x^2 + y^2)$, for $|x| \le 2$, $|y| \le 2$.

21. Sketch the graph of $f(x, y) = \cos(x^2+y^2)/(1+x^2+y^2)$, for $x^2 + y^2 \le 2$.

22. Sketch the graph of $z = \sqrt{5 - x^2 - y^2}$ when $1 \le z \le 2$.

23. Plot the part of the plane $z-x-y = 1$ in $\mathbb{R}^3$ for $1 \le x \le 3$ and $1 \le y \le 2$.

*24. Sketch the part of the sphere of radius 1 centered at the origin that lies above the part of the first quadrant in the xy-plane that lies between the y-axis and the line $y = x$.

*25. Sketch the part of the graph of $f(x, y) = 2 - x^2 - y^3$ that lies above the first and second quadrants in the xy-plane.

2D   Quadric Surfaces

Quadric surfaces are level sets in $\mathbb{R}^3$ of second-degree polynomials in three variables x, y, z; they fall into six distinct types illustrated in Figure 4.18, plus some degenerate cases in which the polynomial depends on only two variables. We'll be returning to all of these surfaces later in Section 4, where we represent them parametrically in a way similar to what we used to represent curves in space.

The elliptic cone in Figure 4.18(2) is a limit of hyperboloids in two ways: (i) As the waist of the hyperboloid of one sheet pinches in while k *decreases* to 0 through

FIGURE 4.18

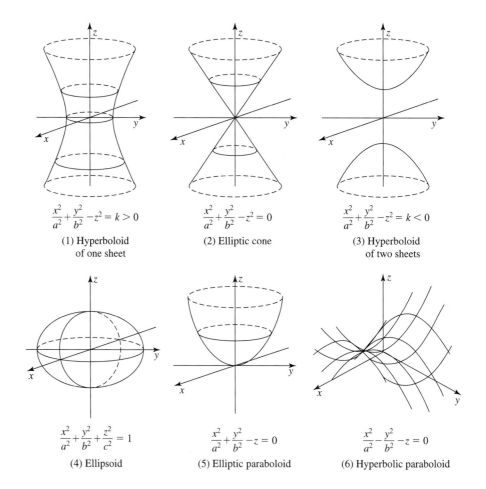

$$\frac{x^2}{a^2}+\frac{y^2}{b^2}-z^2 = k > 0$$

(1) Hyperboloid of one sheet

$$\frac{x^2}{a^2}+\frac{y^2}{b^2}-z^2 = 0$$

(2) Elliptic cone

$$\frac{x^2}{a^2}+\frac{y^2}{b^2}-z^2 = k < 0$$

(3) Hyperboloid of two sheets

$$\frac{x^2}{a^2}+\frac{y^2}{b^2}+\frac{z^2}{c^2} = 1$$

(4) Ellipsoid

$$\frac{x^2}{a^2}+\frac{y^2}{b^2}-z = 0$$

(5) Elliptic paraboloid

$$\frac{x^2}{a^2}-\frac{y^2}{b^2}-z = 0$$

(6) Hyperbolic paraboloid

positive values it tends to the cone. (ii) As the two separate pieces of the hyperboloid of two sheets get closer while k *increases* to 0 through negative values, the two pieces of the hyperboloid become more pointed and come together to form the cone.

As well as being level sets at level 0 of functions of three variables, the two paraboloids are also graphs in $\mathbb{R}^3$ of the respective functions $(x/a)^2 \pm (y/b)^2$ defined on $\mathbb{R}^2$. The surface of Example 5 is an elliptic paraboloid in which $a = b = 1$. The spheres of Example 6 are a special case of the ellipsoid in which $a = b = c$.

The degenerate cases mentioned previously are cylinders, which may be level sets of functions on $\mathbb{R}^3$ that really depend on only two variables, say x and y. For example, the equation $x^2 + y^2 = 1$ that determines a circle in $\mathbb{R}^2$ also determines a circular cylinder in $\mathbb{R}^3$. Since the equation places no restriction on z, a level set satisfying $x^2 + y^2 = k > 0$ in $\mathbb{R}^3$ contains all lines perpendicular to the xy-plane and passing through the circle of radius $\sqrt{k}$.

To make highly accurate pictures of surfaces, in particular quadric surfaces, we use computer graphics. On the other hand, rough sketches are often based on the observation that well-known curves such as lines, parabolas, ellipses, and hyperbolas lie on these surfaces and are useful guides in making a drawing.

EXAMPLE 9   The hyperboloid of one sheet defined by $x^2/a^2 + y^2/b^2 - z^2 = k$, with $k > 0$, contains the hyperbola $x^2/a^2 - z^2 = k$ in the xz-plane. (Just set $y = 0$ to restrict to the xz-plane.) The same hyperboloid contains the hyperbola $y^2/b^2 - z^2 = k$. (Set $x = 0$ to see this.) Cross sections of the hyperboloid by planes $z = l$ parallel to the xz-plane are the ellipses $x^2/a^2 + y^2/b^2 = k + l^2, z = l$ that lie above the elliptical level sets $x^2/a^2 + y^2/b^2 = k + l^2$ in the xz-plane. These cross-sectional curves, together with the two hyperbolas identified previously, form a framework for the surface. Such a framework appears in the generic picture shown of the hyperboloid of one sheet. If $a = b$, this surface arises by rotating one of the hyperbolas about the z-axis. Interchanging x and z, or y and z, in the given equation gives hyperboloids with the x-axis and y-axis respectively as principal axis of symmetry.

EXAMPLE 10   The second picture in our catalogue is the cone $x^2/a^2 + y^2/b^2 = z^2$. This cone contains the lines $z = \pm x/a$ in the xy-plane, where $y = 0$, and the lines $z = \pm y/b$ in the yz-plane, where $x = 0$. Cross sections of the cone by planes $z = l$ parallel to the xy-plane are the ellipses $x^2/a^2 + y^2/b^2 = l^2, z = l$ that lie above the elliptical level sets $x^2/a^2 + y^2/b^2 = l^2$ in the xy-plane. If $a = b$ we get a circular cone generated by rotating one of the lines about the z-axis. Note that the upper half of this circular cone is the graph of $z = (1/a)\sqrt{x^2 + y^2}$ and the lower half is the graph of $z = (-1/a)\sqrt{x^2 + y^2}$.

EXAMPLE 11   The hyperboloid of two sheets $x^2/a^2 + y^2/b^2 - z^2 = k$, with $k < 0$, comes in two pieces, namely the graphs of $z = \pm\sqrt{x^2/a^2 + y^2/b^2 - k}$. Since $k < 0$, the two graphs contain the points $(0, 0, \pm\sqrt{-k})$, but there are no points of the graphs at z-levels between these numbers. Cross sections at z-levels outside these numbers are ellipses as in the previous two examples.

EXERCISES

1. Use the picture of the generic elliptic cone in Figure 4.18 as a guide in making sketches of the circular cones (a) $x^2+y^2-z^2 = 0$. (b) $x^2+z^2-y^2 = 0$. (c) $y^2+z^2-x^2 = 0$.

2. Make sketches in $\mathbb{R}^3$ of (a) circular cylinder $x^2 + y^2 = 1$. (b) parabolic cylinder $x^2 - y = 0$. (c) hyperbolic cylinder $x^2 - y^2 = 1$.

3. The ellipsoid, the elliptic paraboloid, and hyperbolic paraboloid are shown as (4), (5) and (6) in Figure 4.18 as generic level surfaces of quadratic polynomials at respective levels 1, 0, and 0. What, if anything, would be altered in the pictures if we had chosen levels 2 in (4), 1 in (5), and 2 in (6)?

4. (a) Show that the intersection of the hyperboloid H_1 of one sheet

$$(x/a)^2 + (y/b)^2 - (z/c)^2 = 1$$

   with a plane perpendicular to the z-axis is an ellipse.
   (b) Show that the intersection of the hyperboloid H_1 with a plane perpendicular to the x-axis or to the y-axis is a hyperbola.

5. (a) Show that if a plane perpendicular to the z-axis intersects the hyperboloid H_2 of two sheets

$$(x/a)^2 + (y/b)^2 - (z/c)^2 = -1,$$

   then the intersection is an ellipse.
   (b) Show that the intersection of the hyperboloid H_2 with a plane perpendicular to the x-axis or to the y-axis is a hyperbola.

6. Identify the curves of intersection of the hyperbolic paraboloid $(x/a)^2 - (y/b)^2 = z$ with (a) planes perpendicular to the z-axis. (b) planes perpendicular to the x-axis or the y-axis.

7. Consider two distances in $\mathbb{R}^3$: (i) from (x, y, z) to the plane $z = -1$, (ii) from (x, y, z) to the point $(0, 0, 1)$. The points (x, y, z) for which these distances are equal constitute a quadric surface Q. Identify Q and make a sketch of it.

8. The paraboloids $x^2 + y^2 = z$ and $x^2 + y^2 = 8 - z$ intersect in a curve in $\mathbb{R}^3$. Identify the curve and make a sketch of it.

Each of the following quadratic equations describes an example of one of the quadric surface types illustrated in the text. In each case identify the type by name and make a sketch of the surface.

9. $4x^2 - y^2 + 4z^2 = 16$ 10. $x^2/4 + y^2/4 - z^2/9 = 1$

11. $x^2/4 - y^2/4 + z^2/9 = 1$ 12. $4x^2 + y^2 + 4z^2 = 16$

13. $x^2/4 + y^2/4 + z^2/9 = 1$ 14. $x^2/4 - y^2/4 - z^2/9 = 0$

15. $4x^2 - y^2 - z = 0$ 16. $x^2/4 + y^2/9 - z = 1$

17. $x^2/4 - y^2/4 + z = 1$ 18. $4y^2 + z^2 - x = 0$

19. $x^2/4 + z^2/9 - y = 0$ 20. $x^2/4 + y^2/4 + z = 0$

Each of the quadratic equations 21 to 28 in two variables describes a curve in $\mathbb{R}^2$. Each one also describes a quadric surface of cylindrical type in $\mathbb{R}^3$. In each exercise make a perspective sketch of the underlying curve in the appropriate 2-dimensional coordinate plane in $\mathbb{R}^3$. Then sketch the cylinder parallel to the remaining axis in $\mathbb{R}^3$.

21. $x^2 + y = 0$ 22. $z^2 - x = 0$

23. $y^2 + z^2 = 1$ 24. $x^2 - y^2 = 1$

25. $x^2 + z = 1$ 26. $z^2 + y = 2$

27. $x^2 + 4z^2 = 4$ 28. $x^2 - 2y^2 = 1$

SECTION 3   PARTIAL DERIVATIVES

3A   Definition

Partial derivatives are the straightforward generalization to functions from $\mathbb{R}^n$ to $\mathbb{R}$ of the ordinary derivative of a real-valued function of a real variable. For example, if f is defined on $\mathbb{R}^2$ we define the **partial derivatives** $\partial f/\partial x$ and $\partial f/\partial y$ by

$$\frac{\partial f}{\partial x}(x, y) = \lim_{t \to 0} \frac{f(x + t, y) - f(x, y)}{t},$$

$$\frac{\partial f}{\partial y}(x, y) = \lim_{t \to 0} \frac{f(x, y + t) - f(x, y)}{t}.$$

Thus a partial derivative is the result of differentiating with respect to just one variable at a time with the others held fixed. If the derivatives $\partial f/\partial x$ and $\partial f/\partial y$ exist they are also functions from $\mathbb{R}^2$ to $\mathbb{R}$.

A similar definition works for functions defined on $\mathbb{R}^n$. For each $i = 1, \dots, n$, we define a new real-valued function called the **partial derivative** of f with respect to the ith variable, denoted by $\partial f/\partial x_i$. For each $\mathbf{x} = (x_1, \dots, x_n)$ in the domain of f, the number $(\partial f/\partial x_i)(\mathbf{x})$ is by definition

3.1 $$\frac{\partial f}{\partial x_i}(\mathbf{x}) = \lim_{t \to 0} \frac{f(x_1, \dots, x_i + t, \dots, x_n) - f(x_1, \dots, x_i, \dots, x_n)}{t}.$$

The domain space of $\partial f/\partial x_i$ is $\mathbb{R}^n$, and the domain of $\partial f/\partial x_i$ is the subset of the domain of f consisting of all $\mathbf{x}$ for which the preceding limit exists. Thus the domain of $\partial f/\partial x_i$ could conceivably be the empty set. The number $(\partial f/\partial x_i)(\mathbf{x})$ is simply the derivative at x_i of the function of one variable obtained by holding $x_1, \dots, x_{i-1}, x_{i+1}, \dots, x_n$ fixed and by considering f to be a function of the ith variable only. As a result, the differentiation formulas of one-variable calculus apply directly.

It's important to realize that we do not call a function "differentiable" just because it has partial derivatives. For functions of more than one variable, the concept of differentiability is a little more complicated than that; the matter is taken up in Chapter 5.

EXAMPLE 1   Let $f(x, y, z) = x^2 y + y^2 z + z^2 x$. Then

$$\frac{\partial f}{\partial x}(x, y, z) = 2xy + z^2,$$

$$\frac{\partial f}{\partial y}(x, y, z) = x^2 + 2yz,$$

$$\frac{\partial f}{\partial z}(x, y, z) = y^2 + 2zx.$$

The partial derivatives at $\mathbf{x} = (1, 2, 3)$ are

$$\frac{\partial f}{\partial x}(1, 2, 3) = 4 + 9 = 13,$$

$$\frac{\partial f}{\partial y}(1, 2, 3) = 1 + 12 = 13,$$

$$\frac{\partial f}{\partial z}(1, 2, 3) = 4 + 6 = 10.$$

EXAMPLE 2   Let $f(u, v) = \sin u \cos v$. Then

$$\frac{\partial f}{\partial u} = \frac{\partial \sin u \cos v}{\partial u} = \cos u \cos v,$$

$$\frac{\partial f}{\partial v} = \frac{\partial \sin u \cos v}{\partial v} = -\sin u \sin v,$$

$$\frac{\partial f}{\partial v}(\pi/2, \pi/2) = \frac{\partial \sin u \cos v}{\partial v}(\pi/2, \pi/2) = -\sin\frac{\pi}{2}\sin\frac{\pi}{2} = -1.$$

We can repeat the operation of taking partial derivatives. The partial derivative of $\partial f / \partial x_i$ with respect to the jth variable is $\partial / \partial x_j (\partial f / \partial x_i)$ and is denoted by $\partial^2 f / \partial x_j \partial x_i$. We may repeat this indefinitely, provided the derivatives exist. An alternative notation for higher-order partial derivatives is illustrated as follows, in which each variable of differentiation is denoted by a subscript:

$$\frac{\partial f}{\partial x_i} = f_{x_i}$$

$$\frac{\partial}{\partial x_j} \left(\frac{\partial f}{\partial x_i} \right) = \frac{\partial^2 f}{\partial x_j \, \partial x_i} = f_{x_i x_j}$$

$$\frac{\partial}{\partial x_i} \left(\frac{\partial f}{\partial x_i} \right) = \frac{\partial^2 f}{\partial x_i^2} = f_{x_i x_i}$$

$$\frac{\partial}{\partial x_k} \left(\frac{\partial^2 f}{\partial x_j \, \partial x_i} \right) = \frac{\partial^3 f}{\partial x_k \, \partial x_j \, \partial x_i} = f_{x_i x_j x_k}.$$

Note that the order of variables in the subscript notation is the opposite of that in the ∂-notation, since for example f_{xy} means $(f_x)_y$.

EXAMPLE 3   Consider $f(x, y) = xy - x^2$.

$$f_x = \frac{\partial f}{\partial x} = y - 2x$$

$$f_{xy} = \frac{\partial^2 f}{\partial y \, \partial x} = 1$$

$$f_{xx} = \frac{\partial^2 f}{\partial x^2} = -2$$

$$f_{yxx} = \frac{\partial^3 f}{\partial x^2 \, \partial y} = 0$$

3B   Geometric Interpretation

To interpret partial derivatives geometrically, we rely on something we know about real-valued functions of a single variable, namely that the value of the derivative at a point is the slope of the tangent line to the graph of the function at that point. For illustrative purposes it will be enough to consider the graph of a function $\mathbb{R}^2 \xrightarrow{f} \mathbb{R}$, namely, the set of points $(x, y, f(x, y))$ in $\mathbb{R}^3$ where (x, y) is in the domain of f. Such a graph is in Figure 4.19 as a surface lying over a rectangle in the xy-plane. The intersection of the surface with the vertical plane determined by the condition $y = b$ is a curve satisfying the conditions

$$z = f(x, y), \quad y = b.$$

Consider the curve defined by the function $g(x) = f(x, b)$ as a subset of 2-dimensional space. Its slope at $x = a$ is

$$g'(a) = \frac{\partial f}{\partial x}(a, b).$$

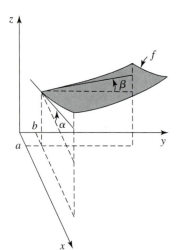

FIGURE 4.19

Similarly, at $y = b$ the curve defined by $h(y) = f(a, y)$ has slope equal to

$$h'(b) = \frac{\partial f}{\partial y}(a, b).$$

The angles α and β shown in Figure 4.19 therefore satisfy

$$\tan \alpha = \frac{\partial f}{\partial x}(a, b), \quad \tan \beta = \frac{\partial f}{\partial y}(a, b).$$

The numbers $\tan \alpha$ and $\tan \beta$ are slopes of tangent lines to two curves contained in the graph of the function f. For this reason it's natural to try to define a **tangent plane** to the graph of f just to be the plane containing these two lines. If f satisfies the condition of differentiability defined in Chapter 5, then that turns out to be consistent with our ultimate definition. We see that the set of points (x, y, z) satisfying

3.2
$$z = f(a, b) + (x - a)\frac{\partial f}{\partial x}(a, b) + (y - b)\frac{\partial f}{\partial y}(a, b)$$

is a plane containing the tangent lines found previously. To see this, specify in turn $y = b$ and $x = a$ in the previous equation to determine the respective tangent lines.

EXAMPLE 4   A sketch of the part of the graph of

$$f(x, y) = 1 - 2x^2 - y^2$$

corresponding to $x \geq 0$, $y \geq 0$ is in Figure 4.20. The function f has partial derivatives at $\left(\frac{1}{2}, \frac{1}{2}\right)$ given by

$$\frac{\partial f}{\partial x}\left(\frac{1}{2}, \frac{1}{2}\right) = -2, \quad \frac{\partial f}{\partial y}\left(\frac{1}{2}, \frac{1}{2}\right) = -1.$$

Since $f\left(\frac{1}{2}, \frac{1}{2}\right) = \frac{1}{4}$, the tangent plane to the graph of f at $\left(\frac{1}{2}, \frac{1}{2}\right)$ is, by Equation 3.2,

$$z = \frac{1}{4} - 2\left(x - \frac{1}{2}\right) - \left(y - \frac{1}{2}\right)$$
$$= \frac{7}{4} - 2x - y.$$

We can sketch the tangent plane by drawing the two tangent lines in it determined at $(x, y) = (\frac{1}{2}, \frac{1}{2})$. It's somewhat easier to locate three points on the plane, for simplicity

$$\left(\tfrac{7}{8}, 0, 0\right), \left(0, \tfrac{7}{4}, 0\right) \left(0, 0, \tfrac{7}{4}\right),$$

and then sketch the plane containing these points. The point of tangency on the graph of f is $\left(\frac{1}{2}, \frac{1}{2}, \frac{1}{4}\right)$. See Figure 4.20.

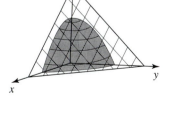

FIGURE 4.20

3C   Continuity

We discuss continuity for functions of more than one variable extensively in Chapter 5. At this point, we'll consider briefly the case $\mathbb{R}^2 \xrightarrow{f} \mathbb{R}$. To allow $\mathbf{z}$ to approach $\mathbf{x}$ from an arbitrary direction we assume, for each $\mathbf{x} = (x, y)$ in the domain of f, that $f(\mathbf{z})$ is defined for all vectors $\mathbf{z} = (z, w)$ satisfying $|\mathbf{x} - \mathbf{z}| < \delta$, where δ is some positive number. We then say that f is **continuous** if for each point $\mathbf{x}$ in the domain of f

$$\lim_{\mathbf{z} \to \mathbf{x}} f(\mathbf{z}) = f(\mathbf{x}).$$

The limit relation means that we can make $f(\mathbf{z})$ arbitrarily close to $f(\mathbf{x})$ if the distance $|\mathbf{x} - \mathbf{z}|$ from $\mathbf{x}$ to $\mathbf{z}$, is small enough. As usual, the intuitive idea of continuity is that the values of the function f should not change abruptly, resulting, for example, in breaks in the graph of f. The graphs shown in Figures 4.19 and 4.20 are those of continuous functions, whereas Figure 4.21 shows a simple example of the graph of a discontinuous function.

　　If we assume certain continuity conditions on f and its partial derivatives then the higher-order partial derivatives of $\mathbb{R}^2 \xrightarrow{f} \mathbb{R}$ are independent of the order of differentiation. The precise statement follows, though we remark that a slightly stronger theorem is true. (See Exercise 13 of Chapter 7, Section 3.)

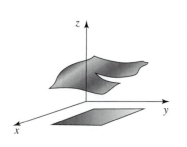

FIGURE 4.21

3.3　Clairaut's Theorem.　　Let $\mathbb{R}^2 \xrightarrow{f} \mathbb{R}$ be continuous and such that f_x, f_y, f_{xy}, and f_{yx} are also continuous on the same domain as f. Then $f_{xy} = f_{yx}$.

Proof.　　Choose x, y, $h \neq 0$, $k \neq 0$ and $\delta > 0$ so the difference

$$F(h, k) = [f(x + h, y + k) - f(x + h, y)] - [f(x, y + k) - f(x, y)]$$

is defined if $\sqrt{h^2 + k^2} < \delta$. We now apply the mean-value theorem in the variable x to the function

$$G(x) = f(x, y + k) - f(x, y)$$

on the interval with endpoints x and $x + h$. We find

$$G(x + h) - G(x) = hG'(x_1),$$

where x_1 is between x and $x + h$. In terms of F and f, this last equation is

$$F(h, k) = h[f_x(x_1, y + k) - f_x(x_1, y)].$$

Now apply the mean-value theorem again, this time to the function $H(y) = f_x(x_1, y)$ on the y-interval with endpoints y and $y + k$. We find

$$F(h, k) = hk f_{xy}(x_1, y_1),$$

where y_1 is between y and $y + k$. Rewriting F in the form

$$F(h, k) = [f(x + h, y + k) - f(x, y + k)] - [f(x + h, y) - f(x, y)]$$

allows us to follow the same general procedure, this time differentiating with respect to y, then x. We find

$$F(h, k) = hk f_{yx}(x_2, y_2),$$

where x_2 and y_2 lie between x, $x + h$ and y, $y + k$ respectively. Equating the two expressions found for $F(h, k)$, and canceling the factor hk, gives

$$f_{xy}(x_1, y_1) = f_{yx}(x_2, y_2).$$

Now let both h and k tend to zero. It follows from the positions of the x_i and y_i that the distances

$$\sqrt{(x_1 - x)^2 + (y_1 - y)^2} \quad \text{and} \quad \sqrt{(x_2 - x)^2 + (y_2 - y)^2}$$

both tend to zero. Therefore, by the continuity of f_{xy} and f_{yx}, we get $f_{xy}(x, y) = f_{yx}(x, y)$. The point (x, y) was arbitrary, so $f_{xy} = f_{yx}$ on the domain of f.   ∎

We may apply Theorem 3.3 successively to still higher-order partial derivatives, provided the analogous differentiability and continuity requirements are satisfied. Moreover, by considering only two variables at a time, we can apply the theorem to functions $\mathbb{R}^n \xrightarrow{f} \mathbb{R}$ where $n > 2$. Thus for the commonly encountered functions that have continuous partial derivatives of arbitrarily high order, we have typically

$$\frac{\partial^2 f}{\partial x \partial y} = \frac{\partial^2 f}{\partial y \partial x}$$

$$\frac{\partial^3 g}{\partial x \partial y \partial x} = \frac{\partial^3 g}{\partial x^2 \partial y}$$

$$\frac{\partial^4 h}{\partial z \partial x \partial y \partial z} = \frac{\partial^4 h}{\partial x \partial y \partial z^2}, \text{ etc.}$$

The last two formulas follow from repeated application of the two-variable formula by interchanging two differentiations at a time.

EXERCISES

In Exercises 1 to 6, find $\dfrac{\partial f}{\partial x}$ and $\dfrac{\partial f}{\partial y}$, where $f(x, y)$ is the given function.

1. $x^2 + x \sin(x + y)$

2. $\sin x \cos(x + y)$

3. e^{x+y+1}

4. $\arctan(y/x)$

5. x^y

6. $\log_x y$

In Exercises 7 to 10, find the general formulas for $\partial f/\partial x$ and $\partial f/\partial y$, then evaluate them at the indicated point $(x, y) = (a, b)$. Then for each function f use Equation 3.2 to write the equation of the tangent plane at the point $(a, b, f(a, b))$ on the graph of f. Simplify the equation of the tangent.

7. $f(x, y) = x^2 y + xy^2, (a, b) = (1, -1)$

8. $f(x, y) = x^2 - y^2, (a, b) = (2, 1)$

9. $f(x, y) = \dfrac{1}{x^2 + y^2}, (a, b) = (1, 1)$

10. $f(x, y) = x(y^2 + 1), (a, b) = (0, 2)$

In Exercises 11 to 14, find $\dfrac{\partial^2 f}{\partial y \partial x}$ and $\dfrac{\partial^2 f}{\partial x \partial y}$, where f is as given.

11. $xy + x^2 y^3$

12. $\sin(x^2 + y^2)$

13. $\dfrac{1}{x^2 + y^2}$

14. $\dfrac{e^{x+y}}{x + y}$

In Exercises 15 to 20, find all first-order partial derivatives of the given functions.

15. $f(x, y, z) = x^2 e^{x+y+z} \cos y$

16. $f(x, y) = x^2 \cos xy$

17. $f(x, y, z, w) = \dfrac{x^2 - y^2}{z^2 + w^2}$

18. $f(x, y, z) = xyz$

19. $f(x, y, z) = x + 2yz$

20. $f(x_1, x_2, x_3) = x_1 x_2 - x_3$

21. Find $\dfrac{\partial^3 f(x, y)}{\partial x^2 \partial y}$ if $f(x, y) = \ln(x + y)$

In Exercises 22 to 25, show that the **Laplace equation** $f_{xx} + f_{yy} = 0$ is satisfied by the given function.

22. $\ln(x^2 + y^2)$

23. $x^3 - 3xy^2$

24. $x/(x^2 + y^2)$

25. $e^x \cos y$

26. If $f(x, y, z) = 1/(x^2 + y^2 + z^2)^{1/2}$, show that

$$f_{xx} + f_{yy} + f_{zz} = 0.$$

27. If $f(x_1, x_2, \dots, x_n) = 1/(x_1^2 + x_2^2 + \cdots + x_n^2)^{(n-2)/2}$, show that
$$f_{x_1 x_1} + f_{x_2 x_2} + \cdots + f_{x_n x_n} = 0.$$

28. Prove directly, without using Theorem 3.3, the general statement that if $f(x,y)$ is a polynomial in x and y, that is, a sum of constant multiples of functions of the form $x^k y^l$, where k and l are nonnegative integers, then $f_{xy}(x, y) = f_{yx}(x, y)$.

For each of the functions 29 to 32, use Equation 3.2 to find a function whose graph is the tangent plane to the graph at the indicated point. Sketch the graph of the function and the tangent plane near the point of tangency.

29. $\sqrt{1 - x^2 - y^2}$ at $(1/2, 1/2, 1/\sqrt{2})$

30. e^{x+y} at $(0, 0, 1)$

31. $e^{-x^2 - y^2}$ at $(0, 0, 1)$

32. $\sqrt{1 - x^3 - y^3}$ at $(0, 0, 1)$

33. Find a parametric representation for the line perpendicular to the tangent plane found in Exercise 29 and passing through the point of tangency.

Verify that the following functions satisfy the **diffusion equation** $u_{xx} = 4u_t$.

34. $u(x, t) = e^{-a^2 t/4} \cos ax$, where a is constant.

35. $u(x, t) = t^{-1/2} e^{-x^2/t}$ for $t > 0$.

Verify that the functions 36 and 37 satisfy the **wave equation** $y_{xx} = y_{tt}$.

36. $y(x, t) = \sin(x - t)$

37. $y(x, t) = \cosh(x + t)$

Assume in Exercises 38 and 39 that a pair $u(x, y)$ and $v(x, y)$ of functions satisfy the equations $u_x(x, y) = v_y(x, y)$ and $u_y(x, y) = -v(x, y)$, called the **Cauchy-Riemann equations**, and assume that u and v have continuous derivatives of order two on some domain D.

38. Show that $u_{xx}(x, y) + u_{yy}(x, y) = 0$ and $v_{xx}(x, y) + v_{yy}(x, y) = 0$ on D.

39. Show that $u(x, y) = e^x \cos y$ and $v(x, y) = e^x \sin y$ satisfy the two Cauchy-Riemann equations.

Harmonic functions on $\mathbb{R}^2$ are real-valued functions $u(x, y)$ that satisfy $u_{xx} + u_{yy} = 0$ for all (x, y) in the domain of u. In Exercises 40 to 49, each formula defines a function on some domain D in $\mathbb{R}^2$. In each case describe D, and state whether the function is harmonic on D or not.

40. $u(x, y) = x^2 - y^2$

41. $u(x, y) = x^3 - y^3$

42. $u(x, y) = e^x \cos y$

43. $u(x, y) = x^3 - 3xy^2$

44. $u(x, y) = x/y$

45. $u(x, y) = \sin(x - y)$

46. $u(x, y) = \ln(x^2 + y^2)$

47. $u(x, y) = \arctan(y/x)$

48. $u(x, y) = \arctan(x/y)$

49. $u(x, y) = \sin(x + y)$

Concavity. A harmonic function u as defined in the preamble to the previous exercises has the property that at every point of its domain either $u_{xx} = u_{yy} = 0$ or else the graph of u exhibits concavity, up or down, along one or both of the lines through the point and parallel to the x- and y-axes. If $u_{xx} \neq 0$ and $u_{yy} \neq 0$, then the concavities have opposite directions.

50. Illustrate the concavity properties using the specific example $u(x, y) = x^2 - y^2$.

*51. Prove the concavity properties for harmonic functions in general.

*52. If

$$f(x, y) = \begin{cases} 2xy \dfrac{x^2 - y^2}{x^2 + y^2}, & \text{for } x^2 + y^2 \neq 0 \\ 0, & \text{for } x = y = 0, \end{cases}$$

show that $f_{xy}(0, 0) = -2$ and $f_{yx}(0, 0) = 2$. [*Hint:* You need to use the definition of partial derivative.] Why does Theorem 3.3 not apply here?

SECTION 4   PARAMETRIZED SURFACES

Parametrization is both useful and fundamental for the representation of curves, and the same statement is true for surfaces. More than one parameter is needed in a vector-valued function to represent differentiable surfaces, as for example planes in Chapter 1, Section 5. Consequently the derivatives we use for computation will be partial derivatives with respect to more than one variable.

4A   Vector Partial Derivatives

In the previous section we considered partial derivatives of real-valued functions only. If $f: \mathbb{R}^n \to \mathbb{R}^m$ is a vector-valued function of n variables, it's natural and useful to define their partial derivatives $\partial f / \partial x_i$ by Equation 3.1 that was used for real-valued functions. The difference is that the quotient in that definition now becomes a vector rather than a number, so the limit $\partial f / \partial x_i(\mathbf{x})$ is a vector also. What we have here is a combination of the ideas of Section 1, on vector-valued functions, and Section 3, where the domain variable is a vector. Since vector limits are computed by taking the limit of each coordinate function, it follows immediately that

4.1   If $f(\mathbf{x}) = \begin{pmatrix} f_1(\mathbf{x}) \\ \vdots \\ f_m(\mathbf{x}) \end{pmatrix}$, then $\dfrac{\partial f}{\partial x_i}(\mathbf{x}) = \begin{pmatrix} \dfrac{\partial f_1}{\partial x_i}(\mathbf{x}) \\ \vdots \\ \dfrac{\partial f_m}{\partial x_i}(\mathbf{x}) \end{pmatrix}$.

EXAMPLE 1   Writing $g(x, y)$ as a column vector, suppose that $\mathbb{R}^2 \xrightarrow{\;g\;} \mathbb{R}^2$ is

$$g(x, y) = \begin{pmatrix} x^2 y \\ xy^2 \end{pmatrix}.$$

Then

$$\frac{\partial g}{\partial x}(x, y) = \begin{pmatrix} 2xy \\ y^2 \end{pmatrix} \quad \text{and} \quad \frac{\partial g}{\partial y}(x, y) = \begin{pmatrix} x^2 \\ 2xy \end{pmatrix}.$$

EXAMPLE 2   If $\mathbb{R}^2 \xrightarrow{\;f\;} \mathbb{R}^3$ is defined by

$$f(u, v) = \begin{pmatrix} u \cos v \\ u \sin v \\ v \end{pmatrix},$$

then

$$\frac{\partial f}{\partial u}(u, v) = \begin{pmatrix} \cos v \\ \sin v \\ 0 \end{pmatrix} \quad \text{and} \quad \frac{\partial f}{\partial v}(u, v) = \begin{pmatrix} -u \sin v \\ u \cos v \\ 1 \end{pmatrix}.$$

EXAMPLE 3   If $\mathbf{x}$ and $\mathbf{y}$ are constant vectors, and $h(u, v) = u\mathbf{x} + v\mathbf{y}$, then $\partial h / \partial u(u, v) = \mathbf{x}$ and $\partial h / \partial v(u, v) = \mathbf{y}$.

The geometric significance of the vector partial derivative is as follows. If all coordinates but one are held fixed, and the remaining one, say x_i, is allowed to vary, then $f(\mathbf{x}) = f(x_1, \ldots, x_i, \ldots, x_n)$ traces an image curve in $\mathbb{R}^m$, sometimes called a **coordinate curve** if $m \geq n$. Hence by the interpretation of Section 1, the vector $\partial f / \partial x_i(\mathbf{x})$ is a tangent vector to this coordinate curve at the image point $f(\mathbf{x})$.

EXAMPLE 4

Consider the very simple situation in which $f: \mathbb{R}^2 \to \mathbb{R}^3$ is given by

$$f(u, v) = u\mathbf{x}_1 + v\mathbf{x}_2,$$

which is a parametric representation of a plane containing $\mathbf{x}_1$ and $\mathbf{x}_2$ and also passing through the origin. If $v = v_0$ is held fixed, then as u varies, $f(u, v_0)$ traces a line through $v_0\mathbf{x}_2$ and parallel to $\mathbf{x}_1$. See Figure 4.22(a). The vector partial derivative with respect to u is

$$\frac{\partial f}{\partial u}(u, v_0) = \frac{\partial(u\mathbf{x}_1 + v_0\mathbf{x}_2)}{\partial u} = \mathbf{x}_1.$$

Similarly,

$$\frac{\partial f}{\partial u}(u_0, v) = \frac{\partial(u_0\mathbf{x}_1 + v\mathbf{x}_2)}{\partial v} = \mathbf{x}_2.$$

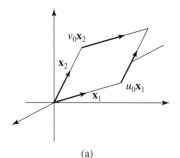

Thus $\mathbf{x}_1$ and $\mathbf{x}_2$ each plays the role of tangent vector to a line in the plane parametrized by f.

(a)

Just as a parametric representation of a plane is a linear function image, perhaps shifted by a constant vector, so curved surfaces are parametrized images of nonlinear vector-valued functions. If f is differentiable in a sense to be made precise in Chapter 5, and the vectors $\mathbf{x}_u(u_0, v_0) = \partial f/\partial u(u_0, v_0)$ and $\mathbf{x}_v(u_0, v_0) = \partial f/\partial v(u_0, v_0)$ are linearly independent, we define the **tangent plane** at $\mathbf{x}(u_0, v_0)$ to a surface parametrized by $\mathbf{x}(u, v) = f(u, v)$ to be the plane passing through $\mathbf{x}(u_0, v_0)$ and parallel to $\mathbf{x}_u(u_0, v_0)$ and $\mathbf{x}_v(u_0, v_0)$. A parametric representation for the tangent plane gives a picture such as Figure 4.22(a) for the image of

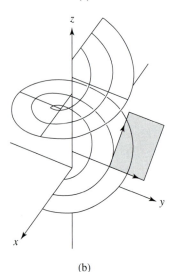

4.2 $$\mathbf{x} = u\mathbf{x}_u(u_0, v_0) + v\mathbf{x}_v(u_0, v_0) + \mathbf{x}(u_0, v_0).$$

If you like, you can make the point of tangency on the tangent plane correspond to the parameter values $(u, v) = (u_0, v_0)$ by replacing the scalar factors u and v in Equation 4.2 by $(u - u_0)$ and $(v - v_0)$. Alternatively, you have the option of using altogether different letters for the parameters, for example using s and t:

(b)

$$\mathbf{x} = s\mathbf{x}_u(u_0, v_0) + t\mathbf{x}_v(u_0, v_0) + \mathbf{x}(u_0, v_0).$$

FIGURE 4.22

EXAMPLE 5

Figure 4.22(b) shows a helical surface that is generated by helical curves. or alternatively by line segments perpendicular to the z-axis. A parametric representation for the surface is

$$f(u, v) = \begin{pmatrix} u \cos v \\ u \sin v \\ v \end{pmatrix},$$

where in the picture we have restricted the parameters u and v so that $0 \le u \le 4$ and $0 \le v \le 3\pi$. If $u = u_0$ is held fixed and v varies, we get a helical curve (see Example 8 in Section 1) winding one and one-half times around the z-axis on a cylinder of radius u_0. With $v = v_0$ and u varying, we get a line segment v units above the xy-plane. The vector partial derivatives of f were computed in Example 2.

At $(u_0, v_0) = (1, \pi/4)$, we get the tangent vectors

$$\frac{\partial f}{\partial u}(1, \pi/4) = \begin{pmatrix} 1/\sqrt{2} \\ 1/\sqrt{2} \\ 0 \end{pmatrix}, \quad \frac{\partial f}{\partial v}(1, \pi/4) = \begin{pmatrix} -1/\sqrt{2} \\ 1/\sqrt{2} \\ 1 \end{pmatrix},$$

to the two parameter curves through the point $f(1, \pi/4) = (1/\sqrt{2}, 1/\sqrt{2}, \pi/4)$ on the surface. The tangent plane at this point is represented parametrically by

$$u \begin{pmatrix} 1/\sqrt{2} \\ 1/\sqrt{2} \\ 0 \end{pmatrix} + v \begin{pmatrix} -1/\sqrt{2} \\ 1/\sqrt{2} \\ 1 \end{pmatrix} + \begin{pmatrix} 1/\sqrt{2} \\ 1/\sqrt{2} \\ \pi/4 \end{pmatrix},$$

as (u, v) ranges over $\mathbb{R}^2$. The plane appears in Figure 4.22(b). Note that the curved surface is *not* the graph of a function of (x, y), because there are z-values differing by 2π corresponding to some pairs (x, y). The image surface is called a **helicoid**.

EXAMPLE 6   The function $f(u, v) = (u \cos v, u \sin v, v)$ with domain altered from the previous example to the rectangle $-1 \leq u \leq 1$, $-3\pi \leq v \leq 3\pi$ has for its image a helicoid H that winds around the central axis. Figure 4.23(a) shows a sketch of a portion of the surface. The tangent plane to H at a point $f(0, v_0) = (0, 0, v_0)$ on the vertical axis is generated by two tangent vectors, $f_u(0, v_0) = (\cos v_0, \sin v_0, 0)$ and $f_v(0, v_0) = (0, 0, 1)$. Since the second of these two vectors is parallel to the central axis, the tangent plane at a point on the axis always contains the axis. Since the dot product of $f_u(0, v_0)$ and $f_v(0, v_0)$ equals zero the tangent plane also contains segments perpendicular to the central axis and lying in the surface. The surface is pictured as a kind of "skeleton," reminiscent of models of the DNA molecule and their linking of points on pairs of helices shown in Figure 4.23(a).

EXAMPLE 7   A cone C represented parametrically by $\mathbf{x}(u, v) = (u \cos v, u \sin v, u)$ is sketched in Figure 4.23(b). (Note that the parametrization differs only in the third coordinate

FIGURE 4.23

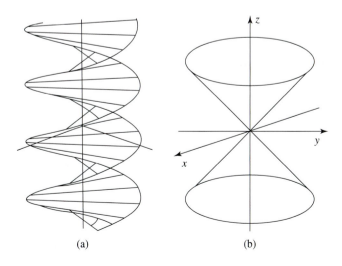

(a)   (b)

function from that of the helicoid.) The parameter values $(u, v) = (0, v_0)$ all correspond to the single point $(0, 0, 0)$ on C. Furthermore, the attempt to find a pair of independent vectors $\mathbf{x}_u(0, v_0) = (\cos v_0, \sin v_0, 0)$ and $\mathbf{x}_v(0, v_0) = (0, 0, 0)$ at $(0, 0, 0)$ fails to produce a tangent plane, since the second vector fails to provide a tangent direction. At every other point C has a well-defined tangent plane, as you are asked to show in Exercise 19.

The sharp point at the ends of the two symmetric halves of the cone in the previous example is called "singular" because the surface lacks a certain smoothness that we like to associate with a typical point of a surface. The official definition is as follows. Recall the related definition of smooth curve in Section 1 of this chapter.

Definition   A surface S parametrized by a function $\mathbb{R}^2 \xrightarrow{g} \mathbb{R}^3$ is **smooth** at points $g(u, v)$ on S if there is a rectangle $a < u < b$, $c < v < d$ such that (i) $g_u(u, v)$ and $g_v(u, v)$ have continuous coordinate functions on the rectangle, and (ii) the vector partial derivatives $g_u(u, v)$ and $g_v(u, v)$ are linearly independent, that is, neither is a constant multiple of the other. The intuitive content of the definition is that smoothness is equivalent to having a continuously varying tangent plane at the $g(u, v)$. A point where a surface fails to be smooth is called a **singular point**.

EXAMPLE 8   Here we consider the transformation $h: \mathbb{R}^2 \to \mathbb{R}^2$ given by

$$h(u, v) = (u \cos v, u \sin v)$$

restricted by $0 \le u \le 2$ and $0 \le v \le \pi$. This example comes from the previous one by elimination of the third coordinate in the range, so the image surface is restricted to the image plane. Figure 4.24(a) shows the domain of h and Figure 4.24(b) shows the image. To get an idea of how the transformation behaves, we look at the images under h of the lines parallel to the axes in the domain. The resulting curves in $\mathbb{R}^2$, given by

$$x = u \cos v, \quad y = u \sin v,$$

are semicircles if u is held fixed and line segments if v is held fixed. The image of f restricted to the rectangle in Figure 4.24(a) is the half-disk in Figure 4.24(b). In this example the image "surface" lies in $\mathbb{R}^2$.

Graphs of $z = f(x, y)$. In Section 2 we considered the graphs of real-valued functions $f(x, y)$ as surfaces S in $\mathbb{R}^3$ lying over the domain D of f in the xy-space $\mathbb{R}^2$. There is a simple way to absorb such surfaces into our present discussion of parametrized surfaces. The idea is just to regard x and y as parameters in a parametric representation $(x, y, z) = (x, y, f(x, y))$, where (x, y) varies over D. A glance at Figure 4.23(a) shows that finding a single-valued function $f(x, y)$ whose graph is the entire helical surface is impossible, so the parametric representation is really more general. You can verify that letting $f(x, y) = \arctan(y/x)$ is a partial

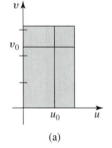

(a)

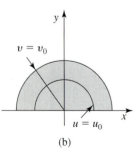

(b)

FIGURE 4.24

solution to finding a function $\mathbb{R}^2 \overset{f}{\longrightarrow} \mathbb{R}^3$ whose graph is a part of the helicoid parametrized by

$$x = u \cos v, \quad y = u \sin v, \quad z = v.$$

The following example illustrates the general proposition that when both representations apply, the tangent planes turn out to be the same.

EXAMPLE 9 The graph of the function $f(x, y) = e^{x-y}$ has partial derivatives $f_x(1, 1) = 1$ and $f_y(1, 1) = -1$. Hence the tangent plane to the graph S of f at the point of tangency $(1, 1, 1)$ is

$$z = 1 + (1)(x - 1) + (-1)(x - 1) \quad \text{or} \quad z = x - y + 1.$$

There is no uniquely determined parametric representation for S, but there is one that is particularly convenient. Letting $x = u$ and $y = v$ to introduce our usual parameters, we write

$$g(u, v) = \begin{pmatrix} u \\ v \\ e^{u-v} \end{pmatrix}, \quad \text{so} \quad g_u(u, v) = \begin{pmatrix} 1 \\ 0 \\ e^{u-v} \end{pmatrix}, \quad g_v(u, v) = \begin{pmatrix} 0 \\ 1 \\ -e^{u-v} \end{pmatrix}$$

as parametric representations for S and its tangent vectors. Parametrically, the tangent at $g(1, 1) = (1, 1, 1)$ is $\mathbf{x} = g(1, 1) + u g_u(1, 1) + v g_v(1, 1)$, or

$$\mathbf{x} = \begin{pmatrix} x \\ y \\ z \end{pmatrix} = \begin{pmatrix} 1 \\ 1 \\ 1 \end{pmatrix} + u \begin{pmatrix} 1 \\ 0 \\ 1 \end{pmatrix} + v \begin{pmatrix} 0 \\ 1 \\ -1 \end{pmatrix} = \begin{pmatrix} 1 + u \\ 1 + v \\ 1 + u - v \end{pmatrix}.$$

Reverting to $u = x$ and $v = y$, we see that $z = 1 + x - y$, which is what we got using the first method.

4B   Quadric Surfaces

The definition of quadric surface given in Section 2D is fundamental for many purposes, but parametric representations provide insight into the structure of some of them, and will also be useful for certain multiple integration problems later on. The elliptic and hyperbolic paraboloids are graphs of real-valued functions, so these two types are only a notational change away from parametrization, as we've just seen. Of the remaining types, we'll treat the elliptic cone and the ellipsoid in detail, leaving details of the hyperboloids as exercises.

EXAMPLE 10 Geometrically an **elliptic cone** is generated by all lines that pass through the points of a fixed ellipse in space and that also pass through a fixed point not in the plane of the ellipse. (In particular, an elliptic cone could be one of the familiar right circular cones.) Using coordinates in $\mathbb{R}^3$, let the fixed point be the origin, and let the fixed ellipse be the one that projects parallel to the z-axis from the plane $z = 1$ onto the ellipse $(x/a)^2 + (y/b)^2 = 1$ in the xy-plane. A typical point on the ellipse has coordinates $(a \cos v, b \sin v, 1)$, so a line joining this point to the origin consists of all points of the form

$$\mathbf{x}(u, v) = u(a \cos v, b \sin v, 1) = (au \cos v, bu \sin v, u).$$

FIGURE 4.25

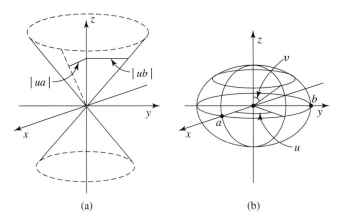

(a)　　　　　　　　(b)

As u varies for fixed v, $\mathbf{x}(u, v)$ traces one of the generating lines of the cone. As v varies for fixed $u \neq 0$, $\mathbf{x}(u, v)$ traces an ellipse with semiaxes $|ua|$ and $|ub|$ in the plane $z = u$. See Figure 4.25(a).

To identify the cone as the quadric surface defined in Section 2, just extract x, y and z from the parametrization, and observe that

$$\frac{x^2}{a^2} + \frac{y^2}{b^2} = u^2 \cos^2 v + u^2 \sin^2 v = z^2.$$

EXAMPLE 11　Geometrically an **ellipsoid** is a closed surface with three perpendicular axes of symmetry such that the plane cross sections perpendicular to these axes are ellipses, possibly circles. Taking the symmetry axes to be the coordinate axes in $\mathbb{R}^3$, we let a, b and c be positive numbers and

$$\mathbf{x}(u, v) = \begin{pmatrix} a \cos u \sin v \\ b \sin u \sin v \\ c \cos v \end{pmatrix}.$$

Identification with the quadric surface of Section 2 comes from checking that

$$(x/a)^2 + (y/b)^2 + (z/c)^2 = \cos^2 u \sin^2 v + \sin^2 u \sin^2 v + \cos^2 v = 1.$$

For a fixed v between 0 and π and varying u, $\mathbf{x}(u, v)$ traces an elliptic latitude curve $(x/a \sin v)^2 + (y/b \sin v)^2 = 1$ in the plane $z = c \cos v$. As v varies for fixed u, $\mathbf{x}(u, v)$ traces longitude curves extending between the "north pole" and the "south pole." Typical curves are shown in Figure 4.25(b).

EXERCISES

Find formulas for the vector partial derivatives $\partial f/\partial x(x, y)$ and $\partial f/\partial y(x, y)$ of the functions in Exercises 1 to 6.

1. $f(x, y) = \begin{pmatrix} x + y \\ x - y \\ x^2 + y^2 \end{pmatrix}$

2. $f(x.y) = (e^x \cos y, e^x \sin y)$

3. $f(x, y) = \begin{pmatrix} xy \\ x + y \end{pmatrix}$

4. $f(x, y) = \begin{pmatrix} x^2 \\ y^2 \\ xy \end{pmatrix}$

5. $f(x, y) = (e^x, e^y, e^{x+y})$

6. $f(x, y) = \begin{pmatrix} x \\ y \end{pmatrix} + \begin{pmatrix} y \\ x \end{pmatrix}$

For each of the functions in Exercises 7 to 10, find all first-order vector partial derivatives at the indicated point. Then for each one, sketch the curves on the image surface passing through the image point $g(u_0, v_0)$ and having the property that only u varies on one curve, and only v varies on the other; that is, sketch the coordinate curves through $g(u_0, v_0)$. Finally, sketch the two tangent vectors given by the partial derivatives.

7. $g(u, v) = (u, v, u^2 + v^2)$ at $(u_0, v_0) = (1, 1)$

8. $g(u, v) = (u, v, uv)$ when $(u_0, v_0) = (1, 1)$

9. $g(u, v) = (\cos u \sin v, \sin u \sin v, \cos v)$ at $(u_0, v_0) = (\pi/4, \pi/4)$

10. $g(u, v) = (u \cos v, u \sin v, v)$ at $(u_0, v_0) = (1, \pi/4)$

For each of Exercises 11 to 14, sketch the image surface corresponding to the domain given here and sketch the tangent plane at the indicated point.

11. $g(u, v) = (u, v, u^2 + v^2)$, $-2 \le u \le 2, -2 \le v \le 2$, tangent where $(u_0, v_0) = (1, 1)$

12. $g(u, v) = (u, v, uv)$, $0 \le u \le 2, 0 \le v \le 2$, tangent where $(u_0, v_0) = (1, 1)$

13. $g(u, v) = (\cos u \sin v, \sin u \sin v, \cos v)$, $0 \le u \le \frac{1}{2}\pi$, $0 \le v \le \frac{1}{2}\pi$, tangent where $(u_0, v_0) = (\pi/4, \pi/4)$ [*Hint:* This surface is a piece of a sphere.]

14. $g(u, v) = (u \cos v, u \sin v, v)$, $0 \le u \le 2$, $0 \le v \le 2\pi$, tangent at $(u_0, v_0) = (1, \pi/4)$

15. Let $f(x, y, t) = (x + t, y + t^2)$, for $t \ge 0$, represent the position at time t of a point starting at (x, y) and tracing a path in $\mathbb{R}^2$.
   (a) Sketch the paths starting at $(x, y) = (0, 0)$, $(1, 0)$, and $(1, 1)$.
   (b) Sketch the path starting at $(-1, 0)$ and give a geometric interpretation for the vector partial derivatives

$$\frac{\partial f}{\partial t}(-1, 0, 0) \quad \text{and} \quad \frac{\partial f}{\partial t}(-1, 0, 1).$$

16. The vector function f is defined by

$$f\begin{pmatrix} x \\ y \end{pmatrix} = \begin{pmatrix} x^2 - y^2 \\ 2xy \end{pmatrix}.$$

Consider the domain space to be the xy-plane and the range space to be the uv-plane.
   (a) What are the coordinate functions of f?

   (b) Find the image of the segment of the line $y = x$ between $(0, 0)$ and $(1, 1)$.
   (c) Find the image of the region defined for positive x and y, and $x^2 + y^2 < 1$.
   (d) Find the angle between the images of the lines $y = 0$ and $y = (1/\sqrt{3})x$.

17. A vector function f from the xy-plane to the uv-plane is defined by

$$f\begin{pmatrix} x \\ y \end{pmatrix} = \begin{pmatrix} u \\ v \end{pmatrix} = \begin{pmatrix} x \\ \dfrac{(x + y)^2}{4x} \end{pmatrix}, \quad x \ne 0.$$

   (a) What are the coordinate functions of f?
   (b) What are the vector partial derivatives of f?
   (c) Describe the image of the region bounded by the four lines

$$x = y, \ y = x - 8, \ x = -y, \ y = 8 - x.$$

18. Let a transformation from the xy-plane to itself be given by

$$f\begin{pmatrix} x \\ y \end{pmatrix} = \begin{pmatrix} x + y \\ -x + y \end{pmatrix}.$$

   (a) Show that f accomplishes an expansion out from the origin by a factor $\sqrt{2}$ combined with a rotation through an angle $\pi/4$.
   (b) Show that the vector partial derivatives of f are constant.

19. Show that the cone C as parametrized in Example 10 of the text is smooth at all points except the pointed tip.

20. Let $f(u, v) = (u^2 \cos v, u^2 \sin v, u)$.
   (a) Show that the image S of f with (u, v) unrestricted coincides with the level set $x^2 + y^2 - z^4 = 0$.
   (b) Make a sketch of S.
   (c) Where are the singular points of S?

21. The top half T of the cone as parametrized in Example 10 of the text appears intuitively to be singular at its pointed end, and is singular in the technical sense because, while the natural tangent vectors there are continuous, they aren't linearly independent since one of them is the zero vector.
   (a) Show that T is also parametrized by $g(u, v) = (u, v, \sqrt{u^2 + v^2})$ by showing that the images of both parametrizations coincide with the level set $x^2 + y^2 - z^2 = 0$ for $z \ge 0$.
   (b) Show that the pointed end of T is singular with respect to the parametrization $g(u, v)$, but for a

different reason than for the parametrization given in Example 10.

(c) Show that, with respect to $g(u, v)$, T is smooth at every point other than the tip.

22. (a) Sketch the surface parametrized by

$$\mathbf{x}(u, v) = ((1 - u^2) \cos v, (1 - u^2) \sin v, u),$$

$$-1 \le u \le 1, -\infty < v < \infty.$$

(b) What are the singular points of the surface in part (a)?

23. Let $a > b > 0$, and consider the set T parametrically represented by

$$\begin{pmatrix} x \\ y \\ z \end{pmatrix} = \begin{pmatrix} (a + b \cos v) \cos u \\ (a + b \cos v) \sin u \\ b \sin v \end{pmatrix},$$

$$0 \le u \le 2\pi, \quad 0 \le v \le 2\pi.$$

(a) Show that T is a torus by identifying two families of circles on it, showing in particular that for fixed $v = v_0$, $\mathbf{x}$ traces a circle of radius $a + b \cos v_0$ parallel to the xy-plane and that for fixed $u = u_0$, $\mathbf{x}$ traces a circle of radius b in a plane containing the z-axis. Note that $a + b$ is the outer radius of the ring, $a - b$ is its inner radius, and b is the radius of the tube.

(b) By eliminating u and v show that the points of T satisfy

$$((x^2 + y^2 + z^2) - (a^2 + b^2))^2 = 4a^2(b^2 - z^2).$$

24. The helical surface represented parametrically by $(x, y, z) = (u \cos v, u \sin v, v)$ for $u > 0$ and $-\pi/2 < v < \pi/2$ is also the graph of a function $f(x, y)$ defined for $x > 0$. Show by eliminating u and v that $f(x, y) = \arctan(y/x)$, where we assume arctan $(0) = 0$.

25. (a) Verify that the parametrizations

$$\mathbf{z}(u, v) = (a \cos v, b \cos u \sin v, c \sin u \sin v)$$

$$\mathbf{x}(u, v) = (a \cos u \sin v, b \cos v, c \sin u \sin v)$$

are different parametrizations of the ellipsoid $(x/a)^2 + (y/b)^2 + (z/c)^2 = 1$.

(b) Use the parametrizations of part (a) to show that the ellipsoid has elliptic cross sections perpendicular to the x-axis and y-axis.

26. Parametrizations for the **hyperboloid of one sheet**, $(x/c)^2 + (y/b)^2 - (z/c)^2 = 1$, and of **two sheets**, $(x/a)^2 + (y/b)^2 - (z/c)^2 = -1$, may involve the hyperbolic functions $\cosh v = \frac{1}{2}(e^v + e^{-v})$ and $\sinh v = \frac{1}{2}(e^v - e^{-v})$.

(a) Verify that $\cosh^2 v - \sinh^2 v = 1$.

(b) Verify that the hyperboloid of one sheet has parametrization

$$\mathbf{x}(u, v) = (a \cos u \cosh v, b \sin u \cosh v, c \sinh v).$$

(c) Verify that each piece of the hyperboloid of two sheets has parametrization

$$\mathbf{x}(u, v) = (a \cos u \sinh v, b \sin u \sinh v, c \cosh v),$$

where $c > 0$ gives the upper half and $c < 0$ gives the lower half.

27. Suppose that $(y, z) = (g(v), h(v))$ parametrizes a curve in the yz-plane of $\mathbb{R}^3$. Rotating this curve about the z-axis generates a **surface of revolution** S. Show that a parametrization for S is given by

$$\mathbf{x}(u, v) = \begin{pmatrix} g(v) \cos u \\ g(v) \sin u, \\ h(v) \end{pmatrix}.$$

28. Show that if the graph S of the real-valued function $f(x, y)$ has a tangent plane at $(x_0, y_0, f(x_0, y_0))$ given by Equation 3.2, then the parametrization of S given by

$$\mathbf{x}(u, v) = \begin{pmatrix} u \\ v \\ f(u, v) \end{pmatrix},$$

together with Equation 4.2, yields the same tangent plane.

4C   Computer Plotting of Image Surfaces

A parametric representation $x = f_1(u, v)$, $y = f_2(u, v)$, $z = f_3(u, v)$ of a surface is one in which the surface is the image of the vector function with coordinate functions f_1, f_2, and f_3 restricted to a domain in $\mathbb{R}^2$ with coordinates (u, v). This is a generalization of the situation considered when we plotted graphs for $z = f(x, y)$.

There we drew pictures in which the domain of a function was itself a part of the picture. In particular, for a positive function, the domain could be thought of as lying in the xy-plane under the graph. In the present situation, the common 2-dimensional domain of the coordinate functions f_1, f_2, f_3 can usually just as well be thought of as lying somewhere in some separate plane. The plotting program is very similar however.

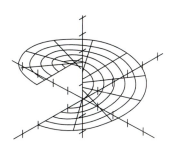

```
DEFINE f₁(u, v)  =  u cos (v)
DEFINE f₂(u, v)  =  u sin (v)
DEFINE f₃(u, v)  =  0.4v
  First plot v-varying curves, for selected u.
FOR u  =  0 TO 3 STEP 0.5
FOR v  =  0 TO 2π STEP 0.01
  PLOT3D (f₁(u, v), f₂(u, v), f₃(u, v))
NEXT v
NEXT u
  Then plot u-varying curves, for selected v.
FOR v  =  0 TO 2π STEP π/5
FOR u  =  0 TO 3 STEP .01
  PLOT3D (f₁(u, v), f₂(u, v), f₃(u, v))
NEXT u
NEXT v
```

FIGURE 4.26

Executing the routine produces Figure 4.26.

EXERCISES

In Exercises 1 to 6, sketch the parametrically defined surfaces. In 5 and 6 you may find it useful to introduce the heaviside function $H(u^2 + v^2)$ as in Section 2C.

1. $f(u, v) = (\cos u \sin v, \sin u \sin v, \cos v), 0 \le u \le \pi/2,$
$0 \le v \le \pi/2$

2. $f(u, v) = (\cos u \sin v, \sin u \sin v, 1 + \cos v), 0 \le u \le 2\pi,$
$0 \le v \le 2\pi$

3. $f(u, v) = (\exp(u - v), u, v), 0 \le u \le 2, 0 \le v \le 2$

4. $f(u, v) = (u + v, u - 2v, 2u + 3v), 0 \le u \le 3, 0 \le v \le 2$

5. $f(u, v) = (u+v, u-2v, 2u+3v), 0 \le u, 0 \le v, u^2 + v^2 \le 2$

6. $f(u, v) = (u, v, u^2 + v^2), 0 \le u, 0 \le v, u^2 + v^2 \le 2$

7. Explain in words the geometric significance of the angle parameters u, v in Exercises 1 and 2.

In Exercises 8 to 11, sketch the parametrically defined surfaces subject to the given conditions.

8. $(x, y, z) = (\cos u \sin v, \sin u \sin v, \cos v), 0 \le u \le \pi/2,$
$0 \le v \le \pi/4$

9. $(x, y, z) = (\cos u \sin v, \sin u \sin v, 1+\cos v), 0 \le u \le 2\pi,$
$0 \le v \le 2\pi$, for $\frac{1}{2} \le y \le 1$

10. $(x, y, z) = (\exp(u - v), u, v), 0 \le u \le 2, 0 \le v \le 2$

11. $(x, y, z) = (u + v, u - 2v, 2u + 3v), |x| \le 1, |y| \le 2,$
$|z| \le 3$

In Exercises 12 to 15, find a parametrization and, using this parametrization, make a computer sketch of an example each of the following.

12. The ellipsoid $x^2 + 2y^2 + z^2 = 1$

13. The hyperboloid of one sheet $x^2 + y^2 - 2z^2 = 1$ for $|z| \le 1$

14. The half of the hyperboloid $z^2 - x^2 - y^2 = 1$ for which $z \ge 1$

15. The half of the hyperboloid $z^2 - x^2 - y^2 = 1$ for which $z \le -1$

Chapter 4 REVIEW

1. Let $\mathbf{x}(t) = (e^t \cos t)\mathbf{i} + (e^t \sin t)\mathbf{j} + e^t\mathbf{k}$ and find the velocity vector $\dot{\mathbf{x}}(t)$ and the vector $\mathbf{t}(t)$ of length 1 pointing in the same direction. Also find the acceleration vector $\ddot{\mathbf{x}}(t)$.

2. The motion of a particle is given by the vector function $\mathbf{x}(t) = (\cos 2t)\mathbf{i} + (\sin 2t)\mathbf{j} + t^2\mathbf{k}$.
   - (a) Sketch the trajectory of the particle when $0 \leq t \leq \pi$.
   - (b) What is the velocity vector when $t = \pi/4$? Add this vector to your sketch.
   - (c) Suppose the particle leaves its prescribed path when $t = \pi/4$ and continues at the constant velocity it has acquired at that time; where will the particle then be at time $t = \pi/2$?

3. The position of a particle at time t is given by

$$\mathbf{x}(t) = \left(\frac{5}{\sqrt{2}} \sin t \right)\mathbf{i} + \left(\frac{5}{\sqrt{2}} \sin t \right)\mathbf{j} + (5\cos t)\mathbf{k}.$$

   - (a) Find the velocity and acceleration vectors when $t = 0$.
   - (b) What distance does the particle travel between $t = 0$ and $t = 2\pi$?
   - (c) Show that the particle always lies on a sphere centered at the origin.
   - (d) Find a plane that contains the path of the particle, and sketch the path from an appropriate point of view.

4. (a) Sketch the image curve of the function given by $f(t) = (t, \cos t, \sin t)$ for $0 \leq t \leq 2\pi$.
   - (b) Find a parametric representation for the line tangent to the curve in part (a) at the point $f(\pi/2) = (\pi/2, 0, 1)$, and add this line to the sketch for part (a).
   - (c) Find the acceleration vector of the curve at $f(\pi/2)$ and add this to the sketch also.

5. Find the maximum and minimum values of the speed of the motion along the curve traced by $\mathbf{x}(t) = (a\cos t, b\sin t, ct)$, where $a > b > 0$ and c are constant.

6. Suppose $\mathbf{x}(t)$ traces a curve in $\mathbb{R}^3$ that lies in a plane $ax + by + cz = d$. Show that if the curve has nonzero tangent and acceleration vectors at each point $\mathbf{x}(t)$, then $\dot{\mathbf{x}}(t)$ and $\ddot{\mathbf{x}}(t)$ are parallel to the plane of the curve.

7. (a) Sketch the curve parametrized by $\mathbf{x}(t) = (e^t \cos t, e^t \sin t)$ for $0 \leq t \leq \pi/2$.
   - (b) Find the length of the curve in part (a).
   - (c) Repeat parts (a) and (b) with the parameter interval replaced by $-\pi/2 \leq t \leq 0$.

8. A particle moves so that at time t its position is $\mathbf{x}(t) = (-\sin t, \cos t, t^2)$.
   - (a) Find the velocity $\mathbf{v}(t) = \dot{\mathbf{x}}(t)$ and the acceleration $\mathbf{a}(t) = \dot{v}(t)$.
   - (b) What is the distance between the positions of the particle at $t = 0$ and at $t = \pi$?
   - (c) Find a parametric expression for the tangent line to the curve traced out by the particle's motion at the point $\mathbf{x}(\pi)$.

9. Find an equation for the tangent plane to the graph of the equation $z = x^2 + y^3$ at $(1, 2, 9)$.

10. Find a parametric representation for the plane tangent to the image surface of the function $f(u, v) = (u+v, u^2, v^2)$ at the point $f(1, 1) = (2, 1, 1)$.

In Exercises 11 to 16, verify that the functions satisfy the **Laplace equation** $u_{xx} + u_{yy} = 0$ in two dimensions.

11. $u(x, y) = x^2 - y^2$

12. $u(x, y) = 2xy$

13. $u(x, y) = x^3 y - xy^3$

14. $u(x, y) = \ln(x^2 + y^2)$

15. $u(x, y) = \arctan(y/x), \; x \neq 0$

16. $u(x, y) = e^{x^2 - y^2} \cos 2xy$

In Exercises 17 and 18, verify that the functions satisfy the 3-dimensional **Laplace equation** $u_{xx} + u_{yy} + u_{zz} = 0$.

17. $u(x, y, z) = xyz + 2x^2 - y^2 - z^2$

18. $u(x, y, z) = (x^2 + y^2 + z^2)^{-1/2}, \; (x, y, z) \neq (0, 0, 0)$

19. Let $u(x, y, z) = (x^2 + y^2 + z^2)^\alpha$, α constant. Show that $u_{xx} + u_{yy} + u_{zz} = 0$ just for $\alpha = 0$ and $\alpha = -1/2$.

20. Let $f(u, v) = (u, v, u^2 + v^2)$.
   - (a) Sketch the image of f.
   - (b) Compute the partial derivatives $f_u(u, v)$ and $f_v(u, v)$. Compute the equation of the tangent plane to the image surface of f when $u = 1, v = 1$.
   - (c) Sketch the u and v coordinate curves of f through $(1, 1, 2)$; that is, the curve where u varies and $v = 1$, and the curve where v varies and $u = 1$. Sketch the tangent vectors to these curves at $(1, 1, 2)$. What is the relationship between these tangent vectors and the partial derivatives computed in (b)?

21. (a) Show that the intersection of the elliptic cone $(x/a)^2 + (y/b)^2 - z^2 = 0$ with a plane perpendicular to the z-axis is an ellipse.

(b) Show that the intersection of the cone with a plane perpendicular to the x-axis or to the y-axis is a hyperbola.

22. Identify the curve of intersection of the ellipsoid $(x/a)^2 + (y/b)^2 + (z/c)^2 = 1$ with a plane perpendicular to one of the coordinate axes.

23. Identify the curves of intersection of the elliptic paraboloid $(x/a)^2 + (y/b)^2 = z$ with (a) planes perpendicular to the z-axis. (b) planes perpendicular to the x-axis or the y-axis.

24. Consider the helicoid H parametrized by
$\mathbf{x}(u, v) = (u \cos v, u \sin v, v)$.
(a) Find a parametrization for the tangent plane to H at the point $(1/\sqrt{2}, 1/\sqrt{2}, \pi/4)$.
(b) Find a normal vector $\mathbf{n}$ of length 1 to H at the same point.

A function $\mathbb{R}^n \xrightarrow{f} \mathbb{R}^m$ represents a set S

(a) **explicitly** if S is the graph of f in $\mathbb{R}^{n+m}$,
(b) **implicitly** if S is a level set of f in $\mathbb{R}^n$,
(c) **parametrically** if S is the image of f in $\mathbb{R}^m$.

For example, the graph of $\mathbb{R}^1 \xrightarrow{f} \mathbb{R}^1$ where $f(x) = x^2$ is an explicit representation of a parabola P in $\mathbb{R}^{1+1} = \mathbb{R}^2$.

25. Find an implicit representation for the parabola P.

26. Find a parametric representation for the parabola P.

27. The function $\mathbb{R}^1 \xrightarrow{g} \mathbb{R}^3$ where $g(t) = (t, t, t^2)$ has as its image a parabola P in $\mathbb{R}^3$, so P is represented parametrically by g. Make a sketch of P, and find an explicit representation of P as the graph of some $\mathbb{R}^1 \xrightarrow{h} \mathbb{R}^2$.

CHAPTER 5

DIFFERENTIABILITY

To say that a function $\mathbb{R} \xrightarrow{f} \mathbb{R}^m$ is **differentiable** on an interval $a < t < b$ means that the m real-valued coordinate functions of $f(t)$ are differentiable on $a < t < b$ and that $f(t)$ has derivative $f'(t) = \left(f_1'(t), \ldots, f_m'(t) \right)$ for every t in the interval. The aim of this chapter is to explain the analogue of that simple definition for functions $\mathbb{R}^n \xrightarrow{f} \mathbb{R}$ and then for $\mathbb{R}^n \xrightarrow{f} \mathbb{R}^m$. The connection between the property of differentiability and the existence of derivatives is more interesting for the case of a higher-dimensional domain even when the range space is 1-dimensional. The reason for this is that with only a 1-dimensional domain there is only one line along which to differentiate. But whenever the domain has dimension $n \geq 2$ we have to deal with infinitely many different directions from which to approach a given point, and hence infinitely many possible directions along which to compute the limit that defines a derivative. Recall that in the previous chapter all the derivatives we computed for functions of more than one variable were *partial* derivatives, each one defined by a limit taken in a direction parallel to one of the coordinate axes; this allowed us to use all the machinery of single-variable calculus in our computations. The only reason we were able to get away with ignoring the infinitely many other possible directions for computing derivatives was that we restricted our examples to functions $\mathbb{R}^n \xrightarrow{f} \mathbb{R}^m$ that were *differentiable* in the specific sense defined below in Section 2 of the present chapter. Once we have clarified this point, we'll see that extending the definition of differentiability to vector-valued functions f is really just a matter of requiring the real-valued coordinate functions of f to be differentiable. Finding the correct analogue of the derivative $f'(t)$ then comes down to using matrix algebra in a very natural way, one advantage of which will become apparent in Section 5 of the present chapter on Newton's method for approximate solution of systems of equations. Another important application appears in Section 3 of Chapter 6 on chain rule computations.

For functions of one real variable, finite or infinite intervals with or without endpoints will be sufficiently general domains for our purposes. To make the ideas mentioned in the previous paragraph precise in higher-dimensional domains we begin with some terminology for describing significant properties of more variously shaped subsets of $\mathbb{R}^n$ such as disks and rectangular regions. We follow these examples by a brief discussion of continuity and then consider differentiability, for real-valued functions in Section 2, and for vector-valued functions in Section 4.

SECTION 1   LIMITS AND CONTINUITY

To discuss differentiability for general vector functions we need two ideas: the *limit* of such a function, and what it means to be an *interior point* of the domain of a function.

We'll also extend the definitions of *limit* and *continuity* from real-valued functions of one variable to real-valued and vector-valued functions of several variables and explain how to construct continuous functions of several variables using the familiar functions of single-variable calculus.

The definition of limit is based on the idea of nearness. The limit relation

$$\lim_{x \to 0} \frac{\sin x}{x} = 1$$

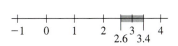

FIGURE 5.1

underlies the usual introduction to the calculus of the trigonometric functions and is most often proved geometrically in that context. The equation says "$(\sin x)/x$ is arbitrarily close to 1 provided x is sufficiently close to 0." We express nearness on the real-number line by inequalities such as $|x - 3| < 0.4$, which says that the distance between the number x and the number 3 is less than 0.4, or equivalently that x lies in the open interval with center 3 and length 0.8. See Figure 5.1. And we translate statements such as "$(\sin x)/x$ is arbitrarily close to 1 provided x is sufficiently close to 0" into statements about inequalities that we can operate on algebraically. Thus the previous displayed formula says the following: For a given positive number ϵ, there is a positive number δ such that if

$$0 < |x - 0| = |x| < \delta, \quad \text{then} \quad \left| \frac{\sin x}{x} - 1 \right| < \epsilon.$$

The condition $0 < |x - 0|$ signifies that the precise value, if any, assigned to the function at $x = 0$ is irrelevant to the existence of the limit; this condition can only make it easier to find the required number $\delta > 0$, because $x = 0$ doesn't have to satisfy the ϵ-inequality.

In the previous chapter we assumed known the properties of elementary functions such as $\sin x/x$, and will continue to do so here, largely avoiding ϵ and δ arguments, and concentrating more on the geometric properties of the natural domain sets in $\mathbb{R}^n$ that play the role of intervals $a < x < b$ and $a \leq x \leq b$ in $\mathbb{R}$.

1A   Neighborhoods

In $\mathbb{R}^n$ a definition of limit also requires the means of asserting that one point is close to another. For a given $\delta > 0$ and point $\mathbf{x}_0$ in $\mathbb{R}^n$, the set of all points $\mathbf{x}$ in $\mathbb{R}^n$ that satisfy the inequality

$$|\mathbf{x} - \mathbf{x}_0| < \delta$$

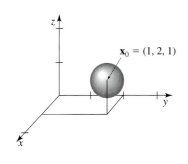

FIGURE 5.2

is called a δ-**ball** with radius δ and center $\mathbf{x}_0$. For example if $\mathbf{x}_0 = (1, 2, 1)$, Figure 5.2 shows the set of all $\mathbf{x}$ in $\mathbb{R}^3$ such that

$$|\mathbf{x} - \mathbf{x}_0| = \sqrt{(x - 1)^2 + (y - 2)^2 + (z - 1)^2} < 0.5.$$

Suppose S is a set of points in $\mathbb{R}^n$ and $\mathbf{x}$ a point in $\mathbb{R}^n$. Then $\mathbf{x}$ is a **limit point** of S if, for a given $\delta > 0$, there exists a point $\mathbf{y}$ in S such that $0 < |\mathbf{x} - \mathbf{y}| < \delta$. Translated into English, the definition says that $\mathbf{x}$ is a limit point of S if there are points in S other than $\mathbf{x}$ that are contained in a ball of arbitrarily small positive radius with center at $\mathbf{x}$. A δ-ball is sometimes called a **neighborhood** of the point at its center. Thus $\mathbf{x}$ is a limit point of S if every neighborhood of $\mathbf{x}$ contains a point of S other than $\mathbf{x}$. Note that a limit point of S need not itself be in S.

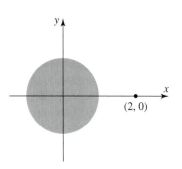

FIGURE 5.3

EXAMPLE 1 The set S in $\mathbb{R}^2$ consisting of all points (x, y) such that

$$x^2 + y^2 < 1,$$

together with the single point $(2, 0)$ appears in Figure 5.3. The set of limit points of S consists of the circular disk together with the circle

$$x^2 + y^2 = 1.$$

Note, however, that the limit points precisely on the circle of radius 1 are not in S. The point $(2, 0)$ is not a limit point of S, even though it is in S, because there is no other point of S within 1 unit of it.

One way for a point $\mathbf{x}$ to be a limit point of a set S is for $\mathbf{x}$ to be an **interior point** of S, that is, a point $\mathbf{x}$ in S such that all points within some neighborhood of $\mathbf{x}$ are also in S. A set S all of whose points are interior points is called **open**. For example $\mathbb{R}^n$ is an open set, and so is the circular disk in Example 1. If something occurs at all points in a neighborhood of a point $\mathbf{x}$ then $\mathbf{x}$ is an interior point of the set where it occurs.

EXAMPLE 2 Consider again the set S shown in Figure 5.3 and described in Example 1. The interior points of S, which are also limit points, are those in the open disk represented by the shaded part of the drawing. A point $\mathbf{x}$ in the disk but not on the circle $x^2 + y^2 = 1$ is an interior point of S, because a disk of small enough radius centered at $\mathbf{x}$ would be contained in S. The point $(2, 0)$ is not an interior point of S, even though it is in S. Even if the circle $x^2 + y^2 = 1$ were included in S, the points of the circle would not be interior points. Figure 5.4(a) shows the interior points of S and Figure 5.4(b) shows the limit points of S.

In the most common examples of functions $f : D \to \mathbb{R}^m$, the domain D is either an open set, or else an open set together with some points called boundary points of D. A **boundary point** of a set D is a point $\mathbf{x}$ such that every neighborhood of $\mathbf{x}$ contains both a point in D and a point not in D. Thus $\mathbf{x}$ may be a boundary point of D without being itself in D. But an interior point of D is never a boundary point of D. The **boundary** of a set D is just the set of all boundary points of D, and a **closed** set is by definition a set that contains all of its boundary points.

EXAMPLE 3 Figure 5.5 shows examples of closed sets D in $\mathbb{R}$ and in $\mathbb{R}^2$, along with the interior of D, which is always open, and the boundary of D. The boundary of the set shown in Figure 5.3 consists of the points on the circle $x^2 + y^2 = 1$ together with the point $(2, 0)$.

1B Limits

Here is the definition of limit for a function $\mathbb{R}^n \xrightarrow{f} \mathbb{R}^m$. Let $\mathbf{y}_0$ be a point in $\mathbb{R}^m$ and $\mathbf{x}_0$ a limit point of the domain of f. Then $\mathbf{y}_0$ is the **limit** of f at $\mathbf{x}_0$ if, for a given $\epsilon > 0$, there is a $\delta > 0$ such that $|f(\mathbf{x}) - \mathbf{y}_0| < \epsilon$ whenever $\mathbf{x}$ is in the domain of f and satisfies $0 < |\mathbf{x} - \mathbf{x}_0| < \delta$. The relation is written

$$\lim_{\mathbf{x} \to \mathbf{x}_0} f(\mathbf{x}) = \mathbf{y}_0.$$

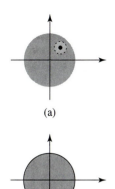

(a)

(b)

FIGURE 5.4

FIGURE 5.5

To put it less formally, the definition says that $f(\mathbf{x})$ is arbitrarily close to $\mathbf{y}_0$ when $\mathbf{x}$ is sufficiently close to $\mathbf{x}_0$ and $\mathbf{x} \neq \mathbf{x}_0$. Geometrically, the idea is this: Given an ϵ-ball B_ϵ centered at $\mathbf{y}_0$, there exists a δ-ball B_δ centered at $\mathbf{x}_0$ whose intersection with the domain of f, except possibly for $\mathbf{x}_0$ itself, is sent by f into B_ϵ. A 2-dimensional example is pictured in Figure 5.6. The statement

$$\lim_{\mathbf{x} \to \mathbf{x}_0} f(\mathbf{x}) = \mathbf{y}_0$$

is also commonly read "The limit of $f(\mathbf{x})$ as $\mathbf{x}$ approaches $\mathbf{x}_0$ is $\mathbf{y}_0$." We always get a **unique limit** if one exists, since if

$$\lim_{bx \to \mathbf{x}_0} f(\mathbf{x}) = \mathbf{y}_1 \quad \text{and} \quad \lim_{bx \to \mathbf{x}_0} f(\mathbf{x}) = \mathbf{y}_2$$

then by the triangle inequality

$$|\mathbf{y}_1 - \mathbf{y}_2| \leq |\mathbf{y}_1 - f(\mathbf{x})| + |f(\mathbf{x}) - \mathbf{y}_2| < \epsilon + \epsilon = 2\epsilon$$

for all $\mathbf{x}$ in a small enough neighborhood of $\mathbf{x}_0$. But we can make 2ϵ as small as we like, so $|\mathbf{y}_1 - \mathbf{y}_2| = 0$ and $\mathbf{y}_1 = \mathbf{y}_2$.

D	Interior of D	Boundary of D
a b	a b	a b
Closed interval $a \leq x \leq b$, denoted $[a, b]$	Open interval $a < x < b$, denoted (a, b)	
(a)	(b)	(c)

Closed rectangle
$a \leq x \leq b$,
$c \leq y \leq d$.

(d)

Open interior
$a < x < b$,
$c < y < d$.

(e)

Boundary rectangle

(f)

FIGURE 5.6

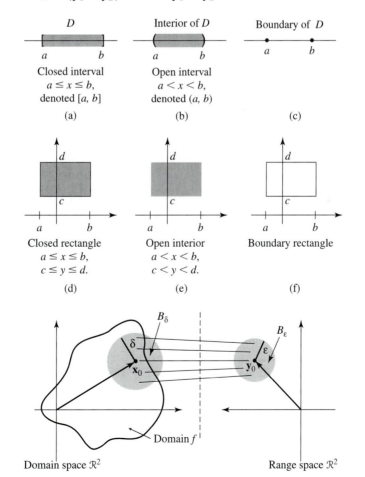

Domain space $\mathcal{R}^2$ Range space $\mathcal{R}^2$

EXAMPLE 4   Consider the function defined by

$$f(t) = (\cos t, \sin t).$$

The domain of f is all of $\mathbb{R}$, and at every point t_0 of $\mathbb{R}$, f has limit $f(t_0)$. To see this we use known facts about $\cos t$ and $\sin t$ and consider

$$|f(t) - f(t_0)| = \sqrt{(\cos t - \cos t_0)^2 + (\sin t - \sin t_0)^2}$$

$$\leq |\cos t - \cos t_0| + |\sin t - \sin t_0|. \qquad (*)$$

The last inequality holds because $\sqrt{a^2 + b^2} \leq |a| + |b|$. (Square both sides.) Using continuity of $\sin t$ and $\cos t$,

$$\lim_{t \to t_0} \cos t = \cos t_0 \quad \text{and} \quad \lim_{t \to t_0} \sin t = \sin t_0.$$

Then we can make both

$$|\cos t - \cos t_0| \quad \text{and} \quad |\sin t - \sin t_0|$$

as small as we like by making $|t - t_0|$ small enough. Hence the inequality $(*)$ shows that $|f(t) - f(t_0)|$ is as small as we like whenever $|t - t_0|$ is small enough.

EXAMPLE 5   Consider the real-valued function defined in all of $\mathbb{R}^2$ except for $(x, y) = (0, 0)$ by

$$f(x, y) = \frac{1}{x^2 + y^2}.$$

In this example we can write

$$\lim_{\mathbf{x} \to 0} f(\mathbf{x}) = +\infty$$

to describe what happens, because as $|(x, y)|$ tends to 0, its square $x^2 + y^2 = |(x, y)|^2$ tends to zero also, so the fraction tends to $+\infty$. By the convention established in our definition of limit, only elements of $\mathbb{R}^n$ are acceptable limits, so we say for this example that the limit fails to exist.

EXAMPLE 6   Let f be real-valued with the same domain as in the preceding example and defined by

$$f(x, y) = \frac{x^2 - y^2}{x^2 + y^2}.$$

There is no limit as $(x, y) \to (0, 0)$. If (x, y) approaches $(0, 0)$ along the line $y = \alpha x$, we obtain

$$\lim_{\mathbf{x} \to 0} \frac{x^2 - y^2}{x^2 + y^2} = \lim_{\mathbf{x} \to 0} \frac{x^2(1 - \alpha^2)}{x^2(1 + \alpha^2)} = \frac{1 - \alpha^2}{1 + \alpha^2}.$$

This limit is not independent of α, because, for example, the limit equals 0 if $\alpha = 1$, and 1 if $\alpha = 0$. But for a unique limit to exist we have to be able to approach $(0, 0)$ from all possible directions, so the overall limit fails to exist.

The functions in Examples 5 and 6 are both real-valued. The following theorem shows that the problem of the existence and evaluation of a limit for a function $\mathbb{R}^n \xrightarrow{f} \mathbb{R}^m$ reduces to the same problem for the real-valued coordinate functions.

1.1   Theorem.   Given $\mathbb{R}^n \xrightarrow{f} \mathbb{R}^m$, with coordinate functions $f_1, \ldots, f_m$, and a point $\mathbf{y}_0 = (y_1, \ldots, y_m)$ in $\mathbb{R}^m$, then

$$\lim_{\mathbf{x} \to \mathbf{x}_0} f(\mathbf{x}) = \mathbf{y}_0 \tag{A}$$

if and only if

$$\lim_{\mathbf{x} \to \mathbf{x}_0} f_i(\mathbf{x}) = y_i, \quad i = 1, \ldots, m. \tag{B}$$

Proof.   To say that Equations (A) and (B) are equivalent is to say that the distance

$$|f(\mathbf{x}) - \mathbf{y}_0| = \sqrt{(f_1(\mathbf{x}) - y_1)^2 + \cdots + (f_m(\mathbf{x}) - y_m)^2}$$

is arbitrarily small for $\mathbf{x}$ in a small enough neighborhood of $\mathbf{x}_0$ if and only if

$$|f_1(\mathbf{x}) - y_1|, \ldots, |f_m(\mathbf{x}) - y_m|$$

also become arbitrarily small. But the equivalence of these last two statements follows at once from the inequalities

$$|f(\mathbf{x}) - \mathbf{y}_0| \geq |f_i(\mathbf{x}) - y_i|, \quad i = 1, \ldots, m, \quad \text{and}$$

$$|f(\mathbf{x}) - \mathbf{y}_0| \leq \sqrt{m} \max_{1 \leq i \leq m} \{|f_i(\mathbf{x}) - y_i|\}.$$

After squaring both sides the first inequality follows from $a_1^2 + \cdots + a_m^2 \geq a_i^2$, $i = 1, \ldots, m$, and the second from $a_1^2 + \cdots + a_m^2 \leq m \left(\max_{1 \leq i \leq m} |a_i| \right)^2$.   ∎

EXAMPLE 7   For vector functions f_1 and f_2 defined by

$$f_1(t) = \left(t, \ t^2, \ \sin t \right), \quad f_2(t) = \left(t, \ t^2, \ \sin(1/t) \right),$$

we have

$$\lim_{t \to 0} f_1(t) = (0, \ 0, \ 0).$$

But $\lim f_2(t)$ doesn't exist because the function $\sin(1/t)$ has no limit at $t = 0$.

1C   Continuity

Roughly speaking, a continuous function f is one whose values do not change abruptly. That is, if $\mathbf{x}$ is close to $\mathbf{x}_0$, then $f(\mathbf{x})$ must be close to $f(\mathbf{x}_0)$. This idea is related to the idea of limit, and the definition of continuity is as follows: A function f is **continuous** at $\mathbf{x}_0$ if

(a) $\mathbf{x}_0$ is in the domain of f.

(b) $\lim_{\mathbf{x} \to \mathbf{x}_0} f(\mathbf{x}) = f(\mathbf{x}_0)$.

At a nonlimit, or **isolated**, point of the domain of f, we can't ask for a limit; instead we extend the definition of continuity simply by defining f to be **continuous** at such a point. A function is **continuous** on a subset S of its domain if it's continuous at every point in S. It's an immediate corollary of Theorem 1.1 that

1.2 Theorem.   A vector function is continuous at a point if and only if its coordinate functions are continuous there.

EXAMPLE 8

Returning to Example 7, the function

$$f_1(t) = (t, \ t^2, \ \sin t)$$

is continuous at every value of t. On the other hand, the function

$$f_2(t) = (t, \ t^2, \ \sin(1/t))$$

is continuous on the set S of all real numbers t with $t = 0$ deleted.

A function is simply called **continuous** if it's continuous at every point of its domain. From Theorem 1.2 we see that a continuous vector-valued function of a single variable, $\mathbb{R} \xrightarrow{f} \mathbb{R}^n$, is precisely one for which the coordinate functions $f_1, \ldots, f_n$ are continuous real-valued functions of a real variable. The latter include most of the functions of ordinary calculus, such as x^2, $\sin x$, and, for $x > 0$, $\ln x$. We use these same functions to construct examples of the continuous coordinate functions that constitute the vector-valued functions $\mathbb{R}^n \xrightarrow{f} \mathbb{R}^m$ of a vector variable. For example, the coordinate functions of

$$f(x, y) = \left(\frac{\sin xy}{e^{x+y}}, \ \frac{\cos xy}{e^{x+y}} \right)$$

turn out to be continuous. The continuity of these and other examples follows from repeated application of the following three theorems, together with Theorem 1.2 on coordinate functions. If you think these theorems are obviously true you're right, but we'll prove them anyway.

1.3 Theorem.   The functions $\mathbb{R}^n \xrightarrow{P_k} \mathbb{R}$, where $P_k(x_1, \ldots, x_n) = x_k$, are continuous for $k = 1, 2, \ldots, n$. P_k is called the kth **coordinate projection**.

1.4 Theorem.   The functions $\mathbb{R}^2 \xrightarrow{S} \mathbb{R}$ and $\mathbb{R}^2 \xrightarrow{M} \mathbb{R}$, defined by $S(x, y) = x + y$ and $M(x, y) = xy$, are continuous.

1.5 Theorem.   If $\mathbb{R}^n \xrightarrow{f} \mathbb{R}^m$ and $\mathbb{R}^m \xrightarrow{g} \mathbb{R}^p$ are continuous, then the composition $g(f(\mathbf{x}))$ is continuous wherever $g(f(\mathbf{x}))$ is defined.

Proof of 1.3.   We have $|P_k(x_1, \ldots, x_n) - P_k(a_1, \ldots, a_n)| = |x_k - a_k| \leq |\mathbf{x} - \mathbf{a}|$, so $|P_k(x_1, \ldots, x_n) - P_k(a_1, \ldots, a_n)|$ is arbitrarily small if $|\mathbf{x} - \mathbf{a}|$ is small enough. ∎

Proof of 1.4.   For $S(x, y) = x + y$, write $|S(x, y) - S(a, b)| = |x - a + y - b| \leq |x - a| + |y - b|$, by the triangle inequality. Hence $|S(x, y) - S(a, b)|$ is small if $|x - a|$ and $|y - b|$ are small enough. Since $|x - a|$ and $|y - b|$ are both at most the distance $\sqrt{(x-a)^2 + (y-b)^2}$ from (x, y) to (a, b), making this distance small enough makes $|S(x, y) - S(a, b)|$ as small as you like.

For $M(x, y) = xy$ use the triangle inequality and factor out $|x|$ and $|b|$ to get $|M(x, y) - M(a, b)| = |xy - xb + xb - ab| \le |x||y - b| + |x - a||b|$. Keeping x within distance 1 of a, makes

$$|M(x, y) - M(a, b)| \le (|a| + 1)|y - b| + |x - a||b|.$$

Hence $|M(x, y) - M(a, b)|$ is as small as you like for a given (a, b) if $|x - a|$ and $|y - b|$ are small enough. From here the argument is the same as for $S(x, y)$.   ∎

Proof of 1.5.   We let $\mathbf{a}$ be a limit point of the domain of $g(f(\mathbf{x}))$ and show that $\lim\limits_{\mathbf{x} \to \mathbf{x}_0} g(f(\mathbf{x})) = g(f(\mathbf{a}))$. Since g is continuous at $f(\mathbf{a})$ there is a neighborhood B_s, $s > 0$, of $f(\mathbf{a})$ such that $|g(\mathbf{y}) - g(f(\mathbf{a}))|$ is a small as we like, say less than $\epsilon > 0$, when $\mathbf{y}$ is in B_s. Similarly, since $f(\mathbf{x})$ is continuous at $\mathbf{x} = \mathbf{a}$ there is a neighborhood B_r, $r > 0$, of $\mathbf{a}$ such that $f(\mathbf{x})$ is in B_s when $\mathbf{x}$ is in B_r. Hence $|g(\mathbf{y}) - g(f(\mathbf{a}))| < \epsilon$ whenever $\mathbf{x}$ is in B_r.   ∎

| **EXAMPLE 9** |

The function $f(x, y) = \sqrt{1 - x^2 - y^2}$, defined for $|(x, y)| \le 1$, is continuous, because we can write it as

$$f(x, y) = \sqrt{1 - (P_1(x, y))^2 - (P_2(x, y))^2},$$

so $f(x, y)$ is a composition of continuous functions. Similarly, $g(x, y) = \ln(x + y)$, defined for $x + y > 0$, is continuous. The product of f and g, given by

$$h(x, y) = \sqrt{1 - x^2 - y^2}\, \ln(x + y),$$

is defined on the half-disk that is the intersection of the domains of f and g, as shown in Figure 5.7. The product is a continuous function because it is the composition of the continuous vector function

$$F(x, y) = (f(x, y), g(x, y))$$

with the function M of Theorem 1.4.

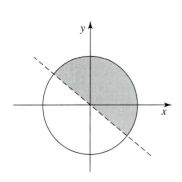

FIGURE 5.7

A function $\mathbb{R}^n \xrightarrow{f} \mathbb{R}^m$ is called **linear** if its coordinate functions have the form

$$f_k(x_1, \ldots, x_n) = a_{k1}x_1 + \cdots + a_{kn}x_n, \quad k = 1, \ldots, m,$$

for some scalars a_{kj}. The linear functions enhance our understanding of the property of *differentiability* taken up in the next section, and are important in their own right in both pure and applied mathematics. Theorems 1.2, 1.3, and 1.4 show that linear functions are continuous, as the next theorem spells out.

1.6   Theorem.   A linear function $\mathbb{R}^n \xrightarrow{f} \mathbb{R}^m$ is continuous.

Proof.   Each scalar a_{kj} in $a_{k1}x_1 + \cdots + a_{kn}x_n$ is continuous because it is constant, and each x_k is continuous by Theorem 1.3, so their product and sum are continuous by Theorem 1.4. Then f is continuous by Theorem 1.2.   ∎

EXAMPLE 10   The pair of equations

$$x = 2u - 3v + w \quad \text{and} \quad y = u + v - w$$

defines a linear, therefore continuous, function from points (u, v, w) in $\mathbb{R}^3$ to points (x, y) in $\mathbb{R}^2$. The examples of the next section will help to illuminate the behavior of functions such as this one.

EXERCISES

1. Assuming $x_0 = (1, 2)$, draw the set of all vectors x in $\mathbb{R}^2$ such that
   (a) $|x - x_0| \leq 3$
   (b) $|x - x_0| = 3$
   (c) $|x - x_0| < 3$

In Exercises 2 to 11, identify (a) the interior and (b) the boundary of the set of points $x = (x, y)$ in $\mathbb{R}^2$. (c) Which sets are open? (d) Which sets are closed?

2. $|x - (1, 2)| \leq 0.5$

3. $|x - (1, 2)| < 0.5$

4. $|x - (1, 2)| < -0.5$

5. $0 < x < 3$ and $0 < y < 2$

6. $2 \leq x < 3$ and $0 < y < 2$

7. $x^2 + 2y^2 < 1$

8. $x \neq (0, 2)$ or $(1, 2)$

9. $x^2 + y^2 > 0$

10. $x > 0$

11. $x > y$

12. Let the set S consist of the points (x, y) in $\mathbb{R}^2$ satisfying $0 < x^2 + y^2 < 1$, together with the interval $1 \leq x < 2$ of the x axis.
    (a) Describe the boundary of S.
    (b) What are the interior points of S?
    (c) Is S open? Closed?

13. Let L be a line and P a plane in $\mathbb{R}^3$. Is either P or L an open subset of $\mathbb{R}^3$?

In Exercises 14 to 19, the formula defines a function f from $\mathbb{R}^n$ to $\mathbb{R}^m$ for some n and m. In each example, state what n and m are, and list the real-valued coordinate functions of f.

14. $f(x, y) = (x - y, x^2 - y^2)$

15. $f(x, y) = \begin{pmatrix} 1 & 3 \\ 0 & 2 \end{pmatrix} \begin{pmatrix} x \\ y \end{pmatrix}$

16. $f(x, y, z) = (x - y)\sqrt{x^2 + y^2 + z^2}$

17. $f(t) = (t, t^2, t^3, t^4)$

18. $f(u, v) = (u + v, u - v, u^2 + v^2)$

19. $f(x, y, z) = (2x, 2y, 2z)$

In Exercises 20 to 25, determine at which points the function fails to have a limit. Use Theorem 1.1. Take the domain of each coordinate function as large as possible. The domain of f is then the part common to the domains of all the coordinate functions.

20. $f\begin{pmatrix} x \\ y \end{pmatrix} = \begin{pmatrix} y + \tan x \\ \ln(x + y) \end{pmatrix}$

21. $f\begin{pmatrix} x \\ y \end{pmatrix} = \begin{pmatrix} \dfrac{y}{x^2 + 1} \\ \dfrac{x}{y^2 - 1} \end{pmatrix}$

22. $f(x, y) = \dfrac{x}{\sin x} + y$

23. $f(x, y) = \begin{cases} \dfrac{x}{\sin x} + y, & \text{if } x \neq 0, \\ 2 + y, & \text{if } x = 0 \end{cases}$

24. $f(t) = \begin{pmatrix} \sin t \\ \cos t \\ \dfrac{1}{\sin t^2} \end{pmatrix}$

25. $f(u, v) = \left(\dfrac{uv}{1 - u^2 - v^2}, \dfrac{1}{2 - u^2 - v^2} \right)$

In Exercises 26 to 31, determine at which points the function fails to be continuous. Take the domain of each coordinate function as large as possible. The domain of f is then the part common to the domains of all the coordinate functions.

26. $f\begin{pmatrix} x \\ y \end{pmatrix} = \begin{pmatrix} \dfrac{1}{x^2} + \dfrac{1}{y^2} \\ x^2 + y^2 \end{pmatrix}$

27. $f\begin{pmatrix} u \\ v \end{pmatrix} = \begin{pmatrix} 3u - 4v \\ u + 8 \end{pmatrix}$

28. $f(x, y) = \begin{cases} \dfrac{x}{\sin x} + y, & \text{if } x \neq 0 \\ 1 + y, & \text{if } x = 0 \end{cases}$

29. $f\begin{pmatrix} x \\ y \end{pmatrix} = \begin{cases} \dfrac{x^2 - y^2}{x^2 + y^2}, & \text{if } x^2 + y^2 \neq 0 \\ 0, & \text{if } x^2 + y^2 = 0 \end{cases}$

30. $f\begin{pmatrix} u \\ v \end{pmatrix} = \begin{pmatrix} v \tan u \\ u \sec v \\ v \end{pmatrix}$

31. $f(\mathbf{x}) = \dfrac{|\mathbf{x}|}{1 - |\mathbf{x}|^2}$, if $\mathbf{x}$ is in $\mathbb{R}^n$

32. A vector function f has a **removable discontinuity** at $\mathbf{x}_0$ if (1) f is not continuous at $\mathbf{x}_0$, and (2) there is a vector $\mathbf{y}_0$ such that $\lim_{\mathbf{x} \to \mathbf{x}_0} f(\mathbf{x}) = \mathbf{y}_0$. Give examples of functions f and g that are discontinuous at a point $\mathbf{x}_0$ such that f has a removable discontinuity at $\mathbf{x}_0$ and g does not.

33. A function $T: \mathbb{R}^n \to \mathbb{R}^n$ is called a **translation** acting on $\mathbb{R}^n$ by $\mathbf{y}_0$ if there is a vector $\mathbf{y}_0$ in $\mathbb{R}^n$ such that $T(\mathbf{x}) = \mathbf{x} + \mathbf{y}_0$ for all $\mathbf{x}$ in $\mathbb{R}^n$.
  (a) Describe in words the effect on $\mathbb{R}^2$ of translation by $\mathbf{y}_0 = (1, 1)$.
  (b) Prove that every translation is a continuous function.

34. Prove that the union of an arbitrary collection of open subsets of $\mathbb{R}^n$ is open.

35. Prove that the intersection of a *finite* collection of open subsets of $\mathbb{R}^n$ is open.

36. Give an example to show that an intersection of *infinitely* many open subsets of $\mathbb{R}^n$ may fail to be open.

37. Prove Theorem 1.3 of the text by first proving the inequalities

$$|x_k| \leq |(x_1, \ldots, x_n)|, \quad k = 1, \ldots, n.$$

38. Prove both parts of Theorem 1.4 of the text. [*Hint:* For the function M, note that by the triangle inequality for absolute value, $|xy - x_0 y_0| \leq |xy - xy_0| + |xy_0 - x_0 y_0|$.]

39. If f and g are vector functions with the same domain and same range space, prove

$$\lim_{\mathbf{x} \to \mathbf{x}_0} (f(\mathbf{x}) + g(\mathbf{x})) = \lim_{\mathbf{x} \to \mathbf{x}_0} f(\mathbf{x}) + \lim_{\mathbf{x} \to \mathbf{x}_0} g(\mathbf{x}),$$

provided that $\lim_{\mathbf{x} \to \mathbf{x}_0} f(\mathbf{x})$ and $\lim_{\mathbf{x} \to \mathbf{x}_0} g(\mathbf{x})$ exist.

40. Let S be a closed subset of $\mathbb{R}^n$. Prove that the complement of S in $\mathbb{R}^n$ is open.

41. If S is an open subset of $\mathbb{R}^n$, show that the complement of S in $\mathbb{R}^n$ is closed.

42. A function of more than one variable can have a limit along every line through a point without having a limit at that point. For example, define $f(x, y) = x^2 y / (x^4 + y^2)$ for $(x, y) \neq (0, 0)$.
  (a) Show that $\lim_{y \to 0} f(0, y) = 0$ and that, for each fixed number α, $\lim_{x \to 0} f(x, \alpha x) = 0$.
  (b) Show that approaching $(0, 0)$ along the parabola $y = \alpha x^2$ you get limit $\alpha / (1 + \alpha^2)$.
  (c) What is the set of possible limits achieved by using the approaches of part (b)?

SECTION 2   REAL-VALUED FUNCTIONS

2A   Differentiability and Continuity

We begin by reviewing the definition for a real-valued function $\mathbb{R} \xrightarrow{f} \mathbb{R}$ of a single real variable x on some interval $a < x < b$. We say that f is **differentiable** at x_0 if there is a number a such that

$$\lim_{x \to x_0} \frac{f(x) - f(x)}{x - x_0} = a, \quad \text{or} \quad \lim_{x \to x_0} \frac{f(x) - f(x_0) - a(x - x_0)}{x - x_0} = 0,$$

in which case f has **derivative** $f'(x_0) = a$. The equivalent form on the right calls attention to an important property of the tangent line equation $y = f(x_0) + a(x - x_0)$, namely that its graph approaches the graph of $f(x)$ as x approaches x_0 more rapidly than $x - x_0$ approaches 0.

Recall that in Chapter 4, Section 3 we used a modification of this same definition to motivate the definition of the partial derivatives of a function of more than one variable. At that point we emphasized that the mere existence of partial derivatives is not enough to provide a reasonable definition of differentiability for a function of more than one variable. The reason for this is that partial derivatives at a point $\mathbf{x}_0$ take account of the behavior of a function near $\mathbf{x}_0$ only along lines through $\mathbf{x}_0$ and parallel to the axes. Thus we have to consider a function's behavior not only along infinitely many other individual lines through $\mathbf{x}_0$ but, even more, throughout an entire open neighborhood of $\mathbf{x}_0$. Thus we formulate the definition as follows.

> **2.1 Definition**   A function $\mathbb{R}^n \xrightarrow{f} \mathbb{R}$ is **differentiable** at $\mathbf{x}_0$ if
> (i) $\mathbf{x}_0$ is an interior point of the domain of f.
> (ii) There is a vector $\mathbf{a}$ such that
> $$\lim_{\mathbf{x} \to \mathbf{x}_0} \frac{f(\mathbf{x}) - f(\mathbf{x}_0) - \mathbf{a} \cdot (\mathbf{x} - \mathbf{x}_0)}{|\mathbf{x} - \mathbf{x}_0|} = 0.$$

To be consistent with the custom in the case of a single real variable, the function f is called simply **differentiable** if it's differentiable at every point of its domain. We'll prove in Theorem 2.2 that there can be only one vector $\mathbf{a}$ for which (ii) is true, and it's called the **gradient** of the differentiable function f at $\mathbf{x}_0$; it's customary to use the notation $\nabla f(\mathbf{x}_0)$ for the vector $\mathbf{a}$. The symbol ∇ is pronounced "grad" here, so $\nabla f(\mathbf{x})$ becomes "grad f at $\mathbf{x}$." If f is a function of a single real variable x we continue to write the customary $f'(x)$ instead of $\nabla f f(x)$.

Remark 1.   Since we can't divide by a vector $\mathbf{x} - \mathbf{x}_0$ we instead multiply by the reciprocal of its length in (ii), the crucial point being to ensure that the numerator tends to zero faster than $|\mathbf{x} - \mathbf{x}_0|$ does.

Remark 2.   Condition (i) of the definition requires the approach of $\mathbf{x}$ to $\mathbf{x}_0$ to be unrestricted, as compared to the restricted 1-dimensional limits used to define partial derivatives.

Remark 3.   According to the definition of differentiability, the domain of a differentiable function is an open set. It's convenient however to extend the definition sufficiently to speak of a **differentiable function** f defined on an arbitrary subset S of the domain space. By such an f we'll mean the restriction to S of a differentiable function whose domain is an open set containing S.

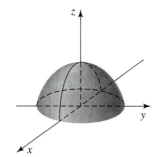

FIGURE 5.8
$z = \sqrt{1 - x^2 - y^2}$.

EXAMPLE 1

The function f defined by $f(x, y) = \sqrt{1 - x^2 - y^2}$ has for its domain the disk $x^2 + y^2 \leq 1$. Its graph appears in Figure 5.8. The interior points of the domain are those (x, y) such that $x^2 + y^2 < 1$. We'll see that $f(x, y)$ is differentiable at these interior points. But it doesn't follow by Remark 3 that f is differentiable at the points of the circle $x^2 + y^2 = 1$; indeed we'll see that f can't be extended to be differentiable on an open set containing the circle. See Examples 5 and 6 and Exercise 26.

The next theorem allows us to compute the vector $\nabla f(\mathbf{x})$ in terms of partial derivatives of a differentiable function. The gradient is the key to putting a firm foundation under the notion of tangent plane introduced in Chapter 4. We'll see in Chapter 6 that the gradient of a real-valued function has several natural interpretations.

2.2   Theorem.   If a function $\mathbb{R}^n \xrightarrow{f} \mathbb{R}$ is differentiable at $\mathbf{x}_0$, then the kth coordinate of the gradient $\nabla f(\mathbf{x}_0)$ of f at $\mathbf{x}_0$ is the kth partial derivative of f at $\mathbf{x}_0$,

$k = 1, 2, \ldots, n$. Thus $\nabla f(\mathbf{x}_0)$ is uniquely determined by the differentiability conditions (i) and (ii), and

$$\nabla f(\mathbf{x}_0) = \left(\frac{\partial f}{\partial x_1}(\mathbf{x}_0), \ldots, \frac{\partial f}{\partial x_k}(\mathbf{x}_0), \ldots, \frac{\partial f}{\partial x_n}(\mathbf{x}_0) \right).$$

Proof. To identify the entries in $\nabla f(\mathbf{x}_0) = \mathbf{a}$ using the definition of this vector, we specialize $\mathbf{x}$ in part (ii) of the definition to vectors of the form $\mathbf{x}_j = \mathbf{x}_0 + t\mathbf{e}_j$. Since $\mathbf{x}_0$ is assumed to be an interior point of the domain of f, then $\mathbf{x}_0 + t\mathbf{e}_j$ is in the domain of f for small enough t. Since now $\mathbf{x} - \mathbf{x}_0 = t\mathbf{e}_j$, these special cases of condition (ii) become

$$\lim_{t \to 0} \frac{f(\mathbf{x}_j) - f(\mathbf{x}_0) - \mathbf{a} \cdot (t\mathbf{e}_j)}{t} = 0, \quad j = 1, \ldots, n.$$

Here we've used $|\mathbf{x} - \mathbf{x}_0| = |t\mathbf{e}_j| = |t||\mathbf{e}_j| = |t|$. We then removed the absolute value since the limit is zero, making the sign of the denominator irrelevant. By the homogeneity of the dot product, $\mathbf{a} \cdot (t\mathbf{e}_j) = t\mathbf{a} \cdot (\mathbf{e}_j)$, so we rewrite these limits as

$$\lim_{t \to 0} \frac{f(\mathbf{x}_j) - f(\mathbf{x}_0)}{t} = \mathbf{a} \cdot \mathbf{e}_j, \quad j = 1, \ldots, n.$$

But $\mathbf{x}_j = \mathbf{x}_0 + t\mathbf{e}_j$ differs from $\mathbf{x}_0$ only in the jth coordinate, and in that coordinate the difference is just t. Hence the limit on the left side of the last equation is just $\partial f / \partial x_j$ at $\mathbf{x}_0$. Since the dot product $\mathbf{a} \cdot \mathbf{e}_j$ is just the jth entry in $\mathbf{a}$, we're done. ∎

EXAMPLE 2

We'll see shortly that the functions $f(x, y) = e^x \sin y + y$ and $g(x, y, z) = xy + yz + zx$ are differentiable at all points (x, y) and (x, y, z), respectively. The two gradient vectors are $\nabla f = (f_x, \ f_y)$ and $\nabla g = (g_x, g_y, g_z)$, so

$$\nabla f(x, y) = (e^x \sin y, \ e^x \cos y + 1) \quad \text{and} \quad \nabla g(x, y, z) = (y + z, \ x + z, \ y + x).$$

How can we tell whether or not a vector function is differentiable? Theorem 2.2 only allows us to conclude that a function is *not* differentiable at a point if one or more of its first-order partial fails to exist at that point, because differentiability of f at $\mathbf{x}$ implies that these partials exist at $\mathbf{x}$. Thus Example 2 is inconclusive to the extent that we have simply assumed that the functions f and g appearing there are differentiable. The converse implication isn't valid, since it's possible for the partials to exist without f being differentiable, as Exercise 24 shows. However, by adding an additional assumption, namely that the partials are themselves continuous on an open set S, we'll deduce in Theorem 2.3 the differentiability of f on the entire set S. The theorem guarantee differentiability for most examples met in practice.

2.3 Theorem. Let the domain of $\mathbb{R}^n \xrightarrow{f} \mathbb{R}$ be an open subset D of $\mathbb{R}^n$ on which all partial derivatives $\partial f_i / \partial x_j$ of the coordinate functions of f are continuous. Then f is differentiable at every point of D.

Proof. Since we can formally write the gradient $\nabla f(\mathbf{x}_0)$ of f at $\mathbf{x}_0$, the theorem will have been proved if we show that $\nabla f(\mathbf{x}_0)$ satisfies

$$\lim_{\mathbf{x} \to \mathbf{x}_0} \frac{f(\mathbf{x}) - f(\mathbf{x}_0) - \nabla f(\mathbf{x}_0) \cdot (\mathbf{x} - \mathbf{x}_0)}{|\mathbf{x} - \mathbf{x}_0|} = 0. \qquad (*)$$

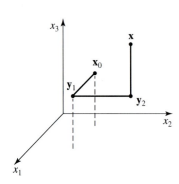

FIGURE 5.9

If $\mathbf{x} = (x_1, \ldots, x_n)$ and $\mathbf{x}_0 = (a_1, \ldots, a_n)$, set

$$\mathbf{y}_k = (x_1, \ldots, x_k, a_{k+1}, \ldots, a_n), \quad k = 0, 1, \ldots, n,$$

so that in particular $\mathbf{y}_0 = \mathbf{x}_0$ and $\mathbf{y}_n = \mathbf{x}$. We show these points with line segments joining them for three dimensions in Figure 5.9. Then we have

$$f(\mathbf{x}) - f(\mathbf{x}_0) = \sum_{k=1}^{n} \left(f(\mathbf{y}_k) - f(\mathbf{y}_{k-1}) \right),$$

since only the first and last terms survive in the sum on the right because of cancellation between successive terms.

Because $\mathbf{y}_k$ and $\mathbf{y}_{k-1}$ differ only in their kth coordinates, we can apply the mean-value theorem for real functions of the real variable x_k to get

$$f(\mathbf{y}_k) - f(\mathbf{y}_{k-1}) = (x_k - a_k) \frac{\partial f}{\partial x_k}(\mathbf{z}_k),$$

where $\mathbf{z}_k$ is a point on the segment joining $\mathbf{y}_k$ and $\mathbf{y}_{k-1}$. Then

$$f(\mathbf{x}) - f(\mathbf{x}_0) = \sum_{k=1}^{n} (x_k - a_k) \frac{\partial f}{\partial x_k}(\mathbf{z}_k).$$

We also have, by the definition of $\nabla f(\mathbf{x}_0)$,

$$\nabla f(\mathbf{x}_0) \bullet (\mathbf{x} - \mathbf{x}_0) = \left(\frac{\partial f}{\partial x_1}(\mathbf{x}_0), \ldots, \frac{\partial f}{\partial x_n}(\mathbf{x}_0) \right) \bullet (x_1 - a_1, \ldots, x_n - a_n)$$

$$= \sum_{k=1}^{n} (x_k - a_k) \frac{\partial f}{\partial x_k}(\mathbf{x}_0).$$

Hence

$$|f(\mathbf{x}) - f(\mathbf{x}_0) - \nabla f(\mathbf{x}_0) \bullet (\mathbf{x} - \mathbf{x}_0)| = \left| \sum_{k=1}^{n} \left(\frac{\partial f}{\partial x_k}(\mathbf{z}_k) - \frac{\partial f}{\partial x_k}(\mathbf{x}_0) \right) (x_k - a_k) \right|$$

$$\leq \sum_{k=1}^{n} \left| \frac{\partial f}{\partial x_k}(\mathbf{z}_k) - \frac{\partial f}{\partial x_k}(\mathbf{x}_0) \right| |\mathbf{x} - \mathbf{x}_0|,$$

where we have used the triangle inequality and the inequalities

$$|x_k - a_k| \leq |\mathbf{x} - \mathbf{x}_0| \quad \text{for} \quad k = 1, 2, \ldots, n.$$

Now divide by $|\mathbf{x} - \mathbf{x}_0|$. Since the partial derivatives are assumed continuous at $\mathbf{x}_0$, and the $\mathbf{z}_k$ tend to $\mathbf{x}_0$ as $\mathbf{x}$ does, Equation $(*)$ follows from making $|\mathbf{x} - \mathbf{x}_0|$ tend to zero. ∎

EXAMPLE 3   The function $f(x, y) = e^x \sin y + y$ of Example 2 is differentiable at all points (x, y). To conclude this from Theorem 2.3 we calculate that

$$\left(f_x(x, y), f_y(x, y) \right) = (e^x \sin y, \ e^x \cos y + 1)$$

According to Theorems 1.3 and 1.4 of Section 1, these partial derivatives are continuous for all (x, y), so by Theorem 2.3 the function $f(x, y)$ is differentiable for all (x, y).

EXAMPLE 4

The function $g(x, y, z) = xy + yz + zx$ of Example 2 is differentiable at all points (x, y, z). To apply Theorem 2.3 we compute the vector

$$\big(g_x(x, y, z),\ g_y(x, y, z),\ g_z(x, y, z)\big) = (y + z,\ x + z,\ y + x).$$

According to Theorem 1.4 of Section 1, these partial derivatives are continuous for all (x, y, z), so by Theorem 2.3 the function $g(x, y, z)$ is differentiable for all (x, y, z).

EXAMPLE 5

The function $h(x, y) = \sqrt{1 - x^2 - y^2}$ of Example 1 turns out to be differentiable at all points in the open set of (x, y) such that $x^2 + y^2 < 1$, and at no other points. Note that h is not even defined if $x^2 + y^2 > 1$. To apply Theorem 2.3 at the interior points of the circular disk we calculate that

$$\big(f_x(x.y),\ f_y(x, y)\big) = \left(-x/\sqrt{1 - x^2 - y^2},\ -y/\sqrt{1 - x^2 - y^2}\right).$$

According to our Theorems 1.3, 1.4 and 1.5 of Section 1, these partial derivatives are continuous for all (x, y), so by Theorem 2.3 the function $f(x, y)$ is differentiable for all (x, y) in the open unit disk.

EXAMPLE 6

Continuing with the previous example, the points on the circle $x^2 + y^2 = 1$ are more problematic. According to Remark 3 following Definition 2.1 of differentiability, if it were possible to extend the definition of $h(x, y)$ to an open set containing the circle in such a way that the resulting extension $h_e(x, y)$ became differentiable, then we could claim that the original function $h(x, y)$ is differentiable on the circle. However, such an extension of $h(x, y)$ is impossible for this example. For depending on the signs of x and y,

$$h_x(x, y) = -x/\sqrt{1 - x^2 - y^2}\quad\text{and}\quad h_y(x, y) = -y/\sqrt{1 - x^2 - y^2}$$

tend to ∞ or $-\infty$ as (x, y) tends to a point on the circle from inside the circle. Thus there is no point on the circle at which both partial derivatives can exist, as would have to be the case if $h(x, y)$ were differentiable there. See Exercise 26.

A single-variable example given in Exercise 25 shows that continuity of the partial derivatives of a function f is not necessary for differentiability of f. However, the hypotheses of Theorem 2.3 are referred to often enough to be given a special name. A function f is **continuously differentiable** on an open set D if the entries in the gradient of f are continuous on D. Thus each of the functions of Examples 1 through 4 is not only differentiable but also continuously differentiable in the interior of its domain. In practice we deal almost exclusively with continuously differentiable functions.

We conclude by considering a relationship between continuity and differentiability that our intuition suggests should hold:

2.4   Differentiability Implies Continuity.   If $\mathbb{R}^n \xrightarrow{f} \mathbb{R}$ is differentiable at a point $\mathbf{x}_0$ of its domain, then f is continuous at $\mathbf{x}_0$.

Proof.   Differentiability of f at $\mathbf{x}_0$ means that the fraction

$$q(\mathbf{x}) = \frac{f(\mathbf{x}) - f(\mathbf{x}_0) - \nabla f(\mathbf{x}_0) \cdot (\mathbf{x} - \mathbf{x}_0)}{|\mathbf{x} - \mathbf{x}_0|}$$

tends to zero as $\mathbf{x}$ tends to $\mathbf{x}_0$. Multiplying this equation by $|\mathbf{x} - \mathbf{x}_0|$ and rearranging terms gives

$$f(\mathbf{x}) - f(\mathbf{x}_0) = q(\mathbf{x})|\mathbf{x} - \mathbf{x}_0| + \nabla f(\mathbf{x}_0) \cdot (\mathbf{x} - \mathbf{x}_0).$$

The first term on the right is a product of factors that tend to 0 as $\mathbf{x}$ tends to $\mathbf{x}_0$. By the Cauchy–Schwarz inequality

$$|\nabla f(\mathbf{x}_0) \cdot (\mathbf{x} - \mathbf{x}_0)| \leq |\nabla f(\mathbf{x}_0)|\, |\mathbf{x} - \mathbf{x}_0|.$$

Thus the second term on the right also tends to zero as $\mathbf{x}$ tends to $\mathbf{x}_0$. So $f(\mathbf{x})$ tends to $f(\mathbf{x}_0)$ as $\mathbf{x}$ tends to $\mathbf{x}_0$, and f is continuous at $\mathbf{x}_0$.   ∎

2B   Tangent Approximations

Many of the concepts and techniques of calculus have at their foundations the idea of approximating a graph by a tangent line or plane. In particular, the tangent line to the graph of a differentiable function $\mathbb{R} \xrightarrow{f} \mathbb{R}$ at a domain point x_0 is defined to be the graph of the function $\mathbb{R} \xrightarrow{T} \mathbb{R}$ given by $T(x) = f(x_0) + f'(x_0)(x - x_0)$. We refer to the function T as the **tangent approximation** to f at x_0, though the term **first-degree Taylor approximation** is also appropriate. Figure 5.10 shows the graphical relationship between a typical f and T if f is a real-valued function of one real variable.

In the natural generalization to real-valued functions $\mathbb{R}^n \xrightarrow{f} \mathbb{R}$ of several variables, the tangent approximation takes the form

2.5 $$T(\mathbf{x}) = f(\mathbf{x}_0) + \nabla f(\mathbf{x}_0) \cdot (\mathbf{x} - \mathbf{x}_0),$$

where $\nabla f(\mathbf{x}_0)$ is the n-dimensional gradient vector of f at the point $\mathbf{x}_0$. With this formula for T we get the expected result $T(\mathbf{x}_0) = f(\mathbf{x}_0)$, since the dot product of the gradient vector with the zero vector in $\mathbb{R}^n$ is the real number zero. If f is a function of two real variables the picture corresponding to Figure 5.10 is Figure 5.11.

FIGURE 5.10

Tangent for $\mathbb{R} \xrightarrow{f} \mathbb{R}$.

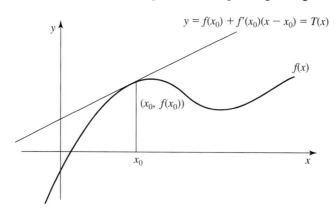

$$y = f(x_0) + f'(x_0)(x - x_0) = T(x)$$

$f(x)$

$(x_0, f(x_0))$

x_0

FIGURE 5.11

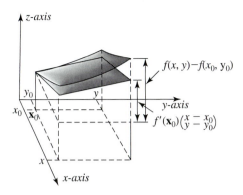

EXAMPLE 7

In Example 4 of Chapter 4, Section 3B we considered the problem of finding the tangent plane to the graph of $f(x) = 1 - 2x^2 - y^2$ where $(x_0, y_0) = (\frac{1}{2}, \frac{1}{2})$. In that example our discussion of tangency was incomplete in that we ignored the behavior of the function except along lines parallel to the axes. Using our thoroughly justified Equation 2.5 we calculate $\nabla f(x, y) = (-4x, -2y)$, so $\nabla f(\frac{1}{2}, \frac{1}{2}) = (-2, -1)$ and $f(\frac{1}{2}, \frac{1}{2}) = \frac{1}{4}$. Thus Equation 2.5 becomes

$$T(x, y) = \tfrac{1}{4} + (-2, -1) \cdot \left(x - \tfrac{1}{2}, y - \tfrac{1}{2}\right) + \tfrac{1}{4} = \tfrac{7}{4} - 2x - y,$$

consistent with the tangent plane $z = \frac{7}{4} - 2x - y$ we found in the earlier example with somewhat less justification.

It's only for functions of two variables that we can draw a graph of a tangent plane in $\mathbb{R}^3$, but the approximation given by Equation 2.5 is valid quite generally as in the next example, where we consider a function of three variables with graph and tangent in $\mathbb{R}^4$.

EXAMPLE 8

If $f(x, y, z) = xy^2z^3$, then $\nabla f(x, y, z) = (y^2z^3, 2xyz^3, 3xy^2z^2)$ To find the tangent approximation at $(1, 1, 1)$ we compute $\nabla f(1, 1, 1) = (1, 2, 3)$ and $f(1, 1, 1) = 1$. Hence the approximation given by Equation 2.5 is

$$T(x, y, z) = 1 + (1, 2, 3) \cdot (x - 1, y - 1, z - 1)$$
$$= -5 + x + 2y + 3z.$$

In the equation that has the tangent plane at $(1, 1, 1, 1)$ as its graph in $\mathbb{R}^4$, we introduce an additional variable w. The desired equation is then

$$w = -5 + x + 2y + 3z.$$

In determining the vector $\nabla f(\mathbf{x}_0)$, we required that the real-valued function

$$f(\mathbf{x}) - T(\mathbf{x}) = f(\mathbf{x}) - f(\mathbf{x}_0) - \nabla f(\mathbf{x}_0) \cdot (\mathbf{x} - \mathbf{x}_0)$$

tend to zero faster than $\mathbf{x} - \mathbf{x}_0$ as $\mathbf{x}$ tends to $\mathbf{x}_0$. This requirement turned out to uniquely determine the entries in the vector $\nabla f(\mathbf{x}_0)$, at the same time ensuring that the graph of $T(\mathbf{x})$ is a good fit to the graph of $f(\mathbf{x})$. To ensure a good geometric approximation near $\mathbf{x}_0$ we also needed to be able to let $\mathbf{x}$ approach $\mathbf{x}_0$ from every possible direction, so we required that some open neighborhood of $\mathbf{x}_0$ be contained in the domain of f. Thus the conditions (i) and (ii) in the definition 2.1 of differentiability are just what are needed to define tangency adequately.

EXERCISES

For each of the functions 1 to 8 find $\nabla f(\mathbf{x})$ at a general point $\mathbf{x}$ in the domain of f.

1. $f(x, y) = x^2 - y^2$

2. $f(x, y) = x^2 - y^2 - \sin xy$

3. $f(x, y) = x + 2y$

4. $f(x, y, z) = (x - y)z$

5. $f(x, y, z) = x + y - z^2$

6. $f(x, y, z) = x^2 + y^2 + z^2$

7. $f(x_1, x_2) = x_1^2 + 2x_2^4$

8. $f(x_1, x_2, x_3) = x_1 x_2 x_3$

Find the tangent approximation $T(x, y)$ or $T(x, y, z)$ as appropriate for each of the functions 9–16 at the indicated point and use it to write the equation for the tangent plane in terms of coordinate variables (x, y) or (x, y, z).

9. $f(x, y) = x^3 - y^3$ at $(1, 1)$

10. $f(x, y) = \sin(x^2 + y^2)$ at (π, π)

11. $f(x, y) = x + 2y$ at $(1, 2)$

12. $f(x, y, z) = (x - y)z$ at $(1, 0, 1)$

13. $f(x, y, z) = x + y - z^2$ at $(0, 0, 1)$

14. $f(x, y, z) = x^2 + y^2 - z^2$ at $(1, 1, 1)$

15. $f(x, y, z) = x + 2yz$ at $(1, 1, 1)$

16. $f(x, y, z) = xy^2z^2$ at $(2, 1, 1)$

At which points do the functions 17 to 20 fail to be differentiable? Give a reason for your answer.

17. $f(x, y) = x^{-2} + y^{-2}$

18. $f(x, y) = \sqrt{x^2 - y^2}$

19. $f(x, y) = |x + y|$

20. $f(x, y) = (x^2 + y^2)^{-1}$

21. Consider the function $f: \mathbb{R}^n \to \mathbb{R}$ defined by $f(\mathbf{x}) = |\mathbf{x}|^2 = \mathbf{x} \cdot \mathbf{x}$. Prove that $\nabla f(\mathbf{x}) = 2\mathbf{x}$ for all $\mathbf{x}$ in $\mathbb{R}^n$.

22. Is the function $g: \mathbb{R}^n \to \mathbb{R}$ defined by $g(\mathbf{x}) = |\mathbf{x}|$ differentiable at every point of its domain? Explain your answer. [*Hint:* $|\mathbf{x}| = \sqrt{\mathbf{x} \cdot \mathbf{x}}$.]

*23. Prove that if the real-valued function f is differentiable at $\mathbf{x}_0$, then

$$\lim_{t \to 0} \frac{f(\mathbf{x}_0 + t\mathbf{x}) - f(\mathbf{x}_0)}{t} = \nabla f(\mathbf{x}_0) \cdot \mathbf{x}.$$

*24. Consider the function

$$f(x, y) = \begin{cases} \dfrac{xy}{x^2 - y^2}, & x \neq \pm y, \\ 0, & x = \pm y. \end{cases}$$

(a) Prove that f has at the origin the partial derivatives $f_x(0, 0) = f_y(0, 0) = 0$.

(b) Prove that f is not differentiable at $(0, 0)$ by assuming it *is* differentiable and contradicting the conclusion of the previous exercise.

(c) Prove that f is not differentiable at $(0, 0)$ by contradicting the conclusion of Theorem 2.4.

*25. Show that the function defined by

$$f(x) = \begin{cases} x^2 \sin \dfrac{1}{x}, & x \neq 0, \\ 0, & x = 0 \end{cases}$$

is differentiable for all x but is not continuously differentiable at $x = 0$.

*26. Referring to Examples 5 and 6 in the text, prove that there is no point on the circle $x^2 + y^2 = 1$ for which the one-sided partial derivatives both exist when the defining limits are taken from within the circle.

SECTION 3   DIRECTIONAL DERIVATIVES

3A   Definition

A partial derivative of a real-valued function measures the rate of change of the function in a particular coordinate direction, parallel to one of the coordinate axes.

To measure the rate of change in an arbitrary direction, we use the *directional deriva-tive*. We first define the derivative with respect to an arbitrary nonzero vector **v**. Let $\mathbb{R}^n \xrightarrow{f} \mathbb{R}$ be a real-valued function, and let **v** be a vector in the domain space $\mathbb{R}^n$. The **derivative with respect to v**, denoted by $\partial f / \partial \mathbf{v}$, is the real-valued function defined by

$$\frac{\partial f}{\partial \mathbf{v}}(\mathbf{x}) = \lim_{t \to 0} \frac{f(\mathbf{x} + t\mathbf{v}) - f(\mathbf{x})}{t}.$$

This is a significant extension of our earlier use of the partial derivative notation, but the derivative is still "partial" since it's computed in a single direction. The domain of $\partial f / \partial \mathbf{v}$ is the subset of the domain of f for which the preceding limit exists. In practice we always assume $\mathbf{v} \neq 0$, since the case $\mathbf{v} = 0$ gives no information. (Why?)

The connection between the derivative with respect to a vector and the gradient is provided in the following theorem.

3.1   Theorem.   If f is differentiable at $\mathbf{x} = (x_1, \ldots, x_n)$ and $\mathbf{v} = (v_1, \ldots, v_n)$, then

$$\frac{\partial f}{\partial \mathbf{v}}(\mathbf{x}) = \nabla f(\mathbf{x}) \cdot \mathbf{v}.$$

We can write this formula in terms of coordinates as

$$\frac{\partial f}{\partial \mathbf{v}}(\mathbf{x}) = v_1 \frac{\partial f}{\partial x_1}(\mathbf{x}) + \cdots + v_n \frac{\partial f}{\partial x_n}(\mathbf{x}).$$

Proof.   First assume $\mathbf{v} \neq 0$ and note that $|t\mathbf{v}| = |t||\mathbf{v}|$. Since f is differentiable at **x**

$$\lim_{t \to 0} \frac{f(\mathbf{x} + t\mathbf{v}) - f(\mathbf{x}) - \nabla f(\mathbf{x}) \cdot (t\mathbf{v})}{|t||\mathbf{v}|} = 0.$$

Since the limit is zero we can remove the vertical bars from $|t|$ to get

$$\lim_{t \to 0} \frac{1}{|\mathbf{v}|} \left(\frac{f(\mathbf{x} + t\mathbf{v}) - f(\mathbf{x})}{t} - \nabla f(\mathbf{x}) \cdot \mathbf{v} \right) = 0.$$

Multiplying by $|\mathbf{v}|$ we get

$$\lim_{t \to 0} \frac{f(\mathbf{x} + t\mathbf{v}) - f(\mathbf{x})}{t} = \nabla f(\mathbf{x}) \cdot \mathbf{v},$$

and the proof is finished for nonzero **v**. When $\mathbf{v} = 0$, both sides of the previous equation are zero.   ∎

Observe that when $\mathbf{v} = \mathbf{e}_j$, a standard basis vector of length 1, the equation in Theorem 3.1 shows that the derivative with respect to that vector is just the partial derivative with respect to x_j, that is,

$$\frac{\partial f}{\partial \mathbf{e}_j} = \frac{\partial f}{\partial x_j}.$$

As in the previous equation we'll most often want to choose the vectors $\mathbf{v}$ in $\partial f/\partial \mathbf{v}$ to have length 1 so these derivatives serve as standardized rates of change in a variety of directions. Nevertheless the more general definition has its uses, as for example in Exercise 21.

For each vector $\mathbf{u}$ in $\mathbb{R}^n$ of length $|\mathbf{u}| = 1$, we define the **directional derivative** of f in the direction of $\mathbf{u}$ to be the function $\partial f/\partial \mathbf{u}$. The reason for the name "directional" derivative is that in $\mathbb{R}^n$ there is a natural way to associate a vector to each direction, namely, take the unit vector in that direction. The number $(\partial f/\partial \mathbf{u})(\mathbf{x})$ is then regarded as a standard measure of the rate of change of the value $f(\mathbf{x})$ in the direction of $\mathbf{u}$.

EXAMPLE 1 Suppose a function $f: \mathbb{R}^3 \to \mathbb{R}$ is $f(x, y, z) = xyz$. We find the directional derivative of f in the direction of the unit vector $\mathbf{u} = (1/2, 1/2, 1/\sqrt{2})$ by letting $\mathbf{x} = (x, y, z)$ and using Theorem 3.1 to get

$$\frac{\partial f}{\partial \mathbf{u}}(\mathbf{x}) = \nabla f(\mathbf{x}) \cdot \mathbf{u} = (yz, xz, xy) \cdot (1/2, 1/2, 1/\sqrt{2})$$

$$= \frac{1}{2}(yx + xz + \sqrt{2}xy).$$

It follows that the directional derivative of f in the direction of $\mathbf{u}$ at $(1, 1, 1)$ has the value $\partial f/\partial \mathbf{u}(1, 1, 1) = 1 + 1/\sqrt{2}$.

Let $\mathbb{R}^2 \xrightarrow{f} \mathbb{R}$ be a function whose graph is a surface in $\mathbb{R}^3$ and let $\mathbf{u}$ be a unit vector in $\mathbb{R}^2$, i.e., $|\mathbf{u}| = 1$. An example appears in Figure 5.12. The value of the directional derivative $\partial f/\partial \mathbf{u}$ at $\mathbf{x} = (x, y)$ is by definition

$$\frac{\partial f}{\partial \mathbf{u}}(\mathbf{x}) = \lim_{t \to 0} \frac{f(\mathbf{x} + t\mathbf{u}) - f(\mathbf{x})}{t}.$$

The distance between the points $\mathbf{x} + t\mathbf{u}$ and $\mathbf{x}$ is given by

$$|(\mathbf{x} + t\mathbf{u}) - \mathbf{x}| = |t\mathbf{u}| = |t|.$$

Hence the ratio

$$\frac{f(\mathbf{x} + t\mathbf{u}) - f(\mathbf{x})}{t}$$

is the slope of the line through the points $f(\mathbf{x} + t\mathbf{u})$ and $f(\mathbf{x})$. It follows that the limit of the ratio, $(\partial f/\partial \mathbf{u})(\mathbf{x})$, is the slope of the tangent line at $(\mathbf{x}, f(\mathbf{x}))$ to the curve formed by the intersection of the graph of f with the plane that contains $\mathbf{x}$ and $\mathbf{x} + \mathbf{u}$, and is parallel to the z axis. This curve appears in dashes in Figure 5.12. The angle γ in Figure 5.12 indicates the inclination of a tangent line to the graph in the vertical plane containing $\mathbf{u}$ and therefore satisfies the equation

$$\tan \gamma = \frac{\partial f}{\partial \mathbf{u}}(\mathbf{x}).$$

The situation here is a generalization of the one shown in Figure 4.19 in Chapter 4, Section 3. If $\mathbf{u} = \mathbf{e}_1$, the angle γ becomes the angle α in the earlier figure and

$$\frac{\partial f}{\partial \mathbf{u}} = \frac{\partial f}{\partial x}, \quad \text{with } \mathbf{u} = (1, 0).$$

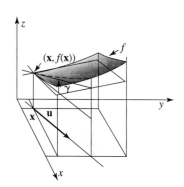

FIGURE 5.12
$\tan \gamma$ is a slope.

If $\mathbf{u} = \mathbf{e}_2$, we get $\gamma = \beta$ in Figure 4.19 of Chapter 4, Section 3, and

$$\frac{\partial f}{\partial \mathbf{u}} = \frac{\partial f}{\partial y}, \text{ with } \mathbf{u} = (0, 1).$$

3B   Mean-Value Theorem

We assume here some acquaintance with the following fundamental theorem from single-variable calculus, and we state it without proof.

3.2   Mean-Value Theorem.   Let $\mathbb{R} \xrightarrow{f} \mathbb{R}$ be continuous on the closed interval $[x, y]$ and differentiable on the open interval (x, y). Then there is a number x_0 strictly between x and y such that $f(x) - f(y) = f'(x_0)(x - y)$.

The function f in Theorem 3.2 is a function of a single real variable, so in that context we can write the mean-value equation with both sides divided by $x - y$; this is a natural way to write it because we can interpret each side of the equation as a slope. For a real-valued differentiable function of a vector variable $\mathbf{x}$ we can't divide by $\mathbf{x} - \mathbf{y}$, but there is still a valid generalization with $f'(x_0)(x - y)$ replaced by $\nabla f(\mathbf{x}_0) \cdot (\mathbf{x} - \mathbf{y})$. The formal statement follows.

3.3   Theorem.   Let $\mathbb{R}^n \xrightarrow{f} \mathbb{R}$ be differentiable on an open set containing the line segment S joining two vectors $\mathbf{x}$ and $\mathbf{y}$ in $\mathbb{R}^n$. Then there is a point $\mathbf{x}_0$ on S such that

$$f(\mathbf{y}) - f(\mathbf{x}) = \nabla f(\mathbf{x}_0) \cdot (\mathbf{y} - \mathbf{x}).$$

Proof.   Consider the function $g(t) = f(t(\mathbf{y} - \mathbf{x}) + \mathbf{x})$, defined for $0 \leq t \leq 1$ and set $\mathbf{m}(t) = t(\mathbf{y} - \mathbf{x}) + \mathbf{x}$. Then if h is a real number,

$$g(t + h) - g(t) = f\big((t + h)(\mathbf{y} - \mathbf{x}) + \mathbf{x}\big) - f\big(t(\mathbf{y} - \mathbf{x}) + \mathbf{x}\big)$$
$$= f\big(h(\mathbf{y} - \mathbf{x}) + \mathbf{m}(t)\big) - f\big(\mathbf{m}(t)\big).$$

Dividing both sides by $h \neq 0$ gives

$$\frac{g(t + h) - g(t)}{h} = \frac{f\big(h(\mathbf{y} - \mathbf{x}) + \mathbf{m}(t)\big) - f\big(\mathbf{m}(t)\big)}{h}.$$

Now let $h \to 0$. On the right side we get $\partial f/\partial(\mathbf{y} - \mathbf{x})$ evaluated at $\mathbf{m}(t)$. Hence the limit of the left side exists also, and is $g'(t)$. Thus

$$g'(t) = \frac{\partial f}{\partial(\mathbf{y} - \mathbf{x})}\big(\mathbf{m}(t)\big) \quad \text{or} \quad g'(t) = \nabla f\big(\mathbf{m}(t)\big) \cdot (\mathbf{y} - \mathbf{x}), \qquad (*)$$

by Theorem 3.1. Now let $t = 0$ in the definition of $g(t)$ to get $g(1) - g(0) = f(\mathbf{y}) - f(\mathbf{x})$. But by the mean-value theorem for functions of one variable, applied to g at $t = 1$ and $t = 0$,

$$\frac{g(1) - g(0)}{1 - 0} = f(\mathbf{y}) - f(\mathbf{x}) = g'(t_0),$$

for some t_0 satisfying $0 < t_0 < 1$. Setting $t = t_0$ and $\mathbf{x}_0 = \mathbf{m}(t_0)$ in Equation $(*)$ and using this last equation gives $f(\mathbf{y}) - f(\mathbf{x}) = \nabla f(\mathbf{x}_0) \cdot (\mathbf{y} - \mathbf{x})$.   ∎

FIGURE 5.13

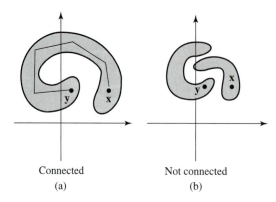

Connected
(a)

Not connected
(b)

One of the most important conclusions we draw from the mean-value theorem for functions of one variable is that a function with zero derivative on an interval is constant. For a function f of a vector variable, we replace the domain interval by an open set D in $\mathbb{R}^n$ that we assume to be **polygonally connected**; a polygonally connected set S is one such that a given pair of points in it can be joined by a finite sequence of line segments lying in S, that is, by a polygonal path. Figure 5.13 shows a set in $\mathbb{R}^2$ that is connected in this way and also one that is not.

3.4   Theorem.   If $\mathbb{R}^n \xrightarrow{f} \mathbb{R}$ is differentiable on a polygonally connected open set D and $\nabla f(\mathbf{x}) = 0$ for every $\mathbf{x}$ in D, then f is constant.

Proof.   If $\mathbf{x}_1$ and $\mathbf{x}_2$ are points of D joined by a single line segment, then Theorem 3.3 and the assumption that $\nabla f(\mathbf{x}) = 0$ in D together imply that $f(\mathbf{x}_1) = f(\mathbf{x}_2)$. Working stepwise, one additional segment at a time, the same conclusion holds for two points $\mathbf{x}_1$ and $\mathbf{x}_p$ joined by a finite sequence of segments. So f is constant on D.   ∎

EXERCISES

In each Exercise 1 to 4, with functions defined on $\mathbb{R}^2$ or $\mathbb{R}^3$, find the directional derivative of f in the direction of the unit vector $\mathbf{u}$ at the point $\mathbf{x}$.

1. $f(x, y, z) = x^2 + y^2 + z^2, \mathbf{u} = (1/\sqrt{3}, 1/\sqrt{3}, 1/\sqrt{3}), \mathbf{x} = (1, 0, 1)$

2. $f(x, y) = x^2 - y^2, \mathbf{u} = (1/\sqrt{2}, 1/\sqrt{2}), \mathbf{x} = (2, 1)$

3. $f(x, y) = x + y, \mathbf{u} = (1, 0), \mathbf{x} = (2, 3)$

4. $f(x, y, z) = xy \sin z, \mathbf{u} = (1/\sqrt{2}, 0, -1/\sqrt{2}), \mathbf{x} = (1, 1, 1)$

In Exercises 5 to 8, for the real-valued function defined in $\mathbb{R}^2$, find the directional derivative at $\mathbf{x}$ in the direction indicated.

5. $f(x, y) = x^2 - y^2$ at $\mathbf{x} = (1, 1)$ and in the direction $(1/\sqrt{5}, 2/\sqrt{5})$

6. $f(x, y) = e^x \sin y$ at $\mathbf{x} = (1, 0)$ and in the direction $(\cos \alpha, \sin \alpha)$

7. $f(x, y) = e^{x+y}$ at $\mathbf{x} = (1, 1)$ and in the direction of the curve defined by $g(t) = (t^2, t^3)$ at $g(2)$ for t increasing

8. $f(x, y) = (x^2 + y^2)^{-1}$ at the point $(1, 3)$ and in the direction of the vector $(1, 2)$

9. Find the directional derivative at $(1, 0, 0)$ of the function $f(x, y, z) = x^2 + ye^z$ in the direction of the tangent vector at $g(0)$ to the curve $\mathbb{R}^3$ defined parametrically by
$$g(t) = (3t^2 + t + 1, 2t, t^2).$$

10. Find the directional derivative at $(1, 0, 0)$ of the function $f(x, y, z) = x^2 + ye^z$ in the direction of increasing t along the curve in $\mathbb{R}^3$ defined by $g(t) = (t^2 - t + 2, t, t + 2)$ at $g(0)$.

11. Find the directional derivative at $(1, 0, 1)$ of the function $f(x, y, z) = 4x^2y + y^2z$ in the direction of the vector $(1, 1, 1)$.

12. Find the directional derivative at $(0, 0)$ of the function $f(x, y) = \sin(x + y)$ in the direction of the vector (a, b).

13. Use the cross product to find the direction of a perpendicular $\mathbf{p}$ at $(1, 2, 1)$ to the surface defined parametrically by $(x, y, z) = (u^2v, u + v, u)$. Then find the directional derivative of $f(x, y, z) = x^3 + y^2 + z$ in the direction of $\mathbf{p}$ at $(1, 2, 1)$.

14. Show that for an arbitrary angle α, the vector $\mathbf{u} = (\cos\alpha, \sin\alpha)$ is a unit vector in $\mathbb{R}^2$ inclined at angle α to the positive x-axis.

15. For the unit vector $\mathbf{u}$ in Exercise 14, show that

$$\frac{\partial f}{\partial \mathbf{u}} = \cos\alpha \frac{\partial f}{\partial x} + \sin\alpha \frac{\partial f}{\partial y}.$$

16. Let $\mathbf{u}$ be a unit vector in $\mathbb{R}^n$ with **direction cosines** $\cos\alpha_i, = \mathbf{u} \cdot \mathbf{e}_i$, relative to the natural basis $\mathbf{e}_1, \dots, \mathbf{e}_n$. Show that $\mathbf{u} = (\cos\alpha_1, \cos\alpha_2, \dots, \cos\alpha_n)$.

17. For the unit vector $\mathbf{u}$ in Exercise 16, show that

$$\frac{\partial f}{\partial \mathbf{u}} = \cos\alpha_1 \frac{\partial f}{\partial x_1} + \dots + \cos\alpha_n \frac{\partial f}{\partial x_n}.$$

18. If $f: \mathbb{R}^n \to \mathbb{R}$ is differentiable and $\mathbf{u}$ is a unit vector in $\mathbb{R}^n$, show that

$$\frac{\partial f}{\partial(-\mathbf{u})}(\mathbf{x}) = -\frac{\partial f}{\partial \mathbf{u}}(\mathbf{x}),$$

for all $\mathbf{x}$ in $\mathbb{R}^n$.

19. Show that the mean-value formula of Theorem 3.3 also takes the form

$$\frac{f(\mathbf{y}) - f(\mathbf{x})}{|\mathbf{y} - \mathbf{x}|} = \frac{\partial f}{\partial \mathbf{u}}(\mathbf{x}_0),$$

where the unit vector is $\mathbf{u} = (\mathbf{y} - \mathbf{x})/|\mathbf{y} - \mathbf{x}|$.

20. Show that the function f defined by

$$f(x, y) = \begin{cases} \dfrac{x|y|}{\sqrt{x^2 + y^2}}, & (x, y) \neq (0, 0) \\ 0, & (x, y) = (0, 0) \end{cases}$$

has a directional derivative in every direction at $(0, 0)$, but that f is not differentiable at $(0, 0)$. [*Hint:* If f were differentiable at $(0, 0)$, then we would have $\nabla f(0, 0) = (0, 0)$.]

21. The mean-value theorem doesn't generalize to vector-valued functions, because for a vector-valued function $f(x)$ of even just one variable there may not be a single point x_0 between x and y at which $\big(f(y) - f(x)\big)/(x - y) = f'(x_0)$. Verify this assertion for the example $f(x) = (\sin x, \sin 2x)$ on the interval $0 \leq x \leq \pi$.

In Exercises 22 and 23, the fundamental derivative approximation $f(\mathbf{x} + \mathbf{h}) \approx f(\mathbf{x}_0) + \nabla f(\mathbf{x}_0)\mathbf{h}$ for differentiable functions $\mathbb{R}^n \xrightarrow{f} \mathbb{R}$ becomes the first-degree part of a higher-degree approximation. We define the **Nth-degree Taylor approximation** to $f(\mathbf{x})$ by

$$f(\mathbf{x} + \mathbf{h}) \approx f(\mathbf{x}) + \frac{\partial f}{\partial \mathbf{h}}(\mathbf{x})$$

$$+ \frac{1}{2!}\frac{\partial^2 f}{\partial \mathbf{h}^2}(\mathbf{x}) + \frac{1}{3!}\frac{\partial^3 f}{\partial \mathbf{h}^3}(\mathbf{x})$$

$$+ \dots + \frac{1}{N!}\frac{\partial^N f}{\partial \mathbf{h}^N}(\mathbf{x}),$$

where we assume that the required Nth-order derivatives are continuous on an open neighborhood of $\mathbf{x}$. The higher-order derivatives with respect to the vector $\mathbf{h}$ are defined recursively by generalizing ordinary higher-order partials as follows:

$$\frac{\partial^2 f}{\partial \mathbf{h}^2}(\mathbf{x}) = \frac{\partial}{\partial \mathbf{h}}\left(\frac{\partial f}{\partial \mathbf{h}}\right)(\mathbf{x}), \quad \frac{\partial^3 f}{\partial \mathbf{h}^3}(\mathbf{x})$$

$$= \frac{\partial}{\partial \mathbf{h}}\left(\frac{\partial^2 f}{\partial \mathbf{h}^2}\right)(\mathbf{x}), \dots, \quad \text{and in general}$$

$$\frac{\partial^N f}{\partial \mathbf{h}^N}(\mathbf{x}) = \frac{\partial}{\partial \mathbf{h}}\left(\frac{\partial^{N-1} f}{\partial \mathbf{h}^{N-1}}\right)(\mathbf{x}).$$

22. Let $f(x, y) = \sin(x^2 + y)$, $\mathbf{x} = (0, 0)$, and $\mathbf{h} = (h, k)$. Compute the second-degree Taylor approximation to $f(h, k)$.

23. Let $f(x, y) = e^{x^2 + y^2 - z^2}$, $\mathbf{x} = (0, 0, 0)$, and $\mathbf{h} = (h, k, l)$. Compute the second-degree Taylor approximation to $f(h, k, l)$.

SECTION 4   VECTOR-VALUED FUNCTIONS

4A   Differentiability

Relying on Definition 2.1 for real-valued functions in Section 2, we define a vector-valued function $\mathbb{R}^n \xrightarrow{f} \mathbb{R}^m$, with domain an open set D, to be **differentiable** if each of its m real-valued coordinate functions $f_1, f_2, \ldots, f_m$ is differentiable. If all n first-order partial derivatives of each of the m coordinate functions of f are continuous on D then f is in addition called **continuously differentiable** on D. In this generality we're looking at m times n real-valued partial derivatives altogether. A natural way to organize these partial derivatives is to compute successively the gradient vectors of the coordinate functions $f_1, f_2, \ldots, f_m$ of f:

$$\nabla f_1(\mathbf{x}) = \left(\frac{\partial f_1}{\partial x_1}(\mathbf{x}), \ \frac{\partial f_1}{\partial x_2}(\mathbf{x}), \ \ldots, \ \frac{\partial f}{\partial x_n}(\mathbf{x}) \right)$$

$$\nabla f_2(\mathbf{x}) = \left(\frac{\partial f_2}{\partial x_1}(\mathbf{x}), \ \frac{\partial f_2}{\partial x_2}(\mathbf{x}), \ \ldots, \ \frac{\partial f_2}{\partial x_n}(\mathbf{x}) \right)$$

$$\vdots$$

$$\nabla f_m(\mathbf{x}) = \left(\frac{\partial f_m}{\partial x_1}(\mathbf{x}), \ \frac{\partial f_m}{\partial x_2}(\mathbf{x}), \ \ldots, \ \frac{\partial f_m}{\partial x_n}(\mathbf{x}) \right).$$

In practice we hope to be able to observe that each of these mn partial derivatives is continuous on the open set D in which case f is not only differentiable but continuously differentiable.

EXAMPLE 1 The function $\mathbb{R}^2 \xrightarrow{f} \mathbb{R}^2$ defined by

$$f(x, y) = \left(\begin{matrix} x^2 - y^2 \\ 2xy \end{matrix} \right)$$

has coordinate functions $f_1(x, y) = x^2 - y^2$ and $f_2(x, y) = 2xy$, so

$$\nabla f_1(x, y) = (2x, \ -2y)$$
$$\nabla f_2(x, y) = (2y, \ 2x).$$

Since the coordinates of $\nabla f_1(x, y)$ and $\nabla f_2(x, y)$ are continuous for all (x, y), the vector-valued function f is continuously differentiable on all of $\mathbb{R}^2$.

An alternative way of displaying the first-order partial derivatives of the coordinate functions of a vector function f is to look directly at the vector partial derivatives of f. We'll see that sometimes there's a distinct advantage to displaying a vector function f and its partial derivatives as column vectors.

EXAMPLE 2 The function $\mathbb{R}^2 \xrightarrow{f} \mathbb{R}^2$ of Example 1 defined by

$$f(x, y) = \left(\begin{matrix} x^2 - y^2 \\ 2xy \end{matrix} \right)$$

has vector partial derivatives

$$\frac{\partial f}{\partial x}(x, y) = \left(\begin{matrix} 2x \\ 2y \end{matrix} \right) \quad \frac{\partial f}{\partial y}(x, y) = \left(\begin{matrix} -2y \\ 2x \end{matrix} \right).$$

We draw the same conclusion as in Example 1 from continuity of the four partial derivatives, namely that f is continuously differentiable on all of $\mathbb{R}^2$.

4B   The Derivative Matrix

Each of the Examples 1 and 2 in Section 4A displays the first-order partial derivatives of a vector-valued function $\mathbb{R}^2 \xrightarrow{f} \mathbb{R}^2$ as the result of computations that are in principle somewhat different. In Example 1 we looked at the gradient vectors of the two coordinate functions, while in Example 2 we looked at the vector partial derivatives of f itself. Both of these interpretations are important to keep in mind, but there is a third way of organizing the results of the computations that is specially important, namely the arrangement of each example's real-valued partial derivatives in the following 2-by-2 matrix:

$$f'(x, y) = \begin{pmatrix} \dfrac{\partial f_1}{\partial x}(x, y) & \dfrac{\partial f_1}{\partial y}(x, y) \\ \dfrac{\partial f_2}{\partial x}(x, y) & \dfrac{\partial f_2}{\partial y}(x, y) \end{pmatrix} = \begin{pmatrix} 2x & -2y \\ 2y & 2x \end{pmatrix}.$$

We have denoted the resulting matrix at a point (x, y) by $f'(x, y)$, a notation that will be particularly useful when we take up the general chain rule in Chapter 5. Notice that the rows of this matrix are the successive coordinates of the gradient vectors of the coordinate functions $f_1(x, y) = x^2 - y^2$ and $f_2(x, y) = 2xy$ of $f(x, y)$. The columns of the matrix consist of the entries in the vector partial derivatives of $f(x, y)$.

In general we define the **derivative matrix** of a differentiable function at $\mathbf{x}$ by

$$f'(\mathbf{x}) = \begin{pmatrix} \dfrac{\partial f_1}{\partial x_1}(\mathbf{x}) & \dfrac{\partial f_1}{\partial x_2}(\mathbf{x}) & \cdots & \dfrac{\partial f_1}{\partial x_n}(\mathbf{x}) \\ \dfrac{\partial f_2}{\partial x_1}(\mathbf{x}) & \dfrac{\partial f_2}{\partial x_2}(\mathbf{x}) & \cdots & \dfrac{\partial f_2}{\partial x_n}(\mathbf{x}) \\ \vdots & \vdots & \ddots & \vdots \\ \dfrac{\partial f_m}{\partial x_1}(\mathbf{x}) & \dfrac{\partial f_m}{\partial x_2}(\mathbf{x}) & \cdots & \dfrac{\partial f_m}{\partial x_n}(\mathbf{x}) \end{pmatrix},$$

where $f_1(\mathbf{x}), f_2(\mathbf{x}), \ldots, f_m(\mathbf{x})$ are the differentiable coordinate functions of $f(\mathbf{x})$. Note that the differentiation variable remains the same in each column, producing a vector partial derivative $\partial f/\partial x_j(\mathbf{x})$, which is a tangent vector at $\mathbf{x}$ to a curve in the image space generated by varying only x_j. Across each row a coordinate function f_i remains the same, producing the coordinate functions of a gradient vector $\nabla f_i(\mathbf{x})$, which we'll see in Section 1 of the next chapter gives the magnitude and direction of maximum increase of f_i at $\mathbf{x}$.

If f is a real-valued function there is only one coordinate function so the matrix has only a single row, whose entries are identical with the coordinates of the gradient vector $\nabla f(\mathbf{x})$. Notationally the only difference between $\nabla f(\mathbf{x})$ and the single-rowed matrix is that $\nabla f(\mathbf{x})$ has its entries separated by commas. So in every case **a derivative matrix $f'(\mathbf{x})$ can be regarded as having tangent vectors for columns and gradient vectors for rows.**

EXAMPLE 3   The function $\mathbb{R}^3 \xrightarrow{f} \mathbb{R}^3$ defined by

$$f(x, y, z) = \begin{pmatrix} x^2 + e^y \\ x + y \sin z \\ x + y \end{pmatrix}$$

has coordinate functions

$$f_1(x, y, z) = x^2 + e^y, \quad f_2(x, y, z) = x + y \sin z, \quad f_3(x, y, z) = x + y.$$

The derivative matrix at (x, y, z) is the matrix whose columns are the three possible vector partial derivatives:

$$f'(x, y, z) = \left(\frac{\partial f}{\partial x}(x, y, z) \quad \frac{\partial f}{\partial y}(x, y, z) \quad \frac{\partial f}{\partial z}(x, y, z) \right).$$

For this example

$$f'(x, y, z) = \begin{pmatrix} 2x & e^y & 0 \\ 1 & \sin z & y \cos z \\ 1 & 1 & 0 \end{pmatrix}.$$

In particular the derivative matrix of f at $(1, 1, \pi)$ is the matrix

$$f'(1, 1, \pi) = \begin{pmatrix} 2 & e & 0 \\ 1 & 0 & -1 \\ 1 & 1 & 3 \end{pmatrix}.$$

should be 0

The three columns are vectors in $\mathbb{R}^3$ tangent to coordinate curves of f through $f(1, 1, \pi) = (1 + e, 1, 2)$, and the three rows are gradient vectors at the same point.

EXAMPLE 4   The function $\mathbb{R}^2 \xrightarrow{f} \mathbb{R}^2$ defined by

$$f(x, y) = (x^2 + 2xy + y^2, \ xy^2 + x^2 y)$$

has coordinate functions $f_1(x, y) = x^2 + 2xy + y^2$ and $f_2(x, y) = xy^2 + x^2 y$. Thinking in terms of gradients this time we find

$$\nabla f_1(x, y) = (2x + 2y, \ 2x + 2y) \quad \text{and} \quad \nabla f_2(x, y) = (y^2 + 2xy, \ 2xy + x^2).$$

Hence the derivative of f at (x, y) is given by the matrix

$$f'(x, y) = \begin{pmatrix} 2x + 2y & 2x + 2y \\ y^2 + 2xy & 2xy + x^2 \end{pmatrix}.$$

EXAMPLE 5   Consider the function $\mathbb{R} \xrightarrow{f} \mathbb{R}^2$

$$f(t) = \begin{pmatrix} \cos t \\ \sin t \end{pmatrix}, \quad -\infty < t < \infty.$$

The derivative $f'(t_0)$ is the 2-by-1 matrix

$$\begin{pmatrix} -\sin t_0 \\ \cos t_0 \end{pmatrix}.$$

It's instructive to consider the matrix as a vector in the range space of f and to draw it with its tail at the image point $f(t_0)$. For $t_0 = 0, \pi/4, \pi/3, \pi/2$, and π, the respective derivative matrices $f'(t_0)$ are

$$\begin{pmatrix} 0 \\ 1 \end{pmatrix}, \quad \begin{pmatrix} -\sqrt{2}/2 \\ \sqrt{2}/2 \end{pmatrix}, \quad \begin{pmatrix} -\sqrt{3}/2 \\ 1/2 \end{pmatrix}, \quad \begin{pmatrix} -1 \\ 0 \end{pmatrix}, \quad \text{and} \quad \begin{pmatrix} 0 \\ -1 \end{pmatrix}.$$

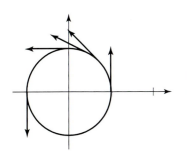

FIGURE 5.14
Tangents to a circular image.

Viewed as vectors, drawn with their tails at their corresponding image points under f, these are shown in Figure 5.14. Evidently, for functions of one real variable, the idea of derivative, introduced here as a matrix, coincides with the vector derivative developed in Chapter 4, Section 4. The first-degree Taylor expansion approximates $f(t)$ near t_0 and is the vector function of t given by

$$T(t) = f'(t_0)(t - t_0) + f(t_0),$$

which in terms of matrices becomes   ~~remove.~~

$$T(t) = (t - t_0) \begin{pmatrix} -\sin t_0 \\ \cos t_0 \end{pmatrix} + \begin{pmatrix} \cos t_0 + t_0 \sin t_0 \\ \sin t_0 - t_0 \cos t_0 \end{pmatrix}. \quad \times$$

This is the parametric representation of the line tangent to the image of f at $f(t_0)$.

4C   Tangent Approximations

We've so far encountered two settings of the tangent approximation, the first being the one appropriate for the case $\mathbb{R} \xrightarrow{f} \mathbb{R}$ of a differentiable real-valued function of a single real variable, namely

$$T(x) = f(x_0) + f'(x_0)(x - x_0).$$

In this familiar setting in single-variable calculus, we use $T(x)$ to write the equation of the tangent line to the graph of $y = f(x)$ at x_0 in the form

$$y = f(x_0) + f'(x_0)(x - x_0).$$

In the second setting we get a geometric picture of the tangent approximation of a real-valued function $f(x, y)$ at (x_0, y_0) by drawing the graph of $z = f(x, y)$ together with the tangent approximation to the surface at the point $(x_0, y_0, f(x_0, y_0))$, namely

$$T(x, y) = f(x_0, y_0) + \nabla f(x_0, y_0) \cdot (x - x_0, y - y_0).$$

Thus the tangent plane is the graph of the equation

$$z = f(x_0, y_0) + \nabla f(x_0, y_0) \cdot (x - x_0, y - y_0).$$

The function $T(x, y)$ is a good first-degree approximation to the function $f(x, y)$, provided (x, y) is close to (x_0, y_0). Figure 5.11 in Section 3 is a picture of the relation between the graphs of $z = f(x, y)$ and $z = T(x, y)$. The more general version of $T(\mathbf{x})$ for real-valued functions is

$$T(\mathbf{x}) = f(\mathbf{x}_0) + \nabla f(\mathbf{x}_0) \cdot (\mathbf{x} - \mathbf{x}_0).$$

We've seen that for general n and m a differentiable function $\mathbb{R}^n \xrightarrow{f} \mathbb{R}^m$ has an m-by-n derivative matrix $f'(\mathbf{x})$ with m rows, one for each coordinate function, and

n columns, one for each coordinate variable in $\mathbf{x}$. It is this matrix that plays the role that the gradient plays in the case of a single real-valued coordinate function. Thus the **first-degree Taylor approximation** to $f: \mathbb{R}^n \longrightarrow \mathbb{R}^m$ at $\mathbf{x}_0$ is

---

4.1 Definition $T(\mathbf{x}) = f(\mathbf{x}_0) + f'(\mathbf{x}_0)(\mathbf{x} - \mathbf{x}_0),$

where the vector $\mathbf{x} - \mathbf{x}_0$ must be interpreted as an n-by-1 column vector so that we can multiply on the left by the m-by-n matrix $f'(\mathbf{x}_0)$ according to the row by column rule for matrix multiplication.

EXAMPLE 6 The function $\mathbb{R}^2 \xrightarrow{f} \mathbb{R}^3$ defined by

$$f(u, v) = \begin{pmatrix} u \cos v \\ u \sin v \\ u \end{pmatrix} \qquad 0 \le u \le 1, \ 0 \le v \le 2\pi$$

has as image one complete turn of a circular helix in $\mathbb{R}^3$. This function f has derivative matrix

$$f'(u, v) = \begin{pmatrix} \cos v & -u \sin v \\ \sin v & u \cos v \\ 1 & 0 \end{pmatrix}.$$

Since the entries in this matrix are continuous real-valued functions on all of $\mathbb{R}^2$, we can conclude that $f(u, v)$ is continuously differentiable on its rectangular domain $0 \le u \le 1, \ 0 \le v \le 2\pi$. Note that the columns of the 3-by-2 matrix $f'(u_0, v_0)$ in a single-variable context are tangent vectors $\mathbf{u}_0$ and $\mathbf{v}_0$ that generate a tangent plane at a point $f(u_0, v_0)$ consisting of all points

$$T(u, v) = f(u_0, v_0) + (u - u_0)\mathbf{u}_0 + (v - v_0)\mathbf{v}_0.$$

This plane is the image of the first-degree tangent approximation at (u_0, v_0). Note also that the inclusion of the terms $-u_0$ and $-v_0$ just has the effect of shifting the parameter values so the point of tangency on the plane corresponds to (u_0, v_0). Deleting those two terms would still produce the same plane as image; with $(u, v) = (0, 0)$ the parameter values indicate the point of tangency on the plane. See Figure 4.22(b) in Chapter 4, Section 4, where the columns of the matrix $f'(u, v)$ were written as the vector partials $\partial f / \partial u$ and $\partial f / \partial v$.

EXAMPLE 7 The function $\mathbb{R}^3 \xrightarrow{f} \mathbb{R}^3$ defined for all (x, y, z) by

$$f(x, y, z) = \begin{pmatrix} -x + y + z \\ 2x - 2y + 2z \\ 3x + 3y - 3z \end{pmatrix}$$

has as image all of $\mathbb{R}^3$. To see this note that f has derivative matrix

$$f'(u, v) = \begin{pmatrix} -1 & 1 & 1 \\ 2 & -2 & 2 \\ 3 & 3 & -3 \end{pmatrix}.$$

This constant 3-by-3 matrix $A = f'(x, y, z)$ has positive determinant 24 and so is invertible with inverse A^{-1} using Theorem 5.7 in Chapter 2, Section 5. Note that $f(\mathbf{x}) = A\mathbf{x}$, where $\mathbf{x}$ is a 3-dimensional column vector. Hence there is an inverse function $f^{-1}(\mathbf{y}) = A^{-1}\mathbf{y}$ that takes each point $\mathbf{y}$ in the image of f back to its corresponding point $\mathbf{x}$ in the domain of f.

The first-degree approximation defined by Equation 4.1 has the same essential character as the two special cases we considered previously, namely that $f(\mathbf{x}) - T(\mathbf{x})$ tends to zero faster than $|\mathbf{x} - \mathbf{x}_0|$ as $\mathbf{x}$ tends to $\mathbf{x}_0$. Here is the complete formal statement and proof.

4.2   Theorem.   If $\mathbb{R}^n \xrightarrow{f} \mathbb{R}^m$ is differentiable at $\mathbf{x}_0$, then

$$\lim_{\mathbf{x} \to \mathbf{x}_0} \frac{f(\mathbf{x}) - f(\mathbf{x}_0) - f'(\mathbf{x}_0)(\mathbf{x} - \mathbf{x}_0)}{|\mathbf{x} - \mathbf{x}_0|} = \mathbf{0},$$

and $f'(\mathbf{x}_0)$ is the unique matrix that satisfies this equation.

Proof.   According to the definition of the matrix-vector product $f'(\mathbf{x}_0)(\mathbf{x} - \mathbf{x}_0)$, the ith coordinate in the product $f'(\mathbf{x}_0)(\mathbf{x} - \mathbf{x}_0)$ is just the dot product of the ith row of the matrix $f'(\mathbf{x}_0)$ with the column vector $\mathbf{x} - \mathbf{x}_0$. But the entries in the ith row of the matrix are just the coordinates of $\nabla f_i(\mathbf{x}_0)$, so the ith coordinate of the desired limit equation is

$$\lim_{\mathbf{x} \to \mathbf{x}_0} \frac{f_i(\mathbf{x}) - f_i(\mathbf{x}_0) - \nabla f_i(\mathbf{x}_0) \cdot (\mathbf{x} - \mathbf{x}_0)}{|\mathbf{x} - \mathbf{x}_0|} = 0.$$

We assumed f was differentiable at $\mathbf{x}_0$, which means that each real-valued function f_i is differentiable there, so the limit is valid in each coordinate. Theorem 1.1 in Section 1 of this chapter says that the vector limit is valid if, and only if, the limit is valid in each coordinate. So the vector limit is valid. Finally, the matrix $f'(\mathbf{x}_0)$ is the unique matrix that has the gradients of the coordinate functions as rows.   ∎

4.3   Corollary.   If A is a constant m-by-n matrix, then the function $\mathbb{R}^n \xrightarrow{f} \mathbb{R}^m$ defined by $f(\mathbf{x}) = A\mathbf{x}$ has A for its derivative matrix, that is, $f'(\mathbf{x}) = A$.

Proof.   The derivative matrix $f'(\mathbf{x}_0)$ is the unique matrix that satisfies the equation of Theorem 4.2. Observe that since $A(\mathbf{x} - \mathbf{x}_0) = A\mathbf{x} - A\mathbf{x}_0$, then

$$\frac{A\mathbf{x} - A\mathbf{x}_0 - A(\mathbf{x} - \mathbf{x}_0)}{|\mathbf{x} - \mathbf{x}_0|} = \frac{A\mathbf{x} - A\mathbf{x}_0 - A\mathbf{x} - A\mathbf{x}_0}{|\mathbf{x} - \mathbf{x}_0|} = \mathbf{0}.$$

The limit as $\mathbf{x}$ tends to $\mathbf{x}_0$ is the $\mathbf{0}$-vector, so A is the unique derivative matrix.   ∎

We saw in Chapter 2, Section 2C that a function of the form $\mathbf{F}(\mathbf{x}) = A\mathbf{x}$ is *linear* in the sense that $\mathbf{F}(s\mathbf{x} + t\mathbf{y}) = s\mathbf{F}(\mathbf{x}) + t\mathbf{F}(\mathbf{y})$ for all scalars s, t and n-dimensional vectors $\mathbf{x}$, $\mathbf{y}$. Thus it's a linear function that's the crucial part of the first-degree Taylor approximation $\mathbf{T}(\mathbf{x}) = f'(\mathbf{x}_0)(\mathbf{x} - \mathbf{x}_0) + f(\mathbf{x}_0)$ to $f(\mathbf{x})$ near $\mathbf{x} = \mathbf{x}_0$.

EXERCISES

In Exercises 1 to 10, find the derivative matrix f' at a general point of the domain of the function f from $\mathbb{R}^n$ to $\mathbb{R}^m$.

1. $f(x, y) = \begin{pmatrix} xy \\ x + y \end{pmatrix}$

2. $f(u, v) = \begin{pmatrix} u \cos v \\ u \sin v \end{pmatrix}$

3. $f(x, y, z) = \begin{pmatrix} x + \sin y \\ y + \cos z \\ x + y + z \end{pmatrix}$

4. $f(t, u, v) = \begin{pmatrix} t\cos u \sin v \\ t \sin u \sin v \\ t \cos v \end{pmatrix}$

5. $f(x) = x^2 e^x$

6. $f(x, y) = (e^{xy}, xy)$

7. $f(u, v, w) = \begin{pmatrix} uv \\ vw \\ wu \end{pmatrix}$

8. $f(t) = \begin{pmatrix} t^2 \\ t^3 \end{pmatrix}$

9. $f(x, y) = \begin{pmatrix} x^2 + y^2 \\ x^2 - y^2 \\ xy \end{pmatrix}$

10. $f(u, v) = (u, v, u^2 + v^2)$

In Exercises 11 to 14, let f be the vector function defined by

$$f\begin{pmatrix} x \\ y \end{pmatrix} = \begin{pmatrix} x^2 - y^2 \\ 2xy \end{pmatrix}$$

Find the derivative matrix of f at the following points:

11. $\begin{pmatrix} x \\ y \end{pmatrix}$

12. $\begin{pmatrix} a \\ b \end{pmatrix}$

13. $\begin{pmatrix} 1 \\ 0 \end{pmatrix}$

14. $\begin{pmatrix} 1/\sqrt{2} \\ 1/\sqrt{2} \end{pmatrix}$

In Exercises 15 to 22, find the derivative matrix of the function at the indicated point.

15. $f\begin{pmatrix} x \\ y \end{pmatrix} = x^2 + y^2$ at $\begin{pmatrix} x \\ y \end{pmatrix} = \begin{pmatrix} 1 \\ 1 \end{pmatrix}$

16. $g(x, y, z) = xyz$ at $(x, y, z) = (1, 0, 0)$

17. $f(t) = \begin{pmatrix} \sin t \\ \cos t \end{pmatrix}$ at $t = \frac{\pi}{4}$

18. $f(t) = \begin{pmatrix} e^t \\ t \\ t^2 \end{pmatrix}$ at $t = 1$

19. $g(x, y) = \begin{pmatrix} x + y \\ x^2 + y^2 \end{pmatrix}$ at $(x, y) = (1, 2)$

20. $A\begin{pmatrix} u \\ v \end{pmatrix} = \begin{pmatrix} u + v \\ u - v \\ 1 \end{pmatrix}$ at $\begin{pmatrix} u \\ v \end{pmatrix} = \begin{pmatrix} 1 \\ 0 \end{pmatrix}$

21. $T\begin{pmatrix} u \\ v \end{pmatrix} = \begin{pmatrix} u \cos v \\ u \sin v \\ v \end{pmatrix}$ at $\begin{pmatrix} u \\ v \end{pmatrix} = \begin{pmatrix} 1 \\ \pi \end{pmatrix}$

22. $f(x, y, z) = (x + y + z, xy + yz + zx, xyz)$ at (x, y, z)

23. Let P be the function from $\mathbb{R}^3$ to $\mathbb{R}^2$ defined by $P(x, y, z) = (x, y)$.
  (a) What is the geometric interpretation of this transformation?
  (b) Show that P is differentiable at all points and find the derivative matrix of P at $(1, 1, 1)$.

24. (a) Draw the curve in $\mathbb{R}^2$ defined parametrically by the function

$$g(t) = (t - 1, t^2 - 3t + 2), \quad -\infty < t < \infty.$$

  (b) Find formulas describing the tangent approximations to g near $t = 0$ and near $t = 2$.
  (c) Draw the lines defined parametrically by the tangent approximation.

25. Let f be the function given in Exercise 1, and let

$$\mathbf{x}_0 = \begin{pmatrix} 1 \\ 0 \end{pmatrix}, \quad \mathbf{y}_1 = \begin{pmatrix} 0.1 \\ 0 \end{pmatrix},$$

$$\mathbf{y}_2 = \begin{pmatrix} 0 \\ 0.1 \end{pmatrix}, \quad \mathbf{y}_3 = \begin{pmatrix} 0.1 \\ 0.1 \end{pmatrix}.$$

  (a) Compute $f(\mathbf{x}_0 + \mathbf{y}_i)$ for $i = 1, 2, 3$.
  (b) Find the tangent approximation to $f(\mathbf{x}_0 + \mathbf{y})$ for an arbitrary vector $\mathbf{y}$.
  (c) Use part (b) to find approximations to the vectors $f(\mathbf{x}_0 + \mathbf{y}_i)$, $i = 1, 2, 3$.

26. (a) Sketch the graph in $\mathbb{R}^3$ defined explicitly by the function

$$f(x, y) = 4 - x^2 - y^2.$$

  (b) Find the tangent approximation to f (i) near $(x, y) = (0, 0)$ and (ii) near $(x, y) = (2, 0)$.
  (c) Draw the graphs of the approximations in (b).

27. What is the derivative matrix $f'(x, y, z)$ of the function

$$f(x, y, z) = \begin{pmatrix} a_1 & a_2 & a_3 \\ b_1 & b_2 & b_3 \\ c_1 & c_2 & c_3 \end{pmatrix} \begin{pmatrix} x \\ y \\ z \end{pmatrix} + \begin{pmatrix} a_0 \\ b_0 \\ c_0 \end{pmatrix}?$$

28. A **translation** is a function of the form $T(\mathbf{x}) = \mathbf{x} + \mathbf{b}$, for a fixed vector $\mathbf{b}$. What is the derivative matrix of a translation from $\mathbb{R}^n$ to $\mathbb{R}^n$?

*29. Show that if $f:\mathbb{R}^m \longrightarrow \mathbb{R}^n$ and $g:\mathbb{R}^m \longrightarrow \mathbb{R}^n$ are differentiable at $\mathbf{x}_0$ and a is a real number, then $f + g$ and af are differentiable at $\mathbf{x}_0$ and

(a) $(f + g)'(\mathbf{x}_0) = f'(\mathbf{x}_0) + g'(\mathbf{x}_0)$

(b) $(af)'(\mathbf{x}_0) = af'(\mathbf{x}_0)$

SECTION 5   NEWTON'S METHOD

In this section we treat Newton's method for approximating a solution of an equation $f(\mathbf{x}) = 0$, where $\mathbb{R}^n \xrightarrow{f} \mathbb{R}^n$ is a nonlinear function. We begin by looking at the idea of approximating a vector in $\mathbb{R}^n$ by a sequence of vectors in $\mathbb{R}^n$. We are used to thinking of a real number like $\sqrt{2}$ as being approximated by a sequence of rational numbers, say, 1, 1.4, 1.41, 1.414, ... The idea extends immediately to vectors.

First we define the *limit of a sequence* in $\mathbb{R}^n$. Let $\mathbf{x}_1, \mathbf{x}_2, \mathbf{x}_3, \ldots$ be an infinite sequence of vectors in $\mathbb{R}^n$. Suppose there is a vector $\mathbf{x}$ in $\mathbb{R}^n$ such that, for a given $\epsilon > 0$, there is an integer N for which

$$|\mathbf{x}_k - \mathbf{x}| < \epsilon$$

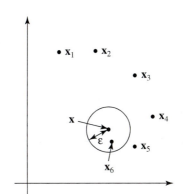

FIGURE 5.15

whenever $k \geq N$. Then we say that the given sequence **converges** to the **limit x**, and we write

$$\lim_{k\to\infty} \mathbf{x}_k = \mathbf{x}.$$

We can summarize by saying that the sequence $\mathbf{x}_1, \mathbf{x}_2, \mathbf{x}_3, \ldots$ converges to $\mathbf{x}$ if $|\mathbf{x}_k - \mathbf{x}|$ is arbitrarily small for all sufficiently large k. Figure 5.15 shows a sequence with entries lying within ϵ of $\mathbf{x}$ whenever $k \geq 6$.

| EXAMPLE 1 | Consider the vector $(\sqrt{2}, \pi)$ in $\mathbb{R}^2$. Suppose that $\sqrt{2}$ is approximated by the decimal expansion sequence 1, 1.4, 1.41, 1.414, ... and that π is approximated by 3, 3.1, 3.14, 3.141, ... Then we can form the sequence of vectors $(1, 3), (1.4, 3.1), (1.41, 3.14), (1.414, 3.141), \ldots$ to approximate the vector $(\sqrt{2}, \pi)$. We leave as an exercise showing that if $x_1, x_2, x_3, \ldots$ and $y_1, y_2, y_3, \ldots$ are the sequences approximating $\sqrt{2}$ and π respectively, then $\lim_{k\to\infty} x_k = \sqrt{2}$ and $\lim_{k\to\infty} y_k = \pi$ implies that $\lim_{k\to\infty}(x_k, y_k) = (\sqrt{2}, \pi)$. |

We look first at Newton's method for approximating a solution of an equation $f(x) = 0$ where f is real-valued and x is a real variable. We assume that f is continuously differentiable. If the graph of f should happen to be convex as shown in Figure 5.16, then it's geometrically apparent that the tangent line to the graph at $(x_0, f(x_0))$ crosses the x-axis at a point x_1 that is a better approximation to the solution $\bar{x}$ than x_0 is. Having chosen x_0 somewhat arbitrarily, and having found x_1, we can repeat the process. This time we use the tangent line at $(x_1, f(x_1))$ and call its intersection with the x-axis x_2. Thus we can generate a sequence of numbers $x_0, x_1, x_2, \ldots$ approximating $\bar{x}$.

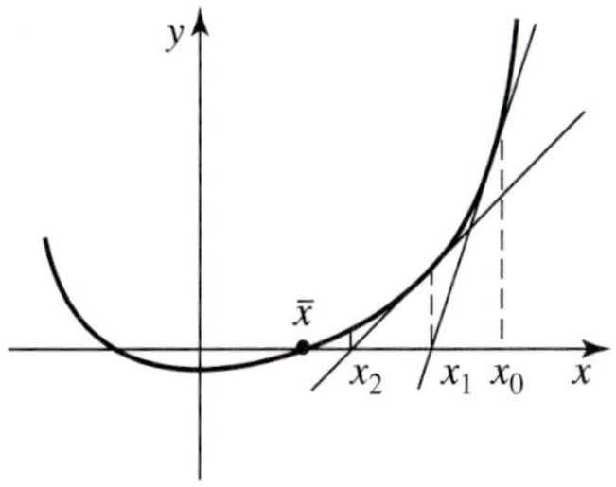

In practice, we need a formula for computing the sequence $x_1, x_2, \ldots$. We observe first that the tangent line at $(x_0, f(x_0))$ has the equation

FIGURE 5.16

5.1 $$y = f'(x_0)(x - x_0) + f(x_0).$$

Since the approximation x_1 is found by intersecting the tangent with the x-axis, we set $y = 0$ in the above equation and solve for x_1. The result is

$$0 = f'(x_0)(x_1 - x_0) + f(x_0), \quad \text{or} \quad f'(x_0)(x_1 - x_0) = -f(x_0).$$

If $f'(x_0) \neq 0$,

$$x_1 - x_0 = -\frac{f(x_0)}{f'(x_0)}. \quad \text{or} \quad x_1 = x_0 - \frac{f(x_0)}{f'(x_0)}.$$

Having found x_1, to find x_2 we replace x_0 by x_1 in the last formula to get

$$x_2 = x_1 - \frac{f(x_1)}{f'(x_1)}.$$

In general, we compute x_{k+1} by

5.2
$$x_{k+1} = x_k - \frac{f(x_k)}{f'(x_k)}.$$

EXAMPLE 2   The equation $x^2 - 3 = 0$ has two solutions, $\sqrt{3}$ and $-\sqrt{3}$. To approximate $\sqrt{3}$ we choose $x_0 = 2$ and compute x_{k+1} from x_k by the Formula (1), which in this case is

$$x_{k+1} = x_k - \frac{(x_k^2 - 3)}{2x_k}$$

$$= \frac{(x_k^2 + 3)}{2x_k}.$$

Thus we get $x_1 = \frac{7}{4} = 1.75$. Substituting this value in the preceding formula for $k = 1$ gives $x_2 = \frac{97}{56} \approx 1.732142857$. This approximation to $\sqrt{3}$ is correct to three decimal places. Calculating one more step gives $x_3 \approx 1.7320508$, which is correct to the number of displayed digits.

We follow a similar procedure to the one just described if f is a function from $\mathbb{R}^n$ to $\mathbb{R}^n$. The difference is that, in this case, $\mathbf{x}_1, \mathbf{x}_0$, and $f(\mathbf{x}_0)$ are vectors in $\mathbb{R}^n$ and $f'(\mathbf{x}_0)$ is an n-by-n derivative matrix. To approximate a solution of a vector equation $f(\mathbf{x}) = 0$, we consider equation generalizing Equation 5.1 defines the value of the tangent approximation to f near $\mathbf{x}_0$, that is,

5.3
$$\mathbf{y} = f'(\mathbf{x}_0)(\mathbf{x} - \mathbf{x}_0) + f(\mathbf{x}_0),$$

where $\mathbf{x}_0$ is chosen as an initial approximation to the desired solution $\bar{\mathbf{x}}$. In Figure 5.16, Equation 5.3 is the equation of the tangent to the graph of f at $(\mathbf{x}_0, f(\mathbf{x}_0))$. As before, we set $\mathbf{y} = \mathbf{0}$ in Equation 5.3 to get

$$\mathbf{0} = f'(\mathbf{x}_0)(\mathbf{x}_1 - \mathbf{x}_0) + f(\mathbf{x}_0), \quad \text{or} \quad f'(\mathbf{x}_0)(\mathbf{x}_1 - \mathbf{x}_0) = -f(\mathbf{x}_0).$$

If $f'(\mathbf{x}_0)$ has an inverse matrix $[f'(\mathbf{x}_0)]^{-1}$, we apply the inverse to both sides to get

$$\mathbf{x}_1 - \mathbf{x}_0 = -[f'(\mathbf{x}_0)]^{-1} f(\mathbf{x}_0), \quad \text{or} \quad \mathbf{x}_1 = \mathbf{x}_0 - [f'(\mathbf{x}_0)]^{-1} f(\mathbf{x}_0).$$

In this equation $[f'(\mathbf{x}_0)]^{-1} f(\mathbf{x}_0)$ is the vector we get by applying the inverse of the matrix $f'(\mathbf{x}_0)$ to the vector $f(\mathbf{x}_0)$. The vector $\mathbf{x}_1$ is the first improvement on the initial approximation $\mathbf{x}_0$ to the solution $\bar{\mathbf{x}}$.

We can repeat what we have just done, replacing $\mathbf{x}_0$ by $\mathbf{x}_1$ to get

$$\mathbf{x}_2 = \mathbf{x}_1 - [f'(\mathbf{x}_1)]^{-1} f(\mathbf{x}_0).$$

After $k + 1$ steps we have the general formula for

5.4   Newton's Method. $\mathbf{x}_{k+1} = \mathbf{x}_k - [f'(\mathbf{x}_k)]^{-1} f(\mathbf{x}_k).$

EXAMPLE 3   Consider the pair of equations

$$x^2 + y^2 = 2$$
$$x^2 - y^2 = 1,$$

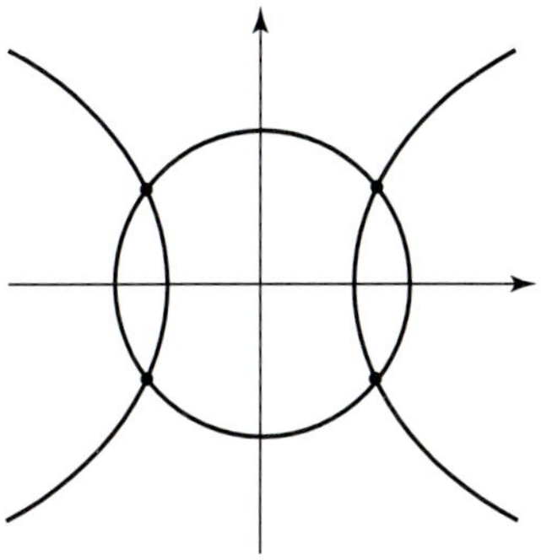

the intersecting graphs of which appear in Figure 5.17. There are four solutions to the pair of equations. To find approximate solutions by Newton's method we define

$$f(x, y) = \begin{pmatrix} x^2 + y^2 - 2 \\ x^2 - y^2 - 1 \end{pmatrix}$$

FIGURE 5.17

and solve the equation $f(x, y) = (0, 0)$. Since f is a function from $\mathbb{R}^2$ to $\mathbb{R}^2$, we require both $x^2 + y^2 - 2 = 0$ and $x^2 - y^2 - 1 = 0$, and it's helpful to sketch the curves defined by these two equations. The exact solutions are represented by the four points of intersection of the circle $x^2 + y^2 - 2 = 0$ and the hyperbola $x^2 - y^2 - 1 = 0$ shown in Figure 5.17. The choice of an initial approximation depends on which solution we want to approximate. To look for the solution in the first quadrant, we try $\mathbf{x}_0 = (1, 1)$. Since

$$f(x, y) = \begin{pmatrix} x^2 + y^2 - 2 \\ x^2 - y^2 - 1 \end{pmatrix}, \quad \text{we have} \quad f'(x, y) = \begin{pmatrix} 2x & 2y \\ 2x & -2y \end{pmatrix}$$

and

$$[f'(x, y)]^{-1} = \begin{pmatrix} \dfrac{1}{4}x^{-1} & \dfrac{1}{4}x^{-1} \\ \dfrac{1}{4}y^{-1} & -\dfrac{1}{4}y^{-1} \end{pmatrix}.$$

Then the right side of Equation 5.4 becomes

$$\begin{pmatrix} x \\ y \end{pmatrix} - [f'(x, y)]^{-1} f(x, y) = \begin{pmatrix} x \\ y \end{pmatrix} - \begin{pmatrix} \dfrac{1}{4}x^{-1} & \dfrac{1}{4}x^{-1} \\ \dfrac{1}{4}y^{-1} & -\dfrac{1}{4}y^{-1} \end{pmatrix} \begin{pmatrix} x^2 + y^2 - 2 \\ x^2 - y^2 - 1 \end{pmatrix}$$

$$= \begin{pmatrix} x \\ y \end{pmatrix} - \begin{pmatrix} \dfrac{2x^2 - 3}{4x} \\ \dfrac{2y^2 - 1}{4y} \end{pmatrix} = \begin{pmatrix} \dfrac{2x^2 + 3}{4x} \\ \dfrac{2y^2 + 1}{4y} \end{pmatrix}.$$

This vector is the analog of the expression $(x^2 + 3)/2x$ in the previous example and is the formula by which the sequence of approximations is actually computed. Setting $\mathbf{x}_0 = (x_0, y_0) = (1, 1)$, we get

$$\mathbf{x}_1 = \begin{pmatrix} \dfrac{2x_0^2 + 3}{4x_0} \\ \dfrac{2y_0^2 + 1}{4y_0} \end{pmatrix} = \begin{pmatrix} \dfrac{5}{4} \\ \dfrac{3}{4} \end{pmatrix} = \begin{pmatrix} 1.25 \\ 0.75 \end{pmatrix}.$$

Substituting $\mathbf{x}_1$ into Equation 5.3 gives

$$\mathbf{x}_2 = \begin{pmatrix} \dfrac{2(1.25)^2 + 3}{4(1.25)} \\ \dfrac{2(0.75) + 3}{4(0.75)} \end{pmatrix} \approx \begin{pmatrix} 1.225 \\ 1.70833 \end{pmatrix}.$$

Substituting our approximate value for x_2 gives

$$\mathbf{x}_3 = \begin{pmatrix} \dfrac{2(1.225)^2 + 3}{4(1.225)} \\ \dfrac{2(1.70833)^2 + 3}{4(1.70833)} \end{pmatrix} \approx \begin{pmatrix} 1.22574 \\ 0.707108 \end{pmatrix}.$$

Similarly, we get

$$\mathbf{x}_4 \approx \begin{pmatrix} 1.22474 \\ 0.707107 \end{pmatrix} \quad \text{and} \quad \mathbf{x}_5 \approx \begin{pmatrix} 1.22474 \\ 0.707107 \end{pmatrix}.$$

As in the previous example, you can check that further iteration using only five places after the decimal point doesn't produce a change.

In this example, the two simultaneous equations can actually be solved by elimination to yield $\mathbf{x} = (\sqrt{1.5}, \sqrt{0.5})$. The approximation $x_4 = (1.22474, 0.707107)$ happens to be correct to that many decimal places. We get the other three vector solutions by symmetry. Referring to Figure 5.17, we get these vectors by changing one or both signs of the coordinates to minus. The numerical procedure could have been applied by taking as initial estimate $\mathbf{x}_0$ one of the vectors $(-1, -1)$, $(-1, 1)$, or $(1, -1)$.

In choosing an initial approximation $\mathbf{x}_0$, some care must be used in getting a sufficiently close approximation. For instance, if in Example 3 we wanted the solution in the first quadrant, then making too gross an error in choosing $\mathbf{x}_0$ could lead to approximating the wrong solution. In many examples a sketch or similar geometric analysis of the function f will show how $\mathbf{x}_0$ should be chosen.

In using Newton's method in $\mathbb{R}^n$ for large n, it may be very time-consuming to invert the matrix $f'(\mathbf{x}_n)$ at each step of the iteration. In such cases we can use the

5.5   Modified Newton Method: $\mathbf{x}_{k+1} = \mathbf{x}_k - [f'(\mathbf{x}_0)]^{-1} f(\mathbf{x}_k).$

As with Newton's method we derive a sequence of approximations to a solution of $f(\mathbf{x}) = \mathbf{0}$. For $k = 0$, the formula defining $\mathbf{x}_1$ is the same as the Newton formula.

For $k \geq 1$, $\mathbf{x}_{k+1}$ as defined by Equation 5.5 will in general be different from the corresponding value determined by the Newton formula in Equation 5.4, because the matrix $[f'(\mathbf{x}_0)]^{-1}$ remains the same at each step in Equation 5.5.

EXAMPLE 4   Returning to the equation of Example 3, namely,

$$\begin{pmatrix} x^2 + y^2 - 2 \\ x^2 - y^2 - 1 \end{pmatrix} = \begin{pmatrix} 0 \\ 0 \end{pmatrix},$$

we apply Equation 5.5 with $\mathbf{x}_0 = (1, 1)$. Then

$$[f'(\mathbf{x}_0)]^{-1} = \begin{pmatrix} \dfrac{1}{4} & \dfrac{1}{4} \\ \dfrac{1}{4} & -\dfrac{1}{4} \end{pmatrix},$$

so

$$\mathbf{x} - [f'(\mathbf{x}_0)]^{-1} f(\mathbf{x}) = \begin{pmatrix} x \\ y \end{pmatrix} - \begin{pmatrix} \dfrac{1}{4} & \dfrac{1}{4} \\ \dfrac{1}{4} & -\dfrac{1}{4} \end{pmatrix} \begin{pmatrix} x^2 + y^2 - 2 \\ x^2 - y^2 - 1 \end{pmatrix}$$

$$= \begin{pmatrix} x \\ y \end{pmatrix} - \begin{pmatrix} \dfrac{2x^2 - 3}{4} \\ \dfrac{2y^2 - 1}{4} \end{pmatrix} = \begin{pmatrix} \dfrac{-2x^2 + 4x + 3}{4} \\ \dfrac{-2y^2 + 4y + 1}{4} \end{pmatrix}.$$

Then $\mathbf{x}_1$, defined by $\mathbf{x}_1 = \mathbf{x}_0 - [f'(\mathbf{x}_0)]^{-1} f(\mathbf{x}_0)$, is

$$\mathbf{x}_1 = \begin{pmatrix} \dfrac{-2x_0^2 + 4x_0 + 3}{4} \\ \dfrac{-2y_0^2 + 4y_0 + 1}{4} \end{pmatrix} = \begin{pmatrix} 1.25 \\ 0.75 \end{pmatrix}.$$

In the next step

$$\mathbf{x}_2 = \begin{pmatrix} \dfrac{-2(1.25)^2 + 4(1.25) + 3}{4} \\ \dfrac{-2(0.75)^2 + 4(0.75) + 1}{4} \end{pmatrix} = \begin{pmatrix} 1.21875 \\ 0.71875 \end{pmatrix}.$$

Continuing, we arrive at

$$\mathbf{x}_{10} = \begin{pmatrix} 1.22474 \\ 0.707107 \end{pmatrix},$$

which agrees with the result obtained in Example 3, though it takes more steps.

In deciding whether to use the Newton Formula 5.4 or its modification 5.5, note that Formula 5.4 produces faster convergence than 5.5, that is, it achieves a smaller error in a given number of steps; on the other hand Formula 5.5 has the advantage that it requires calculation of the derivative matrix f' and its inverse at only one point. Thus if computing the inverse matrices $[f'(x_k)]^{-1}$ is going to be particularly time-consuming, it may be worth taking the extra iteration steps that Equation 5.5 may require to achieve the desired accuracy.

Note.   A Java applet NEWTON is available at http://math.dartmouth.edu/~rewn/ for implementing Newton's method.

EXERCISES

1. (a)   Sketch the graph of $f(x) = \sqrt[3]{x} - x$ for $-2 < x < 2$.
   (b)   Sketch the tangent lines to the graph of f at $x_0 = \frac{3}{4}$, $x_0 = -\frac{3}{4}$, and $x_0 = -\frac{1}{4}$.
   (c)   For each of the three choices for x_0 in part (b), what solution of $f(x) = 0$ can the Newton iteration be expected to converge to?
   (d)   Discuss the choice $x_0 = \sqrt[3]{3}/9$ for an initial approximation to a solution of $f(x) = 0$.

2. To get some idea of how Newton's method can fail, observe that the equation $\sqrt[3]{x} = 0$ has the unique solution $x = 0$. Show that applying Newton's method to this equation with initial guess x_0 for the solution produces the subsequent approximations $x_n = (-2)^n x_0$. What happens if $x_0 \neq 0$ as n increases? [Strictly speaking, the method doesn't apply if we start with $x_0 = 0$, since $f'(0)$ fails to exist.]

3. (a)   To approximate the solution of $\cos x - x = 0$ by Newton's method, show that when x_0 is chosen, then for $k \geq 0$,

$$x_{k+1} = \frac{x_k \sin x_k + \cos x_k}{1 + \sin x_k}.$$

   (b)   Assuming $x_0 = 1$ find x_4.

4. Find approximate solutions to the pair of equations

$$x^2 + y - 1 = 0$$

$$x + y^2 - 2 = 0$$

by following these steps:

(a)   Sketch the curves satisfying each of the two equations.
(b)   Defining f by

$$f(x, y) = \begin{pmatrix} x^2 + y - 1 \\ x + y^2 - 2 \end{pmatrix},$$

find $f'(x, y)$, $[f'(x, y)]^{-1}$, and $(x, y) = [f'(x, y)]^{-1} f(x, y)$.
(c)   Using the sketch in part (a), choose an initial approximation $\mathbf{x}_0 = (x_0, y_0)$ to the solution of $f(x, y) = 0$ that lies in the fourth quadrant of the xy-plane.
(d)   Compute $\mathbf{x}_1 = (x_1, y_1)$ by Formula 5.1.
(e)   Compute $\mathbf{x}_5 = (x_5, y_5)$.

5. Let

$$f(x, y, z) = \begin{pmatrix} x + y + z \\ x^2 + y^2 + z^2 \\ x^3 + y^3 + z^3 \end{pmatrix}$$

and find $f'(1, 2, -1)$. Taking $\mathbf{x}_0 = (1, 2, -1)$, apply the modified Newton Formula 5.2 to approximate a solution to $f(x, y, z) = (2.1, 5.7, 8.2)$.

6. Let

$$g(u, v) = \begin{pmatrix} u^2 + uv^2 \\ u + v^3 \end{pmatrix}.$$

Noting that $g(1, 1) = (2, 2)$, use Newton's method 5.1 or its modification 5.2 to approximate a solution to $g(u, v) = (1.9, 2.1)$.

Chapter 5 REVIEW

State whether each of the following sets 1 to 8 is (a) open, (b) closed, or (c) neither. Also describe each set's (d) interior, and (e) boundary.

1. The positive quadrant in $\mathbb{R}^2$, that is, the set of points $\{(x, y) \mid x > 0, y > 0\}$

2. All (x, y) in $\mathbb{R}^2$ such that $|x| < 1$ and $|y| \leq 1$

3. All (x, y, z) in $\mathbb{R}^3$ such that $x^2 + y^2 + z^2 \leq 1$ and $x < 0$

4. All (x, y, z) in $\mathbb{R}^3$ such that $x > 1$, $y > 0$, and $z > -1$

5. The xy-plane in $\mathbb{R}^3$

6. The set in $\mathbb{R}^2$ consisting of all (x, y) with $|(x, y)| < 1$, along with the point $(2, 2)$

7. The set of all (x, y, z) in $\mathbb{R}^3$ such that $x > 0$, $y > 1$. and $z > 3$

8. The set of vectors $\mathbf{x}$ in $\mathbb{R}^4$ such that $|\mathbf{x} - (1, 1, 1, 1)| \leq 2$

9. Consider the set S in $\mathbb{R}^3$ consisting of the points (x, y, z) such that $x^2 + y^2 + z^2 \leq 1$ when $y \leq 0$, and such that $x^2 + y^2 + z^2 < 1$ when $y > 0$. Describe (a) the interior of S and (b) the boundary of S. What is the smallest closed set containing S?

In Exercises 10 to 15, suppose the function $\mathbb{R} \xrightarrow{\text{H}} \mathbb{R}$ is defined by

$$H(x) = \begin{cases} 1, & \text{if } x \geq 0, \\ 0, & \text{if } x < 0. \end{cases}$$

What points of discontinuity, if any, does the composition $H\big(f(x, y)\big)$ have in its domain of definition, given each of the following definitions for $f(x, y)$?

10. $f(x, y) = x^2 + y^2$

11. $f(x, y, z) = x^2 + y^2 + z^2$

12. $f(x, y) = \ln(x^2 + y^2)$

13. $f(x, y) = (x - y)/(1 + (x - y)^2)$

14. $f(x, y) = y$

15. $f(x, y) = 1/(1 + x^2 + y^2)$

In Exercises 16 to 21, compute $\nabla f(\mathbf{x})$ at a general point $\mathbf{x}$ in the domain of f.

16. $f(x, y) = e^x \sin y$

17. $f(x, y) = (x^2 + y^2)^{-1}$

18. $f(x, y, z) = xy + xyz^2$

19. $f(x, y, z) = x - y$

20. $f(x, y, z) = x/(y^2 + z^2)$

21. $f(x, y, z, w) = xy + yz + zw + wx$

In Exercises 22 to 27, compute all second-order partial derivatives of the given function, including mixed partial derivatives such as $\partial^2 f/\partial x \, \partial y$.

22. $f(x, y) = x^3 - y^3$

23. $f(x, y) = e^x \sin y$

24. $f(x, y) = x/(x^2 + y^2)$

25. $f(x, y, z) = yze^x$

26. $f(x, y, z) = \cos(x + y + z)$

27. $f(x, y, z) = x^4 + y^3 + z^2$

In Exercises 28 to 33, find the derivative matrix for the given function.

28. $f(x, y) = (x + y, x^2 + y^2)$

29. $f(u, v) = (u, v, u + v, u - 1)$

30. $f(t) = (t, t^2, t^3)$

31. $f(x, y) = (\frac{1}{2}x^2 - \frac{1}{2}y^2, xy)$

32. $f(x, y, z) = (x + y, y + z, z + x)$

33. $f(u, v, w) = (uv, vw, wu, uvw)$

34. Assuming $|\mathbf{u}| = 1$, find the rate of increase $\partial f/\partial \mathbf{u}(1, 2, 3)$ of $f(x, y, z) = xyz$ in the direction of the vector $(100, 200, 500)$.

35. Let $f(x, y) = xy^2$ and let $\mathbf{u} = (\cos\theta, \sin\theta)$.
   (a) Compute $(\partial f/\partial \mathbf{u})(-1, 2)$ in terms of θ.
   (b) In what direction does this function increase most rapidly from the value $f(-1, 2)$?

*36. Let two fixed unit vectors in $\mathbb{R}^2$ satisfy $\mathbf{u}_1 \neq \pm\mathbf{u}_2$. Suppose the directional derivatives $\dfrac{\partial f}{\partial \mathbf{u}_1}(\mathbf{x})$ and $\dfrac{\partial f}{\partial \mathbf{u}_2}(\mathbf{x})$ are continuous functions of $\mathbf{x}$ on all of $\mathbb{R}^2$. Is f continuously differentiable? Does your answer change if the unit vectors $\mathbf{u}_1 \neq \pm\mathbf{u}_2$ vary continuously from one point $\mathbf{x}$ to another?

CHAPTER 6

VECTOR DIFFERENTIAL CALCULUS

The previous chapter stresses the fundamentals of multivariable differentiation together with some geometric interpretations. The present chapter has to do with extended techniques and interpretations that have proved to be particularly useful in applications. Each section contains a mixture of the geometric settings introduced in Chapter 4.

SECTION 1   GRADIENT FIELDS

1A   Basic Properties

If f is a differentiable real-valued function $\mathbb{R}^n \xrightarrow{f} \mathbb{R}$, then we saw in Section 2 of Chapter 5 that we could calculate the function ∇f in rectangular coordinates $(x_1, x_2, \ldots, x_n)$ using

$$\nabla f(\mathbf{x}) = \left(\frac{\partial f}{\partial x_1}(\mathbf{x}), \ldots, \frac{\partial f}{\partial x_n}(\mathbf{x}) \right).$$

In Chapter 5 we concentrated on isolated values of the gradient of a real-valued function $f(\mathbf{x})$ and paid relatively little attention to it in a context in which it was important to regard it as a function from $\mathbb{R}^n$ to $\mathbb{R}^n$. As a function of $\mathbf{x}$, the image of $\nabla f(\mathbf{x})$ is most often pictured as a **vector field**, which we'll define here. We'll then refer to a vector field generated by ∇f for some differentiable function f as a **gradient field.** (In Chapter 12 we'll study vector fields that aren't necessarily gradient fields.) Plotting a vector field $\mathbf{F}(x, y) = \big(f(x, y), g(x, y) \big)$ in $\mathbb{R}^2$ is in principle a very simple matter. Associated with each *point* (x, y) in the domain of $\mathbf{F}$ is an *arrow* from $(0, 0)$ to the *point* $\mathbf{F}(x, y)$. We translate this arrow parallel to itself so that it starts at (x, y) and ends at $(x, y) + \mathbf{F}(x, y)$. Carrying out this routine for a suitably chosen selection of points in the plane will produce a sketch of the vector field. Carrying out this procedure by hand for a large number of points is extremely tedious for all but the simplest vector fields, and computer plotting is the preferred alternative, as described in Subsection 1D. Figure 6.1 shows some sketches that include a few such arrows in $\mathbb{R}^2$ and $\mathbb{R}^3$. To make sense of these pictures physically, you can imagine that the direction and length of an arrow $\nabla f(\mathbf{x})$ are respectively the direction and speed of a fluid flow at the point $\mathbf{x}$ to which the arrow is attached. (See Subsection 1D for a discussion of flow lines.)

EXAMPLE 1   The real-valued function $f(x, y) = \frac{1}{2}x^2 + y$ is differentiable in all of $\mathbb{R}^2$. Therefore $\nabla f(x, y) = \big(\partial f/\partial x(x, y), \partial f/\partial y(x, y) \big)$ is also defined on $\mathbb{R}^2$ and necessarily takes

FIGURE 6.1

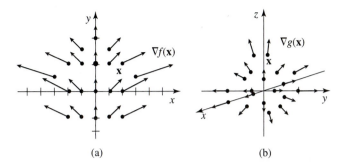

(a)   (b)

its values in $\mathbb{R}^2$. We have

$$\nabla f(x, y) = (x, 1).$$

Figure 6.1(a) shows a sketch with just 20 vectors of the vector field. For instance,

$$\nabla f(1, 0) = (1, 1), \quad \nabla f(-1, 0) = (-1, 1), \quad \nabla f(0, 2) = (0, 1).$$

Notice that the length $|\nabla f(x, y)| = \sqrt{x^2 + 1}$ is independent of y, but increases as $|x|$ does. Also, the arrows starting from a single vertical line, with the same x-coordinate, are parallel, with the same length because, as we remarked, $\nabla f(x, y)$ is independent of y in this example.

EXAMPLE 2   The function $g(x, y, z) = \frac{1}{4}(x^2 + y^2 + z^2)$ has a gradient in $\mathbb{R}^3$,

$$\nabla g(x, y, z) = \left(\tfrac{1}{2}x, \tfrac{1}{2}y, \tfrac{1}{2}z \right).$$

The direction of the field is directly away from the origin at each point, as shown in Figure 6.1(b).

The gradient of a function is important for several reasons, one of which is that it appears often in applications, for example to the concept of energy in Chapter 9, Section 2. Also, we proved in Chapter 5, Section 3, Theorem 3.1 that we can write the directional derivative of a real-valued function f with respect to a unit vector $\mathbf{u}$ in terms of the gradient of f. Thus if f is differentiable, then

1.1
$$\frac{\partial f}{\partial \mathbf{u}}(\mathbf{x}) = \nabla f(\mathbf{x}) \cdot \mathbf{u}.$$

The following theorem is the origin of the mathematical use of the term gradient. The term appears in several other areas, for example road construction, where it refers to the slope of a road.

1.2   Theorem.   Let $\mathbb{R}^n \xrightarrow{f} \mathbb{R}$ be differentiable in an open set D in $\mathbb{R}^n$. Then at each point $\mathbf{x}$ in D for which $\nabla f(\mathbf{x}) \neq 0$, the vector $\nabla f(\mathbf{x})$ points in the direction of maximum increase for f. The number $|\nabla f(\mathbf{x})|$ is the maximum rate of increase of f at $\mathbf{x}$.

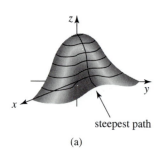

steepest path

(a)

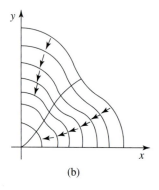

(b)

FIGURE 6.2

Proof.　Recall that the dot product of two vectors equals the product of their lengths multiplied by the cosine of the angle between them. Given a unit vector $\mathbf{u}$, we have, by Equation 1.1 and the assumption that $|\mathbf{u}| = 1$,

$$\frac{\partial f}{\partial \mathbf{u}}(\mathbf{x}) = \nabla f(\mathbf{x}) \cdot \mathbf{u} = |\nabla f(\mathbf{x})||\mathbf{u}| \cos \theta$$

$$= |\nabla f(\mathbf{x})| \cos \theta,$$

where θ is the angle between $\mathbf{u}$ and $\nabla f(\mathbf{x})$. Hence the directional derivative assumes its maximum value when $\cos \theta = 1$ and $\theta = 0$, that is to say when $\mathbf{u}$ has the same direction as $\nabla f(\mathbf{x})$. Thus $\nabla f(\mathbf{x})$ points in the direction of maximum increase of f at the point $\mathbf{x}$, and $|\nabla f(\mathbf{x})|$ is the maximum rate of increase there.　∎

Figure 6.2(a) shows the graph of a function $\mathbb{R}^2 \xrightarrow{f} \mathbb{R}$ along with some level curves and gradient arrows in the domain of f, shown in Figure 6.2(b). The arrows represent unit vectors in the direction of maximum increase at their tails.

Remark.　If the maximum rate of increase of f at $\mathbf{x}_0$ is positive in the direction $\mathbf{u}$, then the maximum rate of *decrease* has the opposite direction, namely the direction $-\mathbf{u} = -\nabla f(\mathbf{x}_0)$, but the same absolute magnitude. The reason is that

$$\frac{\partial f}{\partial(-\mathbf{u})}(\mathbf{x}_0) = \nabla f(\mathbf{x}_0) \cdot (-\mathbf{u}) = -\nabla f(\mathbf{x}_0) \cdot (\mathbf{u}) = -\frac{\partial f}{\partial \mathbf{u}}(\mathbf{x}_0).$$

EXAMPLE 3　Let $f(x, y) = e^{xy}$. Then $\nabla f(x, y) = (ye^{xy}, xe^{xy})$; thus at $(1, 2)$ the function f increases most rapidly in the direction $\nabla f(1, 2) = (2e^2, e^2)$, which has the same direction as the unit vector $(2/\sqrt{5}, 1/\sqrt{5})$. The rate of increase in this direction is $|\nabla f(1, 2)| = \sqrt{5}e^2$. Similarly,

$$\nabla f(-1, 2) = (2e^{-2}, -e^{-2})$$

and has direction $(2/\sqrt{5}, -1/\sqrt{5})$, with maximum rate of increase at $(-1, 2)$ equal to $\sqrt{5}e^{-2}$. The maximum rate of decrease occurs in the opposite direction.

1B　Chain Rule

Next we'll prove a chain rule for differentiating the composition $g(f(t))$, of a function $\mathbb{R} \xrightarrow{f} \mathbb{R}^n$ and a function $\mathbb{R}^n \xrightarrow{g} \mathbb{R}$. For example, if f and g are

$$f(t) = (t, t^2, t) \quad \text{and} \quad g(x, y, z) = x \cos(y + z),$$

then

$$g(f(t)) = t \cos(t^2 + t).$$

In this example the composition defines a new function from $\mathbb{R}$ to $\mathbb{R}$. For example, with $g(x, y, z)$ denoting temperature at a point (x, y, z) in a region D of $\mathbb{R}^3$ and f describing the motion of a point along a path lying in D, we may be interested in finding the rate of change of temperature with respect to t along the path. The theorem gives a formula for doing this in terms of the gradient of g and the vector derivative of f.

1.3   Theorem.   Let g be real-valued and continuously differentiable on an open set D in $\mathbb{R}^n$ and let $f(t)$ be defined and differentiable for $a < t < b$, taking its values in D. Then the composite function $F(t) = g\big(f(t)\big)$ is differentiable for $a < t < b$ and

$$F'(t) = \nabla g\big(f(t)\big) \bullet f'(t).$$

Proof.   By definition,

$$F'(t) = \lim_{h \to 0} \frac{F(t+h) - F(t)}{h}$$

$$= \lim_{h \to 0} \frac{g\big(f(t+h)\big) - g\big(f(t)\big)}{h},$$

if the limit exists. Since f is differentiable, it is continuous. Then we can choose $\delta > 0$ such that, whenever $|h| < \delta$, $f(t+h)$ is always inside an open ball centered at $f(t)$ and contained in D. We now apply the Mean-Value Theorem 3.2 of Section 3, Chapter 5 to g, getting

$$g(\mathbf{y}) - g(\mathbf{x}) = \nabla g(\mathbf{x}_0) \bullet (\mathbf{y} - \mathbf{x}),$$

where $\mathbf{x}_0$ is some point on the segment joining $\mathbf{y}$ and $\mathbf{x}$. Letting $\mathbf{x} = f(t)$ and $\mathbf{y} = f(t + h)$, with $|h| < \delta$, we have

$$\frac{F(t+h) - F(t)}{h} = \nabla g(\mathbf{x}_0) \bullet \frac{f(t+h) - f(t)}{h}.$$

The vector $\mathbf{x}_0$ is now some point on the segment joining $f(t)$ and $f(t + h)$. (Note that $\mathbf{x}_0$ is in the domain D of g because it lies on a radius of a ball contained in D.) Since g was assumed continuously differentiable, $\nabla g(\mathbf{x})$ is continuous, and so $\nabla g(\mathbf{x}_0)$ tends to $\nabla g\big(f(t)\big)$ as h tends to zero. The dot product is continuous, so $F'(t)$ exists, with

$$F'(t) = \lim_{h \to 0} \nabla g(\mathbf{x}_0) \bullet \frac{f(t+h) - f(t)}{h}$$

$$= \nabla g\big(f(t)\big) \bullet f'(t). \qquad \blacksquare$$

Remark.   Theorem 1.3 is true under the weaker assumption that g is differentiable. The previous proof avoids some technical work needed to prove the stronger theorem.

EXAMPLE 4   Let $g(x, y) = x^2 y + xy^3$ for (x, y) in $\mathbb{R}^2$. Let $f(t)$ be differentiable in some neighborhood of $t = t_0$ and take its values in $\mathbb{R}^2$. If it is known only that $f(t_0) = (-1, 1)$ and $f'(t_0) = (2, 3)$, then the composition, $F(t) = g\big(f(t)\big)$, is known only at $t = t_0$, and $F'(t_0)$ cannot be computed by direct differentiation. However, by the previous theorem we have

$$F'(t_0) = \nabla g\big(f(t_0)\big) \bullet f'(t_0).$$

We find that $\nabla g(x, y) = (2xy + y^3, x^2 + 3xy^2)$ so $\nabla g\big(f(t_0)\big) = (-1, -2)$. Then $F'(t_0) = (-1, -2) \cdot (2, 3) = -8$.

We'll extend Theorem 1.3 to vector-valued functions F in Section 2.

1C   Normal Vectors

We'll now see how gradient fields are connected with the level sets of real-valued functions. Level sets were defined and illustrated in Section 2B of Chapter 4. Recall that the level set S at level k of f is the set of points $\mathbf{x}$ in the domain of f such that $f(\mathbf{x}) = k$. For $\mathbb{R}^2 \xrightarrow{f} \mathbb{R}$ we're usually interested in S when S is a curve, and for $\mathbb{R}^3 \xrightarrow{f} \mathbb{R}$ when S is a surface. These two cases are illustrated in Figure 6.3.

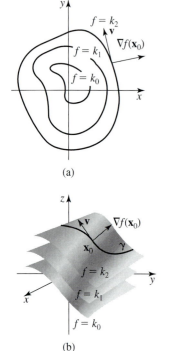

(a)

(b)

FIGURE 6.3

One possible approach to finding a tangent to a level set S of a function f is to first parametrize S and then use the parametrization to determine the tangent as in Sections 1 and 4 of Chapter 4. However we can also proceed more directly as follows. Define a **normal vector** to S at a point $\mathbf{x}_0$ on S to be a vector $\mathbf{n} \neq 0$ that is perpendicular to every smooth curve on S that passes through $\mathbf{x}_0$. If such a vector $\mathbf{n}$ exists, it's then natural to define the **tangent** to S at $\mathbf{x}_0$ to be the unique line or plane that contains $\mathbf{x}_0$ and is perpendicular to $\mathbf{n}$. We show below that under appropriate hypotheses on f we can take $\mathbf{n}$ to be the gradient vector $\nabla f(\mathbf{x}_0)$. These ideas are illustrated in Figure 6.3(a) for level curves and Figure 6.3(b) for level surfaces. The precise statement follows.

1.4   Theorem.   Let $\mathbb{R}^n \xrightarrow{f} \mathbb{R}$, $n \geq 2$, be continuously differentiable at $\mathbf{x}_0$, and let S be a level set of f containing $\mathbf{x}_0$. If $\nabla f(\mathbf{x}_0) \neq \mathbf{0}$, then $\nabla f(\mathbf{x}_0)$ is a normal vector to S at $\mathbf{x}_0$, and all points $\mathbf{x}$ in a tangent line or plane to S at $\mathbf{x}_0$ satisfy the equation

1.5
$$\nabla f(\mathbf{x}_0) \cdot (\mathbf{x} - \mathbf{x}_0) = 0.$$

Proof.   Suppose $g(t)$ parametrizes some smooth curve γ on S and that $g(t_0) = \mathbf{x}_0$. We need to know that $g'(t_0)$ is perpendicular to $\nabla f(\mathbf{x}_0)$, that is $\nabla f(\mathbf{x}_0) \cdot g'(t_0) = 0$. To see this apply the chain rule to the function $h(t) = f\big(g(t)\big)$ to get

$$h'(t_0) = \nabla f\big(g(t_0)\big) \cdot g'(t_0) = \nabla f(\mathbf{x}_0) \cdot g'(t_0).$$

Since γ lies on a level set of f at some level k, the function $h(t)$ is constantly equal to k. Consequently, $h'(t_0) = 0$, so $\nabla f(\mathbf{x}_0)$ is perpendicular to the tangent vector $g'(t_0)$. If $\mathbf{x}$ is a point on a tangent to S at $\mathbf{x}_0$, then the vector $\mathbf{x} - \mathbf{x}_0$ is parallel to the tangent vector $g'(t_0)$. It follows that $\mathbf{x} - \mathbf{x}_0$ also is perpendicular to $\nabla f(\mathbf{x}_0)$, so Equation 1.5 holds. ∎

EXAMPLE 5   The function $f(x, y) = x^3 + y^2$ has a level curve passing through the point $(1, 2)$ and having the equation $x^3 + y^2 = 5$. Then $\nabla f(1, 2) = \big(3(1)^2, 2(2)\big) = (3, 4)$. According to Theorem 1.4 the tangent line to this curve at $(x, y) = (1, 2)$ has equation

$$(3, 4) \cdot (x - 1, y - 2) = 0, \quad \text{or} \quad 3(x - 1) + 2(y - 2) = 0, \quad \text{or} \quad 3x + 2y = 7.$$

EXAMPLE 6   The function $f(x, y, z) = x^2 + y^2 - z^2$ has for one of its level surfaces a cone C consisting of all points satisfying $x^2 + y^2 - z^2 = 0$. The point $\mathbf{x}_0 = (1, 1, \sqrt{2})$ lies

FIGURE 6.4

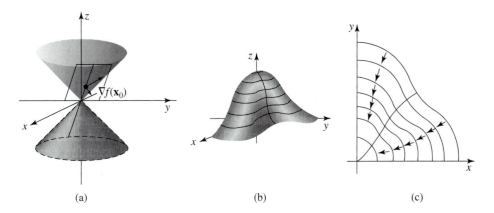

(a)   (b)   (c)

on C, and to find the tangent plane to C at $\mathbf{x}_0$ we compute $\nabla f(\mathbf{x}_0) = (2, 2, -2\sqrt{2})$. Then

$$\nabla f(\mathbf{x}_0) \cdot (\mathbf{x} - \mathbf{x}_0) = (2, 2, -2\sqrt{2}) \cdot (x - 1, y - 1, z - \sqrt{2}),$$

and according to Theorem 1.4, the tangent plane is given by $(x - 1) + (y - 1) - \sqrt{2}(z - \sqrt{2}) = 0$, or $x + y - \sqrt{2}z = 0$. This plane is shown in Figure 6.4(a). Notice that both C and its tangent contain a common line with direction $(1, 1, \sqrt{2})$, and the normal vector to the tangent is perpendicular to that line.

Putting together Theorem 1.2 with 1.4, we get the following theorem.

1.6   Theorem.   The direction of maximum increase of a differentiable function f at $\mathbf{x}_0$ is perpendicular to the level set of f containing $\mathbf{x}_0$, assuming $\nabla f(\mathbf{x}_0) \neq \mathbf{0}$.

Proof.   The reason is that $\nabla f(\mathbf{x}_0)$ is the direction of maximum increase of f at $\mathbf{x}_0$ and at the same time is perpendicular to the level set through $\mathbf{x}_0$ determined by $f(\mathbf{x}) = k$.   ∎

Figure 6.4(b) shows the graph of a differentiable function $f(x, y)$, and Figure 6.4(c) shows some level curves with perpendicular vectors. The curve running from bottom to top on the graph in Figure 6.4(b) has the property that the horizontal component of a tangent vector at $\big(x, y, f(x, y)\big)$ always points in the direction of maximum increase for f, that is, in the direction of $\nabla f(x, y)$. Such a path is called a **path of steepest ascent**, and in Figure 6.4(b) this path appears to lead directly to the point at which the maximum value of f is attained; indeed such paths are sometimes used to locate maxima in practice. See Section 4F.

EXERCISES

In Exercises 1 to 6 find $\nabla f(\mathbf{x})$ for each of the following functions at a general point $\mathbf{x}$ in the domain of f.

1. $f(x, y) = x^2 - y^2$.

2. $f(x, y) = x^2 - y^2 - \sin xy$.

3. $f(x, y) = x + 2y$.

4. $f(x, y, z) = (x - y)z$.

5. $f(x, y, z) = x + y - z^2$.

6. $f(\mathbf{x}) = |\mathbf{x}|^2$, for $\mathbf{x}$ in $\mathbb{R}^n$.

In Exercises 7 to 10 find the gradient of each of the following functions at the indicated point $\mathbf{x}_0$. Then find the unit vector $\mathbf{u}$ that points in the direction of maximum increase of the function at $\mathbf{x}_0$, and also find the rate of maximum increase at $\mathbf{x}$.

7. $f(x, y) = x^2 - y^3$ at $(x_0, y_0) = (1, 1)$

8. $g(x, y) = xy^2$ at $(x_0, y_0) = (-1, 2)$

9. $h(x, y, z) = xy \sin z$ at $(x_0, y_0, z_0) = (1, 2, \pi)$

10. $p(x, y, z, w) = (x^2+y^2+z^2+w^2)^{1/2}$ at $(x_0, y_0, z_0, w_0) = (1, 1, 1, 2)$

In Exercises 11 to 14, sketch the vector fields described by the functions from $\mathbb{R}^2$ to $\mathbb{R}^2$ or $\mathbb{R}^3$ to $\mathbb{R}^3$. To do this pick a few points $\mathbf{x}$ in the indicated domain and draw the arrow for $\mathbf{F}(\mathbf{x})$ with its tail located at the point $\mathbf{x}$.

11. $\mathbf{F}(x, y) = (1, x)$ for $-1 \leq x \leq 2, 0 \leq y \leq 2$

12. $\mathbf{F}(x, y) = (-y, x)$ for $x^2 + y^2 \leq 4$

13. $\mathbf{F}(x, y) = (y, x)$ for $x^2 + y^2 \leq 4$

14. $\mathbf{F}(x, y, z) = -\frac{1}{2}(x, y, z)$ for $x^2 + y^2 + z^2 \leq 4$

In Exercises 15 to 17, first compute ∇f, and then sketch the vector field $\mathbf{F} = \nabla f$.

15. $f(x, y) = xy + y^2$

16. $f(x, y, z) = x^2 + y^2 + z^2$

17. $f(x, y, z) = x^2 + y^2$

18. $f(x, y, z) = x - y + z$

In Exercises 19 to 24, find, if possible, a normal vector and the tangent line or plane to each of the following level curves or surfaces at the indicated points.

19. $x^2 + y^2 - z^2 = 2$ at $(x, y, z) = (1, 1, 0)$

20. $x \sin y = 0$ at $(x, y) = (0, \pi/2)$ and at $(x, y) = (0, 0)$

21. $|\mathbf{x}| = 1$ at $\mathbf{x} = \mathbf{e}_1$, the first natural basis vector in $\mathbb{R}^n$

22. $x^2y + yz + w = 3$ at $(x, y, z, w) = (1, 1, 1, 1)$

23. $xyz = 1$ at $(x, y, z) = (1, 1, 1)$

24. $xyz = 0$ at $(x, y, z) = (1, 2, 0)$

25. If $\mathbb{R}^2 \xrightarrow{f} \mathbb{R}$ is continuously differentiable, its graph is defined implicitly in $\mathbb{R}^3$ as the level surface S of the function $F(x, y, z) = z - f(x, y)$ given by $F(x, y, z) = 0$.
   (a) Show that $\nabla F = (-\partial f/\partial x, -\partial f/\partial y, 1)$, which is never the zero vector.
   (b) Find a normal vector and the tangent plane to the graph of $f(x, y) = xy + ye^x$ at $(x, y) = (1, 1)$.

26. (a) The function $f(x, y) = x^2 + y^2$ has $\nabla f(0, 0) = (0, 0)$, which fails to indicate that there is a direction of maximum increase for f at $(x, y) = (0, 0)$. Is this reasonable? What happens at $(0, 0)$?

(b) Are there directions of maximum increase for $f(x, y) = xy$ and $g(x, y) = x^2 - y^2$ at $(x, y) = (0, 0)$? Does Theorem 1.2 apply?

27. If $g(x, y) = e^{x+y}$ and $f'(0) = (1, 2)$, use the chain rule to find $F'(0)$, where $F(t) = g(f(t))$ and $f(0) = (1, -1)$.

28. Let γ be a curve in $\mathbb{R}^3$ being traversed at time $t = 1$ with speed 2 and in the direction of $(1, -1, 2)$. If $t = 1$ corresponds to the point $(1, 1, 1)$ on γ, find the rate of change of the function $x + y + z + xyz$ along γ at $t = 1$.

29. If $f(x, y, z) = \sin x$ and $F(t) = (\cos t, \sin t, t)$, find $g'(\pi)$, where $g(t) = f(F(t))$.

30. Let $\mathbb{R} \xrightarrow{F} \mathbb{R}^n$ be differentiable. Let $\mathbb{R}^n \xrightarrow{f} \mathbb{R}$ be continuously differentiable, and such that the composition $g(t) = f(F(t))$ exists. If $F'(t_0)$ is tangent to the level surface of f at $F(t_0)$, show that $g'(t_0) = 0$.

31. A spaceship is traveling in $\mathbb{R}^2$ along a path such that at time $t \geq 0$ the ship is at $g(t) = (3t^2, t^3)$. The intensity of gamma radiation at the point (x, y) in $\mathbb{R}^2$ is $I(x, y) = x^2 - y^2$ wherever $I(x, y) \geq 0$. Describe fully, using a labeled sketch where appropriate, the following:
   (a) The level curve of I that the ship is on at $t = 1$.
   (b) The path of the ship for $t \geq 0$.
   (c) The gradient vector of I at the ship's position when $t = 1$.
   (d) The ship's velocity vector at $t = 1$.
   (e) The time, if there is one, when the ship stops increasing its radiation risk and begins its race to safety. Does its course become more dangerous later on?

32. If $T(x, y, z)$ represents the temperature at a point (x, y, z) of a region R in $\mathbb{R}^3$, the vector field ∇T is called the **temperature gradient**. Under certain physical assumptions $\nabla T(x, y, z)$ is negatively proportional to the vector that represents the direction and rate per unit of area of heat flow at (x, y, z). The sets on which T is constant are called isotherms. If the isotherms of a temperature function are concentric spheres, prove that the temperature gradient points either toward or away from the center of the spheres.

33. Show that the vector field defined on $\mathbb{R}^2$ by $\mathbf{F}(x, y) = (-y, x)$ is not of the form $\nabla f(x, y)$ for a function f. [*Hint*: Suppose that $\partial f/\partial x$ $(x, y) = -y$ and $\partial f/\partial y$ $(x, y) = x$. Then differentiate the first equation with respect to y and the second with respect to x.]

For each of the following fields 33 to 36, find an f such that $\nabla f = \mathbf{F}$. The previous exercise established, by specific example, that some vectors fields are not gradient fields. The distinction between gradient fields and nongradient fields is investigated in some detail in Section 2 of Chapter 9. But when a vector field $\mathbf{F}$ is a gradient field, a little guesswork based on experience with indefinite integrals will sometimes yield a real-valued function such that $\nabla f = \mathbf{F}$. For example, if $\mathbf{F}(x, y) = (x, y)$, a little thought leads to the guess that $\nabla(x^2/2 + y^2/2) = \mathbf{F}(x, y)$. It follows from Theorem 3.4 of the previous chapter that every two solutions to the problem differ by at most an additive constant.

34. $\mathbf{F}(x, y) = (y, x)$

35. $\mathbf{F}(x, y, z) = (x, y, z)$

36. $\mathbf{F}(x, y) = (e^{x+y}, e^{x+y})$

37. $\mathbf{F}(x, y) = (x^2, y^2)$

38. The level surfaces of a function $\mathbb{R}^n \xrightarrow{f} \mathbb{R}$ are called the **equipotential surfaces** of the vector field ∇f, and f is called the **potential function** of the field.

   (a) Show that the equipotential surfaces are perpendicular to the field.

   (b) Find the equipotential surfaces of the field $\nabla f(x, y, z) = (x, y, z)$.

(c) Find the field of which $f(x, y, z) = (x^2 + y^2 + z^2)^{-1/2}$, the **Newtonian potential**, is the potential function.

(d) Find the field of which $f(x, y) = -\frac{1}{2} \log(x^2 + y^2)$, the **logarithmic potential**, is the potential function.

(e) Show that the generalized Newtonian potential $f(\mathbf{x}) = |\mathbf{x}|^{2-n}$ in $\mathbb{R}^n$, $n \geq 3$, satisfies $\nabla f(\mathbf{x}) = (2 - n)|\mathbf{x}|^{-n}\mathbf{x}$.

39. The vector equation of motion for the position $\mathbf{x}(t)$ at time t of a single planet relative to a star fixed at the origin has the form $\ddot{\mathbf{x}} = -k|\mathbf{x}|^{-3}\mathbf{x}$, where k is a positive constant depending on the gravitational constant and the masses of the two bodies. See Equations 3.2 in Chapter 12, Section 3.

   (a) Show that the magnitude of the acceleration vector obeys an **inverse-square law:** $|\ddot{\mathbf{x}}| = k/|\mathbf{x}|^2$.

   (b) Show that the vector equation is equivalent to a pair of equations, where $\mathbf{x} = (x, y)$:

   $$\ddot{x} = -kx(x^2 + y^2)^{-3/2}, \quad \ddot{y} = -ky(x^2 + y^2)^{-3/2}.$$

   (c) Show that the vector field $\mathbf{F}(x, y) = \left(-kx(x^2 + y^2)^{-3/2}, -ky(x^2 + y^2)^{-3/2}\right)$ is equal to $\nabla f(x, y)$, where $f(x, y) = k(x^2 + y^2)^{-1/2}$.

1D   Plotting Vector Fields; Flow Lines

As described in Subsection 1A, a vector field is often pictured by drawing an arrow representing $\nabla f(\mathbf{x})$, but shifted to a parallel arrow with its tail at $\mathbf{x}$ instead of at the origin. When plotting by computer the initial points for the arrows are located at the points of a rectangular lattice spaced s units apart. A good routine puts arrow points at the tips of the arrows and little dots at the initial points. The basic routine is as follows for $(f(x + y), g(x + y)) = \left(\frac{1}{10}(x + y), \frac{1}{10}(x - y)\right)$ in a rectangle $x_1 < x < x_2, y_1 < y < y_2$.

FIGURE 6.5

```
DEFINE f(x, y) = 1/10 (x + y)
DEFINE g(x, y) = 1/10 (x - y)
FOR x = x1 TO x2 STEP s
    FOR y = y1 TO y2 STEP s
       PLOT ARROW:
          (x, y) TO (x + f(x, y), y + g(x, y))
       NEXT y
NEXT x
```

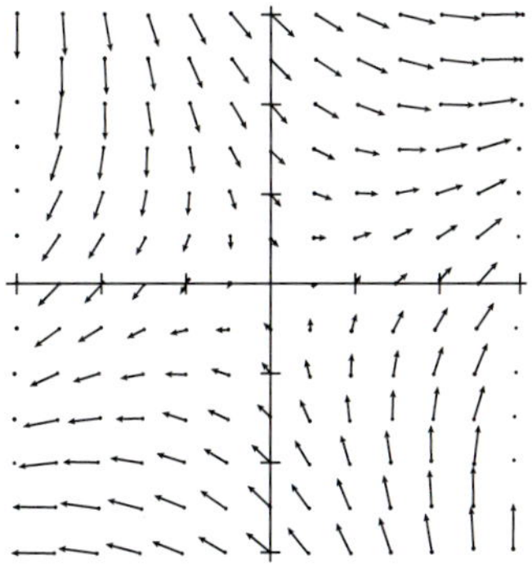

Using such a routine produces Figure 6.5 for $-3 < x < 3, -3 < y < 3$. Note particularly the factor $\frac{1}{10}$ in the vector field $\left(\frac{1}{10}(x + y), \frac{1}{10}(x - y)\right)$ sketched in Figure 6.5. Without that factor the arrows in the graphic representation would be 10 times as long, and the resulting overlap of arrows would lead to a confusing picture. (Try it, leaving out the scale factor!) Thus we often scale the vector lengths in a field, either down or up, to get a better picture, bearing in mind what specific scaling has taken place. What we're often interested in anyway is the relative strength of a field as it varies from point to point, and that information comes across very well in a properly scaled sketch.

Recall that a *gradient field* is a vector field having the form $\mathbf{F}(x, y) = \nabla f(x, y)$, where the real-valued function f is called a **potential function** of $\mathbf{F}$. In $\mathbb{R}^2$, for example,

$$\nabla f(x, y) = \left(\frac{\partial f(x, y)}{\partial x}, \frac{\partial f(x, y)}{\partial y} \right).$$

Considerable insight is needed to distinguish a gradient field from one that isn't a gradient just by looking at sketches such as Figure 6.6. For example, the one in Figure 6.6(a) is a scaled version of the gradient field $\nabla xy = (y, x)$ while the one in Figure 6.6(b) is a scaled version of $\mathbf{F}(x, y) = (y, -x)$, which is not a gradient field. We'll see later in Chapters 8 and 9 that the "rotational" character of the field on the right is a visual clue that it isn't a gradient field. Theorem 2.5 in Chapter 9, Section 2 provides a computational criterion.

Often a potential function f has an **infinite singularity**, that is, a point of its domain where f tends to infinity. At such a point ∇f will not only fail to be defined but at nearby points will have gradient vectors that become arbitrarily long. If such a singularity occurs in the course of making a sketch, we need to make allowances for it in our algorithm. The simplest thing to do is arrange it so that the plotting steps give the troublesome point a wide berth.

In principle, we can make a perspective sketch of a vector field in $\mathbb{R}^3$. The trouble is that the tendency of the arrows to overlap each other often makes the picture hard to interpret.

Flow Lines.   Suppose a continuously differentiable vector field $\mathbf{F}$ is defined on some open subset S of $\mathbb{R}^n$. The image of a parametrized curve $\mathbf{x} = g(t)$, with image in S, is called a **flow line** of $\mathbf{F}$ if the velocity vector of the curve at a point $\mathbf{x} = g(t)$

FIGURE 6.6

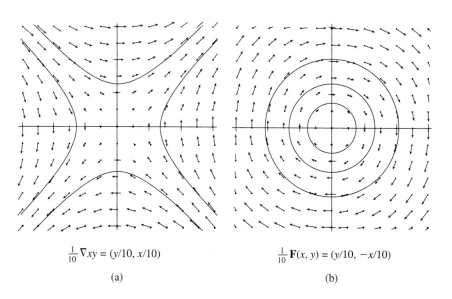

$$\tfrac{1}{10} \nabla xy = (y/10, x/10)$$

(a)

$$\tfrac{1}{10} \mathbf{F}(x, y) = (y/10, -x/10)$$

(b)

in S coincides with the vector $\mathbf{F}(\mathbf{x})$, that is,

$$g'(t) = \mathbf{F}(g(t)).$$

If $\mathbf{F}$ is a velocity field, its flow lines represent the paths of particles moving with velocities given by $\mathbf{F}$. There is a detailed discussion of this relationship between vector fields and curves in Chapter 12 on systems of differential equations. For now we simply observe that given a reasonably accurate sketch of a vector field, we can sketch in some typical flow lines, as in Figure 6.6. The idea is to draw curves that appear to be tangent to the arrows of the field. The speed and direction of traversal of the curve are determined by the length and direction of the arrows of the field, though if the field arrows are scaled as previously described, then only direction and relative speed are apparent directly from the sketch.

EXERCISES

Plot the vector fields 1 to 6 in $\mathbb{R}^2$. Use scaling of the vector lengths as it seems appropriate. Add some sketches of typical flow lines to each of the vector field sketches.

1. $\mathbf{F}(x, y) = (x, y), -2 \le x \le 2, -2 \le y \le 2$

2. $\mathbf{F}(x, y) = (x, 0), -4 \le x \le 4, -4 \le y \le 4$

3. $\mathbf{F}(x, y) = (2x, x), -4 \le x \le 4, -4 \le y \le 4$

4. $\mathbf{F}(x, y) = \nabla(x^2 y^2), -2 \le x \le 2, -2 \le y \le 2$

5. $\mathbf{F}(x, y) = \nabla e^{x+y}, -4 \le x \le 4, -4 \le y \le 4$

6. $\mathbf{F}(x, y) = \nabla \frac{1}{2} \log(x^2 + y^2), -4 \le x \le 4,$
   $-4 \le y \le 4, (x, y) \ne (0, 0)$

7. (a)   Verify that if a and b are constants, not both zero, then the image of the curve parametrized by

   $$(x, y) = (a \cos t + b \sin t, b \cos t - a \sin t)$$

   is a flow line of the vector field $\mathbf{F}(x, y) = (y, -x)$.

   (b)   Show that the flow lines of part (a) are circles.

8. (a)   Verify that if a and b are constants, not both zero, then the image of the curve parametrized by

   $$(x, y) = (a \cosh t + b \sinh t, b \cosh t + a \sinh t)$$

   is a flow line of vector field $\mathbf{F}(x, y) = (y, x)$.

   (b)   Show that the flow lines of part (a) are either hyperbolas or lines.

SECTION 2   THE CHAIN RULE

One of the most useful one-variable calculus formulas is the chain rule, used to compute the derivative of the composition of one function with another:

$$\frac{dg}{dx}(f(x)) = g'(f(x)) f'(x), \quad \text{or} \quad \frac{dz}{dx} = \frac{dz}{dy} \frac{dy}{dx},$$

in the compressed Leibniz notation with $z = g(y)$ and $y = f(x)$. The generalization of both formulas to several variables is just as valuable and, properly formulated, is just as easy to state. We proved the special case where g is real-valued and f is a function of one real variable t in Theorem 1.3 of Section 1.

   If two functions f and g are related so that the range space of f is the same as the domain space of g, we may form the **composite function** $g \circ f$ by first applying f and then g. Thus we define

$$g \circ f(\mathbf{x}) = g(f(\mathbf{x}))$$

FIGURE 6.7

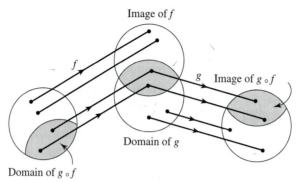

Image of f

f

g Image of $g \circ f$

Domain of g

Domain of $g \circ f$

The image of f and the domain
of g must overlap for $g \circ f$ to be defined

for every vector $\mathbf{x}$ such that $\mathbf{x}$ is in the domain of f and $f(\mathbf{x})$ is in the domain of g. The domain of $g \circ f$ consists of those vectors $\mathbf{x}$ that are carried by f into the domain of g. An abstract picture of the composition of two functions is shown in Figure 6.7.

EXAMPLE 1   Suppose that we are given a two-dimensional region in which the points move about according to some specified law. It may be known that, for a given position with coordinates (u, v), a point is always to be found at some definite later time in a position (x, y). Then (x, y) and (u, v) are related by equations of the form

$$x = g_1(u, v)$$
$$y = g_2(u, v).$$

In vector notation these equations might be written

$$\mathbf{x} = g(\mathbf{u}),$$

where $\mathbf{x} = (x, y)$, $\mathbf{u} = (u, v)$, and g has coordinate functions g_1, g_2. Now suppose that the position $\mathbf{u} = (u, v)$ of a point is itself determined as a function of other variables (s, t) by equations

$$u = f_1(s, t)$$
$$v = f_2(s, t).$$

These may be written in vector form as

$$\mathbf{u} = f(\mathbf{s}),$$

where $\mathbf{s} = (s, t)$ and f has coordinate functions f_1, f_2. Then (x, y) and (s, t) are related by

$$x = g_1\big(f_1(s, t), f_2(s, t)\big)$$
$$y = g_2\big(f_1(s, t), f_2(s, t)\big),$$

or

$$\mathbf{x} = g\big(f(\mathbf{s})\big).$$

2A   General Formula and Examples

To see what the derivative $(g \circ f)'$ should be in terms of the derivatives g' and f', suppose that we are given

$$\mathbb{R}^n \xrightarrow{f} \mathbb{R}^m \xrightarrow{g} \mathbb{R}^p,$$

and suppose first that f and g are the linear functions

$$f(\mathbf{x}) = A\mathbf{x}, \quad g(\mathbf{y}) = B\mathbf{y},$$

where A and B are constant matrices having respective shapes m by n and p by m. Then by associativity of matrix multiplication,

$$g \circ f(\mathbf{x}) = B(A\mathbf{x}) = BA\mathbf{x}.$$

But it follows from Corollary 4.3 of Chapter 5 that for a function $f(\mathbf{x}) = A\mathbf{x}$ or $g(\mathbf{x}) = B\mathbf{x}$ generated by multiplying a vector by a constant matrix, the derivative matrix is just the constant matrix. Thus

$$g' = B, \qquad f' = A \quad \text{and} \quad (g \circ f)' = BA.$$

Hence for functions f and g defined by matrix-vector products we have

2.1
$$(g \circ f)' = g'f'.$$

It's a remarkable fact that for differentiable f and g the previous formula is the correct extension of the chain rule if we take care to evaluate the derivatives at the proper points, as in the next example.

EXAMPLE 2   Consider the special case in which f is a function of a single real variable and g is real-valued. Then $g \circ f$ is a real function of a real variable. Theorem 1.3 shows that if f and g are continuously differentiable, then

$$(g \circ f)'(t) = \nabla g\big(f(t)\big) \bullet f'(t).$$

That is, in terms of coordinate functions,

$$(g \circ f)'(t) = \left(\frac{\partial g}{\partial y_1}\big(f(t)\big), \dots, \frac{\partial g}{\partial y_m}\big(f(t)\big) \right) \bullet \big(f_1'(t), \dots, f_m'(t)\big).$$

The right side of this last equation equals a matrix product in terms of derivative matrices

$$g'\big(f(t)\big) = \left(\frac{\partial g}{\partial y_1}\big(f(t)\big), \dots, \frac{\partial g}{\partial y_m}\big(f(t)\big) \right),$$

and

$$f'(t) = \begin{pmatrix} f_1'(t) \\ \vdots \\ f_m'(t) \end{pmatrix}$$

as $(g \circ f)'(t) = g'(f(t))f'(t)$. The product of $g'(f(t))$ and $f'(t)$ is defined by multiplication of matrices of size 1-by-m and m-by-1 and is equivalent to the dot product of the two matrices looked at as vectors in $\mathbb{R}^m$. Thus for the case where the domain of f and the range of g are both 1-dimensional, we can regard the formulas

$$\nabla g(f(t)) \cdot f'(t) \quad \text{and} \quad g'(f(t))f'(t)$$

as different notations for the same thing.

The chain rule is valid under the assumption that f and g are differentiable, but for a proof requiring less detailed analysis, we make the stronger assumption of continuous differentiability.

2.2 Chain Rule. Let f be continuously differentiable near $\mathbf{x}$, and let g be continuously differentiable near $f(\mathbf{x})$, with

$$\mathbb{R}^n \xrightarrow{f} \mathbb{R}^m \xrightarrow{g} \mathbb{R}^p.$$

If $g \circ f$ is defined on an open set containing $\mathbf{x}$, then $g \circ f$ is continuously differentiable at $\mathbf{x}$, and

$$(g \circ f)'(\mathbf{x}) = g'(f(\mathbf{x}))f'(\mathbf{x}).$$

Proof. We need only show that the derivative matrix of $g \circ f$ at $\mathbf{x}$ has continuous entries given by the entries in the product of $g'(f(\mathbf{x}))$ and $f'(\mathbf{x})$. These matrices have the respective forms

$$\begin{pmatrix} \frac{\partial g_1}{\partial y_1}(f(\mathbf{x})) & \cdots & \frac{\partial g_1}{\partial y_m}(f(\mathbf{x})) \\ \vdots & & \vdots \\ \frac{\partial g_p}{\partial y_1}(f(\mathbf{x})) & \cdots & \frac{\partial g_p}{\partial y_m}(f(\mathbf{x})) \end{pmatrix} \quad \text{and} \quad \begin{pmatrix} \frac{\partial f_1}{\partial x_1}(\mathbf{x}) & \cdots & \frac{\partial f_1}{\partial x_n}(\mathbf{x}) \\ \vdots & & \vdots \\ \frac{\partial f_m}{\partial x_1}(\mathbf{x}) & \cdots & \frac{\partial f_m}{\partial x_n}(\mathbf{x}) \end{pmatrix}.$$

The product of the matrices has for its ijth entry the sum of products

$$\sum_{k=1}^m \frac{\partial g_i}{\partial y_k}(f(\mathbf{x})) \frac{\partial f_k}{\partial x_j}(\mathbf{x}). \tag{1}$$

But this expression is just the dot product of two vectors $\nabla g_i(f(\mathbf{x}))$ and $(\partial f/\partial x_j)(\mathbf{x})$. It follows from Theorem 1.3 that

$$\nabla g_i(f(\mathbf{x})) \cdot \frac{\partial f}{\partial x_j}(\mathbf{x}) = \frac{\partial(g_i \circ f)}{\partial x_j}(\mathbf{x}), \tag{2}$$

because we are differentiating with respect to the single variable x_j. This establishes the matrix relation, because the entries in $(g \circ f)'(\mathbf{x})$ are by definition given by the right side of Equation 2. Since g and f are continuously differentiable, Formula (1) represents a continuous function of x for each i and j. Hence $g \circ f$ is continuously differentiable. ∎

EXAMPLE 3   Let

$$f(x, y) = (x^2 + y^2, x^2 - y^2)$$

and let

$$g(u, v) = (uv, u + v).$$

We find

$$g'(u, v) = \begin{pmatrix} v & u \\ 1 & 1 \end{pmatrix} \quad \text{and} \quad f'(x, y) = \begin{pmatrix} 2x & 2y \\ 2x & -2y \end{pmatrix}.$$

To find $(g \circ f)'(2, 1)$, we note that $f(2, 1) = (5, 3)$ and compute

$$g'(5, 3) = \begin{pmatrix} 3 & 5 \\ 1 & 1 \end{pmatrix} \quad \text{and} \quad f'(2, 1) = \begin{pmatrix} 4 & 2 \\ 4 & -2 \end{pmatrix}.$$

Then the product of these last two matrices gives

$$(g \circ f)'(2, 1) = \begin{pmatrix} 32 & -4 \\ 8 & 0 \end{pmatrix}.$$

EXAMPLE 4   It's common practice in calculus to denote a function by the same symbol as a typical element of its range. Thus the derivative of a function $\mathbb{R} \xrightarrow{f} \mathbb{R}$ is often denoted, in conjunction with the equation $y = f(x)$, by dy/dx. Similarly, the partial derivatives of a function $\mathbb{R}^3 \xrightarrow{f} \mathbb{R}$ are commonly written as

$$\frac{\partial w}{\partial x}, \frac{\partial w}{\partial y}, \quad \text{and} \quad \frac{\partial w}{\partial z},$$

along with the explanatory equation $w = f(x, y, z)$. For example, if $w = f(x, y, z) = xy^2 e^{x+3z}$, then

$$\frac{\partial w}{\partial x} = y^2 e^{x+3z} + xy^2 e^{x+3z},$$

$$\frac{\partial w}{\partial y} = 2xy e^{x+3z},$$

$$\frac{\partial w}{\partial z} = 3xy^2 e^{x+3z}.$$

This notation has the disadvantage that it doesn't contain specific reference to the function being differentiated, but it's convenient and is the traditional language of calculus. To illustrate its convenience, suppose that the functions g and f are given by real-valued coordinate functions

$$w = g(x, y, z), \quad x = f_1(s, t), \quad y = f_2(s, t), \quad z = f_3(s, t).$$

Then by the chain rule,

$$\left(\begin{array}{cc} \dfrac{\partial w}{\partial s} & \dfrac{\partial w}{\partial t} \end{array} \right) = \left(\begin{array}{ccc} \dfrac{\partial g}{\partial x} & \dfrac{\partial g}{\partial y} & \dfrac{\partial g}{\partial z} \end{array} \right) \left(\begin{array}{cc} \dfrac{\partial x}{\partial s} & \dfrac{\partial x}{\partial t} \\[2mm] \dfrac{\partial y}{\partial s} & \dfrac{\partial y}{\partial t} \\[2mm] \dfrac{\partial z}{\partial s} & \dfrac{\partial z}{\partial t} \end{array} \right).$$

Matrix multiplication yields

$$\left. \begin{array}{l} \dfrac{\partial w}{\partial s} = \dfrac{\partial g}{\partial x}\dfrac{\partial x}{\partial s} + \dfrac{\partial g}{\partial y}\dfrac{\partial y}{\partial s} + \dfrac{\partial g}{\partial z}\dfrac{\partial z}{\partial s} \\[3mm] \dfrac{\partial w}{\partial t} = \dfrac{\partial g}{\partial x}\dfrac{\partial x}{\partial t} + \dfrac{\partial g}{\partial y}\dfrac{\partial y}{\partial t} + \dfrac{\partial g}{\partial z}\dfrac{\partial z}{\partial t} \end{array} \right\}. \tag{A}$$

EXAMPLE 5   We get a slightly different-looking application of the chain rule if the domain space of f is one-dimensional, that is, if f is a function of one variable. Consider, for example,

$$w = g(u, v), \quad \left(\begin{array}{c} u \\ v \end{array} \right) = f(t) = \left(\begin{array}{c} f_1(t) \\ f_2(t) \end{array} \right).$$

The composition of $g \circ f$ is in this case a real-valued function of one variable. Its derivative is the 1-by-1 matrix whose entry is the ordinary derivative

$$\frac{d(g \circ f)}{dt} = \frac{dw}{dt}.$$

The derivatives of g and f are defined, respectively, by the derivative matrices

$$\left(\begin{array}{cc} \dfrac{\partial w}{\partial u} & \dfrac{\partial w}{\partial v} \end{array} \right) \quad \text{and} \quad \left(\begin{array}{c} \dfrac{du}{dt} \\[2mm] \dfrac{dv}{dt} \end{array} \right).$$

Hence the chain rule implies that

$$\frac{dw}{dt} = \left(\begin{array}{cc} \dfrac{\partial w}{\partial u} & \dfrac{\partial w}{\partial v} \end{array} \right) \left(\begin{array}{c} \dfrac{du}{dt} \\[2mm] \dfrac{dv}{dt} \end{array} \right) \quad \text{or} \quad \frac{dw}{dt} = \frac{\partial w}{\partial u}\frac{du}{dt} + \frac{\partial w}{\partial v}\frac{dv}{dt}. \tag{B}$$

This is the case treated in Section 1, using the gradient, where we would have written

$$\frac{dw}{dt} = \nabla g \cdot f'.$$

EXAMPLE 6   Let us suppose that both f and g are real-valued functions of one variable, the situation we meet in one-variable calculus. The derivatives of f at t, of g at $s = f(t)$, and of $g \circ f$ at t are represented by the three 1-by-1 derivative matrices $f'(t)$, $g'(s)$, and $(g \circ f)'(t)$, respectively. The chain rule implies that

$$(g \circ f)'(t) = g'(s) f'(t).$$

If the functions are presented in the form

$$x = g(s), s = f(t),$$

the chain rule appears as the familiar equation

$$\frac{dx}{dt} = \frac{dx}{ds} \frac{ds}{dt}. \tag{C}$$

EXAMPLE 7   Given that

$$\begin{cases} x = u^2 + v^3 \\ y = e^{uv}, \end{cases} \quad \text{and} \quad \begin{cases} u = t + 1 \\ v = e^t, \end{cases}$$

find dx/dt and dy/dt at $t = 0$. Let $\mathbb{R} \xrightarrow{f} \mathbb{R}^2$ and $\mathbb{R}^2 \xrightarrow{g} \mathbb{R}^2$ be the functions defined by

$$f(t) = \begin{pmatrix} t+1 \\ e^t \end{pmatrix} = \begin{pmatrix} u \\ v \end{pmatrix}, \quad -\infty < t < \infty,$$

$$g \begin{pmatrix} u \\ v \end{pmatrix} = \begin{pmatrix} u^2 + v^3 \\ e^{uv} \end{pmatrix} = \begin{pmatrix} x \\ y \end{pmatrix}, \quad \begin{cases} -\infty < u < \infty. \\ -\infty < v < \infty. \end{cases}$$

The derivative $f'(t)$ is defined by the 2-by-1 derivative matrix

$$\begin{pmatrix} \dfrac{du}{dt} \\ \dfrac{dv}{dt} \end{pmatrix} = \begin{pmatrix} 1 \\ e^t \end{pmatrix}.$$

The derivative $g'(u, v)$ is

$$\begin{pmatrix} \dfrac{\partial x}{\partial u} & \dfrac{\partial x}{\partial v} \\ \dfrac{\partial y}{\partial u} & \dfrac{\partial y}{\partial v} \end{pmatrix} = \begin{pmatrix} 2u & 3v^2 \\ ve^{uv} & ue^{uv} \end{pmatrix}.$$

The dependence of x and y on t is given by

$$\begin{pmatrix} x \\ y \end{pmatrix} = (g \circ f)(t), \quad -\infty < t < \infty.$$

Hence the two derivatives dx/dt and dy/dt are the entries in the derivative matrix of the composite function $g \circ f$. The chain rule therefore implies that

$$\begin{pmatrix} \dfrac{dx}{dt} \\ \dfrac{dy}{dt} \end{pmatrix} = \begin{pmatrix} \dfrac{\partial x}{\partial u} & \dfrac{\partial x}{\partial v} \\ \dfrac{\partial y}{\partial u} & \dfrac{\partial y}{\partial v} \end{pmatrix} \begin{pmatrix} \dfrac{du}{dt} \\ \dfrac{dv}{dt} \end{pmatrix}.$$

That is,

$$
\left.
\begin{array}{l}
\dfrac{dx}{dt} = \dfrac{\partial x}{\partial u}\dfrac{du}{dt} + \dfrac{\partial x}{\partial v}\dfrac{dv}{dt} \\[2mm]
\dfrac{dy}{dt} = \dfrac{\partial y}{\partial u}\dfrac{du}{dt} + \dfrac{\partial y}{\partial v}\dfrac{dv}{dt}
\end{array}
\right\}.
\qquad \text{(D)}
$$

Substitution of the specific entries in $f'(t)$ and $g'(u, v)$ gives

$$
\left.
\begin{array}{l}
\dfrac{dx}{dt} = 2u + 3v^2 e^t \\[2mm]
\dfrac{dy}{dt} = v e^{uv} + u e^{uv+t}
\end{array}
\right\}.
\qquad \text{(*)}
$$

If $t = 0$, then $\begin{pmatrix} u \\ v \end{pmatrix} = f(0) = \begin{pmatrix} 1 \\ 1 \end{pmatrix}$, and we get $u = v = 1$. It follows that

$$
\frac{dx}{dt}(0) = 2 + 3 = 5,
$$

$$
\frac{dy}{dt}(0) = e + e = 2e.
$$

The definition of matrix multiplication gives the derivative formulas resulting from applications of the chain rule a formal pattern that helps the memory. The pattern is particularly evident when the coordinate functions are denoted by scalar variables, as in Formulas (A), (B), (C), and (D). All formulas of the general form

$$
\cdots + \frac{\partial z}{\partial x}\frac{\partial x}{\partial t} + \frac{\partial z}{\partial y}\frac{\partial y}{\partial t} + \cdots
$$

have the disadvantage of not containing explicit reference to the points at which the various derivatives are evaluated. It's essential to know this information, and we can find it by going to the formula

$$
(g \circ f)'(\mathbf{x}) = g'\big(f(\mathbf{x})\big) f'(\mathbf{x}).
$$

It follows that derivatives appearing in the matrix $f'(\mathbf{x})$ are evaluated at $\mathbf{x}$, and those in the matrix $g'\big(f(\mathbf{x})\big)$ are evaluated at $f(\mathbf{x})$. This is the reason for setting $t = 0$ and $u = v = 1$ in Formula (*) to obtain the final answers in Example 7.

EXAMPLE 8   Let

$$
z = xy \quad \text{and} \quad \begin{cases} x = f(u, v). \\ y = g(u, v). \end{cases}
$$

Suppose that when $u = 1$ and $v = 2$, we have

$$
\frac{\partial x}{\partial u} = -1, \quad \frac{\partial x}{\partial v} = 3, \quad \frac{\partial y}{\partial u} = 5, \quad \frac{\partial y}{\partial v} = 0.
$$

Suppose also that $f(1, 2) = 2$ and $g(1, 2) = -2$. What is $\partial z/\partial u(1, 2)$? The chain rule implies that

$$
\frac{\partial z}{\partial u} = \frac{\partial z}{\partial x}\frac{\partial x}{\partial u} + \frac{\partial z}{\partial y}\frac{\partial y}{\partial u}.
\qquad \text{(E)}
$$

When $u = 1$ and $v = 2$, we are given that

$$x = f(1, 2) = 2 \quad \text{and} \quad y = g(1, 2) = -2.$$

Hence

$$\frac{\partial z}{\partial x}(2, -2) = y \bigg|_{x=2, y=-2} = -2$$

$$\frac{\partial z}{\partial y}(2, -2) = x \bigg|_{x=2, y=-2} = 2.$$

To obtain $\partial z/\partial u$ at $(u, v) = (1, 2)$, it's necessary to know at what points to evaluate the partial derivatives that appear in Equation (E). In greater detail, the chain rule implies that

$$\frac{\partial z}{\partial u}(1, 2) = \frac{\partial z}{\partial x}(2, -2)\frac{\partial x}{\partial u}(1, 2) + \frac{\partial z}{\partial y}(2, -2)\frac{\partial y}{\partial u}(1, 2).$$

Hence

$$\frac{\partial z}{\partial u}(1, 2) = (-2)(-1) + (2)(5) = 12.$$

EXAMPLE 9   If $w = f(ax^2 + bxy + cy^2)$ and $y = x^2 + x + 1$, we may want dw/dx at $x = 1$. The solution relies on formulas that follow from the chain rule such as (A), (B), (C), (D), and (E). Let z be defined by

$$z = ax^2 + bxy + cy^2.$$

Then $w = f(z)$, and since $\partial x/\partial x = 1$,

$$\frac{dz}{dx} = \frac{\partial z}{\partial x}\frac{\partial x}{\partial x} + \frac{\partial z}{\partial y}\frac{dy}{dx} = \frac{\partial z}{\partial x} + \frac{\partial z}{\partial y}\frac{dy}{dx}.$$

Hence

$$\frac{dw}{dx} = \frac{df}{dz}\frac{dz}{dx} = \frac{df}{dz}\left(\frac{\partial z}{\partial x} + \frac{\partial z}{\partial y}\frac{dy}{dx}\right)$$

$$= f'(z)\big(2ax + by + (bx + 2cy)(2x + 1)\big).$$

If $x = -1$, then $y = 1$, and so $z = a - b + c$. Thus

$$\frac{dw}{dx}(-1) = f'(a - b + c)(-2a + 2b - 2c).$$

EXERCISES

1. Assume
$$f\begin{pmatrix} x \\ y \end{pmatrix} = \begin{pmatrix} x^2 + xy + 1 \\ y^2 + 2 \end{pmatrix},$$

$$g\begin{pmatrix} u \\ v \end{pmatrix} = \begin{pmatrix} u + v \\ 2u \\ v^2 \end{pmatrix}.$$

(a) Find the matrices $g'\big(f(x, y)\big)$ and $f'(x, y)$.
(b) Use part (a) to find the matrices $(g \circ f)'(1, 1)$ and $(g \circ f)'(0, 0)$.

2. Assume

$$f(t) = \begin{pmatrix} t \\ t+1 \\ t^2 \end{pmatrix} = \begin{pmatrix} x \\ y \\ z \end{pmatrix}$$

and

$$g \begin{pmatrix} x \\ y \\ z \end{pmatrix} = \begin{pmatrix} x + 2y + z^2 \\ x^2 - y \end{pmatrix} = \begin{pmatrix} u \\ v \end{pmatrix}.$$

(a) Find the matrices $g'(f(t))$ and $f'(t)$.
(b) Use part (a) to find the matrices $(g \circ f)'(1)$ and $(g \circ f)'(0)$.

3. Consider the curve defined parametrically by

$$f(t) = \begin{pmatrix} t \\ t^2 - 4 \\ e^{t-2} \end{pmatrix}, \quad -\infty < t < \infty.$$

Let g be a real-valued differentiable function with domain $\mathbb{R}^3$. If $\mathbf{x}_0 = (2, 0, 1)$, and

$$\frac{\partial g}{\partial x}(\mathbf{x}_0) = 4, \quad \frac{\partial g}{\partial y}(\mathbf{x}_0) = 2, \quad \frac{\partial g}{\partial z}(\mathbf{x}_0) = 2,$$

find $d(g \circ f)/dt$ at $t = 2$.

4. Let $z = xy^2$ and suppose that $x = 2u + 3v$. Assume also that y is a function of u and v with the properties that when $(u, v) = (2, 1)$ then $y = -1$, $\partial y/\partial u = 5$ and $\partial y/\partial v = -2$. Find $\partial z/\partial u$ and $\partial z/\partial v$ when $(u, v) = (2, 1)$.

5. Consider the functions

$$f \begin{pmatrix} u \\ v \end{pmatrix} = \begin{pmatrix} u + v \\ u - v \\ u^2 - v^2 \end{pmatrix} = \begin{pmatrix} x \\ y \\ z \end{pmatrix}$$

and

$$F(x, y, z) = x^2 + y^2 + z^2 = w.$$

(a) Find the derivative matrix of $F \circ f$ at (u, v).
(b) Find $\partial w/\partial u$ and $\partial w/\partial v$.

6. Let $u = f(x, y)$. Make the change of variables $x = r\cos\theta$, $y = r\sin\theta$. Given that

$$\frac{\partial f}{\partial x} = x^2 + 2xy - y^2 \quad \text{and} \quad \frac{\partial f}{\partial y} = x^2 - 2xy + 2,$$

find $\partial f/\partial\theta$, when $r = 2$ and $\theta = \pi/2$.

7. If $g(x, y, z) = \sqrt{x^2 + y^2 + z^2}$ and

$$f(r, \theta) = \begin{pmatrix} r\cos\theta \\ r\sin\theta \\ r \end{pmatrix},$$

find $g'(f(r, \theta))$ and $f'(r, \theta)$; then multiply these together and find $(g \circ f)'(2, \pi)$.

8. Vector functions f and g are defined by

$$f \begin{pmatrix} u \\ v \end{pmatrix} = \begin{pmatrix} u\cos v \\ u\sin v \end{pmatrix}, \quad \begin{cases} 0 < u < \infty, \\ -\pi/2 < v < \pi/2, \end{cases}$$

$$g \begin{pmatrix} x \\ y \end{pmatrix} = \begin{pmatrix} \sqrt{x^2 + y^2} \\ \arctan\frac{y}{x} \end{pmatrix}, \quad 0 < x < \infty.$$

(a) Find the derivative matrix of $g \circ f$ at $\begin{pmatrix} u \\ v \end{pmatrix}$.
(b) Find the derivative matrix of $f \circ g$ at $\begin{pmatrix} x \\ y \end{pmatrix}$.
(c) Are the following statements true or false?
   (i) Domain of f = domain of $g \circ f$.
   (ii) Domain of g = domain of $f \circ g$.

9. Let $\mathbf{v}$ be a tangent vector at $\mathbf{x}_0$ to a curve defined parametrically by a differentiable vector function g. If $\mathbf{x}_0$ is in the domain of a differentiable vector function F, prove that $F'(\mathbf{x}_0)\mathbf{v}$, if not zero, is a tangent vector at $F(\mathbf{x}_0)$ to the curve defined parametrically by $F \circ g$.

10. The convention of denoting coordinate functions by real variables has its pitfalls. Resolve the following paradox: Let $w = f(x, y, z)$ and $z = g(x, y)$. By the chain rule

$$\frac{\partial w}{\partial x} = \frac{\partial w}{\partial x}\frac{\partial x}{\partial x} + \frac{\partial w}{\partial y}\frac{\partial y}{\partial x} + \frac{\partial w}{\partial z}\frac{\partial z}{\partial x}.$$

The quantities x and y are unrelated, so that $\partial y/\partial x = 0$. However $\partial x/\partial x = 1$. Hence

$$\frac{\partial w}{\partial x} = \frac{\partial w}{\partial x} + \frac{\partial w}{\partial z}\frac{\partial z}{\partial x},$$

and so, subtracting $\partial w/\partial x$ from both sides,

$$0 = \frac{\partial w}{\partial z}\frac{\partial z}{\partial x}.$$

In particular, take $w = 2x + y + 3z$ and $z = 5x + 18$. Then

$$\frac{\partial w}{\partial z} = 3 \quad \text{and} \quad \frac{\partial z}{\partial x} = 5.$$

It follows that $0 = 15$.

11. If $y = f(x - at) + g(x + at)$, where a is constant and f and g are twice differentiable, show that

$$a^2 \frac{\partial^2 y}{\partial x^2} = \frac{\partial^2 y}{\partial t^2} \quad \text{(wave equation).}$$

12. Let $U(x, y) = f(x + iy) + f(x - iy)$, where $i^2 = -1$. Show that $U_{xx} + U_{yy} = 0$.

13. If $f(tx, ty) = t^n f(x, y)$ for some integer n, and for all x, y, and t, show that

$$x \frac{\partial f}{\partial x} + y \frac{\partial f}{\partial y} = nf(x, y).$$

14. (a) If

$$w = f(x, y, z, t), \quad x = g(u, z, t), \quad \text{and}$$

$$z = h(u, t),$$

write a formula for dw/dt, where by this symbol is meant the rate of change of w with respect to t, and where all the interrelations of w, x, z, t are taken into account.

(b) If

$$w = f(x, y, z, t) = 2xy + 3z + t^2,$$

$$g(u, z, t) = ut \sin z,$$

$$h(u, t) = 2u + t,$$

evaluate dw/dt at the point $u = 1, t = 2, y = 3$, by using the formula you derived in part (a) and also by substituting in the functions for x and z and then differentiating.

15. Consider a real-valued function $f(x, y)$ such that

$$f_x(2, 1) = 3, \quad f_y(2, 1) = -2, \quad f_{xx}(2, 1) = 0,$$

$$f_{xy}(2, 1) = f_{yx}(2, 1) = 1, \quad f_{yy}(2, 1) = 2.$$

Let $\mathbb{R}^2 \xrightarrow{g} \mathbb{R}^2$ be defined by

$$g(u, v) = (u + v, uv).$$

Find $\partial^2 (f \circ g)/\partial v\, \partial u$ at $(1, 1)$. $\quad (2)$

In Exercises 16 to 19, let f be real-valued and differentiable.

16. If $u(x, y) = f(ax + by)$, show that $b\, \partial u/\partial x = a\, \partial u/\partial y$.

17. If $u(x, y) = f(xy)$, show that $x\, \partial u/\partial x = y\, \partial u/\partial y$.

18. If $u(x, y) = f(x/y)$, show that $x\, \partial u/\partial x = -y\, \partial u/\partial y$, $y \neq 0$.

19. If $u(x, y) = f(x^2 + y^2)$, show that $y\, \partial u/\partial x = x\, \partial u/\partial y$.

If in Exercises 20 to 25 f and g are of the following types, decide whether $g \circ f$, or $f \circ g$, or neither one, can possibly be defined.

20. $f: \mathbb{R}^2 \to \mathbb{R}^2, \ g: \mathbb{R}^2 \to \mathbb{R}^3$

21. $f: \mathbb{R}^3 \to \mathbb{R}^2, \ g: \mathbb{R}^2 \to \mathbb{R}$

22. $f: \mathbb{R} \to \mathbb{R}^2, \ g: \mathbb{R} \to \mathbb{R}^2$

23. $f: \mathbb{R}^3 \to \mathbb{R}^2, \ g: \mathbb{R}^3 \to \mathbb{R}^3$

24. $f: \mathbb{R} \to \mathbb{R}^2, \ g: \mathbb{R}^3 \to \mathbb{R}^2$

25. $f: \mathbb{R} \to \mathbb{R}^3, \ g: \mathbb{R}^3 \to \mathbb{R}^3$

*26. A 2-dimensional **Hamiltonian system** is a pair of equations of the form

$$dx/dt = H_y(x, y, t), \quad dy/dt = -H_x(x, y, t).$$

The function H of three variables that determines the system is called its Hamiltonian. Suppose that the pair $(x(t), y(t))$ satisfies the system, and consider two functions of t:

(i) $\dfrac{d}{dt}[H(x(t), y(t), t)]$, **(ii)** $H_t(x(t), y(t), t)$,

where the partial derivative of the Hamiltonian in (ii) is computed before substituting $x(t)$ and $y(t)$ for x and y.

(a) Show that (i) and (ii) are equal as functions of t.

(b) Show that if H is independent of t, then the curve parametrized by $(x(t), y(t))$ lies on a level curve of H.

2B   Changing Variables

One of the most important uses of the chain rule is computing the effect of a change of variable on the form of important expressions such as $|\nabla f|$, the length of a gradient. In doing computations of this kind it's often simpler and clearer to use the subscript notation for partial derivatives, as in the next example.

| EXAMPLE 10 | If $u = u(x, y)$ is a differentiable real-valued function, then the length of the gradient $|\nabla u| = (u_x^2 + u_y^2)^{1/2}$ limits how fast u can change in an arbitrary direction. Suppose new variables z, w are introduced by setting $z = x + y$, $w = x - y$, with inverse relations $x = (z + w)/2$, $y = (z - w)/2$. Because of the way it arose in the |

definition of differentiability, ∇f has an intrinsic meaning independent of the choice of coordinates. But so far we've calculated ∇f only in terms of standard rectangular coordinates in $\mathbb{R}^n$, so it's important to know how to compute it directly from the vector $(\overline{u}_z, \overline{u}_w)$, where $\overline{u}(z, w) = u\big((z + w)/2, (z - w)/2\big)$. To do this we compute u_x and u_y in terms of $\overline{u}_z$ and $\overline{u}_w$ using the chain rule. Noting from $z = x + y$ and $w = x - y$ that $z_x = w_x = z_y = 1$ and $w_y = -1$, we find

$$u_x = \overline{u}_z z_x + \overline{u}_w w_x = \overline{u}_z + \overline{u}_w$$

$$u_y = \overline{u}_z z_y + \overline{u}_w w_y = \overline{u}_z - \overline{u}_w.$$

Hence $\nabla \overline{u}(z, w) = (\overline{u}_z + \overline{u}_w, \overline{u}_z - \overline{u}_w)$. Squaring and adding the coordinates of $\nabla \overline{u}(z, w)$ gives

$$|\nabla u| = (u_x^2 + u_y^2)^{1/2} = \left((\overline{u}_z + \overline{u}_w)^2 + (\overline{u}_z - \overline{u}_w)^2 \right)^{1/2} = \sqrt{2}(\overline{u}_z^2 + \overline{u}_w^2)^{1/2}.$$

This equation tells us that in this case we can compute $|\nabla u|$ directly from the length of $(\overline{u}_z, \overline{u}_w)$ if we just multiply this length by $\sqrt{2}$.

EXAMPLE 11

For functions of a point (x, y) in $\mathbb{R}^2$, the **Laplace operator** Δ acting on twice continuously differentiable functions $u = u(x, y)$ produces a continuous function Δu:

$$\Delta u(x, y) = u_{xx}(x, y) + u_{yy}(x, y).$$

The Laplace operator Δ is an extension to higher dimensions of the second-order differential operator d^2/dx^2 that occurs naturally in many problems in pure and applied mathematics. It is important to know the form that Δ takes after a change of variable. For example, suppose new variables are introduced by setting $z = x + y$ and $w = x - y$, with inverse relations $x = (z + w)/2$, $y = (z - w)/2$. We need to find out how Δ is expressed in terms of partial derivatives of $\overline{u}$ with respect to z and w, where $\overline{u}(z, w) = u\big((z + w)/2, (z - w)/2\big)$. First compute u_x and u_y as in the previous example to get

$$u_x = \overline{u}_z z_x + \overline{u}_w w_x = \overline{u}_z + \overline{u}_w$$

$$u_y = \overline{u}_z z_y + \overline{u}_w w_y = \overline{u}_z - \overline{u}_w.$$

Then

$$u_{xx} = (\overline{u}_z)_x + (\overline{u}_w)_x = (\overline{u}_{zz} z_x + \overline{u}_{zw} w_x) + (\overline{u}_{wz} z_x + \overline{u}_{ww} w_x)$$

$$= (\overline{u}_{zz} + \overline{u}_{zw}) + (\overline{u}_{wz} + \overline{u}_{ww}),$$

$$u_{yy} = (\overline{u}_z)_y - (\overline{u}_w)_y = (\overline{u}_{zz} z_y + \overline{u}_{zw} w_y) - (\overline{u}_{wz} z_y + \overline{u}_{ww} w_y)$$

$$= (\overline{u}_{zz} - \overline{u}_{zw}) - (\overline{u}_{wz} - \overline{u}_{ww}).$$

Adding the results of these two computations cancels $\overline{u}_{zw}$ and $\overline{u}_{wz}$ to give

$$u_{xx} + u_{yy} = 2(\overline{u}_{zz} + \overline{u}_{ww}).$$

In terms of Δ this equation is $\Delta_{(x,y)}u = 2\Delta_{(z,w)}\bar{u}$. When interpreting this equation it's important to remember that while u and $\bar{u}$ take the same value at corresponding coordinate pairs (x, y) and (z, w), the formal expression of the Laplace operator may change in going over to the (z, w) coordinates.

The term **transformation** is sometimes used instead of function, particularly when the domain and range spaces are the same. When using a change of coordinates we use the term **coordinate transformation**. It's important to know to what extent each choice of coordinates in one system corresponds to a unique choice in the other. In the two previous examples, the coordinate transformations were made by invertible linear transformations. In Example 10 we had

$$\begin{pmatrix} x \\ y \end{pmatrix} = \begin{pmatrix} \frac{1}{2} & \frac{1}{2} \\ \frac{1}{2} & -\frac{1}{2} \end{pmatrix} \begin{pmatrix} z \\ w \end{pmatrix} \quad \text{and} \quad \begin{pmatrix} z \\ w \end{pmatrix} = \begin{pmatrix} 1 & 1 \\ 1 & -1 \end{pmatrix} \begin{pmatrix} x \\ y \end{pmatrix}.$$

These matrix equations establish a one-to-one correspondence between pairs (x, y) in one copy of $\mathbb{R}^2$ and pairs (z, w) in another copy. We've seen in Section 5 on determinants in Chapter 2 that a linear coordinate change $\mathbf{z} = A\mathbf{x}$ with square matrix A is one-to-one precisely when $\det A \neq 0$. Establishing an analogous criterion for nonlinear transformations is possible only **locally**, meaning in some neighborhood of a point. The general result, too technical to prove here, is as follows.

2.3   Inverse Function Theorem.   Let $\mathbb{R}^n \xrightarrow{F} \mathbb{R}^n$ be continuously differentiable on open subset S of $\mathbb{R}^n$, and let $\mathbf{x}_0$ be a point in S with invertible derivative matrix $F'(\mathbf{x}_0)$. Then there is an open neighborhood N of $\mathbf{x}_0$ such that F has a continuously differentiable inverse function F^{-1} defined on the image set $F(N)$. The derivative matrix of F^{-1} is related to $F'(\mathbf{x})$ by $(F^{-1})'(F(\mathbf{x})) = F'(\mathbf{x})^{-1}$ for $\mathbf{x}$ in N.

Briefly the theorem says that F has a continuously differentiable **local inverse** in some neighborhood of a point $\mathbf{x_0}$ where $F'(\mathbf{x_0})$ is invertible, or equivalently where $\det F'(\mathbf{x_0}) \neq 0$. The scalar $\det F'(\mathbf{x})$ is called the **Jacobian determinant** of $F(\mathbf{x})$, and it's crucial for changing variables in multiple integrals in Chapter 7.

EXAMPLE 12

The pair of equations $x = u + v$, $y = u^2 - v^2$ determines a transformation F from points (u, v) in $\mathbb{R}^2$ to points (x, y) in $\mathbb{R}^2$. But note that F sends every point (u, v) for which $u = -v$ onto the single point $(x, y) = (0, 0)$. To put it another way, the entire line $u + v = 0$ gets sent by F into the single point $(0, 0)$, so without some restriction on its domain the transformation F can't be one-to-one. On the other hand, the derivative matrix of F at (u, v) is

$$F'(u, v) = \begin{pmatrix} 1 & 1 \\ 2u & -2v \end{pmatrix}, \quad \text{so } \det F'(u, v) = -2(u + v).$$

The inverse function theorem implies that F has a continuously differentiable inverse defined in a neighborhood of every point $F(u, v)$ for which $\det F'(u, v) = -2(u + v) \neq 0$. Note that these are exactly the points not on the line $u + v = 0$, which is collapsed by F into a single point. For this particular transformation we can actually compute the coordinate functions for the inverse where it exists:

$$u = \frac{1}{2}\left(x + \frac{y}{x}\right), \quad v = \frac{1}{2}\left(x - \frac{y}{x}\right), \quad x = u + v \neq 0.$$

Figure 6.8(b) shows some image curves of vertical and horizontal line segments in the (u, v)-plane. The points on the diagonal in Figure 6.8(a) all have $(x, y) = (0, 0)$ as their image.

We treat more examples in detail in Section 5 where both sets of variables have standard geometric interpretations.

EXERCISES

1. Let $u(x, y)$ be differentiable for all (x, y) in $\mathbb{R}^2$. Let $x = s + t$, $y = s - t$ and $\bar{u}(s, t) = u(s + t, s - t)$. Use the chain-rule to show that

$$\left(\frac{\partial \bar{u}}{\partial s}\right)^2 + \left(\frac{\partial \bar{u}}{\partial t}\right)^2 = 2\left(\frac{\partial u}{\partial x}\right)^2 + 2\left(\frac{\partial u}{\partial y}\right)^2.$$

2. Let $x = u^2 - v^2$ and $y = 2uv$, and suppose that $z = f(x, y)$ is differentiable. Show that

$$\left(\frac{\partial z}{\partial u}\right)^2 + \left(\frac{\partial z}{\partial v}\right)^2 = 4(u^2 + v^2)\left(\left(\frac{\partial z}{\partial x}\right)^2 + \left(\frac{\partial z}{\partial y}\right)^2\right).$$

3. Let $z = f(x, y)$ be differentiable, and let $x = u \cos v$ and $y = u \sin v$. Show that for $u \neq 0$

$$\left(\frac{\partial z}{\partial x}\right)^2 + \left(\frac{\partial z}{\partial y}\right)^2 = \left(\frac{\partial z}{\partial u}\right)^2 + \frac{1}{u^2}\left(\frac{\partial z}{\partial v}\right)^2.$$

4. Suppose that $z = f(x, y)$ is differentiable for all (x, y) and that $x = e^{u+v} + e^{u-v}$ and $y = e^{u+v} - e^{u-v}$.
 (a) Compute the first-order partial derivatives of x and y with respect to u and with respect to v. Then express these four derivatives in terms of x and y.
 (b) Show that

$$\frac{\partial^2 z}{\partial u^2} - \frac{\partial^2 z}{\partial v^2} = (x^2 - y^2)\left(\frac{\partial^2 z}{\partial x^2} - \frac{\partial^2 z}{\partial y^2}\right).$$

5. Let $f : \mathbb{R}^2 \to \mathbb{R}$ be continuously differentiable, let $x(u, v) = u + v$ and $y(u, v) = u^2 - v^2$, and set $z = f(x(u, v), y(u, v))$. Show that $z_u^2 - z_v^2 = 4xf_xf_y + 4yf_y^2$.

6. Let $f : \mathbb{R}^2 \to \mathbb{R}$ be continuously differentiable, let $x(u, v) = u + v$ and $y(u, v) = u^2 + v^2$, and set $z = f(x(u, v), y(u, v))$. Show that $z_u^2 + z_v^2 = 2f_x^2 + 4xf_xf_y + 4yf_y^2$.

7. As in Example 10 of the text, show that if (x, y) and (z, w) are related by

$$z = (x + y)/\sqrt{2}, \quad w = (x - y)/\sqrt{2},$$

then $u_{xx} + u_{yy} = \bar{u}_{zz} + \bar{u}_{ww}$.

8. $F(u, v)$ is defined by the equations, $x = u + v$, $y = u^2 - v^2$ in Example 12 of the text.
 (a) Show that for fixed $v = v_0$ and varying u the image curves are upward-pointing parabolas passing through $(x, y) = (0, 0)$ as indicated in Figure 6.8(b).
 (b) Show that for fixed $u = u_0$ and varying v the image curves are downward-pointing parabolas passing through $(x, y) = (0, 0)$ as indicated in Figure 6.8(b).
 (c) Derive the inverse relations $u = \frac{1}{2}(x + y/x)$, $v = \frac{1}{2}(x - y/x)$, $x \neq 0$. [*Hint*: $y = x(u - v)$.]

9. Define $\mathbb{R}^2 \xrightarrow{f} \mathbb{R}^2$ by the equations $x = 2uv$, $y = u^2 - v^2$.
 (a) Show that the images of lines through the origin in the uv-plane are half-lines emanating from $(x, y) = (0, 0)$ and that each point of a half-line except for $(0, 0)$ is the image of two points.
 (b) Show that image curves of circles $u^2 + v^2 = a^2$ in the uv-plane are circles $x^2 + y^2 = a^4$ in the xy-plane.
 (c) Compute $\det f'(u, v)$ and show that, if $(u_0, v_0) \neq (0, 0)$, the inverse function theorem implies the existence of a local inverse in a neighborhood of $f(u_0, v_0)$.

10. Define $\mathbb{R}^2 \xrightarrow{P} \mathbb{R}^2$ by the equations $x = u \cos v$, $y = u \sin v$ for $u > 0$.
 (a) Show that for fixed $v = v_0$ and varying $u > 0$ the image curves in the xy-plane are half-lines emanating from $(x, y) = (0, 0)$.
 (b) Show that for fixed $u = u_0$ and varying v the image curves in the xy-plane are circles of radius u_0 each one traced infinitely often.
 (c) Compute $\det P'(u, v)$ and show that if $u_0 \neq 0$, the inverse function theorem implies the existence of a local inverse in a neighborhood of $P(u_0, v_0)$.

*11. For a continuously differentiable function $\mathbb{R} \xrightarrow{f} \mathbb{R}$ the inverse function theorem sharpens slightly to assert that if $f'(x_0) \neq 0$, then f is either strictly increasing or else strictly decreasing in some neighborhood of x_0. Prove this using the mean-value theorem for derivatives, and without appealing to the statement of the inverse function theorem. Make sure to make use of the continuity of f'; the conclusion is false without that assumption.

12. The conditions of the inverse function theorem guarantee the existence of a *continuously differentiable* inverse

function. If one of the main hypotheses, $F'(\mathbf{x}_0) \neq 0$, fails there may still be a merely *continuous* inverse. Verify this last assertion using the function $\mathbb{R} \xrightarrow{f} \mathbb{R}$ defined by $f(x) = x^3$.

13. Use the chain rule to show that under the assumptions of the inverse function theorem 2.3, $(F^{-1})'\big(F(\mathbf{x}_0)\big) = \big(F'(\mathbf{x}_0)\big)^{-1}$, that is, the derivative matrix at $F(\mathbf{x}_0)$ of the inverse mapping is equal to the inverse of the matrix $F'(\mathbf{x}_0)$.

SECTION 3   IMPLICIT DIFFERENTIATION

It may happen that two vectors are related by a formula that doesn't express either one directly as a function of the other. For example, the formula

$$\frac{pv}{t} = k_0$$

expresses the relationship between pressure p, volume v, and temperature t, of the gas in some container. Or the equations

$$x^2 + y^2 + z^2 = 1$$

$$x + y + z = 0$$

may be interpreted as a relation between the three coordinates of a point on both the sphere of radius 1 centered at $(0, 0, 0)$ in $\mathbb{R}^3$ and a plane through the origin. In neither example do the equations give an explicit formula for any of the coordinates in terms of others. In this section we study the application of calculus to such relations.

For two functions $\mathbb{R}^2 \xrightarrow{F} \mathbb{R}$ and $\mathbb{R} \xrightarrow{f} \mathbb{R}$, the equation

$$F(x, y) = 0$$

defines f **implicitly** if $F\big((x, f(x)\big) = 0$ for every x in the domain of f. The zero on the right side of the equation could in practice be an arbitrary constant c. But since $F(x, y) = c$ is equivalent to $G(x, y) = F(x, y) - c = 0$, it's customary to absorb the constant into the function F in a generic context.

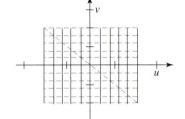

(a)

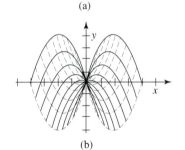

(b)

FIGURE 6.8

| EXAMPLE 1 | Let $F(x, y) = x^2 + y^2 - 1$. Then the condition that $F\big(x, f(x)\big) = x^2 + (f(x))^2 - 1 = 0$, for every x in the domain of f, is satisfied by each of the following choices for f

$$f_1(x) = \sqrt{1 - x^2}, \quad -1 \leq x \leq 1.$$

$$f_2(x) = -\sqrt{1 - x^2}, \quad -1 \leq x \leq 1.$$

$$f_3(x) = \begin{cases} \sqrt{1 - x^2}, & -\frac{1}{2} \leq x \leq 0. \\ -\sqrt{1 - x^2}, & 0 < x \leq 1. \end{cases}$$

Their graphs are shown in Figure 6.9. It follows from the definition of an implicitly defined function that all three functions f_1, f_2, f_3 are defined implicitly by the equation $x^2 + y^2 - 1 = 0$.

FIGURE 6.9

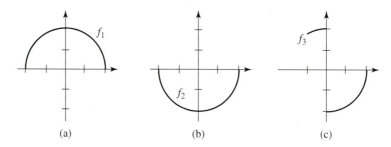

(a)   (b)   (c)

Consider a function $\mathbb{R}^{n+m} \xrightarrow{F} \mathbb{R}^m$. We can write an arbitrary element in $\mathbb{R}^{n+m}$ as $(x_1, \ldots, x_n, y_1, \ldots, y_m)$, or as a pair $(\mathbf{x}, \mathbf{y})$, where $\mathbf{x} = (x_1, \ldots, x_n)$ and $\mathbf{y} = (y_1, \ldots, y_m)$. In this way F looks like either a function of the two vector variables, $\mathbf{x}$ in $\mathbb{R}^n$ and $\mathbf{y}$ in $\mathbb{R}^m$, or a function of the single vector variable $(\mathbf{x}, \mathbf{y})$ in $\mathbb{R}^{n+m}$. The function $\mathbb{R}^n \xrightarrow{G} \mathbb{R}^m$ is defined **implicitly** by the equation

$$F(\mathbf{x}, \mathbf{y}) = 0$$

if $F\big(\mathbf{x}, G(\mathbf{x})\big) = 0$ for every $\mathbf{x}$ in the domain of G.

EXAMPLE 2   The equations

$$x + y + z - 1 = 0$$
$$2x \quad\;\; + z + 2 = 0 \qquad\qquad (*)$$

determine y and z as functions of x. We get

$$y = x + 3, \; z = -2x - 2.$$

Notice that the number of equations is the same as the number of variables that we solve for, two in this example.

In terms of a function $\mathbb{R}^2 \xrightarrow{f} \mathbb{R}^2$ Equations (*) are

$$F\left(x, \begin{pmatrix} y \\ z \end{pmatrix}\right) = \begin{pmatrix} x & + & y & + & z & - & 1 \\ 2x & & & + & z & + & 2 \end{pmatrix} = \begin{pmatrix} 0 \\ 0 \end{pmatrix}$$

$$= \begin{pmatrix} 1 \\ 2 \end{pmatrix} x + \begin{pmatrix} 1 & 1 \\ 0 & 1 \end{pmatrix} \begin{pmatrix} y \\ z \end{pmatrix} + \begin{pmatrix} -1 \\ 2 \end{pmatrix} = \begin{pmatrix} 0 \\ 0 \end{pmatrix}.$$

The implicitly defined function $\mathbb{R} \xrightarrow{G} \mathbb{R}^2$ is

$$G(x) = \begin{pmatrix} y \\ z \end{pmatrix} = \begin{pmatrix} x + 3 \\ -2x - 2 \end{pmatrix}.$$

Although Example 1 shows that an implicitly defined function need not be continuous, we'll be primarily concerned in this section with functions that are not only continuous but also differentiable. The *implicit function theorem* described in Theorem 3.4 gives conditions for the existence of a differentiable G defined by an equation

$F(\mathbf{x}, G(\mathbf{x})) = 0$. However, we consider here the problem of finding the derivative of G only when G and G' are both assumed to exist. Suppose the functions $\mathbb{R}^2 \xrightarrow{f} \mathbb{R}$ and $\mathbb{R} \xrightarrow{G} \mathbb{R}$ are differentiable and that

$$F(x, G(x)) = 0$$

for every x in the domain of g. Then the chain rule applied to $F(x, G(x))$ yields, in terms of the partial derivatives F_x and F_y,

$$F_x(x, G(x)) + F_y(x, G(x))G'(x) = 0.$$

Since in a typical application we don't have an explicit formula for $y = G(x)$, we rewrite the previous equation as

$$F_x(x, y) + F_y(x, y)\frac{dy}{dx} = 0.$$

Solving the last equation for dy/dx gives

3.1
$$\frac{dy}{dx} = -\frac{F_x(x, y)}{F_y(x, y)}, \quad \text{if } F_y(x, y) \neq 0.$$

EXAMPLE 3   If $F(x, y) = x^2 + y^2 - 1 = 0$ is thought of as defining y implicitly as a function $y = y(x)$, we can differentiate both sides with respect to x to get

$$2x + 2y\frac{dy}{dx} = 0.$$

Solving for dy/dx gives

$$\frac{dy}{dx} = -\frac{2x}{2y}, \quad \text{if } y \neq 0.$$

For example, at the point $(x_0, y_0) = (1/\sqrt{2}, 1/\sqrt{2})$, we have $F(x_0, y_0) = 0$, and

$$\frac{dy}{dx}(x_0, y_0) = -\frac{2x_0}{2y_0}$$
$$= -1.$$

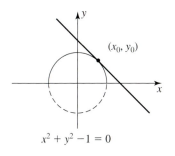

$x^2 + y^2 - 1 = 0$

FIGURE 6.10

Thus the graph of the implicitly defined function has slope -1 at (x_0, y_0). Figure 6.10 shows the tangent line there.

The process just described is called **implicit differentiation**, and it extends to vector-valued functions of several variables too.

EXAMPLE 4   Given the equations

$$x^2 + y^2 + z^2 - 6 = 0, \quad xyz + 2 = 0,$$

suppose that x and y are differentiable functions of z, that is, the function defined implicitly by the equations is of the form $(x, y) = G(z)$. To compute dx/dz and dy/dz, we apply the chain rule to the given equations to get

$$2x\frac{dx}{dz} + 2y\frac{dy}{dz} + 2z = 0,$$

$$yz\frac{dx}{dz} + xz\frac{dy}{dz} + xy = 0.$$

We can solve these new equations for dx/dz and dy/dz. The solution is

$$\begin{pmatrix} \dfrac{dx}{dz} \\[2mm] \dfrac{dy}{dz} \end{pmatrix} = \begin{pmatrix} \dfrac{x(y^2 - z^2)}{z(x^2 - y^2)} \\[2mm] \dfrac{y(z^2 - x^2)}{z(x^2 - y^2)} \end{pmatrix},$$

which is the matrix $G'(z)$. Notice that the corresponding values for x and y have to be known to make the formula completely explicit. That is, from the information given so far, there is no possible way of evaluating dx/dz at $z = 1$. On the other hand, given the point $(x, y, z) = (1, -2, 1)$ satisfying both equations, we have $(dx/dz)(1) = -1$. The reason is that, just as in Example 1, there is more than one function f defined implicitly by the given equations. By specifying a particular point on its graph, we determine f uniquely in the vicinity of the point.

EXAMPLE 5   Consider

$$xu + yv + zw = 1,$$

$$x + y + z + u + v + w = 0,$$

$$xy + zuv + w = 1.$$

Suppose that each of x, y, and z is a function of u, v, and w. To find the partial derivatives of x, y, and z with respect to w, we differentiate the three equations using the chain rule.

$$u\frac{\partial x}{\partial w} + v\frac{\partial y}{\partial w} + w\frac{\partial z}{\partial w} + z = 0,$$

$$\frac{\partial x}{\partial w} + \frac{\partial y}{\partial w} + \frac{\partial z}{\partial w} + 1 = 0,$$

$$y\frac{\partial x}{\partial w} + x\frac{\partial y}{\partial w} + uv\frac{\partial z}{\partial w} + 1 = 0$$

The linear system for x_w, y_w, z_w in matrix form is

$$\begin{pmatrix} u & v & w \\ 1 & 1 & 1 \\ y & x & uv \end{pmatrix} \begin{pmatrix} x_w \\ y_w \\ z_w \end{pmatrix} = \begin{pmatrix} -z \\ -1 \\ -1 \end{pmatrix}.$$

Solving the this linear system is simplest using Cramer's rule, giving, for example, x_w as

$$\frac{\partial x}{\partial w} = \frac{uv^2 + xz + w - zuv - xw - v}{u^2v + vy + wx - yw - ux - uv^2}.$$

Similarly, we could solve for $\partial y/\partial w$ and $\partial z/\partial w$. To find partials with respect to u, differentiate the original equations with respect to u and solve for $\partial x/\partial u$, $\partial y/\partial u$, and $\partial z/\partial u$. Partials with respect to v are found by the same method.

The computation indicated in Example 5 leads to the nine entries in the derivative matrix of an implicitly defined vector function. For the computation to work it's necessary to have the number of given equations equal the number of implicitly defined coordinate functions, just as in Example 2. To get more insight into the reason for this requirement, suppose we are given a differentiable vector function

$$F(u, v, x, y) = \left(\begin{array}{c} F_1(u, v, x, y) \\ F_2(u, v, x, y) \end{array} \right)$$

and that the equations

$$F_1(u, v, x, y) = 0, \quad F_2(u, v, x, y) = 0$$

implicitly define a differentiable function $(x, y) = G(u, v)$. Differentiating these equations with respect to u and v using the chain rule, we get

$$\frac{\partial F_1}{\partial u} + \frac{\partial F_1}{\partial x}\frac{\partial x}{\partial u} + \frac{\partial F_1}{\partial y}\frac{\partial y}{\partial u} = 0, \quad \frac{\partial F_1}{\partial v} + \frac{\partial F_1}{\partial x}\frac{\partial x}{\partial v} + \frac{\partial F_1}{\partial y}\frac{\partial y}{\partial v} = 0,$$

$$\frac{\partial F_2}{\partial u} + \frac{\partial F_2}{\partial x}\frac{\partial x}{\partial u} + \frac{\partial F_2}{\partial y}\frac{\partial y}{\partial u} = 0, \quad \frac{\partial F_2}{\partial v} + \frac{\partial F_2}{\partial x}\frac{\partial x}{\partial v} + \frac{\partial F_2}{\partial y}\frac{\partial y}{\partial v} = 0.$$

These equations written in matrix form are

$$\left(\begin{array}{cc} \dfrac{\partial F_1}{\partial u} & \dfrac{\partial F_1}{\partial v} \\ \dfrac{\partial F_2}{\partial u} & \dfrac{\partial F_2}{\partial v} \end{array} \right) + \left(\begin{array}{cc} \dfrac{\partial F_1}{\partial x} & \dfrac{\partial F_1}{\partial y} \\ \dfrac{\partial F_2}{\partial x} & \dfrac{\partial F_2}{\partial y} \end{array} \right) \left(\begin{array}{cc} \dfrac{\partial x}{\partial u} & \dfrac{\partial x}{\partial v} \\ \dfrac{\partial y}{\partial u} & \dfrac{\partial y}{\partial v} \end{array} \right) = 0.$$

The last matrix on the right is the derivative matrix $G'(u, v)$. Solving for it, we get

3.2 $\quad G'(u, v) = \left(\begin{array}{cc} \dfrac{\partial x}{\partial u} & \dfrac{\partial x}{\partial v} \\ \dfrac{\partial y}{\partial u} & \dfrac{\partial y}{\partial v} \end{array} \right) = - \left(\begin{array}{cc} \dfrac{\partial F_1}{\partial x} & \dfrac{\partial F_1}{\partial y} \\ \dfrac{\partial F_2}{\partial x} & \dfrac{\partial F_2}{\partial y} \end{array} \right)^{-1} \left(\begin{array}{cc} \dfrac{\partial F_1}{\partial u} & \dfrac{\partial F_1}{\partial v} \\ \dfrac{\partial F_2}{\partial u} & \dfrac{\partial F_2}{\partial v} \end{array} \right).$

To be able to solve uniquely for the matrix $G'(u, v)$ it's essential that the inverse matrix appearing in Equation 3.2 should exist. In particular, this requires that the matrix to be inverted be square, in other words that the number of equations originally given must equal the number of implicitly determined variables; equivalently, the number of variables you solve for must be the same as the number of equations that determine them, just as for linear systems for which you may expect a unique solution.

The analog of Equation 3.2 holds for an arbitrary number of equations under suitable hypotheses and the proof follows similar lines. We summarize the generalization of Equations 3.1 and 3.2 as follows.

3.3   Theorem.   Suppose $\mathbb{R}^{n+m} \xrightarrow{F} \mathbb{R}^m$ and $\mathbb{R}^n \xrightarrow{G} \mathbb{R}^m$ are differentiable and that $\mathbf{y} = G(\mathbf{x})$ satisfies $F(\mathbf{x}, \mathbf{y}) = 0$ for all $\mathbf{x}$ in some open subset of $\mathbb{R}^n$. Then

$$G'(\mathbf{x}) = -F_{\mathbf{y}}^{-1}\big(\mathbf{x}, G(\mathbf{x})\big) F_{\mathbf{x}}\big(\mathbf{x}, G(\mathbf{x})\big),$$

provided that the m by m matrix $F_{\mathbf{y}}$ is invertible. The derivative matrix $F_{\mathbf{y}}$ is computed with $\mathbf{x}$ held fixed and $F_{\mathbf{x}}$ is computed with $\mathbf{y}$ held fixed.

The subscript notation used in the theorem is illustrated in the next example.

EXAMPLE 6   Suppose that

$$F(x, y, z) = \begin{pmatrix} x^2 y + xz \\ xz + yz \end{pmatrix}$$

and that we choose $\mathbf{x} = x$, $\mathbf{y} = (y, z)$. Then

$$F_x(x, y, z) = \begin{pmatrix} 2xy + z \\ z \end{pmatrix} \quad \text{and} \quad F_{(y,z)}(x, y, z) = \begin{pmatrix} x^2 & x \\ z & x + y \end{pmatrix}.$$

Note that the vector $\mathbf{y}$ must be chosen so that $F_{\mathbf{y}}$ is a square matrix and that the implicit differentiation formula works only when that matrix is invertible. Thus we must choose $(\mathbf{x}, \mathbf{y})$ so that

$$\det F_{\mathbf{y}}(\mathbf{x}, \mathbf{y}) \neq 0.$$

For the choice made in Example 6, we have

$$\det \begin{pmatrix} x^2 & x \\ z & x + y \end{pmatrix} = x^3 + x^2 y - xz,$$

so the formula fails at any point (x, y, z) for which $x^3 + x^2 y - xz = 0$.

Theorem 3.3 is very general as far at it goes, but it does nothing to guarantee the existence of a function G such that $F\big(\mathbf{x}, G(\mathbf{x})\big) = 0$; this is done in theory at least, by Theorem 3.4. The proof is too complicated to give here, but there's a straightforward equivalence with the inverse function theorem of Section 2B.

3.4   Implicit Function Theorem.   Let $\mathbb{R}^{n+m} \xrightarrow{F} \mathbb{R}^m$ be a continuously differentiable function. Suppose for some $\mathbf{x}_0$ in $\mathbb{R}^n$ and some $\mathbf{y}_0$ in $\mathbb{R}^m$ that

(i) $F(\mathbf{x}_0, \mathbf{y}_0) = \mathbf{0}$ and   (ii) $F_{\mathbf{y}}(\mathbf{x}_0, \mathbf{y}_0)$ is an invertible m-by-m matrix.

Then there is a unique continuously differentiable function $\mathbb{R}^n \xrightarrow{G} \mathbb{R}^m$ defined on an open neighborhood N of $\mathbf{x}_0$ in $\mathbb{R}^n$ such that $F(\mathbf{x}, G(\mathbf{x})) = 0$ for all $\mathbf{x}$ in N and $G(\mathbf{x}_0) = \mathbf{y}_0$.

Note that condition (ii) of the statement is equivalent to $\det F_{\mathbf{y}}(\mathbf{x}_0, \mathbf{y}_0) \neq 0$ and is just what is needed to make sense of the formula for computing $G'(\mathbf{x})$ in Theorem 3.3. Theorem 3.4 is useful for identifying points at which the level sets S of a function are smooth. This identification typically has to be done piecemeal, treating one "patch" of a curve or surface at a time.

EXAMPLE 7

The circular cone $F(x, y, z) = x^2 + y^2 - z^2 = 0$ has two conical parts, symmetric about the z-axis, each with a sharp point at $(0, 0, 0)$. Applying the implicit function theorem, we can ask where z is represented locally as the graph of a smooth function $z = z(x, y)$. We compute $F_z(x, y, z) = -2z$. Hence near any point (x, y, z) where $z \neq 0$ the desired function exists. This example is atypical in that we can actually find such functions. But here $z = \pm\sqrt{x^2 + y^2}$ meet the requirements for the top and bottom halves of the cone respectively. Conceivably, the origin might not be singular with respect to representation as either $x = x(y, z)$ or $y = y(x, z)$. But $F_y(0, 0, 0) = F_z(0, 0, 0) = 0$ also, so the origin is singular with respect to those possibilities also.

EXERCISES

---

1. The equation $x^2 + y^2 - 1 = 0$ is satisfied by many values of (x, y), including $(1, 0)$, $(0, 1)$, and $(1/\sqrt{2}, 1/\sqrt{2})$. Use implicit differentiation in both parts (a) and (b).

(a) Express dy/dx in terms of x and y, and evaluate at $(x, y) = (1/\sqrt{2}, 1/\sqrt{2})$. Does it make sense to evaluate at $(x, y) = (0, 1)$ or $(x, y) = (-1, 1)$?

(b) Express dx/dy in terms of x and y, and evaluate at $(x, y) = (1/\sqrt{2}, -1/\sqrt{2})$. Does it make sense to evaluate at $(x, y) = (1, 0)$ or $(x, y) = (0, 1)$?

(c) Solve the given equation explicitly for y in terms of x and interpret the results of part (a) graphically.

(d) Solve the given equation explicitly for x in terms of y and interpret the results of part (b) graphically.

2. The equation $x^2 - y^2 - 1 = 0$ is satisfied by many points including $(x, y) = (\sqrt{3}, \sqrt{2})$, $(x, y) = (1, 0)$ and $(x, y) = (-1, 0)$. Use implicit differentiation in both parts (a) and (b).

(a) Express dy/dx in terms of x and y, and evaluate at $(x, y) = (\sqrt{3}, \sqrt{2})$. Does it make sense to evaluate at $(x, y) = (1, 0)$ or $(x, y) = (1, 1)$?

(b) Express dx/dy in terms of x and y, and evaluate at $(x, y) = (\sqrt{2}, 1)$. Does it make sense to evaluate at $(x, y) = (1, 0)$ or $(x, y) = (0, 1)$?

(c) Solve the given equation explicitly for y in terms of x and interpret the results of part (a) graphically.

(d) Solve the given equation explicitly for x in terms of y and interpret the results of part (b) graphically.

In Exercises 3 to 6, use implicit differentiation to find dy/dx, and, if possible, dx/dy at the indicated point.

3. $xy + 1 = 0$ at $(x, y) = (-1, 1)$

4. $xe^y + ye^x = 0$ at $(x, y) = (0, 0)$

5. $x + y(x^2 + 1) + \frac{1}{2} = 0$ at $(x, y) = (-1, \frac{1}{4})$

6. $x^2 + y^2 = 1$ at $(x, y) = (1/\sqrt{2}, 1/\sqrt{2})$

7. Suppose that $x^2 y + yz = 0$ and $xyz + 1 = 0$.

(a) Find dx/dz and dy/dz at $(x, y, z) = (1, 1, -1)$.

(b) Find dy/dx and dz/dx at $(x, y, z) = (1, 1, -1)$.

(c) Find dx/dy and dz/dy at $(x, y, z) = (1, 1, -1)$.

8. If $x + y - u - v = 0$ and $x - y + 2u + v = 0$, find $\partial x/\partial u$, $\partial y/\partial u$, $\partial x/\partial v$, and $\partial y/\partial v$ by

(a) first solving for x and y in terms of u and v

(b) implicit differentiation

9. If Exercise 7 is expressed in the general vector notation of Theorem 3.3, what are F, $\mathbf{x}$, $\mathbf{y}$, $F_{\mathbf{x}}$, and $F_{\mathbf{y}}$ for part (a)? Part (b)? Part (c)?

10. If Exercise 8 is expressed in the vector notation of Theorem 3.3, what is the matrix $G'(\mathbf{x})$?

11. If $x^2 + yu + xv + w = 0$, $x + y + uvw + 1 = 0$, then, regarding x and y as functions of u, v, and w, find

$$\frac{\partial x}{\partial u} \quad \text{and} \quad \frac{\partial y}{\partial u} \text{ at } (x, y, u, v, w) = (1, -1, 1, 1, -1).$$

12. The equations $2x^3 y + yx^2 + t^2 = 0$, $x + y + t - 1 = 0$ implicitly define a curve

$$f(t) = \begin{pmatrix} x(t) \\ y(t) \end{pmatrix} \text{ that satisfies } f(1) = \begin{pmatrix} -1 \\ 1 \end{pmatrix}.$$

Find the tangent line to the curve when $t = 1$.

13. Let the equation $x^2/4 + y^2 + z^2/9 - 1 = 0$ define z implicitly as a function $z = f(x, y)$ near the point $x = 1$, $y = \sqrt{11}/6$, $z = 2$. The graph of the function f is a surface. Find its tangent plane at $(1, \sqrt{11}/6, 2)$.

14. Suppose the equation $F(x, y, z) = 0$ implicitly defines $z = f(x, y)$ and that $z_0 = f(x_0, y_0)$. Suppose further that the surface that is the graph of $z = f(x, y)$ has a tangent plane at (x_0, y_0), as defined in Chapter 4, Section 3B. Show that

$$(x - x_0)\frac{\partial F}{\partial x}(x_0, y_0, z_0) + (y - y_0)\frac{\partial F}{\partial y}(x_0, y_0, z_0) +$$

$$(z - z_0)\frac{\partial F}{\partial z}(x_0, y_0, z_0) = 0$$

is an equation for this tangent plane.

15. The equations

$$2x + y + 2z + u - v - 1 = 0$$

$$xy + z - u + 2v - 1 = 0$$

$$yz + xz + u^2 + v = 0$$

define x, y, and z as functions of u and v near $(x, y, z, u, v) = (1, 1, -1, 1, 1)$.

(a) Find the derivative matrix of the implicitly defined function

$$\begin{pmatrix} x \\ y \\ z \end{pmatrix} = \begin{pmatrix} x(u, v) \\ y(u, v) \\ z(u, v) \end{pmatrix}$$

$$= f(u, v) \quad \text{at} \quad (u, v) = (1, 1).$$

(b) The function f parametrically defines a surface in the (x, y, z) space. Find the tangent plane to it at the point $(1, 1, -1)$.

16. Show that the hyperboloid of two sheets $x^2 + y^2 - z^2 + 1 = 0$ has two pieces, not intersecting, that are graphs of smooth functions of the form $z = z(x, y)$.

(a) Do this by finding explicit representations for the two graphs.

(b) Apply the implicit function theorem to show this for a neighborhood N of every point $\mathbf{x}_0$ in $\mathcal{R}^2$.

17. It's intuitively evident that a sphere of radius a centered at the origin is a smooth surface S near all of its points.

(a) Explain how this follows from the implicit function theorem, applied with each of x, y and z as dependent variable, to the function $F(x, y, z) = x^2 + y^2 + z^2 - a^2$.

(b) Find representations for parts of S in six pieces, so showing explicitly the smoothness of S.

18. The sphere $x^2 + y^2 + z^2 - 4 = 0$ and the plane $x + y + z - 2 = 0$ have points in common, for example $(2, 0, 0)$, and the two surfaces appear to intersect in a circle.

(a) Find the center $\mathbf{c}$ and radius a of the circle.

(b) The implicit function theorem doesn't actually find an explicit parametrization of the circle for you, but it does show that this can in principle be done in overlapping pieces. Explain how.

19. The equation $xyz - yz^2 + x^2 y = 1$ is satisfied by the points on a level set S in $\mathcal{R}^3$ of $F(x, y, z) = xyz - yz^2 + x^2 y$.

(a) Find all points (x, y, z) such that $F_x(x, y, z) = 0$, where the implicit function theorem fails to guarantee the existence of $x = x(y, z)$.

(b) Some of the points found in part (a) may not lie on S. Show that the only points on S at which it's impossible to apply the implicit function theorem to guarantee existence of $x = x(y, z)$ are the solutions of $2x + z = 0$ and $5yz^2 + 4 = 0$.

(c) Solve the x-quadratic equation $xyz - yz^2 + x^2 y = 1$ explicitly for $x = x(y, z)$. The points where this solution fails to define a continuously differentiable function should be the same as the points found in part (b).

20. The equation $xyz - yz^2 + x^2 y = 1$ in Exercise 19 is satisfied by the points on a level set S in $\mathcal{R}^3$ of $F(x, y, z) = xyz - yz^2 + x^2 y$.

(a) Find all points (x, y, z) such that $F_z(x, y, z) = 0$, as required by the implicit function theorem for the existence of $z = z(x, y)$.

(b) Some of the points found in part (a) may not lie on S. Show that all points on S near which it's impossible to apply the implicit function theorem to guarantee existence of $z = z(x, y)$ are the solutions of $x - 2z = 0$ and $5x^2y - 4 = 0$.

(c) Solve the z-quadratic equation $xyz - yz^2 + x^2y = 1$ explicitly for $z = z(x, y)$. The points where this solution fails to define a continuously differentiable function should be the same as the points found in part (b).

(d) The results of parts (b) of Exercises 19 and 20 together imply that S is a smooth surface at all of its points. Explain why.

*21. Show that the inverse function theorem (Theorem 2.3) follows from the implicit function theorem (Theorem 3.4) by setting $F(\mathbf{x}, \mathbf{y}) = \mathbf{x} - f(\mathbf{y})$.

SECTION 4   EXTREME VALUES

4A   Critical Points

The problem of finding the maximum and minimum values of a real-valued function of several variables is important in many branches of applied mathematics, as well as in pure mathematics. Familiar examples are extremes of temperature, speed, or economic profit, each of which may be a function of more than one variable in a practical problem.

A real-valued function f has an **absolute maximum value** at $\mathbf{x}_0$ if, for all $\mathbf{x}$ in the domain of f,

$$f(\mathbf{x}) \leq f(\mathbf{x}_0),$$

and an **absolute minimum value** if, instead,

$$f(\mathbf{x}_0) \leq f(\mathbf{x}).$$

The number $f(\mathbf{x}_0)$ is called a **local maximum** value or a **local minimum value** if there is a neighborhood N of $\mathbf{x}_0$ such that, respectively,

$$f(\mathbf{x}) \leq f(\mathbf{x}_0) \quad \text{or} \quad f(\mathbf{x}_0) \leq f(\mathbf{x}),$$

for all $\mathbf{x}$ in N. A maximum or minimum value of f is called an **extreme value**. A point $\mathbf{x}_0$ at which an extreme value occurs is called an **extreme point**. The routines of single-variable and multivariable calculus just identify extreme points, and the corresponding extreme values, of real-valued differentiable functions under the assumption that these extreme values do indeed exist. The fundamental theorem that guarantees existence of extreme values makes no reference to differentiability of a function $f: S \to \mathbb{R}$ but only to continuity on a set S in $\mathbb{R}^n$ that is both **closed**, so contains all its boundary points, and **bounded**, so is contained in a ball of finite radius. For functions of a real variable x, the typical closed, bounded set encountered in this context is an interval $a \leq x \leq b$, often denoted $[a, b]$. We state the theorem without its fairly technical proof. The conditions of the theorem are designed to specifically exclude two possibilities: (i) that f is unbounded on S, and (ii) that f might approach a limit on S that is not attained at a point in S.

4.1   Theorem.   For a function $f: S \to \mathbb{R}$ assume (i) that S is a closed, bounded subset of $\mathbb{R}^n$, and (ii) that f is continuous on S. Then f assumes its absolute maximum and absolute minimum values on S.

EXAMPLE 1   Consider the function defined by $f(x, y) = x^2 + y^2$ for points (x, y) in the set S of points that lie inside or on the ellipse $x^2 + 2y^2 = 1$. Note that S is closed and

FIGURE 6.11

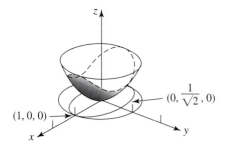

bounded and that f is continuous on S. The graph of f is shown in Figure 6.11. Suppose that f has an extreme value (i.e., maximum or minimum) at a point (x_0, y_0) in the interior of the ellipse. Then both functions f_1 and f_2 defined by

$$f_1(x) = f(x, y_0), \quad f_2(y) = f(x_0, y)$$

must also have extreme values at x_0 and y_0, respectively. Applying the familiar criterion for differentiable functions of one variable, we have

$$f_1'(x_0) = f_2'(y_0) = 0.$$

Since

$$f_1'(x_0) = \frac{\partial f}{\partial x}(x_0, y_0) \quad \text{and} \quad f_2'(y_0) = \frac{\partial f}{\partial y}(x_0, y_0),$$

a necessary condition for f to have an extreme value at (x_0, y_0) is

$$\frac{\partial f}{\partial x}(x_0, y_0) = \frac{\partial f}{\partial y}(x_0, y_0) = 0.$$

In this example, with $f(x, y) = x^2 + y^2$,

$$\frac{\partial f}{\partial x}(x, y) = 2x \quad \text{and} \quad \frac{\partial f}{\partial y}(x, y) = 2y,$$

and so the only extreme value of f in the interior of the ellipse occurs at $(x_0, y_0) = (0, 0)$. From the graph of f, shown in Figure 6.11, we see that the value 0 there is a local minimum. We next consider the values of f on the boundary curve itself. The ellipse is defined parametrically by the function

$$g(t) = (x, y) = \left(\cos t, \frac{1}{\sqrt{2}} \sin t\right), \quad 0 \le t < 2\pi.$$

Thus the values of f on the ellipse are given as the values of the composition $f \circ g$. Any extreme values of f on the ellipse will be extreme for $f \circ g$. The latter is a real-valued function of one variable, and we treat it in the usual way, that is, by setting its derivative equal to zero. By the chain rule, we obtain

$$\frac{d}{dt}(f \circ g) = \nabla f\big(g(t)\big) \cdot g'(t)$$

$$= (2\cos t, 2/\sqrt{2}\sin t) \cdot (-\sin t, 1/\sqrt{2}\cos t)$$

$$= -2\cos t \sin t + \sin t \cos t$$

$$= -\tfrac{1}{2}\sin 2t.$$

Extreme values therefore may occur at $t = 0$, $\pi/2$, π, and $3\pi/2$. The corresponding values of (x, y) are $(1, 0)$, $(0, 1/\sqrt{2})$, $(-1, 0)$, and $(0, -1/\sqrt{2})$, and those of f are $1, \tfrac{1}{2}, 1$, and $\tfrac{1}{2}$, respectively. We see that the absolute minimum of f is 0 at $(0, 0)$ and that the absolute maximum of f occurs at the two points $(1, 0)$ and $(-1, 0)$. Notice that the two extreme values of $f \circ g$ that occur at $t = \pi/2$ and $3\pi/2$ are not extreme for f, as we see by looking at Figure 6.11.

The methods used in the preceding example are valid in any number of dimensions. The next theorem is the principal criterion used in this extension, and although we can prove it by reducing it to the single-variable method, we give a proof that contains the single-variable situation as a special case.

4.2   Theorem.   If a differentiable function $\mathbb{R}^n \xrightarrow{f} \mathbb{R}$ has a local extreme value at a point $\mathbf{x}_0$ interior to its domain, then $\nabla f(\mathbf{x}_0) = \mathbf{0}$.

Proof.   Suppose f has a local minimum at $\mathbf{x}_0$. For any unit vector $\mathbf{u}$ in $\mathbb{R}^n$, there is an $\epsilon > 0$ such that if $-\epsilon < t < \epsilon$, then $f(\mathbf{x}_0) \leq f(\mathbf{x}_0 + t\mathbf{u})$. Hence, for $0 < t < \epsilon$,

$$0 \leq \frac{f(\mathbf{x}_0 + t\mathbf{u}) - f(\mathbf{x}_0)}{t}, \text{ and}$$

$$0 \leq \frac{f(\mathbf{x}_0 - t\mathbf{u}) - f(\mathbf{x}_0)}{t}.$$

It follows from Theorem 4.1 of Chapter 5 that

$$\frac{\partial f}{\partial \mathbf{u}}(\mathbf{x}_0) = f'(\mathbf{x}_0)\mathbf{u}.$$

Therefore

$$0 \leq \lim_{t \to 0+} \frac{f(\mathbf{x}_0 + t\mathbf{u}) - f(\mathbf{x}_0)}{t} = f'(\mathbf{x}_0)\mathbf{u},$$

$$0 \leq \lim_{t \to 0+} \frac{f(\mathbf{x}_0 - t\mathbf{u}) - f(\mathbf{x}_0)}{t} = f'(\mathbf{x}_0)(-\mathbf{u}) = -f'(\mathbf{x}_0)\mathbf{u}.$$

We conclude that $f'(\mathbf{x}_0)\mathbf{u} = 0$. Because $\mathbf{u}$ is an arbitrary unit vector, $f'(\mathbf{x}_0) = \mathbf{0}$. The argument for a maximum value is similar.   ∎

The previous theorem is what we should expect. Recall that

$$\frac{\partial f}{\partial \mathbf{u}}(\mathbf{x}_0) = f'(\mathbf{x}_0)\mathbf{u},$$

and that the derivative with respect to $\mathbf{u}$ measures the rate of change of f in the direction of $\mathbf{u}$. At an extreme point in the interior of the domain of f, this rate should be zero in every direction. The importance of the theorem is that of all the interior points $\mathbf{x}$ of the domain of f we need to look for extreme points only among those for which $f'(\mathbf{x}) = 0$. Points $\mathbf{x}$ for which $f'(\mathbf{x}) = 0$ are called **critical points** of f.

4B   Constraints

As we did in Example 1 we'll consider in more detail real-valued functions f on open sets D, trying to find the extreme points of f when f has its domain restricted to some subset S of D. Two possibilities that we must look out for are

1. A point $\mathbf{x}$ where $\nabla f(\mathbf{x}) = 0$ is not necessarily an extreme point for f.
2. f may have an extreme point $\mathbf{x}$ on a set S without having $\nabla f(\mathbf{x}) = 0$.

EXAMPLE 2   Let $f(x, y, z) = xyz$ in the set defined by $|x| \leq 1, |y| \leq 1, |z| \leq 1$. Thus the domain of f is the cube C with edges of length 2 illustrated in Figure 6.12(a). The condition $\nabla f(\mathbf{x}) = 0$ for critical points amounts to $(yz, xz, xy) = (0, 0, 0)$. The solutions of this equation are the points satisfying $x = y = 0$, or $x = z = 0$, or $y = z = 0$; in other words, the coordinate axes. Since f has the value zero at all of its critical points, and since f has both positive and negative values in the neighborhood of each of these points, no critical point can be an extreme point. Furthermore, a little thought shows that f has maximum value 1 and minimum value -1 on C. These values occur at the eight corners of the cube, none of which is a critical point for f.

The boundary set S of a region R in $\mathbb{R}^n$ is itself never an open subset of $\mathbb{R}^n$, so examining critical points of f is of no use in finding whatever extreme points of f may lie on S, as on the boundary of the cube in the previous example. More generally, we may be interested in maximizing or minimizing a function f whose domain is restricted to a lower-dimensional set, say a curve or a surface, that we may not necessarily regard as the boundary of some region.

EXAMPLE 3   The function $f(x, y, z) = y^2 - z - x$ has as its gradient the vector

$$\nabla f(x, y, z) = (-1, 2y, -1),$$

so f has no critical points as a function defined on $\mathbb{R}^3$. However, suppose that f is restricted to the curve γ defined parametrically by

$$(x, y, z) = (t, t^2, t^3), \quad -\infty < t < \infty.$$

On γ, f takes the values $F(t) = f(t, t^2, t^3) = t^4 - t^3 - t$, while t varies over $(-\infty, \infty)$. We have

$$F'(t) = 4t^3 - 3t^2 - 1 = (t - 1)(4t^2 + t + 1).$$

Then $F'(t)$ is zero only at $t = 1$. Furthermore, since $F''(t) = 12t^2 - 6t$, we have $F''(1) > 0$. It follows that f has a relative minimum at the point $(1, 1, 1)$ while restricted to the curve γ. The minimum value of f on γ is $f(1, 1, 1) = -1$, and there are no other extreme values.

EXAMPLE 4

Suppose the function $f(x, y, z) = x + y + z$ is restricted to the intersection of the two surfaces

$$x^2 + y^2 = 1, \ z = 2$$

shown in Figure 6.12(b). The curve C of intersection is parametrized by

$$\begin{pmatrix} x \\ y \\ z \end{pmatrix} = \begin{pmatrix} \cos t \\ \sin t \\ 2 \end{pmatrix}, \quad 0 \leq t < 2\pi.$$

The function f on C takes the value $F(t) = \cos t + \sin t + 2$. We have $F'(t) = -\sin t + \cos t$, So $F'(t) = 0$ at $t = \pi/4$ and $t = 5\pi/4$. Since $F''(\pi/4) < 0$ and $F''(5\pi/4) > 0$,

$$f\left(\frac{\sqrt{2}}{2}, \frac{\sqrt{2}}{2}, 2\right) = \sqrt{2} + 2$$

is the maximum and

$$f\left(-\frac{\sqrt{2}}{2}, \frac{\sqrt{2}}{2}, 2\right) = -\sqrt{2} + 2$$

is the minimum value for f on C.

   This problem can also be done by setting $y = \pm\sqrt{1 - x^2}$ and $z = 2$ in $f(x, y, z)$ and then finding the maximum and minimum values of the two functions

$$y = x \pm \sqrt{1 - x^2} + 2$$

on the interval $-1 \leq x \leq 1$.

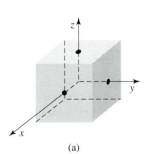

(a)

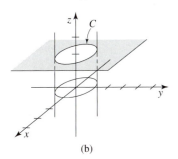

(b)

FIGURE 6.12

4C   Lagrange Multipliers

The solution of the previous problem depended on our being able to find a concrete parametric representation for the curve of intersection of the plane $z - 2 = 0$ and the cylinder $x^2 + y^2 - 1 = 0$. When a specific parametrization is not readily available, we can still sometimes apply the method of **Lagrange multipliers**, to be described next. The method consists in verifying the pure existence of a parametric representation and then deriving necessary conditions for there to be an extreme point for a function f when restricted to the curve or surface.

4.3   Lagrange Multiplier Method.   Suppose that a function $\mathbb{R}^n \xrightarrow{f} \mathbb{R}$ is differentiable and is restricted to a set S and that f has a local extreme point at $\mathbf{x}_0$ on S. Suppose that near $\mathbf{x}_0$, the set S is a smooth level set of a function $\mathbb{R}^n \xrightarrow{G} \mathbb{R}^m$ with $m < n$ and coordinate functions $G_1, G_2, \dots, G_m$. Then there are constants $\lambda_1, \lambda_2, \dots, \lambda_m$ such that $\mathbf{x}_0$ is a critical point of the real-valued function $f + \lambda_1 G_1 + \cdots + \lambda_m G_m$, that is,

$$\nabla f(\mathbf{x}_0) + \lambda_1 \nabla G_1(\mathbf{x}_0) + \cdots + \lambda_m \nabla G_m(\mathbf{x}_0) = \mathbf{0}.$$

FIGURE 6.13

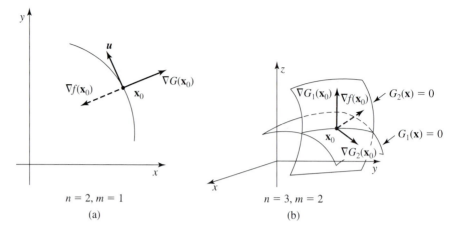

$n = 2, m = 1$

(a)

$n = 3, m = 2$

(b)

Why Does the method work?   What the Lagrange method does for us in practice is allow us to restrict attention to solutions $\mathbf{x}_0$ of the previous equation that also lie on S. A complete proof of the method's correctness is fairly complicated, but the geometric idea behind it is quite plausible. Since $\mathbf{x}_0$ is a local extreme point for f on S, derivatives of f in directions parallel to S at $\mathbf{x}_0$ should be zero; in other words,

$$\frac{\partial f}{\partial \mathbf{u}}(\mathbf{x}_0) = \nabla f(\mathbf{x}_0) \cdot \mathbf{u} = 0$$

for every unit vector $\mathbf{u}$ tangent to S at $\mathbf{x}_0$. But vectors $\mathbf{u}$ tangent to S are perpendicular to the m normal vectors $\nabla G_1(\mathbf{x}_0), \nabla G_2(\mathbf{x}_0), \ldots, \nabla G_m(\mathbf{x}_0)$, as illustrated in Figure 6.13 for the cases $n = 2, m = 1$ and $n = 3, m = 2$. The previous displayed equation shows that $\mathbf{u}$ is also perpendicular to $\nabla f(\mathbf{x}_0)$. The pictures suggest, and it can be proved, that $\nabla f(\mathbf{x}_0)$ is then a linear combination of the vectors $\nabla G_k(\mathbf{x}_0)$, a combination that we choose to write here in the form

$$\nabla f(\mathbf{x}_0) = -\lambda_1 \nabla G_1(\mathbf{x}_0) - \cdots - \lambda_m \nabla G_m(\mathbf{x}_0).$$

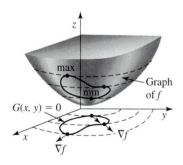

FIGURE 6.14

If $m = 1$, then $\nabla f(\mathbf{x}_0)$ will typically be parallel to $\nabla G_1(\mathbf{x}_0)$, as shown in Figure 6.13(a), but in any case there are constants λ_k such that the Lagrange condition is satisfied at $\mathbf{x}_0$. Figure 6.14 is a picture that includes the graph of a function $\mathbb{R}^2 \xrightarrow{f} \mathbb{R}$ and shows some critical vectors ∇f perpendicular to the set S determined by $G(x, y) = 0$.

Remark 1.   It's important to understand that the Lagrange condition is only a *necessary* condition that must hold at a local extreme point, which is why we can use it to exclude many other points from consideration. The Lagrange condition may hold at some points that are not extreme points, just as the gradient of a function f may be zero at points that are not extreme points of f.

Remark 2.   In all max-min problems it's important to have some grounds for believing that the desired extreme points do indeed exist. In addition we need to be able to distinguish among the critical points for those that provide relative maxima and minima. Section 4E gives a second derivative test that's sometimes helpful for doing this. For the Lagrange method to be effective, we also need to assure ourselves that the set S to which f is restricted is not only closed and bounded but also sufficiently smooth, so sharp corners on curves and sharp edges in surfaces have to be examined separately. We deal with these issues in the Exercises.

EXAMPLE 5 The problem of Example 4 is that of finding the extreme points of $f(x, y, z) = x + y + z$ subject to the conditions

$$G_1(x, y, z) = x^2 + y^2 - 1 = 0, \quad G_2(x, y, z) = z - 2 = 0.$$

According to Theorem 4.3 we write

$$(x + y + z) + \lambda_1(x^2 + y^2 - 1) + \lambda_2(z - 2).$$

The critical points of this function of x, y, and z occur when

$$1 + 2\lambda_1 x = 0, \quad 1 + 2\lambda_1 y = 0, \quad 1 + \lambda_2 = 0.$$

In addition, we have to satisfy the two given conditions on x, y, and z, making in all five equations in the five variables x, y, z, λ_1, λ_2. Solution of such a system is often best carried out by looking for simplifying substitutions among the equations, aimed at reducing the number of variables in any one equation. The first condition in this problem says that $z = 2$, so we're already down to four variables. Subtracting the second Lagrange condition from the first gives

$$2\lambda_1(x - y) = 0, \quad \text{so either} \quad \lambda_1 = 0 \quad \text{or} \quad x = y.$$

The choice $\lambda_1 = 0$ is inconsistent with $1 + \lambda_1 x = 0$, so we reject that possibility and accept $x = y$. Setting $x = y$ in $x^2 + y^2 - 1 = 0$ gives $2y^2 = 1$, so $(x, y) = \pm(1/\sqrt{2}, 1/\sqrt{2})$. Thus the only values of f that we need to check are at $(x, y, z) = (\pm 1/\sqrt{2}, \pm 1/\sqrt{2}, 2)$. There we find respectively $2 + \sqrt{2}$ for a maximum and $2 - \sqrt{2}$ for a minimum as in Example 4. In some problems it's convenient to find explicit values for one or more of the λ's along the way, though that wasn't necessary in this example. (It was nevertheless important not to neglect any possible λ-values.)

EXAMPLE 6 Find the maximum value of $f(x, y, z) = x - y + z$, subject to the condition $x^2 + y^2 + z^2 = 1$. The function

$$x - y + z + \lambda(x^2 + y^2 + z^2 - 1)$$

has critical points satisfying

$$1 - 2\lambda x = 0, \quad -1 + 2\lambda y = 0, \quad 1 + 2\lambda z = 0,$$

and

$$x^2 + y^2 + z^2 = 1.$$

The solutions of these four equations are found as in the previous example:

$$\lambda = \pm \frac{\sqrt{3}}{2}, \quad x = -y = z = \mp \frac{1}{\sqrt{3}}.$$

The maximum of f occurs at $(1/\sqrt{3}, -1/\sqrt{3}, 1/\sqrt{3})$. The maximum value is $\sqrt{3}$. What is the minimum value?

EXAMPLE 7   Let $g(x_1, x_2, \ldots, x_n) = 0$ implicitly define a surface S in $\mathbb{R}^n$ and let $\mathbf{a} = (a_1, a_2, \ldots, a_n)$ be a fixed point not on S. Suppose we want to minimize locally the distance from $\mathbf{a}$ to S. Minimizing the distance from $\mathbf{a}$ to S is the same as minimizing the square of the distance, which is easier to differentiate. Applying the Lagrange method, we look for points $\mathbf{p}$ on S that are critical points of

$$\sum_{k=1}^{n}(x_k - a_k)^2 + \lambda g(x_1, \ldots, x_n)$$

for some λ. The critical points satisfy, in addition to $g(x_1, \ldots, x_n) = 0$, the equations

$$2(x_1 - a_1) + \lambda \frac{\partial g}{\partial x_1}(x_1, \ldots, x_n) = 0$$

$$\vdots$$

$$2(x_n - a_n) + \lambda \frac{\partial g}{\partial x_n}(x_1, \ldots, x_n) = 0.$$

In vector form, these equations reduce at the critical point $\mathbf{p}$ to

$$\begin{pmatrix} p_1 - a_1 \\ \vdots \\ p_n - a_n \end{pmatrix} = -\frac{\lambda}{2}\begin{pmatrix} \dfrac{\partial g}{\partial x_1}(\mathbf{p}) \\ \vdots \\ \dfrac{\partial g}{\partial x_n}(\mathbf{p}) \end{pmatrix},$$

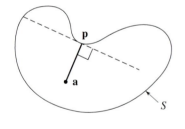

FIGURE 6.15

where $\mathbf{p} = (p_1, \ldots, p_n)$. The vector $\mathbf{p} - \mathbf{a}$ on the left is then either zero or parallel to the normal vector to S at $\mathbf{p}$, which appears on the right side of the equation. In other words, we have shown that $\mathbf{p} - \mathbf{a}$ is perpendicular to S, or else $\mathbf{p} = \mathbf{a}$. A two-dimensional example is illustrated in Figure 6.15, where $\mathbf{p}$ provides a local, but not a global, minimum.

EXAMPLE 8   Suppose that a cylindrical can is to contain a fixed volume V and that its surface area, with top and bottom, is to be as small as possible. If the radius of the can is x, and its height is y, then $V = \pi x^2 y$. We want to minimize the total area $2\pi x^2 + 2\pi xy$ of the top, bottom, and sides. We write

$$F(x, y) = 2\pi x^2 + 2\pi xy + \lambda(\pi x^2 y - V)$$

and look for critical points of F. We find that $F_x = 0$, $F_y = 0$ reduce to

$$2x + y + \lambda xy = 0, \quad 2x + \lambda x^2 = 0.$$

The second equation is satisfied if $x = 0$ or if $\lambda x = -2$. But $x = 0$ would require $V = 0$, so we substitute $\lambda x = -2$ into the first equation to get $2x = y$. Thus height y

must equal diameter $2x$. The value of x for a given volume V can then be determined from the equation $2\pi x^3 = \pi x^2 y = V$.

EXAMPLE 9

The planes

$$x + y + z - 1 = 0 \quad \text{and} \quad x + y - z = 0$$

intersect in a line S as shown in Figure 6.16(a). Let $f(x, y, z) = xy$, and restrict f to the line S. Using the Lagrange method to minimize f on S, we consider

$$xy + \lambda(x + y + z - 1) + \mu(x + y - z).$$

Its critical points occur when

$$y + \lambda + \mu = 0, \quad x + \lambda + \mu = 0, \quad \lambda - \mu = 0.$$

The only point that satisfies these conditions, together with the condition that it lie on S, is $\mathbf{x}_0 = (\frac{1}{4}, \frac{1}{4}, \frac{1}{2})$. The minimum value is thus $\frac{1}{16} = f(\mathbf{x}_0)$. Note that f has no maximum on S. We have also that $\nabla f(\mathbf{x}_0) = (\frac{1}{4}, \frac{1}{4}, 0)$, which is perpendicular to S. The unit vector $\mathbf{u}$ in the direction of $\nabla f(\mathbf{x}_0)$ is shown in Figure 6.16(a) with its initial point moved to $\mathbf{x}_0$.

4D   Saddle Points

A critical point $\mathbf{x}_0$ of a function f such that $f(\mathbf{x}_0)$ is neither a local maximum nor a local minimum value for f is called a **saddle point** for f.

EXAMPLE 10

Let $f(x, y) = y^2 - x^2$. Then $f_x(0, 0) = f_y(0, 0) = 0$, so $\mathbf{x}_0 = (0, 0)$ is a critical point. Since $f(x, 0) = -x^2 < 0$ for all $x \neq 0$, and $f(0, y) = y^2 > 0$ for all $y \neq 0$, $f(0, 0) = 0$ is neither a local maximum nor a local minimum value and so $(0, 0)$ is a saddle point. The graph of f is shown in Figure 6.16(b), a hyperbolic paraboloid.

FIGURE 6.16

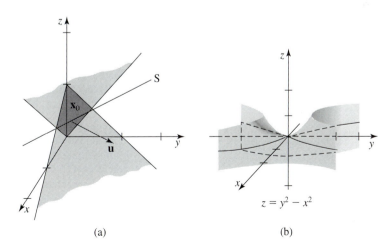

$$z = y^2 - x^2$$

(a)   (b)

Summary of Methods. To find the maximum and minimum values of a differentiable function f defined on a region R in $\mathbb{R}^n$, compare the values of f at the following points:

(a) Critical points of f in the interior of R

(b) Points on the boundary of R

In case (b), or in the case where R has no interior points, we can do either of the following:

1. Find a parametric representation g for the boundary of R, in which case we have a new problem with the function $f \circ g$ defined on a set of one lower dimension.
2. Use the Lagrange method.

EXERCISES

In Exercises 1 to 8, find all critical points, if any, of the given function.

1. $f(x, y) = x^2 + 4xy - y^2 - 8x - 6y$

2. $f(x, y) = x^2 + y^2 - 2xy - y$

3. $f(x, y) = x^2 - y^2 - 2xy - y$

4. $f(x, y) = x^2 - y^2 + x + y$

5. $f(x, y, z) = x^2 + y^2 - z^2 - xz + x$

6. $f(x, y, z) = x^2 + 4xy - y^2 + z^2 - 8x - 6y + z$

7. $f(x, y, z, w) = xy + yz + zw + wx + x + y - z + w$

8. $f(x, y, z, w) = x^2 - z^2 - w^2 - 2xy + 2zw$

In Exercises 9 to 14, find the points at which the largest and smallest values are attained by the function in the region.

9. $x + y$ inside and on the square with corners $(\pm 1, \pm 1)$

10. $x + y \sin x$ in the rectangular region $0 \le x \le 2\pi$, $-1 \le y \le 1$

11. $x^2 + 24xy + 8y^2$ in the circular region $x^2 + y^2 \le 25$

12. $1/(x^2 + y^2)$ in the circular region $(x - 2)^2 + y^2 \le 1$

13. $x^2 + y^2 + (2\sqrt{2}/3)xy$ in the elliptical region $x^2 + 2y^2 \le 1$

14. $x^2 + 2y^2$ in the circular region $x^2 + y^2 \le 1$

15. Find the points that are farthest from the origin on the closed curve in $\mathbb{R}^3$ parametrized by

$$g(t) = \big(\cos t, \sin t, \sin(t/2) \big).$$

In 16 to 19, find all critical points of the following functions in the given region.

16. $x + y + z$ where $x^2 + y^2 + z^2 \le 1$

17. $xy - xz$ where $|x| \le 1$, $|y| \le 1$, $|z| \le 1$

18. $x^2 + y^2 - z^2 + z$ where $x^2 + y^2 + z^2 \le 1$

19. $x^3 - y^3 + z^3 - x + y - z$ where $0 \le x \le 1$, $0 \le y \le 1$, $0 \le z \le 1$

20. Find the maximum value of the function $x(y + z)$, given that $x^2 + y^2 = 1$ and $xz = 1$.

21. Find the minimum value of $x + y^2$, subject to the condition $2x^2 + y^2 = 1$.

22. Let $f(x, y)$ and $g(x, y)$ be continuously differentiable, and suppose that subject to the condition $g(x, y) = 0$, $f(x, y)$ attains its maximum value M at (x_0, y_0). Show that the level curve $f(x, y) = M$ is tangent to the curve $g(x, y) = 0$ at (x_0, y_0).

23. A rectangular box with a square base and no top is to contain exactly 108 cubic inches. Find the dimensions that yield the minimum surface area.

24. A rectangular box with a square base and no top is to contain exactly 54 cubic inches. Find the dimensions that yield the minimum cost if base material costs four times as much as side material.

25. A rectangular box with a square base and no top is to contain volume V. Find the dimensions that yield the minimum cost if material for one side costs twice as much as for the other three sides and material for the base costs three times as much as for the less expensive sides.

26. (a) Find the minimum distance in $\mathbb{R}^2$ from the circle $x^2 + y^2 = 1$ to the line $x + y = 4$. [*Hint*: Treat the square of the distance as a function of four variables.]

(b) Solve part (a) geometrically and compare answers.

27. (a) Find the maximum value of $x^2 + xy + y^2 + yz + z^2$, subject to the condition $x^2 + y^2 + z^2 = 1$.

(b) Find the maximum value of the same function subject to the conditions $x^2 + y^2 + z^2 = 1$ and $x + \sqrt{2}y + z = 0$.

28. (a) Find the points $\mathbf{x}_0$ at which $f(x, y) = x^2 - y^2 - y$ attains its maximum on the circle $x^2 + y^2 = 1$.

(b) Find the directions in which f increases most rapidly at $\mathbf{x}_0$.

29. The planes $x + y - z - 2w = 1$ and $x - y + z + w = 2$ intersect in a set $\mathcal{F}$ in $\mathbb{R}^4$. Find the point on $\mathcal{F}$ that is nearest to the origin.

30. Let $\mathbf{x}_1, \ldots, \mathbf{x}_N$ be points in $\mathbb{R}^n$, and let

$$f(\mathbf{x}) = \sum_{k=1}^{N} |\mathbf{x} - \mathbf{x}_k|^2.$$

Find the point at which f attains its minimum and find the minimum value.

*31. Prove by solving an appropriate minimum problem that if $a_k > 0, k = 1, \ldots, n$, then

$$(a_1 a_2 \cdots a_n)^{1/n} \leq \frac{a_1 + a_2 + \cdots + a_n}{n}.$$

32. Let $f(x, y) = 2x^2 + y^2$ be restricted to the set S satisfying $x^{1/3} + y^{1/3} = 1$ as well as $|x| \leq 1$ and $|y| \leq 1$.

(a) What conclusion can you draw from applying the Lagrange multiplier method to f on S?

(b) Sketch the set S, and explain why f attains both maximum and minimum values on S. What are these values, and where are they attained?

33. Consider the extreme-value problem for $f(x, y, z) = x$ restricted to the set C_h satisfying $x^2 + y^2 + z^2 - 1 = 0$ and $z = h$ for constant $h \neq \pm 1$.

(a) For what values of h is C_h a closed, bounded, nonempty set in $\mathbb{R}^3$?

(b) What conclusion can you draw from applying the Lagrange multiplier method to this problem for the values of h found in part (a)?

(c) What are the maximum and minimum values attained by f on C_1?

34. The extreme-value problem for $f(x, y) = x^2 - y^2$ subject to the condition $x + y = 1$ has no solution for either a maximum or a minimum.

(a) Explain why this is so.

(b) What is the result of trying to apply the Lagrange method to this problem? Does the Lagrange theorem really apply here? Explain why or why not.

35. The extreme-value problem for $f(x, y, z) = x^2 - y^2 + z^2$ subject to the conditions $x+y+z = 3$ and $x^2+y^2+z^2 = 3$ has a unique solution. Find the unique solution to the problem.

36. A rectangular box with no top is to have surface area 48 square units. Our problem is to choose dimensions that maximize the volume $V = xyz$.

(a) Use the Lagrange method to eliminate from consideration those (x, y, z) that can't maximize V.

(b) Show by using the constraint on (x, y, z) that V tends to zero if x or y or z tends to $+\infty$.

(c) Use the results of parts (a) and (b) to find the constrained values of x, y, and z that maximize V.

37. (a) A rectangular shed with an open front and no flooring is to be built to shelter 108 cubic feet. If the roof material costs twice as much as the material for the three walls, what dimensions will be the least expensive?

(b) How would the answer change if roofing costs the same as walls?

38. A rectangular building is to be built to contain a fixed volume V. Heat loss through the roof and walls is proportional to area, and heat loss through the floor is negligible. Heat loss through the roof material is 3 times as rapid as through the wall material. What dimensions will minimize heat loss?

39. Let $\mathbb{R}^n \xrightarrow{F} \mathbb{R}^m$ be differentiable on $\mathbb{R}^n$. Prove that a smooth curve or surface S defined implicitly by $F(\mathbf{x}) = \mathbf{k}$ is a closed set in $\mathbb{R}^n$. [*Hint:* To show that S contains a given boundary point $\mathbf{x}_0$, let $\{\mathbf{x}_j\}$ be a sequence of points in S such that $\lim_{j \to \infty} \mathbf{x}_j = \mathbf{x}_0$ and consider $\lim_{j \to \infty} F(\mathbf{x}_j)$.]

4E   Second-Derivative Criterion

In this section we'll identify **strict** local extreme points, that is, points $\mathbf{x}_0$ for which a strict inequality $f(\mathbf{x}_0) > f(\mathbf{x})$ or $f(\mathbf{x}_0) < f(\mathbf{x})$ holds for $\mathbf{x} \neq \mathbf{x}_0$ in some neighborhood of $\mathbf{x}_0$. For functions of one real variable with two continuous derivatives the **second-derivative test** says that at a critical point x_0 interior to its domain f has (i) a local minimum if $f''(x_0) > 0$, (ii) a local maximum if $f''(x_0) < 0$, or (iii) neither of these if f'' changes sign at x_0. The intuitive geometric content of these alternatives is as follows: near x_0 the graph of f (i) is concave up and so stays above the horizontal tangent through $(x_0, f(x_0))$ if $f''(x_0) > 0$, (ii) is concave down and stays below the tangent if $f''(x_0) < 0$, or (iii) crosses the tangent in case f'' changes sign at x_0. We'll see that the alternatives for functions $\mathbb{R}^n \xrightarrow{f} \mathbb{R}$ are very similar, the main technical difference being the criteria that we use to decide about concavity. The essence of the extension to higher dimensions consists of examining the **second-order directional derivative**, defined naturally by

$$\frac{\partial^2 f}{\partial \mathbf{u}^2}(\mathbf{x}) = \frac{\partial}{\partial \mathbf{u}}\left(\frac{\partial f}{\partial \mathbf{u}}\right)(\mathbf{x}).$$

The basic analysis is in the following theorem, which contains the 1-dimensional case and so bears out the intuitive review we started with.

4.4 Theorem.   Let $\mathbb{R}^n \xrightarrow{f} \mathbb{R}$ be twice continuously differentiable on an open subset of $\mathbb{R}^n$ that contains a critical point $\mathbf{x}_0$ of f.

(i) If $\dfrac{\partial^2 f}{\partial \mathbf{u}^2}(\mathbf{x}_0) > 0$ for all unit vectors $\mathbf{u}$, then $f(\mathbf{x}_0)$ is a strict local minimum value.

(ii) If $\dfrac{\partial^2 f}{\partial \mathbf{u}^2}(\mathbf{x}_0) < 0$ for all unit vectors $\mathbf{u}$, then $f(\mathbf{x}_0)$ is a strict local maximum value.

(iii) If $\dfrac{\partial^2 f}{\partial \mathbf{u}^2}(\mathbf{x}_0)$ is positive for some $\mathbf{u}$ and negative for others, then $\mathbf{x}_0$ is a saddle point.

Proof.   We first observe that if $g(x)$ is real-valued and twice continuously differentiable on an interval containing 0 in its interior, then

$$g(x) = g(0) + g'(0)x + \int_0^x (x - t)g''(t)\,dt. \qquad (*)$$

To see this just compute the integral by parts and apply the fundamental theorem of calculus to the remaining integral. Now let $g(x) = f(\mathbf{x}_0 + x\mathbf{u})$, where $\mathbf{u}$ is a unit vector in $\mathbb{R}^n$. By the chain rule $g'(x) = \nabla f(\mathbf{x}_0 + x\mathbf{u}) \cdot \mathbf{u}$; applying the chain rule again we get

$$g''(x) = \nabla(\nabla f(\mathbf{x}_0 + x\mathbf{u}) \cdot \mathbf{u}) \cdot \mathbf{u} = \frac{\partial}{\partial \mathbf{u}}\left(\frac{\partial f}{\partial \mathbf{u}}\right)(\mathbf{x}_0 + x\mathbf{u}).$$

Since $\mathbf{x}_0$ is a critical point for f, it follows that $g'(0) = \nabla f(\mathbf{x}_0) \cdot \mathbf{u} = 0$. Replacing g by the corresponding expressions in f everywhere in Equation $(*)$, we get

$$f(\mathbf{x}_0 + x\mathbf{u}) = f(\mathbf{x}_0) + \int_0^x (x - t)\frac{\partial^2 f}{\partial \mathbf{u}^2}(\mathbf{x}_0 + t\mathbf{u})\,dt.$$

But $\partial^2 f/\partial \mathbf{u}^2$ is continuous, so for small enough t the sign of $\partial^2 f/\partial \mathbf{u}^2(\mathbf{x}_0 + t\mathbf{u})$ is the same as the sign of $\partial^2 f/\partial \mathbf{u}^2(\mathbf{x}_0)$. Case (i) now follows by checking that the inequality $f(\mathbf{x}_0 + x\mathbf{u}) - f(\mathbf{x}_0) > 0$ holds at all points $\mathbf{x}$ for which $|x\mathbf{u}| = |x| < \delta$, for some positive $\delta > 0$; cases (ii) and (iii) are similar. ∎

Remark. Just as in the 1-dimensional case, the three cases of Theorem 4.4 don't cover all possibilities. (See Exercise 1.) For example, $\partial^2 f/\partial \mathbf{u}^2(\mathbf{x}_0)$ could be zero for all unit vectors $\mathbf{u}$, in which case the statement yields no information.

Theorem 4.4 is in a way a straightforward generalization of the second-derivative criterion of single-variable calculus. But in practice the transition from dimension 1 to dimension 2 is a distinctive one, because in $\mathbb{R}^2$ there are infinitely many ways to approach a critical point, while in $\mathbb{R}$ there are at most two approaches, from the right or the left. For dimension 3 or more the additional distinctions are mainly technical, so we'll concentrate on functions $\mathbb{R}^2 \xrightarrow{f} \mathbb{R}$.

4.5 Theorem. Suppose that $f(x, y)$ has continuous second-order partials f_{xx}, $f_{xy} = f_{yx}$ and f_{yy} defined on an open set containing $\mathbf{x}_0$. Let $\mathbf{u} = (u, v)$. Then

$$\frac{\partial^2 f}{\partial \mathbf{u}^2}(\mathbf{x}_0) = f_{xx}(\mathbf{x}_0)u^2 + 2f_{xy}(\mathbf{x}_0)uv + f_{yy}(\mathbf{x}_0)v^2.$$

Proof. We know that $\partial f/\partial \mathbf{u} = \nabla f \bullet \mathbf{u} = f_x u + f_y v$, so the second-order derivative $\partial^2 f/\partial \mathbf{u}^2$ is

$$\nabla(\nabla f \bullet \mathbf{u}) \bullet \mathbf{u} = \nabla(f_x u + f_y v) \bullet (u, v)$$
$$= (f_x u + f_y v)_x u + (f_x u + f_y v)_y v$$
$$= f_{xx} u^2 + 2f_{xy} uv + f_{yy} v^2. \qquad ∎$$

In applications the second partials in Theorem 4.5 are to be evaluated at some critical point $\mathbf{x}_0 = (x_0, y_0)$. Often the decision as to whether we're in case (i), (ii), (iii), or none of these, in Theorem 4.4 will follow from very elementary observations.

EXAMPLE 11

Suppose $f_{xx}(x_0, y_0) = p > 0$, $f_{yy}(x_0, y_0) = q > 0$ and $f_{xy}(x_0, y_0) = 0$. Then $(\partial^2 f/\partial \mathbf{u}^2)(x_0, y_0) = pu^2 + qv^2 > 0$, since u and v can't both be zero if $|(u, v)| = 1$. Thus we're in case (i) of Theorem 4.3, and $f(x_0, y_0)$ would be a strict minimum at a critical point. On the other hand, if $f_{xx}(x_0, y_0) = f_{yy}(x_0, y_0) = 0$ and $f_{xy}(x_0, y_0) = r \neq 0$, then $(\partial^2 f/\partial \mathbf{u}^2)(x_0, y_0) = ruv$. Since ruv can be both positive and negative, a critical point at (x_0, y_0) would be a saddle point.

EXAMPLE 12

Let $f(x, y) = x^2 y + y^3 - y$. Then $f'(x, y) = (2xy, x^2 + 3y^2 - 1) = (0, 0)$ has solutions $(0, \pm 1/\sqrt{3})$ and $(\pm 1, 0)$, so these four points are the critical points. We have $f_{xx}(x, y) = 2y$, $f_{xy}(x, y) = 2x$, $f_{yy}(x, y) = 6y$, so $f_{xx}(0, \pm 1/\sqrt{3}) = \pm 2/\sqrt{3}$, $f_{xy}(0, \pm 1/\sqrt{3}) = 0$, $f_{yy}(0, \pm 1/\sqrt{3}) = \pm 6/\sqrt{3}$. Then the second directional derivative reduces to

$$\frac{\partial^2 f}{\partial \mathbf{u}^2} = \pm \frac{2}{\sqrt{3}} u^2 \pm \frac{6}{\sqrt{3}} v^2.$$

It follows that $(0, 1/\sqrt{3})$ is a strict minimum point and $(0, -1/\sqrt{3})$ is a strict maximum. At $(1, 0)$ we find $f_{xx}(1, 0) = 0$, $f_{xy}(1, 0) = 2$, $f_{yy}(1, 0) = 0$, so the second derivative is $2uv$, which exhibits different signs depending on whether u and v have the same or opposite sign. Hence $(1, 0)$ is a saddle point. Similarly, $(-1, 0)$ is a saddle point.

If the quadratic polynomial of Theorem 4.5 doesn't yield to the quick sign analysis that applied in the two previous examples, we have a simple test based on the discriminant of a quadratic equation.

4.6 Theorem. Let $D = f_{xx}(x_0, y_0)f_{yy}(x_0, y_0) - f_{xy}^2(x_0, y_0)$, where $\mathbb{R}^2 \xrightarrow{f} \mathbb{R}$ is twice continuously differentiable. Assume (x_0, y_0) is a critical point of $f(x, y)$.

(i) If $D > 0$ and $f_{xx}(x_0, y_0) > 0$ or $f_{yy}(x_0, y_0) > 0$, then $f(x_0, y_0)$ is a strict local minimum.

(ii) If $D > 0$ and $f_{xx}(x_0, y_0) < 0$ or $f_{yy}(x_0, y_0) < 0$, then $f(x_0, y_0)$ is a strict local maximum.

(iii) If $D < 0$, then $f(x, y)$ has a saddle point at (x_0, y_0).

Proof. Since $\mathbf{u} = (u, v)$ is a unit vector, u and v can't both be zero. Thus for any choice of u and v, the quadratic polynomial $\partial^2 f/\partial \mathbf{u}^2$ of Theorem 4.5 can be written in either of the two forms

$$u^2 \left[f_{xx} + 2(f_{xy})(v/u) + (f_{yy})(v/u)^2 \right] \text{ or } v^2 \left[(f_{xx})(u/v)^2 + 2(f_{xy})(u/v) + f_{yy} \right].$$

Deciding whether either of these two functions changes sign or not comes down to deciding whether either of the quadratic polynomials in v/u or u/v has a real root or not; if not, there is no sign change. But "no real root" is equivalent to

$$4f_{xy}^2(x_0, y_0) - 4f_{xx}(x_0, y_0)f_{yy}(x_0, y_0) < 0,$$

in other words to $D > 0$. To decide in that event whether the quadratic polynomial is negative or positive, all we have to do is check one of the coefficients $f_{xx}(x_0, y_0)$ or $f_{yy}(x_0, y_0)$. This covers (i) and (ii). Case (iii) is now quite simple. If $D < 0$, there are two distinct real root values for v/u or u/v, so there are two definite sign changes as (u, v) varies. Hence (x_0, y_0) is a saddle point. ∎

If $D = 0$ no conclusions follow from Theorem 4.6. Keeping the next example in mind along with Figure 6.17 makes it easier to recall the alternatives of Theorem 4.6.

EXAMPLE 13   The general quadratic polynomial $f(x, y) = ax^2 + 2bxy + cy^2$ has critical points satisfying

$$\begin{cases} 2ax + 2by = 0 \\ 2bx + 2cy = 0 \end{cases} \text{ or } \begin{pmatrix} 2a & 2b \\ 2b & 2c \end{pmatrix} \begin{pmatrix} x \\ y \end{pmatrix} = \begin{pmatrix} 0 \\ 0 \end{pmatrix}.$$

Hence there is a single critical point at $(x_0, y_0) = (0, 0)$ unless the determinant $D = 4ac - 4b^2 = 0$. Note that $f_{xx} = 2a$, $f_{yy} = 2c$ and $f_{xy} = f_{yx} = 2b$. If

FIGURE 6.17

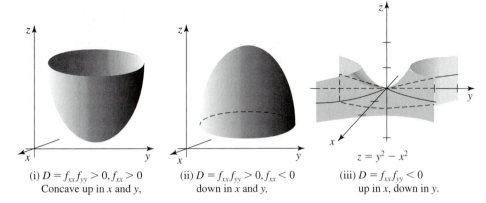

(i) $D = f_{xx}f_{yy} > 0, f_{xx} > 0$
Concave up in x and y,

(ii) $D = f_{xx}f_{yy} > 0, f_{xx} < 0$
down in x and y,

$z = y^2 - x^2$
(iii) $D = f_{xx}f_{yy} < 0$
up in x, down in y.

$a = c = 1$ and $b = 0$, then $D = 2^2 > 0$, and the graph of $f(x, y)$ is an elliptic paraboloid with minimum at $(0, 0)$ [Case (i)]. If $a = c = -1$, then $D = (-2)^2 > 0$, and we get an elliptic paraboloid with maximum at $(0, 0)$ [Case (ii)]. If a and c have opposite signs with $b = 0$, or if $a = c = 0$ with $b \neq 0$, then $D < 0$ and the graph of $f(x, y)$ is a hyperbolic paraboloid with a saddle point at $(0, 0)$ [Case (iii)]. See Figure 6.17.

EXAMPLE 14   Let $f(x, y) = 3x^2 - 6xy + 5y^2 + y^3$. Since $f_x(x, y) = 6x - 6y$ and $f_y(x, y) = (-6x + 10y + 3y^2)$, the equation $f_x = 0$ shows that a critical point must satisfy $x = y$. From the equation $f_y = 0$ we then get $3y^2 + 4y = y(3y + 4) = 0$. Hence the critical points are at $(0, 0)$ and $(-\frac{4}{3}, -\frac{4}{3})$. To classify these points we compute

$$f_{xx}(x, y) = 6, \quad f_{xy}(x, y) = -6, \quad f_{yy}(x, y) = 6y + 10.$$

At $(0, 0)$ we find $D = 60 - 36 = 24$, so $(0, 0)$ is a local extreme point. Because $f_{xx}(0, 0) = 6 > 0$, $f(0, 0) = 0$ is a strict minimum. At $(-\frac{4}{3}, -\frac{4}{3})$, we find $D = (6)(2) - 36 = -24$, so this point is a saddle.

EXERCISES

1. (a)   Show that the functions $f(x) = x^3$ and $g(x) = x^4$ behave differently at their critical points, but that the second-derivative criterion of Theorem 4.4 fails to distinguish between their behaviors.

(b)   Find the critical points of the functions $f(x, y) = (x + y)^3$ and $g(x, y) = (x + y)^4$, and describe the behavior of f and g near these points. Show also that Theorem 4.4 fails to distinguish between the two behaviors.

2. (a)   For twice-differentiable functions $\mathbb{R} \xrightarrow{f} \mathbb{R}$ there are in principle two values for the second-order derivative $\dfrac{\partial^2 f}{\partial \mathbf{u}^2}(x)$, because we can let

$\mathbf{u} = \pm 1$. Show that both second partials are equal to $f''(x)$.

(b)   Show more generally that for twice-differentiable functions $\mathbb{R}^n \xrightarrow{f} \mathbb{R}$

$$\frac{\partial^2 f}{\partial \mathbf{u}^2}(\mathbf{x}) = \frac{\partial^2 f}{\partial(-\mathbf{u})^2}(\mathbf{x}).$$

Note that it's not correct to show this by simply replacing $(-\mathbf{u})^2$ by $\mathbf{u}^2$ in the previous equation.

Find the critical points of each of the functions 3 to 13, and try to apply the second-derivative test to determine whether each critical point is a maximum, a minimum or a saddle point. Note, however, that if the conditions of Theorem 4.4 don't apply you may have to make a decision based on special features of the problem.

3. $f(x, y) = x^2 - 2x + y^2 + 4y$

4. $f(x, y) = x^2 + 4xy - y^2 - 8x - 6y$

5. $f(x, y) = x^2 - xy - y^2 + 5y$

6. $f(x, y) = x^2 - 2y^2 - x$

7. $f(x, y) = x^4 + y^4$

8. $f(x, y) = (x - y)^4$

9. $f(x, y) = x^2 + 2xy$

10. $f(x, y) = x^3 - y^3 - 2xy$

11. $f(x, y) = x^{-1} + xy - 8y^{-1}$

12. $f(x, y) = e^{-x^2 - y^2}$

13. $f(x, y) = e^{x^2 - y^2}$

14. **(a)** Integrate by parts to show that $\int_{x_0}^x (x - t)g''(t)dt = g(x) - g(x_0) - g'(x_0)x$, as in the proof of Theorem 4.4.

   (b) Use the equation of part (a) to establish directly the second-derivative criterion for functions $\mathbb{R} \xrightarrow{g} \mathbb{R}$: At a critical point x_0, (i) $g''(x_0) > 0$ implies $g(x_0)$ is a minimum, (ii) $g''(x_0) < 0$ implies $g(x_0)$ is a maximum, (iii) $g''(x)$ changing sign at x_0 implies $g(x_0)$ is neither a minimum nor a maximum.

   (c) Give three examples to show that if $g''(x_0) = 0$, then $g(x_0)$ can be a maximum, a minimum, or neither.

15. For second-degree polynomials, the second-order derivative is closely related to the quadratic part of the polynomial.

   (a) Show that if $p(x, y) = ax^2 + bxy + cy^2$, then for $\mathbf{u} = (u, v)$,

$$\frac{1}{2}\frac{\partial^2 p}{\partial \mathbf{u}^2}(x, y) = p(u, v), \quad \text{independent of } (x, y).$$

   (b) Show that if $q(x, y, z) = ax^2 + by^2 + cz^2 + lyz + mzx + nxy$, then for $\mathbf{u} = (u, v, w)$,

$$\frac{1}{2}\frac{\partial^2 q}{\partial \mathbf{u}^2}(x, y, z) = q(u, v, w),$$

$$\text{independent of } (x, y, z).$$

16. For twice continuously differentiable functions $\mathbb{R}^n \xrightarrow{f} \mathbb{R}$, the n-by-n matrix

$$H_f(\mathbf{x}) = \left(f_{x_j x_k}(\mathbf{x}) \right)_{\substack{j=1,\ldots,n \\ k=1,\ldots,n}}$$

is called the **Hessian matrix** of f. Note that $H_f(\mathbf{x})$ is *symmetric* relative to its main diagonal since $f_{x_j x_k} = f_{x_k x_j}$.

   (a) Show that the discriminant D of Theorem 4.6 is the determinant of a 2-by-2 Hessian matrix.

   (b) If $\mathbf{u} = (u_1, u_2, \ldots, u_n)$ is a unit vector, show that the second-order directional derivative of f can be written using the matrix-vector product $H_f\mathbf{u}$ as either

   (i) $\dfrac{\partial^2 f}{\partial \mathbf{u}^2} = \displaystyle\sum_{j,k=1}^n f_{x_j x_k} u_j u_k$ or **(ii)** $\dfrac{\partial^2 f}{\partial \mathbf{u}^2} = (H_f\mathbf{u}) \cdot \mathbf{u}$.

   (c) Write out the analogue of the formula in Theorem 4.5 for a generic twice continuously differentiable function $f(\mathbf{x}) = f(x, y, z)$ with $\mathbf{u} = (u, v, w)$.

*17. The roots $\lambda_1, \ldots, \lambda_n$ of the polynomial equation det $(H_f(\mathbf{x}_0) - \lambda I) = 0$ are called the **eigenvalues** of $H_f(\mathbf{x}_0)$, and we can use them to characterize a critical point $\mathbf{x}_0$ of f. It's possible to prove that if all $\lambda_k > 0$, then $f(\mathbf{x}_0)$ is a strict local minimum, and if all $\lambda_k < 0$, then $f(\mathbf{x}_0)$. is a strict local maximum. If roots of both signs occur, then $\mathbf{x}_0$ s a saddle point.

   (a) Show that criteria (i), (ii), and (iii) of Theorem 4.6 are implied by this statement in the two-dimensional case.

   (b) Apply the statement to $f(x, y, z) = x^2 + y^2 + z^2 - 2xy - 4yz - 6xz$.

*18. The **principal subminors** $H_f^{(m)} = \left(f_{x_j x_k}(\mathbf{x}) \right)_{\substack{j=1,\ldots,m \\ k=1,\ldots,m}}$ of the Hessian matrix play a role in the generalization of Theorem 4.6 to higher dimensions. It can be shown that if $\mathbf{x}_0$ is a critical point of f and the determinants det $H_f^{(m)}, m = 1, \ldots, n$ are positive then f has a strict local minimum at $\mathbf{x}_0$. Furthermore, if these determinants alternate in sign so that $(-1)^m$ det $H_f^{(m)}(\mathbf{x}_0)$ are positive then f has a strict local maximum at $\mathbf{x}_0$. Show that this formulation has the criteria (i) and (ii) of Theorem 4.6 as special cases.

4F   Steepest Ascent Method

This is an alternative to the standard calculus method for finding the maximum or minimum value of a continuously differentiable function $\mathbb{R}^n \xrightarrow{f} \mathbb{R}$ when the extreme point is in an open subset of the domain of f. Our classic strategy so far is to restrict attention to the critical points of f, that is, the solutions of the n simultaneous equations embodied in the single vector equation $f'(\mathbf{x}) = 0$, but finding those solutions can be problematic, even with Newton's method.

Steepest Ascent Method.   Though the method described here applies in any number of dimensions, it's easiest to visualize in the special case $\mathbb{R}^2 \xrightarrow{f} \mathbb{R}$. Imagine that the graph of f represents the topography of some mountainous terrain, and that an ambitious climber is determined always to head up the steepest way from a given point. Figures 6.4(b) and (c) in Section 1 illustrate the setting. This strategy determines the climber's path, and it's natural to call that path a **path of steepest ascent**. To make mathematics out of these remarks, all we have to do is to recall that at a point $\mathbf{x} = (x, y)$ in the plane, the direction of maximum increase of f is the same as the direction of the gradient vector $\nabla f(\mathbf{x})$, provided this vector is not zero. (This effectively tells the climber what horizontal compass heading gives the steepest way up at each point of the path.) It seems reasonable that a path of steepest ascent will in general lead to the "top," at which point we would have reached a local maximum of f where $\nabla f = 0$. (It might only be the summit of one of the foothills.) To reach a local minimum we would always head in the direction $-\nabla f(\mathbf{x})$ opposite to that of the gradient. We'd then call what we're doing the **steepest descent method**.

The numerical implementation of steepest ascent amounts to taking a succession of small steps along the direction of the gradient, at each point $\mathbf{x}$ along the way observing the value $f(\mathbf{x})$. At each step we make a decision about whether to continue or not and record the value of f at the end of the last step as our estimate for a local maximum value for f.

Each step in the process has the same general form as the very first step. Most of the computation takes place in the domain of f, which in the case of $\mathbb{R}^2$ we can think of as a topographic map of the graph of f. Having decided on $\mathbf{x}_0$, a starting point, we move in the direction of the gradient vector at $\mathbf{x}_0$ by a certain distance to a new point $\mathbf{x}_1$:

$$\mathbf{x}_1 = \mathbf{x}_0 + h_0 \nabla f(\mathbf{x}_0), \quad \text{where } h_0 > 0.$$

In general we go from $\mathbf{x}_n$ to $\mathbf{x}_{n+1}$ as follows:

$$\mathbf{x}_{n+1} = \mathbf{x}_n + h_n \nabla f(\mathbf{x}_n), \quad \text{where } h_n > 0, n = 1, 2, \ldots$$

By following the gradient in small steps we move at each step from a point $\mathbf{x}$ in a direction determined by the vector $\mathbf{u} = \nabla f(\mathbf{x}_n)$ having two properties:

   (i) $\mathbf{u}$ points in the direction of maximum increase of f.
   (ii) $\mathbf{u}$ is perpendicular to the level set of f containing $\mathbf{x}$.

To search for a local minimum value, we think of pursuing the path of **steepest descent** by making $h_n < 0$. This choice will tend to move us downhill in the direction opposite to the gradient direction.

The remaining question is how the numerical factors h_n are to be chosen. The simplest choice is to make all h_n the same; this choice has the consequence that as we approach the location of a maximum value for f, where the gradient is zero, continuity of ∇f will cause the vectors $h\nabla f(\mathbf{x}_n)$ to get shorter and shorter. This is desirable, for otherwise we face the danger of taking a big step right past the extreme point. The following routine summarizes the method. From a different point of view what we're doing here is finding approximate solutions to a system of two differential equations, namely

$$\frac{dx}{dt} = f_x(x, y), \quad \frac{dy}{dt} = f_y(x, y).$$

At the end of the section we describe a method for improving the accuracy of the process by varying h.

```
DEFINE f(x, y)  = 1 - x² - 2y²
DEFINE fx(x, y) = -2x (x-coordinate of ∇ f(x, y).)
DEFINE fy(x, y) = -4y (y-coordinate of ∇ f(x, y).)
SET h = 0.4
INPUT (x, y) (Starting point.)
DO
    SET (x₁, u₁) = (x, y) (Keep for later use.)
       (Next compute new position.)
    SET (x, y) = (x + hfx(x, y), y + hfy(x, y))
    PRINT x, y, f(x, y)
LOOP UNTIL |f(x, y) - f(x₁, y₁)| < ε
    (Stop if change in f is < ε.)
```

EXAMPLE 15

The function used to illustrate the preceding routine is $f(x, y) = 1 - x^2 - 2y^2$, with a single critical point at $(x, y) = (0, 0)$ and maximum value 1 at that point. Figure 6.18 shows some curves on the graph of f along with their projections into the xy-plane. Starting with $x_0 = y_0 = 0.5$ and $\varepsilon = 0.001$ the output of numbers x_n, y_n and $f(x_n, y_n)$ from the next to last line of the program would be as follows:

x	y	$f(x, y)$
0.1	−0.3	0.81
0.02	0.18	0.9348
0.004	−0.108	0.976656
0.0008	0.0648	0.991601
0.00016	−0.03888	0.996977
0.000032	0.023328	0.999608

Note that the step size $h = 0.4$ is fairly large; this choice results in repeatedly overshooting the origin, as evidenced by the alternating sign in the y-coordinates. A smaller choice, say $h = 0.1$, avoids this effect but requires more steps to achieve the same accuracy. With a very small, and very inefficient, step size the points (x, y) would approximate a flow line of the gradient field that cuts perpendicularly across level curves of f as shown in the picture that accompanies the routine.

FIGURE 6.18

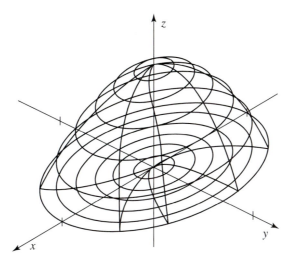

EXAMPLE 16

The function chosen for this example, $f(x, y) = \sin^2 x + \sin^2 y$, has its critical points at the solutions of the equations

$$f_x(x, y) = 2 \sin x \cos x = \sin 2x$$

$$f_y(x, y) = 2 \sin y \cos y = \sin 2y.$$

The solutions are all of the form $(x, y) = (j\pi/2, k\pi/2)$, where j and k are integers. If $j = 2l$, $k = 2m$ are both even we get $f(l\pi, m\pi) = 0$ for a local minimum, and if $j = 2l + 1$, $k = 2m + 1$ we get $f((2l + 1)\pi/2, (2m + 1)\pi/2) = 2$ for a local maximum. Having this information before doing the computing allows us to experiment intelligently with different starting points and different sizes for h to see what the results are. You're asked to do this in the exercises. The trick is to coordinate the choice for h with the starting point; too small or too large a value for h can require many steps to reach an acceptable approximation, or may lead to no convergence at all. In our example, starting at $(x, y) = (1.5, 1.5)$, computation with step size $h = 0.4$ and $\varepsilon = 0.0001$ ends after three steps; note that $\pi/2 \approx 1.57079$:

x	y	$f(x, y)$
1.55645	1.555645	1.99959
1.51341	1.51341	1.9998
1.57022	1.57022	2.0000

In the preceding example we were able to start reasonably close to the extreme point at $(x, y) = (\pi/2, \pi/2)$ because we already knew the exact location of the point. If we always had that kind of information there would be no need for numerical methods at all. In practice, we need to make educated guesses about the location of extreme points. If the domain of the real-valued function f under consideration is two-dimensional, making a computer-aided graph of f may be helpful. Failing that, Newton's method for approximate root location may be helpful, as described in Chapter 5, Section 5.

The preceding program outline is crude in that it forces you to stick rigidly with a single step size h throughout the computation. This rigidity is clearly a defect when the approximations $\mathbf{x}_n$ are getting close to a critical point, causing repeated overshooting from side to side. What we need is a method for automatically choosing the step size and adjusting it as the process proceeds. One way to do this is to think of replacing $f(\mathbf{x} + \mathbf{u})$, where $\mathbf{u} = h\nabla f(\mathbf{x})$, by its second-degree Taylor expansion

$$f(\mathbf{x}) + \frac{\partial f}{\partial \mathbf{u}}(\mathbf{x}) + \frac{1}{2}\frac{\partial^2 f}{\partial \mathbf{u}^2}(\mathbf{x}).$$

In dimension 2, with $\mathbf{u} = (u, v)$, the middle term is

$$\nabla f(x, y) \bullet (u, v) = f_x(x, y)u + f_y(x, y)v.$$

From Theorem 4.5 in Section 4E we have for the third term

$$f_{xx}(x, y)u^2 + 2f_{xy}(x, y)uv + f_{yy}(x, y)v^2.$$

Since in our application $u = hf_x(x, y)$ and $v = hf_y(x, y)$, the expressions get a bit cluttered unless we suppress the evaluations at (x, y), so with that in mind we have the Taylor approximation

$$f + h\left(f_x^2 + f_y^2\right) + \frac{h^2}{2}\left(f_{xx}f_x^2 + 2f_{xy}f_xf_y + f_{yy}f_y^2\right).$$

We now maximize this function, which is possible since it's parabolic with respect to the variable h, and has a maximum if the coefficient of h^2 is negative. Setting the derivative with respect to h equal to zero, we find the critical value of h to be

$$h = -\frac{f_x^2(x, y) + f_y^2(x, y)}{f_{xx}(x, y)f_x^2(x, y) + 2f_{xy}(x, y)f_x(x, y)f_y(x, y) + f_{yy}(x, y)f_y^2(x, y)}.$$

Replacing the line that sets $h = 0.4$ in the earlier program by this more complicated expression requires earlier insertion into the program of definitions for the second derivative functions $f_{xx}(x, y)$, $f_{xy}(x, y)$, and $f_{yy}(x, y)$.

Note that the second derivative that appears in the denominator of the expression for h will typically be negative if we're looking for a maximum, so h will be positive. Correspondingly, in looking for a minimum h will be negative. The applet ASCENT/DESCENT implements the method for fixed h, and ASCENT/DESCENT+ does the same for the variable h method. Both applets are available at http://math.dartmouth.edu/~rewn/. The former program is simpler to apply because it requires the user to calculate only first partial derivatives, so you'd resort to the automatic step-size version only if the simpler version fails.

EXERCISES

1. The function $\sin^2 x + \sin^2 y$ has infinitely many local maxima, with maximum value 2, and local minima, with minimum value 0. Find numerical approximations for the coordinates of those extreme points that lie in the rectangle $0 \le x \le 5, 0 \le y \le 5$.

2. Verify the details in the derivation of the variable step factor h.

In Exercises 3 to 6, find approximate values for the coordinates of the local extreme (both max. and min.) points and the local extreme values of the following functions.

3. $f(x, y) = (x + 2y)e^{-x^2-2y^2}$, for (x, y) in $\mathbb{R}^2$

4. $f(x, y) = (x^2 + 2y)e^{-3x^2-2y^2}$, for $x^2 + y^2 \leq 9$

5. $f(x, y) = ((x - 1)^2 + y^2)^{-1} + ((x + 1)^2 + y^2)^{-1}$, for (x, y) in $\mathbb{R}^2$

6. $f(x, y) = \exp(-x^2 - 2y^2)\sin(x^2 + y^2)$, $x^2 + y^2 \leq 1$

The method of steepest ascent (or descent) applied to a real-valued function $f(x)$ generates a sequence of numbers $x_{n+1} = x_n + hf'(x_n)$. Apply this approach to Exercises 7 and 8 using a hand calculator.

7. $f(x) = \exp(-x^2)\sin x$, $0 \leq x \leq 4$

8. $f(x) = (x^3 + 2x + 1)e^{-x^2}$, for x in $\mathbb{R}$

9. A rectangular box has two faces contained in the positive xz-plane and yz-planes, with their common edge on the z-axis. The two corners farthest from the origin lie on the graphs of $e^{-0.1x^2-0.2y^2}$ and $-e^{-0.4x^2-0.1y^2}$, respectively.

 (a) Use steepest ascent to estimate the dimensions that will maximize the volume of the box.

 (b) Solve a similar problem by hand if the two functions are replaced by $e^{-x^2-y^2}$ and $-e^{-x^2-y^2}$.

SECTION 5   CURVILINEAR COORDINATES

We can often simplify formulas that occur in mathematics and its applications by finding good descriptions of the quantities to be singled out for special attention. Since in practice these quantities are usually represented by vectors whose entries are real-number coordinates, our problem here is one of choosing the most useful system of coordinates. Thus we consider introducing coordinates in $\mathbb{R}^n$ different from the natural coordinates x_k that appear in the designation of a typical point $(x_1, \ldots, x_n)$. Specifically, to each point $(x_1, \ldots, x_n)$ there will be assigned a new n-tuple $(u_1, \ldots, u_n)$. If we are to be able to switch back and forth from one set of coordinates to the other, the assignment described must be one to one, that is, for each $(x_1, \ldots, x_n)$ there will be just one n-tuple $(u_1, \ldots, u_n)$ and vice versa. In practice we sometimes make the new coordinate assignment for some specific subregion of $\mathbb{R}^n$ rather than for the whole space. In what follows we'll denote the space of new coordinate vectors $(u_1, \ldots, u_n)$ by $\mathbb{U}^n$ to help us change variables coherently from the standard coordinates $(x_1, \ldots, x_n)$ in $\mathbb{R}^n$.

5A   Polar Coordinates

Consider two copies of 2-dimensional space: the xy-plane $\mathbb{R}^2$. One of these we'll rename the $r\theta$-plane, denoting it by $\mathbb{U}^2$. The function $\mathbb{U}^2 \xrightarrow{P} \mathbb{R}^2$ defined by

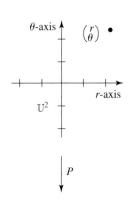

$$\begin{pmatrix} x \\ y \end{pmatrix} = P\begin{pmatrix} r \\ \theta \end{pmatrix} = \begin{pmatrix} r\cos\theta \\ r\sin\theta \end{pmatrix}, \quad \begin{cases} 0 < r < \infty \\ 0 \leq \theta < 2\pi \end{cases}$$

has a simple geometric description. The image under P of a point (r, θ) is the point $\mathbf{x} = (x, y)$ whose distance from the origin is r and such that the angle from the positive x axis to $\mathbf{x}$ in the counterclockwise direction is θ. See Figure 6.19.

The image of P consists of all of $\mathbb{R}^2$ except for the origin, so for any point (x, y) in $\mathbb{R}^2$ there are numbers r and θ, called **polar coordinates** of $\mathbf{x}$, such that

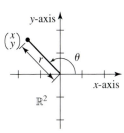

FIGURE 6.19

5.1 $x = r\cos\theta$ and $y = r\sin\theta$.

For two points (r_1, θ_1) and (r_2, θ_2) in the domain of P, the equations

$$r_1\cos\theta_1 = r_2\cos\theta_2,$$

$$r_1\sin\theta_1 = r_2\sin\theta_2$$

hold whenever $r_1 = r_2$ and $\theta_1 = \theta_2 + 2\pi m$ for some integer m. Hence the polar coordinates of a point (x, y) in $\mathbb{R}^2$ are not uniquely specified without some restrictions on r and θ. However if $(x, y) \neq (0, 0)$ the polar coordinates of (x, y) are uniquely specified up to an integer multiple of 2π in the θ-coordinate. To see this square both sides of the two displayed equations and add to get $r_1^2 = r_2^2$. Assuming $r > 0$ we conclude that $r_1 = r_2$. But then $\cos \theta_1 = \cos \theta_2$ and $\sin \theta_1 = \sin \theta_2$, so $(\cos \theta_1, \sin \theta_1)$ and $(\cos \theta_2, \sin \theta_2)$ represent the same point on a circle of radius 1 centered at the origin in $\mathbb{R}^2$. Hence $\theta_1 = \theta_2 + 2\pi m$ for some integer m.

The preceding paragraph says that P is not one-to-one, but that it becomes so if its domain is restricted to be a subset of a rectangular half-strip in the $r\theta$-plane defined by inequalities

$$0 < r < \infty, \quad \theta_0 \leq \theta < \theta_0 + 2\pi.$$

So restricted, P does have an inverse function, and we can find some partial formulas for the inverse by solving the equations $x = r \cos \theta$, $y = r \sin \theta$ for r and θ. We obtain, for $x \neq 0$,

$$r = \sqrt{x^2 + y^2}, \quad \theta = \arctan \frac{y}{x} + k\pi.$$

We have used the common convention of restricting an inverse trigonometric function to the principal branch of the corresponding multiple-valued function. Hence the image of arctan is the interval $-\pi/2 < \theta < \pi/2$. If follows that the function defined by

$$\begin{pmatrix} r \\ \theta \end{pmatrix} = \begin{pmatrix} \sqrt{x^2 + y^2} \\ \arctan \dfrac{y}{x} \end{pmatrix}, \quad x > 0,$$

is the inverse of the restriction of P by $0 < r < \infty$ and $-\pi/2 < \theta < \pi/2$. Similarly the function defined by

$$\begin{pmatrix} r \\ \theta \end{pmatrix} = \begin{pmatrix} \sqrt{x^2 + y^2} \\ \operatorname{arccot} \dfrac{x}{y} \end{pmatrix}, \quad y > 0,$$

is the inverse of the restriction of P by $0 < r < \infty$ and $0 < \theta < \pi$.

EXAMPLE 1   We have not defined polar coordinates for the origin of the xy-plane simply because

$$\begin{pmatrix} 0 \cos \theta \\ 0 \sin \theta \end{pmatrix} = \begin{pmatrix} 0 \\ 0 \end{pmatrix} \quad \text{for all } \theta,$$

so the one-to-one requirement fails at the origin. This failure causes no real difficulty; for example, the equation in rectangular coordinates of the **lemniscate**,

$$(x^2 + y^2)^2 = 2(x^2 - y^2), \tag{1}$$

becomes, upon introduction of polar coordinates,

$$r^2 = 2 \cos 2\theta, \quad r > 0. \tag{2}$$

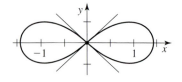

FIGURE 6.20

The image under P of the set of pairs (r, θ) that satisfy Equation 2 is precisely the set of pairs (x, y) that satisfy Equation 1, except for the origin. We may simply fill in this one point. See Figure 6.20.

5B   Spherical Coordinates

Consider the function $\mathbb{U}^3 \xrightarrow{S} \mathbb{R}^3$, defined by

5.2
$$S\begin{pmatrix} r \\ \phi \\ \theta \end{pmatrix} = \begin{pmatrix} r\sin\phi\cos\theta \\ r\sin\phi\sin\theta \\ r\cos\phi \end{pmatrix}, \quad \begin{cases} 0 < r < \infty \\ 0 \le \phi < \pi \\ 0 \le \theta < 2\pi. \end{cases}$$

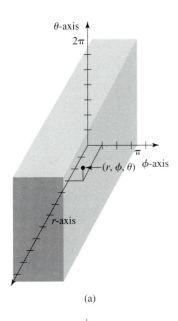

(a)

$\downarrow T$

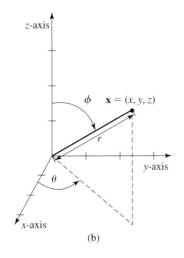

(b)

FIGURE 6.21

Here, for simplicity, we have restricted the domain of S from the outset so that S is one-to-one. Its range is all of $\mathbb{R}^3$ with the exception of the z-axis. Hence it assigns **spherical coordinates** (r, ϕ, θ) to every point of $\mathbb{R}^3$ except those on the z-axis. As with polar coordinates in the plane, the spherical coordinates (r, ϕ, θ) of a point $\mathbf{x} = (x, y, z)$ have a simple geometric interpretation. See Figure 6.21(b). The number r is the distance from $\mathbf{x}$ to the origin. The coordinate ϕ is the angle in radians between the vector $\mathbf{x}$ and the positive z-axis. Finally, θ is the angle in radians from the positive x-axis to the projected image $(x, y, 0)$ of $\mathbf{x}$ on the xy-plane. The symbols ϕ and θ are sometimes interchanged, particularly in physical applications.

We can compute an explicit expression for the inverse function, which we denote by S^{-1}, by solving the equations

$$x = r\sin\phi\cos\theta,$$
$$y = r\sin\phi\sin\theta,$$
$$z = r\cos\phi,$$

for r, θ, and ϕ. We get, for $y \ge 0$,

$$\begin{pmatrix} r \\ \phi \\ \theta \end{pmatrix} = S^{-1}\begin{pmatrix} x \\ y \\ z \end{pmatrix} = \begin{pmatrix} \sqrt{x^2+y^2+z^2} \\ \arccos\frac{z}{\sqrt{x^2+y^2+z^2}} \\ \arccos\frac{x}{\sqrt{x^2+y^2}} \end{pmatrix}, \quad x^2 + y^2 > 0.$$

Since the image of the principal branch the arccosine function is the interval $0 \le \theta \le \pi$, this function is actually the inverse of the function obtained by restricting the domain of S by the further condition $0 \le \theta \le \pi$. To get values of θ in the interval $\pi < \theta < 2\pi$, corresponding to $y < 0$, we add π to the third coordinate in the preceding formula. Note that when $\phi = 0$ the spherical coordinate transformation reduces to $x = r\cos\theta$, $y = r\sin\theta$, $z = 0$; this amounts to changing to polar coordinates in the plane $z = 0$, that is in the xy-plane in $\mathbb{R}^3$.

EXAMPLE 2   Three surfaces in $\mathbb{R}^3$ defined by spherical coordinate equations $r = 1$, $\phi = \pi/4$, and $\theta = \pi/3$, respectively, are shown in Figure 6.22. The corresponding rectangular coordinate equations derived from the preceding expressions for S^{-1} are respectively

$$x^2 + y^2 + z^2 = 1, \text{ with } x^2 + y^2 > 0$$

$$z = \frac{\sqrt{2}}{2}\sqrt{x^2+y^2+z^2}, \text{ with } z > 0$$

$$y = \sqrt{3}x, \text{ with } x > 0,$$

FIGURE 6.22

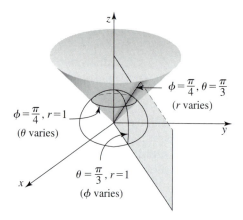

$\phi = \frac{\pi}{4}, \theta = \frac{\pi}{3}$
(r varies)

$\phi = \frac{\pi}{4}, r = 1$
(θ varies)

$\theta = \frac{\pi}{3}, r = 1$
(ϕ varies)

FIGURE 6.23

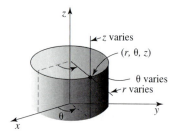

z varies
(r, θ, z)

θ varies
r varies

5C   Cylindrical Coordinates

The coordinate transformation is defined by

5.3

$$\begin{pmatrix} x \\ y \\ z \end{pmatrix} = \begin{pmatrix} r \cos \theta \\ r \sin \theta \\ z \end{pmatrix}, \quad \begin{cases} 0 < r < \infty \\ -\pi < \theta \leq \pi \\ -\infty < z < \infty. \end{cases}$$

The coordinates (r, θ, z) are obtained by a straightforward extension to $\mathbb{R}^3$ of polar coordinates in $\mathbb{R}^2$. Figure 6.23 shows the effect of varying each of the three coordinates.

5D   Jacobian Matrices

The name "curvilinear" is applied to coordinates for the reason that if all but one of the nonrectangular coordinates are held fixed and the remaining one is varied, the coordinate transformation defines a curve in $\mathbb{R}^n$. Thus in plane polar coordinates the coordinate curves are circles and straight lines, as shown in Figure 6.24(b). For spherical coordinates, typical coordinate curves are the circle, semi-circle, and half-line obtained as intersections of the pairs of surfaces shown in Figure 6.22. The curves and surfaces obtained by varying one or more curvilinear coordinate variables play the same role that the natural coordinate lines and planes of $\mathbb{R}^n$ do. For example, to say that a point in $\mathbb{R}^3$ has rectangular coordinates $(x, y, z) = (1, 2, 1)$ is to say that it lies at the intersection of the coordinate planes $x = 1$, $y = 2$, and $z = 1$. Similarly,

FIGURE 6.24

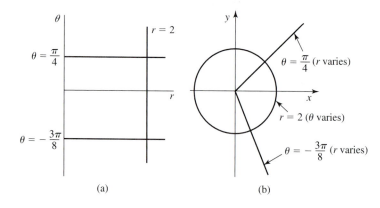

(a)   (b)

saying that a point in $\mathbb{R}^3$ has spherical coordinates $(r, \phi, \theta) = (1, \pi/4, \pi/3)$ is to say that the point lies at the intersection of the surfaces shown in Figure 6.22.

Generalizing from the preceding examples, we see that a system of curvilinear coordinates in $\mathbb{R}^n$ is determined by a function $\mathbb{U}^n \xrightarrow{T} \mathbb{R}^n$. It's assumed that for some open subset N in the domain of T, the restriction of T to N is one-to-one and therefore has an inverse T^{-1}. The **curvilinear coordinates** of a point $\mathbf{x}$ lying in the image set $T(N)$ are

$$\begin{pmatrix} u_1 \\ \vdots \\ u_n \end{pmatrix} = T^{-1} \begin{pmatrix} x_1 \\ \vdots \\ x_n \end{pmatrix}.$$

We impose fairly stringent regularity conditions on a coordinate transformation; specifically we'll assume that at every point $\mathbf{u}$ of N the function T is continuously differentiable and that $T'(\mathbf{u})$ is invertible. Thus the inverse function theorem (Theorem 2.3 will apply to T locally.

The polar, spherical, and cylindrical coordinate changes, represented by Equations 5.1, 5.2, and 5.3, have derivative matrices

5.4
$$\begin{pmatrix} \cos\theta & -r\sin\theta \\ \sin\theta & r\cos\theta \end{pmatrix},$$

5.5
$$\begin{pmatrix} \sin\phi\cos\theta & r\cos\phi\cos\theta & -r\sin\phi\sin\theta \\ \sin\phi\sin\theta & r\cos\phi\sin\theta & r\sin\phi\cos\theta \\ \cos\phi & -r\sin\phi & 0 \end{pmatrix},$$

5.6
$$\begin{pmatrix} \cos\theta & -r\sin\theta & 0 \\ \sin\theta & r\cos\theta & 0 \\ 0 & 0 & 1 \end{pmatrix},$$

respectively. These matrices, and, more generally, the derivative matrices of differentiable coordinate transformations, are called **Jacobian matrices**. We've seen in Chapter 5, Section 4 that the columns of these matrices have simple geometric interpretations; each column of a Jacobian matrix is obtained by differentiation of the coordinate functions with respect to a single variable, while holding the other variables fixed. This means that the jth column of the matrix represents a tangent vector to the curvilinear coordinate curve for which the jth coordinate is allowed to vary. That is, let the coordinate transformation be given by $\mathbb{U}^n \xrightarrow{T} \mathbb{R}^n$. Then the jth column of the matrix of the derivative $T'(\mathbf{u}_0)$ is a tangent vector, which we'll denote by $\mathbf{c}_j$, at $\mathbf{x}_0 = T(\mathbf{u}_0)$, to the curvilinear coordinate curve formed by allowing only the jth coordinate of $\mathbf{u}_0$ to vary. Tangent vectors are shown (with their initial points translated to the point $\mathbf{x}_0$) in Figure 6.25 for some polar, spherical, and cylindrical coordinate curves. The coordinates of the tangent vectors $\mathbf{c}_1, \ldots, \mathbf{c}_n$ are rectangular coordinates, not curvilinear coordinates.

Remark.   We can now see that the Jacobian matrix itself of a coordinate transformation is the matrix of a certain first-degree change of coordinates at each point. To see this, consider curvilinear coordinates in $\mathbb{R}^n$ given by $\mathbf{x} = T(\mathbf{u})$, where $\mathbf{u}$

FIGURE 6.25

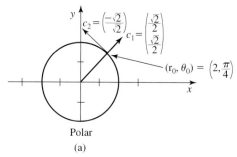

Polar
(a)

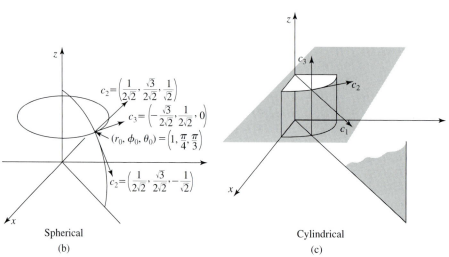

Spherical   Cylindrical
(b)   (c)

is the curvilinear coordinate variable. Fix a point $\mathbf{x}_0$ having curvilinear coordinates $\mathbf{u}_0$. At $\mathbf{x}_0$ we can introduce a new origin and new unit vectors $\mathbf{e}_1, \ldots, \mathbf{e}_n$ with the same directions as the natural basis vectors for $\mathbb{R}^n$. Then multiplication by the matrix $T'(\mathbf{u}_0)$ transforms the vectors $\mathbf{e}_1, \ldots, \mathbf{e}_n$ into the vectors $\mathbf{c}_1, \ldots, \mathbf{c}_n$ that are the tangent vectors to the curvilinear coordinate curves. Figure 6.25 illustrates the relation between the $\mathbf{e}_i$ and the $\mathbf{c}_i$. Notice that the vectors $\mathbf{c}_1, \ldots, \mathbf{c}_n$ will be linearly independent if and only if $T'(\mathbf{u}_0)$ is invertible. This is one reason for requiring not only that a coordinate transformation be one-to-one in a neighborhood of a point, but also that its derivative be invertible.

EXERCISES

In Exercises 1 to 4, make a sketch using xy coordinates in $\mathbb{R}^2$ of the curves given in polar coordinates.

1. $r = 1, \pi \le \theta \le 3\pi/2$

2. $r = \theta, 0 \le \theta \le \pi/2$

3. $r(\sin\theta - \cos\theta) = \pi/2, r > 0, \pi/2 \le \theta \le \pi$

4. $r = \pi/2\cos\theta, \pi \le \theta \le 3\pi/2$

In Exercises 5 to 8, make a sketch using xyz coordinates in $\mathbb{R}^3$ of the curves and surfaces given in spherical coordinates.

5. $r = 2, 0 \le \theta \le \pi/4, \pi/4 \le \phi \le \pi/2$

6. $1 \le r \le 2, \theta = \pi/2, \phi = \pi/4$

7. $0 \le r \le 1, 0 \le \theta \le \pi/2, \phi = \pi/4$

8. $0 \le r \le 1, \theta = \pi/4, 0 \le \phi \le \pi/4$

9. Use cylindrical coordinates in $\mathbb{R}^3$ to describe the region defined in rectangular coordinates by $0 \le x,\, x^2 + y^2 \le 1$.

10. Let (r, θ) be polar coordinates in $\mathbb{R}^2$. The equation

$$\begin{pmatrix} r \\ \theta \end{pmatrix} = \begin{pmatrix} 2\cos t \\ t \end{pmatrix}, \quad 0 \le t \le \frac{\pi}{2},$$

describes a curve in $\mathbb{U}^2$. Sketch this curve, and sketch its image in $\mathbb{R}^2$ under the polar coordinate transformation.

11. Let (r, ϕ, θ) be spherical coordinates in $\mathbb{R}^3$. The equation

$$\begin{pmatrix} r \\ \phi \\ \theta \end{pmatrix} = \begin{pmatrix} 1 \\ t \\ t \end{pmatrix}, \quad 0 \le t \le \frac{\pi}{2},$$

determines a curve in $\mathbb{R}^3$ (as well as in the $r\phi\theta$ space $\mathbb{U}^3$). Sketch the curve in $\mathbb{R}^3$. [*Suggestion*: The curve lies on a sphere.]

12. Compute the determinants of the matrices 5.4, 5.5, and 5.6, and show that they are

(a) $\dfrac{\partial(x, y)}{\partial(r, \theta)} = r$

(b) $\dfrac{\partial(x, y, z)}{\partial(r, \phi, \theta)} = r^2 \sin \phi$

(c) $\dfrac{\partial(x, y, z)}{\partial(r, \theta, z)} = r$

For which points (x, y) in $\mathbb{R}^2$, or (x, y, z) in $\mathbb{R}^3$, do the matrices 5.4, 5.5, and 5.6 fail to be invertible?

13. With a, b, c positive, the equations

$$x = ar \sin \phi \cos \theta$$
$$y = br \sin \phi \sin \theta \quad ,\, 0 < \phi < \pi,\, 0 \le \theta < 2\pi,$$
$$z = cr \cos \phi$$

define ellipsoidal coordinates in $\mathbb{R}^3$. For $a = 1, b = c = 2$, sketch a typical example of each of the three kinds of coordinate surface.

14. Compute the coordinates of tangent vectors to the coordinate curves for the general ellipsoidal coordinates given in Exercise 7, when $a = b = 1$, $c = 2$, and $r = \frac{1}{2}$, $\phi = \theta = \pi/2$.

15. Let r, ϕ, and θ be spherical coordinates in $\mathbb{R}^3$. The equation

$$\begin{pmatrix} r \\ \phi \\ \theta \end{pmatrix} = \begin{pmatrix} 1 \\ t \\ t^2 \end{pmatrix}$$

determines a curve in $\mathbb{R}^3$. Compute the coordinates of a tangent vector to the curve.

16. Prove that in 3-dimensional spherical coordinates, the sphere $x_1^2 + x_2^2 + x_3^2 = 1$ has the equation $r = 1$.

Chapter 6 REVIEW

In Exercises 1 to 6, let $f(x, y, z) = x^2 + y^2 - z^2$, $g(u, v) = (\cos u, \sin u, v)$, $h(u, v) = (u + v, uv)$, and $K(x, y) = xy$. Find

1. $f'(x, y, z)$

2. $h'(u, v)$

3. $g'(\pi/3, \pi^2/36)$

4. $(g \circ h)'(\pi/6, \pi/6)$

5. $\partial K / \partial \mathbf{u}$ at $(1, \sqrt{3})$, where $\mathbf{u}$ is the unit vector in the direction of the gradient of K at $(1, \sqrt{3})$

6. $\dfrac{\partial}{\partial x} f\big(K(x, y),\ K(x + y, x - y),\ x^2 + y^2\big)$

7. If the temperature at a point (x, y, z) of a solid ball of radius 3 centered at $(0, 0, 0)$ is given by $T(x, y, z) = yz + zx + xy$, find the direction in which T is increasing most rapidly at $(1, 1, 2)$.

8. Suppose $T(x, y, x) = Ke^{-c(x^2 + y^2 + z^2)}$ is the temperature at a point (x, y, z) inside a solid ball $x^2 + y^2 + z^2 \le a^2$, where K and c are positive constants. Show that the surfaces of equal temperature (i.e., the isotherms) are spheres, and that the temperature gradients point toward the center of the ball. What is the magnitude of the temperature gradient on a sphere of radius $b < a$?

9. Define $\mathbf{F}(x, y) = (x/\sqrt{x^2 + y^2},\ y/\sqrt{x^2 + y^2})$ at points $(x, y) \ne (0, 0)$.
  (a) Sketch the vector field defined by $\mathbf{F}(x, y)$.
  (b) Find the maximum rate of increase of $\sqrt{x^2 + y^2}$ at $(x, y) \ne (0, 0)$.

10. Let $x = u^2 - v^2$ and $y = u^2 + v^2$. Assume $z = f(x, y)$ has partial derivatives $f_x(0, 2) = 3$ and $f_y(0, 2) = 4$. Find $\partial z / \partial u$ and $\partial z / \partial v$ at $(u, v) = (1, 1)$.

11. A function $\mathbb{R}^n \xrightarrow{f} \mathbb{R}$ is said to be **homogeneous of degree** m if $f(t\mathbf{x}) = t^m f(\mathbf{x})$ for all $t > 0$ and all $\mathbf{x}$ in the domain of f. For example $f(x, y) = (x^3 + y^3) \cos(y/x)$ is homogeneous of degree 3. Show that if f is differentiable and homogeneous of degree m, then $f(\mathbf{x}) = \frac{1}{m}\mathbf{x} \cdot \nabla f(\mathbf{x})$.

12. Let $f(z)$ be differentiable and $w = f(ax + by)$, a, b constant. Show that $bw_x = aw_y$.

13. Let $h(z)$ be a real-valued function, differentiable for all real z. Define $u(x, t) = h(x - at)$, where a is a constant. Show that

$$a^2 \frac{\partial^2 u}{\partial x^2} - \frac{\partial^2 u}{\partial t^2} = 0 \quad \text{for all pairs } (x, t).$$

14. Suppose $w = f(x, y)$ is differentiable and $x = u + v$, $y = u - v$. Show that

$$\frac{\partial w}{\partial u} \frac{\partial w}{\partial v} = \left(\frac{\partial f}{\partial x}\right)^2 - \left(\frac{\partial f}{\partial y}\right)^2.$$

15. Suppose that $f(u, v) = u^2 v + uv^3$ and that u and v are differentiable functions of s and t with $u(2, 1) = -2$, $u_s(2, 1) = 3$, $v(2, 1) = 2$ and $v_s(2, 1) = -4$. Find $\dfrac{\partial f}{\partial s}(2, 1)$.

16. Define F from the uv plane to the xy plane by

$$x = e^{u+v} + e^{u-v}, \quad y = e^{u+v} - e^{u-v}.$$

(a) Show that an image point (x, y) necessarily satisfies $x > |y|$. Sketch the region R in the xy plane determined by $x > |y|$.

(b) Show that F satisfies the hypotheses of the Inverse Function Theorem 2.3 in Section 2, and so has a continuously differentiable inverse defined in some neighborhood of each image point (x, y) in R.

(c) Show that F has a global inverse defined for all (x, y) with $x > |y|$ by

$$u = \tfrac{1}{2}\ln\left(\tfrac{1}{2}(x + y)\right) + \tfrac{1}{2}\ln\left(\tfrac{1}{2}(x - y)\right)$$

$$v = \tfrac{1}{2}\ln\left(\tfrac{1}{2}(x + y)\right) - \tfrac{1}{2}\ln\left(\tfrac{1}{2}(x - y)\right).$$

17. The pressure P, volume V and temperature T of a gas are related by $PV = 6T$.

(a) Show that it's impossible to have P, V, and T all increase simultaneously with each one increasing at its own constant rate.

(b) Assume T increases steadily at $2°$ per minute and V at 3 cubic centimeters per minute. At $t = 0$, $T = 10°$ and $V = 8$ cubic centimeters. Find dP/dt at $t = 0$.

(c) Using the same data as in part (b), find dP/dt at $t = 30$.

18. The graphs of $x + yz = 0$ and $y + xz = 0$ intersect in a curve containing the point $(x, y, z) = (1, -1, 1)$ and parametrized by functions of the form $y = y(x), z = z(x)$.

(a) Find $y'(1)$ and $z'(1)$ by implicit differentiation.

(b) Find a tangent vector to the curve at $(1, -1, 1)$.

19. A function $z = f(x, y)$ is defined implicitly by $x + y^2 + z^2 = 3z$. Find z_x and z_y at $(1, 1, 1)$.

20. The equation $e^x + e^y + e^z - 3xyz = 3$ defines $x = x(y, z)$, $y = y(x, z)$ and $z = z(x, y)$ near $(x, y, z) = (0, 0, 0)$.

(a) Find $\partial x/\partial y$ at this point.

(b) Find the other possible partial derivatives at $(0, 0, 0)$ without any additional computation, and explain your reasoning.

21. Consider the curve γ of intersection of the two level surfaces

$$x^2 + 2y^2 + 3z^2 = 6$$

$$x^2 + y^2 - z^2 = 1.$$

You are asked to find a tangent vector to γ at $(1, 1, 1)$ in each of two different ways:

(a) Solve explicitly for $x = x(z)$ and $y = y(z)$ to find a parametric representation $(x, y, z) = (x(z), y(z), z)$. Then find a tangent vector.

(b) Use implicit differentiation to find dx/dz and dy/dz near $z = 1$. Then find a tangent vector.

22. The equations $x^2 + y^2 + z^2 = 6$, $x + y + z = 4$ define $z = z(x)$ and $y = y(x)$ as differentiable functions near $(x, y, z) = (1, 2, 1)$.

(a) Sketch the curve defined by the intersection of the two surfaces.

(b) Find dz/dx and dy/dx at the point $(x, y, z) = (1, 2, 1)$.

(c) Find a vector parallel to the line tangent to the curve in part (a) at the point $(1, 2, 1)$. [*Hint*: A parametric representation for the curve near $(1, 2, 1)$ is given by $(x, y(x), z(x))$.]

23. Let $f(x, y)$ be differentiable, and let $\mathbf{u} = (\cos\theta, \sin\theta)$.

(a) Show that the directional derivative $\partial f/\partial\mathbf{u}$ has a critical point as a function of θ whenever $\mathbf{u}$ and ∇f are parallel.

(b) How does the result of part (a) relate to Theorem 1.2 in Section 1?

24. Find a point on the sphere $x^2 + y^2 + z^2 = 28$ at which the tangent plane is parallel to the tangent plane to $xy + \ln z = 6$ at the point $(2, 3, 1)$.

25. The equations

$$\sin(x + u) - \cos(y + v) + xy - u + uv = 1$$

$$u + v + x + y - ux - vy = 5$$

implicitly define x and y as functions of u and v. Find x_u and y_u at the point $(u, v, x, y) = (2, 1, -2, -1)$.

26. Let $f(x, y, z) = 3x^2z + 3xy - 6y^2 - 3z + 3z^3$.
   (a) Find $\nabla f(x, y, z)$.
   (b) Let S be the level set of f at 0. Find a unit vector that is perpendicular to S at the point $(1, 1, 1)$.
   (c) Find the tangent plane to S at $(1, 1, 1)$.

27. The equations $xy + zt = -4, x^2 + y + z + t^2 = 8$ define x and z as differentiable functions of y and t near $(x, y, z, t) = (-2, 1, -1, 2)$. Find $\partial x / \partial y$ and $\partial z / \partial y$ at this point.

28. A point moves on a differentiable curve in the xy-plane in $\mathbb{R}^3$ so that its position at time t is $(x(t), y(t), 0)$. A corresponding point on the graph of a differentiable function $z = f(x, y)$ is then at $(x(t), y(t), f(x(t), y(t)))$. Suppose that the velocity vector of the plane curve is given by the gradient field of $f(x, y)$, so that $(dx/dt, dy/dt) = (f_x(x, y), f_y(x, y))$ for a point (x, y) on the curve.
   (a) Show that, if $z(t) = f(x(t), y(t))$, then $dz/dt = |\nabla f(x, y)|^2$.
   (b) Show that the point on the graph of $f(x, y)$ corresponding to $(x(t), y(t))$ has speed
   $$v = |\nabla f(x, y)| \sqrt{1 + |\nabla f(x, y)|^2}.$$

29. Find the points where $f(x, y) = x^2 + y$ attains it maximum and minimum values on the subset of $\mathbb{R}^2$ where $x^2 + 2y^2 \le 1$.

30. Find the maximum value of $x + 2y$ subject to $x^2 + y^2 = 1$.

31. A rectangular box is to be constructed so that the length of one of its internal diagonals is 3 units. What is the maximum possible volume?

32. (a) It's clear that the minimum value of $f(x, y, z) = xyz$ subject to the conditions $0 \le x, 0 \le y, 0 \le z$ and $z + y + z = 1$ is zero. Find the maximum value.
   (b) What if xyz is replaced by xy^2z^3?

33. Let $u = f(r)$ and $r = \sqrt{x^2 + y^2}$. Show that $\dfrac{\partial^2 u}{\partial x^2} + \dfrac{\partial^2 u}{\partial y^2} = \dfrac{d^2 f}{dr^2} + \dfrac{1}{r}\dfrac{df}{dr}$ if $r \ne 0$.

34. Let $f(x, y) = (x^2 - y^2, 2xy) = (u, v)$ and $g(u, v) = (e^u \cos v, e^u \sin v) = (s, t)$. Using the chain rule, show that $\partial s / \partial x = \partial t / \partial y$ and $\partial s / \partial y = -\partial t / \partial x$.

35. Find all critical points in $\mathbb{R}^3$ of the function $f(x, y, z) = xy + yz + xz$.

36. Find the maximum and minimum values of the function $f(x, y) = x^2 + 4xy + y^2$ on the region in $\mathbb{R}^2$ defined by $x^2 + y^2 \le 1$.

37. Check all critical points of $2xy + y^2 + 4y + 2x$ for local maximality and minimality.

38. Use Lagrange multipliers to find the point or points on the parabola $y = x^2$ closest to the point $(0, b)$, where $(0, b)$ is a fixed point on the y-axis. Note that the answer will depend on b.

39. Let $\mathbf{x}_0$ be a nonzero vector in $\mathbb{R}^n$. Define a real-valued function on $\mathbb{R}^n$ by $f(\mathbf{x}) = \mathbf{x}_0 \cdot \mathbf{x}$. Without calculating any derivatives, show that the maximum value of $f(\mathbf{x})$ for $\mathbf{x}$ restricted by $|\mathbf{x}| = 1$ is $|\mathbf{x}_0|$. What is the minimum value? Do these answers change if the restriction is replaced by $|\mathbf{x}| \le 1$?

40. Find the minimum value of $f(x, y) = 3x^2 + 2y^2 - 6x - 4y$ subject to the condition $x + 2y = 4$. What can you say about the maximum? Answer both questions if the condition is replaced by $x + 2y \le 4$.

41. Let $f(x, y) = x^2y - 2x - y$.
   (a) Find all critical points of $f(x, y)$ in $\mathbb{R}^2$.
   (b) Find the maximum and minimum values of $f(x, y)$ when (x, y) is restricted to lie on the line segment joining the points $(0, 1)$ and $(1, 0)$.

42. Find the points on $x^2 + 2xy + 3y^2 = 14$ which are closest to and farthest from the origin.

43. (a) Find all critical points of the function $x^3 + 3xy + y^2$.
   (b) Find the maximum value of the function in part (a) given that (x, y) is restricted to the closed square $-1 \le x \le 1, -1 \le y \le 1$.

44. Find the maximum and minimum values of the function $g(x, y) = x^2 + x + y + y^2$ over the closed region $x^2 + y^2 \le 1$.

45. Suppose (x_0, y_0) is a point in $\mathbb{R}^2$ with polar coordinates (r_0, θ_0). Show that the polar coordinate equation $r^2 - 2rr_0 \cos(\theta - \theta_0) + r_0^2 = a^2$ represents the circle with radius a and center (x_0, y_0).

46. A cone with vertex at the origin in $\mathbb{R}^3$, and that intersects planes through the z-axis in two perpendicular lines, has a simple representation in terms of spherical (r, ϕ, θ) coordinates. Find such a representation. Then find an equation in terms of rectangular (x, y, z) coordinates for the same cone.

CHAPTER 7

MULTIPLE INTEGRATION

This chapter is devoted to the study of integrals of functions with domains in $\mathbb{R}^n$. Such integrals occur in many branches of pure and applied mathematics, with interpretations such as volume, mass, probability, and flux. In Section 1 we start with iterated integrals because they have a direct interpretation in terms of volume, and because they are then immediately available for computing the multiple integrals introduced in Section 2.

SECTION 1   ITERATED INTEGRALS

1A   Integration over a Rectangle

Recall that we can interpret the definite integral

$$\int_a^b f(x)\,dx$$

as the signed area between the x-axis and the graph of f. We want to extend this idea to the integral of a function $\mathbb{R}^2 \xrightarrow{f} \mathbb{R}$. Suppose that $f(x, y)$ is a function defined on a rectangle $a \leq x \leq b$, $c \leq y \leq d$. By

$$\int_c^d f(x, y)\,dy$$

is meant simply the definite integral of the function of one variable obtained by holding x fixed; for example,

$$\int_0^2 x^3 y^2\,dy = \left[\frac{x^3 y^3}{3}\right]_{y=0}^{y=2} = \frac{8}{3}x^3.$$

As this example shows, if the integral exists, it depends on x. Thus we may set

$$F(x) = \int_c^d f(x, y)\,dy$$

and form the **iterated integral** in the order first y, then x:

$$\int_a^b F(x)\,dx = \int_a^b \left[\int_c^d f(x, y)\,dy\right] dx.$$

To interpret the iterated integral, look at Figure 7.1(a). For a fixed value of x, the integral with respect to y is the area of the shaded region, which we have called $F(x)$. Then we can interpret the iterated integral, which is the integral of the area

FIGURE 7.1

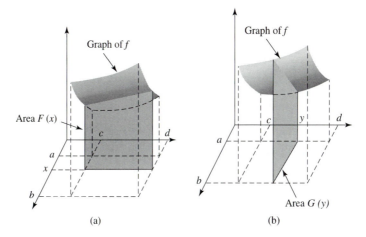

(a)   (b)

function $F(x)$ with respect to x, as the volume of the region between the graph of f and the rectangle $a \le x \le b$, $c \le y \le d$. If f assumes negative values there, we interpret the integral as a signed volume.

We can also define an iterated integral in the opposite order. We set

$$G(y) = \int_a^b f(x, y)\, dx,$$

and form the iterated integral in the order first x, then y:

$$\int_c^d G(y)\, dy = \int_c^d \left[\int_a^b f(x, y)\, dx \right] dy.$$

We may also interpret this last integral as the signed volume under the graph of f, as suggested by Figure 7.1(b). Intuitively we expect the two iterated integrals to be equal, and this will follow from Theorem 2.2 in the next section.

A common notational convention, which we'll sometimes use, is to omit the brackets and write the previous iterated integral as

$$\int_c^d \left[\int_a^b f(x, y)\, dx \right] dy = \int_a^b dx \int_c^d f(x, y)\, dy.$$

This alternative notation has the advantage of emphasizing which variable goes with which integral sign, namely, x with $\int_a^b$ and y with $\int_c^d$.

EXAMPLE 1   Consider $f(x, y) = x^2 + y$, defined on the rectangular region

$$0 \le x \le 1, \qquad 1 \le y \le 2.$$

$$\int_0^1 dx \int_1^2 (x^2 + y)\, dy = \int_0^1 \left[x^2 y + \frac{y^2}{2} \right]_{y=1}^{y=2} dx$$

$$= \int_0^1 \left[(2x^2 + 2) - \left(x^2 + \frac{1}{2}\right) \right] dx$$

$$= \int_0^1 \left(x^2 + \frac{3}{2}\right) dx = \frac{1}{3} + \frac{3}{2} = \frac{11}{6}.$$

To interpret this example geometrically, look at the surface defined by $z = x^2 + y$ shown in Figure 7.2(a). For each x in the interval between 0 and 1, the integral

$$\int_1^2 (x^2 + y)\, dy = x^2 + \frac{3}{2}$$

is the area of the shaded cross section. It is customary to interpret the definite integral of an area-valued function as volume. Thus we can regard the iterated integral

$$\int_0^1 dx \int_1^2 (x^2 + y)\, dy = \frac{11}{6}$$

as the volume of the 3-dimensional region lying below the surface and above the rectangle $0 \le x \le 1$, $1 \le y \le 2$.

EXAMPLE 2   We can perform the integration in Example 1 in the opposite order.

$$\int_1^2 dy \int_0^1 (x^2 + y)\, dx = \int_1^2 \left[\frac{x^3}{3} + yx \right]_{x=0}^{x=1} dy$$

$$= \int_1^2 \left(\frac{1}{3} + y\right) dy = \left[\frac{y}{3} + \frac{y^2}{2} \right]_1^2$$

$$= \left(\frac{2}{3} + 2\right) - \left(\frac{1}{3} + \frac{1}{2}\right) = \frac{11}{6}.$$

This time

$$\int_0^1 (x^2 + y)\, dx = \frac{1}{3} + y$$

FIGURE 7.2

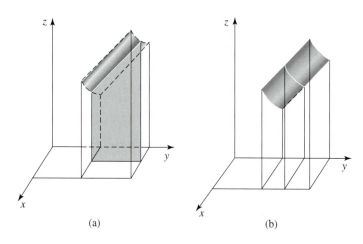

(a)   (b)

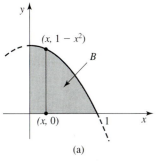

(a)

is the area of a cross section parallel to the xz-plane. See Figure 7.2(b). The second integral again gives the volume of the 3-dimensional region lying below the surface $z = x^2 + y$ and above the rectangle $0 \le x \le 1$, $1 \le y \le 2$. So it isn't surprising that the two iterated integrals of Examples 1 and 2 are equal.

1B   Nonrectangular Regions

It is important to be able to integrate over subsets of the plane that are more general than rectangles. In such problems, the limits in the first integration will depend on the remaining variable.

EXAMPLE 3

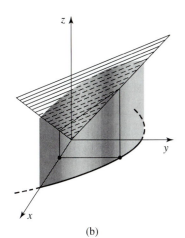

(b)

Consider the iterated integral

$$\int_0^1 dx \int_0^{1-x^2} (x + y)\, dy = \int_0^1 \left[xy + \frac{y^2}{2} \right]_0^{1-x^2} dx$$

$$= \int_0^1 \left[x(1 - x^2) + \frac{(1 - x^2)^2}{2} \right] dx$$

$$= \int_0^1 \left[x - x^3 + \frac{1 - 2x^2 + x^4}{2} \right] dx = \frac{31}{60}.$$

For each x between 0 and 1, the number y is between 0 and $1 - x^2$. In other words, the point (x, y) runs along the line segment joining $(x, 0)$ and $(x, 1 - x^2)$. As x varies between 0 and 1, this line segment sweeps out the shaded region B as shown in Figure 7.3(a). The integrand $f(x, y) = x + y$ has the graph shown in Figure 7.3(b), and the iterated integral is the volume under the graph and above the region B, shown in Figure 7.3(b).

EXAMPLE 4

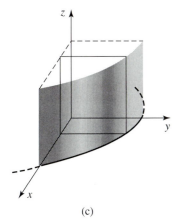

(c)

FIGURE 7.3

Suppose we are given an iterated integral over a plane region B in which the integrand is the constant function f defined by $f(x, y) = 1$, for all (x, y) in B. The integral may then be interpreted either as the volume of the slab of unit thickness and with base B or simply as the area of B. For example,

$$\int_0^1 dx \int_0^{1-x^2} dy = \frac{2}{3}$$

is the area of the region B shown in Figure 7.3(a). The integral also represents the volume of the solid region of height 1 based on the plane region B and shown in Figure 7.3(c).

A region of integration is often described by inequalities, or else by specifying its boundary curves using equations. The usual procedure for setting up an iterated integral over such a region is as follows:

1. Sketch the region of integration B using the given information.
2. If they are not already given, find equations whose graphs make up the boundary of B.
3. Using the equations for the boundary curves in the limits of integration, write the iterated integral in the order that seems simplest.

FIGURE 7.4

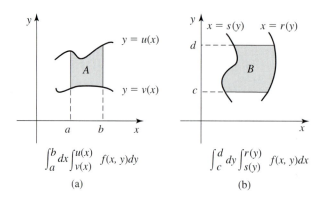

$$\int_a^b dx \int_{v(x)}^{u(x)} f(x,y)dy$$

(a)

$$\int_c^d dy \int_{s(y)}^{r(y)} f(x,y)dx$$

(b)

For example, to integrate over a region A between the graphs of $y = u(x)$ and $y = v(x)$ for x between a and b, shown in Figure 7.4(a), we would choose the order of integration to be first with respect to y, then x. On the other hand, the region B in Figure 7.4(b) would naturally lead us to the opposite order: first x, then y.

EXAMPLE 5

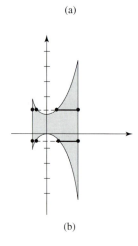

(a)

Let f be defined by $f(x, y) = xy$ over the region B bounded by the vertical lines $x = -1$ and $x = 2$ and by the graphs $y = 1 + x^2$ and $y = -x^2$, shown in Figure 7.5(a). To find the iterated integral of f over B in the order first y, then x, we think of holding x fixed somewhere between -1 and 2 and letting y vary between $y = -x^2$ and $y = 1 + x^2$. Thus we get the single integral

$$\int_{-x^2}^{1+x^2} xy\,dy.$$

Then integrate with respect to x between $x = -1$ and $x = 2$ to get

$$\int_{-1}^{2} \left[\int_{-x^2}^{1+x^2} xy\,dy \right] dx.$$

To compute the value of the integral we write it as

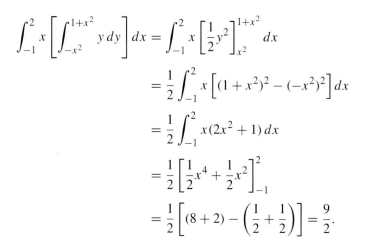

FIGURE 7.5

(b)

The choice of the order of integration was dictated by noticing that integrating the other way, first with respect to x, then y, leads to two separate integrals when y is between 1 and 5 or between -4 and 0. See Figure 7.5(b).

EXAMPLE 6   Consider the region D in $\mathbb{R}^2$ defined by the inequalities

$$0 \le x \quad \text{and} \quad x^2 + y^2 \le 1.$$

This region is shown in Figure 7.6(a) and consists of the half-disk that is the intersection of the circular disk $x^2 + y^2 \le 1$ and the right half-plane $x \ge 0$. The boundary of D consists of the line segment $x = 0$ for $-1 \le y \le 1$ on the left and the graph of $x = \sqrt{1 - y^2}$ for $-1 \le y \le 1$ on the right. To integrate the function $f(x, y) = x$ over D, we integrate first with respect to x and then with respect to y to get

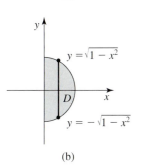

(a)

(b)

FIGURE 7.6

$$\int_{-1}^{1} dy \int_{0}^{\sqrt{1-y^2}} x \, dx = \int_{-1}^{1} \left[\frac{1}{2} x^2 \right]_{0}^{\sqrt{1-y^2}} dy$$

$$= \frac{1}{2} \int_{-1}^{1} (1 - y^2) \, dy$$

$$= \frac{1}{2} \left[y - \frac{1}{3} y^3 \right]_{-1}^{1} = \frac{1}{2} \left[\frac{2}{3} - \left(-\frac{2}{3} \right) \right] = \frac{2}{3}.$$

We can also integrate in the other order. We think of D as bounded above by the graph of $y = \sqrt{1 - x^2}$ and below by the graph of $y = -\sqrt{1 - x^2}$. We first integrate with respect to y between $-\sqrt{1 - x^2}$ and $\sqrt{1 - x^2}$, then with respect to x between 0 and 1, as indicated in Figure 7.6(b). The result is

$$\int_{0}^{1} dx \int_{-\sqrt{1-x^2}}^{\sqrt{1-x^2}} x \, dy = \int_{0}^{1} x[y]_{-\sqrt{1-x^2}}^{\sqrt{1-x^2}} dx$$

$$= \int_{0}^{1} 2x \sqrt{1 - x^2} \, dx$$

$$= \left[-\frac{2}{3} (1 - x^2)^{3/2} \right]_{0}^{1} = \frac{2}{3}.$$

In this example the two orders of integration lead to computations of about equal complexity. In practice it may happen that you get stuck using one order, so you try the other.

EXAMPLE 7   Let f be defined by $f(x, y) = x^2 y + xy^2$ over the region bounded by $y = |x|$, $y = 0$, $x = -1$ and $x = 1$. See Figure 7.7. The two iterated integrals over the region are

$$\int_{-1}^{1} dx \int_{0}^{|x|} (x^2 y + xy^2) \, dy$$

and

$$\int_{0}^{1} dy \left[\int_{-1}^{-y} (x^2 y + xy^2) \, dx + \int_{y}^{1} (x^2 y + xy^2) \, dx \right].$$

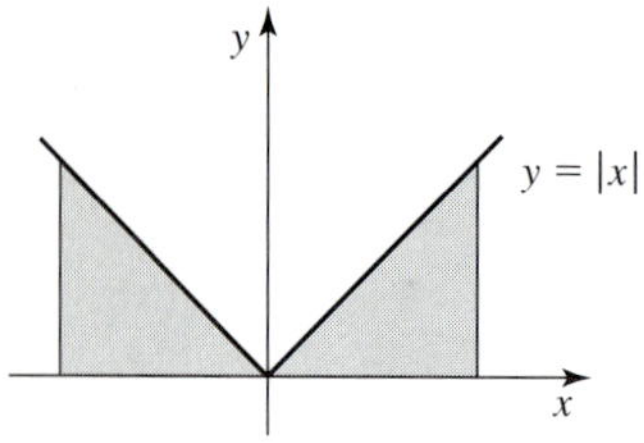

y = |x|

FIGURE 7.7

The second integral breaks into two pieces because, for fixed y between 0 and 1, the integration with respect to x is carried out over two separate intervals. Computation of the integral is straightforward. We get

$$\int_0^1 \left[\frac{x^3 y}{3} + \frac{x^2 y^2}{2}\right]_{-1}^{-y} + \left[\frac{x^3 y}{3} + \frac{x^2 y^2}{2}\right]_y^1 dy = \int_0^1 \frac{2}{3}(y - y^4)\, dy = \frac{1}{3} - \frac{2}{15} = \frac{1}{5}.$$

The iterated integral in the other order is

$$\int_{-1}^1 \left[\frac{x^2 y^2}{2} + \frac{xy^3}{3}\right]_0^{|x|} dx = \int_{-1}^1 \left(\frac{x^4}{2} + \frac{x|x|^3}{3}\right) dx$$

$$= \int_{-1}^1 \frac{x^4}{2}dx + \int_{-1}^1 \frac{x|x|^3}{3}dx.$$

The functions $x^4/2$ and $x|x|^3/3$ are even and odd, respectively. It follows that the sum of the two integrals is

$$\frac{1}{2}\int_{-1}^0 x^4 dx - \frac{1}{3}\int_{-1}^0 x^4 dx + \frac{1}{2}\int_0^1 x^4 dx + \frac{1}{3}\int_0^1 x^4 dx = \int_0^1 x^4 dx = \frac{1}{5}.$$

1C   Higher Dimensions

Iterated integrals for functions defined on sets of dimension greater than 2 can also be computed by repeated 1-dimensional integration.

EXAMPLE 8   We compute the value of the iterated integral

$$\int_0^1 dx \int_0^{1-x} dy \int_0^{1-x-y} x\, dz.$$

Since x is held fixed during the integration with respect to z and y, the integral is

$$\int_0^1 x\, dx \int_0^{1-x} dy \int_0^{1-x-y} dz = \int_0^1 x\, dx \int_0^{1-x} [z]_0^{1-x-y} dy$$

$$= \int_0^1 x\, dx \int_0^{1-x} (1 - x - y)\, dy$$

$$= \int_0^1 x \left[y - xy - \frac{1}{2}y^2\right]_0^{1-x} dx$$

$$= \int_0^1 x \left[(1 - x) - x(1 - x) - \frac{1}{2}(1 - x)^2\right] dx$$

$$= \int_0^1 \left(\frac{1}{2}x^3 - x^2 + \frac{1}{2}x\right) dx = \frac{1}{24}.$$

The region of integration is shown in Figure 7.8(a). It is bounded on top by the graph of $z = 1 - x - y$ and on three other faces by the coordinate planes $x = 0$,

$y = 0$, $z = 0$. The first integration, with respect to z, is along a vertical segment depending on x and y. The second integration, with respect to y, is sweeping out a triangular region depending on x. The third integration with respect to x, pushes the triangle across the entire solid. Note that the graph of the function $f(x, y, z) = x$ cannot be pictured in $\mathbb{R}^3$.

If the integrand $f(x, y, z) = x$ is replaced by the constant function with value 1, then we interpret the resulting iterated integral as the volume of the solid region shown in Figure 7.8(a). We compute the volume as follows:

$$
\int_0^1 dx \int_0^{1-x} dy \int_0^{1-x-y} dz = \int_0^1 dx \int_0^{1-x} [z]_0^{1-x-y} dy
$$

$$
= \int_0^1 dx \int_0^{1-x} (1 - x - y)\, dy
$$

$$
= \int_0^1 \left[y - xy - \frac{1}{2}y^2 \right]_0^{1-x} dx
$$

$$
= \int_0^1 \left(\frac{1}{2}x^2 - x + \frac{1}{2} \right) dx
$$

$$
= \left[\frac{1}{6}x^3 - \frac{1}{2}x^2 + \frac{1}{2}x \right]_0^1 = \frac{1}{6}.
$$

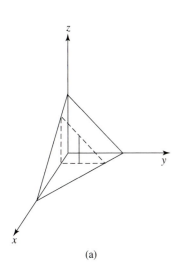

(a)

We may interpret the integrand $f(x, y, z) = x$ in the previous integral as a variable density that increases from 0 as we move away from the yz-plane in the positive x-direction. The integral would then be the total mass of the solid.

EXAMPLE 9

In this example, the function $f(x, y, z) = x + y$ is integrated over the region in $\mathbb{R}^3$ shown in Figure 7.8(b). We integrate first with respect to z, then y, and then x to get

$$
\int_0^1 dx \int_{x^2}^x dy \int_0^x (x + y)\, dz = \int_0^1 dx \int_{x^2}^x [xz + yz]_{z=0}^{z=x} dy
$$

$$
= \int_0^1 dx \int_{x^2}^x (x^2 + xy)\, dy
$$

$$
= \int_0^1 \left[x^2 y + \frac{1}{2}xy^2 \right]_{y=x^2}^{y=x} dx
$$

$$
= \int_0^1 \left(\frac{3}{2}x^3 - x^4 - \frac{1}{2}x^5 \right) dx
$$

$$
= \left[\frac{3}{8}x^4 - \frac{1}{5}x^5 - \frac{1}{12}x^6 \right]_0^1
$$

$$
= \frac{3}{8} - \frac{1}{5} - \frac{1}{12} = \frac{11}{120}.
$$

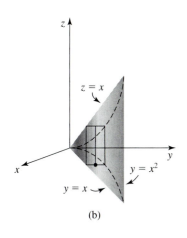

(b)

FIGURE 7.8

We can interpret the first integration with respect to z as taking place along a vertical segment joining the points $(x, y, 0)$ and (x, y, x). The integration with respect to

(a)

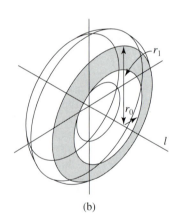

(b)

FIGURE 7.9

y takes place for each fixed x between 0 and 1, and sweeps out a vertical rect-angle between the xy-plane and the plane $z = x$. [The graph of the integrand $f(x, y, z) = x + y$ cannot be drawn in Figure 7.8(b); the sketch shown there is the domain of f for the purposes of integration, but the graph of f would be in $\mathbb{R}^4$.]

In the integration with respect to x, the vertical rectangle sweeps out the 3-dimen-sional region between the planar top and bottom. We can interpret the value of the integral as the total mass of a solid region with variable density equal to $f(x, y, z) = x + y$ at the point (x, y, z).

1D   Solids with Known Sectional Areas

Integration over solid regions B in $\mathbb{R}^3$ can sometimes be simplified if the coordinates used to describe B are chosen so that plane cross-sections perpendicular to one of the axes have area that you already know or can routinely compute in advance. Examples are shown in Figure 7.9.

EXAMPLE 10   A solid column C of height h has circular cross-sections with radii that decrease linearly from b at the base to a at the top. Our problem is to find its volume $V(C)$. We choose an x-interval from $x = 0$ to $x = h$ to coincide with the column's axis of symmetry as shown in Figure 7.9(a). The radius and area of a cross-sectional disk at distance x from the base are given, consistently with $r(0) = b$ and $r(h) = a$, by

$$r(x) = \left(\frac{a - b}{h}\right) x + b \quad \text{and} \quad A(x) = \pi r^2(x).$$

This formula for $A(x)$ could have been the result of an additional, but unnecessary, iterated integration with respect to y and z. In any case the volume of the column is

$$V(C) = \int_0^h \pi r^2(x)\, dx = \int_0^h \pi \left[\left(\frac{a - b}{h}\right) x + b\right]^2 dx.$$

We can compute this integral either by making the substitution $u = r(x)$ or by squaring the bracketed expression for $r(x)$. The result is

$$V(C) = (\pi h/3)(b^3 - a^3)/(b - a) = (\pi h/3)(b^2 + ab + a^2).$$

Note that when $a = 0$ we get a formula for the volume of a right circular cone with base radius b and height h: $V = \frac{1}{3}\pi b^2 h$. If $a = b$ we get the volume of a cylinder.

EXAMPLE 11   A solid torus B is generated by rotating a circular disk of radius a about a line l in the same plane and at distance $c > a$ from the center of the disk. See Figure 7.9(b). If S is sliced by a plane perpendicular to l the resulting intersection is a flat annular region with inner radius of the form $r_0 = c - p$ and outer radius $r_1 = c + p$. Thus the annulus will have area

$$A = \pi r_1^2 - \pi r_0^2 = \pi((c + p)^2 - (c - p)^2) = 4\pi cp.$$

The increments $\pm p$ depend on the level at which the slice is made. At level z measured along the horizontal axis having label l, we see that $p = \sqrt{a^2 - z^2}$. Hence $A(z) = 4\pi c\sqrt{a^2 - z^2}$. To find $V(B)$ we integrate $A(z)$ from $-a$ to a:

$$V(B) = \int_{-a}^{a} A(z)\,dz = 4\pi c \int_{-a}^{a} \sqrt{a^2 - z^2}\,dz.$$

This last integral is usually computed using the substitution $z = a\sin u$. However a moment's thought shows that the integral represents the area of a semicircle of radius a, namely $\frac{1}{2}\pi a^2$. Hence $V(B) = (4\pi c)(\frac{1}{2}\pi a^2) = 2\pi^2 ca^2$, $0 < a < c$.

EXERCISES

In Exercise 1 to 14 evaluate the following iterated integrals and sketch the region of integration for each.

1. $\displaystyle\int_{-1}^{0}\left[\int_{1}^{2}(x^2y^2 + xy^3)\,dy\right]dx$

2. $\displaystyle\int_{0}^{2}\left[\int_{1}^{3}|x-2|\sin y\,dx\right]dy$

3. $\displaystyle\int_{1}^{0}\left[\int_{2}^{0}(x+y^2)\,dy\right]dx$

4. $\displaystyle\int_{0}^{\pi/2}\left[\int_{-y}^{y}\sin x\,dx\right]dy$

5. $\displaystyle\int_{-2}^{1}dy\int_{0}^{y^2}(x^2+y)\,dx$

6. $\displaystyle\int_{-1}^{1}dx\int_{0}^{|x|}dy$

7. $\displaystyle\int_{0}^{1}dx\int_{0}^{\sqrt{1-x}}dy$

8. $\displaystyle\int_{1}^{-1}dx\int_{x}^{2x}e^{x+y}dy$

9. $\displaystyle\int_{0}^{\pi/2}dy\int_{0}^{\cos y}x\sin y\,dx$

10. $\displaystyle\int_{1}^{2}dx\int_{x^2}^{x^3}x\,dy$

11. $\displaystyle\int_{0}^{1}\left[\int_{0}^{z}\left[\int_{0}^{y}dx\right]dy\right]dz$

12. $\displaystyle\int_{0}^{1}\left[\int_{1}^{x}\left[\int_{0}^{x+y}y\,dz\right]dy\right]dx$

13. $\displaystyle\int_{1}^{2}dy\int_{0}^{1}dx\int_{x}^{y}dz$

14. $\displaystyle\int_{-1}^{1}dx\int_{0}^{|x|}dy\int_{0}^{1}(x+y+z)\,dz$

15. Evaluate the integral $\displaystyle\int_{0}^{\pi}\sin x\,dx\int_{0}^{1}dy\int_{0}^{2}(x+y+z)\,dz$.

16. Evaluate the integral $\displaystyle\int_{0}^{1}dx\int_{-x}^{x}dy\int_{-x-y}^{x+y}dz\int_{-z}^{x}dw$.

17. Sketch the subset B of $\mathbb{R}^2$ defined by $0 \le x \le 1$, $0 \le y \le x$, and write down the integral over B in each of the two possible orders of $f(x, y) = x\sin y$. Evaluate both integrals.

18. Sketch the region defined by $x \ge 0$, $x^2 + y^2 \le 2$, and $x^2 + y^2 \ge 1$. Write down the integral over the region in each of the two possible orders of $f(x, y) = x^2$. Evaluate both integrals.

19. Consider two real valued functions $c(x)$ and $d(x)$ of a real variable x. Suppose that, for all x in the interval $a \le x \le b$, we have $c(x) \le d(x)$.
    (a) Make a sketch of two such functions and of the subset B of the xy-plane consisting of all (x, y) such that $a \le x \le b$ and $c(x) \le y \le d(x)$.
    (b) Express the area of B as an iterated integral.
    (c) Set up the iterated integral of $f(x, y)$ over B.

20. Sketch the subset B of $\mathbb{R}^3$, defined by $0 \le x \le 1$, $0 \le y \le 1$, and $0 \le z \le 2$. Write down the iterated integral with order of integration z, then y, and then x, of the function $f(x, y, z) = x^2 + z$ over the subset B. Compute the integral.

21. Sketch the region defined by $0 \le x \le 1$, $x^2 \le y \le \sqrt{x}$, and $0 \le z \le x + y$, and evaluate the iterated integral, in some order, of $f(x, y, z) = x + y + z$ over the region.

22. Let f be defined by $f(x, y, z) = 1$ on the hemisphere bounded by the plane $z = 0$ and the surface $z = \sqrt{1 - x^2 - y^2}$. Evaluate an iterated integral of f in some order over the region.

23. A solid region B has a circular base of radius a. Cross-sections by planes perpendicular to a fixed diameter of the circle are squares. Sketch B and find its volume.

24. Derive the formula $V = \frac{4}{3}\pi a^3$ for the volume of a sphere of radius a by computing an integral of areas of parallel slices of the sphere.

25. (a) Two planes intersect at angle $\pi/4$. A cylinder of radius a with central axis perpendicular to one of the planes intersects that plane in a disk of radius a tangent to the line l of intersection of the planes. Sketch the wedge-shaped solid W bounded by the planes and the cylinder.

  (b) Find the volume of W by integrating sectional areas perpendicular to the axis of the cylinder.

  (c) Find the volume of W by integrating sectional areas perpendicular to l.

26. Let f be defined by $f(x_1, \dots, x_n) = x_1 x_2 \dots x_n$ on the cube $0 \le x_1 \le 1, 0 \le x_2 \le 1, \dots, 0 \le x_n \le 1$. Evaluate

$$\int_0^1 dx_1 \int_0^1 dx_2 \dots \int_0^1 x_1 x_2 \dots x_n \, dx_n.$$

27. Evaluate

$$\int_0^1 dx_1 \int_0^1 dx_2 \dots \int_0^1 dx_{n-1} \int_0^{x_1} (x_1 + x_2) \, dx_n.$$

28. (a) Evaluate $I_N = \int_0^N dy \int_0^N e^{-x-y} \, dx$.

  (b) Evaluate the improper integral $\int_0^\infty dy \int_0^\infty e^{-x-y} \, dx$ by finding $\lim_{N \to \infty} I_N$.

29. (a) Evaluate $J_\delta = \int_\delta^1 dy \int_\delta^1 \frac{1}{\sqrt{xy}} \, dx$.

  (b) Evaluate $\int_0^1 dy \int_0^1 \frac{1}{\sqrt{xy}} \, dx$ by finding $\lim_{\delta \to 0+} J_\delta$.

SECTION 2   MULTIPLE INTEGRALS

2A   Definition

Recall that the integral of a function with values $f(x)$ assumed on some interval $a \le x \le b$ is defined by

$$\lim_{\max(\Delta x_k) \to 0} \sum_{k=0}^K f(x_k) \Delta x_k = \int_a^b f(x) \, dx$$

where Δx_k is the distance between points of subdivision shown in Figure 7.10(a) and x_k is some point in the kth interval. The sum on the left is most simply interpreted as the total area of a collection of rectangles. The purpose of this section is to extend the definition of integral to functions with values $f(\mathbf{x})$ where $\mathbf{x}$ is in some subset B of $\mathbb{R}^n$ for $n \ge 2$.

The extension proceeds very naturally if we keep in mind the analog of Figure 7.10(a) shown in Figure 7.10(b) for the case $n = 2$. A segment with length Δx is replaced by a rectangle in the xy-plane with dimensions Δx and Δy, and the rectangle area $f(x)\Delta x$ is replaced by a volume $f(\mathbf{x})\Delta x \Delta y$. The integral will then be defined as a limit of sums of volumes.

Although integration over intervals is adequate for practically all purposes in dimension 1, we need more general sets in $\mathbb{R}^n$. We first consider some simple sets in $\mathbb{R}^n$. A closed **coordinate rectangle** is a subset of $\mathbb{R}^n$ consisting of all points $\mathbf{x} = (x_1, \dots, x_n)$ that satisfy a set of inequalities

$$a_i \le x_i \le b_i, \qquad i = 1, \dots, n. \tag{1}$$

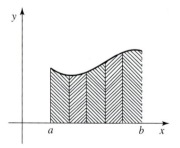

(a)

(b)

FIGURE 7.10

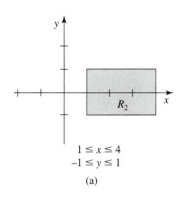

$1 \leq x \leq 4$
$-1 \leq y \leq 1$

(a)

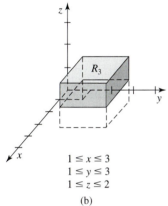

$1 \leq x \leq 3$
$1 \leq y \leq 3$
$1 \leq z \leq 2$

(b)

FIGURE 7.11

FIGURE 7.12

If in Formula (1) some of the symbols "$\leq$" are replaced by "$<$," the resulting set is still called a coordinate rectangle. In particular, if all the inequalities are of the form $a_i < x_i < b_i$, the set is open and is an open coordinate rectangle. A coordinate rectangle has its edges parallel to the coordinate axes. Throughout this section the word *rectangle* will be understood to mean *coordinate rectangle*. Figure 7.11 illustrates rectangles in $\mathbb{R}^2$ and $\mathbb{R}^3$. A "rectangle" in $\mathbb{R}$ is just an interval.

Let R be a rectangle (open, closed, or neither) defined by Formula (1), and with replacement of some or all symbols "$\leq$" by "$<$" permitted. The **volume** or **content** of R, written $V(R)$, is by definition the product of the lengths of the edges of R, and so equals

$$V(R) = (b_1 - a_1)(b_2 - a_2) \ldots (b_n - a_n). \tag{2}$$

In the examples shown in Figure 7.11,

$$V(R_2) = (4 - 1)(1 - (-1)) = 6 \quad \text{and} \quad V(R_3) = (3 - 1)(3 - 1)(2 - 1) = 4.$$

If, for some i in Formula (1), $a_i = b_i$, then R is called **degenerate** and $V(R) = 0$. For rectangles in $\mathbb{R}^2$, content is the same thing as area, and we often write $A(R)$ instead of $V(R)$ to have the notation remind us of area rather than volume.

A subset B of $\mathbb{R}^n$ is called **bounded** if there is a real number k such that $|\mathbf{x}| < k$ for all $\mathbf{x}$ in B. A finite set of $(n-1)$-dimensional planes in $\mathbb{R}^n$ (lines in $\mathbb{R}^2$) parallel to the coordinate planes will be called a **grid**. A grid separates $\mathbb{R}^n$ into a finite number of closed, bounded rectangles $R_1, \ldots, R_r$ and a finite number of unbounded regions. A grid **covers** a subset B of $\mathbb{R}^n$ if B is contained in the union of the bounded rectangles $R_1, \ldots, R_r$, so a set B can be covered by a grid if and only if B is bounded. As a measure of the fineness of a grid, we take the maximum of the lengths of the edges of the rectangles $R_1, \ldots, R_r$. This number is called the **mesh** of the grid. In Figure 7.12 the shadings are parts of planes that cut B_3.

Remark. When $n \geq 4$ a note is in order about the planes that form a grid. In the space $\mathbb{R}^n$ each of the planes of dimension $n - 1$ that we use to form a grid will be perpendicular to one of the coordinate axes. For example, planes with equations $x_1 = c$ consist of all points $(c, x_2, x_3, \ldots, x_n)$. Thus vectors of the form $(0, x_2, x_3, \ldots, x_n)$

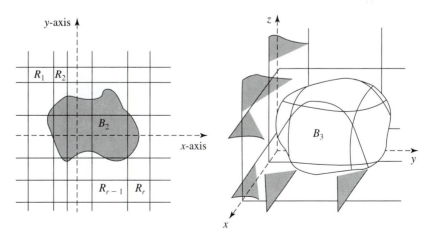

joining two points in this plane will be perpendicular to a vector $(d, 0, 0, \ldots, 0)$ with $d \neq 0$ that is parallel to the x_1 axis. Similar remarks apply to each of the other $n - 1$ types of plane with respective equation types $x_2 = c$, $x_3 = c, \ldots, x_n = c$. Each of these planes has content zero in $\mathbb{R}^n$, and the volume of a box bounded by n pairs of parallel planes is the product of the distances between the two planes in each pair.

We now give a definition of the multiple integral, called the Riemann integral after Bernhard Riemann. Consider a function $\mathbb{R}^n \xrightarrow{f} \mathbb{R}$ and a set B such that

(a) B is a bounded subset of the domain of f.
(b) f is bounded on B.

Assertion (b) means that there exists a real number K such that $|f(\mathbf{x})| \leq K$, for all $\mathbf{x}$ in B. The multiple integral of f over B will be defined in terms of the function f_B, which is f altered to be zero outside B, that is,

$$f_B(\mathbf{x}) = \begin{cases} f(\mathbf{x}), & \text{if } \mathbf{x} \text{ is in } B \\ 0, & \text{if } \mathbf{x} \text{ is not in } B \end{cases}$$

Figure 7.13 shows the shaded graph of a function f_B cut from a graph over the first quadrant in $\mathbb{R}^2$. Let G be a grid that covers B and has mesh equal to $m(G)$. In each of the bounded rectangles R_i formed by G, with $i = 1, \ldots, r$, choose an arbitrary point $\mathbf{x}_i$. The sum

$$\sum_{i=1}^{r} f_B(\mathbf{x}_i) V(R_i)$$

is called a **Riemann sum** for f over B. Its value, for given f and B, depends on G and $\mathbf{x}_1, \ldots, \mathbf{x}_r$. If, no matter how we choose grids G with mesh $m(G)$ tending to zero, it happens that

$$\lim_{m(G) \longrightarrow 0} \sum_{i=1}^{r} f_B(\mathbf{x}_i) V(R_i)$$

exists and is always the same number, then this limit is the **integral** of f over B and is denoted by $\int_B f \, dV$. If the integral exists, f is said to be **integrable** over B.

FIGURE 7.13

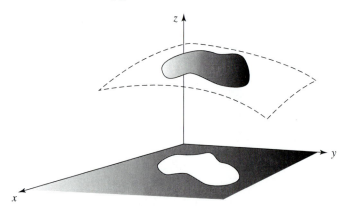

The limit that defines the multiple integral is somewhat different from the limit of a vector function defined in Chapter 5, Section 1, although the idea behind it is similar. The defining equation

$$\lim_{m(G) \longrightarrow 0} \sum_{i=1}^{r} f_B(\mathbf{x}_i) V(R_i) = \int_B f \, dV$$

means that, for any $\varepsilon > 0$, there exists $\delta > 0$ such that if G is any grid that covers B and has mesh less than δ, and S is an arbitrary Riemann sum for f_B formed from G, then

$$\left| S - \int_B f \, dV \right| < \varepsilon.$$

It should be emphasized that the integral is not defined for functions f and sets B unless the boundedness conditions on f and on B are satisfied. Without these conditions, even the Riemann sums may not be defined.

If f is a real-valued function of one real variable, that is, if $n = 1$, and if B is an interval $a \leq x \leq b$, the Riemann integral of f over B is the familiar definite integral

$$\int_a^b f(x) \, dx.$$

Other common notations for the integral of $\mathbb{R}^n \xrightarrow{f} \mathbb{R}$ over B are

$$\int_B f \, dA \text{ and } \int_B f(x, y) \, dx \, dy, \qquad\qquad \text{if } n = 2,$$

$$\int_B f(x, y, z) \, dx \, dy \, dz, \qquad\qquad \text{if } n = 3,$$

$$\int_B f \, dx_1 \ldots dx_n, \qquad\qquad \text{for arbitrary } n.$$

2B   Existence

Multiple integrals are often computed by first rewriting them as iterated integrals, which are then evaluated by repeated application of 1-dimensional integration techniques. Even though they are too technical to prove here, it is nevertheless important to have criteria for the existence of an integral $\int_B f(\mathbf{x}) \, dV$. The criteria provided below in Theorem 2.1 impose conditions (i) on the set of boundary points of B and (ii) on the set of discontinuity points of f. Both (i) and (ii) require that the respective sets be negligible in the following sense: A set S has **zero content** if $\int_S dV = 0$. For example, finite sets of points, finite collections of smooth curves in $\mathbb{R}^2$ and $\mathbb{R}^3$, and finite collections of smooth curves and surfaces in $\mathbb{R}^3$ all have zero content, though we won't prove this.

2.1   Theorem.   Let $\mathbb{R}^n \xrightarrow{f} \mathbb{R}$ be defined and bounded on a bounded set B such that (i) the boundary of B has zero content and (ii) f is continuous except possibly on a set of zero content. Then f is Riemann integrable over B.

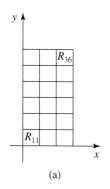

(a)

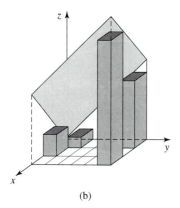

(b)

FIGURE 7.14

EXAMPLE 1   We'll evaluate $\int_B (2x+y)\,dx\,dy$ directly from its definition, where B is the rectangle $0 \le x \le 1, 0 \le y \le 2$. Note that (i) the boundary of B consists of four line segments having total content zero, and (ii) $f(x, y) = 2x + y$ is continuous everywhere. Thus we know that f is integrable on B, so we can use an arbitrary sequence of Riemann sums with mesh tending to 0 to evaluate the integral. For each $n = 1, 2, \ldots$, consider the grid G_n consisting of the lines

$$x = \frac{i}{n}, \quad i = 0, \ldots, n \quad \text{and} \quad y = \frac{j}{n}, \quad j = 0, \ldots, 2n.$$

See Figure 7.14(a). The mesh of G_n is $1/n$, and the area of the rectangles R_{ij} is $1/n^2$. Setting

$$\mathbf{x}_{ij} = (x_i, y_j) = \left(\frac{i}{n}, \frac{j}{n}\right),$$

we form the Riemann sum, partly illustrated in Figure 7.14(b),

$$\sum_{i=1}^{n}\sum_{j=1}^{2n}(2x_i + y_j)A(R_{ij}) = \sum_{i=1}^{n}\sum_{j=1}^{2n}\left(\frac{2i}{n} + \frac{j}{n}\right)\frac{1}{n^2}$$

$$= \frac{1}{n^3}\left(2\sum_{i=1}^{n}\sum_{j=1}^{2n}i + \sum_{i=1}^{n}\sum_{j=1}^{2n}j\right)$$

$$= \frac{1}{n^3}\left(4n\sum_{i=1}^{n}i + n\sum_{j=1}^{2n}j\right)$$

$$= \frac{1}{n^2}\left(\frac{4n^2 + 4n}{2} + \frac{4n^2 + 2n}{2}\right)$$

$$= \frac{4n^2 + 3n}{n^2} = 4 + \frac{3}{n}.$$

Hence

$$\int_B (2x + y)\,dx\,dy = \lim_{n\to\infty}\left(4 + \frac{3}{n}\right) = 4.$$

Direct evaluation of a multiple integral would be very difficult for most functions we want to integrate. Fortunately in many instances we can evaluate the multiple integral by repeated application of ordinary 1-dimensional integration instead of by finding the limits of Riemann sums. The pertinent theorem, which we don't prove, is the following.

2.2   Theorem.   Let B be a subset of $\mathbb{R}^n$ such that the iterated integral

$$\int dx_1 \int dx_2 \ldots \int f\,dx_n$$

exists over B. If, in addition, the multiple integral

$$\int_B f\, dV$$

exists, then the two integrals are equal.

Since the argument that proves Theorem 2.2 applies equally well to any order of iterated integration, we have an immediate corollary:

2.3   Theorem.   If $\int_B f\, dV$ exists and iterated integrals exist for some orders of integration, then all these integrals are equal.

We won't prove this theorem, but looking at Example 1 gives us an intuitive justification for interchanging the order of integration. In a Riemann sum we can add first with respect to one index, then with respect to the other, to get

$$\sum_{i,j} f(x_i, y_j) A(R_{ij}) = \sum_{i=1}^{n} \left[\sum_{j=1}^{m} f(x_i, y_j)\, \Delta y_j \right] \Delta x_i$$

$$= \sum_{j=1}^{m} \left[\sum_{i=1}^{n} f(x_i, y_j)\, \Delta x_i \right] \Delta y_j,$$

where Δx_i and Δy_j are the dimensions of the rectangle R_{ij}. Thus we can expect these sums to tend to the respective integrals as the mesh of the grid tends to zero:

$$\int_B f(x, y)\, dA = \int \left[\int f(x, y)\, dy \right] dx$$

$$= \int \left[\int f(x, y)\, dx \right] dy.$$

2C   Double Integrals

Computing multiple integrals by iterated integration often requires us to describe the region over which are to integrate so we can make reasonable choices for the order of integration and the limits of integration. We start with 2-dimensional examples. A double integral is usually written

$$\iint_B f(x, y)\, dx\, dy \quad \text{or} \quad \int_B f(x, y)\, dx\, dy,$$

where B is a subset of $\mathbb{R}^2$ on which f is defined. If f is nonnegative, we may interpret the integral as the volume between B and the graph of f in $\mathbb{R}^3$.

EXAMPLE 2   Compute $\int_B (2x + y)\, dx\, dy$ where B is the rectangle

$$0 \le x \le 1, \quad 0 \le y \le 2.$$

This is the same integral that occurs in Example 1, but we evaluate it here by iterated integration as follows.

$$\int_B (2x + y)\,dx\,dy = \int_0^1 dx \int_0^2 (2x + y)\,dy$$

$$= \int_0^1 \left[2xy + \frac{1}{2}y^2\right]_{y=0}^{y=2} dx$$

$$= \int_0^1 (4x + 2)\,dx$$

$$= [2x^2 + 2x]_0^1 = 4.$$

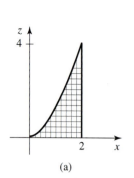

(a)

Integration in the other order, first with respect to x and then with respect to y, would produce the same final result.

EXAMPLE 3   Consider the 2-dimensional region B satisfying $0 \le x \le 2$ and $0 \le y \le x^2$ shown in Figure 7.15(a). The double integral of a function with values $f(x, y)$ defined on B equals an iterated integral in either of two orders. Integrating first with respect to y, we would have

$$\int_B f(x, y)\,dx\,dy = \int_0^2 dx \int_0^{x^2} f(x, y)\,dy.$$

The integral with respect to y, namely

$$\int_0^{x^2} f(x, y)\,dy,$$

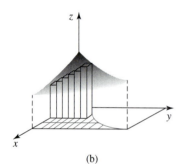

(b)

depends on x and represents the area of the shaded vertical slice shown in Figure 7.15(b). If we want to integrate first with respect to x, perhaps to make the indefinite integrals easier to find, we hold y fixed with $0 \le y \le 4$ and note that then x satisfies $\sqrt{y} \le x \le 2$. The double integral is then given by [see Figure 7.15(c)]

$$\int_B f(x, y)\,dx\,dy = \int_0^4 dy \int_{\sqrt{y}}^2 f(x, y)\,dx.$$

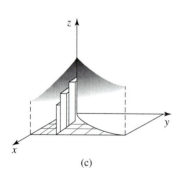

(c)

FIGURE 7.15

For example, if $f(x, y) = xy$, we would have

$$\int_B xy\,dx\,dy = \int_0^4 dy \int_{\sqrt{y}}^2 xy\,dx$$

$$= \int_0^4 \left[\frac{1}{2}x^2 y\right]_{x=\sqrt{y}}^{x=2} dy$$

$$= \int_0^4 \left(2y - \frac{1}{2}y^2\right) dy$$

$$= \left[y^2 - \frac{1}{6}y^3\right]_0^4 = 16 - \frac{32}{3} = \frac{16}{3}.$$

EXAMPLE 4

The inequality $x^2 + y^2 \leq 1$ defines a disk D in the xy-plane shown in Figure 7.16(a). The volume above D and under the graph of $f(x, y) = x^2 + y^2$ is shown in Figure 7.16(b). For each fixed x satisfying $-1 \leq x \leq 1$, we have y restricted so that

$$-\sqrt{1 - x^2} \leq y \leq \sqrt{1 - x^2}.$$

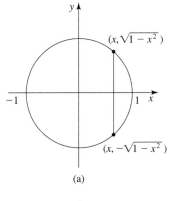

(a)

We compute the double integral of f over D as an iterated integral as follows.

$$\int_D (x^2 + y^2)\, dx\, dy = \int_{-1}^{1} dx \int_{-\sqrt{1-x^2}}^{\sqrt{1-x^2}} (x^2 + y^2)\, dy$$

$$= \int_{-1}^{1} \left[x^2 y + \frac{1}{3} y^3 \right]_{-\sqrt{1-x^2}}^{\sqrt{1-x^2}} dx$$

$$= \int_{-1}^{1} \left(2x^2 \sqrt{1 - x^2} + \frac{2}{3}(1 - x^2)\sqrt{1 - x^2} \right) dx$$

$$= \frac{4}{3} \int_{0}^{1} \left(\sqrt{1 - x^2} + 2x^2 \sqrt{1 - x^2} \right) dx$$

$$= \frac{4}{3} \left(\frac{\pi}{4} + \frac{\pi}{8} \right) = \frac{\pi}{2}.$$

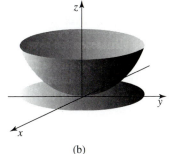

(b)

The indefinite integrals needed in the last step are in the Appendix and most standard tables; they're evaluated by making the substitution $x = \sin \theta$.

FIGURE 7.16

More complicated regions like those shown in Figure 7.17 are handled by cutting them up into disjoint regions C_k over each of which it is possible to compute an iterated integral. Then use as often as necessary the equation

$$\int_{C_1 \cup C_2} f(x, y)\, dx\, dy = \int_{C_1} f(x, y)\, dx\, dy + \int_{C_2} f(x, y)\, dx\, dy;$$

this is proved as Theorem 3.6 in the next section.

FIGURE 7.17

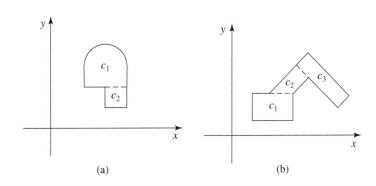

(a)   (b)

2D   Triple Integrals

A triple integral is usually written

$$\iiint_B f(x, y, z)\, dx\, dy\, dz \quad \text{or} \quad \int_B f(x, y, z)\, dx\, dy\, dz,$$

where B is a subset of $\mathbb{R}^3$ on which f is defined. If f is nonnegative, the number $f(x, y, z)$ can be interpreted as the density of material at the point (x, y, z), so under this interpretation the value of the integral becomes the total mass of material in B.

EXAMPLE 5

If R is the 3-dimensional rectangle defined by

$$0 \le x \le 2$$
$$0 \le y \le 1$$
$$0 \le z \le 2,$$

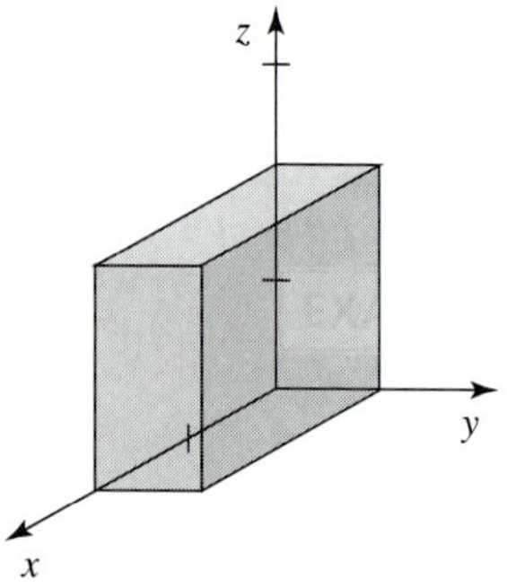

we can sketch it as in Figure 7.18. The integral of $f(x, y, z) = xyz$ over R is then computed as an iterated integral in any of the possible orders.

$$\int_R xyz\, dx\, dy\, dz = \int_0^2 dx \int_0^1 dy \int_1^2 xyz\, dz$$
$$= \int_0^2 x\, dx \int_0^1 y\, dy \int_1^2 z\, dz$$
$$= \left[\frac{1}{2}x^2\right]_0^2 \left[\frac{1}{2}y^2\right]_0^1 \left[\frac{1}{2}z^2\right]_1^2 = \frac{3}{2}$$

FIGURE 7.18

We have factored the x or y out of each integral in which it's not the integration variable, being careful not to do this when x or y is the integration variable.

EXAMPLE 6

Let $f(x, y, z) = xyz$, and let the subset B of $\mathbb{R}^3$ be defined by $x^2 + y^2 + z^2 \le 4$, $x \ge 0, y \ge 0, z \ge 0$. B is the interior and boundary of one-eighth of the spherical ball of radius 2 with center at the origin, shown in Figure 7.19. The integral $\int_B f\, dV$ equals the triple iterated integral of the function $f(x, y, z) = xyz$ over B. For fixed x and y, the variable z runs from 0 to $\sqrt{4 - x^2 - y^2}$, which are the limits of the first integration with respect to z. The result of this integration is a function of x and y that must be integrated over the 2-dimensional subset obtained by projecting B on the xy-plane, that is, over the region

$$x^2 + y^2 \le 4, \quad x \ge 0, y \ge 0.$$

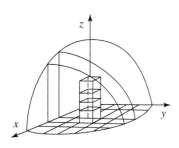

FIGURE 7.19

For fixed x, the variable y runs from 0 to $\sqrt{4 - x^2}$; hence these are the limits on the integration with respect to y. Finally, x runs from 0 to 2, so we conclude that

$$\int_B f\, dV = \int_0^2 dx \int_0^{\sqrt{4-x^2}} dy \int_0^{\sqrt{4-x^2-y^2}} xyz\, dz.$$

Then

$$\int_B f \, dV = \frac{1}{2} \int_0^2 x \, dx \int_0^{\sqrt{4-x^2}} y(4 - x^2 - y^2) \, dy$$

$$= \frac{1}{2} \int_0^2 x \left(2(4 - x^2) - \frac{x^2}{2}(4 - x^2) - \frac{(4 - x^2)^2}{4} \right) dx.$$

The last integral simplifies to

$$\int_0^2 \left(2x - x^3 + \frac{1}{8}x^5 \right) dx = \frac{4}{3}.$$

2E   Content and Mass

If an integrand $f(x, y)$ is constantly 1 on a region R in $\mathbb{R}^2$, then the integral of f over R can be interpreted as the volume of the solid B with base R and height 1. An alternative interpretation is as the area of R. More generally, and more precisely, if a set B in $\mathbb{R}^n$ satisfies the conditions of Theorem 2.1, we define the **content** of B to be

$$V(B) = \int_B dV.$$

In case $n = 2$ we call this number the **area** of B and denote it by $A(B)$. When $n \geq 3$ we speak of n-**dimensional volume** and retain the notation $V(B)$, or else use $V_n(B)$ if the dimension isn't clear from the context. One virtue of this definition is that it allows us to remove the ambiguity inherent in the multiple ways of computing via iterated integrals. However, the definition is dependent on the choice of coordinates used to describe B; this point is addressed in Section 4.

EXAMPLE 7   Referring back to the previous Example 4, the volume of the region B that lies vertically between the disk D and the graph of $f(x, y) = x^2 + y^2$ is equal, by Theorem 2.2 and our definition of $V(B)$, to

$$V(B) = \int_B dV = \int_D \left[\int_0^{f(x,y)} dz \right] dx \, dy.$$

The integral with respect to z works out immediately to $f(x, y)$, so the value $\pi/2$ of the integral computed in Example 4 is assigned to $V(B)$ without concern about order of integration.

If $\mu(\mathbf{x})$ is nonnegative and integrable for over B, then $\mu(\mathbf{x})$ can be interpreted as the **density** at $\mathbf{x}$ of a mass distribution μ. Then, assuming $V(B) > 0$, the integral

$$M(B) = \int_B \mu(\mathbf{x}) \, dV$$

is called the **total mass** of the distribution μ on B. The type of density considered here describes mass per volume unit, and in case $V(B) = 0$, the integral of an integrable density μ over B will always be zero. Some alternative densities for curves and surfaces with zero volume are described in Chapter 8, Section 2 and Chapter 9, Section 3.

EXAMPLE 8   Referring again to Example 4, we take B to be the 2-dimensional disk D and let $\mu(x, y) = x^2 + y^2$. Think of D as being made of material that increases in density from 0 at the center to 1 at the edge as the square of the distance from the center. According to the computation in Example 4, the total mass is $M(D) = \pi/2$.

The importance of the multiple integral is due partly to the variety of interpretations that stem from it. Content and total mass are conceptually two of the simplest; others are discussed in subsequent sections.

EXERCISES

In Exercises 1 to 4, make a drawing of the set B and compute $\int_B f \, dx \, dy$.

1. $f(x, y) = x^2 + 3y^2$ and B is the disk $x^2 + y^2 \le 1$.

2. $f(x, y) = 1/(x + y)$ and B is the region bounded by the lines $y = x$, $x = 1$, $x = 2$, $y = 0$.

3. $f(x, y) = x \sin xy$ and B is the rectangle $0 \le x \le \pi$, $0 \le y \le 1$.

4. $f(x, y) = x^2 - y^2$ and B consists of all (x, y) such that $0 \le x \le 1$ and $x^2 - y^2 \ge 0$.

In Exercises 5 and 6, use the definition of the double integral as a limit of Riemann sums to compute the integrals $\int_B f(x, y) \, dx \, dy$ with given f and B. Then verify your answer by computing an appropriate iterated integral. The following formulas will be useful.

$$\sum_{i=1}^{n} i = \frac{n(n + 1)}{2}, \quad \sum_{i=1}^{n} i^2 = \frac{n(n + 1)(2n + 1)}{6},$$

$$\sum_{i=1}^{n} i^3 = \left(\sum_{i=1}^{n} i \right)^2$$

5. $f(x, y) = x + 4y$ and B is the rectangle $0 \le x \le 2$, $0 \le y \le 1$.

6. $f(x, y) = 3x^3 + 2y$ and B is the rectangle $0 \le x \le 2$, $0 \le y \le 1$.

In Exercises 7 to 10, find the volume under the graph of f and above the set B, where

7. $f(x, y) = x + y^2$ and B is the rectangle with corners $(1, 1)$, $(1, 3)$, $(2, 3)$, and $(2, 1)$.

8. $f(x, y) = x + y + 2$ and B is the region bounded by the curves $y^2 = x$, and $x = 2$.

9. $f(x, y) = |x + y|$ and B is the disk $x^2 + y^2 \le 1$.

10. $f(x, y) = x^2 + y^2$ and B is the square with corners at $(x, y) = (\pm 1, \pm 1)$.

11. Find by integration the area of the subset of $\mathbb{R}^2$ bounded by the curve

$$x^2 - 2x + 4y^2 - 8y + 1 = 0.$$

12. Given that $f(x, y, z) = xyz$ and that

$$\int_B f(x, y, z) \, dx \, dy \, dz = \int_0^2 dx \int_0^x dy \int_0^{x+y} xyz \, dz,$$

sketch the region B and evaluate the integral.

13. Sketch the region B in $\mathbb{R}^3$ bounded by the surface $z = 4 - 4x^2 - y^2$ and the xy-plane. Set up the volume of B as a triple integral and also as a double integral. Compute the volume.

14. Write an expression for the volume of the ball $x^2 + y^2 + z^2 \le a^2$
   (a)   as a triple integral.
   (b)   as a double integral.

15. Sketch in $\mathbb{R}^3$ the two cylindrical solids defined by $x^2 + z^2 \le 1$ and $y^2 + z^2 \le 1$, respectively. Find the volume of their intersection.

16. The 4-dimensional ball B of radius 1 and with center at the origin is the subset of $\mathbb{R}^4$ defined by $x_1^2 + x_2^2 + x_3^2 + x_4^2 \le 1$. Set up an expression for the volume $V(B)$ as a fourfold iterated integral.

17. A hemispherical bowl of radius a contains liquid with maximum depth h. Find the volume of the liquid.

18. Cavalieri's principle as originally formulated in the 17th century states that two solids that have equal cross-sectional areas at the same height will have equal volumes.

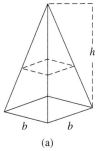 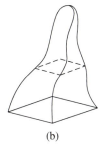

      (a)   (b)

(a) Assuming that the hypotheses of the principle hold for the two solids with square cross sections that follow, find the volume of the one below.

(b) Explain how Cavalieri's principle follows from Theorem 2.2 and the definition of area and volume.

19. A semicircular steel plate with a two foot radius has a concentric semicircle of 6-inch radius removed from its straight edge. If the steel has uniform density $\mu = 12$ pounds per square foot, find the total mass of the plate.

20. A rectangular 2-by-3 foot steel plate has been machined so that its density varies linearly from 10 pounds per square foot to 12 pounds per square foot as measured in the long direction. Find the total mass of the plate.

21. A column with circular cross section varying from diameter 12 inches to diameter 8 inches is 10 feet long. The density μ of the material in the column varies linearly along the length of the column from 50 pounds per cubic foot at the thick end to 40 pounds per cubic foot at the thin end. Find the total mass. See Figure 7.9(a).

SECTION 3   INTEGRATION THEOREMS

The emphasis in the two previous sections is on computational technique and on interpretation of the integral. Here we look at four characteristic properties of integrals and show how some other important properties follow from them.

3.1   Theorem.   Linearity: If f and g are integrable over B and a and b are any two real numbers, then $af + bg$ is integrable over B and

$$\int_B (af + bg)\, dV = a \int_B f\, dV + b \int_B g\, dV.$$

3.2   Theorem.   Positivity: If f is nonnegative and integrable over B, then

$$\int_B f\, dV \geq 0.$$

3.3   Theorem. If R is a rectangle, then $\int_R dV = V(R)$, where the content $V(R)$ is defined as the product of the lengths of the edges of R.

In the next theorem recall that $f_B(\mathbf{x})$ is defined to equal $f(\mathbf{x})$ for $\mathbf{x}$ in B and to equal 0 for $\mathbf{x}$ not in B.

3.4   Theorem. If B is a subset of a bounded set C, then $\int_B f\, dV$ exists if and only if $\int_C f_B\, dV$ exists. Whenever both integrals exist, they are equal.

Proof of 3.1.   Let $\epsilon > 0$ be given, and choose $\delta > 0$ so that if S_1 and S_2 are Riemann sums for f_B and g_B respectively whose grids have mesh less that δ, then

$$|a|\left|S_1 - \int_B f\,dV\right| < \frac{\epsilon}{2} \quad \text{and} \quad |b|\left|S_2 - \int_B g\,dV\right| < \frac{\epsilon}{2}.$$

Let S be any Riemann sum for $(af + bg)_B$ whose grid has mesh less than δ. Then

$$S = \sum_i (af + bg)_B(\mathbf{x}_i)V(R_i)$$

$$= a\sum_i f_B(\mathbf{x}_i)V(R_i) + b\sum_i g_B(\mathbf{x}_i)V(R_i)$$

$$= aS_1 + bS_2.$$

Hence

$$\left|S - a\int_B f\,dV - b\int_B g\,dV\right| = \left|aS_1 - a\int_B f\,dV + bS_2 - b\int_B g\,dV\right|$$

$$\leq |a|\left|S_1 - \int_B f\,dV\right| + |b|\left|S_2 - \int_B g\,dV\right|$$

$$< \frac{\epsilon}{2} + \frac{\epsilon}{2} = \epsilon.$$

Thus

$$\lim_{m(G) \to 0} \sum_i (af + bg)B(\mathbf{x}_i)V(R_i) = a\int_B f\,dV + b\int_B g\,dV,$$

and the proof is complete.   ∎

Proof of 3.2.   Since all the Riemann sums are nonnegative, the limit must also be nonnegative.   ∎

Proof of 3.3.   This follows immediately from Theorems 2.1 and 2.2.   ∎

Proof of 3.4.   The existence and the value of the integral $\int_B f\,dV = \int f_B\,dV$ depends only on the function f_B. Similarly, $\int_C f_B\,dV$ is defined by using $(f_B)_C$, which is equal to f_B.   ∎

We can now prove the next two theorems directly.

3.5   Theorem.   If f and g are integrable over B, and $f \leq g$ on B, then

$$\int_B f\,dV \leq \int_B g\,dV.$$

If in addition $|f|$ is integrable over B then

$$\left|\int_B f\,dV\right| \leq \int_B |f|\,dV.$$

Proof.   The function $g - f$ is nonnegative and, by Theorem 3.1, is integrable over B. Hence, by Theorems 3.1 and 3.2,

$$0 \leq \int_B (g - f)\, dV = \int_B g\, dV - \int_B f\, dV,$$

from which the conclusion follows. The second part is left as Exercise 2.   ∎

The next theorem establishes an analog for the equation

$$\int_a^c f(x)\, dx = \int_a^b f(x)\, dx + \int_b^c f(x)\, dx$$

that holds for functions of one variable.

3.6   Theorem.   If f is integrable over each of two disjoint sets B_1 and B_2, then f is integrable over their union and

$$\int_{B_1 \cup B_2} f\, dV = \int_{B_1} f\, dV + \int_{B_2} f\, dV.$$

Proof.   By Theorem 3.4,

$$\int_{B_1} f\, dV + \int_{B_2} f\, dV = \int_{B_1 \cup B_2} f_{B_1}\, dV + \int_{B_1 \cup B_2} f_{B_2}\, dV.$$

Since B_1 and B_2 are disjoint, $f_{B_1 \cup B_2} = f_{B_1} + f_{B_2}$. Hence, by Theorem 3.1, the function $f_{B_1 \cup B_2}$ is integrable over $B_1 \cup B_2$, and

$$\int_{B_1 \cup B_2} f_{B_1}\, dV + \int_{B_1 \cup B_2} f_{B_2}\, dV = \int_{B_1 \cup B_2} f_{B_1 \cup B_2}\, dV.$$

Finally, by Theorem 3.4 again, f is integrable over $B_1 \cup B_2$ and

$$\int_{B_1 \cup B_2} f_{B_1 \cup B_2}\, dV = \int_{B_1 \cup B_2} f\, dV,$$

which completes the proof.   ∎

The possibility of changing order of integration has a number of consequences other than its convenience for computing multiple integrals. One of these is the theorem for change of order in partial differentiation, proved in Section 3 of Chapter 4 by other means, and in a slightly stronger form in Exercise 13 of this section. Another consequence is the Leibniz Rule for interchanging differentiation and integration.

3.7   Leibniz Rule.   If $(\partial g / \partial y)(x, y)$ is continuous for $a \leq x \leq b$ and $c \leq y \leq d$, then

$$\frac{d}{dy} \int_a^b g(x, y)\, dx = \int_a^b \frac{\partial g}{\partial y}(x, y)\, dx.$$

Proof.   The trick is to start with the following change in order of integration:

$$\int_c^y \left[\int_a^b g_y(t, y)\, dt \right] dy = \int_a^b \left[\int_c^y g_y(t, y)\, dy \right] dt.$$

For each t the integral in square brackets on the right evaluates to $g(t, y) - g(t, c)$. (Use the version of the Fundamental Theorem of Calculus that tells you how to integrate a derivative.) Thus the previous equation becomes

$$\int_c^y \left[\int_a^b g_y(t, y)\, dt \right] dy = \int_a^b g(t, y)\, dt - \int_a^b g(t, c)\, dt.$$

Note that the subtracted term is a constant. Now apply the other version of the fundamental theorem of calculus to both sides. On the left we undo the y-integration, so we get the desired result

$$\int_a^b g_y(t, y)\, dt = \frac{d}{dy} \int_a^b g(t, y)\, dt. \qquad \blacksquare$$

EXAMPLE 1   Let $G(y) = \int_0^1 \sin(y e^x)\, dx$. There seems to be no way to evaluate the integral in terms of elementary functions, but we can find $G'(y)$ using the Leibniz rule. We find

$$G'(y) = \int_0^1 \cos(y e^x) e^x\, dx = \left[\frac{1}{y} \sin(y e^x) \right]_0^1 = \frac{1}{y}(\sin(ey) - \sin y), \ y \neq 0.$$

EXAMPLE 2   If u and v are fixed numbers, both positive or both negative, the formula

$$F(y, u, v) = \int_u^v \frac{1}{x} e^{yx}\, dx$$

defines a function of y. To find the derivative of F with respect to y, we can write, using Theorem 3.7,

$$\frac{d}{dy} F(y, u, v) = \int_u^v \frac{\partial}{\partial y} \left[\frac{1}{x} e^{yx} \right] dx$$

$$= \int_u^v e^{yx}\, dx$$

$$= \left[\frac{1}{y} e^{yx} \right]_{x=u}^{x=v} = \frac{e^{vy} - e^{uy}}{y}.$$

EXERCISES

1. Consider the rectangles   and the function

B_1 defined by $0 < x \leq 1, \ 0 \leq y < 1,$
B_2 defined by $1 \leq x \leq 2, \ -1 \leq y \leq 1$

$$f(x, y) = \begin{cases} 2x - y, & \text{if} \quad x < 1, \\ x^2 + y, & \text{if} \quad x \geq 1. \end{cases}$$

Compute

$$\int_{B_1 \cup B_2} f(x, y) \, dx \, dy.$$

2. Use the first part of Theorem 3.5 to show that if f and $|f|$ are integrable over B, then

$$\left| \int_B f \, dV \right| \le \int_B |f| \, dV. \quad [\textit{Hint:} - |f| \le f \le |f|.]$$

3. Use the result of Exercise 2 to show that if $\mathbb{R}^n \xrightarrow{f} \mathbb{R}$ is continuous on a set B, and $\mathbf{x}_0$ is interior to B, then

$$\lim_{r \to 0} \frac{1}{V(B_r)} \int_{B_r} f \, dV = f(\mathbf{x}_0),$$

where B_r is a ball of radius r centered at $\mathbf{x}_0$.

4. Let B be the subset of $\mathbb{R}^2$ consisting of all points (x, y) such that $0 \le y \le 1$, and x is rational, $0 \le x \le 1$. Does the area exist?

5. On the rectangle $0 \le x \le 1$ and $0 \le y \le 1$, let $f(x, y) = 1$, if x is rational, and $f(x, y) = 2y$ if x is irrational. Show that

$$\int_0^1 dx \int_0^1 f(x, y) \, dy = 1,$$

but that f is not Riemann integrable over the rectangle.

6. Interchange of order may not always work for improper integrals. Prove that

$$\int_0^1 dy \int_1^\infty (e^{-xy} - 2e^{-2xy}) \, dx$$

$$\ne \int_1^\infty dx \int_0^1 (e^{-xy} - 2e^{-2xy}) \, dy.$$

7. Let $\mathbb{R}^n \xrightarrow{f} \mathbb{R}^m$ be defined on a set B in $\mathbb{R}^n$. We define

$$\int_B f \, dV = \left(\int_B f_1 \, dV, \dots, \int_B f_m \, dV \right),$$

provided that the integrals of the coordinate functions $f_1, \dots, f_m$ of f all exist.

(a) Show that if $\mathbb{R}^n \xrightarrow{f} \mathbb{R}^m$ and $\mathbb{R}^n \xrightarrow{g} \mathbb{R}^m$ are both integrable over B, then

$$\int_B (af + bg) \, dV = a \int_B f \, dV + b \int_B g \, dV,$$

where a and b are constants.

(b) If $\mathbf{k}$ is some fixed vector in $\mathbb{R}^m$, and $\mathbb{R}^n \xrightarrow{f} \mathbb{R}^m$ is integrable over B, show that

$$\int_B \mathbf{k} \cdot f \, dV = \mathbf{k} \cdot \int_B f \, dV.$$

(c) Show that if $\mathbb{R}^n \xrightarrow{f} \mathbb{R}^m$ and $\mathbb{R}^n \xrightarrow{|f|} \mathbb{R}$ are integrable over B, then $\left| \int_B f \, dV \right| \le \int_B |f| \, dV$ by the Cauchy-Schwarz inequality.

[*Hint:* $f(\mathbf{x}) \cdot \int_B f \, dV \le |f(\mathbf{x})| \left| \int_B f \, dV \right|$, for all $\mathbf{x}$ in B. Integrate with respect to $\mathbf{x}$ and apply the result of part (b).]

8. (a) Use the Leibniz rule together with the chain rule to prove that if $g_y(x, y)$ is continuous, and $h_1(y)$ and $h_2(y)$ are differentiable, then

$$\frac{d}{dy} \int_{h_1(y)}^{h_2(y)} g(t, y) \, dt$$

$$= \int_{h_1(y)}^{h_2(y)} g_y(t, y) \, dt + h_2'(y) g(h_2(y), y)$$

$$- h_1'(y) g(h_1(y), y).$$

[*Hint:* The integral on the left has the form $G\big(y, h_1(y), h_2(y)\big)$.]

(b) Use part (a) to compute $F'(y)$, where $F(y) = \int_{-y}^{y} (1/x)(1 - e^{-xy}) \, dx$.

In Exercises 9 to 12, use the Leibniz rule to find the indicated derivative of the given function.

9. $f(y) = \int_0^1 (y^2 + t^2) \, dt$. Find $f'(y)$.

10. $g(t) = \int_1^2 \frac{1}{x} e^{tx} \, dx$. Find $g'(t)$.

11. $h(x) = \int_0^x (x - u) e^{u^2} \, du$. Find $h'(x)$ and $h''(x)$.

12. $k(s) = \int_0^1 \frac{u^s - 1}{\ln u} \, du$. Assuming $s > -1$, find $k'(s)$ and then find $k(s)$.

*13. Prove the following stronger version of Theorem 3.3 of Chapter 4: If f_x, f_y, f_{yx}, and f_{xy} are continuous on an open set, then $f_{xy} = f_{yx}$. [*Hint:* Apply the Leibniz rule to the equation

$$f(x, y) - f(a, y) = \int_a^x f_x(t, y) \, dt,$$

and then differentiate both sides with respect to x.]

SECTION 4   CHANGE OF VARIABLE

A change of variable is often used in a 1-dimensional integral to simplify the integrand. For example, we can make the substitution $x = u + 1$ and $dx = du$ to find

$$\int_1^2 x\sqrt{x-1}\,dx = \int_0^1 (u+1)u^{1/2}du$$

$$= \left[\frac{2}{5}u^{5/2} + \frac{2}{3}u^{3/2}\right]_0^1 = \frac{16}{15}.$$

The aim of this change of variable was to simplify the integrand, and the change of interval of integration from $1 \le x \le 2$ to $0 \le u \le 1$ makes very little difference in the computation. In computing multiple integrals it's more often the corresponding change in the region of integration that we're concerned with, the point being that we can use a change of variable to simplify the region. We first consider some simple examples of multivariable coordinate changes.

4A   Polar Coordinates

In a double integral $\int_D f(x, y)\,dx\,dy$ over a circular region, it is often helpful to introduce polar coordinates by the transformation

$$x = r\cos\theta$$

$$y = r\sin\theta.$$

Corresponding regions in the xy-plane and $r\theta$-plane are shown in Figure 7.20. As r and θ vary within the limits $r_0 \le r \le r_1$ and $\theta_0 \le \theta \le \theta_1$, the values of x and y give the coordinates of the points in the shaded part D of the disk shown in Figure 7.20(c). Rather than approximate the value of an integral by decomposing D using a rectangular grid, we can think of using a polar coordinate grid. A typical subdivision S of such a grid is shown in Figure 7.20(b). Using elementary geometry, we can compute the exact area of S as a certain fraction, namely, $(\Delta\theta)/2\pi$ of the region between circles of radius r and $r + \Delta r$. We find

$$A(S) = \frac{\Delta\theta}{2\pi}[\pi(r + \Delta r)^2 - \pi r^2]$$

$$= \frac{\Delta\theta}{2}[2r\,\Delta r + (\Delta r)^2] = r\,\Delta r\,\Delta\theta + \frac{1}{2}(\Delta r)^2\Delta\theta.$$

If Δr and $\Delta\theta$ are small, the second term is relatively small compared with the first term, and we make the approximation $A(S) \approx r\,\Delta r\,\Delta\theta$. Because $x = r\cos\theta$ and $y = r\sin\theta$, we can approximate the integral of f over D as follows:

$$\int_D f(x, y)\,dA \approx \sum_{k=1}^N f(r_k\cos\theta_k, r_k\sin\theta_k)A(S_k)$$

$$\approx \sum_{k=1}^N f(r_k\cos\theta_k, r_k\sin\theta_k)r_k\,\Delta r_k\,\Delta\theta_k.$$

It follows from Equation 4.4 in Section 4D that

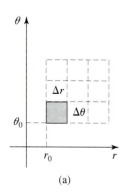

(a)

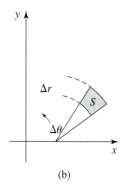

(b)

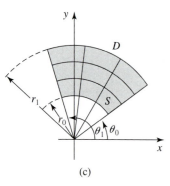

(c)

FIGURE 7.20

4.1
$$\int_D f(x, y)\, dA = \int_{\theta_0}^{\theta_1} d\theta \int_{r_0}^{r_1} f(r\cos\theta, r\sin\theta) r\, dr.$$

The computations in the next two examples are considerably simpler than direct use of iterated integrals with respect to x and y.

EXAMPLE 1   If f is constantly 1 on a quarter-circle Q of radius 1, we get

$$\int_Q dA = \int_0^{\pi/2} d\theta \int_0^1 r\, dr$$

$$= [\theta]_0^{\pi/2} \left[\frac{1}{2} r^2 \right]_0^1 = \frac{\pi}{2} \cdot \frac{1}{2} = \frac{\pi}{4},$$

which is the area of a quarter-circle of radius 1.

EXAMPLE 2   If $f(x, y) = x^2 + y^2$ on a half-circle H, then $f(r\cos\theta, r\sin\theta) = r^2$, so we would have

$$\int_H (x^2 + y^2)\, dA = \int_0^\pi d\theta \int_0^1 r^2 r\, dr$$

$$= [\theta]_0^\pi \left[\frac{1}{4} r^4 \right]_0^1 = \pi \cdot \frac{1}{4} = \frac{\pi}{4}.$$

4B   Spherical Coordinates

We introduce spherical coordinates in $\mathbb{R}^3$ by the transformation

$$\begin{pmatrix} x \\ y \\ z \end{pmatrix} = \begin{pmatrix} r\sin\phi\cos\theta \\ r\sin\phi\sin\theta \\ r\cos\phi \end{pmatrix}.$$

Corresponding regions are shown in Figure 7.21. The spherical coordinate "cube" C has by a direct calculation a volume approximated by

$$V(C) \approx r^2 \sin\phi\, \Delta r\, \Delta\phi\, \Delta\theta.$$

It will follow from Equation 4.3 that the equation

4.2
$$\int_B f(x, y, z)\, dV = \int_{\theta_0}^{\theta_1} d\theta \int_{\phi_0}^{\phi_1} d\phi \int_{r_0}^{r_1} \overline{f}(r, \phi, \theta) r^2 \sin\phi\, dr$$

FIGURE 7.21   is valid, where $\overline{f}(r, \phi, \theta) = f(r\sin\phi\cos\theta, r\sin\phi\sin\theta, r\cos\phi)$.

EXAMPLE 3   A solid ball B of radius a is described by the spherical coordinate inequalities

$$0 \le r \le a, \quad 0 \le \phi \le \pi, \quad 0 \le \theta < 2\pi.$$

We can compute the volume of the ball by

$$\int_B dV = \int_0^{2\pi} d\theta \int_0^{\pi} d\phi \int_0^a r^2 \sin\phi \, dr$$

$$= \int_0^{2\pi} d\theta \int_0^{\pi} \sin\phi \, d\phi \int_0^a r^2 dr$$

$$= [\theta]_0^{2\pi} \, [-\cos\phi]_0^{\pi} \left[\frac{1}{3}r^3\right]_0^a$$

$$= (2\pi)(2)\left(\frac{1}{3}a^3\right) = \frac{4}{3}\pi a^3,$$

the formula for the volume of a sphere of radius a.

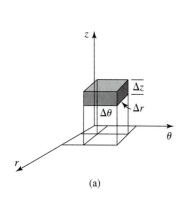

(a)

(b)

FIGURE 7.22

4C   Cylindrical Coordinates
The transformation

$$\begin{pmatrix} x \\ y \\ z \end{pmatrix} = \begin{pmatrix} r\cos\theta \\ r\sin\theta \\ z \end{pmatrix}$$

is used to introduce cylindrical coordinates in $\mathbb{R}^3$. Corresponding regions are shown in Figure 7.22. Note the close connection with plane polar coordinates. Using the result of the similar calculation for polar coordinates, we can see that the volume of a cylindrical coordinate "cube" C is given approximately by

$$V(C) \approx r\Delta r \Delta\theta \Delta z.$$

It will follow from Equation 4.3 that the equation

$$\int_B f(x, y, z) \, dV = \int_{z_0}^{z_1} dz \int_{\theta_0}^{\theta_1} d\theta \int_{r_0}^{r_1} \overline{f}(r, \theta, z) r \, dr$$

is valid, where $\overline{f}(r, \theta, z) = f(r\cos\theta, r\sin\theta, z)$.

EXAMPLE 4   A longitudinal wedge cut from a cylinder C of height h and radius a is described by the inequalities

$$0 \le r \le a, \quad 0 \le \theta \le w, \quad 0 \le z \le h.$$

To integrate the function $f(x, y, z) = x^2 + y^2 + z^2$ over C, we compute as follows.

$$\int_C (x^2 + y^2 + z^2) \, dV = \int_0^w d\theta \int_0^h dz \int_0^a (r^2 + z^2) r \, dr$$

$$= w \int_0^h \left[\frac{1}{4}r^4 + \frac{1}{2}z^2\right]_{r=0}^{r=a} dz$$

$$= w \int_0^h \left(\frac{1}{4}a^4 + \frac{1}{2}a^2 z^2 \right) dz$$

$$= w \left[\frac{1}{4}a^4 z + \frac{1}{6}a^2 z^3 \right]_0^h$$

$$= wa^2 h \left(\frac{1}{4}a^2 + \frac{1}{6}h^2 \right).$$

The formulas for integration in polar, spherical, and cylindrical coordinates can all be derived in a uniform way. The computation involves the determinants of the Jacobian matrices of the coordinate transformations. For polar coordinates, we have

$$\begin{pmatrix} x \\ y \end{pmatrix} = \begin{pmatrix} r\cos\theta \\ r\sin\theta \end{pmatrix} \quad \text{and} \quad \det \begin{pmatrix} \cos\theta & -r\sin\theta \\ \sin\theta & r\cos\theta \end{pmatrix} = r.$$

For spherical coordinates, we have, correspondingly,

$$\det \begin{pmatrix} \sin\phi\cos\theta & r\cos\phi\cos\theta & -r\sin\phi\sin\theta \\ \sin\phi\sin\theta & r\cos\phi\sin\theta & r\sin\phi\cos\theta \\ \cos\phi & -r\sin\phi & 0 \end{pmatrix} = r^2 \sin\phi.$$

And for cylindrical coordinates

$$\det \begin{pmatrix} \cos\theta & -r\sin\theta & 0 \\ \sin\theta & r\cos\theta & 0 \\ 0 & 0 & 1 \end{pmatrix} = r.$$

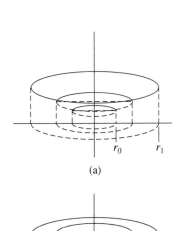

(a)

(b)

FIGURE 7.23

Thus we see that the extra factor in the integrand on the right side of Equation 4.1 and Equation 4.2 is in each case supplied by the Jacobian determinant of the coordinate transformation. The expression $r\Delta r\Delta\theta$ is called the **area element** in polar coordinates. Similarly $r^2 \sin\phi\Delta r\Delta\phi\Delta\theta$ is called the **volume element** in spherical coordinates, whereas in cylindrical coordinates the volume element is $r\Delta r\Delta\theta\Delta z$. These formulas will be generalized in Section 4D.

Cylindrical Shells.   Suppose a solid figure B is generated by rotating a plane region R about a line l in the same plane. If R doesn't intersect l and is bounded above and below by graphs of $u(r)$ and $v(r)$, we can imagine that B is composed of coaxial cylindrical shells, each one of height $h(r) = u(r) - v(r)$ depending on the distance r of a point on the shell from the line. See Figure 7.23(a). When such a shell is slit vertically and rolled out flat, it has surface area $S(r) = 2\pi rh(r)$, which is the circumference $2\pi r$ of the shell multiplied by its height $h(r)$. It seems plausible that we can find the volume of B by computing the integral of $S(r)$ over the relevant interval $r_0 \leq r \leq r_1$:

4.3
$$V(B) = \int_a^b S(r)\, dr = \int_{r_0}^{r_1} 2\pi rh(r)\, dr.$$

To make a rigorous reconciliation of Equation 4.3 with the fundamental definition $V(B) = \int_B dx\, dy\, dz$, we simply introduce cylindrical coordinates as follows

$$V(B) = \int_0^{2\pi} d\theta \int_{r_0}^{r_1} \left[\int_{v(r)}^{u(r)} dz \right] r\, dr = 2\pi \int \left(u(r) - v(r) \right) r\, dr.$$

Since $u(r) - v(r) = h(r)$, Equation 4.3 follows immediately.

EXAMPLE 5

Rotate a square with side length a about a line l in the same plane, where l is parallel to an edge of the square and lies at distance $c > a/2$ from the center of the square. The resulting solid B is a ring with right-angled corners. See Figure 7.23(b).

To use Equation 4.3 to compute $V(B)$, we note that r should run from $r_0 = c - \frac{1}{2}a$ to $r_1 = c + \frac{1}{2}a$ and that $h(r) = a$ for all r in the interval. Then

$$V(B) = \int_{c-\frac{1}{2}a}^{c+\frac{1}{2}a} 2\pi a r\, dr = \left[\pi a r^2 \right]_{c-\frac{1}{2}a}^{c+\frac{1}{2}a} = \pi a \left[(c + \tfrac{1}{2}a)^2 - (c - \tfrac{1}{2}a)^2 \right] = 2\pi c a^2.$$

This result follows in an interesting way using Pappus's theorem, discussed in Section 5.

4D   Jacobi's Theorem

The foregoing discussion shows that the Jacobian determinant of a coordinate transformation is related to the volume of the natural curvilinear coordinate subdivisions. The Jacobian determinants of arbitrary one-to-one continuously differentiable transformations discussed in Chapter 6, Section 2B have similar interpretations. In what follows it will usually be more convenient to consider the domain space and range space of T as distinct. We therefore regard T as a transformation from one copy of $\mathbb{R}^n$, which we label $\mathcal{U}^n$, to another copy, which we continue to label $\mathbb{R}^n$, writing typically $T(\mathbf{u}) = \mathbf{x}$ where $\mathbf{u}$ is in $\mathcal{U}^n$ and $\mathbf{x}$ is in $\mathbb{R}^n$. The statement of the n-dimensional change-of-variable theorem follows. We won't give the proof because it's quite complicated. Typical regions R and $T(R)$ are shown in Figure 7.24.

4.4   Jacobi's Theorem.   Let $\mathcal{U}^n \xrightarrow{T} \mathbb{R}^n$ be a continuously differentiable transformation. Let R be a set in $\mathcal{U}^n$ having a boundary consisting of finitely many smooth sets. Suppose that R and its boundary are contained in the interior of the domain of T and that

   (i) T is one-to-one on the interior of R.
   (ii) $\det T'$, the Jacobian determinant of T, is not zero in the interior of R.

If the function f is bounded and continuous on the image of R under T, denoted by $T(R)$, then we have

$$\int_{T(R)} f(\mathbf{x})\, dV_{\mathbf{x}} = \int_R f\big(T(\mathbf{u})\big) |\det T'(\mathbf{u})|\, dV_{\mathbf{u}}.$$

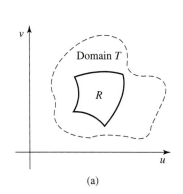

Domain T

R

v

u

(a)

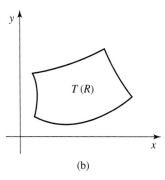

y

$T(R)$

x

(b)

FIGURE 7.24

Using **Leibniz notation** for the Jacobian determinant of

$$T(u, v) = \big(F(u, v), G(u, v)\big),$$

we have

$$\det T' = \frac{\partial(x, y)}{\partial(u, v)} = \det \begin{pmatrix} \dfrac{\partial F}{\partial u} & \dfrac{\partial F}{\partial v} \\[2ex] \dfrac{\partial G}{\partial u} & \dfrac{\partial G}{\partial v} \end{pmatrix}.$$

Then Jacobi's formula becomes

$$\int_{T(R)} f(x, y)\, dx\, dy = \int_R f\big(F(u, v), G(u, v)\big) \left| \frac{\partial(x, y)}{\partial(u, v)} \right| du\, dv.$$

In three dimensions, we have, with

$$T(u, v, x) = \big(F(u, v, w), G(u, v, w), H(u, v, w)\big),$$

the formulas

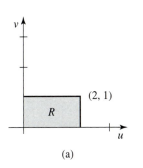

(a)

$$\det T' = \frac{\partial(x, y, z)}{\partial(u, v, w)} = \det \begin{pmatrix} \dfrac{\partial F}{\partial u} & \dfrac{\partial F}{\partial v} & \dfrac{\partial F}{\partial w} \\[2ex] \dfrac{\partial G}{\partial u} & \dfrac{\partial G}{\partial v} & \dfrac{\partial G}{\partial w} \\[2ex] \dfrac{\partial H}{\partial u} & \dfrac{\partial H}{\partial v} & \dfrac{\partial H}{\partial w} \end{pmatrix}$$

and

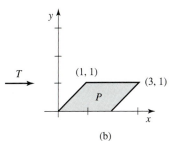

(b)

FIGURE 7.25

$$\int_{T(R)} f(x, y, z)\, dx\, dy\, dz$$

$$= \int_R f\big(F(u, v, w), G(u, v, w), H(u, v, w)\big) \left| \frac{\partial(x, y, z)}{\partial(u, v, w)} \right| du\, dv\, dw.$$

Aside from the computation of $\det T'$, the application of the transformation formula is a matter of finding the geometric relationship between the subset R and its image $T(R)$ for a transformation T.

EXAMPLE 6   The integral $\displaystyle\int_P (x + y)\, dx\, dy$, in which P is the parallelogram shown in Figure 7.25, transforms into an integral over a rectangle. This is done using the transformation

$$\begin{pmatrix} x \\ y \end{pmatrix} = T\begin{pmatrix} u \\ v \end{pmatrix} = \begin{pmatrix} u + v \\ v \end{pmatrix}.$$

The Jacobian determinant of T is

$$\det T' = \begin{vmatrix} 1 & 1 \\ 0 & 1 \end{vmatrix} = 1.$$

344   Chapter 7 Multiple Integration

By the change-of-variable theorem,

$$\int_P (x+y)\,dx\,dy = \int_R [(u+v)+v][1]\,du\,dv$$

$$= \int_0^2 du \int_0^1 (u+2v)\,dv = 4.$$

The transformation T is one-to-one because it is a linear transformation with nonzero determinant. Notice that the region of integration in the given integral is in the range of the transformation rather than in its domain.

EXAMPLE 7 The polar coordinate transformation

$$\begin{pmatrix} x \\ y \end{pmatrix} = \begin{pmatrix} u\cos v \\ u\sin v \end{pmatrix}$$

goes between the regions shown in Figure 7.26. The Jacobian is

$$\det T' = \begin{vmatrix} \cos v & -u\sin v \\ \sin v & u\cos v \end{vmatrix} = u.$$

The transformation is one-to-one between R and $T(R)$. We can see this geometrically, because of the interpretation of v and u as angle and radius, respectively, or directly from the relations

$$u = \sqrt{x^2 + y^2}, \qquad \cos v = \frac{x}{\sqrt{x^2 + y^2}},$$

together with the fact that $\cos v$ is one-to-one for $0 \le v \le \pi/2$. Given the integral of $x^2 + y^2$ over $T(R)$, we can transform as follows:

$$\int_{T(R)} (x^2 + y^2)\,dA = \int_R u^2 u\,dA = \int_1^2 u^3\,du \int_0^{\pi/2} dv = \frac{15\pi}{8}.$$

FIGURE 7.26

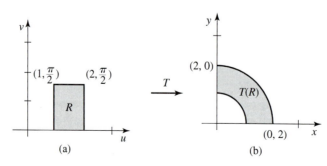

(a)   (b)

EXAMPLE 8 Let B be the positive octant in 3-dimensional space $\mathbb{R}^3$ defined by the inequalities

$$x^2 + y^2 + z^2 \le 1, \quad x \ge 0, y \ge 0, z \ge 0.$$

FIGURE 7.27

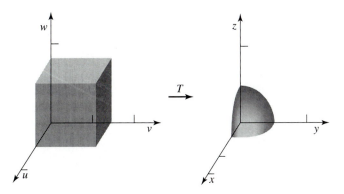

To transform the integral $\int_B (x^2 + y^2)\, dx\, dy\, dz$, we can define T by

$$\begin{pmatrix} x \\ y \\ z \end{pmatrix} = T \begin{pmatrix} u \\ v \\ w \end{pmatrix} = \begin{pmatrix} u \sin v \cos w \\ u \sin v \sin w \\ u \cos v \end{pmatrix}.$$

Restricting (u, v, w) to the rectangle R in $\mathcal{U}^3$ defined by

$$0 \le y \le 1,\ 0 \le v \le \frac{\pi}{2},\ 0 \le w \le \frac{\pi}{2},$$

we get $T(R) = B$. The corresponding regions are shown in Figure 7.27. Since

$$u = \sqrt{x^2 + y^2 + z^2}$$

$$\cos v = \frac{z}{\sqrt{x^2 + y^2 + z^2}},$$

$$\cos w = \frac{x}{\sqrt{x^2 + y^2}},$$

we conclude that the transformation T is one-to-one from R to B except on the boundary planes $u = 0$ and $v = 0$. The Jacobian determinant is

$$\det T' = \begin{vmatrix} \sin v \cos w & u \cos v \cos w & -u \sin v \sin w \\ \sin v \sin w & u \cos v \sin w & u \sin v \cos w \\ \cos v & -u \sin v & 0 \end{vmatrix}$$

$$= u^2 \sin v.$$

The transformed integral is

$$\int_B (x^2 + y^2)\, dx\, dy\, dz = \int_R (u^2 \sin^2 v \cos^2 w + u^2 \sin^2 v \sin^2 w)\, u^2 \sin v\, du\, dv\, dw$$

$$= \int_0^1 u^4\, du \int_0^{\pi/2} \sin^3 v\, dv \int_0^{\pi/2} dw$$

$$= \frac{1}{5} \cdot \frac{2}{3} \cdot \frac{\pi}{2} = \frac{\pi}{15}.$$

EXERCISES

In the definite integrals 1 and 2, make the indicated change of variable together with the appropriate change in the limits of integration. Then compute the resulting integral.

1. Let $x = \sqrt{u}$ in $\int_0^2 x e^{x^2} \, dx$.

2. Let $x = \sin\theta$ in $\int_0^1 \sqrt{1 - x^2} \, dx$.

3. Let B be the region in $\mathbb{R}^2$ described by the inequalities

$$0 \le x, \qquad 0 \le y, \qquad \text{and} \qquad x^2 + y^2 \le 4.$$

  (a) Sketch the region B and describe it by using polar coordinates.

  (b) Use the polar coordinates and Equation 4.1 to evaluate the double integral

$$\int_B \sqrt{x^2 + y^2} \, dx \, dy.$$

  (c) Use polar coordinates to evaluate the double integral

$$\int_B x \, dx \, dy.$$

4. (a) Let R be the region in $\mathbb{R}^2$ bounded by the x-axis and the polar coordinate curve $r = 1 + \cos\theta$ for $0 \le \theta \le \pi$. Sketch R and find its area.

  (b) Compute the integral

$$\int_R (x^2 + y^2) \, dx \, dy,$$

  where R is the region in part (a).

5. Let A be the annular region in $\mathbb{R}^2$ consisting of points (x, y) that satisfy $1 \le x^2 + y^2 \le 4$. Suppose that A is used as a pattern for a flat annular piece of plastic that has density at each point inversely proportional to the distance of the point from the center of the hole. What is the total mass of the piece of plastic if the density is 10 grams per unit area at the inner edge of the region?

6. Compute $\int_0^1 dx \int_0^{\sqrt{1-x^2}} (x^2 + y^2)^3 \, dy$.

7. Compute $\int_D \cos(x^2 + y^2) \, dx \, dy$, where D is the disk of radius $\sqrt{\pi/2}$ centered at $(0, 0)$.

8. Compute the area bounded by the polar coordinate curves $\theta = 0$, $\theta = \pi/4$, and $r = \theta^2$.

9. Find the area bounded by the **lemniscate** $(x^2 + y^2)^2 = 2a^2(x^2 - y^2)$ by changing to polar coordinates.

10. Let B be the region in $\mathbb{R}^3$ described by the inequalities $0 \le x$, $0 \le y$, $0 \le z$, and $x^2 + y^2 + z^2 \le 4$.

  (a) Sketch the region B, and describe it by using spherical coordinates.

  (b) Use spherical coordinates and Equation 4.2 to evaluate the triple integral

$$\int_B \sqrt{x^2 + y^2 + z^2} \, dx \, dy \, dz.$$

  (c) Use spherical coordinates to evaluate the triple integral

$$\int_B z \, dx \, dy \, dz.$$

Use spherical coordinates to compute the triple integrals 11 and 12.

11. $\int_B \sqrt{x^2 + y^2} \, dx \, dy \, dz$, where B is the solid ball of radius 1 centered at the origin in $\mathbb{R}^3$.

12. $\int_C z^2 \, dx \, dy \, dz$, where C is the region in $\mathbb{R}^3$ described by

$$1 \le x^2 + y^2 + z^2 \le 4.$$

13. Find the total mass of a solid ball of radius a with density at each point equal to the distance from the point to the center of the ball.

Use cylindrical coordinates to compute the integrals 14 and 15.

14. $\int_B z \, dx \, dy \, dz$, where B satisfies $1 \le z \le 2$, $x^2 + y^2 \le 1$.

15. $\int_C (x^2 + y^2) \, dx \, dy \, dz$, where C is the region in $\mathbb{R}^3$ described by $0 \le x$, $0 \le y$, $0 \le z \le 1$, and $x^2 + y^2 \le 2$.

16. Prove the 1-dimensional change-of-variable formula

$$\int_{\phi(a)}^{\phi(b)} f(x) \, dx = \int_a^b f(\phi(u)) \phi'(u) \, du,$$

under the assumptions that f is continuous and ϕ is continuously differentiable. [*Hint:* Differentiate both sides with respect to b.]

For Exercises 17 to 20 use multiple integration to prove the geometric volume formulas.

17. S is a sphere of radius a. Show $V(S) = \frac{4}{3}\pi a^3$.

18. C is a cone of height h and base radius a. Show $V(C) = \frac{\pi}{3}a^2 h$.

19. L is a right circular cylinder of height k and radius a. Show $V(L) = \pi a^2 k$.

20. R is a slice of thickness k perpendicular to the axis of a right circular cone having maximum radius b and minimum radius a. Show that its volume is $V(R) = \frac{\pi}{3}(a^2 + ab + b^2)k$. Explain how Exercises 18 and 19 are essentially special cases of this.

21. Consider the transformation T defined by

$$\begin{pmatrix} x \\ y \end{pmatrix} = T\begin{pmatrix} u \\ v \end{pmatrix} = \begin{pmatrix} u^2 - v^2 \\ 2uv \end{pmatrix}.$$

Let R_{uv} be the region $1 \le u^2 + v^2 \le 4$, $u \ge 0$, $v \ge 0$.
(a) Sketch the image region $R_{xy} = T(R_{uv})$.
(b) Compute $\displaystyle\int_{R_{xy}} \frac{dx\,dy}{\sqrt{x^2 + y^2}}$.

22. Define a transformation from the uv-plane to the xy-plane by $x = u + v$, $y = u^2 - v$. Let R_{uv} be the region bounded by (1) u-axis, (2) v-axis, and (3) the line $u + v = 2$.
(a) Find and sketch the image region R_{xy}.
(b) Compute the integral $\displaystyle\int_{R_{xy}} \frac{dx\,dy}{\sqrt{1 + 4x + 4y}}$.

23. Let a transformation of the uv-plane to the xy-plane be given by

$$x = u, \qquad y = v(1 + u^2),$$

and let R_{uv} be the rectangular region given by $0 \le u \le 3$ and $0 \le v \le 2$.
(a) Find and sketch the image region R_{xy}.
(b) Find $\dfrac{\partial(x, y)}{\partial(u, v)}$.
(c) Transform $\displaystyle\int_{R_{xy}} x\,dx\,dy$ to an integral over R_{uv} and compute either one of them.

24. Rotate a circular disk of radius a about a line in the same plane at distance $c > a$ from the center of the circle. This generates a solid torus B. Find the volume of B using the cylindrical shell approach to setting up an integral for $V(B)$, as in Equation 4.3. (A way to verify the correctness of the answer is to use the Pappus theorem of Section 5.)

25. If $0 < c < a$ the solid figure obtained by rotation in the previous exercise is no longer a torus but rather a kind of dimpled sphere. Make a sketch of such a solid for the case $c = 1$, $a = 2$, and find its volume.

26. A cylindrical hole of radius $\frac{1}{2}a$ is bored through the center of a sphere of radius a. Find the volume of the remaining solid.

27. A cylindrical hole of radius b is to be bored through the center of a solid ball of radius $a > b$. If you want a

proportion $k < 1$ of the volume of the whole sphere to remain as a ring, how should you choose b?

*28. A solid ball B_a of radius a is **spherically homogeneous** if its density is constant on every spherical shell with center at the center of B_a. The purpose of this exercise is to establish Newton's result that according to the inverse-square law the gravitational attraction of a spherically homogeneous ball acting at a point $\mathbf{p}$ is the same as it would be if all the mass of the ball were concentrated at its center. If $\mathbf{p}$ is inside the ball, the part of the ball at distance from the center greater than $|\mathbf{p}|$ is irrelevant. By definition, if B_a is centered at the origin and has density $\mu(|\mathbf{x}|)$ at $\mathbf{x}$, the attracting force vector on a particle of mass 1 at $\mathbf{p}$ is given by G times the 3-dimensional vector integral

$$\int_{B_a} \mu(|\mathbf{x}|) \frac{\mathbf{x} - \mathbf{p}}{|\mathbf{x} - \mathbf{p}|^3}\, dV_{\mathbf{x}} = -\frac{M_{\mathbf{p}}}{|\mathbf{p}|^3}\mathbf{p},$$

where $M_{\mathbf{p}}$ is the mass of the part of B_a that lies within distance $|\mathbf{p}|$ of its center, and G is the gravitational constant. Newton showed without using our techniques that the integral equals the expression on the right.
(a) Choose perpendicular (x, y, z)-axes with origin at the center of B_a and positive z-axis passing through $\mathbf{p} = (0, 0, p)$. Show, without computing any integrals, that the x and y coordinates of the vector integral are zero and that the z coordinate is given in spherical coordinates by

$$2\pi \int_0^a r^2 \mu(r)$$
$$\left[\int_0^\pi \frac{r\cos\phi - p}{(r^2 - 2pr\cos\phi + p^2)^{3/2}} \sin\phi\,d\phi \right] dr.$$

(b) Let $\cos\phi = u$ and integrate by parts to show that the inner integral in part (a) is

$$\int_{-1}^1 (ru - p)(r^2 + p^2 - 2pru)^{-3/2}\,du$$
$$= \begin{cases} -2/p^2, & p > r, \\ 0, & p < r. \end{cases}$$

(c) Show that the mass of B_b is $4\pi \int_0^b r^2\mu(r)\,dr$, and then use the previous results to prove Newton's formula for the attracting force.
(d) Use the results of parts (a) and (b) to show that the attracting force of matter distributed in a spherically homogeneous way between two concentric spheres cancels out, and so paradoxically exerts zero gravitational attraction at points inside the inner sphere.

(e) Specialize the results of parts (a) and (b) to the case of a homogeneous ball with *constant* density μ to show that the gravitational attraction of the ball on a unit point-mass inside the ball and r units from the center has magnitude $\frac{4}{3}\pi\mu Gr$.

SECTION 5   CENTROIDS AND MOMENTS

If positive masses $m_1, \ldots, m_N$ are concentrated at the respective points $\mathbf{x}_1, \ldots, \mathbf{x}_N$ in space, then the **center of mass** of the system is the point

5.1
$$\overline{\mathbf{x}} = \frac{1}{M}\sum_{k=1}^{N} m_k\mathbf{x}_k, \qquad \text{where} \qquad M = \sum_{k=1}^{N} m_k.$$

Thus the center of mass $\overline{\mathbf{x}}$ is a weighted average of the position vectors $\mathbf{x}_k$ in the system. We'll see that $\overline{\mathbf{x}}$ is the unique point at which a physical system consisting of masses m_k at points $\mathbf{x}_k$ would "balance" under the influence of constant gravity if the mutual distances $|\mathbf{x}_k - \mathbf{x}_l|$ are held fixed. The meaning of the term "balance" is expressed by saying that the "moment" of the system about an arbitrary plane P through $\overline{\mathbf{x}}$ is the zero vector. To make these ideas precise, we define the **moment** M_P of the mass system about a plane P to be the weighted algebraic sum of the distances from the points to P. See Figure 7.28. If $\mathbf{n} \cdot (\mathbf{x} - \mathbf{x}_0) = 0$ is the equation of a plane through $\mathbf{x}_0$, normalized so that $|\mathbf{n}| = 1$, then the distance from $\mathbf{x}_k$ to the plane is $\mathbf{n} \cdot (\mathbf{x}_k - \mathbf{x}_0)$ if $\mathbf{x}_k$ is on the side of the plane toward which $\mathbf{n}$ points and is minus that number if $\mathbf{x}_k$ is on the other side. (See Section 5 of Chapter 1.) A formula suitable for computation of M_P is then

$$M_P = \sum_{k=1}^{N} m_k\mathbf{n} \cdot (\mathbf{x}_k - \mathbf{x}_0).$$

The moment M_P is independent of the point $\mathbf{x}_0$ on the plane P, since M_P is just a weighted sum of distances to P. However, the sign of M_P does depend on an arbitrary choice for the direction of the unit vector $\mathbf{n}$.

5.2   Theorem.   Let P be an arbitrary plane containing the center of mass $\overline{\mathbf{x}}$ of the system of masses $m_1, m_2, \ldots, m_N$ at the respective points $\mathbf{x}_1, \mathbf{x}_2, \ldots, \mathbf{x}_N$. Then the moment M_P of the system about P is zero.

FIGURE 7.28

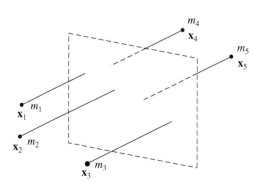

Proof.   To verify that the moment about a plane P_0 containing the center of mass $\bar{\mathbf{x}}$ is 0, we replace the generic point $\mathbf{x}_0$ by $\bar{\mathbf{x}}$ and use the distributive law for the dot product. We get

$$M_{P_0} = \sum_{k=1}^{N} m_k \mathbf{n} \cdot (\mathbf{x}_k - \bar{\mathbf{x}}) = \mathbf{n} \cdot \left[\sum_{k=1}^{N} m_k \mathbf{x}_k - \left(\sum_{k=1}^{N} m_k \right) \bar{\mathbf{x}} \right].$$

The vector in square brackets is **0** by the definition of $\bar{\mathbf{x}}$, so M_{P_0} is the **0**-vector.   ∎

For mass that is distributed according to a continuous, nonnegative density $\mu(\mathbf{x}) \geq 0$ over a body B in space, by analogy with Equation 5.1, we define the **center of mass** of the distribution μ over B to be the point

5.3 $\bar{\mathbf{x}} = \dfrac{1}{M(B)} \int_B \mu(\mathbf{x})\mathbf{x}\, dV$, if the **total mass** $M(B) = \int_B \mu(\mathbf{x})\, dV$ is positive.

The term **centroid** is used for $\bar{\mathbf{x}}$ if the distribution $\mu(\mathbf{x})$ is uniformly equal to 1, in which case we're talking about a purely geometric property of B rather than a physical property associated with mass. From this perspective it's appropriate to write $V(B)$ for volume instead of $M(B)$.

Just as with Equation 5.1 for discrete masses, the center of mass of a continuous density $\mu(\mathbf{x})$ should be a vector that we picture as a point in space. This is just what we get from Equation 5.3. The reason is that the **vector-valued integral** of a vector-valued function is defined to be the vector whose coordinates are the integrals of the individual coordinate functions x, y and z of the vector $\mathbf{x} = (x, y, z)$. In applying Equation 5.3 in $\mathbb{R}^3$, we have

$$\mu(\mathbf{x})\mathbf{x} = \big(x\mu(x, y, z), \, y\mu(x, y, z), \, z\mu(x, y, z) \big).$$

Since M is a positive number, we can compute $\bar{\mathbf{x}} = (\bar{x}, \bar{y}, \bar{z})$ by

5.4

$$\bar{x} = \frac{1}{M} \int_B x\mu(x, y, z)\, dx\, dy\, dz$$

$$\bar{y} = \frac{1}{M} \int_B y\mu(x, y, z)\, dx\, dy\, dz$$

$$\bar{z} = \frac{1}{M} \int_B z\mu(x, y, z)\, dx\, dy\, dz.$$

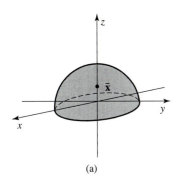

(a)

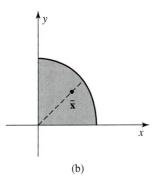

(b)

FIGURE 7.29

EXAMPLE 1

To find the center of mass of a solid hemisphere H with radius a and constant density μ, we'll take as given the formula $\frac{4}{3}\pi a^3$ for the volume of a sphere. The total mass of H is then $M = \frac{1}{2}(\frac{4}{3}\pi a^3)\mu = 2\pi\mu a^3/3$. To compute $\bar{\mathbf{x}}$, we introduce coordinates as shown in Figure 7.29(a). We note that $\bar{x} = \bar{y} = 0$, because H is symmetric about the yz-plane and the xz-plane. To find $\bar{z}$ we change to spherical coordinates, getting

$$\int_H \mu z \, dx\, dy\, dz = \mu \int_0^{2\pi} d\theta \int_0^{\pi/2} d\phi \int_0^a (r\cos\phi)r^2 \sin\phi\, dr$$

$$= \mu \int_0^{2\pi} d\theta \int_0^{\pi/2} \sin\phi\cos\phi\, d\phi \int_0^a r^3\, dr$$

$$= [\theta]_0^{2\pi} \, [\tfrac{1}{2}\sin^2\phi]_0^{\pi/2} \, [\tfrac{1}{4}r^4]_0^a$$

$$= \mu(2\pi)(\tfrac{1}{2})(\tfrac{1}{4}a^4) = \pi\mu a^4/4.$$

Hence $\bar{z} = \pi\mu a^4/4M = (\pi\mu a^4)(8\pi\mu a^3/3) = 3a/8$. In other words, the center of mass is $\frac{3}{8}$ of the way along the axis of symmetry of H, measured from the flat surface of H.

In mechanical problems it's often convenient to idealize a flat piece of material, treating it as a plane region R that carries with it a density function $\mu(x,\ y)$ that may be constant or may vary from point to point of R. Choosing the plane in which R sits to be the xy-plane and assuming $\mu(x,\ y)$ independent of z forces the moment about this plane to be zero, so we automatically get $\bar{z} = 0$. The remaining two coordinates are then computed from

5.5
$$\bar{x} = \frac{1}{M} \int_R x\mu(x,\ y)\,dx\,dy$$

$$\bar{y} = \frac{1}{M} \int_R y\mu(x,\ y)\,dx\,dy, \quad \text{where } M = \int_R \mu(x,\ y)\,dx\,dy.$$

EXAMPLE 2   Let Q be a quarter-disk of radius a, weighted so that the density at a point is equal to the square of the distance from the center of the full disk. We can start by introducing rectangular coordinates, placing Q in the first quadrant with vertex at $(0, 0)$, as in Figure 7.29(b). Then $\mu(x,\ y) = x^2 + y^2$. Because Q is symmetric about the line $y = x$ it's apparent from Equations 5.5 that $\bar{x} = \bar{y}$, so we'll just compute $\bar{x}$. Changing to polar coordinates is not required, but is helpful. We find

$$M = \int_Q \mu(x,\ y)\,dx\,dy = \int_0^{\pi/2} d\theta \int_0^a r^2\,r\,dr$$

$$= \left(\frac{\pi}{2}\right)\left(\frac{a^4}{4}\right) = \frac{\pi a^4}{8}.$$

Similarly,

$$\int_Q x\mu(x,\ y)\,dx\,dy = \int_0^{\pi/2} d\theta \int_0^a (r\cos\theta)r^2\,r\,dr$$

$$= \int_0^{\pi/2} \cos\theta\,d\theta \int_0^a r^4\,dr = (1)\left(\frac{a^5}{5}\right) = \frac{a^5}{5}.$$

Hence $\bar{x} = \bar{y} = (a^5/5)/(\pi a^4/8) = 8a/(5\pi)$. It follows that the center of mass is $8\sqrt{2}/(5\pi) \approx 72\%$ of the way out from the vertex at the origin along the line of symmetry of the quarter circle. This is reasonable; the weighting is heavier near the circular edge of Q, so we expect a value more than $\frac{1}{2}$.

We define the **moment** M_P of the mass density μ distributed over B to be the number

$$M_P = \int_B \mu(\mathbf{x})\mathbf{n} \cdot (\mathbf{x} - \mathbf{x}_0)\,dV,$$

where $\mathbf{n} \cdot (\mathbf{x} - \mathbf{x}_0) = 0$ is a normalized equation for the plane P. As in the case of a discrete distribution, M_P changes sign if n changes direction, and in many applications there is a natural choice for this direction.

5.6   Theorem.   The mass and moment of the union of disjoint regions B_1 and B_2 about a plane P is the sum of their respective masses and moments about P:

$$M(B_1 \cup B_2) = M(B_1) + M(B_2) \quad \text{and} \quad M_P(B_1 \cup B_2) = M_P(B_1) + M_P(B_2).$$

Proof.   Both equations are immediate consequences of Theorem 3.6 of Section 3, which expresses additivity of the integral as a function of the domain of integration.   ∎

Denoting the moments of B about the yz-plane, the xz-plane and the xy-plane in $\mathbb{R}^3$ by M_{yz}, M_{xz} and M_{xy} respectively, we summarize Equations 5.4 as

$$\bar{x} = M_{yz}/M, \quad \bar{y} = M_{xz}/M, \quad \bar{z} = M_{xy}/M, \quad \text{where } M \text{ is the mass of } B.$$

By Theorem 5.6, to compute the center of mass of the union of two disjoint bodies B_1, B_2, we can write

$$\bar{x} = \frac{M_{yz}(B_1) + M_{yz}(B_2)}{M(B_1) + M(B_2)},$$

$$\bar{y} = \frac{M_{zx}(B_1) + M_{xz}(B_2)}{M(B_1) + M(B_2)},$$

$$\bar{z} = \frac{M_{xy}(B_1) + M_{xy}(B_2)}{M(B_1) + M(B_2)}.$$

These formulas extend by successive application to an arbitrary finite union of distinct bodies.

EXAMPLE 3   Consider the body B consisting of a hemispherical region H_a of radius a with a concentric hemisphere H_b of radius $b < a$ removed. Assuming uniform density μ, the mass of B is $M = \frac{2}{3}\pi\mu(a^3 - b^3)$. Choose coordinates in $\mathbb{R}^3$ so that the flat base of B rests in the xy-plane, with the axis of symmetry along the positive z-axis. Then $\bar{x} = \bar{y} = 0$. To compute $\bar{z}$ we borrow the information from Example 1 that H_a has moment about the xy-plane $M_{xy}(H_a) = \frac{1}{4}\pi\mu a^4$. Since moments can be added, we find

$$M_{xy}(H_a) = M_{xy}(B) + M_{xy}(H_b) \quad \text{or} \quad M_{xy}(B) = M_{xy}(H_a) - M_{xy}(H_b).$$

Hence $M_{xy}(B) = \frac{1}{4}\pi\mu a^4 - \frac{1}{4}\pi\mu b^4 = \frac{1}{4}\pi\mu(a^4 - b^4)$. For the center of mass of B we have

$$\bar{z} = \tfrac{1}{4}\pi\mu(a^4 - b^4)/\tfrac{2}{3}\pi\mu(a^3 - b^3) = \tfrac{3}{8}(a^4 - b^4)/(a^3 - b^3).$$

Recall that the location of the *centroid* of a geometric object is a geometric property of the object, though it coincides with *center of mass* if we think of area or volume as a uniform mass distribution of density 1. The next theorem invokes the idea of centroid and is an intuitively appealing consequence of our ability to give two different geometric interpretations to the same mathematical object. In this case we're rotating a plane region R about a line L in the same plane as R but not intersecting R.

5.7 Pappus's Theorem.   The volume of a solid of revolution B about a line L is equal to the area of the rotated region R times the circumference $2\pi\bar{r}$ of the circle traced by its centroid during rotation: $V(B) = 2\pi\bar{r}A(R)$.

Proof.   According to the "cylindrical shell" analysis for volume of revolution discussed in Section 4C,

$$V(B) = 2\pi \int_a^b rh(r)\, dr, \qquad 0 < a \le r \le b,$$

where $z = h(r)$ defines the height of the region R measured parallel to L and at distance r from the line. Dividing both sides of this equation by $2\pi A(R)$ shows that $V(B)/(2\pi A(R)) = \bar{r}$, the distance of the centroid of R from L. Now just multiply this last equation by $2\pi A(R)$ to get Pappus's formula. ∎

Pappus's theorem is particularly simple to apply when we know on geometric grounds where the centroid R is located relative to the axis of rotation, as in the next example.

EXAMPLE 4   Rotate an a-by-b rectangle R about a line l in the same plane and parallel to an edge of length a and lying at distance d from the nearest edge. Figure 7.23(b) in the previous Section 4C shows the solid for the case $a = b$. The resulting solid B is a ring with sharp corners. Since the centroid is $d + b/2$ units from l, the circle it traces has circumference $2\pi(d + b/2) = \pi(2d + b)$. Since R is a rectangle $A(R) = ab$, so $V(B) = \pi(2d + b)ab$.

EXERCISES

In Exercises 1–4, find the center of mass $\bar{x}$ of each of the following discrete mass distributions. Sketch the given points and the center of mass $\bar{x}$.

1. In $\mathbb{R}$, $m_1 = 1$ at $x_1 = 1$, $m_2 = 3$ at $x_2 = 2$, $m_3 = 2$ at $x_3 = -4$.

2. In $\mathbb{R}^2$, $m_1 = 1$ at $x_1 = (1, 1)$, $m_2 = 2$ at $x_2 = (1, 0)$, $m_3 = 3$ at $x_3 = (0, 1)$.

3. In $\mathbb{R}^2$, $m_1 = 1$ at $x_1 = (1, -1)$, $m_2 = 2$ at $x_2 = (1, 2)$, $m_3 = 3$ at $x_3 = (-1, 1)$.

4. In $\mathbb{R}^3$, $m_1 = \frac{1}{2}$ at $x_1 = (1, -1, 2)$, $m_2 = \frac{1}{4}$ at $x_2 = (0, 1, 2)$, $m_3 = \frac{1}{4}$ at $x_3 = (1, 1, 1)$.

In Exercises 5 to 8, find the center of mass $\bar{x}$ of each of the following continuous distributions with density μ defined on the set described. Sketch the set and show the location of the center of mass $\bar{x}$.

5. Let I be the interval $0 \le x \le 2$ in $\mathbb{R}$, and let $\mu(x) = 1 - \frac{1}{2}x$.

6. Let D be the disk of radius 1 centered at the origin in $\mathbb{R}^2$, and let $\mu(x, y) = |x + y|$.

7. Let Q be the quarter-disk of radius 1 in the first quadrant with edges on the axes in $\mathbb{R}^2$, and let $\mu(x, y) = x + y$.

8. Let C be the unit cube in $\mathbb{R}^3$ defined by $0 \le x \le 1$, $0 \le y \le 1$, $0 \le z \le 1$, and let $\mu(x, y, z) = xyz$.

9. Find the centroid of the region R in the first quadrant of $\mathbb{R}^2$ bounded by the graph of $y = x^2$, the line $y = 4$ and the y axis.

10. Find the centroid of a solid right circular cone of base radius a, and height b from base to tip.

11. Show that the center of mass of a homogeneous solid ball is at the center of the ball.

12. Use the result of Example 3 of the text to find the center of mass of a hemispherical surface of radius a and constant surface density μ. (The result will be corroborated later with an approach tailored to more general surfaces.)

13. (a) Find the centroid of the part of the annulus with radii $a > b$ and center $(0, 0)$ that lies in the first quadrant.
    (b) Use part (a) to locate the centroid of a quarter circle of radius a.

14. A square region R of side length a is rotated about a line parallel to a diagonal and containing a vertex not on that diagonal. Find the volume of the solid generated this way.

15. Use Pappus's theorem and the volume $\frac{4}{3}\pi a^3$ of a ball of radius a to find the centroid of a plane semicircular region.

16. A right circular cone of height h and base area A has volume $V = \frac{1}{3}hA$. Use Pappus's theorem to find the centroid of a right triangle with perpendicular side lengths a and b.

17. Let B be a set in $\mathbb{R}^n$ and $\mu(\mathbf{x})$ the density at $\mathbf{x}$ in B. If $\mathbf{x}_0$ is a fixed point in $\mathbb{R}^n$ and $\mathbf{n} \cdot (\mathbf{x} - \mathbf{x}_0) = 0$ is the normalized equation of a plane P through $\mathbf{x}_0$, we've defined the *moment* of B about P by

$$M_P(B) = \int_B \mu(\mathbf{x})\mathbf{n} \cdot (\mathbf{x} - \mathbf{x}_0) \, dV.$$

(a) Show that M_P is independent of $\mathbf{x}_0$ as long as $\mathbf{x}_0$ is on the same plane P. [*Hint:* If $\mathbf{x}_1$ is another point on P, then $\mathbf{n} \cdot (\mathbf{x}_1 - \mathbf{x}_0) = 0$.]
(b) Show that if P passes through the center of mass $\bar{\mathbf{x}}$ of B, then $M_P = 0$, an extension of Theorem 5.2.

Exercises 18 to 22 refer to the **moment of inertia** $I(\mathbf{z}_0)$ of a mass density $\mu(\mathbf{x})$ over a set R in $\mathbb{R}^2$ about $\mathbf{z}_0$,

where $\mathbf{z}_0$ is a fixed point in $\mathbb{R}^2$; the number $I(\mathbf{z}_0)$ is

$$I(\mathbf{z}_0) = \int_R |\mathbf{x} - \mathbf{z}_0|^2 \, \mu(\mathbf{x}) \, dA.$$

18. Find the moment of inertia of a disk R_a of radius a about its center if (i) R_a has constant density $\mu(\mathbf{x}) = 1$ and (ii) R_a has density $\mu(\mathbf{x}) = |\mathbf{x}|^p$, $p > 0$.

19. Find the moment of inertia of a square S of side b about one corner if S has uniform density μ.

20. Find the moment of inertia of a square S of side b about its center if S has uniform density μ.

21. Show that the moment of inertia $I(\mathbf{z}_0)$ of R as defined previously satisfies

$$I(\mathbf{z}_0) = M(R)|\mathbf{z}_0 - \bar{\mathbf{x}}|^2 + I(\bar{\mathbf{x}}),$$

where $\bar{\mathbf{x}}$ is the center of mass of the weighted region R. [*Hint:* In the definition of $I(\mathbf{z}_0)$, $|\mathbf{x} - \mathbf{z}_0|^2$ can be replaced by the dot product of $(\mathbf{x} - \bar{\mathbf{x}}) + (\bar{\mathbf{x}} - \mathbf{z}_0)$ with itself.]

22. Use the previous exercise to show that $I(\mathbf{z}_0)$ is minimized by taking $\mathbf{z}_0$ to be the center of mass $\bar{\mathbf{x}}$ of R.

23. Let $B_1, \ldots, B_N$ be nonoverlapping regions with union B, having respective masses $M(B_k)$, $M(B)$ and centers of mass $\bar{\mathbf{x}}_k$, $\bar{\mathbf{x}}$.
    (a) Prove that $\bar{\mathbf{x}}$ is given by a weighted sum of the $\bar{\mathbf{x}}_k$ as

$$\bar{\mathbf{x}} = \frac{1}{M(B)} \sum_{k=1}^{N} M(B_k)\bar{\mathbf{x}}_k.$$

This reduces finding $\bar{\mathbf{x}}$ to the case for point masses in Equation 5.1.
    (b) Illustrate part (a) with the example in which B_1 and B_2 are the rectangles in $\mathbb{R}^2$ having the same uniform density 1 and respective corners at $(1, 3), (3, 3), (3, 2), (1, 2)$ and $(3, -1), (4, -1), (4, -2), (3, -2)$.

SECTION 6   IMPROPER INTEGRALS

The underlying definition of the Riemann integral as a limit of weighted sums of function values requires the integrand to be a bounded function $f(\mathbf{x})$ that is defined on a bounded domain B. The definition extends to some functions that are unbounded or that aren't necessarily zero outside a bounded set. Such an integral is called "improper" and is defined as a limit of "proper" integrals of bounded functions defined on bounded sets B_δ of a family $\{B_\delta\}$ that expands to cover all of B. The indices δ are chosen at our convenience to tend, either increasingly or decreasingly, to some finite number or to infinity.

EXAMPLE 1   Let $f(x) = x^{-1/3}$ for $0 < x \le 1$. See Figure 7.30(a). The integral of f over this interval I is not an ordinary Riemann integral because $f(x)$ tends to infinity as $x \longrightarrow 0$. To assign a value to $\int_0^1 f(x)\,dx$, we let I_δ be the interval $[\delta, 1]$ and first compute

$$\int_{I_\delta} f(x)\,dx = \int_\delta^1 x^{-1/3}\,dx = \left[\frac{3}{2}x^{2/3}\right]_\delta^1 = \frac{3}{2}(1 - \delta^{2/3}).$$

Thus a value of the integral is determined by

$$\int_I f(x)\,dx = \lim_{\delta \to 0+} \int_{I_\delta} f(x)\,dx = \lim_{\delta \to 0+} \tfrac{3}{2}(1 - \delta^{2/3}) = \tfrac{3}{2}.$$

EXAMPLE 2   Let $g(x) = e^{-x}$ for $0 \le x$. See Figure 7.30(b). The integral of f over this interval J is not an ordinary Riemann integral because the domain of the integrand $f(x)$ is an unbounded interval. To assign a value to $\int_0^\infty g(x)\,dx$, we let J_δ be the interval $[0, \delta]$ and first compute

$$\int_{J_\delta} f(x)\,dx = \int_0^\delta e^{-x}\,dx = [-e^{-x}]_0^\delta = 1 - e^{-\delta}.$$

Thus a value of the integral is determined by

$$\int_J f(x)\,dx = \lim_{\delta \to \infty} \int_{J_\delta} f(x)\,dx = \lim_{\delta \to \infty} (1 - e^{-\delta}) = 1.$$

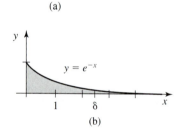

(a)

(b)

FIGURE 7.30

We say that $\int_B f(\mathbf{x})\,dV$ is defined as an **improper integral** if a limit $\int_B f(\mathbf{x})\,dV = \lim \int_{B_\delta} f(\mathbf{x})\,dV$ is finite and independent of the family $\{B_\delta\}$ used to define it. It's possible to show that if either $f(\mathbf{x}) \ge 0$ on B, or $f(\mathbf{x}) \le 0$ on B, then the choice of expanding sets B_δ that cover all of B doesn't affect the final outcome; the limit value assigned to the improper integral will either be finite, in which case we speak of **convergence** of the integral to that value, or else we have **divergence** of the integral, perhaps to $+\infty$ or to $-\infty$.

EXAMPLE 3   Since $\ln z \le 0$ for $0 < z \le 1$, the function $f(x, y) = -\ln(xy)$ is non-negative on the square S: $0 < x \le 1$, $0 < y \le 1$. See Figure 7.31(a). Noting that f is bounded on the square S_δ determined by $\delta \le x \le 1$, $\delta \le y \le 1$, we integrate f over S_δ and compute the limit as δ tends to 0.

$$\int_{S_\delta} -\ln(xy)\,dx\,dy = \int_{S_\delta} -(\ln x + \ln y)\,dx\,dy$$

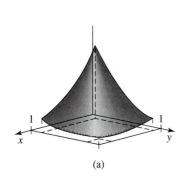

(a)

$$= -\int_\delta^1 dy \int_\delta^1 \ln x \, dx - \int_\delta^1 dx \int_\delta^1 \ln y \, dy$$

$$= -2 \int_\delta^1 dy \int_\delta^1 \ln x \, dx$$

$$= -2(1-\delta)[x \ln x - x]_\delta^1 = -2(1-\delta)(-1 - \delta \ln \delta + \delta)$$

An elementary limit calculation, $\lim_{\delta \to 0+} \delta \ln \delta = 0$, shows that the improper integral is convergent and we can write $\int_S - \ln(xy) \, dx \, dy = 2$.

EXAMPLE 4

The integral $\int_D 1/(x^2 + y^2)^p \, dx \, dy$, where D is the disk $x^2 + y^2 \leq 1$ in $\mathbb{R}^2$, is improper if $p > 0$. To check for convergence, first compute an ordinary Riemann integral over the annulus D_δ determined by $\delta^2 \leq x^2 + y^2 \leq 1$. See Figure 7.31(b). This computation is best done by changing to polar coordinates. The case $p = 1$ turns out to be special, so we'll assume $p \neq 1$.

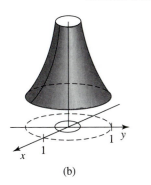

(b)

FIGURE 7.31

$$\int_{D_\delta} 1/(x^2 + y^2)^p \, dx \, dy = \int_0^{2\pi} d\theta \int_\delta^1 r^{-2p} r \, dr = 2\pi \int_\delta^1 r^{1-2p} \, dr$$

$$= 2\pi [(2 - 2p)^{-1} r^{2(1-p)}]_\delta^1 = \left(\pi/(1-p)\right)(1 - \delta^{2(1-p)})$$

Now let $\delta \longrightarrow 0+$. If $0 < p < 1$ the limit of the integrals over D_δ is $\pi/(1-p)$, so the improper integral converges to that value. If $p > 1$ the integrals tend to $+\infty$, so the improper integral diverges for $p > 0$. The case $p = 1$ is left as an exercise.

EXAMPLE 5

The function $f(x, y) = 1/x^2 y^2$, defined for $x \geq 1$ and $y \geq 1$, has the graph shown in Figure 7.32. If B is the set of points (x, y) for which $x \geq 1$ and $y \geq 1$, it is natural to define $\int_B f \, dA$ in such a way that it stands for the volume under the graph of f. We can approximate this volume by computing the volume lying above bounded subrectangles of B. To be specific, let B_N be the rectangle with corners at $(1, 1)$ and (N, N) and with edges parallel to the edges of B. For $N > 1$ we have

FIGURE 7.32

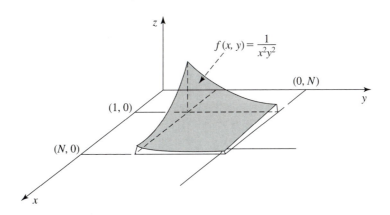

$$\int_{B_N} f \, dA = \int_1^N dx \int_1^N \frac{1}{x^2 y^2} \, dy$$

$$= \left(\int_1^N \frac{dx}{x^2} \right)^2 = \left(1 - \frac{1}{N} \right)^2.$$

As N tends to infinity, the rectangles B_N eventually cover every point of B, and the regions above the B_N fill out the region under the graph of f. Then we define

$$\int_B f \, dA = \lim_{N \to \infty} \int_{B_N} f \, dA = 1.$$

Probability Densities.   Integrals over unbounded regions occur often in statistics and statistical mechanics. If $p(\mathbf{x}) \geq 0$ for $\mathbf{x}$ in some subset S of $\mathbb{R}^n$, and the function p is normalized so that

6.1 $$\int_S p(\mathbf{x}) \, dV = 1,$$

then $p(\mathbf{x})$ can be interpreted as the **density** of a statistical outcome. To be more specific, suppose E is an experiment with possible outcomes in S. Then the **probability** that the outcome of the experiment lies in a subset B of S can sometimes be expressed in the form

$$\Pr[E \text{ in } B] = \int_B p(\mathbf{x}) \, dV,$$

for some density function $p(\mathbf{x})$. For example, the coordinates of a vector outcome $\mathbf{x}$ might be the results of measuring simultaneously several distinct properties of some physical object. In analogy with the center of mass of a mass distribution, we define the **mean** of a probability distribution as the vector

$$\mathbf{m}[p] = \int_B \mathbf{x} p(\mathbf{x}) \, dV.$$

| EXAMPLE 6 | The **symmetric normal probability density** in $\mathbb{R}^2$ is the function defined in all of $\mathbb{R}^2$ by

$$N(x, y) = \frac{1}{2\pi \sigma^2} e^{-[(x^2 + y^2)/2\sigma^2]}.$$

The number σ^2 is a positive constant. To verify that N is a probability density, we need to check the normalization condition 5.4. The integral in question is an improper double integral, which we can evaluate using polar coordinates as follows:

$$\frac{1}{2\pi \sigma^2} \int_{\mathbb{R}^2} e^{-(x^2 + y^2)/2\sigma^2} dx \, dy = \frac{1}{2\pi \sigma^2} \int_0^{2\pi} d\theta \int_0^\infty r e^{-(r^2/2\sigma^2)} \, dr$$

$$= \frac{1}{\sigma^2} \lim_{R \to \infty} \int_0^R r e^{-(r^2/2\sigma^2)} \, dr$$

$$= \frac{1}{\sigma^2} \lim_{R \to \infty} \left[-\sigma^2 e^{-(r^2/2\sigma^2)} \right]_0^R$$

$$= \lim_{R \to \infty} \left[-e^{-(R^2/2\sigma^2)} + 1 \right] = 1.$$

Thus the integral of N has value 1 as we wanted to show and so is independent of the constant σ^2, called the **variance** of the normal density in $\mathbb{R}^2$. The variance is a measure of the dispersion away from the origin of the density. The mean of N is zero in both its coordinates, because the integrals

$$\int_{\mathbb{R}^2} x e^{-(x^2+y^2)/2\sigma^2} dx\, dy \quad \text{and} \quad \int_{\mathbb{R}^2} y e^{-(x^2+y^2)/2\sigma^2} dx\, dy$$

are both zero. For example,

$$\int_{\mathbb{R}^2} x e^{-(x^2+y^2)/2\sigma^2} dx\, dy = \int_{-\infty}^{\infty} x e^{-(x^2/2\sigma^2)} dx \int_{-\infty}^{\infty} e^{-(y^2/2\sigma^2)} dy.$$

The first integral on the right is zero because the integrand is an odd function.

EXAMPLE 7

The function

$$p(x, y) = e^{-x-y}, \quad x \geq 0, \quad y \geq 0,$$

defines a probability density in the first quadrant Q of $\mathbb{R}^2$. All we have to check is that the integral of p over Q is equal to 1. We compute

$$\int_Q e^{-x-y} dx\, dy = \int_0^{\infty} e^{-x} dx \int_0^{\infty} e^{-y} dy$$

$$= \left(\int_0^{\infty} e^{-x} dx \right)^2$$

$$= \left(\lim_{N \to \infty} \left[-e^{-x} \right]_0^N \right)^2$$

$$= \left(\lim_{N \to \infty} (-e^{-N} + 1) \right)^2 = 1.$$

The probability that the outcome of the experiment E is in some rectangle R: $a \leq x \leq b$, $c \leq y \leq d$, where a and c are nonnegative is

$$\Pr[E \text{ in } R] = \int_R e^{-x-y} dx\, dy$$

$$= \left(\int_a^b e^{-x} dx \right) \left(\int_c^d e^{-y} dy \right)$$

$$= [-e^{-x}]_a^b [-e^{-y}]_c^d$$

$$= (e^{-a} - e^{-b})(e^{-c} - e^{-d}).$$

EXAMPLE 8   The squaring trick used in the previous example used in combination with a switch to polar coordinates allows us to show that the integral of e^{-x^2} between $-\infty$ and ∞ equals $\sqrt{\pi}$ as follows:

$$\left(\int_{-\infty}^{\infty} e^{-x^2} \, dx \right)^2 = \int_{\mathbb{R}^2} e^{-x^2 - y^2} \, dx \, dy$$

$$= \int_0^{2\pi} d\theta \int_0^{\infty} e^{-r^2} r \, dr$$

$$= 2\pi \left[-\tfrac{1}{2} e^{-r^2} \right]_0^{\infty} = (2\pi)(\tfrac{1}{2}) = \pi.$$

EXERCISES

In Exercises 1 to 9, determine which of the following improper integrals have finite values, and for those that do, compute the value.

1. $\int_0^{\infty} (1+x)^{-4} dx.$

2. $\int_0^{\infty} x^{-3} dx.$

3. $\int_R \dfrac{x}{y^2} \, dx \, dy$, where R is the infinite rectangle $0 \le x \le 1$, $2 \le y$ in $\mathbb{R}^2$.

4. $\int_R \dfrac{x}{y} \, dx \, dy$, where R is the same as in Exercise 3.

5. $\int_C e^{-x-y-z} \, dx \, dy \, dz$, where C is the infinite rectangle $0 \le x \le 1, 0 \le y \le 1, 0 \le z$ in $\mathbb{R}^3$.

6. $\int_C \dfrac{1}{z^2 \sqrt{xy}} \, dx \, dy \, dz$, where C is the same as in Exercise 5.

7. $\int_D \ln(x^2 + y^2) \, dx \, dy$, where D is the disk $x^2 + y^2 \le 1$ in $\mathbb{R}^2$. [*Hint:* Let D_δ be the annulus $\delta^2 \le x^2 + y^2 \le 1$, and change to polar coordinates.]

8. $\int_D (x^2 + y^2)^{-1} dx \, dy$, where D is the disk $x^2 + y^2 \le 1$ in $\mathbb{R}^2$.

9. $\int_Q \dfrac{dx \, dy}{\sqrt{x^2 + y^2}}$, where Q is the quarter-disk $x \ge 0$, $y \ge 0$, $x^2 + y^2 \le 1$ in $\mathbb{R}^2$. [*Hint:* Use polar coordinates as in Exercise 7.]

10. For which values of α does $\int_0^1 x^\alpha \, dx$ have a finite value as an improper or ordinary Riemann integral?

11. For which values of α does $\int_1^{\infty} x^\alpha \, dx$ have a finite value as an improper or as an ordinary Riemann integral?

12. (a) Show that $p(x) = e^{-x}$ is a probability density on the interval $0 \le x < \infty$ in $\mathbb{R}$.

(b) If E is an experiment with probability density p, as given in part (a), find the probability that the outcome of E lies between a and b, where $0 \le a < b$.

13. (a) For what constant k is the function

$$p(x, y) = k(1 - x^2 - y^2), \quad x^2 + y^2 \le 1$$

a probability density on a disk of radius 1?

(b) If the outcomes of an experiment E are distributed according to the density of part (a), find the probability that E has an x-coordinate bigger than $\tfrac{1}{2}$.

(c) What is the mean of p?

14. If an experiment E has as its probability density the symmetric normal density in $\mathbb{R}^2$ with constant σ^2, find the probability that the coordinates of the outcome are both positive.

15. (a) Show that

$$\frac{1}{\sigma \sqrt{2\pi}} \int_{-\infty}^{\infty} e^{-x^2/2\sigma^2} \, dx = 1$$

by using the result of Example 6 in the text that

$$\frac{1}{2\pi\sigma^2} \int_{-\infty}^{\infty} dx \int_{-\infty}^{\infty} e^{-(x^2+y^2)/2\sigma^2} \, dy = 1.$$

(b) Deduce from part (a) that

$$N_m(x) = \frac{1}{\sigma\sqrt{2\pi}} e^{-(x-m)^2/2\sigma^2}$$

is a probability density on the interval $-\infty < x < \infty$.

(c) Show that the mean of N_m is

$$\int_{-\infty}^{\infty} x N_m(x)\, dx = m.$$

(d) Show that

$$\int_{-\infty}^{\infty} (x-m)^2 N_m(x)\, dx = \sigma^2.$$

The number σ^2 is the *variance* of N_m.

16. (a) Show that the vector mean $\mathbf{m}[N_{m.n}]$ of

$$N_{m,n}(x, y) = \frac{1}{2\pi\sigma^2} e^{-[(x-m)^2 + (y-n)^2]/2\sigma^2}$$

is $\mathbf{m}[N_{m,n}] = (m, n)$.

(b) The **variance** of a density $p(\mathbf{x})$ on $\mathbb{R}^2$ with vector mean $\mathbf{m} = (m, n)$ is defined by

$$\int_{\mathbb{R}^2} |\mathbf{x} - \mathbf{m}|^2 p(\mathbf{x})\, dV,$$

Compute the variance of the density $N_{n,m}$ of part (a).

17. The **Maxwell distribution** for a gas molecule of mass m at temperature T assigns a probability that the molecule's velocity vector at a given instant lies in a region B of the 3-dimensional space of possible velocity vectors $\mathbf{v} = (v_1, v_2, v_3)$. A fundamental assumption is that for a given speed $v = |\mathbf{v}|$, all possible directions in the velocity space are equally likely. In other words, we assume that the probability density for $\mathbf{v}$ is spherically symmetric, so it is appropriate to restrict attention to events dependent entirely on statements about the speed v. We write the corresponding probabilities in terms of spherical coordinates with radial variable v in the form

$$\Pr[a \leq v \leq b] = \int_0^\pi \sin\phi\, d\phi \int_0^{2\pi} d\theta \int_a^b f(v)v^2 dv$$

$$= 4\pi \int_a^b f(v)v^2\, dv.$$

The function $f(v)$ has been determined on theoretical and experimental grounds to be

$$f(v) = \left(\frac{m}{2\pi kT}\right)^{3/2} e^{-mv^2/(2kT)}.$$

The **Boltzmann constant** k is a factor relating the mean kinetic energy of the molecule at temperature T, as stated in part (c).

(a) Calculate the integral that verifies $\Pr[0 \leq v < \infty] = 1.$

(b) Show that the **mean speed** $\mathbf{m}[4\pi v^2 f(v)]$ is $\sqrt{8kT/(\pi m)}$.

SECTION 7   NUMERICAL INTEGRATION

A double integral $\int_R f(x, y)\, dx\, dy$ on a rectangle R determined by $a \leq x \leq b$, $c \leq y \leq d$ is an iterated integral in either of two orders:

$$\int_a^b \left[\int_c^d f(x, y)\, dx \right] dx \quad \text{or} \quad \int_c^d \left[\int_a^b f(x, y)\, dx \right] dy.$$

If evaluating either of these two integrals is possible by finding a succession of two indefinite integrals, then that is probably the best method. (For example, that would give us an answer for lots of values of a, b, c, d.) If we can't find the required indefinite integrals, it may still be possible to evaluate the integral for particular choices of limits, perhaps by some clever change of variable. [See Exercise 4(a).] When all else fails, numerical approximations are available, though only for specific numerical choices of the limits. We'll consider two such approximation methods.

7A   Midpoint Approximations

For a 1-dimensional integral $\int_a^b f(x)\,dx$ we partition the interval $a \leq x \leq b$ into p equal subintervals with endpoints $x_j = (a + j(b-a)/p)$, $j = 0, \ldots p)$. The **midpoint approximation** is

$$\int_a^b f(x)\,dx \approx \sum_{j=0}^{p-1} f\big(a + (j + \tfrac{1}{2})(b-a)/p\big),$$

which evaluates $f(x)$ at the midpoint of each of the p subintervals.

For a double integral over a rectangle R we impose a grid on R with intersection points at

$$(x_j, y_k) = (a + j(b-a)/p,\ c + k(d-c)/q);$$

here p and q are the respective numbers of lines in the grid in the x and y directions, while j runs from 0 to p and k runs from 0 to q. Since the dimensions of each rectangle are $(b-a)/p$ by $(d-c)/q$, the midpoint of the grid rectangle with lower left corner (x_j, y_k) is at

$$(\overline{x}_j, \overline{y}_k) = \big(x_j + \tfrac{1}{2}(b-a)/p,\ y_k + \tfrac{1}{2}(d-c)/q\big)$$
$$= \big(a + (j + \tfrac{1}{2})(b-a)/p,\ c + (k + \tfrac{1}{2})(d-c)/q\big).$$

The **midpoint approximation** to the value of the integral is

$$\int_R f(x, y)\,dx\,dy \approx \frac{(b-a)(d-c)}{pq} \sum_{j=0}^{p-1} \sum_{k=0}^{q-1} f(\overline{x}_j, \overline{y}_k).$$

The approximate value we get this way is just the value of a Riemann sum of the type used to define the integral. In particular, if f is a positive function, the approximate value is a sum of volumes of vertical boxes as illustrated in Figure 7.14(b) of Section 2. The routine to implement the midpoint approximation method is a double loop of the form

```
SET s = 0
FOR j = 0  TO  p - 1
FOR k = 0  TO  q - 1
    LET s = s + f(a + (j + .5)(b - a)/p, c + (k + .5)(d - c)/q)
NEXT k
NEXT j
PRINT s(b - a)(d - c)/(pq)
```

The rest of the routine consists of a definition for f, and an assignment of values to the limits a, b, c, d.

To integrate over a region D in $\mathbb{R}^2$ that's more complicated than a rectangle, simply enclose D in a rectangle R. Then define f by its given values for (x, y)

in D and define $f(x, y) = 0$ for (x, y) outside D. The Heaviside function H of Chapter 4, Section 2C is helpful here. Alternatively, it may be simpler first to make a change of variable in the integral that results in integration over a rectangle.

The analogous formula for the midpoint approximation for a function $g(x, y, z)$ integrated over a rectangular region R with extreme corners at (a, c, e) and (b, d, f) is

$$\int_R g(x, y)\, dx\, dy\, dz \approx \frac{(b-a)(d-c)(e-f)}{pqr} \sum_{j=0}^{p-1} \sum_{k=0}^{q-1} \sum_{l=0}^{r-1} g(\overline{x}_j, \overline{y}_k, \overline{z}_l),$$

where

$$(\overline{x}_j, \overline{y}_k, \overline{z}_l) = \left(x_j + \tfrac{1}{2}(b-a)/p,\ y_k + \tfrac{1}{2}(d-c)/q,\ z_k + \tfrac{1}{2}(e-f)/r \right)$$

$$= \left(a + (j + \tfrac{1}{2})(b-a)/p,\ c + (k + \tfrac{1}{2})(d-c)/q,\ e + \tfrac{1}{2}(e-f)/r \right).$$

7B   Simpson Approximations

If the integrand f in a multiple integral is a fairly smooth function we can take advantage of its smoothness by repeated use of the 1-dimensional **Simpson approximation** over an *even* number p of intervals:

$$\int_a^b f(x)\, dx \approx \frac{b-a}{3p} \left(f(x_0) + 4f(x_1) + 2f(x_2) + \cdots + 4f(x_{p-1}) + f(x_p) \right)$$

$$\approx \frac{b-a}{3p} \sum_{j=0}^{p} S_j^{(p)} f(x_j),$$

where $x_j = a + j(b-a)/p$. The pattern for the coefficients $S_j^{(p)}$ is such that the first and the last coefficients are $S_0^{(p)} = S_p^{(p)} = 1$, while the intermediate values are given by the formula $S_j^{(p)} = 3 - (-1)^j$, in other words, alternating 4's and 2's, beginning and ending with 4. The requirement for the even number of subdivision intervals comes from the geometry underlying the method; with just two intervals, the Simpson approximation is precisely the integral over $a \le x \le b$ of the unique quadratic polynomial that interpolates the values of f at a, $\tfrac{1}{2}(a + b)$ and b.

To apply the Simpson formula to a double integral we use a two-stage Simpson approximation to an iterated integral

$$\int_c^d \left[\int_a^b f(x, y)\, dx \right] dy.$$

Thinking of y as held fixed for the moment, start with

$$F(y) = \int_a^b f(x, y)\, dx \approx \frac{b-a}{3p} \sum_{j=0}^{p} S_j^{(p)} f(x_j, y), \quad \text{where} \quad x_j = a + j(b-a)/p.$$

Letting $y_k = c + k(d - c)/q$, with q even, we approximate

$$\int_c^d F(y)\, dy \approx \frac{(d - c)}{3q} \sum_{k=0}^{q} S_k^{(q)} F(y_k).$$

Now replace $F(y_k)$ by the Simpson approximation previously obtained to get

$$\int_c^d \left[\int_a^b f(x, y)\, dx \right] dy \approx \frac{(b - a)(d - c)}{9pq} \sum_{k=0}^{q} S_k^{(q)} \left(\sum_{j=0}^{p} S_j^{(p)} f(x_j, y_k) \right)$$

$$\approx \frac{(b - a)(d - c)}{9pq} \sum_{j=0}^{p} \sum_{k=0}^{q} S_j^{(p)} S_k^{(q)} f(x_j, y_k).$$

Note that in this formula we use the values of f at all grid points in the rectangle R, including those on its boundary. The double loop in the implementing routine has the form

```
SET s = 0
FOR j = 0  TO  p (Odd number of values, with p even.)
FOR k = 0  TO  q (Odd number of values, with q even.)
   LET s = s + S(j,p)S(k,q)f(a + j(b - a)/p,c + k(d - c)/q)
NEXT k
NEXT j
PRINT s(b - a)(d - c)/(9pq)
```

Before the loop we need to include the definition

$$S(j, k) = \begin{cases} 3 - (-1)^j, & \text{if } 0 < j < p, \\ 1, & \text{if } j = 0 \text{ or } j = p. \end{cases}$$

The remarks at the end of Section 7B about defining $f(x, y)$ on nonrectangular regions apply here as well.

EXERCISES

1. Use a program to implement the midpoint approximation for an $f(x, y)$ and test it on the example $f(x, y) = x^2 + y^4$ over the rectangle R: $0 \le x \le 1$, $0 \le y \le 1$. Having computed the correct value via indefinite integrals, you can find out how small p and q can be while still producing four-place accuracy.

2. Use a program to implement Simpson's approximation for $f(x, y)$ and test it on the same example as in the previous problem to find minimal values for p and q for four-place accuracy, such that reducing either p or q fails to yield that degree of accuracy, in particular *increasing* p or q should not change the first four digits.

In Exercises 3 to 8, use the midpoint or Simpson approximations to find an approximate value to four-place accuracy for $\int_R f(x, y)\, dx\, dy$ where f, R are

3. $f(x, y) = 1 - x^2 - y^2$, R: $0 \le x \le 2, 0 \le y \le 3$.

4. $f(x, y) = 1 - x - y$, $R: 0 \le x$, $0 \le y$, $x + y \le 1$.

5. $f(x, y) = 1 - x^2 - y^2$, $R: x^2 + y^2 \le 1$.

6. $f(x, y) = 1 - x^2 - y^2$, $R: x^2 + y^2 \le 1$, $y \ge 0$.

7. $f(x, y) = 1 - x^2 - y^2$, $R: 0 \le x, 0 \le y, x^2 + y^2 \le 1$.

8. $f(x, y) = y$, $R: 0 \le x \le 1$, $0 \le y \le x$.

9. The double integral

$$G(a, b, c, d) = \int_a^b \int_c^d e^{-x^2 - y^2} \, dx \, dy$$

can't be evaluated in terms of elementary functions of a, b, c, d. Nevertheless, we can still find such a value when the rectangular region is replaced by a circular region of radius a centered at the origin. The trick is first to make the change of variable $x = r\cos\theta$, $y = r\sin\theta$, $dx\,dy = r\,dr\,d\theta$, then to compute the resulting double integral by iterated integration over the region $0 \le r \le R$, $0 \le \theta \le 2\pi$.

(a) Compute the value of the integral over $\mathbb{R}^2$ by using polar coordinates over a disk of radius R as R tends to ∞.

(b) Compute approximations to π by finding Simpson approximations to $4G(0, a, 0, a)$ for suitable values of a.

(c) Estimate how large you need to make the positive number a in part (b) to get four-place accuracy.

In Exercises 10 to 13, use Simpson's approximation to find approximate values for the following integrals. The Simpson formula for a triple integral over a 3-dimensional rectangle R is

$$\int\int\int_R g(x, y, z) \, dx \, dy \, dz$$

$$= \int_e^f \left[\int_c^d \left[\int_a^b g(x, y, z) \, dx \right] dy \right] dz$$

$$\approx \frac{(b - a)(d - c)(f - e)}{27pqr}$$

$$\sum_{j=0}^p \sum_{k=0}^q \sum_{l=0}^r S_j^{(p)} S_k^{(q)} S_l^{(r)} g(x_j, y_k, z_l).$$

10. $\displaystyle\int_R \ln(xyz) \, dx \, dy \, dz$, R: $1 \le x \le 2$, $1 \le y \le 3$, $2 \le z \le 3$.

11. $\displaystyle\int_R \sqrt{x^2 + y^2 + z^2} \, dx \, dy \, dz$, R: $x^2 + y^2 + z^2 \le 1$, $0 \le z$.

12. $\displaystyle\int_R e^{(x+y+z)} \, dx \, dy \, dz$, R: $0 \le x \le 1$, $0 \le y \le 2$, $0 \le z \le 3$.

13. $\displaystyle\int_R (x + y + z) \, dx \, dy \, dz$, $R: 0 \le x \le 1$, $0 \le y \le 1$, $0 \le z \le 1$.

14. (a) Sketch the region in $\mathbb{R}^2$ bounded by the four lines $x + y = 1$, $x + 2y = 4$, $x - 2y = -1$ and $x - 3y = 1$.

(b) Find an approximate value for the area of the region in part (a). Can you find the exact value, 187/120, by elementary geometry?

(c) Find an approximate value for the integral of $f(x, y, z) = x^3 + y^3$ over the region described in part (a).

15. (a) Sketch the region in the positive octant of $\mathbb{R}^3$ bounded by the planes $x + y + z = 3$, $x + y + 2z = 6$, $z = 1$ and $z = 2$.

(b) Find an approximate value for the volume of the region in part (a). Can you find the exact value by elementary geometry?

(c) Find an approximate value for the integral of $f(x, y, z) = x^4 + y^4 + z^4$ over the region described in part (a).

Chapter 7 REVIEW

Sketch the region of integration in each of the following iterated integrals, and evaluate the integral as given. Then evaluate the integral in the reversed order as a check.

1. $\displaystyle\int_0^1 \left[\int_0^2 xy^2 \, dx \right] dy$.

2. $\displaystyle\int_{-1}^1 \left[\int_0^2 e^{x-y} \, dy \right] dx$.

3. $\displaystyle\int_0^2 \left[\int_0^x e^{x-y} \, dy \right] dx$.

4. $\displaystyle\int_0^1 dy \int_0^y (x + y)^2 \, dx$.

5. $\displaystyle\int_0^\pi \left[\int_0^{2\pi} \sin(x - y) \, dy \right] dx$.

6. $\displaystyle\int_0^1 \left[\int_{-1}^x x^4 \, dy \right] dx$.

Sketch the region of integration stated for each of the following double integrals. Then evaluate the integral. You may want to pick your order carefully, and you may even want to change coordinates.

7. $\int_T (x+y)\, dx\, dy$; T is the triangle with corners at $(0, 0)$, $(1, 0)$ and $(1, 1)$.

8. $\int_R y\, dA$; R is the half of the disk of radius 1 centered at $(0, 0)$ where $y \geq 0$.

9. $\int_Q x\, dx\, dy$; Q is the part in the first quadrant of the disk of radius 1 centered at $(0, 0)$.

10. $\int_D (x^2 + y^2)^5\, dx\, dy$; D is the disk of radius 4 centered at $(0, 0)$.

11. $\int_D (x^2 - y^2)^2\, dx\, dy$; D is the disk of radius 1 centered at $(0, 0)$.

12. $\int_D (1+x^2+y^2)^{-3/2}\, dV$; D is the disk of radius 2 centered at $(0, 0)$.

13. $\int_S (x^2 + y^2)\, dx\, dy$; S is the square of side 2 centered at $(1, 0)$, edges parallel to the axes.

14. $\int_S x^2 y^2\, dx\, dy$; S is the square $|x| + |y| \leq 1$.

Sketch the 3-dimensional region of integration of each of the following integrals. Then evaluate the integral, possibly after a change of coordinates.

15. $\int_0^1 \left[\int_1^2 \left[\int_2^3 xyz\, dz \right] dy \right] dx$.

16. $\int_0^1 \left[\int_0^x \left[\int_0^y xyz\, dz \right] dy \right] dx$.

17. $\int_C z(x^2 + y^2)\, dV$; C is the solid cylinder $x^2 + y^2 \leq 1$, $0 \leq z \leq 2$.

18. $\int_C (x^2 + y^2 + z^2)\, dV$; C is the solid cylinder $x^2 + y^2 \leq 4$, $0 \leq z \leq 1$.

19. $\int_B (x^2 + y^2 + z^2)\, dV$; B is the solid ball $x^2 + y^2 + z^2 \leq 1$.

20. $\int_B z\, dx\, dy\, dz$; B is the solid ball $x^2 + y^2 + z^2 \leq 1$.

21. $\int_K (x^2 + y^2)\, dV$; K is the solid cone $\sqrt{x^2 + y^2} \leq z \leq 1$.

22. $\int_K z\, dV$; K is the solid cone $\frac{1}{2}\sqrt{x^2 + y^2} \leq z \leq 3$.

23. $\int_K \sqrt{x^2 + y^2}\, dV$; K is the solid cone $\sqrt{x^2 + y^2} \leq z \leq 2$.

24. **(a)** Sketch the region R bounded by the graphs of $y = x^3$ and $x = y^2$.

    (b) The double integral $\int_R x\, dx\, dy$ is equal to each of two iterated integrals over R; write down both of them and evaluate one of them.

25. Let R be the region in the plane between the parabola $y = x^2$ and the line $y = 2x + 3$. Write the integral $\int_R x^2 y\, dA$ as an iterated integral in both possible orders; then evaluate one of them.

Make a sketch of each of the following plane or solid regions numbered 26 through 33, and find the area or volume as the case may be. Use your own reasonable choices for the constants in making the sketches.

26. Elliptic region E in $\mathbb{R}^2$: $x^2/a^2 + y^2/b^2 \leq 1$.

27. Solid B based on the region E of the previous exercise and with square cross sections perpendicular to the x-axis.

28. Solid B based on the region E of the previous two exercises and with semicircular cross sections perpendicular to the x-axis.

29. Ellipsoid B: $x^2/a^2 + y^2/b^2 + z^2/c^2 \leq 1$. Note that B has elliptical cross sections.

30. Region H in $\mathbb{R}^2$: $x^2 \leq y^2 + 1$, $|y| \leq 1$.

31. Solid B generated by rotating the region H of the previous exercise about the y-axis.

32. Solid B generated by rotating the region H of the previous exercises about the x-axis.

33. Solid B bounded by the three coordinate planes in $\mathbb{R}^3$ and the plane $ax + by + cz = 1$, where a, b, c are positive.

34. For the integral

$$\int_0^1 dy \int_y^{\sqrt{y}} 2xy\, dx,$$

    (a) Sketch the region of integration in $\mathbb{R}^2$.
    (b) Evaluate the integral.

35. Given

$$\int_0^1 \int_0^{\sqrt{2-2x^2}} 2x\, dy\, dx.$$

    (a) Sketch the region of integration.
    (b) Write an equivalent integral with the order of integration reversed.
    (c) Evaluate either integral, or both as a check.

36. Let B be the region bounded by the xy-plane, the yz-plane, the xz-plane and the plane $x + 2y + z = 4$. (a) Make a sketch of B. (b) Find the volume of B.

37. Integrate the function $f(x, y) = 3x^2 + 2y$, over the region in the plane bounded by the curves $y = x^2$ and $y = 2 - x$.

38. Compute $\displaystyle\int_R \sin(x - y)\, dx\, dy$, where R is the region bounded by parallelogram with vertices at $(0, 0)$, $(0, \pi/2)$, $(\pi/2, \pi/2)$, and $(\pi/2, \pi)$.

39. Compute $\displaystyle\int_R \sqrt{x^2 + y^2}\, dx\, dy$, where R is the region enclosed by the part of the polar coordinate curve $r = \cos\theta$ traced out as θ varies from $-\pi/2$ to $\pi/2$. Sketch the region R.

40. Let B be the part in the first octant of the solid ball in $\mathbb{R}^3$ of radius 3 centered at the origin. Use spherical coordinates to compute $\displaystyle\int_R yz\, dV$.

41. Let C be a solid cylinder of radius 1 symmetric about the z-axis. Let W be the wedge-shaped subset of C where $0 \leq z \leq x$. Write an iterated integral for $\displaystyle\int_W z\, dV$

   (a) in rectangular coordinates.
   (b) in cylindrical coordinates.
   (c) Evaluate the multiple integral in whichever way you prefer.

42. (a) Set up an iterated integral whose evaluation will yield the volume of the spherical ball B_a of radius a centered at the origin in $\mathbb{R}^3$.
   (b) Use Jacobi's theorem to show that every spherical ball of radius a in $\mathbb{R}^3$ has the same volume.

43. Let R be the region in $\mathbb{R}^3$ determined by the inequalities $0 \leq x$, $0 \leq y$, $x^2 + y^2 \leq 1$ and $0 \leq z \leq x^2 + y^2$. Evaluate $\displaystyle\int_R xyz\, dV$.

44. Compute the value of the integral of $f(x, y) = x^2 + y^2$ over the triangle in $\mathbb{R}^2$ with corners at $(0, 0)$, $(1, 0)$, and $(1, 1)$ by computing iterated integrals in both possible orders.

45. Find limits of integration, some of which may be nonconstant, for this integral if the region of integration is the circular disk of radius 2 centered at $(3, 0)$:

$$\int_a^b \left[\int_c^d f(x, y)\, dx \right] dy.$$

46. The region of integration for $h(x, y, z)$ is bounded above by the plane $z = 1$ and below by the circular paraboloid $z = x^2 + y^2$; find the limits, which may be non-constant:

$$\int_a^b \left[\int_c^d \left[\int_e^f h(x, y, z)\, dz \right] dx \right] dy.$$

47. The equations $x = u + v$, $y = u + 3v$ define a transformation from the uv-plane to the xy-plane that carries the points inside the rectangle with corners at $(0, 0)$, $(2, 0)$, $(2, 1)$ and $(0, 1)$ onto a region R. Compute $\displaystyle\int_R y\, dA$.

48. The equations $x = 2u + 1/(v + 1)$, $y = u + v$ define a transformation from the uv-plane to the xy-plane that carries the points inside the rectangle with corners at $(0, 0)$, $(1, 0)$, $(1, 1)$ and $(0, 1)$ onto a region R. Sketch R and find its area.

49. A vertical cylinder C has a flat base in the xy-plane consisting of the semicircle for which $x^2 + y^2 \leq 9$ and $x \geq 0$. The top of C is part of the graph of $z = 1 + x + y^2$. Find the volume of C.

50. Let R be the region in the first quadrant of the xy-plane where $4 \leq xy \leq 9$ and $1 \leq y/x \leq 4$.
   (a) Solve the equations $u = xy$ and $v = y/x$ uniquely for x and y in R in terms of u and v.
   (b) Express $\displaystyle\int_R x^{-2}\, dx\, dy$ as an integral in u and v, and find its value.

In Exercises 51 to 56, evaluate those improper integrals that have finite values; for those that don't, explain why they fail to converge.

51. $\displaystyle\int_{\mathbb{R}^2} e^{-(x^2 + y^2)}\, dx\, dy$.

52. $\displaystyle\int_{\mathbb{R}^2} (x^2 + y^2)^{-1}\, dx\, dy$.

53. $\displaystyle\int_{x^2 + y^2 \leq 1} (x^2 + y^2)^{-1/3}\, dA$.

54. $\displaystyle\int_{\mathbb{R}^2} e^{-\sqrt{x^2 + y^2}}\, dx\, dy$.

55. $\displaystyle\int_{\mathbb{R}^3} e^{-(x^2 + y^2 + z^2)^{3/2}}\, dx\, dy\, dz$.

56. $\displaystyle\int_{x^2 + y^2 + z^2 \leq 1} (x^2 + y^2 + z^2)^{-2}\, dV$.

In Exercises 57 to 60, decide for what values of α the integrals have finite values.

57. $\displaystyle\int_{x^2 + y^2 \leq 1} \frac{1}{(x^2 + y^2)^\alpha}\, dA$

58. $\displaystyle\int_{x^2 + y^2 \geq 1} \frac{1}{(x^2 + y^2)^\alpha}\, dA$

59. $\displaystyle\int_{x^2 + y^2 + z^2 \leq 1} \frac{1}{(x^2 + y^2 + z^2)^\alpha}\, dV$

60. $\displaystyle\int_{x^2+y^2+z^2\geq 1} \frac{1}{(x^2 + y^2 + z^2)^\alpha}\, dV$

61. $\displaystyle\int_0^{2\pi} d\theta \int_0^1 dr \int_0^{\sqrt{1-r^2}} r\,dz$ is expressed in cylindrical coordinates.

(a) Sketch the region of integration.

(b) Write the integral in terms of rectangular coordinates.

(c) Write the integral in terms of spherical coordinates.

(d) Compute the value of the integral.

CHAPTER 8

INTEGRALS AND DERIVATIVES ON CURVES

One version of the Fundamental Theorem of Calculus is

$$\int_a^b f'(t)\,dt = f(b) - f(a).$$

The present chapter features just one of several ways to extend the Fundamental Theorem, and we postpone further extensions to Chapter 9. Here we first introduce a simple generalization of the integral itself, called the line integral over a parametrized curve. Combining the line integral with the gradient operator we'll arrive at an analogue of the Fundamental Theorem,

$$\int_{\mathbf{a}}^{\mathbf{b}} \nabla f(\mathbf{x}) \bullet d\mathbf{x} = f(\mathbf{b}) - f(\mathbf{a}),$$

which we'll use to express a key relationship between the physical concepts *work* and *energy*. Recall that curves in $\mathbb{R}^2$ and $\mathbb{R}^3$ are in a sense negligible in the multiple integrals of Chapter 7. However, all the integrals in this chapter will reduce to ordinary one-variable integrals for computational purposes, even though the concepts that give rise to them have a distinctly higher-dimensional flavor.

Differentiation along a curve is a concept intrinsic to the geometry of the curve, typically distinct from differentiation of the curve's parametrization. The idea is particularly important in describing motion along curves. In some applications to motion on curves we use the term **particle** to describe a body moving on a curved path, but we often understand *particle* to stand for the position of the center of mass of what is really a large and complex body such as the earth, a simplification that turns out to be adequate for many purposes.

SECTION 1   LINE INTEGRALS

1A   Definition and Examples

The integral $\int_a^b f(x)\,dx$ of a real-valued function of one scalar variable generalizes in several ways. One generalization that has applications in physics is the line integral, which we describe here. Let $\mathbb{R}^3 \xrightarrow{\mathbf{F}} \mathbb{R}^3$ be a continuous function defined in a region D of $\mathbb{R}^3$. We picture $\mathbf{F}$ as a **vector field**, that is, as an assignment of the arrow $\mathbf{F}(\mathbf{x})$ to the point $\mathbf{x}$ for each $\mathbf{x}$ in D. A sketch of a 3-dimensional vector field is shown in Figure 8.1(a). Suppose also that γ is a curve lying in D, and parametrized by a function $g(t)$, continuously differentiable for $a \le t \le b$.

FIGURE 8.1

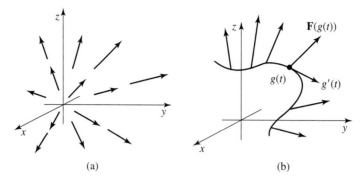

(a)   (b)

We'll be particularly interested in the arrows of the field $\mathbf{F}$ that stem from points on γ, as shown in Figure 8.1(b). These arrows will depend on t in a specific way if we introduce the composition $\mathbf{F}(g(t))$. At each point $g(t)$ on γ there is also a tangent vector $g'(t)$, and the dot product

$$\mathbf{F}(g(t)) \bullet g'(t)$$

is a continuous real-valued function for $a \le t \le b$. The **line integral** of $\mathbf{F}$ over γ is, by definition,

1.1
$$\int_a^b \mathbf{F}(g(t)) \bullet g'(t)\, dt.$$

EXAMPLE 1   If a vector field is given in $\mathbb{R}^3$ by $\mathbf{F}(x, y, z) = (x^2, y^2, z^2)$ and γ is given by $g(t) = (t, t^2, t^3)$ for $0 \le t \le 1$, then the integral of $\mathbf{F}$ over γ is

$$\int_0^1 (t^2, t^4, t^6) \bullet (1, 2t, 3t^2)\, dt = \int_0^1 (t^2 + 2t^5 + 3t^8)\, dt$$

$$= \left[\tfrac{1}{3}t^3 + \tfrac{1}{3}t^6 + \tfrac{1}{3}t^9 \right]_0^1 = 1.$$

We interpret the line integral in qualitative terms as follows. The dot product

$$\mathbf{F}(g(t)) \bullet \frac{g'(t)}{|g'(t)|}$$

is the coordinate of $\mathbf{F}(g(t))$ in the direction of the unit tangent vector to γ at $g(t)$. Then $\mathbf{F}(g(t)) \bullet g'(t)$, the integrand in Formula 1.1, is the tangential coordinate of $\mathbf{F}(g(t))$ times $|g'(t)|$, the speed of traversal of γ at $g(t)$. In particular, if $\mathbf{F}(g(t))$ is always perpendicular to γ at $g(t)$, the integrand, and hence the integral will be zero. At the other extreme, for a given field $\mathbf{F}$, if the speed $|g'(t)|$ is prescribed at each point of the curve, then the integrand will be maximized by choosing a curve γ that at each point has the same direction as the field there. Thus the integrand in the line integral is a local measure of the *circulation* of the vector field along γ. The term *circulation* is justified by the frequent interpretation of $\mathbf{F}$ as the velocity field of a fluid flow.

Equation 1.1 works in any number of dimensions. If $\mathbb{R}^n \overset{\mathbf{F}}{\longrightarrow} \mathbb{R}^n$ is a vector field, and $\mathbb{R} \overset{g}{\longrightarrow} \mathbb{R}^n$ describes for $a \le t \le b$ a smooth curve γ, lying in the domain D of $\mathbf{F}$, the line integral of $\mathbf{F}$ over γ is still defined by Equation 1.1 in which the dot product is now formed in $\mathbb{R}^n$. In general the **circulation** of $\mathbf{F}$ over γ is defined to be the value of the integral of $\mathbf{F}$ over a curve γ, whether γ is a closed curve or not.

EXAMPLE 2

Let $\mathbf{F}(x, y) = (x, y)$ define a vector field in $\mathbb{R}^2$. The curve given by $g(t) = (\cos t, \sin t)$ for $0 \le t \le \pi/2$ is the quarter-circle shown in Figure 8.2(a) together with some tangent vectors and some vectors of the field. Because the field is perpendicular to the curve at each point, we expect the integral to be zero, and we have

$$\int_0^{\pi/2} \mathbf{F}\big(g(t)\big) \cdot g'(t)\, dt = \int_0^{\pi/2} (\cos t, \sin t) \cdot (-\sin t, \cos t)\, dt$$

$$= \int_0^{\pi/2} (-\cos t \sin t + \sin t \cos t)\, dt$$

$$= \int_0^{\pi/2} 0\, dt = 0.$$

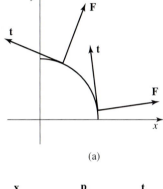

(a)

(b)

FIGURE 8.2

Work.   An important physical interpretation of the line integral arises as follows. Suppose that the function $\mathbb{R}^3 \overset{\mathbf{F}}{\longrightarrow} \mathbb{R}^3$ determines a continuous force field in a region D in $\mathbb{R}^3$. Thus $\mathbf{F}(\mathbf{x})$ represents the magnitude and direction of a force applied at $\mathbf{x}$. To define the work W done in moving a particle along a curve γ in D, we use a preliminary definition for linear motion in a constant field, namely that work is scalar force acting in the direction of motion multiplied by distance covered. Thus **work** is the product

$$W = \mathbf{F_t}\, s,$$

where s is the distance traversed and $\mathbf{F}_t$ is the force in the unit direction $\mathbf{t}$ of motion. In Figure 8.2(b) a particle moves along a line having direction vector $\mathbf{t}$ with $|\mathbf{t}| = 1$, and it is subject at each point $\mathbf{x}$ to the constant force vector $\mathbf{F}$. The coordinate of $\mathbf{F}$ in the direction of motion is $\mathbf{F_t} = \mathbf{F} \cdot \mathbf{t}$, so work is

$$W = (\mathbf{F} \cdot \mathbf{t})\, s = (\text{force coordinate}) \times (\text{distance}).$$

For motion along a continuously differentiable curve, we begin by approximating the curve by tangent vectors. If the curve γ is described parametrically by $\mathbb{R} \overset{g}{\longrightarrow} \mathbb{R}^3$ with $g(t)$ defined for $a \le t \le b$, then the arrows representing the tangent vectors

$$g'(t_{k-1})(t_k - t_{k-1}), \quad t_0 < t_1 < \cdots < t_k,$$

will approximate γ as shown in Figure 8.3, since the number $|g'(t_{k-1})||(t_k - t_{k-1})|$ approximates the distance from $g(t_{k-1})$ to $g(t_k)$. We fix a point $\mathbf{x}_k = g(t_k)$ on γ, and near $\mathbf{x}_k$ approximate $\mathbf{F}$ by the constant field $\mathbf{F}(\mathbf{x}_k)$. That is, near $\mathbf{x}_k$ we approximate $\mathbf{F}(\mathbf{x})$ by the vector field that assigns the constant vector $\mathbf{F}(\mathbf{x}_k)$ to every point. The tangential coordinate of $\mathbf{F}(\mathbf{x}_k)$ is $\mathbf{F}(\mathbf{x}_k) \cdot \mathbf{t}(t_k)$, where $\mathbf{t}(t) = g'(t)/|g'(t)|$. Thus the work done in moving a particle along γ from $\mathbf{x}_k$ to $\mathbf{x}_{k+1}$ is approximately

FIGURE 8.3

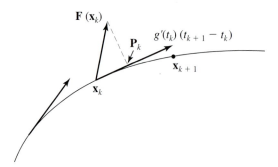

$$W_k = \big(\mathbf{F}(\mathbf{x}_k) \cdot \mathbf{t}(t_k)\big)|g'(t_k)|(t_{k+1} - t_k)$$
$$= \mathbf{F}\big(g(t_k)\big) \cdot g'(t_k)(t_{k+1} - t_k).$$

Letting $m(P) = \max\limits_{1 \le k \le K} (t_k - t_{k-1})$, we get

$$\lim_{m(P) \to 0} \sum_{k=0}^{K-1} W_k = \int_a^b \mathbf{F}\big(g(t)\big) \cdot g'(t)\, dt,$$

an integral formula that we define to be the **work** done by the field $\mathbf{F}$ in moving the particle through the domain of $\mathbf{F}$ along γ.

A suggestive shorthand notation for the general line integral uses the unit tangent vector $\mathbf{t}(t) = g'(t)/|g'(t)|$ to a smooth path of integration on which $g'(t) \ne 0$. Given that arc length along such a path γ is defined by $s(t) = \int_a^t |g'(t)|\, dt$, it's natural to write $ds = |g'(t)|\, dt$ for the so-called **arc length differential**. The line integral for work is then the natural extension of the special case $W = (\mathbf{F} \cdot \mathbf{t})s$:

$$W = \int_\gamma \mathbf{F} \cdot \mathbf{t}\, ds.$$

This way of writing the integral captures the essence of our interpretation of the line integral, since $\mathbf{F} \cdot \mathbf{t}$ is the coordinate of $\mathbf{F}$ along the tangent direction to γ.

The assumptions that $\mathbf{F}$ be continuous and that g' be continuous assured that the integrand $\mathbf{F}\big(g(t)\big) \cdot g'(t)$ would be continuous and hence that the line integral would exist. However these conditions are stronger than necessary. It's enough to assume that the path of integration is piecewise smooth and then that the vector field $\mathbf{F}$ is sufficiently regular so the integral in Formula 1.1 exists. Thus the derivative g' may be discontinuous at finitely many points, allowing γ to have sharp corners at some points.

EXAMPLE 3   Let a vector field be defined in $\mathbb{R}^3$ by $\mathbf{F}(x, y, z) = (x, y, z)$. Let the curve γ in $\mathbb{R}^3$ be given by $g(t) = (\cos t, \sin t, |t - \pi/2|)$ for $0 \le t \le \pi$. Then γ has a corner at $(0, 1, 0)$, where $t = \pi/2$. Indeed, g is not differentiable there, and $\lim\limits_{t \to \pi/2-} g'(t) = (-1, 0, -1)$ and $\lim\limits_{t \to \pi/2+} g'(t) = (-1, 0, 1)$, showing that the direction of the tangent jumps abruptly at $t = \pi/2$. Nevertheless, the integral of **F** over γ exists. To compute it, the interval of integration would ordinarily be broken at $t = \pi/2$. But in this particular case $\mathbf{F}\big(g(t)\big) \cdot g'(t) = t - \pi/2$ unless $t = \pi/2$. It follows that

$$\int_{\gamma} \mathbf{F} \cdot \mathbf{t}\,ds = \int_0^{\pi} \mathbf{F}\big(g(t)\big) \cdot g'(t)\,dt = \int_0^{\pi} \left(t - \frac{\pi}{2} \right) dt = 0.$$

A convenient notation for line integrals denotes the parametrization of γ by $g(t) = \big(x(t), y(t), z(t)\big)$ for $a \le t \le b$. If the coordinate functions of **F** are F_1, F_2, and F_3, and we suppress the variable t in the integrand, we get

$$\int_a^b \mathbf{F}\big(g(t)\big) \cdot g'(t)\,dt = \int_a^b \left[F_1(x, y, z)\frac{dx}{dt} + F_2(x, y, z)\frac{dy}{dt} + F_3(x, y, z)\frac{dz}{dt} \right] dt.$$

The last integral abbreviates to

$$\int_{\gamma} F_1\,dx + F_2\,dy + F_3\,dz.$$

This is still shorter if we write $d\mathbf{x} = (dx, dy, dz)$, giving a convenient shorthand for the line integral of **F** over γ:

$$\int_{\gamma} \mathbf{F} \cdot d\mathbf{x}.$$

EXAMPLE 4   Let a vector field be given in $\mathbb{R}^3$ by

$$\mathbf{F}(x, y, z) = (x - y, y - z, z - x).$$

A curve γ, given by $g(t) = (t, -t, t^2)$ for $0 \le t \le 1$, passes through the field. We compute the integral of **F** over γ as follows. First, the values of **F** on γ are given by

$$\mathbf{F}\big(g(t)\big) = (2t, -t - t^2, t^2 - t).$$

We write

$$d\mathbf{x} = g'(t)\,dt = (1, -1, 2t)\,dt.$$

Then $\mathbf{F} \cdot d\mathbf{x} = \mathbf{F}\big(g(t)\big) \cdot g'(t)\,dt = (2t, -t - t^2, t^2 - t) \cdot (1, -1, 2t)\,dt$, so

$$\int_{\gamma} \mathbf{F} \cdot d\mathbf{x} = \int_0^1 \left[(2t)(1) + (-t - t^2)(-1) + (t^2 - t)(2t) \right] dt$$

$$= \int_0^1 (2t^3 - t^2 + 3t)\,dt = \frac{5}{3}.$$

If we can choose coordinates so that one or more of the sections of a line integral path are parallel to an axis, computing the value of an integral may be substantially simplified, as in the following example.

FIGURE 8.4

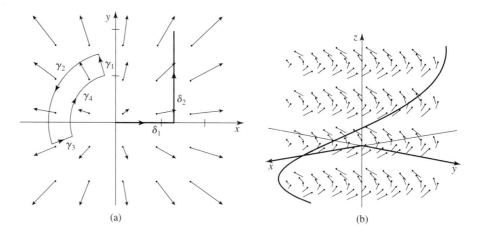

(a)   (b)

EXAMPLE 5 Let $\mathbf{F}(x, y) = (x, y)$ define a 2-dimensional velocity field along the path δ from $(0, 0)$ to $(a, 0)$ along the x-axis and then along the line segment from $(a, 0)$ to (a, b). Thus δ consists of a path δ_1 along the x-axis followed by a path δ_2 parallel to the y-axis. See Figure 8.4(a), where it's assumed that $a > 0$ and $b > 0$. No matter how the path is parametrized by a function $g(t) = \big(g_1(t), g_2(t)\big)$, we see that $g_2'(t) = 0$ along δ_1. In other words, $dy = 0$ along δ_1. Similarly, $dx = 0$ along δ_2. Since $\mathbf{F} \cdot d\mathbf{x} = x\,dx + y\,dy$, we can compute the circulation of $\mathbf{F}$ along δ using x and y as parameters on δ_1 and δ_2 respectively. For arbitrary a and b we have

$$\int_\delta x\,dx + y\,dy = \int_{\delta_1} x\,dx + \int_{\delta_2} y\,dy$$

$$= \int_0^a x\,dx + \int_0^b y\,dy = \frac{1}{2}a^2 + \frac{1}{2}b^2.$$

EXAMPLE 6 Let $\mathbf{F}(x, y) = (x, y)$ as in the previous example. This time we integrate $\mathbf{F}$ over the closed path, shown in Figure 8.4(a), consisting of two circular arcs with their ends joined by radial segments. The entire path γ is to be traced counterclockwise. Over the circular arcs the tangent vectors $\mathbf{t}$ are perpendicular to the vector field arrows, so $\mathbf{F} \cdot \mathbf{t} = 0$ there. Thus the integral of $\mathbf{F}$ is zero over γ_2 and γ_4. Let $\mathbf{t}$ be the unit tangent at a typical point of the segment γ_1. Since $\mathbf{F}$ and $\mathbf{t}$ point in the same direction along γ_1, $\mathbf{F} \cdot \mathbf{t} = |\mathbf{F}| = \sqrt{x^2 + y^2}$ at a point (x, y) of γ_1. In contrast, the tangent vectors to γ_3 all point toward the origin. Since $\mathbf{F}$ and $\mathbf{t}$ point in opposite directions along γ_3, $\mathbf{F} \cdot \mathbf{t} = -|\mathbf{F}| = -\sqrt{x^2 + y^2}$ at a point (x, y) of γ_3. The integrals of $\mathbf{F}$ are thus negatives of each other:

$$\int_{\gamma_3} \mathbf{F} \cdot \mathbf{t}\,ds = -\int_{\gamma_1} \mathbf{F} \cdot \mathbf{t}\,ds.$$

Thus the two remaining integrals cancel in the computation of the integral over γ, giving zero net circulation over the complete circuit.

EXAMPLE 7

Figure 8.4(b) shows the 3-dimensional vector field $\mathbf{F}(x, y, z) = -y\mathbf{i} + x\mathbf{j} + \mathbf{k}$ along with the helix $\mathbf{x}(t) = (2\cos t)\mathbf{i} + (2\sin t)\mathbf{j} + (t)\mathbf{k}$. To integrate $\mathbf{F}$ over the helix we compute $\mathbf{F} \cdot d\mathbf{x}$ along the curve. Note that

$$\mathbf{F}(\mathbf{x}(t)) = (-2\sin t)\mathbf{i} + (2\cos t)\mathbf{j} + \mathbf{k}, \text{ and } \dot{\mathbf{x}}(t) = (-2\sin t)\mathbf{i} + (2\cos t)\mathbf{j} + \mathbf{k}.$$

Hence the velocity vector to the helix at a point $\mathbf{x}$ coincides with the vector $\mathbf{F}(\mathbf{x})$, so $\mathbf{F}(\mathbf{x}) \cdot \dot{\mathbf{x}} = |\dot{\mathbf{x}}|^2 = 5$. It follows that the circulation of $\mathbf{F}$ along the helix from $\mathbf{x}(a)$ to $\mathbf{x}(b)$ is

$$\int_a^b \mathbf{F}(\mathbf{x}(t)) \cdot \dot{\mathbf{x}}(t)\, dt = \int_a^b 5\, dt = 5(b - a).$$

Equivalent Parametrizations.   Different parametrizations may describe the same image curve, for example $f(t) = (t, t^2)$, $0 \le t \le 1$ and $g(u) = (u^2, u^4)$, $0 \le u \le 1$ both describe the same segment of a parabola extending between (0, 0) and (1, 1). It's conceivable that line integrals of the same field using the two functions f and g might be different. However, these parametrizations are related in two significant ways. One is that each segment of the image curve is traced the same number of times (once in this example) by either representation. The other is that there is a correspondence between points t and u of the parameter domain such that corresponding points on the curve have the tangent vectors $f'(t)$ and $g'(u)$ pointing in the same direction. These conditions should be met if we expect to get the same value for a line integral using either parametrization. The conditions will be satisfied in general by imposing the requirement that parametrizations

$$\mathbf{x} = f(t), \quad a \le t \le b \quad \text{and} \quad \mathbf{x} = g(u), \quad \alpha \le u \le \beta$$

be **equivalent**, meaning that there is a continuously differentiable function ϕ with $\phi' > 0$ from $[\alpha, \beta]$ onto $[a, b]$ such that $f(\phi(u)) = g(u)$. (Note that $\phi' > 0$ implies that ϕ is strictly increasing, so changing to the new parametrization won't produce a zero tangent where there wasn't one before.) The following theorem allows us some latitude in the choice of a convenient parametrization for a curve γ. The proof is a direct application of the chain rule and the change-of-variable theorem for integrals.

1.2   Theorem.   Equivalent parametrizations of a curve γ yield the same value for

$$\int_\gamma \mathbf{F} \cdot d\mathbf{x}.$$

Proof.   Let the parameter correspondence be $t = \phi(u)$, so that $f(\phi(u)) = g(u)$. By the change-of-variable theorem

$$\int_\gamma F \cdot d\mathbf{x} = \int_a^b F(f(t)) \cdot f'(t)\, dt = \int_\alpha^\beta F(f(\phi(u))) \cdot f'(\phi(u))\phi'(u)\, du.$$

By the chain rule, $g'(u) = [f(\phi(u))]' = f'(\phi(u))\phi'(u)$, so

$$\int_\gamma F \cdot d\mathbf{x} = \int_\alpha^\beta F(g(u)) \cdot g'(u)\, du. \qquad \blacksquare$$

EXAMPLE 8   Consider three parametrizations for the line segment joining **a** and **b**:

$$f(t) = t\mathbf{b} + (1-t)\mathbf{a}, \quad g(u) = u^2\mathbf{b} + (1-u^2)\mathbf{a}, \quad h(v) = (1-v)\mathbf{b} + v\mathbf{a},$$

with parameter interval $[0, 1]$ for all three. The first two are equivalent via the function $\phi(u) = u^2$, and indeed the tangent vectors for both of them have the same direction as $\mathbf{b} - \mathbf{a}$, pointing in the direction of traversal of the segment. With $h(v)$ we have $t = \phi(v) = 1 - v$ so $\phi'(v) < 0$, and the curve is traversed in the opposite direction, from **b** to **a**; this changes the sign of the integral as compared with the other two parametrizations.

1B   Fundamental Theorem of Calculus

Recall that the gradient ∇f of a differentiable function f from $\mathbb{R}^n$ to $\mathbb{R}$ is the vector field defined by

$$\nabla f(\mathbf{x}) = \left(\frac{\partial f}{\partial x_1}(\mathbf{x}), \dots, \frac{\partial f}{\partial x_n}(\mathbf{x}) \right).$$

If ∇f is continuous, it generalizes the derivative in the formula for the fundamental theorem of calculus for one variable:

$$\int_a^b f'(t)\, dt = f(b) - f(a). \tag{$*$}$$

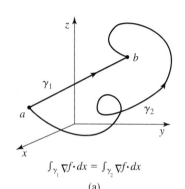

$\int_{\gamma_1} \nabla f \cdot dx = \int_{\gamma_2} \nabla f \cdot dx$

(a)

1.3   Theorem.   Let f be a continuously differentiable real-valued function defined in an open set D of $\mathbb{R}^n$. (Thus ∇f is a continuous vector field in D.) If γ is a smooth curve in D with initial and terminal points **a** and **b**, then

$$\int_\gamma \nabla f \cdot d\mathbf{x} = f(\mathbf{b}) - f(\mathbf{a}).$$

In particular, the value of the line integral of a *gradient* field over a curve depends only on the endpoints of the curve; thus in this case, the notation

$$\int_{\mathbf{a}}^{\mathbf{b}} \nabla f \cdot d\mathbf{x} = f(\mathbf{b}) - f(\mathbf{a})$$

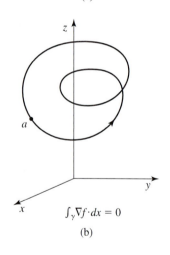

$\int_{\gamma} \nabla f \cdot dx = 0$

(b)

FIGURE 8.5

is justified. If **a** equals **b**, $\displaystyle\int_{\mathbf{a}}^{\mathbf{b}} \nabla f \cdot d\mathbf{x} = 0$, so the integral of a gradient field over a path starting and ending at the same point is zero. (See Figure 8.5.)

Proof.   Suppose γ is parametrized by $g(t)$ with $a \leq t \leq b$, and $g(a) = \mathbf{a}$, $g(b) = \mathbf{b}$. Using first the definition of the line integral we have

$$\int_\gamma \nabla f \cdot d\mathbf{x} = \int_a^b \nabla f\big(g(t)\big) \cdot g'(t)\, dt$$

$$= \int_a^b \frac{d}{dt} f\big(g(t)\big)\, dt,$$

where in the second step we used the chain rule, Theorem 1.3 of Chapter 6. But by Equation $(*)$, the fundamental theorem for one variable, the last integral is equal to
$$f\big(g(b)\big) - f\big(g(a)\big) = f(\mathbf{b}) - f(\mathbf{a}).$$ ∎

EXAMPLE 9   Consider the vector field $\nabla f(x, y)$ in $\mathbb{R}^2$, where $f(x, y) = \frac{1}{2}(x^2 + y^2)$. Then $\nabla f(x, y) = (x, y)$. If γ is some continuously differentiable curve with respective initial and final endpoints $\mathbf{x}_1 = (x_1, y_1)$, and $\mathbf{x}_2 = (x_2, y_2)$, then

$$\int_\gamma \nabla f(\mathbf{x}) \bullet d\mathbf{x} = \int_{(x_1, y_1)}^{(x_2, y_2)} x \, dx + y \, dy = f(x_2, y_2) - f(x_1, y_1)$$

$$= \tfrac{1}{2}(x_2^2 + y_2^2) - \tfrac{1}{2}(x_1^2 + y_1^2)$$

$$= \tfrac{1}{2}(x_2^2 - x_1^2) + \tfrac{1}{2}(y_2^2 - y_1^2).$$

This is what we would expect formally from the fundamental theorem.

EXAMPLE 10   If we let $f(x, y) = xy$, we have $\nabla f(x, y) = (y, x)$, and for a continuously differentiable curve γ we have

$$\int_\gamma y \, dx + x \, dy = x_2 y_2 - x_1 y_1.$$

In particular, if the path starts at $(x_0, y_0) = (1, 2)$, and ends at (x, y), we find that

$$\int_{(1,2)}^{(x,y)} y \, dx + x \, dy = f(x, y) - f(1, 2) = xy - 2.$$

Examples 9 and 10 show in detail how line integrals solve a vector equation

$$\nabla f(x, y) = \big(F_1(x, y), F_2(x, y)\big)$$

for $f(x, y)$ subject to a condition of the form

$$f(x_0, y_0) = 0,$$

provided that $\big(F_1(x, y), F_2(x, y)\big)$ is a given gradient field. The solution is then

$$f(x, y) = \int_{(x_0, y_0)}^{(x,y)} F_1(x, y) \, dx + F_2(x, y) \, dy.$$

More generally the line integral

$$f(\mathbf{x}) = \int_{\mathbf{x}_0}^{\mathbf{x}} \mathbf{F}(\mathbf{x}) \bullet d\mathbf{x} \ \ \text{solves} \ \ \nabla f(\mathbf{x}) = \mathbf{F}(\mathbf{x}), \ \text{with} \ f(\mathbf{x}_0) = 0,$$

assuming again that $\mathbf{F}(\mathbf{x})$ is indeed a gradient field.

The vector differential equation $\nabla f = \mathbf{F}$ is discussed in more detail in Section 2 of Chapter 9. The examples in Exercises 9 to 12 show that if $\mathbf{F}$ is not a gradient field, the value of a line integral of f over a path joining two points may depend on the path and not just on its endpoints.

EXERCISES

In Exercises 1 to 8 compute the line integrals.

1. $\displaystyle\int_L x\,dx + x^2\,dy + y\,dz$, where L is given by $g(t) = (t, t, t)$, for $0 \le t \le 1$.

2. $\displaystyle\int_P (x + y)\,dx + dy$, where P is given by $g(t) = (t, t^2)$, $0 \le t \le 1$.

3. $\displaystyle\int_{\gamma_1} x\,dy$ and $\displaystyle\int_{\gamma_2} x\,dy$, where γ_1 is given by $g(t) = (\cos t, \sin t)$ for $0 \le t \le 2\pi$, and where γ_2 is given by $h(t) = (\cos t, \sin t)$ for $0 \le t \le 4\pi$.

4. $\displaystyle\int_{\gamma_1} (dx + dy)$, where γ_1 is given parametrically by $(x, y) = (\cos t, \sin t)$, $0 \le t \le 2\pi$.

5. $\displaystyle\int_{\gamma_1} \frac{dx + dy}{x^2 + y^2}$, where γ_1 is the curve in Exercise 4.

6. $\displaystyle\int_{\gamma} (e^x\,dx + z\,dy + \sin z\,dz)$, where γ is given by $(x, y, z) = (t, t^2, t^3)$, $0 \le t \le 1$.

7. $\displaystyle\int_{\gamma} \mathbf{F} \cdot d\mathbf{x}$, where $\mathbf{F}(x, y, z) = (z, x, y)$ and γ is given parametrically by $(x, y, z) = (\cos t, \sin t, t)$, $0 \le t \le 2\pi$.

8. $\displaystyle\int_{\gamma} \mathbf{F} \cdot d\mathbf{x}$, where $\mathbf{F}(x, y, z, w) = (x, x, y, xw)$ and γ is given by $(x, y, z, w) = (t, 1, t, t)$, $0 \le t \le 2$.

In Exercises 9 to 12, let γ_1 be given by $(x, y) = (\cos t, \sin t)$, $0 \le t \le \pi/2$ and γ_2 by $(x, y) = (1 - u, u)$, $0 \le u \le 1$. Compute $\displaystyle\int_{\gamma_1} (f\,dx + g\,dy)$ and $\displaystyle\int_{\gamma_2} (f\,dx + g\,dy)$ for the given choices of f and g.

9. $f(x, y) = x$, $g(x, y) = x + 1$.

10. $f(x, y) = x + y$, $g(x, y) = 1$.

11. $f(x, y) = \dfrac{1}{x^2 + y^2}$, $g(x, y) = \dfrac{1}{x^2 + y^2}$.

12. $f(x, y) = xy$, $g(x, y) = x + 1$.

In Exercises 13 to 16, let γ_2 be given on $0 \le u \le 1$ by $(x, y) = (1-u, u)$. The vector fields $\big(F_1(x, y), F_2(x, y)\big)$

are gradient fields. Use Theorem 1.3 to compute $\displaystyle\int_{\gamma_2} (F_1\,dx + F_2\,dy)$ for the given choices of F_1 and F_2.

13. $F_1(x, y) = x^2$, $F_2(x, y) = y^2$.

14. $F_1(x, y) = xy^2$, $F_2(x, y) = x^2 y$.

15. $F_1(x, y) = \sin y$, $F_2(x, y) = x \cos y$.

16. $F_1(x, y) = e^{x-y}$, $F_2(x, y) = -e^{x-y}$.

17. Find the work done in moving a particle along the curve $(x, y, z) = (t, t, t^2)$, $0 \le t \le 2$, under the influence of the field $\mathbf{F}(x, y, z) = (x + y, y, y)$.

18. (a) Find the work done by the force field $\mathbf{F}(x, y) = y\mathbf{i} - x\mathbf{j}$ in moving a particle clockwise once around the circle of radius 1 centered at the origin in $\mathbb{R}^2$.
    (b) How does the answer to part (a) change if the circle is moved so that its center is at an arbitrary point (a, b)?

19. Consider the vector field $\mathbf{F}(x, y) = (y, x)$ and the curve $g(t) = (e^t, e^{-t})$ for $0 \le t \le 1$.
    (a) Sketch $\mathbf{F}$ and g in the same picture.
    (b) Compute the integral of $\mathbf{F}$ over the curve.

20. Show that

$$f(t) = (\cos t, \sin t), \quad 0 \le t \le \pi/2,$$

$$\text{and } g(u) = \left(\frac{1 - u^2}{1 + u^2}, \frac{2u}{1 + u^2}\right), \quad 0 \le u \le 1$$

are equivalent parametrizations of a quarter-circle. (The relevant definition of equivalence is given in the preamble to Theorem 1.2.)

21. Show that $f(t) = (t^{1/2}, t^{3/2})$, $1 \le t \le 2$, and $g(u) = (u, u^3)$, $1 \le u \le \sqrt{2}$ are equivalent parametrizations of a cubic curve. (The relevant definition of equivalence is given in the preamble to Theorem 1.2.)

22. Show that if $\displaystyle\int_{\gamma} \mathbf{F} \cdot d\mathbf{x}$ and $\displaystyle\int_{\gamma} \mathbf{G} \cdot d\mathbf{x}$ exist, then

$$\int_{\gamma} (a\mathbf{F} + b\mathbf{G}) \cdot d\mathbf{x} = a \int_{\gamma} \mathbf{F} \cdot d\mathbf{x} + b \int_{\gamma} \mathbf{G} \cdot d\mathbf{x},$$

where a and b are constants.

23. Let a function $g(t)$ represent the position of a particle of varying mass $m(t)$ in $\mathbb{R}^3$ at time t. Then the velocity vector of the particle is $\mathbf{v}(t) = g'(t)$, and the force acting on the particle at $g(t)$ is $\mathbf{F}(g(t)) = [m(t)\mathbf{v}(t)]'$.

    (a) Show that
$$\mathbf{F}(g(t)) \cdot g'(t) = m'(t)v^2(t) + m(t)v(t)v'(t),$$
where v is the speed of the particle.

    (b) Show that if $m(t)$ is constant, then the work done in moving the particle over its path between times $t = a$ and $t = b$ is $w = (m/2)(v^2(b) - v^2(a))$. $[(\frac{1}{2})mv^2(t)$ is the kinetic energy of the particle.]

24. Sketch the vector field $\mathbf{F}(x, y) = (x, y)$. Explain on geometric grounds why $\int_\gamma \mathbf{F} \cdot d\mathbf{x} = 0$ if the path γ is confined to a circle centered at the origin.

25. Sketch the vector field $\mathbf{F}(x, y) = (-y, x)$. Explain on geometric grounds why $\int_\gamma \mathbf{F} \cdot d\mathbf{x} \neq 0$ if the path γ traces an ellipse centered at the origin and with major and minor axes not necessarily parallel to the (x, y)-axes.

26. The purpose of this exercise is to display a pattern in the results of integrating the vector field $\mathbf{F}(x, y) = x\mathbf{j} = (0, x)$ over some closed paths in $\mathbb{R}^2$.

    (a) Make a sketch of the vector field $\mathbf{F}$.

    (b) Compute $\int_c \mathbf{F} \cdot d\mathbf{x}$, where c is a circular path of radius a, centered at (α, β) and traced counterclockwise.

    (c) Compute $\int_r \mathbf{F} \cdot d\mathbf{x}$, where r is a rectangle with sides parallel to the axes, traced counterclockwise. [*Hint:* Only the vertical sides of the rectangle make a nonzero contribution.]

    (d) Do a computation analogous to the ones in parts (b) and (c) for a triangle with vertices at $(0, 0)$, $(a, 0)$ and $(0, b)$ where a and b are positive. What is the pattern in the answers?

27. The purpose of this exercise is to display a pattern in the results of integrating the vector field $\mathbf{G}(x, y) = -\frac{1}{2}y\mathbf{i} + \frac{1}{2}x\mathbf{j} = (-\frac{1}{2}y, \frac{1}{2}x)$ over some closed paths in $\mathbb{R}^2$.

    (a) Make a sketch of the vector field $\mathbf{G}$.

    (b) Compute $\int_c \mathbf{G} \cdot d\mathbf{x}$, where c is a circular path of radius a, centered at (α, β) and traced counterclockwise.

    (c) Compute $\int_r \mathbf{G} \cdot d\mathbf{x}$, where r is a rectangle with sides parallel to the axes, traced counterclockwise.

    (d) Do a computation analogous to the ones in parts (b) and (c) for a triangle with vertices at $(0, 0)$, $(a, 0)$ and $(0, b)$ where a and b are positive. What is the pattern in the answers?

In Exercises 28 to 31, compute $\int_\gamma \nabla f \cdot d\mathbf{x}$ for the indicated choices of f and γ.

28. $f(x, y) = x^2 + y^2; \gamma: g(t) = (1+t^2, 1-t^2), -1 \leq t \leq 2$.

29. $f(x, y, z) = x - y^2 + z; \gamma: g(t) = (t, t^2, -t^2), 0 \leq t \leq 2$.

30. $f(x, y) = x^3 - 2y^3; \gamma:$ line segment from $(1, 1)$ to $(5, -1)$.

31. $f(x, y, z) = (x - y + z)^2; \gamma:$ one turn of a helix from $(1, 0, 0)$ to $(1, 0, 4)$.

32. (a) Sketch the vector field $\mathbf{F}(x, y) = (y, 0)$.

    (b) Show that the vector field of part (a) can't be the gradient of a real-valued function f by finding distinct paths from some point $\mathbf{a}$ to another point $\mathbf{b} \neq \mathbf{a}$ such that the integrals of $\mathbf{F}$ over these paths have different values.

    (c) Show that the vector field of part (a) can't be the gradient of a real-valued function f by finding a closed path γ starting and ending at the same point such that $\int_\gamma \mathbf{F} \cdot d\mathbf{x} \neq 0$.

33. (a) Sketch the vector field $\mathbf{F}(x, y) = (-y, x)$.

    (b) Show that the vector field of part (a) can't be the gradient of a real-valued function f by finding distinct paths from some point $\mathbf{a}$ to another point $\mathbf{b} \neq \mathbf{a}$ such that the integrals of $\mathbf{F}$ over these paths have different values.

    (c) Show that the vector field of part (a) can't be the gradient of a real-valued function f by finding a closed path γ starting and ending at the same point such that $\int_\gamma \mathbf{F} \cdot d\mathbf{x} \neq 0$.

34. Assume a continuous vector field $\mathbf{F}(\mathbf{x})$ satisfies (i) $|\mathbf{F}(\mathbf{x})| = k$ for a constant $k > 0$, and (ii) $\mathbf{F}(\mathbf{x})$ is tangent at each point $\mathbf{x}$ to a continuously differentiable curve γ of finite length. Prove that $\int_\gamma \mathbf{F} \cdot d\mathbf{x}$ equals $\pm k$ times the length of γ.

SECTION 2   WEIGHTED CURVES AND SURFACES OF REVOLUTION

In Chapter 4, Section 1 the **arc length** function $s = s(t)$ of a curve parametrized by $\mathbf{x} = g(t)$ on $t_0 \leq t \leq t_1$ is defined by

$$s(t) = \int_{t_0}^{t} |\dot{\mathbf{x}}(u)| \, du.$$

The definition is a natural one, because the length $|\dot{\mathbf{x}}(t)|$ is the speed at which the curve is being traced at time t. But a given path in space can be traced at many different varying speeds, including possible multiple back-and-forth tracings. Hence it's useful to have a standard parametrization for a curve that depends only on the intrinsic geometry of the image set of the curve. If points in the image correspond one-to-one with values of arc length measured from a specific point $\mathbf{x}_0$ on the image curve, we can use arc length s as the parameter in a representation of the curve by a function $g(s)$, where $g(s_0) = \mathbf{x}_0$. It would then follow that $s = \int_{s_0}^{s} |\dot{g}(u)| \, du$. Differentiating both sides of the equation with respect to s gives $1 = |\dot{g}(s)|$. For this reason we say that a curve $g(t)$, $t_0 \leq t \leq t_1$ is **parametrized by arc length** if $|\dot{g}(t)| = 1$ for all t in the parameter interval; in other words the curve is traced with constant speed 1. The expression $|\dot{\mathbf{x}}(t)| \, dt = |g'(t)| \, dt$ is traditionally called the **arc length element** of the curve.

EXAMPLE 1   Let $a > 0$ be the radius of a circle centered at the origin in $\mathbb{R}^2$. The simplest parametrization for this circle is $g(t) = (a \cos t, a \sin t)$, $0 \leq t \leq 2\pi$. Since $|\dot{g}(t)| = |(-a \sin t, a \cos t)| = a$, the circle has been parametrized by arc length just when $a = 1$. For other values of a, note that the arc length of the part of the circle corresponding to the parameter interval $0 \leq u \leq t$ is

$$s = \int_0^t |\dot{g}(u)| \, du = \int_0^t a \, du = at.$$

This equation $s = at$ suggests the idea of introducing arc length s as parameter by substituting s/a for t in the given parametrization; we get

$$g(s/a) = \big(a \cos(s/a), a \sin(s/a)\big),$$

so we now have an arc length parametrization, since

$$\left| \frac{dg(s/a)}{ds} \right| = \big|\big(-\sin(s/a), \cos(s/a)\big)\big| = 1.$$

Whether we denote the parameter by s or some other letter t is irrelevant; what is important for applications is that the point $g(t/a)$ is t units along the curve from $g(0)$, or equivalently, that the curve is traced with uniform speed 1.

EXAMPLE 2   The plane curve $g(t) = (t, \frac{2}{3}t^{3/2})$ has velocity vector $\dot{g}(t) = (1, t^{1/2})$, so the speed is $|\dot{g}(t)| = \sqrt{1+t}$. Arc length in terms of the parameter t measured from 0 is then

$$s(t) = \int_0^t \sqrt{1+u} \, du = \frac{2}{3}(1+u)^{3/2} \Big|_0^t = \frac{2}{3}(1+t)^{3/2} - \frac{2}{3}.$$

Solving for t in terms of s gives $t = (1 + \frac{3}{2}s)^{2/3} - 1$. Exercise 14 shows that if we use arc length as parameter, letting $h(s) = g((1 + \frac{3}{2}s)^{3/2} - 1)$, then $|h'(s)| = 1$, so the curve is now traced by $h(s)$ with uniform speed 1 for $s > 0$.

Weighted Curves.   Thinking in terms of arc length parametrization is particularly appropriate for integration over a curve for which points in the image of the curve correspond one-to-one to parameter values. In particular, suppose we have a scalar-valued weight function μ that assigns a number $\mu(h(s))$ to each point $h(s)$ at distance s along the curve. For example, μ might be the mass density of a wire bent into the shape of some curve. Then the **total mass** M of the weighted curve is naturally defined to be

2.1
$$M = \int_{s_0}^{s_1} \mu(h(s))\, ds.$$

If the curve happens to be parametrized by something other than arc length, using instead a function $g(t)$ on $t_0 \leq t \leq t_1$, the arc length function s is given by

$$s(t) = \int_{t_0}^{t} |\dot{g}(u)|\, du, \quad t_0 \leq t \leq t_1.$$

The significance of this equation for the two vector functions $h(s)$ and $g(t)$ that parametrize the curve is that $h(s(t)) = g(t)$ for $t_0 \leq t \leq t_1$. Since $ds/dt = |\dot{g}(t)|$, the one-dimensional change-of-variable theorem for integrals applied to the integral for M gives

$$M = \int_{s_0}^{s_1} \mu(h(s))\, ds = \int_{t_0}^{t_1} \mu(h(s(t))) \frac{ds(t)}{dt}\, dt = \int_{t_0}^{t_1} \mu(g(t)) |\dot{g}(t)|\, dt.$$

The differential $ds = |\dot{g}(t)|dt$ is called the **arc length differential** for a curve parametrized by $g(t)$.

EXAMPLE 3   Consider a full turn of the helix described by

$$g(t) = (a\cos t, a\sin t, t), \quad a > 0, 0 \leq t \leq 2\pi.$$

Suppose that the density of the helix at a point $\mathbf{x}$ is equal to the square of the distance from $\mathbf{x}$ to the midpoint $\mathbf{q} = (0, 0, \pi)$ of the helix's axis. Thus the density at $g(t)$ is

$$|g(t) - \mathbf{q}|^2 = a^2 \cos^2 t + a^2 \sin^2 t + (t - \pi)^2 = a^2 + (t - \pi)^2.$$

Since $|\dot{g}(t)| = \sqrt{(-a\sin t)^2 + (a\cos t)^2 + 1} = \sqrt{a^2 + 1}$, the total mass is

$$M_a = \int_0^{2\pi} (a^2 + (t - \pi)^2)\sqrt{a^2 + 1}\, dt$$

$$= \sqrt{a^2 + 1} \int_{-\pi}^{\pi} (a^2 + t^2)\, dt = \sqrt{a^2 + 1}\, (2\pi a^2 + 2\pi^3/3).$$

FIGURE 8.6

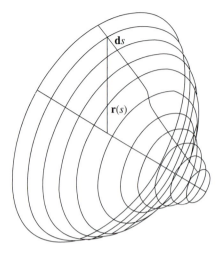

Surfaces of Revolution.   Certain weightings of a curve provide a way to find some surface areas. By a **surface of revolution** we'll mean a surface S in $\mathbb{R}^3$ that is generated by rotating a plane curve about a fixed line lying in the plane of the curve. For example, a curve in $\mathbb{R}^2$ parametrized by $\big(x(s), y(s)\big)$, $s_0 \le s \le s_1$, can be rotated about a variety of lines including, but not restricted to, the x-axis and the y-axis, to produce a surface. See Figure 8.6 where the rotation is around the y-axis. During rotation a point at distance s along the curve traces out a circle of radius $r(s)$ and circumference $2\pi r(s)$ centered on the axis of rotation. It's plausible that the area swept out by a short segment of length ds is about $2\pi r(s)\, ds$, so it's natural to define the **surface area** $\sigma(S)$ of S by an integral with respect to arc length

2.2
$$\sigma(S) = \int_{s_1}^{s_2} 2\pi r(s)\, ds.$$

This definition gives the expected result for commonly met surfaces such as cylinders, spheres, and cones, and is a special case of a more comprehensive definition given in the next chapter that includes surfaces that aren't necessarily surfaces of revolution.

In a typical application we're given a curve parametrized in $\mathbb{R}^2$ by $g(t) = \big(x(t), y(t)\big)$, $t_0 \le t \le t_1$. The arc length element is $ds = \sqrt{\dot{x}(t)^2 + \dot{y}(t)^2}\, dt$. If $y(t) \ge 0$ and we rotate the curve about the x-axis then $r\big(s(t)\big) = y(t)$. Hence

2.3
$$\sigma(S) = \int_{t_0}^{t_1} 2\pi y(t)\sqrt{\dot{x}(t)^2 + \dot{y}(t)^2}\, dt.$$

EXAMPLE 4   The circle parametrized by $\big(y(t), z(t)\big) = (a\cos t, b + a\sin t)$, $0 \le t \le 2\pi$ has radius a and center at $(0, b)$ in the yz plane. If $0 < a < b$, rotating about the y-axis generates a torus or "donut" surface T, shown in Figure 8.7(a). The arc length element is $|\dot{g}(t)|\, dt = \sqrt{(-a\sin t)^2 + (a\cos t)^2}\, dt = a\, dt$, and $r\big(s(t)\big) = (b + a\sin t)$. Then

$$\sigma(T) = \int_0^{2\pi} 2\pi(b + a\sin t)a\, dt$$

$$= 2\pi ab \int_0^{2\pi} dt + 2\pi a^2 \int_0^{2\pi} \sin t\, dt = 4\pi^2 ab.$$

FIGURE 8.7

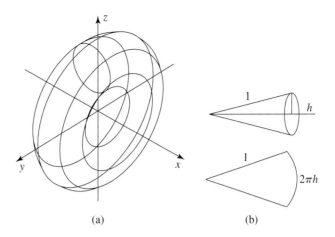

(a)   (b)

If $0 < b < a$, the surface of revolution is more complicated than a torus; to make the surface area computation valid we would have to replace the integrand by its absolute value.

If we rotate the graph $y = f(x)$ for a differentiable function $f(x)$, $a \leq x \leq b$, we can use x as the parameter in $g(x) = (x, f(x))$. Then $ds = \sqrt{1 + f'(x)^2}\, dx$, and $r(s(x)) = f(x)$. If $f(x) \geq 0$, the area of the surface generated by rotating the graph about the x-axis is

$$\sigma(S) = \int_a^b 2\pi f(x)\sqrt{1 + f'(x)^2}\, dx.$$

EXAMPLE 5 A line segment of length l extends from the origin in $\mathbb{R}^2$ to a point h units above the positive x-axis and is then rotated about the x-axis to produce a cone C. Slitting the cone along the segment allows it to be rolled out flat as a sector of a circle, as shown in Figure 8.7(b). The circular arc of the sector has length $2\pi h$, which is the circumference of the cone's base. The area of the sector is thus $2\pi h/(2\pi l) = h/l$ times the area πl^2 of a full circle of radius l. Hence the cone should have area $\sigma(C) = (h/l)\pi l^2 = \pi hl$.

Using instead the previous displayed formula, we'll represent the segment as the graph of $f(x) = (h/\sqrt{l^2 - h^2})x$ for $0 \leq x \leq \sqrt{l^2 - h^2}$. We find $ds = (l/\sqrt{l^2 - h^2})\, dx$. In agreement with our purely geometric computation we get

$$\sigma(C) = \int_0^{\sqrt{l^2-h^2}} 2\pi (h/\sqrt{l^2 - h^2})x(l/\sqrt{l^2 - h^2})\, dx$$

$$= \frac{2\pi hl}{l^2 - h^2} \int_0^{\sqrt{l^2-h^2}} x\, dx = \pi hl.$$

This example provides some supporting evidence for the correctness of the definition of $\sigma(S)$.

EXERCISES

In Exercises 1 to 4, find the length $l(\gamma)$ of the indicated curves.

1. $(x, y) = (t, \ln \cos t)$, $0 \le t \le 1$.

2. $(x, y) = (t^2, \frac{2}{3}t^3 - \frac{1}{2}t)$, $0 \le t \le 2$.

3. $y = x^{3/2}$, $0 \le x \le 5$.

4. $g(t) = (6t^2, 4\sqrt{2}t^3, 3t^4)$, $-1 \le t \le 2$.

5. If a curve is described in plane polar coordinates by a function $r = f(\theta)$ for $a \le \theta \le b$, then in rectangular coordinates the curve may be parametrized by

$$(x, y) = (r \cos \theta, r \sin \theta)$$

$$= (f(\theta) \cos \theta, f(\theta) \sin \theta), \quad a \le \theta \le b.$$

  (a) Show that the arc length formula for a curve $f = f(\theta)$ in polar coordinates is

$$\int_a^b \sqrt{f(\theta)^2 + f'(\theta)^2}\, d\theta.$$

  (b) Sketch the curve given by $r = (1 + \cos \theta)$ for $0 \le \theta \le \pi$ and find its length.

6. A 5-foot piece of wire is coiled in a uniform spiral 3 inches in diameter. Find the height of the coil if it contains six complete turns.

7. Find the total mass of the helix $g(t) = (a \cos t, a \sin t, bt)$, $0 \le t \le 2\pi$, if its density per unit length at (x, y, z) is equal to $x^2 + y^2 + z^2$.

8. Let γ be a continuously differentiable curve with endpoints $\mathbf{p}_1$ and $\mathbf{p}_2$. Let λ be the line segment $\mathbf{p}_1 + t(\mathbf{p}_2 - \mathbf{p}_1)$, $0 \le t \le 1$. Prove that $l(\lambda) \le l(\gamma)$. Thus the shortest distance between two points is a straight line. [*Hint:* Use the result of Exercise 7(c) of Chapter 7, Section 3.]

9. Find the total mass of the wire with shape $(x, y, z) = (6t^2, 4\sqrt{2}t^3, 3t^4)$, $0 \le t \le 1$,

  (a) if the density at the point corresponding to t is t^2.

  (b) if the density at a point is equal to the square of its distance from the yz-plane.

10. Suppose γ is given by $g(t)$ for $a \le t \le b$ and γ is then reparametrized by are length s so that $t = t(s)$. Show that the line integral equation

$$\int_a^b \mathbf{F}\big(g(t)\big) \cdot g'(t)\, dt = \int_0^{l(\gamma)} \mathbf{F}\big(h(s)\big) \cdot \mathbf{t}(s)\, ds$$

holds, where $h(s) = g(t(s))$ and $\mathbf{t}(s) = (dh/ds)(s)$. [*Hint:* Use the change of variable theorem for integrals.]

11. Compute the surface area of a sphere S_a of radius a in two ways using Equation 2.2.

  (a) Parametrize a semicircle in $\mathbb{R}^2$ by $g(t) = (a \cos t, a \sin t)$, $0 \le t \le \pi$. Show that $ds = a\, dt$ and $r(s(t)) = a \sin t$. Then rotate the semicircle about the horizontal axis to get $\sigma(S_a) = 4\pi a^2$.

  (b) Parametrize a semicircle by $g(t) = (t, \sqrt{a^2 - t^2})$, $-a \le t \le a$. Show that $ds = a(a^2 - t^2)^{-1/2}dt$ and $r(s(t)) = (a^2 - t^2)^{1/2}$. Then rotate the semicircle about the horizontal axis to get $\sigma(S_a) = 4\pi a^2$.

12. The graph of $y = a \cosh(x/a)$, $0 \le x \le b$, is rotated about the x-axis in $\mathbb{R}^2$ to generate a surface S. Find $\sigma(S)$.

13. (a) Set up an integral for the arc length of the ellipse in $\mathbb{R}^2$ parametrized by

$$(x, y) = (a \cos t, b \sin t), \quad 0 \le t \le 2\pi.$$

  (b) Assume $a > b$ and show that the arc length integral found in part (a) is equal to

$$4a \int_0^{\pi/2} \sqrt{1 - k^2 \sin^2 t}\, dt, \quad k^2 = (1 - b^2/a^2).$$

This integral is a standard form of an **elliptic integral**; it can't be evaluated using elementary functions if $0 < k^2 < 1$.

  (c) Approximate the length of the ellipse if $a = 2$ and $b = 1$, either by direct numerical approximation of the integral using Simpson's rule or by finding the value of the elliptic integral in a table.

14. If γ is given by $g(t) = (t, \frac{2}{3}t^{3/2})$, and $h(s) = g\big((1 + \frac{3}{2}s)^{2/3} - 1\big)$, show that $|h'(s)| = 1$ for $s > 0$, and that the parametrization $h(s)$ is an arc length parametrization for γ.

15. Compute $|\dot{g}(t)|$ for the helix $g(t) = (\cos t, \sin t, t)$ and use the result to find an arc length parametrization for this helix.

16. In Example 3 of the text, if the radius a of the helix tends to zero the total mass M_a tends to $2\pi^3/3$. What geometric interpretation does this number have in the present context?

17. The **centroid** of a curve γ of finite length $l(\gamma)$ is the average position $\mathbf{p}_0$ of the points on the curve. Thus the vector $\mathbf{p}_0$ is given in terms of an arc length parametrization $h(s)$ or more general parametrizations $g(t)$ in a vector-valued integral by

$$\mathbf{p}_0 = \frac{1}{l(\gamma)} \int_{s_0}^{s_1} h(s)\, ds = \frac{1}{l(\gamma)} \int_{t_0}^{t_1} g(t)|\dot{g}(t)|\, dt.$$

(a) Let γ be an arc of a circle of radius a such that the ends of the arc subtend angle θ at the center of the circle. Show that the centroid of γ lies at distance $(2a/\theta)\sin(\theta/2)$ from the center of the circle, measured along the line from the center of the circle to the midpoint of the arc.

(b) Show that the half-turn of a helix parametrized by $g(t)=(a\cos t, a\sin t, bt), 0 \le t \le \pi$ has its centroid at $\mathbf{p}_0 = (0, 2a/\pi, b\pi/2)$.

18. Using the definition of centroid of a curve γ in the previous exercise, prove **Pappus's theorem**: Rotating a plane curve about a line in the same plane generates a surface S of area $\sigma(S)$ equal to $l(\gamma)$ times the circumference of the circle traced by rotating the centroid of γ about the line. [*Hint:* The distance in $\mathbb{R}^2$ from a point $\mathbf{y}$ to a line is $|n \cdot (\mathbf{y} - \mathbf{x}_0)|$, where $n \cdot (\mathbf{y} - \mathbf{x}_0) = 0$ is a normalized equation for the line.]

SECTION 3   NORMAL VECTORS AND CURVATURE

The purpose of this section is to analyze the connection between the shape of a smooth curve in space and the variety of possible motions that can occur along the path of the curve. We'll denote position on a curve as a function of time by $\mathbf{x} = \mathbf{x}(t)$ and assume $\mathbf{x}(t)$ is twice continuously differentiable. Since arc length $s(t)$ along a curve parametrized by $\mathbf{x}(t)$ is an integral of speed $|\dot{\mathbf{x}}(t)|$, it follows that the speed is $v = ds/dt$, or sometimes more conveniently $v = \dot{s}$. Thus

$$\frac{ds}{dt}(t) = \dot{s}(t) = |\dot{\mathbf{x}}(t)|.$$

It's customary to denote the vector of length 1 having the same direction as the velocity or tangent vector $\mathbf{v}(t) = \dot{\mathbf{x}}(t)$ by $\mathbf{t}(t)$. Recall that, by definition, $\dot{s} \neq 0$ on a smooth curve. Thus we can write $\dot{\mathbf{x}}(t) = \dot{s}(t)\mathbf{t}(t)$ or $\mathbf{t} = (1/\dot{s})\dot{\mathbf{x}}$, with $|\mathbf{t}| = 1$.

Turning to the acceleration vector along the curve, we have by definition $\ddot{\mathbf{x}} = \dot{\mathbf{v}} = d(\dot{s}\mathbf{t})/dt$. Now apply the product rule for a scalar times a vector, Formula 1.3 in Chapter 4, Section 1, to get

3.1 $$\ddot{\mathbf{x}} = \ddot{s}\mathbf{t} + \dot{s}\dot{\mathbf{t}}.$$

As a first step in interpreting Equation 3.1, we'll verify in the following proof that the vector $\dot{\mathbf{t}}(t)$ is **orthogonal** to $\mathbf{t}(t)$, that is $\mathbf{t}(t) \cdot \dot{\mathbf{t}}(t) = 0$. Since $\mathbf{t} \neq 0$, this means that either (i) $\dot{\mathbf{t}} = 0$ or else (ii) $\dot{\mathbf{t}}$ is perpendicular to $\mathbf{t}$. In case $\dot{\mathbf{t}}(t) \neq 0$, we define a unit vector $\mathbf{n}(t)$ called the **principal normal** to the curve at a point by $\mathbf{n}(t) = (|\dot{\mathbf{t}}|^{-1})\dot{\mathbf{t}}$. Thus $\dot{\mathbf{t}} = |\dot{\mathbf{t}}|\mathbf{n}$. This observation allows us to refine Equation 3.1 as follows.

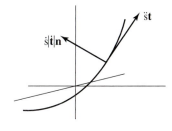

FIGURE 8.8

3.2 Theorem. For a twice-differentiable smooth curve $\mathbf{x}(t)$, the acceleration vector is expressible as a sum of orthogonal components as

$$\ddot{\mathbf{x}} = \ddot{s}\mathbf{t} + \dot{s}|\dot{\mathbf{t}}|\mathbf{n}.$$

Proof. The orthogonality of $\mathbf{t}$ and $\mathbf{n}$ follows from having $|\mathbf{t}|^2 = \mathbf{t} \cdot \mathbf{t}$ equal to a constant, namely 1 in this case; just differentiate $\mathbf{t} \cdot \mathbf{t} = 1$ with respect to time t using the product rule. On the left we get $(d/dt)\mathbf{t} \cdot \mathbf{t} = \mathbf{t} \cdot \dot{\mathbf{t}} + \dot{\mathbf{t}} \cdot \mathbf{t} = 2\mathbf{t} \cdot \dot{\mathbf{t}}$. On the right side the derivative of 1 is 0, so $\mathbf{t} \cdot \dot{\mathbf{t}} = 0$. Hence $\mathbf{t}$ and $\dot{\mathbf{t}}$ are orthogonal. Since $\dot{\mathbf{t}} = |\dot{\mathbf{t}}|\mathbf{n}$, we can rewrite Equation 3.1 as claimed. ∎

Figure 8.8 is a typical picture of how $\mathbf{t}$ and $\mathbf{n}$ relate to the path followed by a curve. The two orthogonal components, $\mathbf{a_t} = \ddot{s}\mathbf{t}$ and $\mathbf{a_n} = \dot{s}|\dot{\mathbf{t}}|\mathbf{n}$ are called respectively the

tangential acceleration and the **centripetal acceleration** of motion along the curve. The *tangential* acceleration measures the rate of change of speed along the curve. The *centripetal* acceleration measures the rate at which the motion bends away from a straight-line path. The force that bends the path of a particle of mass m generates the centripetal acceleration, so the centripetal component of that force is $m\mathbf{a_n}$, and the total force is $m\mathbf{a_t} + m\mathbf{a_n}$.

EXAMPLE 1

If a path is traversed with constant speed, i.e., $\dot{s} = v_0 = \text{const.}$, then $\ddot{s} = 0$, and Equation 3.1 reduces to $\ddot{\mathbf{x}} = v_0|\dot{\mathbf{t}}|\mathbf{n}$. Thus any acceleration vectors $\ddot{\mathbf{x}}$ are perpendicular to the path of motion, i.e., are centripetal. This is true, for example for the helical motion $\mathbf{x}(t) = (a\cos t, a\sin t, bt)$, since $\dot{s} = \sqrt{a^2 + b^2}$ is constant.

We can get more insight into centripetal acceleration by introducing a measure of the shape of a curve at each point called curvature. If $\mathbf{x} = \mathbf{x}(t)$ traces a twice-differentiable smooth curve then the **curvature** at $\mathbf{x}(t)$ is the scalar

3.3
$$\kappa(t) = \left|\frac{d\mathbf{t}}{ds}\right|.$$

The unit tangent vector $\mathbf{t}$ and the arc length s are intrinsic to the geometry of the image path, so the derivative $d\mathbf{t}/ds$ is intrinsic also. Therefore the scalar curvature $\kappa(t)$ is an intrinsic measure of the turning rate of the unit tangent vector $\mathbf{t}$, viewed as a function of arc length. In particular, if the tangent vector doesn't turn at all, as for a straight line when $\mathbf{t}$ is constant, we get $\kappa = 0$.

The definition of curvature involves a derivative with respect to arc length s, so a direct approach to computing curvature would start with introducing arc length as parameter. The following theorem allows us to avoid this step, often awkward in practice; it also allows us to identify the role that curvature plays in the centripetal term $\mathbf{a}_n = \dot{s}|\dot{\mathbf{t}}|\mathbf{n}$ of Theorem 3.2.

3.4 Theorem. For a twice-differentiable smooth curve $\mathbf{x}(t)$, the curvature function is determined by $|\dot{\mathbf{t}}| = \dot{s}\kappa$ so $\kappa = |\dot{\mathbf{t}}|/\dot{s}$. Hence the acceleration vector of Theorem 3.2 equals $\ddot{\mathbf{x}} = \ddot{s}\mathbf{t} + \dot{s}^2\kappa\mathbf{n}$.

Proof. By the chain rule for vector functions of a scalar, $d\mathbf{t}/dt = \dot{s}(d\mathbf{t}/ds)$. Hence, since $\dot{s} \geq 0$, $|\dot{\mathbf{t}}| = |\dot{s}(d\mathbf{t}/ds)| = \dot{s}|d\mathbf{t}/ds| = \dot{s}\kappa$. Since $\dot{s} > 0$ for smooth curves we can write $\kappa = |\dot{\mathbf{t}}|/\dot{s}$. ∎

EXAMPLE 2

The helix $\mathbf{x}(t) = (a\cos t, a\sin t, bt)$ has radius $a > 0$ and vertical climb rate $b > 0$. To find its curvature κ we first compute

$$\dot{\mathbf{x}}(t) = (-a\sin t, a\cos t, b).$$

Hence $\dot{s} = \sqrt{a^2\sin^2 t + a^2\cos^2 t + b^2} = \sqrt{a^2 + b^2}$. Hence

$$\mathbf{t}(t) = (1/\dot{s})\dot{\mathbf{x}}(t) = (a^2 + b^2)^{-1/2}(-a\sin t, a\cos t, b),$$

so $\dot{\mathbf{t}}(t) = (a^2 + b^2)^{-1/2}(-a\cos t, -a\sin t, 0)$. Then $|\dot{\mathbf{t}}| = a(a^2 + b^2)^{-1/2}$, so

$$\kappa = (1/\dot{s})|\dot{\mathbf{t}}| = a/(a^2 + b^2).$$

As b tends to 0 with a fixed, a single turn of the helix approaches a circle of radius a, while κ approaches $1/a$. Decreasing a then gives greater curvature, so qualitatively tighter curling goes with greater curvature. On the other hand, as b gets large a turn of the helix stretches out in the direction of its axis, making κ tend to 0. If $b = 0$, the curve we get is a circle of radius a with constant curvature $\kappa = 1/a$.

Some special formulas for computing curvature are taken up in Exercises 12 to 15. Here are two, one written directly in terms of the time derivatives $\dot{\mathbf{x}}$ and $\ddot{\mathbf{x}}$ of the position vector $\mathbf{x}(t)$ that traces the curve, another for the graph of $y = f(x)$.

3.5
$$\kappa(t) = \frac{\sqrt{|\dot{\mathbf{x}}|^2|\ddot{\mathbf{x}}|^2 - (\dot{\mathbf{x}} \cdot \ddot{\mathbf{x}})^2}}{|\dot{\mathbf{x}}|^3}, \quad \text{assuming } \dot{\mathbf{x}} \neq 0.$$

3.6
$$\kappa(x) = \frac{y''}{(1 + (y')^2)^{3/2}}.$$

EXERCISES

1. Show that the curvature of a plane circular path of radius $a > 0$ is $1/a$.

2. Find the curvature $\kappa(t)$ of the parabola $\mathbf{x}(t) = (t, t^2)$ for $-\infty < t < \infty$.

3. Centripetal acceleration $\mathbf{a_n}$ increases in magnitude if either speed $\dot{s}$ or curvature κ is increased by a factor $\rho > 1$. Which does more to increase $|\mathbf{a_n}|$?

4. For the circular helix motion $\mathbf{x}(t) = (a \cos t, a \sin t, bt)$, show that the tangential component of the acceleration is always zero and that the centripetal component at a point of the path is directed toward the axis of the helix.

5. Equation 3.5 expresses curvature κ in terms of the square root of an expression of the form $|\mathbf{a}|^2|\mathbf{b}|^2 - (\mathbf{a} \cdot \mathbf{b})^2$; is this expression always nonnegative? Explain.

6. Motion along a linear path can be described by $\mathbf{x}(t) = \phi(t)\mathbf{c} + \mathbf{d}$, $\mathbf{c} \neq 0$, where we assume that the real-valued function $\phi(t)$ is strictly increasing and has two continuous derivatives. Show that the acceleration vector has centripetal component identically zero.

7. Here is the converse to the statement in the previous exercise: Suppose $\mathbf{x}(t)$ has the centripetal component of its acceleration identically equal to zero, but that $\dot{s}(t) \neq 0$. Then the path of motion is a straight line. Prove this as follows.
   (a) Show that $\dot{\mathbf{t}} = 0$ and hence that $\dot{\mathbf{x}} = \dot{s}\mathbf{c}$ for some constant vector $\mathbf{c} \neq 0$.
   (b) Integrate the result of part (a) with respect to time t to show that $\mathbf{x}(t) = s(t)\mathbf{c} + \mathbf{d}$ for some constant vector $\mathbf{d}$, so that the path of motion is a line.

8. Let a, b and ω be positive constants. Let $g(t) = (a \cos \omega t, a \sin \omega t, bt)$, $t \geq 0$.
   (a) Find explicitly the arc length parametrization $h(s)$ of the curve.
   (b) Find the unit tangent and principle normal vectors at an arbitrary point $h(s)$.
   (c) Find the curvature $\kappa(s)$.

9. Show that the curve $(x, y) = (\cos s, \sin s)$, $0 \leq s \leq 2\pi$ is parametrized by arc length. Sketch the curve together with its velocity and acceleration vectors at $s = \pi/2$.

10. (a) Show that for a line given by $g(t) = t\mathbf{x}_1 + \mathbf{x}_0$, the curvature is identically zero.
    (b) Show that if a curve γ, parametrized by arc length and given by a function $f(s)$, has a tangent at every point and has curvature identically zero, then γ is a straight line.

11. Use Theorem 3.2 to show how that if a particle of constant mass m moves so that at time t it is at $\mathbf{x}(t)$, then the work done in traversing a part of the path having length s_0 is equal to an integral with respect to arc length s along $\mathbf{x}(t)$ in the form
$$W = \int_0^{s_0} m\frac{d^2s}{dt^2} \, ds.$$

12. Here are two different ways to prove Equation 3.5 for curvature.
    (a) Verify the equation using the substitutions $\dot{\mathbf{x}} = \dot{s}\mathbf{t}$ and $\ddot{\mathbf{x}} = \ddot{s}\mathbf{t} + \dot{s}^2\kappa\mathbf{n}$. Then expand the dot products

and use the relations $\mathbf{t} \cdot \mathbf{t} = \mathbf{n} \cdot \mathbf{n} = 1$ and $\mathbf{t} \cdot \mathbf{n} = 0$.

(b) Derive the equation by first computing the time derivative of the vector $\mathbf{t} = (\dot{\mathbf{x}} \cdot \dot{\mathbf{x}})^{-1/2}\dot{\mathbf{x}}$. Then use the formula $\kappa = |\dot{\mathbf{t}}|/|\dot{\mathbf{x}}|$.

13. A twice differentiable real-valued function $f(x)$, has as its graph a smooth curve in $\mathbb{R}^2$.

(a) Show that the curvature of the graph of f is

$$\kappa(x) = \frac{|f''(x)|}{(1 + f'(x)^2)^{3/2}}.$$

This follows from Equation 3.5 if we use x as parameter in $(x, y) = (x, f(x))$.

(b) If $y = \cos x$ on $[-\pi/2, \pi/2]$, find the maximum and minimum of $\kappa(x)$.

(c) Find all the points of maximum and minimum $\kappa(x)$ on the graph of $y = x^4$.

14. (a) Use Equation 3.5 to show that if θ is the angle between $\dot{\mathbf{x}}$ and $\ddot{\mathbf{x}}$, then

$$\kappa = \frac{|\ddot{\mathbf{x}}|\, |\sin \theta|}{|\dot{\mathbf{x}}|^2}.$$

(b) If the speed and magnitude of acceleration are given, what does part (a) tell you about the dependence of κ on θ? Explain why your answer to this question agrees, or fails to agree, with your physical intuition.

15. (a) Show that for a special case of a plane curve parametrized by two scalar functions $x(t)$ and $y(t)$ Equation 3.5 reduces to

$$\kappa = \frac{|\dot{x}\ddot{y} - \dot{y}\ddot{x}|}{(\dot{x}^2 + \dot{y}^2)^{3/2}}.$$

(b) Find the curvature $\kappa(t)$ for $x = t^2$, $y = t^3$ when $t \neq 0$. This is not a smooth curve at $t = 0$; what is the limit of $\kappa(t)$ as t tends to 0?

16. Position on a helical curve is $\mathbf{x}(t) = (a \cos \phi(t), a \sin \phi(t), b\phi(t))$, where $\phi(t)$ is a twice-differentiable real-valued function of time t. Imagine a bead of mass 1 that starts sliding from rest at $t = 0$, with negligible friction, on a helical wire whose central axis is vertical, and so parallel to the acceleration $-g$ of gravity. In that case $\phi(t) = -\frac{1}{2}bgt^2/(a^2 + b^2)$.

(a) Show that for $t > 0$, $\dot{s}(t) = bgt/\sqrt{a^2 + b^2}$.

(b) Since curvature depends only on the shape of the helix, which depends on a and b but not $\phi(t)$, the curvature $\kappa(t)$ is given by Example 2 of the text. Use this information to show that the tangential and normal components of the acceleration vector have respective magnitudes $bg/\sqrt{a^2 + b^2}$ and $ab^2g^2t^2/(a^2 + b^2)^2$.

*17. If a smooth curve is parametrized by arc length, its curvature is $\kappa(s) = |(d/ds)\mathbf{t}(s)|$. Show that if $\theta(s, h)$ is the angle between $\mathbf{t}(s)$ and $\mathbf{t}(s + h)$, which tends to zero as h tends to zero, then

$$\kappa(s) = \lim_{h \to 0} \left| \frac{\theta(s, h)}{h} \right|.$$

[*Hint:* Show that $|\mathbf{t}(s + h) - \mathbf{t}(s)| = \sqrt{2 - 2\cos\theta(s, h)}$.]

SECTION 4   FLOW LINES, DIVERGENCE, AND CURL

Suppose $\mathbb{R}^n \xrightarrow{\mathbf{F}} \mathbb{R}^n$ is a vector field. A differentiable curve $\mathbf{x} = \mathbf{x}(t)$ with the property that at each of its points $\mathbf{x}$ the corresponding velocity vector $\dot{\mathbf{x}}$ coincides with the field vector $\mathbf{F}(\mathbf{x})$ is called a **flow line** of the field $\mathbf{F}$; thus $\dot{\mathbf{x}}(t) = \mathbf{F}(\mathbf{x}(t))$ for each t in some interval $a < t < b$. If the field $\mathbf{F}$ represents the velocity field of a fluid flow, a flow line models the path followed by a particular molecule of the fluid, or alternatively the path traced by a small foreign object dropped into the fluid at some point. Finding explicit formulas for flow lines is generally not possible except under some special assumptions discussed in Chapter 12, Section 2. Indeed we often have to resort to the approximate numerical methods of Chapter 12, Section 5, to make accurate pictures of flow lines. However, given a reasonably accurate sketch of a vector field, some flow lines can usually be sketched in roughly by hand; the idea is to draw a flow line so it appears to be tangent to an arrow of the field if the arrow's tail is on the curve. (The computer methods for drawing flow lines described in Chapter 12 use essentially this idea.) Figure 8.9(a) shows such a sketch. The lengths

FIGURE 8.9

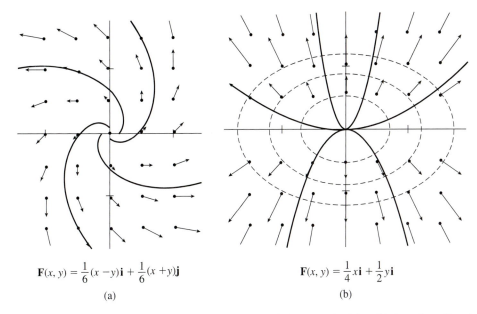

$$\mathbf{F}(x,\,y) = \tfrac{1}{6}(x - y)\mathbf{i} + \tfrac{1}{6}(x + y)\mathbf{j}$$

(a)

$$\mathbf{F}(x,\,y) = \tfrac{1}{4}x\mathbf{i} + \tfrac{1}{2}y\mathbf{i}$$

(b)

of the nearby field arrows give an indication of the speed with which a flow line is traversed at a given point.

EXAMPLE 1 A sketch of the 2-dimensional vector field $\mathbf{F}$ defined by the vector equation $\mathbf{F}(x,\,y) = \tfrac{1}{6}(x - y)\mathbf{i} + \tfrac{1}{6}(x + y)\mathbf{j}$ appears in Figure 8.9(a). For example, $\mathbf{F}(1,\,1)$ is represented by a vertical arrow of length $\tfrac{1}{3}$ with its tail at the point $F(1,\,1)$. Four flow lines have been sketched in with attention paid to their tangency relation to the arrows in the field sketch.

EXAMPLE 2 If in the previous example we had wanted instead a sketch of the field $6\mathbf{F}(x,\,y) = (x - y)\mathbf{i} + (x + y)\mathbf{j}$ we might have preferred to settle for the picture shown in Figure 8.9(a) anyway. The point is that making the arrows 6 times as long for $6\mathbf{F}$ makes for a more cluttered picture, particularly if the domain is enlarged beyond the one shown in the figure. In accepting the temporary convention that the arrow lengths be scaled down by the factor $\tfrac{1}{6}$, we make a clearer picture at the small expense of asking our minds to interpret the picture as if the arrows were 6 times as long. The flow lines will appear to be the same in either case, though they will be traced with 6 times the velocity in $6\mathbf{F}$ as in $\mathbf{F}$. There is nothing about the image of the flow lines themselves that shows their velocities, so we rely on our interpretation of the field arrows for that information.

    If a field $\mathbf{F}$ is known to be a *gradient* field, with $\dot{\mathbf{x}} = \mathbf{F}(\mathbf{x}) = \nabla f(\mathbf{x})$, then Theorem 1.4 of Section 1C tells us that the flow lines are perpendicular to the level sets of f, as shown in Figure 8.9(b). Thus for a gradient field ∇f we have the option of drawing flow lines by first drawing level sets of f and using these as a guide in drawing flow lines.

EXAMPLE 3 Suppose $f(x,\,y) = \tfrac{1}{8}x^2 + \tfrac{1}{4}y^2$. The gradient field $\mathbf{F} = \nabla f$ is then given by

$$\mathbf{F}(x,\,y) = \tfrac{1}{4}x\mathbf{i} + \tfrac{1}{2}y\mathbf{j}.$$

Figure 8.9(b) shows a sketch of the field together with some level curves of the function f. Some flow lines have also been sketched in. These flow lines have the property that, as well as being tangent to vectors of the field, they are perpendicular to the level curves that they cross. To sketch the flow lines we can use either tangency to the field arrows or perpendicularity to the level curves as a guide, whichever seems simpler. In this example the level curves are the family of ellipses $\frac{1}{8}x^2 + \frac{1}{4}y^2 = k$ or equivalently, with $c = 8k$, $x^2 + 2y^2 = c$; the ellipses are fairly easy to draw, so we might prefer using these as an aid in sketching the perpendicular flow lines instead of first sketching the vector field.

Divergence of a Vector Field.   Important aspects of the behavior of a differentiable vector field $\mathbf{F}$ and its flow lines are characterized by the **divergence** of $\mathbf{F}$, abbreviated div $\mathbf{F}(\mathbf{x})$ and defined as the sum of the main diagonal elements of the derivative matrix $\mathbf{F}'$. Thus the divergence of $\mathbf{F}$ is a real-valued function wherever $\mathbf{F}$ is differentiable. For example, if $\mathbf{F}$ has real-valued coordinate functions $F_1, F_2, \ldots, F_n$, then

4.1 $\operatorname{div}\mathbf{F}(x, y) = \dfrac{\partial F_1}{\partial x}(x, y) + \dfrac{\partial F_2}{\partial y}(x, y), \quad \text{for } \mathbf{F}: \mathbb{R}^2 \to \mathbb{R}^2.$

4.2 $\operatorname{div}\mathbf{F}(x, y, z) = \dfrac{\partial F_1}{\partial x}(x, y, z) + \dfrac{\partial F_2}{\partial y}(x, y, z) + \dfrac{\partial F_3}{\partial z}(x, y, z),$

for $\mathbf{F}: \mathbb{R}^3 \to \mathbb{R}^3$. If we write $\nabla = \left(\dfrac{\partial}{\partial x}, \dfrac{\partial}{\partial y}, \dfrac{\partial}{\partial z}\right)$ and $\mathbf{F} = (F_1, F_2, F_3)$ then it makes sense to write $\operatorname{div}\mathbf{F} = \nabla \cdot \mathbf{F}$.

EXAMPLE 4   The 2-dimensional field $\mathbf{F}(x, y) = xy\mathbf{i} + (x - y^2)\mathbf{j}$ has

$$\operatorname{div}\mathbf{F}(x, y) = \frac{\partial xy}{\partial x} + \frac{\partial(x - y^2)}{\partial y} = y + (-2y) = -y.$$

EXAMPLE 5   The 3-dimensional field $\mathbf{F}(x, y, z) = x\mathbf{i} + y\mathbf{j} + z\mathbf{k}$ has

$$\operatorname{div}\mathbf{F}(x, y, z) = \frac{\partial x}{\partial x} + \frac{\partial y}{\partial y} + \frac{\partial z}{\partial z} = 1 + 1 + 1 = 3.$$

To attach some meaning to div $\mathbf{F}$, think of $\mathbf{F}$ as the velocity field of a 2 or 3-dimensional fluid flow. Using Gauss's theorem in Sections 1C and 4B of the following Chapter 9 we show that div $\mathbf{F}(\mathbf{x})$ is the expansion rate of fluid at $\mathbf{x}$ per unit of area and volume respectively; in terms of the density $\rho(\mathbf{x})$ of the fluid at $\mathbf{x}$ this property can be expressed as the **continuity equation**

4.3 $$\operatorname{div}\mathbf{F}(\mathbf{x}) = -\frac{\partial \rho}{\partial t}(\mathbf{x}).$$

In particular, if div $\mathbf{F}(\mathbf{x}) < 0$ in a region then the fluid is contracting there and becoming more dense, because $\rho_t > 0$. If div $\mathbf{F}(\mathbf{x}) = 0$ the fluid volume and density remain constant, and if div $\mathbf{F}(\mathbf{x}) > 0$ the fluid is expanding and becoming less dense. We can even use div $\mathbf{F}$ to measure the compression or expansion of a fluid such as a gas in another way. Indeed a special case of Theorem 1.6 in Section 1D of Chapter 12 shows that if div $\mathbf{F}$ is constant, then in time $t > 0$ a region of volume V will flow into a region of volume $e^{t \, \mathrm{div} \, \mathbf{F}}$ times V; such pairs of regions appear in Figure 8.10. For $t > 0$. the factor $e^{t \, \mathrm{div} \, \mathbf{F}} > 1$ if div $\mathbf{F} > 0$, is equal to 1 if div $\mathbf{F} = 0$, and less than 1 if div $\mathbf{F} < 0$. A basic assumption in this discussion is that no fluid is being created or destroyed near $\mathbf{x}$.

EXAMPLE 6   The field $\mathbf{F}(x, y) = \frac{1}{8}(-x + y)\mathbf{i} + \frac{1}{8}(-x - y)\mathbf{j}$ shown in Figure 8.10(a) has div $\mathbf{F}(x, y) = -\frac{1}{8} - \frac{1}{8} = -\frac{1}{4}$ in all of $\mathbb{R}^2$, so the fluid is being compressed at a constant rate per unit of area; in other words the fluid density is increasing. Following the points of the shaded region in the second quadrant along the flow lines of the field for some fixed time unit, we see the initial area compressed into a smaller region in the first quadrant. If the flow direction is reversed, the fluid expands.

EXAMPLE 7   If we were to reverse the direction of the field in the previous example and consider instead the field $-\mathbf{F}$, the arrows in the sketch would all reverse direction and the flow lines would spiral outward from the origin as in Figure 8.9(a). We would have div$(-\mathbf{F})(x, y) = \frac{1}{4}$, and conclude that we have expansion of areas along flow lines of $-\mathbf{F}$.

EXAMPLE 8   The vector field $\mathbf{G}(x, y) = -\frac{1}{4}(y/\sqrt{x^2 + y^2})\mathbf{i} + \frac{1}{4}(x/\sqrt{x^2 + y^2})\mathbf{j}$ is shown in Figure 8.10(b) with two flow lines enclosing a pair of shaded regions. We find

FIGURE 8.10

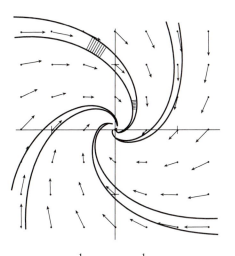

$$\mathbf{F}(x, y) = \frac{1}{8}(-x + y)\mathbf{i} + \frac{1}{8}(-x - y)\mathbf{j}$$

Area decreased: div $\mathbf{F}(x, y) = -\frac{1}{4}$

(a)

$$\mathbf{G}(x, y) = -\frac{1}{4}(y/\sqrt{x^2 + y^2})\mathbf{i} + \frac{1}{4}(x/\sqrt{x^2 + y^2})\mathbf{j}$$

Area preserved: div $\mathbf{G}(x, y) = 0$

(b)

$$\text{div }\mathbf{G}(x, y) = \frac{\partial}{\partial x}\left(\frac{-y}{4\sqrt{x^2 + y^2}}\right) + \frac{\partial}{\partial y}\left(\frac{x}{4\sqrt{x^2 + y^2}}\right)$$

$$= \frac{xy}{4(x^2 + y^2)^{3/2}} - \frac{xy}{4(x^2 + y^2)^{3/2}} = 0.$$

Since div $\mathbf{G}$ is identically 0 the points in the shaded region in the first quadrant move along their flow lines during a fixed time interval into a region of the same area.

EXAMPLE 9   The 3-dimensional vector field $\mathbf{F}(x, y, z) = xy^2\mathbf{i} - yz^2\mathbf{j} + x^2z\mathbf{k}$ would be fairly complicated to sketch. However we can get some significant information about its behavior by computing div $\mathbf{F}$. We find div $\mathbf{F}(x, y, z) = y^2 - z^2 + x^2$, so div $\mathbf{F}(\mathbf{x}) < 0$ whenever $x^2 + y^2 < z^2$, that is, precisely when $\mathbf{x} = (x, y, z)$ is inside a right circular cone symmetric about the z-axis. Thus a fluid flow with velocity field $\mathbf{F}$ would be compressive inside the cone and expansive outside the cone.

The Curl of a Vector Field.   The term *curl* is meant to suggest that we're trying to measure the local tendency of a vector field and its flow lines to circulate around some axis. In dimension 3 the **curl** of a differentiable vector field $\mathbf{F} = F_1\mathbf{i} + F_2\mathbf{j} + F_3\mathbf{k}$ is the 3-dimensional field

4.4 $$\text{curl }\mathbf{F} = \left(\frac{\partial F_3}{\partial y} - \frac{\partial F_2}{\partial z}\right)\mathbf{i} + \left(\frac{\partial F_1}{\partial z} - \frac{\partial F_3}{\partial x}\right)\mathbf{j} + \left(\frac{\partial F_2}{\partial x} - \frac{\partial F_1}{\partial y}\right)\mathbf{k}.$$

To help in recalling the formula, we can express curl $\mathbf{F}$ as a kind of cross-product of the gradient operator $(\partial/\partial x, \partial/\partial y, \partial/\partial z)$ and $\mathbf{F} = (F_1, F_2, F_3)$. Thus

$$\text{curl }\mathbf{F} = \det\begin{pmatrix} \mathbf{i} & \mathbf{j} & \mathbf{k} \\ \partial/\partial x & \partial/\partial y & \partial/\partial z \\ F_1 & F_2 & F_3 \end{pmatrix} = \nabla \times \mathbf{F}.$$

EXAMPLE 10   If $\mathbf{F}(x, y, z) = (y, z^2, x^3)$, then

$$\text{curl }\mathbf{F} = \det\begin{pmatrix} \mathbf{i} & \mathbf{j} & \mathbf{k} \\ \partial/\partial x & \partial/\partial y & \partial/\partial z \\ y & z^2 & x^3 \end{pmatrix} = -2z\mathbf{i} - 3x^2\mathbf{j} - 1\mathbf{k}.$$

We can conclude that along the y-axis, where $z = x = 0$, the vectors of the field curl $\mathbf{F}$ all have length 1 and point in the direction of the negative z-axis.

EXAMPLE 11   Sometimes we can choose coordinates so that the third coordinate function of the given vector field is identically zero and the other two coordinate functions are independent of z, that is,

$$\mathbf{F}(x, y, z) = F_1(x, y)\mathbf{i} + F_2(x, y)\mathbf{j}.$$

Then all partials with respect to z are zero, and

$$\text{curl } \mathbf{F} = \det \begin{pmatrix} \mathbf{i} & \mathbf{j} & \mathbf{k} \\ \partial/\partial x & \partial/\partial y & \partial/\partial z \\ F_1(x, y) & F_2(x, y) & 0 \end{pmatrix} = 0\mathbf{i} + 0\mathbf{j} + \left(\frac{\partial F_2}{\partial x} - \frac{\partial F_1}{\partial y} \right) \mathbf{k}.$$

We conclude that the arrows representing the curl of a field of this special kind either have length zero or else are parallel to the z-axis.

We can look at a 2-dimensional vector field as a horizontal slice of the very special kind of 3-dimensional field described in the previous example: $\mathbf{F}(x, y) = F_1(x, y)\mathbf{i} + F_2(x, y)\mathbf{j}$. As we see from the example,

$$\text{curl } \mathbf{F}(x, y) = \left(\partial F_2/\partial x - \partial F_1/\partial y \right)\mathbf{k},$$

so we define the **curl**; sometimes called the **scalar curl**, of a 2-dimensional vector field to be the real-valued function $\text{curl } \mathbf{F} = (\partial F_2/\partial x - \partial F_1/\partial y)$. The scalar curl plays an important part in Green's theorem taken up in Section 1 of the next chapter, and it's particularly helpful in conveying an intuitive feeling for the significance of $\text{curl } \mathbf{F}$ in general. It will follow from Green's theorem that if the scalar curl of a 2-dimensional field is continuous and positive at a point (x, y), then the line integral of $\mathbf{F}$ over a small enough counterclockwise oriented circle centered at (x, y) will be positive; thus a field with positive scalar curl will have positive counterclockwise circulation near (x, y). If $\text{curl } \mathbf{F}(x, y) < 0$ the circulation will be clockwise near (x, y). If $\text{curl } \mathbf{F} = 0$ identically, the circulation will be zero near every point.

The statements in the previous paragraph can't be interpreted as predictions about how a fluid particle would move at a given time and place; that information is given by the vector values of $\mathbf{F}$. Indeed, without some external constraint, a fluid particle will simply follow a flow line with its velocity at each point $\mathbf{x}$ determined by $\mathbf{F}(\mathbf{x})$. Circulation, as defined by a line integral in Section 1, is just a cumulative measure of the effect of the field along a particular path.

EXAMPLE 12   The 2-dimensional field $\mathbf{F}(x, y) = \frac{1}{8}(-x + y)\mathbf{i} + \frac{1}{8}(-x - y)\mathbf{j}$ of Example 6, shown in Figure 8.10(a), has

$$\text{curl } \mathbf{F}(x, y) = \frac{1}{8}\frac{\partial(-x - y)}{\partial x} - \frac{1}{8}\frac{\partial(-x + y)}{\partial y} = -\frac{1}{8} - \frac{1}{8} = -\frac{1}{4}.$$

This tells us that near each point (x, y) the circulation around a counterclockwise oriented circle is negative, or alternatively that the circulation around a clockwise circle is positive.

EXAMPLE 13 The vector field $\mathbf{G}(x, y) = -\frac{1}{4}(y/\sqrt{x^2 + y^2})\mathbf{i} + \frac{1}{4}(x/\sqrt{x^2 + y^2})\mathbf{j}$ of Example 8 shown in Figure 8.10(b) has a scalar curl given for $(x, y) \neq (0, 0)$ by

$$\text{curl}\,\mathbf{G}(x, y) = \frac{\partial}{\partial x}\left(\frac{x}{4\sqrt{x^2 + y^2}}\right) - \frac{\partial}{\partial y}\left(\frac{-y}{4\sqrt{x^2 + y^2}}\right)$$

$$= \frac{(x^2 + y^2)^{1/2} - x^2(x^2 + y^2)^{-1/2}}{4(x^2 + y^2)}$$

$$+ \frac{(x^2 + y^2)^{1/2} - y^2(x^2 + y^2)^{-1/2}}{4(x^2 + y^2)}$$

$$= \frac{1}{4\sqrt{x^2 + y^2}} > 0.$$

Since curl $\mathbf{G}$ is positive everywhere on the domain of $\mathbf{G}$ we can conclude that the circulation of $\mathbf{G}$ around small enough counterclockwise circles in the domain of $\mathbf{G}$ will be positive. It's tempting to think that by some reasoning we can conclude that the circulation around the closed flow lines shown in Figure 8.10(b) will also be positive; the conclusion is true, but for a somewhat different reason explained in Section 1 on line integrals, namely that the tangent vectors to the curve coincide with the vectors of the field. (See also Exercise 8 of the present section.)

EXAMPLE 14 The vectors of the field $\mathbf{G}$ in the previous example all have length $\frac{1}{4}$. By varying the arrow-lengths but leaving the directions alone we find a field having the same flow lines as $\mathbf{G}$ but traced with different speeds. For example, consider the vector field

$$\mathbf{H}(x, y) = \frac{4}{\sqrt{x^2 + y^2}}\mathbf{G}(x, y) = \frac{-y}{x^2 + y^2}\mathbf{i} + \frac{x}{x^2 + y^2}\mathbf{j}.$$

Only on the circle $x^2 + y^2 = 16$ do the two vector fields coincide; outside this circle the arrows of $\mathbf{H}$ are shorter, while inside the circle they are longer. The scalar curl of $\mathbf{H}$ is

$$\text{curl}\,\mathbf{H}(x, y) = \frac{\partial}{\partial x}\left(\frac{x}{x^2 + y^2}\right) - \frac{\partial}{\partial y}\left(\frac{-y}{x^2 + y^2}\right)$$

$$= \frac{(x^2 + y^2) - 2x^2}{(x^2 + y^2)^2} + \frac{(x^2 + y^2) - 2y^2}{(x^2 + y^2)^2} = 0, \quad (x, y) \neq (0, 0).$$

We conclude that the circulation of $\mathbf{H}$ is zero near every point. Nevertheless, it's intuitively evident that the circulation along the flow lines will be nonzero. (See Exercise 8.)

We'll see in Chapter 9 that the scalar curl measures the local tendency of a 2-dimensional vector field to have a nonzero circulation about a point, as defined by line integrals of the field over small closed paths around the point. However, such a tendency by no means implies that a particle acted upon by a velocity field with

FIGURE 8.11

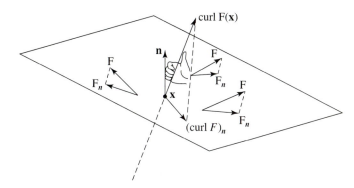

nonzero scalar curl will exhibit vortex motion locally. Indeed Figure 8.10(b) shows that the circular flow lines would cut right across a small circular path centered at a point other than the origin. Similar remarks apply to the 3-dimensional vector field **F** of a 3-dimensional field **F**.

In $\mathbb{R}^3$ we can ask if there is an interpretation not only for the magnitude of curl **F** but also for its direction, assuming curl $\mathbf{F(x)} \neq 0$. The answer is yes, and we interpret the vector curl as follows. Let P be a plane through **x** with unit normal vector **n**. For each point of P that is also in the domain of **F**, project the vector $\mathbf{F(x)}$ perpendicularly onto P as shown in Figure 8.11 to get a 2-dimensional vector field $\mathbf{F_n}$ in P having a *scalar curl*, namely curl $\mathbf{F_n}$.

The following three observations show that understanding the special case of the scalar curl is a help in understanding the 3-dimensional vector field curl **F**.

(i) $|\operatorname{curl} \mathbf{F(x)}|^2 = |\operatorname{curl} \mathbf{F_n(x)}|^2 + \big(\mathbf{n} \cdot \operatorname{curl} \mathbf{F(x)}\big)^2$.

(ii) Choosing **n** perpendicular to curl $\mathbf{F(x)}$, which means the vector curl $\mathbf{F(x)}$ is parallel to P, maximizes the absolute value of curl $\mathbf{F_n(x)}$ among all choices for a plane through **x**.

(iii) If curl $\mathbf{F_n(x)} > 0$ in a neighborhood of a point $\mathbf{x}_0$. then the circulation of $\mathbf{F_n}$ around nearby circular paths centered at $\mathbf{x}_0$ in P, is positive, following the fingers of the right hand rule, with thumb pointing in the direction of curl $\mathbf{F_n(x)}$, as shown in Figure 8.11.

Statement (i) is just the Theorem of Pythagoras for the large triangle in Figure 8.11. Statement (ii) follows from the first statement, since curl $\mathbf{F_n(x)}$ is maximized by making $\mathbf{n} \cdot \operatorname{curl} \mathbf{F(x)} = 0$. Statement (iii) will follow from Example 5 in Section 1 of Chapter 9.

EXAMPLE 15 For a 3-dimensional vector field of the form

$$\mathbf{F}(x, y, z) = -y\mathbf{i} + x\mathbf{j} + \phi(z)\mathbf{k}$$

we find curl $\mathbf{F}(x, y, z) = 0\mathbf{i} + 0\mathbf{j} + 2\mathbf{k} = 2\mathbf{k}$ regardless of our choice for $\phi(z)$. The third coordinate of curl **F**, namely 2, is equal to the scalar curl of the 2-dimensional field $\mathbf{F_n}(x, y) = -y\mathbf{i} + x\mathbf{j}$ that we get if we project the arrows of $\mathbf{F}(x, y, z)$ onto the xy-plane, with unit normal $\mathbf{n} = \mathbf{k}$; this projection is the 2-dimensional field $\mathbf{F_n}(x, y) = -y\mathbf{i} + x\mathbf{j}$ with scalar curl equal to 2.

EXERCISES

In Exercises 1 to 6, compute div $\mathbf{F}$ and, as appropriate, the scalar or vector curl for each of the indicated real-valued functions.

1. $\mathbf{F}(x, y) = \cos(xy)\mathbf{i} + \sin(xy)\mathbf{j}$.

2. $\mathbf{F}(x, y, z) = xy\mathbf{i} + yz\mathbf{j} + zx\mathbf{k}$.

3. $\mathbf{F}(x, y) = (2x - y)\mathbf{i} + (x - 3y)\mathbf{j}$.

4. $\mathbf{F}(x, y, z) = yz\mathbf{i} + xz\mathbf{j} + xy\mathbf{k}$.

5. $\mathbf{F}(x, y) = (x^2 + y^2)^2\mathbf{i} + (x^2 - y^2)^2\mathbf{j}$.

6. $\mathbf{F}(x, y, z) = (x^2y, y^2z^2, xz^3)$.

In Exercises 7 to 10 describe the region in $\mathbb{R}^2$ or $\mathbb{R}^3$ in which div $\mathbf{F}$ is positive, and hence in which the corresponding flow is expanding.

7. $\mathbf{F}(x, y) = (x^2 + y^2)\mathbf{i} + (x^2 - y^2)\mathbf{j}$.

8. $\mathbf{F}(x, y, z) = xy\mathbf{i} + yz\mathbf{j} + zx\mathbf{k}$.

9. $\mathbf{F}(x, y) = e^{x+y}\mathbf{i} + e^{x-y}\mathbf{j}$.

10. $\mathbf{F}(x, y, z) = x^3\mathbf{i} - y^3\mathbf{j} + z^3\mathbf{k}$.

 Note. General methods for deriving the parametrizations of flow lines in the next three exercises are taken up in Chapter 12.

11. (a) Verify that the gradient field $\mathbf{F}(x, y) = \frac{1}{4}x\mathbf{i} + \frac{1}{2}y\mathbf{j}$ in Example 3 of the text has flow lines parametrized by $\mathbf{x}(t) = c_1e^{t/4}\mathbf{i} + c_2e^{t/2}\mathbf{j}$, where c_1 and c_2 are arbitrary real constants.

 (b) Show that the flow lines in part (a) usually follow parabolic paths, degenerating in some cases into straight lines heading away from the origin.

12. (a) Verify that the vector field $6\mathbf{F}(x, y) = (x - y)\mathbf{i} + (x+y)\mathbf{j}$ in Example 2 of the text has flow lines given by $\mathbf{x}(t) = Ae^t \cos(t + \alpha)\mathbf{i} + Ae^t \sin(t + \alpha)\mathbf{j}$, where A and α are arbitrary real constants.

 (b) Show that the flow lines described in part (a) follow generally spiral paths, in one case degenerating into a point.

13. (a) Verify that the vector field $4\mathbf{G}(x, y) = -\frac{1}{4}(y/\sqrt{x^2 + y^2})\mathbf{i} + \frac{1}{4}(x/\sqrt{x^2 + y^2})\mathbf{j}$ in Example 8 of the text has flow lines parametrized by

$$\mathbf{x}(t) = A \cos(\tfrac{1}{4}t/A + \alpha)\mathbf{i} + A \sin(\tfrac{1}{4}t/A + \alpha)\mathbf{j},$$

 where A and α are real constants with $A > 0$.

 (b) Show that the flow lines described in part (a) are counterclockwise circles centered at the origin in $\mathbb{R}^2$.

14. Let $\mathbf{F}$ be a continuously differentiable vector field, suppose $\mathbf{x} = \mathbf{x}(t)$, $a \leq t \leq b$, parametrizes a flow line γ of $\mathbf{F}$, and consider the circulation integral $\int_\gamma \mathbf{F} \cdot d\mathbf{x}$ of $\mathbf{F}$ on γ.

 (a) Show that if $\dot{s}(t)$ is the speed along γ at time t, then

$$\int_\gamma \mathbf{F} \cdot d\mathbf{x} = \int_a^b \dot{s}^2(t)\, dt$$

 (b) Use part (a) to show that the circulation of $\mathbf{F}$ is always positive along a flow line of positive length.

15. Suppose that a 3-dimensional vector field $\mathbf{F}$ is continuously differentiable and that $\mathbf{F} \cdot \text{curl } \mathbf{F}$ is identically zero. Show that the line integral of curl $\mathbf{F}$ along a flow line of $\mathbf{F}$ is equal to zero.

16. Suppose that a 3-dimensional vector field $\mathbf{F}$ is differentiable and that g is a differentiable real-valued function defined on the domain of $\mathbf{F}$. Show that

$$(g\mathbf{F}) \cdot \text{curl}(g\mathbf{F}) = g^2\mathbf{F} \cdot \text{curl } \mathbf{F}$$

 holds identically on the domain of $\mathbf{F}$. (This is easy to show if g is constant, but otherwise is a little more work.)

*17. (a) Show that if $\mathbf{F} = \nabla f$ is a gradient field with $\mathbb{R}^3 \xrightarrow{f} \mathbb{R}$ twice continuously differentiable, then curl $\mathbf{F}$ is identically zero.

 (b) Use part (a) and the result of the previous exercise to find a 3-dimensional vector field $\mathbf{G}$ that isn't a gradient field, but such that $\mathbf{G} \cdot \text{curl } \mathbf{G}$ is identically zero.

 (c) Find a 3-dimensional differentiable vector field $\mathbf{F}$ such that $\mathbf{F} \cdot \text{curl } \mathbf{F}$ is not always zero.

18. The scalar curl of a continuously differentiable 2-dimensional gradient field $\mathbf{F} = \nabla f$ is always zero.

 (a) Show this by computing the second-order derivatives in curl (∇f).

 (b) Show this by calculating the circulation integral $\int_c \nabla f \cdot d\mathbf{x}$ over a closed curve c.

19. The divergence of the curl of a twice continuously differentiable vector field $\mathbb{R}^3 \xrightarrow{\mathbf{F}} \mathbb{R}^3$ is identically zero.

 (a) Prove this by direct computation of the required mixed partial derivatives.

 (b) What can you conclude about the effect of motion along flow lines of curl $\mathbf{F}$ on volume?

20. The curl of the gradient of a twice continuously differentiable function $\mathbb{R}^3 \xrightarrow{f} \mathbb{R}$ is identically zero.

    (a) Prove this by direct computation of the required mixed partial derivatives.

    (b) What can you conclude about the effect on local circulation of ∇f?

21. The divergence of the gradient of a twice continuously differentiable real-valued function $\mathbb{R}^2 \xrightarrow{f} \mathbb{R}$ is the **Laplacian** of f, denoted by Δf; thus $\Delta f = \mathrm{div}(\nabla f)$. What can you conclude about the effect of flow with velocity ∇f on area from the sign of the Laplacian of f in various regions? In particular, suppose that f is a **harmonic function**, which by definition is a function with the property that $\Delta f = 0$; what can you then conclude about the effect of flow with velocity $\mathbf{F} = \nabla f$?

22. (a) Show that the flow line of the 2-dimensional field $\mathbf{F}(x, y) = y\mathbf{i} + x\mathbf{j}$ passing through (u, v) is parametrized by

$$x(t) = \tfrac{1}{2}(u + v)e^t + \tfrac{1}{2}(u - v)e^{-t},$$

$$y(t) = \tfrac{1}{2}(u + v)e^t - \tfrac{1}{2}(u - v)e^{-t}.$$

    (b) Define the transformations $T_t(u, v) = \big(x(t), y(t)\big)$ from $\mathbb{R}^2$ to $\mathbb{R}^2$ using the definitions of $x(t)$ and $y(t)$ in part (a). Show that the Jacobian determinants $\partial(x, y)/\partial(u, v)$ of the transformations T_t are identically equal to 1 for all t.

    (c) Show that $\mathrm{div}\,\mathbf{F}(x, y)$ is identically zero. How does this relate to part (b)?

Chapter 8 REVIEW

Compute the following line integrals by whatever correct method seems simplest.

1. $\int_s xy\,dx + (x^2 + y^2)\,dy$, where s is the closed counterclockwise oriented square with corners at $(0, 0)$, $(1, 0)$, $(1, 1)$ and $(0, 1)$.

2. $\int_q xy\,dx + (x^2 + y^2)\,dy$, where q is the part in the first quadrant of the counterclockwise oriented circle $x^2 + y^2 = 1$.

3. $\int_{\mathbf{a}}^{\mathbf{b}} (y^2 z^3\,dx + 2xyz^3 dy + 3xy^2 z^2\,dz)$, where $\mathbf{a} = (1, 1, 1)$ and $\mathbf{b} = (2, 2, 2)$. The field is the gradient field of a function that's fairly easy to guess.

4. $\int_{\lambda} y^2 z^3\,dx + xyz^3\,dy + xy^2 z^2\,dz$, where λ is the line from $(1, 1, 1)$ to $(2, 2, 2)$.

5. $\int_c y\,dx$, where c is the counterclockwise oriented circle $x^2 + y^2 = 1$.

In Exercises 6 to 9, consider a line integral $I(\gamma) = \int_\gamma x^2 y\,dy$, where γ starts at $(0, 0)$ and ends at $(1, 1)$.

6. Compute $I(\gamma_1)$ if γ_1 is parametrized by $g(t) = (t^2, t^3)$ for $0 \le t \le 1$.

7. Compute $I(\gamma_2)$ if γ_2 is parametrized by $g(t) = (t, t^2)$ for $0 \le t \le 1$.

8. Compute $I(\gamma_3)$ if γ_3 is parametrized by $g(t) = (t^2, t^4)$ for $0 \le t \le 1$.

9. Explain in general terms why $I(\gamma_2) = I(\gamma_3)$, while $I(\gamma_1)$ has a different value.

10. (a) Which, if any, of the parametrizations in the previous exercise are equivalent, so that the line integrals of an arbitrary continuous vector field $\mathbf{F}$ over them will be equal?

    (b) Suppose $F = \nabla f$ is a continuous gradient field on $\mathbb{R}^2$. Which of the integrals of f along the three curves in the previous exercise will be equal?

11. Find the work done by the force field $F(x, y, z) = (x, 2y, z)$ on a particle moving from $(1, 0, 0)$ to $(-1, 2, 1)$ on the straight line segment joining these two points.

12. Find the work done by the force field $F(x, y, z) = (0, 0, mg)$ on a particle moving up through exactly two full turns of the elliptic helix $h(u) = (\cos u, 2\sin u, u)$.

13. Let γ be a smooth parametrized curve, and let $\mathbf{F}$ be a vector field that assigns to each point $\mathbf{x}$ on γ the unit tangent vector to the curve, pointing in the direction of traversal. Show that $\int_\gamma F(\mathbf{x}) \cdot d\mathbf{x} = l(\gamma)$, the length of γ.

14. Define a vector field $F(\mathbf{x}) = \mathbf{x}$ and parametrize the segment joining point $\mathbf{a}$ and point $\mathbf{b}$ by $\mathbf{x}(t) = t\mathbf{b} + (1-t)\mathbf{a}$, with $0 \le t \le 1$.

    (a) Show by direct computation of the line integral that

$$\int_\gamma F \cdot d\mathbf{x} = \tfrac{1}{2}(|\mathbf{b}|^2 - |\mathbf{a}|^2).$$

    (b) Do the computation in part (a) by finding a real-valued function f such that $\nabla f = F$ and applying

Theorem 1.3, the Fundamental Theorem of Calculus for line integrals.

15. Let t stand for time and consider the time-dependent plane vector field

$$F(t, x, y) = ((1 - t)x - ty, tx + (1 - t)y).$$

(a) Find the work done by this field on a particle at $g(t) = (\cos t, \sin t)$ in the time interval $0 \leq t \leq 2\pi$.

(b) How does the answer to part (a) change if, instead of varying with time, the vectors of the field are constantly equal to their values at some fixed time t_0? Explain why the answer is geometrically evident for $t = 0$ and for $t = 1$.

16. Let γ be a smooth curve parametrized by $\mathbf{x} = g(s)$, $0 \leq s \leq l$, where s stands for arc length measured along the curve starting at $g(0)$ and ending at $g(l)$. Show that if F is a continuous vector field defined along γ, then

$$\int_\gamma F \cdot d\mathbf{x} = \int_0^l F\big(g(s)\big) \cdot \mathbf{t}(s)\, ds, \text{ where } \mathbf{t}(s) \text{ is the unit}$$

tangent vector to the curve at $g(s)$ pointing in its direction of traversal.

17. An outdoor sculpture consists of a vertical wall of heavy sheet steel. The base of the wall follows the curve $x = t^3 - 3t$, $y = 3t^2$ for $1 \leq t \leq 2$ where x and y are measured in meters and the height of the wall at (x, y) is y. If the steel weighs 30 kilograms per square meter, what is the total weight of the steel used? [*Hint:* Consider a weighted curve.]

18. (a) The graph of the parabola $y = x^2$ for $0 < a \leq x \leq b$ is rotated about the y-axis. Find the surface area generated.

(b) What is the area of the surface generated by rotation about the x-axis?

19. Show that a parabola has its maximum curvature at its *vertex*, where the parabola's line of symmetry intersects the curve.

20. Show that the curvature of the graph of $y = e^x$ tends to zero as $x \to \pm\infty$. Where is the point of maximum curvature?

21. Let $\mathbf{x}(t) = (a \cos t, a \sin t, bt)$ where a and b are nonnegative constants.

(a) For a fixed positive value of b, what values of a yield the maximum and minimum values for the curvature κ?

(b) For a fixed value of a, what values of b yield the maximum and minimum values for the curvature κ?

22. Let $\mathbf{x}(t)$ trace a smooth curve with speed $\dot{s}(t)$ along the curve.

(a) Show that $|\ddot{\mathbf{x}}(t)|^2 \geq |\ddot{s}(t)|^2$.

(b) Show that the magnitude $\dot{s}^2\kappa$ of the centripetal component of acceleration along the curve is equal to $\sqrt{|\ddot{\mathbf{x}}(t)|^2 - |\ddot{s}(t)|^2}$ at $\mathbf{x}(t)$, so that the discrepancy between $|\ddot{\mathbf{x}}|$ and $|\ddot{s}|$ is zero just when $\kappa = 0$.

23. Using the cross-product, show that curvature of a smooth curve $\mathbf{x}(t)$ is

$$\kappa(t) = \frac{|\dot{\mathbf{x}}(t) \times \ddot{\mathbf{x}}(t)|}{|\dot{\mathbf{x}}(t)|^3}.$$

24. Consider the decomposition $\ddot{\mathbf{x}}(t) = \ddot{s}(t)\mathbf{t}(t) + \dot{s}^2(t)\kappa(t)\mathbf{n}(t)$ of acceleration of motion along $\mathbf{x} = g(t)$ into perpendicular components. Along with $\ddot{\mathbf{x}}$, the tangential and normal components of $\ddot{\mathbf{x}}$ define vector fields on the path of motion, so we can integrate them along the path.

(a) Show by direct computation that the line integral of the normal component $\dot{s}^2(t)\kappa(t)\mathbf{n}(t)$ along a given part γ of the path is always zero.

(b) Show by direct computation that the line integral of the tangential component $\ddot{s}(t)\mathbf{t}(t)$ along a given part γ of the path from $\mathbf{x}(a)$ to $\mathbf{x}(b)$ is equal to $\frac{1}{2}\dot{s}^2(b) - \frac{1}{2}\dot{s}^2(a)$.

(c) Use the results of parts (a) and (b) to compute

$$\int_\gamma \ddot{\mathbf{x}} \cdot d\mathbf{x}.$$

CHAPTER 9

VECTOR FIELD THEORY

The fundamental theorem of calculus for one variable says that if f' is integrable for $a \le t \le b$, then

$$\int_a^b f'(t)\,dt = f(b) - f(a). \tag{1}$$

In Section 1 of the previous chapter, the theorem was extended to line integrals of a gradient ∇f by the equation

$$\int_a^b \nabla f(\mathbf{x}) \cdot d\mathbf{x} = f(\mathbf{b}) - f(\mathbf{a}). \tag{2}$$

The main theorems of the present chapter are also variations on the idea that an integral of some kind of derivative of a function can be evaluated by using only the values of the function itself on a boundary set, for example the endpoints a and b in the fundamental theorem stated above. We begin with the version known as Green's theorem.

SECTION 1 GREEN'S THEOREM

1A Statement and Examples

Let D be a plane region whose boundary is a single curve γ, parametrized by a function g in such a way that, as t increases from a to b, $g(t)$ traces γ once in the *counterclockwise* direction as indicated by the oriented circle on the integral sign. An example is shown in Figure 9.1. If F and G are real-valued functions defined on D, including its boundary, then the formula for **Green's Theorem** says that

$$\int_D \left(\frac{\partial G}{\partial x} - \frac{\partial F}{\partial y} \right) dx\,dy = \oint_\gamma F\,dx + G\,dy, \tag{3}$$

under appropriate differentiability conditions on F and G, and on the boundary curve γ. It's enough to assume additionally about γ that it's **piecewise smooth**, which means that γ consists of finitely many curves each of which is *smooth* as defined in Section 1A of Chapter 4, though the condition that the parametrization have nonzero derivative isn't really necessary here. The requirement that γ be traced counterclockwise is the analog of the requirement that in Equations (1) and (2), the differences on the right have to be taken in the proper order. We can further strengthen the analogy of Equation (3) with Equations (1) and (2) if we think of the integrand $(\partial G/\partial x) - (\partial F/\partial y)$ as a kind of derivative of the vector field $\mathbf{F} = (F, G)$.

EXAMPLE 1

Suppose that D is the square defined by $-1 \le x \le 1$, $-1 \le y \le 1$, and let F and G be defined on D by $F(x, y) = -ye^x$ and $G(x, y) = xe^y$. Then

$$\frac{\partial G}{\partial x}(x, y) - \frac{\partial F}{\partial y}(x, y) = e^y + e^x$$

so

$$\begin{aligned}\int_D \left(\frac{\partial G}{\partial x} - \frac{\partial F}{\partial y}\right) dx\,dy &= \int_{-1}^{1} dx \int_{-1}^{1} (e^y + e^x)\,dy \\ &= \int_{-1}^{1} (e + 2e^x - e^{-1})\,dx \\ &= 4(e - e^{-1}).\end{aligned}$$

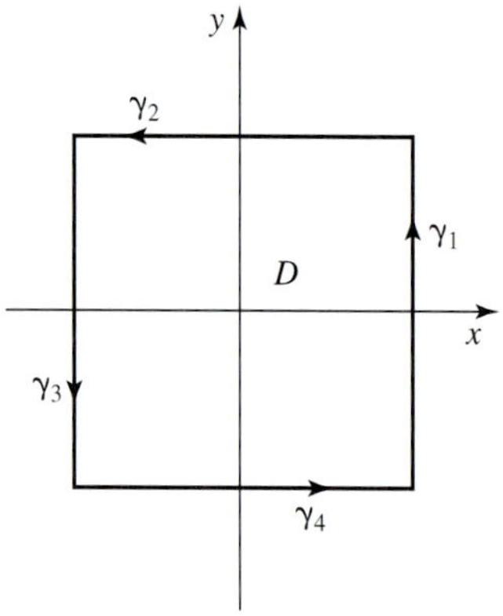

FIGURE 9.1

We parametrize the boundary curve γ in four pieces γ_i, $i = 1, 2, 3, 4$, by

$$\begin{pmatrix} x \\ y \end{pmatrix} = \begin{cases} \begin{pmatrix} 1 \\ t \end{pmatrix} \\ \begin{pmatrix} -t \\ 1 \end{pmatrix} \\ \begin{pmatrix} -1 \\ -t \end{pmatrix} \\ \begin{pmatrix} t \\ -1 \end{pmatrix} \end{cases}, \quad -1 \le t \le 1.$$

Notice that the traversal of γ is counterclockwise, as is shown in Figure 9.1. On the first side of the square we have

$$\begin{aligned}\int_{\gamma_1} F\,dx + G\,dy &= \int_{\gamma} -ye^x\,dx + xe^y\,dy \\ &= \int_{-1}^{1} \left[(-te)\frac{dx}{dt} + e^t \frac{dy}{dt}\right] dt \\ &= \int_{-1}^{1} e^t\,dt = e - \frac{1}{e}.\end{aligned}$$

Similarly, the integrals over the other three sides are also equal to $(e - 1/e)$, so

$$\int_{\gamma} F\,dx + G\,dy = 4\left(e - \frac{1}{e}\right).$$

Equation (3) is thus verified for this particular example.

In computing a line integral, a given parametrization can always be replaced by an equivalent one for which the line integral will have the same value. In the previous example the boundary curve γ was given what appears to be the simplest parametrization, although an equivalent one would do. The question becomes more important if the boundary is presented without a parametrization but merely as a set. It may be necessary to choose a parametrization, and if Green's theorem is to be applied, we'll see that this must be done so that the boundary is traced just once, and in the proper counterclockwise direction.

The need for a counterclockwise instead of clockwise traversal of the boundary curve is apparent when we observe that reversal of direction changes the sign of a nonzero line integral, whether over a closed path or not. In other words, we have the following theorem.

1.1   Theorem.   Let γ be a smooth curve and let $\mathbf{F}$ be a continuous vector field defined on γ. Denote by γ^- the curve γ traced in the opposite direction. Then

$$\int_{\gamma^-} \mathbf{F} \cdot d\mathbf{x} = -\int_{\gamma} \mathbf{F} \cdot d\mathbf{x}.$$

Proof.   If γ is parametrized by $\mathbf{x}(t) = g(t)$ for $a \leq t \leq b$, we parametrize γ^- by $\mathbf{x}(t) = g(a+b-t)$ over the same interval; this change reverses direction, going from $g(b)$ to $g(a)$ instead of the other way around. Since $dg(a+b-t)/dt = -g'(a+b-t)$, we have

$$\int_{\gamma^-} \mathbf{F} \cdot d\mathbf{x} = -\int_a^b \mathbf{F}\big(g(a+b-t)\big) \cdot g'(a+b-t)\, dt.$$

Now change the variable of integration by $t = a+b-u, dt = -du$ to get

$$\int_{\gamma^-} \mathbf{F} \cdot d\mathbf{x} = \int_b^a \mathbf{F}\big(g(u)\big) \cdot g'(u)\, du$$

$$= -\int_a^b \mathbf{F}\big(g(u)\big) \cdot g'(u)\, du = -\int_{\gamma} \mathbf{F} \cdot d\mathbf{x}. \qquad \blacksquare$$

We can prove Green's Theorem most easily for regions D such that γ, the boundary of D, is crossed at most twice by a line parallel to a coordinate axis. Such a region is called **simple**. Thus a coordinate line intersects the boundary of a simple region either in a line segment or else in at most two points. Using Theorem 1.1 we can extend the theorem to finite unions of simple regions. A few such are shown in Figure 9.2, where only D_1 is simple. As shown for D_2, when the boundary of the region is not a single curve, only the outer boundary is traced counterclockwise, while the inner boundary is traced clockwise. A rule that covers all cases is to trace each piece of the boundary so that the region is always to the left as a point traces the boundary. Line integrals around a path that begins and ends at the same point, called a **closed path** or **circuit**, are important enough that they are often distinguished from other integrals by means of an integral sign like $\oint$, or perhaps $\oint$ to indicate a direction of traversal.

FIGURE 9.2

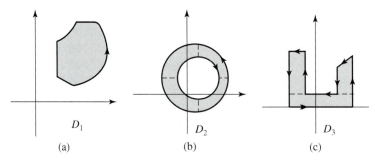

D_1 D_2 D_3

(a)   (b)   (c)

1.2   Green's Theorem.   Let D be a bounded plane region that is a finite union of simple regions, each with a boundary consisting of a piecewise smooth curve. Let F and G be continuously differentiable real-valued functions defined on D together with γ, the boundary of D. Then

$$\int_D \left(\frac{\partial G}{\partial x} - \frac{\partial F}{\partial y} \right) dx\, dy = \oint_\gamma F\, dx + G\, dy,$$

where γ is parametrized so that it's traced once, with D on the left.

Proof.   Consider first the case in which D is a simple region, with boundary γ parametrized by

$$\left(\begin{array}{c} x \\ y \end{array} \right) = \left(\begin{array}{c} g_1(t) \\ g_2(t) \end{array} \right), \quad a \le t \le b.$$

Since

$$\int_\gamma F\, dx + G\, dy = \int_\gamma F\, dx + \int_\gamma G\, dy,$$

we can work with each of the terms on the right separately. We have

$$\int_\gamma F(x, y)\, dx = \int_a^b F\big(g_1(t), g_2(t)\big) g_1'(t)\, dt.$$

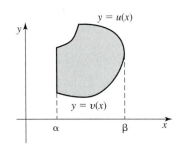

$y = u(x)$

$y = v(x)$

α β

FIGURE 9.3

The curve γ consists of the graphs of two functions $u(x)$ and $v(x)$, perhaps together with one or two vertical segments, as shown in Figure 9.3. On a vertical segment, g_1 is constant, so $g_1' = 0$ there. On the remaining parts of γ we apply the change of variable $x = g_1(t)$ so that, on the top curve, $g_2(t) = y = u(x)$, whereas on the bottom, $g_2(t) = y = v(x)$. It follows that

$$\int_\gamma F(x, y)\, dx = \int_\beta^\alpha F\big(x, u(x)\big)\, dx + \int_\alpha^\beta F\big(x, v(x)\big)\, dx,$$

where the integration from β to α occurs because the graph of u is traced from right to left. Reversing the limits in the first integral, we get

$$\int_\gamma F(x, y)\, dx = \int_\alpha^\beta [-F\big(x, u(x)\big) + F\big(x, v(x)\big)]\, dx$$

$$= \int_{\alpha}^{\beta} \left[- \int_{v(x)}^{u(x)} \frac{\partial F}{\partial x}(x, y) \, dy \right] dx$$

$$= \int_{D} - \frac{\partial F}{\partial y} \, dx \, dy.$$

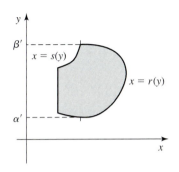

A similar proof, referring to Figure 9.4, shows that

$$\int_{\gamma} G(x, y) \, dy = \int_{D} \frac{\partial G}{\partial x} \, dx \, dy.$$

Combining this equation with the previous one gives Green's Theorem for the special class of simple regions.

We now extend the theorem to a finite union, $D = D_1 \cup \cdots \cup D_K$, of simple regions each with a piecewise smooth boundary curve γ_k, $k = 1, \ldots, K$. Applying Green's Theorem to each simple region D_k we get

FIGURE 9.4

$$\int_{D_k} \left(\frac{\partial G}{\partial x} - \frac{\partial F}{\partial y} \right) dx \, dy = \int_{\gamma_k} F \, dx + G \, dy.$$

The sum of integrals over D_k is an integral over D; so

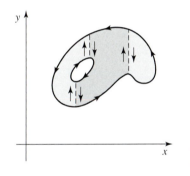

$$\int_{D} \left(\frac{\partial G}{\partial x} - \frac{\partial F}{\partial y} \right) dx \, dy = \int_{\gamma_1} F \, dx + G \, dy + \cdots + \int_{\gamma_k} F \, dx + G \, dy.$$

Now the boundary of D consists of pieces taken from several of the curves γ_k. In addition, there may be parts of curves γ_k that are not a part of γ but that act as a common boundary to two simple regions. The effect is illustrated in Figure 9.5.

A piece δ of common boundary will be traced in one direction or the opposite depending on which simple region it's associated with. But for a line integral we always have, by Theorem 1.1,

FIGURE 9.5

$$\int_{\delta} F \, dx + G \, dy + \int_{\delta^-} F \, dx + G \, dy = 0,$$

where δ^- is δ traced in reverse order. Thus although the parts of the curves γ_k that make up γ contribute to $\int_{\gamma} F \, dx + G \, dy$, the other parts cancel, leaving

$$\int_{D} \left(\frac{\partial G}{\partial x} - \frac{\partial F}{\partial y} \right) dx \, dy = \int_{\gamma} F \, dx + G \, dy.$$

This completes the proof of Green's Theorem.   ∎

FIGURE 9.6

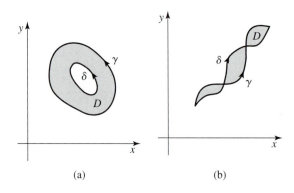

(a)   (b)

1B Changing Paths

The last part of the proof just given extends Green's Theorem from simple regions to those such as are shown in Figure 9.6. The extension has an important consequence for line integrals $\int F\,dx + G\,dy$ over two closed curves γ and δ, when the functions F and G are defined in the region D between γ and δ. In Figure 9.6(a), the curves are traced in the same direction (counterclockwise in the figure), and in Figure 9.6(b), the curves go from one point to another in the same direction. Given this relative orientation of the two curves, if the equation

$$\frac{\partial G}{\partial x} - \frac{\partial F}{\partial y} = 0 \tag{4}$$

holds throughout D, then we can conclude that

$$\int_{\gamma} F\,dx + G\,dy = \int_{\delta} F\,dx + G\,dy.$$

We'll show the validity of this principle in the next two examples.

| EXAMPLE 2 |

Let F and G be defined by

$$F(x, y) = \frac{-y}{x^2 + y^2}, \qquad G(x, y) = \frac{x}{x^2 + y^2},$$

for $(x, y) \neq (0, 0)$. Direct computations show that these functions satisfy Equation (4). If γ is the ellipse shown in Figure 9.7 and defined by

$$\begin{pmatrix} x \\ y \end{pmatrix} = \begin{pmatrix} 2\cos t \\ 3\sin t \end{pmatrix}, \qquad 0 \le t \le 2\pi,$$

then the integral $\int_{\gamma} F\,dx + G\,dy$ would be troublesome to compute directly, even using tables. However we can apply Green's Theorem to the region D between γ

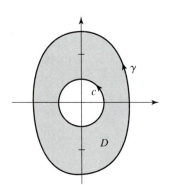

FIGURE 9.7

and the circle c of radius 1 about the origin, parametrized by $(x, y) = (\cos t, \sin t)$, for $0 \le t \le 2\pi$. Because Equation (4) is satisfied, Green's Theorem yields

$$\int_{\gamma \cup c^-} F\,dx + G\,dy = 0,$$

where c^- is c traced clockwise, so that D is on its left. By Equation 1 the last equation is equivalent to

$$\int_\gamma F\,dx + G\,dy = \int_c F\,dx + G\,dy.$$

But on c we have $x^2 + y^2 = 1$, so

$$\int_\gamma F\,dx + G\,dy = \int_c -y\,dx + x\,dy$$

$$= \int_0^{2\pi} (\sin^2 t + \cos^2 t)\,dt = 2\pi.$$

It's important to observe that Green's Theorem could not have been applied directly to the entire interior of the ellipse because $(\partial G/\partial x)$ and $(\partial F/\partial y)$ fail to exist at the origin.

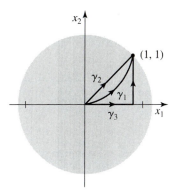

FIGURE 9.8

EXAMPLE 3   The curve γ_1 given by $g(t) = (t, t^2), 0 \le t \le 1$, is shown in Figure 9.8. Suppose that $\mathbf{F}(x, y) = \big(F(x, y), G(x, y)\big)$ is a continuously differentiable vector field for $x^2 + y^2 < 4$ and satisfies Equation (4), namely $G_x(x, y) - F_y(x, y) = 0$ in the disk of radius 2. The line integral of F over γ_1 could perhaps be computed directly in the form

$$\int_{\gamma_1} \mathbf{F} \cdot d\mathbf{x} = \int_0^1 [F(t, t^2) + G(t, t^2)(2t)]\,dt.$$

But there are other possibilities. For example, the curve γ_2 can be parametrized by $g_2(t) = (t, t), 0 \le t \le 1$. Since we can apply Green's Theorem to the region between γ_1 and γ_2, Equation (4) implies that

$$\int_D \left(\frac{\partial G}{\partial x} - \frac{\partial F}{\partial y} \right) dx\,dy = 0,$$

and hence

$$\int_{\gamma_1} \mathbf{F} \cdot d\mathbf{x} + \int_{\gamma_2^-} \mathbf{F} \cdot d\mathbf{x} = 0.$$

Here γ_2^- is given by $g_2^-(t) = (1 - t, 1 - t)$ for $0 \le t \le 1$. Then the line integrals over γ_1 and γ_2 are equal by Equation 1, and the latter integral is then

$$\int_{\gamma_2} \mathbf{F} \cdot d\mathbf{x} = \int_0^1 [F(t, t) + G(t, t)] \, dt.$$

Another alternative would be to replace γ_1 by γ_3, where γ_3 is parametrized in two pieces, one horizontal and one vertical, by

$$g_3(t) = \begin{cases} (t, 0), & 0 \le t \le 1, \\ (1, t), & 0 \le t \le 1. \end{cases}$$

Thus

$$\int_{\gamma_3} \mathbf{F} \cdot d\mathbf{x} = \int_0^1 F(t, 0) \, dt + \int_0^1 G(1, t) \, dt.$$

This may be easier to compute than either of the integrals over γ_1 and γ_2, although all three are equal. In Exercise 5 you are asked to compute the value of a specific example using each of these three paths.

The previous examples are typical applications of Green's Theorem; in summary we have the

Path independence principle.   If a plane vector field $\mathbf{F}(x, y) = \big(F(x, y), G(x, y)\big)$ satisfies $\partial G/\partial x - \partial F/\partial y = 0$ in a region whose boundary is the union of two paths γ_1 and γ_2 with common initial and terminal points, then

$$\int_{\gamma_1} \mathbf{F} \cdot d\mathbf{x} = \int_{\gamma_2} \mathbf{F} \cdot d\mathbf{x}.$$

The paths may even intersect at points other than their common endpoints, as indicated in Figure 9.6(b); the only requirement is that we be able to apply Green's Theorem to the region or regions bounded by the curves.

1C   Physical Interpretations

Green's Theorem has two distinct but closely related physical interpretations. We assume D to be a region in $\mathbb{R}^2$ whose boundary is a single counterclockwise-oriented curve γ. If γ has a smooth parametrization $g(t) = \big(g_1(t), g_2(t)\big)$, $a \le t \le b$, and has a nonzero tangent at each point, we can form the unit tangent and normal vectors

$$\mathbf{t}(t) = \frac{g'(t)}{|g'(t)|} = \left(\frac{g_1'(t)}{|g'(t)|}, \frac{g_2'(t)}{|g'(t)|} \right)$$

and

$$\mathbf{n}(t) = \left(\frac{g_2'(t)}{|g'(t)|}, \frac{-g_1'(t)}{|g'(t)|} \right).$$

An example is shown in Figure 9.9. Note that this normal vector isn't related to the curvature of γ and doesn't necessarily have the same direction as the principal normal to the curve, as defined in Chapter 8, Section 3.

Stokes's Theorem in the Plane.   In Section 4 of Chapter 8 we defined the scalar curl of a vector field as a real-valued function. Stokes's Theorem in that context equates an area integral over a region D to a line integral over the boundary of D.

| EXAMPLE 4 |

If $\mathbf{F} = (F, G)$ is a continuously differentiable vector field defined on a region containing D and γ, then using the abbreviation, $|g'(t)|\,dt = ds$, the line integral in Green's Theorem is

$$\oint_\gamma F\,dx + G\,dy = \int_a^b \mathbf{F}\big(g(t)\big) \cdot \mathbf{t}(t)|g'(t)|\,dt$$

$$= \oint_\gamma \mathbf{F} \cdot \mathbf{t}\,ds.$$

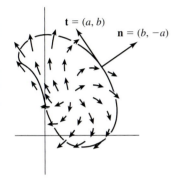

FIGURE 9.9

We define a real-valued function, curl $\mathbf{F}$, called the **scalar curl** of $\mathbf{F}$, by

$$\operatorname{curl} \mathbf{F}(\mathbf{x}) = \frac{\partial G}{\partial x}(\mathbf{x}) - \frac{\partial F}{\partial y}(\mathbf{x}).$$

Green's Theorem then becomes

$$\int_D \operatorname{curl} \mathbf{F}\,dA = \oint_\gamma \mathbf{F} \cdot \mathbf{t}\,ds,$$

sometimes called **Stokes's Theorem** for the plane. Now interpret $\mathbf{F}$ as the velocity field of a fluid flow in the plane, which means that at each point $\mathbf{x}$ the arrow representing $\mathbf{F}(\mathbf{x})$ has the direction of the flow at $\mathbf{x}$, with the speed of the flow there equal to the length of the arrow. The line integral represents the **circulation** of the flow around γ in the counterclockwise direction. (Recall that *circulation* of a vector field over a smooth curve, closed or not, was defined in Chapter 8, Section 1.) Stokes's Theorem says that this circulation is equal to the integral of curl $\mathbf{F}$ over D. In particular, if curl $\mathbf{F}$ is identically zero in D, then the circulation is zero for every smooth circuit γ contained in D, whether γ is oriented counterclockwise or not. For this conclusion to hold, it's necessary that curl $\mathbf{F}$ be defined throughout the inside of every circuit in D to which Stokes's Theorem is applied. Conversely, we can show that if the circulation is zero over every smooth circuit, then the function curl $\mathbf{F}$ must be identically zero. See Exercise 10 and Section 2 for an alternative approach. In Section 5, we treat the scalar curl generalized to a vector field in $\mathbb{R}^3$.

| EXAMPLE 5 |

In Chapter 8 we defined the 3-dimensional curl of a 3-dimensional vector field $\mathbf{F}(x, y, z) = \big(F(x, y, z), F(x, y, z), H(x, y, z)\big)$ by

$$\operatorname{curl} \mathbf{F}(x, y, z) = \left(\frac{\partial H}{\partial y} - \frac{\partial G}{\partial z}\right)\mathbf{i} + \left(\frac{\partial F}{\partial z} - \frac{\partial H}{\partial x}\right)\mathbf{j} + \left(\frac{\partial G}{\partial x} - \frac{\partial F}{\partial y}\right)\mathbf{k}.$$

This vector field curl $\mathbf{F}$ reduces to curl $\mathbf{F}(x, y) = \left(\dfrac{\partial G}{\partial x} - \dfrac{\partial F}{\partial y} \right) \mathbf{k}$ when $\mathbf{F}$ is independent of z and H is identically zero, because all partials by z are zero and all partials of H are zero. This 3-dimensional field curl $\mathbf{F}(x, y)$ is in all respects but one the same as the scalar curl that occurs in Green's Theorem, the difference being that the vector curl has a direction, a direction that's always perpendicular to the xy-plane.

Now suppose the scalar $G_x(x, y) - F_y(x, y)$ is positive in a neighborhood of a point $\mathbf{x}_0 = (x_0, y_0)$ in $\mathbb{R}^2$. Applying Green's Theorem over a disk D centered at $\mathbf{x}_0$, and making D small enough that $G_x(x, y) - F_y(x, y) > 0$ there, we get

$$0 < \int_D \left(G_x(x, y) - F_y(x, y) \right) dA = \int_D \operatorname{curl} \mathbf{F} \, dA = \oint_\gamma \mathbf{F} \cdot \mathbf{t} \, ds,$$

where γ is the counterclockwise oriented boundary circle of D. It follows that the predominant circulation of the flow, taken counterclockwise around the boundary of D, and given by the line integral on the right, is positive. Thus the orientation of the vector $\mathbf{k}$ relative to the circulation follows the right-hand rule, with the fingers curling in the direction of the flow and the thumb pointing in the direction of $\mathbf{k}$. See Figure 8.11 in Chapter 8.

In the previous example we interpreted the field $\mathbf{F}$ as the velocity field of a fluid flow in D. That is, the vector field $\mathbf{F}$ at each point of D represents the speed and direction of the flow at that point. In this case the line integral in Stokes's Theorem is called the **circulation** of $\mathbf{F}$ around γ, and Stokes's Theorem says that circulation of $\mathbf{F}$ along γ is the integral of curl $\mathbf{F}$ over D. Thus if curl $\mathbf{F}$ is identically zero in D, then the circulation is zero around every smooth closed curve with its interior contained in D. A field $\mathbf{F}$ for which curl $\mathbf{F}$ is zero is called **irrotational** for this reason.

Gauss's Theorem in the Plane.   Using the **divergence** of a vector field introduced in Section 4 of Chapter 8, we can rewrite Green's Theorem in another way. Instead of applying the fundamental Equation (3) to the field $\mathbf{F} = (F, G)$, here we instead apply it to the related vector field $\mathbf{H} = (-G, F)$. If $\mathbf{t} = (a, b)$ is a unit tangent vector pointing so that the region is on the left, then the perpendicular vector $\mathbf{n} = (b, -a)$ is a unit vector that points away from the region as shown in Figure 9.9. Since $\mathbf{F} = (F, G)$, we have

$$\mathbf{H} \cdot \mathbf{t} = (-G, F) \cdot (a, b) = -aG + bF = (F, G) \cdot (b, -a) = \mathbf{F} \cdot \mathbf{n}.$$

Hence the line integral of $\mathbf{H}$ over γ becomes

$$\oint_\gamma \mathbf{H} \cdot d\mathbf{x} = \int_\gamma \mathbf{H} \cdot \mathbf{t} \, ds$$

$$= \oint_\gamma \mathbf{F} \cdot \mathbf{n} \, ds.$$

On the other hand, the area integral for Green's Theorem applied to $\mathbf{H}$ is

$$\int_D \left(\frac{\partial F}{\partial x} + \frac{\partial G}{\partial y} \right) dx \, dy.$$

We define a real-valued function div $\mathbf{F}$ called the **divergence** of $\mathbf{F}$ by

$$\operatorname{div}\mathbf{F}(x, y) = \frac{\partial F}{\partial x}(x, y) + \frac{\partial G}{\partial y}(x, y).$$

In terms of the divergence, Green's Theorem is

$$\int_D \operatorname{div}\mathbf{F}\, dA = \oint_\gamma \mathbf{F}\cdot\mathbf{n}\, ds.$$

This version of Green's Theorem is called the 2-dimensional **Gauss's Theorem**, or the 2-dimensional **divergence theorem**.

Using the fluid flow interpretation, in which $\mathbf{F}$ represents the velocity field of a fluid flow, the line integral in the divergence theorem is the integral of the outward normal coordinate $\mathbf{F}\cdot\mathbf{n}$ of $\mathbf{F}$ over γ and gives the rate at which fluid is flowing out of the region D bounded by γ. The value of this line integral is called the total flow rate, called **flux**, of $\mathbf{F}$ across γ in the outward direction. Gauss's Theorem shows that the flux across γ, denoted $\Phi(\gamma)$, is equal to the integral of the divergence of $\mathbf{F}$ over the region bounded by γ. Thus $\operatorname{div}\mathbf{F}(\mathbf{x})$ measures the rate of change of the density of the fluid at the point $\mathbf{x}$. If $\operatorname{div}\mathbf{F}(\mathbf{x})$ is predominantly positive in D, then $\Phi(\gamma)$, the outward flow, will be positive, while a negative $\Phi(\gamma)$ indicates that more fluid is going into D than is going out. If $\operatorname{div}\mathbf{F}$ is identically zero, then $\mathbf{F}$ is said to represent an **incompressible** flow, since the flow into and out of arbitrarily small neighborhoods of every point will be exactly balanced.

EXAMPLE 6   A flow in the plane determined by the vector field $\mathbf{F}(x, y) = y\mathbf{i}+x\mathbf{j}$ is incompressible since $\operatorname{div}\mathbf{F}(x, y) = 0$. Hence

$$0 = \int_D \operatorname{div}\mathbf{F}\, dA = \oint_\gamma y\, dx + x\, dy$$

for every circular disk D with counterclockwise oriented boundary circle γ. Indeed, if we were to parametrize γ by $x = x_0 + r\cos t$, $y = y_0 + r\sin t$ for $0 \le t \le 2\pi$ we would discover that the total flux of $\mathbf{F}$ across γ is 0 for all choices of the point (x_0, y_0) and radius r.

In Chapter 8, Section 4 we introduced without proof the **continuity equation** for fluid flow, namely

1.3 $$\frac{\partial \rho}{\partial t}(\mathbf{x}) = -\operatorname{div}\mathbf{F}(\mathbf{x}),$$

where $\mathbf{F}$ is the velocity field of a fluid flow and $\rho(\mathbf{x})$ is the density of the fluid at $\mathbf{x}$. Proving the 2-dimensional continuity equation is an interesting application of Gauss's Theorem in the next example.

EXAMPLE 7   Let $\mathbf{F}(\mathbf{x})$ be the continuously differentiable velocity field of a fluid flow in 2-dimensional space, with continuously differentiable fluid density $\rho(\mathbf{x})$, and let D be an arbitrary region of finite area in the domain of $\mathbf{F}(\mathbf{x})$. We assume that no fluid is created or destroyed in D, so any change in density is due to compression or expansion of the fluid. Then

$$\frac{d}{dt} \int_D \rho \, dx \, dy = - \int_\delta \mathbf{F} \cdot \mathbf{n} \, ds,$$

where $\mathbf{n}$ is the outward-pointing unit normal to the boundary curve δ of D. The left side is the rate of change of mass with respect to t, which because we assume no fluid is created or destroyed in D, must be due to flux across δ, as measured by the right side. We need the minus sign because the integral by itself measures total outward flux, which would be positive if the density ρ were decreasing, making the left side negative. Now apply the Leibniz differentiation rule of Chapter 7, Section 3 on the left and Gauss's Theorem on the right to get

$$\int_D \frac{\partial \rho}{\partial t} \, dx \, dy = - \int_D \operatorname{div} \mathbf{F} \, dx \, dy, \quad \text{or} \quad \int_D \left(\frac{\partial \rho}{\partial t} + \operatorname{div} \mathbf{F} \right) dx \, dy = 0.$$

Since D is arbitrary we can choose D to be a disk $(\mathbf{x}_0)$ of radius r centered at an arbitrary point $\mathbf{x}_0$ in the domain of $\mathbf{F}(\mathbf{x})$. Since the integrand in the right-hand equation is continuous it must be identically zero, otherwise a nonzero value for it at $\mathbf{x}_0$ would, for small enough positive r, give a nonzero value for the integral over D_r. This establishes Equation 3.

EXERCISES

In Exercises 1 to 4, use Green's theorem to compute the value of the line integral $\oint_\gamma y \, dx + x^2 dy$, where γ is the indicated closed path.

1. The circle given by $g(t) = (\cos t, \sin t), 0 \le t \le 2\pi$

2. The square with corners at $(\pm 1, \pm 1)$, traced counterclockwise

3. The square with corners at $(0, 0)$, $(1, 0)$, $(1, 1)$, and $(0, 1)$, traced counterclockwise

4. The ellipse $x^2 + 4y^2 = 4$, traced clockwise

5. Use each of the three paths γ_1, γ_2, or γ_3 in Example 3 of the text to compute this path-independent line integral from $(0, 0)$ to $(1, 1)$: $\int_{\gamma_k} y \, dx + x \, dy.$

6. Let γ be the curve parametrized by $g(t) = (2 \cos t, 3 \sin t)$, $0 \le t \le 2\pi$. Compute $\int_\gamma (2x + y) \, dx + (x + 3y) \, dy.$

In Exercises 7 to 10, evaluate the following line integrals by whatever method seems simplest.

7. $\int_\gamma (x - y) \, dx + (x + y) \, dy$, where γ is a triangle traced counterclockwise and having for its three vertices $(0, 0)$, $(1, 0)$, and $(1, 1)$

8. Use the same integrand as in Exercise 7, but change the path to the square with corners at $(0, 0)$, $(1, 0)$, $(1, 1)$, and $(0, 1)$, traced counterclockwise.

9. $\int_c (x^2 - y^2) \, dx + (x^2 + y^2) \, dy$, where c is the circle of radius 1 centered at the origin and traced *clockwise*

10. $\int_\gamma (x^2 - y^2) \, dx + (x^2 + y^2) \, dy$, where γ is the circle of radius 1 centered at the origin traced *clockwise*, together with the circle of radius 2 traced *counterclockwise*

11. Show that if D is a simple region bounded by a piecewise smooth curve γ, traced counterclockwise, then the area of D is given by

$$A(D) = \frac{1}{2} \oint_\gamma (-y \, dx + x \, dy).$$

12. Let f be a real-valued function with continuous second-order derivatives in an open set D in $\mathbb{R}^2$. Let $\mathbf{F}$ be the

vector field defined in D by $\mathbf{F}(\mathbf{x}) = \nabla f(\mathbf{x})$, the gradient of f. Show that if $\mathbf{F}(\mathbf{x}) = (F(\mathbf{x}), G(\mathbf{x}))$, then the equation $(\partial G/\partial x) - (\partial F/\partial y) = 0$ is satisfied in D.

13. (a) If $f(x, y) = \arctan(y/x)$ for $x > 0$, compute $\nabla f(x, y)$.

(b) Show that the formulas for the coordinate functions of ∇f found in part (a) define a continuous vector field $\mathbf{F}(x, y) = (F(x, y), G(x, y))$ for all $(x, y) \neq (0, 0)$.

(c) Show that there is no function g such that $\nabla g(x, y) = \mathbf{F}(x, y)$ for all $(x, y) \neq (0, 0)$. [Hint: If g existed, then the line integral of ∇g would be independent of the path as long as the path avoided $(0, 0)$.]

14. (a) Consider a particle moving in a plane vertical to the surface of the earth and subject to the gravitational field $\mathbf{N}(x, y) = (0, mg)$, where m is the mass of the particle and g is the acceleration of gravity. Show that as the particle moves in the plane, the amount of work done is independent of the path between two points and depends only on the initial and final points. In particular, the work done in moving along a closed path is zero.

(b) Replace the field $\mathbf{N}$ by a field $\mathbf{F} = (F, G)$ satisfying $(\partial G/\partial x) = (\partial F/\partial y)$ throughout the plane. Show that the same conclusions hold.

15. Assume that the vector field $\mathbf{F} = (F, G)$ is a gradient field, that is, $\mathbf{F} = \nabla f$ for some real-valued f. Show that Green's Formula can be written in the form

$$\int_D \Delta f \, dA = \int_\gamma \nabla f \cdot \mathbf{n} \, ds,$$

where $\Delta f = (\partial^2 f/\partial x^2) + (\partial^2 f/\partial y^2)$, the **Laplacian** of f.

16. (a) Show that if $f(x, y)$ is a continuous real-valued function defined in an open set B of $\mathbb{R}^2$, and $\int_D f(x, y) \, dx \, dy = 0$ for every circular disk D in B, then $f(x, y)$ is identically zero in B. [Hint:

Show that if $f(\mathbf{x}_0) \neq 0$ for some $\mathbf{x}_0$ in B, then there is a disk D centered at $\mathbf{x}_0$ such that $|f(\mathbf{x})| \geq \delta$ for some $\delta > 0$, and all $\mathbf{x}$ in D.]

(b) Use part (a) and Stokes's Theorem to show that if curl $\mathbf{F}$ is continuous in an open set D, and the circulation of $\mathbf{F}$ is zero around every smooth circuit in D, then $\mathbf{F}$ is irrotational in D; that is, curl $\mathbf{F}$ is identically zero in D.

(c) Use part (a) and Gauss's Theorem to show that if $\mathbf{F}$ is continuous in D and the flux $\Phi(\gamma) = 0$ for every smooth circuit γ in D, then $\mathbf{F}$ is incompressible.

17. Define

$$\mathbf{F}(x, y) = \frac{-y}{x^2 + y^2}\mathbf{i} + \frac{x}{x^2 + y^2}\mathbf{j}, \quad \text{for} \quad (x, y) \neq (0, 0).$$

(a) Show that div $\mathbf{F}$ is identically zero. What implication does this have for areas of regions under the influence of the flow generated by $\mathbf{F}$?

(b) Show that curl $\mathbf{F}$ is identically zero. What implication does this fact have for the circulation of $\mathbf{F}$ around circular paths that don't go around the origin?

(c) What is the circulation of $\mathbf{F}$ along a counterclockwise-oriented circle of radius a centered at the origin? Does this result contradict part (b)? Explain your answer.

The equations curl $\mathbf{F} = \mathbf{0}$ and div $\mathbf{F} = 0$ occur in complex variable theory in a slightly different form as the Cauchy-Riemann equations. In Exercises 17 to 20 show that if $u(x, y)$ and $v(x, y)$ are the real and imaginary parts, respectively, of the following complex-valued functions, then the vector field given by $\mathbf{F}(x, y) = (u(x, y), -v(x, y))$ is irrotational and incompressible.

18. $(x + iy)^2$

19. $(x + iy)^3$

20. e^{x+iy}

21. $\frac{1}{2}\ln(x^2 + y^2) + i \arctan y/x, x > 0$

SECTION 2   CONSERVATIVE VECTOR FIELDS

2A   Potentials

The examples of the previous section show that, under certain conditions, it's possible to alter the path of integration in a line integral in the plane without affecting the value of the integral. Not all line integrals have this property, but those that do are particularly important, not only for the computational reasons already illustrated, but also because of their relation to the gradient. We have the following theorem, valid in $\mathbb{R}^n$, which is a converse to Theorem 1.3 of Chapter 6, Section 1.

2.1   Theorem.   Let **F** be a continuous vector field defined in a polygonally connected open subset D of $\mathbb{R}^n$. If the line integral

$$\int_\gamma \mathbf{F} \cdot d\mathbf{x}$$

is independent of the piecewise smooth path γ from $\mathbf{x}_0$ to $\mathbf{x}$ in D, then the real-valued function defined by

$$f(\mathbf{x}) = \int_{\mathbf{x}_0}^{\mathbf{x}} \mathbf{F} \cdot d\mathbf{x}$$

is continuously differentiable and satisfies the vector equation $\nabla f = \mathbf{F}$ throughout D.

Proof.   We have to show that, for each $\mathbf{x}$ in D, $\nabla f(\mathbf{x}) = \mathbf{F}(\mathbf{x})$. Since $\mathbf{x}$ is an interior point of D, there is a ball of radius δ centered at $\mathbf{x}$ and contained in D. This implies that, for any unit vector $\mathbf{u}$ and for all real numbers t satisfying $|t| < \delta$, the vectors $\mathbf{x} + t\mathbf{u}$ are contained in D. Since the line integral is independent of the path, we choose an arbitrary piecewise smooth path from $\mathbf{x}_0$ to $\mathbf{x}$, lying in D, and extend it by a linear segment to the vector $\mathbf{x} + t\mathbf{u}$, $|t| < \delta$, as shown in Figure 9.10. Then

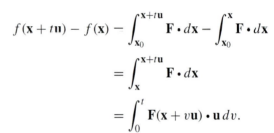

$$f(\mathbf{x} + t\mathbf{u}) - f(\mathbf{x}) = \int_{\mathbf{x}_0}^{\mathbf{x}+t\mathbf{u}} \mathbf{F} \cdot d\mathbf{x} - \int_{\mathbf{x}_0}^{\mathbf{x}} \mathbf{F} \cdot d\mathbf{x}$$

$$= \int_{\mathbf{x}}^{\mathbf{x}+t\mathbf{u}} \mathbf{F} \cdot d\mathbf{x}$$

$$= \int_0^t \mathbf{F}(\mathbf{x} + v\mathbf{u}) \cdot \mathbf{u} \, dv.$$

FIGURE 9.10

In the result of this computation we let $\mathbf{u} = \mathbf{e}_j$, the jth standard basis vector in $\mathbb{R}^n$. Then

$$\frac{\partial f}{\partial x_j}(\mathbf{x}) = \lim_{t \to 0} \frac{f(\mathbf{x} + t\mathbf{e}_j) - f(\mathbf{x})}{t}$$

$$= \lim_{t \to 0} \frac{1}{t} \int_0^t \mathbf{F}(\mathbf{x} + v\mathbf{e}_j) \cdot \mathbf{e}_j \, dv.$$

Since the integral in this last limit is zero when $t = 0$, the limit is the derivative with respect to t of the integral, evaluated at $t = 0$. By the fundamental theorem of calculus, this is just the integrand evaluated at $v = 0$, so

$$\frac{\partial f}{\partial x_j}(\mathbf{x}) = \mathbf{F}(\mathbf{x}) \cdot \mathbf{e}_j = F_j(\mathbf{x}),$$

where F_j is the jth coordinate function of $\mathbf{F}$. Since $\mathbf{F}$ was assumed continuous, so are the partial derivatives $\partial f / \partial x_j$; therefore f is continuously differentiable on D. Finally, the equations $(\partial f / \partial x_j)(\mathbf{x}) = F_j(\mathbf{x})$, $j = 1, \dots, n$, taken all together mean that $\nabla f = \mathbf{F}$ in D.   ∎

A vector field $\mathbf{F}$ for which there is a real-valued function f such that $\mathbf{F} = \nabla f$ is called a **conservative field** or **gradient field**. In that case f is called a **field potential** of $\mathbf{F}$. The next example motivates this terminology.

EXAMPLE 1   Suppose that a continuous force field $\mathbf{F}$ is defined in a region D of $\mathbb{R}^3$. Suppose also that the work done in moving a particle from one point to another under the influence of the field is independent of whatever path it's constrained to take between the two points. Thus if $\mathbf{x}_1$ and $\mathbf{x}_2$ are two points in the field and $W(\mathbf{x}_1, \mathbf{x}_2)$ represents the work done in going from $\mathbf{x}_1$ to $\mathbf{x}_2$, we can write

$$W(\mathbf{x}_1, \mathbf{x}_2) = \int_{\mathbf{x}_1}^{\mathbf{x}_2} \mathbf{F} \cdot d\mathbf{x}.$$

If the particle follows a particular path given by $\mathbf{x}(t) = g(t)$, then the velocity and acceleration vectors are $\mathbf{v}(t) = g'(t)$ and $\mathbf{a}(t) = g''(t)$, and we have $F(g(t)) = m\mathbf{a}(t)$, where m is the mass of the particle. We write $v = |\mathbf{v}|$, so that $v^2 = \mathbf{v} \cdot \mathbf{v}$. Hence if $\mathbf{x}_1 = g(t_1)$ and $\mathbf{x}_2 = g(t_2)$, then

$$W(\mathbf{x}_1, \mathbf{x}_2) = \int_{t_1}^{t_2} m\mathbf{a}(t) \cdot \mathbf{v}(t)\, dt.$$

But since $\mathbf{a}(t) = \dot{\mathbf{v}}(t)$, and $(d/dt)v^2(t) = 2\mathbf{v}(t) \cdot \dot{\mathbf{v}}(t)$, we have

$$W(\mathbf{x}_1, \mathbf{x}_2) = \frac{m}{2} \int_{t_1}^{t_2} \frac{d}{dt}[v^2(t)]\, dt,$$

$$= \frac{m}{2}\left(v^2(t_2) - v^2(t_1)\right). \tag{1}$$

The function $T(t) = (m/2)v^2(t)$ is called the **kinetic energy** of the particle at time t.

On the other hand, if we fix a point $\mathbf{x}_0$ in D, then by Theorem 2.1, the equation

$$U(\mathbf{x}) = -\int_{\mathbf{x}_0}^{\mathbf{x}} \mathbf{F} \cdot d\mathbf{x}$$

defines a continuously differentiable function U in D. Using independence of path to integrate from $\mathbf{x}_1$ to $\mathbf{x}_2$ via $\mathbf{x}_0$, we get

$$W(\mathbf{x}_1, \mathbf{x}_2) = \int_{\mathbf{x}_1}^{\mathbf{x}_2} \mathbf{F} \cdot d\mathbf{x}$$

$$= \int_{\mathbf{x}_0}^{\mathbf{x}_2} \mathbf{F} \cdot d\mathbf{x} - \int_{\mathbf{x}_0}^{\mathbf{x}_1} \mathbf{F} \cdot d\mathbf{x} \tag{2}$$

$$= -U(\mathbf{x}_2) + U(\mathbf{x}_1).$$

Comparison of Equations (1) and (2) shows that

$$U(\mathbf{x}_2) + \frac{m}{2}v^2(t_2) = U(\mathbf{x}_1) + \frac{m}{2}v^2(t_1).$$

In other words, along the path traced by $g(t)$, the sum $U\big(g(t)\big) + T(t)$ is a constant, independent of t, called the **total energy** of the particle. For this reason, the function $U(\mathbf{x})$, which is a function of position in D, is called the **potential energy** of the field $\mathbf{F}$. Thus the potential energy is *minus* the field potential. Notice that there is an arbitrary choice made in defining the potential in that the point $\mathbf{x}_0$ was picked to have zero potential. The choice of some other point $\mathbf{x}_0$ would change the function U by at most an additive constant equal to $W(\mathbf{x}_0, \mathbf{x}_1)$. It is the constant total energy that's "conserved" and that gives rise to the term conservative field.

2B   Path Independence

For a vector field $\mathbf{F}$ defined in a region D of $\mathbb{R}^n$, **independence of path** in the line integral $\int \mathbf{F} \cdot d\mathbf{x}$ means that

2.2
$$\int_{\gamma[\mathbf{x}_1, \mathbf{x}_2]} \mathbf{F} \cdot d\mathbf{x} = \int_{\delta[\mathbf{x}_1, \mathbf{x}_2]} \mathbf{F} \cdot d\mathbf{x},$$

where $\gamma[\mathbf{x}_1, \mathbf{x}_2]$ and $\delta[\mathbf{x}_1, \mathbf{x}_2]$ are any two piecewise smooth curves in D having initial point $\mathbf{x}_1$ and terminal point $\mathbf{x}_2$. An alternative formulation of the independence property is that

2.3
$$\int_{\gamma} \mathbf{F} \cdot d\mathbf{x} = 0$$

for every piecewise smooth *closed* curve γ lying in D. The equivalence of the two properties follows from the observations that $\gamma[\mathbf{x}_1, \mathbf{x}_2]$ followed by $\delta[\mathbf{x}_1, \mathbf{x}_2]$ in reverse direction is a closed path, and that a closed path may be regarded as two paths joining $\mathbf{x}_1$ and $\mathbf{x}_2$, but traced in opposite directions.

The following theorem is a formal summary of three equivalent characteristics of gradient fields, the first of which is just our original definition.

2.4   Theorem.   Let $\mathbf{F}$ be a continuous vector field defined in a polygonally connected open set D in $\mathbb{R}^n$. Then each of the following three statements implies the others.

(a) The integral of $\mathbf{F}$ over every piecewise smooth path from $\mathbf{x}_1$ to $\mathbf{x}_2$ in D has the same value, so we can write the integral as $\int_{\mathbf{x}_1}^{\mathbf{x}_2} \mathbf{F}(\mathbf{x}) \cdot d\mathbf{x}$.

(b) The integral over every piecewise smooth closed path γ in D is zero, that is, $\int_{\gamma} \mathbf{F}(\mathbf{x}) \cdot d\mathbf{x} = 0$.

(c) There is a continuously differentiable function $f : D \to \mathbb{R}$ such that $\mathbf{F}$ is the gradient of f, that is, $\mathbf{F}(\mathbf{x}) = \nabla f(\mathbf{x})$ for all $\mathbf{x}$ in D.

Proof.   To see that (a) implies (b), let the two points $\mathbf{x}_1$ and $\mathbf{x}_2$ on the closed path γ separate γ into two paths, p from $\mathbf{x}_1$ and $\mathbf{x}_2$ and another, q, from $\mathbf{x}_2$ to $\mathbf{x}_1$. Reversing direction on the second of these paths gives another path q^- from $\mathbf{x}_1$ to $\mathbf{x}_2$. Assuming (a), we have $\int_p \mathbf{F} \cdot d\mathbf{x} = \int_{q^-} \mathbf{F} \cdot d\mathbf{x}$. Hence $\int_p \mathbf{F} \cdot d\mathbf{x} - \int_{q^-} \mathbf{F} \cdot d\mathbf{x} = 0$. Thus we obtain (b) from

$$
\int_\gamma \mathbf{F} \cdot d\mathbf{x} = \int_p \mathbf{F} \cdot d\mathbf{x} + \int_q \mathbf{F} \cdot d\mathbf{x}
$$

$$
= \int_p \mathbf{F} \cdot d\mathbf{x} - \int_{q^-} \mathbf{F} \cdot d\mathbf{x} = 0.
$$

To see that (b) implies (a), we reverse the previous argument, letting r and s be two given piecewise smooth paths from $\mathbf{x}_1$ to $\mathbf{x}_2$. Then r together with the reversed curve s^- make up a closed path γ over which $\mathbf{F}$ has integral zero. It follows that

$$
\int_r \mathbf{F} \cdot d\mathbf{x} + \int_{s^-} \mathbf{F} \cdot d\mathbf{x} = 0, \quad \text{so} \quad \int_r \mathbf{F} \cdot d\mathbf{x} - \int_s \mathbf{F} \cdot d\mathbf{x} = 0.
$$

Finally, Theorem 1.3 of Chapter 8, Section 1 states that (c) implies (a), while Theorem 2.1 of the present section states that (a) implies (c). It then follows from the previous implications that (b) and (c) imply each other. ∎

2C   Derivative Criterion

A more intrinsic criterion for deciding whether a continuous vector field is a gradient field arises as follows. Suppose first that $\mathbb{R}^2 \xrightarrow{\ \mathbf{F}\ } \mathbb{R}^2$ is continuous on an open set D, and that $\mathbf{F}$ is a gradient field, that is, there is a real-valued function f defined on D such that $\nabla f = \mathbf{F}$. In terms of coordinate functions F_1 and F_2 of $\mathbf{F}$, this means that

$$
\frac{\partial f}{\partial x_1} = F_1 \quad \text{and} \quad \frac{\partial f}{\partial x_2} = F_2.
$$

If $\mathbf{F}$ itself is continuously differentiable, we can form the second partials,

$$
\frac{\partial^2 f}{\partial x_2 \partial x_1} = \frac{\partial F_1}{\partial x_2} \quad \text{and} \quad \frac{\partial^2 f}{\partial x_1 \partial x_2} = \frac{\partial F_2}{\partial x_1},
$$

and conclude from their equality that

$$
\frac{\partial F_1}{\partial x_2} = \frac{\partial F_2}{\partial x_1} \tag{3}
$$

throughout D. By the definition of curl $\mathbf{F}$, Equation (3) says curl $\mathbf{F} = \mathbf{0}$. This equation has an extended consequence: We consider a more general vector field $\mathbb{R}^n \xrightarrow{\ \mathbf{F}\ } \mathbb{R}^n$, which we assume continuously differentiable in an open subset D of $\mathbb{R}^n$. If $\mathbf{F}$ is a gradient field, there is an f such that $\nabla f = \mathbf{F}$, or, in terms of coordinate functions

$$
\frac{\partial f}{\partial x_j} = F_j, \quad j = 1, \dots, n.
$$

Differentiating with respect to x_i, we get

$$\frac{\partial F_j}{\partial x_i} = \frac{\partial^2 f}{\partial x_i \partial x_j} = \frac{\partial^2 f}{\partial x_j \partial x_i} = \frac{\partial F_i}{\partial x_j}. \tag{4}$$

The functions $\partial F_i/\partial x_j$ are the entries in the n-by-n Jacobian matrix of $\mathbb{R}^n \xrightarrow{\mathbf{F}} \mathbb{R}^n$, and Equation (4) expresses its **symmetry**, which means that $\mathbf{F}'$ equals its transpose across its main diagonal.

2.5 Theorem. If $\mathbb{R}^n \xrightarrow{\mathbf{F}} \mathbb{R}^n$ is a continuously differentiable gradient field, then $\mathbf{F}'$, the Jacobian matrix of $\mathbf{F}$, is symmetric.

EXAMPLE 2

The converse of Theorem 2.5 is false, as we see by looking at an example in $\mathbb{R}^2$. The vector field

$$\mathbf{F}(x, y) = \left(\frac{-y}{x^2 + y^2}, \frac{x}{x^2 + y^2} \right)$$

is defined for all $(x, y) \neq (0, 0)$. You can check that $\partial F_1/\partial y = \partial F_2/\partial x$, but there is no continuously differentiable f such that $\nabla f(x, y) = \mathbf{F}(x, y)$ for all $(x, y) \neq (0, 0)$. The underlying reason is that for $x > 0$ the function $f(x, y) = \arctan(y/x)$ satisfies $\nabla f = \mathbf{F}$, but this f cannot be extended to be a single-valued solution of the equation in the entire plane with the origin deleted. (See Exercise 13 of the previous section.) If $f(x, y)$ could be so extended to a function $g(x, y)$, we would have

$$\oint_c \mathbf{F}(\mathbf{x}) \cdot d\mathbf{x} = \oint_c \nabla g(\mathbf{x}) \cdot d\mathbf{x} = g(1, 0) - g(1, 0) = 0,$$

where c lies on $x^2 + y^2 = 1$, starting and ending at $(1, 0)$. This is impossible, since explicit calculation along c, traced once counterclockwise with $(x, y) = (\cos t, \sin t)$, shows

$$\oint_c \mathbf{F}(\mathbf{x}) \cdot d\mathbf{x} = \oint_c -y\,dx + x\,dy = \int_0^{2\pi} (\sin^2 t + \cos^2 t)\,dt = 2\pi \neq 0.$$

Example 2 shows that the nature of the region D on which $\mathbf{F}$ is defined is significant in determining whether $\mathbf{F}$ is a gradient field. By making a special assumption about D we can obtain a partial converse to Theorem 2.5.

2.6 Theorem. Let R be an open coordinate rectangle in $\mathbb{R}^n$, and let $\mathbf{F}$ be a continuously differentiable vector field on R. If $\mathbf{F}'(\mathbf{x})$, the Jacobian matrix of $\mathbf{F}$, satisfies $\partial F_i/\partial x_j = \partial F_j/\partial x_i$, then $\mathbf{F}$ is a gradient field.

Proof. Pick a fixed point $\mathbf{x}_0$ in R and let $\mathbf{x}$ be any other point of R. We consider paths from $\mathbf{x}_0$ to $\mathbf{x}$, each consisting of a sequence of line segments parallel to the axes and such that each coordinate variable varies on at most one such segment. Three-dimensional examples are shown in Figure 9.11. The reason for looking at

FIGURE 9.11

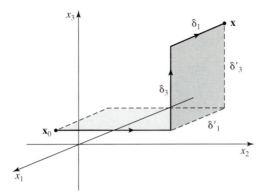

such paths is to be able to approach $\mathbf{x}$ from any coordinate direction for the purpose of taking partial derivatives at $\mathbf{x}$. Choosing one of these paths, call it $\gamma_\mathbf{x}$, define a real-valued function f by

$$f(\mathbf{x}) = \int_{\gamma_\mathbf{x}} \mathbf{F} \cdot d\mathbf{x}. \tag{5}$$

Although the particular path $\gamma_\mathbf{x}$ is only one of several of the same type, we'll see that any of the other possible choices would lead to the same value for $f(\mathbf{x})$. The reason is that any one of these paths can be altered step by step into any one of the others by changes, each of which leaves the value of the integral (5) unaltered. Each path can be described as a sequence of segments, along which only one coordinate variable varies. (For example, the dashed path in Figure 9.11 corresponds to x_1, x_2, x_3, and the solid one to x_2, x_3, x_1.) We can change one such sequence into another by successively interchanging adjacent variables in pairs until the desired order is reached. But each interchange replaces a pair of segments (δ_i, δ_j) by another pair (δ_i', δ_j') lying in the same 2-dimensional plane. To see that the replacement leaves the value of the integral invariant, we form the circuit δ consisting of the segments δ_i and δ_j, followed by δ_i' and δ_j' in the reverse of their original directions. On these segments, x_i and x_j are the only variables that vary, so we can write the circuit integral as

$$\oint_\delta \mathbf{F} \cdot d\mathbf{x} = \oint_\delta F_i \, dx_i + F_j \, dx_j.$$

We apply Green's Theorem to the 2-dimensional rectangle R_δ bounded by δ and get

$$\oint_\delta \mathbf{F} \cdot d\mathbf{x} = \int_{R_\delta} \left(\frac{\partial F_j}{\partial x_i} - \frac{\partial F_i}{\partial x_j} \right) dx_i \, dx_j = 0,$$

since by the symmetry assumption, $\partial F_j / \partial x_i - \partial F_i / \partial x_j = 0$ in R. Thus

$$\oint_\delta \mathbf{F} \cdot d\mathbf{x} = \int_{(\delta_i, \delta_j)} \mathbf{F} \cdot d\mathbf{x} - \int_{(\delta_j', \delta_i')} \mathbf{F} \cdot d\mathbf{x} = 0,$$

and so the change of path leaves the value of the integral invariant.   ∎

Once it has been established that $\mathbf{x}$ can be approached along a path of integration that varies only in an arbitrary coordinate, say the kth, we have, as in the proof of Theorem 2.1, the equation $\partial f/\partial x_k(\mathbf{x}) = F_k(\mathbf{x})$, for all k. Thus $\nabla f(\mathbf{x}) = \mathbf{F}(\mathbf{x})$ for all x in R.

EXAMPLE 3　　Applying Theorem 2.6 to the field

$$\mathbf{F}(x, y) = \left(\frac{-y}{x^2 + y^2}, \frac{x}{x^2 + y^2} \right), \quad (x, y) \neq (0, 0),$$

of Example 2, we conclude that $\mathbf{F}$, when restricted to any coordinate rectangle not containing the origin, is a gradient field. This is true, for example, in any of the four half-planes bounded by a coordinate axis. A potential function f for the half-plane $x > 0$ can be computed by the line integral

$$f(x, y) = \int_{(1,0)}^{(x,y)} \frac{-y\,dx}{x^2 + y^2} + \frac{x\,dy}{x^2 + y^2},$$

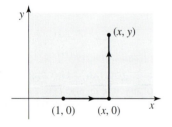

FIGURE 9.12

where the path of integration is any piecewise smooth curve from $(1, 0)$ to (x, y). A polygonal path from $(1, 0)$ to $(x, 0)$ and from $(x, 0)$ to (x, y) is shown in Figure 9.12. On the first segment, the entire integral is zero because y is identically zero, and on the second segment, with x constant, the integral reduces to

$$\int_0^y \frac{x\,dy}{x^2 + y^2} = \arctan \left(\frac{y}{x} \right).$$

The most general potential of $\mathbf{F}$ in the right half-plane differs from this one by at most a constant. (Why?) The general solution of $\nabla f = \mathbf{F}$ in the half-plane is therefore

$$f(x, y) = \arctan \frac{y}{x} + C.$$

EXAMPLE 4　　In this example we look at the vector field

$$\mathbf{F}(x, y) = \left(\frac{x}{x^2 + y^2}, \frac{y}{x^2 + y^2} \right), \quad (x, y) \neq (0, 0).$$

Like the field in Example 3 this one satisfies the hypotheses of Theorem 2.6, so we conclude from the theorem that there is a potential function f such that $\nabla f = \mathbf{F}$ on any half-plane bounded by a coordinate axis. But in contrast to Example 3, for this field there is a continuous potential function defined everywhere in $\mathbb{R}^2$ except at the origin. This information can't be obtained simply by verifying the symmetry condition, but requires checking one of the conditions of Theorem 2.4. To actually find a potential we calculate the line integral of $\mathbf{F}(x, y)$ from $(1, 0)$ to (x, y) along the polygonal path from $(1, 0)$ to $(x, 0)$ and then from $(x, 0)$ to (x, y) as in Example 3. After some cancellation the result is

$$f(x, y) = \int_0^y \frac{y\,dy}{x^2 + y^2} = \tfrac{1}{2}\ln(x^2 + y^2), \quad (x, y) \neq (0, 0).$$

Thus we have a potential function $f(x, y)$ for the vector field $\mathbf{F}(x, y)$, usually called the **logarithmic potential**, valid everywhere in the plane except at the origin.

2D   Indefinite Integration

Given a vector field $\mathbb{R}^n \xrightarrow{\mathbf{F}} \mathbb{R}^n$, finding a real-valued function f such that $\nabla f = \mathbf{F}$ amounts to solving for the function F in the system of partial differential equations

$$\frac{\partial f}{\partial x_1} = F_1, \ \ \frac{\partial f}{\partial x_2} = F_2, \dots, \ \frac{\partial f}{\partial x_n} = F_n,$$

where $F_1, F_2, \dots, F_n$ are the given coordinate functions of $\mathbf{F}$. It's sometimes simpler to avoid working with definite integrals along explicit paths of integration as in the previous example and instead use indefinite integrals. Assuming equality of mixed partials, the **consistency conditions** $\dfrac{\partial F_i}{\partial x_j} = \dfrac{\partial F_j}{\partial x_i}$ must be satisfied, so it usually makes sense to verify them before proceeding further; failure of even one of these equations means that there is no solution f to the system of equations.

EXAMPLE 5

Suppose $\mathbf{F}(x, y) = (y^2, 2xy + 1)$, and we're looking for a function $\mathbb{R}^2 \xrightarrow{f} \mathbb{R}$ such that $f_x(x, y) = y^2$ and $f_y(x, y) = 2xy + 1$. Since $\partial(y^2)/\partial y = \partial(2xy)/\partial x = 2y$ for all (x, y), Theorem 2.6 guarantees that the desired function is defined for all (x, y). To find f, start for example with $f_x(x, y) = y^2$ and integrate with respect to x while holding y fixed. We get

$$\int f_x(x, y)\, dx = f(x, y) = \int y^2\, dx = xy^2 + C(y).$$

It's important in principle, and often in practice, to allow the "constant" of integration $C(y)$ to depend on the temporarily fixed, but arbitrary, value y. Now apply $\partial/\partial y$ to this partly determined expression for $f(x, y)$ and compare the result with the given expression $f_y(x, y)$. We find

$$\frac{\partial\left(xy^2 + C(y)\right)}{\partial y} = 2xy + C'(y) = 2xy + 1.$$

Canceling $2xy$, we see that we need to have $C'(y) = 1$, so $C(y) = y + c$ where c is a real constant. (It's only at this point of the process that we have concrete evidence for the existence of solutions; they are $f(x, y) = xy^2 + y + c$, now directly verifiable as solutions.) As a final payoff, we see that a line integral of $\mathbf{F}$ from, for example $\mathbf{a} = (1, 1)$ to $\mathbf{b} = (2, 3)$, is really a line integral of a gradient field. We can choose the constant $c = 0$. Hence $\int_a^b \mathbf{F} \cdot d\mathbf{x} = f(2, 3) - f(1, 1) = 21 - 2 = 19$.

Theorem 3.4 of Chapter 5 guarantees that any two solutions f to the equation $\nabla f = \mathbf{F}$ differ by at most a constant, so we know we have the most general solution in the previous example. Note also that the "partial integrations" in the example are essentially integrations along paths parallel to the axes.

EXAMPLE 6

Let

$$\mathbf{F}(x, y, z) = \left(F_1(x, y, z), F_2(x, y, z), F_3(x, y, z)\right) = e^{y+z}(y, x(y+1), xy).$$

We want to solve

$$f_x(x, y, z) = ye^{y+z}, \; f_y(x, y, z) = x(y+1)e^{y+z}, \; f_z(x, y, z) = xye^{y+z},$$

that is, find f such that $\nabla f = (F_1, F_2, F_3)$. The three consistency conditions, for example, $\partial F_1 / \partial y = (y+1)e^{y+z} = \partial F_2 / \partial x$, all hold, so we go ahead and integrate, choosing to start with the first equation $f_x(x, y, z) = (y+1)e^{y+z}$. We find

$$f(x, y, z) = \int f_x(x, y, z)\, dx = \int ye^{y+z}\, dx = xye^{y+z} + C(y, z).$$

The constant of integration may depend on the two variables not involved in the integration. Now apply $\partial/\partial z$ to this last expression for f to get $f_z(x, y, z) = xye^{y+z} + C_z(y, z)$. The third equation, $f_z(x, y, z) = xye^{y+z}$, of our given system shows by comparison that $C_z(y, z) = 0$. This says $C(y, z) = C(y)$ is independent of z, so $f(x, y, z) = xye^{y+z} + C(y)$. Now compute $f_y(x, y, z) = x(y+1)e^{y+z} + C'(y)$; comparison with the second equation of the system shows that $C'(y) = 0$, so $C(y)$ is constant. Thus $f(x, y, z) = xye^{y+z} + c$.

EXERCISES

1. Consider the approximation to the earth's gravitational field acting on a particle of mass 1 represented by the vector field $\mathbf{F}(x, y, z) = (0, 0, -g)$.
   (a) Find for $\mathbf{F}$ the potential energy function $U(x, y, z)$ that is zero when $(x, y, z) = (0, 0, 0)$.
   (b) If a particle of mass 1 has at $(0, 0, 0)$ a velocity vector (v_1, v_2, v_3) with $v_3 > 0$, and no force but $\mathbf{F}$ acts on the particle, find the path of the particle.
   (c) Verify that the sum of potential energy and kinetic energy remains constant for the path of part (b).

2. Show that if $\mathbf{F}$ and $\mathbf{G}$ are gradient fields defined on the same domain D, then $\mathbf{F} + \mathbf{G}$ and $c\mathbf{F}$ are gradient fields, where c is a constant.

In Exercises 3 to 6, use Theorem 2.4, Theorem 2.5, or 2.6 to decide whether the vector field is a gradient field.

3. $\mathbf{F}(x, y) = (x - y, x + y)$, for (x, y) in $\mathbb{R}^2$

4. $\mathbf{G}(x, y, z) = (y, z, x)$, for (x, y, z) in $\mathbb{R}^3$

5. $\mathbf{H}(x, y) = \left(\dfrac{-y}{x^2 + y^2}, \dfrac{x}{x^2 + y^2} \right)$, $(x, y) \neq (0, 0)$

6. $\mathbf{K}(x, y) = \left(\dfrac{x}{x^2 + y^2}, \dfrac{y}{x^2 + y^2} \right)$, $(x, y) \neq (0, 0)$

7. Use Theorem 2.5 to show that the vector fields in Exercises 3 and 4 are not gradient fields in any open subset at all of $\mathbb{R}^2$ or $\mathbb{R}^3$ respectively.

8. Show that the vector field $\mathbf{H}$ of Exercise 5 is a gradient field in the region $y > 0$ of $\mathbb{R}^2$ and find an explicit representation for its potential.

9. Consider the vector field defined in $\mathbb{R}^3$, with the z-axis deleted, by

$$\mathbf{F}(x, y, z) = \left(\dfrac{-y}{x^2 + y^2}, \dfrac{x}{x^2 + y^2}, 0 \right).$$

Is $\mathbf{F}$ a gradient field?

In Exercises 10 to 13, find a field potential for the given field.

10. $\mathbf{F}(x, y, z) = (2xy, x^2 + z^2, 2yz)$

11. $\mathbf{G}(x, y) = (y \cos xy, x \cos xy)$

12. $\mathbf{H}(x, y) = \left(\dfrac{-y}{x^2 + y^2}, \dfrac{x}{x^2 + y^2} \right)$, $(x, y) \neq (0, 0)$

13. $\mathbf{K}(x, y) = \left(\dfrac{x}{x^2 + y^2}, \dfrac{y}{x^2 + y^2} \right)$, $(x, y) \neq (0, 0)$

14. Consider the vector field $\mathbf{F}$ which is the gradient of the **Newtonian potential** $f(\mathbf{x}) = -|\mathbf{x}|^{-1}$ for nonzero $\mathbf{x}$ in $\mathbb{R}^3$. Find the work done in moving a particle from $(1, 1, 1)$ to $(-2, -2, -2)$ along a smooth curve lying in the domain of $\mathbf{F}$.

15. Give a detailed proof of the equivalence of Relations 2.2 and 2.3 of the text.

16. In $\mathbb{R}^n$, how many paths can there be from $\mathbf{x}_0$ to $\mathbf{x}$ of the special kind described in the proof of Theorem 2.6?

17. Apply the method of indefinite integration to find a potential of the field
$$\mathbf{F}(x, y) = \left(2x/(x^2 + y^2), 2y/(x^2 + y^2) \right).$$

18. Redo Example 5 of the text by first integrating the equation $f_y(x, y) = 2xy + 1$ with respect to y.

In Exercises 19 to 22, find a potential f following the method of Examples 5 and 6 of the text.

19. Find f, if $\nabla f(x, y) = (e^y, xe^y)$.

20. Find f, if $\nabla f(x, y) = (y^2 + 2xy, 2xy + x^2)$.

21. Find f, if $\nabla f(x, y, z) = (y + z, z + x, x + y)$.

22. Find f, if $\nabla f(x, y, z) = (yz + z, xz, xy + x)$.

23. Find the function f of Exercise 20 by direct computation of a line integral of ∇f from $(0, 0)$ to (x, y).

24. Find the function f of Exercise 21 by direct computation of a line integral from $(0, 0, 0)$ to (x, y, z).

25. Verify that the line integral of the field $\mathbf{F}(x, y) = 2x/(x^2 + y^2)\mathbf{i} + 2y/(x^2 + y^2)\mathbf{j}$ is zero around every closed path that avoids the origin.

26. Suppose $\mathbf{x} = \mathbf{x}(t)$ satisfies the vector equation $\ddot{\mathbf{x}} + \nabla U(\mathbf{x}) = \mathbf{0}$ on a t-interval I.

   (a) Show that the scalar equation $\ddot{\mathbf{x}} \cdot \dot{\mathbf{x}} + \nabla U(\mathbf{x}) \cdot \dot{\mathbf{x}} = 0$ holds on I.

   (b) Show that $d|\dot{\mathbf{x}}|^2/dt = 2(\ddot{\mathbf{x}} \cdot \dot{\mathbf{x}})$.

   (c) Apply part (b) to part (a) and integrate to show that $\frac{1}{2}|\dot{\mathbf{x}}|^2 + U(\mathbf{x}) = C$, where C is constant. This equation is called a **first integral** of the vector differ-

ential equation, because the order of differentiation has been reduced from two to one.

27. A particle of unit mass moves with constant angular velocity ω on a circle of radius a about the origin in $\mathbb{R}^2$. The **centripetal force field** that constrains the position $\mathbf{x}$ of the particle to remain on its circular path is $\mathbf{F}(\mathbf{x}) = -a\omega^2\mathbf{x}$.

   (a) Show that $f(\mathbf{x}) = -\frac{1}{2}a\omega^2|\mathbf{x}|^2$ is a field potential for $\mathbf{F}$, that is, show that $\nabla f(\mathbf{x}) = \mathbf{F}(\mathbf{x})$.

   (b) Show that the total work done by the field during one complete traversal of the circle is equal to zero.

   (c) Compute the work done by the field if, instead of traversing the circle, the particle moves along a smooth path from a circle of radius a to a circle of radius $b > a$.

28. The location $\mathbf{x}$ of a single satellite relative to a fixed earth at the origin in $\mathbb{R}^2$ is governed by the force field $\mathbf{F}(\mathbf{x}) = -k|\mathbf{x}|^{-3}\mathbf{x}$, where k is a positive constant.

   (a) Show that $f(\mathbf{x}) = k/|\mathbf{x}|$ is a field potential for $\mathbf{F}$, that is, show that $\nabla f(\mathbf{x}) = \mathbf{F}(\mathbf{x})$.

   (b) Show that the total work done by the field during one complete satellite orbit is equal to zero.

   (c) Compute the work done by the field if, instead of moving on a circular orbit of radius a, the satellite moves along a smooth path to a circular orbit of radius $b > a$.

SECTION 3   SURFACE INTEGRALS

3A   Normal Vectors

In Chapter 8, we defined integrals both of a real-valued function and of a vector field over a smooth curve. Defining an integral over a surface S leads to a different geometric situation with a close analogy with the line integral. As described in Chapter 4, Section 4, we assume as our primary representation for a surface S a parametrization by a continuously differentiable function $\mathbb{R}^2 \overset{g}{\to} \mathbb{R}^3$. We'll write g as

$$g(u, v) = \begin{pmatrix} g_1(u, v) \\ g_2(u, v) \\ g_3(u, v) \end{pmatrix}, \tag{1}$$

with $\mathbf{u} = (u, v)$ in the interior of some set D in $\mathbb{R}^2$, which we assume bounded by finitely many smooth curves. We further assume that, at each point $g(u, v)$ of S, the tangent vectors defined by the vector partial derivatives

$$\frac{\partial g}{\partial u}(u, v), \quad \frac{\partial g}{\partial v}(u, v)$$

determine a 2-dimensional tangent plane to S; in other words, that the two tangents are linearly independent. If S satisfies all these conditions, we'll refer to it as a piece of **smooth surface**.

FIGURE 9.13

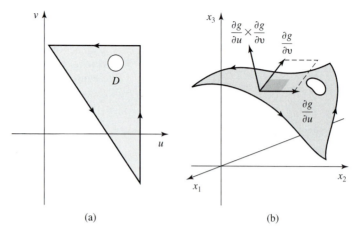

(a)   (b)

On a smooth curve, the choice of a parametrization going from one endpoint to the other establishes an orientation for the curve. Analogously, on a piece of smooth surface, a one-to-one parametrization determines a **standard normal vector**

$$\frac{\partial g}{\partial u}(u, v) \times \frac{\partial g}{\partial v}(u, v) \tag{2}$$

pointing out of one side or the other of the surface at $g(u, v)$. See Figure 9.13(b).

3B   Area and Mass

We recall that the length of the cross product of two vectors $\mathbf{a}$ and $\mathbf{b}$ is the area of the parallelogram spanned by $\mathbf{a}$ and $\mathbf{b}$. In particular,

$$\left| \frac{\partial g}{\partial u}(u, v) \times \frac{\partial g}{\partial v}(u, v) \right|$$

represents the area of the outlined tangent parallelogram shown in Figure 9.13(b). If we think of scaling down such parallelograms by factors du and dv at the points $g(u_k, v_k)$ corresponding to the corner points (u_k, v_k) of a grid over D, then it's natural to define the **area** of S by integrating the parallelogram area over D:

3.1
$$\sigma(S) = \int_D \left| \frac{\partial g}{\partial u}(u, v) \times \frac{\partial g}{\partial v}(u, v) \right| \, du \, dv.$$

We assume that $\mathbb{R}^2 \overset{g}{\to} \mathbb{R}^3$ is one-to-one so that each part of the image surface is covered just once. The integral over D will exist as a finite Riemann integral because g is assumed to be continuously differentiable. The expression

$$d\sigma = |g_u(u, v) \times g_v(u, v)| \, du \, dv$$

is called the **area element differential** for the parametrized surface S. In addition, if $\mu(\mathbf{x})$ is a continuous scalar-valued function defined for $\mathbf{x}$ on S, then

3.2
$$\int_D \mu \, d\sigma = \int_D \mu\big(g(u, v)\big) \left| \frac{\partial g}{\partial u}(u, v) \times \frac{\partial g}{\partial v}(u, v) \right| \, du \, dv$$

exists. If $\mu(\mathbf{x}) \geq 0$, then Equation 2 defines the **total mass** due to the **density** μ.

FIGURE 9.14

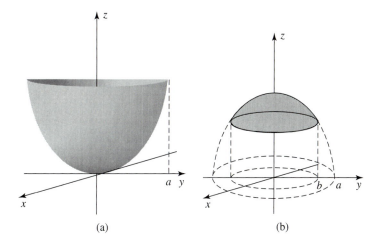

(a)   (b)

EXAMPLE 1

Let S be parametrized by

$$g(u, v) = \begin{pmatrix} u \\ v \\ u^2 + v^2 \end{pmatrix}, \quad u^2 + v^2 \le a^2;$$

thus S is actually the graph of $z = x^2 + y^2$ for $x^2 + y^2 \le a^2$. The surface is shown in Figure 9.14(a). Then

$$\frac{\partial g}{\partial u}(u, v) = \begin{pmatrix} 1 \\ 0 \\ 2u \end{pmatrix}, \quad \frac{\partial g}{\partial v}(u, v) = \begin{pmatrix} 0 \\ 1 \\ 2v \end{pmatrix}.$$

We have

$$\frac{\partial g}{\partial u}(u, v) \times \frac{\partial g}{\partial v}(u, v) = \left(\begin{vmatrix} 0 & 1 \\ 2u & 2v \end{vmatrix}, \begin{vmatrix} 2u & 2v \\ 1 & 0 \end{vmatrix}, \begin{vmatrix} 1 & 0 \\ 0 & 1 \end{vmatrix} \right).$$

The length of this vector is

$$|(-2u, -2v, 1)| = \sqrt{4u^2 + 4v^2 + 1}.$$

Changing to polar coordinates gives

$$\sigma(S) = \int_{u^2 + v^2 \le a^2} \sqrt{4u^2 + 4v^2 + 1} \, du \, dv$$

$$= \int_0^{2\pi} d\theta \int_0^a \sqrt{4r^2 + 1} \, r \, dr$$

$$= 2\pi \left[\frac{1}{12}(4r^2 + 1)^{3/2} \right]_0^a = \frac{\pi}{6}((4a^2 + 1)^{3/2} - 1).$$

The surface in the previous example can be thought of as a piece of the graph of an equation $z = f(x, y)$, where in the example $f(x, y) = x^2 + y^2$, with the domain of integration the disk $x^2 + y^2 \leq a^2$. In general, such a graph can be parametrized by a function $\mathbb{R}^2 \xrightarrow{g} \mathbb{R}^3$ of the form

$$g(x, y) = \begin{pmatrix} x \\ y \\ f(x, y) \end{pmatrix}, \quad (x, y) \text{ in } D.$$

Then $g_x(x, y) \times g_y(x, y) = (-f_x(x, y), -f_y(x, y), 1)$ so the area differential becomes

3.3 $d\sigma = \sqrt{f_x^2(x, y) + f_y^2(x, y) + 1}\, dx\, dy = \sqrt{|\nabla f(x, y)|^2 + 1}\, dx\, dy.$

Using this approach in the previous example would have led to the same integral that we computed before for the surface area, namely

$$\sigma(S) = \int_{x^2+y^2 \leq a^2} \sqrt{4x^2 + 4y^2 + 1}\, dx\, dy.$$

EXAMPLE 2

To compute the area of a sphere S_a of radius a using Equation 3.3, we start with the hemispherical graph H_a of $z = \sqrt{a^2 - x^2 - y^2}$ over the disk $x^2 + y^2 < a^2$. See Figure 9.14(b). Then we have $f_x(x, y) = -x(a^2 - x^2 - y^2)^{-1/2}$ and $f_y(x, y) = -y(a^2 - x^2 - y^2)^{-1/2}$, so Equation 3.3 becomes

$$\sigma(H_a) = \int_{x^2+y^2<a^2} \sqrt{x^2/(a^2 - x^2 - y^2) + y^2/(a^2 - x^2 - y^2) + 1}\, dx\, dy$$

$$= \int_{x^2+y^2<a^2} a/\sqrt{a^2 - x^2 - y^2}\, dx\, dy.$$

This is an improper integral, because the integrand tends to infinity as (x, y) tends from within the disk to an arbitrary point on the boundary. To compute the integral, integrate first over a smaller disk of radius b, and then let $b \to a$. This can be done by changing to polar coordinates and then letting $b \to a$. We get

$$\sigma(H_a) = \lim_{b \to a} \int_{x^2+y^2<b^2} a/\sqrt{a^2 - x^2 - y^2}\, dx\, dy$$

$$= \lim_{b \to a} a \int_0^{2\pi} d\theta \int_0^b r/\sqrt{a^2 - r^2}\, dr$$

$$= 2\pi a \lim_{b \to a} [-(a^2 - r^2)^{1/2}]_0^b$$

$$= 2\pi a \lim_{b \to a} [-(a^2 - b^2)^{1/2} + a] = 2\pi a^2.$$

Hence $\sigma(S_a) = 2\sigma(H_a) = 4\pi a^2$, the formula for the area of a sphere of radius a.

EXAMPLE 3

We parametrize a complete turn of a **helicoid** surface of width a by

$$g(u, v) = \begin{pmatrix} u \cos v \\ u \sin v \\ v \end{pmatrix}, \quad 0 \le u \le a, 0 \le v \le 2\pi.$$

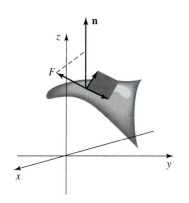

Suppose the helicoid is weighted with density $\mu(x, y, z) = \sqrt{x^2 + y^2}$ at (x, y, z). In other words, the density at a point of the helicoid equals its distance from the central axis of the surface, which is just u in terms of the parameter pair (u, v). To find the total mass distributed this way, we compute

$$g_u(u, v) \times g_v(u, v) = \begin{pmatrix} \cos v \\ \sin v \\ 0 \end{pmatrix} \times \begin{pmatrix} -u \sin v \\ u \cos v \\ 1 \end{pmatrix} = \begin{pmatrix} \sin v \\ -\cos v \\ u \end{pmatrix}.$$

FIGURE 9.15

The weighted area differential is thus

$$\mu \, d\sigma = u |g_u(u, v) \times g_v(u, v)| \, du \, dv$$

$$= u |(\sin v, -\cos v, u)| \, du \, dv = u\sqrt{1 + u^2} \, du \, dv.$$

The total mass is

$$M_a = \int_0^{2\pi} dv \int_0^a u\sqrt{1 + u^2} \, du$$

$$= 2\pi \left[\tfrac{1}{3}(1 + u^2)^{3/2} \right]_0^a = (2\pi/3)[(1 + a^2)^{3/2} - 1].$$

3C   Integrating Vector Fields

The main purpose of this section is the definition of the integral of a continuous vector field $\mathbb{R}^3 \xrightarrow{\mathbf{F}} \mathbb{R}^3$ over a surface S. Continuing with the assumption that S is a piece of smooth surface represented by the function of Equation (1), we compare the standard normal vector $\partial g/\partial u \times \partial g/\partial v$ with the vector field $\mathbf{F}$ at a point $g(u, v)$ of S. These are shown in Figure 9.15 at one point. If $\mathbf{n}$ is a *unit* normal to S at $g(u, v)$, then the dot product $\mathbf{F} \cdot \mathbf{n}$ at $g(u, v)$ is the coordinate of $\mathbf{F}$ in the direction of $\mathbf{n}$. But since

$$\mathbf{n} = \frac{\dfrac{\partial g}{\partial u} \times \dfrac{\partial g}{\partial v}}{\left| \dfrac{\partial g}{\partial u} \times \dfrac{\partial g}{\partial v} \right|},$$

it follows that

$$\mathbf{F}\big(g(u, v)\big) \bullet \left(\frac{\partial g}{\partial u}(u, v) \times \frac{\partial g}{\partial v}(u, v) \right)$$

is equal to the coordinate of $\mathbf{F}\big(g(u, v)\big)$ in the direction of $\mathbf{n}$, multiplied by the area of the tangent parallelogram spanned by $\partial g/\partial u$ and $\partial g/\partial v$ at $g(u, v)$. We define the **surface integral** of $\mathbf{F}$ over S by

3.4
$$\int_D \mathbf{F}\big(g(u, v)\big) \cdot \left(\frac{\partial g}{\partial u}(u, v) \times \frac{\partial g}{\partial v}(u, v)\right) du\, dv,$$

and denote it by $\int_S \mathbf{F} \cdot d\mathbf{S}$ or $\int_S \mathbf{F} \cdot \mathbf{n}\, d\sigma$.

Suppose that a continuous vector field $\mathbb{R}^3 \xrightarrow{\mathbf{F}} \mathbb{R}^3$ describes the speed and direction of a fluid flow at each point of a region R in which it's defined. We'll define, using a surface integral, the **flux**, or rate of flow across a piece of smooth surface S lying in the region R. If S is perfectly flat and $\mathbf{F}$ is a constant field, then the *flux* is equal to $F_{\mathbf{n}}\sigma(S)$, where $F_{\mathbf{n}}$ is the coordinate $\mathbf{F} \cdot \mathbf{n}$ of $\mathbf{F}$ in the direction of a unit normal $\mathbf{n}$ to S. Thus, for a flat S, the flux is equal to the volume of the tube of fluid illustrated in Figure 9.16, which shows the amount of fluid passing through its base in one time unit. Because $F_{\mathbf{n}} = \mathbf{F} \cdot \mathbf{n}$, we define, for a flat S with area $\sigma(S)$, the **flux** to be the rate of flow of $\mathbf{F}$ across S given by the formula

$$\Phi(\mathbf{F}, S) = \mathbf{F} \cdot \mathbf{n}\, \sigma(S).$$

If S is a piece of smooth surface in a region R, we partition S along coordinate curves of the form $u = $ constant and $v = $ constant and assume that, within each part of S so formed, the field $\mathbf{F}$ is constant. Approximating S by tangent parallelograms S_k having for adjacent edges the vectors $\Delta u\, g_u(\mathbf{u}_k)$ and $\Delta v\, g_v(\mathbf{u}_k)$ gives S_k the area

$$\sigma(S_k) = |g_u(\mathbf{u}_k) \times g_v(\mathbf{u}_k)|\, \Delta u\, \Delta v.$$

See Figure 9.17. The approximate flux across a typical subdivision S_k of S will have the form

$$\Phi_k = \mathbf{F}\big(g(\mathbf{u}_k)\big) \cdot \mathbf{n}_k\, \sigma(S_k)$$
$$= \mathbf{F}\big(g(\mathbf{u}_k)\big) \cdot \left(\frac{\partial g}{\partial u}(\mathbf{u}_k) \times \frac{\partial g}{\partial v}(\mathbf{u}_k)\right) \Delta u\, \Delta v,$$

since the length $|g_u(\mathbf{u}_k) \times g_v(\mathbf{u}_k)|$ cancels from the area $\sigma(S_k)$ and the denominator of $\mathbf{n}_k$. The sum

$$\sum_{k=1}^N \Phi_k = \sum_{k=1}^N \mathbf{F}\big(g(\mathbf{u}_k)\big) \cdot \left(\frac{\partial g}{\partial u}(\mathbf{u}_k) \times \frac{\partial g}{\partial v}(\mathbf{u}_k)\right) \Delta u\, \Delta v$$

becomes a better approximation to what we would like to call the flux of $\mathbf{F}$ across S as the subdivision of S is refined by making the corresponding grid G finer in the parameter domain D. On the other hand, if $\mathbf{F}$ is continuous on S and g is continuously differentiable on D, then

$$\lim_{m(G)\to 0} \sum_{k=1}^N \Phi_k = \int_D \mathbf{F}\big(g(\mathbf{u})\big) \cdot \left(\frac{\partial g}{\partial u}(\mathbf{u}) \times \frac{\partial g}{\partial v}(\mathbf{u})\right) du\, dv$$
$$= \int_S \mathbf{F} \cdot d\mathbf{S},$$

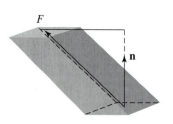

FIGURE 9.16

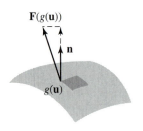

FIGURE 9.17

which is the previously defined integral of $\mathbf{F}$ over S. Consequently, we define the **flux** of $\mathbf{F}$ across S to be the rate of flow given by

$$\Phi(\mathbf{F}, S) = \int_S \mathbf{F} \cdot d\mathbf{S} = \int_S \mathbf{F} \cdot \mathbf{n}\, d\sigma.$$

We remark that the sign of Φ would change if S were reparametrized so that the unit normal vector $\mathbf{n}$ determined by the parametrization pointed in the opposite direction.

EXAMPLE 4

Suppose the vector field $\mathbf{F}(\mathbf{x}) = (F_1(\mathbf{x}), F_2(\mathbf{x}), F_3(\mathbf{x}))$ is tangent to the surface S at every point $\mathbf{x}$ of S. Then $\mathbf{F} \cdot \mathbf{n} = 0$ at every point of S, so the surface integral $\int_S \mathbf{F} \cdot \mathbf{n}\, d\sigma$ is zero. At the other extreme, if the continuous vector field $\mathbf{F}(\mathbf{x})$ is perpendicular to S at every point $\mathbf{x}$ of S we expect that the integral of $\mathbf{F}$ over S will be different from zero. For example, if F coincides with the standard unit normal vector $\mathbf{n}$ at each point of S, then $\int_S \mathbf{F} \cdot \mathbf{n}\, d\sigma = \int_S \mathbf{n} \cdot \mathbf{n}\, d\sigma = \int_S d\sigma$, which is just the area of S.

The motivation for the definition of flux given previously is stated in terms of the velocity field of a fluid flow, because that is the physical setting for flux measurements across surfaces that we most easily visualize. However some of the most important applications of surface integrals concern the flux of the more abstract fields: gravitational, electric, and magnetic. The next example is fundamental for these areas.

EXAMPLE 5

A body of mass M concentrated at the origin in $\mathbb{R}^3$ generates a gravitational field at $\mathbf{x}$ that attracts a body of mass m toward the origin with force

$$\mathbf{F}(\mathbf{x}) = -\frac{GMm}{|\mathbf{x}|^3}\mathbf{x},$$

where G is the universal gravitational constant. Note that the magnitude of the field at $\mathbf{x}$ is $|\mathbf{F}(\mathbf{x})| = GMm|\mathbf{x}|^{-2}$; in other words, this is an inverse-square law of attraction. To compute the flux of this field across a sphere S_a of radius a centered at the origin, we make the simplifying observation that if $\mathbf{x}$ is on S_a, then we have $\mathbf{F}(\mathbf{x}) = -GMma^{-2}\mathbf{n}$, where $\mathbf{n}$ is an outward-pointing unit vector directed from $\mathbf{x}$ away from the origin. Since $\mathbf{n} \cdot \mathbf{n} = 1$, the flux of the field is

$$\Phi = \int_{S_a} \mathbf{F}(\mathbf{x}) \cdot \mathbf{n}\, d\sigma = -(GMm/a^2)\int_{S_a} \mathbf{n} \cdot \mathbf{n}\, d\sigma = -(GMm/a^2)\int_{S_a} d\sigma.$$

This last integral is the area of the sphere S_a, namely $4\pi a^2$, so flux $\Phi = -4\pi GMm$. The most significant feature of this result is that Φ is independent of a, the radius of the sphere. We'll use Gauss's Theorem in the next section to show that the same phenomenon holds for closed surfaces other than spheres.

The coordinates of the standard normal vector to a surface parametrized by $\mathbf{x}(u, v) = g(u, v)$ can be written in terms of 2-by-2 Jacobian determinants as follows.

Let $\mathbf{x} = (x, y, z)$. Then the coordinates of the cross-product $\partial g / \partial u \times \partial g / \partial v$ have the form

$$\frac{\partial(y, z)}{\partial(u, v)} = \begin{vmatrix} y_u & y_v \\ z_u & z_v \end{vmatrix}, \qquad \frac{\partial(z, x)}{\partial(u, v)} = \begin{vmatrix} z_u & z_v \\ x_u & x_v \end{vmatrix}, \qquad \frac{\partial(x, y)}{\partial(u, v)} = \begin{vmatrix} x_u & x_v \\ y_u & y_v \end{vmatrix}.$$

Thus we can write general surface integrals of a vector field $\mathbf{F} = (F_1, F_2, F_3)$ in either of the successively more abbreviated forms

$$\int_S \mathbf{F} \cdot dS = \int_D \left(F_1 \frac{\partial(y, z)}{\partial(u, v)} + F_2 \frac{\partial(z, x)}{\partial(u, v)} + F_3 \frac{\partial(x, y)}{\partial(u, v)} \right) du \, dv$$

$$= \int_S (F_1 dy \, dz + F_2 dz \, dx + F_3 dx \, dy).$$

The last abbreviation is analogous to our abbreviation for a line integral:

$$\int_\gamma \mathbf{F} \cdot d\mathbf{x} = \int_\gamma (F_1 dx + F_2 dy + F_3 dz).$$

3D   Orientation

In computing a line integral over a piecewise smooth curve, it's customary to orient the smooth pieces of the curve *coherently* so that the terminal point of one piece is the same as the initial point of the one that follows it. To integrate a vector field over a piecewise smooth surface, we need a notion of orientation for pieces of smooth surfaces S. If $\mathbb{R}^2 \xrightarrow{g} \mathbb{R}^3$ represents S parametrically with g defined on D, then Figure 9.13 shows how D and S might possibly be related. The edge of S, corresponding under g to the boundary of D, we'll call the **border** of S. As a point $\mathbf{u}$ moves around the piecewise smooth boundary of D in the *counterclockwise* direction, its image $g(\mathbf{u})$ traces the border of S with what we'll call its **positive orientation**. It will be convenient later to use the notation ∂S to denote the **positively oriented border** of S.

An alternative way to describe the positive orientation is as follows. Define the **positive side** of S by saying that "positive" is the side of S out from which the normal vector $\partial g / \partial u \times \partial g / \partial v$ points. If you then walk on the positive side of S keeping S on your left as you follow the border around, you are going in its **positive direction**. See Figure 9.18(b) for a picture. The equivalence of these two notions

FIGURE 9.18

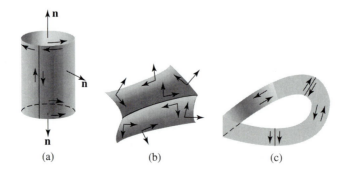

(a)   (b)   (c)

of positivity can be stated and proved as a formal theorem, but we won't attempt that here.

A **piecewise smooth surface** is defined to be a finite union of pieces of smooth surface that are joined along common border curves. Figure 9.18 shows some examples. The border curve of each piece of surface has a positive orientation that comes from some parametrization of that piece. The parametrizations of two adjacent pieces are *coherent* if they give opposite orientations to common border curves, as in parts (a) and (b) of Figure 9.18. A piecewise smooth surface is said to be **orientable** if its adjacent pieces can be parametrized coherently. The border orientation of a single piece can always be reversed to accommodate a neighbor by interchanging the roles of its two parameters, for example replacing (u, v) by (v, u) throughout. Parts (a) and (b) of Figure 9.18 show orientable surfaces. However part (c) of Figure 9.18 shows two rectangular strips joined together, one of them with a twist. This surface is not orientable, because no matter how the orientations of the pieces are changed there will be some part of the common border traced in the same direction. The resulting surface is called a **Möbius strip**.

We define the integral of a continuous vector field over a piecewise smooth surface to be the sum of the integrals over each of its smooth pieces. Thus if $S = S_1 \cup S_2$,

$$\int_S \mathbf{F} \cdot d\mathbf{S} = \int_{S_1} \mathbf{F} \cdot d\mathbf{S} + \int_{S_2} \mathbf{F} \cdot d\mathbf{S}.$$

This definition holds even if S is not orientable, but in practice it's of little interest to integrate a vector field over a nonorientable surface. On the other hand, we can compute the integral of a real-valued function over a surface without regard to orientation. The reason is that in Formulas 3.1 and 3.2 the area differential of surface area,

$$d\sigma = \left| \frac{\partial g}{\partial u} \times \frac{\partial g}{\partial v} \right| du \, dv,$$

doesn't change when the orientation is reversed by interchanging the roles of u and v. But in Formula 3.3, the **vector surface differential**,

$$d\mathbf{S} = \left(\frac{\partial g}{\partial u} \times \frac{\partial g}{\partial v} \right) du \, dv,$$

does change sign when u and v are interchanged. We observe that the surface differential $d\mathbf{S}$ can also be written in the form

$$d\mathbf{S} = \mathbf{n} \, d\sigma,$$

where $\mathbf{n}$ is a unit normal to the surface. Possible choices for this vector are considered in the following examples.

EXAMPLE 6

For a flat surface S parallel to the xy-plane, there are two possible choices for the unit normal; $\mathbf{n}$ must be either $(0, 0, 1)$ or $(0, 0, -1)$. In Figure 9.19(a) either choice would in principle be appropriate for the rectangle S_1 in the xy-plane, but having chosen one, say $\mathbf{n_1} = (0, 0, 1)$, there is only one possible choice of $\mathbf{n}$ for the

FIGURE 9.19

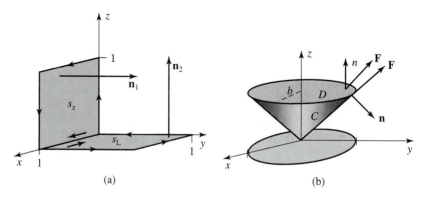

(a)   (b)

rectangle S_2 in the xz-plane that will lead to a coherent orientation. Following the conventions illustrated in Figure 9.19(a), we choose $\mathbf{n_2} = (0, 1, 0)$. (Note that with these choices the two normals point out from the same side of the two-piece surface.) To compute the total flux of the vector field $\mathbf{F}(x, y, z) = (y, z, x)$ over $S_1 \cup S_2$ with this orientation, we first note that $\mathbf{F}(x, y, z) \cdot \mathbf{n_1} = (y, z, x) \cdot (0, 0, 1) = x$ on S_1. Also on S_1 we have $d\sigma = dx\, dy$. Hence

$$\int_{S_1} \mathbf{F} \cdot \mathbf{n_1}\, d\sigma = \int_{S_1} x\, dx\, dy = \int_0^1 dy \int_0^1 x\, dx = \frac{1}{2}.$$

On S_2, we have $\mathbf{F}(x, y, z) \cdot \mathbf{n_2} = (y, z, x) \cdot (0, 1, 0) = z$, so

$$\int_{S_2} \mathbf{F} \cdot \mathbf{n_2}\, d\sigma = \int_{S_2} z\, dx\, dz = \int_0^1 dx \int_0^1 z\, dz = \frac{1}{2}.$$

The total flux across the oriented surface is then equal to $\frac{1}{2} + \frac{1}{2} = 1$.

The orientation of a surface is sometimes most naturally determined so that all normal vectors are "outward-pointing" or "inward-pointing." This way of describing coherent orientation is particularly appropriate for a closed surface that comprises the boundary of a solid 3-dimensional region, as in the next example.

EXAMPLE 7   A vector field is defined in $\mathbb{R}^3$ by $\mathbf{F}(x, y, z) = (x, y, z)$. Let C denote the part of the cone $z = \sqrt{x^2 + y^2}$ between $z = 0$ and $z = b$. We close the top of the cone with a flat disk D of radius b in the plane $z = b$. See Figure 9.19(b). If we choose $\mathbf{n}$ to be an outward-pointing unit normal at each point of C or D, this will be consistent with Figure 9.19(b). To compute the flux of $\mathbf{F}$ across C, note that on C the arrows representing $\mathbf{F}$ lie along C, and so they are perpendicular to $\mathbf{n}$. Thus $\mathbf{F} \cdot \mathbf{n} = 0$ on C, so the flux of $\mathbf{F}$ across C is zero. At a point (x, y, b) of D we have $\mathbf{F}(x, y, b) \cdot \mathbf{n} = (x, y, b) \cdot (0, 0, 1) = b$. Hence the flux of $\mathbf{F}$ out across D is $\int_D \mathbf{F} \cdot \mathbf{n}\, d\sigma = \int_D b\, dx\, dy = \pi b^3$, which is then the total flux across $C \cup D$.

EXAMPLE 8   Suppose the vector field in the previous example is replaced by the field $\mathbf{G}(x, y, z) = (x, y, 0)$, but we retain the closed surface $C \cup D$ shown in Figure 9.19(b). This time the normal vector $\mathbf{n}$ to D is perpendicular to the field, so $\int_D \mathbf{G} \cdot \mathbf{n}\, d\sigma = 0$.

The conical surface is parametrized either by $(x, y, z) = (v \cos u, v \sin u, v)$ or else by $g(x, y) = (x, y, \sqrt{x^2 + y^2})$ for (x, y) in the disk R_b defined by $x^2 + y^2 \le b^2$. We choose the latter, in which case

$$g_x \times g_y = (-x/\sqrt{x^2 + y^2}, -y/\sqrt{x^2 + y^2}, 1) \quad \text{and} \quad |g_x \times g_y| = \sqrt{2}.$$

The area element on C is $d\sigma = |g_x \times g_y| \, dx \, dy = \sqrt{2} \, dx \, dy$. Since the third coordinate of $g_x \times g_y$ is positive, this vector points in, so we must change sign to get an outward-pointing normal. However we'll compute $\mathbf{G} \cdot \mathbf{n}$ just by observing that $\mathbf{G}$ is parallel to the xy-plane and $\mathbf{n}$ is perpendicular to the lines on C. Thus the angle between $\mathbf{G}$ and $\mathbf{n}$ is $\pi/4$ at every point of C, and

$$\mathbf{G} \cdot \mathbf{n} = |\mathbf{G}| \cos\left(\frac{\pi}{4}\right) = \sqrt{x^2 + y^2}/\sqrt{2}.$$

It follows, changing the integral to polar coordinates, that

$$\int_C \mathbf{G} \cdot \mathbf{n} \, d\sigma = \frac{1}{\sqrt{2}} \int_{R_b} \sqrt{x^2 + y^2} \, dx \, dy = \frac{1}{\sqrt{2}} \int_0^{2\pi} d\theta \int_0^b r^2 \, dr = \frac{\sqrt{2}}{3} \pi b^3.$$

Thus the total flux of the field $\mathbf{G}$ out across $C \cup D$ is $\frac{\sqrt{2}}{3} \pi b^3$.

EXERCISES

1. (a) Sketch the plane triangle T in $\mathbb{R}^3$ parametrized by $g(u, v) = (2u + v, v, 3u + v)$ for $0 \le u, 0 \le v, u + v \le 1$.
   (b) Find the area of T.

2. (a) Sketch the plane elliptical region E in the part of the plane $z = 4 - x - 2y$ that lies above the disk $x^2 + y^2 \le 1$ in the xy-plane.
   (b) Find the area of E.

3. (a) Sketch the part of the graph of the hyperbolic paraboloid $z = y^2 - x^2$ that lies above the disk $x^2 + y^2 \le 1$ in the xy-plane.
   (b) Find the area of the part of the graph described in part (a).

4. Let $\mathbf{a}$ and $\mathbf{b}$ be vectors in $\mathbb{R}^3$. Let P be the part of a plane parametrized by $\mathbf{x}(u, v) = u\mathbf{a} + v\mathbf{b}$ for parameter variables (u, v) in a region R with area $A(R)$. Show that the area of P is $|\mathbf{a} \times \mathbf{b}| A(R)$.

5. Let P be the part of the graph of $z = x^2 + y$ lying vertically above the square $0 \le x \le 1, 0 \le y \le 1$ in the xy-plane.
   (a) Assume that P is weighted by density $\mu(x, y, z) = x$. Find the total mass of the weighted surface.

(b) Find the flux across P of the vector field $\mathbf{F}(x, y, z) = -x\mathbf{i} + y\mathbf{j} + z\mathbf{k}$.

6. Use the parametrization

$$g(u, v) = \begin{pmatrix} a \cos u \sin v \\ a \sin u \sin v \\ a \cos v \end{pmatrix}, \quad 0 \le u \le 2\pi, \quad 0 \le v \le \pi,$$

for a sphere of radius a to show that the area of the sphere is $4\pi a^2$.

7. A repelling electric field $\mathbf{E}(\mathbf{x}) = |\mathbf{x}|^{-3}\mathbf{x}$ has flux Φ across the sphere of radius a centered at the origin. Find Φ.

8. (a) Find the area of the spiral ramp represented parametrically by

$$g(u, v) = \begin{pmatrix} u \cos v \\ u \sin v \\ v \end{pmatrix}, \quad 0 \le u \le 1, \quad 0 \le v \le 3\pi.$$

(b) Let the surface of part (a) have a density per unit of area at each point equal to the distance of that point from the central axis of the surface. Find the total mass of the weighted surface.

9. Compute $\displaystyle\int_S \mathbf{F} \cdot d\mathbf{S}$, where

(a) $\mathbf{F}(x, y, z) = (x, y, z)$ and S is given by

$$g(u, v) = \begin{pmatrix} u - v \\ u + v \\ uv \end{pmatrix}, \qquad \begin{matrix} 0 \le u \le 1, \\ 0 \le v \le 2. \end{matrix}$$

(b) $\mathbf{F}(x, y, z) = (x^2, 0, 0)$ and S is given by

$$g(u, v) = \begin{pmatrix} u \cos v \\ u \sin v \\ v \end{pmatrix}, \qquad \begin{matrix} 0 \le u \le 1, \\ 0 \le v \le 2\pi. \end{matrix}$$

10. Find the total mass of a spherical film having density at each point equal to the linear distance of the point from a single fixed point on the sphere.

*11. Let $\mathbf{x} = g(u, v)$, for (u, v) in D, and $\mathbf{x} = h(s, t)$, for (s, t) in B, be parametrizations for the same piece of smooth surface S in $\mathbb{R}^3$. If there is a one-to-one transformation T, continuously differentiable both ways between D and B, such that the Jacobian determinant of T is positive, and such that $g(u, v) = h(T(u, v))$ for (u, v) in D, then g and h are called **equivalent** parametrizations of S.
   (a)   Show that equivalent parametrizations assign the same surface area to S. (*Hint:* Use the change-of-variable theorem.)
   (b)   Show that the equivalent parametrizations assign the same value to the surface integral of a vector field over S.

12. Let the temperature at a point (x, y, z) of a region R be given by a continuously differentiable function $T(x, y, z)$. Then the vector field ∇T is called the **temperature gradient**, and under some reasonable assumptions about the region, $\nabla T(x, y, z)$ is proportional to the direction and rate of flow of heat per unit of area at (x, y, z).
   (a)   If $T(x, y, z) = x^2 + y^2$ for $x^2 + y^2 \le 4$, find the total rate of flow of heat across the cylindrical surface $x^2 + y^2 = 1, 0 \le z \le 1$.
   (b)   Give an example of a continuously differentiable vector field that cannot be a temperature gradient.

13. The Newtonian potential function $(x^2 + y^2 + z^2)^{-1/2}$ has as its gradient the attractive force field $\mathbf{F}$ of a charged particle at the origin acting on an oppositely charged particle at (x, y, z). The flux of the field across a piece of smooth surface is defined to be the surface integral of $\mathbf{F}$ over S. Show that the flux of $\mathbf{F}$ across a sphere of radius a centered at the origin is independent of a.

14. (a)   If $\mathbb{R}^2 \xrightarrow{f} \mathbb{R}$ is continuously differentiable on a set D bounded by a piecewise smooth curve, show that the area of the graph of f is

$$\sigma(S) = \int_D \sqrt{1 + (f_x)^2 + (f_y)^2} \, dx \, dy.$$

(b)   Find the area of the graph of $f(x, y) = x^2 + y$ for $0 \le x \le 1, 0 \le y \le 1$.

15. (a)   Show that if $\mathbb{R}^3 \xrightarrow{G} \mathbb{R}$ is continuously differentiable and implicitly determines a piece of smooth surface S on which $\partial G / \partial z \ne 0$, and which lies over a region D of the xy-plane, then

$$\sigma(S) = \int_D \sqrt{\left(\frac{\partial G}{\partial x}\right)^2 + \left(\frac{\partial G}{\partial y}\right)^2 + \left(\frac{\partial G}{\partial z}\right)^2}$$

$$\times \left|\frac{\partial G}{\partial z}\right|^{-1} dx \, dy.$$

Assume that just one point of S lies over each point of D.
   (b)   Compute the surface area of the hemisphere

$$x^2 + y^2 + z^2 = a^2, \quad z \ge 0,$$

using part (a).

16. (a)   Show that if a surface S is the graph of $z = f(x, y)$ for (x, y) in D, then the surface integral of $\mathbf{F} = (F_1, F_2, F_3)$ over S is

$$\int_D \left(-F_1 \frac{\partial f}{\partial x} - F_2 \frac{\partial f}{\partial y} + F_3\right) dx \, dy.$$

(b)   Use part (a) to compute the integral of $\mathbf{F}(x, y, z) = (x, y, z)$ over the graph of $z = x^2 + y$ for $0 \le x \le 1, 0 \le y \le 1$.

In Exercises 17 to 20, find a parametrization as a piecewise smooth orientable surface with outward-pointing normal for the given surface.

17. The cylindrical can with bottom and no top given by $x^2 + y^2 = 1, 0 \le z \le 1$ and $x^2 + y^2 \le 1, z = 0$

18. The funnel given by $x^2 + y^2 - z^2 = 0, 1 \le z \le 4$ and $x^2 + y^2 = 1, 0 \le z \le 1$

19. The trough given by

$$y - z = 0, 0 \le x \le 1, 0 \le z \le 1, \text{ and}$$

$$y + z = 0, 0 \le x \le 1, 0 \le z \le 1$$

20. The top half of the sphere of radius 1 centered at the origin in $\mathbb{R}^3$, together with the disk of radius 1 centered at the origin of the xy plane in $\mathbb{R}^3$

21. Let $\mathbf{F}$ be the vector field in $\mathbb{R}^3$ given by $\mathbf{F}(x, y, z) = (x, y, 2z - x - y)$. Find the integral of $\mathbf{F}$ over the oriented surface of Exercise 17

22. Let F be a continuous fluid flow field and let M be a piecewise smooth Möbius strip lying in the domain of F. Is it possible to define the flux of F across M?

23. Parametrize the set of Exercise 17 so it is reoriented, with normals pointing out on the bottom and in on the sides. Compute the integral of $\mathbf{F}(x, y, z) = (x, y, 2z - x - y)$ over this surface.

24. Prove that if $\mathbf{F}$ and $\mathbf{G}$ are continuous vector fields on a piece of smooth surface S, then

$$\int_S (a\mathbf{F} + b\mathbf{G}) \cdot d\mathbf{S} = a \int_S \mathbf{F} \cdot d\mathbf{S} + b \int_S \mathbf{G} \cdot d\mathbf{S},$$

where a and b are constants.

*25. (a) Let $\mathbf{F}$ be a continuous vector field on a piece of smooth surface S. Show that

$$\left| \int_S \mathbf{F} \cdot d\mathbf{S} \right| \le M\sigma(S),$$

where M is the maximum of $|\mathbf{F}(\mathbf{x})|$ for $\mathbf{x}$ on S. [*Hint:* Write $\int \mathbf{F} \cdot d\mathbf{S}$ in the form $\int \mathbf{F} \cdot \mathbf{n} \, d\sigma$ and use Theorem 3.5 in Chapter 7, Section 3.]

(b) Show that if the piece of S shrinks to a point $\mathbf{x}_0$ in such a way that $\sigma(S)$ tends to zero, then $\left(1/\sigma(S)\right) \int_S \mathbf{F} \cdot d\mathbf{S}$ tends to $\mathbf{F}(\mathbf{x}_0) \cdot \mathbf{n}_0$, where $\mathbf{n}_0$ is a unit normal to S at $\mathbf{x}_0$.

26. Let f be a real-valued continuously differentiable function of one variable, nonnegative for $a \le x \le b$. The graph of f, rotated around the x-axis, generates a **surface of revolution** in $\mathbb{R}^3$.

(a) Find a parametric representation of S in terms of f.

(b) Prove that $\sigma(S) = 2\pi \int_a^b f(x)\sqrt{1 + (f'(x))^2} \, dx$.

27. The **solid angle** determined by a solid cone $\mathcal{C}$ with vertex at the origin in $\mathbb{R}^3$ is defined to be the surface area of the intersection of $\mathcal{C}$ with the unit sphere $|\mathbf{x}| = 1$.

(a) Show that 2-dimensional reduction of this definition leads to the usual definition of the angle between two lines.

(b) Compute the solid angle determined by the cone $x^2 + y^2 \le 2z^2, 0 \le z$.

28. Let $\mathbf{G}(x, y, z) = y\mathbf{j} + z\mathbf{k}$ define a vector field in $\mathbb{R}^3$ and let $S_1 \cup S_2$ be the oriented two-piece surface described in Example 6 of the text. Compute the flux of $\mathbf{G}$ across $S_1 \cup S_2$ as oriented in the example.

29. Let $\mathbf{H}(x, y, z) = (x, 2y, 3z)$ define a vector field in $\mathbb{R}^3$ and let $C \cup D$ be the closed surface described in Example 7 of the text. Compute the flux of $\mathbf{H}$ out across $C \cup D$.

30. Let $f(x, y, z)$ be a continuously differentiable function defined on a smooth surface S in $\mathbb{R}^3$. Suppose that every level surface of f is perpendicular to S wherever the two surfaces intersect. (This just means that their normal vectors are perpendicular at each point of intersection.) Prove that $\int_S \nabla f \cdot d\mathbf{S} = 0$.

SECTION 4   GAUSS'S THEOREM

4A   Statement and Examples

Gauss's Theorem is a fairly straightforward generalization of Green's Theorem from $\mathbb{R}^2$ to $\mathbb{R}^3$. Both of these theorems are generalizations of the Fundamental Theorem of Calculus, so we can expect them to play a fundamental role in relating multivariable integrals to multivariable derivatives. We begin with a region R in $\mathbb{R}^3$ having as boundary a piecewise smooth surface S. Each piece of S will be parametrized by a continuously differentiable function $\mathbb{R}^2 \xrightarrow{g} \mathbb{R}^3$ with the normal vector $\partial g/\partial u \times \partial g/\partial v$ pointing away from R at each point of S. The boundary surface S is then said to have **positive orientation**, and we denote the positively oriented boundary of R by ∂R. To state the theorem, we consider a vector field $\mathbf{F}$, continuously differentiable on R and its boundary. We define the **divergence** of $\mathbf{F}$ to be the real-valued function div $\mathbf{F}$ defined on R by

$$\text{div } \mathbf{F}(\mathbf{x}) = \frac{\partial F_1}{\partial x}(\mathbf{x}) + \frac{\partial F_2}{\partial y}(\mathbf{x}) + \frac{\partial F_3}{\partial z}(\mathbf{x}),$$

where F_1, F_2, F_3 are the coordinate functions of $\mathbf{F}$. If $\mathbf{F}$ is the velocity field of a fluid flow, we can interpret div $\mathbf{F}(\mathbf{x})$ as the rate at $\mathbf{x}$ of expansion or contraction of the fluid per volume unit. In particular, if div $\mathbf{F}(\mathbf{x}) > 0$ the fluid is expanding at $\mathbf{x}$ and if div $\mathbf{F}(\mathbf{x}) < 0$ the fluid is contracting at $\mathbf{x}$. This interpretation is justified in Section 4B. (See also Chapter 8, Section 4 and Chapter 12, Section 1D.)

EXAMPLE 1   Let $\mathbf{F}(x, y, z) = (x^3, y^2, z)$. Then div $\mathbf{F}(x, y, z) = 3x^2 + 2y + 1$.

The formula for **Gauss's Theorem**, or the **divergence theorem**, is

$$\int_R \operatorname{div} \mathbf{F} \, dV = \int_{\partial R} \mathbf{F} \cdot d\mathbf{S}.$$

Gauss's Theorem is like Green's Theorem and the formula

$$\int_a^b \nabla f \cdot d\mathbf{x} = f(\mathbf{b}) - f(\mathbf{a}),$$

in that it relates an integral of some kind of derivative of a function to the behavior of that function on a boundary. In each case the orientation of the boundary is important. For example, if we apply Gauss's Theorem to the region R in $\mathbb{R}^3$ given by $1 \le |\mathbf{x}| \le 2$, then the **oriented boundary**, denoted ∂R, must be such that its normal vectors on the outer sphere point out from the sphere, and on the inner sphere point in from the sphere, as shown in Figure 9.20. We'll say in general that ∂R is **positively oriented** with respect to R if the normal vectors given by the parametrization of ∂R point away from R.

We'll prove Gauss's Theorem for the case in which R is a finite union of simple regions, where a **simple region** in $\mathbb{R}^3$ is one whose boundary is crossed by a line parallel to a coordinate axis at most twice. For example, the non-simple region between two spheres, shown in Figure 9.20, splits into a union of eight simple regions, one in each coordinate octant.

4.1   Gauss's Theorem.   Let R be a finite union of simple regions in $\mathbb{R}^3$, having a positively oriented piecewise smooth boundary ∂R. If $\mathbf{F}$ is a continuously differentiable vector field on R and ∂R, then

$$\int_R \operatorname{div} \mathbf{F} \, dV = \int_{\partial R} \mathbf{F} \cdot d\mathbf{S}.$$

Proof.   In terms of coordinate functions of $\mathbf{F}$, Gauss's formula reads

$$\int_R \left(\frac{\partial F_1}{\partial x} + \frac{\partial F_2}{\partial y} + \frac{\partial F_3}{\partial z} \right) dx \, dy \, dz = \int_{\partial R} F_1 \, dy \, dz + F_2 \, dz \, dx + F_3 \, dx \, dy.$$

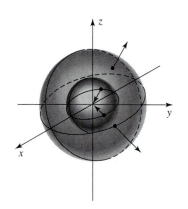

FIGURE 9.20

We assume first that R is a simple region and prove only the equation

$$\int_R \frac{\partial F_2}{\partial y} dx \, dy \, dz = \int_{\partial R} F_2 \, dz \, dx,$$

the proofs for the terms containing F_1 and F_3 being similar. Addition of the resulting equations will then prove the theorem for simple regions. Because R is simple, ∂R

FIGURE 9.21

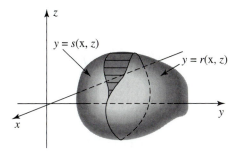

consists of the graphs of two functions, $s(x, z)$ and $r(x, z)$, perhaps together with pieces consisting of lines parallel to the y-axis as shown in Figure 9.21. Let

$$g(u, v) = \begin{pmatrix} g_1(u, v) \\ g_2(u, v) \\ g_3(u, v) \end{pmatrix}, \quad (u, v) \text{ in } D,$$

be a parametrization for ∂R that orients it positively. Then by the definition of the surface integral,

$$\int_{\partial R} F_2 \, dz \, dx = \int_D F_2(g_1, g_2, g_3) \frac{\partial(g_3, g_1)}{\partial(u, v)} du \, dv, \tag{1}$$

and, on the sections of ∂R that are parallel to the y axis, the normal vector to ∂R is perpendicular to the y axis. Hence $\partial(g_3, g_1)/\partial(u, v)$, the second coordinate of the normal, is equal to zero, thus eliminating the part of the integral that is not on the graph of r or s. We now apply the change-of-variable theorem to the two remaining parts of the integral in Equation (1). The appropriate transformations are

$$\begin{pmatrix} z \\ x \end{pmatrix} = \begin{pmatrix} g_3(u, v) \\ g_1(u, v) \end{pmatrix},$$

with (u, v) in either D_r or D_s, where D_r and D_s are the parts of D corresponding to the graphs of r and s. The Jacobian determinant $\partial(g_3, g_1)/\partial(u, v)$ is positive on the graph of r and negative on the graph of s, because it represents the x_2 coordinate of the outward normal. On D_r we have $g_2(u, v) = r(x, z)$, whereas on D_s we have $g_2(u, v) = s(x, z)$. Using these facts, we get from the change-of-variable theorem and Equation (1),

$$\int_{\partial R} F_2 \, dz \, dx = \int_{R_2} F_2(x, s(x, z), z)(-1) \, dx \, dz$$

$$+ \int_{R_2} F_2(x, r(x, z), z) \, dx \, dz,$$

where R_2 is the plane region we get by projecting R onto the xz-plane. These last two integrals are not surface integrals, but rather 2-dimensional multiple integrals.

Then by the fundamental theorem of calculus,

$$\int_{\partial R} F_2 \, dz \, dx = \int_{R_2} \left[\int_{s(x,z)}^{r(x,z)} \frac{\partial F_2}{\partial y}(x, y, z) \, dy \right] dx \, dz$$

$$= \int_R \frac{\partial F_2}{\partial y} \, dx \, dy \, dz.$$

Similar arguments involving F_1 and F_3 complete the proof for simple regions, since the addition of the three resulting equations gives

$$\int_{\partial R} F_1 \, dy \, dz + F_2 \, dz \, dx + F_3 \, dx \, dy = \int_R \left(\frac{\partial F_1}{\partial x} + \frac{\partial F_2}{\partial y} + \frac{\partial F_3}{\partial z} \right) dx \, dy \, dz.$$

This is the Gauss formula in coordinate form.

   The extension of Gauss's Theorem to a finite union R of simple regions is essentially the same as the analogous extension of Green's Theorem. In the present case, when two simple regions have a common boundary surface, the respective outward normals will be negatives of one another. The corresponding surface integrals are then negatives of one another, and so cancel out. The remaining surface integrals add up to the integral over the surface ∂R.   ∎

EXAMPLE 2   Example 5 of Section 3 consists of showing that the flux of the gradient field **F** of the potential function

$$f(x, y, z) = (x^2 + y^2 + z^2)^{-1/2}$$

across a sphere of radius a, centered at the origin, is independent of a. Using Gauss's Theorem we can prove something more general, and with a minimum of calculation. Let S_1 and S_2 be any two piecewise smooth closed surfaces, one contained in the other, both containing the origin, and bounding a region R between them; for example, R might be the region between two spheres, as shown in Figure 9.20. A routine calculation shows that the gradient is

$$\mathbf{F}(x, y, z) = (x^2 + y^2 + z^2)^{-3/2}(-x\mathbf{i} - y\mathbf{j} - z\mathbf{k}),$$

and then that the divergence of this field is zero (i.e., div **F** $= 0$ everywhere except at the origin). In particular, div **F** $= 0$ throughout R. Applying Gauss's Theorem to R gives

$$\int_{\partial R} \mathbf{F} \cdot d\mathbf{S} = \int_R \text{div} \, \mathbf{F} \, dV = 0.$$

But ∂R consists of S_1 with inward pointing normal and S_2 with outward pointing normal; so, with the understanding that S_1^- stands for the inner surface with reversed normal, we get

$$\int_{\partial R} \mathbf{F} \cdot d\mathbf{S} = \int_{S_1^-} \mathbf{F} \cdot d\mathbf{S} + \int_{S_2} \mathbf{F} \cdot d\mathbf{S} = 0.$$

Thus the integrals over the outward-oriented surfaces are equal. To find the actual value, it's enough to compute it for one surface, say a sphere. The result is -4π, as shown in Example 5 of Section 3 with $GMm = 1$. This result is a special case of one version of **Gauss's Law**, which says that the gravitational flux out across a surface S containing a mass distribution of total mass M on R is $-4\pi M$.

Example 2 is a typical application of Gauss's Theorem. The same argument implies the following general statement.

Surface independence principle.   If a vector field $\mathbf{F}$ satisfies div $\mathbf{F} = 0$ in a region R whose entire boundary consists of two surfaces S_1 and S_2, one with inward-oriented normal and the other with outward-oriented normal, then

4.2
$$\int_{S_1} \mathbf{F} \cdot d\mathbf{S} = \int_{S_2} \mathbf{F} \cdot d\mathbf{S}.$$

In other words, $\mathbf{F}$ has the same flux across the two surfaces S_1 and S_2.

Two examples of appropriately oriented pairs S_1, S_2 of surfaces are in Figures 9.22(a) and (b). We choose an orientation for each surface so we can apply Gauss's Theorem to the region bounded by S_1 and S_2. The principle is analogous to the Path Independence Principle for line integrals in the plane, one difference being that in the plane case the derivative condition was stated in terms of scalar curl as curl $\mathbf{F} = 0$ or $\partial F_2/\partial x_1 = \partial F_1/\partial x_2$. Stokes's Theorem in Section 5 will add further insight into these ideas.

EXAMPLE 3   It's easy to check that the field $\mathbf{F}(x, y, z) = (xz, yz, -z^2)$ satisfies div $\mathbf{F}(\mathbf{x}) = 0$ everywhere. It follows from the Surface Independence Principle that if two surfaces S_1 and S_2 bound a region R, one with inward-oriented normal, the other with outward-oriented normal, then $\mathbf{F}$ has the same flux across the two surfaces. If one of the two surfaces, say S_1, is contained in the xy-plane, as in Figure 9.22(a), then the flux across S_1 is zero, because $\mathbf{F}(x, y, z) = \mathbf{0}$ when $z = 0$. Hence the flux across S_2 is also zero.

4B   Interpretation of Divergence
The divergence of a vector field $\mathbf{F}$ at a point $\mathbf{x}$ is a measure of the tendency of the field to radiate away from $\mathbf{x}$, hence the term divergence. To justify this interpretation, consider a solid ball B_a of radius a centered at a point $\mathbf{x}_0$ in the interior of the set

FIGURE 9.22

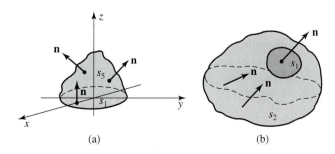

(a)   (b)

on which $\mathbf{F}$ is continuously differentiable. Apply Gauss's formula to $\mathbf{F}$ on B_a and divide both sides of the equation by the volume $V(B_a)$ to get

$$\frac{1}{V(B_a)} \int_{B_a} \operatorname{div} \mathbf{F} \, dV = \frac{1}{V(B_a)} \int_{\partial B_a} \mathbf{F} \cdot \mathbf{n} \, d\sigma,$$

where $\mathbf{n}$ is the outward-pointing unit normal vector to the spherical boundary surface $S_a = \partial B_a$. The ratio on the left is the average value of $\operatorname{div} \mathbf{F}$ in a neighborhood of $\mathbf{x}_0$, and so tends to $\operatorname{div} \mathbf{F}(\mathbf{x}_0)$ as a tends to zero. (See Exercise 3 of Chapter 7, Section 3.) The integral on the right is the average flux, per unit of volume, of $\mathbf{F}$ directed out across S_a. Hence this average flux tends to the limit of the left side, namely $\operatorname{div} \mathbf{F}(\mathbf{x}_0)$, as a tends to zero. The number $\operatorname{div} \mathbf{F}$ is the **expansion rate** of $\mathbf{F}$ at $\mathbf{x}_0$. Gauss's Theorem itself is often called the **divergence theorem** because the theorem is a statement about $\operatorname{div} \mathbf{F}$. In particular, if $\operatorname{div} \mathbf{F}(\mathbf{x}) > 0$ the flow generated by $\mathbf{F}$ is expanding near $\mathbf{x}$ and if $\operatorname{div} \mathbf{F}(\mathbf{x}) < 0$ the flow generated by $\mathbf{F}$ is contracting near $\mathbf{x}$. (See also Chapter 8, Section 4 and Chapter 12, Section 1D.)

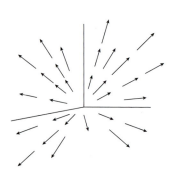

FIGURE 9.23

EXAMPLE 4   Consider the vector field $\mathbf{F}(x, y, z) = x^3 \mathbf{i} + y^3 \mathbf{j} + z^3 \mathbf{k}$. A glance at the sketch of $\mathbf{F}$ in Figure 9.23 shows the field radiating away from the origin with increasing strength as distance from the origin increases. So it follows from the definition of flux that the average flux of $\mathbf{F}$ over a sphere centered at the origin will be positive. A little additional thought shows on similar grounds that the average flux of $\mathbf{F}$ out from a sphere that doesn't enclose the origin will also be positive. The reason is that the part of the flow more distant from the origin is both outward-directed and stronger than the parts nearer to the origin. Beyond these simple remarks, we can conclude from Gauss's Theorem that the average flux of $\mathbf{F}$ out across an arbitrary smooth surface bounding a solid 3-dimensional region R is positive and equal to

$$\int_R \operatorname{div} \mathbf{F}(x, y, z) \, dx \, dy \, dz = \int_R 3(x^2 + y^2 + z^2) \, dx \, dy \, dz.$$

In Chapter 8, Section 4 we introduced without proof the **continuity equation** for fluid flow, namely

4.3
$$\frac{\partial \rho}{\partial t}(\mathbf{x}) = -\operatorname{div} \mathbf{F}(\mathbf{x}),$$

where $\mathbf{F}$ is the velocity field of a fluid flow and $\rho(\mathbf{x})$ is the density of the fluid at $\mathbf{x}$. Proving the continuity equation is a nice application of Gauss's Theorem, as follows.

EXAMPLE 5   Let $\mathbf{F}(\mathbf{x})$ be the continuously differentiable velocity field of a fluid flow in 3-dimensional space, with continuously differentiable fluid density $\rho(\mathbf{x})$, and let B be an arbitrary region of finite volume in the domain of $\mathbf{F}(\mathbf{x})$. We assume that no fluid is created or destroyed in B so any change in density is due to compression or expansion of the fluid. Then

$$\frac{d}{dt} \int_B \rho \, dV = -\int_{\partial B} \mathbf{F} \cdot \mathbf{n} \, dS,$$

where $\mathbf{n}$ is the outward-pointing unit normal to the surface ∂B. The left side is the rate of change of mass with respect to t, which because of the absence of creation or destruction of fluid must be due to flux across ∂B, as measured by the right side. We need the minus sign because the integral by itself measures total outward flux, which would be positive if the left side were negative, and vice versa. Now apply the Leibniz rule on the left and Gauss's theorem on the right to get

$$\int_B \frac{\partial \rho}{\partial t}\, dV = -\int_B \operatorname{div} \mathbf{F}\, dV, \quad \text{or} \quad \int_B \left(\frac{\partial \rho}{\partial t} + \operatorname{div} \mathbf{F} \right) dV = 0.$$

Since B is arbitrary we can choose B to be a ball $B_r(\mathbf{x}_0)$ of radius r centered at an arbitrary point $\mathbf{x}_0$ in the domain of $\mathbf{F}(\mathbf{x})$. Since the integrand in the right hand integral is continuous it must be identically zero, otherwise a nonzero value for it would, for small enough positive r, give a nonzero value for the integral over B_r. This establishes Equation 4.3.

EXERCISES

In Exercises 1 to 4, compute the divergence of the vector field $\mathbf{F}$.

1. $\mathbf{F}(x, y, z) = (x^2, y^2, z^2)$

2. $\mathbf{F}(x, y, z) = (\sin xy, 0, 0)$

3. $\mathbf{F}(x, y, z) = (y, z, x)$

4. $\mathbf{F}(x, y, z) = (xy, yz, zx)$

In Exercises 5 to 8, verify Gauss's Theorem for the vector field $\mathbf{F}$ and regions R in $\mathbb{R}^3$. Sketch R, together with a few outward-pointing normal vectors.

5. $\mathbf{F}(x, y, z) = (x^2, y^2, z^2); R : x^2 + y^2 \leq 1, 0 \leq z \leq 1$

6. $\mathbf{F}(x, y, z) = (y, -x, 0); R : x^2 + y^2 + z^2 \leq 4$

7. $\mathbf{F}(x, y, z) = (0, 0, z); R : x^2 + y^2 \leq 1, 0 \leq z \leq 1$

8. $\mathbf{F}(x, y, z) = (x, y, z); R : 0 \leq x \leq 1, 0 \leq y \leq 1, 0 \leq z \leq 1$

In Exercise 9 to 12, sketch the closed surface S, and compute $\displaystyle\int_S \mathbf{F} \cdot d\mathbf{S}$ over S by using Gauss's Theorem. Assume that the normal vectors to S point out.

9. $\mathbf{F}(x, y, z) = (x, y, z); S : x^2 + y^2 + z^2 = 4$

10. $\mathbf{F}(x, y, z) = (x, x, x); S :$ cylindrical surface $x^2 + y^2 = 1$, $0 \leq z \leq 1$; bottom $x^2 + y^2 \leq 1$, $z = 0$; top $x^2 + y^2 \leq 1, z = 1$

11. $\mathbf{F}(x, y, z) = (xz, -yz, xy); S : x^2 + 2y^2 + 3z^2 = 1$

12. $\mathbf{F}(x, y, z) = (x, y, z); S :$ bottom $x^2 + y^2 \leq 1$, $z = 0$; top $z = 1 - x^2 - y^2$

In Exercises 13 and 14, prove the identity for a twice continuously differentiable vector field $\mathbf{F}$ or real-valued function f.

13. $\operatorname{div}(\operatorname{curl} \mathbf{F})(\mathbf{x}) = 0$

14. $\operatorname{curl}(\nabla f)(\mathbf{x}) = \mathbf{0}$

15. (a) Show that for $f(x, y, z) = (x^2 + y^2 + z^2)^{-1/2}$ the equation $\operatorname{div}(\nabla f)(\mathbf{x}) = 0$ holds for all $\mathbf{x} \neq 0$.

(b) Show by example that $\operatorname{div}(\nabla f)(\mathbf{x}) \neq 0$ may hold for some twice continuously differentiable function f.

(c) If the operator Δ is defined by $\Delta f = \operatorname{div}(\nabla f)$, find a formula for Δf in terms of partial derivatives of f. A function such that $\Delta f(\mathbf{x}) = 0$ for all $\mathbf{x}$ in the domain of f is called **harmonic function**, and Δ is called the **Laplace operator**.

16. The **trace** of a square matrix is defined as the sum of the elements on its main diagonal. If $\mathbb{R}^n \xrightarrow{\mathbf{F}} \mathbb{R}^n$ is a differentiable vector field, we define $\operatorname{div} \mathbf{F}$ to be the real-valued function given by

$$\operatorname{div} \mathbf{F}(\mathbf{x}) = \operatorname{tr} \mathbf{F}'(\mathbf{x}),$$

where $\operatorname{tr} A$ stands for the trace of A. Show that in the 2- and 3-dimensional cases this definition agrees with those previously given.

In Exercises 17 and 18, use Gauss's Theorem to compute $\displaystyle\int_S \mathbf{F} \cdot d\mathbf{S}$ over the sphere of radius 1 centered at the origin in $\mathbb{R}^3$ and with outward-pointing normal.

17. $\mathbf{F}(x, y, z) = (x^2, y^2, z^2)$

18. $\mathbf{F}(x, y, z) = (xz^2, 0, z^3)$

19. Show that for a region R to which Gauss's Theorem applies, the volume of R is given by

$$V(R) = \frac{1}{3} \int_{\partial R} x \, dy \, dz + y \, dz \, dx + z \, dx \, dy.$$

20. (a) Use Gauss's theorem to prove that if $\mathbf{F}$ is a continuously differentiable vector field with zero divergence in a region R, then the integral of $\mathbf{F}$ over ∂R is zero.

(b) Write an intuitive argument, based on the interpretation of the divergence, for the assertion in part (a).

21. Let S be the ellipsoid $\dfrac{x^2}{a^2} + \dfrac{y^2}{b^2} + \dfrac{z^2}{c^2} = 1$, and let $D(x, y, z)$ be the distance from the origin to the tangent plane to S at (x, y, z).

(a) Let

$$\mathbf{F}(x, y, z) = \tfrac{1}{2} \nabla \left(\frac{x^2}{a^2} + \frac{y^2}{b^2} + \frac{z^2}{c^2} \right) = \left(\frac{x}{a^2}, \frac{y}{b^2}, \frac{z}{c^2} \right).$$

Show that $\mathbf{F} \cdot \mathbf{n} = D^{-1}$, where $\mathbf{n}$ is the outward unit normal to S at (x, y, z).

(b) Show that $\displaystyle \int_S D^{-1} d\sigma = \frac{4\pi}{3} \left(\frac{bc}{a} + \frac{ca}{b} + \frac{ab}{c} \right)$.

22. A vector field $\mathbb{R}^3 \xrightarrow{\mathbf{F}} \mathbb{R}^3$ defined in a region R is called **incompressible** in R if $\operatorname{div} \mathbf{F}(\mathbf{x}) = 0$ for all $\mathbf{x}$ in R. If $\mathbf{F}$ is continuously differentiable and incompressible in R, show that the flux of $\mathbf{F}$ is zero across every sufficiently small sphere with its interior in R.

23. Suppose that $u(x, y, z)$ is twice continuously differentiable in a region R and that u is a **harmonic function**

in R, that is $u_{xx} + u_{yy} + u_{zz} = 0$ in R. Show that if the boundary of R consists of finitely many smooth surfaces, then the outward flux of the field ∇u across ∂R is zero.

24. Use the surface independence principle, Equation 4.2, to compute the flux of the constant field $\mathbf{F}(x, y, z) = (0, 0, 1)$ across the hemisphere $z = \sqrt{1 - x^2 - y^2}$, where $x^2 + y^2 \leq 1$.

25. A field $\mathbf{F}$ for which $\operatorname{div} \mathbf{F}(\mathbf{x}) = 0$ everywhere is called **divergence free**. Show that the flux of a divergence-free field across a smooth closed surface is zero.

26. Define the vector field $\mathbf{F}(x, y, z) = (ax, by, cz)$, where a, b, and c are constants.

(a) Find the flux of $\mathbf{F}$ across a sphere of radius $\rho > 0$, oriented so that its normal vector points out from the sphere.

(b) Answer the question of part (a) with $\mathbf{F}$ defined instead by $\mathbf{F}(x, y, z) = (yz, zx, xy)$.

27. Gauss's Law. The **gravitational field** generated by an integrable mass density μ defined on a region R is

$$\mathbf{F}(\mathbf{x}) = G \int_R \frac{\mu(\mathbf{y})(\mathbf{y} - \mathbf{x})}{|\mathbf{y} - \mathbf{x}|^3} \, dV_{\mathbf{y}}.$$

(a) Show that the flux of this field across a smooth closed surface S with no points of R inside or on S is zero. Do this by interchanging the order of surface and volume integrals.

(b) Show that the flux of this field across a smooth closed surface S containing all points of R in its interior is $-4\pi G \displaystyle\int_R \mu \, dV$.

SECTION 5   STOKES'S THEOREM

5A   Statement and Examples

An important extension of Green's theorem is as follows. Instead of considering a plane region D bounded by a curve, we can think of lifting such a region, together with its boundary curve, into a 2-dimensional surface S in $\mathbb{R}^3$. Then S will have as its border a space curve γ corresponding to the boundary of D. The lifting is made precise by defining on D and its piecewise smooth boundary a function $\mathbb{R}^2 \xrightarrow{g} \mathbb{R}^3$ having S as the image of D. A typical picture is shown in Figure 9.24. The region D has its boundary oriented counterclockwise, and γ, the border curve of S, inherits what we'll call the **positive orientation** with respect to S. If we parametrize the boundary of D by $\mathbb{R} \xrightarrow{h} \mathbb{R}^2$, for $a \leq t \leq b$, then the composition $g(h(t))$ will describe the border of S. We'll denote the **positively oriented border** of S by ∂S.

FIGURE 9.24

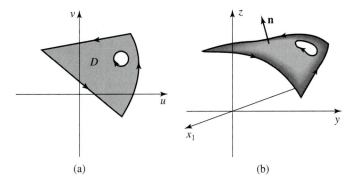

(a)   (b)

(The term **border** instead of boundary is used to avoid confusion with what we earlier called the boundary of S in Chapter 5, Section 1.)

We can now relate the line integral of a vector field $\mathbf{F}$ around ∂S to the surface integral of an associated vector field over S. We assume that $\mathbb{R}^3 \xrightarrow{\mathbf{F}} \mathbb{R}^3$ is a continuously differentiable vector field whose domain contains S. In Chapter 8, Section 4 we defined the vector field $\operatorname{curl} \mathbf{F}$ by

5.1 $$\operatorname{curl} \mathbf{F}(\mathbf{x}) = \left(\frac{\partial F_3}{\partial y}(\mathbf{x}) - \frac{\partial F_2}{\partial z}(\mathbf{x}), \ \frac{\partial F_1}{\partial z}(\mathbf{x}) - \frac{\partial F_3}{\partial x}(\mathbf{x}), \ \frac{\partial F_2}{\partial x}(\mathbf{x}) - \frac{\partial F_1}{\partial y}(\mathbf{x}) \right),$$

where F_1, F_2, and F_3 are the coordinate functions of $\mathbf{F}$. If the domain of $\mathbf{F}$ is an open set, then the domain of $\operatorname{curl} \mathbf{F}$ is the same set. As a memory aid, we express $\operatorname{curl} \mathbf{F}$ as a formal cross-product of the gradient operator $(\partial/\partial x, \partial/\partial y, \partial/\partial z)$ and $\mathbf{F} = (F_1, F_2, F_3)$, namely

$$\operatorname{curl} \mathbf{F} = \det \begin{pmatrix} \mathbf{i} & \mathbf{j} & \mathbf{k} \\ \partial/\partial x & \partial/\partial y & \partial/\partial z \\ F_1 & F_2 & F_3 \end{pmatrix}.$$

EXAMPLE 1   If $\mathbf{F}(x, y, z) = (y^2, z^2, x^2)$, then

$$\operatorname{curl} \mathbf{F} = \det \begin{pmatrix} \mathbf{i} & \mathbf{j} & \mathbf{k} \\ \partial/\partial x & \partial/\partial y & \partial/\partial z \\ y^2 & z^2 & x^2 \end{pmatrix} = -2z\mathbf{i} - 2x\mathbf{j} - 2y\mathbf{k}.$$

The vector field $\operatorname{curl} \mathbf{F}$ is a kind of derivative of the field $\mathbf{F}$, and it plays a central role in another extension of the Fundamental Theorem of Calculus called **Stokes's Theorem**, to be proved later as Theorem 4.2:

$$\int_S \operatorname{curl} \mathbf{F} \cdot d\mathbf{S} = \int_{\partial S} \mathbf{F} \cdot d\mathbf{x}. \qquad (1)$$

The way in which the positively oriented border curve ∂S in Equation (1) inherits its orientation from a parametrization of S is crucial to the validity of Equation (1). For example, an incorrect choice will produce the wrong sign on the right side. Even worse, failure to orient the two border curves coherently in Figure 9.24(b) can lead to a result having no significant relation to an integral over the surface.

If **F** were essentially a 2-dimensional vector field, with $F_3 = 0$ and F_1 and F_2 independent of z, then only the third coordinate of curl **F**, namely $\partial F_2/\partial x - \partial F_1/\partial y$, would be different from zero. In addition we could write $dS = dx\,dy$, so Stokes's Theorem would reduce to Green's Theorem of Section 1.

EXAMPLE 2

Let S be the helicoid parametrized by

$$\begin{pmatrix} x \\ y \\ z \end{pmatrix} = \begin{pmatrix} u\cos v \\ u\sin v \\ v \end{pmatrix}, \quad \text{for} \quad 0 \le u \le 1, \quad 0 \le v \le \frac{\pi}{2}.$$

Then the border of S consists of three line segments and a spiral curve shown in Figure 9.25 together with the domain D of the parametrization. Restricting the parametrization of S to the boundary of D gives the following parametrizations of the smooth pieces of the border of S:

$$\gamma_1 : \begin{pmatrix} x \\ y \\ z \end{pmatrix} = \begin{pmatrix} 0 \\ 1 - t \\ \dfrac{\pi}{2} \end{pmatrix}, 0 \le t \le 1$$

$$\gamma_2 : \begin{pmatrix} x \\ y \\ z \end{pmatrix} = \begin{pmatrix} 0 \\ 0 \\ \dfrac{\pi}{2} - t \end{pmatrix}, \quad 0 \le t \le \frac{\pi}{2}$$

$$\gamma_3 : \begin{pmatrix} x \\ y \\ z \end{pmatrix} = \begin{pmatrix} t \\ 0 \\ 0 \end{pmatrix}, \quad 0 \le t \le 1$$

$$\gamma_4 : \begin{pmatrix} x \\ y \\ z \end{pmatrix} = \begin{pmatrix} \cos t \\ \sin t \\ t \end{pmatrix}, \quad 0 \le t \le \frac{\pi}{2}.$$

Now let **F** be the vector field $\mathbf{F}(x, y, z) = (z, x, y)$. The line integrals of **F** over the γ_i are all of the form

$$\int_{\gamma_i} z\,dx + x\,dy + y\,dz.$$

FIGURE 9.25

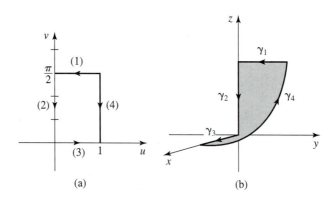

(a)   (b)

It's easy to see that the integrals over γ_1, γ_2, and γ_3 are all zero, whereas over γ_4 we get

$$\int_{\gamma_4} \mathbf{F} \cdot d\mathbf{x} = \int_0^{\pi/2} (\cos^2 t + \sin t - t \sin t)\, dt = \frac{\pi}{4}.$$

On the other hand $\operatorname{curl} \mathbf{F}(x_1, x_2, x_3) = (1, 1, 1)$; so the integral of $\operatorname{curl} \mathbf{F}$ over S is

$$\int_S \operatorname{curl} \mathbf{F} \cdot dS = \int_D \left(\frac{\partial(y, z)}{\partial(u, v)} + \frac{\partial(z, x)}{\partial(u, v)} + \frac{\partial(x, y)}{\partial(u, v)} \right) du\, dv$$

$$= \int_0^1 du \int_0^{\pi/2} (\sin v - \cos v + u)\, du = \frac{\pi}{4}.$$

This verifies Equation (1) for our special example.

The proof that we give of Stokes's Theorem depends on an application of Green's Theorem to the region D on which the parametrization of S is defined. For this reason we need to assume enough about D to make Green's Theorem hold on it. Also, if $\mathbb{R}^2 \xrightarrow{g} \mathbb{R}^3$ is the parametrization of S, we'll want the second-order partial derivatives of g to be continuous, that is, g should be twice continuously differentiable on D. We can relax these conditions, but to do so makes the proof much more difficult.

5.2 Stokes's Theorem.   Let S be a piece of smooth surface in $\mathbb{R}^3$, parametrized by a twice continuously differentiable function g. Assume that D, the parameter domain of g, is a finite union of simple regions bounded by a piecewise smooth curve. If $\mathbf{F}$ is a continuously differentiable vector field defined on S, then

$$\int_S \operatorname{curl} \mathbf{F} \cdot d\mathbf{S} = \oint_{\partial S} \mathbf{F} \cdot d\mathbf{x},$$

where ∂S is the positively oriented border of S.

Proof.   Let F_1, F_2, F_3 be coordinate functions of $\mathbf{F}$. We'll prove that

$$\oint_{\partial S} F_1 dx = \int_S -\frac{\partial F_1}{\partial y} dx\, dy + \frac{\partial F_1}{\partial z} dz\, dx. \qquad (2)$$

The proofs that

$$\oint_{\partial S} F_2 dy = \int_S -\frac{\partial F_2}{\partial y} dy\, dz + \frac{\partial F_2}{\partial x} dx\, dy$$

and

$$\oint_{\partial S} F_3 dz = \int_S -\frac{\partial F_3}{\partial x} dz\, dx + \frac{\partial F_3}{\partial y} dy\, dz$$

are similar, and addition of the three equations gives Stokes's formula. To prove the Equation (2), suppose that $h(t) = \big(u(t), v(t)\big)$ is a counterclockwise-oriented parametrization of δ, the boundary of D. Then $g(h(t))$ is a piecewise smooth parametrization of the border of S, which by definition is then positively oriented. Using g_1, g_2, g_3 for the coordinate functions of g, we can write the differential dx as

$$dx = \frac{d}{dt}\left[g_1\big(u(t), v(t)\big)\right]dt.$$

This substitution and the chain rule give

$$\oint_{\partial S} F_1 \, dx = \int F_1\big(g(u, v)\big)\frac{d}{dt}g_1(u, v)\, dt$$

$$= \int F_1\big(g(u, v)\big)\left[\frac{\partial g_1}{\partial u}(u, v)\frac{du}{dt} + \frac{\partial g_1}{\partial v}(u, v)\frac{dv}{dt}\right]dt$$

$$= \oint_{\delta} F_1(g)\frac{\partial g_1}{\partial u}du + F_1(g)\frac{\partial g_1}{\partial v}\, dv.$$

This last integral is a line integral around the region D in $\mathbb{R}^2$, and we can apply Green's Theorem to it, getting

$$\oint_{\partial S} F_1 \, dx = \int_D \left[\frac{\partial}{\partial u}\left(F_1(g)\frac{\partial g_1}{\partial v}\right) - \frac{\partial}{\partial v}\left(F_1(g)\frac{\partial g_1}{\partial u}\right)\right]du\, dv. \qquad (3)$$

The assumption that g is twice continuously differentiable ensures that the integral over D will exist. The same assumption allows us to interchange the order of partial differentiation in a computation which shows that

$$\frac{\partial}{\partial u}\left(F_1(g)\frac{\partial g_1}{\partial v}\right) - \frac{\partial}{\partial v}\left(F_1(g)\frac{\partial g_1}{\partial u}\right) = -\frac{\partial F_1}{\partial y}\frac{\partial(g_1, g_2)}{\partial(u, v)} + \frac{\partial F_1}{\partial z}\frac{\partial(g_3, g_1)}{\partial(u, v)}. \qquad (4)$$

Substitution of this identity into Equation (3) gives Equation (2), thus completing the proof. A suggestion for deriving Equation (4) is given in Exercise 17.   ∎

5B   Interpretation of Curl

Using Stokes's Theorem we can derive an interpretation for the vector field curl $\mathbf{F}$ that gives some information about $\mathbf{F}$ itself. Let $\mathbf{x}_0$ be a point of an open set on which $\mathbf{F}$ is continuously differentiable. Let $\mathbf{n}_0$ be an arbitrary unit vector pointing away from $\mathbf{x}_0$, and construct a disk S_r of radius r centered at $\mathbf{x}_0$ and perpendicular to $\mathbf{n}_0$. This is shown in Figure 9.26. Applying Stokes's Theorem to $\mathbf{F}$ on the surface S_r and its border γ_r gives

$$\oint_{\gamma_r} \mathbf{F}\cdot d\mathbf{x} = \int_{S_r} \text{curl}\,\mathbf{F}\cdot d\mathbf{S}.$$

The value of the line integral was defined more generally in Chapter 8, Section 1A to be the *circulation* of $\mathbf{F}$ around γ_r. For small r, the circulation around γ_r is a measure of the tendency of the field near $\mathbf{x}_0$ to rotate around the axis determined by $\mathbf{n}_0$. On the other hand, the surface integral is, for small enough r, nearly equal to the dot product curl $\mathbf{F}(\mathbf{x}_0)\cdot\mathbf{n}_0$ multiplied by the area of S_r. See Exercise 22 of Section 3.

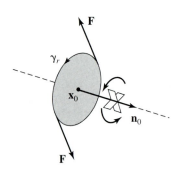

FIGURE 9.26

It follows that the circulation around γ_r tends to be larger if $\mathbf{n}_0$ points in the same direction as curl $\mathbf{F}(\mathbf{x}_0)$. Thus we can think of curl $\mathbf{F}(\mathbf{x}_0)$ as determining the axis about which the circulation of $\mathbf{F}$ is greatest near $\mathbf{x}_0$. Similarly, $|\text{curl}\,\mathbf{F}(\mathbf{x}_0)|$ measures the magnitude of the circulation around this axis near $\mathbf{x}_0$. Mechanically speaking, if the vanes of a paddle-wheel were attached to an arrow $\mathbf{n}_0$ (see Figure 9.26.) and inserted in a velocity field $\mathbf{F}$ at $\mathbf{x}_0$, the wheel would be expected to rotate most rapidly with $\mathbf{n}_0$ held parallel to the vector curl $\mathbf{F}(\mathbf{x}_0)$ and not at all with $\mathbf{n}_0$ perpendicular to curl $\mathbf{F}(\mathbf{x}_0)$. In summary, we can think intuitively about the curl of a field as follows:

    (i) The direction of curl $\mathbf{F}(\mathbf{x})$ is the axis about which $\mathbf{F}$ rotates most rapidly at $\mathbf{x}$.

    (ii) The length of curl $\mathbf{F}(\mathbf{x})$ determines the maximum rate of rotation at $\mathbf{x}$.

The extension of Stokes's Theorem to piecewise smooth orientable surfaces is very simple, though we need to be careful in orienting the border of such a surface. Figure 9.27 illustrates the method. The surfaces S_1 and S_2 have their borders joined so as to produce a piecewise smooth positively oriented surface, which we denote by $S_1 \cup S_2$. Recall that the surface integral of a vector field $\mathbf{F}$ over $S_1 \cup S_2$ has already been defined by

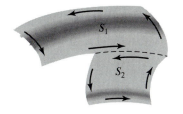

FIGURE 9.27

$$\int_{S_1 \cup S_2} \mathbf{F} \cdot d\mathbf{S} = \int_{S_1} \mathbf{F} \cdot d\mathbf{S} + \int_{S_2} \mathbf{F} \cdot d\mathbf{S}.$$

The piece of common border curve, indicated by a dashed line in Figure 9.27, will be traced in opposite directions, depending on whether the parametrization induced by S_1 or by S_2 is used. Hence the respective line integrals of $\mathbf{F}$ over the common border will have opposite sign, and when the line integrals over ∂S_1 and ∂S_2 are added, the integrals over the common part will cancel, leaving a line integral over the rest of the borders of S_1 and S_2. It is this remaining part that we call the positively oriented border of $S_1 \cup S_2$, and denote by $\partial(S_1 \cup S_2)$. With this understanding, we write Stokes's Theorem in the form

$$\int_S \text{curl}\,\mathbf{F} \cdot d\mathbf{S} = \oint_{\partial S} \mathbf{F} \cdot d\mathbf{x},$$

for a piecewise smooth surface S.

    EXAMPLE 3   We can regard a sphere as a piecewise smooth surface on which all of the border curves cancel one another. Indeed if we parametrize a sphere S_a in $\mathbb{R}^3$ by

$$g(u, v) = \begin{pmatrix} a \sin v \cos u \\ a \sin v \sin u \\ a \cos v \end{pmatrix}, \qquad \begin{array}{l} 0 \le u \le 2\pi \\ 0 \le v \le \pi, \end{array}$$

then the positively oriented "border" of the sphere consists of the half-circle shown in Figure 9.28 traced once in each direction. Thus the half-circle corresponds to the segments $u = 0$ and $u = 2\pi$ in the parameter domain. (What happens to the

FIGURE 9.28

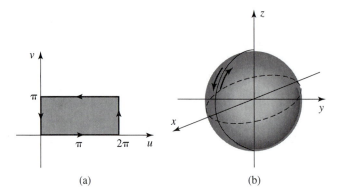

(a)   (b)

segments $v = 0$ and $v = \pi$?) The result is that a line integral over ∂S_a will be zero, and Stokes's Theorem applied to a vector field $\mathbf{F}$ on S_a gives

$$\int_{S_a} \text{curl } \mathbf{F} \cdot d\mathbf{S} = 0.$$

A surface like that in Example 3, in which the border is effectively nonexistent for the purpose of line integration over ∂S, is called a **closed surface**.

EXAMPLE 4   According to the electromagnetic theory embodied in **Maxwell's equations**, the vector current flow $\mathbf{I}$ in an electrical conductor is related to the magnetic field $\mathbf{B}$ that the current flow induces in the surrounding space by the equation $\text{curl } \mathbf{B} = \mathbf{I}$. To apply Stokes's Theorem to this equation, let a bordered surface S cut the conductor cross-sectionally. Then

$$\int_S \mathbf{I} \cdot d\mathbf{S} = \int_S \text{curl } \mathbf{B} \cdot d\mathbf{S} = \int_{\partial S} \mathbf{B} \cdot d\mathbf{x}.$$

The first integral is the total current flux across S, and the last one is the circulation of the magnetic field around the border curve ∂S that encircles the conductor. The equality of these two quantities is called **Ampère's law**.

A vector field for which $\text{div } \mathbf{F} = 0$, is called a **divergence-free field**; for a quick way to find one, start with an arbitrary twice continuously differentiable vector field $\mathbf{G} = (G_1, G_2, G_3)$ and set $\mathbf{F} = \text{curl } \mathbf{G}$. Then $\text{div } \mathbf{F} = 0$, since

$$\text{div } \mathbf{F} = \frac{\partial}{\partial x}\left(\frac{\partial G_3}{\partial y} - \frac{\partial G_2}{\partial z}\right) + \frac{\partial}{\partial y}\left(\frac{\partial G_1}{\partial z} - \frac{\partial G_3}{\partial x}\right) + \frac{\partial}{\partial z}\left(\frac{\partial G_2}{\partial x} - \frac{\partial G_1}{\partial y}\right) = 0$$

by equality of mixed partials. This way of generating fields that are locally divergence-free exhausts all possibilities, since locally every divergence-free field is the curl of some other field G. Though we won't prove this, note that it's analogous to the result of Theorem 2.6 of Section 2C which implies in $\mathbb{R}^3$ that locally every **curl-free field**, that is, a field $\mathbf{F}$ for which $\text{curl } \mathbf{F} = \mathbf{0}$, is the gradient of some scalar-valued function f, that is, $\nabla f = \mathbf{F}$.

EXAMPLE 5   Let $\mathbf{G}(x, y, z) = (y \sin z, x \cos z, z \sin x)$. Define a divergence-free vector field $\mathbf{F}$ by $\mathbf{F} = \operatorname{curl} \mathbf{G}$. Thus

$$\mathbf{F}(x, y, z) = (x \sin z, y \cos z - z \cos x, \cos z - \sin z),$$

and $\operatorname{div} \mathbf{F}(x, y, z) = \sin z + \cos z + (-\sin z - \cos z) = 0$.

EXAMPLE 6   To simplify the pictures, we'll consider here some vector fields in which the z-coordinate is zero. Our pictures will show a horizontal slice of the field. Figure 9.29 shows three snapshots of a time-dependent field

$$\mathbf{F}_t(x, y, z) = \big((1 - t)x - ty, \ tx + (1 - t)y, \ 0\big)$$

taken at times $t = 0$, $\frac{1}{2}$ and 1. With t fixed we compute $\operatorname{div} \mathbf{F}_t(x, y, z) = 2 - 2t$ and $\operatorname{curl} \mathbf{F}_t(x, y, z) = (0, 0, 2t)$. The field varies from curl-free at $t = 0$ to divergence-free at $t = 1$. Note carefully the geometric character of these extreme states. Intermediate states have both curl and divergence nonzero.

FIGURE 9.29

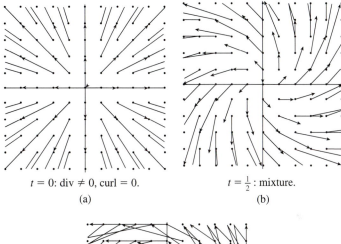

$t = 0$: div $\neq 0$, curl $= 0$.

(a)

$t = \frac{1}{2}$: mixture.

(b)

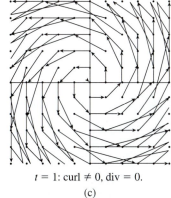

$t = 1$: curl $\neq 0$, div $= 0$.

(c)

$$\mathbf{F}_t(x, y, z)$$
$$= ((1 - t) x - ty, \ tx + (1 - t) y, 0)$$
$$0 \le t \le 1.$$

5C   Simple Connectedness

Stokes's Theorem has many applications and in particular gives information about gradient fields, that is, fields $\mathbf{F}$ such that $\mathbf{F} = \nabla f$, or, using an alternative notation, $\mathbf{F} = \operatorname{grad} f$. If we assume that $\mathbf{F}$ is the continuously differentiable gradient field of f, we can form the vector field $\operatorname{curl} \mathbf{F}$. But $\mathbf{F} = (\partial f / \partial x, \partial f / \partial y, \partial f / \partial z)$, so we get immediately from the definition of $\operatorname{curl} \mathbf{F}$ and the equality of mixed partials that

5.3
$$\operatorname{curl}(\operatorname{grad} f)(\mathbf{x}) = \mathbf{0},$$

for all $\mathbf{x}$ in the domain of f. We have already met the condition $\operatorname{curl} \mathbf{F} = \mathbf{0}$ in Theorems 2.5 and 2.6, where, for the 3-dimensional case that we consider here, it was stated in terms of the Jacobian matrix

$$\mathbf{F}' = \begin{pmatrix} \dfrac{\partial F_1}{\partial x} & \dfrac{\partial F_1}{\partial y} & \dfrac{\partial F_1}{\partial z} \\[2mm] \dfrac{\partial F_2}{\partial x} & \dfrac{\partial F_2}{\partial y} & \dfrac{\partial F_2}{\partial z} \\[2mm] \dfrac{\partial F_3}{\partial x} & \dfrac{\partial F_3}{\partial y} & \dfrac{\partial F_3}{\partial z} \end{pmatrix}.$$

The symmetry of $\mathbf{F}'$ about its main diagonal is equivalent to $\operatorname{curl} \mathbf{F} = \mathbf{0}$. Theorem 2.5 says, in particular, that if $\mathbf{F}$ is a gradient field, then $\operatorname{curl} \mathbf{F}$ is identically zero. Theorem 2.6 gives only a partial converse, to the effect that if $\operatorname{curl} \mathbf{F}$ is identically zero, then there is some rectangle in which $\mathbf{F}$ equals a gradient field. This is sometimes paraphrased by saying that $\mathbf{F}$ is *locally* a gradient field. Example 2 of Section 2 shows that the strict converse is false. Using Stokes's Theorem, we can prove another partial converse, in which the local condition is replaced by a different kind of restriction on the domain of the given field.

For this purpose we'll define a **simply connected** open set B in $\mathbb{R}^n$. Roughly, a set B is simply connected if every closed curve γ in B can be continuously contracted to a point in such a way as to stay within B during the contraction. As γ contracts to a point, it sweeps out a surface S lying in B, and γ is the border of S. The region between two spheres shown in Figure 9.30(a) is simply connected, because a closed curve can slip past the inner ball and then shrink to a point. However the open ball with a hole bored all the way through it isn't simply connected, because a surface whose border encircles the hole must lie at least partly in the hole, and so outside B. See Figure 9.30(b). In $\mathbb{R}^2$, the typical simply connected region is the inside of

FIGURE 9.30

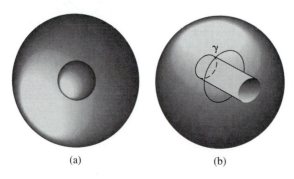

(a)   (b)

FIGURE 9.31

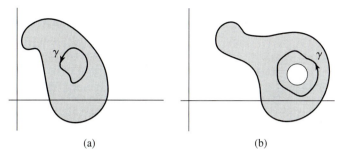

(a)   (b)

a closed curve, whereas the outside of such a curve is not simply connected. In Figure 9.31(a), the curve γ is the border of the surface consisting of the part of the plane lying inside γ. However, the presence of the hole in Figure 9.31(b) prevents a similar construction. More precisely, we'll say that an open set is simply connected if every piecewise smooth closed curve γ lying in B is the border of some piecewise smooth orientable surface S lying in B, and with parameter domain a disk in $\mathbb{R}^2$. We assume for applications that S is parametrized by twice continuously differentiable functions.

Now we can prove the following.

5.4   Theorem.   Let $\mathbf{F}$ be a continuously differentiable vector field defined on an open set B in $\mathbb{R}^2$ or $\mathbb{R}^3$. If

(a) B is simply connected, and
(b) curl $\mathbf{F}$ is identically zero in B,

then $\mathbf{F}$ is a gradient field in B, that is, there is a real-valued function f such that $\mathbf{F} = \nabla f$.

Proof.   By Theorem 2.4 it is enough to show that $\oint_\gamma \mathbf{F} \cdot d\mathbf{x} = 0$ for every piecewise smooth curve γ lying in B. Because B is simply connected, there is a piecewise smooth surface S of which γ is the border and to which we can apply Stokes's theorem in either two or three dimensions. Thus

$$\oint_{\partial S} \mathbf{F} \cdot d\mathbf{x} = \int_S \text{curl}\, \mathbf{F} \cdot d\mathbf{S} = 0,$$

as we wanted to show.   ∎

EXERCISES

In Exercises 1 to 4, compute curl $\mathbf{F}$.

1. $\mathbf{F}(x, y, z) = (y - z^2, z - x^2, x - y^2)$

2. $\mathbf{F}(x, y, z) = (z, 2y, 3z)$

3. $\mathbf{F}(x, y, z) = (x - y, z - x, y - z)$

4. $\mathbf{F}(x, y, z) = (x, y, z)$

In Exercises 5 to 8, verify by computing both integrals

that Stokes's Theorem holds for the vector field $\mathbf{F}$ and surface S. Sketch S and its border, showing orientation.

5. $\mathbf{F}(x, y, z) = (x, y, z)$;
     $S: g(u, v) = (u, v, \sqrt{1 - u^2 - v^2}), u^2 + v^2 \le 1$

6. $\mathbf{F}(x, y, z) = (z, x, y)$;
     $S: g(u, v) = (u, v, 1 - u^2 - v^2), u^2 + v^2 \le 1$

7. $\mathbf{F}(x, y, z) = (x, y, 0)$;
$S : g(u, v) = (u, v, u^2 + v^2), u^2 + v^2 \le 4$

8. $\mathbf{F}(x, y, z) = (x, y, z)$;
$S : g(u, v) = (\cos u, \sin u, v), 0 \le u \le 2\pi, 0 \le v \le 2$

In Exercises 9 and 10, compute $\displaystyle\int_S \text{curl}\,\mathbf{F} \cdot d\mathbf{S}$ by using Stokes's Theorem. In other words, choose a properly oriented parametrization for the border curve γ of S, and compute $\displaystyle\int_\gamma \mathbf{F} \cdot d\mathbf{x}$.

9. $\mathbf{F}(x, y, z) = (y, z, x)$;
$S : g(u, v) = (u, v, \sqrt{1 - u^2 - v^2}), u^2 + v^2 \le 1$

10. $\mathbf{F}(x, y, z) = (z^2, x^2, y^2)$;
$S : g(u, v) = (u, v, u^2 + v^2), u^2 + v^2 \le 4$

11. (a) Verify that if $\mathbf{F}(x, y, z)$ is independent of z and the third coordinate function of $\mathbf{F}$ is identically zero, then Stokes's Theorem, applied to a planar surface in the xy plane, becomes Green's Theorem.

(b) Consider the function $\mathbb{R}^2 \overset{g}{\to} \mathbb{R}^3$ defined by

$$g(u, v) = \begin{pmatrix} u \cos v \\ u \sin v \\ 0 \end{pmatrix}, \qquad \begin{array}{l} 1 \le u \le 2, \\ 0 \le v \le 4\pi. \end{array}$$

If S is the image in $\mathbb{R}^3$ of g, give a precise description of the oriented border of S.

(c) Use Stokes's Theorem to compute the integral of $\mathbf{F}(x, y, z) = (x, x, 0)$ over the border of S as oriented by the parametrization in part (b).

12. Show that Stokes's formula can be written in the form

$$\int_S \text{curl}\,\mathbf{F} \cdot \mathbf{n}\, d\sigma = \oint_{\partial S} \mathbf{F} \cdot \mathbf{t}\, ds,$$

where $\mathbf{n}$ is a unit normal to S and $\mathbf{t}$ is a unit tangent to ∂S.

*13. Use the result of Exercise 25 of Section 3 and Stokes's Theorem to prove that if $\mathbf{F}$ is a continuously differentiable vector field at $\mathbf{x}_0$, then

$$\lim_{r \to 0} \frac{1}{A(D_r)} \oint_c \mathbf{F} \cdot \mathbf{t}\, ds = \text{curl}\,\mathbf{F}(\mathbf{x}_0) \cdot \mathbf{n}_0,$$

where D_r is a disk of radius r centered at $\mathbf{x}_0$, $\mathbf{n}_0$ is a unit normal to the disk, and c is the boundary of D_r.

14. Prove that if $\mathbf{F}$ is a continuously differentiable vector field such that at each point $\mathbf{x}$ of a piece of smooth surface S, the vector curl $\mathbf{F}(\mathbf{x})$ is tangent to S, then the integral of $\mathbf{F}$ around the border of S is zero.

15. Let $\mathbf{F}$ be a differentiable vector field defined in an open subset B of $\mathbb{R}^3$. Use the decomposition of a square matrix A into symmetric and skew-symmetric parts given by $A = \frac{1}{2}(A + A^t) + \frac{1}{2}(A - A^t)$ to show that for all $\mathbf{y}$ in $\mathbb{R}^3$

$$\mathbf{F}'(\mathbf{x})\mathbf{y} = S(\mathbf{x})\mathbf{y} + \tfrac{1}{2}\,\text{curl}\,\mathbf{F}(\mathbf{x}) \times \mathbf{y},$$

where $S(\mathbf{x})$ is a symmetric matrix.

16. Let $\mathbf{F}(x, y, z)$ be the gradient field of the real-valued **Newtonian potential**

$$f(x, y, z) = \nabla(x^2 + y^2 + z^2)^{-1/2}.$$

Show that the circulation of $\mathbf{F}$ is zero around a smooth closed curve that is sufficiently close to a point of the domain of $\mathbf{F}$.

17. Carry out the computation of the identity in Equation (4) of the proof of Stokes's Theorem. For the first term on the left of Equation (4) we find

$$(F_1 \circ g)\frac{\partial^2 g_1}{\partial u \partial v} + \frac{\partial}{\partial u}(F_1 \circ g)\frac{\partial g_1}{\partial v} = (F_1 \circ g)\frac{\partial^2 g_1}{\partial u \partial v}$$
$$+ \left(\frac{\partial F_1}{\partial x}\frac{\partial g_1}{\partial u} + \frac{\partial F_1}{\partial y}\frac{\partial g_2}{\partial u} + \frac{\partial F_1}{\partial z}\frac{\partial g_3}{\partial u} \right) \frac{\partial g_1}{\partial v}.$$

The second term works out similarly. Then subtract, using equality of mixed partials.

18. A vector field $\mathbb{R}^3 \overset{\mathbf{F}}{\to} \mathbb{R}^3$ defined in a region R is called **irrotational** in R if curl $\mathbf{F}(\mathbf{x}) = \mathbf{0}$ for all $\mathbf{x}$ in R. If $\mathbf{F}$ is continuously differentiable and irrotational in R, show that the circulation of $\mathbf{F}$ is zero around every sufficiently small circular path in R.

19. Consider a cylindrical can C of radius 1 having a closed flat bottom and open top with an unspecified smooth border ∂C, oriented as shown in Figure 9.32. Let $\mathbf{F}(x, y, z) = (x, x, 0)$. What is the value of the line integral of $\mathbf{F}$ over the border of C?

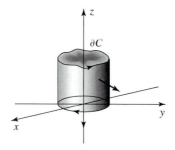

FIGURE 9.32

20. Compute the integral of curl $\mathbf{F}$, where $\mathbf{F}(x, y, z) = (y^3, -x^3, z^3)$, over the hemisphere $x^2 + y^2 + z^2 = 1$, $z \geq 0$, by considering an integral over the disk that closes the bottom of the hemisphere.

21. Show that the open subset of $\mathbb{R}^2$ consisting of $\mathbb{R}^2$ with the origin deleted is not simply connected by finding a vector field $\mathbf{F}$ for which curl $\mathbf{F}$ is identically zero, but such that $\mathbf{F}$ is not a gradient field. [*Hint:* See Exercise 7 of Section 1.]

22. The open set in $\mathbb{R}^2$ consisting of $\mathbb{R}^2$ with two points $\mathbf{x}_1$ and $\mathbf{x}_2$ deleted is not simply connected. However, you're asked here to show that if $\mathbf{F}$ is any continuously differentiable vector field in such a region such that curl $\mathbf{F} = \mathbf{0}$ there, then the integral of $\mathbf{F}$ over the smooth curve shown in Figure 9.33 is equal to zero.

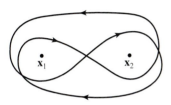

FIGURE 9.33

23. Finding a vector field $\mathbf{G} = (G_1, G_2, G_3)$ such that curl $\mathbf{G} = \mathbf{F}$, where $\mathbf{F} = (F_1, F_2, F_3)$ is (necessarily) divergence free, is in principle a rather complicated-looking problem that in practice may have a fairly straightforward solution.

(a) Show that finding $\mathbf{G}$ amounts to solving this system of partial differential equations for G_1, G_2, and G_3:

$$\frac{\partial G_3}{\partial y} - \frac{\partial G_2}{\partial z} = F_1, \qquad \frac{\partial G_1}{\partial z} - \frac{\partial G_3}{\partial x} = F_2,$$

$$\frac{\partial G_2}{\partial x} - \frac{\partial G_1}{\partial y} = F_3.$$

(b) Show that if $\mathbf{F}$ is continuously differentiable on all of $\mathbb{R}^3$, the system of equations in part (a) always has solutions in which, for example, G_3 is constant and

G_1 and G_2 are given by

$$G_1(x, y, z) = \int_0^z F_2(x, y, t)\, dt - \int_0^y F_3(x, t, 0)\, dt,$$

$$G_2(x, y, z) = -\int_0^z F_1(x, y, t)\, dt.$$

To verify this you need to use the Leibniz rule for differentiation under the integral sign and the assumption that div $\mathbf{F} = 0$.

In Exercises 23 to 26, use the result of part (b) of the previous exercise to find a vector field $\mathbf{G}$ such that curl $\mathbf{G} = \mathbf{F}$, and check that the equation is satisfied for each of the following fields, in which div $\mathbf{F} = 0$.

24. $\mathbf{F}(x, y, z) = (2x, -y, -z)$

25. $\mathbf{F}(x, y, z) = (y, z, x)$

26. $\mathbf{F}(x, y, z) = (yz, xz, xy)$

27. $\mathbf{F}(x, y, z) = (x, -y, 3x)$

The vector fields $\mathbf{G}$ found in the previous exercise are not the only ones for which curl $\mathbf{G} = \mathbf{F}$ when div $\mathbf{F} = 0$.

28. Show that adding a gradient field will also work, that is, curl$(\mathbf{G} + \nabla f) = \mathbf{F}$.

29. Show that if $\mathbf{G}$ and $\mathbf{H}$ satisfy curl $\mathbf{G}(\mathbf{x}) =$ curl $\mathbf{H}(\mathbf{x}) = \mathbf{F}(\mathbf{x})$ for all $\mathbf{x}$ in $\mathbb{R}^3$, then $\mathbf{G} - \mathbf{H} = \nabla f$ for some $f \colon \mathbb{R}^3 \to \mathbb{R}$.

It's possible to prove that if div $\mathbf{F}(\mathbf{x}) = 0$ for all $\mathbf{x}$ in $\mathbb{R}^3$, then

$$\mathbf{G}(\mathbf{x}) = \int_0^1 [\mathbf{F}(t\mathbf{x}) \times (t\mathbf{x})]\, dt$$

defines a vector field such that curl $\mathbf{G} = \mathbf{F}$. In Exercises 30 to 33, use this formula to find curl $\mathbf{G}$ given $\mathbf{F}$, and check that curl $\mathbf{G} = \mathbf{F}$, for each of the following fields for which the divergence is zero.

30. $\mathbf{F}(x, y, z) = (2x, -y, -z)$

31. $\mathbf{F}(x, y, z) = (y, z, x)$

32. $\mathbf{F}(x, y, z) = (yz, xz, xy)$

33. $\mathbf{F}(x, y, z) = (x, -y, 3x)$

SECTION 6   THE OPERATORS ∇, $\nabla\times$ AND $\nabla\cdot$

6A   Derivative Formulas

To facilitate the application of the Gauss and Stokes theorems, it's helpful to extend the use of the symbol ∇, called "del," that is used in denoting the gradient field of a real-valued function. In terms of the natural basis $\mathbf{i}, \mathbf{j}, \mathbf{k}$ for $\mathbb{R}^3$, we recall that

6.1
$$\nabla f = \frac{\partial f}{\partial x}\mathbf{i} + \frac{\partial f}{\partial y}\mathbf{j} + \frac{\partial f}{\partial z}\mathbf{k}.$$

This equation defines ∇ as an operator from real-valued differentiable functions $\mathbb{R}^3 \xrightarrow{f} \mathbb{R}$, to vector fields $\mathbb{R}^3 \xrightarrow{\mathbf{F}} \mathbb{R}^3$. If we write

6.2
$$\nabla = \frac{\partial}{\partial x}\mathbf{i} + \frac{\partial}{\partial y}\mathbf{j} + \frac{\partial}{\partial z}\mathbf{k},$$

then Equation 6.1 follows by application of both sides of Equation 6.2 to f.

The formalism just described makes the following definitions natural. If $\mathbf{F}$ is a differentiable vector field given by

$$\mathbf{F}(\mathbf{x}) = F_1(\mathbf{x})\mathbf{i} + F_2(\mathbf{x})\mathbf{j} + F_3(\mathbf{x})\mathbf{k},$$

then the operator $\nabla \times$ is defined by taking the formal cross product of ∇ and $\mathbf{F}$ to get

6.3
$$\nabla \times \mathbf{F} = \begin{vmatrix} \frac{\partial}{\partial y} & \frac{\partial}{\partial z} \\ F_2 & F_3 \end{vmatrix} \mathbf{i} + \begin{vmatrix} \frac{\partial}{\partial z} & \frac{\partial}{\partial x} \\ F_3 & F_1 \end{vmatrix} \mathbf{j} + \begin{vmatrix} \frac{\partial}{\partial x} & \frac{\partial}{\partial y} \\ F_1 & F_2 \end{vmatrix} \mathbf{k}$$

$$= \left(\frac{\partial F_3}{\partial y} - \frac{\partial F_2}{\partial z} \right)\mathbf{i} + \left(\frac{\partial F_1}{\partial z} - \frac{\partial F_3}{\partial x} \right)\mathbf{j} + \left(\frac{\partial F_2}{\partial x} - \frac{\partial F_1}{\partial y} \right)\mathbf{k}.$$

Thus $\nabla \times \mathbf{F}$ is the vector field that we have called the curl of $\mathbf{F}$ and written $\operatorname{curl} \mathbf{F}$. Similarly, for a differentiable vector field $\mathbf{F}$, we define the operator $\nabla \cdot$ by taking the formal dot product of ∇ and $\mathbf{F}$ to get

6.4
$$\nabla \cdot \mathbf{F} = \frac{\partial F_1}{\partial x} + \frac{\partial F_2}{\partial y} + \frac{\partial F_3}{\partial z},$$

This real-valued function we have called the divergence of $\mathbf{F}$ and have written $\operatorname{div} \mathbf{F}$. The meaning of the notation just introduced is easy to remember if Equation 6.2 is kept in mind.

Using the ∇ notation, Stokes's formula becomes

6.5
$$\int_S (\nabla \times \mathbf{F}) \cdot \mathbf{n}\, d\sigma = \int_{\partial S} \mathbf{F} \cdot \mathbf{t}\, ds,$$

and Gauss's formula becomes

6.6
$$\int_R \nabla \cdot \mathbf{F}\, dV = \int_{\partial R} \mathbf{F} \cdot \mathbf{n}\, d\sigma.$$

To exploit these formulas fully, we need some identities involving ∇. In the following formulas, f and g are real-valued differentiable functions, $\mathbf{F}$ and $\mathbf{G}$ are differentiable vector fields, and a and b are constants.

$$\nabla(af + bg) = a\nabla f + b\nabla g \tag{1}$$

$$\nabla(fg) = f\nabla g + g\nabla f \tag{2}$$

$$\nabla \times (a\mathbf{F} + b\mathbf{G}) = a\nabla \times \mathbf{F} + b\nabla \times \mathbf{G} \tag{3}$$

$$\nabla \times (f\mathbf{F}) = f\nabla \times \mathbf{F} + \nabla f \times \mathbf{F} \tag{4}$$

$$\nabla \cdot (a\mathbf{F} + b\mathbf{G}) = a\nabla \cdot \mathbf{F} + b\nabla \cdot \mathbf{G} \tag{5}$$

$$\nabla \cdot (f\mathbf{F}) = f\nabla \cdot \mathbf{F} + \nabla f \cdot \mathbf{F} \tag{6}$$

$$\nabla \cdot (\mathbf{F} \times \mathbf{G}) = (\nabla \times \mathbf{F}) \cdot \mathbf{G} - \mathbf{F} \cdot (\nabla \times \mathbf{G}). \tag{7}$$

Checking each of these formulas is a matter of writing out the expressions using the coordinate definitions of the operators. All of these formulas compress into a useful form the extraordinary amount of clutter that writing the corresponding coordinate expressions requires. Note also that with the possible exception of Formula 7 the formulas appear to be natural extensions of familiar calculus formulas and so are easy to remember.

Using the same kind of verification used for Formulas 1 to 7 establishes that if f and $\mathbf{F}$ are *twice* differentiable, then

$$\nabla \cdot (\nabla \times \mathbf{F}) = 0, \tag{8}$$

$$\nabla \times (\nabla f) = 0, \tag{9}$$

$$\nabla \cdot \nabla f = \nabla^2 f, \tag{10}$$

where $\nabla^2 f$ is just shorthand for $\nabla \cdot (\nabla f)$ and so denotes the Laplace operator

$$\nabla^2 f = \frac{\partial^2 f}{\partial x^2} + \frac{\partial^2 f}{\partial y^2} + \frac{\partial^2 f}{\partial z^2}.$$

Equations (8) and (9), are the same as those in Exercises 13 and 14 of Section 4, where we used the notations div and curl.

6B   Green's Identities

The preceding formulas imply many special cases of the Gauss and Stokes theorems. A particularly important kind arises if the vector field $\mathbf{F}$ is assumed to be a gradient ∇f, or a multiple $f\nabla g$ of a gradient. If we set $\mathbf{F} = \nabla f$ in Equation 6, the result is

$$\int_R \nabla \cdot \nabla f\, dV = \int_{\partial R} \nabla f \cdot \mathbf{n}\, d\sigma. \tag{11}$$

But by Formula (10), $\nabla \cdot \nabla f = \nabla^2 f$, and by Equation 1 of Chapter 6, Section 1, $\nabla f \cdot \mathbf{n} = (\partial/\partial\mathbf{n})f$. Thus we have

$$\int_R \nabla^2 f\, dV = \int_S \frac{\partial f}{\partial\mathbf{n}}\, d\sigma. \tag{12}$$

If we replace $\mathbf{F}$ in Formula 6.6 by $f\nabla g$, instead of by ∇f, we have, from Equation (6),

$$\nabla \cdot (f\nabla g) = f\nabla \cdot \nabla g + \nabla f \cdot \nabla g,$$

and so Gauss's formula yields

6.7
$$\int_R f\nabla^2 g\, dV + \int_R \nabla f \cdot \nabla g\, dV = \int_S f\frac{\partial g}{\partial \mathbf{n}}\, d\sigma.$$

This is called **Green's first identity**. Because of the symmetry in the middle term, interchange of f and g and subtraction of the corresponding terms gives **Green's second identity**:

6.8
$$\int_R (f\nabla^2 g - g\nabla^2 f)\, dV = \int_S \left(f\frac{\partial g}{\partial \mathbf{n}} - g\frac{\partial f}{\partial \mathbf{n}} \right) d\sigma.$$

EXAMPLE 1 Let R be a polygonally connected region in $\mathbb{R}^3$ with a piecewise smooth boundary surface S. If h is a real-valued function defined in R, we consider the **Poisson equation**

$$\nabla^2 u = h,$$

subject to a preassigned boundary condition, $u(\mathbf{x}) = \phi(\mathbf{x})$ for $\mathbf{x}$ on S. We suppose that there is at least one solution $u(\mathbf{x})$ defined in R and satisfying the boundary condition. We can prove, using Green's first identity, that *such a solution must be unique*. Let us suppose that there were two solutions u_1 and u_2; then the function u defined by $u(\mathbf{x}) = u_1(\mathbf{x}) - u_2(\mathbf{x})$ would satisfy the **Laplace equation** $\nabla^2 u = 0$ in R, together with the boundary condition $u(\mathbf{x}) = 0$ on S. Setting $f = g = u$ in Formula 6.7 gives

$$\int_R u\nabla^2 u\, dV + \int_R |\nabla u|^2\, dV = \int_S u\frac{\partial u}{\partial \mathbf{n}}\, d\sigma.$$

But the first and last terms are zero because $\nabla^2 u = 0$ in R and $u = 0$ on S. It follows from $\int_R |\nabla u|^2\, dV = 0$ that $\nabla u = 0$ identically on R and S. Hence u must be a constant in the polygonally connected region R. Finally, u must be identically zero because $u(\mathbf{x}) = 0$ for $\mathbf{x}$ on S. We remark that the Laplace equation is the special case of the Poisson equation obtained by taking h identically zero; thus we have proved a uniqueness theorem for the Laplace equation also. The Laplace and Poisson equations are important in various physical problems. For example, the steady-state temperature in a homogeneous solid satisfies Laplace's equation. If h is the density of electric charge in a region of space, then the electrostatic field is proportional to the gradient of a solution u of the Poisson equation $\nabla^2 u = h$.

EXAMPLE 2 Green's second identity, Equation 8, illuminates many features of the Laplace operator ∇^2. In particular, suppose u and v are harmonic functions on a region R of $\mathbb{R}^3$ and its smooth boundary surface ∂R, that is, suppose $\nabla^2 u = 0$ and $\nabla^2 v = 0$ on

$R \cup \partial R$. Setting $f = u$ and $g = v$ in Equation 8 makes the left side zero, so

$$\int_{\partial R} \left(u \frac{\partial v}{\partial \mathbf{n}} - v \frac{\partial u}{\partial \mathbf{n}} \right) d\sigma = 0 \quad \text{or} \quad \int_{\partial R} u \frac{\partial v}{\partial \mathbf{n}} d\sigma = \int_{\partial R} v \frac{\partial u}{\partial \mathbf{n}} d\sigma.$$

This last equation expresses a symmetry that holds for an arbitrary pair of harmonic functions on $R \cup \partial R$. In particular, with $v = 1$ as constant harmonic function, $\partial v / \partial \mathbf{n} = 0$ so we conclude that

$$\int_{\partial R} \frac{\partial u}{\partial \mathbf{n}} d\sigma = 0.$$

In words this last equation says that the average value over ∂R of the normal derivative of a harmonic function must be zero. This last result can also be obtained directly from Green's first identity. (See Exercise 20.)

From Equation 6 we can derive some equations for vector-valued integrals. Let $\mathbf{v}$ be an arbitrary constant vector and let $\mathbf{F}(\mathbf{x}) = f(\mathbf{x})\mathbf{v}$, where f is real-valued and continuously differentiable on a region R. Then because $\nabla \cdot f\mathbf{v} = \nabla f \cdot \mathbf{v}$ (verify!), Formula 6.6 becomes

$$\int_R \nabla f \cdot \mathbf{v} \, dV = \int_{\partial R} f\mathbf{v} \cdot \mathbf{n} \, d\sigma.$$

Since $\mathbf{v}$ is constant,

$$\mathbf{v} \cdot \int_R \nabla f \, dV = \mathbf{v} \cdot \int_{\partial R} f\mathbf{n} \, d\sigma,$$

Because $\mathbf{v}$ is arbitrary, we can successively set $\mathbf{v}$ equal to $\mathbf{e}_1$, $\mathbf{e}_2$, $\mathbf{e}_3$, and conclude that the two vector integrals have the same coordinates in $\mathbb{R}^3$. Hence

$$\int_R \nabla f \, dV = \int_{\partial R} f\mathbf{n} \, d\sigma. \tag{13}$$

Similarly replacing $\mathbf{F}$ in Equation 6 by $\mathbf{v} \times \mathbf{F}$, where $\mathbf{v}$ is a constant vector, we can conclude that

$$\int_R \nabla \times \mathbf{F} \, dV = \int_{\partial R} \mathbf{n} \times \mathbf{F} \, d\sigma. \tag{14}$$

6C   Changing Coordinates

We've already dealt with this issue in Chapter 6, Section 2B and Section 5, and in Chapter 7, Section 4. In the first instance we looked mainly at fairly simple linear examples, in the second mainly at geometry, and in the third mainly just at Jacobian determinants in multiple integrals. Here we take up nonlinear changes of variable as they affect first and second order differential operators. The calculations can be fairly messy, so to understand them it's important to follow a general principle.

Changing from rectangular coordinates $\mathbf{x}$ in $\mathbb{R}^n$ to curvilinear coordinates $\mathbf{u}$ we use a **coordinate transformation** $\mathbf{x} = T(\mathbf{u})$ that's both continuously differentiable and invertible with continuously differentiable inverse T^{-1}. Suppose $\mathbf{v}(\mathbf{x})$ is a scalar or vector function, and for reasons of symmetry in the field we'd like to compute $\nabla \cdot \mathbf{v}$, not in $\mathbf{x}$-coordinates but in $\mathbf{u}$-coordinates. But we defined the divergence as a sum of partial derivatives of the $\mathbf{x}$-variables, and just adding up partials with respect to the $\mathbf{u}$-variables isn't correct. To clarify the notation we introduce a new function $\bar{\mathbf{v}}(\mathbf{u}) = \mathbf{v}(T(\mathbf{u}))$. Thus we can differentiate and integrate $\bar{\mathbf{v}}(\mathbf{u})$ directly in terms of $\mathbf{u}$-variables. Using the derivative matrix $T'(\mathbf{x})$ and its inverse $(T')^{-1}(\mathbf{u})$, the switch to new coordinates sorts out as follows.

6.9   Theorem.   If $\mathbf{v}(\mathbf{x})$ is continuously differentiable then $\mathbf{v}'(\mathbf{x}) = \bar{\mathbf{v}}'(\mathbf{u})(T')^{-1}(\mathbf{u})$, where $\mathbf{x} = T(\mathbf{u})$ is a coordinate transformation and $\bar{\mathbf{v}}(\mathbf{u}) = \mathbf{v}(T(\mathbf{u}))$.

Proof.   Since $\mathbf{x} = T(\mathbf{u})$, and we assume T^{-1} exists and is differentiable, the chain-rule gives us

$$\mathbf{v}'(\mathbf{x}) = \left(\bar{\mathbf{v}}(T^{-1}(\mathbf{x}))\right)' = \bar{\mathbf{v}}'(T^{-1}(\mathbf{x}))(T^{-1})'(\mathbf{x}).$$

By Theorem 2.3 of Chapter 6 the derivative of an inverse is the inverse of the derivative, so $(T^{-1})'(\mathbf{x}) = (T')^{-1}(\mathbf{u})$. Finally note that $\bar{\mathbf{v}}'(T^{-1}(\mathbf{x})) = \bar{\mathbf{v}}'(\mathbf{u})$.   ∎

We'll show how this theorem works with plane polar coordinates.

EXAMPLE 3

Divergence and gradient in polar coordinates. We first compute the inverse of the derivative matrix for the polar coordinate transformation

$$\left. \begin{array}{l} x = r\cos\theta \\ y = r\sin\theta \end{array} \right\} : \quad \begin{pmatrix} \cos\theta & -r\sin\theta \\ \sin\theta & r\cos\theta \end{pmatrix}^{-1} = \begin{pmatrix} \cos\theta & \sin\theta \\ -(1/r)\sin\theta & (1/r)\cos\theta \end{pmatrix}.$$

To find $\nabla u(x, y)$ in polar coordinates we use Theorem 6.9 and multiply this inverse matrix by the derivative matrix of $\bar{u}(r, \theta)$: $\left(\dfrac{\partial \bar{u}}{\partial r} \quad \dfrac{\partial \bar{u}}{\partial \theta}\right) \begin{pmatrix} \cos\theta & \sin\theta \\ -(1/r)\sin\theta & (1/r)\cos\theta \end{pmatrix}.$
Thus

$$\nabla \bar{u} = \left(\cos\theta\,\frac{\partial \bar{u}}{\partial r} - \frac{1}{r}\sin\theta\,\frac{\partial \bar{u}}{\partial \theta}\right)\mathbf{i} + \left(\sin\theta\,\frac{\partial \bar{u}}{\partial r} + \frac{1}{r}\cos\theta\,\frac{\partial \bar{u}}{\partial \theta}\right)\mathbf{j}, \text{ for } r > 0. \quad (15)$$

In particular we see that we can replace partial differentiation of u with respect to x and y by action on $\bar{u}$ according to

$$\frac{\partial u}{\partial x} = \left(\cos\theta\,\frac{\partial \bar{u}}{\partial r} - \frac{1}{r}\sin\theta\,\frac{\partial \bar{u}}{\partial \theta}\right) \text{ and } \frac{\partial u}{\partial y} = \left(\sin\theta\,\frac{\partial \bar{u}}{\partial r} + \frac{1}{r}\sin\theta\,\frac{\partial \bar{u}}{\partial \theta}\right) \quad (16)$$

Note that the coefficients of the partials with respect to r and θ are just the columns of the inverse matrix. Thus the divergence of a vector field $\mathbf{F}(x, y) = F_1(x, y)\mathbf{i} + F_2(x, y)\mathbf{j}$ becomes in polar coordinates

$$\nabla \cdot \mathbf{F} = \left(\cos\theta\,\frac{\partial \bar{F}_1}{\partial r} - \frac{1}{r}\sin\theta\,\frac{\partial \bar{F}_1}{\partial \theta}\right) + \left(\sin\theta\,\frac{\partial \bar{F}_2}{\partial r} + \frac{1}{r}\sin\theta\,\frac{\partial \bar{F}_2}{\partial \theta}\right). \quad (17)$$

If $\bar{u}(r)$ and $\bar{\mathbf{F}}(r)$ are functions of r alone then Equations (15) and (17) simplify to

$$\nabla\bar{u} = \cos\theta\frac{\partial\bar{u}}{\partial r}\mathbf{i} + \sin\theta\frac{\partial\bar{u}}{\partial r}\mathbf{j} \quad \text{and} \quad \nabla\bullet\bar{\mathbf{F}} = \cos\theta\frac{\partial\bar{F}_1}{\partial r} + \sin\theta\frac{\partial\bar{F}_2}{\partial r}.$$

Note that while the vector coordinates are computed using polar coordinates, the vectors $\nabla\bar{u}$ are represented using standard basis vectors $\mathbf{i}$ and $\mathbf{j}$ in $\mathbb{R}^2$.

An operator of the form $\nabla^2 = \nabla\bullet\nabla$ is called a **Laplacian**, and it operates on functions defined in spaces of various dimensions and expressed in a variety of coordinate systems. Here's the simplest example in non-rectangular coordinates.

EXAMPLE 4

Laplacian in polar coordinates. In plane rectangular coordinates ∇^2 acts on a scalar-valued function by $\nabla^2 u(x, y) = u_{xx}(x, y) + u_{yy}(x, y)$. Switching to polar coordinates we let $\bar{u}(r, \theta) = u(r\cos\theta, r\sin\theta)$. In operator form Equations 16 lead to the operator replacements

$$\frac{\partial}{\partial x} \text{ by } \left(\cos\theta\frac{\partial}{\partial r} - \frac{1}{r}\sin\theta\frac{\partial}{\partial\theta}\right) \quad \text{and} \quad \frac{\partial}{\partial y} \text{ by } \left(\sin\theta\frac{\partial\bar{u}}{\partial r} + \frac{1}{r}\cos\theta\frac{\partial}{\partial\theta}\right). \quad (18)$$

Applying the first of these twice to the first equation in (16) gives

$$\begin{aligned}
\frac{\partial^2 u}{\partial x^2} &= \left(\cos\theta\frac{\partial}{\partial r} - \frac{1}{r}\sin\theta\frac{\partial}{\partial\theta}\right)\left(\cos\theta\frac{\partial\bar{u}}{\partial r} - \frac{1}{r}\sin\theta\frac{\partial\bar{u}}{\partial\theta}\right) \\
&= \cos\theta\frac{\partial}{\partial r}\left(\cos\theta\frac{\partial\bar{u}}{\partial r}\right) - \cos\theta\frac{\partial}{\partial r}\left(\frac{1}{r}\sin\theta\frac{\partial\bar{u}}{\partial\theta}\right) \\
&\quad - \frac{1}{r}\sin\theta\frac{\partial}{\partial\theta}\left(\cos\theta\frac{\partial\bar{u}}{\partial r}\right) + \frac{1}{r^2}\sin\theta\frac{\partial}{\partial\theta}\left(\sin\theta\frac{\partial\bar{u}}{\partial\theta}\right).
\end{aligned}$$

We get the polar expression for $\dfrac{\partial}{\partial y}$ from $\dfrac{\partial}{\partial x}$ by interchanging $\cos\theta$ and $\sin\theta$ and replacing $-$ by $+$, so

$$\begin{aligned}
\frac{\partial^2 u}{\partial y^2} &= \left(\sin\theta\frac{\partial}{\partial r} + \frac{1}{r}\cos\theta\frac{\partial}{\partial\theta}\right)\left(\sin\theta\frac{\partial\bar{u}}{\partial r} + \frac{1}{r}\cos\theta\frac{\partial\bar{u}}{\partial\theta}\right) \\
&= \sin\theta\frac{\partial}{\partial r}\left(\sin\theta\frac{\partial\bar{u}}{\partial r}\right) + \sin\theta\frac{\partial}{\partial r}\left(\frac{1}{r}\cos\theta\frac{\partial\bar{u}}{\partial\theta}\right) \\
&\quad + \frac{1}{r}\cos\theta\frac{\partial}{\partial\theta}\left(\sin\theta\frac{\partial\bar{u}}{\partial r}\right) + \frac{1}{r^2}\cos\theta\frac{\partial}{\partial\theta}\left(\cos\theta\frac{\partial\bar{u}}{\partial\theta}\right).
\end{aligned}$$

With the help of $\sin^2\theta + \cos^2\theta = 1$ we can extract from the first and last terms of the sum $u_{xx} + u_{yy}$ the terms $\bar{u}_{rr}$ and $r^{-2}\bar{u}_{\theta\theta}$. Everything else but $r^{-1}\bar{u}_r$ cancels, so in polar coordinates

$$\nabla^2\bar{u} = \frac{\partial^2\bar{u}}{\partial r^2} + \frac{1}{r}\frac{\partial\bar{u}}{\partial r} + \frac{1}{r^2}\frac{\partial^2\bar{u}}{\partial\theta^2}, \quad 0 < r. \quad (19)$$

Exercise 20 shows it's easier to verify this equation than derive it from $u_{xx} + u_{yy}$.

EXAMPLE 5

Laplacian in cylindrical coordinates. An argument analogous to the ones in Examples 3 and 4 is based on the inverse of the derivative matrix Formula 5.6 of Chapter 6, Section 5, namely

$$\left.\begin{array}{l} x = r\cos\theta \\ y = r\sin\theta \\ z = z \end{array}\right\} : \quad \begin{pmatrix} \cos\theta & -r\sin\theta & 0 \\ \sin\theta & r\cos\theta & 0 \\ 0 & 0 & 1 \end{pmatrix}^{-1} = \begin{pmatrix} \cos\theta & \sin\theta & 0 \\ -(1/r)\sin\theta & (1/r)\cos\theta & 0 \\ 0 & 0 & 1 \end{pmatrix}.$$

Using the polar coordinate result, we find

$$\nabla^2 \bar{u} = \frac{\partial \bar{u}}{\partial r^2} + \frac{1}{r}\frac{\partial \bar{u}}{\partial r} + \frac{1}{r^2}\frac{\partial \bar{u}}{\partial \theta^2} + \frac{\partial^2 \bar{u}}{\partial z^2}, \quad r > 0. \tag{20}$$

EXAMPLE 6

Laplacian in spherical coordinates. An argument analogous to the ones in Examples 3 and 4 is based on the inverse of the derivative matrix Formula 5.5 of Chapter 6, Section 5, namely

$$\begin{pmatrix} \sin\phi\cos\theta & r\cos\phi\cos\theta & -r\sin\phi\sin\theta \\ \sin\phi\sin\theta & r\cos\phi\sin\theta & r\sin\phi\cos\theta \\ \cos\phi & -r\sin\phi & 0 \end{pmatrix}^{-1} = \begin{pmatrix} \sin\phi\cos\theta & \sin\phi\sin\theta & \cos\phi \\ \dfrac{\cos\phi\cos\theta}{r} & \dfrac{\cos\phi\sin\theta}{r} & -\dfrac{\sin\phi}{r} \\ -\dfrac{\sin\theta}{r\sin\phi} & \dfrac{\cos\theta}{r\sin\phi} & 0 \end{pmatrix}.$$

The replacements of the partials with respect to x, y, and z are, respectively,

$$\frac{\partial}{\partial x} \text{ by } \left(\sin\phi\cos\theta\frac{\partial}{\partial r} + \frac{\cos\phi\cos\theta}{r}\frac{\partial}{\partial\phi} - \frac{\sin\theta}{r\sin\phi}\frac{\partial}{\partial\theta} \right),$$

$$\frac{\partial}{\partial y} \text{ by } \left(\sin\phi\sin\theta\frac{\partial}{\partial r} + \frac{\cos\phi\sin\theta}{r}\frac{\partial}{\partial\phi} + \frac{\cos\theta}{r\sin\phi}\frac{\partial}{\partial\theta} \right),$$

$$\frac{\partial}{\partial z} \text{ by } \left(\cos\phi\frac{\partial}{\partial r} - \frac{\sin\phi}{r}\frac{\partial}{\partial\phi} \right).$$

Applying these twice in succession to a function $\bar{u}(r, \phi, \theta)$, and then adding the results, gives for $0 < r$, $0 < \phi < \pi$,

$$\nabla^2 \bar{u} = \frac{\partial^2 \bar{u}}{\partial r^2} + \frac{2}{r}\frac{\partial \bar{u}}{\partial r} + \frac{1}{r^2}\frac{\partial^2 \bar{u}}{\partial \phi^2} + \frac{\cos\phi}{r^2\sin\phi}\frac{\partial \bar{u}}{\partial \phi} + \frac{1}{r^2\sin^2\phi}\frac{\partial^2 \bar{u}}{\partial \theta^2}.$$

EXERCISES

For Exercises 1 to 7, verify the corresponding identity (1)–(7) in Section 6A. For Exercises 8 to 10, verify the corresponding identity (8)–(10) in Section 6A.

11. Prove that if $\mathbf{v}$ is a constant vector and $\mathbf{x}$ is not zero, then

$$\nabla \times \frac{\mathbf{v} \times \mathbf{x}}{|\mathbf{x}|} = \frac{\mathbf{v}}{|\mathbf{x}|} + \frac{\mathbf{v} \cdot \mathbf{x}}{|\mathbf{x}|^3}\mathbf{x}.$$

In Exercises 12 and 13, assume $\mathbf{x}$ in $\mathbb{R}^3$, and prove the equation for $\mathbf{x} \neq 0$.

12. $\nabla\left(\dfrac{1}{|\mathbf{x}|}\right) = \dfrac{-\mathbf{x}}{|\mathbf{x}|^3}$

13. $\nabla^2\left(\dfrac{1}{|\mathbf{x}|}\right) = 0$

14. Replace $\mathbf{F}$ in Equation 6 of the text by $\mathbf{v} \times \mathbf{F}$ where $\mathbf{v}$ is a constant vector in $\mathbb{R}^3$. Use this to prove Formula (14) at the end of Section 6B. [*Hint:* Use Formula (7) in Section 6A, and Equation 8 in Section 4C of Chapter 1 to show that the dot product of the two sides of Formula (14) with $\mathbf{v}$ are equal as in the proof of Formula (13).]

15. If $T(\mathbf{x})$ is the steady-state temperature at a point $\mathbf{x}$ of an open set R in $\mathbb{R}^3$, then the flux of the temperature gradient across any smooth surface in R is zero. Use this fact and Equation (12) to prove that a steady-state temperature function that is twice continuously differentiable is harmonic, i.e., $\nabla^2 T \equiv 0$. [*Hint:* Suppose that $\nabla^2 T(\mathbf{x}_0) > 0$. Prove that $\nabla^2 T(\mathbf{x}) > 0$ in some ball centered at $\mathbf{x}_0$.]

16. Use Green's first identity to prove that if u is a harmonic function on a region R together with its smooth boundary ∂R in $\mathbb{R}^3$, then the average value over ∂R of the normal derivative of u must be zero.

17. Consider the Newtonian potential function $N(\mathbf{x}) = |\mathbf{x}|^{-1}$ and its associated gradient field $\nabla N(\mathbf{x})$. (See Exercise 16.) Prove that $N(\mathbf{x})$ can be interpreted as the work done in

moving a particle from ∞ to $\mathbf{x}$ along some smooth path through the field ∇N.

18. Show that if $f(x, y)$ equals a function $\overline{f}\left(\sqrt{x^2 + y^2}\right) = \overline{f}(r)$, then
$$\nabla^2 f(x, y) = \frac{\partial^2 \overline{f}(r)}{\partial r^2} + \frac{1}{r}\frac{\partial \overline{f}(r)}{\partial r}.$$

19. Show that if $f(x, y, z)$ equals a function $\overline{f}$ $\left(\sqrt{x^2 + y^2 + z^2}\right) = \overline{f}(r)$, then
$$\nabla^2 f(x, y, z) = \frac{\partial^2 \overline{f}(r)}{\partial r^2} + \frac{2}{r}\frac{\partial \overline{f}(r)}{\partial r}.$$

20. Verify the formula for $\nabla^2 \overline{u}(r, \theta)$ in text Example 4 by computing $\overline{u}_r$, $\overline{u}_{rr}$, and $\overline{u}_{\theta\theta}$ from $\overline{u}(r, \theta) = u(r\cos\theta, r\sin\theta)$. This computation doesn't qualify as a derivation of the polar form of $\nabla^2 u(x, y)$.

21. Verify that the cancellations claimed at the end of text Example 4 do occur.

Chapter 9 REVIEW

1. (a) Find a function f such that
$$\nabla f(x, y) = (3x^2 y, x^3 + 3y^2).$$
   (b) Use your answer to part (a) to prove that $\int_\gamma 3x^2 y \, dx + (x^3 + 3y^2) \, dy = 8$ for any path γ from $(1, 1)$ to $(1, 2)$.

2. Let γ be the closed curve consisting of line segments from $(0, 0)$ to $(1, 0)$, from $(1, 0)$ to $(1, 2)$ and from $(1, 2)$ back to $(0, 0)$. Show that
$$\int_\gamma (-xy + \sin x^2) \, dx + \cos^2 y \, dy = \tfrac{2}{3}.$$

3. Let R be the region in $\mathbb{R}^2$ above the x-axis and below the curve γ parametrized by $g(t) = (1 + t^2, t - t^2)$ with $0 \le t \le 1$. Use Green's Theorem to prove that the area of R is $\tfrac{1}{6}$.

4. (a) Find the unique choice β_0 of the constant β for which the line integral $I(\gamma; \beta)$ is independent of the path γ in the first quadrant, where
$$I(\gamma; \beta) = \int_\gamma \frac{1 + \beta y^2}{(1 + xy)^2} dx + \frac{1 + \beta x^2}{(1 + xy)^2} dy.$$

   (b) Prove that $I(\gamma; \beta_0) = (x_0 + y_0)/(1 + x_0 y_0)$ for all piecewise smooth paths in the first quadrant from $(0, 0)$ to (x_0, y_0), using whatever method seems most convenient.
   (c) Explain to what extent the results of parts (a) and (b) extend to other quadrants.

5. Let S be the closed surface that is the boundary of the solid region inside the cylinder $x^2 + y^2 = 4$ and between the planes $z = 0$ and $z = 2$. Suppose S is positively oriented, with normal vector pointing out at each point. Find the flux of the vector field $\mathbf{F}(x, y, z) = (x^3, y^3 + x, xy)$ across S.

6. Consider the vector field in $\mathbb{R}^3$ given by $\mathbf{F}(\mathbf{x}) = \frac{1}{|\mathbf{x}|^\beta}\mathbf{x}$ for $\mathbf{x} \ne 0$.
   (a) Prove that for every real constant β, curl $\mathbf{F}(\mathbf{x}) = \mathbf{0}$.
   (b) What does part (a) tell you about the circulation of $\mathbf{F}$ around closed paths in $\mathbb{R}^3$?
   (c) Prove that div $\mathbf{F}(\mathbf{x}) = 0$ if and only if $\beta = 3$, that is, just when $|\mathbf{F}(\mathbf{x})| = |\mathbf{x}|^{-2}$.
   (d) For $\beta = 3$ what does part (c) tell you about the flux of $\mathbf{F}$ across closed surfaces in $\mathbb{R}^3$?
   (e) Prove that the flux of $\mathbf{F}$ across the sphere of radius a centered at the origin is $4\pi a^{3-\beta}$; is this consistent with your answer to part (d)?

7. The circulation and flux of the vector field $\mathbf{F}(x, y) = (-x, -y)$ relative to the circle $x^2 + y^2 = 1$ can be computed in several ways, some easier than others.
   (a) Find the total circulation of $\mathbf{F}(x, y)$ around the circle $x^2 + y^2 = 1$ (relative to the counterclockwise direction).
   (b) Find the total flux of $\mathbf{F}$ across the circle $x^2 + y^2 = 1$, relative to outward-pointing unit normal vectors.

8. Let $\mathbf{F}$ be given by $\mathbf{F}(x, y, z) = (x + y, y + z, z + x)$, and let $g(u, v) = (u \cos v, u \sin v, v)$ parametrize a helicoidal surface H.
   (a) Compute curl $\mathbf{F}(x, y, z)$.
   (b) Compute the normal vector $g_u(u, v) \times g_v(u, v)$.
   (c) Compute the integral $\int_{H_1}$ curl $\mathbf{F} \cdot d\mathbf{S}$, where H_1 is the part of the helicoid corresponding to $0 \le u \le 1, 0 \le v \le 4\pi$, representing two complete turns of a helicoid of width 1.

9. Consider the family of 2-dimensional vector fields defined by

$$\mathbf{F}_\alpha(x, y) = \frac{-y}{(x^2 + y^2)^\alpha}\mathbf{i} + \frac{x}{(x^2 + y^2)^\alpha}\mathbf{j},$$

$$(x, y) \ne (0, 0),$$

where α is a positive constant. Note that curl $\mathbf{F}_0(x, y) = 2\mathbf{k}$.
   (a) Compute the scalar curl $\mathbf{F}_\alpha(x, y)$ and prove that it's zero if and only if $\alpha = 1$, in which case it's identically zero.
   (b) What can you say, depending on $\alpha > 0$, about the circulation of $\mathbf{F}_\alpha$ around a smooth closed curve that doesn't contain the origin?
   (c) What can you say, depending on α, about the circulation of $\mathbf{F}_\alpha$ around a circle centered at the origin that encircles the origin once, counterclockwise?

10. Find a function $f(x, y)$ such that $\nabla f(x, y) = (2xy + y^3 + 1, x^2 + 3xy^2)$. Explain why you cannot find an $f(x, y)$ such that $\nabla f(x, y) = (x^2 + 3xy^2, 2xy + y^3 + 1)$.

11. Let R be a plane region with piecewise smooth boundary curve ∂R, oriented counterclockwise. Prove that
$$\int_{\partial R} x \, dy = -\int_{\partial R} y \, dx = A(R),$$
and $\int_{\partial R} x \, dx = \int_{\partial R} y \, dy = 0.$

12. Find $\frac{1}{2}\int_c -y \, dx + x \, dy$, where c is the boundary of the region between the curves $y = x^2$ and $y = 8 - x^2$, oriented counterclockwise. Sketch the region, its boundary, and the field.

13. The conical graph of $z = 2\sqrt{x^2 + y^2}, x^2 + y^2 \le 1$ has the hemispherical graph of $z = 2 + \sqrt{1 - x^2 - y^2}, x^2 + y^2 \le 1$ placed over its top to form a surface S.
   (a) Find parametrizations that orient each of the two parts of S such that the normals point outward over the whole surface.
   (b) Compute the total surface area of S and its enclosed volume.
   (c) Find the flux of $\mathbf{F}(x, y, z) = (x, y, z)$ across S.

14. Let S be the portion of the sphere $x^2 + y^2 + z^2 = 1$ above the xy plane. Calculate $\iint_S z \, d\sigma$ in two ways:
   (a) Directly, as the integral of a function over a surface.
   (b) By noting that $\iint_S z \, d\sigma =$
   $$\iint_S (0, 0, 1) \cdot (x, y, z) \, d\sigma = \iint_S \text{curl } \mathbf{F} \cdot \mathbf{n} \, d\sigma,$$
   where the field $\mathbf{F}$ is $\mathbf{F}(x, y, z) = (-y/2, x/2, 0)$ and S is given an upward orientation, and then applying Stokes's Theorem, either to change the surface of integration or to convert to a line integral.

15. Let f and g be scalar functions of three variables whose second partial derivatives are all continuous.
   (a) Prove $(\nabla f) \times (\nabla g) = \text{curl}(f \nabla g)$.
   (b) Let $f(x, y, z) = x + y + z$ and $g(x, y, z) = x^2 + y^2 - z^2$. Compute
   $$\iint_S ((\nabla f) \times (\nabla g)) \cdot \mathbf{n} \, d\sigma,$$
   where S is the hemisphere $x^2 + y^2 + z^2 = 1, z \ge 0$ and $\mathbf{n}$ is the upward directed normal.
   [Hint: Think about the change of surface or line integral using Stokes's Theorem.]

16. (a) Find a formula for the function $f \colon \mathbb{R}^2 \to \mathbb{R}$ such that $\nabla f(x, y) = (\sin y, x \cos y)$.
   (b) Use your answer to part (a) to compute
   $$\int_\gamma \sin y \, dx + x \cos y \, dy, \text{ where } \gamma \text{ is any path from } (1, 1) \text{ to } (1, 2).$$

17. Let γ be the counterclockwise path consisting of the part of the circle $x^2 + y^2 = 1$ lying in the first quadrant together with the segments $0 \le x \le 1$ and $0 \le y \le 1$ on the x- and y-axes. Use Green's Theorem to compute $\int_\gamma xy \, dx + y \, dy$.

18. Let $\mathbf{F} \colon \mathbb{R}^3 \to \mathbb{R}^3$ be twice continuously differentiable.
   (a) Prove that $\text{div}(\text{curl } \mathbf{F})$ is identically zero.
   (b) Use the result of part (a) to prove that if S is the oriented boundary surface of a region R in $\mathbb{R}^3$, then the surface integral $\int_S$ curl $\mathbf{F} \cdot d\mathbf{S}$ is 0.

19. Let S be a part of a plane parametrized by $g(u, v) = (u + v, 2u + v, u - v)$ for $0 \le u \le 1$ and $0 \le v \le 1$. Let ∂S denote the border of S with the positive orientation. With $\mathbf{F}(x, y, z) = (y, z, x)$, use Stokes's Theorem to find a surface integral equal to $\displaystyle\int_{\partial S} \mathbf{F}\cdot d\mathbf{x}$. Evaluate the resulting surface integral.

20. Let $\mathbf{F}(\mathbf{x})$ be a continuously differentiable vector field on a region B, including its piecewise smooth boundary surface ∂B. Suppose that at each point $\mathbf{x}$ of ∂B the vector $\mathbf{F}(\mathbf{x})$ is tangent to ∂B. Explain why $\displaystyle\int_{\partial B} \operatorname{div} \mathbf{F}\, dV = 0$.

21. Let B be a region in $\mathbb{R}^3$ with a piecewise smooth boundary ∂B.

(a)   Prove that $V(B) = \frac{1}{3}\displaystyle\int_{\partial B} \mathbf{x} \cdot d\mathbf{S}$.

(b)   Prove that $V(B)$ is also equal to each of the following three (equal) surface integrals:
$$\int_{\partial B} x\, dy\, dz, \quad \int_{\partial B} y\, dz\, dx, \quad \int_{\partial B} z\, dx\, dy.$$

22. Define a vector field $\mathbf{F}(\mathbf{x}) = f(\mathbf{x})\mathbf{v}$, where $\mathbf{v}$ is a constant vector and f is a continuously differentiable function $\mathbb{R}^3 \xrightarrow{f} \mathbb{R}$ defined on a region B with a piecewise smooth boundary surface ∂B.

(a)   Apply Gauss's Theorem to prove that
$$\int_B \mathbf{v} \cdot \nabla f(\mathbf{x})\, dV = \int_{\partial B} f(\mathbf{x})\mathbf{v} \cdot d\mathbf{S}.$$

(b)   Use part (a) to establish an equation for vector-valued integrals: $\displaystyle\int_B \nabla f(\mathbf{x})\, dV = \int_{\partial B} f(\mathbf{x})\, d\mathbf{S}$.

(c)   What conclusion can you draw from part (b) if f is identically zero on ∂B?

23. Suppose $\mathbf{F}$ is a continuously differentiable vector field on a simply connected region B in $\mathbb{R}^3$ in which $\operatorname{div}\mathbf{F} = 0$. Let S_1 and S_2 be smooth surfaces in B with the same border. Explain why the flux of $\mathbf{F}$ is the same for both surfaces.

24. Let $u(\mathbf{x})$ be an **harmonic function**, that is, $u_{xx} + u_{yy} + u_{zz} = 0$, in a region B of $\mathbb{R}^3$ having a smooth boundary ∂B.

(a)   Prove that $\displaystyle\int_B |\nabla u|^2\, dV = \int_{\partial B} u\frac{\partial u}{\partial \mathbf{n}}\, d\sigma$, where $\mathbf{n}$ is the outward-pointing unit normal vector to ∂B.

(b)   Assume that in addition to being harmonic u is homogeneous of degree m (i.e., $u(t\mathbf{x}) = t^m u(\mathbf{x})$ for $t > 0$). Prove that if ∂B is a sphere of radius a centered at the origin, then $\displaystyle\int_B |\nabla u|^2\, dV = \frac{m}{a}\int_{\partial B} u^2\, d\sigma$. [*Hint:* Prove that $\mathbf{x}\cdot\nabla u(\mathbf{x}) = mu(\mathbf{x})$.]

25. Use Gauss's Theorem and the definition of centroid to prove that
$$\int_S x^2\, dy\, dz + y^2\, dz\, dx + z^2\, dx\, dy$$
$$= \tfrac{8}{3}\pi a^3(x_0 + y_0 + z_0),$$
where S is the sphere of radius a centered at (x_0, y_0, z_0).

26. Let f and g be twice continuously differentiable real-valued functions defined on $\mathbb{R}^3$, and let Δ denote the Laplace operator $\Delta u = u_{xx} + u_{yy} + u_{zz}$.

(a)   Prove that $\Delta(fg) = f\Delta g + g\Delta f + 2\nabla f \cdot \nabla g$.

(b)   Prove that $\Delta(fg) = f\Delta g + g\Delta f$ if the level surfaces of f and g intersect only at right angles.

CHAPTER 10

FIRST-ORDER DIFFERENTIAL EQUATIONS

This chapter is a brief introduction to differential equations, and it makes no assumption about prior knowledge of the subject. However, the use of derivatives in studying functions of a single variable is assumed to be familiar from elementary calculus. One of Newton's many great discoveries was that while we may lack precise information about the values $y(x)$ of a function or its derivatives $y'(x)$ and $y''(x)$ we can sometimes find an equation, called a **differential equation**, relating the function and its derivatives. For example,

$$y'(x) + y(x) = x \quad \text{and} \quad y''(x) + y'(x) = 0$$

are differential equations, usually abbreviated

$$y' + y = x \quad \text{and} \quad y'' + y' = 0.$$

A **solution** of a differential equation on an interval $a < x < b$ is a function $y(x)$, which, when substituted along with its relevant derivatives into the differential equation, satisfies the equation for all x in some subinterval. For example, it's a routine check that

$$y(x) = x - 1 \quad \text{and} \quad y(x) = e^{-x}$$

are respective solutions of the preceding two equations. As with algebraic equations, we'll want find all solutions of a differential equation. We'll also consider the geometric interpretation of an equation and its solutions and the derivation of the equation from scientific principles. This chapter takes up these matters for **first-order** equations, that is, differential equations in which the derivatives that occur are of first order.

SECTION 1   DIRECTION FIELDS

We can interpret a differential equation of the form

$$y' = F(x, y)$$

as assigning a slope y' to a point (x, y). The assignment is usually represented geometrically by drawing through the point with coordinates (x, y) a line segment with slope $y' = F(x, y)$. Just such an array of points and segments is shown in Figure 10.1(b). A collection of points with directions attached is called a **direction**

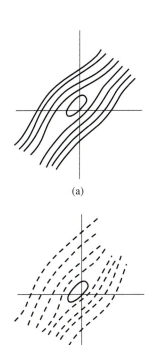

(a)

(b)

FIGURE 10.1

field or **slope field**, and geometrically speaking the assignment of slopes to points is the essence of the equation $y' = F(x, y)$.

It's important to be clear about the distinction between direction fields and the vector fields introduced in Chapter 6, Section 1. A picture of a direction field always contains a 1-dimensional domain axis whose positive direction determines a direction for the solution curves, while speed is equal to the slope of segments relative to this axis. In a vector field we need some device such as an arrow point to indicate direction, and we indicate speed by the length of a segment. A 1-dimensional vector field has its arrows all on the same line, not a helpful picture, which is one reason for resorting to direction fields here instead.

1A   Plotting Direction Fields

When a thin film of fluid flows steadily over a plane surface, the particles of fluid trace in the plane paths called **flowlines**. Figure 10.1(a) illustrates such a flow by showing some of its flowlines. In practice, we might try to describe the flow by giving even less information, namely, just some short line segments tangent to the flowlines at a selection of points. Figure 10.1(b) shows some tangent segments, chosen from among the tangents to the paths in Figure 10.1(a). Visually it's fairly easy to reconstruct the significant features of Figure 10.1(a) from Figure 10.1(b); to do this graphically, we can sketch curves through the selected points, making them appear to be tangent to the segment through each point. A study of such a reconstruction is the geometric theme of this chapter.

There are two natural ways to produce a sketch of a direction field. One way is draw tangents to flowlines. The other is to make the sketch associated with the first-order differential equation $y' = F(x, y)$ by drawing a short segment with slope $F(x, y)$ through the point (x, y). These two ways of looking at a direction field blend together when we solve the differential equation. The reason is that a solution $y(x)$ satisfies

$$y'(x) = F\big(x, y(x)\big);$$

therefore the graph of $y(x)$ has a slope equal to the slope specified by the differential equation

$$y' = F(x, y)$$

at the point $(x, y) = \big(x, y(x)\big)$. In particular, the curves in Figure 10.1(a) are the graphs of solutions coming from the direction field in Figure 10.1(b).

EXAMPLE 1   Suppose that by physical measurement we decide that the directions in a flow of particles are determined according to the differential equation

$$y' = -\frac{y}{x}, \quad \text{for } x \neq 0.$$

At each point in the xy-plane, except for points on the y-axis, where $x = 0$, the equation specifies a numerical slope y'. We can make a table of some sample points and slopes:

(x, y)	$Y' = -\dfrac{Y}{X}$
$(1, 1)$	-1
$(1, 2)$	-2
$(2, 1)$	$-\frac{1}{2}$
$(-1, 2)$	2
$(-2, 2)$	1

By plotting some points and at each drawing a short segment with the specified slope, we get the picture of the direction field shown in Figure 10.2(a). The shape of the curves tangent to the segments in Figure 10.2(a) is fairly easy to sketch, and some are in the figure. Note in particular that the positive and negative x-axes, where $y = 0$, are such curves, but that the vertical y-axis is excluded because of the restriction that $x \neq 0$.

We can use calculus in this example to find formulas for the solution curves. We multiply the given equation by x and rearrange to give

$$xy' + y = 0.$$

Treating y as a function of x, the product rule for differentiation shows that our equation is the same as

$$(xy)' = 0.$$

But this means that the product xy must be a constant: $xy = c$. In other words,

$$y = \frac{c}{x}.$$

The graphs of $y = c/x$, for various choices of c are just the curves tangent to the segments of the direction field in Figure 10.2(a). In particular, $c = 0$ corresponds to the x-axis except for $x = 0$, and $c = 1$ corresponds to the curves sketched in the first and third quadrants. Finally, we can verify directly that given a constant c,

$$y = \frac{c}{x} \quad \text{satisfies} \quad y' = -\frac{y}{x}, \quad \text{for } x \neq 0.$$

FIGURE 10.2

Direction fields and solution graphs.

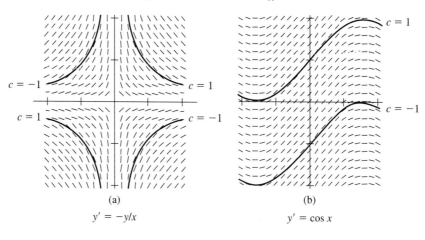

(a)

$y' = -y/x$

(b)

$y' = \cos x$

The reason is that $y' = -c/x^2$, and on the other hand, $-y/x = -c/x^2$, also. Hence

$$y' = -\frac{c}{x^2} = -\frac{y}{x}.$$

EXAMPLE 2   The special case of the equation $y' = F(x, y)$ in which F is independent of y takes the form

$$y' = G(x).$$

We assume that G is a continuous function on some interval. To solve the equation, we integrate both sides with respect to x, getting formally,

$$y = \int G(x)\, dx + C.$$

Here the indefinite integral stands for any function whose derivative is G. We know that any two such integrals differ by at most an additive constant C. Each different constant C gives a graph parallel to the others because the function

$$F(x, y) = G(x)$$

is independent of y: that is, direction segments lying on the same vertical line are all parallel.

For example, the differential equation

$$y' = \cos x$$

has solutions

$$y = \int \cos x\, dx + c$$

$$= \sin x + c.$$

The direction field generated by

$$y' = G(x) = \cos x$$

is sketched in Figure 10.2(b) together with the particular solutions $y = \sin x + 1$ and $y = \sin x - 1$.

The previous example suggests that solving first-order differential equations is something like finding indefinite integrals. In particular, we should expect solutions of a first-order differential equation to be distinguished from one another by specifying an arbitrary constant as in the preceding examples. The usual way to single out a particular solution is to specify that its graph should pass through some preassigned point (x_0, y_0). An **initial condition** for a solution $y(x)$ of a first-order differential equation requires $y(x)$ to satisfy a condition of the form $y(x_0) = y_0$. The problem of satisfying

$$y' = F(x, y) \quad \text{and} \quad y(x_0) = y_0,$$

a differential equation and an initial condition, is called an **initial-value problem**.

Sometimes people refer loosely to a solution formula that contains an arbitrary constant as a "general solution formula" for a first-order differential equation, even though the formula may not contain all possible solutions as special cases. We'll avoid the term "general solution" except when we can actually show that the formula really does contain all solutions.

EXAMPLE 3   We return here to the differential equation

$$y' = -\frac{y}{x}$$

of Example 1, with solutions

$$y = \frac{c}{x}, \quad x \neq 0, \quad c \text{ constant.}$$

We find the particular solution whose graph passes through the point $(x_0, y_0) = (\frac{1}{2}, 2)$ by substituting these values into the solution formula. The constant c is determined by the equation

$$2 = \frac{c}{\frac{1}{2}},$$

so that $c = 1$. Thus the solution to the initial-value problem

$$y' = -\frac{y}{x}, \quad y(\tfrac{1}{2}) = 2$$

is given by the formula

$$y = \frac{1}{x}, \quad x > 0.$$

The graph of this solution is shown in Figure 10.2(a). To find the solution curve through an arbitrary preassigned point (x_0, y_0), with $x_0 \neq 0$, we make the substitution

$$y_0 = \frac{c}{x_0}$$

and find $c = x_0 y_0$. The solution through (x_0, y_0) is evidently

$$y = \frac{x_0 y_0}{x}, \quad x \neq 0.$$

The preceding examples all suggest that, with minor exceptions, through each point of a direction field there is a unique solution curve for $y' = F(x, y)$ and that the solution extends without hindrance wherever the field is defined. The following examples show that neither of these statements is true in general.

EXAMPLE 4   The differential equation

$$y' = \begin{cases} \sqrt{y}, & y \geq 0 \\ 0, & y < 0 \end{cases}$$

FIGURE 10.3

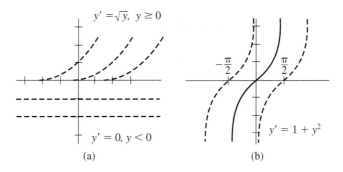

$y' = \sqrt{y},\ y \geq 0$

$y' = 0,\ y < 0$

(a)

$-\dfrac{\pi}{2}$ $\dfrac{\pi}{2}$

$y' = 1 + y^2$

(b)

has two distinct solutions passing through $(x_0, y_0) = (0, 0)$:

$$y(x) = 0, \quad \text{for} \quad -\infty < x < \infty$$

and

$$y(x) = \begin{cases} 0, & -\infty < x < 0, \\ \dfrac{x^2}{4}, & 0 \leq x < \infty. \end{cases}$$

The first solution has its graph coinciding with the x-axis for all x, whereas the graph of the second coincides with the x-axis for $x \leq 0$ and then assumes a parabolic shape, shown in Figure 10.3(a). Thus a solution to the differential equation is not uniquely determined by the requirement that its graph pass through $(0, 0)$. For still more solutions of this differential equation see Exercise 6.

EXAMPLE 5   The differential equation

$$y' = 1 + y^2$$

has the solution $y = \tan x$ with its graph passing through $(x_0, y_0) = (0, 0)$. But the solution tends to infinity discontinuously at $x = \pm\pi/2$, despite having $F(x, y) = 1 + y^2$ well-behaved throughout the entire xy-plane. Figure 10.3(b) makes it clear graphically why the solution can't be carried on continuously outside the interval $-\pi/2 < x < \pi/2$.

The question of existence and uniqueness of solutions is taken care of for large classes of differential equations by using the methods of the next three chapters. We'll state without proof a theorem for the initial-value problem $y' = F(x, y)$, $y(x_0) = y_0$.

1.1   Existence and Uniqueness Theorem.   Suppose the function $F(x, y)$ and its partial derivative $F_y(x, y)$ are both continuous for $a < x < b$ and $\alpha < y < \beta$, and that $a < x_0 < b$ and $\alpha < y_0 < \beta$. Then the initial-value problem $y' = F(x, y)$, $y(x_0) = y_0$ has a unique solution defined on some subinterval of $a < x < b$; if in addition there is a constant B such that $|F_y(x, y)| < B$ for all x in the interval and for all real numbers y, then this solution will exist on the entire interval $a < x < b$.

EXERCISES

By substituting into the given differential equation in Exercises 1 to 6, verify that the corresponding formula to the right gives one or more solutions to the differential equation. Then determine the arbitrary constant so that the differentiable function $y(x)$ satisfies the given initial condition of the form $y(a) = b$ and satisfies the given differential equation on an interval containing a.

1. $y' = y + 1$; $y = Ce^x - 1$, $y(0) = 2$

2. $\dfrac{dy}{dx} = -\dfrac{x}{y}$; $y = \sqrt{a^2 - x^2}$, $|x| < a$, $y(1) = 4$

3. $y' + y = 0$; $y = Ke^{-x}$. $y(5) = 6$

4. $y' = \dfrac{1}{x}$, $x \neq 0$; $y = \log|x| + C$, $y(-1) = 3$

5. $y' = y^2$; $y = (C - x)^{-1}$, $y(3) = 2$

6. $y' = 1 + y^2$; $y = \tan(x + c)$, $y(1) = 1$

For each of the differential equations 7 to 10 of the form $y' = F(x, y)$, sketch the associated direction field, locating a short segment with slope $F(x, y)$ at enough points (x, y) so that a geometric pattern begins to appear. Then sketch into the same picture a solution graph containing the given point (x_0, y_0).

7. $y' = \dfrac{y}{x}$, $(x_0, y_0) = (1, 2)$

8. $\dfrac{dy}{dx} = -\dfrac{x}{y}$, $(x_0, y_0) = (1, 1)$

9. $\dfrac{dy}{dx} = y + x$, $(x_0, y_0) = (1, -1)$

10. $y' = x^2$, $(x_0, y_0) = (1, 0)$

Isoclines An **isocline** in a direction field is a curve along which the directions of the field are all the same. Finding the isoclines of a field is helpful in sketching the field because the direction segments on an isocline are all parallel. For the direction field determined by a differential equation $y' = F(x, y)$, the isoclines satisfy equations of the form $F(x, y) = m$, where m is some constant slope. In each of Exercises 11 to 16, sketch several isoclines, and then sketch the direction field by drawing parallel segments crossing the isocline curves $F(x, y) = m$ with slope m.

11. $y' = -\dfrac{y}{x}$

12. $y' = x + y$

13. $y' = x^2 + y^2$

14. $y' = x^2$

15. $y' = \dfrac{1 - y}{x}$, $x \neq 0$

16. $y' = y$

The differential equations 17 to 22 are of the special form $y' = f(x)$, having isoclines that are lines parallel to the y-axis. Thus to sketch the direction field you need to determine only one slope on each such line, making all slope-segments centered on that line parallel to the first one. Sketch the direction field for each of the following differential equations and then use the field to sketch in a few solution graphs.

17. $y' = x^3$

18. $y' = \sqrt{x^2 + 1}$

19. $y' = 1/(1 + x^4)$

20. $y' = x^4$

21. $y' = \sqrt[3]{1 - x^3}$

22. $y' = x/(1 + x^4)$

23. (a) Verify that a differential equation of the form $y' = F(x)$, where F is continuous on an interval containing x_0, has solutions on that interval of the form $y(x) = y_0 + \int_{x_0}^{x} F(t)\, dt$.

  (b) Prove that the solution in part (a) is uniquely determined by the requirement that $y(x_0) = y_0$. [*Hint:* Suppose there are two solutions, $y_1(x)$ and $y_2(x)$. What is $(y_1 - y_2)'$?]

24. (a) For the differential equation $y' = \sqrt{y}$, $y \geq 0$, show that there are infinitely many different solutions passing through the point $(x_0, y_0) = (0, 0)$. [*Hint:* Consider $y = (x - a)^2/4$ for $x \geq a$ and $y = 0$ for $x < a$.]

  (b) Verify that the equation $y' = \sqrt{y}$ for $y \geq 0$ has the identically zero solution $y(x) = 0$. Explain why the uniqueness part of Theorem 1.1 is not contradicted by this example.

  (c) Prove that there is no value of the number a such that the formula for solutions through $(x_0, y_0) = (0, 0)$ found in part (a) yields the identically zero solution as a special case.

25. The differential equation $y' = \sqrt{1 - y^2}$ is satisfied by $y(x) = \sin(x + a)$ on any interval on which $y'(x) \geq 0$. The differential equation is also satisfied by $y(x) \equiv 1$ and $y(x) \equiv -1$ (identically 1 and identically -1). Show that on the interval $-\pi/2 < x < \pi/2$ there are infinitely many different solutions passing through $(0, 1)$ and also infinitely many different solutions passing through $(0, -1)$. Explain why the uniqueness part of Theorem 1.1 is not contradicted by this example.

1B   Numerical Methods

Hand-plotting direction fields and solution graphs is good practice for understanding the concepts, but computer graphics programs are much better for producing accurate pictures, particularly when the geometry is complicated. Figure 10.4 shows an example that would be difficult to deal with by hand. We can use commercially available software to sketch direction fields that most people would consider too tiresome to sketch by hand, for example Maple, Matlab and Mathematica. The Web site **http://math.dartmouth.edu/~rewn** contains the Java program DFDEM that will plot a direction field and allow you to draw solution graphs tangent to the field and starting at a graphically determined initial point.

A straightforward way to implement a numerical routine for making approximations to the solution of the initial-value problem

$$y' = F(x, y), \quad y(x_0) = y_0$$

is to start with equally spaced x values

$$x_0, x_1 = x_0 + h, x_2 = x_1 + h, \ldots, x_{m+1} = x_m + h$$

and use the tangent line approximation

$$y - y_k = F(x_k, y_k)(x - x_k) \quad \text{at the point} \quad (x, y) = (x_k, y_k)$$

to get an approximate value y_{k+1} at x_{k+1}. Setting $x = x_{k+1}$ in the tangent line equation and noting that $(x_{k+1} - x_k) = h$, gives

$$y_{k+1} - y_k = F(x_k, y_k)(x - x_k). \quad \text{or} \quad y_{k+1} = y_k + hF(x_k, y_k).$$

The value y_{k+1} is called the kth **Euler approximation** at x_{k+1}, and the entire process is called **Euler's method**.

A computing routine to print approximate values for the solution of the initial-value problem

$$y' = xy, \quad y(0) = 1$$

for x values between 0 and 1, with step size $h = 0.01$, might look like this:

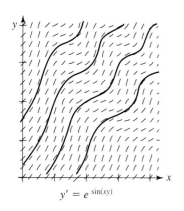

$$y' = e^{\sin(xy)}$$

FIGURE 10.4

```
DEFINE F(X, Y)  =  X * Y
SET X  =  0
SET Y  =  1
SET H  =   0.01
DO
SET Y  =  Y  +  H * F(X, Y)
SET X  =  X  +  H
PRINT X, Y
LOOP WHILE X<1
```

To improve accuracy, we think first of increasing m, the number of subdivisions of the interval from $x_0 = 0$ to $x_m = 1$, thus making h smaller. But increasing the number of subdivisions increases the likelihood of significant round-off error accumulation in the arithmetic. Rather than carelessly increasing m, we prefer to use a simple improvement of the method. The improvement produces a smaller error at each step without increasing the number of steps, so the error increases more slowly.

The **improved Euler method** uses a process called **prediction-correction**. The method assigns a corrected slope to each approximating segment that is the average of the Euler slope at x_k and what the predicted Euler slope would have been at x_{k+1}.

We'll now use y_{k+1} to denote our improved approximate value at step $k + 1$, and use p_{k+1} for the corresponding simple Euler prediction, based on a previously computed value y_k. We follow these steps to approximate the solution to the initial-value problem $y' = F(x, y)$, $y(x_0) = y_0$:

1. Compute the slope $F(x_k, y_k)$
2. Determine a predictor estimate p_{k+1} by $p_{k+1} = y_k + hF(x_k, y_k)$.
3. Compute the average of the two slopes $F(x_k, y_k)$ and a predicted slope $F(x_{k+1}, p_{k+1})$ and use it to determine y_{k+1} by

$$y_{k+1} = y_k + h\left(\frac{F(x_k, y_k) + F(x_{k+1}, p_{k+1})}{2}\right)$$

$$= y_k + \tfrac{1}{2}h\big(F(x_k, y_k) + F(x_{k+1}, p_{k+1})\big).$$

The formula for y_{k+1} displayed above comes from writing the equation for the line through (x_k, y_k) with the average slope and then setting $x = x_{k+1}$ to get the corresponding value $y = y_{k+1}$. A computing routine to implement the method for the initial-value problem $y' = xy$, $y(0) = 1$ might look like this, with step size $h = 0.01$, printing values for x between 0 and 1:

```
DEFINE  F(X, Y)  =  X * Y
SET X  =  0
SET Y  =  1
SET H  =   0.01
DO WHILE X  <  1
SET P  =  Y  +  H * F(X, Y)
SET Y  =  Y  +  (H/2) * (F(X, Y)  +  F(X + H,  P))
SET X  =  X  +  H
PRINT X, Y
LOOP
```

At each stage the value p_k is the *prediction* and y_k is the *correction*. If $h < 0$ the approximations move from larger to smaller x values. For graphic output use some form of PLOT instead of PRINT. Matlab, Maple, and Mathematica software is available for doing the following exercises. Also Java applets DFDEM, 1ORDPLOT, and 1ORD are at the Web site **http://math.dartmouth.edu/~rewn/**.

EXERCISES

In Exercises 1 to 12, make computer-aided direction field sketches for each equation $-2 \le x \le 2$, $-2 \le y \le 2$, and then add a few solution graphs.

1. $y' = \sin(x - y)$

2. $y' = e^{\sin(x+y)}$

3. $y' = \sqrt{9 - y^3}$

4. $y' = x^4 - y^3$

5. $4y' = \sin(x^2 + y^2)$

6. $y' = (1 + y^4)^{-1}$

7. $y' = \cos(x^2 + y^2)$

8. $y' = y/(x^2 + 1)$

9. $y' = (1 + x^4)^{-1}$

10. $y' = \cos(xy)$

11. $y' = e^{-y^2}$

12. $y' = (1 + y^2)^{\frac{3}{2}}$

SECTION 2   APPLICATIONS

One of Newton's many contributions to science was that it's useful to formulate a differential equation to be solved for a physically interesting unknown and then solve the equation. This apparently simple observation has had a profound influence on science.

2A   Direct Integration

We'll consider first some problems that are reducible to solving differential equations of the form $dy/dx = F(x)$. If $F(x)$ is continuous on some interval, then all solutions are

$$y(x) = \int F(x)\, dx + C \quad \text{or} \quad y(x) = G(x) + C,$$

where C is an arbitrary constant and $G'(x) = F(x)$ on the interval in question. To satisfy an initial condition $y(x_0) = y_0$, solve for the constant C in the equation $y(x_0) = G(x_0) + C$, getting $C = y(x_0) - G(x_0)$. This routine for solving the initial-value problem proves the special case of Theorem 1.1 of the previous section in which $F(x, y) = F(x)$ is a function of x alone. In geometric terms, we see that the slopes of a continuously varying direction field generated by an equation $y' = F(x)$ determine a function $y(x)$ on the interval of definition of $F(x)$ whose graph satisfies two conditions: (i) It passes through a given point (x_0, y_0) if x_0 is in the interval. (ii) It is tangent to a direction segment at each of its points.

EXAMPLE 1   If $F(x) = (1 + x^2)^{-1}$ for all real x, then all solutions of $y' = F(x)$ have the form

$$y = G(x) = \int F(x)\, dx + C$$

$$= \int \frac{dx}{1 + x^2} + C = \arctan x + C.$$

We'll assume that the arctangent function is the principal branch, the branch for which $\arctan 0 = 0$. To satisfy the initial condition $y(1) = \pi/2$ we need

$$\frac{\pi}{2} = \arctan 1 + C = \frac{\pi}{4} + C$$

so we take $C = \pi/4$, making the unique solution to the initial-value problem $y = \arctan x + \pi/4$.

FIGURE 10.5

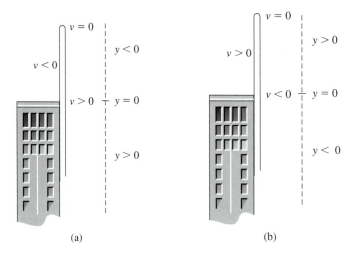

(a)   (b)

EXAMPLE 2

Let g be a constant approximation to the acceleration of gravity near the surface of the earth. A projectile is fired straight up from the top of a building ($y_0 = 0$) with velocity $v(0) = 1000$ feet per second. If we choose to measure distance up from the top of the building, as in Figure 10.5(b), then at time $t \geq 0$ we have $dv/dt = -g$ since gravity acts to decrease velocity. Integrating $dv/dt = -g$ with respect to t using initial condition $v(0) = 1000$ and the estimate $g = 32.2$, gives

$$v = dy/dt = -gt + C_1,$$
$$\approx -32.2t + 1000.$$

Integrating the velocity with initial condition $y(0) = 0$ to find $y(t)$ gives

$$y(t) = -\tfrac{1}{2}gt^2 + v(0)t + C_2$$
$$\approx -16.1t^2 + 1000t.$$

The maximum height is reached when $v = 0$, at $t_{max} \approx 1000/32.2 \approx 31$ seconds. The maximum height is $y_{max} \approx -16.1(t_{max})^2 + 1000t_{max} \approx 15,528$ feet. Note that if we had measured y down from the top of the building our original differential equation would then have been $dv/dt = g$, and the first initial condition would have been $v(0) = -1000$. See Figure 10.5(a).

2B   Separation of Variables

We'll now consider more examples where first-order differential equations arise from geometric or scientific assumptions. The equations are chosen so that we can solve them by a method called separation of variables, illustrated in the next example.

EXAMPLE 3

It's often observed in biological studies that the rate of change dP/dt of the size $P(t)$ of a bacteria population at time t is very nearly proportional to $P(t)$. Expressing this proportionality in the form

$$\frac{dP}{dt} = kP, \qquad (1)$$

where k is a constant, gives a first-order differential equation for P. Because the derivation of the differential equation depends on assumptions that may not be precisely true, we can expect a solution $P(t)$ to be at best an approximation to the true situation. It's our purpose here to study this approximation. Experience with the exponential function allows us to guess one solution. If we let

$$P(t) = Ke^{kt},$$

we see that $P'(t) = kP(t)$ for all real numbers t. In other words, $P = Ke^{kt}$ is a solution. If we had not been able to guess a solution, or if we wanted to try to find still other solutions, we would have proceeded as follows. Assuming that the population size is always positive, we can divide Equation (1) by P, getting

$$\frac{1}{P}\frac{dP}{dt} = k.$$

Next we integrate both sides of the equation with respect to t:

$$\int \frac{1}{P}\frac{dP}{dt}\, dt = \int k\, dt.$$

The integral on the left is $\ln P$; that on the right is kt. Both integrals are determined only to within an additive constant. Hence we can lump the constants together and write

$$\ln P = kt + c,$$

where c is the constant of integration, as yet undermined, Taking the exponential of both sides and recalling that $\exp(\ln P) = P$, we have

$$P(t) = e^{kt+c}$$
$$= e^c e^{kt}.$$

Now set the positive constant $e^c = P_0$, so

$$P(t) = P_0 e^{kt}.$$

Figure 10.6 shows the graph of P for, rather arbitrarily, $k = 2$ and various choices of P_0. The constant P_0 is usually determined by observing that, for $t = 0$, we have $P_0 = P(0)$, which is the size of the population at $t = 0$. If instead of $P(0)$ we happen to know $P(t_1)$ for some $t_1 > 0$, then the equation

$$P(t_1) = P_0 e^{kt_1}$$

leads to

$$P_0 = e^{-kt_1} P(t_1).$$

Hence

$$P(t) = P(t_1) e^{k(t-t_1)}$$

for all $t > 0$.

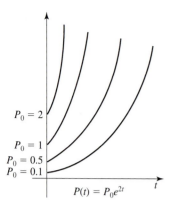

$P_0 = 2$

$P_0 = 1$

$P_0 = 0.5$
$P_0 = 0.1$

$P(t) = P_0 e^{2t}$

FIGURE 10.6

A condition that requires a solution $P(t)$ to satisfy an equation of the form $P(t_0) = P_0$ is called an **initial condition**. The term comes from the interpretation of t_0 in applications as a starting time for an evolving process.

We could solve the differential equation in the previous example because we could rewrite it as an equation between two functions, for each of which we could find an indefinite integral. The typical equation of this type looks like

$$g(y)\frac{dy}{dx} = f(x), \tag{2}$$

though this form isn't possible for all first-order differential equations. (See Exercise 9.) By assuming that y is some differentiable function of x, we can try to find an indefinite integral with respect to x for each side, and so write

$$\int g(y)\frac{dy}{dx}\,dx = \int f(x)\,dx. \tag{3}$$

If there is an indefinite integral G, of g, such that $G'(y) = g(y)$, then we have, by the chain rule, $dG(y)/dx = g(y)\,dy/dx$. If we can also find an indefinite integral F such that $F'(x) = f(x)$, then we can integrate Equation (3) to get

$$G(y) = F(x) + C, \tag{4}$$

relating y and x. There still remains the problem of solving this last equation for y in terms of x.

The process outlined is usually called **separation of variables** because it involves getting the x's on one side of the equation and the y's on the other. The whole matter becomes simpler notationally if we cancel the dx's on the left side of Equation (3). The resulting formal equation

$$\int g(y)\,dy = \int f(x)\,dx \tag{5}$$

still leads to Equation (4) for the solution. The original Equation (2) is sometimes written in the symmetric form

$$g(y)\,dy = f(x)\,dx,$$

which can be interpreted as either

$$g(y)\frac{dy}{dx} = f(x) \quad \text{or} \quad g(y) = f(x)\frac{dx}{dy}.$$

Analogously, we can try to find y as a function of x, or x as a function of y, from Equation (4), whichever suits our purpose better

EXAMPLE 4   The differential equation

$$\frac{dy}{dx} = \frac{y}{x} \tag{6}$$

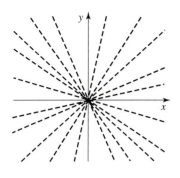

FIGURE 10.7

has associated with it the direction field with slope y/x at the point (x, y). Since the slope y/x is just the same as the slope of the line from $(0, 0)$ to (x, y), the direction field looks like the sketch shown in Figure 10.7. It appears that the solution curves are radial lines from the origin. To prove this, we write the equation in either of the two forms

$$\frac{1}{y} \frac{dy}{dx} = \frac{1}{x} \quad \text{or} \quad \frac{dy}{y} = \frac{dx}{x},$$

assuming both $y \neq 0$ and $x \neq 0$. Integration with respect to x on the left of the first equation gives

$$\int \frac{1}{y} \frac{dy}{dx} dx = \int \frac{1}{x} dx,$$

or, formally,

$$\int \frac{dy}{y} = \int \frac{dx}{x}.$$

In either formulation, we find

$$\ln |y| = \ln |x| + C.$$

Taking the exponential of both sides gives

$$e^{\ln |y|} = e^{\ln |x|} e^c$$

or

$$|y| = e^c |x|.$$

Removing the absolute value symbols, we get

$$y = \pm e^c x.$$

Since e^c is always positive, and since $y = 0$ is a solution of Equation (6), a solution formula for Equation (6) is

$$y = kx,$$

where k is any real number. In other words, the graphs of the solutions are lines through the origin.

EXAMPLE 5   Suppose that a tank containing a chemical in solution is divided into two compartments by a porous membrane. Suppose that the chemical in one compartment is maintained at a fixed concentration C (e.g., in grams per liter) and let $u(t)$ be the concentration of the chemical in the other compartment at time t. It may sometimes be determined experimentally that diffusion takes place across the dividing membrane in such a way that the rate of change of the concentration $u(t)$ is proportional to the difference in concentrations.

Then

$$\frac{du}{dt} = k(C - u),$$

where k is the constant of proportionality. To solve the differential equation, we write it in the form

$$\frac{du}{C - u} = k\,dt.$$

Carrying out the integration gives

$$-\ln|C - u| = kt + c.$$

We can now take exponentials of both sides to get

$$|C - u| = e^{-c}e^{-kt}.$$

Removing absolute values gives

$$C - u = Ke^{-kt},$$

where K is now any nonzero constant. Finally,

$$u(t) = C - Ke^{-kt}.$$

To determine the constant K (remember that C is given at the start), we could, for example, measure $u(0)$. Setting $t = 0$ in the preceding solution formula then gives

$$u(0) = C - K \quad \text{or} \quad K = C - u(0).$$

Hence

$$u(t) = C - \big(C - u(0)\big)e^{-kt}.$$

The constant k also could be determined experimentally by measuring $u(t_1)$ for some $t_1 > 0$ (see Exercise 19). The shape of the graph of $u(t)$ is shown in Figure 10.8 for a single arbitrary choice of C and k and for values of $u(0)$ that are relatively larger and smaller than C. Indeed, the original differential equation $u' = k(C - u)$, with $k > 0$, shows that whenever $C > u(t)$, then $u'(t) > 0$, so that u is increasing. Similarly whenever $C < u(t)$, we must have $u'(t) < 0$, so that u is decreasing. We assumed above that $u \neq C$; if $u(0) = C$ then $u(t) = C$ is a solution, so u is constant.

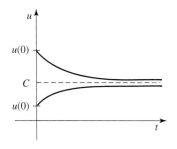

FIGURE 10.8

EXAMPLE 6 Chemical solutions in a tank, for example solutions of salt in water, are often subject to inflow and outflow of a particular chemical at different rates. If $S = S(t)$ is the amount of chemical in the tank at time t, then

$$\frac{dS}{dt} = (\text{rate of inflow}) - (\text{rate of outflow}). \tag{7}$$

As an example, suppose that a full 100-gallon tank contains 150 pounds of salt in solution at time $t = 0$, that a salt solution with a concentration of 2 pounds per gallon is being added at a rate of 2 gallons per minute, and that thoroughly mixed salt solution is flowing out of the tank at a rate of 2 gallons per minute. Thus salt is flowing in at a constant rate of 4 pounds per minute and is overflowing at a rate of $2S(t)/100$ pounds per minute at time t. The differential equation

$$\frac{dS}{dt} = 4 - \frac{2S}{100}$$

then expresses the general relation of Equation (7). To solve the equation for $S(t)$, we write it as

$$\frac{dS}{S - 200} = -\frac{2\,dt}{100}.$$

Integration gives

$$\ln |S - 200| = -\tfrac{1}{50}t + c$$

or

$$|S - 200| = e^c e^{-t/50}.$$

Removing the absolute value, we get

$$S(t) = 200 \pm e^c e^{-t/50}.$$

But $S(0) = 150$ by assumption, so $\pm e^c$ must be equal to -50. The amount of salt at time t is then

$$S(t) = 200 - 50e^{-t/50},$$

and the graph of S is shown in Figure 10.9.

    If instead of salt solution pouring in at a steady rate, the rate is allowed to vary with time, then the method is the same. Suppose, for example, that salt solution at a concentration of 2 pounds per gallon is poured into the same tank at a rate $r(t) = 2 - t/2$ for the first four hours and then no more solution is introduced, so there's no overflow after $t = 4$. The graph of r is shown in Figure 10.10(a). The differential equation for S is now $dS/dt = 0$ when $t > 4$, and when $0 \le t \le 4$,

$$\frac{dS}{dt} = 2\left(2 - \frac{t}{2}\right) - \left(2 - \frac{t}{2}\right)\frac{S}{100}$$

$$= \left(2 - \frac{t}{2}\right)\left(2 - \frac{S}{100}\right).$$

Written in the form

$$\frac{dS}{S - 200} = -\frac{\left(2 - (t/2)\right)dt}{100},$$

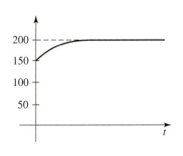

FIGURE 10.9

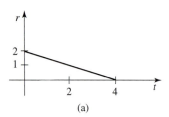

(a)

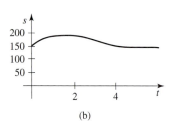

(b)

FIGURE 10.10

the last equation is still easy to solve. We get

$$\ln|S - 200| = -\frac{1}{100}\left(2t - \frac{t^2}{4}\right) + C$$

or

$$|S - 200| = e^c e^{-(t/50)+(t^2/400)}.$$

Removing the absolute value gives

$$S(t) = 200 \pm e^c e^{-(t/50)+(t^2/400)},$$

and, as before, the assumption that $S(0) = 150$ shows that $\pm e^c = -50$ is correct. Thus

$$S(t) = 200 - 50e^{-(t/50)+(t^2/400)}$$

is the solution to the problem for $0 \le t \le 4$. For $t > 4$, the simple differential equation

$$\frac{dS}{dt} = 0$$

means that there is no further change in the amount of salt. The correct value of the solution

$$S(t) = C, \quad 4 < t,$$

comes from setting $t = 4$ in the formula that holds for $0 \le t \le 4$. We find that

$$S(10) = 200 - 50e^{-0.04}$$

$$\approx 152.$$

The graph of S, for both $t \le 4$ and $t > 4$, is shown in Figure 10.10(b). There is no single elementary formula to represent the function, so we write

$$S(t) = \begin{cases} 200 - 50e^{-(t/50)+(t^2/400)}, & 0 \le t \le 4, \\ 200 - 50e^{-0.04}, & 4 < t. \end{cases}$$

EXAMPLE 7   A satellite moving on a radial line away from a planet is subject to a force of magnitude

$$F = -\frac{GMm}{r^2},$$

where r is the distance between the two bodies, m is the mass of the satellite, M is the mass of the planet, and G is a constant depending on the units of measurement (see Figure 10.11). The acceleration of the satellite has magnitude $a = d^2r/dt^2$, so we also have

$$F = ma = m\frac{d^2r}{dt^2}.$$

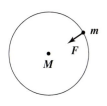

FIGURE 10.11

Equating the two expressions for F and canceling m gives the second-order differential equation

$$\frac{d^2r}{dt^2} = -\frac{GM}{r^2}.$$

If we let $v = dr/dt$ denote the radial speed, then by the chain rule,

$$\frac{d^2r}{dt^2} = \frac{dv}{dt} = \frac{dv}{dr}\frac{dr}{dt}$$

$$= v\frac{dv}{dr}.$$

The differential equation then becomes

$$v\frac{dv}{dr} = -\frac{GM}{r^2}.$$

Integration of both sides with respect to r gives

$$\int_{r_0}^{v} v\,dv = -GM \int_{r_0}^{r} \frac{dr}{r^2}$$

or

$$\frac{v^2}{2} - \frac{v_0^2}{2} = \frac{GM}{r} - \frac{GM}{r_0},$$

where v_0 is the radial speed at distance r_0 from the planet. This relation between speed and distance enables us to determine the **escape speed** of the satellite, namely the speed v_0 that must be attained at distance r_0 so that the speed v always remains positive thereafter. We must have

$$v^2 = v_0^2 + \frac{2GM}{r} - \frac{2GM}{r_0} > 0.$$

Since $GM/r \to 0$ as $r \to \infty$, the only way the inequality can hold is to have

$$v_0^2 - \frac{2GM}{r_0} > 0.$$

The critical escape speed that must be exceeded at distance r_0 is thus

$$v_0 = \sqrt{\frac{2GM}{r_0}}.$$

The analysis presented here ignores the possibility that the planet moves under the influence of the satellite; practically speaking this is fine if the satellite has negligible mass compared with the planet's mass. Also, considerations involving kinetic and potential energy show that a formula of this type for escape speed is correct even if the satellite is on a nonradial path. We treat both of these issues in Chapter 12, Section 3, where we allow for two bodies having commensurate masses as would be the case in a double star system.

EXERCISES

In Exercises 1 to 10, solve each differential equation by direct integration, and find the particular solution that satisfies the associated initial condition by determining one or more constants of integration.

1. $y' = x(1 - x)$, $y(0) = 1$

2. $ds/dt = (t + 1)^2$, $s(1) = 2$

3. $y' = x/(1 - x^2)$, $y(0) = 1$

4. $du/dv = v^2 + 1$, $u(-1) = 1$

5. $y'' = \sin x$, $y(0) = 1$, $y'(0) = 1$

6. $y''' = 1$, $y(0) = y'(0) = y''(0) = 0$

7. $dz/dt = te^t$, $z(0) = 1$

8. $y' = \arctan x$, $y(0) = 0$

9. $dx^2/dt^2 = e^t$, $x(0) = 1$, $dx/dt(0) = 0$

10. $y'''' = x$, $y(0) = y''(0) = 0$, $y'(1) = y'''(1) = 1$

11. A projectile is fired up from ground level with an initial velocity of 5000 feet per second. What is the maximum altitude attained, and how long does it take to get there, assuming $g = 32$ feet per second per second?

12. A weight is dropped from 5000 feet above ground. How long does it take to reach the ground, and with what final velocity does it hit? Assume $g = 32$ ft./sec².

13. Suppose the two objects described in Exercises 11 and 12 are released at the same time and are aimed directly at each other.
(a) How long after release do they meet, and at what height above ground?
(b) What initial velocity should the projectile be given so that the two objects meet 2500 feet above ground?

14. A projectile is fired up from ground level so that its maximum height will be 5000 feet. What is its initial velocity?

15. A weight is thrown down from 5000 feet above ground so as to reach the ground in 10 seconds. What is the velocity of the throw?

16. Suppose the objects described in the previous two exercises are sent on their way at the same time and are aimed directly at each other.
(a) About how long after release do they meet, and at what height above ground?
(b) What initial velocity should the projectile be given so that the two objects meet 2500 feet above ground?

Find a solution formula for each of the following differential equations, and then find a particular solution that satisfies the given additional condition. Verify by substitution that your solution does satisfy the differential equation.

17. $\dfrac{dy}{dt} = 2y$, $y(0) = 2$

18. $\dfrac{dy}{dt} = 2ty$, $y(0) = 2$

19. $y' = \dfrac{x}{y^2}$, $y(1) = 0$

20. $\dfrac{dy}{dx} = -\dfrac{x}{y}$, $y(1) = 1$

21. (a) Suppose that a spherical ball of dry ice evaporates in such a way that the rate of evaporation dV/dt is always proportional to the radius r of the ball. Use $V = (\frac{4}{3})\pi r^3$ to show that the first-order differential equation satisfied by r as a function of t is of the form
$$\frac{dr}{dt} = \frac{k}{r},$$
where k is a negative constant.
(b) Solve the differential equation in part (a), and use the observed measurements that at time $t = 0$ the radius of the ball is 1 inch, whereas 1 hour later the radius is $\frac{1}{2}$ inch, to determine a particular solution as well as the constant k.
(c) How long does it take the ball to evaporate completely starting with a radius of 1 inch?

22. Psychological studies of stimulus and response often attempt to treat these as numerical variables s and r related by an equation of the form $r = f(s)$. It's sometimes hypothesized that f satisfies a differential equation of the form
$$\frac{dr}{ds} = k\frac{r^n}{s}, \quad \text{with } k > 0.$$

Which of the two hypotheses on the exponent n, $n = 0$ or $n = 1$, is consistent with the following table of experimental values?

r	s
0.5	1
1	2
3	6

23. **(a)** Suppose that a 100-gallon tank containing 150 pounds of salt in solution at $t = 0$ has pure water added at a rate of 2 gallons per minute and that the resulting mixture is drawn off at a rate of 2 gallons per minute also. Find a differential equation satisfied by $S(t)$, the amount of salt in the tank at time t, and solve the equation to find $S(t)$. Is S increasing or decreasing? What is $\lim_{t \to \infty} S(t)$?

   (b) Suppose that the process described in part (a) is modified as follows: (i) each 2 gallons flowing in per minute contains 1 pound of salt; (ii) only 1 gallon of solution is drawn off per minute; (iii) 1 gallon of water per minute is boiled away as steam. Answer the same questions as in part (a).

24. Assume that a membrane separating a vat into two components has a porosity that is variable with time, so that the equation

$$\frac{du}{dt} = k(t)(C - u)$$

is satisfied by $u(t)$, the concentration of some chemical in one of the compartments. Suppose that by measuring $u(t)$ we find that

$$u(t) = C(1 - e^{-t^2}).$$

What is the corresponding porosity factor $k(t)$?

25. If the solution u found in Example 5 of the text has the form

$$u(t) = 10 - 5e^{-kt},$$

and $u(2) = 5$, what is the constant k?

26. Show that the differential equation

$$\frac{dy}{dx} = y + x$$

cannot be written in the form

$$g(y)\frac{dy}{dx} = f(x),$$

and therefore cannot be solved by separating variables.

27. **(a)** Sketch the direction fields for the two differential equations

$$\frac{dy}{dx} = \frac{y}{x} \quad \text{and} \quad \frac{dy}{dx} = -\frac{x}{y},$$

and show that at points (x, y), for which both are defined, the direction fields are perpendicular.

(b) What are the solution curves of the two differential equations in part (a)?

28. A function F of (x, y) is called **homogeneous** of degree n if $F(tx, ty) = t^n F(x, y)$ for all x, y, and t.

   (a) Show that if F is homogeneous of degree zero, then the substitution $y = xu$ transforms the differential equation

$$\frac{dy}{dx} = F(x, y)$$

into

$$\frac{du}{dx} = \frac{F(1, u) - u}{x},$$

so that this equation is of the form that's solvable by separation of variables.

   (b) Show that $F(x, y) = (x^2 + y^2)/2xy$ is homogeneous, and use the substitution of part (a) to change the equation

$$\frac{dy}{dx} = \frac{x^2 + y^2}{2xy}$$

into an equation of the form

$$\frac{du}{dx} = G(x, u).$$

   (c) Solve the last differential equation of part (b), and replace u by y/x in the resulting solution. Then check to see that you have found a solution to the equation $y' = (x^2 + y^2)/2xy$.

29. **(a)** Referring to Example 7 of the text, suppose that the planet is the earth, having a radius of 4000 miles, and that the acceleration of gravity at the surface of the earth is -0.006 miles per second (-32.2 feet per second, approximately). Find the constant GM.

   (b) Find the initial velocity that a projectile fired from the surface of the earth would need in order not to fall back to the surface of the earth.

   (c) Find the velocity that a projectile would need 1000 miles above the surface of the earth in order not to fall back to the surface of the earth.

30. Let g be the acceleration of gravity near the surface of the earth. ($g \approx 32.2$ feet per second per second.) By Newton's law an object falling with negligible air resistance has acceleration

$$\frac{dv}{dt} = g,$$

where $v = v(t)$ is the velocity of the object at time t. Use integration to derive the following relations.

(a) $v(t) = gt + v_0$, where v_0 is the velocity at time 0.

(b) $s(t) = \frac{1}{2}gt^2 + v_0 t + s_0$, where $s = s(t)$ is the distance at time t of the object from the reference point $s = 0$.

31. An object dropped near the earth's surface falls distance $s(t) = \frac{1}{2}gt^2$ in time t. In particular, $s_0 = s(0) = 0$ and $v_0 = s'(0) = 0$.

(a) Show that $s = s(t)$ satisfies the first-order differential equation

$$\frac{ds}{dt} = \sqrt{2gs}.$$

(b) Show that the differential equation in part (a) has each member of the one-parameter family

$$s = \left(\sqrt{\frac{g}{2}}\, t + c \right)^2 = \frac{1}{2}gt^2 + \sqrt{2g}\, ct + c^2$$

as a solution.

(c) Show that the solution in part (b) satisfies $v_0^2 = 2gs_0$.

32. The general solution to the falling object problem treated in the two previous problems is

$$s = \frac{1}{2}gt^2 + v_0 t + s_0,$$

where v_0 is initial velocity and s_0 is initial displacement.

(a) Show that $s = s(t)$ satisfies the first-order differential equation

$$\frac{ds}{dt} = \sqrt{v_0^2 + 2g(s - s_0)}.$$

[*Hint:* Solve for t in terms of both s and ds/dt.]

(b) Show that the expression $v_0^2 + 2g(s - s_0)$ under the radical is always nonnegative, given our assumptions on s. (*Hint:* When does that expression reach its minimum as a function of t?)

33. Flow of liquid from a tank. A cylindrical tank with cross-sectional area A has an outlet hole in its side near the bottom. If $h = h(t)$ is the height of an ideal fluid above the outlet at time t, and a is the area of the outlet hole, then $V(t)$, the remaining fluid volume at time t satisfies **Torricelli's equation**

$$\frac{dV}{dt} = -a\sqrt{2gh}.$$

An intuitive justification for the equation is to note that it depends on having the outlet velocity equal to the free-fall velocity of a drop of fluid from height h, as derived in the previous exercise; thus $-dV/dt$ equals area a times outlet velocity $\sqrt{2gh}$. (A thoroughly scientific justification depends on principles of fluid mechanics.) Thus for an ideal fluid, the equation takes the form

$$\frac{dh}{dt} = -\frac{a}{A}\sqrt{2gh}.$$

(a) Show that the Torricelli equation has a solution of the form $h(t) = (bt + c)^2$. Then determine what the constants b and c must be.

(b) Use your answer to part (a) to find out how long it would take for the fluid height above the outlet to drop from h_0 to 0. In particular, estimate how long it would take to empty a full cylindrical tank with diameter 10 feet, height 20 feet, and circular outlet at the bottom with diameter 6 inches.

34. The first-order nonlinear equation $dy/dx = e^{-x^2 - y^2}$ can in principle be solved by using separation of variables.

(a) Try to find an effective solution formula for the initial-value problem with initial condition $y(0) = -1$, and explain the difficulty you encounter.

(b) Make a computer graphics plot of the solution to the initial-value problem in part (a).

SECTION 3   LINEAR EQUATIONS

The first-order differential equation of the form

$$y' = F(x, y)$$

has a particularly important special case, namely the one in which

$$F(x, y) = -g(x)y + f(x)$$

for some functions g and f. The resulting differential equation is usually written in the **normalized form**

$$y' + g(x)y = f(x). \tag{1}$$

For reasons explained at the end of the section the equation is called a first-order **linear** differential equation.

EXAMPLE 1

If f happens to be identically zero, we can find solutions y to Equation (1) by assuming $y \neq 0$ and writing

$$\frac{y'}{y} = -g(x).$$

Integrating with respect to x, we get

$$\ln |y| = -G(x) + c,$$

where G is an indefinite integral of g and c is a constant. Taking the exponential of both sides gives

$$|y| = e^c e^{-G(x)},$$

and removing the absolute value allows us to replace the positive constant e^c by an arbitrary nonzero constant K:

$$y = K e^{-G(x)}$$
$$= K e^{-\int g(x)\,dx}.$$

3A   Exponential Integrating Factors

The method of solution used in Example 1 fails if the function f in Equation (1) is not zero; it also has the technical defect that it forces us to assume $y \neq 0$ (conceivably there are solutions that take on the value zero). We avoid both objections at once if we use the following method, suggested by the form of the solution found in Example 1. For the differential equation

$$y' + g(x)y = f(x),$$

written in normalized form, we define an **exponential integrating factor** to be

$$M(x) = e^{\int g(x)\,dx},$$

where $\int g(x)\,dx$ is an indefinite integral of g. The trick is next to multiply the differential equation by M to get

$$e^{\int g(x)\,dx} y' + g(x)e^{\int g(x)\,dx} y = f(x)e^{\int g(x)\,dx}.$$

The whole point is that the left side can now be written as the derivative of $e^{\int g(x)\,dx}\,y$, because, by the product rule, applied to the factors $e^{\int g(x)\,dx}$ and y,

$$\frac{d}{dx}(e^{\int g(x)\,dx}\,y) = e^{\int g(x)\,dx}\,y' + g(x)e^{\int g(x)\,dx}\,y.$$

Thus we have rewritten the standard linear differential equation in the form

$$\frac{d}{dx}(e^{\int g(x)\,dx}\,y) = e^{\int g(x)\,dx}\,f(x);$$

it remains only to integrate both sides with respect to x and then solve for y. The integrating factor $M(x)$ is sometimes called an **exponential multiplier**.

EXAMPLE 2

To find all solutions of the linear differential equation

$$y' = xy + x,$$

we first rewrite the equation in the standard form

$$y' - xy = x.$$

The exponential multiplier is then found by identifying the coefficient function $g(x) = -x$ and computing

$$M(x) = e^{\int g(x)\,dx}$$

$$= e^{-\int x\,dx}$$

$$= e^{-(1/2)x^2}.$$

Multiplying the differential equation by M gives

$$e^{-(1/2)x^2}\,y' - xe^{-(1/2)x^2}\,y = xe^{-(1/2)x^2}.$$

But we know from the preceding discussion, or we could verify directly, that this last equation is the same as

$$\frac{d}{dx}(e^{-(1/2)x^2}\,y) = xe^{-(1/2)x^2}.$$

Integrating both sides with respect to x gives

$$e^{-(1/2)x^2}\,y = \int xe^{-(1/2)x^2}\,dx + C$$

$$= -e^{-(1/2)x^2} + C.$$

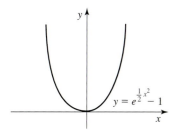

FIGURE 10.12

Then multiplying by $e^{+(1/2)x^2}$ gives

$$y = -1 + Ce^{(1/2)x^2}$$

for the solution. Figure 10.12 shows the graph of the particular solution satisfying $y(0) = 0$.

Two points should be emphasized about applying the exponential multiplier method:

1. The linear differential equation must be in standard form

$$y' + g(x)y = f(x)$$

before identification of the coefficient function g for the purpose of computing $M(x) = e^{\int g(x)\,dx}$.

2. The differential equation

$$y' + g(x)y = f(x)$$

and its multiplied form

$$M(x)y' + M(x)g(x)y = M(x)f(x)$$

are completely equivalent to one another in the sense that any solution of one equation is also a solution of the other. The reason is that the multiplier M, being an exponential function, is never equal to zero, so we can multiply and divide by it as we please.

3B   Applications

EXAMPLE 3   Suppose that a 100-gallon vat contains 10 pounds of a certain chemical dissolved in water and that a solution of the same chemical is being run into the vat at a rate of 3 gallons per minute. The solution being run into the vat has a concentration that increases slowly with time according to the formula

$$C(t) = 1 - e^{-t/100}.$$

The solution is kept thoroughly mixed and the excess is drawn off, also at a rate of 3 gallons per minute. Let $S(t)$ stand for the amount of chemical in the tank at time $t \geq 0$. We have rate of inflow minus rate of outflow equal to

$$\frac{dS}{dt}(t) = 3C(t) - 3\frac{S(t)}{100}$$

$$= 3(1 - e^{-t/100}) - \frac{3}{100}S(t).$$

The resulting first-order equation can't be solved by separation of variables unless $C(t)$ were to be replaced by a constant. But the equation is linear in any case, and in standard form is

$$\frac{dS}{dt} - \frac{3}{100}S = 3(1 - e^{-t/100}).$$

An exponential multiplier is given by

$$M(t) = e^{\int (3/100)dt} = e^{3t/100}.$$

Multiplying the equation by M puts it in the form

$$\frac{d}{dt}(e^{3t/100}S) = 3(e^{3t/100} - e^{2t/100}),$$

and integration with respect to t gives

$$e^{3t/100}S(t) = \int 3(e^{3t/100} - e^{2t/100})dt + K$$

$$= 100e^{3t/100} - 150e^{2t/100} + K.$$

Then multiplication by $e^{-3t/100}$ gives

$$S(t) = 100 - 150e^{-t/100} + Ke^{-3t/100}.$$

To determine the constant K we recall that the vat initially contains 10 pounds of the chemical, so that $S(0) = 10$. Then setting $t = 0$ in the formula for $S(t)$ gives

$$10 = -50 + K, \quad \text{or} \quad K = 60.$$

Thus the desired particular solution is

$$S(t) = 100 - 150e^{-t/100} + 60e^{-3t/100}.$$

Notice that

$$\lim_{t \to \infty} C(t) = 1,$$

so that the concentration of the solution being added approaches 1 pound per gallon. From this information we could conclude on physical grounds that the total amount of salt in the 100-gallon tank should approach 100 pounds; indeed the formula for $S(t)$ shows that

$$\lim_{t \to \infty} S(t) = 100.$$

EXAMPLE 4   Let $f(t)$ be the concentration of a chemical solution on one side of a porous membrane, and let $u(t)$ be the concentration on the other side. Suppose that diffusion takes place through the membrane in such a way that

$$\frac{du}{dt} = 2\big(f(t) - u\big),$$

that is, so that the rate of change of u is proportional to the difference in concentrations. If $u(0) = 3$, and f is maintained so that

$$f(t) = \begin{cases} 4, & 0 \le t < 10 \\ 1, & 10 \le t, \end{cases}$$

then we can most easily solve the equation by writing it as

$$\frac{du}{dt} + 2u = 2f(t).$$

An exponential multiplier M is given by

$$M(t) = e^{\int 2\,dt} = e^{2t}.$$

Hence the differential equation is

$$\frac{d}{dt}(e^{2t}u) = 2e^{2t}f(t).$$

Integration of both sides from $t = 0$ to $t = s$ gives

$$\int_0^s \frac{d}{dt}\left(e^{2t}u(t)\right) dt = \int_0^s 2e^{2t}f(t)\,dt$$

or

$$e^{2s}u(s) - u(0) = \int_0^s 2e^{2t}f(t)\,dt.$$

Then

$$u(s) = u(0)e^{-2s} + 2e^{-2s}\int_0^s e^{2t}f(t)\,dt. \tag{2}$$

Using the integral with limits is convenient here because we can write, according to the definition of f,

$$\int_0^s e^{2t}f(t)\,dt = \begin{cases} \displaystyle\int_0^s 4e^{2t}\,dt, & 0 \le s \le 10, \\[2ex] \displaystyle\int_0^{10} 4e^{2t}\,dt + \int_{10}^s e^{2t}\,dt, & 10 \le s, \end{cases}$$

$$= \begin{cases} 2(e^{2s} - 1), & 0 \le s \le 10, \\[1ex] 2(e^{20} - 1), + \frac{1}{2}(e^{2s} - e^{20}), & 10 \le s. \end{cases}$$

$$= \begin{cases} 2(e^{2s} - 1), & 0 \le s \le 10, \\[1ex] \frac{3}{2}e^{20} - 2 + \frac{1}{2}e^{2s}, & 10 \le s. \end{cases}$$

Then returning to Equation (2),

$$u(s) = 3e^{-2s} + \begin{cases} 4 - 4e^{-2s}, & 0 \le s \le 10, \\ (3e^{20} - 4)e^{-2s} + 1, & 10 \le s. \end{cases}$$

$$= \begin{cases} 4 - e^{-2s}, & 0 \le s \le 10, \\ (3e^{20} - 1)e^{-2s} + 1, & 10 \le s. \end{cases}$$

Sketching the graph of the solution u is left as an exercise.

EXAMPLE 5   **Newton's law of cooling** asserts that the rate of change of the surface temperature $u(t)$ of an object is proportional to the difference between $u(t)$ and $f(t)$, the
temperature of the surrounding medium. Thus

$$\frac{du}{dt} = k(f - u), \quad k > 0.$$

The constant k must be positive to be consistent with knowing that if $f(t) > u(t)$
then $du/dt > 0$. We can't solve this differential equation by separation of variables
unless f is constant, but in any case the equation is linear, with the form

$$\frac{du}{dt} + ku = kf(t).$$

One important problem is to figure out how to control the temperature $u(t)$ in some
desired way by choosing $f(t)$ properly. Our solution method leads to

$$\frac{d}{dt}(e^{kt}u) = ke^{kt} f(t).$$

Hence

$$u(t) = ke^{-kt} \int e^{kt} f(t)\, dt + Ce^{-kt}.$$

It's convenient to choose the indefinite integral to have the value 0 when $t = 0$ so
we can write it as a definite integral. We get

$$u(t) = ke^{-kt} \int_0^t e^{ks} f(s)\, ds + u(0)e^{-kt},$$

where we have replaced C by $u(0)$. If we now take $f(t)$ to be a constant and call it
f_0, the solution becomes

$$u(t) = f_0(1 - e^{-kt}) + u(0)e^{-kt}$$

$$= f_0 + (u(0) - f_0)e^{-kt}.$$

Since $\lim_{t \to \infty} e^{-kt} = 0$, we find that $\lim_{t \to \infty} u(t) = f_0$, which is reasonable from physical
intuition. Other choices for $f(t)$ are considered in the Exercises.

EXERCISES

In Exercises 1 to 4, assume that y represents some differentiable function of x. Find an exponential multiplier M for each combination such that the product has the form $(d/dx)(M(x)y)$.

1. $y' + 2y$

2. $\dfrac{dy}{dx} + xy$

3. $\dfrac{dy}{dx} + \dfrac{2}{x}y$

4. $y' + e^x y$

In Exercises 5 to 8, find the general solution of each of the following linear equations, and then find a particular solution that satisfies the given initial condition.

5. $\dfrac{ds}{dt} + ts = t$, $s(0) = 0$

6. $y' = y + 1$, $y(0) = 1$

7. $2\dfrac{dy}{dx} = xy$, $y(1) = 0$

8. $t\dfrac{dP}{dt} + P = t^3$, $P(1) = 0$

9. Salt solution enters a 100-gallon tank of initially pure water from two different sources. One source provides water containing 1 pound of salt per gallon at a rate of 2 gallons per minute. A second source provides 3 gallons of salt solution per minute at a varying concentration $C(t) = 2e^{-2t}$, measured in pounds of salt per gallon. Assume that the contents of the tank are kept thoroughly mixed at all times and that solution is drawn off at a rate of 5 gallons per minute. Find the amount of salt in the tank at an arbitrary time $t > 0$.

10. The current $i(t)$ in an electric circuit satisfies the differential equation

$$L\frac{di}{dt} + Ri = E(t),$$

where L and R are positive constants, called inductance and resistance, respectively, and $E(t)$ is an applied voltage. Show that

$$i(t) = \frac{1}{L}e^{-Rt/L}\int_0^t E(u)e^{Ru/L}\,du + i(0)e^{-Rt/L}.$$

11. A pellet of mass m failing under the influence of gravity through a viscous medium has a velocity $v(t)$ at time t,

satisfying

$$m\frac{dv}{dt} = mg - kv.$$

Here, g is the acceleration of gravity and k is a positive constant depending on the viscosity. Show that

$$v(t) = \left(v(0) - \frac{mg}{k}\right)e^{-kt/m} + \frac{mg}{k}.$$

12. Choosing an appropriate scale, sketch the graph of a typical function u found at the conclusion of Example 5 of the text, if $0 < u(0) < f_0$. What is the maximum value of u, and what is $\lim_{s \to \infty} u(s)$?

In Exercises 13 and 14, use **Newton's law of cooling** to find the result of choosing $f(t)$ in Example 5 of the text in two ways.

13. $f(t) = e^{-2t}$, for $t \geq 0$, with $u(0) = 10$

14. $f(t) = \begin{cases} f_0(\text{constant}), & \text{for } 0 \leq t \leq 1, \text{ with } u(0) = 5, \\ 0, & \text{for } 1 < t \end{cases}$

15. A container of milk at $70°$ F is placed in a mixture of ice and brine constantly at $30°$ F. Assume the validity of Newton's law described in Example 5 and that the milk has reached $40°$ after 15 minutes.
    (a) Find an approximate value for the constant k in Newton's law.
    (b) When will the milk reach $35°$?

16. Suppose that a metal bar initially at $300°$ F is immersed in a water bath at $100°$ F. for 30 minutes and then is transferred to another water bath at $50°$ F. Assume the validity of Newton's law described in Example 5 of the text.
    (a) What will the temperature of the bar be after an additional 30 minutes, assuming the cooling coefficient for the iron in water is $k = 0.1$?
    (b) Suppose that initially the bar is cooled for 30 minutes in air at $100°$, for which the cooling coefficient is only $k = 0.07$ and is then immersed in water for 30 minutes. What will the temperature of the bar be at the end of the hour?

17. Verify directly that if $y_1(x)$ is a solution of

$$y' + gy = 0,$$

then $cy_1(x)$ is a solution, for every constant c.

18. Verify directly that if $y_1(x)$ and $y_2(x)$ are solutions of the respective equations

$$y' + gy = f_1 \quad \text{and} \quad y' + gy = f_2,$$

then $c_1 y_1 + c_2 y_2$ is, for every pair of constants c_1, c_2, a solution of

$$y' + gy = c_1 f_1 + c_2 f_2.$$

19. An initially full 100 cubic-foot tank starts with 10 pounds of salt dissolved in water. At a certain time additional salt solution begins to enter the tank at a rate of 1 cubic foot per hour, while thoroughly mixed solution runs out a drain at the same rate. However, the amount of salt in the added solution decreases at a constant rate from 1 pound per cubic foot initially all the way down to zero pounds per cubic foot at the end of one hour.

    (a) Find the amount of salt in the tank at a given time during the first hour. In particular, about how much salt will be in the tank at the end of one hour?

    (b) If pure water continues to run into the tank after the first hour at the rate of 1 cubic foot per hour, how much more time will it take for the total amount of salt in the tank to reach 5 pounds?

20. Two 100-gallon mixing tanks are initially full of pure water. A solution containing one pound of salt per gallon of water pours into the first tank at the rate of one gallon per minute. Thoroughly mixed solution runs from the first tank to the second at the rate of one gallon per minute, where it too is thoroughly mixed in before draining away at 1 gallon per minute. There will always be at least as much salt in the first tank as in the second; find the maximum amount of this excess.

21. A 100-gallon tank is initially full of pure water. Salt solution is added for 10 minutes at the rate of 1 gallon per minute with salt content of the added solution increasing linearly over the 10 minutes from 1 pound per gallon to 2 pounds per gallon. Thoroughly mixed salt solution is drawn off at the rate of one gallon per minute. Estimate the amount of salt in the tank at the end of the 10 minutes.

22. A 100-gallon mixing vat is initially half-full of pure water. Two gallons of salt solution per minute at a concentration of one pound of salt per gallon begin to flow in, while one gallon per minute of mixed solution flows out. Estimate the amount of salt in the vat at the moment it begins to overflow.

23. A 100-gallon tank initially contains 50 gallons of water with a total of 10 pounds of salt dissolved in it. A drain is opened in the bottom that is regulated so as to let out 1 gallon of solution per minute. Simultaneously, salt solution begins to be added at 2 gallons per minute with a concentration of 2 pounds per gallon.

    (a) How much salt is in the tank when it first becomes full and starts to overflow?

    (b) If the process is allowed to continue with overflow at an additional outflow of 1 gallon per minute, what is the upper limit for the total amount of salt in the tank? Estimate the additional time after the start of overflow for the amount of salt in the tank to reach 175 pounds.

24. The first-order linear equation $dy/dx - (\sin x)y = e^{-x^2}$ can in principle be solved by using an exponential integrating factor.

    (a) Try to find an effective solution formula for the initial-value problem with initial condition $y(0) = -1$, and explain the difficulty you encounter.

    (b) Make a computer graphics plot of the solution to the initial-value problem of part (a).

Chapter 10 REVIEW

In Exercises 1 to 14, find all functions that satisfy the differential equation.

1. $x(dy/dx) + y - x = 0$

2. $dy/dx = 1/(y(1-x)^2)$

3. $dx/dt = tx + e^t$

4. $(1+x)y' + y = \cos x$

5. $y^3 y' = (y^4 + 1)e^x$

6. $dy/dx = 4x^3 y - y$, $y(1) = 1$

7. $xy' + (2x - 3)y = x^4$

8. $y' = xy + y$

9. $t(dx/dt) = -2x + t^3$, $x(2) = 1$

10. $t(dx/dt) = 1$

11. $dx/dt = -3x^2$

12. $dy/dt + ty = 1$

13. $dx/dt = (x + t)^2$ [*Hint:* Let $x + t = y$.]

14. $dy/dt = \cos^2 y$

15. Consider the differential equation $dy/dx = e^{x-y}$.

(a) In what region of the (x, y)-plane are all solutions strictly increasing?

(b) In what region of the (x, y)-plane are all solutions concave up?

(c) Is the line $y = x$ a solution graph?

(d) Is the line $y = x$ an isocline?

(e) Solve the differential equation by separation of variables. Can you get the information asked for above directly from your solution formula? Which approach seems simpler?

16. Consider the differential equation $dy/dx = e^{x-y}$.

(a) What conclusions can you draw from Theorem 1.1 on existence and uniqueness about solutions of this equation?

(b) Can a solution graph passing through the point $(x, y) = (0, 1)$ cross the line $y = x$? Explain your reasoning.

17. Consider the family of linear equations $y' + ay = c$, with a, c constant, $a \neq 0$.

(a) Show (a) that the isoclines of the direction field of this equation are horizontal lines and (b) that every such line is an isocline.

(b) Sketch the direction field associated with the differential equation $y' + 2y = 1$.

18. Early experiments with objects dropped from rest above the earth led to the conjecture that after an object had fallen distance s its velocity would be proportional to s. Under the contemporary assumption that the acceleration of gravity is constant, the velocity is proportional to $\sqrt{s}$.

(a) Is the early conjecture consistent with initial velocity zero? Explain your reasoning.

(b) Is the early conjecture consistent with positive initial velocity? How would acceleration be related to s under this assumption?

19. A 100-gallon mixing vat is initially full of pure water, whereupon two gallons of salt solution per minute is added, each gallon containing 1 pound of salt. Water evaporates from the tank at the rate of one gallon per minute, and the excess solution overflows into a drain. Find the amount of salt in the tank at time t under the given assumptions and also under the altered assumption that the tank initially contains 50 pounds of salt in solution.

20. Coffee cooling. We are presented two choices for cooling one cup of coffee over a period of 10 minutes: (i) let the coffee cool by itself for 10 minutes and then add cream, or (ii) add the same amount of cream right away and then allow the mixture to cool for 10 minutes. Assume that mixing quantity p of liquid at temperature T_0 and quantity q at temperature T_1 instantly results in quantity $p+q$ with average temperature given by $(pT_0+qT_1)/(p+q)$. Which method will end up with cooler coffee?

CHAPTER 11

SECOND-ORDER EQUATIONS

Most of this chapter is about *linear* differential equations such as

$$y'' - 2y = x, \tag{1}$$

$$y'' - 3y = e^x, \tag{2}$$

$$y'' - 3y' + 2y = f(x), \tag{3}$$

in which the left-hand side is a sum of multiples, with constant coefficients, of functions $y(x)$, $y'(x)$, $y''(x)$. Such equations have several important applications that are taken up in Section 4 and are called **linear constant-coefficient equations.** The reason for focusing on second-order equations is that it's the first and second order derivatives of a function that have the most important interpretations, velocity and acceleration, matched geometrically with slope and concavity. Quite apart from any applications, these differential equations are interesting because we can describe their solutions very explicitly in terms of solution formulas. Furthermore the set of all solutions of a linear equation has a form that enables us to take a geometric view of the sets of all solutions similar to the form of solutions of an algebraic system $A\mathbf{x} = \mathbf{b}$. The solution techniques and concepts of this chapter apply also to many of the systems of equations in Chapter 12.

In Section 3C we'll relax somewhat the requirement that the coefficients of combination in the linear equation should be constants. And finally in Section 7 we'll remove the restriction to linear equations and discuss some features of *non*linear equations such as $y'' + (y')^2 = 0$.

SECTION 1   DIFFERENTIAL OPERATORS

Before beginning a systematic treatment of constant-coefficient linear differential operators, we'll look at some simple examples of linear differential equations.

1A   Examples

EXAMPLE 1

We can write the differential equation $y' - ry = 0$, where r is a constant, as

$$y' = ry, \tag{4}$$

which specifies that the rate of change of y is proportional to the value of y for every value of the variable x. This type of equation appears in Chapter 10, Section 2 for describing population growth. To find solutions we use repeatedly the formula $y' = re^{rx}$ for the derivative of $y = e^{rx}$. It follows that Equation (4) is satisfied if we take $y = e^{rx}$. More generally, if c is an arbitrary constant, then Equation (4) is

490

satisfied if we take

$$y = ce^{rx}, \tag{5}$$

because the c will cancel on both sides. Equation (5) gives the most general solution to (4); observe that we can write (5) in the form

$$e^{-rx} y = c.$$

Differentiating with respect to x gives

$$(e^{-rx} y)' = 0$$

or, using the product rule for derivatives,

$$e^{-rx} y' - re^{-ry} y = e^{-rx}(y' - ry) = 0.$$

Dividing by e^{-rx} leaves $y' - ry = 0$, which is the given Equation (4) rewritten. But now we can reverse these steps, supposing that y is *some* solution. We start with

$$y' - ry = 0$$

and then multiply by e^{-rx} to get

$$e^{-rx} y' - re^{-rx} y = 0.$$

By the product rule, this last equation is

$$(e^{-rx} y)' = 0.$$

Integrating both sides with respect to x gives

$$e^{-rx} y = c,$$

where c is a constant of integration. Multiplying both sides by e^{rx} shows that y must be of the form

$$y = ce^{rx}.$$

Thus we have shown that ce^{rx} is the most general solution of $y' = ry$ in the sense that all particular solutions arise from specifying the value of c.

The method used in the preceding example consists of multiplying the expression $y' + ay$ by e^{ax} and then recognizing the result as the derivative $(e^{ax} y)' = e^{ax} y' + ae^{ax} y$. We'll use this **exponential multiplier** e^{ax} repeatedly in what follows.

EXAMPLE 2   To solve the differential equation

$$y' - 3y = e^x,$$

we multiply by e^{-3x} and get

$$e^{-3x} y' - 3e^{-3x} y = e^{-2x},$$

which is the same as

$$(e^{-3x}y)' = e^{-2x}.$$

Now we integrate both sides with respect to x, getting

$$e^{-3x}y = -\tfrac{1}{2}e^{-2x} + c,$$

where c is some constant of integration. Then multiplying by e^{3x} we obtain

$$y = -\tfrac{1}{2}e^x + ce^{3x}$$

for the most general solution. We can verify directly that we have indeed found *some* solutions, one for each value of c. What we have shown additionally is that *all* solutions must be of the form $-\tfrac{1}{2}e^x + ce^{3x}$.

   Before considering more complicated examples, it will be useful to describe some notation that is often used in solving differential equations. We let D stand for differentiation with respect to some agreed-on variable, say x, and interpret $D + 2$, $D^2 - 1$, and similar expressions as operations acting on suitably differentiable functions y. For example,

$$(D + 2)y = Dy + 2y$$
$$= y' + 2y,$$
$$(D^2 - 1)y = D^2y - y$$
$$= D(Dy) - y = y'' - y.$$

An important observation is that D acts *linearly* on y; the term **linear operator** is sometimes used to avoid possible confusion over y itself being a function of x, though not necessarily a linear function. To see that D acts linearly all we have to do is recall two familiar properties of differentiation:

$$D(y_1 + y_2) = Dy_1 + Dy_2$$
$$D(cy) = cDy, \qquad c \text{ constant.}$$

These two equations express the **linearity** of D. Repeated application of linear operators is linear, so it follows that the operators D^2, D^3, and in general D^n are also linear. Because scalar multiplication is a linear operation and because the sum of linear operations is linear, the operator $(D + a)$ is linear for all constants a. Putting these ideas together allows us to conclude that expressions such as

$$D^2 + a, \quad D^2 + aD + b, \quad (D + s)(D + t)$$

are all linear operators, with the respective interpretations

$$(D^2 + a)y = y'' + ay,$$
$$(D^2 + aD + b)y = y'' + ay' + by,$$
$$(D + s)(D + t)y = (D + s)(y' + ty)$$
$$= D(y' + ty) + s(y' + ty)$$

$$= y'' + ty' + sy' + sty$$
$$= y'' + (t + s)y' + sty$$
$$= (D^2 + (s + t)D + st)y.$$

The last computation shows that for constants s and t

$$(D + s)(D + t) = D^2 + (s + t)D + st,$$

and also that

$$(D + t)(D + s) = D^2 + (s + t)D + st.$$

Thus if a is constant we can formally multiply operators of the form $D - a$ as we do ordinary polynomials with variable D. Conversely it's also sometimes important to be able to factor an operator, for example, $D^2 - 1$. We see immediately that for this example

$$D^2 - 1 = (D - 1)(D + 1)$$
$$= (D + 1)(D - 1).$$

Returning to differential equations, suppose we are given one of the form

$$y'' + ay' + by = 0;$$

Equation (3) at the beginning of this chapter is similar to this, with $a = -3, b = 2$. Writing the equation using differential operators gives

$$(D^2 + aD + b)y = 0.$$

If we try to find a solution of the form $y = e^{rx}$, then $Dy = re^{rx}$ and $D^2y = r^2e^{rx}$, so e^{rx} is a solution if and only if $r^2e^{rx} + are^{rx} + be^{rx} = 0$. Then dividing by e^{rx} gives the condition on r

$$r^2 + ar + b = 0,$$

called the **characteristic equation** of the given differential equation.

EXAMPLE 3   The differential equation

$$y'' - 3y' + 2y = 0$$

has characteristic equation

$$r^2 - 3r + 2 = 0 \quad \text{or} \quad (r - 1)(r - 2) = 0.$$

The roots are $r_1 = 1$ and $r_2 = 2$, so there are solutions

$$y_1(x) = e^x, \quad y_2(x) = e^{2x}.$$

The operator $L = D^2 - 3D + 2$ is linear, so if both $L(y_1) = 0$ and $L(y_2) = 0$ then we also have $L(c_1y_1 + c_2y_2) = 0$, and additional solutions are given by

$$y(x) = c_1e^x + c_2e^{2x}$$

for each pair of constants c_1, c_2. This formula gives all the solutions, but to prove that, we must proceed differently as shown next.

1B   Factoring Operators

Our general method of solution will be to factor an operator into factors of the form $(D+s)$ and $(D+t)$, and then apply the exponential multiplier method of Examples 1 and 2 repeatedly.

EXAMPLE 4 Suppose we want to find all functions $y = y(x)$ that satisfy

$$y'' + 5y' + 6y = 0.$$

We write the equation in operator form as

$$(D^2 + 5D + 6)y = 0.$$

Next we try to factor the operator. We see that

$$(D^2 + 5D + 6) = (D + 3)(D + 2);$$

thus we need to solve

$$(D + 3)(D + 2)y = 0.$$

To find all solutions, we suppose that y is *some* solution. Letting

$$(D + 2)y = u$$

for the moment, we substitute u into the previous equation and arrive at

$$(D + 3)u = 0.$$

But we can solve this first-order linear equation for u if we multiply through by e^{3x}. We get

$$e^{3x}Du + 3e^{3x}u = 0$$

or

$$D(e^{3x}u) = 0.$$

Therefore

$$e^{3x}u = c_1,$$

for some constant c_1, and so

$$u = c_1 e^{-3x}.$$

Recall now that we have temporarily set $(D + 2)y = u$. We then have

$$(D + 2)y = c_1 e^{-3x}.$$

Multiply this first-order linear equation by e^{2x} to get

$$e^{2x}Dy + 2e^{2x}y = c_1 e^{-x} \quad \text{or} \quad D(e^{2x}y) = c_1 e^{-x}.$$

Integrating with respect to x gives

$$e^{2x} y = -c_1 e^{-x} + c_2 \quad \text{or} \quad y = -c_1 e^{-3x} + c_2 e^{-2x}.$$

Since the constants c_1 and c_2 are arbitrary anyway, we can change the sign on the first one to get

$$y = c_1 e^{-3x} + c_2 e^{-2x}$$

for the form of the most general solution.

We find the exponential multiplier used in the previous examples as follows: multiply $(D + a)y$ by e^{ax} to get $D(e^{ax} y)$, that is,

$$e^{ax}(D + a)y = D(e^{ax} y).$$

Repeated application of this formula to constant-coefficient equations leads to the following general theorem that lets us write the solutions knowing only the equation's **characteristic roots**, that is, the roots of the characteristic equation. We consider first the case of second-order equations of the form

$$y'' + ay' + by = (D - r_1)(D - r_2)y = 0,$$

where r_1 and r_2 are real numbers, and the values of the arbitrary constants c_1 and c_2 are determined by **initial conditions** that prescribe the values $y(x_0)$ and $y'(x_0)$ of the solution and its derivative at a point x_0.

1.1 Theorem.   The constant-coefficient equation

$$y'' + ay' + by = 0,$$

with unequal characteristic roots r_1, r_2 has its most general solution of the form

$$y = c_1 e^{r_1 x} + c_2 e^{r_2 x}.$$

If $r_1 = r_2$, then $e^{r_2 x}$ is replaced in the general solution formula by $x e^{r_1 x}$ to get

$$y = c_1 e^{r_1 x} + c_2 x e^{r_1 x}.$$

The constants c_1, c_2 are uniquely determined by prescribing initial conditions $y(x_0) = y_0$, $y'(x_0) = z_0$.

Proof.   In operator form, the differential equation is

$$(D - r_1)(D - r_2)y = 0.$$

We assume $y(x)$ is a solution and show that it has the form claimed in the theorem. Set $z(x) = (D - r_2)y(x)$ and substitute z for $(D - r_2)y$ in the previous equation. Now solve the resulting equation $(D - r_1)z = 0$ to get

$$z = c_1 e^{r_1 x}.$$

Note that c_1 is determined by $c_1 = z(x_0)e^{-r_1 x_0} = \big(y'(x_0) - r_2 y(x_0)\big)e^{-r_1 x_0}$. Given the relation between y and z, the solution y then satisfies

$$(D - r_2)y = c_1 e^{r_1 x},$$

and multiplication by $e^{-r_2 x}$ gives

$$D(e^{-r_2 x} y) = c_1 e^{(r_1 - r_2)x}.$$

If $r_1 \neq r_2$, we integrate to get

$$e^{-r_2 x} y = \frac{c_1}{r_1 - r_2} e^{(r_1 - r_2)x} + c_2.$$

Now multiply by $e^{r_2 x}$ to get

$$y = \frac{c_1}{r_1 - r_2} e^{r_1 x} + c_2 e^{r_2 x}.$$

For neatness, we can rename the constant $c_1/(r_1 - r_2)$ and call it c_1 to get

$$y(x) = c_1 e^{r_1 x} + c_2 e^{r_2 x}. \tag{$*$}$$

If $r_1 = r_2$, we have $D(e^{-r_1 x} y) = c_1$, so integrating both sides gives

$$e^{-r_1 x} y = c_1 x + c_2.$$

In that case,

$$y(x) = c_1 x e^{r_1 x} + c_2 e^{r_1 x}. \tag{$**$}$$

Finally, note that once c_1 is determined from $y(x_0)$ and $y'(x_0)$ as noted previously, the constant c_2 can in any case be determined from the value $y(x_0)$ alone; just solve the appropriate equation $(*)$ or $(**)$ for c_2 with $x = x_0$.   ∎

The problem of finding a solution to a differential equation that also satisfies given initial conditions is called an **initial-value problem**. Geometrically, the initial conditions described in Theorem 1.1 require the graph of a solution to go through a given point (x_0, y_0) with given slope $y'(x_0) = z_0$. It's possible to extract from the proof of Theorem 1.1 some formulas for determining coefficients c_1 and c_2 from initial conditions in a linear combination of two solutions. Since such formulas aren't particularly memorable, it's usually just as efficient to work directly, as in the next example.

EXAMPLE 5   Suppose we want a solution graph of $y'' + 5y' + 6y = 0$ that passes through $(0, 1)$ with slope 2. In other words we want the solution for which $y(0) = 1$ and $y'(0) = 2$. Since the characteristic equation of the differential equation is $r^2 + 5r + 6 = 0$, the characteristic roots are $r = -2$ and $r = -3$. All solutions thus have the form

$$y(x) = c_1 e^{-2x} + c_2 e^{-3x}.$$

FIGURE 11.1

$y(x) = 5e^{-2x} - 4e^{-3x}$, $y'(x) = -10e^{-2x} + 12e^{-3x}$.

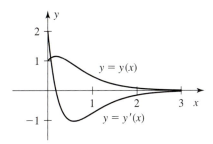

The corresponding derivative formula is

$$y'(x) = -2c_1 e^{-2x} - 3c_2 e^{-3x}.$$

Setting $x = 0$, in $y(x)$ and $y'(x)$ we get the two equations

$$y(0) = c_1 + c_2 = 1 \quad \text{and} \quad y'(0) = -2c_1 - 3c_2 = 2.$$

Solving these for c_1 and c_2 gives the unique solution $c_1 = 5, c_2 = -4$, so our solution to the initial-value problem is $y(x) = 5e^{-2x} - 4e^{-3x}$. Figure 11.1 shows the graphs of $y(x)$ and $y'(x)$. Note that $y(x)$ attains its maximum when $y'(x) = 0$ at $x = \ln \frac{6}{5}$.

EXAMPLE 6   The equation $y'' + 2y' + y = 0$ has for its characteristic equation $r^2 + 2r + 1 = 0$ with a repeated characteristic root $r_1 = r_2 = -1$. Theorem 1.1 says that the general solution is $y(x) = c_1 e^{-x} + c_2 x e^{-x}$. Initial conditions $y(0) = 1$, $y'(0) = 0$ require first that $y(0) = c_1 = 1$. Since $y'(x) = -c_1 e^{-x} + c_2 e^{-x} - c_2 x e^{-x}$, the second condition requires $y'(0) = -c_1 + c_2 = 0$. Hence $c_1 = c_2 = 1$ and the initial-value problem has solution $y = e^{-x} + x e^{-x}$.

Suppose that in the previous example we didn't want to satisfy initial conditions at a single point, but wanted instead a solution graph passing through two given points in the xy-plane. Such conditions applied to a single solution at more than one point are called **boundary conditions**. The problem of finding a solution of a differential equation that also satisfies boundary conditions is called a **boundary-value problem**.

Boundary-value problems are theoretically more complicated than initial-value problems and don't always have solutions. (See Exercises 10 and 11 in the next section.) Nevertheless some boundary-value problems are quite simple computationally. The next example is of this kind.

EXAMPLE 7   Our aim is to find the solution to $y'' - 4y = 0$ that satisfies boundary conditions $y(0) = 1$ and $y(1) = 2$. Thus in this example we need work only with the expression

$$y(x) = c_1 e^{2x} + c_2 e^{-2x}$$

for the general solution itself, since values of $y'(x)$ aren't involved. The resulting equations for the coefficients are

$$y(0) = c_1 + c_2 = 1 \quad \text{and} \quad y(1) = e^2 c_1 + e^{-2} c_2 = 2.$$

Though it's not guaranteed by Theorem 1.1, these equations turn out here to have a unique solution, namely

$$c_1 = (2 - e^{-2})/(e^2 - e^{-2}) \approx 0.26, c_2 = (e^2 - 2)/(e^2 - e^{-2}) \approx 0.74.$$

The desired solution is

$$y(x) = \left(\frac{2 - e^{-2}}{e^2 - e^{-2}}\right)e^{2x} + \left(\frac{e^2 - 2}{e^2 - e^{-2}}\right)e^{-2x}.$$

EXERCISES

In Exercises 1 to 6, with $D = d/dx$, compute

1. $(D + 1)e^{-2x}$

2. $(D^2 + 1)e^x$

3. $D^3 e^{3x}$

4. $(D^2 + D - 1)\sin x$

5. $(D^2 + 1)x\cos x$

6. $(D^2 - 1)xe^{-2x}$

In Exercises 7 to 14, find the characteristic equation of each of the following differential equations. Then solve the characteristic equation and use the roots to write the general solution of the differential equation. Finally determine the arbitrary constants in the general solution to produce the solution to the initial-value or boundary-value problem.

7. $y'' + y' - 6y = 0$, $y(0) = 2$, $y'(0) = 2$

8. $2y'' - y = 0$, $y(0) = 1$, $y'(0) = 0$

9. $y'' + 2y' + y = 0$, $y(0) = 1$, $y'(0) = 2$

10. $y'' + 3y' + ys = 0$, $y(1) = 1$, $y'(1) = 1$

11. $y'' - y' = 0$, $y(0) = 1$, $y(1) = 0$

12. $y'' - 3y' - y = 0$, $y(0) = 0$, $y(1) = 0$

13. $2y'' - 3y' + y = 0$, $y(0) = 0$, $y'(0) = 0$

14. $3y'' + 3y' = 0$, $y(1) = 1$, $y(2) = 2$

In Exercises 15 to 20, sketch the graph of the given function of x. Then find a differential equation of the form

$$y'' + ay' + by = 0$$

of which each is a solution; write the general solution of the differential equation and verify that the given function is a special case of your general solution.

15. xe^{-x}

16. $e^x + e^{-x}$

17. $1 + x$

18. $2e^{2x} - 3e^{3x}$

19. $xe^{-x} - e^{-x}$

20. $e^{-3x} + e^{5x}$

[*Hint:* What characteristic roots go with each solution?]

In Exercises 21 to 26, put each of the linear differential equations in the operator form $(aD^2 + bD + c)y = 0$. Then factor the operator, e.g. $D^2 - 1 = (D - 1)(D + 1)$.

21. $y'' + 2y' + y = 0$

22. $y'' - 2y = 0$

23. $2y'' - y = 0$

24. $y'' + 3y' = 0$

25. $y'' = 0$

26. $y'' - y' = 0$

Each of the equations 27 to 30 has a factored operator form:

$$(D - r_1)(D - r_2)y = f(x).$$

In each case let $(D - r_2)y = z$ and solve

$$(D - r_1)z = f(x)$$

for the most general possible z. Having found z, solve $(D - r_2)y = z$.

27. $D(D - 3)y = 0$

28. $D^2 y = 1$

29. $y'' - y = 1$

30. $y'' + 2y' + y = x$

31. The differential equation $y'' + (1/x)y' - (1/x^2)y = 0$, $x > 0$, has operator form as $(D^2 + (1/x)D - 1/x^2)y = 0$.
   (a) Show that the equation can also be written $D(D + 1/x)y = 0$.
   (b) Solve the equation in part (a) by letting $z = (D + 1/x)y$ and solving a succession of first-order equations.
   (c) Show that $D(D + 1/x) \neq (D + 1/x)D$.
   (d) Solve $(D + 1/x)Dy = 0$.

32. The hyperbolic cosine and hyperbolic sine are defined by

$$\cosh x = \tfrac{1}{2}(e^x + e^{-x}), \quad \sinh x = \tfrac{1}{2}(e^x - e^{-x}).$$

(a) Show that, if constants d_1 and d_2 are suitably chosen in terms of c_1 and c_2, then

$$c_1 e^{rx} + c_2 e^{-rx} = d_1 \cosh rx + d_2 \sinh rx.$$

(b) Express the general solution of

$$y'' - k^2 y = 0$$

in terms of hyperbolic functions.

33. (a) Show that the characteristic equation of

$$Ay'' + By' + Cy = 0,$$

with A, B, C constant, $A \neq 0$, has real roots if and only if $B^2 \geq 4AC$.

(b) Show that when $B^2 > 4AC$, the general solution of the differential equation in part (a) also has the form

$$y = e^{\alpha x}(d_1 \cosh \beta x + d_2 \sinh \beta x),$$

where $\alpha = -B/2A$, $\beta = \sqrt{B^2 - 4AC}/2A$.

34. Assume $|A| < \frac{1}{4}$ in the equation

$$Ay'' + y' + y = 0$$

and show that, as A tends to 0, and with proper choice of arbitrary constants, there are solutions of this equation tending, for each fixed x, to solutions of $y' + y = 0$.

35. The differential equation $y'' - 2y' + y = 0$ has infinitely many solutions $y(x)$ with graphs passing through the point $(0, 1)$. Find the three that have slopes -1, 0 and 1 at that point and sketch their graphs.

36. Initial conditions $y(x_0) = a_0$ and $y'(x_0) = a_1$, imposed at a single point x_0, will always be satisfied by some solution of $y'' - 3y' + 2y = 0$; show that the boundary conditions $y'(0) = a_0$ and $y(\ln 2) = a_1$, at two different points $x_0 = 0$ and $x_1 = \ln 2$, are satisfied only if $a_1 = 2a_0$, and even then not uniquely.

37. A chain of length l and mass density δ per unit of length lies unattached and in a straight line on the deck of a ship.

(a) If the chain runs out over the side with no force acting on it but gravity, in particular without friction, at constant acceleration g, show that the amount y hanging over the side satisfies $d^2y/dt^2 = (g/l)y$ as long as $0 \leq y \leq l$. (Assume the deck is more than height l above the water.)

(b) How fast is the chain accelerating as the last link goes over the side?

(c) Find $y(t)$ if $y(0) = y_0$ and $dy/dt(0) = v_0$.

(d) Show that if the chain starts from rest with length $y_0 > 0$ hanging over the side, then the last link goes over the side at time $t_1 = \sqrt{l/g}\ln((l + \sqrt{l^2 - y_0^2})/y_0)$.

38. Here is an alternative way to arrive at the modified exponential solution xe^{mx} when the characteristic equation of $y'' + ay' + by = 0$ has m as a double root. First write the equation in operator form as $(D - m)^2 y = 0$, or as

$$y'' - 2my' + m^2 y = 0.$$

Now try to find a solution of the form $y = e^{mx}u(x)$ by substitution into the displayed equation. (This technique is used in an essential way in Section 3C for dealing with linear differential equations with nonconstant coefficients.) [*Hint:* Show that $u''(x) = 0$.]

39. This exercise gives a clue as to why the factor x occurs in the "equal-root" case.

(a) Show that $(D - r)(D - (r + h))y = 0$, or $(D^2 - (2r + h)D + r(r + h))y = 0$ can also be written $y'' - (2r + h)y' + r(r + h)y = 0$.

(b) Show that if $h \neq 0$ the general solution is $y_h = c_1 e^{(r+h)x} + c_2 e^{rx}$.

(c) Let $c_1 = 1/h, c_2 = -1/h$, and show that with these choices $\lim_{h \to 0} y_h(x) = xe^{rx}$ for all x.

(d) Show that the limit in part (c) is, by definition, the derivative of the solution e^{rx} with respect to r.

*40. Assume that the characteristic roots of the constant-coefficient differential equation $y'' + ay' + by = 0$ are real numbers r_1, r_2. Show that if $x_1 \neq x_2$, the **boundary-value problem**

$$y'' + ay' + by = 0, \quad y(x_1) = y_1, \quad y(x_2) = y_2$$

always has a unique solution for given numbers y_1 and y_2. [*Hint:* Consider the cases $r_1 \neq r_2$ and $r_1 = r_2$ separately, and show that you can always solve for the desired constants c_1 and c_2 in the general solution.]

41. Two functions $y_1(x)$ and $y_2(x)$ are **linearly independent** on an x-interval if and only if neither one is a constant multiple of the other.

(a) Show that e^{rx} and e^{sx} are linearly independent on a given interval $a < x < b$ if $r \neq s$.

(b) Show that e^{rx} and xe^{rx} are linearly independent on a given interval $a < x < b$.

SECTION 2   COMPLEX SOLUTIONS

2A   Complex Exponentials

Complex numbers arise just as naturally in the solution of constant-coefficient equations of the form

$$ay'' + by' + cy = 0,$$

as they do in the solution of the related algebraic equation

$$ax^2 + bx + c = 0.$$

But to exploit fully the analogy between these two kinds of equation we need the **complex exponential** function, defined for purely imaginary numbers ix, with x real, by

$$e^{ix} = \cos x + i \sin x.$$

Our motivation for this definition comes from Equations 2.1 and 2.2 below. See also Exercise 40.

The **absolute value** of a complex number $\alpha + i\beta$ is $|\alpha + i\beta| = \sqrt{\alpha^2 + \beta^2}$, and equals its distance from the complex number 0. Figure 11.2 shows that in e^{ix} we can interpret x as an angle. The absolute value $|e^{ix}|$ equals 1 for all x because

$$|e^{ix}| = |\cos x + i \sin x|$$

$$= \sqrt{\cos^2 x + \sin^2 x} = 1.$$

Using the addition formulas for sine and cosine shows that

$$(\cos x + i \sin x)(\cos x' + i \sin x')$$

$$= (\cos x \cos x' - \sin x \sin x') + i(\cos x \sin x' + \sin x \cos x')$$

$$= \cos(x + x') + i \sin(x + x').$$

It follows that

$$e^{ix} e^{ix'} = e^{i(x+x')}.$$

FIGURE 11.2

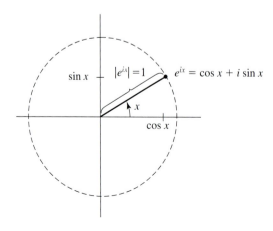

In particular, when $x' = -x$, we get $e^{ix}e^{-ix} = 1$, so that

$$\frac{1}{e^{ix}} = e^{-ix}.$$

These equations justify using the exponential notation: The function e^{ix} behaves very much like the real-valued exponential e^x, for which $e^x e^{x'} = e^{x+x'}$ and $1/e^x = e^{-x}$.

In addition to its algebraic simplicity, another reason for using the complex exponential function is the simplicity of the formulas for its derivative and integral. To differentiate or integrate a complex-valued function $u(x) + iv(x)$ with respect to the real variable x, we simply differentiate or integrate the real and imaginary parts. By definition,

$$\frac{d}{dx}\big(u(x) + iv(x)\big) = \frac{du}{dx}(x) + i\frac{dv}{dx}(x),$$

and

$$\int \big(u(x) + iv(x)\big)\, dx = \int u(x)\, dx + i \int v(x)\, dx.$$

Then the derivative of e^{ix} with respect to x is given by

$$\frac{d}{dx}e^{ix} = \frac{d}{dx}(\cos x + i \sin x)$$

$$= -\sin x + i \cos x$$

$$= i(\cos x + i \sin x) = i e^{ix}.$$

In short, we have

$$\frac{d}{dx}e^{ix} = i e^{ix}.$$

Similarly,

$$\int e^{ix}\, dx = \frac{1}{i}e^{ix} + c,$$

where c may be a real or complex constant. These are analogous to the formulas for the derivative and integral of e^{ax} when a is real. More generally, we can define

$$e^{(\alpha + i\beta)x} = e^{\alpha x}e^{i\beta x}$$

and compute

2.1
$$\frac{d}{dx}e^{(\alpha + i\beta)x} = (\alpha + i\beta)e^{(\alpha + i\beta)x}$$

and

2.2
$$\int e^{(\alpha + i\beta)x}\, dx = \frac{1}{\alpha + i\beta}e^{(\alpha + i\beta)x} + c, \quad \alpha + i\beta \neq 0.$$

These computations are left as exercises. We are now in a position to discuss the differential equation

$$(D^2 + aD + b)y = 0$$

when the factored operator

$$(D - r_1)(D - r_2) = D^2 - (r_1 + r_2)D + r_1 r_2$$

contains complex numbers r_1 and r_2. We'll see that the usual techniques, as discussed in Section 1, still apply. The exponential multiplier method goes over formally unchanged because of Equation 2.1; we have

$$D(e^{rx} y) = e^{rx}(D + r)y,$$

whether r is real or complex.

EXAMPLE 1   Consider the differential equation $y'' + y = 0$. We write the equation in operator form,

$$(D^2 + 1)y = 0,$$

and factor $D^2 + 1$ to get

$$(D - i)(D + i)y = 0.$$

Then set

$$(D + i)y = u, \tag{1}$$

and try to solve

$$(D - i)u = 0$$

for u. As in the real case, we multiply by a factor designed to make the left side the derivative of a product. The same multiplier rule suggests that the correct factor is e^{-ix}, so we write

$$e^{-ix}(D - i)u = 0.$$

Since $D(e^{-ix}u) = e^{-ix}(D - i)u$, we can write

$$D(e^{-ix}u) = 0.$$

We integrate both sides with respect to x to get

$$e^{-ix}u = c_1 \quad \text{or} \quad u = c_1 e^{ix}.$$

Substituting this result for u into Equation (1) gives

$$(D + i)y = c_1 e^{ix},$$

which must now be solved for y. We do it by multiplying through by e^{ix} to get

$$e^{ix}(D + i)y = c_1 e^{2ix}$$

or, since $D(e^{ix}y) = e^{ix}(D + i)y$,

$$D(e^{ix}y) = c_1 e^{2ix}.$$

Integrating gives

$$e^{ix} y = \frac{1}{2i} c_1 e^{2ix} + c_2$$

or

$$y = \frac{1}{2i} c_1 e^{ix} + c_2 e^{-ix}.$$

On replacing the arbitrary constant c_1 by $2i c_1$, we have

$$y = c_1 e^{ix} + c_2 e^{-ix}$$
$$= c_1(\cos x + i \sin x) + c_2(\cos x - i \sin x)$$
$$= (c_1 + c_2) \cos x + i(c_1 - c_2) \sin x.$$

To simplify the solution we can set $d_1 = (c_1 + c_2)$ and $d_2 = i(c_1 - c_2)$. This involves no change in generality in the constants because for given d_1 and d_2 we can solve for c_1 and c_2. Solving for c_1 and c_2, we find

$$c_1 = \tfrac{1}{2}(d_1 - i d_2) \quad \text{and} \quad c_2 = \tfrac{1}{2}(d_1 + i d_2).$$

Whenever the equation $y'' + a y' + b y = 0$ has coefficients that are real numbers, if the characteristic equation

$$r^2 + ar + b = 0$$

has complex roots as in Example 2, these complex roots will be **conjugate** to each other, that is, of the form $r_1 = \alpha + i\beta$ and $r_2 = \alpha - i\beta$ where α and β are real numbers. This follows from the quadratic formula

$$r = \frac{-a \pm \sqrt{a^2 - 4b}}{2},$$

and the assumption that $a^2 - 4b < 0$. Thus

$$r_1 = -\frac{a}{2} + \frac{i}{2}\sqrt{4b - a^2}, \quad r_2 = -\frac{a}{2} - \frac{i}{2}\sqrt{4b - a^2}.$$

It follows that the complex solutions

$$y = c_1 e^{(\alpha + i\beta)x} + c_2 e^{(\alpha - i\beta)x}, \quad \alpha, \beta \text{ real},$$

can always be written

$$y = e^{\alpha x}(c_1 e^{i\beta x} + c_2 e^{-i\beta x})$$
$$= e^{\alpha x}[c_1(\cos \beta x + i \sin \beta x) + c_2(\cos \beta x - i \sin \beta x)]$$
$$= e^{\alpha x}[(c_1 + c_2) \cos \beta x + i(c_1 - c_2) \sin \beta x]$$
$$= d_1 e^{\alpha x} \cos \beta x + d_2 e^{\alpha x} \sin \beta x.$$

This is a form of the solution that is often used in practice, so we include it in the statement of the following Theorem 2.3. The proof is formally the same as that of

Theorem 1.1 in the previous section, so we omit it. The only difference here is that we can interpret the solutions as being complex-valued, though the case of real roots is automatically included also.

2.3   Theorem.   The differential equation

$$y'' + ay' + by = 0, \quad a, b \text{ constant},$$

has for its general solution

$$y = c_1 e^{r_1 x} + c_2 e^{r_2 x}, \quad r_1 \neq r_2$$
$$y = c_1 x e^{r_1 x} + c_2 e^{r_1 x}, \quad r_1 = r_2,$$

where r_1, r_2 are the roots of $r^2 + ar + b = 0$. If a and b are real numbers with $a^2 - 4b < 0$, we can write $r_1 = \alpha + i\beta, r_2 = \alpha - i\beta$. The general solution can then be written

$$y = c_1 e^{\alpha x} \cos \beta x + c_2 e^{\alpha x} \sin \beta x.$$

Initial conditions $y(x_0) = y_0, \ y'(x_0) = z_0$ can always be satisfied by a unique choice of c_1 and c_2.

EXAMPLE 2   The equation $y'' + \omega^2 y = 0$, with $\omega^2 > 0$, has characteristic equation $r^2 + \omega^2 = 0$ with characteristic roots $r = \pm i\omega$. The associated solutions to the differential equation are

$$y = c_1 e^{i\omega x} + c_2 e^{-i\omega x} = d_1 \cos \omega x + d_2 \sin \omega x.$$

Functions of this form are called **harmonic oscillations** because of the role they play in the analysis of sound waves, and the differential equation is called a **harmonic oscillator equation**.

EXAMPLE 3   We classify the solutions of

$$Ay'' + By' + Cy = 0$$

according to relations among the constants A, B, and C. The characteristic equation is

$$Ar^2 + Br + C = 0$$

with roots

$$r_1, r_2 = \frac{-B \pm \sqrt{B^2 - 4AC}}{2A}.$$

If $B^2 - 4AC > 0$, the roots are real and unequal, so the solutions are

$$y = c_1 e^{r_1 x} + c_2 e^{r_2 x}.$$

If $B^2 - 4AC = 0$, the roots are equal and real, so solutions are all of the form

$$y = c_1 x e^{r_1 x} + c_2 e^{r_1 x}.$$

Finally, with $B^2 - 4AC < 0$, the solutions have the form

$$y = c_1 e^{\alpha x} \cos \beta x + c_2 e^{\alpha x} \sin \beta x,$$

where $\alpha = -B/2A$ and $\beta = \sqrt{4AC - B^2}/2A$. The case $B^2 - 4AC = 0$ is critical in that it represents the division between oscillatory solutions, for which $B^2 - 4AC < 0$ and nonoscillatory solutions $(B^2 - 4AC > 0)$. The physical significance of this classification is explained in Section 4. It's apparent in any case that the presence of sine and cosine functions in a solution produces oscillatory behavior.

| EXAMPLE 4 | An alternative way to find solutions of the differential equation

$$y'' + 2y' + 5y = 0$$

is as follows. This approach is technically simpler than the method of successive integration, but fails to prove that we've found all solutions, as the integration method does. We let $y(x) = e^{rx}$ be a trial solution and try to determine what values of r will indeed yield a solution. Since

$$y(x) = e^{rx}, \quad y'(x) = re^{rx}, \quad y''(x) = r^2 e^{rx},$$

we must have, after substitution into the differential equation,

$$r^2 e^{rx} + 2re^{rx} + 5e^{rx} = 0.$$

But since $e^{rx} \neq 0$ for all complex numbers rx, we can divide by e^{rx} to get

$$r^2 + 2r + 5 = 0.$$

The polynomial on the left is just the characteristic polynomial of the differential equation, and its roots are

$$r_1 = -1 + 2i, \quad r_2 = -1 - 2i.$$

It follows that
$$y_1(x) = e^{(-1+2i)x} = e^{-x}(\cos 2x + i \sin 2x)$$

and
$$y_2(x) = e^{(-1-2i)x} = e^{-x}(\cos 2x - i \sin 2x)$$

are complex-valued solutions of the differential equation. Hence the linear combination

$$y(x) = c_1 e^{-x}(\cos 2x + i \sin 2x) + c_2 e^{-x}(\cos 2x - i \sin 2x)$$
$$= (c_1 + c_2)e^{-x} \cos 2x + i(c_1 - c_2)e^{-x} \sin 2x$$

is also a complex-valued solution. Here it's understood that c_1 and c_2 may be complex numbers. But as we saw at the end of Example 1, we can choose c_1 and c_2 so that the combinations $c_1 + c_2$ and $i(c_1 - c_2)$ assume arbitrary values, in particular, real values. Thus the solutions we have found have the form

$$y(x) = d_1 e^{-x} \cos 2x + d_2 e^{-x} \sin 2x.$$

EXERCISES

In Exercises 1 to 4, show that each of the given complex numbers has absolute value 1. Then find a real number x such that the complex number has the form $e^{ix} = \cos x + i \sin x$; for example,

$$\frac{\sqrt{3} + i}{2} = \cos \frac{\pi}{6} + i \sin \frac{\pi}{6} = e^{i\pi/6}.$$

1. i **2.** $(1+i)/\sqrt{2}$ **3.** $(1-i)/\sqrt{2}$ **4.** $(\sqrt{3} - i)/2$

In Exercises 5 to 8, for each pair d_1, d_2 of real numbers given, find complex numbers c_1, c_2 such that

$$c_1 e^{(\alpha + i\beta)x} + c_2 e^{(\alpha - i\beta)x} = e^{\alpha x}(d_1 \cos \beta x + d_2 \sin \beta x).$$

5. $d_1 = 1, d_2 = 0$ **6.** $d_1 = 4, d_2 = -2$

7. $d_1 = 0, d_2 = \pi$ **8.** $d_1 = 1, d_2 = 1$

9. Recall that a function is periodic, with period p, if $f(x + p) = f(x)$ for all x in the domain of f.
  (a) Show that e^{ix} has period $2k\pi$ if k is an integer.
  (b) Show that $e^{i\beta x}$ is periodic for β real, and find the smallest positive period if $\beta \neq 0$.

10. Show that $e^{-i\beta x} = \cos x - i \sin \beta x$. What properties of cos and sin are used here?

Solve each of the differential equations 11 to 14 by factoring the differential operator associated with it and then successively solving a pair of first-order linear equations.

11. $y'' + y = 1$ **12.** $y'' + 2y' + 2y = 0$

13. $y'' + 2y = 0$ **14.** $y'' + y' = x$

Find the roots of the characteristic equation of each of the differential equations 15 to 22. Then write the general solution of the differential equation, replacing complex exponentials by $e^{\alpha x} \cos \beta x$ and $e^{\alpha x} \sin \beta x$ where it's appropriate. Finally determine the constants of integration so the given initial conditions will be satisfied.

15. $y'' + 2y = 0$, $y(0) = 0$, $y'(0) = 1$

16. $2y'' + 3y' = 0$, $y(0) = 1$, $y'(0) = 0$

17. $y'' - 2y' + 2y = 0$, $y(\pi) = 0$, $y'(\pi) = 0$

18. $y'' - y' + y = 0$, $y(0) = 2$, $y'(0) = -1$

19. $2y'' + y' - y = 0$, $y(0) = 0$, $y'(0) = 2$

20. $y'' + y' = 2y$, $y(0) = 0$, $y(1) = 1$

21. $2y'' + y' + y = 0$, $y(0) = 0$, $y'(0) = 0$

22. $3y'' - y' + y = 0$, $y(0) = 0$, $y(1) = 0$

In Exercise 23 to 30, find a second-order differential equation $y'' + ay' + by = 0$ that has the given solution. [*Hint:* What are the characteristic roots associated with each solution?]

23. $\sin 2x$ **24.** $e^{2x} \sin 2x$

25. $e^x \cos 2x$ **26.** $\cos 2x$

27. $\cos(x/2)$ **27.** xe^{2x}

29. $\sin 3x - \cos 3x$ **30.** $x - 7$

In Exercises 31 to 33, we deal with the issue that a constant-coefficient differential equation with complex characteristic roots may fail to have a unique solution if we impose certain critically chosen boundary conditions.

31. Show that the boundary-value problem $y'' + y = 0$, $y(0) = 1$, $y(\pi) = 1$ has no solution. [*Hint:* You know all solutions of $y'' + y = 0$.]

32. Show that the boundary-value problem $y'' + y = 0$, $y(0) = 1$, $y(2\pi) = 1$ has infinitely many solutions.

33. Show that if the constant-coefficient equation $y'' + ay' + by = 0$ has real characteristic roots, then the associated boundary-value problem $y(x_1) = y_1$, $y(x_2) = y_2$ always has a unique solution if $x_1 \neq x_2$.

34. (a) Show that $c_1 \cos \beta x + c_2 \sin \beta x$ also has the form $A \cos(\beta x - \phi)$, where $A = \sqrt{c_1^2 + c_2^2}$ and ϕ satisfies $\cos \phi = c_1/A$ and $\sin \phi = c_2/A$.
  (b) The result of part (a) is useful because it shows that

$$c_1 \cos \beta x + c_2 \sin \beta x$$

has a graph that is the same as that of $\cos \beta x$ shifted by a **phase angle** ϕ and multiplied by an

amplitude A. Sketch the graph of

$$\cos 2x + \sqrt{3}\sin 2x$$

by first finding ϕ and A.

35. (a) Let c_1 and c_2 be real numbers. Show that $y(x) = c_1 \cos \beta x + c_2 \sin \beta x$ can also be written as $A \sin(\beta x + \theta)$ where $A = \sqrt{c_1^2 + c_2^2}$ is the **amplitude** of $y(x)$ and θ satisfies $\sin \theta = c_1/A$ and $\cos \theta = c_2/A$.

(b) The result of part (a) says that

$$c_1 \cos \beta x + c_2 \sin \beta x = A \sin(\beta x + \theta)$$

for appropriately chosen real numbers A and θ. Use this result to sketch the graph of $y(x) = \cos 2x + \sqrt{3}\sin 2x$, by first finding A and θ.

(c) Show that

$$A \sin(\beta x + \theta) = A \cos(\beta x - \phi),$$

where $\phi = \pi/2 - \theta$. The number ϕ (or sometimes $-\theta$) is called a **phase angle**, and A is the **amplitude** of the trigonometric function $A \sin(\beta x + \theta)$.

36. Verify Equation 2.1 in the text.

37. Verify Equation 2.2 in the text.

*38. Find all real or complex solutions of $y'' + iy' = 0$.

*39. Find all real or complex solutions of $y'' + iy = 0$

40. Separate the real and imaginary terms in the infinite series

$$\sum_{k=1}^{\infty} \frac{(ix)^k}{k!}$$

into two power series. Then use the result as another justification of the definition $e^{ix} = \cos x + i \sin x$.

Complex-valued differentiable functions $f(x) = u(x) + iv(x)$ and $g(x) = s(x) + it(x)$ obey the same basic rules relative to differentiation that real-valued functions do. In Exercises 41 to 44, use the corresponding relations for real-valued functions to show that the following formulas hold on an interval $a < x < b$ on which both f, g and f/g are differentiable complex-valued functions.

41. $(f + g)' = f' + g'$

42. $(cf)' = cf'$, c constant

43. $(fg)' = fg' + f'g$

44. $(f/g)' = (f'g - fg')/g^2$, $g \neq 0$

45. (a) Show that if $y = y(x)$ is a solution to the constant-coefficient equation $y'' + ay' + by = 0$ and c is constant, then the function y_c defined by $y_c(x) = y(x + c)$ is also a solution. [*Hint:* It's not necessary to know the form of $y(x)$ in terms of elementary functions.]

(b) Generalize the result of part (a) to a solution $y(x)$ of the nth order constant-coefficient equation

$$y^{(n)} + a_{n-1}y^{(n-1)} + \cdots + a_1 y' + a_0 y = 0.$$

(c) Generalize the result of part (a) to a solution $y(x)$, $a < x < b$, of a second-order equation, linear or nonlinear, of the form $y'' = F(y, y')$. [*Hint:* $y_c(x)$ will in general be defined on an interval different from $a < x < b$.]

46. Let $y(x)$ be a solution of a second-order equation $y'' + ay' + by = 0$. Can constants a and b be chosen so that $y(x) = x \cos x$? What about the function $\sin x + \cos 2x$? Justify your answers.

47. For fixed α and fixed $\beta \neq 0$, show directly that $e^{\alpha x}\cos \beta x$ and $e^{\alpha x}\sin \beta x$ are linearly independent. What if $\beta = 0$?

2B   Higher-Order Equations

While most of the differential equations that arise directly in applications have order 1 or 2, understanding higher-order equations is technically useful for solving certain second-order equations in Section 3 and for solving systems of equations in Chapter 12. Therefore it's useful to record an extension of Theorem 1.1 as follows. The proof continues step-by-step as in the proof of Theorem 1.1 and uses no new ideas so we omit it.

2.4   Theorem.   The differential equation

$$(D - r_1)(D - r_2) \cdots (D - r_n)y = 0,$$

with characteristic roots r_k *all different* has its most general solution of the form

$$y = c_1 e^{r_1 x} + c_2 e^{r_2 x} + \cdots + c_n e^{r_n x}.$$

If some r_k are equal, say $r_1 = r_2 = \cdots = r_m$, then $e^{r_2 x}, e^{r_3 x}, \ldots, e^{r_m x}$ are replaced in the general solution formula by $xe^{r_1 x}, x^2 e^{r_1 x}, \ldots, x^{m-1}e^{r_1 x}$ respectively. The constants $c_1, c_2, \ldots, c_n$ are uniquely determined by prescribing values $y(x_0), y'(x_0), \ldots, y^{n-1}(x_0)$ of the solution and its derivatives at a single point x_0.

The key to applying Theorem 2.4 is finding the **characteristic roots** of a differential equation

$$y^{(n)} + a_{n-1}y^{(n-1)} + \cdots + a_1 y' + a_0 y = 0,$$

that is, finding the roots of the purely algebraic **characteristic equation**

$$r^n + a_{n-1}r^{n-1} + \cdots + a_1 r + a_0 = 0.$$

Although there are general formulas for the roots if n is 3 or 4, these are awkward to use and we'll rely in our examples on equations that reduce to solving quadratic equations.

EXAMPLE 5   To solve $y''' - 4y'' + 4y' = 0$, we solve the characteristic equation $r^3 - 4r^2 + 4r = 0$. We observe that we can factor the right side:

$$r(r^2 - 4r + 4) = r(r - 2)^2 = 0.$$

The roots are 0 and 2, where 2 is a double root. According to Theorem 2.4 the general solution to the differential equation is a linear combination of the solutions $e^{0x} = 1$, e^{2x} and xe^{2x}. Thus the solution is $y = c_1 + c_2 e^{2x} + c_3 xe^{2x}$.

It will be useful for us to apply the idea behind Theorem 2.4 in reverse order, starting with solutions and arriving at a differential equation that has those solutions.

EXAMPLE 6   To find a constant-coefficient equation $L(y) = 0$ of least possible order having e^x, e^{2x} and e^{-2x} for solutions, we write the linear operator equation

$$(D - 1)(D - 2)(D + 2)y = 0.$$

The function e^{-2x} is a solution, because $(D + 2)e^{-2x} = 0$. Since the operator factors may appear in any order, and $(D - 1)e^{2x} = 0$ and $(D - 2)e^x = 0$, all three functions are solutions of the differential equation. Linear combinations $y = c_1 e^x + c_2 e^{2x} + c_3 xe^{-2x}$ constitute the general solution, as in Theorem 2.4

2C   Independent Solutions

The substitution method used in the preceding example doesn't by itself show that the solution formula obtained gives the most general solution; this is true and it follows from Theorem 2.3. Knowing about the solutions to the nth-order constant-coefficient equation is useful in Section 3 for understanding some special types of second-order nonhomogeneous equations $y'' + ay' + by = f(x)$. Otherwise it's mainly 4th-order equations that arise directly in applications. We restate Theorem 2.4 to take account of the trigonometric form that solutions may take if the constant coefficients in the differential equation are all real.

2.5   Theorem.   The nth-order constant-coefficient equation $L(y) = 0$ has for its general solution a sum of constant multiples

$$c_1 y_1(x) + \cdots + c_n y_n(x)$$

of solutions $y_k(x)$, where the c_k are arbitrary constants; these constants are uniquely prescribed by initial conditions:

$$y(x_0) = z_0, \ y'(x_0) = z_1, \ldots, \ y^{(n-1)}(x_0) = z_{n-1}.$$

If $r_1, \ldots, r_n$ are the roots of the characteristic equation $r^n + a_{n-1}r^{n-1} + \cdots + a_1 r + a_0 = 0$, the terms $y_k(x)$ in the solution can each be written in the form $x^l e^{r_k x}$, $l = 0, 1, \ldots, m-1$, where m is the multiplicity of the root r_k. If roots $\alpha + i\beta$ and $\alpha - i\beta$ occur in complex conjugate pairs, then the corresponding pairs of exponential solutions are equivalent to

$$x^l e^{\alpha x} \cos \beta x, \quad x^l e^{\alpha x} \sin \beta x.$$

The constants of integration in the solution formulas we've just been dealing with appear frequently in the next section. An expression of the form

$$c_1 y_1 + c_2 y_2 + \cdots + c_n y_n$$

is called a **linear combination** of $y_1, y_2, \ldots, y_n$ with **coefficients** $c_1, c_2, \ldots, c_n$. Alternatively, the numbers c_k may be regarded as parameters, and the linear combination displayed above is an example of an n-**parameter family** of functions.

EXAMPLE 7

The differential equation $y^{(4)} - y = 0$ has characteristic equation $r^4 - 1 = 0$. To find the roots, we note that since $r^4 = 1$ then either $r^2 = 1$ or else $r^2 = -1$. Hence the roots are $r_1 = 1, r_2 = -1, r_3 = i$ and $r_4 = -i$. The first two roots provide the solutions e^x and e^{-x}. The second pair provides e^{ix} and e^{-ix} or the alternative form $\cos x$ and $\sin x$. The complete solution is then

$$y(x) = c_1 e^x + c_2 e^{-x} + c_3 \cos x + c_4 \sin x.$$

Initial conditions $y(0) = 0$, $y'(0) = 1$, $y''(0) = 2$, $y'''(0) = 1$ impose conditions on the constants c_k. For example,

$$y'(x) = c_1 e^x - c_2 e^{-x} c_3 \sin x + c_4 \cos x,$$

so $y'(0) = c_1 - c_2 + c_4 = 1$. The complete set of conditions reduces to

$$c_1 + c_2 + c_3 = 0$$
$$c_1 - c_2 + c_4 = 1$$
$$c_1 + c_2 - c_3 = 2$$
$$c_1 - c_2 - c_4 = 1.$$

Straightforward elimination shows that $c_1 = 1, c_2 = 0, c_3 = -1$ and $c_4 = 0$. Thus the particular solution that satisfies the initial conditions is $y(x) = e^x - \cos \ x$.

EXAMPLE 8 The third-order differential equation $(D - 1)^3 y = 0$, which looks like $y''' - 3y'' + 3y' - y = 0$ when written without operator notation, has the single characteristic root $r = 1$ with multiplicity 3. The general solution is, by Theorem 2.4,

$$y = c_1 e^x + c_2 x e^x + c_3 x^2 e^x.$$

EXAMPLE 9 The equation $y^{(4)} + \lambda y'' = 0$ is satisfied under certain conditions by a function $y(x)$ that describes the lateral deflection, measured x units from one end, of a uniform column under a vertical compressive force. (The constant $\lambda = P/\rho$ depends on the structure of the column and on the vertical load P applied to it.) The characteristic equation is $r^4 + \lambda r^2 = 0$, or $r^2(r^2 + \lambda) = 0$. With $\lambda > 0$, the roots are $r_1 = r_2 = 0$ and $r_3 = \sqrt{\lambda} i, r_4 = -\sqrt{\lambda} i$. The general solution is

$$y(x) = c_1 + c_2 x + c_3 \cos \sqrt{\lambda} x + c_4 \sin \sqrt{\lambda} x.$$

Initial conditions at a single point x_0 are physically uninteresting in this problem. What is usually done is to impose "boundary conditions" on $y(x)$ and $y'(x)$, or else on $y(x)$ and $y''(x)$ at points corresponding to the two ends of the column, say at $x = 0$ and $x = L$. The existence of a unique solution depends critically on the value of λ. These matters are taken up in the Exercises.

Finding the roots of an nth degree characteristic equation is in general a difficult problem. The following examples illustrate some fairly simple special cases.

EXAMPLE 10 We can factor the cubic in $r^3 + 2r^2 + 5r = 0$ to get $r(r^2 + 2r + 5) = 0$. Apart from the root $r_1 = 0$ due to the first factor, there are the roots $r_2 = -1 + 2i$ and $r_3 = -1 - 2i$ of the quadratic equation $r^2 + 2r + 5 = 0$, so the corresponding solutions of the differential equation

$$y''' + 2y'' + 5y' = 0$$

are

$$y = c_1 + c_2 e^{-x} \cos 2x + c_3 e^{-x} \sin 2x.$$

EXAMPLE 11 The fourth degree, or quartic, equation $r^4 - 13r^2 + 36 = 0$ is also a quadratic equation in r^2, with solutions $r^2 = (13 \pm 5)/2$. Since $r^2 = 4$ or $r^2 = 9$, the four distinct roots are $r = \pm 2, \pm 3$. Hence the solutions to

$$y^{(4)} - 13y'' + 36y = 0$$

consist of all members of the four-parameter family

$$y = c_1 e^{2x} + c_2 e^{-2x} + c_3 e^{3x} + c_4 e^{-3x}.$$

It will be useful in Section 3 to apply the ideas behind Theorem 2.5 in reverse order, starting with solutions and arriving at a differential equation that has those solutions.

EXAMPLE 12   To find a constant-coefficient equation $L(y) = 0$ of least possible order having e^x, e^{2x} and e^{-2x} for solutions, we write the linear operator equation

$$(D - 1)(D - 2)(D + 2)y = 0.$$

The function e^{-2x} is a solution, because $(D + 2)e^{-2x} = 0$. Since the operator factors may appear in any order, and $(D - 2)e^{2x} = 0$ and $(D - 1)e^x = 0$, all three exponentials are solutions of the differential equation. (By the linearity of the equation, linear combinations $y = c_1 e^x + c_2 e^{2x} + c_3 e^{-2x}$ constitute the general solution.)

EXAMPLE 13   Let us find a constant-coefficient equation $L(y) = 0$ of least possible order having $\cos x$ and $\sin 2x$ for solutions. These solutions would have arisen from characteristic roots $\pm i$ and $\pm 2i$ respectively. We write the linear operator equation

$$(D^2 + 1)(D^2 + 4)y = 0.$$

The function $y = \sin 2x$ is a solution, because $(D^2 + 4)\sin 2x = 0$. Since the operator factors may appear in any order, and $(D^2 + 1)\cos x = 0$, both functions are solutions of the differential equation. (Linear combinations $y = c_1 \cos x + c_2 \sin x + c_3 \cos 2x + c_4 \sin 4x$ constitute the general solution.)

Independence of basic solutions.   A set of n functions $y_1(x), y_2(x), \ldots, y_n(x)$ defined on an interval is **linearly independent** on that interval if the identity

$$c_1 y_1(c) + c_2 y_2(x) + \cdots + c_n y_n(x) = 0$$

holds there only with the choice $c_1 = c_2 = \cdots = c_n = 0$ for the constants. Exercise 48 asks you to show that a set of two or more functions is linearly independent if and only if no one of them is expressible as a linear combination of the others. Thus there is no ambiguity about the coefficients in the expression of a function as a linear combination of elements from a given independent set of functions. A set B of linearly independent functions whose linear combinations constitute *all* solutions of a linear differential equation is called a **basis** for the solution set. That the solutions $x^l e^{\alpha x} \cos \beta x$, $x^l e^{\alpha x} \sin \beta x$ listed at the conclusion of Theorem 2.6 form a linearly independent basis is a simple consequence of the theorem, as follows.

2.6 Corollary.   The solutions $x^l e^{r_k x}$ listed in Theorem 2.5, including those with real form $x^l e^{\alpha x} \cos \beta x$ and $x^l e^{\alpha x} \sin \beta x$, are linearly independent.

Proof.   Suppose a linear combination $y(x)$ of these n solutions is identically zero:

$$c_1 y_1(x) + \cdots + c_n y_n(x) = 0.$$

To prove independence we need to show that all $c_k = 0$. The linear combination is a solution of a linear homogeneous differential equation $L(y) = 0$ with possibly multiple characteristic roots $\alpha + i\beta$. Since this solution is identically zero, it satisfies initial conditions $y(x_0) = y'(x_0) = \cdots = y^{(n-1)}(x_0) = 0$ at a given point x_0 of the interval. Since the choice $c_k = 0$ produces this solution, and since Theorem 2.5 guarantees uniqueness of the coefficients c_k, the numbers c_k must all be zero.   ∎

EXERCISES

In Exercises 1 to 10, find the general solution to the differential equation.

1. $y''' + y = 0$

2. $y^{(4)} + 2y'' + y = 0$

3. $y''' - 2y' = 0$

4. $y^{(4)} - y'' = 0$

5. $y''' - 16y' = 0$

6. $y^{(4)} - 4y'' + 4y = 0$

7. $(D^2 + 4)(D^2 - 1)y = 0$

8. $D(D - 1)^3 y = 0$

9. $y''' - 2y'' = 0$

10. $y^{(4)} = 0$

In Exercises 11 to 28, find constant-coefficient linear differential equation of the smallest possible order that has the function $y(x)$ as solution. [*Hint:* What are the characteristic roots associated with each function?]

11. $y(x) = e^{5x}$

12. $y(x) = xe^{5x}$

13. $y(x) = x^2$

14. $y(x) = x^2 e^{-x}$

15. $y(x) = x + e^x$

16. $y(x) = x^3$

17. $y(x) = x^5 + x^2 e^x$

18. $y(x) = -1$

19. $y(x) = \cos 4x$

20. $y(x) = x \cos 4x$

21. $y(x) = x^2 \cos 4x$

22. $y(x) = x \sin 4x$

23. $y(x) = xe^x \sin x$

24. $y(x) = x^3 \cos 4x$

25. $y(x) = x^5$

26. $y(x) = \cos 4x + \sin 3x$

27. $y(x) = e^{-x} \cos x$

28. $y(x) = x \cos x + \cos 2x$

In Exercise 29 to 38, there are given families of solutions to some linear constant-coefficient differential equations of order more than 2. In each case find a differential equation of least possible order satisfied by the family.

29. $c_1 + c_2 x + c_3 e^x$

30. $c_1 \cos x + c_2 \sin x + c_3 + c_4 x$

31. $c_1 \cos x + c_2 \sin x + c_3 e^x$

32. $c_1 \cos x + c_2 \sin x + c_3 \cos 3x + c_4 \sin 3x$

33. $c_1 \cos 2x + c_2 \sin 2x + c_3 e^{-x}$

34. $c_1 + c_2 x + c_3 x^2$

35. $c_1 \cos 3x + c_2 \sin 3x + c_3$

36. $c_1 e^x + c_2 e^{-x} + c_3 \cos x + c_4 \sin x$

37. $c_1 e^x \cos 2x + c_2 e^x \sin 2x + c_3$

38. $c_1 \cos x + c_2 \sin x$

Each of the solution families in Exercises 29 to 32 is listed again in Exercises 39 to 42 along with a set of initial conditions. Find the correct values for the constants c_k so the conditions will be satisfied.

39. $c_1 + c_2 x + c_3 e^x$; $y(0) = 1$, $y'(0) = 2$, $y''(0) = 1$

40. $c_1 \cos x + c_2 \sin x + c_3 + c_4 x$; $y(0) = 2$, $y'(0) = y''(0) = y'''(0) = 0$

41. $c_1 \cos x + c_2 \sin x + c_3 e^x$; $y(0) = 2$, $y'(0) = y''(0) = -3$

42. $c_1 \cos x + c_2 \sin x + c_3 \cos 3x + c_4 \sin 3x$; $y(0) = y'(0) = 1$, $y''(0) = -1$, $y'''(0) = 3$

In Exercises 43 to 48, find the general solution to each equation.

43. $y^{(4)} - y = 0$

44. $y^{(4)} - 2y'' + y = 0$

45. $y^{(4)} - 2y' = 0$

46. $y^{(4)} + y = 0$

47. $(D^2 - 4)(D^2 - 1)y = 0$

48. $D^2(D - 1)^2 y = 0$

[*Hint:* For $r^4 + 1 = 0$, note that $r^2 = \pm i = \pm e^{i(\pi/2)}$. Then $r = \pm i e^{i(\pi/4)}, \pm e^{i(\pi/4)}$.]

49. Explain why $y(x) = \cos x + \sin 2x$ can't be the solution to a constant-coefficient equation of the form $y'' + ay' + by = 0$. Find an equation of higher order that $y(x)$ does satisfy.

50. The general solution $y(x) = c_1 + c_2 x + c_3 \cos \sqrt{\lambda} x + c_4 \sin \sqrt{\lambda} x$ to $y'''' + \lambda y'' = 0$ is derived in Example 7 of the text; use it to do the following.

(a) If $\lambda = 4\pi^2$, find the infinitely many solutions that satisfy the boundary conditions $y(0) = y(1) = 0$, $y'(0) = y'(1) = 0$.

(b) Under the assumption that $y(x)$ represents the horizontal deflection of a column under a vertical compressing force, we can interpret the boundary conditions in part (a) to mean that the ends of the column are rigidly embedded in floor and ceiling. Sketch some typical solutions, assuming small displacements.

(c) Show that if $\lambda = \pi^2$ then the only solution satisfying the boundary conditions in part (a) is the identically zero solution.

51. Euler beam equation. Suppose a uniform horizontal beam has profile shape $y = y(x)$, with x measured from the left end. For a rather rigid beam with uniform loading, $y(x)$ typically satisfies the fourth order differential equation $y'''' = -P$, where $P > 0$ is a constant depending on the characteristics of the beam. If the left end of the beam is embedded horizontally in a wall at $x = 0$, called a *cantilever* support, then $y'(0) = 0$. indicating that the beam is flat there. If the beam is just supported from below at $x = L$, but at the same level, say level

0, then we'll use boundary conditions $y(0) = y(L) = 0$, $y'(0) = 0$, and $y''(L) = 0$. Imagine the beam extended beyond $x = L$ and bending with an inflection at $x = L$.

(a) Solve the differential equation by four successive integrations, and use the boundary conditions to show that the beam's shape is described by the graph of

$$y(x) = -\frac{P}{48}(2x^4 - 5Lx^3 + 3L^2x^2).$$

(b) Show that the graph of $y(x)$ has an inflection on $0 < x < L$. What is the maximum downward vertical deflection from level 0 on $L < x < 0$?

(c) Make a sketch that shows the qualitative features of the graph of $y(x)$ for $L \le x \le 0$.

52. Rotating shaft. A differential equation for the lateral displacement $y = y(x)$ at distance x from one end of a uniform rotating shaft is

$$y^{(4)} - \lambda y = 0,$$

where the constant $\lambda > 0$ is proportional to the speed of rotation.

(a) Show that the most general solution to the differential equation may be written

$$y = c_1 \cosh \sqrt[4]{\lambda}x + c_2 \sinh \sqrt[4]{\lambda}x$$
$$+ c_3 \cos \sqrt[4]{\lambda}x + c_4 \sin \sqrt[4]{\lambda}x,$$

where $\cosh u = (e^u + e^{-u})/2$ and $\sinh u = (e^u - e^{-u})/2$.

(b) Let $\lambda = n^4\pi^4$ where n is a positive integer, and find solutions of the differential equation subject to boundary conditions $y(0) = y''(0) = 0$, $y(1) = y''(1) = 0$.

(c) Sketch the graphs of the solutions found in part (b) for $n = 1, 2, 3$.

(d) Show that if λ is not of the form prescribed in part (b) then the only solution to the problem posed there is the identically zero solution.

53. Prove that a set $\{y_1(x), y_2(x), \ldots, y_n(x)\}$ of functions defined on an interval is linearly independent if and only if no one of them is a linear combination, using constant coefficients, of any remaining functions in the set.

SECTION 3   NONHOMOGENEOUS EQUATIONS

3A   Superposition

An operator, for example, $L = D^2 + D + 1$, is **linear** if for functions y_1, y_2, and constants c, both

$$\text{(i) } L(y_1 + y_2) = L(y_1) + L(y_2), \text{ and}$$

$$\text{(ii) } L(cy) = cL(y).$$

To a given linear operator L we can associate the **homogeneous equation**

$$L(y) = 0,$$

and for a given function f we can also consider the **nonhomogeneous** equation

$$L(y) = f.$$

The associated homogeneous equation is the special case of the nonhomogeneous equation obtained by letting f be the identically zero function, and this special case is fundamental to understanding the more general case. For constant-coefficient linear differential operators, the theorems of the two preceding sections give a complete description of the set of all solutions to the homogeneous equation. Furthermore, the exponential multiplier method developed there provides a practical method for solving many nonhomogeneous equations. The next example illustrates the method.

EXAMPLE 1   Given

$$y'' + 2y' + y = e^{3x},$$

we write the characteristic polynomial in the form $D^2 + 2D + 1$ and factor it, putting the equation in the form

$$(D + 1)^2 y = e^{3x}.$$

Letting $(D + 1)y = u$, we try to solve

$$(D + 1)u = e^{3x}.$$

Multiplication by e^x gives

$$e^x Du + e^x u = e^{4x}$$

or

$$D(e^x u) = e^{4x}.$$

Then integration gives

$$e^x u = \tfrac{1}{4} e^{4x} + c_1,$$

or

$$u = \tfrac{1}{4} e^{3x} + c_1 e^{-x}.$$

Since $(D + 1)y = u$, we have

$$(D + 1)y = \tfrac{1}{4} e^{3x} + c_1 e^{-x}.$$

Again multiplying by e^x, we get

$$e^x Dy + e^x y = \tfrac{1}{4} e^{4x} + c_1$$

or

$$D(e^x y) = \tfrac{1}{4} e^{4x} + c_1.$$

Then

$$e^x y = \tfrac{1}{16} e^{4x} + c_1 x + c_2$$

or

$$y = \tfrac{1}{16} e^{3x} + c_1 x e^{-x} + c_2 e^{-x}.$$

In the preceding example, the solution breaks naturally into a sum of two parts y_h and y_p:

$$y_h = c_1 x e^{-x} + c_2 e^{-x},$$

$$y_p = \tfrac{1}{16} e^{3x}.$$

The function y_h is called the **homogeneous** part of the solution because it's a solution of the **homogeneous equation**

$$L(y) = 0$$

associated with $L(y) = f$. The function y_p is called a **particular solution** of

$$L(y) = f$$

because it's just that: a particular solution, though not the most general one. We sometimes refer to the homogeneous part of a general solution as the **homogeneous solution**. We get y_p by setting $c_1 = c_2 = 0$ in the general solution. The breakup of the solution into two parts is an example of a general property of linear operators discussed in Chapter 2, Section 2C on systems of linear algebraic equations. The principle is important enough, and at the same time simple enough, that we state it here also.

3.1   Theorem.   Let L be a linear operator. Let f be a function, and let y_p be a function in the domain of L such that $L(y_p) = f$. Then every solution y of

$$L(y) = f \text{ is a sum } y = y_h + y_p,$$

where y_h is a solution to $L(y) = 0$.

Proof.   Suppose that $L(y) = f$ and that also $L(y_p) = f$, Then since L is linear,

$$L(y - y_p) = L(y) - L(y_p)$$
$$= f - f = 0.$$

It follows that $y - y_p = y_h$ for some homogeneous solution y_h. But then $y = y_h + y_p$ as we wanted to show.   ∎

The method of Example 1 can always be used to find the most general solution to an equation $L(y) = f$ of the form

$$(D - r_1) \cdots (D - r_n)y = f.$$

In second-order examples we can use Theorems 1.1 or 2.3, since y_h follows immediately from the roots of the characteristic polynomial. Theorem 3.1 then says that if we find the general homogeneous part of the solution y_h using Theorems 1.1 or 2.3, and somehow find a particular solution y_p, then the general solution of the given equation is $y_h + y_p$.

To find y_p it's often convenient to take advantage of the linearity of L in case the right-hand side f is a sum of two or more terms. If we want to solve

$$L(y) = a_1 f_1 + a_2 f_2 \tag{1}$$

and we can find solutions y_1 and y_2 such that

$$L(y_1) = f_1, \quad L(y_2) = f_2,$$

then because L is linear, the function

$$y = a_1 y_1 + a_2 y_2$$

is a solution of Equation (1). In this context, the property of linearity is sometimes called the **superposition principle** because the desired solution is found by superposition (i.e., addition) of solutions of more than one equation.

EXAMPLE 2   In Example 1 we found that the differential equation

$$(D+1)^2 y = e^{3x}$$

had the general solution

$$\tfrac{1}{16} e^{3x} + c_1 x e^{-x} + c_2 e^{-x}.$$

When $c_1 = c_2 = 0$ we get the particular solution $y_1 = \tfrac{1}{16} e^{3x}$. If we now wanted to solve

$$(D+1)^2 y = e^{3x} + 1, \tag{2}$$

we would not have to start all over again, but would only have to find a particular solution for

$$(D+1)^2 y = 1.$$

This could be solved by using exponential multipliers, but in this case the differential equation is so simple that we can guess a solution, namely, $y_2 = 1$. Then a particular solution of Equation (2) is $y_p = \tfrac{1}{16} e^{3x} + 1$, and the general solution is

$$y = \tfrac{1}{16} e^{3x} + 1 + c_1 x e^{-x} + c_2 e^{-x}.$$

Solving

$$(D+1)^2 y = e^{3x} + e^{-x},$$

with an extra term on the right requires us to find a particular solution to

$$(D+1)^2 y = e^{-x}.$$

To do this we could return to the exponential multiplier method, and let

$$(D+1)y = u.$$

Then we solve

$$(D+1)u = e^{-x}$$

by using the multiplier e^x to get

$$e^x (D+1)u = 1$$

or

$$D(e^x u) = 1.$$

Hence

$$e^x u = x + c_1 \quad \text{or} \quad u = x e^{-x} + c_1 e^{-x}.$$

Solving $(D+1)y = u$ for y, we use again the multiplier e^x to get

$$e^x (D+1)y = e^x u \quad \text{or} \quad D(e^x y) = x + c_1.$$

Integration then gives
$$e^x y = \tfrac{1}{2}x^2 + c_1 x + c_2$$

or

$$y = \tfrac{1}{2}x^2 e^{-x} + c_1 x e^{-x} + c_2 e^{-x}.$$

Using the linearity of $(D + 1)^2$, we conclude that the general solution of

$$(D + 1)^2 y = e^{3x} + e^{-x}$$

is

$$y = c_1 x e^{-x} + c_2 e^{-x} + \tfrac{1}{16}e^{3x} + \tfrac{1}{2}x^2 e^{-x}.$$

The exponential multiplier method shown previously for finding particular solutions of $L(y) = f(x)$ will provide a solution if we can perform the integrations involving $f(x)$. The method of *undetermined coefficients* explained next has more restricted applicability, but is often more efficient when it does apply.

3B   Undetermined Coefficients

The method depends on the observation that if we want to solve

$$L(y) = f(x),$$

where $f(x)$ is itself a solution of a homogeneous equation $My = 0$, then

$$M\big(L(y)\big) = M\big(f(x)\big) = 0.$$

Then the desired solution $y(x)$ must be among the solutions of

$$M\big(L(y)\big) = 0.$$

If M and L are linear constant-coefficient operators then the solutions of the preceding equation are linear combinations of functions of the form $x^k e^{rx}$, where r may be real or complex. Thus the only computational problem is to determine the so far "undetermined coefficients" of combination that will actually give a solution of the original equation $L(y) = f(x)$. Guessing the operator M is based on experience solving homogeneous equations.

EXAMPLE 3   The differential equation
$$y'' - y = e^x$$

in operator form is
$$(D^2 - 1)y = e^x.$$

Since $(D - 1)e^x = 0$, we have for any solution $y(x)$,

$$(D - 1)(D^2 - 1)y = (D - 1)e^x = 0.$$

Hence any particular solution y must have the form

$$y(x) = c_1 e^{-x} + c_2 e^x + c_3 x e^x.$$

Since the first two terms are solutions of the associated homogeneous equation $y'' - y = 0$, we can concentrate on the remaining term $c_3 x e^x$. To find c_3, we compute

$$y_p(x) = c_3 x e^x$$
$$y_p'(x) = c_3(x e^x + e^x)$$
$$y_p''(x) = c_3(x e^x + 2 e^x).$$

Thus for $y_p'' - y_p = e^x$ to hold, we must have

$$c_3(x e^x + 2 e^x) - c_3 x e^x = e^x,$$

or

$$(2 c_3 - 1) e^x = 0.$$

Hence $2 c_3 = 1$, so that $c_3 = \frac{1}{2}$. The general solution is thus

$$y = c_1 e^{-x} + c_2 e^x + \frac{1}{2} x e^x.$$

EXAMPLE 4 We can write the differential equation

$$y'' + y = 2 \sin x$$

in operator form as

$$(D^2 + 1) y = 2 \sin x.$$

Since $(D^2 + 1) \sin x = 0$, it follows that any solution $y(x)$ to the given equation must satisfy

$$(D^2 + 1)^2 y = 0.$$

Hence y must have the form

$$y(x) = c_1 x \cos x + c_2 x \sin x + c_3 \cos x + c_4 \sin x.$$

The last two terms satisfy the homogeneous equation $y'' + y = 0$, so we try to determine c_1 and c_2 so that

$$y_p(x) = c_1 x \cos x + c_2 x \sin x$$

satisfies $y'' + y = 2 \sin x$. We compute

$$y_p'(x) = c_1 \cos x - c_1 x \sin x + c_2 \sin x + c_2 x \cos x,$$
$$y_p''(x) = -2 c_1 \sin x - c_1 x \cos x + 2 c_2 \cos x - c_2 x \sin x.$$

Then to satisfy the given differential equation we substitute y_p and its derivatives to get

$$-2 c_1 \sin x + 2 c_2 \cos x = 2 \sin x.$$

Because $\cos x$ and $\sin x$ are linearly independent on any interval we must have $c_1 = -1$ and $c_2 = 0$. Thus $y_p(x) = -x \cos x$ is a particular solution, and

$$y(x) = c_3 \cos x + c_4 \sin x - x \cos x$$

is the general solution.

Here is an outline of the routine for finding the terms in a linear combination for a trial particular solution y_p to $L(y) = f(x)$, where f is a constant multiple of

$$x^n e^{\alpha x} \cos \beta x \quad \text{or} \quad x^n e^{\alpha x} \sin \beta x,$$

perhaps with n, α or β equal to zero.

(i) Include in the linear combination y_p the function f itself and all terms in its **derivative set**, consisting of the linearly independent sets of functions of which f and its successive derivatives are linear combinations. For example, the derivative set of $x^2 + x \sin x$ consists of the two sets $\{x^2, x, 1\}$ and $\{x \sin x, x \cos x, \sin x, \cos x\}$.

(ii) If a term included in step (i) happens to be a solution of the homogeneous equation, multiply that term and all terms in its derivative set by the single lowest power x^k such that the resulting terms are no longer homogeneous solutions.

(iii) Form a linear combination with undetermined constant coefficients of the terms from (ii), and determine the values of the coefficients by substitution into $L(y) = f$.

Here are some examples of functions $f(x)$ and corresponding trial solutions $y_p(x)$, assuming no term in $y_p(x)$ satisfies the homogeneous equation.

$$
\begin{aligned}
f(x) &= ce^{rx}; & y_p(x) &= Ae^{rx} \\
f(x) &= cx^2; & y_p(x) &= Ax^2 + Bx + C \\
f(x) &= cx^2 e^{rx}; & y_p(x) &= (Ax^2 + Bx + C)e^{rx} \\
f(x) &= c \cos \beta x; & y_p(x) &= A \cos \beta x + B \sin \beta x \\
f(x) &= cx \sin \beta x; & y_p(x) &= (Ax + B) \sin \beta x + (Cx + D) \cos \beta x \\
f(x) &= ce^{\alpha x} \cos \beta x; & y_p(x) &= e^{\alpha x}(A \cos \beta x + B \sin \beta x)
\end{aligned}
$$

EXAMPLE 5 The differential equation

$$y'' + 2y' + y = 3e^{-x}$$

has the homogeneous solution

$$y_h(x) = c_1 e^{-x} + c_2 x e^{-x}.$$

For a nonhomogeneous solution, we try functions of the form

$$y(x) = Ax^2 e^{-x}.$$

Since

$$y'(x) = A(2x - x^2)e^{-x},$$
$$y''(x) = A(2 - 4x + x^2)e^{-x},$$

substitution into the nonhomogeneous differential equation gives

$$A(2 - 4x + x^2)e^{-x} + 2A(2x - x^2)e^{-x} + Ax^2 e^{-x} = 3e^{-x}.$$

The terms with x and x^2 as factors all cancel, and we are left with

$$2Ae^{-x} = 3e^{-x},$$

so $A = \frac{3}{2}$. Thus $y_p(x) = \frac{3}{2}x^2e^{-x}$ is a particular solution, and the general solution is

$$y = c_1e^{-x} + c_2xe^{-x} + \frac{3}{2}x^2e^{-x}.$$

To find the form of a trial solution $y_p(x)$ for a nonhomogeneous equation it's often simpler, and just as effective, to make an educated guess at $y_p(x)$ rather than methodically following the three rules listed above. For example, a little experience shows that $y_p = Axe^{-x}$ is a good choice for the equation $y'' - y = e^{-x}$.

EXERCISES

In Exercises 1 to 10, find the general solution of the differential equations and then find the particular solution satisfying $y(0) = 0$ and $y'(0) = 1$.

1. $y'' - y = e^{2x}$ **2.** $y'' - y = 3e^x$

3. $y'' + 2y' + y = e^x$ **4.** $y'' - y = x$

5. $y'' - y = e^x + x$ **6.** $y'' - 2y = \cos 2x$

7. $y'' + y = \cos x$ **8.** $y'' = \cos x + \sin x$

9. $y'' + y = x \cos x$ **10.** $y'' - y = xe^x$

In Exercises 11 to 14, use factored operators and the exponential multiplier method to find the general solution for the differential equation.

11. $y'' + y' - 2y = e^x$ **12.** $y'' - y = e^{2x}$

13. $y'' + y = e^{ix}$ **14.** $y'' = x$

In Exercise 15 to 22, find a homogeneous differential equation of least possible order for which the given function is a solution.

15. $e^x + 2e^{2x}$ **16.** $e^x \cos x - e^x \sin x$

17. $x + 1$ **18.** $xe^x - 2e^x$

19. $x \sin 3x$ **20.** $x^2 \cos 4x$

21. $xe^x \sin x$ **22.** $x^3 e^{-x} \cos 2x$

In Exercises 23 to 28, find the appropriate form for a trial solution for the equation. For example you would use $y_p = A \cos 2x + B \sin 2x$ for $y'' - y = \sin 2x$.

23. $y'' - y = \cos x$ **24.** $y'' + y = \cos x$

25. $y'' - y = e^x$ **26.** $y'' - y = xe^x$

27. $y'' - 2y' + y = xe^x$ **28.** $y'' = x^5$

In Exercises 29 to 32, find the general solution of the equation by first finding the general solution of the associated homogeneous equation and then adding to it a particular solution found by the undetermined coefficient method. Then find the particular solution satisfying $y(0) = 0$, $y'(0) = 1$ and sketch the graph of that solution.

29. $y'' + 4y' + 4y = 3x$ **30.** $y'' - y' - 12y = 2e^{4x}$

31. $y'' + 2y' + 2y = e^x$ **32.** $y'' - y' = x$

In Exercise 33 to 42, find the general form for a trial solution y_p for each of the following. (For example, if $y'' - y = e^x$, choose $y_p = Axe^x$.) You need not determine the coefficient values.

33. $y'' - 4y = xe^{2x} + e^{2x}$

34. $y'' + y = x^2 \cos x$

35. $y'' - 5y' + 6y = xe^{2x} + e^{3x}$

36. $y'' + 4y = x^2 \cos 2x - 2 \sin 2x$

37. $y'' - 4y = e^{2x} + 5 \cos x$

38. $y'' + y = 3x \sin(x - 3)$

39. $y'' - y' = x^2 + 2e^x$

40. $y''' - y = e^{x/2} \sin \sqrt{3}x$

41. $y''' = 1 + x + x^3$

42. $y'' + y = x^{99} \cos x$

Falling body in a resisting medium. The distance $y(t)$ covered in time t by a falling body of mass m under the sole influence of a constant gravitational field satisfies a differential equation of the form $d^2y/dt^2 = g$. (If distance is measured up from the surface of the earth, then instead we would use $d^2y/dt^2 = -g$.) To take atmosphere resistance into account in a simple way, we write the differential equation in terms of force rather than resistance in the form

$$m\frac{d^2y}{dt^2} = gm - k\frac{dy}{dt} \quad \text{or} \quad \frac{d^2y}{dt^2} + \frac{k}{m}\frac{dy}{dt} = g.$$

Here k is a positive constant that is used to express a retarding force proportional to velocity dy/dt. This equation applies to Exercises 43 to 46.

43. (a) Show that the general solution to the retarded falling body equation is

$$y = c_1 + c_2 e^{-kt/m} + \frac{mg}{k}t.$$

(b) Show that if initial conditions $y(0) = y_0$ and $y'(0) = v_0$ are observed, then $c_1 = y_0 - (m/k)(mg/k - v_0)$ and $c_2 = (m/k)(mg/k - v_0)$.

(c) Show that regardless of the choice of initial conditions, $\lim_{t\to\infty} y'(t) = mg/k$; this limit is called the **terminal velocity** of the falling body.

(d) If the initial velocity $y'(0) = v_0$ is negative, show that the velocity reaches zero at time $t_1 = (m/k)\ln\left(1 - kv_0/(mg)\right)$. (Remember that y is measured down toward the attracting body, so negative initial velocity is directed up.)

44. Suppose a falling body is subject to a linear friction force $ky'(t)$ and has mass m.

(a) If the initial velocity is 0, show that the distance covered in time t is

$$y(t) = \frac{mg}{k}t - \frac{m^2 g}{k^2}(1 - e^{-kt/m}).$$

(b) Show that the formula for $y(t)$ in part (a) satisfies $y(t) \leq \frac{1}{2}gt^2$ for $t \geq 0$. [*Hint:* $y'' = g - (k/m)y' \leq g$. Now integrate from 0 to t.]

(c) Find the analogue of the formula given in part (a) for the case of initial velocity $y'(0) = v_0$.

45. (a) For a body of mass m subject to friction constant k, show that initial velocity v_0, leads to velocity at time t given by

$$y'(t) = \frac{mg}{k} + \left(v_0 - \frac{mg}{k}\right)e^{-kt/m}.$$

(b) Find the limit as k tends to 0 of the formula for $y'(t)$ in part (a). Does this agree with the free-fall formula $v_0 + gt$?

46. We can estimate the friction constant k by using an observed value of the terminal velocity $v_\infty = mg/k$ of a falling body dropped from rest.

(a) Find k if a body of weight $w = 100$ pounds achieves terminal velocity $v_\infty = 180$ feet per second.

(b) How far must the body in part (a) fall to attain the velocity of 150 feet per second?

*47. Use successive integration by the exponential multiplier method to show that if f is continuous on an interval containing x_0, then the equation

$$(D - r_1)(D - r_2)y = f(x)$$

has solution

$$y_p(x) = e^{r_2 x}\int_{x_0}^{x} e^{(r_1 - r_2)t}\left[\int_{x_0}^{t} e^{-r_1 s} f(s)\,ds\right]dt$$

satisfying $y(x_0) = y'(x_0) = 0$.

3C   Variation of Parameters

The undetermined coefficient method is inadequate if the nonhomogeneity involves a term that is not itself a solution of some homogeneous equation. If we know a nontrivial solution y_1 of a homogeneous equation, we can try to find a function $u(x)$ such that $y(x) = y_1(x)u(x)$ will be a solution of the associated *nonhomogeneous* equation. We can find the complete solution this way, and the procedure is called **variation of parameters**. The substitution $y(x) = y_1(x)u(x)$ will leave us with a linear differential equation that we can solve for $u(x)$; then solve this equation to find the "variable parameter" $u(x)$. This method doesn't require that the coefficients be constant.

| EXAMPLE 6 |

We know that the associated *homogeneous* equation for

$$y'' + 2y' + y = e^{-x}$$

has characteristic equation $(r + 1)^2 = 0$ and so has $y_1(x) = e^{-x}$ for a solution. Letting $y = y_1(x)u(x) = e^{-x}u(x)$, we compute

$$y = e^{-x}u, \quad y' = e^{-x}u' - e^{-x}u, \quad y'' = e^{-x}u'' - 2e^{-x}u' + e^{-x}u.$$

Substitution into the given differential equation followed by division by e^{-x} yields

$$(u'' - 2u' + u) + 2(u' - u) + u = 1, \quad \text{simplifying to} \quad u'' = 1.$$

We integrate $u'' = 1$ twice to get $u = \frac{1}{2}x^2 + c_1x + c_2$. So our solution is

$$y = y_1(x)u(x) = e^{-x}u(x)$$
$$= \tfrac{1}{2}x^2e^{-x} + c_1xe^{-x} + c_2e^{-x},$$

in agreement with what we would have found by the method of Section 3B.

EXAMPLE 7   It's routine to check that the equation

$$x^2y'' - 2xy' + 2y = x^4$$

has $y_1(x) = x$ for a solution of the associated *homogeneous* equation. (We just accept this as a guess for now.) Letting $y = y_1(x)u(x) = xu(x)$, we compute

$$y = xu, \quad y' = xu' + u, \quad y'' = xu'' + 2u'.$$

Substitution followed by simplification of the differential equation yields

$$x^2(xu'' + 2u') - 2x(xu' + u) + 2xu = x^4 \quad \text{or} \quad x^3u'' = x^4.$$

We divide by x^3 and integrate $u'' = x$ twice to get $u = \frac{1}{6}x^3 + c_1x + c_2$. So our solution is

$$y = y_1(x)u(x) = xu(x)$$
$$= \tfrac{1}{6}x^4 + c_1x^2 + c_2x.$$

Note that what we just did would have given us the second independent solution $y_2(x) = x^2$ regardless of what function of x we had on the right side of the equation.

In the previous example we found the homogeneous solution $y_1(x) = x$ by guessing, but lacking a correct guess we could have found a solution by a method described in Exercise 26 at the end of this section.

In applying variation of parameters, we can reduce by one the number of integrations required if we already know two independent solutions y_1, y_2 of the associated homogeneous equation. While it's similar in principle to what we did previously, the routine using two solutions is lengthy enough that rather than repeat it for every application we'll standardize it, along with the final result, as follows. Suppose $a(x)$ and $b(x)$ are continuous on some interval and that $f(x)$ is continuous. Then consider the **normalized equation**

$$y'' + a(x)y' + b(x)y = f(x),$$

in which the coefficient of y'' equals 1. If y_1, y_2 are homogeneous solutions, we form the linear combination

$$y(x) = y_1(x)u_1(x) + y_2(x)u_2(x),$$

where $u_1(x)$ and $u_2(x)$ are to be determined so that $y(x)$ is a solution of the nonhomogeneous equation. What we do now is what we did in the previous example: compute the derivatives y' and y'' and substitute into the nonhomogeneous equations. Then rearrange the terms as follows:

$$(y_1'' + ay_1' + by_1)u_1 + (y_2'' + ay_2' + by_2)u_2 + (y_1u_1' + y_2u_2')'$$
$$+ a(y_1u_1' + y_2u_2') + (y_1'u_1' + y_2'u_2') = f.$$

The first two collections of terms are zero because y_1 and y_2 are homogeneous solutions. What remains of the equation will be satisfied if we can choose u_1 and u_2 so that the two equations

3.2
$$y_1u_1' + y_2u_2' = 0,$$
$$y_1'u_1' + y_2'u_2' = f$$

hold identically for all x on the interval in question. We solve this system of equations for u_1' and u_2', with the result that

$$u_1'(x) = \frac{-y_2(x)f(x)}{y_1(x)y_2'(x) - y_2(x)y_1'(x)},$$

$$u_2'(x) = \frac{y_1(x)f(x)}{y_1(x)y_2'(x) - y_2(x)y_1'(x)}.$$

The expression in the denominators is the same in both formulas and we can write them as the 2-by-2 determinant

$$w(x) = \begin{vmatrix} y_1(x) & y_2(x) \\ y_1'(x) & y_2'(x) \end{vmatrix} = y_1(x)y_2'(x) - y_2(x)y_1'(x),$$

called the **Wronskian determinant** of the pair y_1, y_2. It's possible to prove that $w(x)$ is never zero if y_1, y_2 are linearly independent solutions, so the examples we consider here will have that property. To complete the solution, integrate the formulas for u_1', u_2' to find u_1 and u_2, and then combine with y_1, y_2 to get a particular solution

3.3
$$y_p(x) = y_1(x)u_1(x) + y_2(x)u_2(x),$$
$$= y_1(x) \int \frac{-y_2(x)f(x)}{w(x)} \, dx + y_2(x) \int \frac{y_1(x)f(x)}{w(x)} \, dx.$$

Then the solutions of the original normalized equation is

$$y(x) = c_1y_1(x) + c_2y_2(x) + y_p(x).$$

Because Equations 3.2 are easier to remember than Equation 3.3, people sometimes prefer to start with them in each problem and carry out the rest of the computation to arrive at Equation 3.3. The next example will be done that way. Because of the way that f enters Equations 3.2 and 3.3, the equation to be solved must be in normalized form to make these formulas valid.

EXAMPLE 8   We normalize $x^2 y'' - 2xy' + 2y = x^3$, say for positive x, to get

$$y'' - \frac{2}{x} y' + \frac{2}{x^2} y = x, \quad x > 0.$$

It's routine to check that the equation has homogeneous solutions $y_1(x) = x$, $y_2(x) = x^2$. For this example Equations 3.2 are

$$x u_1' + x^2 u_2' = 0,$$
$$u_1' + 2x u_2' = x.$$

Multiplying the second equation by x and then subtracting the first equation from it gives $x^2 u_2' = x^2$, or $u_2' = 1$. It then follows from the first equation that $u_1' = -x$. Integrating to find u_1 and u_2 gives

$$u_1(x) = -\tfrac{1}{2} x^2, \qquad u_2(x) = x.$$

A particular solution is

$$y_p = y_1 u_1 + y_2 u_2$$
$$= x \cdot (-\tfrac{1}{2} x^2) + x^2 \cdot x = \tfrac{1}{2} x^3.$$

Adding constants of integration to u_1 and u_2 would only add linear combination of homogeneous solutions to y_p. In any case, we have the solution

$$y = c_1 x + c_2 x^2 + \tfrac{1}{2} x^3.$$

We'll now solve a generalization of Example 1, but using Equation 3.3 instead of Equation 3.2.

EXAMPLE 9   In normalized form, we consider

$$y'' - \frac{2}{x} y' + \frac{2}{x^2} y = f(x), \quad x > 0.$$

Homogeneous solutions are $y_1(x) = x$, $y_2(x) = x^2$. The Wronskian determinant of y_1, y_2 is

$$w(x) = \begin{vmatrix} x & x^2 \\ 1 & 2x \end{vmatrix} = 2x^2 - x^2 = x^2.$$

Equation 3.3 reduces to

$$y_p(x) = x \int \frac{-x^2 f(x)}{x^2} \, dx + x^2 \int \frac{x f(x)}{x^2} \, dx$$
$$= -x \int f(x) \, dx + x^2 \int \frac{f(x)}{x} \, dx.$$

To make the integration fairly easy, we can use the example $f(x) = x \cos x$. For this choice, we get

$$y_p = -x \int x \cos x \, dx + x^2 \int \cos x \, dx$$

$$= -x \left(x \sin x - \int \sin x \, dx \right) + x^2 \int \cos x \, dx$$

$$= -x(x \sin x + \cos x) + x^2 \sin x$$

$$= -x \cos x.$$

3D   Green's Functions

A slight modification of Equation 3.3 yields formulas for the initial-value problem $y(x_0) = y'(x_0) = 0$ associated with a second-order constant-coefficient operator. The first step is to convert the integrals in Equation 3.3 into definite integrals over the interval from x_0 to x:

$$y_p(x) = y_1(x) \int_{x_0}^{x} \frac{-y_2(t) f(t)}{w(t)} \, dt + y_2(x) \int_{x_0}^{x} \frac{y_1(t) f(t)}{w(t)} \, dt,$$

3.4

$$= \int_{x_0}^{x} \frac{y_1(t) y_2(x) - y_2(t) y_1(x)}{w(t)} f(t) \, dt,$$

where $w(t) = y_1(t) y_2'(t) - y_2(t) y_1'(t)$ is the Wronskian determinant of $y_1(t)$, $y_2(t)$.

Setting $x = x_0$ in Equation 3.4 gives $y_p(x_0) = 0$. It's left as Exercise 19 or 20 to show that $y_p'(x_0) = 0$ also. Thus we define the **Green's function** to be

3.5

$$G(x, t) = \frac{y_1(t) y_2(x) - y_2(t) y_1(x)}{w(t)},$$

and this generates the solution

$$y_p(x) = \int_{x_0}^{x} G(x, t) f(t) \, dt$$

to the initial-value problem

$$y'' + a(x) y' + b(x) y = f(x), \quad y(x_0) = y'(x_0) = 0.$$

For equations $y'' + ay' + by = f(x)$, where a and b are real constants, the homogeneous solutions reduce to three distinct types, identified in Theorem 2.3 of Section 2A, depending on the nature of the characteristic roots of $r^2 + ar + b = 0$. Since the Green's functions are constructed from the solutions of constant-coefficient equations, $G(x, t)$ also has just three types:

(i) $G(x, t) = \dfrac{1}{r_1 - r_2} (e^{r_1(x-t)} - e^{r_2(x-t)})$, $r_1 \neq r_2$

(ii) $G(x, t) = (x - t) e^{r_1(x-t)}$, $r_1 = r_2$

(iii) $G(x, t) = \dfrac{1}{\beta} e^{\alpha(x-t)} \sin \beta(x - t)$, $r_1 = \alpha + i\beta$, $r_2 = \alpha - i\beta$

Using what we know about solving constant-coefficient equations, these formulas are fairly easy to remember, and it's straightforward in Exercise 21 to derive them from Equation 3.5.

Quite apart from the neatness with which Equation 3.4 displays a solution, it's convenient for calculating solutions when the forcing function $f(x)$ is discontinuous, as in the next example.

EXAMPLE 10   The normalized constant-coefficient equation $y'' - 3y' + 2y = f(x)$ has characteristic equation $r^2 - 3r + 2 = 0$ with roots $r_1 = 2$ and $r_2 = 1$ and independent solutions $y_1 = e^{2x}$, $y_2 = e^x$. Then $w(t) = -e^{3t}$ and the Green's function solution is

$$y_p(x) = \int_{x_0}^x \frac{e^{2t}e^x - e^t e^{2x}}{-e^{3t}} f(t)\, dt = \int_{x_0}^x (e^{2(x-t)} - e^{(x-t)}) f(t)\, dt.$$

Suppose $f(x) = -1$ if $x < 0$ and $f(x) = 2$ if $0 \le x$. In a purely formal sense the two cases differ only by a constant factor. Assume first that $x < 0$, where $f(x) = -1$. In that case, with $x_0 = 0$,

$$y_p(x) = -\int_0^x (e^{2(x-t)} - e^{(x-t)})\, dt = -e^{2x} \int_0^x e^{-2t}\, dt + e^x \int_0^x e^{-t}\, dt$$

$$= -e^{2x} \left[-\frac{1}{2} e^{-2t} \right]_0^x + e^x [-e^{-t}]_0^x = \frac{1}{2} e^{2x}(e^{-2x} - 1) + e^x(-e^{-x} + 1)$$

$$= e^x - \frac{1}{2} e^{2x} - \frac{1}{2} \quad \text{for } x < 0.$$

For the case $x \ge 0$, just replace the factor -1 by 2 in the integral to get altogether

$$y_p(x) = \begin{cases} e^x - \frac{1}{2} e^{2x} - \frac{1}{2}, & x < 0, \\ 1 + e^{2x} - 2e^x, & x \ge 0. \end{cases}$$

Note that the behavior of the two parts is quite different: $\lim_{x \to -\infty} y_p(x) = -\frac{1}{2}$, but $\lim_{x \to \infty} y_p(x) = \infty$.

EXAMPLE 11   Recall from earlier examples that $x^2 y'' - 2xy' + 2y = 0$ has independent solutions $y_1(x) = x$, $y_2(x) = x^2$. The Wronskian is $w(x) = x^2 \ne 0$ except at $x = 0$. The Green's function is $G(x, t) = (x^2/t) - x$. We have the solution to the normalized nonhomogeneous equation $y'' - (2/x)y' + (2/x^2)y = f(x)$ given by

$$y(x) = c_1 x + c_2 x^2 + \int_{x_0}^x \left(\frac{x^2}{t} - x \right) f(t)\, dt, \quad x_0 \ne 0.$$

If $f(x) = x^3$ the integral term is

$$y_p(x) = \int_{x_0}^x ((x^2/t) - x)t^3\, dt$$

$$= x^2 \int_{x_0}^x t^2\, dt - x \int_{x_0}^x t^3\, dt = \frac{1}{3} x^2 (x^3 - x_0^3) - \frac{1}{4} x(x^4 - x_0^4)$$

$$= \frac{1}{12} x^5 - \frac{1}{3} x_0^3 x^2 + \frac{1}{4} x_0^4 x.$$

This is the solution that satisfies $y(x_0) = y'(x_0) = 0$. The terms containing x_0 combine with the homogeneous solution to give the solution $y = c_1 x + c_2 x^2 + \frac{1}{12} x^5$.

EXAMPLE 12

The equation $y'' - \left(2x/(x^2 - 1)\right)y' + \left(2/(x^2 - 1)\right)y = f(x)$ has homogeneous solutions $y_1(x) = x$ and $y_2(x) = x^2 + 1$, with Wronskian

$$w(x) = \begin{vmatrix} x & x^2 + 1 \\ 1 & 2x \end{vmatrix} = x^2 - 1, \ x \neq \pm 1.$$

The Green's function for an initial-value problem is then

$$G(x, t) = \frac{t(x^2 + 1) - (t^2 + 1)x}{t^2 - 1}.$$

Suppose we want the solution with

$$f(x) = \begin{cases} 0, & -1 < x < 0, \\ 1, & 0 \leq x \leq \frac{1}{2}, \\ 0, & \frac{1}{2} < x < 1, \end{cases}$$

and satisfying $y(0) = y'(0) = 0$. This is

$$y_p(x) = \int_0^x G(x, t) f(t) \, dt = \begin{cases} 0, & -1 < x < 0, \\ \int_0^x G(x, t) \, dt, & 0 \leq x \leq \frac{1}{2}, \\ \int_0^{1/2} G(x, t) \, dt, & \frac{1}{2} < x < 1. \end{cases}$$

We compute, for $0 \leq x \leq \frac{1}{2}$,

$$\int_0^x G(x, t) \, dt = (x^2 + 1) \int_0^x \frac{t}{t^2 - 1} \, dt - x \int_0^x \frac{t^2 + 1}{t^2 - 1} \, dt$$

$$= (x^2 + 1)\left[\tfrac{1}{2} \ln(1 - x^2)\right] - x\left[x + \ln\left(\frac{1 - x}{1 + x}\right)\right].$$

The complete solution is then

$$y_p(x) = \begin{cases} 0, & -1 < x < 0, \\ \frac{1}{2}(x^2 + 1) \ln(1 - x^2) - x \ln((1 - x)/(1 + x)) - x^2, & 0 \leq x \leq 1/2, \\ \frac{1}{2} \ln(\frac{3}{4})(x^2 + 1) - (\frac{1}{2} - \ln 3)x, & \frac{1}{2} < x < 1. \end{cases}$$

The third line comes from evaluating the bracketed expressions in the previous integral evaluation at $x = \frac{1}{2}$. The first and third lines are solutions of the homogeneous equation on their respective intervals, because $f(x) = 0$ there. Finding the solution that satisfies more general initial conditions, $y(0) = y_0$, $y'(0) = z_0$, is just a matter of solving for c_1 and c_2 from

$$y(x) = c_1 x + c_2(x^2 + 1) + y_p(x).$$

We have

$$y'(x) = c_1 + 2c_2 x + y_p'(x).$$

But $y_p(0) = y_p'(0) = 0$, so we find c_1, c_2 from equations $c_2 = y_0$, $c_1 = z_0$.

Summary.   What we have seen in this section is a collection of methods for finding explicit solution formulas of the form

3.6 $$y(x) = c_1 y_1(x) + c_2 y_2(x) + y_p(x)$$

for differential equations of the form

3.7 $$y'' + a(x)y' + b(x)y = f(x).$$

It will follow from Theorem 1.1 in Chapter 12 that Equation 3.7 always has solutions of the form in Equation 3.6 on an interval $x_0 \leq x \leq x_1$ if $a(x)$ and $b(x)$ are continuous on that interval. In the constant-coefficient case we've seen that by choosing c_1 and c_2 properly we could satisfy arbitrary initial conditions of the form $y(x_0) = y_0$, $y'(x_0) = z_0$. This last possibility follows simply from the meaning of c_1 and c_2 as arbitrary constants of integration. If $a(x)$ and $b(x)$ are continuous functions, the analogous theorem still holds, though we won't prove it. The next example illustrates how this works.

| EXAMPLE 13 | The equation $(x - 1)y'' - xy' + y = 1$ has a particular solution $y_p(x) = 1$, and the associated homogeneous equation has solutions $y_1(x) = x$ and $y_2(x) = e^x$. Initial conditions $y(x_0) = y_0$, $y'(x_0) = z_0$ are satisfied by solving for c_1, c_2 in

$$y_0 = c_1 y_1(x_0) + c_2 y_2(x_0) + 1$$
$$z_0 = c_1 y_1'(x_0) + c_2 y_2'(x_0).$$

If $x_0 = 0$ it turns that $c_1 = z_0 - y_0 + 1$ and $c_2 = y_0 - 1$. But our method fails to apply at $x_0 = 1$, because the coefficient $(x - 1)$ of y'' is zero there so $y''(1)$ is not determined by $y(1)$ and $y'(1)$. You're asked to show in Exercise 27 that the initial-value problem at $x_0 = 1$ has a solution only if $z_0 = y_0 - 1$ and that in that case the solution is not unique.

EXERCISES

For each equation in Exercises 1 to 4, find or guess a solution y_1 of the associated homogeneous equation. Then determine $u(x)$ so that $y(x) = y_1(x)u(x)$ is a solution of the given equation containing two arbitrary constants.

1. $y'' - 4y' + 4y = e^x$

2. $y'' + (1/x)y' = x, x > 0$

3. $x^2 y'' - 3xy' + 3y = x^4, x > 0$ [*Hint*: Try $y_1 = x^n$, for some n.]

4. $xy'' - (2x + 1)y' + (x + 1)y = 3x^2 e^x, x > 0$ [*Hint*: Try $y_1(x) = e^{rx}$.]

In Exercises 5 to 10, find a particular solution y_p by solving Equation 3.2 for $u_1(x), u_2(x)$ to get $y_p(x) = y_1(x)u_1(x) + y_2(x)u_2(x)$. If suitable y_1 and y_2 are not given, find them first. Then find a solution that contains two arbitrary constants to the nonhomogeneous equation.

5. $y'' + y' - 2y = e^{2x}$

6. $y'' + y = \tan x, \ -\pi/2 < x < \pi/2$

7. $y'' + y = \sec x, \ -\pi/2 < x < \pi/2$

8. $y'' - y = xe^x$

9. $y'' = x^2 e^x$

10. $x^2 y'' - 2xy' + 2y = 1; \ y_1(x) = x, \ y_2(x) = x^2, \ x > 0$

In Exercises 11 to 14, use Equation 3.3 to find a formula for a solution $y_p(x)$. Complete the required integration if you can. Don't forget to make sure that the equation is normalized.

11. $y'' - 2y' + y = e^x$

12. $y'' + 3y' + 2y = 1 + e^x$

13. $y'' + 3y' + 2y = (a + e^x)^{-1}$

14. $2y'' + 8y = e^x$

In Exercises 15 to 18, find a pair of independent solutions to the associated homogeneous equation and use them to write the Green's function for the normalized equation. Then normalize the equation and solve the given initial-value problem.

15. $y'' + 3y' + 2y = \begin{cases} 0, & x < 1, \\ 1, & 1 \le x; \ y(0) = 1, \ y'(0) = 2 \end{cases}$

16. $2y'' + 4y = \begin{cases} 1, & x < 1 \\ 0, & 1 \le x; \ y(0) = -1, \ y'(0) = 1 \end{cases}$

17. $y'' + (1/x)y' = 1/x; \ y(1) = 0, \ y'(1) = 2$

18. $x^2 y'' - 2xy' + 2y = \begin{cases} 0, & 0 < x < 2, \\ 3, & 2 \le x \le 4, \\ 1, & 4 < x; \ y(3) = 0, \ y'(3) = 0 \end{cases}$

19. Show that the derivative of the Green's function Formula 3.4 is

$$y_p'(x) = \int_{x_0}^x \frac{y_1(t)y_2'(x) - y_2(t)y_1'(x)}{w(t)} f(t) \, dt.$$

Show then that $y_p'(x_0) = 0$. [*Hint:* Use Equation 3.4 separated into two integrals, and then apply the product rule for differentiation.]

20. The **Leibniz rule** for differentiating an integral states that

$$\frac{d}{dx} \int_{a(x)}^{b(x)} F(x, t) \, dt = \int_{a(x)}^{b(x)} \frac{\partial F}{\partial x}(x, t) \, dt$$
$$+ b'(x)F(x, b(x)) - a'(x)F(x, a(x)),$$

if $\partial F / \partial x$ is continuous. Use this result to establish the formula for $y_p'(x)$ in Exercise 19.

21. The constant-coefficient equation $y'' + ay' + by = f(x)$ has homogeneous solutions $y_1(x) = e^{r_1 x}, y_2(x) = e^{r_2 x}$. These solutions are independent if $r_1 \ne r_2$; otherwise we consider $y_1(x) = e^{r_1 x}, y_2(x) = xe^{r_1 x}$. Find the Green's function for the equation in the following cases by using Equation 3.5 of the text.

(a) $r_1 \ne r_2$
(b) $r_1 = r_2$
(c) $r_1 = \alpha + i\beta, r_2 = \alpha - i\beta, \beta \ne 0$

22. Sketch the graph of the solution found in Example 10 of the text.

Do the calculation of the Green's function integral in Example 11 of the text for the choices

23. $f(x) = x$

24. $f(x) = -1; x < 0; \ f(x) = 1, x \ge 0$

Do the calculation of the Green's function integral in Example 10 of the text for the choices

25. $f(x) = e^x$

26. $f(x) = e^x, x < 0; \ f(x) = e^{-x}, x \ge 0$

27. Consider the differential equation $(x-1)y'' - xy' + y = 1$ of Example 13 of the text. Show that the initial-value problem $y(1) = y_0, \ y'(1) = z_0$ has a solution only if $z_0 = y_0 - 1$, and that in that case there are infinitely many solutions. [*Hint:* x and e^x are homogeneous solutions.]

28. (a) Show that we can solve the **Euler differential equation**

$$x^2 y'' + axy' + by = 0, \quad a, b \text{ real constants},$$

as follows, assuming $x > 0$. Let $y = x^\mu$, so that $y' = \mu x^{\mu-1}$, and $y'' = \mu(\mu-1)x^{\mu-2}$. Show that for y to solve the differential equation, μ must satisfy the **indicial equation**
$$\mu^2 + (a-1)\mu + b = 0.$$

(b) Show that if the indicial equation has real roots $\mu_1 \ne \mu_2$ then $y_1 = x^{\mu_1}, y_2 = x^{\mu_2}$ are solutions.

(c) Show that if the indicial equation has complex conjugate roots $\mu_1 = \alpha + i\beta, \ \mu_2 = \alpha - i\beta$ then $y_1 = x^\alpha \cos(\beta \ln x), \ y_2 = x^\alpha \sin(\beta \ln x)$ are solutions. Note that, by definition, $x^{\alpha+i\beta} = x^\alpha e^{i\beta \ln x}$ for $x > 0$.

(d) Show that if μ_1 is a double root of the indicial equation then $y_1 = x^{\mu_1}, y_2 = x^{\mu_1} \ln x$ are solutions. Use Theorem 2.4 for this.

In Exercises 29 to 32, use the results of the previous exercise to solve the following Euler Equations.

29. $x^2 y'' + xy' - y = 0$ **30.** $x^2 y'' + 4xy' + y = 0$

31. $x^2 y'' + 3xy' + y = 0$ **32.** $x^2 y'' + xy' + y = 0$

33. Let $f(x)$ and $g(x)$ be twice differentiable functions on an interval $a < x < b$ on which $f(x)g'(x) - f'(x)g(x) \neq 0$.

   (a) Show that the 3-by-3 determinant equation

$$\begin{vmatrix} y & f & g \\ y' & f' & g' \\ y'' & f'' & g'' \end{vmatrix} = 0$$

      is a second-order homogeneous linear equation on the interval (a, b) having $f(x)$ and $g(x)$ as solutions.

   (b) Find a second-order homogeneous linear equation having $f(x) = \sin x$ and $g(x) = x \sin x$ as solutions for $0 < x < \pi$.

   (c) Find a second-order homogeneous linear equation having $f(x) = x$ and $g(x) = e^x$ as solutions for all $x \neq 1$.

34. It's sometimes erroneously inferred from insufficient evidence that the **Wronskian determinant**

$$W[f, g](x) = \begin{vmatrix} f(x) & g(x) \\ f'(x) & g'(x) \end{vmatrix}$$

of two linearly independent functions f and g can't equal 0. Put this idea to rest by computing $W[f, g](x)$ for all real x when $f(x) = x^2$ and $g(x) = x|x|$. You need to use the definition of derivative to compute $g'(0)$. What's true is that $W[y_1, y_2](x) \neq 0$ for two independent solutions of a second-order equation.

SECTION 4   OSCILLATIONS

A second-order constant-coefficient linear differential equation

$$a\frac{d^2 x}{dt^2} + b\frac{dx}{dt} + cx = f(t)$$

often has a physical interpretation that allows a neat classification of the solutions and equations into distinct types, depending on the relations between the constants a, b, c and the function f. Equations of this kind are important not only because of their direct physical applicability, but also because of the insight they yield about related nonlinear phenomena. A typical mechanism that we can analyze using a constant-coefficient equation is shown in Figure 11.3, in cross section. Automobile shock absorbers and artillery recoil mechanisms are designed using the principles illustrated here. The working parts consist of a piston that travels in a cylinder containing fluid, and a spring that can expand and compress. A spring usually exerts a force roughly proportional to its extension or compression from its equilibrium position, denoted by 0 on the fixed scale. Thus if x is the amount of displacement from 0, then the force f_s exerted by the spring is representable, according to **Hooke's law**, by

$$f_s = -hx, \quad h > 0,$$

where for small enough displacements h is constant. We also assume that the frictional force f_F in the mechanism due to the viscosity of the fluid is proportional to the velocity:

$$f_F = -k\frac{dx}{dt}, \quad k > 0.$$

The time-dependent external force $f_E = f(t)$ acts independently of f_s and f_F, which are, in turn, assumed to act independently of one another, so that the total force acting parallel to the scale in the figure is

$$f_s + f_F + f_E = -hx - k\frac{dx}{dt} + f(t).$$

FIGURE 11.3

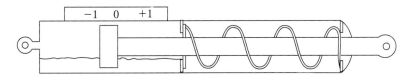

On the other hand, general physical principles assert that this force must also be equal to the mass m of the moving parts times the acceleration d^2x/dt^2. Thus

4.1
$$m\frac{d^2x}{dt^2} + k\frac{dx}{dt} + hx = f(t).$$

We'll investigate various assumptions about k, h, and $f(t)$. The mass m will be a fixed positive constant. We consider first **harmonic oscillation**, also called **free oscillation**, with external force f identically zero.

4A   Harmonic Oscillation

We assume that $k = 0$ and $f = 0$. These assumptions represent an ideal situation that can only be approximated by the mechanism shown in Figure 11.3. Under these assumptions the differential equation becomes

$$\frac{d^2x}{dt^2} + \frac{h}{m}x = 0.$$

The solutions all have the form

$$x(t) = c_1 \cos\sqrt{\frac{h}{m}}\, t + c_2 \sin\sqrt{\frac{h}{m}}\, t.$$

To interpret the solution easily it's good to rewrite it as follows. We choose an α such that

$$\cos\alpha = \frac{c_1}{\sqrt{c_1^2 + c_2^2}}, \quad \sin\alpha = \frac{c_2}{\sqrt{c_1^2 + c_2^2}};$$

such an α can always be found if c_1 and c_2 are not both zero. Then letting $A = \sqrt{c_1^2 + c_2^2}$ gives

$$x(t) = A\left(\cos\alpha \cos\sqrt{\frac{h}{m}}\, t + \sin\alpha \sin\sqrt{\frac{h}{m}}\, t\right)$$

$$= A\cos\left(\sqrt{\frac{h}{m}}\, t - \alpha\right).$$

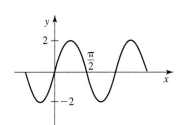

FIGURE 11.4

This last formula shows that A is the **amplitude** of an oscillation about the equilibrium position. The number $\omega_0 = \sqrt{h/m}$ is is the number of complete oscillations in time 2π, called the **circular frequency** of the oscillation. Thus if $\omega_0 = 2$, both $\cos(\omega_0 t)$ and $\sin(\omega_0 t)$ go through 2 oscillation periods in time 2π. The graph of the displacement as a function of time is shown in Figure 11.4 for the choice $\alpha = \pi/2$,

$A = 2$, and $h = m = 1$. Changing the number α, called the **phase angle** shifts the graph to the right or left. The frequency of the oscillation depends only on the ratio h/m; increasing h or decreasing m increases the frequency of the oscillation. That this ideal oscillation is periodic is a direct consequence of the assumption that there is no friction.

EXAMPLE 1   If in Equation 4.1 we take $m = h = 1$, $k = 0$, and $f = 0$, then the resulting differential equation

$$\frac{d^2x}{dt^2} + x = 0$$

has the solution

$$x(t) = A \cos(t - \alpha).$$

The initial conditions

$$x(0) = x_0, \quad \frac{dx}{dt}(0) = v_0,$$

require that

$$A \cos(-\alpha) = x_0, \quad -A \sin(-\alpha) = v_0.$$

We can solve these equations for A and α in terms of x_0 and v_0 to get

$$A = \sqrt{x_0^2 + v_0^2}, \quad \alpha = \arctan \frac{v_0}{x_0}.$$

In the special case when $x_0 = 0$, the initial displacement from equilibrium is zero, and we find

$$A = |v_0|, \quad \alpha = \pm\frac{\pi}{2}.$$

Thus the solution is

$$x(t) = |v_0| \cos\left(t \mp \frac{\pi}{2}\right)$$

$$= \pm|v_0| \sin t.$$

The sign has to be chosen so that $\pm|v_0| = v_0$. Other special cases are treated in the exercises.

4B   Damped Oscillation

The piston in the mechanism shown in Figure 11.3 exerts a damping force that depends on the viscosity of the medium where the piston moves. If we continue to assume that $f = 0$ in Equation 4.1, then we have to deal with the differential equation

$$\frac{d^2x}{dt^2} + \frac{k}{m}\frac{dx}{dt} + \frac{h}{m}x = 0.$$

The characteristic equation is

$$r^2 + \frac{k}{m}r + \frac{h}{m} = 0,$$

which has the roots

$$r_1 = \frac{1}{2m}(-k + \sqrt{k^2 - 4mh}), \quad r_2 = \frac{1}{2m}(-k - \sqrt{k^2 - 4mh}).$$

We distinguish three distinct cases depending on the discriminant $k^2 - 4mh$.

Overdamping: $k^2 - 4mh > 0$. Because k and h are both positive this case occurs when $k > 2\sqrt{mh}$. Physically this inequality means that the friction constant k exceeds the constant $\sqrt{mh}$, depending on the spring stiffness h and the mass m, by a factor of more than 2. The effect of the assumption $k > 2\sqrt{mh}$ is to make the roots of the characteristic equation satisfy

$$r_2 < r_1 < 0.$$

As a result, the general solution has the form

$$x(t) = c_1 e^{r_1 t} + c_2 e^{r_2 t},$$

where the exponentials decrease as t increases.

EXAMPLE 2   We assume $m = 2$, $h = 1$, and $k = 3$, so that $k > 2\sqrt{mh}$. Then

$$r_1 = -\tfrac{1}{2}, \quad r_2 = -1,$$

so that the displacement from equilibrium at time t is

$$x(t) = c_1 e^{-t/2} + c_2 e^{-t}.$$

A typical graph is shown in Figure 11.5. The maximum displacement occurs at just one point, after which the displacement tends steadily to 0 as t increases.

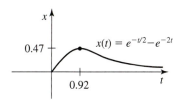

FIGURE 11.5

Underdamping, $k^2 - 4mh < 0$. This case occurs when $k < 2\sqrt{mh}$, so that, relative to $\sqrt{mh}$, the friction constant k is small. The characteristic roots are now complex conjugates of one another:

$$r_1 = \frac{1}{2m}(-k + i\sqrt{4mh - k^2}), \quad r_2 = \frac{1}{2m}(-k - i\sqrt{4mh - k^2}).$$

The general form of the displacement function is then

$$x(t) = e^{-kt/2m} \left(c_1 \cos \frac{\sqrt{4mh - k^2}}{2m} t + c_2 \sin \frac{\sqrt{4mh - k^2}}{2m} t \right)$$

$$= A e^{-kt/2m} \cos \left(\frac{\sqrt{4mh - k^2}}{2m} t - \alpha \right),$$

where $A = \sqrt{c_1^2 + c_2^2}$, and $\alpha = \arctan(c_2/c_1)$.

FIGURE 11.6

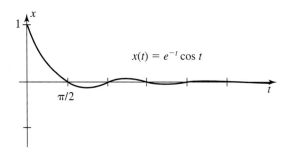

EXAMPLE 3

Take $h = 2$, $m = 1$, and $k = 2$. Then $k < 2\sqrt{mh}$, and the displacement at time t is

$$x(t) = Ae^{-t}\cos(t - \alpha).$$

Figure 11.6 shows the graph of such a function with $A = 1$ and $\alpha = 0$. It's easy to check that this choice for the constants A and α gives a solution satisfying the initial conditions

$$x(0) = 1, \quad \frac{dx}{dt}(0) = -1.$$

Critical Damping, $k = 2\sqrt{mh}$. This case lies between overdamping and under-damping, and it is critical in the sense that an arbitrarily small change in one of the parameters k, m, or h will disturb the equality $k = 2\sqrt{mh}$ and produce one of the other two cases. Numerically, the case of critical damping is distinguished by the equality of the characteristic roots: $r_1 = r_2 = -k/2m$. It follows that the displacement function is given by

$$x(t) = c_1 t e^{-kt/2m} + c_2 e^{-kt/2m}.$$

EXAMPLE 4

Take $m = h = 1$ and $k = 2$. Then

$$x(t) = (c_1 t + c_2)e^{-t}.$$

If $x(0) = x_0$ and $dx/dt(0) = v_0$, then

$$c_1 = (v_0 + x_0), \quad c_2 = x_0,$$

so that

$$x(t) = [(v_0 + x_0)t + x_0]e^{-t}.$$

Figure 11.7 shows four possibilities, depending on the size of v_0, the initial velocity.

A critically damped displacement is like an overdamped one in that there is no oscillation from one side to the other of the equilibrium position. But with fixed mass m and spring constant h, the critical viscosity value $k = 2\sqrt{mh}$ produces the most rapid return toward equilibrium from an initial position in which $v_0 = 0$. The physical reason is that a higher viscosity produces a more sluggish return, and lower viscosity allows oscillation.

FIGURE 11.7

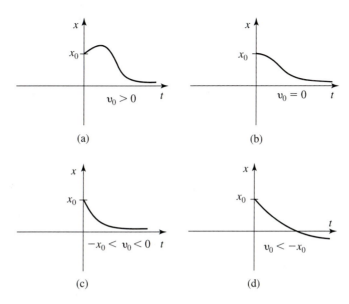

(a)   (b)

(c)   (d)

4C   Forced Oscillation

In the specific instances considered so far, the differential equation

$$m\frac{d^2x}{dt^2} + k\frac{dx}{dt} + hx = f(t)$$

has been subject to initial conditions $x(0) = x_0$, $dx/dt(0) = v_0$, but the external force function f has been assumed to be identically zero. The resulting free oscillation is described by a solution of a homogeneous differential equation. (Note that *free* doesn't imply *undamped*.) When f is not identically zero, we speak of **forced oscillation**. From a purely mathematical point of view, there is no reason why the function f on the right-hand side of the preceding differential equation cannot be chosen to be an arbitrary continuous function for $t \geq 0$. However a force function f that assumes large values could easily drive the oscillations outside the range in which we can maintain the original assumptions used to derive the differential equation. (For example stretching a spring too far might change its characteristics to the point of destroying its elasticity altogether.) For this reason, the function f is chosen in the examples and exercises to have a rather restricted range of values. In every example we can use the decomposition of a solution $x(t)$ into homogeneous and particular parts,

$$x(t) = x_h(t) + x_p(t).$$

The solution $x_h(t)$ has already been discussed earlier in this section for various choices of m, k, and h in the homogeneous differential equation. What remains to be done is to discuss the effect of adding a solution of the nonhomogeneous equation. If $k > 0$, the analysis given in the earlier examples shows that every homogeneous solution tends to zero like an exponential of the form $e^{-kt/2m}$. Thus for values of t that make $(kt/2m)$ moderately large, the addition of the homogeneous solution has a negligible effect. Such an effect is called **transient**, and the complementary particular solution is called the **steady-state** solution.

EXAMPLE 5   If $f(t) = a_0 \cos \omega t$, then the differential equation

$$m\frac{d^2x}{dt^2} + k\frac{dx}{dt} + hx = a_0 \cos \omega t, \quad k > 0,$$

has a particular solution of the form

$$x_p(t) = A \cos \omega t + B \sin \omega t$$

that we can find by the undetermined coefficient method of Section 3B. Substitution of x_p into the equation yields

$$(h - \omega^2 m)A + \omega k B = a_0,$$

$$\omega k A - (h - \omega^2 m)B = 0.$$

It follows that

$$A = \frac{(h - \omega^2 m)a_0}{(h - \omega^2 m)^2 + \omega^2 k^2}, \quad B = \frac{\omega k a_0}{(h - \omega^2 m)^2 + \omega^2 k^2},$$

so the particular solution we get is

$$x_p(t) = \frac{a_0}{(h - \omega^2 m)^2 + \omega^2 k^2}((h - \omega^2 m) \cos \omega t + \omega k \sin \omega t)$$

$$= \frac{a_0}{\sqrt{(h - \omega^2 m)^2 + \omega^2 k^2}} \cos(\omega t - \alpha),$$

where $\alpha = \arctan(\omega k/(h - \omega^2 m))$. What we have found is just one of many solutions, each satisfying different initial conditions. But since $k > 0$, the exponential in $x_h(t)$ decreases to zero so our particular solution is the steady-state solution. Notice that $x_p(t)$ is much like the external force $f(t)$. Since $f(t) = a_0 \cos \omega\, t$,

$$x_p(t) = \frac{1}{\sqrt{(h - \omega^2 m)^2 + \omega^2 k^2}} f\left(t - \frac{\alpha}{\omega}\right).$$

The choice $\omega^2 = h/m$ makes the maximum amplitude of x_p equal to $a_0\sqrt{m}/(k\sqrt{h})$. This choice of the frequency ω is called **resonant** because it produces a response of large amplitude for values of the system parameter k that are small relative to $\sqrt{m/h}$. Notice that in this example of resonance we have $\alpha = \arctan(\infty) = \pi/2$, so that

$$x_p(t) = \frac{a_0}{\omega k} \cos\left(\omega t - \frac{\pi}{2}\right)$$

$$= \sqrt{\frac{m}{h}} \cdot \frac{a_0}{k} \cos\left(\sqrt{\frac{h}{m}}\, t - \frac{\pi}{2}\right).$$

Thus in a system with $\sqrt{m}$ large relative to $k\sqrt{h}$, a small external force may produce vibrations of large amplitude if the external force oscillates at the resonant frequency. For this reason, resonance can completely upset an operating system, even though the external force remains small in magnitude.

EXERCISES

Note. Some of the following exercises use the overdot notation $\dot{x} = dx/dt$ and $\ddot{x} = d^2x/dt^2$ for first and second time derivatives, used also in Chapter 4, Section 1.

Each of the differential equations 1 to 6 generates a free (i.e., undriven) oscillation. (a) Without solving it first, classify each equation according to type: harmonic, over-damped, underdamped, or critically damped. (b) Find the general solution formula for each equation.

1. $d^2x/dt^2 + 2dx/dt + x = 0$

2. $d^2x/dt^2 + 2dx/dt + 2x = 0$

3. $\ddot{x} + 9x = 0$

4. $\ddot{x} + 3\dot{x} + x = 0$

5. $\ddot{x} + a\dot{x} + a^2x = 0, a > 0$

6. $d^2x/dt^2 + \frac{1}{4}dx/dt + \frac{1}{8}x = 0$

For each of the following general solutions of a second-order, constant-coefficient equation, find the choice of the arbitrary constants that satisfies the corresponding initial conditions. Sketch the solution that you find. Also find the differential equation of lowest order that the solution satisfies.

7. $x(t) = c_1 \cos 2t + c_2 \sin 2t$; $x(0) = 0$, $dx/dt(0) = 1$

8. $x(t) = A \cos(3t - \phi)$; $x(0) = 1$, $dx/dt(0) = -1$

9. $x(t) = c_1 t e^{-2t} + c_2 e^{-2t}$; $x(0) = 0$, $dx/dt(0) = -1$

10. $x(t) = A e^{-t} \cos(2t - \phi)$; $x(0) = 2$, $dx/dt(0) = 2$

11. $x(t) = c_1 e^{-2t} + c_2 e^{-4t}$; $x(0) = -1$, $dx/dt(0) = -1$

Find the steady-state solution to each of the differential equations 12 and 13. Also estimate the earliest time beyond which the transient solution remains less than 0.01, assuming initial conditions $x(0) = 1$, $\dot{x}(0) = 0$.

12. $d^2x/dt^2 + 2dx/dt + 2x = 2 \cos 3t$

    [*Hint:* Show $|x(t)| \le \sqrt{2}e^{-t}$.]

13. $d^2x/dt^2 + 3dx/dt + 2x = \cos t$

14. There is a well-known analogy between the behavior of a damped mass-spring system and of an *RLC* electrical circuit. Here L is the inductance (analogue of mass) of a coil, R is the resistance (analogue of friction constant) in the circuit, and C is the capacitance (analogue of reciprocal of spring stiffness) or ability of a capacitor to store a charge. The differential equation satisfied by the charge $Q(t)$ on the capacitor at time t is

$$L\frac{d^2Q}{dt^2} + R\frac{dQ}{dt} + \frac{1}{C}Q = E(t),$$

where $E(t)$ is the voltage impressed on the circuit from an external source. The charge $Q(t)$ is related to the current flow $I(t)$ by $I = dQ/dt$.

(a) Derive the relations that must hold between R, L, and C in order that the response of $Q(t)$ should be respectively underdamped, critically damped, and overdamped.

(b) Show that if $C = \infty$ (capacitor is absent) the equation for $I(t)$ is $L dI/dt + RI = E(t)$.

(c) Solve the equation in part (b) when $E(t) = E \sin \omega t$ and $I(0) = 0$, and show that, if t is large enough, the current response differs negligibly from

$$\frac{E}{Z} \sin(\omega t - \theta),$$

where $Z = \sqrt{R^2 + \omega^2 L^2}$ and $\cos\theta = R/Z$. The function $Z(\omega)$ is called the **impedance** of the circuit in response to the sinusoidal input of frequency ω.

*(d) Show that, if $0 < C < \infty$, a long-term response to input voltage $E(t) = E \sin \omega t$ is $E(t + \alpha)/Z(\omega)$, where $Z(\omega) = \sqrt{R^2 + [\omega L - 1/(\omega C)]^2}$, $\tan\alpha = (C^{-1} - \omega^2 L)/(\omega R)$.

15. We can determine the range of validity of Hooke's law for a given spring, and the corresponding value of a Hooke constant h, as follows. Hang the spring with known weights $W_j = m_j g$, $j = 1, \dots, n$, attached to the free end. If the additional extension is always proportional to the additional weight, then Hooke's law is valid for this range of extensions, and h is the constant of proportionality. A similar procedure applies to compression.

(a) Assume distance units are in feet. A spring with a 5-pound weight appended has length 6 inches, but with an 8-pound weight appended has length 1 foot. If the spring satisfies Hooke's law with constant h, find h.

(b) What if distance is measured in meters and force in kilograms in part (a)? (There are about 3.28 feet in a meter and 2.2 pounds in a kilogram.)

(c) Suppose we know Hooke's constant to be $h = 120$ for a certain spring. We observe that between hanging a 20-pound weight from it and then a larger weight we get an additional extension of 6 inches. How big is the larger weight?

(d) A spring is compressed to length 20 cm by a force of 5 kg, to length 10 cm by a force of 6 kg, and to 5 cm by a force of 7 kg. Discuss the possible validity of Hooke's law given this information.

16. Answer the following questions about solutions $x(t)$ of the forced harmonic oscillator $\ddot{x} + hx = \cos \omega t$, if $x(0) = (h - \omega^2)^{-1}$ and $\dot{x}(0) = 0$.

(a) If $h = 2$, what value should ω have to make the response amplitude equal 4?

(b) If $\omega = 2$, what value should h have to make the response amplitude equal to 5?

(c) If $h = 2$, what is the unique positive value of ω for which the response $x(t)$ becomes unbounded as $t \to \infty$?

(d) If $\omega = 10$, for what range of h-values will the response amplitude remain between 3 and 4?

17. Answer the following questions about solutions $x(t)$ to the damped and unforced equation $m\ddot{x} + k\dot{x} + hx = 0$.

(a) If $m = 2$, how should h and k be related so that the solutions will be oscillatory?

(b) If $h = k = 1$, how should m be chosen so that all nontrivial solutions will oscillate?

(c) If $m = h = 1$, how should k be chosen so that $x(t)$ has circular frequency $\omega = \frac{1}{2}$?

(d) If $m = h = 1$ how should k be chosen so that $x(t)$ is oscillatory?

18. Answer the following questions about solutions $x(t)$ of the damped and forced equation $m\ddot{x} + k\dot{x} + hx = \cos \omega t$.

(a) If $m = k = h = 1$, how should ω be chosen so that the amplitude of the steady-state solution will be 1?

(b) If $k = \omega = 1$, what relation must hold between m and h so that the steady-state amplitude will be 1?

(c) If $m = h = \omega = 1$, how should k be chosen so that the amplitude of the steady-state solution will be 2?

(d) What polynomial relation must hold among m, h, k, and ω if the frequency of the transient solution is to be the same as the steady-state frequency? As a special case, show that if the homogeneous solutions have frequency ω, and also $h = m\omega^2$, then there is no transient solution. [*Hint*: Show $k = 0$.]

Find the amplitude A, the frequency $\omega/2\pi$ and a phase angle ϕ for each of the following periodic functions.

19. $2\cos t + 3\sin t$

20. $-2\cos 2t + 3\sin 2t$

21. $\sin \pi t + 2\cos \pi t$

22. $\sin(3t)$

By how much are each of the following pairs of oscillations out of phase? You can decide this by expressing each pair in the general form $A\cos(\omega t - \phi)$, $\cos(\omega t - \psi)$.

23. $(\sqrt{3}/2)\cos t + (1/2)\sin t$, $\cos t$

24. $(1/2)\cos t + (\sqrt{3}/2)\sin t$, $\cos t$

25. $(1/\sqrt{2})\cos t + (1/\sqrt{2})\sin t$, $\sin t$

26. $\cos t - \sin t$, $\sin t$

27. A weight of mass $m = 1$ is attached by springs with Hooke constants h_1, h_2 to two fixed vertical supports. The weight oscillates along a horizontal line with negligible friction. By analyzing the force due to each spring, show that the displacement $x = x(t)$ of the weight from equilibrium satisfies $\ddot{x} = -(h_1 + h_2)x$.

28. A weight of mass m is attached by springs with Hooke constants h_1, h_2 to two fixed vertical supports. The weight oscillates along a horizontal line with negligible friction. Let the respective unstressed lengths of the two springs be l_1, l_2, and let b denote the distance between the supports.

(a) By analyzing the force due to each spring, show that the displacement $x = x(t)$ of the weight from the support attached to the first spring satisfies

$$\ddot{x} = -(h_1 + h_2)x + h_1 l_1 + h_2(b - l_2).$$

(b) Use the result of part (a) to show that the equilibrium value for $x(t)$ is

$$x_e = \big(h_1 l_1 + h_2(b - l_2)\big)/(h_1 + h_2).$$

(c) Show that the constant value x_e of part (b) is the solution to the differential equation that satisfies the initial conditions $x(0) = x_e$, $\dot{x}(0) = 0$.

29. The recoil mechanism of an artillery piece is designed containing a linearly damped spring mechanism. The spring stiffness h and the damping factor k should be chosen so that after firing the gun barrel will tend to its original position before firing without additional oscillation. We'll assume a given initial velocity V_0 and mass m for the gun barrel during recoil, and also a fixed maximum recoil distance E, always attained by the barrel.

(a) How should h and k be chosen so that under these conditions the gun barrel undergoes critical damping after firing?

(b) Write the differential equation for displacement $x(t)$ so as to display dependence on the parameters V_0 and E instead of h and k.

30. During construction of a suspension bridge, two towers have been erected, and a 10-ton weight is suspended between the towers by a cable anchored to both towers. Because of the elastic properties of the cable and the towers, it takes a $\frac{1}{2}$-ton force to move the weight sideways by 0.1 feet. An earthquake moves the base of each tower sideways with identical displacements of the form $0.25\cos 6t$ feet in t seconds. Assume a linear model for the lateral force on the weight and that damping is negligible. Find Hooke's constant h and the natural unforced frequency of oscillation for the weight.

31. The differential equation

$$m\ddot{x} + k\dot{x} + hx = a_0 \sin \omega t$$

determines the displacement $x(t)$ of a damped spring with external forcing $f(t) = a_0 \sin \omega t$ as a function of time t. A frequency ω that maximizes the amplitude of $x(t)$ is called a **maximum resonance frequency**. This exercise asks you to investigate the maximum resonance frequency for fixed ω, and under various assumptions about the mechanism.

(a) Consider the mechanically ideal case where $k = 0$. Show that choosing ω to equal the natural circular frequency $\omega_0 = \sqrt{h/m}$ produces a response $x(t)$ that contains the factor t and hence has deviations from the equilibrium position $x = 0$ that become arbitrarily large as t increases. (Thus there is no theoretical maximum resonance frequency in this case, though in practice the maximum response will be limited by the structural capacity of the mechanism to accommodate wide deviations from equilibrium.)

(b) k is a fixed positive number in Example 5 of the text. Show that the steady-state displacement $x_p(t)$ has maximum amplitude when h and m are chosen so that $\sqrt{h/m} = \omega$ and that the maximum amplitude is $a_0(\omega k)^{-1}$. (Making such choices for h and m constitute **tuning** of the mechanism for maximum response.)

(c) Continuing with the ideas of part (b), show that we get a small response amplitude in the steady-state solution to a given forcing frequency ω by making $|h - \omega^2 m|$ large.

32. In the previous exercise we assumed the input frequency in Example 5 of the text to be fixed and considered the effect of varying h and m in the differential equation. Suppose now that h, m and k are fixed positive numbers and that we want to choose ω so as to maximize the amplitude of the response $x_p(t)$.

(a) Show that the amplitude factor

$$\rho(\omega) = \left((h - \omega^2 m)^2 + k^2\omega^2\right)^{-1/2}$$

of $x_p(t)$ is maximized when the function $F(\omega) = (h - \omega^2 m)^2 + k^2\omega^2$ is minimized.

(b) Show that if $k^2 \le 2hm$, then $F'(\omega) = 0$ when $\omega^2 = (2hm - k^2)/(2m^2)$. Conclude from this that in this case the maximum response amplitude

$$\frac{2m|a_0|}{k\sqrt{4hm - k^2}} \quad \text{occurs for } \omega = \omega_0\sqrt{1 - \frac{k^2}{2hm}},$$

where $\omega_0 = \sqrt{h/m}$ is the natural circular frequency of the undamped ($k = 0$), unforced ($a_0 = 0$) mechanism.

(c) Show that $4mh - k^2 > 0$ is equivalent to assuming that the transient response $x_h(t)$ is oscillatory.

(d) Show that if $k^2 \ge 2hm$ then $F'(\omega) = 0$ only when $\omega = 0$ and that $F(\omega)$ is strictly increasing for $\omega \ge 0$. Hence conclude that in this case the maximum response is $|a_0|/h$, and occurs only for the constant $f(t) = a_0$.

33. The purpose of this exercise is to show that if m, k and h are positive constants, then for large enough t each particular solution of $m\ddot{x} + k\dot{x} + hx = b_0 \sin \omega t$ is bounded by a number proportional to $|b_0|$.

(a) Show that the solutions of the associated homogeneous equation all tend to zero as $t \to \infty$. (These are the transient solutions.)

(b) Show that

$$x_p(t) = \frac{-b_0}{k^2\omega^2 + (h - m\omega^2)^2}$$
$$\times (k\omega \cos \omega t - (h - m\omega^2) \sin \omega t)$$

is a particular solution and that it satisfies $|x_p(t)| \le |b_0|/\sqrt{k^2\omega^2 + (h - m\omega^2)^2}$.

[*Hint:* See Example 5 of the text.]

(c) Show how to conclude from the results of (a) and (b) that every solution is bounded for $t \ge 0$.

34. The purpose of this exercise is to observe the effect on the individual solutions of the initial-value problem

$$\ddot{x} + hx = \sin \omega t, \quad x(0) = \dot{x}(0) = 0$$

of letting the parameter ω approach the positive constant $\sqrt{h}$. (The differential equation represents a highly idealized situation from a physical point of view, because there is no damping term.)

(a) Show that the unique solution to the initial-value problem with $\omega \ne \sqrt{h}$ is

$$x(t) = \frac{1}{h - \omega^2}\left(\sin \omega t - \frac{\omega}{\sqrt{h}} \sin \sqrt{h}t\right),$$

and that the solution satisfies

$$|x(t)| \le \frac{1 + \omega/\sqrt{h}}{|h - \omega^2|} \quad \text{for all values of } t.$$

(b) Show that as ω approaches $\sqrt{h}$, the solution values found in part (a) approach

$$\frac{1}{2h}(-\sqrt{h}t \cos \sqrt{h}t + \sin \sqrt{h}t).$$

Show also that in contrast to the inequality in part (a), this function oscillates with arbitrarily large amplitude as t tends to infinity.

(c) Find an initial-value problem that has the function obtained in part (b) as a solution.

[*Hint:* What happens to the original differential equation as $\omega \to \sqrt{h}$?]

35. Suppose that an undamped, but forced, oscillator has the form

$$\ddot{x} + 2x = \sum_{k=0}^{n} a_k \cos kt.$$

(a) Use the linearity of the differential equation to show that it has the particular solution

$$x_p(t) = \sum_{k=0}^{n} \frac{a_k}{2 - k^2} \cos kt.$$

The trigonometric sum on the right in the differential equation is an example of a *Fourier series*, discussed in general in Chapter 14. An extension of such a sum to an infinite series can represent a very general class of functions.

(b) How does the solution in part (a) change if the left side of the differential equation is replaced by $\ddot{x} + 4x$ and the right side remains the same?

36. Consider the differential equation

$$\ddot{x} + 25x = 16 \cos 3t.$$

(a) Show that the equation has general solution

$$x(t) = c_1 \cos 5t + c_2 \sin 5t + \cos 3t.$$

(b) Show that the particular solution satisfying $x(0) = 0$, $\dot{x}(0) = 0$ is $x_p(t) = \cos 3t - \cos 5t$.

(c) Show that $\cos 3t - \cos 5t = 2 \sin 4t \sin t$.

(d) Use the result of part (c) to sketch the graph of the particular solution found in part (b) for $0 \le t \le 2\pi$.

37. (a) The phase difference between $\cos(\omega t - \alpha)$ and $\cos(\omega t - \beta)$ is $\alpha - \beta$. What is the *time*-shift required to put the two oscillations in phase?

(b) What is the phase difference between $\cos(\omega t - \alpha)$ and $\sin(\omega t - \beta)$? [*Hint:* Express the second one in terms of cosine.]

38. Let $f(t) = \sin \alpha t + \sin \beta t$ where α and β are positive numbers.

(a) Show that if $\beta = r\alpha$ for some rational number r then $f(t)$ is periodic for some period $p > 0$, i.e., $f(t + p) = f(t)$ for all t. Show also that p can be expressed as (possibly different) integer multiples of both π/α and π/β.

***(b)** Prove that if an $f(t)$ of the form given above is periodic with period $p > 0$ then $\beta = r\alpha$ for some rational number r. Thus for example $\sin t + \sin \sqrt{2}t$ can't be periodic. [*Hint:* Check that $f(p) = 0$ and $f''(p) = 0$. Then conclude that $(\alpha^2 - \beta^2) \sin \alpha p = (\alpha^2 - \beta^2) \sin \beta p = 0$, so that either $a = \pm\beta$ or else αp and βp are integer multiples of π.]

39. Suppose we want to construct a damped harmonic oscillator with Hooke constant $h = 2$ and damping constant $k = 3$. What is the lower limit m_0 for the mass m such that oscillatory solutions are possible? Does oscillation occur for $m = m_0$?

In Exercises 40 to 43, suppose a physical process is accurately modeled by a differential equation of the form

$$m\frac{d^2x}{dt^2} + k\frac{dx}{dt} + hx = 0,$$

with m, k and h positive constants. It may be possible by observation to draw conclusions about the parameters in the underlying process. Given each of the following sets of information about the constants and a solution, find the implications for the other constants.

40. $m = 1$, $x(t) = e^{-3t} \cos 6t$

41. $h = 1$, $x(t) = e^{-t} \sin 5t$

42. $k = 1$, $x(t) = e^{-t/2} \cos(t/2)$

43. $k = 3$, $h = 2$, $x(t) = e^{-4t} \sin 4t$

SECTION 5   LAPLACE TRANSFORMS

The techniques described earlier in the chapter are used to find formulas for the solution of initial-value problems such as

$$y'' + ay' + by = f(t), \quad y(0) = y_0, y'(0) = y_1.$$

Recall that the routine so far has been to solve the homogeneous equation by finding the roots of the characteristic equation, then solve the general nonhomogeneous equation by an integration that involves not only $f(t)$ but two independent homogeneous solutions. If we are willing to assume that $f(t)$ is defined for all $t \geq 0$, and that $f(t)$ and the solution $y(t)$ don't grow too rapidly as $t \to \infty$, we can use an alternative method that incorporates all of these steps and has for historical reasons achieved considerable popularity in electrical engineering and control theory. Experience with exponential integrating factors shows that it's natural to multiply a solution $y(t)$ by a factor e^{-st}, where s is some real or complex number, to get a product $e^{-st} y(t)$. What seems less natural, but nevertheless turns out to be effective, is to integrate with respect to t between 0 and ∞. This gives us an improper integral, leading for example to the calculation

5.1
$$\int_0^\infty e^{-st} e^{at}\, dt = \frac{1}{s-a}$$

which holds when $(s - a) > 0$. We'll use this result repeatedly, and to verify it we first compute the partial integral

$$\int_0^T e^{-(s-a)t}\, dt = \left[-\frac{1}{s-a} e^{-(s-a)t} \right]_0^T$$

$$= -\frac{1}{s-a} e^{-(s-a)T} + \frac{1}{s-a}.$$

When $T \to \infty$, the exponential factor tends to zero, so letting $T \to \infty$ on both sides proves the formula correct. If for some real or complex numbers s, an integral of the form

$$\mathcal{L}[f](s) = \int_0^\infty e^{-st} f(t)\, dt$$

converges to a function $\mathcal{L}[f]$ depending on s, then $\mathcal{L}[f]$ is called the **Laplace transform** of the function f. The key to using the Laplace transform to solve differential equations is the following formula:

5.2
$$\int_0^\infty e^{-st} y'(t)\, dt = -y(0) + s \int_0^\infty e^{-st} y(t)\, dt,$$

which we prove under the assumptions (i) the improper integrals are convergent, and (ii) $\lim_{t \to \infty} e^{-st} y(t) = 0$. Indeed, integration by parts of the partial integral on the left gives

$$\int_0^T e^{-st} y'(t)dt = [e^{-st} y(t)]_0^T + s \int_0^T e^{-st} y(t)\, dt$$

$$= e^{-sT} y(T) - y(0) + s \int_0^T e^{-st} y(t)\, dt.$$

Because of assumption (ii), the first term on the right tends to 0 as $T \to \infty$. Equation 5.2 follows by letting $T \to \infty$ in the preceding equation.

EXAMPLE 1	The example

$$y' + 2y = 0, \quad y(0) = 3,$$

is too simple to show the real advantages of using Laplace transforms, but it does illustrate the general principles involved. We form the Laplace transform of both sides of the differential equation by multiplying both sides by e^{-st} and then integrating from 0 to ∞ with respect to t. The result is

$$\int_0^\infty e^{-st} y' \, dt + 2 \int_0^\infty e^{-st} y \, dt = 0.$$

Equation 5.2 allows us to rewrite the first integral, obtaining

$$-3 + s \int_0^\infty e^{-st} y \, dt + 2 \int_0^\infty e^{-st} y \, dt = 0.$$

Here we have used the assumption that $y(0) = 3$. We also rely on our knowledge of the exponential nature of the solution to justify the assumptions (i) and (ii) needed for the application of Equation 5.2. The previous equation can now be solved for the Laplace transform of the solution $y(t)$ in the form

$$\int_0^\infty e^{-st} y \, dt = \frac{3}{s+2}. \tag{1}$$

Thus we have found not the solution $y(t)$, but its Laplace transform. However if we set $a = -2$ in Equation 5.1, and then multiply by 3, we get

$$\int_0^\infty e^{-st} 3e^{-2t} dt = \frac{3}{s+2}. \tag{2}$$

Since we already know from the general theory of this chapter that $y(t)$ must be an exponential solution, there remains only the one question of the constants involved, and we see by comparing Equations (1) and (2) that the solution

$$y(t) = 3e^{-2t}$$

satisfies our requirements.

To apply the Laplace transform to differential equations with order higher than one, we need a simple extension of Equation 5.2. To simplify the notation, we write 5.2 in the form

$$\mathcal{L}[y'](s) = -y(0) + s\mathcal{L}[y](s).$$

Applying this equation to $\mathcal{L}[y''](s)$, the Laplace transform of y'', gives

$$\mathcal{L}[y''](s) = -y'(0) + s\mathcal{L}[y'].$$

Applying the equation again gives

$$\mathcal{L}[y''](s) = -y'(0) + s\{-y(0) + s\mathcal{L}[y](s)\}$$
$$= -y'(0) - sy(0) + s^2\mathcal{L}[y](s).$$

The same routine leads after n steps to the formula

5.3
$$\mathcal{L}[y^{(n)}](s) = -y^{(n-1)}(0) - sy^{(n-2)}(0) - \cdots$$
$$- s^{n-2}y(0) + s^n\mathcal{L}[y](s).$$

The assumptions (i) and (ii) needed for 5.2 have to be increased to

(a) the integrals $\displaystyle\int_0^\infty e^{-st}y^{(k)}(t)\,dt$ are convergent for $k = 1, 2, \ldots, n$.

(b) $\displaystyle\lim_{t\to\infty} e^{-st}y^{(k)}(t) = 0$ for $k = 0, 1, \ldots, n-1$.

EXAMPLE 2   We can solve the differential equation

$$y'' - y' - 2y = 3e^t,$$

with initial conditions

$$y(0) = 1, \quad y'(0) = 0,$$

by applying the Laplace transform to both sides. For simplicity, we denote the Laplace transform of y by Y, that is, $Y(s) = \mathcal{L}[y](s)$. Using Equation 5.3 for $n = 1$ and 2, together with the initial conditions, we find

$$\mathcal{L}[y] = Y(s),$$
$$\mathcal{L}[y'] = -y(0) + sY(s)$$
$$= -1 + sY(s),$$
$$\mathcal{L}[y''] = -y'(0) - sy(0) + s^2Y(s)$$
$$= -s + s^2Y(s).$$

Because integration from 0 to ∞ and multiplication by e^{-st} are both linear operations, the equation

$$\mathcal{L}[y'' - y' - 2y] = \mathcal{L}[3e^t]$$

simplifies to

$$\mathcal{L}[y''] - \mathcal{L}[y'] - 2\mathcal{L}[y] = 3\mathcal{L}[e^t].$$

The expressions found for $\mathcal{L}[y]$, $\mathcal{L}[y']$, and $\mathcal{L}[y'']$, together with 5.1, allow us to write the equation as

$$[-s + s^2Y(s)] - [-1 + sY(s)] - 2[Y(s)] = 3\frac{1}{s-1}.$$

Rearrangement gives

$$(s^2 - s - 2)Y(s) = \frac{3}{s-1} + s - 1 = \frac{s^2 - 2s + 4}{s-1}.$$

or

$$Y(s) = \frac{s^2 - 2s + 4}{(s - 1)(s^2 - s - 2)}.$$

Having found an expression for $Y(s)$, our problem is now to identify precisely the solution $y(t)$ that satisfies $\mathcal{L}[y](s) = Y(s)$. Because $Y(s)$ is a rational function, it can theoretically always be broken down according to the partial fraction decomposition usually associated with the computation of indefinite integrals. In our example the decomposition works because the denominator of Y factors. We need to determine the coefficients A, B, and C in

$$\frac{s^2 - 2s + 4}{(s - 1)(s + 1)(s - 2)} = \frac{A}{s - 1} + \frac{B}{s + 1} + \frac{C}{s - 2}.$$

Multiplying through by $(s - 1)$ with $s \neq 1$, and then letting s go to 1, gives $A = -\frac{3}{2}$. Similarly, we multiply by $(s + 1)$ and then set $s = -1$ to get $B = \frac{7}{6}$. Finally, we multiply by $(s - 2)$ and then set $s = 2$ to get $C = \frac{4}{3}$. As a result, we have

$$Y(s) = -\frac{\frac{3}{2}}{s - 1} + \frac{\frac{7}{6}}{s + 1} + \frac{\frac{4}{3}}{s - 2}.$$

Equation 5.1 now allows us to identify $y(t)$ as

$$y(t) = -\tfrac{3}{2}e^t + \tfrac{7}{6}e^{-t} + \tfrac{4}{3}e^{2t}.$$

As a check on the computation, we can verify that $y(0) = 1$ and $y'(0) = 0$.

In the examples given previously, we have used the linearity of $\mathcal{L}$, the Laplace transform operator. The property is formally expressed by the two equations

5.4
$$\begin{aligned} \mathcal{L}[y_1 + y_2] &= \mathcal{L}[y_1] + \mathcal{L}[y_2] \\ \mathcal{L}[cy] &= c\mathcal{L}[y], \quad c = \text{constant.} \end{aligned}$$

These equations, together with Equation 5.3 for the Laplace transform of a derivative, need to be supplemented in practice by the calculation of Laplace transforms of specific functions as for example in Equation 5.1, which asserts that $\mathcal{L}[e^{at}](s) = 1/(s-a)$. Table 11.1 contains more than enough entries to do all the problems in this section, although more elaborate tables may contain several hundred entries. Such tables are meant to be used in both directions, so that while the entry

$$\mathcal{L}[t^n] = \frac{n!}{s^{n+1}}, \quad n = 0, 1, 2, \ldots$$

provides the transform of $f(t) = t^n$, it also provides, after division by $n!$, the inverse transform

$$\mathcal{L}^{-1}\left[\frac{1}{s^{n+1}}\right] = \frac{t^n}{n!}, \quad n = 0, 1, 2, \ldots$$

For the proof that for every Laplace transform Y there is a *unique* function y such that $\mathcal{L}[y] = Y$, we can refer to more theoretical accounts of the subject. All the entries in Table 11.1 are computed using elementary integration techniques.

TABLE 11.1 Table of Laplace transforms.

$f(t)$	$\mathcal{L}[f](s) = \int_0^\infty e^{-st} f(t)\, dt$
1. 1	$\dfrac{1}{s}$
2. t	$\dfrac{1}{s^2}$
3. t^n	$\dfrac{n!}{s^{n+1}}, \quad n = 0, 1, 2, \ldots$
4. e^{at}	$\dfrac{1}{s - a}$
5. $t e^{at}$	$\dfrac{1}{(s - a)^2}$
6. $t^n e^{at}$	$\dfrac{n!}{(s - a)^{n+1}}, \quad n = 0, 1, 2, \ldots$
7. $\sin bt$	$\dfrac{b}{s^2 + b^2}$
8. $\cos bt$	$\dfrac{s}{s^2 + b^2}$
9. $t \sin bt$	$\dfrac{2bs}{(s^2 + b^2)^2}$
10. $t \cos bt$	$\dfrac{s^2 - b^2}{(s^2 + b^2)^2}$
11. $e^{at} \sin bt$	$\dfrac{b}{(s - a)^2 + b^2}$
12. $e^{at} \cos bt$	$\dfrac{s - a}{(s - a)^2 + b^2}$
13. $e^{at} - e^{bt}$	$\dfrac{(a - b)}{(s - a)(s - b)}$
14. $a e^{at} - b e^{bt}$	$\dfrac{(a - b)s}{(s - a)(s - b)}$
15. $(b - c)e^{at} + (c - a)e^{bt} + (a - b)e^{ct}$	$\dfrac{(a - b)(b - c)(a - c)}{(s - a)(s - b)(s - c)}$
16. $\sin bt - bt \cos bt$	$\dfrac{2b^3}{(s^2 + b^2)^2}$

Partial fraction decomposition.   For finding inverse transforms it's sometimes essential to decompose a rational function $P(s)/Q(s)$, with degree of P less than the degree of Q, according to the following two rules; otherwise, long division shows that $P(s)/Q(s) = T(s) + R(s)/Q(s)$, where the remainder $R(s)$ has lower degree.

1. If the denominator $Q(s)$ has the factor $(s - a)^m$ as the highest power of $s - a$ that divides $Q(s)$, then include in the decomposition of $P(s)/Q(s)$ the fractions of the form

$$\frac{A_j}{(s - a)^j}, \quad j = 1, 2, \ldots, m.$$

2. If the denominator $Q(s)$ has the factor $(s^2 + ps + q)^n$ as the highest power of $s^2 + ps + q$ that divides $Q(s)$, then include in the decomposition of $P(s)/Q(s)$ the fractions of the form

$$\frac{B_k s + C_k}{(s^2 + ps + q)^k}, k = 1, 2, \ldots, n.$$

EXAMPLE 3   To find the function $f(t)$ having Laplace transform

$$F(s) = \frac{s + 1}{(s - 1)^2 (s^2 + 1)},$$

we decompose the function into a sum of fractions as follows:

$$\frac{s + 1}{(s - 1)^2 (s^2 + 1)} = \frac{A}{s - 1} + \frac{B}{(s - 1)^2} + \frac{Cs + D}{s^2 + 1}.$$

To compute B, we can multiply through by $(s - 1)^2$ and then set $s = 1$. We get $B = 1$. The same kind of trick doesn't apply directly to the other coefficients, but if we subtract $1/(s - 1)^2$ from both sides we find we can cancel $(s - 1)$ on the left to get

$$\frac{-s^2 + s}{(s - 1)^2 (s^2 + 1)} = \frac{-s}{(s - 1)(s^2 + 1)} = \frac{A}{s - 1} + \frac{Cs + D}{s^2 + 1}.$$

Now multiply by $(s - 1)$ and then set $s = 1$ to get $A = -\frac{1}{2}$. As a result,

$$\frac{-s}{(s - 1)(s^2 + 1)} = \frac{-\frac{1}{2}}{s - 1} + \frac{Cs + D}{s^2 + 1}.$$

To find C and D, we can multiply through by $(s - 1)(s^2 + 1)$ to get

$$-s = -\tfrac{1}{2}(s^2 + 1) + (s - 1)(Cs + D).$$

Rearranging the powers of s gives

$$\tfrac{1}{2}s^2 - s + \tfrac{1}{2} = Cs^2 + (D - C)s - D.$$

We equate coefficients of like powers on both sides and find that $C = \frac{1}{2}$ whereas $D = -\frac{1}{2}$. The result is that

$$\frac{s + 1}{(s - 1)^2 (s^2 + 1)} = \frac{-\frac{1}{2}}{s - 1} + \frac{1}{(s - 1)^2} + \frac{\frac{1}{2}s}{s^2 + 1} - \frac{\frac{1}{2}}{s^2 + 1}.$$

From the table of transforms, we conclude that

$$\mathcal{L}^{-1}[F(s)] = -\tfrac{1}{2}e^t + te^t + \tfrac{1}{2}\cos t - \tfrac{1}{2}\sin t.$$

The coefficients A, B, C, and D can also be computed by multiplying through by $(s - 1)^2 (s^2 + 1)$ at the first step and then equating coefficients of like powers of s

on both sides of the equation. The resulting linear equations can then be solved for the coefficients.

EXAMPLE 4   The differential equation

$$y'' + 4y = 3\sin 2t$$

with initial conditions $y(0) = 1$, $y'(0) = -1$, transforms into

$$-y'(0) - sy(0) + s^2 Y(s) + 4Y(s) = 3\frac{2}{s^2 + 4},$$

by using Equation 5.3 on the left side and Entry 7 of the transform table for the right side. Using the given initial values, we get

$$(s^2 + 4)Y(s) = \frac{6}{s^2 + 4} + s - 1$$

or

$$Y(s) = \frac{6}{(s^2 + 4)^2} + \frac{s}{(s^2 + 4)} - \frac{1}{(s^2 + 4)}.$$

To use entries 16, 8, and 7 in the table, we first write

$$Y(s) = \frac{3}{8} \cdot \frac{16}{(s^2 + 4)^2} + \frac{s}{s^2 + 4} - \frac{1}{2} \cdot \frac{2}{s^2 + 4}.$$

From the table, we read directly

$$y(t) = \tfrac{3}{8}(\sin 2t - 2t\cos 2t) + \cos 2t - \tfrac{1}{2}\sin 2t$$
$$= -\tfrac{1}{8}\sin 2t + \cos 2t - \tfrac{3}{4}t\cos 2t.$$

The amount of arithmetic required is less than if we had used any of the methods described earlier in the chapter; the reason is that we have relied heavily on the already-computed table of Laplace transforms.

EXERCISES

In Exercises 1 to 4, compute directly, assuming that $y(t)$ has a transform $Y(s)$ and is such that all required integrals and limits exist and are finite.

1. Integrate to verify that

$$\mathcal{L}[e^{at}](s) = \int_0^\infty e^{-st}e^{at}\,dt = \frac{1}{s-a}, \quad \text{for } s > a.$$

2. Use integration by parts to verify that

$$\mathcal{L}[t](s) = \int_0^\infty te^{-st}\,dt = \frac{1}{s^2}, \quad \text{for } s > 0.$$

3. Integrate once by parts to verify that if $y(0) = 1$ then

$$\int_0^\infty e^{-st}y'(t)\,dt = sY(s) - 1.$$

4. Integrate twice by parts to verify that if $y(0) = 2$ and $y'(0) = 3$, then

$$\int_0^\infty e^{-st}y''(t)\,dt = s^2Y(s) - 2s - 3.$$

In Exercises 5 to 8, by computing the appropriate integral, or by using Table 11.1 of Laplace transforms, compute $\mathcal{L}[f](s)$ where $f(t)$ is as follows:

5. $t \sin 2t$ **6.** $\cos t + 2 \sin t$ **7.** $t^2 + 2t - 1$

8. $\cos(t + a)$ **9.** $(2t + 1)e^{3t}$ **10.** $e^t + e^{-t}$

Use Table 11.1 to find the *inverse* Laplace transforms of the following functions $f(s)$—that is, to find $y(t)$ such that $\mathcal{L}[y](s) = F(s)$.

11. $\dfrac{1}{s^2 - 1}$ **12.** $\dfrac{2}{s^2 + 4}$

13. $\dfrac{1}{(s - 2)^2 + 9}$ **14.** $\dfrac{s}{s^2 - 4}$

15. $\dfrac{4s}{(s^2 + 4)^2}$ **16.** $\dfrac{1}{s^2} - \dfrac{1}{(s - 1)^2}$

In Exercises 17 to 22, use the Laplace transform to solve the following initial-value problems. Check by substitution.

17. $y' - y = t$, $y(0) = 2$

18. $y' + 2y = 1$, $y(0) = 1$

19. $y' + 3y = \cos 2t$, $y(0) = 0$

20. $y'' + y = e^{-t} + 1$, $y(0) = -1$, $y'(0) = 1$

21. $2y'' - y' = 2 \cos 3t$, $y(0) = 0$, $y'(0) = 2$

22. $y'' + y' + y = 1$, $y(0) = y'(0) = 0$

23. (a) Define the Heaviside function

$$H(t) = \begin{cases} 0, & \text{if } t < 0 \\ 1, & \text{if } 0 \le t. \end{cases}$$

Show that $\mathcal{L}[H(t - a)](s) = (1/s)e^{-as}$.

(b) Show that if $g(t) = H(t - a)f(t)$ for $0 \le t$ and $a \ge 0$, then

$$\mathcal{L}[g](s) = e^{-as}\mathcal{L}[f(t + a)](s).$$

(c) Sketch the graph of $H(t) - H(t - 1)$.

(d) Solve the differential equation $y'' = H(t - a)$, $0 < a$, with initial conditions $y(0) = 1$, $y'(0) = 0$.

24. Let $P(D)$ be an nth-order, linear, constant-coefficient, differential operator. Show that

$$\mathcal{L}[P(D)y](s) = P(s)\mathcal{L}[y](s) + Q(s),$$

for some polynomial Q of degree $n - 1$. Use induction.

25. (a) Show that if $f(t)$ is defined and differentiable only for $t > 0$ (instead of $t \ge 0$), then

$$\mathcal{L}[f'](s) = -f(0+) + s\mathcal{L}[f](s),$$

where $f(0+) = \lim_{t \to 0+} f(t)$.

(b) Show that if the limits $f^{(k)}(0+)$ all exist, then Equation 5.3 generalize to

$$\mathcal{L}[f^{(n)}](s) = -f^{(n-1)}(0+) - \cdots - s^{(n-1)}f(0+) + s^n \mathcal{L}[f](s).$$

26. The assumption that $\displaystyle\int_0^\infty |f'(t)|\, dt < \infty$ implies that

$$\lim_{s \to 0+} \int_0^\infty e^{-st} f'(t)\, dt = \int_0^\infty f'(t)\, dt.$$

Show that, under the additional assumption that $\lim_{t \to \infty} f(t)$ exists,

$$\lim_{t \to \infty} f(t) = \lim_{s \to 0+} s\mathcal{L}[f](s).$$

This formula makes it possible to determine something about the long-run behavior of f from the behavior of $\mathcal{L}[f](s)$ near $s = 0$, without finding f.

SECTION 6   CONVOLUTION

Let us review the solution of the second-order differential equation

$$y'' + py' + qy = f(t), \quad y(0) = y_0, \, y'(0) = y_1.$$

Taking the Laplace transform of both sides gives

$$(s^2 + ps + q)Y(s) = F(s) + y'(0) + sy(0) + py(0).$$

The polynomial factor $P(s) = s^2 + ps + q$ on the left is the **characteristic polynomial** of the operator $D^2 + pD + q$. The reciprocal $Q(s) = 1/P(s)$ is called the **transfer**

function of the operator, and if we multiply by $Q(s)$, or divide by $P(s)$, we get the formula

$$Y(s) = \frac{F(s) + y'(0) + sy(0) + py(0)}{P(s)}$$

for the Laplace transform $Y = \mathcal{L}[y]$. The remaining step is to find the inverse transform $y(t) = \mathcal{L}^{-1}[Y](t)$. The essence of the method is to use the Laplace transform to reduce the solution of the problem to some routine algebraic manipulations.

In addition to the table of specific Laplace transforms in the previous section (Table 11.1), there are a number of general formulas, such as Formula 5.3, that are useful in solving problems. The most important of these answers the following question: If $F(s)$ and $G(s)$ are the Laplace transforms of $f(t)$ and $g(t)$, respectively, what function has Laplace transform equal to the product $F(s)G(s)$? It turns out that under rather general hypotheses there is an answer, given by the **convolution integral**

$$f * g(t) = \int_0^t f(u)g(t-u)\,du.$$

The function $f * g(t)$ is called the **convolution** of the functions $f(t)$ and $g(t)$ and is defined for $t \geq 0$, provided that f and g are integrable on every finite interval. The convolution $f * g$ is to be thought of as a kind of product of f and g and it turns out that $f * g = g * f$, although this is not obvious from the definition. The basic information about convolutions is summarized as follows.

6.1   Theorem.   Let $f(u)$ and $g(u)$ be integrable on $0 \leq u \leq t$ for every positive t; then $f * g$ and $g * f$ both exist and are equal, that is, convolution is commutative:

$$f * g = g * f.$$

If $|f(t)|$ and $|g(t)|$ are such that $\mathcal{L}[|f|](s)$ and $\mathcal{L}[|g|](s)$ are both finite, then

$$\mathcal{L}[f * g](s) = \bigl(\mathcal{L}[f](s)\bigr)\bigl(\mathcal{L}[g](s)\bigr).$$

Proof.   The first statement follows from changing variable in the definition of $f * g$. We have, on replacing u by $t - v$,

$$f * g(t) = \int_0^t f(u)g(t-u)\,du = -\int_t^0 f(t-v)g(v)\,dv$$

$$= \int_0^t g(v)f(t-v)\,dv = g * f(t).$$

Proving the second statement involves an important technical point for which we won't give a proof, but the rest of the argument is complete. To simplify the writing of limits of integration, we can extend both $f(t)$ and $g(t)$ to have the value 0 for $t < 0$. Since e^{-su} is independent of v, we can write

$$\mathcal{L}[f](s)\mathcal{L}[g](s) = \int_{-\infty}^{\infty} e^{-su} f(u)\,du \int_{-\infty}^{\infty} e^{-sv} g(v)\,dv$$

$$= \int_{-\infty}^{\infty} f(u)\left[\int_{-\infty}^{\infty} e^{-s(u+v)} g(v)\,dv\right] du.$$

We next make the change of variable $v = t - u$ in the inner integral to get

$$\mathcal{L}[f](s)\mathcal{L}[g](s) = \int_{-\infty}^{\infty} f(u)\left[\int_{-\infty}^{\infty} e^{-st} g(t-u)\, dt\right] du.$$

Under our assumptions we can interchange the order of integration using a theorem called Fubini's theorem. Then we have

$$\mathcal{L}[f](s)\mathcal{L}[g](s) = \int_{-\infty}^{\infty} e^{-st}\left[\int_{-\infty}^{\infty} f(u)g(t-u)\, du\right] dt.$$

Because we have assumed $f(t)$ and $g(t)$ are zero for $t < 0$, the inner integral is zero for $t < 0$. It follows that we need the t integration only for $0 \leq t < \infty$. Similarly we need the u integration only for $0 \leq u \leq t$. Hence

$$\mathcal{L}[f](s)\mathcal{L}[g](s) = \int_{0}^{\infty} e^{-st}\left[\int_{0}^{t} f(u)g(t-u)\, du\right] dt$$
$$= \mathcal{L}[f * g](s),$$

which is what we wanted to prove. ∎

EXAMPLE 1   From Table 11.1, we see that $\mathcal{L}[t](s) = 1/s^2$ and $\mathcal{L}[\sin t](s) = 1/(s^2+1)$. It follows from Theorem 6.1 that

$$\frac{1}{s^2}\cdot\frac{1}{s^2+1} = \mathcal{L}\left[\int_{0}^{t} (t-u)\sin u\, du\right](s).$$

Holding t fixed, we can use integration by parts to show that

$$\int_{0}^{t} (t-u)\sin u\, du = [-(t-u)\cos u]_{0}^{t} - \int_{0}^{t} \cos u\, du$$
$$= -t - \sin t.$$

We could have obtained the same result by computing a partial fraction decomposition of the form

$$\frac{1}{s^2}\cdot\frac{1}{s^2+1} = \frac{A}{s} + \frac{B}{s^2} + \frac{Cs+D}{s^2+1},$$

and then finding the inverse transform of each term.

Table 11.2 lists the most frequently used general properties of the Laplace transform. The entries that haven't already been discussed follow from elementary calculus techniques. The table omits the precise conditions under which each formula holds. The distinction between Formulas 2 and 3 and Equations 5.2 and 5.3 of the previous section occurs because in 5.2 and 5.3 we assumed that we were dealing with solutions of differential equations and that these solutions had continuous derivatives at $t = 0$. The corresponding formulas in Table 11.2 are valid under the weaker assumption that $\lim_{t\to 0+} f^{(k)}(t)$ exists, but is not necessarily equal to $f^{(k)}(0)$. (See Exercise 8 of Section 5.)

TABLE 11.2 General properties of the Laplace transform.

1. $\mathcal{L}[af + bg] = a\mathcal{L}[f] + b\mathcal{L}[g], \quad a, b$ constant

2. $\mathcal{L}[f'](s) = s\mathcal{L}[f](s) - f(0+), \quad f(0+) = \lim_{t \to 0+} f(t)$

3. $\mathcal{L}[f^{(n)}](s) = s^n \mathcal{L}[f](s) - s^{n-1} f(0+) - s^{n-2} f'(0+) - \cdots - f^{(n-1)}(0+)$

4. $\mathcal{L}\left[\int_0^t f(u) \, du\right](s) = \dfrac{1}{s}\mathcal{L}[f](s)$

5. $\mathcal{L}[e^{at} f(t)](s) = \mathcal{L}[f](s - a)$

6. $\mathcal{L}[f(t - a)](s) = e^{-as} \mathcal{L}[f](s), \quad a > 0, \quad f(t) = 0$ if $t < 0$

7. $\mathcal{L}[tf(t)](s) = -\dfrac{d}{ds}\mathcal{L}[f](s)$

8. $\mathcal{L}[f * g](s) = \mathcal{L}[f](s)\mathcal{L}[g](s)$

EXAMPLE 2

From Table 11.1, we find that

$$\mathcal{L}[\sin t](s) = \frac{1}{s^2 + 1}.$$

Taking $f(t) = \sin t$ in Formula 4 of Table 11.2 gives

$$\frac{1}{s(s^2 + 1)} = \frac{1}{s}\mathcal{L}[\sin t](s)$$

$$= \mathcal{L}\left[\int_0^t \sin u \, du\right](s) = \mathcal{L}[-\cos t + 1].$$

Hence

$$\mathcal{L}^{-1}\left[\frac{1}{s(s^2 + 1)}\right] = -\cos t + 1.$$

Repeating the application of Formula 4 gives

$$\frac{1}{s^2(s^2 + 1)} = \mathcal{L}\left[\int_0^t (-\cos u + 1) \, du\right]$$

$$= \mathcal{L}[-\sin t + t].$$

This establishes the formula

$$\mathcal{L}^{-1}\left[\frac{1}{s^2(s^2 + 1)}\right] = -\sin t + t,$$

which was derived in Example 1 using convolution.

EXAMPLE 3

Starting with the formula

$$\mathcal{L}[\cos t](s) = \frac{s}{s^2 + 1},$$

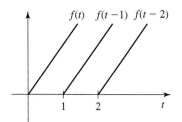

FIGURE 11.8

we can apply Formula 7 in Table 11.2 to get

$$\mathcal{L}[t\cos t](s) = -\frac{d}{ds}\frac{s}{s^2+1}$$

$$= \frac{s^2-1}{(s^2+1)^2}.$$

Another application of the same formula gives

$$\mathcal{L}[t^2\cos t](s) = -\frac{d}{ds}\frac{s^2-1}{(s^2+1)^2}$$

$$= \frac{2s^3-6s}{(s^2+1)^3}.$$

EXAMPLE 4   To apply Formula 6 of Table 11.2 to the function $f(t) = t$, we define $f(t) = 0$ for $t < 0$. The graphs of $f(t)$ and $f(t-a)$ are in Figure 11.8 for $a = 1$ and $a = 2$. Each function is zero where it's not positive. From Formula 6 we find

$$\mathcal{L}[f(t-1)](s) = e^{-s}\mathcal{L}[f(t)](s)$$

$$= e^{-s}\mathcal{L}[t](s) = e^{-s}\frac{1}{s^2}.$$

Similarly,

$$\mathcal{L}[f(t-2)](s) = e^{-2s}\mathcal{L}[f(t)](s)$$

$$= e^{-2s}\mathcal{L}[t](s) = e^{-2s}\frac{1}{s^2}.$$

EXERCISES

In Exercises 1 to 4, find the convolution $f * g$ of the given pair of functions.

1. $f(t) = t$, $g(t) = e^{-t}$, $t \geq 0$

2. $f(t) = t^2$, $g(t) = (t^2 + 1)$, $t \geq 0$

3. $f(t) = 1$, $g(t) = 1$, $t \geq 0$

4. $f(t) = t$, $g(t) = \cos t$, $t \geq 0$

In Exercises 5 to 8, use the convolution of two functions to find the inverse Laplace transform of each of the given products of Laplace transforms.

5. $\dfrac{1}{s^2(s+1)}$ **6.** $\dfrac{1}{(s-1)(s-2)}$

7. $\dfrac{1}{s^3}e^{-2s}$ **8.** $\dfrac{1}{(s^2+1)(s-1)}$

In Exercises 9 to 14, use the formulas in Tables 11.1 and 11.2, find the inverse Laplace transform of the given function.

9. $\dfrac{1}{s(s+3)^2}$ **10.** $\dfrac{e^{-2s}}{s(s^2+4)}$

11. $\dfrac{1}{s^2+2s+2}$ **12.** $\dfrac{1}{s^2+1}$

13. $(e^{-s}+1)/s$ **14.** $\dfrac{s}{s^2+10}$

In Exercises 15 to 18, solve the given initial-value problem and check by substitution.

15. $y'' - y = \sin 2t + 1$, $y(0) = 1$, $y'(0) = -1$

16. $y'' + 2y = t$, $y(0) = 0$, $y'(0) = 1$

17. $y'' + y' = t + e^{-t}$, $y(0) = 2$, $y'(0) = 1$

18. $y'' + y = \sin t$, $y(0) = 0$, $y'(0) = 1$

19. Solve the equation $y' + y = \int_0^t y(u)\,du + t$, given that $y(0) = 1$

20. Solve the equation $y' - y = \int_0^t y(u)\,du$, given that $y(0) = 1$

21. (a) Use Formula 4 in Table 11.2 repeatedly to show that

$$\mathcal{L}\left[\int_0^t \left(\int_0^{t_1} \left(\cdots \int_0^{t_n} f(t_{n+1})\,dt_{n+1}\right)\cdots dt_2\right) dt_1\right]$$
$$= \frac{1}{s^{n+1}} F(s).$$

(b) Use the Convolution Theorem 6.1 to show that the $(n+1)$-fold iterated integral equals

$$\frac{1}{n!} \int_0^t (t-u)^n f(u)\,du.$$

22. One possible definition of the **gamma function**, denoted by $\Gamma(z)$, is

$$\Gamma(z) = \int_0^\infty t^{z-1} e^{-t}\,dt, \quad z > 0.$$

(a) Use integration by parts to show that

$$\Gamma(z+1) = z\Gamma(z).$$

(b) Deduce from part (a) that $\Gamma(n+1) = n!$, for $n = 0, 1, 2, \ldots$.

(c) Show that if $a > -1$, then

$$\mathcal{L}[t^a](s) = \frac{\Gamma(a+1)}{s^{a+1}}.$$

*23. The conditions on the absolute values of f and g in Theorem 2.1 that $\mathcal{L}[|f|](s)$ and $\mathcal{L}[|g|](s)$ both be finite is needed to allow interchange of integration order in the second part of the proof. These hypotheses are usually routine to verify in ordinary practice, but to see that it really is a restriction we need an example of an f that has a transform but such that $|f|$ doesn't.

(a) Let

$$f(t) = e^{(t+e^t)} \sin e^{(e^t)}.$$

Show that

$$\int_0^\infty e^{-st} f(t)\,dt = \int_{e^e}^\infty \frac{\sin u}{(\ln u)^s}\,du,$$

and that this is finite for all $s > 0$. [*Hint:* Express the second integral as an alternating infinite series.]

(b) Show that $\int_0^\infty e^{-st}|f(t)|\,dt = +\infty$ for all s. [*Hint:* Compare the analogue of the second integral above with a smaller, but divergent, infinite series.]

(c) Prove that if $\mathcal{L}[|f(t)|](s)$ is finite then so is $F(s) = \mathcal{L}[f(t)](s)$. [*Hint:* Compare $|\int_N^M e^{-st} f(t)\,dt|$ and $\int_N^\infty e^{-st}|f(t)|\,dt$.]

SECTION 7   NONLINEAR EQUATIONS

In the earlier sections of this chapter we've seen a complete treatment of the initial-value problem

$$\ddot{y} + a\dot{y} + by = f(t), \quad y(t_0) = y_0, \quad \dot{y}(t_0) = z_0,$$

where a and b are real constants, and $f(t)$ is continuous on an interval I containing t_0. In particular we proved the existence of a unique solution for the problem and exhibited a general solution formula of the form

$$y(t) = c_1 y_1(t) + c_2 y_2(t) + y_p(t),$$

where $y_1(t)$ and $y_2(t)$ are solutions of the associated homogeneous equation with $f(t) = 0$ on I, and the constants c_1, c_2 are chosen to satisfy the initial conditions. The analogous problem where coefficients $a = a(t)$ and $b = b(t)$ in the differential equation are continuous functions on the interval I has solutions of the same form, and this will follow from Theorem 7.1. However, Theorem 7.1 covers much more than linear equations; also included in its scope are nonlinear equations such as

$$\ddot{y} = y\dot{y}, \quad \text{or} \quad \ddot{y} = ty^2, \quad \text{or} \quad \ddot{y} = -\sin y.$$

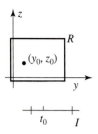

7.1   Existence and Uniqueness Theorem.

Assume the function of three variables $f(t, y, z)$ and its two first-order partial derivatives $f_y(t, y, z)$ and $f_z(t, y, z)$ are continuous for t in an interval I and for all (y, z) in an open rectangle R containing (y_0, z_0) shown in the figure. Then the initial-value problem

$$\ddot{y} = f(t, y, \dot{y}), \quad y(t_0) = y_0, \quad \dot{y}(t_0) = z_0$$

has a unique solution on some subinterval J of I containing t_0. If in addition there is constant B such that $|f_y(t, y, z)| \le B$ and $|f_z(t, y, z)| \le B$ for all t in I and all (y, z) in $\mathbb{R}^2$, then the unique solution is defined on the entire interval I.

EXAMPLE 1   Of the three equations preceding Theorem 7.1 $\ddot{y} = -\sin y$ is the only one that satisfies the boundedness condition. In this example $|f_y(t, y, z)| = |\cos y| \le 1$, and $f_z(t, y, z) = 0$, so the initial-value problem has a solution for all real t, starting with arbitrary t_0, y_0, and z_0.

We'll consider in Sections 7A and 7B two special cases of the equation $\ddot{y} = f(t, y, \dot{y})$ for which we can sometimes find explicit solutions. The trick in each case is to do what we did with constant-coefficient linear equations, reduce the problem to a pair of successive integrations. These methods apply also to some linear equations, including the ones discussed earlier in this chapter.

7A   Dependent Variable Absent: $\ddot{y} = f(t, \dot{y})$

Given that $\ddot{y} = f(t, \dot{y})$ the natural thing to do is first let $z = \dot{y}$, giving us a system of two first-order equations

$$\dot{z} = f(t, z)$$

$$\dot{y} = z$$

to solve first for z and then for y. In the next example we solve a linear equation with a variable coefficient, one that we can't solve using the constant-coefficient methods of the previous sections in this chapter.

EXAMPLE 2   Suppose we want to solve the initial-value problem

$$\ddot{y} = \frac{1}{t}\dot{y}, \quad y(t_0) = y_0, \quad \dot{y}(t_0) = z_0,$$

with $t_0 \ne 0$. Letting $z = \dot{y}$, this problem is equivalent to

$$\dot{z} = \frac{1}{t}z, \quad z(t_0) = z_0$$

$$\dot{y} = z, \quad y(t_0) = y_0.$$

The top equation doesn't contain y and is first-order linear with integrating factor $\exp\left(\int(-1/t)\,dt\right) = 1/t$, so the integrable form of the equation is

$$(1/t)\dot{z} - (1/t^2)z = 0 \quad \text{or} \quad \frac{d}{dt}(z/t) = 0, \quad \text{with solution} \quad z = c_1 t.$$

The initial condition on z requires $c_1 = z_0/t_0$, so $z(t) = z_0 t/t_0$. To find y we integrate $\dot{y} = z$ to get $y = \frac{1}{2}z_0 t^2/t_0 + c_2$. Finally, the initial condition $y(t_0) = y_0$ requires $c_2 = y_0 - \frac{1}{2}z_0 t_0$ so $y = \frac{1}{2}z_0 t^2/t_0 + y_0 - \frac{1}{2}z_0 t_0$. A routine check shows this to be the solution to the initial-value problem for $t > 0$ when $t_0 > 0$, or for $t < 0$ when $t_0 < 0$.

7B   Independent Variable Absent: $\ddot{y} = f(y, \dot{y})$

As usual we reduce the problem to two first-order problems by letting $dy/dt = z$ and considering the system

$$\frac{dz}{dt} = f(y, z)$$

$$\frac{dy}{dt} = z.$$

At this point we make an assumption, namely that z is expressible as a differentiable function of y, an assumption that will be verifiable in practice. Under that assumption we apply the chain rule for functions of a single variable as follows:

$$\frac{dz}{dt} = \frac{dz}{dy}\frac{dy}{dt}, \quad \text{or since } \frac{dy}{dt} = z, \quad \frac{dz}{dt} = z\frac{dz}{dy}.$$

Thus the shortcut to remember is the replacement of $\dfrac{dz}{dt}$ by $z\dfrac{dz}{dy}$ in the top equation of our first-order system, yielding

$$z\frac{dz}{dy} = f(y, z)$$

$$\frac{dy}{dt} = z.$$

If we can solve the top equation for z as a function of y then we can put the result in the bottom equation with some hope of solving for $y = y(t)$. Note that we would then have $\dot{y} = z(y(t))$ as a check on the accuracy of our computations.

EXAMPLE 3

The initial-value problem $\ddot{y} = y\dot{y}$, $y(0) = 0$, $\dot{y}(0) = \frac{1}{2}$ breaks down, according to the outline above, into

$$z\frac{dz}{dy} = zy, \quad z(0) = \tfrac{1}{2},$$

$$\frac{dy}{dt} = z, \quad y(0) = 0.$$

Assuming $z \neq 0$, we can divide by z in the top equation to get the separable equation

$$\frac{dz}{dy} = y, \quad \text{with solutions } z = \tfrac{1}{2}y^2 + c_1.$$

Using the two initial conditions $z(0) = \frac{1}{2}$ and $y(0) = 0$ we see that $c_1 = \frac{1}{2}$, so $z = \frac{1}{2}y^2 + \frac{1}{2}$. (Note that z is never zero.) Putting this expression for z into the bottom equation of the system gives the equation

$$\frac{dy}{dt} = \tfrac{1}{2}y^2 + \tfrac{1}{2}. \quad \text{or, in separated form,} \quad \frac{dy}{y^2 + 1} = \tfrac{1}{2}\, dt.$$

Integration gives $\arctan y = \frac{1}{2}t + c_2$, or since $y(0) = 0$, $\arctan y = \frac{1}{2}t$. Hence the solution to the second-order initial-value problem is $y = \tan\left(\frac{1}{2}t\right)$.

Note. The solution $y = \tan\left(\frac{1}{2}t\right)$ found in the previous example has its domain restricted by the behavior of $\tan x$ near $x = \pm\pi/2$. Thus the solution is valid only when $-\pi/2 < t/2 < \pi/2$, or $\pi < t < \pi$. This is an instance of Theorem 7.1 where the domain interval J of a solution is actually smaller than the domain interval I in which the differential equation makes sense. There is no way to tell from looking at the differential equation $\ddot{y} = y\dot{y}$ just what the interval J will be, because J depends critically on the initial conditions.

EXAMPLE 4

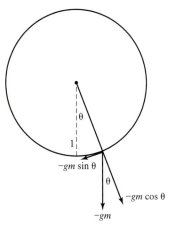

FIGURE 11.9
Pendulum analysis.

Initially we best describe the motion of a pendulum with no force but gravity acting on it in terms of the angle $\theta = \theta(t)$ that the pendulum makes with a vertical line as shown in Figure 11.9. Here we'll assume an *ideal* pendulum with a rod of negligible mass attaching a weight of mass m to the pivot, and with all mass concentrated at the weight's center of gravity at distance l from the pivot. Figure 11.9 shows the set-up, where the possible positions for the center of mass describe a circle of radius l. The typical motion is a back and forth oscillation, such as we considered in Section 4.

The downward-directed gravitational force **F** of magnitude mg has a radial component $\mathbf{F}_R$ of magnitude $|gm\cos\theta|$ and a component $\mathbf{F}_T$ of magnitude $|gm\sin\theta|$ tangential to the circle. At any given position the gravitational force component in the direction of motion is perpendicular to the component directed along the length of the pendulum. The coordinates relative to these directions are $F_T = -gm\sin\theta$ and $F_R = -gm\cos\theta$. It follows by the Pythagorean relation that the sum of the squares of the perpendicular component magnitudes must be $g^2 m^2$. The force $\mathbf{F}_R$ acting along the length of the pendulum toward its end must be exactly balanced by an opposite force at the pivot, and these forces play no other role in our description of the motion. If θ is measured in radians, distance along the circular path of motion is $y = l\theta$, so we can express the force coordinate F_T in the direction of motion as mass times acceleration: $F_T = md^2(l\theta)/dt^2$. Equating our two expressions for $\mathbf{F}_T$ gives the differential equation satisfied by $\theta = \theta(t)$:

$$m\frac{d^2(l\theta)}{dt^2} = -gm\sin\theta.$$

The minus sign signifies that the signed velocity $d(l\theta)/dt$ is decreasing if $0 < \theta < \pi$ and increasing if $-\pi < \theta < 0$. If $y(t)$ stands for distance measured along the circle then $\theta(t) = y(t)/l$, leading to the alternative form

$$\frac{d^2 y}{dt^2} = -g\sin\frac{y}{l}.$$

Theorem 7.1 guarantees the existence of a unique solution $y(t)$ to an initial-value problem with $y(t_0) = y_0$, $\dot{y}(t_0) = z_0$, but this nonlinear equation has no solutions in terms of elementary functions.

If y remains small, say $|y| < 0.1$, we may find approximate solutions that are acceptable for some purposes by using the tangent-line replacement for the graph of $\sin y$ near $y = 0$, namely $\sin y \approx y$. This approximation leads to the linear equation

$$\frac{d^2 y}{dt^2} = -\frac{g}{l}y.$$

It's routine to check that this differential equation has among its solutions

$$y = \cos\sqrt{g/l}\,t \quad \text{and} \quad y = \sin\sqrt{g/l}\,t.$$

The solutions of the linear version of the pendulum equation are examples of *harmonic oscillation*, studied in detail in Section 4A. The solutions of the nonlinear pendulum equation are oscillatory also, but have their own distinct character, which we'll take up in the Exercises and in Section 8 on numerical methods. In particular, it's not at all obvious from looking at the differential equation $\ddot{y} = -\sin y$ that it has periodic solutions, but this is nevertheless true and doesn't depend on the periodicity of the sine function.

EXERCISES

In Exercises 1 to 6, use the method of Section 7A to solve the initial value problem, stating explicitly for what values of the independent variable your solution is defined.

1. $t\ddot{y} + \dot{y} = 0$; $y(1) = 0$, $\dot{y}(1) = 1$

2. $t^2\ddot{y} + \dot{y}^2 = 0$; $y(1) = \dot{y}(1) = -1$

3. $\ddot{y} + \dot{y}^2 = 0$; $y(0) = 0$, $\dot{y}(0) = 1$

4. $t\ddot{y} + \dot{y} = 0$; $y(1) = \dot{y}(1) = 1$

5. $t\ddot{y} + \dot{y} = t^3$; $y(1) = \dot{y}(1) = 1$

6. $\ddot{y} + \dot{y} = 0$; $y(0) = \dot{y}(0) = 1$

In Exercises 7 to 13, use the method of Section 7B to solve the initial value problem, stating explicitly for what values of the independent variable your solution is defined.

7. $y\ddot{y} - \dot{y}^2 = 0$; $y(0) = \dot{y}(0) = 1$

8. $y^2\ddot{y} + \dot{y}^3 = 0$; $y(0) = \dot{y}(0) = 1$

9. $\ddot{y} - \dot{y}^3 = 0$; $y(0) = \dot{y}(0) = 1$

10. $\ddot{y} + \dot{y}^2 = 0$ $y(0) = 0$, $\dot{y}(0) = 1$

11. $\ddot{y} + \dot{y}^2 = 1$; $y(0) = \dot{y}(0) = 0$

12. $\ddot{y} - y^3 = 0$; $y(0) = \dot{y}(0) = \sqrt{2}$

13. Show that the differential equation $t^2\ddot{y} + \dot{y}^2 = 0$ has more than one solution satisfying $y(0) = \dot{y}(0) = 0$. Explain why this doesn't contradict the uniqueness part of Theorem 7.1.

14. Show that the differential equation $t\ddot{y} + \dot{y} = t^3$ has more than one solution satisfying $y(0) = \dot{y}(0) = 0$. Explain why this doesn't contradict the uniqueness part of Theorem 7.1.

15. (a) Solve the initial value problems for $\ddot{y} = -y$ and $\ddot{y} = y$, using the same initial conditions $y(0) = 0$, $\dot{y}(0) = 1$ for both equations.
    (b) Sketch the graphs of both solutions of part (a).

(c) Use the complex exponential e^{it} to find a relation between the solutions of part (a).

16. (a) Solve the initial-value problem $\ddot{y} = y\dot{y}$, $y(0) = 0$, $\dot{y}(0) = -\frac{1}{2}$.
    (b) Sketch the graph of the solution to part (a).
    (c) Use the complex exponential $e^{it/2}$ to find a relation between the solution to part (a) and the solution to the problem in Example 3 of the text.

17. (a) Apply the method of Section 7B to the initial-value problem

    $$\ddot{y} = -\sin y, \quad y(0) = y_0, \quad \dot{y}(0) = z_0$$

    for a nonlinear pendulum equation to establish the equation

    $$\tfrac{1}{2}\dot{y}^2 = \cos y - \cos y_0 + \tfrac{1}{2}z_0^2.$$

    (b) Use the result of part (a) to show that if $-\cos y_0 + \tfrac{1}{2}z_0^2 > 1$ the pendulum will rotate "over the top" repeatedly.
    (c) Use the result of part (a) to show that to have oscillatory motion, with angle y strictly between $-\pi$ and π, we must have $-\cos y_0 + \tfrac{1}{2}z_0^2 < 1$.

18. (a) Apply the method of Section 7B to the initial-value problem

    $$\ddot{y} = -\sin y, \quad y(0) = \eta, \quad \dot{y}(0) = 0$$

    for a nonlinear pendulum equation to establish the equation

    $$\tfrac{1}{2}\dot{y}^2 = \cos y - \cos \eta,$$

    where $-\pi < \eta < \pi$.
    (b) Prove that the time $T(\eta)$ it takes for the pendulum in part (a) to fall from angle $y = \eta$ to the vertical

position at angle $y = 0$ is

$$T(\eta) = \int_0^\eta \frac{dy}{\sqrt{2(\cos y - \cos \eta)}}.$$

(c) Make a change of variable in the integral in part (b) chosen to show that $T(\eta) = K(k)$, where $k =$

$\sin(\eta/2)$, with $0 < \eta < \pi$, and

$$K(k) = \int_0^{\pi/2} \frac{d\phi}{\sqrt{1 - k^2 \sin \phi}}, \quad k^2 < 1,$$

which is an **elliptic integral** and isn't computable using elementary functions.

7C   Phase Space

If we can't find an explicit formula for the solution of a second-order equation and we want to study a particular solution satisfying given initial conditions, one option is to apply numerical methods as discussed in Section 8 and another is to display solutions in what is called *phase space*, described here. To do either of these we first convert second-order equations $\ddot{y} = f(t, y, \dot{y})$, with initial conditions $y(t_0) = y_0$, $\dot{y}(t_0) = z_0$, into first-order systems as follows. Let $\dot{y} = z$ so that $\dot{z} = \ddot{y}$. Then $\dot{z} = f(t, y, \dot{y})$, and we can write the original second order initial-value problem as

$$\begin{aligned} \dot{y} &= z, & y(t_0) &= y_0, \\ \dot{z} &= f(t, y, z), & z(t_0) &= z_0. \end{aligned}$$

There are two advantages to this reformulation. A purely technical advantage is that it allows us to apply the first-order numerical methods of Chapter 10, Section 2 to second-order problems. The other advantage is conceptual, in that the first-order system gives equal weight to displaying the two fundamental quantities, position y and velocity $z = \dot{y}$. The 2-dimensional (y, z)-space is the **phase space** of the second-order equation. For the purpose of plotting curves in phase space we restrict attention to equations $\ddot{y} = f(y, \dot{y})$ in which the function f is not explicitly dependent on time t; such equations are called **autonomous**.

EXAMPLE 5

The harmonic oscillator equation $\ddot{y} = -\omega^2 y$ is equivalent to the system

$$\begin{aligned} \dot{y} &= z, \\ \dot{z} &= -\omega^2 y. \end{aligned}$$

The solutions $y = A \cos(\omega t - \alpha)$ of the second-order equation correspond to solutions

$$y = A \cos(\omega t - \alpha), \quad z = -\omega A \sin(\omega t - \alpha)$$

of the first-order system. Since

$$y^2 + z^2/\omega^2 = A^2 \cos^2(\omega t - \alpha) + A^2 \sin^2(\omega t - \alpha) = A^2,$$

the vector functions $(y(t), z(t))$ trace out ellipses in the phase space. Figure 11.10(a) shows some curves in phase space and Figure 11.10(b) shows the corresponding solution graphs, which relate time and position. The phase curves relate the fundamental quantities position and velocity. In Figure 11.10 half the *width* of an ellipse corresponds to the *amplitude* of a graph.

FIGURE 11.10

(a) $y^2 + z^2/4 = A^2$
(b) $y = A \sin(2t)$.

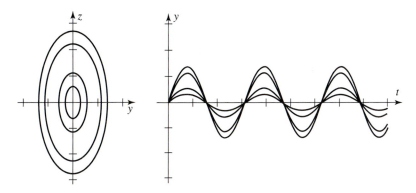

The previous example illustrates an important point about phase curves and periodic solutions $y = y(t)$ of second-order differential equations $\ddot{y} = f(y, \dot{y})$.

7.2   Periodicity Theorem.   Assume $f_y(y, \dot{y})$ and $f_z(y, \dot{y})$ are bounded continuous functions of y and that $y(t)$ is a solution of $\ddot{y} = f(y, \dot{y})$. Then $y(t)$ is periodic if and only if the corresponding phase space curve traced by $\big(y(t), \dot{y}(t)\big)$ is a closed loop.

Proof.   If $y(t)$ is periodic with period $P > 0$, that is $y(t + P) = y(t)$ for all t, then differentiation shows that also $\dot{y}(t + P) = \dot{y}(t)$. Thus the vector function defined by $(y, z) = \big(y(t), \dot{y}(t)\big)$ traces a closed loop in the yz phase space whenever t traverses an interval of length P. Conversely, if this same vector function traces a closed loop when t traverses an arbitrary interval $t_0 \leq t \leq t_0 + P$ of length p, then applying the Existence and Uniqueness Theorem 7.1 with the initial-values $\big(y(t_0), \dot{y}(t_0)\big) = \big(y(t_0 + P), \dot{y}(t_0 + P)\big)$ implies that the solution $y(t)$ repeats over intervals $t_0 + P \leq t \leq t_0 + (k + 1)P$ for integer k and so is periodic.   ∎

EXAMPLE 6

Example 5 is about a linear equation that has explicit solutions in terms of the periodic functions sine and cosine. Since the corresponding phase curves are ellipses, Theorem 7.2 implies that a solution must be periodic with period P equal to the time it takes to traverse the corresponding ellipse.

The next example is the nonlinear pendulum equation, for which the solutions are well understood but for which there are no simple formulas.

EXAMPLE 7

The simplest form of the pendulum equation is $\ddot{y} = -\sin y$. Letting $z = \dot{y}$ we arrive at the equivalent 1-dimensional system

$$\dot{y} = z$$
$$\dot{z} = -\sin y$$

for the phase space variables y and z. We interpret y as a displacement angle in radians and z as its angular velocity. Applying the method of Section 7B, we solve

$$z\frac{dz}{dy} = -\sin y \quad \text{to get} \quad \tfrac{1}{2}z^2 = \cos y + c_1 \quad \text{or} \quad z = \pm\sqrt{2\cos y - 2\cos y_0 + z_0^2},$$

where y_0 and z_0 are initial values for y and z_0. To get a real value for $\dot{y}$ we must have $-2 \le -2 \cos y_0 + z_0^2$. Furthermore, if $-2 \cos y_0 + z_0^2 > 2$, then z is either a positive or negative periodic function of y whose graph can't cross the y axis to form the closed loop that goes with a solution that's periodic as a function of t. Thus the solutions $y(t)$ that are *periodic* are generated by initial-conditions satisfying

$$-2 \le -2 \cos y_0 + z_0^2 < 2;$$

we can think of forming the corresponding phase curves by joining at the y axis the two graphs we get by choosing opposite sign for the square-root. See Figure 11.11(a). The periodic rotational curves above and below the y axis go with "over-the-top" motions of the pendulum, motions with increasing angle for $z = \dot{y} > 0$ and decreasing angle for $z = \dot{y} < 0$. On the closed loops the motions are "back-and-forth" periodic, with top part of the loop representing increasing angle y and the bottom part decreasing angle y. The periodicity of the periodic solutions $y = y(t)$ of $\ddot{y} = -\sin y$ depends in no way on the periodicity of $\sin y$. For comparison we include a phase portrait for a somewhat more realistic linearly damped pendulum; the top two curves in Figure 11.11(b) represent motions that start out rotating over the top but after a while are damped down to swinging back and forth with decreasing amplitude. We take up the plotting of these curves in Section 8.

Note. Other than closed loops and rotational traces, we list three types of special points and curves in Figure 11.11(a). The unstable equilibrium points and separating curves listed as types 2 and 3 are highly theoretical, and are impossible to realize mechanically.

1. Single points $(y_1, 0)$ on the y axis at the center of closed loops represent vertical stable equilibrium positions of the pendulum, hanging down, and with velocity $z = 0$.
2. Single points $(y_2, 0)$ midway between the points of type 1 represent vertical unstable equilibrium positions of the pendulum, balanced up, with velocity $z = 0$.

FIGURE 11.11

Phase portraits for
(a) $\ddot{y} = -\sin y$ and
(b) $\ddot{y} = -\frac{1}{10}\dot{y} - \sin y$.

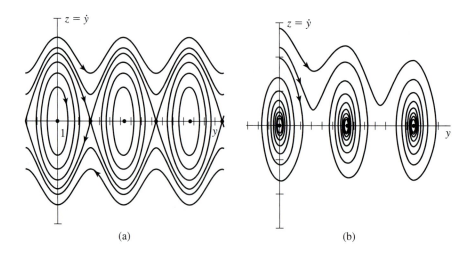

(a)   (b)

3. Curves extending between two points of type 2 separate the rotational motions from the back-and-forth motions and are the phase-space traces of motions that tend away from and toward unstable equilibrium without ever attaining it.

The single points referred to as type 2 are traces in phase space of constant solutions $y(t) = C$ of the differential equation $\ddot{y} = -\sin y$. Note that $\dot{y}(t) = 0$ so the corresponding phase space plot of a constant solution is a point $(y, z) = (C, 0)$ on the y-axis. In general a constant solution of a differential equation must satisfy $\dot{y}(t) = 0$, and is called an **equilibrium solution** because the position $y(t)$ doesn't change over time. The pendulum example shows that an equilibrium may be very unstable, as in the upward vertical position $(y, z) = (\pi, 0)$ of a pendulum. In general an equilibrium point $(y, z) = (C, 0)$ is called **stable** if all phase curves starting sufficiently close to it remain close to it. Otherwise the point $(y, z) = (C, 0)$ is called unstable.

EXAMPLE 8

The pendulum equation $\ddot{y} = -\sin y$, with phase portrait shown in Figure 11.1(a) has equilibrium solutions satisfying $\dot{y}(t) = 0$, so these solution also satisfy $\ddot{y}(t) = 0$. For the pendulum equation this amounts to asking for the solutions of $-\sin y = 0$. The solutions of this equation are $y = k\pi$ for integer values of k. Thus the equilibrium points are at $(y, z) = (k\pi, 0)$. When k is an even integer, these points lie at the centers of the closed loops in Figure 11.1(a), and these points are stable equilibrium points, because a closed loop starting close enough to such a point remains close to the point. When k is an odd integer, a loop starting near the point loops nearly 2π units away while going around a stable point that is nearly π units away.

EXERCISES

In Exercises 1 to 6, solve the initial-value problem and sketch the graph of the solution $y = y(t)$. Then sketch the trace of the solution in yz phase space, indicating the direction of traversal as t increases.

1. $\ddot{y} + y = 1$, $y(0) = 2$, $\dot{y}(0) = 0$

2. $\ddot{y} - y = 0$, $y(0) = 1$, $\dot{y}(0) = 0$

3. $\ddot{y} = 1$, $y(0) = \dot{y}(0) = 0$

4. $\ddot{y} + y = 0$, $y(0) = 2$, $\dot{y}(0) = -1$

5. $\ddot{y} + \dot{y} = 0$, $y(0) = 2$, $\dot{y}(0) = 0$

6. $\ddot{y} - \dot{y} = 0$, $y(0) = 1$, $\dot{y}(0) = 0$

In Exercises 7 to 14, use the method of Section 7B to find a relation depending on a constant C between y and $z = \dot{y}$; then use this to sketch a phase portrait of the differential equation containing at least three curves.

7. $\ddot{y} + y = 1$

8. $\ddot{y} - y = 0$

9. $\ddot{y} = 1$

10. $\ddot{y} = 0$

11. $\ddot{y} + y = 0$

12. $\ddot{y} + 2y^3 = 0$

13. $\ddot{y} - 2y^3 = 0$

14. $\ddot{y} = 1 + y$

15. The differential equation $\ddot{y} - \dot{y} = 0$ has phase plots that decompose into three distinct parts, some of which correspond to constant equilibrium solutions.

(a) Make a phase portrait of the differential equation in $(y, z) = (y, \dot{y})$ space, making clear where the equilibrium points and what the directions of traversal are for the other phase curves.

(b) On the basis of your sketch for part (a) do the equilibrium points appear to be stable or unstable? Explain your answer.

16. The differential equation $\ddot{y} + \dot{y} = 0$ has phase plots that decompose into three distinct parts, some of which correspond to constant equilibrium solutions.

(a) Make a phase portrait of the differential equation in $(y, z) = (y, \dot{y})$ space, making clear where the equilibrium points and what the directions of traversal are for the other phase curves.

(b) On the basis your sketch for part (a) do the equilibrium points appear to be stable or unstable? Explain your answer.

SECTION 8   NUMERICAL METHODS

EXAMPLE 1 We saw in Section 4 that the differential equation

$$m\ddot{y} + k\dot{y} + hy = f(t)$$

is satisfied by the displacement function $y(t)$ of a vibrating spring. There, the factors m (mass), k (frictional constant), and h (spring stiffness) were constants, whereas $f(t)$, the externally applied force, was allowed to be a nonconstant function. Using the method of characteristic equations, we were able to make a fairly complete analysis of the solutions of the differential equation. But suppose that some or all of $m = m(t)$, $k = k(t)$, and $h = h(t)$ vary with time. For example, the spring may weaken, or the friction may increase because of heating, or the weight of the mechanism might increase or decrease for some reason. Assuming that $m(t) > 0$, we can divide by it to get an equation of the form

$$\ddot{y} = a(t)\dot{y} + b(t)y + f(t),$$

where $a = -k/m$ and $b = -h/m$. In exceptional circumstances, it may be possible to solve this differential equation explicitly, but usually we have to settle for an approximate solution.

Second-order differential equations are either linear, of the form

$$\frac{d^2y}{dt^2} + a(t)\frac{dy}{dt} + b(t)y = f(t),$$

or nonlinear, for example, the damped pendulum equation

$$\frac{d^2y}{dt^2} + k\frac{dy}{dt} + \frac{g}{l}\sin y = 0.$$

Equations of both types appear in the very general form

$$\frac{d^2y}{dt^2} = F\left(t, y, \frac{dy}{dt}\right).$$

It is this form that we'll treat here along with **initial conditions** of the form

$$y(t_0) = y_0, \quad \frac{dy}{dt}(t_0) = z_0,$$

where t_0, y_0, and z_0 are given constants. The Existence and Uniqueness Theorem 1.1 described in the introduction to this chapter guarantees a unique solution to this problem if $F(t, y, z)$ and its partial derivatives with respect to y and z are continuous on some interval containing t_0 and in some rectangle containing the point (y_0, z_0). It's important to realize that the only type of problem for which we so far have a universally effective method of actually displaying such solutions is the linear equation with $a(t)$ and $b(t)$ both constant. Even then, if $f(t)$ is not a function we can integrate explicitly, we may have trouble finding a formula for a particular solution. What we may then settle for is a numerical approximation y_k to the value $y(t_k)$ of the true solution at a discrete set of points t_k. Such approximations were treated in Chapter 2 for first-order equations, and the methods used here for second-order equations are simple modifications of the first-order methods.

If the purpose in solving a differential equation is just to obtain numerical values or a graph for some particular solution, then a purely numerical approach may be more efficient than first finding a solution formula and then finding the desired graphical or numerical results from the formula. On the other hand, if what you want is to display the nature of a solution's dependence on certain parameters in the differential equation, or on initial conditions, then solution by formula is preferable if at all possible. Beyond that, detailed properties of a solution, such as whether it's periodic or only approximately so, can be hard to get from a numerical approximation and easy to get from a formula such as $y(t) = 2\cos 3t$.

8A   Euler's Method

Numerical methods for first-order equations are motivated by using the interpretation of the first derivative as a slope. Rather than trying to make something of the interpretation of the second derivative in a second-order equation, what we'll do is find a pair of first-order equations equivalent to a given second-order equation and then apply first-order methods to the simultaneous solution of the pair of equations. The principle is easiest to understand in the general case

$$y'' = F(t, y, y'), \quad y(t_0) = y_0, \quad y'(t_0) = z_0.$$

There are many ways to find an equivalent pair of first-order equations, but the most natural is usually to introduce the first derivative y' as a new unknown function z. We write $z = y'$, $z' = y''$, so we can replace $y'' = F(t, y, y')$ by $z' = F(t, y, z)$. The pair to be solved numerically is then

$$y' = z, \quad y(t_0) = y_0,$$

$$z' = F(t, y, z), \quad z(t_0) = z_0.$$

Since $y'(t_0) = z(t_0) = z_0$, there is an initial condition that goes naturally with each equation. To find an approximate solution, we can do what we would do with a single first-order equation, except that at each step we find new approximate values for both unknown functions y and z, and then use these values to compute new approximations in the next step. The iterative formulas are as follows for the simple **Euler method**, with step size h:

$$y_{k+1} = y_k + hz_k,$$

$$z_{k+1} = z_k + hF(t_k, y_k, z_k),$$

where $t_k = t_0 + kh$. The starting values y_0, z_0 come from the initial conditions.

The initial-value problem

$$\ddot{y} = -\sin y, \quad y(0) = \dot{y}(0) = 1, \quad \text{for} \ \ 0 \le t < 20$$

describes the motion of a pendulum with fairly large amplitude, so large that the linear approximation $\ddot{y} = -y$ would be inadequate. We go ahead to solve numerically the system

$$\dot{y} = z, \quad y(0) = 1,$$

$$\dot{z} = -\sin y, \quad z(0) = 1,$$

We can express the computation as follows.

```
DEFINE F(T,Y,Z) = -SIN(Y)
SET T=0
SET Y=1
SET Z=1
SET H=0.01
DO WHILE T <  20
  SET S=Y
  SET Y=Y+H*Z
  SET Z=Z+H*F(T,S,Z)
  SET T=T+H
  PRINT T,Y
LOOP
```

Note the command SET S=Y, saving the current value of y for use two lines later; without this precaution, the advanced value $y + hz$ would be used, which is not correct. The printout results in 2000 values of t from 0.01 to 20 by steps of size 0.01 along with the corresponding y-values. We could also print the z-values, which are approximations to the values of the derivative $\dot{y}$. This is useful in making a phase-space plot of the solution, plotting approximations to the points $(y(t), \dot{y}(t))$. In the formal routine listed above, the line PRINT T,Y would be replaced by something like PLOT Y,Z for a phase plot.

Rather than displaying a table with 2000 entries, Figure 11.12 shows the graph of y as an unbroken curve for the initial conditions $y(0) = \dot{y}(0) = 1$. The replacement of $\sin y$ by y in the differential equation is inappropriate in this instance, because the values of y that occur are too large to make the approximation a good one. The linearized initial-value problem $\ddot{y} = -y$, $y(0) = y'(0) = 1$ has solution $\cos t + \sin t = \sqrt{2} \sin\left(t - \frac{\pi}{4}\right)$, with graph is shown in Figure 11.12 as a dotted curve. The graphs show that the solution to the nonlinear pendulum equation differs substantially in amplitude and period from the solution to the linear equation. At very small amplitudes the discrepancy is much less, because the approximation of $\sin y$ by y is better the closer y is to zero.

FIGURE 11.12

Solutions $y(t)$ to $\ddot{y} = -\sin y$ and $\ddot{y} = -y$ $(\cdots)$, both with $y(0) = \dot{y}(0) = 1$.

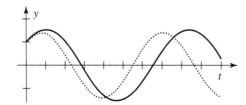

8B   Improved Euler Method

The **improved Euler method** results from applying the single-variable version of the method to each of the equations $y' = z$, $z' = F(t, y, z)$, as we did for the Euler method:

$$p_{k+1} = y_k + hz_k$$

$$q_{k+1} = z_k + hF(t_k, y_k, z_k)$$

$$y_{k+1} = y_k + \frac{h}{2}(z_k + q_{k+1})$$

$$z_{k+1} = z_k + \frac{h}{2}[F(t_k, y_k, z_k) + F(t_{k+1}, p_{k+1}, q_{k+1})].$$

Here p_k and q_k provide the simple Euler estimates that are then used to compute the final estimates for y_k and $z_k = \dot{y}_k$; $t_k = t_0 + kh$ as before. (As with the simple Euler method, the value y_k has to be kept for use in computing z_{k+1} and cannot be replaced by y_{k+1} without significant error.)

The advantage of the modification is that the error in the final estimates is substantially reduced without adding much complexity to the computation.

EXAMPLE 2   The initial-value problem

$$y'' + y = 0, \quad y(0) = 0, \quad y'(0) = 1$$

has solution $y = \sin x$. A numerical table of values compares the Euler method, the improved Euler method, and the correct value rounded to six decimal places (Table 11.3). The value of h used is 0.001, but only every hundredth value is given. The only discrepancies in the last two columns are between the final digits in the sixth, eleventh and last entries.

An algorithm to produce the first, third, and fourth columns in the previous table might look like this. The routine produces only every hundredth row of the computed values.

```
DEFINE F(T,Y,Z) = -Y
SET T=0
SET Y=0
SET Z=1
SET H=.001
FOR J=1 TO 30
FOR K=1 TO 100
 SET P=Y+H*Z
 SET Q=Z+H*F(T,Y,Z)
 SET S=Y
 SET Y=Y+.5*H*(Z+Q)
 SET Z=Z+.5*H*(F(T,S,Z)+F(T+H,P,Q))
 SET T=T+H
NEXT K
PRINT T, Y, SIN(T)
NEXT J
```

Recall that in this algorithm we are dealing with a pair of equations of the form

$$\dot{y} = z \quad \dot{z} = F(t, y, z).$$

The Web site *http://math.dartmouth.edu/~rewn/* has Java applets 2ORD and 2ORD-PLOT at use this routine. Decreasing the step size h often improves accuracy in the Euler methods. This requires more steps to reach a given value of t and may produce

TABLE 11.3

x	Euler y	Imp-Euler y	y ≈ sin x
0.1	0.099838	0.099834	0.099834
0.2	0.198689	0.198669	0.198669
0.3	0.295564	0.295520	0.295520
0.4	0.389496	0.389418	0.389418
0.5	0.479545	0.479426	0.479426
0.6	0.564812	0.564643	0.564642
0.7	0.644443	0.644218	0.644218
0.8	0.717643	0.717356	0.717356
0.9	0.783679	0.783327	0.783327
1.0	0.841892	0.841471	0.841471
1.1	0.891697	0.891208	0.891207
1.2	0.932598	0.932039	0.932039
1.3	0.964185	0.963558	0.963558
1.4	0.986140	0.985450	0.985450
1.5	0.998243	0.997495	0.997495
1.6	1.000370	0.999574	0.999574
1.7	0.992508	0.991665	0.991665
1.8	0.974725	0.973848	0.973848
1.9	0.947200	0.946300	0.946300
2.0	0.910297	0.909297	0.909297
2.1	0.864117	0.863209	0.863209
2.2	0.809387	0.808496	0.808496
2.3	0.746564	0.745705	0.745705
2.4	0.676275	0.675463	0.675463
2.5	0.599221	0.598472	0.598472
2.6	0.516173	0.515501	0.515501
2.7	0.427958	0.427379	0.427379
2.8	0.335458	0.334988	0.334988
2.9	0.239597	0.239249	0.239249
3.0	0.141333	0.141119	0.141120

more approximate solution values than is convenient. Thus we'd print results only after m steps of calculation. For example, $h = 0.001$ and $m = 10$ would produce approximate values with argument differences of 0.01. The applets just referred to allow for this feature.

EXERCISES

MATLAB, Maple, and Mathematica are widely available for doing these exercises. In addition there are Java applets 2ORD, 2ORDPLOT, and PHASEPLOT at the Web site http://math.dartmouth.edu/~rewn/ and the Heaviside function $H(t)$ is available for use in these applets.

1. The **Airy equation** $\ddot{y} + ty = 0$ has solutions for $t > 0$ somewhat similar to solutions of $\ddot{y} + y = 0$.

    (a) Estimate the location of positive t-values for which $y(t) = 0$ for the solution satisfying $y(0) = 0$ and

$\dot{y}(0) = 2$. What changes if you replace the condition $\dot{y}(0) = 0$ by $\dot{y}(0) = a$ for various choices of a? [*Hint:* Look for successive approximate values for y with opposite sign.]

    (b) What can you say about the questions in part (a) on an interval $n \leq t \leq 0$ with $n < 0$?

2. The **Bessel equation** of order zero is

$$x^2 y'' + xy' + x^2 y = 0.$$

For $1 \leq x \leq 40$, estimate the location of the zero values of a solution satisfying $y(1) = 1$, $y'(1) = 0$.

3. Make a numerical comparison of the solution of $\ddot{y} = -\sin y$, $y(0) = 0$, $\dot{y}(0) = 1$, with the solution of $\ddot{y} = -y$ using the same initial conditions. In particular, estimate the discrepancies between the location of successive zero values for the two solutions, one of which is $y(t) = \sin t$.

4. The nonlinear equation

$$\frac{d^2 y}{dx^2} + k\frac{dy}{dx} + hy^2 = 0, \quad h, k, \text{ constant,}$$

has solutions defined near $x = 0$. Compare the behavior of numerical solutions of the initial value problem $y(0) = 0$, $y'(0) = 1$ with the corresponding behavior when the nonlinear term hy^2 is replaced by a linear term hy. To do this you should investigate the result of choosing several different values of $k > 0$ and $h > 0$.

5. The linear equation

$$\frac{d^2 y}{dx^2} + a(x)\frac{dy}{dx} + b(x)y = 0$$

with continuous coefficients $a(x)$ and $b(x)$ occurs often with nonconstant coefficients. For other choices of the coefficients, we use numerical methods. Study the behavior of numerical solutions of the initial value problem $y(0) = 0$, $y'(0) = 1$ as follows.:

(a) Let $a(x) = \sin x$ and $b(x) = \cos x$ for $0 \le x \le 2\pi$.
(b) Let $a(x) = e^{-x/2}$ and $b(x) = e^{-x/3}$ for $0 \le x \le 1$.

6. Nonuniqueness. The initial-value problem $\ddot{y} = 3\sqrt[3]{y}$, $y(0) = 0$, $\dot{y}(0) = 0$ has the identically zero solution.

(a) Verify that $y(t) = \frac{1}{16}t^4$ is also a solution.
(b) Investigate the application of the Euler methods to the problem.

FALLING OBJECTS

7. Suppose that the displacement $y(t)$ of a falling object is subject to a nonlinear friction force

$$m\ddot{y} = -k\dot{y}^{\alpha} + mg.$$

(a) Find numerical approximations to $y(t)$ in the range $0 \le t \le 20$, with $g = 32.2$, $y(0) = 0$, $\dot{y}(0) = 0$, $m = 1$ and $\alpha = 1.5$. Use the values $k = 0, 0.1, 0.5, 1$, and sketch the graphs of $y = y(t)$ using an appropriate scale.
(b) Estimate the values of k that, along with $\dot{y}(0) = 0$ and the other parameter values in part (a), produce approximately the values $y(0) = 0$, $y(5) = 66$, $y(10) = 137$ and $y(15) = 208$.

8. A nonlinear model for an object of mass 1 dropped from rest has frictional force $\ddot{y} = -k\dot{y}^{\alpha} + g$, $y(0) = \dot{y}(0) = 0$; the model is linear if $\alpha = 1$. In numerical work assume $g = 32$ ft/sec^2.

(a) If the terminal velocity for the linear model is 36 feet per second, what is k?
(b) Estimate the time it takes for the linear model to reach velocity 35.99 ft/sec.
(c) If the terminal velocity for the nonlinear model with $\alpha = 1.1$ is 36 feet per second, estimate k. This requires trial and error.
(d) Estimate the time it takes for the nonlinear model of part (c) to reach velocity 35.99 ft/sec.

9. Bead on a wire. Consider a bead sliding without friction under constant vertical gravity along a wire bent into the shape of the twice continuously differentiable graph of $y = f(x)$. It turns out that $x = x(t)$ satisfies the equation $\ddot{x} = -(g + f''(x)\dot{x}^2)f'(x)/(1 + f'(x)^2)$. With $f(x) = -x^3 + 4x^2 - 3x$, $g = 32$ and $x(0) = 0$, estimate how large $\dot{x}(0) > 0$ should be for the bead to overcome periodic oscillation and go over the hump in the wire.

PENDULUM

10. (a) Use the Euler method for

$$y'' = F(t, y, y'), \quad y(t_0) = y_0, y'(t_0) = z_0,$$

and apply it to the pendulum equation with $F(t, y, y') = -16\sin y$, $y(0) = 0$, $y'(0) = 0.5$.
(b) Do part (a) using the improved Euler method. For small oscillations of y, the approximation $\sin y \approx y$ is fairly good, leading to the replacement of the pendulum equation $y'' + (g/l)\sin y = 0$ by the linearized equation $y'' + (g/l)y = 0$, with solutions of the form

$$y(t) = c_1 \cos\sqrt{g/l}\,t + c_2 \sin\sqrt{g/l}\,t.$$

Assuming $g/l = 16$, compare $y(t)$ with the improved Euler approximation to the solution of the nonlinear equation under initial conditions.
(c) $y(0) = 0$, $y'(0) = 0.1$
(d) $y(0) = 0$, $y'(0) = 4$

11. Consider the damped pendulum equation $\ddot{\theta} = -(g/l)\sin\theta - (k/m)\dot{\theta}$, with $g = 32.2$, $l = 20$, $k = 0.03$, and $m = 5$. For the solution with $\theta(0) = 0$, $\dot{\theta}(0) = 0.2$.

(a) Estimate the maximum angles θ for $0 \le t \le 15$.
(b) Estimate the successive times between occurrences of the value $\theta = 0$ for $0 \le t \le 15$.
(c) Repeat part (b), but with initial conditions $\theta(0) = 0$, $\dot{\theta}(0) = 2.0$

12. Consider the following modification of the pendulum equation $ml\ddot{\theta} = -gm\sin\theta$, written here in terms of forces. If the pendulum pivot is moved vertically from its usual fixed position at level 0, so that at time t it is at $f(t)$, with $f(0)=0$, the additional vertical force component is $m\ddot{f}(t)$. Thus the vertical force due to gravity alone is replaced by $\left(-gm + m\ddot{f}(t) \right)\sin\theta$. It follows that the equation for displacement angle $\theta = \theta(t)$ becomes

$$ml\ddot{\theta} = -gm\sin\theta + m\ddot{f}(t)\sin\theta \quad \text{or}$$

$$\ddot{\theta} = \frac{1}{l}\left(-g + \ddot{f}(t) \right)\sin\theta.$$

(a) Show that if $f(t) = at$, with a constant, then there is no change in acceleration as compared with the fixed-pivot case and hence that the equation for the displacement angle θ remains the same: $\ddot{\theta} = -(g/l)\sin\theta$.

(b) Show that in the case of a general twice-differentiable $f(t)$, the position of the pendulum weight at time t is $\mathbf{x}(t) = \left(l\sin\theta(t), f(t) - l\cos\theta \right)$, where θ satisfies either of the differential equations displayed above.

(c) Let $g = 32, l = 5, m = 1$ and $f(t) = 2\sin 4t$. Plot $\left(t, \theta(t) \right)$ for t-values 0.01 apart between 0 and 100, assuming $\theta(0) = 0$ and $\dot{\theta}(0) = 0.01, 0.001, 0.0001$. Note the long term deviation in behavior as compared with the identically zero solution corresponding to $\dot{\theta}(0) = 0$.

(d) Plot the path of the pendulum weight under the assumptions in part (c).

OSCILLATORS AND PHASE SPACE

13. An unforced oscillator displacement $x = x(t)$ satisfies $m\ddot{x} + k\dot{x} + hx = 0$, where k and h may depend on time t.
(a) Suppose that $k(t) = 0.2(1 - e^{-0.1t})$, $h = 5$, $m = 1$ and that $x(0) = 0$, $\dot{x}(0) = 5$. Compute a numerical approximation to $x(t)$ on the range of $0 \le t \le 20$. Then sketch the graph of $x = x(t)$.
(b) Do part (a) using instead $k = 0$ and $h(t) = 5(1 - e^{-0.2t})$.

14. Make phase-plots of the **soft spring oscillator equation** $\ddot{y} = -\gamma y^3 + \delta y$ under each of the following assumptions.
(a) $\gamma = \delta = 1$
(b) $\gamma = 1, \delta = 2$
(c) $\gamma = 2, \delta = 1$

15. Make solution graphs and $(y, \dot{y})$ phase plots for the periodically driven **hard spring oscillator equation** $\ddot{y} = -y - y^3 + k\dot{y} + \frac{3}{10}\cos t$ and use the results to

detect long-term approach to periodic behavior for some $k < 0$.

16. Plot closed periodic phase paths for the **Morse model** of displacement y from equilibrium of the distance between the two atoms of a diatomic molecule:

$$\ddot{y} = K(e^{-2ay} - e^{-ay}), \quad K = a = 1.$$

17. Consider the nonlinear oscillator equation $m\ddot{x} + k\dot{x}|\dot{x}|^\beta + hx = 0$, $0 \le \beta = $ const. Let $m = 1$, $k = 0.2$, $h = 5$ and suppose that $x(0) = 0$, $\dot{x}(0) = 5$. Compute numerical approximations to $x(t)$ on the range $0 \le t \le 20$ for $\beta = 0, 0.5, 1$. Sketch the resulting graphs using computer graphics.

18. **Chaos.** The nonlinear **Duffing oscillator** models a periodically-driven, damped initial-value process:

$$\ddot{y} + k\dot{y} - y + y^3 = A\cos\omega t, \quad y(0) = \tfrac{1}{2}, \; \dot{y}(0) = 0.$$

(a) Make a computer plot of the solution $y(t)$ for $0 \le t \le 300$ for the parameter choices $k = 0.2$, $A = 0.3$, $\omega = 1$ and initial conditions $y(0) = \dot{y}(0) = 0$. The behavior of the damped, periodically driven Duffing equation is often described as "chaotic," which in practical terms means unpredictable. In particular, the specific output that you get will depend significantly not only on the parameter and initial values, but on the choice of numerical method and step size, and even on the internal arithmetic of the machine used to generate the output. For this reason it seems impossible to describe accurately the global shape of the output from the damped, periodically driven Duffing oscillator.

(b) Change just the damping constant in part (a) to $k = 0$, and make a plot of the solution. Comment on the qualitative changes that you see as compared with the output in part (a).

(c) Make several phase plane plots, starting at different points in the $(y, \dot{y})$-plane, of solutions of the Duffing equation. Use the parameter values $k = 0.2$, $A = 0.3$, and $\omega = 1$.

(d) Experiment with part (c) by trying your own choices for the three parameter values.

Chapter 11 REVIEW

In Exercises 1 to 6, find all solutions that satisfy the given equation.

1. $y'' + 2y' + y = e^{-x} + 3e^x$

2. $y'' + y = x\sin x$
3. $y'' - y = \sin x$
4. $y'' - y' - y = 1, \; y(0) = y'(0) = 1$

5. $y'' + 2y' + 3y = 1$

6. $(D-1)^2 y = x^3 - x$

7. $y''' = x$

8. $y'''' = 81y$

9. $y'' + 9y = \sin 3x$, $y(0) = 1$, $y'(0) = 0$

10. $(D^2 + 4)y = \cos 3x$, using Section 3C

11. $y'' + y = 0$, $y(0) = -1$, $y(\pi) = 1$

12. $y'' + y = 0$, $y(0) = 0$, $y(\pi/2) = 2$

The basic real solution forms for $y'' + ay' + by = 0$ are $\{e^{r_1 x}, e^{r_2 x}\}$, $\{e^{r_1 x}, xe^{r_1 x}\}$ and $\{e^{\alpha x} \cos \beta x, e^{\alpha x} \sin \beta x\}$ and are prototypes for Exercises 13 and 14.

13. Make a corresponding list of triples for the equation $y''' + ay'' + by' + cy = 0$.

14. Make a corresponding list of quadruples for $y'''' + ay''' + by'' + cy' + dy = 0$.

15. Derive from scratch the fundamental sinusoidal solutions to the harmonic oscillator problem $\ddot{y} + y = 0$, $y(0) = 0$, $\dot{y}(0) = 1$. Do this in the following steps: (Our earlier derivations were made using complex exponentials.)

   (a) Multiply the equation by $\dot{y}$, and then integrate with respect to t to get $\frac{1}{2}\dot{y}^2 + \frac{1}{2}y^2 = C$.

   (b) Find C, solve for $\dot{y}$, and solve the resulting first order equation to get $y(t) = \sin(t + c)$.

16. Suppose that $y_1(x)$ and $y_2(x)$ are real-valued functions defined for all real x and you know that $y_1(x)$ is not a constant multiple of $y_2(x)$. Are the two functions necessarily linearly independent? Explain your answer, using an example if necessary.

17. Let the functions y_1, y_2 be defined for all real x by $y_1(x) = e^{rx}$ and $y_2(x) = e^{sx}$, where r and s are unequal complex numbers. Show that y_1 and y_2 are linearly independent even if complex constants c_1, c_2 are allowed in $c_1 y_1(x) + c_2 y_2(x)$.

18. For what values of the constant b do the nonidentically zero solutions of $y'' + y' + by = 0$ oscillate as functions of x?

19. The current $I(t)$ flowing through a certain electric circuit at time t satisfies

$$I'' + RI' + I = \sin t,$$

where $R > 0$ is a constant resistance. The equation has a solution of the form $I(t) = A \sin(t - \alpha)$ for certain constants $A > 0$ and α. Find A and α.

20. Show that the functions e^x and e^{-x} are linearly independent on an interval $a < x < b$ by showing directly that

neither function is a constant multiple of the other on such an interval.

21. (a) Show that the functions e^x and e^{-x} are linearly independent on an interval $a < x < b$ by showing directly that the equation

$$c_1 e^x + c_2 e^{-x} = 0$$

   can't be satisfied for all real x unless the constants c_1 and c_2 are both zero.

   (b) Show that the equation $2e^x - 3e^{-x} = 0$ is satisfied for exactly one real x and that $2e^x + 3e^{-x} = 0$ is satisfied for no real x.

   (c) For what *complex* values of x is $2e^x + 3e^{-x} = 0$ satisfied?

22. Suppose that

$$\frac{d^2 y}{dt^2} = -y.$$

   (a) Find the general solution $y(t)$ of this equation.

   (b) Let

$$z(t) = \frac{dy}{dt}(t).$$

   Show that the parametrized curve $(y(t), z(t))$ traces clockwise a circular path or else reduces to a single point.

23. Theorem 2.4 implies that there is a one-to-one correspondence between the set of all n- tuples of initial values $(z_0, z_1, \dots, z_{n-1})$ and all solutions to the nth-order homogeneous equation $L(y) = 0$. Explain how this conclusion follows.

24. Theorem 2.4 implies that there is a one-to-one correspondence between the set of all n-tuples of initial values $(z_0, z_1, \dots, z_{n-1})$ and all n-tuples $(c_1, c_2, \dots, c_n)$ of coefficients of linear combination. Explain how this conclusion follows.

25. (a) Show that the family of differential equations $y'' - (2r + h)y' + r(r + h)y = 0$, depending on the parameter h, can also be written $(D - r)\big(D - (r + h)\big)y = 0$.

   (b) Show that the equations of part (a) have solutions $y = c_1 e^{(r+h)x} + c_2 e^{rx}$ if $h \neq 0$.

   (c) Let $c_1 = 1/h, c_2 = -1/h$, and show that for each fixed x, the resulting solution $y_h(x)$, tends as $h \to 0$ to

$$y = \frac{d}{dr} e^{rx} = xe^{rx}.$$

In Exercises 26 to 29, assume that L is a linear operator such that $L(y_1) = w_1$, $L(y_2) = w_2$, and $L(y_3) = w_3$.

Using just this information, find linear combinations z of y_1, y_2 and y_3 so that the given equation holds.

26. $L(z) = 2w_1 - 3w_2$

27. $L(z) = w_1 + 2w_2 - 4w_3$

28. $L(z) = L(2z) + w_1 + w_2$

29. $L(z) = 0$

The use of the Laplace transform is limited to those functions $f(t)$, defined for $t > 0$ for which the improper integral $\int_0^\infty e^{-st} f(t)\, dt$ exists for some values of s. Disregarding the problem of actually finding a formula in Exercises 40 to 47, decide whether the given function has a Laplace transform for some values of s.

30. $t^2 e^t$

31. $t \ln t$

32. $\ln t$

33. e^{t^2}

34. e^{-t^2}

35. $\sin(\ln t)$

36. $\sin(1/t)$

37. t^{-1}

CHAPTER 12

INTRODUCTION TO SYSTEMS

In the two previous chapters we have considered differential equations whose solutions are real-valued functions of a real variable. It turns out that a natural and useful generalization is to consider vector differential equations (or, equivalently, systems of real differential equations) whose solutions are vector-valued functions of a real variable. There are two main reasons for making this generalization: one is that many phenomena in applied mathematics can most naturally be expressed in vector form; another is that real differential equations of order higher than one can often be reduced advantageously to vector equations of order one. Both of these statements will be explained in this chapter.

In dealing with vector equations it will be convenient to use the letter t to denote the variable with respect to which derivatives are taken. This choice has the advantage that applications most frequently involve time-dependent phenomena, and also that the letters x, y, z, etc., are left free to denote space coordinates, as usual. To write general systems of differential equations more compactly we'll use the notation $\mathbf{x} = (x, y)$ for 2-dimensional systems, $\mathbf{x} = (x, y, z)$ for 3-dimensional systems, and $\mathbf{x} = (x_1, \dots, x_n)$ for n-dimensional systems.

SECTION 1   VECTOR FIELDS

1A   Geometric Interpretation
A first-order differential equation of the form

$$\frac{dx}{dt} = F(t, x)$$

has a real-valued solution of the form $x = x(t)$. For example, the equation

$$\frac{dx}{dt} = x + t$$

has the general solution $x(t) = Ce^t - t - 1$, defined for all real numbers t. Similarly a *pair* of equations

$$\frac{dx}{dt} = F(t, x, y)$$

$$\frac{dy}{dt} = G(t, x, y),$$

called a *system* of dimension 2, has as a solution a pair of functions

$$x = x(t)$$

$$y = y(t),$$

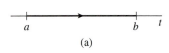

(a)

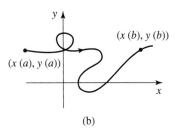

(b)

FIGURE 12.1

defined on some interval $a < t < b$ and satisfying the differential equations. As t increases from a to b, the point in the xy-plane with coordinates $(x(t), y(t))$ will trace a path, perhaps like the one in Figure 12.1(b). Such a path is called a **trajectory** of the system. It's important to remember that a *solution* of a system is a function of t and that its trajectory is the image of this function, containing only a part of the information contained in the solution; by itself, the trajectory fails to show explicitly the correspondence between values of t and values $(x(t), y(t))$ of the solution, nor does the trajectory display the speed of traversal. Nevertheless a sketch of several judiciously chosen trajectories of a system, called for historical reasons a **phase-portrait** of the system, is often enough to convey important information about the system, particularly if the directions of traversal are shown by inserting appropriate arrow points. Note however that a "trajectory" may be a single point $\mathbf{x}_0$ arising from a constant solution $\mathbf{x}(t) = \mathbf{x}_0$; in that case we refer respectively to an **equilibrium point** and an **equilibrium solution**.

In the description of scientific problems, the variable t often represents time. Thus the derivatives x' and y' may stand for the rates of change of x and y with respect to time; if t is to be interpreted as time, the derivatives are often written with dots instead of primes: $\dot{x}$, $\dot{y}$.

EXAMPLE 1   Consider the system

$$\dot{x} = x$$
$$\dot{y} = 2y.$$

This system is particularly simple because each unknown function occurs in just one equation. Such a system is called **uncoupled**, We can therefore solve each equation separately to get the general solution

$$x(t) = c_1 e^t$$
$$y(t) = c_2 e^{2t}.$$

We see that $x(0) = c_1$ and $y(0) = c_2$, so that imposing initial conditions such as

$$x(0) = 1$$
$$y(0) = 2$$

will determine the values of c_1 and c_2 and will single out the particular solution

$$x(t) = e^t > 0,$$
$$y(t) = 2e^{2t} > 0.$$

The trajectory of this solution satisfies

$$y = 2x^2,$$

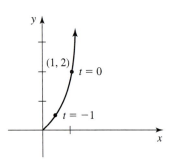

FIGURE 12.2

with the restriction that $x(t) > 0$ and $y(t) > 0$. Figure 12.2 shows a part of the trajectory for $-\infty < t < \infty$. Because $x(t) = e^t$ can't be negative, the trajectory consists of only half of the parabola $y = 2x^2$.

We can write the system of differential equations in Example 1 in vector notation by letting $\mathbf{x} = (x, y)$, $d\mathbf{x}/dt = (dx/dt, dy/dt)$, and $\mathbf{F}(x, y) = (x, 2y)$. Then the system becomes

$$\frac{d\mathbf{x}}{dt} = \mathbf{F}(\mathbf{x}).$$

Similarly, the system

$$\frac{dx}{dt} = x + y + t$$

$$\frac{dy}{dt} = x - y - t$$

would be written $d\mathbf{x}/dt = \mathbf{F}(t, \mathbf{x})$, where

$$\mathbf{F}(t, x, y) = (x + y + t, x - y - t).$$

When we speak of a general first-order system of dimension n in **normal form**, we'll mean a system of the form

$$\frac{dx_1}{dt} = F_1(t, x_1, x_2, \dots, x_n)$$

$$\frac{dx_2}{dt} = F_2(t, x_1, x_2, \dots, x_n)$$

$$\vdots$$

$$\frac{dx_n}{dt} = F_n(t, x_1, x_2, \dots, x_n).$$

Using the vector notations

$$\mathbf{x} = (x_1, \dots, x_n) \quad \text{and} \quad \frac{d\mathbf{x}}{dt} = \left(\frac{dx_1}{dt}, \dots, \frac{dx_n}{dt} \right)$$

we can write the system more compactly as

$$\frac{d\mathbf{x}}{dt} = \mathbf{F}(t, \mathbf{x}),$$

where $\mathbf{F}(t, \mathbf{x}) = \big(F_1(t, \mathbf{x}), \dots, F_n(t, \mathbf{x}) \big)$. A solution $\mathbf{x} = \mathbf{x}(t)$ is a vector-valued function of a real variable t on an interval $a < t < b$, such that substitution into the differential equation satisfies the system of equations on the interval, as $\big(x(t), y(t) \big) = (e^t, 3e^{2t})$ satisfies $(\dot{x}, \dot{y}) = (x, 2y)$ in Example 1. The image of the interval under this function is a trajectory curve in $\mathbb{R}^n$. Such a trajectory in $\mathbb{R}^3$ is shown in Figure 12.3.

One advantage of the vector interpretation of a system is that the derivative $d\mathbf{x}/dt$ has a geometric meaning that the derivatives dx_i/dt do not have when taken separately: $d\mathbf{x}/dt$ is a tangent vector to a trajectory, and if t is time, then $d\mathbf{x}/dt$ is a velocity vector. Formally, tangent and velocity are matters of definition. The following discussion shows how the formal definitions are suggested by the intuitive ideas behind tangent and velocity.

FIGURE 12.3

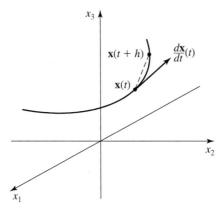

Reviewing the vector derivative, Figure 12.3 shows the points $\mathbf{x}(t)$, $\mathbf{x}(t + h)$, and the chord $\mathbf{x}(t + h) - \mathbf{x}(t)$ joining them. If we multiply the vector by $1/h$, then

$$\frac{\mathbf{x}(t + h) - \mathbf{x}(t)}{h}$$

will be parallel to the chord, and if it has a limit as $h \to 0$, this limit vector can reasonably be defined to be a tangent vector at $\mathbf{x}(t)$. The limit is defined as in Chapter 4 so that

$$\frac{d\mathbf{x}}{dt}(t) = \lim_{h \to 0} \frac{\mathbf{x}(t + h) - \mathbf{x}(t)}{h}$$

$$= \left(\lim_{h \to 0} \frac{x_1(t + h) - x_1(t)}{h}, \dots, \lim_{h \to 0} \frac{x_n(t + h) - x_n(t)}{h} \right)$$

$$= \left(\frac{dx_1}{dt}(t), \frac{dx_2}{dt}(t), \dots, \frac{dx_n}{dt}(t) \right) = \frac{d\mathbf{x}}{dt}(t).$$

Thus $\dfrac{d\mathbf{x}}{dt}(t)$ is a tangent vector at $\mathbf{x}(t)$. The velocity interpretation is valid when t is time because the Euclidean length $|(d\mathbf{x}/dt)(t)|$ is defined to be the **speed** of traversal of the trajectory at $\mathbf{x}(t)$. The reason is that for small values of h, the Euclidean length

$$\left| \frac{\mathbf{x}(t + h) - \mathbf{x}(t)}{h} \right|$$

is nearly the average rate of traversal of the trajectory over the interval from t to $t + h$. This approximation is really good only if $\mathbf{x}(t)$ is differentiable, in which case, because length is continuous,

$$\lim_{h \to 0} \left| \frac{\mathbf{x}(t + h) - \mathbf{x}(t)}{h} \right| = \left| \frac{d\mathbf{x}}{dt}(t) \right|.$$

EXAMPLE 2   The pair of equations

$$\frac{dx}{dt} = -y$$

$$\frac{dy}{dt} = x$$

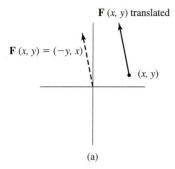

F (x, y) = (−y, x)

F (x, y) translated

(x, y)

(a)

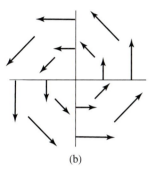

(b)

FIGURE 12.4

has the form $d\mathbf{x}/dt = \mathbf{F}(\mathbf{x})$, where $\mathbf{x} = (x, y)$ and $\mathbf{F}(\mathbf{x}) = (-y, x)$. We interpret the system as saying that the tangent vector to a solution trajectory passing through (x, y) has coordinates $(-y, x)$. Figure 12.4(b) shows a sketch of some tangent vectors at a few points, made by first locating the arrow that represents $(-y, x)$ and then moving the arrow parallel to itself so that its tail coincides with the point (x, y) as in Figure 12.4(a). The picture suggests that the trajectories have a circular shape. Notice also that the arrows get longer as they get farther from the origin, which indicates that the speed along a trajectory is greater when the trajectory is farther from the origin. We postpone actually solving the equations until we develop a systematic procedure in the next section.

Figure 12.4(b) plays somewhat the same role that a sketch of a direction field does for a single first-order equation, but the present picture contains more information: here, the lengths and orientations of the tangents are significant, whereas in a direction field all the important information is conveyed by the slopes of the line segments. The reason is that representing the slope of a graph by a single number is possible only for real functions of a real variable.

1B   Autonomous Systems

If, in the general first-order system

$$\frac{d\mathbf{x}}{dt} = \mathbf{F}(t, \mathbf{x}),$$

the function $\mathbf{F}(t, \mathbf{x})$ is independent of t, then the system is called **autonomous** and so has the form

$$\frac{d\mathbf{x}}{dt} = \mathbf{F}(\mathbf{x}).$$

Example 2 describes an autonomous system. For an autonomous system, the tangent vector $\mathbf{F}(\mathbf{x})$ located at the point $\mathbf{x}$ is always the same regardless of what time it is when the trajectory passes through $\mathbf{x}$. Such an assignment of vectors $\mathbf{F}(\mathbf{x})$ to points $\mathbf{x}$ is called a **vector field** and Figure 12.4(b) shows a sketch of one. For the general nonautonomous system, the function $\mathbf{F}(t, \mathbf{x})$ specifies a tangent vector (to a trajectory through $\mathbf{x}$) that may be different for each t, in this way producing a time-dependent vector field. The pictorial analogue of the static vector field shown in Figure 12.4 for a time-dependent vector field would be a sequence of "snapshots" taken at different times. Each snapshot would have the same general form as Figure 12.4, but would show changes in the individual arrows as time t varies.

EXAMPLE 3   The 2-dimensional system

$$\frac{dx}{dt} = (1 - t)x - ty$$

$$\frac{dy}{dt} = tx + (1 - t)y$$

determined by the time-dependent vector field

$$\mathbf{F}(t, x, y) = \left(\begin{array}{c} (1-t)x - ty \\ tx + (1-t)y \end{array} \right)$$

$$= (1-t)\left(\begin{array}{c} x \\ y \end{array} \right) + t \left(\begin{array}{c} -y \\ x \end{array} \right)$$

is not autonomous. Figure 12.5 shows some sketches of the vector fields for $t = 0, \frac{1}{2}, 1$.

EXAMPLE 4   Suppose the 2-dimensional system $dx/dt = F(t, x, y)$, $dy/dt = G(t, x, y)$ is such that the ratio $G(t, x, y)/F(t, x, y) = R(x, y)$ happens to be independent of t. This would occur in particular if neither F nor G depended explicitly on t. Since the chain rule allows us to write

$$\frac{dy}{dx} = \frac{dy/dt}{dx/dt}$$

under fairly general conditions, we can conclude that there are trajectory curves of the system satisfying the differential equation

$$\frac{dy}{dx} = R(x, y).$$

If we can solve this equation, we have a way to plot trajectories without finding solutions $x(t)$, $y(t)$. For example, $dx/dt = ty$, $dy/dt = -tx$ leads us to consider

$$\frac{dy}{dx} = -\frac{x}{y},$$

which has solutions $x^2 + y^2 = c$, representing circular trajectories if $c > 0$. Using this method, you are asked sketch some trajectories for the systems in Exercises 24 to 27. Looking at the vector field will tell you roughly how a trajectory is traced. The trajectory of a constant solution, for which $\dot{x} = \dot{y} = 0$ for all t is just a single point.

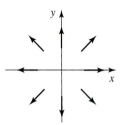

$\mathbf{F}(0, x, y) = \begin{pmatrix} x \\ y \end{pmatrix}$

(a)

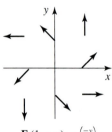

$\mathbf{F}(1, x, y) = \begin{pmatrix} -y \\ x \end{pmatrix}$

(b)

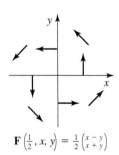

$\mathbf{F}\left(\frac{1}{2}, x, y\right) = \frac{1}{2}\begin{pmatrix} x-y \\ x+y \end{pmatrix}$

(c)

FIGURE 12.5

1C   Second-Order Equations

Systems of first-order differential equations arise very naturally in the study of higher-order equations. We restrict ourselves here to the most important case, namely second order, a context we discussed also in Chapter 11, Section 7. In a second-order 1-dimensional equation $\ddot{y} = f(t, y, \dot{y})$ we reduce the order to 1 and simultaneously raise the dimension to 2 by introducing a new dependent variable z. We let $\dot{y} = z$. Then $\ddot{y} = \dot{z}$. Thus an initial-value problem

$$\ddot{y} = f(t, y, \dot{y}), \quad y(t_0) = y_0, \quad \dot{y}(t_0) = z_0$$

is equivalent to a 2-dimensional system of the form

$$\dot{y} = z, \quad y(t_0) = y_0,$$
$$\dot{z} = f(t, y, z), \quad z(t_0) = z_0,$$

a first-order system $\dot{\mathbf{x}} = \mathbf{F}(t, \mathbf{x})$, with $\mathbf{x} = (y, z)$ and time-dependent vector field

$$\mathbf{F}(t, y, z) = \begin{pmatrix} z \\ f(t, y, z) \end{pmatrix}.$$

The yz-space is the **phase space** or **state space** of the second-order equation. The state-space is important in part because it explicitly displays all the initial-value information necessary for determining the system's evolution. Note, for example, that the solution graphs of the original equation of order 2 don't display the vital velocity component $\dot{y}(t)$ at times $t > t_0$. The image plot of the 2-dimensional system's typical solutions is a **phase portrait** of the original 1-dimensional equation.

EXAMPLE 5   The second-order differential equation

$$\ddot{y} + y = 0$$

converts into a first-order system if we let $\dot{y} = z$. With $\ddot{y} = \dot{z}$ we get

$$\dot{z} = -y$$

$$\dot{y} = z.$$

This is the system considered in Example 2, with its vector field sketched in Figure 12.4. We're familiar with the general solution to $\ddot{y} + y = 0$ and its derivative $z(t) = \dot{y}(t)$, namely

$$y(t) = c_1 \cos t + c_2 \sin t, \quad z(t) = -c_1 \sin t + c_2 \cos t.$$

We determine particular solutions most naturally by choosing initial conditions. For example, $y(0) = 1$, $\dot{y}(0) = 0$ are equivalent to $y(0) = 1$, $z(0) = 0$, which implies

$$y(0) = c_1 = 1,$$

$$z(0) = c_2 = 0.$$

The vector solution determined by these values of c_1 and c_2 is

$$\begin{pmatrix} y(t) \\ z(t) \end{pmatrix} = \begin{pmatrix} \cos t \\ -\sin t \end{pmatrix}.$$

The trajectory of this solution is a circle of radius 1, shown in Figure 12.6(a), which taken altogether is a phase portrait for the original equation $\ddot{y} + y = 0$, consisting of the circles

$$y^2 + z^2 = (c_1 \cos t + c_2 \sin t)^2 + (-c_1 \sin t + c_2 \cos t)^2$$

$$= c_1^2 + c_2^2 = r^2.$$

These circles are traced repeatedly as t tends to infinity.

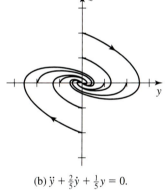

(a) $\ddot{y} + y = 0$;

(b) $\ddot{y} + \frac{2}{5}\dot{y} + \frac{1}{5}y = 0.$

FIGURE 12.6
(a) Circles; $\ddot{y} + y = 0$
(b) Spirals; $\ddot{y} + \frac{2}{5}\dot{y} + \frac{1}{5}y = 0.$

The curves in Figure 12.6 are trajectories of vector solutions $y = y(t)$, $z = z(t)$ and as such have directions as indicated by arrow points in the figures. If a phase curve has parts on both sides of the y axis, these directions in a phase portrait always have a clockwise orientation relative to the usual orientation of the y-axis and z-axis. The general principle is that when $z = \dot{y}$ is positive then y is increasing, and when $z = \dot{y}$ is negative then y is decreasing. Since $z = \dot{y} = 0$ on the y axis, it follows that $dz/dy = \dot{z}/\dot{y} = 0$, so a trajectory that crosses the y axis crosses vertically, as displayed in both parts of Figure 12.6.

EXAMPLE 6   The differential equation $\ddot{y} + \frac{2}{5}\dot{y} + \frac{1}{5}y = 0$ has general solution $y = e^{-\frac{1}{5}t}\left(c_1 \cos \frac{2}{5}t + c_2 \sin \frac{2}{5}t\right)$ and is equivalent to the first-order system

$$\dot{y} = z$$

$$\dot{z} = -\frac{1}{5}y - \frac{2}{5}z.$$

We solve the second-order equation for y and then find $z = \dot{y}$ to solve the system:

$$y = e^{-\frac{1}{5}t}\left(c_1 \cos \tfrac{2}{5}t + c_2 \sin \tfrac{2}{5}t\right)$$

$$z = \tfrac{1}{5}e^{-\frac{1}{5}t}\left((2c_2 - c_1)\cos \tfrac{2}{5}t - (c_2 + 2c_1)\sin \tfrac{2}{5}t\right).$$

Figure 12.6(b) shows some phase space plots starting at eight different points; the trajectories all spiral in toward the origin as t increases.

A major advantage of plotting a phase curve as compared with plotting graphs is that a phase curve typically compresses the long-term behavior of the functions $y(t)$ and $\dot{y}(t) = z(t)$ into a relatively small picture, as compared with a vain attempt to depict graphs extended over an infinite t-axis. A disadvantage of a phase curve is that it fails to associate points on the curve with specific values of t. These ideas appear in more detail in Section 7C of the previous chapter.

EXERCISES

The uncoupled systems 1 to 4 are solvable by treating each equation separately. Find the general solution, and then find the particular solution that satisfies the given initial conditions.

1. $dx/dt = x + 1$, $x(0) = 1$,
   $dy/dt = y$, $y(0) = 2$

2. $dx/dt = t$, $x(1) = 0$,
   $dy/dt = y$, $y(1) = 0$

3. $dx/dt = x$, $x(0) = 0$,
   $dy/dt = \tfrac{1}{2}y$, $y(0) = 1$,
   $dz/dt = \tfrac{1}{3}z$, $z(0) = -1$

4. $dx/dt = x + t$, $x(0) = 0$,
   $dy/dt = y - t$, $y(0) = 0$,
   $dz/dt = z$, $z(0) = 1$

5. For the system in Exercise 2, there is a vector-valued function $\mathbf{F}(t, \mathbf{x})$, with $\mathbf{x} = (x, y)$ such that the system has the form $d\mathbf{x}/dt = \mathbf{F}(t, \mathbf{x})$.
   (a) Find $\mathbf{F}$.
   (b) Find the speed of a trajectory through $\mathbf{x}$ at time t.

6. For the system in Exercise 4, there is a vector-valued function $\mathbf{F}(t, \mathbf{x})$, with $\mathbf{x} = (x, y, z)$ such that the system has the form $d\mathbf{x}/dt = \mathbf{F}(t, \mathbf{x})$.
   (a) Find $\mathbf{F}$.
   (b) Find the speed of a trajectory through $\mathbf{x}$ at time t.

Sketch the vector fields 7 to 10 by drawing a few arrows for $\mathbf{F}(\mathbf{x})$ or $\mathbf{F}(t, \mathbf{x})$ with their tails at selected points $\mathbf{x}$ of the form (x, y) or (x, y, z). In Exercises 8 and 10, make separate sketches for $t = -1$, $t = 0$, and $t = 1$.

7. $\mathbf{F}(x, y) = (x + 1, y)$

8. $\mathbf{F}(t, x, y) = (t, y)$

9. $\mathbf{F}(x, y, z) = (x, \tfrac{1}{2}y, \tfrac{1}{3}z)$

10. $\mathbf{F}(t, x, y, z) = (x + t, y - t, z)$

11. Sketch the vector field $\mathbf{F}(x, y) = (-y, x)$. Then sketch the trajectory curve tangent to arrows in the field sketch, starting at $(x, y) = (1, 0)$.

12. Show that the system $\dot{x} = -ty$, $\dot{y} = tx$ has circular solution trajectories of radius $r > 0$, traced with increasing speed rt as time increases. [*Hint:* Show that $x\dot{x} + y\dot{y} = 0$.]

In Exercises 13 and 14, by letting $\dot{y} = z$, express each second-order differential equation as a first-order system of dimension 2. Also find the corresponding initial conditions for $y(0)$ and $z(0)$, and solve the initial-value problem for y and z.

13. $\ddot{y} + \dot{y} + y = 0$, $y(0) = 1$, $\dot{y}(0) = 1$

14. $\ddot{y} + t\dot{y} = t$, $y(0) = 0$, $\dot{y}(0) = 1$

Find first-order systems equivalent to the differential equations 15 to 18 by setting $dy/dt = z$ and, if appropriate, $dz/dt = w$.

15. $d^2y/dt^2 + (dy/dt)^2 + y^2 = e^t$

16. $d^2y/dt^2 = y\,(dy/dt)$

17. $d^3y/dt^3 = (d^2y/dt^2)^2 - y\,(dy/dt) - t$

18. $d^3y/dt^3 = 12x\,dy/dt$

In Exercises 19 to 22, reduce the system to normal form, with each first derivative by itself on the left side.

19. $dx/dt + dy/dt = t$,
$dx/dt - dy/dt = y$

20. $dx/dt + dy/dt = y$,
$dx/dt + 2dy/dt = x$

21. $2dx/dt + dy/dt + x + 5y = t$,
$dx/dt + dy/dt + 2x + 2y = 0$

22. $dx/dt - dy/dt = e^{-t}$,
$dx/dt + dy/dt = e^{t}$

23. Sketch a phase portrait for the second-order equation $\ddot{y} = \dot{y}$, indicating the directions of traversal where appropriate. Note however that since one of the equations in the system will be $\dot{y} = z$ these directions are always left to right when $z > 0$ and right to left when $z < 0$. [*Hint:* One set of trajectories consists of the individual points on the y axis.]

24. Sketch a phase portrait for the second-order equation $\ddot{y} = y$, indicating the directions of traversal. [*Hint:* Two families of hyperbolas make up the trajectories.]

25. Consider the 2-dimensional coupled system

$$\dot{x} = x + y, \quad \dot{y} = 4x + y.$$

(a) Change the coordinates (x, y) to (z, w) with the relations
$$x = z + w, \quad y = 2z - 2w,$$

and show that this change results in the uncoupled system
$$\dot{z} = 3z, \quad \dot{w} = -w.$$

(b) Solve the uncoupled system in part (a) for z and w, and use the coordinate change to solve for x and y. Then verify by substitution that your solution for x and y satisfies the given system.

In Exercises 26 to 29, convert the problem of solving the system into solving an equation of the form $dy/dx = R(x, y)$ as in Example 4 of the text and then solve the first-order equation and sketch some trajectories of the system.

26. $dx/dt = x - y,\ dy/dt = x^2 - y^2$

27. $dx/dt = e^{2y},\ dy/dt = e^{x+y}$

28. $dx/dt = e^{t}y,\ dy/dt = e^{t}x$

29. $dx/dt = xy + y^2,\ dy/dt = x + y$

30. The 1-dimensional equation $\ddot{y} = -g$ is used to determine the motion of an object moving perpendicularly to the surface of a large attracting body and subject to no other forces. Suppose the range of motion is extended to a vertical plane with horizontal coordinate x. If there are still no forces acting horizontally, the single equation is replaced by the 2-dimensional uncoupled system

$$\ddot{x} = 0,$$

$$\ddot{y} = -g.$$

(a) Solve the 2-dimensional system, subject to the four initial conditions

$$x(0) = 0, \quad y(0) = 0,$$

$$\dot{x}(0) = z_0 > 0, \quad \dot{y}(0) = w_0 > 0.$$

(b) Show that the trajectory of the solution found in part (a) follows a parabolic path.

(c) Show that the maximum height is attained when the horizontal displacement is $z_0 w_0/g$ and that the maximum height is $w_0^2/(2g)$.

(d) Show that the horizontal distance traversed before returning to height $y(0) = 0$ is $2z_0 w_0/g$. Show also that for a given initial speed $v_0 = \sqrt{z_0^2 + w_0^2}$, this horizontal distance is maximized by having $z_0 = w_0$.

31. A projectile fired against **air resistance** proportional to velocity satisfies the uncoupled system

$$\ddot{x} = -k\dot{x},$$

$$\ddot{y} = -k\dot{y} - g, \quad k > 0.$$

(a) Solve the 2-dimensional system, subject to the four initial conditions

$$x(0) = 0, \quad y(0) = 0,$$

$$\dot{x}(0) = z_0 > 0, \quad \dot{y}(0) = w_0 > 0.$$

The equations are linear with constant coefficients so you can use the methods of Chapter 11.

(b) Show that the trajectory of the solution found in part (a) rises to a unique maximum at time $t_{\max} = (1/k)\ln(1 + kw_0/g)$.

(c) Show that the position of maximum height has coordinates

$$x_{\max} = z_0 w_0/(g + kw_0),$$

$$y_{\max} = w_0/k - (g/k^2)\ln(1 + kw_0/g).$$

(d) Show that as k tends to zero the maximum height tends to $\frac{1}{2}w_0^2/g$.

32. Here is an outline of a derivation of the **pendulum equation** $\ddot{\theta} = -(g/l)\sin\theta$ using a system of differential equations.

(a) Show that if x, y are rectangular coordinates and θ is the angle formed by the vector (x, y), measured counterclockwise from the downward vertical direction, then $x = l \sin \theta$ and $y = -l \cos \theta$. [*Hint:* These equations are a slight modification of the usual polar coordinate relations.]

(b) Show that $\ddot{x} = -l \sin \theta \dot{\theta}^2 + l \cos \theta \ddot{\theta}$ and $\ddot{y} = l \cos \theta \dot{\theta}^2 + l \sin \theta \ddot{\theta}$.

(c) Use the representation $\ddot{x} = 0$, $\ddot{y} = -g$ for the coordinates of the acceleration of gravity together with the result of part (b), to derive the pendulum equation. (This derivation safely ignores the lengthwise, or radial, force on the pendulum, since that force is always perpendicular to the path of motion.) [*Hint:* Eliminate terms containing $\dot{\theta}^2$.]

33. Heat exchange. The temperatures $u(t) \geq v(t)$ of two bodies in thermal contact with each other may be governed for the warmer body by Newton's law of cooling and for the cooler body by the analogous heating law:

$$\frac{du}{dt} = -p(u - v), \quad \frac{dv}{dt} = q(u - v),$$

where p and q are positive constants and the equations are subject to initial conditions $u(0) = u_0, v(0) = v_0$. (Note that p may be different from q if the two bodies have different capacities to absorb heat.) This is a coupled system, but because of its simple form we can solve it as follows.

(a) Show that $q\,du/dt + p\,dv/dt = 0$. Then integrate with respect to t to show that $qu(t) + pv(t) = c_0$, where $c_0 = qu_0 + pv_0$.

(b) Use the relation between u and v derived in part (a) together with the given system to derive a single differential equation satisfied by $u(t)$. Then solve this equation using the initial conditions.

(c) Find a formula for $v(t)$.

(d) Find out how long it takes for the initial temperature difference between the two bodies to be cut in half.

34. Bugs in mutual pursuit. Four identical bugs are on a flat table, each moving at the same constant speed v. Use (x, y)-coordinates on the table, and locate bugs 1 through 4 initially in respective quadrants 1 through 4, each at one of the points $(\pm 1, \pm 1)$. Bug 1 always heads directly toward bug 2, bug 2 toward bug 3, bug 3 toward bug 4, and bug 4 toward bug 1, so their paths are mutually congruent.

(a) Use the symmetry of the paths to show that the bugs are at all times at the corners of a square, and in particular, if bug 1 is at (x, y), then 2, 3 and 4 are respectively at $(-y, x)$, $(-x, -y)$ and $(y, -x)$.

(b) Show for a bug at (x, y) that $\dot{y}/\dot{x} = (y - x)/(x + y)$ and that $\dot{x}^2 + \dot{y}^2 = v^2$.

(c) Use part (b) to show that the path $(x, y) = \big(x(t), y(t)\big)$ followed by bug 1 satisfies the nonlinear autonomous system

$$\frac{dx}{dt} = \frac{-v}{\sqrt{2}} \frac{x + y}{\sqrt{x^2 + y^2}}, \quad \frac{dy}{dt} = \frac{v}{\sqrt{2}} \frac{x - y}{\sqrt{x^2 + y^2}}.$$

1D   Existence, Uniqueness, and Flows Optional

Reduction of a system to first-order normal form isn't always possible, but when it is, and certain differentiability conditions are met, it's possible to apply the following theorem, the proof of which we omit. See M. W. Hirsch and S. Smale, *Differential Equations Dynamical Systems and Linear Algebra*, Ch. 15, Academic Press (1974).

1.1   Existence and Uniqueness Theorem.   Suppose that $\mathbf{F}(t, \mathbf{x})$ and the entries $\partial F_j / \partial x_i$ in its derivative matrix $\mathbf{F_x}(t, \mathbf{x})$ with respect to $\mathbf{x}$ are continuous for t in an interval I containing t_0, and for $\mathbf{x}$ in an open rectangle R in $\mathbb{R}^n$ containing $\mathbf{x}_0$. Then the system $\dot{\mathbf{x}} = \mathbf{F}(t, \mathbf{x})$ has a unique solution satisfying $\mathbf{x}(t_0) = \mathbf{x}_0$ on some subinterval J of I containing t_0, and $\mathbf{x}(t)$ is a continuously differentiable function of the vector $\mathbf{x}_0$. If in addition the entries in $\mathbf{F_x}(t, \mathbf{x})$ are bounded for all $\mathbf{x}$ in $\mathbb{R}^n$, then the solutions exist for all t in I, as, for example, in the case of a nonhomogeneous linear system $\dot{\mathbf{x}} = A(t)\mathbf{x} + \mathbf{b}(t)$.

The uniqueness part of the theorem tells us that for an autonomous system $\dot{\mathbf{x}} = \mathbf{F}(\mathbf{x})$ that satisfies the hypotheses of the theorem, two different solution trajectories can never have a point $\mathbf{x}_0$ in common, that is "trajectories of an autonomous system can't cross." Put more formally what happens is this.

1.2 Corollary. If the autonomous system $\dot{\mathbf{x}} = \mathbf{F}(\mathbf{x})$ satisfies the conditions of the Uniqueness Theorem, and two solution trajectories of the system have a point $\mathbf{x}_0$ in common, then on either side of $\mathbf{x}_0$ one trajectory is contained in the other.

Proof. If two trajectories did agree at $\mathbf{x}_0$, this common value taken as initial value would dictate that the trajectories are the same from that time on until one of them terminates. Similarly the reverse trajectory that satisfies $\dot{\mathbf{x}} = -\mathbf{F}(\mathbf{x})$, and that coincides as a curve with a trajectory approaching $\mathbf{x}_0$ from the other side, would also be uniquely determined until termination of one of them. ∎

EXAMPLE 7

It's routine to verify that the system $\dot{x} = -y$, $\dot{y} = x$ has solutions that are given by $x(t) = A\cos(t + \alpha)$, $y(t) = A\sin(t + \alpha)$, where A and α are constants, $A > 0$. These solutions trace circular trajectories of radius A shown in Figure 12.7(a).

A trajectory of a **nonautonomous** system can cross itself, and distinct trajectories can cross each other, as the next example explains.

EXAMPLE 8

Figure 12.7(b) shows some computer plots of trajectories for the nonautonomous system $\dot{x} = (1 - t)x - ty$, $\dot{y} = tx + (1 - t)y$. Each one of the four trajectories is shown crossing one of the others. One trajectory of a nonautonomous system may very well cross another one, or even intersect itself at a nonzero angle, because on arrival at the same point in phase space at a different time there may have been a change of direction in the vector field $\mathbf{F}(t, \mathbf{x})$. Snapshots of the vector field of this system are in Figure 12.5. The graphs of different solutions in (t, x, y)-space will have no points in common, because t varies from point to point.

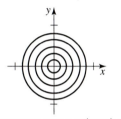

Autonomous system trajectories

(a)

Flows. The trajectories of an autonomous system $\dot{\mathbf{x}} = \mathbf{F}(\mathbf{x})$ are called the **flow lines** of the vector field $\mathbf{F}$, and we can picture them, as shown for example in Figure 12.7(a), as the possible paths followed by fluid particles in a steady fluid flow with velocity vector $\mathbf{F}(\mathbf{x})$ at $\mathbf{x}$. These ideas are also discussed in Chapter 8, Section 4. In what follows we'll assume that the autonomous vector field $\mathbf{F}$ satisfies the conditions of Theorem 1.1 in some region B in $\mathbb{R}^n$, thus guaranteeing (i) that there is a unique flow line through each $\mathbf{x}$ in B and (ii) that distinct flow lines have no points in common. We associate with each such n-dimensional vector field $\mathbf{F}$ a family of **flow transformations** T_t from B to B defined by

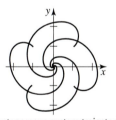

Nonautonomous system trajectories

(b)

FIGURE 12.7

1.3 $T_t(\mathbf{x}) = \mathbf{y}(t)$,   where $\mathbf{y}(t)$ solves $\dot{\mathbf{y}} = \mathbf{F}(\mathbf{y})$ with initial value $\mathbf{y}(0) = \mathbf{x}$.

In words, $T_t(\mathbf{x})$ is the point on the flow line of $\mathbf{F}$ starting at $\mathbf{x}$ that the flow reaches after time t.

EXAMPLE 9

The system $\dot{x} = -y$, $\dot{y} = x$ has circular trajectories, as in Figure 12.7(a). Thus a flow line of radius A for the vector field $\mathbf{F}(x, y) = (-y, x)$ is parametrized by $x(t) = A\cos(t + \alpha)$, $y(t) = A\sin(t + \alpha)$. To start one of these flow lines at a fixed point (u, v) when $t = 0$, we note that $A = \sqrt{u^2 + v^2}$ and write

$$T_t(u, v) = \left(\sqrt{u^2 + v^2}\cos(t + \alpha), \sqrt{u^2 + v^2}\sin(t + \alpha)\right),$$

where α is an angle that the radius from the origin to (u, v) makes with the positive x-axis. (Thus $\alpha(u, v) = \arctan(v/u)$, extended for $u = 0$ and $v \neq 0$ to be the odd multiple of $\pi/2$ that makes $\alpha(u, v)$ continuous.)

Since sine and cosine are periodic functions, the vector-valued function $\phi : \mathbb{R}^3 \to \mathbb{R}^2$ defined by $\phi(t, u, v) = T_t(u, v)$ in the previous example is not one-to-one as a function of t and (u, v) unless t is somehow restricted. However for fixed $t = t_0$, the function $T_{t_0} : \mathbb{R}^2 \to \mathbb{R}^2$ turns out not only to be one-to-one but to have a nice inverse, namely T_{-t_0}. It's a straightforward exercise to show that T_{t_0} is just a rotation about the origin through angle t_0. Hence the inverse of T_{t_0} is T_{-t_0}. We'll see that this simple relationship between T_t and its inverse holds very generally.

The flow transformations T_t defined above have the **composition property**.

1.4
$$T_t T_s = T_{t+s}; \quad \text{in other words,} \quad T_t\big(T_s(\mathbf{x})\big) = T_{t+s}(\mathbf{x}),$$

whenever all three transformations are defined. Equation 1.4 holds because the system

$$\frac{d\mathbf{x}}{dt} = \mathbf{F}(\mathbf{x}),$$

of which $\mathbf{y}(t) = T_t(\mathbf{x})$ is a solution, has a unique solution starting at $T_t(\mathbf{x}_0)$ whose value at s time units later must coincide with the unique solution value achieved by starting at $\mathbf{x}_0$ and running for time $t + s$. Furthermore, the *reversed system*,

$$\frac{d\mathbf{x}}{dt} = -\mathbf{F}(\mathbf{x}),$$

has solution trajectories traced in the direction $-\mathbf{F}(\mathbf{x})$ exactly opposite to that of the solutions of $d\mathbf{x}/dt = \mathbf{F}(\mathbf{x})$. We can use solutions of the reversed system to define T_t for $t < 0$ by $T_t(\mathbf{x}_0) = \mathbf{z}(t)$, where $\mathbf{z}(t)$ satisfies

$$\frac{d\mathbf{z}}{dt} = -\mathbf{F}(\mathbf{z}), \quad \mathbf{z}(0) = \mathbf{x}_0.$$

It follows that each of T_{-t} and T_t is an **inverse operator** to the other, so

1.5
$$T_{-t} T_t = T_t T_{-t} = I,$$

where I is an identity operator that leaves points fixed.

We now consider the effect of T_t on area in $\mathbb{R}^2$ and volume in higher dimensions. This effect is measured locally for each fixed t by the Jacobian determinant J_t of the transformation T_t. Section 4 of Chapter 8 contains a detailed examination of the consequences of the next theorem in the context of interpreting the divergence of a vector field.

1.6   Theorem.   Let $F(\mathbf{x})$ be a continuously differentiable vector field on $\mathbb{R}^n$ such that the derivative matrix $\mathbf{F}'(\mathbf{x})$ has bounded entries. The system $\dot{\mathbf{x}} = F(\mathbf{x})$ defines for $t \geq 0$ a family of one-to-one transformations $\mathbb{R}^n \xrightarrow{T_t} \mathbb{R}^n$ by $T_t(\mathbf{x}) = \mathbf{y}(t)$, where $\mathbf{y}(t)$ represents the flow line of the vector field $\mathbf{F}$ starting at $\mathbf{x}$. For fixed t the flow transformation T_t is volume-preserving in its action on a region B of $\mathbb{R}^n$ if and only if $\operatorname{div}\mathbf{F}$ is identically zero in B. If $\operatorname{div}\mathbf{F} < 0$ in B, then T_t is volume decreasing in B, and if $\operatorname{div}\mathbf{F} > 0$ in B, then T_t is volume-increasing in B. In the special case that $\operatorname{div}\mathbf{F}(\mathbf{x})$ is constant, the Jacobian determinant of T_t is $J_t(\mathbf{x}) = e^{t\operatorname{div}\mathbf{F}(\mathbf{x})}$.

Proof.   The existence of a *unique* solution curve passing through each $\mathbf{x}_0$ ensures that T_t is a well-defined transformation. Furthermore T_t is one-to-one because if

$T_t(\mathbf{x}_0) = T_t(\mathbf{x}_1)$ for some $t > 0$, and for $\mathbf{x}_0 \neq \mathbf{x}_1$, then there would be two distinct solution curves to the system

$$\frac{d\mathbf{x}}{dt} = -\mathbf{F}(\mathbf{x})$$

starting at $\mathbf{y}_0 = T_t(\mathbf{x}_0) = T_t(\mathbf{x}_1)$ and passing back through $\mathbf{x}_0$ and $\mathbf{x}_1$, respectively. But this is impossible, again by the uniqueness theorem.

To see what the transformation T_t does to volumes, we use the following.

1.7 Lemma.   The Jacobian determinant $J_t(\mathbf{x})$ satisfies

$$(d/dt)J_t(\mathbf{x}) = \operatorname{div}\mathbf{F}\big(\mathbf{y}(t)\big)J_t(\mathbf{x}).$$

Lemma Proof.   (Exercise 10 is the 2-dimensional case of this proof.) Let

$$T_t(\mathbf{x}) = \mathbf{y}(t) = \big(y_1(t), \dots, y_n(t)\big),$$

where $\dot{\mathbf{y}}(t) = \mathbf{F}\big(\mathbf{y}(t)\big)$ and $\mathbf{y}(0) = \mathbf{x}$. By Theorem 1.1 $T_t(\mathbf{x})$ is a continuously differentiable function of $\mathbf{x}$. We use the Leibniz notation of Chapter 7, Section 4D to write

$$J_t = \frac{\partial\big(y_1(t), \dots, y_n(t)\big)}{\partial(x_1, \dots, x_n)}.$$

The derivative of a determinant is the sum of the determinants obtained by differentiating one row at a time, as shown in Exercise 9. With rows indexed by i, we then have

$$\frac{dJ_t}{dt} = \sum_{i=1}^{n} \frac{\partial\big(y_1(t), \dots, \dot{y}_i(t), \dots, y_n(t)\big)}{\partial(x_1, \dots, x_k, \dots, x_n)}.$$

By the chain rule, the ikth determinant entry in the ith term above is

$$\frac{\partial \dot{y}_i}{\partial x_k} = \sum_{j=1}^{n} \frac{\partial \dot{y}_i}{\partial y_j}\frac{\partial y_j}{\partial x_k}.$$

By row-linearity of the determinant, the ith term in the sum for dJ_t/dt is then

$$\sum_{j=1}^{n} \frac{\partial \dot{y}_i}{\partial y_j}\frac{\partial(y_1, \dots, y_j, \dots, y_n)}{\partial(x_1, \dots, x_k, \dots, x_n)}.$$

But the determinants in this last sum are 0 (two rows equal) unless $j = i$, in which case the determinant is J_t. To finish proving the lemma, we note that the remaining multiplier of $\partial \dot{y}_i/\partial y_i$ is just J_t. To finish proving the lemma we have

$$\frac{dJ_t}{dt} = J_t \sum_{i=1}^{n} \frac{\partial \dot{y}_i(t)}{\partial y_i}$$

$$= J_t \sum_{i=1}^{n} \frac{\partial F_i\big(\mathbf{y}(t)\big)}{\partial y_i} = J_t \operatorname{div}\mathbf{F}\big(\mathbf{y}(t)\big).$$

We'll now finish proving the theorem. If $\operatorname{div} \mathbf{F} = 0$ then by the lemma, J_t is constant as a function of t. But T_0 is an identity transformation, so $J_0 = 1$ and $J_t = 1$ for $t > 0$ also. Hence the transformation T_t is volume-preserving by the Jacobi change-of-variable theorem for multiple integrals. Conversely, volume-preservation implies $J_t = 1$, so $\operatorname{div} \mathbf{F} = 0$. Finally, the first-order linear differential equation for J_t with initial condition $J_0 = 1$ has the solution

$$J_t(\mathbf{x}) = e^{\int_0^t \operatorname{div} \mathbf{F}(\mathbf{y}(u))\, du}, \quad \text{where} \quad \mathbf{y}(u) = T_u(\mathbf{x}).$$

(Note that if $\operatorname{div} \mathbf{F}$ is constant, then the exponent is just $t \operatorname{div} \mathbf{F}$.) The statements about volume-decreasing and volume-increasing follow as previously by Jacobi's theorem. ∎

EXAMPLE 10 You can see directly that the uncoupled system $\dot{x} = x$, $\dot{y} = 2y$, $\dot{z} = 3z$ has the solution $x(t) = ue^t$, $y(t) = ve^{2t}$, $z(t) = we^{3t}$ with initial values $x(0) = u$, $y(0) = v$. $z(0) = w$. The flow generated by the vector field $\mathbf{F}(x, y, z) = (x, 2y, 3z)$ is therefore

$$\phi(t, u, v, w) = T_t(u, v, w) = (ue^t, ve^{2t}, we^{3t}).$$

Since $\operatorname{div} \mathbf{F}(x, y, z) = 1 + 2 + 3 = 6$, either Theorem 1.6 or direct computation tells us that the Jacobian determinant of T_t is $J_t(u, v, w) = e^{6t}$. It then follows from Jacobi's theorem 4.4 of Chapter 7, Section 4D that in a time interval of length t the flow transformation T_t sends a set B of volume $V(B)$ into a set $T_t(B)$ of volume $e^{6t} V(B)$. For this example, T_t expands length in each coordinate direction separately, giving again the volume expansion factor $e^t e^{2t} e^{3t} = e^{6t}$.

EXERCISES

1. The domain of t-values for which the solution of even an autonomous system exists may be quite restricted. Illustrate this point by deriving the explicit solution to the 1-dimensional initial-value problem $\dot{x} = ax^2$, $x(0) = 1$, where $a > 0$ is constant.

2. If $x(0) = 0$ Theorem 1.1 on existence and uniqueness of solutions fails to apply to the 1-dimensional equation
$$\dot{x} = \begin{cases} \sqrt{x}, & x \geq 0, \\ 0, & x < 0. \end{cases}$$
   (a) Explain why the theorem doesn't apply if $x(0) = 0$.
   (b) Find two distinct solutions to the equation, both satisfying $x(0) = 0$.

3. Can the flow of a continuously differentiable 2-dimensional vector field send a region of positive area into a region of area zero in finite time? Explain your answer.

4. What is the flow of the identically zero vector field on $\mathbb{R}^3$?

5. A 2-dimensional **Hamiltonian system** has the form
$$\dot{x} = \frac{\partial H}{\partial y}, \quad \dot{y} = -\frac{\partial H}{\partial x},$$
   where the real-valued **Hamiltonian function** $H(x, y)$ is assumed to be twice continuously differentiable.
   (a) Show that the flow of a 2-dimensional Hamiltonian system preserves areas.
   (b) Show that the system $\dot{x} = -y$, $\dot{y} = x$ is Hamiltonian, and find a Hamiltonian function $H(x, y)$ for this system.
   (c) Show that the flow lines of a 2-dimensional Hamiltonian system follow level curves of the associated Hamiltonian function.

6. Show that the second-order equation $\ddot{x} = -f(x)$ is equivalent to a first-order system if we set $y = \dot{x}$. Then show that the first-order system is a Hamiltonian system, as defined in the previous exercise, with Hamiltonian
$$H(x, y) = \tfrac{1}{2}y^2 + U(x), \quad \text{where} \quad U'(x) = f(x).$$

The function $U(x)$ is the **potential energy** of the system, and $H(x, y)$ is the **total energy**.

7. A 2-dimensional **gradient system** has the form

$$\dot{x} = \frac{\partial U}{\partial x}, \quad \dot{y} = \frac{\partial U}{\partial y},$$

where the real-valued **potential function** $U(x, y)$ is assumed to be twice continuously differentiable. Show that the flow of such a system preserves areas if and only if $U_{xx} + U_{yy}$ is identically zero that is, if and only if $U(x, y)$ is a **harmonic function**.

8. Consider the 2-dimensional uncoupled system $\dot{x} = x^3$, $\dot{y} = y^3$.
   (a) Sketch the vector field of the system near the origin.
   (b) Compute the Jacobian determinant J_t of the flow transformation T_t of the system.
   (c) Let B be a region of positive area in $\mathbb{R}^2$. Use the result of part (b) to show that the area of the image of B under T_t is bigger than the area of B if $t > 0$ and less than the area of B if $t < 0$.
   (d) Can you draw the same conclusion as in part (c) if the original system is replaced by $\dot{x} = x^2$, $\dot{y} = y^2$? Explain your reasoning.

9. Show that if the entries in an n-by-n matrix $A(t) = \left(a_{ij}(t)\right)$ are differentiable functions of a real variable t, then the derivative of $\det A(t)$ is computed by differentiating the entries in one row of $A(t)$ at a time and then adding the resulting n determinants.
   (a) Carry out the proof for 2-by-2 matrices by first expanding the determinant and then applying the product rule for differentiation.

(b) Carry out the proof for n-by-n matrices by induction, first expanding the determinant by one row, for example the first row, and then applying the induction hypothesis to the cofactors.

10. If the proof of the lemma for Theorem 1.6 is restricted to dimension 2, the computation is in principle no simpler but we avoid using so many subscripts. In particular, we deal with the flow transformation $T_t(u, v) = \left(x(t), y(t)\right)$ generated by the system $\dot{x} = F(x, y)$, $\dot{y} = G(x, y)$ with initial conditions $x(0) = u$, $y(0) = v$.
   (a) For fixed t let $J_t(u, v)$ be the Jacobian determinant of $T_t(u, v)$ with respect to u and v. Use the system to show that

$$\frac{d}{dt} J_t(u, v) = F_u y_v - F_v y_u + G_v x_u - G_u x_v.$$

   (b) Apply the chain rule, for example $F_u = F_x x_u + F_y y_u$, to the partials of F and G in part (a) to show that $(d/dt)J_t = (F_x + G_y)J_t = \text{div}(F, G)J_t$. Here F and G are evaluated at $(x(t), y(t))$.
   (c) Noting that T_0 is an identity transformation, so that $J_0 = \det T_0 = 1$, solve the first order linear differential equation for J_t in part (b) to show that $J_t = \exp\left(\int_0^t \text{div}(F, G)\, dt\right)$, where F and G are evaluated at $(x(t), y(t))$.

11. The last part of the proof of Theorem 1.6 shows that the Jacobian determinant of a flow transformation T_t is $J_t(\mathbf{x}) = \exp\left(\int_0^t \text{div}\, \mathbf{F}(T_u(\mathbf{x}))\, du\right)$.
   (a) Show that if $\text{div}\,\mathbf{F}$ is constant, then the exponent is $t\,\text{div}\,\mathbf{F}$.
   (b) Show generally that the exponent is t times the time-average of $\text{div}\,\mathbf{F}$ over the part of the flow line starting at $\mathbf{x}$ traced between time 0 and time t.

SECTION 2   LINEAR SYSTEMS

2A   Elimination Method

In this section we define what it means for a *system* of first-order differential equations to be linear. As in the case of a single differential equation, the methods of solution are explicit for *constant-coefficient* linear systems, and we'll concentrate on such systems. Unfortunately there is no generally applicable analogue of the method for solving a single linear differential equation with nonconstant coefficients.

Recall that a single first-order differential equation is called linear if it is equivalent to an equation of the form

$$\frac{dx}{dt} = a(t)x + b(t),$$

where a and b are real-valued functions defined on some interval. Similarly an n-dimensional first-order system of differential equations is called a **linear system** if it has the **normal form**

$$\frac{d\mathbf{x}}{dt} = A(t)\mathbf{x} + \mathbf{b}(t),$$

where $A(t)$ is an n-by-n matrix

$$A(t) = \begin{pmatrix} a_{11}(t) & a_{12}(t) & \cdots & a_{1n}(t) \\ a_{21}(t) & a_{22}(t) & \cdots & a_{2n}(t) \\ \vdots & \vdots & \ddots & \vdots \\ a_{n1}(t) & a_{n2}(t) & \cdots & a_{nn}(t) \end{pmatrix}$$

and $\mathbf{b}(t)$ is an n-dimensional vector

$$\mathbf{b}(t) = \begin{pmatrix} b_1(t) \\ b_2(t) \\ \vdots \\ b_n(t) \end{pmatrix},$$

both with coordinate functions defined on an interval.

EXAMPLE 1   The first term in the right side of the vector differential equation

$$\begin{pmatrix} dx/dt \\ dy/dt \end{pmatrix} = \begin{pmatrix} 2 & 4 \\ 1 & -1 \end{pmatrix}\begin{pmatrix} x \\ y \end{pmatrix} + \begin{pmatrix} 2 \\ 4 \end{pmatrix}$$

in terms of a matrix product is

$$\begin{pmatrix} 2 \\ 4 \end{pmatrix} + \begin{pmatrix} 2 & 4 \\ 1 & -1 \end{pmatrix}\begin{pmatrix} x \\ y \end{pmatrix} = \begin{pmatrix} 2x + 4y + 2 \\ x - y + 4 \end{pmatrix}.$$

Hence the vector equation can also be written as a system,

$$\frac{dx}{dt} = 2x + 4y + 2$$
$$\frac{dy}{dt} = x - y + 4.$$

It's convenient to denote differentiation with respect to t by D; that is, let $D = d/dt$. Then regrouping the terms involving x and y on the left side gives

$$(D - 2)x - 4y = 2$$
$$-x + (D + 1)y = 4.$$

At this point, we follow a routine similar to the row reduction method for solving linear algebraic equations. For example, we can operate on the second equation with the differential operator $(D - 2)$ to eliminate x when we add the first equation to the second:

$$(D - 2)x - 4y = 2$$
$$-(D - 2)x + (D - 2)(D + 1)y = (D - 2)4.$$

Addition gives
$$(D-2)(D+1)y - 4y = (D-2)4 + 2$$

or
$$D^2 y - Dy - 6y = -6.$$

We can solve this equation by the methods of the previous chapter, because it contains only one unknown function, namely $y(t)$. The characteristic equation is
$$r^2 - r - 6 = (r+2)(r-3) = 0,$$

which has roots $r_1 = -2$ and $r_2 = 3$. Hence the general solution of the associated homogeneous equation is
$$y_h = c_1 e^{-2t} + c_2 e^{3t}.$$

By inspection, we see that $y_p = 1$ is a particular solution, so what we have shown is that if $(x(t), y(t))$ is an arbitrary solution of the original system, then $y(t)$ must be of the form
$$y(t) = c_1 e^{-2t} + c_2 e^{3t} + 1.$$

We use the second equation of the system to express $x(t)$ directly in terms of $y(t)$:
$$x(t) = (D+1)y - 4$$
$$= -c_1 e^{-2t} + 4c_2 e^{3t} - 3.$$

If we put $x(t)$ and $y(t)$ together in vector form, we get
$$\begin{pmatrix} x(t) \\ y(t) \end{pmatrix} = c_1 e^{-2t} \begin{pmatrix} -1 \\ 1 \end{pmatrix} + c_2 e^{3t} \begin{pmatrix} 4 \\ 1 \end{pmatrix} + \begin{pmatrix} -3 \\ 1 \end{pmatrix}.$$

Our method of solution guaranteed only that every solution of the original system must be a special case of the general formula just obtained. Therefore we should substitute the formula into the system to see if it really provides a solution for every choice of c_1 and c_2. The general theory to be developed in Chapter 13, Section 3 applies, showing that for systems of the form $dx/dt = Ax + b(t)$, the general solution of an n-dimensional system contains n arbitrary constants. Therefore substitution is necessary in such an example only if the number of arbitrary constants present is greater than the dimension of the system, in which case the substitution leads to relations between the constants. Substitution is always a useful check on the accuracy of a computation.

The previous example was misleadingly simple, because after solving for $y(t)$ it was not necessary to solve another differential equation to find $x(t)$; as a result, no extra arbitrary constants were introduced, so there was no need to find relations among the constants so the number of constants would equal the dimension of the system. We'll content ourselves here with such simple examples, leaving the more complicated ones for Chapter 13 where we use more efficient methods.

The general theory of linear operators applies to linear systems of differential equations. If we let $D = d/dt$, then the equation

$$\frac{d\mathbf{x}}{dt} = A(t)\mathbf{x} + \mathbf{b}(t)$$

takes the form

$$(D - A(t))\mathbf{x} = \mathbf{b}(t)$$

since both D and $A(t)$ act as operators on vector functions $\mathbf{x}(t)$. The solutions of the system can then be derived in two parts: the homogeneous solutions (i.e., solutions of the homogeneous equation) and a particular solution. By Theorem 3.1 of Chapter 11, Section 3, each solution of the nonhomogeneous equation is the sum of a fixed particular solution and some solution of the homogeneous linear equation

$$\frac{d\mathbf{x}}{dt} = A(t)\mathbf{x}.$$

For instance, in the previous example

$$\mathbf{x}_p(t) = \begin{pmatrix} -3 \\ 1 \end{pmatrix}$$

is a particular solution that happens to be constant, and

$$\mathbf{x}_h(t) = c_1 \begin{pmatrix} -e^{-2t} \\ e^{-2t} \end{pmatrix} + c_2 \begin{pmatrix} 4e^{3t} \\ e^{3t} \end{pmatrix}$$

represents all the homogeneous solutions.

2B   Nonstandard Forms

Two somewhat more general looking types of constant-coefficient linear systems arise in applications. Both reduce to the normal form we have already considered, namely systems of the form $d\mathbf{x}/dt = A\mathbf{x} + \mathbf{b}$. One advantage of this reduction is that it enables us to apply the geometric intuition associated with tangent vectors and vector fields to equations to which these interpretations are not directly applicable. Another advantage is that the general theory of linear equations develops most naturally in standard form. Finally, standard form is used for application of numerical methods, where in practice we rely most heavily on existence and uniqueness theory.

The next two examples illustrate two types of reduction to standard form and show how to solve them by the elimination method we have already used in Example 1.

EXAMPLE 2   The second-order system

$$\frac{d^2x}{dt^2} = x + 2y + t$$

$$\frac{d^2y}{dt^2} = 3x + 2y$$

reduces to a first-order system of dimension 4 by letting

$$u = \frac{dx}{dt}, \quad v = \frac{dy}{dt}.$$

The system then becomes

$$\frac{du}{dt} = x + 2y + t$$

$$\frac{dv}{dt} = 3x + 2y$$

$$\frac{dx}{dt} = u$$

$$\frac{dy}{dt} = v.$$

This system is of the standard form $d\mathbf{x}/dt = A\mathbf{x} + \mathbf{b}(t)$, where

$$A = \begin{pmatrix} 0 & 0 & 1 & 2 \\ 0 & 0 & 3 & 2 \\ 1 & 0 & 0 & 0 \\ 0 & 1 & 0 & 0 \end{pmatrix} \quad \text{and} \quad \mathbf{b}(t) = \begin{pmatrix} t \\ 0 \\ 0 \\ 0 \end{pmatrix}.$$

The order u, v, x, y has been used in forming the matrix. Alternatively, the system takes the form

$$\begin{aligned} Du \quad\quad - \; x \; - \; 2y &= t \\ Dv - \; 3x \; - \; 2y &= 0 \\ -u \quad\quad + \; Dx \quad\quad &= 0. \\ -v \quad\quad + \; Dy &= 0. \end{aligned}$$

The numerical methods in Section 4 will apply directly to the system in standard form. We take up matrix methods for solving the first-order system in Chapter 13, Sections 1 to 3, but here we use just the techniques of Chapter 11. A moment's thought shows that the elimination method applied to the first-order system will simply take us back to the original second-order system, or one equivalent to it. Therefore we might as well try to solve the second-order system directly by elimination. We first solve the associated homogeneous system, and, as usual, we write $D = d/dt$ to get

$$(D^2 - 1)x - 2y = 0$$

$$-3x + (D^2 - 2)y = 0.$$

If we multiply the second equation by 2 and operate on the first with $(D^2 - 2)$, then addition of the resulting equations eliminates y:

$$(D^2 - 2)(D^2 - 1)x - 6x = 0.$$

Multiplying out the operators gives

$$(D^4 - 3D^2 - 4)x = 0.$$

We solve the equation by finding its characteristic roots from the equation

$$r^4 - 3r^2 - 4 = (r^2 + 1)(r^2 - 4) = 0;$$

they are evidently $r_1 = i$, $r_2 = -i$, $r_3 = 2$, $r_4 = -2$. Hence the homogeneous solution for x is

$$x(t) = c_1 \cos t + c_2 \sin t + c_3 e^{2t} + c_4 e^{-2t}.$$

The first of the two homogeneous equations allows us to solve for y directly

$$y(t) = \tfrac{1}{2}(D^2 - 1)x(t).$$

A straightforward calculation shows that

$$y(t) = -c_1 \cos t - c_2 \sin t + \tfrac{3}{2}c_3 e^{2t} + \tfrac{3}{2}c_4 e^{-2t}.$$

In vector form the homogeneous solution looks like

$$\begin{pmatrix} x_h(t) \\ y_h(t) \end{pmatrix} = c_1 \begin{pmatrix} \cos t \\ -\cos t \end{pmatrix} + c_2 \begin{pmatrix} \sin t \\ -\sin t \end{pmatrix} + c_3 \begin{pmatrix} e^{2t} \\ \tfrac{3}{2}e^{2t} \end{pmatrix} + c_4 \begin{pmatrix} e^{-2t} \\ \tfrac{3}{2}e^{-2t} \end{pmatrix}.$$

To find a particular solution, we try

$$\begin{pmatrix} x_p(t) \\ y_p(t) \end{pmatrix} = \begin{pmatrix} at + b \\ ct + d \end{pmatrix},$$

which we substitute into the given system, getting

$$0 = (at + b) + 2(ct + d) + t$$
$$0 = 3(at + b) + 2(ct + d),$$

or

$$0 = (a + 2c + 1)t + (b + 2d)$$
$$0 = (3a + 2c)t + (3b + 2d).$$

It follows that

$$b + 2d = 0, \quad a + 2c = -1,$$
$$3b + 2d = 0, \quad 3a + 2c = 0.$$

The solutions are $b = d = 0$, and $a = \tfrac{1}{2}$, $c = -\tfrac{3}{4}$. The particular solution is then

$$\begin{pmatrix} x_p(t) \\ y_p(t) \end{pmatrix} = \begin{pmatrix} \tfrac{1}{2}t \\ -\tfrac{3}{4}t \end{pmatrix}.$$

We could have computed the particular solutions along with the homogeneous solution, by applying elimination to the nonhomogeneous system.

A typical set of initial conditions for the system might take the form

$$\begin{pmatrix} x(0) \\ y(0) \end{pmatrix} = \begin{pmatrix} 0 \\ 0 \end{pmatrix}, \quad \begin{pmatrix} dx/dt(0) \\ dy/dt(0) \end{pmatrix} = \begin{pmatrix} 0 \\ 1 \end{pmatrix}.$$

Applying these to the general solution, consisting of homogeneous solution plus particular solution, we get

$$c_1 \begin{pmatrix} 1 \\ -1 \end{pmatrix} + c_2 \begin{pmatrix} 0 \\ 0 \end{pmatrix} + c_3 \begin{pmatrix} 1 \\ \frac{3}{2} \end{pmatrix} + c_4 \begin{pmatrix} 1 \\ \frac{3}{2} \end{pmatrix} = \begin{pmatrix} 0 \\ 0 \end{pmatrix},$$

$$c_1 \begin{pmatrix} 0 \\ 0 \end{pmatrix} + c_2 \begin{pmatrix} 1 \\ -1 \end{pmatrix} + c_3 \begin{pmatrix} 2 \\ 3 \end{pmatrix} + c_4 \begin{pmatrix} -2 \\ -3 \end{pmatrix} + \begin{pmatrix} \frac{1}{2} \\ -\frac{3}{4} \end{pmatrix} = \begin{pmatrix} 0 \\ 1 \end{pmatrix}.$$

These two vector equations are equivalent to the equations

$$
\begin{aligned}
c_1 \quad\quad\quad + c_3 + c_4 &= 0 \\
-c_1 \quad\quad\quad + \tfrac{3}{2}c_3 + \tfrac{3}{2}c_4 &= 0 \\
c_2 + 2c_3 - 2c_4 &= -\tfrac{1}{2} \\
-c_2 + 3c_3 - 3c_4 &= \tfrac{7}{4}.
\end{aligned}
$$

These equations have the unique solution $c_1 = 0$, $c_2 = -1$, $c_3 = \frac{1}{8}$, $c_4 = -\frac{1}{8}$, so the particular solution we are looking for is

$$\begin{pmatrix} x(t) \\ y(t) \end{pmatrix} = \begin{pmatrix} -\sin t + \frac{1}{8}e^{2t} - \frac{1}{8}e^{-2t} + \frac{1}{2}t \\ \sin t + \frac{3}{16}e^{2t} - \frac{3}{16}e^{-2t} - \frac{3}{4}t \end{pmatrix}.$$

Next is an example of a first-order system that is not presented in standard form.

EXAMPLE 3   The system of differential equations

$$\frac{dx}{dt} + \frac{dy}{dt} = 2x + 4y$$

$$2\frac{dx}{dt} + 3\frac{dy}{dt} = 2x + 6y$$

takes the form $d\mathbf{x}/dt = A\mathbf{x}$ if we apply the elimination method to the left side. We multiply the first equation by 2 and subtract from the second to get

$$\frac{dy}{dt} = -2x - 2y.$$

Now we subtract this equation from the first one to get

$$\frac{dx}{dt} = 4x + 6y.$$

As a result we can rewrite the equation as

$$(D - 4)x - 6y = 0$$

$$2x + (D + 2)y = 0.$$

We can now proceed as in Example 1 to eliminate x. We multiply the first equation by 2 and operate on the second with $(D - 4)$. Subtracting the first from the second gives

$$(D - 4)(D + 2)y + 12y = 0$$

or

$$(D^2 - 2D + 4)y = 0.$$

The roots of the characteristic equation $r^2 - 2r + 4 = 0$ are $r_1 = 1 + i\sqrt{3}$, $r_2 = 1 - i\sqrt{3}$, so $y(t)$ has the form

$$y(t) = c_1 e^{(1+i\sqrt{3}t)} + c_2 e^{(1-i\sqrt{3}t)}$$

$$= d_1 e^t \cos \sqrt{3}\, t + d_2 e^t \sin \sqrt{3}\, t.$$

Using the second equation of the modified system, we can express $x(t)$ directly in terms of $y(t)$:

$$x(t) = -\frac{1}{2}(D+2)y(t)$$

$$= -\frac{1}{2}(D+2)(d_1 e^t \cos \sqrt{3}\, t + d_2 e^t \sin \sqrt{3}\, t)$$

$$= d_1 e^t \left(-\frac{3}{2} \cos \sqrt{3}\, t + \frac{\sqrt{3}}{2} \sin \sqrt{3}\, t \right)$$

$$+ d_2 e^t \left(-\frac{3}{2} \sin \sqrt{3}\, t - \frac{\sqrt{3}}{2} \cos \sqrt{3}\, t \right).$$

In vector form we can write

$$\begin{pmatrix} x(t) \\ y(t) \end{pmatrix} = d_1 e^t \begin{pmatrix} -\dfrac{3}{2} \cos \sqrt{3}\, t + \dfrac{\sqrt{3}}{2} \sin \sqrt{3}\, t \\ \cos \sqrt{3}\, t \end{pmatrix}$$

$$+ d_2 e^t \begin{pmatrix} -\dfrac{3}{2} \sin \sqrt{3}\, t - \dfrac{\sqrt{3}}{2} \cos \sqrt{3}\, t \\ \sin \sqrt{3}\, t \end{pmatrix}.$$

EXERCISES

In Exercises 1 to 4, classify the first-order system as linear or nonlinear:

1. $\dfrac{dx}{dt} = t + x^2 + y$

$\dfrac{dy}{dt} = t^2 + x + y$

2. $\dfrac{dy}{dt} = t^2 + z$

$\dfrac{dz}{dt} = t^3 + y$

3. $\dfrac{dx}{dt} = t^2 x + y + e^t$

$\dfrac{dy}{dt} = 1$

4. $\dfrac{dx}{dt} = tx + y$

$\dfrac{dy}{dt} = x + y$

In Exercises 5 and 6, solve by first eliminating one of the unknown functions. Then determine the arbitrary constants so that the initial conditions are satisfied.

5. $\dfrac{dx}{dt} = 6x + 8y,\ \ x(0) = 1$

$\dfrac{dy}{dt} = -4x - 6y,\ \ y(0) = 0$

6. $\dfrac{dx}{dt} = x + 2y,\ \ x(0) = 0$

$\dfrac{dy}{dt} = -2x + y,\ \ y(0) = -1$

In Exercises 7 and 8, find a particular solution for the system. Then noting that the homogeneous equations are the same as the ones in Exercises 5 and 6, write the most general solution.

7. $\begin{pmatrix} dx/dt \\ dy/dt \end{pmatrix} = \begin{pmatrix} 6 & 8 \\ -4 & -6 \end{pmatrix} \begin{pmatrix} x \\ y \end{pmatrix} + \begin{pmatrix} 1 \\ t \end{pmatrix}$

[*Hint:* Try $x = at + b$; $y = ct + d$.]

8. $\begin{pmatrix} dx/dt \\ dy/dt \end{pmatrix} = \begin{pmatrix} 1 & 2 \\ -2 & 1 \end{pmatrix} \begin{pmatrix} x \\ y \end{pmatrix} + \begin{pmatrix} e^t \\ t \end{pmatrix}$

[*Hint:* Use undetermined coefficients as in part (a).]

9. Solve by elimination

$$\frac{dx}{dt} = x + z.$$

$$\frac{dy}{dt} = x + 2y.$$

$$\frac{dz}{dt} = -z.$$

Then satisfy the initial condition $x(0) = 1$, $y(0) = -1$, $z(0) = 2$.

10. (a) Find a first-order system of dimension 4 equivalent to the second-order system

$$\ddot{x} - 3x - 2\ddot{y} = 0.$$

$$\ddot{x} - \ddot{y} + 2x = 0.$$

[*Hint:* Let $\dot{x} = u$, $\dot{y} = v$.]

(b) By solving for $\dot{x}$, $\dot{y}$, $\dot{u}$, and $\dot{v}$ in the first-order system obtained in part (a), write the equivalent 4-dimensional system in the form $d\mathbf{x}/dt = A\mathbf{x}$, where A is a 4-by-4 matrix.

(c) By collecting terms properly, write the system in part (a) in the form

$$L_1(D)x + L_2(D)y = 0$$

$$L_3(D)x + L_4(D)y = 0,$$

where each $L_k(D)$ is a second-order constant-coefficient operator.

(d) Use the method of elimination to solve the system found in part (c).

In Exercises 11 and 12, apply row operations to the terms containing first derivatives to reduce the system to the standard form $d\mathbf{x}/dt = A(t)\mathbf{x} + \mathbf{b}(t)$, where $A(t)$ is a square matrix.

11. $\dfrac{dx}{dt} + \dfrac{dy}{dt} = x + ty.$

$\dfrac{dx}{dt} - 2\dfrac{dy}{dt} = x + t^2.$

12. $\dfrac{dx}{dt} + e^t\dfrac{dy}{dt} = x + e^t.$

$\dfrac{dx}{dt} + \dfrac{dy}{dt} = y + e^{-t}.$

13. (a) Verify that if $\mathbf{x}_1(t)$ and $\mathbf{x}_2(t)$ are solutions of the system $d\mathbf{x}/dt = A(t)\mathbf{x}$, where $A(t)$ is an n-by-n matrix then $c_1\mathbf{x}_1(t) + c_2\mathbf{x}_2(t)$ is also a solution for arbitrary constants c_1 and c_2.

(b) If $d\mathbf{x}/dt = A\mathbf{x}$, where A is an m-by-n matrix, can you have $n \neq m$? Give an example or explain why not.

(c) Show that for a system of the form $d\mathbf{x}/dt = A(t)\mathbf{x} + \mathbf{b}(t)$, the conclusion of part (a) follows only if $\mathbf{b}(t)$ is identically zero.

In Exercises 14 to 17, classify the system as linear or nonlinear.

14. $dx/dt + dz/dt = 1,$

$dx/dt - t(dz/dt) = x$

15. $d^2x/dt^2 + dy/dt = 0,$

$y^2 + t(d^2y/dt^2) = t$

16. $dx/dt + dz/dt = 1,$

$dx/dt - t^2(dz/dt) = 0$

17. $d^2x/dt^2 - dy/dt = x^2,$

$y + t^2(d^2y/dt^2) = 0$

In Exercises 18 and 19, use elimination by operator multiplication to get rid of one of the dependent variables. Solve the resulting equation for the remaining variable, and then determine the general solution $(x(t), y(t))$.

18. $dx/dt = x + 2y,$

$dy/dt = x + y + t$

19. $dx/dt = -y - t,$

$dy/dt = x + t$

In Exercises 20 to 23, reduce the system to the standard form with just one first derivative on the left side of each equation.

20. $dx/dt + dy/dt = t,$

$dx/dt - dy/dt = x$

21. $dx/dt + dy/dt = y,$

$dx/dt + 2dy/dt = x$

22. $2dx/dt + dy/dt + x + 5y = t,$

$dx/dt + dy/dt + 2x + 2y = 0$

23. $dx/dt + dy/dt = \sin t,$

$dx/dt - dy/dt = \cos t$

In Exercises 24 to 27, use elimination by operator multiplication to get rid of one of the dependent variables. Solve the resulting equation for the remaining variable and then determine the general solution of the system. Substitution may be necessary to find relations among constants. Then determine the constants so that the initial conditions are satisfied.

24. $d^2x/dt^2 - x + dy/dt + y = 0$,

$dx/dt - x + d^2y/dt^2 + y = 0$, $x(0) = y(0) = 0$,
$\dot{x}(0) = 0$, $\dot{y}(0) = 1$

25. $d^2x/dt^2 - dy/dt = 0$,

$dx/dt + d^2y/dt^2 = 0$, $x(0) = 1$, $y(0) = 0$,
$\dot{x}(0) = \dot{y}(0) = 0$

26. $d^2x/dt^2 - dy/dt = t$,

$dx/dt + dy/dt = x + y$, $x(0) = y(0) = 0$, $\dot{x}(0) = 1$

27. $d^2x/dt^2 - y = e^t$,

$d^2y/dt^2 + x = 0$, $x(0) = y(0) = \dot{x}(0) = \dot{y}(0) = 0$

In Exercises 28 to 31, introduce new independent variables, $u = \dot{x}$, $v = \dot{y}$, and reduce the system to first-order standard form in u, v, x, and y. (Section 2 of Chapter 13 develops an efficient method for solving these systems in first-order standard form.)

28. $d^2x/dt^2 - x + dy/dt + y = 0$,
$dx/dt - x + d^2y/dt^2 + y = 0$

29. $d^2x/dt^2 - dy/dt = 0$,
$dx/dt + d^2y/dt^2 = 0$

30. $d^2x/dt^2 - dy/dt = t$,
$dx/dt + dy/dt = x + y$

31. $d^2x/dt^2 - y = e^t$,
$d^2y/dt^2 + x = 0$

None of the linear systems in Exercises 32 to 35 is equivalent to a first-order system in standard form. Discuss the solutions, or lack thereof.

32. $dx/dt + dy/dt = 0$,

$dx/dt + dy/dt = 1$

33. $dx/dt + dy/dt = 0$,

$dx/dt + dy/dt = x$

34. $dx/dt + dy/dt = t$,

$dx/dt + dy/dt = x$

35. $dx/dt + dy/dt = y$,

$dx/dt + dy/dt = x$

36. (a) Find a first-order system of dimension 4 equivalent to the second-order system

$$\ddot{y} - 3x - 2y = 0,$$

$$\ddot{x} - y + 2x = 0.$$

(b) Write the system found in part (a) in standard form.
(c) Solve the system in part (a).

37. (a) Find a 2-dimensional system of order 2 satisfied by the x and y coordinates of the solutions to

$$\dot{x} = z + w,$$

$$\dot{y} = z - w,$$

$$\dot{z} = x - y,$$

$$\dot{w} = x + y.$$

Then solve the second-order system and use its solution to solve the given system.
(b) Find a 2-dimensional system of order 2 satisfied by the z and w coordinates of the solutions to the system in part (a).

38. It's generally not possible to find closed-form solutions for linear systems with nonconstant coefficient functions. Here is one that can nevertheless be solved readily by solving a second-order equation for y. Find the general solution.

$$\dot{x} = (t^{-1} - t)x - t^2 y, \quad t > 0,$$

$$\dot{y} = x + ty.$$

SECTION 3   APPLICATIONS

The examples in this section are all of a type that arise frequently in applied mathematics. For some we'll be able to give complete solutions, whereas the others are examples for which we need the numerical methods described in the next section.

EXAMPLE 1   Figure 12.8 shows two 50-gallon tanks connected by flow pipes and with inlets and outlets all having the rates of flow as marked in gallons per minute (g/m). The flow rates are arranged so that each tank is maintained at its capacity at all times. We suppose that each tank initially contains salt solution at a concentration in pounds per gallon that we leave unspecified for the moment, that the left-hand tank is receiving salt solution at a concentration of 1 pound per gallon, and that the right-hand tank is receiving pure water. The problem is to find out what happens to the amount of salt,

FIGURE 12.8
Fluid exchange.

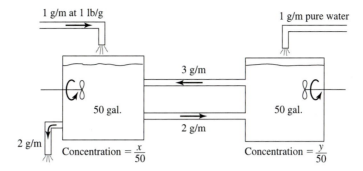

1 g/m at 1 lb/g   1 g/m pure water

3 g/m

50 gal.   50 gal.

2 g/m

2 g/m   Concentration $= \frac{x}{50}$ Concentration $= \frac{y}{50}$

in pounds, as time goes on. We assume that each tank is kept thoroughly mixed at all times, so that the concentration of salt is always the same throughout the whole tank. In the left-hand tank, with salt content $x(t)$, the rate of change of the amount of salt is dx/dt. On the other hand, because of the various flow rates, we can break this rate of change into three parts:

$$\frac{dx}{dt} = -4\left(\frac{x}{50}\right) + 3\left(\frac{y}{50}\right) + 1,$$

where $x/50$ is the concentration of salt in the left tank and $y/50$ the concentration in the right tank, both in pounds per gallon. The term $-4(x/50)$ is the rate of outflow of salt, and the other two terms represent the rate of inflow. Similarly,

$$\frac{dy}{dt} = 2\left(\frac{x}{50}\right) - 3\left(\frac{y}{50}\right).$$

Thus we have a system of differential equations that we can write as

$$\frac{dx}{dt} = -\frac{4}{50}x + \frac{3}{50}y + 1$$

$$\frac{dy}{dt} = \frac{2}{50}x - \frac{3}{50}y.$$

To solve it, we can use the elimination method, first writing the system in the form

$$\left(D + \frac{4}{50}\right)x - \frac{3}{50}y = 1$$

$$-\frac{2}{50}x + \left(D + \frac{3}{50}\right)y = 0.$$

We multiply the first equation by $\frac{2}{50}$ and operate on the second by $(D + \frac{4}{50})$. Addition of the two equations then gives

$$\left(D + \frac{4}{50}\right)\left(D + \frac{3}{50}\right)y - \frac{6}{(50)^2}y = \frac{2}{50},$$

or

$$\left(D^2 + \frac{7}{50}D + \frac{6}{(50)^2}\right)y = \frac{2}{50}.$$

The roots of the characteristic equation come from the factorization

$$r^2 + \frac{7}{50}r + \frac{6}{(50)^2} = \left(r + \frac{1}{50}\right)\left(r + \frac{6}{50}\right)$$

and are $r_1 = -\frac{1}{50}$ and $r_2 = -\frac{6}{50}$. A particular solution is evidently $y_p(t) = \frac{50}{3}$, a constant. Thus in general,

$$y(t) = c_1 e^{-(1/50)t} + c_2 e^{-(6/50)t} + \frac{50}{3}.$$

Using the second equation of the system to write $x(t)$ in terms of $y(t)$, we find

$$x(t) = \frac{50}{2}\left(D + \frac{3}{50}\right)y(t)$$

$$= c_1 e^{-(1/50)t} - \frac{3}{2}c_2 e^{-(6/50)t} + \frac{50}{2}.$$

Thus the general solution is

$$x(t) = c_1 e^{-(1/50)t} - \frac{3}{2}c_2 e^{-(6/50)t} + \frac{50}{2}$$

$$y(t) = c_1 e^{-(1/50)t} + c_2 e^{-(6/50)t} + \frac{50}{3}.$$

From these equations, we see immediately that

$$\lim_{t\to\infty} x(t) = \frac{50}{2},$$

$$\lim_{t\to\infty} y(t) = \frac{50}{3}.$$

In other words, the concentration, in pounds per gallon, in the left tank approaches $\frac{1}{2}$, and in the right tank approaches $\frac{1}{3}$.

The constants c_1 and c_2 depend on the initial values $x(0)$ and $y(0)$. Thus the equations

$$x(0) = c_1 - \frac{3}{2}c_2 + \frac{50}{2}$$

$$y(0) = c_1 + c_2 + \frac{50}{3}$$

determine c_1 and c_2 when $x(0)$ and $y(0)$ are known. The values $x(t_1)$ and $y(t_1)$ at a time t_1 also determine the constants. We leave these details as an exercise.

EXAMPLE 2 Consider two weights of mass m_1 and m_2 separated by springs from each other and from fixed walls. Suppose the springs have stiffness constants k_1, k_2, k_3 as shown in Figure 12.9; thus the restoring force toward the motionless equilibrium position for the ith spring is proportional to k_i. Let x and y be the displacements from equilibrium of the first and second weights. The force acting on the first weight is equal to $m_1(d^2x/dt^2)$, but we also have

$$m_1 \frac{d^2x}{dt^2} = -k_1 x + k_2(y - x).$$

FIGURE 12.9

Mass-spring system.

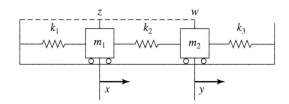

The choice of signs is dictated by whether a positive displacement causes an increase or decrease in velocity. Similarly,

$$m_2 \frac{d^2 y}{dt^2} = -k_2(y - x) - k_3 y.$$

In deriving both equations, we have neglected frictional forces. We can rewrite the system in the form

$$\frac{d^2 x}{dt^2} = -\frac{(k_1 + k_2)}{m_1} x + \frac{k_2}{m_1} y,$$

$$\frac{d^2 y}{dt} = \frac{k_2}{m_2} x - \frac{k_2 + k_3}{m_2} y.$$

For example, if the weights are equal, say $m_1 = m_2 = 1$, and $k_1 = k_2 = k_3 = 1$, then we write the system as

$$(D^2 + 2)x - y = 0$$

$$-x + (D^2 + 2)y = 0.$$

Operating on the second equation with $(D^2 + 2)$ and adding gives

$$(D^4 + 4D^2 + 3)y = (D^2 + 1)(D^2 + 3)y = 0.$$

The general solution of this equation is

$$y(t) = c_1 \cos t + c_2 \sin t + c_3 \cos \sqrt{3}\, t + c_4 \sin \sqrt{3}\, t.$$

Using the second of the pair of equations to find x gives

$$x(t) = (D^2 + 2)y(t)$$

$$= c_1 \cos t + c_2 \sin t - c_3 \cos \sqrt{3}\, t - c_4 \sin \sqrt{3}\, t.$$

The constants c_1, c_2, c_3, and c_4 would be determined by initial displacements and velocities, namely $x(0)$, $y(0)$, $\dot{x}(0)$, $\dot{y}(0)$.

The oscillation $\cos \sqrt{3}\, t$ in the previous example is called a **normal mode** of the oscillation, and it is determined by its **circular frequency** $\sqrt{3}$. The other normal mode, $\cos t$, with circular frequency $\mu = 1$ appearing in the same example arises from different initial conditions. In identifying a normal mode, it's customary to focus attention on the circular frequency itself. Thus the typical normal mode looks like $\cos \mu t$ with circular frequency μ. The normal modes are important characteristics of an oscillatory system, and particularly efficient routes to their computation are in Chapter 13 using eigenvalue methods and exponential matrices.

EXAMPLE 3   Suppose the third spring is removed altogether from our previous example, so that $k_3 = 0$. We're left with

$$\frac{d^2x}{dt^2} = -2x + y,$$

$$\frac{d^2y}{dt^2} = x - y,$$

or, in operator form

$$(D^2 + 2)x - y = 0$$

$$-x + (D^2 + 1)y = 0.$$

Elimination of x proceeds as in the previous example, but this time the differential equation for y is $(D^4 + 3D^2 + 1)y = 0$, with characteristic equation

$$r^4 + 3r^2 + 1 = 0.$$

Regarding the left side quadratic as a function of r^2, we find $r^2 = (-3 \pm \sqrt{5})/2$. Both values are negative, so the characteristic roots are $\pm i\sqrt{(3 + \sqrt{5})/2} \approx \pm 1.62i$ and $\pm i\sqrt{(3 - \sqrt{5})/2} \approx \pm 0.62i$. The normal modes from which solutions are constructed have circular frequencies $\mu_1 = \sqrt{(3 + \sqrt{5})/2}$ and $\mu_2 = \sqrt{(3 - \sqrt{5})/2}$.

EXAMPLE 4   A typical autonomous second-order system has the form

$$\ddot{x} = f(x, y, \dot{x}, \dot{y})$$

$$\ddot{y} = g(x, y, \dot{x}, \dot{y}),$$

with initial conditions $x(t_0) = x_0$, $y(t_0) = y_0$, $\dot{x}(t_0) = u_0$, $\dot{y}(t_0) = v_0$. A particularly important special case is that of **Newton's equations** of planetary motion,

$$\ddot{x} = \frac{-kx}{(x^2 + y^2)^{3/2}} \quad \ddot{y} = \frac{-ky}{(x^2 + y^2)^{3/2}},$$

in which k is a positive constant. In these equations, x and y stand for the rectangular coordinates of a planet in a planar orbit relative to a fixed sun at the origin.

We'll derive the planetary motion equations in a more general vector form that allows for the motion of both bodies, which could be applied to a double star interaction for example. Let $\mathbf{x}_1 = \mathbf{x}_1(t)$ and $\mathbf{x}_2 = \mathbf{x}_2(t)$ represent the positions at time t of two bodies in space such that each acts on the other by the inverse square law of gravitational attraction, with no other forces considered. If m_1 and m_2 are the

FIGURE 12.10

Equal but opposite forces:
$F/m_1 < F/m_2, m_1 > m_2$.

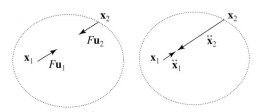

respective masses of the two bodies, the magnitude of the mutually attractive force is then

$$F = \frac{Gm_1m_2}{r^2},$$

where $r = |\mathbf{x}_1 - \mathbf{x}_2|$ is the distance between $\mathbf{x}_1$ and $\mathbf{x}_2$ (i.e., the length of the vector between them). The **gravitational constant** G is about $6.673 \cdot 10^{-11}$ if the relevant units are meters, kilograms, and seconds. The normalized vectors

$$\mathbf{u_2} = \frac{\mathbf{x}_1 - \mathbf{x}_2}{|\mathbf{x}_1 - \mathbf{x}_2|}, \quad \mathbf{u_1} = -\frac{\mathbf{x}_1 - \mathbf{x}_2}{|\mathbf{x}_1 - \mathbf{x}_2|}$$

have length 1 and point respectively from the second body to the first, and vice versa. Thus the vectors that describe the force acting on each body are the product of magnitude F and a normalized direction unit vector $\mathbf{u}$; the vector $F\mathbf{u_1}$ acts on the first body and $F\mathbf{u_2}$ acts on the second body. Since these forces can also be described by Newton's second law as mass times acceleration, we have

$$m_1\ddot{\mathbf{x}}_1 = F\mathbf{u_1}, \quad m_2\ddot{\mathbf{x}}_2 = F\mathbf{u_2}.$$

Figure 12.10 shows the positions and force vectors. The acceleration vectors, which actually govern the motion, are depicted as if m_1 is much larger than m_2. Written out in more detail these **Newton equations** are

3.1 $$\ddot{\mathbf{x}}_1 = -\frac{Gm_2}{|\mathbf{x}_1 - \mathbf{x}_2|^3}(\mathbf{x}_1 - \mathbf{x}_2), \quad \ddot{\mathbf{x}}_2 = -\frac{Gm_1}{|\mathbf{x}_1 - \mathbf{x}_2|^3}(\mathbf{x}_2 - \mathbf{x}_1),$$

where m_1 has been canceled from the first equation and m_2 from the second. Subtracting the second equation from the first gives

$$\ddot{\mathbf{x}}_1 - \ddot{\mathbf{x}}_2 = -\frac{G(m_1 + m_2)(\mathbf{x}_1 - \mathbf{x}_2)}{|\mathbf{x}_1 - \mathbf{x}_2|^3}.$$

Equations 3.1 form a system of vector equations for the motions of the two bodies relative to some coordinate system. If a moving coordinate system has its origin maintained at the center of mass of one of the bodies, say the second, we can let $\mathbf{x} = \mathbf{x_1} - \mathbf{x_2}$ and consider only the equation of relative motion for the first body:

3.2 $$\ddot{\mathbf{x}} = -\frac{G(m_1 + m_2)}{|\mathbf{x}|^3}\mathbf{x}.$$

Writing $\mathbf{x} = (x, y, z)$ and $G(m_1 + m_2) = k$, we get three scalar-valued equations:

3.3 $\quad \ddot{x} = \dfrac{-kx}{(x^2 + y^2 + z^2)^{3/2}}, \quad \ddot{y} = \dfrac{-ky}{(x^2 + y^2 + z^2)^{3/2}}, \quad \ddot{z} = \dfrac{-kz}{(x^2 + y^2 + z^2)^{3/2}}.$

A solution of this nonlinear system will describe a trajectory of the first body relative to the second, or vice versa. We can eliminate the third equation from consideration by choosing (x, y, z) coordinates so that initial conditions on z are $z(0) = \dot{z}(0) = 0$. By Theorem 1.1 of Section 1D, if $\mathbf{x}(0) \neq 0$ the system has a unique solution with its third coordinate $z = z(t)$ identically zero. Thus the system is 2-dimensional and has order 2:

$$\ddot{x} = \dfrac{-kx}{(x^2 + y^2)^{3/2}}, \quad \ddot{y} = \dfrac{-ky}{(x^2 + y^2)^{3/2}}.$$

There are no simple formulas for the solutions $x = x(t)$, $y = y(t)$ of these equations. The classical approach to the problem is to derive certain significant properties of the solutions without actually finding the solutions explicitly. For example, the trajectories have reasonably simple equations. These properties are usually stated as **Kepler's laws** of planetary motion, laws that were discovered empirically from astronomical observation before the work of Newton. Kepler's laws hold for solutions to Newton's equations that have closed paths for trajectories.

1. The path described by a solution $(x(t), y(t))$ is an ellipse with one focus at the sun.
2. The radius from the sun to the planet sweeps out equal areas in equal periods of time.
3. If T is the time required to complete one orbit and a is half the major axis of the orbit, then

$$T^2 = \frac{4\pi^2}{G(m_1 + m_2)} a^3.$$

The derivation of these beautiful laws from Newton's differential equations is given in many physics texts and in some calculus texts. (An outline of the derivation is given in a series of exercises at the end of this section.)

Although the results just described tell us a great deal about planetary motion, if what we want to know is the position or velocity of a planet at a given time, then we may resort to numerical methods of the kind described in the next section. These methods apply directly to a first-order system of arbitrary dimension, and to apply them to Newton's equations we consider an equivalent system of four first-order equations. Let $\dot{x} = u$ and $\dot{y} = v$. Because $\ddot{x} = \dot{u}$ and $\ddot{y} = \dot{v}$, the system takes the first-order form

$$\dot{x} = u$$

$$\dot{y} = v$$

$$\dot{u} = -\frac{G(m_1 + m_2)x}{(x^2 + y^2)^{3/2}}$$

$$\dot{v} = -\frac{G(m_1 + m_2)y}{(x^2 + y^2)^{3/2}}.$$

The prescription for initial position $(x(t_0), y(t_0))$ and velocity $(\dot{x}(t_0), \dot{y}(t_0))$ takes the form

$$x(t_0) = x_0, \quad u(t_0) = u_0,$$

$$y(t_0) = y_0, \quad v(t_0) = v_0.$$

Thus (x_0, y_0) represents the position of the planet at time $t = t_0$, whereas u_0 and v_0 are the rates of change of x and y at the same time.

In analyzing orbits it's important to understand that while an ordinary planetary orbit is an ellipse, which is a closed curve, increasing the orbital speed sufficiently gives an unbounded hyperbolic orbit of a type that is observed for some nonreturning comets. The critical speed, called the escape speed, is given by the formula

$$v_e = \sqrt{\frac{2G(m_1 + m_2)}{r}}.$$

In other words, at distance r from the sun, a speed greater than v_e implies a hyperbolic trajectory, and a speed less than v_e implies an elliptic trajectory. For a derivation of the formula for v_e under the assumption that the less massive body has a negligible effect on the other one and that the motion is radial, see Example 6 of Chapter 10, Section 2. Exercise 28 in the present section shows how to eliminate the radial-motion assumption.

EXERCISES

1. Suppose that two 100-gallon tanks of salt solution contain amounts of salt $y(t)$ and $z(t)$ at time t. Suppose that the solution in the y tank is flowing to the z tank at a rate of 1 gallon per minute, and that the solution in the z tank is flowing to the y tank at the rate of 4 gallons per minute. Suppose also that the overflow from the y tank goes down the drain, whereas the z tank is kept full by the addition of fresh water. Assume that each tank is kept thoroughly mixed at all times.
   (a) Find a linear system satisfied by y and z.
   (b) Find the general solution of the system in part (a) and then determine the constants in it so that the initial values will be $y(0) = 10$ and $z(0) = 20$.
   (c) Draw the graphs of the particular solutions found in part (b) and interpret the results.

2. In Example 1 of the text, the general solution to a system of differential equations is found to be

   $$x(t) = c_1 e^{-(1/50)t} - \tfrac{3}{2} c_2 e^{-(6/50)t} + \tfrac{50}{2},$$

   $$y(t) = c_1 e^{-(1/50)t} + c_2 e^{-(6/50)t} + \tfrac{50}{3}.$$

   (a) Find values for the constants c_1 and c_2 so that the initial conditions $x(0) = 25$, $y(0) = \tfrac{2}{3}$ are satisfied.

   (b) Show that it's possible to choose c_1 and c_2 so that an arbitrary initial condition $(x(0), y(0)) = (x_0, y_0)$ is satisfied. Is this a reasonable state of affairs from a physical standpoint?

3. In Example 2 of the text, the system

   $$(D^2 + 2)x - y = 0$$

   $$-x + (D^2 + 2)y = 0$$

   are shown to have the general solution

   $$x(t) = c_1 \cos t + c_2 \sin t - c_3 \cos \sqrt{3}\, t - c_4 \sin \sqrt{3}\, t,$$

   $$y(t) = c_1 \cos t + c_2 \sin t + c_3 \cos \sqrt{3}\, t + c_4 \sin \sqrt{3}\, t,$$

   where $x(t)$ and $y(t)$ are interpreted as the displacements at time t of two masses in a mass-spring physical system.
   (a) Show the initial conditions

   $$x(0) = 0, \quad \dot{x}(0) = 1$$

   $$y(0) = 1, \quad \dot{y}(0) = 0$$

   are satisfied by choosing the constants properly in the general solution.

(b) Show that general initial conditions of the form $x(0) = x_0$, $y(0) = y_0$, $\dot{x}(0) = u_0$, $\dot{y}(0) = v_0$ can always be satisfied.

4. Two points start from $x_1 = 0$ and $x_2 = 1$ on a line and move with positions $x_1(t)$ and $x_2(t)$ at time $t \geq 0$. Suppose that the x_1-point always maintains its velocity at exactly 10 units per second greater than that of the x_2-point. Suppose also that the sum of the two velocities is e^{-t} for $t \geq 0$.

   (a) Express the relation between the velocities as a first-order system.

   (b) Describe the motion of the two points. Are they ever at the same position at the same time?

5. **(a)** Show that under initial conditions of the special form

$$x(0) = x_0 > 0, \quad \dot{x}(0) = u_0$$
$$y(0) = 0, \quad \dot{y}(0) = 0,$$

Newton's equations of planetary motion reduce to

$$\frac{d^2x}{dt^2} = -\frac{GM}{x^2},$$

together with the condition that $y(t)$ is identically zero.

   (b) Taking the physical situation into account, what can you say about the behavior of a solution $x(t)$ of the reduced system in part (a) if $u_0 = 0$?

6. The **Lotka-Volterra equations**

$$\frac{dH}{dt} = (a - bP)H,$$

$$\frac{dP}{dt} = (cH - d)P,$$

with $a, b, c, d > 0$, model the size relationship of parasite $P(t)$ and host $H(t)$ populations at time t.

   (a) Show that if $P(t) > a/b$, then $H(t)$ decreases, and that if $H(t) < d/c$, then $P(t)$ decreases. Show also that the equilibrium points (H_e, P_e) are $(0, 0)$ and $(a/b, d/c)$.

   (b) Show that the parameterized solution curves $(H, P) = (H(t), P(t))$ satisfy

$$\frac{dH}{dP} = \frac{(a - bP)H}{(cH - d)P}$$

and solve this equation by separation of variables to get

$$H^d P^a = k e^{cH + bP},$$

where k is constant.

   (c) Show that for $k > 0$ the curves found in part (b) are closed circuits in the HP-plane. Thus the Lotka-Volterra theory models the cyclic variation in the sizes of certain populations. [*Hint:* There are at most two positive x-values for which $f(x) = x^a/e^{bx}$ has a given value.]

7. Two 100-gallon tanks X and Y contain initially 50 and 100 gallons, respectively, of pure water. From an external source, salt solution is added to Y at 1 gallon per minute (gpm), each gallon containing 1 pound of salt. Mixed solution flows from Y to X at 2 gpm and from X to Y at 1 gpm. Let $x = x(t)$ and $y = y(t)$ be the respective amounts of salt in X and Y at time $t \geq 0$. *Note.* You're not asked to solve any differential equations for this question.

   (a) At what time t_1 will X begin to overflow? Express the total amount of salt in the two tanks as a function of t while $0 \leq t \leq t_1$.

   (b) Find a system of differential equations satisfied by $x(t)$ and $y(t)$ for $0 \leq t \leq t_1$.

   (c) Find a system of differential equations satisfied by $x(t)$ and $y(t)$ for $t_1 \leq t$, while X is overflowing.

8. Two 100-gallon tanks X and Y are initially full of salt solution, with x_0 pounds of salt in X and y_0 pounds of salt in Y. Mixed solution is pumped from X to Y at 2 gallons per minute and from Y to X at 3 gallons per minute. Pure water evaporates from X at 2 gallons per minute. Let $x = x(t)$ and $y = y(t)$ be the respective amounts of salt in X and Y at time $t \geq 0$.

   (a) At what time t_1 will one of the tanks first overflow or become empty?

   (b) Find, but don't solve, a system of differential equations satisfied by $x(t)$ and $y(t)$ for $0 \leq t \leq t_1$.

   (c) Show that $x(t) + y(t)$ remains constant and that the amount of salt in each tank separately is constant whenever $x_0 = \frac{3}{2} y_0$.

   (d) Assume that $x(0) = 10$ and $y(0) = 20$. Use the equation $x(t) + y(t) = 30$ to solve the system you found in part (b).

9. Two tanks, one of capacity 100 gallons, the other of capacity 200 gallons are each initially half-full of liquid. The 100-gallon tank starts with nothing but pure water, but the other tank starts out with 10 pounds of salt dissolved in the water. Solution flows from the 100-gallon tank to the other tank at 2 gallons per minute. Solution flows in the opposite direction at 1 gallon per minute. Pure water is added to the 100-gallon tank at 1 gallon per min. The entire process is stopped if either tank becomes empty or either tank overflows.

   (a) How long does it take for the process to stop?

   (b) Write down the system of differential equations and initial conditions whose solutions describes the process as a function of time. As a check, notice that the total amount of salt present in the system remains unchanged.

(c) Use the check in part (b) to find a first-order linear initial-value problem for $x(t)$ alone.

(d) Solve the initial-value problem in part (c) for $x(t)$. Then find $y(t)$, and estimate the amount of salt in each tank when the process stops.

Normal modes. In Exercises 10 to 13, calculate the circular frequencies of the various constituent oscillations associated with the system of Example 2 of the text under the following assumptions.

10. $m_1 = m_2 = 1$, $k_2 = 2$, $k_1 = k_3 = 1$

11. $m_1 = m_2 = 1$, $k_1 = k_2 = 1$, $k_3 = 2$

12. $m_1 = 1$, $m_2 = 2$, $k_1 = 1$, $k_2 = k_3 = 4$

13. $m_1 = 1$, $m_2 = 2$, $k_1 = 2$, $k_2 = k_3 = 3$

14. Suppose that the middle spring is removed from the system governed by the equations of Example 2 of the text.
   (a) Show that the system becomes uncoupled.
   (b) What are the normal modes?

15. A 2-dimensional mechanical system $m_1\ddot{x} = f(x, y)$, $m_1\ddot{y} = g(x, y)$ is called **conservative** if there is a **potential function** $U(x, y)$ such that

$$\frac{\partial U(x, y)}{\partial x} = -f(x, y) \quad \text{and} \quad \frac{\partial U(x, y)}{\partial y} = -g(x, y).$$

   (a) Show that this two-body system is conservative by computing a potential:

$$m_1\ddot{x} = -(k_1 + k_2)x + k_2 y,$$
$$m_2\ddot{y} = k_2 x - (k_2 + k_3)y.$$

   (b) The **kinetic energy** of the system is $T = \frac{1}{2}(m_1\dot{x}^2 + m_2\dot{y}^2)$. Show that for a general conservative system of the type considered here, the **total energy** $T + U$ is constant. [*Hint:* Multiply the first equation by $\dot{x}$, the second by $\dot{y}$, add the two equations and integrate.]

*16. In deriving the equations of Example 2 of the text to establish the precise location of each mass relative to the other, and to the spring supports, we needed to know in advance the equilibrium positions of the two masses. This is a problem of finding an **equilibrium solution** to the appropriate equations of motion, that is, finding a constant solution, for which all time derivatives are zero. For the equations derived in Example 2, it's routine to check that the unique equilibrium solution is $x(t) = 0$, $y(t) = 0$. Indeed we chose our coordinates so that these would be the equilibrium solutions, so we get no new information. This exercise asks you to derive

a form of the equations of motion that predicts mathematically the relative equilibrium positions of the two masses.

Instead of measuring the locations of the two masses shown in Figure 12.9 from their equilibrium positions we can measure both displacements from the same point at the left-hand support. If we know the unstressed (i.e., relaxed) lengths l_1, l_2, l_3 of the three springs, and the distance b between the supports, this approach allows us to determine the precise location of the equilibrium positions. (This information was assumed known in our earlier analysis.) Let z and w be the respective distances of masses m_1 and m_2 from the left end, as shown in Figure 12.9.

   (a) Show that

$$m_1 d^2 z/dt^2 = -k_1(z - l_1) + k_2((w - z) - l_2),$$
$$m_2 d^2 w/dt^2 = -k_2((w - z) - l_2) + k_3((b - w) - l_3).$$

   (b) Show that the equations are equivalent to

$$m_1 d^2 z/dt^2 = -(k_1 + k_2)z + k_2 w + k_1 l_1 - k_2 l_2,$$
$$m_2 d^2 w/dt^2 = k_2 z - (k_2 + k_3)w + k_2 l_2 - k_3 l_3 + k_3 b.$$

   (c) The equations derived in part (b) are similar to the ones derived in Example 2 of the text except for the presence of additional constant terms on the right side. Thus they constitute a nonhomogeneous system rather than a homogeneous one. Since equilibrium solutions are constant, the second derivatives $\ddot{z}$ and $\ddot{w}$ are identically zero. Consequently, to find the equilibrium positions, all we have to do is set the right sides of the differential equations equal to zero and solve for z and w. Find the equilibrium solutions in terms of the ls and ks.

17. Let g be the acceleration of gravity at the surface of a homogeneous solid spherical body of mass M and radius R. Use the inverse-square law to show that $g = GM/R^2$, where G is the gravitational constant in appropriate units of measurement. Assume the mass of the body is concentrated at its center.

18. (a) Use the equation established in the previous exercise to estimate the gravitational constant using a measured value of 9.8 meters per second for the acceleration of gravity near the surface of the earth. (Use the values for the mass and radius of the earth $m = 6 \cdot 10^{24}$ kg, $R = 6368$ km.)
   (b) Estimate the acceleration of gravity near the surface of the earth using the value $6.67 \cdot 10^{-11}$ for the gravitational constant G.

19. The outer radius (not the thickness!) of the earth's atmospheric shell is about $5600 \cdot 10^3$ meters, and the earth's mass is about $5976 \cdot 10^{24}$ kilograms. With $G = 6.673 \cdot 10^{-11}$, estimate the escape speed required at the outer limit of the atmosphere for a projectile of mass 100 kilograms. How is your answer affected if the projectile mass is instead 1000 kilograms? How about 10^{22} kilograms?

20. Suppose at some time that two bodies subject only to their mutual gravitational attraction are at distance r_0 apart and are receding from each other along a fixed line at a certain fraction q of escape velocity, where $0 < q < 1$. Show that their separation velocity reaches zero, and the bodies start to "fall" toward each other, when their distance apart becomes $r_0/(1 - q^2)$.

21. This exercise is a reminder that there would be no such thing as escape velocity for a body of constant mass if the acceleration of gravity were really constant. For linear motion away from the attracting body, we would have $\ddot{x} = -g$, for some positive constant g. Show that no matter how large $x_0 = x(0) > 0$ and $v_0 = \dot{x}(0) > 0$ are, $x(t)$ has a finite maximum.

22. **(a)** Use Kepler's second law, equal areas swept out in equal times, to show that a planet moving in circular orbit must have constant speed.

(b) Use Kepler's third law, $T^2 = 4\pi^2 a^3/(G(m_1 + m_2))$, together with the result of part (a), to show that a circular orbit of radius a has constant orbital speed $v = \sqrt{G(m_1 + m_2)/a}$. [*Hint:* Express v in terms of the period T.]

23. The **uniform orbital speed** of a satellite of mass m_1 at distance x_0 from an attracting body of mass m_2 is the speed v_1 that the satellite must attain to keep it in a uniform circular orbit.

(a) Show that the orbit

$$\mathbf{x} = x_0(\cos(v/x_0)t, \, \sin(v/x_0)t)$$

represents circular motion of radius x_0 with uniform speed v and acceleration $\ddot{\mathbf{x}}$ toward the origin of magnitude v^2/x_0. This acceleration vector is called **centripetal acceleration**. [*Hint:* Compute $|\mathbf{x}|$, $|\dot{\mathbf{x}}|$ and $|\ddot{\mathbf{x}}|$.]

(b) Show that if gravitational acceleration $G(m_1 + m_2)/x_0^2$ is to provide precisely the centripetal acceleration of the circular orbit found in part (a), then the uniform orbital speed will be $v_1 = \sqrt{G(m_1 + m_2)/x_0}$.

(c) How is uniform orbital speed related to escape speed?

(d) How many days would there be in a month if the earth's moon had a uniform circular orbit of radius equal to 384,404 kilometers, the mean distance of the moon from the earth, and the earth's motion around the sun is ignored? (Because of the earth's motion around the sun, the number of days from full moon on earth to full moon is about two days more than the answer to this exercise.)

24. The **synchronous orbit** of a body of mass m about a uniformly rotating body of mass $M > m$ is the one that maintains the orbiting body directly over one point on the rotating one. Assume the mass of each body is concentrated at its center.

(a) Use the first two Kepler laws to show that a synchronous orbit is necessarily circular, and that it must lie in the plane of the equator of the rotating body.

(b) Use the third Kepler law to show that if T is the period of rotation of the larger body, then the radius of the synchronous orbit is $R = KT^{2/3}$, where $K = \sqrt[3]{G(M + m)/4\pi^2}$.

(c) Show that the synchronous orbit about the earth for a small satellite has radius approximately 6.22846 times the radius of the earth, or 26,246 miles. (Continuing orbital correction of communication satellites is required because of uneven mass concentrations on earth and the influence of other bodies such as the sun and the moon.)

25. The Newton equations for orbits of a single planet of mass m_2 relative to a fixed sun of mass m_1 have the form

$$\ddot{x} = \frac{-kx}{(x^2 + y^2)^{3/2}}, \qquad \ddot{y} = \frac{-ky}{(x^2 + y^2)^{3/2}},$$

where $k = G(m_1 + m_2)$.

(a) Find the relationship that must hold between the positive constants a and ω so that these differential equations will have solutions with circular orbits described by $x(t) = a\cos\omega t$, $y(t) = a\sin\omega t$.

(b) Show that the relationship described in part (a) expresses the third Kepler law.

(c) Show that the orbit

$$\mathbf{x} = (a\cos\omega t, \, a\sin\omega t), \qquad \omega = \text{const.} > 0$$

obeys the second Kepler law.

26. A vector system $\ddot{\mathbf{x}} = -F(\mathbf{x})$ is called **conservative** if there is a real-valued **potential energy** function $U(\mathbf{x})$ such that $F(\mathbf{x}) = \nabla U(\mathbf{x})$. For a 1-dimensional vector field the relation is just $F(x) = U'(x)$; a potential function is determined only up to an additive constant.

(a) Verify that the Newtonian vector field

$$\mathbf{F}(x, y) = \left(\frac{kx}{(x^2 + y^2)^{3/2}}, \, \frac{ky}{(x^2 + y^2)^{3/2}} \right)$$

has $U(x, y) = -k(x^2 + y^2)^{-1/2}$ as potential.

(b) The **kinetic energy** of a body of mass 1 following a path $(x, y) = (x(t), y(t))$ is $T = \frac{1}{2}(\dot{x}^2 + \dot{y}^2)$, and the **total energy** of motion in a conservative field is

$$E = T + U = \frac{1}{2}(\dot{x}^2 + \dot{y}^2) - \frac{k}{\sqrt{x^2 + y^2}}$$

for the Newtonian field. Verify that the total energy E is constant for the motion in the vector field of part (a). *Hint:* Show that $dE/dt = 0$, and use the Newton equations of motion.]

(c) Verify that for motion governed by the equation $\ddot{\mathbf{x}} = F(\mathbf{x})$ the total energy E is constant if the vector field is conservative: $F(\mathbf{x}) = \nabla U(\mathbf{x})$.

The results of the next four exercises establish the validity of **Kepler's laws.**

27. We've seen that the orbit of one body relative to a second always lies in a fixed plane containing both bodies. This is often shown as follows.

(a) Show that if a body of mass m has a path of motion that obeys the inverse-square law $m\ddot{\mathbf{x}} = -(k/|\mathbf{x}|^3)\mathbf{x}$, then the motion is confined to a plane through the center of attraction determined by the initial position and the velocity vectors. [*Hint:* Establish the relation $\frac{d}{dt}(\mathbf{x} \times \dot{\mathbf{x}}) = \mathbf{x} \times \ddot{\mathbf{x}}$ to show that the plane containing $\mathbf{x}$ and $\dot{\mathbf{x}}$ is perpendicular to a fixed vector.]

(b) A **central force** law is one such that motion is governed by an equation of the form $\ddot{\mathbf{x}} = G(\mathbf{x})\mathbf{x}$, where $G(\mathbf{x})$ is a real-valued function. Show that motion subject to a central force law is confined to a plane.

28. The **angular momentum** of a planet at position $\mathbf{x}$ in its plane orbit about the sun is the vector $\mathbf{L} = \mathbf{x} \times m\dot{\mathbf{x}}$, that is, $\mathbf{L}$ is the cross-product of the position vector $\mathbf{x}$ with the linear momentum vector $m\dot{\mathbf{x}}$.

(a) Introduce rectangular coordinates x, y in the plane of motion so that $\mathbf{x} = (x, y, 0)$ to show that the length $L = |\mathbf{L}|$ of angular momentum equals $L = m|x\dot{y} - y\dot{x}|$.

(b) Show that in terms of polar coordinates $x = r\cos\theta$, $y = r\sin\theta$, the angular momentum is $mr^2\dot{\theta}$, if $\dot{\theta} > 0$.

(c) Kepler's second law of planetary motion decrees that the radius joining a planet to the sun sweeps out equal areas in equal times. Use the Kepler law together with the formula

$$A = \frac{1}{2}\int_{\theta_1}^{\theta_2} r^2 \, d\theta$$

for area in polar coordinates to show that the angular momentum $mr^2\dot{\theta}$ is constant on an orbit. [*Hint:*

Express area swept out along an orbit as an integral with respect to time t between t and $t + \tau$.]

29. Kepler's second law (radius vector from sun to planet sweeps out equal areas in equal times) holds for all **central force laws**, that is, force laws expressible in the form $\ddot{\mathbf{x}} = G(\mathbf{x})\mathbf{x}$, where $G(\mathbf{x})$ is some real-valued function. This includes as special cases the inverse-square law of attraction, where $G(\mathbf{x}) = -k|\mathbf{x}|^{-2}, k > 0$, and the **Coulomb repulsion law**, where $G(\mathbf{x}) = k|\mathbf{x}|^{-2}, k > 0$; the latter governs interaction of particles bearing electric charges of the same sign.

(a) Assuming planar motion and using rectangular coordinates (x, y) for $\mathbf{x}$, show that a central force law has the form

$$\ddot{x} = G(x, y)x, \quad \ddot{y} = G(x, y)y,$$

and conclude that $x\ddot{y} - y\ddot{x} = 0$ for a motion governed by a central force law.

(b) Use the conclusion of part (a) to show that $x\dot{y} - y\dot{x} = h$ for some constant h.

(c) Change to polar coordinates by $x = r\cos\theta$, $y = r\sin\theta$ to show that $x\dot{y} - y\dot{x} = r^2\dot{\theta}$ and hence, using part (b), show that $r^2\dot{\theta} = h$ for some constant h. The result says that for motion in a central force field, the angular velocity $\dot{\theta}$ is inversely proportional to the square of the distance from the center of the field.

(d) Use the result of part (c) and a computation of area in polar coordinates to prove Kepler's second law for a central force field by showing that, as a function of time t, area swept out has the form $A = \frac{1}{2}ht + c$. Explain why this proves Kepler's second law under the given assumptions.

*(e) Apply Green's theorem to the equation $x\dot{y} - y\dot{x} = h$ derived in part (b) to show directly, without using polar coordinates, that Kepler's second law holds.

*30. A single planet with position $\mathbf{x} = \mathbf{x}(t)$ obeying $\ddot{\mathbf{x}} = -(k/|\mathbf{x}|^3)\mathbf{x}$ follows an elliptic, parabolic, or hyperbolic path. Here is an outline of a way to show this by deriving a linear differential equation from the vector equation.

(a) Use $\mathbf{x} = (r\cos\theta, r\sin\theta)$ to express the vector equation of motion in the two polar coordinate equations $\ddot{r} - r\dot{\theta}^2 = -k/r^2$, $r\ddot{\theta} + 2\dot{r}\dot{\theta} = 0$.

(b) Show that the second equation derived in part (a) implies $r^2\dot{\theta} = h$ for some constant h, and use this to write the other equation in the form $\ddot{r} = h^2r^{-3} - kr^{-2}$. In particular, show that if $h = 0$ the motion is confined to a line and results either in collision or escape.

(c) Use the results of part (b) to show that if $h \neq 0$,

$$\frac{1}{r^2}\frac{d^2r}{d\theta^2} - 2\frac{1}{r^3}\left(\frac{dr}{d\theta}\right)^2 = \frac{1}{r} - \frac{k}{h^2}.$$

[*Hint:* Use the chain rule to express $\dot{r}$ and $\ddot{r}$ in terms of derivatives with respect to θ.]

(d) Make the change of variable $r = 1/u$ to show that the equation in part (c) becomes the second-order linear equation $d^2u/d\theta^2 + u = k/h^2$.

(e) Show that the solution $u = 1/r = A\cos(\theta + \alpha) + k/h^2$ to the previous equation represents an ellipse, parabola or hyperbola in polar coordinates according as $|A| < k/h^2$, $|A| = k/h^2$, or $|A| > k/h^2$. [*Hint:* Let $x = r\cos(\theta + \alpha)$, $y = r\sin(\theta + \alpha)$, a rotation by α of the original xy-axes.]

(f) Each **focus** of an ellipse lies on the major axis at distance c from the center where $c^2 = a^2 - b^2$ and a and b are the semi-axis. The **eccentricity** is $e = c/a$. Show that for an elliptic orbit, the center of attraction is at one focus and the eccentricity is $|A|h^2/k$. Then show that the polar equation for an orbit is

$$r = \frac{h^2/k}{1 + e\cos\theta}.$$

[*Hint:* For the first part, convert the polar equation, with $\alpha = 0$, to rectangular coordinates.]

(g) Assume that the orbit in part (f) is elliptic, with $0 \leq e < 1$. Show that the time for one complete revolution is $T = 2\pi ab/h$. Then show that $h^2/k = b^2/a$ to derive the third Kepler law $T^2 = 4\pi^2 a^3/k$. [*Hint:* The sum of the maximum and minimum values for r is equal to $2a$.]

*31. We established the formula for escape speed in the text under the assumption that the relative distance separating two bodies, subject only to the forces of mutual gravitational attraction, would always be measured radially along the same fixed line. The purpose of this problem is to show that the fixed-line assumption isn't necessary, and that the relative speed v and distance r are always related by

$$\frac{v^2}{2} - \frac{v_0^2}{2} = \frac{k}{r} - \frac{k}{r_0}.$$

Here the constant is $k = Gm$, where m is the sum of the two masses, G is the gravitational constant, and v_0

and r_0 are initial speed and distance. Starting with the Newton vector equation $\ddot{\mathbf{x}} = -k\mathbf{x}/|\mathbf{x}|^3$, we form the dot product of both sides by $\dot{\mathbf{x}}$ to get the scalar equation $\dot{\mathbf{x}} \cdot \ddot{\mathbf{x}} = -k(\mathbf{x} \cdot \dot{\mathbf{x}})/|\mathbf{x}|^3$.

(a) Show that the left side of the previous equation is equal to $\dfrac{d}{dt}\dfrac{v^2}{2}$, where $v = \sqrt{\dot{\mathbf{x}} \cdot \dot{\mathbf{x}}} = |\dot{\mathbf{x}}|$.

(b) Show that the right side of that same equation is equal to $k(\nabla|\mathbf{x}|^{-1}) \cdot \dot{\mathbf{x}}$, where ∇ is the gradient operator: $\nabla f = (\partial f/\partial x, \partial f/\partial y, \partial f/\partial z)$.

(c) Show that the result of part (b) is also $k(d/dt)|\mathbf{x}|^{-1}$, and conclude that

$$\frac{d}{dt}\frac{v^2}{2} = k\frac{d}{dt}\frac{1}{r}.$$

(d) Integrate the previous equation between 0 and an arbitrary positive time t to get the equation relating v, v_0, r and r_0.

32. Suppose a projectile is fired directly away from and at distance x_0 from the center of mass of a planet with initial speed z_0. If z_0 is less than the escape speed, show that the maximum additional distance attained from the center of mass of the planet is

$$\frac{x_0^2 z_0^2}{2GM - x_0 z_0^2},$$

where M is the sum of the masses of the two bodies.

33. A 2-dimensional **Hamiltonian system** is a pair of differential equations of the form

$$dx/dt = H_y(x, y, t), \quad dy/dt = -H_x(x, y, t).$$

The function H that determines the system is called its Hamiltonian. Suppose that $(x(t), y(t))$ satisfies the system, and consider two functions of t:

(a) $\dfrac{d}{dt}[H(x(t), y(t), t)]$,

(b) $H_t(x(t), y(t), t)$,

where the partial derivative of the Hamiltonian in (ii) is computed before substituting $x(t)$ and $y(t)$ for x and y. Show that these two functions of t are equal.

SECTION 4   NUMERICAL METHODS

Our numerical methods apply to a first-order vector initial-value problem

$$\frac{d\mathbf{x}}{dt} = \mathbf{F}(t, \mathbf{x}), \quad \mathbf{x}(t_0) = \mathbf{x}_0.$$

If the system is linear and has an explicit solution formula, that formula may well be preferred to a numerical approximation, because the approximation may be unable to give a convincing description of a solution's long-term behavior. However, even for a solvable system the numerical approach may be the quickest way to get some short-term qualitative information about solution trajectories.

4A   Euler's Method

We choose a step of size h to find successive approximations $\mathbf{x}_k$ to the true values $\mathbf{x}(t_0 + kh)$ of the solution $\mathbf{x}(t)$. The idea is to use the derivative approximation

$$\frac{\mathbf{x}(t+h) - \mathbf{x}(t)}{h} \approx \mathbf{F}(t, \mathbf{x}),$$

in the form

$$\mathbf{x}(t+h) \approx \mathbf{x}(t) + h\mathbf{F}(t, \mathbf{x}).$$

Thus having found $\mathbf{x}$ corresponding to $t_k = t_0 + kh$, we define the approximation $\mathbf{x}_{k+1}$ at t_{k+1} by

$$\mathbf{x}_{k+1} = \mathbf{x}_k + h\mathbf{F}(t_k, \mathbf{x}_k).$$

Starting at a point $\mathbf{x}_0$, this equation generates a sequence $\mathbf{x}_1, \mathbf{x}_2, \ldots, \mathbf{x}_m$ of arrow tips $\mathbf{x}_{k+1}$ designed to lie close to a trajectory containing $\mathbf{x}_k$. Figure 12.11(a) shows an example of an autonomous vector field $\mathbf{F}(\mathbf{x})$ along with some arrows tangent to points $\mathbf{x}_k$ on a solution trajectory, the latter shown as a dotted curve. If we scale these tangent arrows down in length by a small enough factor $h > 0$, we can expect the tip of each arrow to land at points $\mathbf{x}_k + h\mathbf{F}(t_k, \mathbf{x}_k)$ that are good approximations to points on the trajectory. Having accepted one of these approximations as $\mathbf{x}_{k+1}$, we may then go on similarly to the next approximation by starting at $\mathbf{x}_{k+1}$. Figure 12.11(b) shows how using a small scale factor h can improve the approximation.

For a 2-dimensional system,

$$\dot{x} = F(t, x, y), \quad x(t_0) = x_0,$$

$$\dot{y} = G(t, x, y), \quad y(t_0) = y_0,$$

the 0th step starts with x_0 and y_0. Then

$$x_1 = x_0 + hF(t_0, x_0, y_0)$$

$$y_1 = y_0 + hG(t_0, x_0, y_0).$$

FIGURE 12.11

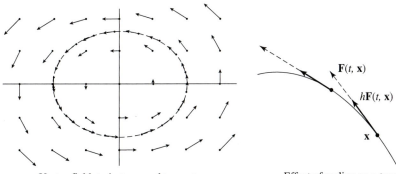

Vector field, trajectory, and tangents

(a)

Effect of scaling on a tangent

(b)

Next, with $t_1 = t_0 + h$,

$$x_2 = x_1 + hF(t_1, x_1, y_1)$$

$$y_2 = y_1 + hG(t_1, x_1, y_1).$$

In general, with $t_k = t_{k-1} + h = t_0 + kh$, we get

$$x_{k+1} = x_k + hF(t_k, x_k, y_k)$$

$$y_{k+1} = y_k + hG(t_k, x_k, y_k).$$

The basic loop for computer implementation is then

$$S = X \quad \text{(Save previous } X.\text{)}$$

$$X = X + H * F(T, X, Y)$$

$$Y = Y + H * G(T, S, Y)$$

$$T = T + H,$$

where the letters on the left represent the new values and the letters on the right represent the values computed in the previous step of the loop.

| **EXAMPLE 1** | The system

$$\dot{x} = ty + 1, \quad x(0) = 1$$
$$\dot{y} = x, \qquad\quad y(0) = -1$$

arose in Example 4 of the previous section. With step size $h = 0.1$, the loop

$$S = X$$

$$X = X + H * (T * Y + 1)$$

$$Y = Y + H * S$$

$$T = T + H$$

produces the values in Table 12.1.

4B   Improved Euler Method

A more accurate numerical method for a vector equation is a modification of the Euler method in which, instead of using the tangent vector $\mathbf{F}(t_k, \mathbf{x}_k)$ to find the next value, we use the vector average

$$\tfrac{1}{2}[\mathbf{F}(t_k, \mathbf{x}_k) + \mathbf{F}(t_k + h, \mathbf{p}_{k+1})],$$

where $\mathbf{p}_{k+1}$ is the value that the Euler method would have predicted, namely

$$\mathbf{p}_{k+1} = \mathbf{x}_k + h\mathbf{F}(t_k, \mathbf{x}_k).$$

Using the predictor value $\mathbf{F}(t_k + h, \mathbf{p}_{k+1})$ to modify the Euler estimate, we define the **improved Euler approximation** to be

$$\mathbf{x}_{k+1} = \mathbf{x}_k + \frac{h}{2}[\mathbf{F}(t_k, \mathbf{x}_k) + \mathbf{F}(t_k + h, \mathbf{p}_{k+1})].$$

TABLE 12.1

t	x	y	$\dot{x} = ty + 1$	$\dot{y} = x$
0	1	-1	1	1
0.1	1.1	-0.89	1	1.1
0.2	1.19	-0.77	0.92	1.19
0.3	1.28	-0.64	0.87	1.28
0.4	1.35	-0.51	0.85	1.35
0.5	1.44	-0.36	0.85	1.44
0.6	1.52	-0.21	0.89	1.52
0.7	1.61	-0.05	0.97	1.61
0.8	1.70	0.12	1.08	1.70
0.9	1.81	0.30	1.24	1.81
1	1.94	0.49	1.44	1.94
1.1	2.09	0.70	1.70	2.09
1.2	2.26	0.93	2.02	2.26
1.3	2.48	1.18	2.41	2.48
1.4	2.73	1.45	2.88	2.73
1.5	3.03	1.75	3.45	3.03
1.6	3.39	2.09	4.14	3.39
1.7	3.83	2.47	4.96	3.83
1.8	4.35	2.91	5.95	4.35
1.9	4.97	3.41	7.13	4.97
2	5.72	3.98	8.56	5.72
2.1	6.62	4.64	10.28	6.62
2.2	7.69	5.41	12.36	7.69
2.3	8.98	6.31	14.88	8.89
2.4	10.53	7.36	17.93	10.53
2.5	12.40	8.60	21.64	12.40

Figure 12.12 shows a geometric rationale for using this particular modification to get an improved estimate. The tip of the arrow $\mathbf{x}_{k+1}$ lies at the midpoint between the tip of the Euler arrow computed at $\mathbf{x}_k$ and time t_k and the tip of the Euler arrow computed at $\mathbf{p}_{k+1}$ and time $t_k + h$, then translated back to $\mathbf{x}_k$. Thus the improved estimate used information not just from the pair $(t_k, \mathbf{x}_k)$ but also estimated future information from $(t_k + h, \mathbf{p}_{k+1})$.

FIGURE 12.12

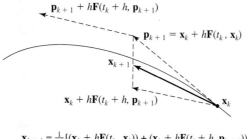

$$\mathbf{x}_{k+1} = \tfrac{1}{2}[(\mathbf{x}_k + h\mathbf{F}(t_k, \mathbf{x}_k)) + (\mathbf{x}_k + h\mathbf{F}(t_k + h, \mathbf{p}_{k+1}))]$$
$$= \mathbf{x}_k + \tfrac{h}{2}[\mathbf{F}(t_k, \mathbf{x}_k) + \mathbf{F}(t_k + h, \mathbf{p}_{k+1})]$$

For a 2-dimensional system

$$\dot{x} = F(t, x, y), \quad x(t_0) = x_0$$
$$\dot{y} = G(t, x, y), \quad y(t_0) = y_0,$$

we start with (x_0, y_0). Then letting $\mathbf{p} = (p, q)$, we compute the Euler approximation

$$p_1 = x_0 + hF(t_0, x_0, y_0)$$

$$q_1 = y_0 + hG(t_0, x_0, y_0),$$

followed by the modified approximation

$$x_1 = x_0 + \frac{h}{2}[F(t_0, x_0, y_0) + F(t_0 + h, p_1, q_1)]$$

$$y_1 = y_0 + \frac{h}{2}[G(t_0, x_0, y_0) + G(t_0 + h, p_1, q_1)].$$

At the $(k + 1)$th step, we compute $t_k = t_{k-1} + h = t_0 + kh$ and

$$p_{k+1} = x_k + hF(t_k, x_k, y_k)$$

$$q_{k+1} = y_k + hG(t_k, x_k, y_k)$$

$$x_{k+1} = x_k + \frac{h}{2}[F(t_k, x_k, y_k) + F(t_{k+1}, p_{k+1}, q_{k+1})]$$

$$y_{k+1} = y_k + \frac{h}{2}[G(t_k, x_k, y_k) + G(t_{k+1}, p_{k+1}, q_{k+1})].$$

EXAMPLE 2   Consider the second-order equation

$$\ddot{y} + y = 0; \quad y(0) = 0, \quad \dot{y}(0) = 1.$$

Reducing the equation to a first-order system by letting $\dot{y} = x$ gives

$$\dot{x} = -y, \quad x(0) = 1$$
$$\dot{y} = x, \quad y(0) = 0.$$

The recursive formulas for the basic loop have the form

$$P = X - H * Y$$

$$Q = Y - H * X$$

$$X = X + \frac{H}{2} * (-Y - Q)$$

$$Y = Y + \frac{H}{2} * (X + P).$$

The results in Table 12.2 use step size $h = 0.01$, but records only every tenth step, including also the values of $\cos t$ and $\sin t$ for comparison, because $(x(t), y(t)) = (\cos t, \sin t)$ is the correct elementary formula for the solution.

If $F(t, \mathbf{x})$ is continuously differentiable in $\mathbf{x}$, then the one-step error using the Euler method is of order h^2, where h is the step size. Using the improved Euler method, the one-step error is of order h^3, provided $F(t, \mathbf{x})$ is twice continuously differentiable in $\mathbf{x}$.

TABLE 12.2

T	X	Y	cos T	sin T
0	1	0	0.1	0.0
0.1	0.99505	0.09983	0.99500	0.09983
0.2	0.98011	0.19868	0.98006	0.19867
0.3	0.95538	0.29553	0.95533	0.29552
0.4	0.92110	0.38944	0.92105	0.38942
0.5	0.87762	0.47945	0.87757	0.47943
0.6	0.82537	0.56467	0.82533	0.56465
0.7	0.76487	0.64425	0.76483	0.64422
0.8	0.69673	0.71740	0.69669	0.71736
0.9	0.62163	0.78337	0.62159	0.78333
1	0.54031	0.84152	0.54028	0.84148
1.1	0.45360	0.89126	0.45358	0.89121
1.2	0.36235	0.93209	0.36233	0.93204
1.3	0.26749	0.96361	0.26747	0.96356
1.4	0.16995	0.98550	0.16994	0.98545
1.5	0.07071	0.99754	0.07071	0.99749
1.6	0.02922	0.99962	0.02922	0.99957

EXAMPLE 3 Recall from Section 3 that Newton's equations of planetary motion for a planet of mass m_1 orbiting a star of mass m_2 are

$$\ddot{x} = -\frac{G(m_1 + m_2)x}{(x^2 + y^2)^{3/2}}, \quad x(0) = x_0, \quad \dot{x}(0) = u_0,$$

$$\ddot{y} = -\frac{G(m_1 + m_2)y}{(x^2 + y^2)^{3/2}}, \quad y(0) = y_0, \quad \dot{y}(0) = v_0.$$

These equations are derived in the previous section, where we remarked that the second-order system is equivalent to the first-order system

$$\begin{aligned}
\dot{x} &= u, & x(0) &= x_0 \\
\dot{y} &= v, & y(0) &= y_0 \\
\dot{u} &= -\frac{G(m_1 + m_2)x}{(x^2 + y^2)^{3/2}}, & u(0) &= u_0 \\
\dot{v} &= -\frac{G(m_1 + m_2)y}{(x^2 + y^2)^{3/2}}, & v(0) &= v_0.
\end{aligned}$$

We often choose units of measurement so that $G(m_1 + m_2) = 4\pi^2$, although for some purposes we could just as well choose them so that $G(m_1 + m_2) = 1$. In choosing initial values for position and velocity, recall that we get an elliptic orbit only if the orbital speed is less than the escape speed, $v_e = \sqrt{2G(m_1 + m_2)/r}$. In other words, we want

$$\sqrt{u_0^2 + v_0^2} < \sqrt{\frac{2G(m_1 + m_2)}{x_0^2 + y_0^2}}.$$

Thus if $G(m_1 + m_2) = 1$ and $(x_0, y_0) = (1, 0)$, we should choose (u_0, v_0) so that

$$u_0^2 + v_0^2 < 2$$

TABLE 12.3

t	x	y
0	1	0
0.5	0.877588	0.479435
1.0	0.540327	0.841496
1.5	0.070792	0.997561
2.0	−0.416073	0.909448
2.5	−0.801107	0.598749
3.0	−0.990088	0.141517
3.5	−0.936777	−0.350347
4.0	−0.654219	−0.756473
4.5	−0.211552	−0.977463
5.0	0.282893	−0.959211
5.5	0.708087	−0.706161
6.0	0.959937	−0.280329

to get a closed trajectory. Table 12.3 for (y, x, y) came from using the improved Euler method, having chosen $G(m_1 + m_2) = 1$, $(x_0, y_0) = (1, 0)$, and $(u_0, v_0) = (0, 1)$. The step size was $h = 0.01$, but the result is printed only for every 50 steps.

The initial conditions we have chosen in these examples are satisfied by the solution $(x(t), y(t)) = (\cos t, \sin t)$, which has a circular orbit for its trajectory. Hence we can use this solution as a check on the accuracy of our method of numerical approximation.

EXERCISES

Software for doing these exercises is widely available; in particular the Web site **http://math.dartmouth.edu/~rewn** contains applicable Java applets, along with some graphical demonstration applets for specific applications.

1. The first order autonomous system

$$\dot{x} = y$$
$$\dot{y} = x$$

is equivalent to the single equation $\ddot{y} = y$, via the relation $\dot{y} = x$.

(a) Show that the system has solutions

$$x(t) = c_1 e^t + c_2 e^{-t},$$
$$y(t) = c_1 e^t - c_2 e^{-t}.$$

(b) Find the particular solution satisfying $x(0) = 1$, $y(0) = 2$.

(c) Compute a table of numerical approximations to the particular solution found in part (b). Do this computation on the interval $0 \le t \le \frac{1}{2}$ in steps

of size $h = 0.1$, both by computing from the explicit exponential solution formula and by a direct numerical solution of the system using either the Euler method or its improved modification.

2. Find a table of approximations to the solution $x(t)$, $y(t)$ of the system

$$\dot{x} = x + y^2$$
$$\dot{y} = x^2 + y + t,$$

with initial condition $x(0) = 1$, $y(0) = 2$. Use a step of size $h = 0.1$ on the interval $0 \le t \le \frac{1}{2}$, and make the approximation with

(a) The Euler method
(b) The improved Euler method.

3. (a) Show that the second-order equation

$$\ddot{y} - 2\dot{y} + y = t$$

with initial conditions $y(0) = 1$, $\dot{y}(0) = 2$, is equivalent to the first order system

$$\dot{x} = 2x - y + t, \quad x(0) = 2,$$
$$\dot{y} = x, \quad y(0) = 1.$$

(b) Find a numerical approximation to the solution of the system in part (a) for the interval $0 \le t \le 1$.

(c) Solve the given second-order equation by using its characteristic equation, and compare the solution with the numerical results of part (b).

4. (a) Find a first-order system equivalent to

$$y'' - ty' - y = t, \quad y(0) = 1, \quad y'(0) = 2.$$

(b) Find a numerical approximation to the solution of the system in part (a) for the interval $0 \le t \le 1$.

Apply the improved Euler method to the 1-dimensional systems in Exercises 5 to 8.

5. $\dfrac{dx}{dt} = 1 - x^{1/3}, x(0) = \frac{1}{2}$

6. $\dfrac{dy}{dt} = t^2 + y^2, y(0) = 1$

7. $\dfrac{dx}{dt} = \sqrt{1 + x^4}, x(0) = 0$

8. $\dfrac{dy}{dt} = \sin y, \; y(0) = 1$

9. Make a computer plot of the solution to the bug-pursuit problem of Exercise 34 in Section 1C.

10. The general **Lorenz system** is $\dot{x} = \sigma(y - x)$, $\dot{y} = \rho x - y - xz$, $\dot{z} = -\beta z + xy$, where β, ρ, σ are positive constants. For certain values of the parameters, in particular $\beta = 8/3$, $\rho = 28$, $\sigma = 10$, solution trajectories exhibit an often-studied type of unpredictable or "chaotic" oscillation. Plot the orbits with these parameter choices and initial value $(x, y, z) = (2, 2, 21)$. A particularly good view is obtained by projecting on the plane through the origin perpendicular to the vector $(-2, 3, 1)$. Note the effect of small changes in the initial vector on the successive numbers of circuits in each spiral configuration.

11. A basic result of multivariable calculus, proved in Chapter 5, Section 1, says that the gradient vector $\nabla f(x, y) = \big(f_x(x, y), f_y(x, y)\big)$ is perpendicular to the level curve of f that contains (x, y); consequently, $\big(- f_y(x, y), f_x(x, y)\big)$ is tangent to a level set, which is then a trajectory of the system $\dot{x} = -f_y(x, y)$, $\dot{y} = f_x(x, y)$. Assume that $f(x, y) = x^2 + \frac{1}{2}y^2$.

(a) Use these ideas to make a computer-graphics plot of the elliptic level curves of f.

(b) Make a computer-graphics plot of some **orthogonal trajectories**, that is, curves perpendicular to the level set of the function f.

(c) Identify the well-known family of orthogonal trajectory curves by solving the relevant uncoupled system analytically.

MIXING

12. Tanks 1 at capacity 100 gallons and 2 at 200 gallons are initially full of salt solution. Tank 1 has 5 gallons per minute of salt solution at 1 pound per gallon running in while mixed solution is drawn off, also at 5 gallons per minute, with an additional 3 gallons per minute flowing out to tank 2. Tank 2 has 2 gallons per minute of pure water running in and 3 gallons per minute being drawn off, while 2 gallons per minute more flow to tank 1.

(a) Find a system of differential equations satisfied by the salt contents of the tanks up to the time when one is empty.

(b) Make a computer plot that compares the graphs of the components of the solution to part (a), assuming tank 1 has initially 10 pounds of salt and tank 2 has 20 pounds. Estimate the maximum amount of salt in each tank and the time when these are attained.

13. Tanks 1 at capacity 100 gallons and 2 at 200 gallons are initially half-full of salt solution. Tank 1 has 5 gallons per minute of pure water running in while mixed solution is drawn off at 4 gallons per minute, with an additional 3 gallons per minute pumped to tank 2. Tank 2 has 2 gallons per minute of salt solution at 1 pound per gallon running in and 1 gallon per minute being drained off, while 3 gallons per minute pour into tank 1.

(a) Find a system of differential equations satisfied by the salt contents of the tanks up to the time when one is empty.

(b) Make a computer plot that compares the graphs of the components of the solution to part (a), assuming tank 1 has initially 10 pounds of salt and tank 2 has 20 pounds. Estimate the minimum amount of salt in tank 1 and the time when this is attained.

PLANETARY ORBITS

14. The system of Newton equations $\ddot{x} = -x(x^2 + y^2)^{-3/2}, \ddot{y} = -y(x^2 + y^2)^{-3/2}$ with initial conditions $x(0) = 1, y(0) = 0, \dot{x}(0) = 0, \dot{y} = v_0$, has a solution with a closed trajectory if $v_0 < \sqrt{2}$. Use the improved Euler method to make an approximate computation of the trajectory of a single orbit if

(a) $v_0 = 0.35$ **(b)** $v_0 = 0.7$ **(c)** $v_0 = 1.4$

15. When the inverse-square law $F = Gm_1m_2r^{-2}$ is replaced by $F = Gm_1m_2r^{-p}$, where $0 < p$, the Newton equations for the orbit of a single planet about a fixed sun take the form

$$\ddot{x} = -kx(x^2 + y^2)^{-(p+1)/2}, \quad \ddot{y} = -ky(x^2 + y^2)^{-(p+1)/2}.$$

Assume $k = 1$ and make a pictorial comparison of the orbits with initial conditions $x(0) = 1$, $y(0) = 0$, $\dot{x}(0) = 0$, $\dot{y}(0) = 0.5$ for the choices $p = 1.9$, $p = 2$ and $p = 2.1$. Discuss the differences among the three cases.

OSCILLATORY SYSTEMS

16. The equations $\ddot{\theta} = -(g/l)\sin\theta + \dot{\phi}^2\sin\theta\cos\theta$, $\ddot{\phi} = -2\dot{\theta}\dot{\phi}\cot\theta$, $\theta \neq k\pi$, k integer, govern the **spherical pendulum**, where ϕ and $0 < \theta < \pi$ are spherical coordinate angles where θ is measured from the *downward-pointing* vertical axis in $\mathbb{R}^3$ and ϕ is the longitudinal angle. Use $g/l = 1$.
    (a) Assuming $g/l = 1$, plot the trajectory in $\theta\phi$-space for a solution if $\theta(0) = \dot{\theta}(0) = 1$, $\phi(0) = 0$, $\dot{\phi}(0) = 1$.
    (b) Assuming $g/l = 1$, make a 3-dimensional perspective plot of the (x, y, z) path of the bob if the initial conditions are as in part (a).

17. The **van der Pol equation** is $\ddot{x} - \alpha(1 - x^2)\dot{x} + x = 0$, where α is a positive constant.
    (a) Let $y = \dot{x}$ and write the first-order system in x and y that is equivalent to the van der Pol equation with $k = 0.1$.
    (b) Use the improved Euler method to plot numerical solutions to the system found in part (a), using initial values $x(0) = 2$, $y(0) = 0$, while successively letting $\alpha = 0.1, 1.0$, and 2.0. Plot for $0 < t < t_1$, where t_1 is in each plot large enough that the trajectory of $(x(t), y(t))$ appears to be a closed loop.
    (c) Repeat the three experiments in part (b) with initial values $x(0) = 1$, $y(0) = 1$, and with $x(0) = 3$, $y(0) = 0$. The closed loops being approximated in part (b) are each an example of a **limit cycle**.

18. The nonlinear **Duffing oscillator** with periodic external driver

$$\ddot{y} + k\dot{y} - y + y^3 = A\cos\omega t, \quad y(0) = \tfrac{1}{2}, \dot{y}(0) = 0$$

is equivalent to the 2-dimensional first-order system obtained by setting $\dot{y} = z$.
    (a) Make a computer plot of the solution $y(t)$ for $0 \leq t \leq 100$ for the parameter choice $k = 0.25$, $A = 0.35$, $\omega = 1$. Compare this result graphically with what you get using $k = 0.25$, $A = 0$, $\omega = 1$ and then $k = 0$, $A = 0.35$, $\omega = 1$.
    (b) Make $(y, z) = (y, \dot{y})$ phase plots using the three sets of parameter values suggested in part (a). Estimate the apparent period τ of the solution you get in the undamped $(k = 0)$ case, indicated by the return of the phase path to its starting point.

19. Make phase plots of the **hard spring oscillator equation** $\ddot{y} = -\gamma y^3 + \delta y$ under each of the following assumptions.
    (a) $\gamma = \delta = 1$
    (b) $\gamma = 1, \delta = 2$
    (c) $\gamma = 2, \delta = 1$

20. Make solution graphs and $(y, \dot{y})$ phase plots for the periodic driven **hard spring oscillator equation** $\ddot{y} = -y - y^3 + k\dot{y} + \frac{3}{10}\cos t$ and use the results to detect long-term approach to periodic behavior.

Time-dependent linear spring mechanism. The equation $\ddot{y} + k(t)\dot{y} + h(t)y = \sin t$ represents an oscillator externally forced by $f(t) = \sin t$, damped by factor $k(t)$ and with stiffness $h(t)$. Use initial conditions $y(0) = 0$, $\dot{y}(0) = 1$ to plot solutions with the following choices for $k(t)$ and $h(t)$.

21. $k(t) = 0$, $h(t) = e^{t/10}$

22. $k(t) = 0$, $h(t) = e^{-t}$

23. $k(t) = \frac{1}{10}$, $h(t) = e^{t/10}$

24. $k(t) = (1 - e^{-t})$, $h(t) = 1$

25. $k(t) = 1$, $h(t) = 1/(1 + t^2)$

26. $k(t) = 0$, $h(t) = t/(1 + t^2)$

INTERACTING POPULATIONS

27. The special Lotka-Volterra system $\dot{H} = (3 - 2P)H$, $\dot{P} = (\frac{1}{2}H - 1)P$ is described in Exercise 6 of the previous section.
    (a) Using the initial conditions $H(0) = 3$, $P(0) = \frac{3}{2}$, compute sufficiently close approximations to the solutions $(H(T), P(t))$ so that the values nearly return to the initial values.
    (b) Sketch the graphs of $H = H(t)$ and $P = P(t)$ using the same vertical axis and the same horizontal t-axis for the time interval found in part (a).
    (c) Sketch an approximate trajectory in the HP-plane for the solution found in part (a).

28. The Lotka-Volterra system $\dot{H} = (3 - 2P)H$, $\dot{P} = (\frac{1}{2}H - 1)P$ has solutions $(H(t), P(t))$ that are periodic, because a given orbit always returns to its initial point in some finite time t_1; thus $H(t + t_1) = H(t)$ and $P(t + t_1) = P(t)$ for all t. The time period t_1 depends on the particular orbit however. Make numerical estimates of t_1 for the system on orbits with (a) $H(0) = 3$, $P(0) = \frac{1}{2}$. (b) $H(0) = 3$, $P(0) = \frac{1}{3}$, (c) $H(0) = 3$, $P(0) = 2$.

A refinement $\dot{H} = (a - bP)H(L - H)$, $\dot{P} = (cH - d)P(M - P)$, a, b, c, d positive constants, of the Lotka-Volterra equations takes account of fixed limits $L > H(t)$ and $M > P(t)$ to the growth of the host and parasite populations. Assume $L > d/c > 1$ and $M > a/b > 1$. The restrictions L and M are typically imposed by lack of sufficient habitat or food supply. Use $a = 3$, $b = d = 2$, $c = 1$ and $L = 4$, $M = 3$ to plot the trajectory of the system for

29. $H(0) = 2$, $P(0) = \frac{1}{2}$

30. $H(0) = 2$, $P(0) = \frac{3}{2}$

31. $H(0) = \frac{5}{2}$, $P(0) = 2$

32. $H(0) = \frac{3}{2}$, $P(0) = \frac{5}{2}$

33. Estimate the time it takes for an orbit to close under each of the three sets of initial conditions proposed in Exercise 28.

34. The three-species system $\dot{x} = x(3 - y)$, $\dot{y} = y(x + z - 3)$, $\dot{z} = z(2 - y)$ has equilibrium solutions at $(3, 3, 0)$ and $(0, 2, 3)$, as well as the trivial one at $(0, 0, 0)$. Use initial conditions $(x, y, z) = (0.001, 3, 2)$ to plot the following.

   (a) The orbit of $(x(t), y(t))$

   (b) The orbit of $(y(t), z(t))$

   (c) The orbit of $(x(t), z(t))$

Chapter 12 REVIEW

Solve the initial-value problems in Exercises 1 to 12.

1. $dx/dt = x^2 + 1$, $x(0) = 1$,

   $dy/dt = y$, $y(0) = 2$

2. $dx/dt = t$, $x(1) = 0$,

   $dy/dt = x + y$, $y(1) = 0$

3. $dx/dt = y + 1$, $x(0) = 1$,

   $dy/dt = x$, $y(0) = 2$

4. $dx/dt = -y$, $x(1) = 0$,

   $dy/dt = -x$, $y(1) = 0$

5. $\dot{x} = 3x - 4y$, $x(0) = 1$,

   $\dot{y} = 4x - 7y$, $y(0) = 2$

6. $\dot{x} = 3x - 5y$, $x(0) = 0$,

   $\dot{y} = x - y$, $y(0) = 1$

7. $\dot{x} = y + t$, $x(0) = 1$,

   $\dot{y} = 4x - 1$, $y(0) = 2$

8. $\dot{x} = 2$, $x(0) = 0$,

   $\dot{y} = x + y$, $y(0) = 1$

9. $\ddot{x} = -3x + 2y$, $x(0) = 3$, $\dot{x}(0) = 0$

   $\ddot{y} = 2x - 2y$, $y(0) = 3$, $\dot{y}(0) = 0$

10. $\ddot{x} = y$, $x(0) = 0$, $\dot{x}(0) = 1$,

   $\ddot{y} = x$, $y(0) = 1$, $\dot{y}(0) = 0$

11. $dx/dt = -y$, $x(0) = 0$,

   $dy/dt = x$, $y(0) = 1$,

   $dz/dt = z$, $z(0) = -1$

12. $dx/dt = t$, $x(0) = 0$,

   $dy/dt = y$, $y(0) = 0$,

   $dz/dt = y + z$, $z(0) = 1$

13. Suppose that the autonomous differential equation $\dot{\mathbf{x}} = F(\mathbf{x})$ has solution $\mathbf{x} = \mathbf{x}(t)$ satisfying $\mathbf{x}(0) = \mathbf{x}_0$ and $\mathbf{x}(1) = \mathbf{x}_1$. What, if anything, can you say about a solution to $\dot{\mathbf{x}} = -F(\mathbf{x})$?

14. Consider the system $(\dot{x}, \dot{y}, \dot{z}) = (-\omega y, \omega x, \sigma)$, where ω and σ are nonzero constants.

   (a) Without solving the system, show that the acceleration vector of a nonzero solution trajectory (i) is always perpendicular to its velocity vector, (ii) is parallel to the xy-plane, with length equal to ω^2 times the length of the corresponding position vector $(x, y, 0)$.

   (b) Find the complete solution of the system. Then sketch the trajectory passing through $(1, 0, 0)$, assuming $\omega = \sigma = 1$.

Let $f(\mathbf{x}) = f(x, y)$ be a continuously differentiable function defined on $\mathbb{R}^2$. The **gradient field** $\nabla f(x, y) = (f_x(x, y), f_y(x, y))$ generates the autonomous vector differential equation $\dot{\mathbf{x}} = \nabla f(\mathbf{x})$, called a **gradient system**.

15. Show that the nonconstant solution trajectories of a gradient system are perpendicular to the level curves $f(x, y) = C$ of f.

16. Illustrate Exercise 15 using the example $f(x, y) = x^2 + y^2$.

*17. Suppose t_0 and t_1 are arbitrary points in the domain of a solution $\mathbf{x}(t)$ of a gradient system with $\mathbf{x}(t_0) = \mathbf{x}_0$ and $\mathbf{x}(t_1) = \mathbf{x}_1$, where $\mathbf{x}_0$ and $\mathbf{x}_1$ lie on the same smooth level set of f, that is $f(\mathbf{x}_0) = f(\mathbf{x}_1)$. Show that $\int_{t_0}^{t_1} |\dot{\mathbf{x}}(t)|^2 \, dt = 0$ and hence that $\mathbf{x}(t)$ must be a constant solution.

Let $H(\mathbf{x}) = H(x, y)$ be a continuously differentiable function of two real variables. The vector field $\mathbf{H}(x, y) = \left(H_y(x, y), -H_x(x, y)\right)$ is called a **Hamiltonian field**, and the autonomous vector differential equation $\dot{\mathbf{x}} = \mathbf{H}(\mathbf{x})$, called a **Hamiltonian system**.

18. Show that the solution trajectories of a Hamiltonian system are level curves of $H(x, y)$. [*Hint:* See Exercise 15.]

19. Illustrate Exercise 18 using the example $H(x, y) = x^2 - y^2$.

20. Illustrate Exercise 18 using the example $H(x, y) = x^2 + y^2$.

CHAPTER 13

MATRIX METHODS

This chapter is about some special techniques for solving linear systems of differential equations, principally those with constant coefficients. The methods all depend on the notion of eigenvalue and eigenvector, introduced in Section 1. The results are analogous to those for a single linear constant-coefficient equation, and the methods developed for them are special cases of the eigenvector analysis in this chapter. Section 4 deals with equilibrium and stability for linear and nonlinear systems.

SECTION 1   EIGENVALUES AND EIGENVECTORS

1A   Exponential Solutions

We've discussed the vector differential equation

$$\frac{d\mathbf{x}}{dt} = A\mathbf{x}$$

in Chapter 12 for a few examples. The examples show that if A is a constant n-by-n matrix, then we can expect to find exponential solutions. For example, in the case $n = 1$, we would have $\dot{x} = ax$, with solutions of the form $x(t) = ce^{at}$. Consequently, we try solutions

$$\mathbf{x}(t) = e^{\lambda t}\mathbf{u},$$

where $\mathbf{u}$ is a constant vector in $\mathbb{R}^n$. Differentiation of $\mathbf{x}(t)$ gives

$$\frac{d\mathbf{x}}{dt} = \lambda e^{\lambda t}\mathbf{u}.$$

Since the matrix A acts linearly, we also have

$$A\mathbf{x} = e^{\lambda t} A\mathbf{u}.$$

Thus to solve the differential equation, we must have, after division by $e^{\lambda t}$,

$$A\mathbf{u} = \lambda\mathbf{u}.$$

The case $\mathbf{u} = 0$ is too trivial to be interesting, so with that possibility ruled out we define the nonzero vector $\mathbf{u}$ to be an **eigenvector** of the matrix A, and the number λ to be the corresponding **eigenvalue**. Going the other way, if $\mathbf{u}$ is an eigenvector with eigenvalue λ, then $\mathbf{x}(t) = e^{\lambda t}\mathbf{u}$ is a solution to the differential equation, and so is an arbitrary scalar multiple $ce^{\lambda t}\mathbf{u}$.

EXAMPLE 1   To solve the system

$$\begin{pmatrix} dx/dt \\ dy/dt \end{pmatrix} = \begin{pmatrix} x + y \\ 4x + y \end{pmatrix}$$

$$= \begin{pmatrix} 1 & 1 \\ 4 & 1 \end{pmatrix} \begin{pmatrix} x \\ y \end{pmatrix},$$

try to find nonzero vectors $\mathbf{u} = (u, v)$ that satisfy the eigenvector equation

$$\begin{pmatrix} 1 & 1 \\ 4 & 1 \end{pmatrix} \begin{pmatrix} u \\ v \end{pmatrix} = \lambda \begin{pmatrix} u \\ v \end{pmatrix}$$

for some number λ. In other words, try to find numbers λ such that the equation

$$\begin{pmatrix} 1 - \lambda & 1 \\ 4 & 1 - \lambda \end{pmatrix} \begin{pmatrix} u \\ v \end{pmatrix} = \begin{pmatrix} 0 \\ 0 \end{pmatrix}$$

has nonzero solutions. If the 2-by-2 matrix is invertible, then only the solution $(u, v) = (0, 0)$ exists, so we assume it *isn't* invertible. Theorem 5.7 of Chapter 2, Section 5 tells us that a square matrix A is invertible if and only if $\det A \neq 0$. Hence we require

$$\det \begin{pmatrix} 1 - \lambda & 1 \\ 4 & 1 - \lambda \end{pmatrix} = 0.$$

In other words,

$$(1 - \lambda)^2 - 4 = \lambda^2 - 2\lambda - 3$$

$$= (\lambda - 3)(\lambda + 1) = 0.$$

The only solutions are $\lambda = 3$ and $\lambda = -1$.

Case (a). $\lambda = 3$. We want nonzero vectors (u, v) such that

$$\begin{pmatrix} -2 & 1 \\ 4 & -2 \end{pmatrix} \begin{pmatrix} u \\ v \end{pmatrix} = \begin{pmatrix} 0 \\ 0 \end{pmatrix}.$$

The two numerical equations

$$-2u + v = 0$$

$$4u - 2v = 0$$

are equivalent to $v = 2u$, so we can choose $u = 1, v = 2$. Thus since $\lambda = 3$,

$$\mathbf{x}_1(t) = e^{3t} \begin{pmatrix} 1 \\ 2 \end{pmatrix}$$

is a solution and so is a numerical multiple $c_1 \mathbf{x}_1(t)$.

Case (b). $\lambda = -1$. We want nonzero vectors (u, v) such that

$$\begin{pmatrix} 2 & 1 \\ 4 & 2 \end{pmatrix} \begin{pmatrix} u \\ v \end{pmatrix} = \begin{pmatrix} 0 \\ 0 \end{pmatrix}.$$

Note that $u = 1$, $v = -2$ will do, so

$$\mathbf{x}_2(t) = e^{-t} \begin{pmatrix} 1 \\ -2 \end{pmatrix}$$

is a solution, and so is a numerical multiple of it. Thus the general solution of the vector differential equation is

$$\mathbf{x}(t) = c_1 e^{3t} \begin{pmatrix} 1 \\ 2 \end{pmatrix} + c_2 e^{-t} \begin{pmatrix} 1 \\ -2 \end{pmatrix}$$

$$= \begin{pmatrix} c_1 e^{3t} + c_2 e^{-t} \\ 2c_1 e^{3t} - 2c_2 e^{-t} \end{pmatrix}.$$

Geometric interpretation. If we take $c_2 = 0$ and make $c_1 > 0$ in the previous example, the resulting solution $c_1 e^{3t} \begin{pmatrix} 1 \\ 2 \end{pmatrix}$ traces a half-line in the xy plane that as t increases extends away from the origin in the direction of the eigenvector $\begin{pmatrix} 1 \\ 2 \end{pmatrix}$. Similarly with $c_1 = 0$ and $c_2 > 0$, we get a half line traced by $c_2 e^{-t} \begin{pmatrix} 1 \\ -2 \end{pmatrix}$ toward the origin and parallel to the eigenvector $\begin{pmatrix} 1 \\ -2 \end{pmatrix}$. More generally each solution of the system traces a curve in the xy plane that results from forming a particular linear combination of points that correspond to the same t-value. We'll now pursue this observation further. To see the geometric significance of the computation in Example 1, observe that neither of the vectors

$$\mathbf{u} = \begin{pmatrix} 1 \\ 2 \end{pmatrix}, \quad \mathbf{v} = \begin{pmatrix} 1 \\ -2 \end{pmatrix}$$

is a multiple of the other, as shown in Figure 13.1. Using the parallelogram law, a vector $\mathbf{x}$ in $\mathbb{R}^2$ is expressible using coordinates z and w relative to these vectors as

$$\mathbf{x} = z \begin{pmatrix} 1 \\ 2 \end{pmatrix} + w \begin{pmatrix} 1 \\ -2 \end{pmatrix}.$$

FIGURE 13.1
Eigenvectors **u**, **v**.

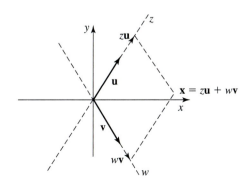

Figure 13.1 shows a typical $\mathbf{x}$ as a linear combination of $\mathbf{u}$ and $\mathbf{v}$. Thus given a vector $\mathbf{x}$, by the linearity of multiplication by A

$$A\mathbf{x} = Az \begin{pmatrix} 1 \\ 2 \end{pmatrix} + Aw \begin{pmatrix} 1 \\ -2 \end{pmatrix}$$

$$= zA \begin{pmatrix} 1 \\ 2 \end{pmatrix} + wA \begin{pmatrix} 1 \\ -2 \end{pmatrix}.$$

Since $\begin{pmatrix} 1 \\ 2 \end{pmatrix}$ and $\begin{pmatrix} 1 \\ -2 \end{pmatrix}$ are eigenvectors, with eigenvalues 3 and -1, respectively,

$$A\mathbf{x} = 3z \begin{pmatrix} 1 \\ 2 \end{pmatrix} - w \begin{pmatrix} 1 \\ -2 \end{pmatrix}.$$

In other words, A has the effect of multiplying the first vector $\begin{pmatrix} 1 \\ 2 \end{pmatrix}$ by 3 and the second vector $\begin{pmatrix} 1 \\ -2 \end{pmatrix}$ by -1. Thus expressing vectors as linear combinations of eigenvectors shows that the action of A is given by the diagonal matrix

$$B = \begin{pmatrix} 3 & 0 \\ 0 & -1 \end{pmatrix}.$$

It follows that relative to the (z, w) coordinates, the vector differential equation is

$$\begin{pmatrix} dz/dt \\ dw/dt \end{pmatrix} = \begin{pmatrix} 3 & 0 \\ 0 & -1 \end{pmatrix} \begin{pmatrix} z \\ w \end{pmatrix}.$$

So expressed in (z, w) coordinates, the system is uncoupled, that is,

$$\frac{dz}{dt} = 3z,$$

$$\frac{dw}{dt} = -w.$$

These two differential equations are particularly simple to solve, because each one involves only one unknown function. We have

$$z(t) = c_1 e^{3t}$$

$$w(t) = c_2 e^{-t},$$

using (z, w) coordinates. This shows geometrically why we can write the general solution as

$$\mathbf{x}(t) = c_1 e^{3t} \begin{pmatrix} 1 \\ 2 \end{pmatrix} + c_2 e^{-t} \begin{pmatrix} 1 \\ -2 \end{pmatrix}.$$

1B   Eigenvector Matrices

The procedure in the previous example generalizes to arbitrary dimensions. We proceed as follows to solve the n-dimensional constant-coefficient equation

$$\frac{d\mathbf{x}}{dt} = A\mathbf{x}:$$

1. Find the eigenvalues of A by finding the roots of the polynomial equation $\det(A - \lambda I) = 0$.
2. For each eigenvalue λ_k, find an eigenvector $\mathbf{u}_k$ by solving

$$(A - \lambda_k I)\mathbf{u} = 0.$$

Theorem 4.3 of Chapter 2 guarantees the existence of an eigenvalue.

3. If the solutions $e^{\lambda_1 t}\mathbf{u}_1, \ldots, e^{\lambda_n t}\mathbf{u}_n$ are linearly independent, so that none is a linear combination of the others, write the general solution

$$\mathbf{x}(t) = c_1 e^{\lambda_1 t}\mathbf{u}_1 + \cdots + c_n e^{\lambda_n t}\mathbf{u}_n.$$

We'll see shortly that if the matrix U with the $\mathbf{u}_k$ as columns is invertible then we do get the most general solution. In particular we prove in Chapter 3, Section 7B that if the eigenvalues of A are all different, then the corresponding solutions will always be linearly independent. If the solutions $e^{\lambda_k t}\mathbf{u}_k$ are linearly dependent, the procedure outlined produces solutions but not the most general one. In this case, we can use the elimination method explained in Chapter 12 or the exponential matrix method described in the next section.

If A has some complex numbers for eigenvalues, then the same method still works, with the complex exponential replacing the real exponential.

EXAMPLE 2   The system

$$\frac{dx}{dt} = x - y$$

$$\frac{dy}{dt} = x + y$$

has matrix

$$A = \begin{pmatrix} 1 & -1 \\ 1 & 1 \end{pmatrix}.$$

The eigenvalues are solutions of

$$\det \begin{pmatrix} 1 - \lambda & -1 \\ 1 & 1 - \lambda \end{pmatrix} = 0,$$

which is the same as

$$(1 - \lambda)^2 + 1 = \lambda^2 - 2\lambda + 2 = 0,$$

which has solutions $\lambda_1 = 1 + i$ and $\lambda_2 = 1 - i$.

Case (a). $\lambda = 1 + i$. The eigenvectors are the nonzero solutions of

$$\begin{pmatrix} -i & -1 \\ 1 & -i \end{pmatrix} \begin{pmatrix} u \\ v \end{pmatrix} = \begin{pmatrix} 0 \\ 0 \end{pmatrix}, \quad \text{that is, of} \quad \begin{aligned} -iu - v &= 0 \\ u - iv &= 0. \end{aligned}$$

One solution is

$$\left(\begin{array}{c} u \\ v \end{array}\right) = \left(\begin{array}{c} 1 \\ -i \end{array}\right).$$

Case (b). $\lambda = 1 - i$. The eigenvectors are the nonzero solutions of

$$\left(\begin{array}{cc} i & -1 \\ 1 & i \end{array}\right)\left(\begin{array}{c} u \\ v \end{array}\right) = \left(\begin{array}{c} 0 \\ 0 \end{array}\right), \quad \text{that is, of} \quad \begin{array}{l} iu - v = 0 \\ u + iv = 0. \end{array}$$

One solution is

$$\left(\begin{array}{c} u \\ v \end{array}\right) = \left(\begin{array}{c} 1 \\ i \end{array}\right).$$

The general solution of the differential equation is then

$$\mathbf{x}(t) = c_1 e^{(1+i)t}\left(\begin{array}{c} 1 \\ -i \end{array}\right) + c_2 e^{(1-i)t}\left(\begin{array}{c} 1 \\ i \end{array}\right)$$

$$= c_1 e^t\left(\begin{array}{c} \cos t + i \sin t \\ -i \cos t - \sin t \end{array}\right) - c_2 e^t\left(\begin{array}{c} \cos t - i \sin t \\ i \cos t + \sin t \end{array}\right)$$

$$= \left(\begin{array}{c} (c_1 + c_2)e^t \cos t + i(c_1 - c_2)e^t \sin t \\ -i(c_1 - c_2)e^t \cos t + (c_1 + c_2)e^t \sin t \end{array}\right).$$

If we rename the constants so that $c_1 + c_2 = d_1$ and $i(c_1 - c_2) = d_2$, then

$$\mathbf{x}(t) = \left(\begin{array}{c} d_1 e^t \cos t + d_2 e^t \sin t \\ -d_2 e^t \cos t + d_1 e^t \sin t \end{array}\right)$$

$$= d_1\left(\begin{array}{c} e^t \cos t \\ e^t \sin t \end{array}\right) + d_2\left(\begin{array}{c} e^t \sin t \\ -e^t \cos t \end{array}\right).$$

Because $c_1 = (d_1 - id_2)/2$ and $c_2 = (d_1 + id_2)/2$, the constants c_1 and c_2 can always be chosen so that d_1 and d_2 have arbitrary preassigned values; in particular, we can choose them to make d_1 and d_2 real numbers.

As usual, we may want to choose arbitrary constants in general solutions such as those preceding so as to satisfy prescribed initial conditions. If we assume that the eigenvectors $\mathbf{u}_1, \ldots, \mathbf{u}_n$ of the n-by-n matrix A are *linearly independent*, so that none of them is a linear combination of the others, this calculation reduces to a routine as follows. We denote by U the matrix with column vectors $\mathbf{u}_1, \ldots, \mathbf{u}_n$ in some fixed order. The matrix $U = (\mathbf{u}_1, \ldots, \mathbf{u}_n)$ is called the **eigenvector matrix** of the system. We denote by Λ_t the diagonal matrix

$$\Lambda_t = \left(\begin{array}{cccc} e^{\lambda_1 t} & 0 & \cdots & 0 \\ 0 & e^{\lambda_2 t} & \cdots & 0 \\ \vdots & \vdots & & \vdots \\ 0 & 0 & \cdots & e^{\lambda_n t} \end{array}\right),$$

with corresponding eigenvalues λ_k in the *same order* as the eigenvectors. If $\mathbf{c}$ is an arbitrary constant column vector with entries $c_1, \ldots, c_n$, we can form the vector-valued function

$$\mathbf{x}(t) = U \Lambda_t \mathbf{c}.$$

The vector $\mathbf{x}(t)$ has the form

$$\mathbf{x}(t) = U \begin{pmatrix} c_1 e^{\lambda_1 t} \\ \vdots \\ c_n e^{\lambda_n t} \end{pmatrix} = c_1 e^{\lambda_1 t} \mathbf{u}_1 + \cdots + c_n e^{\lambda_n t} \mathbf{u}_n,$$

and so is a solution of $d\mathbf{x}/dt = A\mathbf{x}$, though not the most general solution unless the vectors $\mathbf{u}_k$ are linearly independent. But if the $\mathbf{u}_k$ *are* independent, thus making U invertible, we can let $\mathbf{c} = U^{-1}\mathbf{x}_0$ for some $\mathbf{x}_0$ in $\mathbb{R}^n$. Hence

$$\mathbf{x}(t) = U \Lambda_t U^{-1} \mathbf{x}_0.$$

Since $\Lambda_0 = I$, we have $\mathbf{x}(0) = UU^{-1}\mathbf{x}_0 = \mathbf{x}_0$; so $\mathbf{x}(0) = \mathbf{x}_0$. Thus $\mathbf{x}(t)$ is the solution of the vector differential equation that satisfies $\mathbf{x}(0) = \mathbf{x}_0$. To satisfy an initial condition $\mathbf{x}(t_0) = \mathbf{x}_0$ at $t = t_0$, we let $\mathbf{c}$ in $\mathbf{x}(t) = U \Lambda_t \mathbf{c}$ equal

$$\mathbf{c} = \Lambda_{-t_0} U^{-1} \mathbf{x}_0.$$

Since Λ_t is diagonal matrix with entries $e^{\lambda_k t}$, $\Lambda_t \Lambda_{-t_0} = \Lambda_{t-t_0}$ and

$$\mathbf{x}(t) = U \Lambda_{(t-t_0)} U^{-1} \mathbf{x}_0.$$

This formula gives all solutions, because every solution has some value at t_0, and the formula assigns the arbitrary value $\mathbf{x}_0$ there. By Theorem 1.1 in Chapter 12, Section 1D, the solution is uniquely determined by the initial condition, so there is a 1-to-1 correspondence between solutions and initial conditions.

If the eigenvector matrix U isn't invertible we don't get all possible solutions this way, but we'll see in the next section that the role of $U \Lambda_t U^{-1}$ is assumed by a routinely computed *exponential matrix* that provides all solutions in the form $\mathbf{x}(t) = e^{tA}\mathbf{x}_0$.

EXAMPLE 3   In Example 1, we found for

$$A = \begin{pmatrix} 1 & 1 \\ 4 & 1 \end{pmatrix}$$

the eigenvalues $\lambda_1 = 3$, $\lambda_2 = -1$ with eigenvectors

$$\mathbf{u}_1 = \begin{pmatrix} 1 \\ 2 \end{pmatrix}, \quad \mathbf{u}_2 = \begin{pmatrix} 1 \\ -2 \end{pmatrix}.$$

We have

$$\Lambda_t = \begin{pmatrix} e^{3t} & 0 \\ 0 & e^{-t} \end{pmatrix},$$

$$U = \begin{pmatrix} 1 & 1 \\ 2 & -2 \end{pmatrix},$$

$$U^{-1} = \begin{pmatrix} \frac{1}{2} & \frac{1}{4} \\ \frac{1}{2} & -\frac{1}{4} \end{pmatrix}.$$

Thus if $\mathbf{x}_0 = (2, 3)$, the solution

$$\mathbf{x}(t) = \begin{pmatrix} 1 & 1 \\ 2 & -2 \end{pmatrix} \begin{pmatrix} e^{3t} & 0 \\ 0 & e^{-t} \end{pmatrix} \begin{pmatrix} \frac{1}{2} & \frac{1}{4} \\ \frac{1}{2} & -\frac{1}{4} \end{pmatrix} \begin{pmatrix} 2 \\ 3 \end{pmatrix}$$

$$= \begin{pmatrix} \frac{7}{4} e^{3t} + \frac{1}{4} e^{-t} \\ \frac{7}{2} e^{3t} - \frac{1}{2} e^{-t} \end{pmatrix}$$

satisfies the differential equation $d\mathbf{x}/dt = A\mathbf{x}$ and the initial condition $\mathbf{x}(0) = (2, 3)$.

To satisfy instead the initial condition $\mathbf{x}(1) = (3, 4)$, we form

$$\mathbf{x}(t) = \begin{pmatrix} 1 & 1 \\ 2 & -2 \end{pmatrix} \begin{pmatrix} e^{3(t-1)} & 0 \\ 0 & e^{-(t-1)} \end{pmatrix} \begin{pmatrix} \frac{1}{2} & \frac{1}{4} \\ \frac{1}{2} & -\frac{1}{4} \end{pmatrix} \begin{pmatrix} 3 \\ 4 \end{pmatrix}$$

$$= \begin{pmatrix} \frac{5}{2} e^{3(t-1)} + \frac{1}{2} e^{-(t-1)} \\ 5 e^{3(t-1)} - e^{-(t-1)} \end{pmatrix}.$$

EXERCISES

Find the eigenvalues and eigenvectors of the matrices in Exercises 1 to 6.

1. $\begin{pmatrix} 8 & -3 \\ 10 & -3 \end{pmatrix}$ **2.** $\begin{pmatrix} 8 & 9 \\ -4 & -4 \end{pmatrix}$

3. $\begin{pmatrix} -1 & 3 \\ -2 & -4 \end{pmatrix}$ **4.** $\begin{pmatrix} 1 & 1 & 2 \\ 0 & 1 & -1 \\ 0 & 0 & 2 \end{pmatrix}$

5. $\begin{pmatrix} 1 & 1 & 1 \\ 0 & 1 & 0 \\ 0 & 0 & 2 \end{pmatrix}$ **6.** $\begin{pmatrix} 3 & -1 & 0 \\ -1 & 2 & -1 \\ 0 & -1 & 3 \end{pmatrix}$

The 2-dimensional systems of differential equations in Exercises 7 to 10 can all be written $\dfrac{d\mathbf{x}}{dt} = A\mathbf{x}$ in which the matrix A has constant entries. In each example, find the eigenvalues of A, and for each eigenvalue find a corresponding eigenvector. Use the eigenvalues and eigenvectors of the system to write the general solution

of the system in the form $\mathbf{x}(t) = c_1 e^{\lambda_1 t} \mathbf{u}_1 + c_2 e^{\lambda_2 t} \mathbf{u}_2$. In the case of complex eigenvalues, convert the solution to real form. Then use the initial conditions to determine c_1 and c_2.

7. $\begin{pmatrix} \dfrac{dx}{dt} \\ \dfrac{dy}{dt} \end{pmatrix} = \begin{pmatrix} -3 & 2 \\ -4 & 3 \end{pmatrix} \begin{pmatrix} x \\ y \end{pmatrix}, \begin{pmatrix} x(0) \\ y(0) \end{pmatrix} = \begin{pmatrix} 1 \\ 0 \end{pmatrix}$

8. $\dfrac{dx}{dt} = 3x, \ x(0) = 1,$

$\dfrac{dy}{dt} = 2y, \ y(0) = 0$

9. $\dfrac{dx}{dt} = x + 4y, \ x(0) = 1$

$\dfrac{dy}{dt} = 5y, \ y(0) = 1$

10. $\begin{pmatrix} \dfrac{dx}{dt} \\ \dfrac{dy}{dt} \end{pmatrix} = \begin{pmatrix} 2 & -1 \\ 1 & 2 \end{pmatrix} \begin{pmatrix} x \\ y \end{pmatrix}, \begin{pmatrix} x(0) \\ y(0) \end{pmatrix} = \begin{pmatrix} 1 \\ 0 \end{pmatrix}$

11. (a) Find the general *homogeneous* solution of the system

$$\frac{d\mathbf{x}}{dt} = \begin{pmatrix} 1 & -1 & 4 \\ 3 & 2 & -1 \\ 2 & 1 & -1 \end{pmatrix} \mathbf{x} + \begin{pmatrix} 1 \\ 0 \\ 2 \end{pmatrix}.$$

(b) Find a particular solution of the system in part (a).

(c) Find the particular solution $\mathbf{x}$ of the system in part (a) that satisfies the condition $\mathbf{x}(0) = (1, 1, 2)$.

12. (a) Find the general solution of the system

$$\frac{dx}{dt} = y$$

$$\frac{dy}{dt} = -x$$

$$\frac{dz}{dt} = -z$$

by a method of your choice.

(b) What are the eigenvalues of the following matrix?

$$\begin{pmatrix} 0 & 1 & 0 \\ -1 & 0 & 0 \\ 0 & 0 & -1 \end{pmatrix}$$

13. (a) Find the general solution of the system

$$\frac{dx}{dt} = -x + y$$

$$\frac{dy}{dt} = -y.$$

(b) What are the eigenvectors of the following matrix?

$$\begin{pmatrix} -1 & 1 \\ 0 & -1 \end{pmatrix}$$

14. Show that if the eigenvectors of the n-by-n matrix A span $\mathbb{R}^n$ and the eigenvalues of A have negative real parts, then all solutions of $d\mathbf{x}/dt = A\mathbf{x}$ tend to zero as t tends to $+\infty$.

15. The second-order, constant-coefficient, differential equation

$$\frac{d^2 y}{dt^2} + a\frac{dy}{dt} + by = 0$$

is equivalent to the first-order system

$$\frac{dx}{dt} = -ax - by$$

$$\frac{dy}{dt} = x.$$

Show that the eigenvalues of the matrix

$$\begin{pmatrix} -a & -b \\ 1 & 0 \end{pmatrix}$$

are the same as the characteristic roots of the second-order differential equation.

16. Prove that a square matrix U is invertible if and only if its columns are linearly independent. [*Hint:* Row operations reduce U to I.]

17. Let U be an invertible n-by-n matrix with columns $\mathbf{u}_1, \ldots, \mathbf{u}_n$. Let D be the diagonal matrix with diagonal entries $\lambda_1, \ldots, \lambda_n$, and define the n-by-n matrix A by $A = UDU^{-1}$.

(a) Show that $A\mathbf{u}_k = \lambda_k \mathbf{u}_k$, for $k = 1, \ldots, n$.

(b) Find the 2-by-2 matrix that has eigenvalues $\mathbf{u}_1 = (2, 3)$, $\mathbf{u}_2 = (1, 1)$, and corresponding eigenvectors $\lambda_1 = 3, \lambda_2 = 1$.

(c) Show that the system $d\mathbf{x}/dt = A\mathbf{x}$ has solutions $e^{\lambda_k t}\mathbf{u}_k$, for $k = 1, \ldots, n$.

(d) Find a 2-dimensional system having

$$\mathbf{x}(t) = c_1 e^{3t} \begin{pmatrix} 2 \\ 3 \end{pmatrix} + c_2 e^t \begin{pmatrix} 1 \\ 1 \end{pmatrix}$$

as its general solution.

SECTION 2   MATRIX EXPONENTIALS

2A   Definition

To round out the theory of linear systems having constant coefficients, we use the idea of the exponential of a square matrix A, which we define using the familiar power series for the exponential function, namely

$$e^x = 1 + x + \frac{1}{2!}x^2 + \frac{1}{3!}x^3 + \cdots.$$

If A has dimensions n-by-n and I is the n-by-n identity matrix, we consider the finite sum

$$I + A + \frac{1}{2!}A^2 + \cdots + \frac{1}{k!}A^k = \sum_{j=0}^{k} \frac{1}{j!}A^j.$$

This sum of n-by-n matrices is also an n-by-n matrix. We define the **exponential** of A by

$$e^A = \lim_{k \to \infty} \sum_{j=0}^{k} \frac{1}{j!}A^j = \sum_{j=0}^{\infty} \frac{1}{j!}A^j,$$

where the existence of the matrix limit is understood to mean that the limit exists in each of the n^2 entries in the matrix. It's sometimes convenient to use the notation $\exp A$ for e^A. For example, if

$$A = \begin{pmatrix} 2 & 0 \\ 0 & 3 \end{pmatrix}, \quad A^2 = \begin{pmatrix} 2^2 & 0 \\ 0 & 3^2 \end{pmatrix}, \dots, A^j = \begin{pmatrix} 2^j & 0 \\ 0 & 3^j \end{pmatrix},$$

then

$$\exp \begin{pmatrix} 2 & 0 \\ 0 & 3 \end{pmatrix} = \lim_{k \to \infty} \sum_{j=0}^{k} \frac{1}{j!} \begin{pmatrix} 2^j & 0 \\ 0 & 3^j \end{pmatrix}$$

$$= \begin{pmatrix} \sum_{j=0}^{\infty} \dfrac{2^j}{j!} & 0 \\ 0 & \sum_{j=0}^{\infty} \dfrac{3^j}{j!} \end{pmatrix} = \begin{pmatrix} e^2 & 0 \\ 0 & e^3 \end{pmatrix}.$$

It's remarkable that the exponential of a square matrix always exists and has many of the properties of the ordinary real or complex exponential function. In the matrix exponential's most useful form, A is multiplied by a scalar t that we can pull out of the powers $(tA)^j$ to give $t^j A^j$. The most important properties of e^{tA} are as follows.

2.1   Theorem.   If A is an n-by-n real or complex matrix, the matrix series

$$\sum_{j=0}^{\infty} \frac{t^j}{j!}A^j$$

converges to an n-by-n matrix e^{tA} satisfying

(a) $e^{(t+s)A} = e^{tA}e^{sA} = e^{sA}e^{tA}$ for scalars t and s

(b) e^{tA} is invertible, and $e^{-tA}e^{tA} = e^{tA}e^{-tA} = I$

(c) $\dfrac{d}{dt}e^{tA} = Ae^{tA} = e^{tA}A$

Proof.   If $A = (a_{ij})$, choose a positive number b such that $|a_{ij}| \leq b$ for $i, j = 1, \dots, n$. Since the entries in A^2 are of the form

$$a_{i1}a_{1j} + \cdots + a_{in}a_{nj},$$

it follows that they are all at most nb^2 in absolute value. Proceeding inductively, the entries in A^k are at most $n^{k-1}b^k$ in absolute value. It follows that each entry in e^A is defined by an absolutely convergent infinite series dominated by the convergent series

$$1 + b + \frac{nb^2}{2!} + \cdots + \frac{1}{j!}n^{j-1}b^j + \cdots,$$

as discussed in Chapter 14, Section 3D. Hence all the entries exist and e^A is defined. These estimates show that if the entries a_{ij} in A are replaced by the entries ta_{ij} in tA, then the convergence is uniform on every bounded interval $c \le t \le d$. (See Chapter 14, Section 4.)

To prove property (a), we apply the binomial theorem to $(t+s)^j$ to get

$$e^{(t+s)A} = \sum_{j=0}^{\infty} \frac{(t+s)^j A^j}{j!} = \sum_{j=0}^{\infty} \frac{1}{j!} \left[\sum_{l=0}^{j} \binom{j}{l} t^l s^{j-l} A^j \right].$$

Since $\binom{j}{l} = j!/l!(j-l)!$, we can cancel $j!$ to get

$$e^{(t+s)A} = \sum_{j=0}^{\infty} \left[\sum_{l=0}^{j} \frac{t^l A^l s^{j-l} A^{j-l}}{l!(j-l)!} \right].$$

This last sum is just the product of the two absolutely convergent series that represent e^{tA} and e^{sA} respectively. Since s and t commute in the original series, we also get the product in the other order.

Property (b) follows from (a) on taking $t = 1$ and $s = -1$. Since $e^0 = I$, property (a) implies $e^A e^{-A} = e^{-A} e^A = I$.

Formally we can compute the derivative of e^{tA} from the definition by

$$\frac{d}{dt} e^{tA} = \frac{d}{dt} \sum_{j=0}^{\infty} \frac{t^j A^j}{j!}$$

$$= \sum_{j=1}^{\infty} \frac{j t^{j-1} A^j}{j!}$$

$$= A \sum_{j=1}^{\infty} \frac{t^{j-1} A^{j-1}}{(j-1)!} = A e^{tA}.$$

Note that the factor A could be taken out on the right just as well as on the left. This computation using term-by-term differentiation of series is justified because the differentiated series in each entry is uniformly convergent by the estimates made in the first part of the proof. See Chapter 14, Section 4.   ∎

2B   Solving Systems

The simplest justification for introducing e^{tA} is to show that

$$\mathbf{x}(t) = e^{tA} \mathbf{x}_0$$

defines the solution of

$$\frac{d\mathbf{x}}{dt} = A\mathbf{x}, \quad \mathbf{x}(0) = \mathbf{x}_0.$$

First, to differentiate the vector $e^{tA}\mathbf{x}_0$, we need only differentiate each entry in the matrix e^{tA}, because $\mathbf{x}_0$ has constant entries. Hence

$$\frac{d}{dt}e^{tA}\mathbf{x}_0 = Ae^{tA}\mathbf{x}_0$$

by part (c) of the previous theorem. Thus the differential equation is satisfied. Second, to show that the initial condition is satisfied, note that

$$\mathbf{x}(0) = I\mathbf{x}_0 = \mathbf{x}_0.$$

More generally,

$$\mathbf{x}(t) = e^{(t-t_0)A}\mathbf{x}_0$$

satisfies the initial condition $\mathbf{x}(t_0) = \mathbf{x}_0$, since $e^{0A} = I$. It's one of the nice features of e^{tA} that it always exists and that the equation $\mathbf{x}(t) = e^{tA}\mathbf{c}$ provides *all* solutions of $\dfrac{d\mathbf{x}}{dt} = A\mathbf{x}$ as the constant vector $\mathbf{c}$ ranges over all constant vectors with the same dimension as $\mathbf{x}$; this will follow from Theorem 3.2 in Section 3A.

EXAMPLE 1   We can write the system

$$\frac{dx}{dt} = x + y$$
$$\frac{dy}{dt} = y$$

in matrix form $d\mathbf{x}/dt = A\mathbf{x}$ as

$$\begin{pmatrix} dx/dt \\ dy/dt \end{pmatrix} = \begin{pmatrix} 1 & 1 \\ 0 & 1 \end{pmatrix}\begin{pmatrix} x \\ y \end{pmatrix}.$$

We compute

$$A = \begin{pmatrix} 1 & 1 \\ 0 & 1 \end{pmatrix}, \quad A^2 = \begin{pmatrix} 1 & 2 \\ 0 & 1 \end{pmatrix}, \dots, A^k = \begin{pmatrix} 1 & k \\ 0 & 1 \end{pmatrix}, \dots.$$

Then

$$e^{tA} = \sum_{k=0}^{\infty}\frac{t^k}{k!}\begin{pmatrix} 1 & k \\ 0 & 1 \end{pmatrix} = \begin{pmatrix} \sum_{k=0}^{\infty}\frac{t^k}{k!} & \sum_{k=1}^{\infty}\frac{t^k}{(k-1)!} \\ 0 & \sum_{k=0}^{\infty}\frac{t^k}{k!} \end{pmatrix}$$

$$= \begin{pmatrix} e^t & te^t \\ 0 & e^t \end{pmatrix}.$$

Hence the solution with initial conditions $x(0) = x_0$, $y(0) = y_0$ is

$$\begin{pmatrix} x(t) \\ y(t) \end{pmatrix} = \begin{pmatrix} e^t & te^t \\ 0 & e^t \end{pmatrix} \begin{pmatrix} x_0 \\ y_0 \end{pmatrix}$$

$$= \begin{pmatrix} x_0 e^t + y_0 te^t \\ y_0 e^t \end{pmatrix}.$$

To find the solution with initial conditions $x(t_0) = x_0$, $y(t_0) = y_0$, just replace t by $(t - t_0)$ everywhere in the vector solution for $t_0 = 0$; this is simpler than recomputing undetermined coefficients and it shows one of the many advantages of using the exponential matrix e^{tA}.

2C   Relationship to Eigenvectors

The connection between exponential solutions and the eigenvector method of the previous section is as follows. If the eigenvectors of the square matrix A are linearly independent, we form the eigenvector matrix

$$U = (\mathbf{u}_1, \mathbf{u}_2, \dots, \mathbf{u}_n)$$

with these vectors as columns. We also form the diagonal matrix

$$\Lambda^t = \begin{pmatrix} e^{\lambda_1 t} & \cdots & 0 \\ \vdots & & \\ 0 & \cdots & e^{\lambda_n t} \end{pmatrix},$$

where λ_k is the eigenvalue of $\mathbf{u}_k$. Then we have seen that

$$\mathbf{x}(t) = U \Lambda_t U^{-1} \mathbf{x}_0$$

solves the initial value problem for the equation $d\mathbf{x}/dt = A\mathbf{x}$. Since

$$\mathbf{x}(t) = e^{tA} \mathbf{x}_0$$

solves the same problem, we are faced with the question of whether the two solutions are the same. By Theorem 1.1 in Chapter 12, Section 1D, there is only one solution satisfying $\mathbf{x}(0) = \mathbf{x}_0$. (Exercise 18 in Section 3 indicates a direct proof.) Hence $e^{tA}\mathbf{x}_0 = U\Lambda_t U^{-1}\mathbf{x}_0$ for all t. Since $\mathbf{x}_0$ is arbitrary, it follows that

2.2 $$e^{tA} = U \Lambda_t U^{-1}.$$

EXAMPLE 2

In Example 1 of the previous section, we solved the system

$$\begin{pmatrix} dx/dt \\ dy/dt \end{pmatrix} = \begin{pmatrix} 1 & 1 \\ 4 & 1 \end{pmatrix} \begin{pmatrix} x \\ y \end{pmatrix}$$

by finding the eigenvalues $\lambda_1 = 3$, $\lambda_2 = -1$ and corresponding eigenvectors $(1, 2)$, $(1, -2)$ of the 2-by-2 matrix A of the system. Thus

$$U = \begin{pmatrix} 1 & 1 \\ 2 & -2 \end{pmatrix}, \quad \Lambda_t = \begin{pmatrix} e^{3t} & 0 \\ 0 & e^{-t} \end{pmatrix}, \quad U^{-1} = \begin{pmatrix} \frac{1}{2} & \frac{1}{4} \\ \frac{1}{2} & -\frac{1}{4} \end{pmatrix}$$

and

$$e^{tA} = U \Lambda_t U^{-1} = \begin{pmatrix} \frac{1}{2}e^{3t} + \frac{1}{2}e^{-t} & \frac{1}{4}e^{3t} - \frac{1}{4}e^{-t} \\ e^{3t} - e^{-t} & \frac{1}{2}e^{3t} + \frac{1}{2}e^{-t} \end{pmatrix}.$$

As a check on the computation, notice that $e^{tA} = I$ when $t = 0$. This example shows that if the eigenvectors of A are linearly independent, it may well be easier to use them to compute e^{tA} than to use the matrix power series definition.

Equation 2.2 is ineffective for computing e^{tA} if the eigenvector matrix U fails to have an inverse. For example if $A = \begin{pmatrix} 1 & 1 \\ 0 & 1 \end{pmatrix}$ there is only a single repeated eigenvalue $\lambda = 1$, and all corresponding eigenvectors have the form $\begin{pmatrix} u \\ 0 \end{pmatrix}$ with $u \neq 0$. Hence the only possibility for U is a matrix of the form $\begin{pmatrix} u_1 & u_2 \\ 0 & 0 \end{pmatrix}$, which is never invertible. Section 2D provides an alternative method for computing e^{tA} that's often more efficient.

EXERCISES

In Exercises 1 to 6, find the exponential e^{tA} of the matrix A by first computing the successive terms $I, tA, t^2 A^2/2! \ldots$ in the series definition.

1. $A = \begin{pmatrix} -1 & 0 \\ 0 & 1 \end{pmatrix}$ **2.** $A = \begin{pmatrix} 0 & 1 \\ 1 & 0 \end{pmatrix}$

3. $A = \begin{pmatrix} 0 & 1 \\ 0 & 1 \end{pmatrix}$ **4.** $A = \begin{pmatrix} i & 0 \\ 0 & -i \end{pmatrix}$

5. $A = \begin{pmatrix} 1 & 0 & 1 \\ 0 & 1 & 0 \\ 0 & 0 & 1 \end{pmatrix}$ **6.** $A = \begin{pmatrix} 2 & 0 \\ 0 & 3 \end{pmatrix}$

7. In Example 1 of the text, we showed that

$$\exp t \begin{pmatrix} 1 & 1 \\ 0 & 1 \end{pmatrix} = \begin{pmatrix} e^t & te^t \\ 0 & e^t \end{pmatrix}.$$

Verify directly for this example that

(a) $\exp \left(t \begin{pmatrix} 1 & 1 \\ 0 & 1 \end{pmatrix} \right)$ and $\exp \left(-t \begin{pmatrix} 1 & 1 \\ 0 & 1 \end{pmatrix} \right)$ are inverse to one another.

(b) $\exp \left(t \begin{pmatrix} 1 & 1 \\ 0 & 1 \end{pmatrix} \right) \exp \left(s \begin{pmatrix} 1 & 1 \\ 0 & 1 \end{pmatrix} \right)$
$= \exp \left((t+s) \begin{pmatrix} 1 & 1 \\ 0 & 1 \end{pmatrix} \right).$

(c) $\dfrac{d}{dt} \exp \left(t \begin{pmatrix} 1 & 1 \\ 0 & 1 \end{pmatrix} \right) = \begin{pmatrix} 1 & 1 \\ 0 & 1 \end{pmatrix}$
$\exp \left(t \begin{pmatrix} 1 & 1 \\ 0 & 1 \end{pmatrix} \right).$

(d) Find the solution of the initial-value problem

$$\begin{pmatrix} \dfrac{dx}{dt} \\ \dfrac{dy}{dt} \end{pmatrix} = \begin{pmatrix} 1 & 1 \\ 0 & 1 \end{pmatrix} \begin{pmatrix} x \\ y \end{pmatrix},$$

$$\begin{pmatrix} x(0) \\ y(0) \end{pmatrix} = \begin{pmatrix} -1 \\ 2 \end{pmatrix}.$$

In Exercise 8 to 10, find the square matrix e^{tA} in terms of the general identity matrix I.

8. $A = I$ **9.** $A = 2I$ **10.** $A = -I$

In Exercise 11 to 19, compute e^{tA} for each matrix A using Equation 2.2. Then find the inverse matrix e^{-tA}, and check your original computation by showing that the derivative of e^{tA} at $t = 0$ is equal to A.

11. $A = \begin{pmatrix} -3 & 2 \\ -4 & 3 \end{pmatrix}$ **12.** $A = \begin{pmatrix} 1 & 4 \\ 0 & 5 \end{pmatrix}$

13. $A = \begin{pmatrix} 2 & -1 \\ 1 & 2 \end{pmatrix}$ 14. $A = \begin{pmatrix} 5 & 3 \\ 2 & 1 \end{pmatrix}$

15. $A = \begin{pmatrix} 4 & 5 \\ 3 & 4 \end{pmatrix}$ 16. $A = \begin{pmatrix} -1 & -2 & -2 \\ 0 & 1 & -2 \\ 0 & 0 & 2 \end{pmatrix}$

17. $A = \begin{pmatrix} 0 & 1 \\ -6 & 5 \end{pmatrix}$ 18. $A = \begin{pmatrix} -4 & 4 \\ -6 & 6 \end{pmatrix}$

19. $A = \begin{pmatrix} -1 & 0 & 0 \\ 0 & \frac{3}{2} & -\frac{1}{2} \\ 0 & -\frac{1}{2} & \frac{3}{2} \end{pmatrix}$

20. **(a)** Use the method of elimination to find the general solution of the system

$$\begin{pmatrix} \dfrac{dx}{dt} \\ \dfrac{dy}{dt} \end{pmatrix} = \begin{pmatrix} 9 & -4 \\ 4 & 1 \end{pmatrix} \begin{pmatrix} x \\ y \end{pmatrix}.$$

(b) Use the result of part (a) to compute the matrix e^{tA} where

$$A = \begin{pmatrix} 9 & -4 \\ 4 & 1 \end{pmatrix}.$$

[*Hint:* Find solutions such that $\mathbf{x}_1(0) = \mathbf{e}_1$ and $\mathbf{x}_2(0) = \mathbf{e}_2$.]

(c) Show that the eigenvectors of A are linearly dependent.

21. Let A be an n-by-n matrix with real entries. The matrix e^{itA} is defined by

$$e^{itA} = \sum_{k=0}^{\infty} \frac{1}{k!}(i)^k t^k A^k.$$

Define $\cos tA$ to be the real part of the series and $\sin tA$ to be the imaginary part, so that

$$e^{itA} = \cos tA + i \sin tA.$$

Show that the matrices $\cos tA$ and $\sin tA$ satisfy

(a) $\cos(-tA) = \cos tA$, $\sin(-tA) = -\sin tA$

(b) $\dfrac{d}{dt} \cos tA = -A \sin tA$, $\dfrac{d}{dt} \sin tA = A \cos tA$

[*Hint:* Express $\cos tA$ and $\sin tA$ in terms of e^{itA}.]

(c) $(\cos tA)^2 + (\sin tA)^2 = I$, where I is the n-by-n identity matrix

22. Let A be the 2-by-2 matrix

$$\begin{pmatrix} 1 & 1 \\ 0 & 1 \end{pmatrix}.$$

Define $\cos tA$ and $\sin tA$ as in Exercise 21, and verify the formulas given in (a), (b), and (c).

23. Show that if A is an n-by-n matrix, then a system of the form

$$\frac{d^2\mathbf{x}}{dt^2} + A^2\mathbf{x} = 0$$

has solutions of the form

$$\mathbf{x}(t) = (\cos tA)\mathbf{c}_1 + (\sin tA)\mathbf{c}_2,$$

where $\mathbf{c}_1$ and $\mathbf{c}_2$ are constant n-dimensional vectors, and $\cos tA$ and $\sin tA$ are the n-by-n matrices defined in Exercise 21. Is this always the most general solution?

24. Let A be the n-by-n matrix with all entries equal to 1.

(a) Show that $A^2 = nA$, and more generally that $A^k = n^{k-1}A$ for integer $k \geq 1$.

(b) Show that $e^{tA} = I + \dfrac{1}{n}(e^{nt} - 1)A$.

(c) Find the four entries in e^{tA} when $n = 2$.

25. It turns out that if $AB = BA$, then $e^A e^B = e^{A+B}$. Find noncommuting 2-by-2 matrices for which this last equation fails.

2D   Computing e^{tA} in Practice

The method we describe here is usually simpler than appealing directly to the definition of e^{tA} or first finding an eigenvector matrix as described in the previous Section 2C. Furthermore, the method doesn't depend on knowing the eigenvectors of A, and it works whether A has independent eigenvalues or not. Our aim is to show that an n by n exponential matrix always reduces to a polynomial $P(A)$:

2.3

$$e^{tA} = \sum_{j=0}^{n-1} b_j(t)A^j,$$

where the coefficient functions $b_k(t)$ contain the eigenvalues $\lambda_1, \dots, \lambda_n$ of A explicitly. The next theorem shows how to compute the coefficients by solving a system of linear equations. The complete proof of the theorem is complicated, so we'll just sketch it. There is a complete proof in Chapter 7, Section 2 of *Introduction to Differential Equations*, 2nd ed., by Richard Williamson, McGraw-Hill (2001).

2.4   Theorem.   The coefficient functions $b_k(t)$ in the matrix Equation 2.3 satisfy the linear scalar equations

$$\textbf{(i)} \qquad e^{t\lambda_k} = \sum_{j=0}^{n-1} b_j(t)\lambda_k^j, \quad k = 1, \dots, n.$$

If some m of the eigenvalues λ_k are equal, say $\lambda_1 = \dots = \lambda_m$, then the following $m - 1$ additional relations hold:

$$\textbf{(ii)} \qquad \frac{d^k}{d\lambda^k} e^{t\lambda} = \frac{d^k}{d\lambda^k} \sum_{j=0}^{n-1} b_j(t)\lambda^j, \quad \text{at} \quad \lambda = \lambda_1, \quad \text{for } k = 1, \dots, m - 1.$$

Sketch of Proof.   We'll assume Equation 2.3 holds for some choice of the coefficients $b_j(t)$. Let $\mathbf{v}_k$ be an eigenvector of A corresponding to eigenvalue $\lambda_k : A\mathbf{v}_k = \lambda_k\mathbf{v}_k$, $\mathbf{v}_k \neq \mathbf{0}$. Apply the matrix sum on the right side of Equation 2.3 to $\mathbf{v}_k$, noting that $A^j\mathbf{v}_k = \lambda_k^j\mathbf{v}_k$:

$$e^{tA}\mathbf{v}_k = \sum_{j=0}^{n-1} b_j(t)A^j\mathbf{v}_k = \left(\sum_{j=0}^{n-1} b_j(t)\lambda_k^j\right)\mathbf{v}_k.$$

Similarly, apply the matrix e^{tA} to $\mathbf{v}_k$ to get another expression for the same thing:

$$e^{tA}\mathbf{v}_k = \lim_{N\to\infty} \sum_{j=0}^{N} \frac{t^j}{j!} A^j\mathbf{v}_k = \sum_{j=0}^{\infty} \frac{t^j}{j!} \lambda_k^j\mathbf{v}_k = e^{t\lambda_k}\mathbf{v}_k.$$

Since $\mathbf{v}_k$ is an eigenvector, it isn't zero, so the coefficients of $\mathbf{v}_k$ at the ends of the previous two displayed lines must be equal, that is, Equations (i) hold.

   If there are multiple eigenvalues we alter the entries of A slightly to produce matrices A_h with distinct eigenvalues whose limit as $h \to 0$ is A. Equations (ii) follow by calculating an appropriate limit of a difference quotient as $h \to 0$.   ∎

EXAMPLE 3

We saw in Example 1 of Section 1 that the matrix $A = \begin{pmatrix} 1 & 1 \\ 4 & 1 \end{pmatrix}$ had eigenvalues $\lambda_1 = -1, \lambda_2 = 3$. Equations (1) of Theorem 2.4 are then

$$e^{-t} = b_0(t) - b_1(t)$$

$$e^{3t} = b_0(t) + 3b_1(t).$$

Solve for $b_0(t)$ and $b_1(t)$ to get $b_1(t) = -\frac{1}{4}e^{-t} + \frac{1}{4}e^{3t}$, $b_0(t) = \frac{3}{4}e^{-t} + \frac{1}{4}e^{3t}$. Plugging these coefficient functions into Equation 2.3 gives

$$e^{tA} = \left(\tfrac{3}{4}e^{-t} + \tfrac{1}{4}e^{3t}\right) \begin{pmatrix} 1 & 0 \\ 0 & 1 \end{pmatrix} + \left(-\tfrac{1}{4}e^{-t} + \tfrac{1}{4}e^{3t}\right) \begin{pmatrix} 1 & 1 \\ 4 & 1 \end{pmatrix}$$

$$= \begin{pmatrix} \frac{1}{2}e^{-t} + \frac{1}{2}e^{3t} & -\frac{1}{4}e^{-t} + \frac{1}{4}e^{3t} \\ -e^{-t} + e^{3t} & \frac{1}{2}e^{-t} + \frac{1}{2}e^{3t} \end{pmatrix}.$$

As a partial check on the accuracy of our computation we can verify that our expression for e^{tA} equals the identity matrix I when $t = 0$.

EXAMPLE 4   In Example 2 of Section 1 we saw that the matrix $\begin{pmatrix} 1 & -1 \\ 1 & 1 \end{pmatrix}$ has eigenvalues $\lambda_1 = 1 + i$ and $\lambda_2 = 1 - i$. Equations (i) of Theorem 2.4 are then

$$e^{(1+i)t} = b_0(t) + b_1(t)(1 + i)$$

$$e^{(1-i)t} = b_0(t) + b_1(t)(1 - i).$$

Subtracting the second equation from the first gives

$$2ib_1(t) = e^t(e^{it} - e^{-it}), \quad \text{so } b_1(t) = \frac{1}{2i}e^t(e^{it} - e^{-it}) = e^t \sin t.$$

Substituting for $b_1(t)$ in the first of the equations above containing b_0,

$$b_0(t) = -(1 + i)e^t \sin t + e^t(\cos t + i \sin t) = e^t(\cos t - \sin t).$$

Then

$$\exp\left(t\begin{pmatrix} 1 & -1 \\ 1 & 1 \end{pmatrix}\right) = e^t(\cos t - \sin t)\begin{pmatrix} 1 & 0 \\ 0 & 1 \end{pmatrix} + e^t \sin t \begin{pmatrix} 1 & -1 \\ 1 & 1 \end{pmatrix}$$

$$= \begin{pmatrix} e^t \cos t & -e^t \sin t \\ e^t \sin t & e^t \cos t \end{pmatrix}.$$

EXAMPLE 5   The matrix $A = \begin{pmatrix} 0 & 1 \\ -4 & -4 \end{pmatrix}$ has for its characteristic equation $\lambda^2 + 4\lambda + 4 = 0$, with eigenvalues $\lambda_1 = \lambda_2 = -2$. Equations (i) of Theorem 2.4 are identical,

$$e^{t\lambda_1} = b_0(t) + b_1(t)\lambda_1 \quad \text{or} \quad e^{-2t} = b_0(t) - 2b_1(t).$$

Hence we need another equation satisfied by the coefficients. As suggested by Theorem 2.4, we differentiate formally with respect to λ_1 in the first equation above and then set $\lambda_1 = -2$ to get

$$te^{t\lambda_1} = b_1(t) \quad \text{or} \quad te^{-2t} = b_1(t).$$

We see right away that $b_0(t) = e^{-2t} + 2te^{-2t}$. Equation 2.3 then becomes

$$e^{tA} = b_0(t)\begin{pmatrix} 1 & 0 \\ 0 & 1 \end{pmatrix} + b_1(t)\begin{pmatrix} 0 & 1 \\ -4 & -4 \end{pmatrix}$$

$$= (1 + 2t)e^{-2t}\begin{pmatrix} 1 & 0 \\ 0 & 1 \end{pmatrix} + te^{-2t}\begin{pmatrix} 0 & 1 \\ -4 & -4 \end{pmatrix}$$

$$= e^{-2t}\begin{pmatrix} 1 + 2t & t \\ -4t & 1 - 2t \end{pmatrix}.$$

EXAMPLE 6

Let $A = \begin{pmatrix} 2 & 1 & 0 \\ 0 & 2 & 1 \\ 0 & 0 & 2 \end{pmatrix}$. The characteristic equation is $(2 - \lambda)^3 = 0$, so there is a triple root $\lambda_1 = 2$. Equations (i) of Theorem 2.4 reduce to

$$e^{t\lambda_1} = b_0(t) + b_1(t)\lambda_1 + b_2(t)\lambda_1^2 \quad \text{or} \quad e^{2t} = b_0(t) + 2b_1(t) + 4b_2(t).$$

We get two additional relations among the $b_k(t)$ by differentiating the first equation above twice with respect to λ_1 and then setting $\lambda_1 = 2$ after each differentiation:

$$te^{2t} = b_1(t) + 4b_2(t), \quad t^2 e^{2t} = 2b_2(t).$$

Solving the last three displayed equations for the $b_k(t)$ gives

$$b_2(t) = \tfrac{1}{2}t^2 e^{2t}, \quad b_1(t) = (t - 2t^2)e^{2t}, \quad b_0(t) = (1 - 2t + 2t^2)e^{2t}.$$

Equation 2.3 is then

$$e^{tA} = (1 - 2t + 2t^2)e^{2t} \begin{pmatrix} 1 & 0 & 0 \\ 0 & 1 & 0 \\ 0 & 0 & 1 \end{pmatrix}$$

$$+ (t - 2t^2)e^{2t} \begin{pmatrix} 2 & 1 & 0 \\ 0 & 2 & 1 \\ 0 & 0 & 2 \end{pmatrix} + \tfrac{1}{2}t^2 e^{2t} \begin{pmatrix} 4 & 4 & 1 \\ 0 & 4 & 4 \\ 0 & 0 & 4 \end{pmatrix}$$

$$= e^{2t} \begin{pmatrix} 1 & t & \tfrac{1}{2}t^2 \\ 0 & 1 & t \\ 0 & 0 & 1 \end{pmatrix}.$$

The multiple-eigenvalue case in the proof we gave for Theorem 2.4 is incomplete. However the following remarkable algebraic result makes the theorem plausible and leads to a complete proof. This theorem allows in principle for the possibility of collapsing the infinite series for e^{tA} into a finite sum by replacing powers of A higher than $n - 1$ by lower powers as in Equation 2.3. The next example works out the 2-by-2 case.

2.5 Cayley–Hamilton Theorem. If A is an n-by-n matrix with characteristic polynomial

$$P(\lambda) = \det(A - \lambda I),$$

then the matrix polynomial obtained by substituting A^k for λ^k in $P(\lambda)$ satisfies $P(A) = O$, with the understanding that $A^0 = I$ replaces $\lambda^0 = 1$ in the substitution.

The theorem, which we won't prove, is often stated briefly as "a square matrix satisfies its own characteristic equation."

EXAMPLE 7   Suppose $A = \begin{pmatrix} a & b \\ c & d \end{pmatrix}$. The characteristic polynomial of A is

$$\det \begin{pmatrix} a - \lambda & b \\ c & d - \lambda \end{pmatrix} = \lambda^2 - (a+d)\lambda + (ad - bc).$$

The Cayley–Hamilton theorem asserts that $A^2 - (a+d)A + (ad - bc)I = O$, or $A^2 = (a+d)A - (ad - bc)I$. Multiplying this last equation by A gives

$$A^3 = (a+d)\, A^2 - (ad - bc)\, A$$
$$= (a+d)((a+d)\, A - (ad - bc)\, I) - (ad - bc)\, A$$
$$= \big((a+d)^2 - (ad - bc)\big) A - (a+d)(ad - bc)I.$$

Continuing as in the previous example, we see that a power of a 2-by-2 matrix A, and hence a polynomial $p(A)$ of degree $n \geq 2$, equals a first-degree polynomial: $p(A) = \alpha A + \beta I$. The main difficulty in proving Theorem 2.4 is that we're dealing with an infinite sum that defines e^{tA}. Nevertheless Theorem 2.4 implies that the successive additions to coefficient terms in $I, A, \dots, A^{n-1}$ converge to sums $b_0, \dots, b_{n-1}$ that we compute by the routine of Theorem 2.4.

If a square matrix A is invertible, then the Cayley–Hamilton Theorem allows us to write A^{-1} as a polynomial in A. For if

$$a_0 I + a_1 A + \cdots + a_{n-1} A^{n-1} \pm A^n = 0,$$

is the characteristic equation of A with λ replaced by A, we can multiply by A^{-1} to get

$$a_0 A^{-1} + a_1 I + \cdots + a_{n-1} A^{n-2} + A^{n-1} = 0.$$

Now solve for A^{-1}, noting that $a_0 = \det A \neq 0$.

EXAMPLE 8   The matrix $A = \begin{pmatrix} 2 & 3 & 1 \\ 0 & 2 & 2 \\ 0 & 0 & 2 \end{pmatrix}$ has characteristic equation

$$(2 - \lambda)^3 = 8 - 12\lambda + 6\lambda^2 - \lambda^3 = 0,$$

so $8I - 12A + 6A^2 - A^3 = 0$. Multiply by A^{-1} to get

$$8A^{-1} - 12I + 6A - A^2 = 0 \quad \text{or} \quad 8A^{-1} = 12I - 6A + A^2.$$

Hence to find A^{-1} we need only divide by 8 after computing

$$8A^{-1} = 12 \begin{pmatrix} 1 & 0 & 0 \\ 0 & 1 & 0 \\ 0 & 0 & 1 \end{pmatrix} - 6 \begin{pmatrix} 2 & 3 & 1 \\ 0 & 2 & 2 \\ 0 & 0 & 2 \end{pmatrix} + \begin{pmatrix} 4 & 12 & 10 \\ 0 & 4 & 8 \\ 0 & 0 & 4 \end{pmatrix}$$

$$= \begin{pmatrix} 4 & -6 & 4 \\ 0 & 4 & -4 \\ 0 & 0 & 4 \end{pmatrix}.$$

EXERCISES

In Exercises 1 to 4, solve the initial-value problem $\dot{\mathbf{x}} = A\mathbf{x}$ for the given matrix A and corresponding initial condition by first finding e^{tA}.

1. $\begin{pmatrix} 8 & -3 \\ 10 & -3 \end{pmatrix}$; $\mathbf{x}(0) = \begin{pmatrix} 2 \\ -3 \end{pmatrix}$

2. $\begin{pmatrix} 8 & 9 \\ -4 & -4 \end{pmatrix}$; $\mathbf{x}(1) = \begin{pmatrix} 1 \\ 1 \end{pmatrix}$

3. $\begin{pmatrix} 1 & 1 & 2 \\ 0 & 1 & -1 \\ 0 & 0 & 2 \end{pmatrix}$; $\mathbf{x}(0) = \begin{pmatrix} 0 \\ 1 \\ 0 \end{pmatrix}$

4. $\begin{pmatrix} 1 & 1 & 1 \\ 0 & 1 & 0 \\ 0 & 0 & 2 \end{pmatrix}$; $\mathbf{x}(0) = \begin{pmatrix} 2 \\ 1 \\ 1 \end{pmatrix}$

In Exercises 5 and 6, find e^{tA} for the given matrix A.

5. $A = \begin{pmatrix} 1 & 1 & 1 & 0 \\ 0 & 0 & -1 & 0 \\ 1 & 1 & 2 & 0 \\ 1 & 0 & -1 & 1 \end{pmatrix}$

6. $A = \begin{pmatrix} 1 & 0 & 0.5 & -0.5 \\ 1 & 0 & -1 & 0 \\ 0 & 2 & 2.5 & 0.5 \\ -1 & 1 & 0.5 & 1.5 \end{pmatrix}$

7. Find the appropriate exponential matrix and use it to solve

$$\frac{dx}{dt} = 2x \qquad + z,$$
$$\frac{dy}{dt} = -x + 3y \quad + z,$$
$$\frac{dz}{dt} = -x \qquad + 4z.$$

(Note that $\lambda = 3$ is a triple eigenvalue.)

8. Solve by whatever method seems simplest:

$$\frac{dx}{dt} = 2x + z,$$
$$\frac{dy}{dt} = y + w,$$
$$\frac{dz}{dt} = 2z + w,$$
$$\frac{dw}{dt} = -y + w.$$

9. Let $A = \begin{pmatrix} \alpha & 1 \\ 0 & \beta \end{pmatrix}$ where α, β are real numbers.
   (a) Show that if $\alpha \neq \beta$, then A has two linearly independent eigenvectors.
   (b) Show that if $\alpha = \beta$, then the only eigenvectors of A are of the form $\mathbf{u} = \begin{pmatrix} c \\ 0 \end{pmatrix}$, $c \neq 0$.
   (c) Compute e^{tA} in each of the two cases.

10. How should Equation 2.3 be interpreted when $n = 1$?

11. Use the definition of e^{tA} as a matrix power series to show that $e^{tI} = e^t I$.

12. Theorem 2.4 shows that the coefficients $b_k(t)$ in an expansion

$$e^{tA} = \sum_{k=0}^{n-1} b_k(t) A^k$$

are completely determined by the eigenvalues of the n-by-n matrix A. For example, the matrices $\begin{pmatrix} 1 & \beta \\ 0 & 2 \end{pmatrix}$, $\begin{pmatrix} 2 & \beta \\ 0 & 1 \end{pmatrix}$ both have characteristic polynomial with 1 and 2 as roots. Hence the $b_k(t)$ are the same for all these matrices regardless of the value of the entry β.
   (a) Compute $b_0(t)$ and $b_1(t)$ for the two matrices above, and use them to find the corresponding exponential matrices, each depending on the parameter β.
   (b) Compute the exponential matrix for $\begin{pmatrix} \alpha & \beta \\ 0 & \alpha \end{pmatrix}$.

13. Use Theorem 2.5 to compute A^2 and A^3 if $A = \begin{pmatrix} 1 & 2 \\ 3 & 4 \end{pmatrix}$.

14. Verify the Cayley–Hamilton Theorem for the matrix $\begin{pmatrix} 0 & 1 \\ -4 & -4 \end{pmatrix}$.

In Exercise 15 to 23, use the Cayley–Hamilton theorem to find the inverse matrix if the given matrix is invertible.

15. $\begin{pmatrix} 1 & 0 & 0 \\ 3 & 1 & 5 \\ -2 & 0 & 1 \end{pmatrix}$

16. $\begin{pmatrix} 1 & 2 & 3 \\ -1 & 1 & 0 \\ 0 & 3 & 3 \end{pmatrix}$

17. $\begin{pmatrix} 2 & 4 & 8 \\ 1 & 0 & 0 \\ 1 & -3 & -7 \end{pmatrix}$

18. $\begin{pmatrix} t & 0 & 0 \\ 0 & 2 & 0 \\ 0 & 0 & 1 \end{pmatrix}$, t real

19. $\begin{pmatrix} 1 & 2 & 1 \\ 0 & 0 & 1 \\ 0 & 0 & 3 \end{pmatrix}$ **20.** $\begin{pmatrix} 1 & -1 & 1 \\ 0 & -1 & 1 \\ 0 & 0 & 1 \end{pmatrix}$ **23.** $\begin{pmatrix} 1 & 0 & 0 \\ 0 & e^t & te^t \\ 0 & 0 & e^t \end{pmatrix}$, t real

21. $\begin{pmatrix} 1 & 2 & -1 & 3 \\ 0 & 2 & 0 & 1 \\ 0 & 0 & 1 & 1 \\ 0 & 0 & 0 & 4 \end{pmatrix}$ **22.** $\begin{pmatrix} 1 & 0 & 1 & 0 \\ 0 & 2 & 0 & 0 \\ 0 & 0 & 3 & 0 \\ 0 & 0 & 0 & 4 \end{pmatrix}$

2E   Independent Solutions

The discussion of this section shows that there is an exponential solution formula $\mathbf{x}(t) = e^{tA}\mathbf{c}$ for every equation $d\mathbf{x}/dt = A\mathbf{x}$ in which A is a constant square matrix. In the example we had

$$e^{tA}\mathbf{c} = \begin{pmatrix} e^t & te^t \\ 0 & e^t \end{pmatrix} \begin{pmatrix} c_1 \\ c_2 \end{pmatrix}$$

$$= c_1 \begin{pmatrix} e^t \\ 0 \end{pmatrix} + c_2 \begin{pmatrix} te^t \\ e^t \end{pmatrix}.$$

Since we want different solutions for every different choice of c_1, c_2, it's important to avoid the redundancy that would occur in the formula in case one of the two columns in the matrix is a constant multiple of the other. We'll see that this cannot happen in general, but to state the general result we need to look more closely at what is meant by linear independence of vector functions. Let $\mathbf{x}_1(t)$, $\mathbf{x}_2(t)$, ... , $\mathbf{x}_m(t)$ be n-dimensional column vectors whose entries are functions on some common interval $a < t < b$. (It's not ruled out that some or all of the entries may happen to be constant.) Vector functions $\mathbf{x}_k(t)$, $k = 1, \ldots, m$ defined on a t-interval are said to be **linearly independent** if whenever

$$c_1\mathbf{x}_1(t) + c_2\mathbf{x}_2(t) + \cdots + c_m\mathbf{x}_m(t) = 0$$

for all t, then the constant coefficients c_k are all zero. When we have only two functions ($m = 2$), asserting linear independence is the same as saying that neither function is a constant multiple of the other. The reason is that if either c_1 or c_2 is not zero we could divide by it and express one vector as a multiple of the other. Similarly, if we have $m > 2$ vector functions, their linear independence means that none of them is equal to a sum of scalar multiples, or **linear combination**. of the others. The negation of linear independence of a set of vectors is called **linear dependence**, and it means simply that at least one of the vectors is a linear combination of the others.

| EXAMPLE 9 | The vector functions

$$\mathbf{x}_1(t) = \begin{pmatrix} e^t \\ 0 \end{pmatrix}, \quad \mathbf{x}_2 = \begin{pmatrix} te^t \\ e^t \end{pmatrix}$$

that form the exponential matrix in Example 1 are linearly independent. For

$$c_1 \begin{pmatrix} e^t \\ 0 \end{pmatrix} + c_2 \begin{pmatrix} te^t \\ e^t \end{pmatrix} = \begin{pmatrix} 0 \\ 0 \end{pmatrix}, \quad -\infty < t < \infty$$

is the same as

$$c_1 e^t + c_2 t e^t = 0$$
$$c_2 e^t = 0.$$

It follows that $c_2 = 0$. Hence $c_1 = 0$. This conclusion holds for a given value of t, so in particular the constant vectors

$$\begin{pmatrix} 1 \\ 0 \end{pmatrix}, \quad \begin{pmatrix} 0 \\ 1 \end{pmatrix}$$

are linearly independent. Just set $t = 0$.

EXAMPLE 10   Consider the vector functions

$$\mathbf{x}_1(t) = \begin{pmatrix} e^t \\ 0 \\ 0 \end{pmatrix}, \quad \mathbf{x}_2(t) = \begin{pmatrix} 0 \\ e^t \\ te^t \end{pmatrix}, \quad \mathbf{x}_3(t) = \begin{pmatrix} 0 \\ e^t \\ t^2 e^t \end{pmatrix}.$$

The check for independence for $-\infty < t < \infty$ is to solve

$$c_1 \begin{pmatrix} e^t \\ 0 \\ 0 \end{pmatrix} + c_2 \begin{pmatrix} 0 \\ e^t \\ te^t \end{pmatrix} + c_3 \begin{pmatrix} 0 \\ e^t \\ t^2 e^t \end{pmatrix} = \begin{pmatrix} 0 \\ 0 \\ 0 \end{pmatrix}$$

for c_1, c_2, c_3. This is the same as

$$c_1 e^t = 0$$
$$c_2 e^t + c_3 e^t = 0$$
$$c_2 t e^t + c_3 t^2 e^t = 0.$$

The first equation shows that $c_1 = 0$. The middle equation implies $c_2 = -c_3$, so the last equation says $c_2 t - c_2 t^2 = 0$ for all t. Thus $c_2 = c_3 = 0$, so the vector functions are independent as defined on an interval $a < t < b$. Note, however, that when $t = 0$ we get

$$\mathbf{x}_1(0) = \begin{pmatrix} 1 \\ 0 \\ 0 \end{pmatrix}, \quad \mathbf{x}_2(0) = \begin{pmatrix} 0 \\ 1 \\ 0 \end{pmatrix}, \quad \mathbf{x}_3(0) = \begin{pmatrix} 0 \\ 1 \\ 0 \end{pmatrix},$$

and these constant vectors are linearly dependent. This shows that functions may be linearly independent while their restrictions to some smaller domain (in this example a single point) may be linearly dependent.

Here is a theorem that guarantees independence of the columns of an exponential matrix for all values of t, and hence the existence of an independent set of solutions for $\dot{\mathbf{x}} = A\mathbf{x}$ for every constant square matrix A.

2.6   Theorem.   Let A be an n-by-n matrix with constant entries and let $\mathbf{x}_k(t)$ be the kth column of the exponential matrix e^{tA}. Then the vector functions $\mathbf{x}_1(t), \ldots, \mathbf{x}_n(t)$ are linearly independent over an arbitrary set of t-values.

Proof.   Apply the matrix e^{-tA} to both sides of the vector equation

$$c_1\mathbf{x}_1(t) + \cdots + c_n\mathbf{x}_n(t) = 0.$$

Using the distributivity of matrix multiplication, we get

$$c_1 e^{-tA}\mathbf{x}_1(t) + \cdots + c_n e^{-tA}\mathbf{x}_n(t) = 0.$$

But e^{-tA} is the inverse of the matrix whose kth column is $\mathbf{x}_k(t)$. Hence $e^{-tA}\mathbf{x}_k(t)$ is the kth column of the identity matrix I. Thus our equation becomes

$$c_1\mathbf{e}_1 + \cdots + c_k\mathbf{e}_k + \cdots + c_n\mathbf{e}_n = 0,$$

where $\mathbf{e}_k$ is the column vector with 1 in the kth entry and 0 elsewhere. Adding up the linear combination gives

$$\begin{pmatrix} c_1 \\ \vdots \\ c_k \\ \vdots \\ c_n \end{pmatrix} = \begin{pmatrix} 0 \\ \vdots \\ 0 \\ \vdots \\ 0 \end{pmatrix}.$$

So all $c_k = 0$ and the vectors $\mathbf{x}_k(t)$ are linearly independent.   ∎

EXERCISES

Each of the given matrices in Exercises 1 to 4 is the exponential matrix of some constant matrix A. For each matrix, find A by computing the derivative of e^{tA} at $t = 0$. Then express the vector function $\mathbf{x}(t) = e^{tA}\mathbf{c}$ as a linear combination of the columns of e^{tA} and verify that each column of e^{tA} is a solution of $\dot{\mathbf{x}} = A\mathbf{x}$.

1. $e^{tA} = \begin{pmatrix} 1 & 0 \\ 0 & e^t \end{pmatrix}$ **2.** $e^{tA} = \begin{pmatrix} e^t & 0 \\ 0 & e^{2t} \end{pmatrix}$

3. $e^{tA} = \begin{pmatrix} e^t & 0 & 0 \\ 0 & e^{2t} & 0 \\ 0 & 0 & e^{3t} \end{pmatrix}$

4. $e^{tA} = \begin{pmatrix} e^t & te^t & 0 \\ 0 & e^t & 0 \\ 0 & 0 & e^{2t} \end{pmatrix}$

Not every square matrix with linearly independent columns is an exponential matrix. For example, an exponential matrix e^{tA} must equal I when $t = 0$ and properties (a) and (b) of Theorem 2.1 must hold. In Exercises 5 to 8 show that the matrix has linearly independent columns, but is not an exponential matrix.

5. $\begin{pmatrix} e^t & 1 \\ 0 & e^t \end{pmatrix}$ **6.** $\begin{pmatrix} e^{2t} & t^2 e^t \\ 0 & e^t \end{pmatrix}$

7. $\begin{pmatrix} e^t & te^{2t} \\ 0 & e^t \end{pmatrix}$ **8.** $\begin{pmatrix} e^t & t \\ 0 & e^t \end{pmatrix}$

9. Theorem 2.6 is a simple consequence of the following more general theorem: If $A(t)$ is an invertible square matrix for each t in some interval $a < t < b$, then the columns of $A(t)$ are linearly independent vector functions on that interval. Show how to prove this theorem using the ideas in the proof of Theorem 2.6.

10. Let D be the n-by-n diagonal matrix with entries $d_1, \ldots, d_n$ on the main diagonal and zeros elsewhere. Show that e^{tD} is the diagonal matrix with entries $e^{d_1 t}, \ldots, e^{d_n t}$.

11. Prove that a set $\{\mathbf{x}_1(t), \mathbf{x}_2(t), \ldots, \mathbf{x}_m(t)\}$ of vector-valued functions of the same dimension n is linearly independent on an interval $a \leq t \leq b$ if and only if no one of them is a linear combination of the others.

SECTION 3   NONHOMOGENEOUS SYSTEMS

3A   Solution Formula

To develop efficient methods for solving nonhomogeneous systems, we need a formula for the derivative of a matrix product, or, more particularly, the product of a matrix and a vector. The rule is similar to the usual product rule for derivatives:

3.1
$$\frac{d}{dt}[A(t)B(t)] = \left[\frac{d}{dt}A(t)\right]B(t) + A(t)\left[\frac{d}{dt}B(t)\right].$$

However, the order of the factors on the right is important because it involves matrix multiplication, which is not in general commutative. To prove the formula, we differentiate one entry at a time on the left side; the derivative of the ijth entry is

$$\frac{d}{dt}\sum_{k=0}^{n} a_{ik}(t)b_{kj}(t) = \sum_{k=0}^{n} \frac{da_{ik}}{dt}(t)b_{kj}(t) + \sum_{k=0}^{n} a_{ik}(t)\frac{db_{kj}}{dt}(t).$$

But this is just the ijth entry in the sum of products of matrices on the right, so the formula is proved. In our first application $B(t)$ will be a column vector $\mathbf{x}(t)$.

The proof of the next theorem is formally just an application of the exponential multiplier method of Chapter 10, Section 3A.

3.2   Theorem.   The vector differential equation

$$\frac{d\mathbf{x}}{dt} = A\mathbf{x} + \mathbf{b}(t),$$

where A is a constant matrix and $\mathbf{b}(t)$ is a continuous function on some interval has for its general solution

$$\mathbf{x}(t) = e^{tA}\int e^{-tA}\mathbf{b}(t)\,dt + e^{tA}\mathbf{c},$$

where $\mathbf{c}$ is an arbitrary constant vector. In particular, the homogeneous equation $\dfrac{d\mathbf{x}}{dt} = A\mathbf{x}$ has $\mathbf{x}(t) = e^{tA}\mathbf{c}$ for its general solution.

Proof.   We rewrite the differential equation as $\dfrac{d\mathbf{x}}{dt} - A\mathbf{x} = \mathbf{b}(t)$, then multiply through by the matrix e^{-tA} to get

$$e^{-tA}\frac{d\mathbf{x}}{dt} - e^{-tA}A\mathbf{x} = e^{-tA}\mathbf{b}(t).$$

By Equation 3.1, the product rule for differentiation of matrices, this is the same as

$$\frac{d}{dt}(e^{-tA}\mathbf{x}) = e^{-tA}\mathbf{b}(t).$$

Integration of both sides gives

$$e^{-tA}\mathbf{x} = \int e^{-tA}\mathbf{b}(t)\,dt + \mathbf{c}.$$

Since e^{tA} is the inverse of e^{-tA}, we can multiply through by e^{tA} to get

$$\mathbf{x}(t) = e^{tA}\int e^{-tA}\mathbf{b}(t)\,dt + e^{tA}\mathbf{c}. \qquad \blacksquare$$

EXAMPLE 1   In the first example of the previous section, we saw that the system

$$\begin{pmatrix} dx/dt \\ dy/dt \end{pmatrix} = \begin{pmatrix} 1 & 1 \\ 0 & 1 \end{pmatrix}\begin{pmatrix} x \\ y \end{pmatrix}$$

had associated with it the matrix

$$e^{tA} = \begin{pmatrix} e^t & te^t \\ 0 & e^t \end{pmatrix}.$$

Hence to find a particular solution of

$$\begin{pmatrix} dx/dt \\ dy/dt \end{pmatrix} = \begin{pmatrix} 1 & 1 \\ 0 & 1 \end{pmatrix}\begin{pmatrix} x \\ y \end{pmatrix} + \begin{pmatrix} e^t \\ e^{-t} \end{pmatrix},$$

we compute the particular solution

$$
\begin{aligned}
e^{tA}\int e^{-tA}\begin{pmatrix} e^t \\ e^{-t} \end{pmatrix}dt &= \begin{pmatrix} e^t & te^t \\ 0 & e^t \end{pmatrix}\int \begin{pmatrix} e^{-t} & -te^{-t} \\ 0 & e^{-t} \end{pmatrix}\begin{pmatrix} e^t \\ e^{-t} \end{pmatrix}dt \\
&= \begin{pmatrix} e^t & te^t \\ 0 & e^t \end{pmatrix}\int \begin{pmatrix} 1 - te^{-2t} \\ e^{-2t} \end{pmatrix}dt \\
&= \begin{pmatrix} e^t & te^t \\ 0 & e^t \end{pmatrix}\begin{pmatrix} t + \frac{1}{2}te^{-2t} + \frac{1}{4}e^{-2t} \\ -\frac{1}{2}e^{-2t} \end{pmatrix} \\
&= \begin{pmatrix} te^t + \frac{1}{4}e^{-t} \\ -\frac{1}{2}e^{-t} \end{pmatrix}.
\end{aligned}
$$

Adding the particular solution just found to the general homogeneous solution we already had gives

$$
\begin{aligned}
\mathbf{x}(t) &= \begin{pmatrix} e^t & te^t \\ 0 & e^t \end{pmatrix}\begin{pmatrix} c_1 \\ c_2 \end{pmatrix} + \begin{pmatrix} te^t + \frac{1}{4}e^{-t} \\ -\frac{1}{2}e^{-t} \end{pmatrix} \\
&= \begin{pmatrix} c_1 e^t + c_2 t e^t + te^t + \frac{1}{4}e^{-t} \\ c_2 e^t - \frac{1}{2}e^{-t} \end{pmatrix}
\end{aligned}
$$

for the general solution.

3B   Variation of Parameters
Even in the case of an n-by-n matrix $A(t)$ with nonconstant continuous entries there is a formula for a particular solution of

$$\frac{d\mathbf{x}}{dt} = A(t)\mathbf{x} + \mathbf{b}(t)$$

in terms of solutions of the related homogeneous equation. Suppose $\mathbf{x}_1(t), \ldots, \mathbf{x}_n(t)$ is a set of n linearly independent solutions of

$$\frac{d\mathbf{x}}{dt} = A(t)\mathbf{x}.$$

We now form the n-by-n **fundamental matrix**

$$X(t) = \left(\mathbf{x}_1(t) \ldots \mathbf{x}_n(t)\right)$$

whose columns are these independent vector solutions.

EXAMPLE 2   If A is a constant matrix, then the matrix $X(t) = e^{tA}$ is an example of a fundamental matrix, because its columns are linearly independent solutions of $d\mathbf{x}/dt = A\mathbf{x}$. In Example 1, a fundamental matrix is

$$X_1(t) = e^{tA} = \begin{pmatrix} e^t & te^t \\ 0 & e^t \end{pmatrix}.$$

Another fundamental matrix for the same system is

$$X_2(t) = \begin{pmatrix} te^t & e^t \\ e^t & 0 \end{pmatrix},$$

although $X_2(t)$ is not an exponential matrix, because $X_2(0) \neq I$.

Having found a fundamental matrix $X(t)$, we try to find a vector-valued function $\mathbf{v}(t)$ such that

$$\mathbf{x}_p(t) = X(t)\mathbf{v}(t)$$

is a solution of the nonhomogeneous equation. It turns out that this can always be done as follows. Using the product rule for differentiation, we substitute $X(t)\mathbf{v}(t)$ into the nonhomogeneous equation to get

$$\frac{dX(t)}{dt}\mathbf{v}(t) + X(t)\frac{d\mathbf{v}(t)}{dt} = A(t)X(t)\mathbf{v}(t) + \mathbf{b}(t).$$

Since each column of $X(t)$ is a solution of the homogeneous equation, we have

$$\frac{dX(t)}{dt} = A(t)X(t).$$

Therefore, the first term cancels on each side, leaving

$$X(t)\frac{d\mathbf{v}(t)}{dt} = \mathbf{b}(t).$$

Since the columns of $X(t)$ are independent as vector functions, it follows (see Exercise 16.) that these columns are independent vectors for each fixed t. Hence the inverse matrix $X^{-1}(t)$ exists, and multiplying by it gives

$$\frac{d\mathbf{v}(t)}{dt} = X^{-1}(t)\mathbf{b}(t).$$

Integration gives the formula for $\mathbf{v}(t)$:

$$\mathbf{v}(t) = \int X^{-1}(t)\mathbf{b}(t)\,dt.$$

Finally,

3.3
$$\mathbf{x}_p(t) = X(t)\mathbf{v}(t)$$
$$= X(t)\int X^{-1}(t)\mathbf{b}(t)\,dt.$$

Notice that this formula is the same as the one previously derived in the constant-coefficient case, with e^{tA} now replaced by the more general $X(t)$. This process for finding $\mathbf{x}_p$ is sometimes called **variation of parameters**, because to find it we replace the constant vector $\mathbf{v}_0$ in the homogeneous vector solution $X(t)\mathbf{v}_0$ by a function that varies with t.

EXAMPLE 3

It's routine to verify that the homogeneous system associated with

$$\frac{d\mathbf{x}}{dt} = \begin{pmatrix} 1 & e^t \\ 0 & 1 \end{pmatrix}\mathbf{x} + \begin{pmatrix} e^t \\ e^{2t} \end{pmatrix}$$

has independent solutions

$$\mathbf{x}_1(t) = \begin{pmatrix} e^t \\ 0 \end{pmatrix}, \quad \mathbf{x}_2(t) = \begin{pmatrix} e^{2t} \\ e^t \end{pmatrix}.$$

We form a fundamental matrix $X(t)$ and its inverse:

$$X(t) = \begin{pmatrix} e^t & e^{2t} \\ 0 & e^t \end{pmatrix}, \quad X^{-1}(t) = \begin{pmatrix} e^{-t} & -1 \\ 0 & e^{-t} \end{pmatrix}.$$

Formula 3.3 gives the particular solution

$$\mathbf{x}_p(t) = \begin{pmatrix} e^t & e^{2t} \\ 0 & e^t \end{pmatrix}\int \begin{pmatrix} e^{-t} & -1 \\ 0 & e^{-t} \end{pmatrix}\begin{pmatrix} e^t \\ e^{2t} \end{pmatrix} dt$$
$$= \begin{pmatrix} e^t & e^{2t} \\ 0 & e^t \end{pmatrix}\int \begin{pmatrix} 1 - e^{2t} \\ e^t \end{pmatrix} dt$$
$$= \begin{pmatrix} e^t & e^{2t} \\ 0 & e^t \end{pmatrix}\begin{pmatrix} t - \frac{1}{2}e^{2t} \\ e^t \end{pmatrix} = \begin{pmatrix} te^t + \frac{1}{2}e^{3t} \\ e^{2t} \end{pmatrix}.$$

The general solution is then

$$\mathbf{x}(t) = c_1\mathbf{x}_1(t) + c_2\mathbf{x}_2(t) + \mathbf{x}_p(t).$$

3C   Summary of Methods

For linear systems in the standard form $d\mathbf{x}/dt = A\mathbf{x} + \mathbf{b}$, and hence for systems and equations reducible to this form, we usually proceed as follows:

1. Find the general solution of the homogeneous equation $dx/dt = Ax$, either by elimination, by the eigenvector method, if applicable, or by finding e^{tA} directly. In the constant-coefficient case the homogeneous solution is always of the form $x_h(t) = e^{tA}c$, where c is a constant vector.

2. Find a particular solution to the nonhomogeneous equation, either as a by-product of the elimination method, by undetermined coefficients, if applicable, or by Formula 3.2 or 3.3.

3. Write the general solution as $x(t) = x_h(t) + x_p(t)$.

If A isn't constant there is no general method for finding $x_h(t)$, and we'll very likely have to use numerical methods.

EXERCISES

In Exercises 1 to 4, use Equation 3.2 to solve the initial-value problem of the form $dx/dt = Ax + b(t)$, $x(t_0) = x_0$. The associated homogeneous equations $dx/dt = Ax$ were found in Exercises 11, 12, and 13 for Section 2A to 2C to have the exponential matrices e^{tA} needed here in Exercises 2, 3, and 4.

1. $\left(\begin{array}{c} \dfrac{dx}{dt} \\ \dfrac{dy}{dt} \end{array} \right) = \left(\begin{array}{cc} 3 & 0 \\ 0 & 2 \end{array} \right) \left(\begin{array}{c} x \\ y \end{array} \right) + \left(\begin{array}{c} e^t - 1 \\ e^{-t} \end{array} \right)$;

$\left(\begin{array}{c} x(0) \\ y(0) \end{array} \right) = \left(\begin{array}{c} -1 \\ -1 \end{array} \right)$

2. $\left(\begin{array}{c} \dfrac{dx}{dt} \\ \dfrac{dy}{dt} \end{array} \right) = \left(\begin{array}{cc} -3 & 2 \\ -4 & 3 \end{array} \right) \left(\begin{array}{c} x \\ y \end{array} \right) + \left(\begin{array}{c} e^{2t} \\ 1 \end{array} \right)$;

$\left(\begin{array}{c} x(1) \\ y(1) \end{array} \right) = \left(\begin{array}{c} 0 \\ 0 \end{array} \right)$

3. $\left(\begin{array}{c} \dfrac{dx}{dt} \\ \dfrac{dy}{dt} \end{array} \right) = \left(\begin{array}{cc} 1 & 4 \\ 0 & 5 \end{array} \right) \left(\begin{array}{c} x \\ y \end{array} \right) + \left(\begin{array}{c} 1 \\ e^t \end{array} \right)$;

$\left(\begin{array}{c} x(0) \\ y(0) \end{array} \right) = \left(\begin{array}{c} 0 \\ 1 \end{array} \right)$

4. $\left(\begin{array}{c} \dfrac{dx}{dt} \\ \dfrac{dy}{dt} \end{array} \right) = \left(\begin{array}{cc} 2 & -1 \\ 1 & 2 \end{array} \right) \left(\begin{array}{c} x \\ y \end{array} \right) + \left(\begin{array}{c} e^{2t} \\ 2e^{2t} \end{array} \right)$;

$\left(\begin{array}{c} x(0) \\ y(0) \end{array} \right) = \left(\begin{array}{c} -1 \\ -2 \end{array} \right)$

5. (a) Show that for a solution of the form $x(t) = e^{tA}c$, where c is a constant vector, to satisfy the condition $x(t_0) = x_0$, we must have $c = e^{-t_0 A}x_0$.

(b) Show that if $X(t)$ is an n-by-n matrix with linearly independent columns, in particular if $X(t)$ is a fundamental matrix, then in order for

$$x(t) = X(t)c, \quad c \text{ constant}$$

to satisfy $x(t_0) = x_0$, we must have $c = X^{-1}(t_0)x_0$. Exercise 19 shows that $X(t_0)$ is invertible.

6. Let $X(t)$ be a fundamental matrix whose columns span the set of solutions of the homogeneous equation

$$\frac{dx}{dt} = A(t)x.$$

(a) Show that if $x_p(t)$ is a particular solution of the nonhomogeneous system

$$\frac{dx}{dt} = A(t)x + b(t),$$

then the general solution of the nonhomogeneous system is

$$x(t) = x_p(t) + X(t)c,$$

where c is a constant.

(b) Show that if the general solution in part (a) is to satisfy an initial condition $x(t_0) = x_0$, then c should be chosen so that

$$c = X^{-1}(t_0)(x_0 - x_p(t_0)).$$

For Exercises 7 to 10, consider the systems in Exercises 1 to 4 respectively, which are solvable by the method of undetermined coefficients: Form linear combinations of the terms, and their derivatives, that occur in each entry of the nonhomogeneous part of the differential equation, taking care to include appropriate multiples by powers of t for terms that are also homogeneous solutions. Then substitute into the equation to determine the coefficients of combination. In Exercises 7 to 10, use this method on the corresponding system in Exercises 1 to 4.

In Exercises 11 and 12, the system has the *homogeneous* solutions shown. Verify that these are linearly independent solutions. Find a particular solution of the nonhomogeneous equation, using Equation 3.3.

11. $\begin{pmatrix} \dfrac{dx}{dt} \\ \dfrac{dy}{dt} \end{pmatrix} = \begin{pmatrix} \dfrac{3}{2t} & -\dfrac{1}{2} \\ -\dfrac{1}{2t^2} & \dfrac{1}{2t} \end{pmatrix} \begin{pmatrix} x \\ y \end{pmatrix} + \begin{pmatrix} t^3 \\ 2t^2 \end{pmatrix};$

$\begin{pmatrix} t \\ 1 \end{pmatrix}, \begin{pmatrix} -t^2 \\ t \end{pmatrix}$

12. $\begin{pmatrix} \dfrac{dx}{dt} \\ \dfrac{dy}{dt} \end{pmatrix} = \begin{pmatrix} \dfrac{t}{t-1} & \dfrac{-1}{t-1} \\ 1 & 0 \end{pmatrix} \begin{pmatrix} x \\ y \end{pmatrix} + \begin{pmatrix} 1-t \\ 1-t^2 \end{pmatrix};$

$\begin{pmatrix} 1 \\ t \end{pmatrix}, \begin{pmatrix} e^t \\ e^t \end{pmatrix}$

13. (a) Let $x_1(t), \dots, x_n(t)$ be continuously differentiable functions taking values in $\mathbb{R}^n$, and forming a linearly independent set of vectors for each t. Let $X(t)$ be the n-by-n matrix with columns $x_1(t), \dots, x_n(t)$. Show that if we define

$$A(t) = X'(t)X^{-1}(t),$$

then the system $dx/dt = A(t)x$ has $x_1(t), \dots, x_n(t)$ as solutions and thus has $X(t)$ as a fundamental matrix.

(b) Find a first-order homogeneous linear system of the form $dx/dt = A(t)x$ having

$$x_1(t) = \begin{pmatrix} e^t \\ 2e^{2t} \end{pmatrix}, \quad x_2(t) = \begin{pmatrix} 1 \\ e^t \end{pmatrix}$$

as solutions. Are these two solutions linearly independent?

14. Let A be an n-by-n invertible matrix of constants, and let $\mathbf{b}$ be a fixed vector in $\mathbb{R}^n$. Show that the equation

$$\frac{dx}{dt} = Ax + b$$

always has $x_p = -A^{-1}b$ for a particular solution.

15. Show that if A is a constant n-by-n matrix, then the equation

$$\frac{dx}{dt} = Ax$$

has $X(t) = e^{tA}$ for its fundamental matrix of independent column solutions with $X(0) = I$.

In Exercises 16 and 17, $A(t)$ is a square matrix with entries differentiable on $a \le t < b$.

16. (a) Show that if $A(t)$ and $dA(t)/dt$ commute, then
$$\frac{dA^2(t)}{dt} = 2A(t)\frac{dA(t)}{dt}.$$

(b) Generalize part (a) to $\dfrac{dA^k(t)}{dt} = kA^{k-1}(t)\dfrac{dA(t)}{dt}$.

17. Use the previous exercise to show that $\dfrac{de^{A(t)}}{dt} = e^{A(t)}\dfrac{dA(t)}{dt}$.

18. Modify the derivation of Formula 3.2 to show that, for A constant and $\mathbf{b}(t)$ continuous, the initial-value problem

$$\frac{dx}{dt} = Ax + b(t), \quad x(t_0) = x_0,$$

has a unique solution of the form

$$x(t) = e^{tA}\int_{t_0}^{t} e^{-uA}b(u)\,du + e^{(t-t_0)A}x_0.$$

[*Hint:* Integrate from t_0 to t instead of using an indefinite integral.]

*19. Let $X(t)$ be a fundamental matrix of independent solutions of $\dot{x} = A(t)x$ where $A(t)$ has continuous entries on some interval $a < t < b$. Prove for each t_0 in the interval that $X(t_0)$ is invertible as follows.

(a) Assuming the contrary, show that there is a vector $\mathbf{c} \ne 0$ such that $X(t_0)\mathbf{c} = 0$.

(b) Show that the initial-value problem $\dot{x} = A(t)x$, $x(t_0) = 0$ has the solution $x(t) = X(t)\mathbf{c}$ on the interval, and also has the identically zero solution there.

(c) Use the uniqueness part of Theorem 1.1 in Chapter 12, Section 1D to show that $X(t)\mathbf{c}$ is identically zero on the interval, thus contradicting the linear independence of the columns of the fundamental matrix $X(t)$.

SECTION 4   EQUILIBRIUM AND STABILITY

An **equilibrium solution** of a single differential equation involving a time derivative, or of a system of such differential equations, is just a solution that is constant over

FIGURE 13.2

At $(y, z) = (2k\pi, 0)$, stable in (a) and asymptotically stable in (b); unstable at $(y, z) = \big((2k + 1)\pi,\ 0\big)$ in (a) and (b).

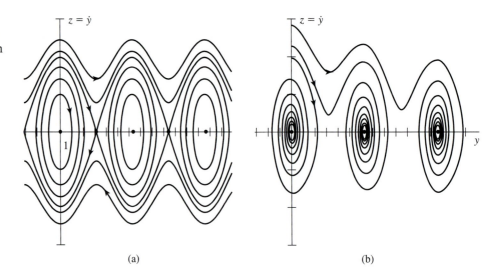

(a)   (b)

time. The reason for using the term *equilibrium solution* rather than *constant solution* is that we're mainly interested in the stability of solutions that result from small perturbations of an equilibrium solution. For the purposes of graphing, an equilibrium solution appears as a single point in the space of the dependent variables, so we often find it natural to refer to an equilibrium solution as an **equilibrium point**.

EXAMPLE 1

The classic example of a stable equilibrium in a mechanical system is a pendulum in a motionless downward position. In the pendulum equation and its equivalent 2-dimensional system, the equilibrium positions are expressed by solutions $y(t) = 2k\pi$ for integer k. A slight perturbation leads to motion that varies only slightly from the equilibrium position, expressed by solutions $y(t)$ having small amplitude. At the other extreme are equilibrium solutions $y(t) = (2k + 1)\pi$ representing precariously balanced upward vertical pendulum positions; a slight perturbation always leads to positions remote from the unstable equilibrium, no matter how slight the perturbation. Thus the stable equilibrium points are at $(y, z) = (2k\pi, 0)$ and the unstable points are at $(y, z) = \big((2k+1)\pi, 0\big)$. See Figure 13.2(a), which we discussed also in Example 7 of Chapter 11, Section 7C.

In general we consider behavior of solutions $\mathbf{x} = \mathbf{x}(t)$ of autonomous systems $\dot{\mathbf{x}} = \mathbf{F}(\mathbf{x})$ near an equilibrium point $\mathbf{x}_0$, that is, a point $\mathbf{x}_0$ such that $\mathbf{F}(\mathbf{x}_0) = 0$. The basic types of behavior are as follows:

1. $\mathbf{x}_0$ is **asymptotically stable** if there is a number $d_0 > 0$ such that every solution starting within distance d_0 of $\mathbf{x}_0$ tends to $\mathbf{x}_0$ as t tends to infinity.
2. $\mathbf{x}_0$ is **stable** if there is a $d_0 > 0$ such that all solutions starting within some distance $d_1 < d_0$ from $\mathbf{x}_0$ remain within distance d_0 of $\mathbf{x}_0$. Note that asymptotic stability near an equilibrium point $\mathbf{x}_0$ implies stability there.
3. An equilibrium point $\mathbf{x}_0$ that is not stable is called **unstable**.

4A   Linear Systems

An n-dimensional autonomous linear system in first-order standard form is

$$\dot{\mathbf{x}} = A\mathbf{x} + \mathbf{b},$$

where A is a constant n by n matrix and $\mathbf{b}$ is a constant n-dimensional vector. An equilibrium solution is a constant vector satisfying

$$A\mathbf{x}_0 + \mathbf{b} = 0 \quad \text{or} \quad A\mathbf{x}_0 = -\mathbf{b}.$$

If A is invertible, there is a unique equilibrium point given by $\mathbf{x}_0 = -A^{-1}\mathbf{b}$. If A^{-1} fails to exist, then either (a) there will be no equilibrium point, or else (b) there will be an entire line or plane consisting entirely of equilibrium points. In either case, if $\mathbf{x} = \mathbf{x}_0$ is an equilibrium solution, the homogeneous plus particular form $\mathbf{x}(t) = \mathbf{x}_h(t) + \mathbf{x}_0$ of the general solution shows that the qualitative behavior of $\mathbf{x}(t)$ near $\mathbf{x}_0$ is the same as the behavior of the homogeneous solution $\mathbf{x}_h(t)$ near $\mathbf{x} = 0$. So to find out about stability of $\mathbf{x}_0$ we just need to find out about the behavior of solution trajectories of $\dot{\mathbf{x}} = A\mathbf{x}$ near $\mathbf{x} = 0$. But we know from Sections 1 and 2 that, for constant square matrices A, the basic solutions of $\dot{\mathbf{x}} = A\mathbf{x}$ are all of the form $e^{\lambda t}$, or $t^k e^{\lambda t}$, where λ is an eigenvalue of A. Thus the qualitative behavior of solutions is entirely determined by the eigenvalues of A along with their multiplicities. The case in which zero is an eigenvalue is atypical and is often called *degenerate*, because it implies the existence of nonzero solutions $\mathbf{x}$ to $A\mathbf{x} = 0$, which as we remarked earlier implies an entire line or plane of equilibrium points.

The following displays I to VIII of types of 2-dimensional trajectory behavior of $\dot{\mathbf{x}} = A\mathbf{x}$ covers all possibilities for which zero is not an eigenvalue, and thus for which the origin is the only equilibrium point. Where there are side-by-side pictures in a category the example on the left has the standard basis vectors as eigenvectors, while the one on the right has an eigenvector that's tilted relative to the axes. When there is at least one eigenvalue with positive real part the equilibrium is necessarily unstable. If one eigenvalue is negative and one is positive the equilibrium is called a **saddle point**, as shown in Display II. An extension to a display for 3-dimensional space would contain more pictures, that take into account whether the third eigenvalue is positive or negative.

I. Unstable Node: $0 < \lambda_1 < \lambda_2$

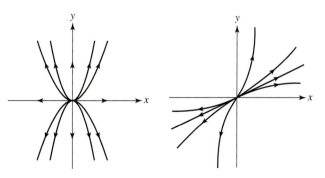

Example: $x = 2e^t,$
$\qquad\qquad y = e^{3t}; 8y = x^3.$

II. Saddle (Unstable): $\lambda_1 < 0 < \lambda_2$

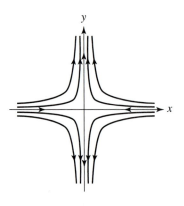
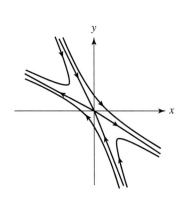

Example: $x = e^{-t}$,
               $y = 3e^{t}; xy = 3$.

III. Asymptotically Stable Node: $\lambda_1 < \lambda_2 < 0$

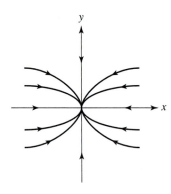
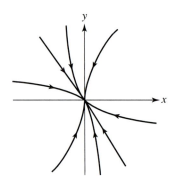

Example: $x = 2e^{-3t}$,
               $y = e^{-t}; x = 2y^{3}$.

IV. Unstable Spiral: $\lambda_1 = p + iq$, $\lambda_2 = p - iq$, $p > 0$, $q \neq 0$

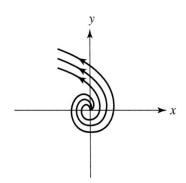

Example: $x = e^{t} \cos 2t$,
               $y = e^{t} \sin 2t; x^{2} + y^{2} = e^{(\arctan y/x)}$.

V. Stable Center: $\lambda_1 = iq$, $\lambda_2 = -iq$, $q \neq 0$.

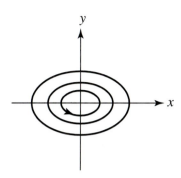

Example: $x = 2\cos t$,
$\qquad\qquad y = \sin t; x^2 + 4y^2 = 4$.

VI. Asymptotically Stable Spiral: $\lambda_1 = -p + iq$, $\lambda_2 = -p - iq$, $p > 0$, $q \neq 0$

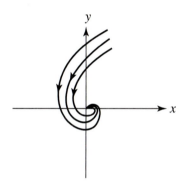

Example: $x = e^{-t}\cos t$,
$\qquad\qquad y = e^{-t}\sin t; x^2 + y^2 = e^{-2(\arctan y/x)}$.

VII. Unstable Star: $\lambda_1 = \lambda_2 > 0$

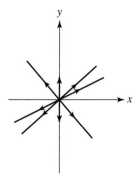

Example: $x = 2e^t$,   Example: $x = e^t$,
$\qquad\qquad y = 3e^t, 3x = 2y$. $\qquad\qquad y = te^t, y = x\ln x$.

VIII. Asymptotically Stable Star: $\lambda_1 = \lambda_2 < 0$

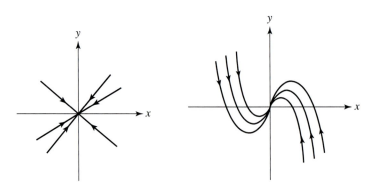

Example: $x = e^{-t}$,
$\qquad\;\; y = 3e^{-t}, 3x = y.$

Example: $x = e^{-t}$,
$\qquad\;\; y = te^{-t}, y = -x \ln x.$

EXAMPLE 2   The unique equilibrium solution $x = x_0$, $y = y_0$ of the system

$$\begin{aligned} \dot{x} &= x + y + 1 \\ \dot{y} &= 4x + y - 1 \end{aligned} \quad \text{satisfies} \quad \begin{aligned} x + y + 1 &= 0 \\ 4x + y - 1 &= 0, \end{aligned}$$

and this solution is $x_0 = \frac{2}{3}$, $y_0 = -\frac{5}{3}$. So for stability of this constant solution we study the homogeneous system

$$\begin{aligned} \dot{x} &= x + y \\ \dot{y} &= 4x + y. \end{aligned} \quad \text{or} \quad \dot{\mathbf{x}} = \begin{pmatrix} 1 & 1 \\ 4 & 1 \end{pmatrix} \mathbf{x}.$$

We saw in Example 1 of Section 1 that the eigenvalues of the 2 by 2 matrix are $\lambda = 3$ and $\lambda = -1$, so there are solutions

$$\mathbf{x}_1(t) = c_1 e^{3t} \mathbf{u} \quad \text{and} \quad \mathbf{x}_2(t) = c_2 e^{-t} \mathbf{v},$$

where $\mathbf{u}$ and $\mathbf{v}$ are respective eigenvectors of 3 and -1. The presence of solutions of the form $\mathbf{x}_1(t)$ is enough to demonstrate that the equilibrium point $(\frac{2}{3}, -\frac{5}{3})$ of the original nonhomogeneous system is unstable. The reason is that by taking $c_2 \neq 0$ small enough in absolute value we can produce solutions starting arbitrarily close to equilibrium that tend arbitrarily far away as t increases. Thus it's just the presence of the positive eigenvalue $\lambda = 3$ that is decisive. Note that the values of the eigenvectors $\mathbf{u}$ and $\mathbf{v}$ are irrelevant in making the decision; these vectors are important in determining the shape and direction of the solution trajectories, but they don't affect stability.

EXAMPLE 3   The 3-dimensional system

$$\begin{aligned} \dot{x} &= y \\ \dot{y} &= -x - y, \\ \dot{z} &= x - z \end{aligned} \quad \text{in matrix form} \quad \dot{\mathbf{x}} = \begin{pmatrix} 0 & 1 & 0 \\ -1 & -1 & 0 \\ 1 & 0 & -1 \end{pmatrix} \mathbf{x},$$

has a unique equilibrium point at the origin $(0, 0, 0)$. The characteristic equation is

$$\det \begin{pmatrix} -\lambda & 1 & 0 \\ -1 & -1-\lambda & 0 \\ 1 & 0 & -1-\lambda \end{pmatrix} = 0,$$

which works out to be $\lambda^3 + 2\lambda^2 + 2\lambda + 1 = 0$. By inspection we see that $\lambda_1 = -1$ is a root. Factoring out $\lambda + 1$ leaves $\lambda^2 + \lambda + 1 = 0$, so the eigenvalues are

$$\lambda_1 = -1, \quad \lambda_2 = \tfrac{1}{2}\left(-1 + \sqrt{3}\,i\right), \quad \lambda_3 = \tfrac{1}{2}\left(-1 - \sqrt{3}\,i\right).$$

Since all roots have negative real part, either -1 or $-\tfrac{1}{2}$, the equilibrium is asymptotically stable, with all solutions tending to the origin as t increases. We could compute the general solution without much trouble, but we don't need that if we're only checking stability near equilibrium. Note that showing the existence of just one eigenvalue with positive real part would have been enough to guarantee instability.

EXAMPLE 4   The 3-dimensional systems

$$\dot{\mathbf{x}} = \begin{pmatrix} -1 & -1 & 0 \\ 1 & -1 & 0 \\ 0 & 0 & -1 \end{pmatrix} \mathbf{x} \quad \text{and} \quad \dot{\mathbf{x}} = \begin{pmatrix} 1 & -1 & 0 \\ 1 & 1 & 0 \\ 0 & 0 & 1 \end{pmatrix} \mathbf{x}$$

have respective characteristic equations

$$(\lambda^2 + 2\lambda + 2)(-\lambda - 1) = 0 \quad \text{and} \quad (\lambda^2 - 2\lambda + 2)(-\lambda + 1) = 0$$

with respective eigenvalues

$$\lambda = -1 \pm i, \lambda = -1 \quad \text{and} \quad \lambda = 1 \pm i, \lambda = 1.$$

The respective solutions are

$$\mathbf{x}(t) = \begin{pmatrix} c_1 e^{-t} \cos t \\ c_2 e^{-t} \sin t \\ c_3 e^{-t} \end{pmatrix} \quad \text{and} \quad \mathbf{x}(t) = \begin{pmatrix} c_1 e^{t} \cos t \\ c_2 e^{t} \sin t \\ c_3 e^{t} \end{pmatrix}.$$

The origin is respectively stable and unstable for these solutions as shown in Figure 13.3. Note that the trajectories *appear* to be very similar, but the ones in (b) necessarily start at a positive distance from the origin, while in theory the ones in (a) approach arbitrarily close to the origin.

The previous examples make the next theorem plausible.

4.1 Linear Stability.   Let A be a real n-by-n constant matrix. The equilibrium solution $\mathbf{x}_0 = 0$ for the homogeneous system $\dot{\mathbf{x}} = A\mathbf{x}$ is

(a) asymptotically stable if every eigenvalue of A has negative real part, and is
(b) unstable if A has at least one eigenvalue with positive real part.

FIGURE 13.3

(a) Stable. (b) Unstable.

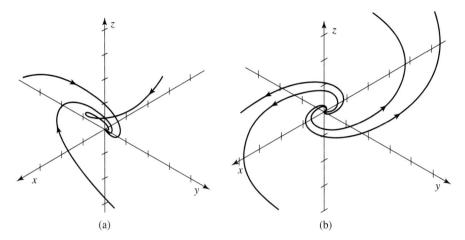

(a)   (b)

The same conclusions hold for stability of an equilibrium solution for $\dot{\mathbf{x}} = A\mathbf{x} + \mathbf{b}$, where $\mathbf{b}$ is an n-dimensional constant vector. If all eigenvalues of A have real part zero, we can draw no immediate conclusion in dimension $n \geq 4$, for then an equilibrium solution $\mathbf{x}_0$ may be stable or may be unstable, but if $n \leq 3$ and the eigenvalues are all distinct then $\mathbf{x}_0$ will be stable.

Proof. The method for computing the exponential matrix described in Section 2D shows that every entry in the exponential matrix e^{tA} has the form $e^{\lambda t} Q(t)$, where $Q(t)$ is a polynomial and λ is an eigenvalue of A. If every such λ has negative real part, then the entries, and hence all solutions, tend to zero as t tends to $+\infty$. On the other hand, if a single eigenvalue λ, with eigenvector $\mathbf{v}$, has positive real part, then the solution $\mathbf{x}(t) = \delta e^{\lambda t} \mathbf{v}$ is unbounded in every nonzero coordinate as t tends to $+\infty$, regardless of how small the positive number δ is chosen.

If the nonhomogeneous equation has $\mathbf{x}_0$ for an equilibrium solution the homogeneous plus particular form $\mathbf{x}(t) = e^{tA}\mathbf{c} + \mathbf{x}_0$ for the general solution shows that the same conclusions hold for the nonhomogeneous equation. The last statement of the theorem is settled by checking out the two examples in Exercise 23 and noting that with real and nonzero parts both zero when $n = 2$ the eigenvalues are $\pm iq$ with q real and nonzero, and when $n = 3$ the eigenvalues are $\pm iq$, and 0. ∎

EXERCISES

Find the eigenvalues λ_1, λ_2 associated with each system 1 through 8. Then classify the behavior of each system near equilibrium by name according to the types categorized in the Displays I to VIII of the text and sketch a phase portrait for each system.

1. $dx/dt = -3x + 2y$,
$dy/dt = -4x + 3y$.

2. $dx/dt = x + 4y$,
$dy/dt = 5y$.

3. $dx/dt = 2x - y$,
$dy/dt = x + 2y$.

4. $dx/dt = x$,
$dy/dt = 3x + y$.

5. $\dot{x} = 2x - 3y$,
$\dot{y} = -2x - 2y$.

6. $\dot{x} = -x + y$,
$\dot{y} = -4x - y$.

7. $\dot{x} = x + y$,
$\dot{y} = 2x$.

8. $\dot{x} = x + y$,
$\dot{y} = x - y$.

Each of the second-order equations 9 to 12 is equivalent to a first-order system obtained by letting $dx/dt = y$. Classify by name, according to the list I–VIII, each equation, or equivalently the system, by finding the characteristic roots of the given second-order equation directly. Sketch a phase-portrait for each equation.

9. $d^2x/dt^2 - x = 0$.

10. $d^2x/dt^2 + dx/dt - x = 0$.

11. $d^2x/dt^2 + dx/dt + x = 0$.

12. $d^2x/dt^2 + x = 0$.

The second-order equations 13 through 16 determine the evolution $x(t)$ of a spring system subject to a frictional force $k(dx/dt)$, where $k \geq 0$ is constant. Find conditions on k under which the equation belongs to any or all of the classes in the standard I–VIII.

13. $d^2x/dt^2 + kdx/dt + x = 0$.

14. $d^2x/dt^2 + kdx/dt + 3x = 0$.

15. $2d^2x/dt^2 + kdx/dt + x = 0$.

16. $md^2x/dt^2 + kdx/dt + 3x = 0, \ m > 0$.

17. Show that the system $\dot{x} = ax + by, \ \dot{y} = cx + dy$ has infinitely many constant equilibrium solutions if and only if $ad - bc = 0$. [*Hint:* Try to solve $ax + by = cx + dy = 0$.]

18. Consider the general solution $x = c_1 e^{\lambda_1 t} + c_2 e^{\lambda_2 t}$, $y = c_3 e^{\lambda_1 t} + c_4 e^{\lambda_2 t}$ of a 2-dimensional linear system in which $\lambda_1 < \lambda_2 < 0$, and for which $(0, 0)$ is an asymptotically stable node.

(a) Show that, if c_2 and c_4 are not both zero, then as $t \to \infty$ the slope $dy/dx = (dy/dt)/(dx/dt)$ of the associated trajectory approaches c_4/c_2, interpreted as a vertical slope if $c_2 = 0$.

(b) Show that if $c_2 = c_4 = 0$, but c_1 and c_3 are not both zero, then all trajectories are straight lines with slopes c_3/c_1, or a vertical line if $c_1 = 0$.

19. What difference, if any, is there between the phase portraits of the system $\dot{x} = f(x, y), \ \dot{y} = g(x, y)$ and the system $\dot{x} = -f(x, y), \ \dot{y} = -g(x, y)$?

20. Show that if the system $\dot{x} = ax + by, \ \dot{y} = cx + dy$ has real characteristic roots then it has an asymptotically stable node at the origin if and only if $a + d < 0$ and $ad - bc > 0$.

21. (a) Solve the system $\dot{x} = y, \ \dot{y} = -x - y, \ \dot{z} = x - z$.
   (b) Show that every solution tends to the equilibrium point $(0, 0, 0)$, which is thus asymptotically stable.
   (c) What changes if the last equation is replaced by $\dot{z} = x + z$?

22. Use the systems $\dot{\mathbf{x}} = A\mathbf{x}$ with these matrices to confirm the last statement in Theorem 4.1:

$$\begin{pmatrix} 0 & 1 & 0 & 0 \\ 0 & 0 & 1 & 0 \\ 0 & 0 & 0 & 1 \\ -1 & 0 & -2 & 0 \end{pmatrix} \text{ and } \begin{pmatrix} 0 & 1 & 0 & 0 \\ 0 & 0 & 1 & 0 \\ 0 & 0 & 0 & 1 \\ -4 & 0 & -5 & 0 \end{pmatrix}.$$

4B   Nonlinear Systems

To extend eigenvalue analysis of equilibrium solutions to autonomous nonlinear systems we start by finding the equilibrium points $\mathbf{x}_0 = (a_1, \ldots, a_n)$ of a nonlinear system $\dot{\mathbf{x}} = \mathbf{F}(\mathbf{x})$, that is, points such that $\mathbf{F}(\mathbf{x}_0) = 0$. To do this for linear systems we had the routine of Chapter 2, Section 2, but if $\mathbf{F}(\mathbf{x})$ is nonlinear we have to resort to ad hoc methods or perhaps numerical approximation using Newton's method as described in Chapter 5, Section 5. After locating an equilibrium point $\mathbf{x}_0$, the next step is to *linearize* the system at $\mathbf{x}_0$, replacing each real-valued equation $\dot{x}_k = F_k(x_1, \ldots, x_n)$ in the system by its linearization

$$\dot{x}_k = \sum_{j=1}^{n} \frac{\partial F_K}{\partial x_j}(\mathbf{x}_0)(x_j - a_j) + F_k(\mathbf{x}_0), \quad k = 1, \ldots, x_n.$$

The resulting linear autonomous system is called the **linearization** of $\dot{\mathbf{x}} = \mathbf{F}(\mathbf{x})$ at $\mathbf{x}_0$, and we can write it as $\dot{\mathbf{x}} = \mathbf{F}'(\mathbf{x}_0)(\mathbf{x} - \mathbf{x}_0) + \mathbf{F}(\mathbf{x}_0)$ using the **derivative matrix**

$$\mathbf{F}'(\mathbf{x}_0) = \begin{pmatrix} \dfrac{\partial F_1}{\partial x_1}(\mathbf{x}_0) & \cdots & \dfrac{\partial F_1}{\partial x_n}(\mathbf{x}_0) \\ \vdots & \ddots & \vdots \\ \dfrac{\partial F_n}{\partial x_1}(\mathbf{x}_0) & \cdots & \dfrac{\partial F_n}{\partial x_n}(\mathbf{x}_0) \end{pmatrix}.$$

For simplicity we work with the homogeneous equation $\dot{\mathbf{x}} = \mathbf{F}'(\mathbf{x}_0)\mathbf{x}$ at $\mathbf{x}_0$, as we did in Section 4A, and we'll see that the eigenvalues of the constant matrix $\mathbf{F}'(\mathbf{x}_0)$ are the key to our criteria.

EXAMPLE 5   A 2-dimensional system and its linearization are respectively

$$\begin{aligned} \dot{x} &= x \\ \dot{y} &= -y + x^2 \end{aligned} \quad \text{and} \quad \begin{aligned} \dot{x} &= x \\ \dot{y} &= -y, \end{aligned} \quad \text{or} \quad \mathbf{x} = \begin{pmatrix} 1 & 0 \\ 0 & -1 \end{pmatrix} \mathbf{x}.$$

For both the nonlinear system and its linearization there is a single equilibrium point at $(x_0, y_0) = (0, 0)$. The characteristic equation of the 2-by-2 matrix is $(1 - \lambda)(-1 - \lambda) = \lambda^2 - 1 = 0$, with roots $\lambda = \pm 1$. Thus the basic solutions are linear combinations of e^t and e^{-t} and so are of the unstable saddle type in Display II in Section 4A. Theorem 4.2 on page 654 guarantees that the solutions of the nonlinear system will be similarly unstable.

Aside from the equilibrium point at the origin, in Figure 13.4(a) there are four other exceptional phase curves in the phase portrait: the positive and negative y axes directed toward the origin, and the left and right halves of the parabola $y = \frac{1}{3}x^2$, both directed away from the origin. To derive the equations that these parabolic trajectories satisfy you can solve the first-order linear equation

$$\frac{\dot{y}}{\dot{x}} = \frac{dy}{dx} = \frac{-y + x^2}{x}, \quad x \neq 0.$$

FIGURE 13.4

(a) Nonlinear saddle. (b) Lorenz trajectory; $\beta = \frac{8}{3}$, $\rho = 28$, $\sigma = 10$.

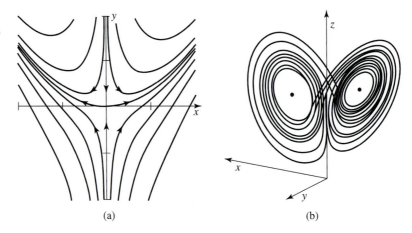

(a)   (b)

EXAMPLE 6   The **general Lorenz system** is

$$\dot{x} = -\sigma x + \sigma y$$
$$\dot{y} = \rho x - y - xz$$
$$\dot{z} = xy - \beta z,$$

and it has been studied extensively in recent years because of the apparently chaotic behavior of its solution trajectories near equilibrium. The equilibrium solutions are just the solutions to the algebraic system we get by setting the right hand sides equal to zero. Looking at the special case $\beta = \sigma = 1$, $\rho = 2$, we solve

$$-x + y = 0$$
$$2x - y - xz = 0$$
$$xy - z = 0.$$

Noting from the first equation that $x = y$, we then see that there are just three solutions: $(1, 1, 1)$, $(0, 0, 0)$ and $(-1, -1, 1)$. The derivative matrix $\mathbf{F}'(x, y, z)$ for the linearization at (x, y, z) is

$$\begin{pmatrix} \dfrac{\partial(-x+y)}{\partial x} & \dfrac{\partial(-x+y)}{\partial y} & \dfrac{\partial(-x+y)}{\partial z} \\ \dfrac{\partial(2x-y-xz)}{\partial x} & \dfrac{\partial(2x-y-xz)}{\partial y} & \dfrac{\partial(2x-y-xz)}{\partial z} \\ \dfrac{\partial(xy-z)}{\partial x} & \dfrac{\partial(xy-z)}{\partial y} & \dfrac{\partial(xy-z)}{\partial z} \end{pmatrix} = \begin{pmatrix} -1 & 1 & 0 \\ 2-z & -1 & -x \\ y & x & -1 \end{pmatrix}.$$

Evaluating this matrix at $(x, y, z) = (1, 1, 1)$ gives

$$\mathbf{F}'(1, 1, 1) = \begin{pmatrix} -1 & 1 & 0 \\ 2-z & -1 & -x \\ y & x & -1 \end{pmatrix}_{(1,1,1)} = \begin{pmatrix} -1 & 1 & 0 \\ 1 & -1 & -1 \\ 1 & 1 & -1 \end{pmatrix}.$$

Similarly, we get the linearization matrices at $(0, 0, 0)$ and $(-1, -1, 1)$ by evaluating the same derivative matrix at these two additional points:

$$\mathbf{F}'(0, 0, 0) = \begin{pmatrix} -1 & 1 & 0 \\ 2 & -1 & 0 \\ 0 & 0 & -1 \end{pmatrix}; \quad \mathbf{F}'(-1, -1, 1) = \begin{pmatrix} -1 & 1 & 0 \\ 1 & -1 & 1 \\ -1 & -1 & -1 \end{pmatrix}.$$

The next theorem draws conclusions about stability from the eigenvalues of derivative matrices. For a general n-dimensional system the number of possibilities is very large, so we list only a few general categories. We omit the detailed proof.

4.2 Linearization Theorem.   Assume that the real-valued coordinate functions F_k of $\mathbf{F}$ are continuously differentiable and that the system $\dot{\mathbf{x}} = \mathbf{F}(\mathbf{x})$ has an equilibrium point at $\mathbf{x}_0$. The equilibrium solution $\mathbf{x}_0$ for the system is asymptotically stable if every eigenvalue of the derivative matrix $\mathbf{F}'(\mathbf{x}_0)$ has negative real part. The point $\mathbf{x}_0$ is unstable if $\mathbf{F}'(\mathbf{x}_0)$ has at least one eigenvalue with positive real part, and is called a **saddle point** if both signs occur. If all eigenvalues of $\mathbf{F}'(\mathbf{x}_0)$ have real part zero, we can draw no definite conclusion, and the equilibrium $\mathbf{x}_0$ may be stable or may be unstable. Exercise 5 has an example of each possibility.

EXAMPLE 7   The special Lorenz system treated in Example 3 has an equilibrium at $(1, 1, 1)$ with characteristic polynomial $P(\lambda) = \det(\mathbf{F}'(1, 1, 1) - \lambda I)$. From Example 3 we see that

$$P(\lambda) = \det \begin{pmatrix} -1-\lambda & 1 & 0 \\ 1 & -1-\lambda & -1 \\ 1 & 1 & -1-\lambda \end{pmatrix}.$$

Computing the determinant, we get $P(\lambda) = -\big((\lambda+1)^3+1\big)$, and we see by inspection that $\lambda_1 = -2$ is a root. Division by $\lambda + 2$ gives $P(\lambda) = -(\lambda + 2)(\lambda^2 + \lambda + 1)$. The roots of the quadratic factor are $\lambda_2 = (-1 + \sqrt{3}\,i)/2$ and $\lambda_3 = (-1 - \sqrt{3}\,i)/2$. Thus the real parts of all three eigenvalues are negative, so we conclude from Theorem 4.2 that $(1, 1, 1)$ is an asymptotically stable equilibrium solution.

EXAMPLE 8   Continuing with the special Lorenz system, we examine the equilibrium at $(0, 0, 0)$. The relevant characteristic polynomial is evaluated at $(0, 0, 0)$ and is $P(\lambda) = \det (\mathbf{F}'(0, 0, 0) - \lambda I)$, or

$$P(\lambda) = \begin{pmatrix} -1-\lambda & 1 & 0 \\ 2 & -1-\lambda & 0 \\ 0 & 0 & -1-\lambda \end{pmatrix}.$$

The determinant is $-(\lambda + 1)^3 + 2\lambda + 2 = -(\lambda + 1)(\lambda^2 + 2\lambda - 1)$. The roots are $\lambda_1 = -1$, $\lambda_2 = -1 - \sqrt{2}$ and $\lambda_3 = -1 + \sqrt{2}$. Since $\lambda_3 > 0$ we conclude from Theorem 4.2 that $(0, 0, 0)$ is an unstable equilibrium. This point is a saddle point, since there are two negative eigenvalues that contribute to making the other basic solutions tend to zero. Checking out the equilibrium at $(-1, -1, 1)$ is left as an exercise.

EXAMPLE 9   The general **Lorenz system**

$$\dot{x} = \sigma(y - x), \quad \dot{y} = \rho x - y - xz, \quad \dot{z} = -\beta z + xy, \quad \beta, \ \rho, \ \sigma \text{ positive constants}$$

has been studied extensively with the aim of understanding trajectories such as the one shown in Figure 13.4(b). With the choice of parameters shown there, the equilibrium points, aside from the one at the origin, are at $(\pm 6\sqrt{2}, \pm 6\sqrt{2}, 27)$. The trajectory shown in the figure has initial point $(2, 2, 21)$. It winds around in the area of one equilibrium an apparently random number of times, then switches toward the other equilibrium with similar behavior, continuing back and forth unpredictably. The eigenvalues of the linearizations are the same at the two equilibrium points; they are approximately as follows: $\lambda_1 \approx -13.85$, λ_2, $\lambda_3 \approx 0.09 \pm 10.19i$. Thus these two points are saddle points, and each one has a surface containing it on which all trajectories gradually spiral away from the point, as well as a trajectory at a positive angle to the surface that converges to the point. The typical trajectory behavior lies somewhere between these extremes, winding away from one equilibrium until it is attracted by the other, then reversing. The number of circuits about each point, and the path taken, is very sensitive to minute changes in the initial conditions.

Edward Lorenz began the study of the Lorenz system by using it to approximate more complicated differential equations in the study of weather patterns; hence the interest in the system's sensitivity to initial conditions, sometimes called the butterfly effect. For more details about the system see Colin Sparrow, *The Lorenz Equations: Bifurcations, Chaos, and Strange Attractors*, Springer-Verlag (1982).

EXERCISES

1. Assume A is not the 0 matrix and let $\dot{\mathbf{x}} = A\mathbf{x}$ be a 2-dimensional autonomous system for which $\det A = 0$. Show that the system has zero for an eigenvalue and that the equilibrium solutions make up an entire line in $\mathbb{R}^2$.

2. The nonautonomous system

$$\dot{x} = (1 - t)x - ty$$

$$\dot{y} = tx + (1 - t)y$$

exhibits a change of character at its lone equilibrium point. This system appears also in Example 3 of Chapter 12, Section 1.

(a) Show that for each real number t the system has a single equilibrium point at $(x, y) = (0, 0)$.

(b) Show that while $t > 1$ solutions behave in a stable manner near the equilibrium point, and that behavior is unstable when $t < 1$.

3. Show that the equilibrium points of the Lorenz system

$$\dot{x} = \sigma(y - x), \quad \dot{y} = \rho x - y - xz,$$

$$\dot{z} = -\beta z + xy, \quad \beta, \ \rho, \ \sigma$$

positive constants, are $(0, 0, 0)$ and, if $\rho > 1$, the two points $\left(\pm\sqrt{\beta(\rho - 1)}, \ \pm\sqrt{\beta(\rho - 1)}, \ \rho - 1 \right)$.

4. (a) According to Theorem 4.2, the nonlinear system $\dot{x} = x$, $\dot{y} = -y + x^2$ of Example 5 has an unstable saddle equilibrium at $(x, y) = (0, 0)$. Solve the system explicitly to show the unstable saddle behavior of solutions near $(0, 0)$.

(b) Show that the trajectories for which $x \neq 0$ satisfy the linear equation $dy/dx = (-y + x^2)/x$, and identify the equations $y = y(x)$ of the two parabolic trajectories.

5. Linearized analysis is inadequate for some systems, as claimed in Theorem 4.2, and we can see this by looking at the family of nonlinear systems

$$\dot{x} = y + \alpha x(x^2 + y^2),$$

$$\dot{y} = -x + \alpha y(x^2 + y^2),$$

where α is constant.

(a) Show that the only equilibrium point is $(x_0, y_0) = (0, 0)$, regardless of the value of α.

(b) Show that the linearized system associated with $(0, 0)$ is $\dot{x} = y$, $\dot{y} = -x$, and that the origin is a stable center as $t \to \infty$. Note that $(0, 0)$ is also a stable center for the given nonlinear system when $\alpha = 0$.

(c) Show that in polar coordinates the given system takes the form $\dot{r} = \alpha r^3$, $\dot{\theta} = -1$.
[*Hint:* Apply d/dt to the equations $x = r \cos\theta$, $y = r \sin\theta$.]

(d) Solve the polar-form system in part (c), and show that if $\alpha > 0$ the equilibrium point is unstable, and that if $\alpha < 0$ it is stable. Thus the parameter value $\alpha = 0$ is called a **bifurcation point** for the system, because the stability of the system at the equilibrium point changes in a fundamental way as α increases through zero.

6. A nonlinear pendulum with frictional damping has a displacement angle $\theta = \theta(t)$ that satisfies $\ddot{\theta} + (k/(lm))\dot{\theta} + \frac{g}{l} \sin\theta = 0$, where $k > 0$ is constant.

(a) Show that the equation for θ is equivalent, with $x = \theta$, $y = \dot{\theta}$, to the first-order system

$$\dot{x} = y, \quad \dot{y} = -\frac{g}{l} \sin x - \frac{k}{lm} y.$$

(b) Show that the equilibrium points of the system in part (a) are independent of k.

(c) Show that the unstable equilibrium points are all saddles.

(d) An equilibrium solution $\mathbf{x}_0$ of a nonlinear system is a **node** or a **spiral point** if $\mathbf{x}_0$ is a node or spiral point for the linearization at $\mathbf{x}_0$. Show that the stable equilibrium points are nodes if $k^2 > 4glm^2$ and asymptotic spirals if $k^2 < 4glm^2$. Why does this distinction make sense physically?

7. Consider the nonlinear system

$$\dot{x} = A - Bx - x + x^2 y, A \neq 0,$$

$$\dot{y} = Bx - x^2 y.$$

(a) Find the system's single equilibrium point.

(b) Assume that $A > 0$ and $B > A^2 + 1$. Trajectories starting sufficiently near, but not at, the equilibrium point exhibit **limit cycle** behavior in that they approach a closed trajectory. Investigate this claim by using a graphical-numerical method.

8. The **Lotka-Volterra equations** that model the interaction between the sizes H and P of certain host and parasite populations are

$$\frac{dH}{dt} = (a - bP)H, \quad \frac{dP}{dt} = (cH - d)P, \ a, b, c, d > 0.$$

(a) Show that for $H > 0$ and $P > 0$, the only equilibrium point is $(H_0, P_0) = (d/c, a/b)$, and find the associated linearized system.

(b) Show that the equilibrium solution of the linearized system is a stable center.

(c) Discuss the equilibrium solution of the nonlinear system at $(H, P) = (0, 0)$. Do your conclusions make sense, given the interpretation of P and H as sizes of parasite and host populations respectively?

9. Find all equilibrium points of $\dot{x} = -y(1 - x^2 - y^2)$, $\dot{y} = x(1 - x^2 - y^2)$. Then show that all other trajectories are circles and that no trajectory converges to an equilibrium point. [*Hint: $dy/dx = -x/y$.*]

10. Find all equilibrium points of $\dot{x} = x(2 - x - y)$, $\dot{y} = y(x - 1)$ and discuss their stability. Then by hand or using computer graphics, make a sketch of some typical trajectories near the equilibrium points.

11. Discuss the stability of the equilibrium solutions of the system

$$\dot{x} = -x(x^2 + y^2 - 1), \quad \dot{y} = -y(x^2 + y^2 + 1).$$

12. The system $\dot{x} = -x(x^2 + y^2 - 1)$, $\dot{y} = -y(x^2 + y^2 + 1)$ reduces to $\dot{x} = -x(x^2 - 1)$ on the x-axis and to $\dot{y} = -y(y^2 + 1)$ on the y-axis.

(a) Solve the y-equation explicitly to show that a trajectory with $y(0) = y_0$ on the y-axis converges to $(0, 0)$ as $t \to \infty$.

(b) Solve the x-equation explicitly to show that if $|x_0| < 1$ a trajectory starting at x_0 on the x-axis converges to either $(1, 0)$ or $(-1, 0)$ as $t \to \infty$ and converges to the origin as $t \to -\infty$.

13. Consider the system $\dot{x} = x(1 - x^2 - y^2) - y$, $\dot{y} = x + y(1 - x^2 - y^2)$.

(a) Show that the system has circular trajectories of radius 1.

(b) Show that all solutions (x, y) satisfy $(d/dt)(y/x) = 1 + (y/x)^2$ when $x \neq 0$.

(c) Use part (b) to show that the polar angle θ of a point on a nonconstant trajectory satisfies $\theta = \arctan(y/x) = t + c$ and hence that all such trajectories wind counterclockwise infinitely often around the origin.

(d) Show that all solutions (x, y) satisfy $x\dot{x} + y\dot{y} = (x^2 + y^2)(1 - x^2 - y^2)$.

(e) Use part (d) to show that the polar radius r of a point on a trajectory satisfies $dr/dt = r(1 - r^2)$, with solutions $r = ke^t/\sqrt{k^2 e^{2t} \pm 1}$, the sign depending on whether $0 < r < 1$ or $r > 1$. Show that the trajectories of these solutions approach the circular trajectory $x^2 + y^2 = 1$ as $t \to \infty$.

14. Consider the system

$$\dot{x} = x(1 - x^2 - y^2)^3 - y(1 - x^2 - y^2)^2 - y^3,$$

$$\dot{y} = x(1 - x^2 - y^2)^2 + y(1 - x^2 - y^2)^3 + xy^2.$$

(a) Show that all solutions satisfy $x\dot{x} + y\dot{y} = (x^2 + y^2)(1 - x^2 - y^2)^3$ and hence that the polar radius r of a point on a trajectory satisfies $dr/dt = r(1 - r^2)^3$.

(b) Show that the system has trajectories on the unit circle satisfying $\dot{x} = -y^3$, $\dot{y} = xy^2$. Explain why the unit semicircle with $y > 0$ has trajectories with $(1, 0)$ as and $(-1, 0)$ as limit points. What can you say about the semicircle with $y < 0$?

15. The **van der Pol equation** $\ddot{x} + \alpha(x^2 - 1)\dot{x} + x = 0$ is equivalent to the system $\dot{x} = y$, $\dot{y} = -x - \alpha(x^2 - 1)y$.

(a) Find the linearization of the system near $(x_0, y_0) = (0, 0)$.

(b) Discuss the behavior of solutions near $(x_0, y_0) = (0, 0)$ and their dependence on the constant $\alpha > 0$. What happens if $\alpha = 0$ or if $\alpha < 0$?

Chapter 13 REVIEW

The initial-value problems in Exercises 1 to 12 are solvable (i) by the elimination method of the previous chapter, (ii) by computing eigenvalues and eigenvectors, (iii) by computing an exponential matrix, or (iv) by an ad hoc approach involving an educated guess. You are asked to use whatever approach to a particular problem seems most efficient.

1. $\dot{\mathbf{x}} = \begin{pmatrix} 2 & -3 \\ 1 & -2 \end{pmatrix} \mathbf{x} + \begin{pmatrix} 1 \\ e^t \end{pmatrix}$, $\mathbf{x}(0) = \begin{pmatrix} 1 \\ 2 \end{pmatrix}$

2. $\dot{\mathbf{x}} = \begin{pmatrix} 2 & 2 \\ 1 & 2 \end{pmatrix} \mathbf{x} + \begin{pmatrix} 1 \\ e^t \end{pmatrix}$, $\mathbf{x}(0) = \begin{pmatrix} 1 \\ 1 \end{pmatrix}$

3. $\dot{\mathbf{x}} = \begin{pmatrix} 1 & -1 \\ 0 & -1 \end{pmatrix} \mathbf{x} + \begin{pmatrix} e^t \\ e^{2t} \end{pmatrix}$, $\mathbf{x}(0) = \begin{pmatrix} 1 \\ 0 \end{pmatrix}$

4. $\dot{\mathbf{x}} = \begin{pmatrix} 0 & -1 \\ 1 & 0 \end{pmatrix} \mathbf{x} + \begin{pmatrix} 1 \\ 2 \end{pmatrix}$, $\mathbf{x}(0) = \begin{pmatrix} 0 \\ 1 \end{pmatrix}$

5. $\dot{\mathbf{x}} = \begin{pmatrix} 2 & -1 \\ 1 & 5 \end{pmatrix} \mathbf{x} + \begin{pmatrix} 1 \\ e^t \end{pmatrix}$, $\mathbf{x}(1) = \begin{pmatrix} 1 \\ -1 \end{pmatrix}$

6. $\dot{\mathbf{x}} = \begin{pmatrix} 0 & 1 \\ -1 & 0 \end{pmatrix} \mathbf{x} + \begin{pmatrix} 1 \\ \sin t \end{pmatrix}$, $\mathbf{x}(0) = \begin{pmatrix} 0 \\ 0 \end{pmatrix}$

7. $\dot{\mathbf{x}} = \begin{pmatrix} 1 & -1 & 1 \\ 0 & 0 & 1 \\ 0 & -1 & 2 \end{pmatrix} \mathbf{x}$, $\mathbf{x}(0) = \begin{pmatrix} 1 \\ 2 \\ 0 \end{pmatrix}$

8. $\dot{\mathbf{x}} = \begin{pmatrix} 2 & 1 & -2 \\ 3 & -2 & 0 \\ 3 & 1 & -3 \end{pmatrix} \mathbf{x} + \begin{pmatrix} 1 \\ 0 \\ t \end{pmatrix}$, $\mathbf{x}(0) = \begin{pmatrix} 1 \\ 1 \\ 0 \end{pmatrix}$

9. $\dot{\mathbf{x}} = \begin{pmatrix} 3 & -1 & -1 \\ 1 & 1 & -1 \\ 1 & -1 & 1 \end{pmatrix} \mathbf{x}$, $\mathbf{x}(0) = \begin{pmatrix} 1 \\ 2 \\ 0 \end{pmatrix}$

10. $\dot{\mathbf{x}} = \begin{pmatrix} 2 & 1 & -2 \\ 3 & -2 & 0 \\ 3 & 1 & -3 \end{pmatrix} \mathbf{x}$, $\mathbf{x}(0) = \begin{pmatrix} 1 \\ 1 \\ 0 \end{pmatrix}$

11. $\dot{\mathbf{x}} = \begin{pmatrix} 1 & 0 & 0 & 1 \\ 0 & 2 & 0 & 0 \\ 0 & 0 & 2 & 0 \\ 0 & 0 & 0 & 1 \end{pmatrix} \mathbf{x} + \begin{pmatrix} 1 \\ 0 \\ 0 \\ e^t \end{pmatrix}$, $\mathbf{x}(0) = \begin{pmatrix} 1 \\ 2 \\ 0 \\ 1 \end{pmatrix}$

12. $\dot{\mathbf{x}} = \mathbf{x} + \mathbf{e}_1$, $\mathbf{x}(0) = \mathbf{e}_n$, $n \geq 2$

13. Find the general solution of the second-order system $\ddot{x} = x + y$, $\ddot{y} = x - y$ by first writing it as a first-order system of dimension 4 in matrix form and finding an exponential matrix in complex form.

14. (a) Find an equivalent first-order system of dimension $2n$ for the n-dimensional second-order initial-value problem $\ddot{\mathbf{x}} = \mathbf{x}$, $\mathbf{x}(0) = \mathbf{x}_0$, $\dot{\mathbf{x}}(0) = \mathbf{z}_0$, and write the resulting $2n$-dimensional system as an uncoupled sequence of 2-dimensional coupled systems.

(b) Solve the given second-order system for the case $n = 1$ and deduce the solution to the general case from part (a) and this special case.

(c) Deduce from the result of part (b) the form of the exponential matrix for the $2n$-dimensional system of part (a).

15. (a) Find an equivalent first-order system of dimension $2n$ for the n-dimensional second-order initial-value problem $\ddot{\mathbf{x}} = -\mathbf{x}$, $\mathbf{x}(0) = \mathbf{x}_0$, $\dot{\mathbf{x}}(0) = \mathbf{z}_0$, and write the resulting $2n$-dimensional system as an uncoupled sequence of 2-dimensional coupled systems.

(b) Solve the given second-order system for the case $n = 1$ and deduce the solution to the general case from part (a) and this special case.

(c) Deduce from the result of part (b) the form of the exponential matrix for the $2n$-dimensional system of part (a).

16. If all eigenvalues of the n-by-n matrix A are real, show that the system $t\dot{\mathbf{x}} = A\mathbf{x}$ has solutions $\mathbf{x}(t) = t^\lambda \mathbf{u}$ for $t > 0$, where $\mathbf{u}$ is an eigenvector of A with eigenvalue λ.

CHAPTER 14

INFINITE SERIES

The study of numerical infinite series is an important branch of the study of numerical approximation. In the first section we'll treat limits of sequences somewhat informally. Later on we'll use calculus technique to deal with sequential limits. Section 1 introduces the idea of convergence of a series, and Section 2 complements it using Taylor expansions. Additional sections deal with the more technical aspects of convergence. The chapter closes with power series solutions of ordinary differential equations and an introduction to Fourier series and the 1-dimensional heat and wave equations.

SECTION 1 EXAMPLES AND DEFINITIONS

Infinite series generalize finite sums to infinite sums of the form

$$a_1 + a_2 + a_3 + \cdots,$$

where the three dots indicate that a term a_k is included for each of the infinitely many positive integers $k = 1, 2, 3, \ldots$. The decimal expansion

$$\frac{1}{3} = 0.3333\cdots$$

is really an example of an infinite series, because suitably rewritten it's an infinite sum in the form

$$\frac{1}{3} = \frac{3}{10} + \frac{3}{100} + \frac{3}{1000} + \frac{3}{10,000} + \cdots = \frac{3}{10} + \frac{3}{10^2} + \frac{3}{10^3} + \frac{3}{10^4} + \cdots.$$

In both decimal and sum form the dots at the end mean that the most obvious pattern is to be carried on indefinitely. In principle this convention could lead to ambiguity, but we'll be more specific when necessary. To write infinite series more briefly, and less ambiguously, recall the Σ-notation for finite sums, defined by

$$\sum_{k=m}^{n} a_k = a_m + a_{m+1} + \cdots + a_n, \quad m \le n.$$

For example,

$$\sum_{k=1}^{n} k = 1 + 2 + 3 + \cdots + n.$$

A natural extension allows us to write a general infinite series in more compressed form like this:

$$a_1 + a_2 + a_3 + \cdots = \sum_{k=1}^{\infty} a_k.$$

Writing formulas like this requires us to have a formula for the general term a_k. Thus, for example,

$$\frac{3}{10} + \frac{3}{10^2} + \frac{3}{10^3} + \frac{3}{10^4} + \cdots = \sum_{k=1}^{\infty} \frac{3}{10^k}.$$

Assigning an appropriate numerical value to an infinite series is usually done by taking the limit as $n \to \infty$ of the nth **partial sum** $s_n = \sum_{k=1}^{n} a_k$. Thus by definition, the **sum** s of a series is

$$\sum_{k=1}^{\infty} a_k = \lim_{n \to \infty} \sum_{k=1}^{n} a_k = s,$$

if the limit exists. If the series has a sum in this sense, then the series is said to **converge** to s, and if the limit fails to exist, the series is said to **diverge**.

EXAMPLE 1

The **geometric series** with ratio x is

$$\sum_{k=0}^{\infty} x^k = 1 + x + x^2 + \cdots.$$

There is a simple formula for the partial sums if $x \neq 1$, given by

$$s_n = 1 + x + x^2 + \cdots + x^{n-1} = \frac{1 - x^n}{1 - x}.$$

To verify the formula, multiply both sides by $1 - x$ and note that all but two terms cancel on the left. If $0 < |x| < 1$, then x^n tends to 0 as $n \to \infty$; to see this take $n > \ln \delta / \ln |x|$ to make $|x|^n < \delta < 1$. Thus for all x for which $|x| < 1$ the sum is

$$\sum_{k=0}^{\infty} x^k = \lim_{n \to \infty} \frac{1 - x^n}{1 - x} = \frac{1}{1 - x}.$$

If $|x| > 1$, then x^n is unbounded as $n \to \infty$, so the sum fails to exist. If $x = -1$, the formula for s_n gives 0 if n is odd and 1 if n is even, so there is no limit then either. Finally, if $x = 1$ the formula for s_n is invalid, but in that case we see directly that $s_n = n$, which tends to ∞ as $n \to \infty$.

We can get some feeling for convergence of a sum by looking at specific numerical examples. In each of the next three examples we can find a simple formula for the partial sums, something that is not possible for many important infinite series.

EXAMPLE 2

Consider the geometric series with ratio $x = \frac{1}{3}$. Using Example 1, we have

$$\sum_{k=0}^{\infty} \frac{1}{3^k} = 1 + \frac{1}{3} + \frac{1}{3^2} + \frac{1}{3^3} + \cdots$$

$$= \lim_{n \to \infty} \sum_{k=0}^{n} \frac{1}{3^k} = \lim_{n \to \infty} \frac{1 - (1/3^{n+1})}{1 - (1/3)}$$

$$= \frac{1 - 0}{1 - (1/3)} = \frac{3}{2}.$$

EXAMPLE 3   Here is a particularly simple series.

$$\sum_{k=1}^{\infty}\left(\frac{1}{k}-\frac{1}{k+1}\right)=\lim_{n\to\infty}\sum_{k=1}^{n}\left(\frac{1}{k}-\frac{1}{k+1}\right)$$

$$=\lim_{n\to\infty}\left(1-\frac{1}{2}\right)+\left(\frac{1}{2}-\frac{1}{3}\right)+\cdots+\left(\frac{1}{n}-\frac{1}{n+1}\right)$$

$$=\lim_{n\to\infty}\left(1-\frac{1}{n+1}\right)=1-\lim_{n\to\infty}\frac{1}{n+1}=1-0=1$$

This is an example of what is called a "telescoping" series, because the interior terms cancel in the partial sum. Since $(1/k)-(1/k+1)=1/k(k+1)$, the series can also be written

$$\sum_{k=1}^{\infty}\frac{1}{k(k+1)}=1.$$

EXAMPLE 4   The infinite series $\sum_{k=1}^{\infty}k=1+2+3+\cdots$ is divergent. To see this, note that the nth partial sum is the sum of the first n integers:

$$s_n=1+2+\cdots+n=\frac{n(n+1)}{2}.$$

Hence $\lim_{n\to\infty}s_n=\infty$, so we agree to say that the series "diverges to $+\infty$."

EXAMPLE 5   The geometric series with $x=-2$ is

$$\sum_{k=0}^{\infty}(-2)^k=1-2+2^2-2^3+\cdots,$$

and the nth partial sum is, according to the formula in Example 1,

$$s_n=\frac{1-(-2)^{n+1}}{1+2}=\frac{1+(-1)^n2^n}{3}.$$

As $n\to\infty$, the numerator oscillates between being large and positive and large and negative. Since the partial sums do not tend to a fixed finite value, the series diverges.

EXERCISES

In Exercises 1 to 4, use the definition $\sum_{k=m}^{n}a_k=$
$a_m+a_{m+1}+\cdots+a_n$ for the Σ-notation to write out and simplify the sums.

1. $\sum_{k=1}^{5}\left(\frac{1}{k}-\frac{1}{k+1}\right)$ 2. $\sum_{k=2}^{5}k-\sum_{k=3}^{6}(k-1)$

3. $\sum_{k=0}^{4}2^{-k}$ 4. $6\sum_{k=1}^{11}(-1)^k$

In Exercises 5 to 10, find a formula for the kth term, $k = 1, 2, 3, \ldots$, of the infinite series that is consistent with the given part of the series.

5. $1 + \dfrac{1}{2} + \dfrac{1}{3} + \dfrac{1}{4} + \cdots$

6. $1 - \dfrac{1}{2} + \dfrac{1}{4} - \dfrac{1}{8} + \cdots$

7. $\dfrac{1}{1\cdot3} + \dfrac{1}{2\cdot4} + \dfrac{1}{3\cdot5} + \cdots$

8. $\dfrac{1}{3} + \dfrac{1}{2\cdot3^2} + \dfrac{1}{3\cdot3^3} + \cdots$

9. $\dfrac{4}{2\cdot3^2} + \dfrac{5}{3\cdot3^3} + \dfrac{6}{4\cdot3^4} + \cdots$

10. $\dfrac{1}{1\cdot2} - \dfrac{1}{2\cdot3} + \dfrac{1}{3\cdot4} - \cdots$

In Exercises 11 to 14, verify that each of the partial sum formulas is correct. Then find the sum of the corresponding infinite series as $n \to \infty$ if it exists.

11. $\displaystyle\sum_{k=1}^{n} \dfrac{1}{2k(2k+2)} = \dfrac{n}{4(n+1)}$

12. $\displaystyle\sum_{k=0}^{n} \dfrac{3}{4^k} = 4 - 4^{-n}$

13. $\displaystyle\sum_{k=1}^{n} 2k = n(n+1)$

14. $\displaystyle\sum_{k=0}^{n} \left(\dfrac{2^k+1}{2^k} - 1\right) = 2 - 2^{-n}$

In Exercise 15 to 20, what is the sum of the geometric series?

15. $\displaystyle\sum_{k=0}^{\infty} \left(\dfrac{1}{6}\right)^k$

16. $\displaystyle\sum_{k=0}^{\infty} \left(-\dfrac{3}{4}\right)^k$

17. $\displaystyle\sum_{k=1}^{\infty} \left(\dfrac{1}{1+\pi}\right)^k$

18. $\displaystyle\sum_{k=0}^{\infty} \left(\dfrac{1}{1+x^2}\right)^k, x \neq 0$

19. $\displaystyle\sum_{k=0}^{\infty} e^{-2k}$

20. $\displaystyle\sum_{k=0}^{\infty} (0.01)^k$

In Exercises 21 to 24, write the repeating decimal expansions as an infinite geometric series and find the sum of each one.

21. $0.8888\underline{8}$

22. 0.10101010

23. $0.123123\underline{123}$

24. $1.23452\underline{345}$

25. Prove that an infinite decimal expansion that repeats periodically from some point on must represent a rational number.

The symbol $k!$, called k-**factorial** is defined by

$$k! = \begin{cases} 1, & k = 0, \\ k(k-1)\cdots 3\cdot2\cdot1, & k \geq 1. \end{cases}$$

Thus $0! = 1$, $1! = 1$, $2! = 2\cdot1 = 2$, $3! = 3\cdot2\cdot1 = 6$, and so on. Rewrite each of the infinite series 26 to 31 using factorial notation. Then write out the first three terms.

26. $\displaystyle\sum_{k=1}^{\infty} \dfrac{3^{-k}}{1\cdot2\cdots k}$

27. $\displaystyle\sum_{k=1}^{\infty} \dfrac{1}{2\cdot4\cdots(2k)}$

28. $\displaystyle\sum_{k=1}^{\infty} \dfrac{2^k}{2\cdot4\cdot6\cdots(2k)}$

29. $\displaystyle\sum_{k=1}^{\infty} \dfrac{1}{k(k+1)(k+2)\cdots(2k-1)(2k)}$

30. $\displaystyle\sum_{k=1}^{\infty} \dfrac{1\cdot3\cdot5\cdots(2k-1)}{2\cdot4\cdot6\cdots(2k)}$

31. $\displaystyle\sum_{k=0}^{\infty} \dfrac{(-1)^k}{1\cdot3\cdots(2k+1)}$

In Exercises 32 to 37, write out s_n for $n = 1, 2$, and 3. Also find $\lim_{n\to\infty} s_n$

32. $s_n = \left(1 + \dfrac{1}{n}\right)$

33. $s_n = \left(2 - \dfrac{1}{n}\right)$

34. $s_n = \dfrac{3n^2-1}{4n^2+n}$

35. $s_n = \dfrac{2^n}{3^{n+1}}$

36. $s_n = \dfrac{3n+4}{n}$

37. $s_n = \dfrac{n^{10}+1}{n(n^9+1)}$

Verify the equations in Exercises 38 to 41.

38. $\displaystyle\sum_{k=1}^{\infty} \dfrac{1}{10^k} = \dfrac{1}{9}$

39. $\displaystyle\sum_{k=2}^{\infty} \dfrac{1}{3^k} = \dfrac{1}{6}$

40. $\displaystyle\sum_{k=1}^{\infty} \left(-\dfrac{1}{4}\right)^k = -\dfrac{1}{5}$

41. $\displaystyle\sum_{k=3}^{\infty} \dfrac{1}{k(k+1)} = \dfrac{1}{3}$

42. Find an infinite series with positive terms that converges to 3.

43. Find an infinite series with alternating positive and negative terms that converges to 3.

44. Assume that the series $\sum_{k=1}^{\infty} a_k = s$ is convergent. Show that $\sum_{k=m}^{\infty} a_k$ is also convergent if $m > 1$. What is the sum of the second series in terms of s and the terms a_k?

45. **Cantor set**. From the interval $[0, 1]$, the open middle third $(\frac{1}{3}, \frac{2}{3})$ is deleted. Then the open middle third is deleted from each of the two remaining closed intervals, then the open middle third from each of the four remaining closed intervals, and so on. The set C remaining after the entire infinite sequence of deletions is called the **Cantor middle-third set**.

(a) Find a formula for c_n, the sum of the lengths of the intervals remaining after n steps.

(b) Show that $\lim_{n \to \infty} c_n = 0$, showing that C has total length zero.

SECTION 2   TAYLOR SERIES

2A   Taylor Polynomials

A polynomial f of degree n is just a sum of numerical multiples of powers x: $f(x) = c_0 + c_1 x + \cdots + c_n x^n$. To examine the behavior of f near a point $x = a$, it's useful to be able to write $f(x)$ as a polynomial in powers of $(x - a)$. The next theorem gives a simple formula for the coefficients a_k in terms of derivatives $f^{(k)}(a)$ of $f(x)$ at a and the **factorial** function defined by $k! = 1 \cdot 2 \cdot 3 \cdots k$ for $k \geq 1$ and $0! = 1$.

2.1   Theorem.   If $f(x)$ is a polynomial of degree n of the form

$$f(x) = a_0 + a_1(x - a) + a_2(x - a)^2 + \cdots + a_n(x - a)^n, \text{ then } a_k = \frac{1}{k!} f^{(k)}(a).$$

Thus $a_0 = f(a)$, $a_1 = f'(a)$, $a_2 = \frac{1}{2} f''(a)$, $a_3 = \frac{1}{6} f''(a)$, $a_4 = \frac{1}{24} f^{(4)}(a)$, and so on.

Proof.   To prove the formula for the coefficients a_k we note first that k successive differentiations of the terms $a_j(x - a)^j$ of $f(x)$ with $0 \leq j \leq k - 1$ give zero, starting with the constant a_0 and going through the term of degree $k - 1$. All the remaining terms but the one of degree k are zero at $x = a$, because each contains a positive power of $(x - a)$. Hence the only surviving term in the computation of $f^{(k)}(a)$ is

$$\frac{d^k}{dx^k} \left[a_k(x - a)^k \right] = 1 \cdot 2 \cdots (k - 1) \cdot k a_k.$$

In other words, $f^{(k)}(a) = k! a_k$. The formula for a_k follows upon division by $k!$.   ∎

EXAMPLE 1   It will follow from the next theorem that every polynomial has the form in Theorem 2.1 using powers of $x - a$ for arbitrary a. Let $f(x) = 1 - x^2 + 2x^3$. To write $f(x)$ in terms of powers of $x - a$ with $a = 1$, we compute

$$f(1) = 2, \quad f'(1) = 4, \quad f''(1) = 10, \quad f'''(1) = 12.$$

Hence

$$f(x) = 2 + \frac{4}{1!}(x - 1) + \frac{10}{2!}(x - 1)^2 + \frac{12}{3!}(x - 1)^3$$

$$= 2 + 4(x - 1) + 5(x - 1)^2 + 2(x - 1)^3.$$

More generally, suppose that $f(x)$ is a function, not necessarily a polynomial, that has n derivatives at $x = a$. We can use the coefficients a_k given by Theorem 2.1 to produce a polynomial

$$T_n(x) = \sum_{k=0}^{n} \frac{1}{k!} f^{(k)}(a)(x - a)^k,$$

called the nth degree **Taylor polynomial** of $f(x)$ at $x = a$. Theorem 2.1 shows that if $f(x)$ is itself a polynomial of degree at most n, then the coefficients a_k are designed so that $T_n(x) = f(x)$ for all real x. When $f(x)$ is not a polynomial, $T_n(x)$ will not usually be equal to $f(x)$ except at $x = a$. The difference $f(x) - T_n(x) = R_n(x)$ is the **Taylor remainder**. If $R_n(x)$ is small $T_n(x)$ will be a good approximation to $f(x)$. Here is a simple estimate for the size of the remainder, under the assumption that $f^{(n+1)}(x)$ is continuous on an interval containing a.

2.2   Taylor Remainder Theorem.   Suppose f has $n + 1$ continuous derivatives on an open interval containing a. Then for all x in the interval,

$$f(x) = \sum_{k=0}^{n} \frac{1}{k!} f^{(k)}(a)(x - a)^k + R_n(x)$$

where

$$R_n(x) = \frac{1}{(n+1)!} f^{(n+1)}(c)(x - a)^{n+1},$$

and c is a number between x and a. If $f(x)$ is a polynomial of degree n, then $R_n(x)$ is identically zero, so $f(x)$ is a polynomial in powers of $x - a$.

Proof.   With x held fixed and $x \neq a$, define the unique number K by

$$f(x) = \sum_{k=0}^{n} \frac{1}{k!} f^{(k)}(a)(x - a)^k + K(x - a)^{n+1}.$$

With K determined this way, define $g(t)$ by

$$g(t) = -f(x) + \sum_{k=0}^{n} \frac{1}{k!} f^{(k)}(t)(x - t)^k + K(x - t)^{n+1}.$$

Note that $g(a) = 0$ because of the way K is defined, and that $g(x) = 0$ no matter what value K has. Applying the product rule to the terms in the summation over k, we find that differentiation with respect to t gives

$$g'(t) = \sum_{k=0}^{n} \frac{1}{k!} f^{(k+1)}(t)(x - t)^k - \sum_{k=1}^{n} \frac{1}{(k-1)!} f^{(k)}(t)(x - t)^{k-1}$$
$$- (n+1)K(x - t)^n.$$

All but two terms cancel, so

$$g'(t) = \frac{1}{n!} f^{(n+1)}(t)(x - t)^n - (n+1)K(x - t)^n.$$

By the Mean-Value Theorem for derivatives, there is a number c between x and a such that $g'(c) = 0$. (Recall that $g(x) = g(a) = 0$.) But the equation $g'(c) = 0$

allows us to solve for K to get

$$K = \frac{1}{(n+1)!} f^{(n+1)}(c),$$

which is what we wanted to show.   ∎

The nth degree Taylor polynomial $T_n(x)$ is also called the **nth degree Taylor approximation** to $f(x)$ about $x = a$. Note that to increase the degree of approximation, all we do is add another term without altering the previous terms. Without worrying about convergence, we can write the *infinite* **Taylor series**.

2.3

$$\sum_{k=0}^{\infty} \frac{1}{k!} f^{(k)}(a)(x-a)^k$$

$$= f(a) + f'(a)(x-a) + \frac{1}{2!} f''(a)(x-a)^2$$

$$+ \frac{1}{3!} f'''(a)(x-a)^3 + \cdots$$

and arrive at the nth degree Taylor approximation by stopping after $(n+1)$ terms. The infinite series is defined only when $f(x)$ has derivatives of all orders at $x = a$, and even then may not converge except at $x = a$.

EXAMPLE 2   Let $f(x) = x^{1/2}$ for $x \geq 0$. Then $f'(x) = (\frac{1}{2})x^{-1/2}$, $f''(x) = -(\frac{1}{4})x^{-3/2}$, $f'''(x) = (\frac{3}{8})x^{-5/2}$, and so on. We find, with $a = 1$ in Taylor's formula

$$\sqrt{x} = 1 + \frac{1}{1!}\left(\frac{1}{2}\right)(x-1) + \frac{1}{2!}\left(-\frac{1}{4}\right)(x-1)^2 + R_2(x)$$

$$= 1 + \frac{1}{2}(x-1) - \frac{1}{8}(x-1)^2 + R_2(x),$$

where $R_2(x) = (1/3!)\frac{3}{8}c^{-5/2}(x-1)^3$, and c is somewhere between x and 1. Note that the first two terms of the approximation give the function T_1 whose graph is the tangent line to the graph of $y = \sqrt{x}$ at $x = 1$. The first three terms describe a quadratic function T_2 that approximates $\sqrt{x}$ near $x = 1$. The graphs of both approximations are shown in Figure 14.1(a). Note that the approximations get better the closer we get to $x = 1$.

EXAMPLE 3   The Taylor approximations of $f(x) = e^x$ are particularly simple to compute, because $f^{(k)}(x) = e^x$ for $k = 1, 2, 3, \ldots$. Since $f^{(k)}(0) = e^0 = 1$, the formal infinite series of Equation 2.3, with $a = 0$, becomes

$$\sum_{k=0}^{\infty} \frac{1}{k!} x^k = 1 + \frac{1}{1!}x + \frac{1}{2!}x^2 + \frac{1}{3!}x^3 + \cdots.$$

The partial sums $T_1(x) = 1 + x$ and $T_2(x) = 1 + x + \frac{1}{2}x^2$ are compared with e^x graphically in Figure 14.1(b). The remainder after $n + 1$ terms is

$$R_n(x) = \frac{1}{(n+1)!} e^c x^{n+1}.$$

FIGURE 14.1

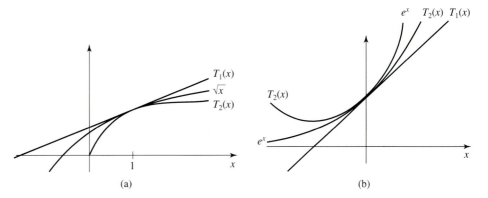

(a)   (b)

We can then estimate that e^x differs from $\sum_{k=1}^{n} \frac{1}{k!}x^k$ by at most

$$e^x - \sum_{k=1}^{n} \frac{1}{k!}x^k = \frac{1}{(n+1)!}e^c x^{n+1}$$

$$\leq \frac{e}{(n+1)!}, \quad \text{if } 0 \leq x \leq 1.$$

2B   Convergence of Taylor Series
Taylor's formula in the form

$$f(x) - \sum_{k=0}^{n} f^{(k)}(a)(x-a)^k = R_n(x)$$

gives us a way of showing that in some cases

$$f(x) = \sum_{k=0}^{\infty} \frac{1}{k!} f^{(k)}(a)(x-a)^k,$$

with the series on the right converging for x in at least some subinterval of the domain of f. All we have to do is show that

$$\lim_{n \to \infty} R_n(x) = 0$$

for the values of x in question. This at once proves convergence of the series and shows that the sum at x is $f(x)$.

2.4   Here is a list of important Taylor expansions that many people find useful to remember.

(a) $e^x = \sum_{k=0}^{\infty} \frac{x^k}{k!} = 1 + \frac{x}{1!} + \frac{x^2}{2!} + \frac{x^3}{3!} + \cdots, \quad -\infty < x < \infty$

(b) $\cos x = \sum_{k=0}^{\infty} \frac{(-1)^k}{(2k)!} x^{2k} = 1 - \frac{x^2}{2!} + \frac{x^4}{4!} + \frac{x^6}{6!} + \cdots, \quad -\infty < x < \infty$

(c) $\sin x = \sum_{k=0}^{\infty} \frac{(-1)^k}{(2k+1)!} x^{2k+1} = x - \frac{x^3}{3!} + \frac{x^5}{5!} - \frac{x^7}{7!} + \cdots, \quad -\infty < x < \infty$

(d) $\ln(1+x) = \sum_{k=1}^{\infty} \frac{(-1)^{k+1}}{k} x^k = x - \frac{x^2}{2} + \frac{x^3}{3} - \frac{x^4}{4} + \cdots, \quad -1 < x < 1$

EXAMPLE 4   As we showed in Example 3, Taylor's formula for e^x with $a = 0$ is

$$e^x = 1 + \frac{x}{1!} + \frac{x^2}{2!} + \frac{x^3}{3!} + \cdots + \frac{x^n}{n!} + R_n(x),$$

where $R_n(x) = \frac{1}{(n+1)!} e^c x^{n+1}$ for some c between x and 0. To prove the infinite series expansion 2.4(a), we have to show that the remainder satisfies

$$\lim_{n \to \infty} \frac{1}{(n+1)!} e^c x^{n+1} = 0$$

for all real numbers x. Pick a fixed value for x. Since $c \le |c| \le |x|$, we see that $e^c < e^{|x|}$ for all relevant values of c. Hence, with x fixed, all we need to show is that

$$\lim_{n \to \infty} \frac{1}{(n+1)!} x^{n+1} = 0.$$

We prove this as follows. Choose $k > l \ge 2x > 0$, hold l fixed, and write

$$\frac{x^k}{k!} = \frac{x}{1} \cdot \frac{x}{2} \cdots \frac{x}{l} \cdot \frac{x}{l+1} \cdots \frac{x}{k}.$$

By the assumption on k, l and x, we have $x/l \le \frac{1}{2}$. Thus for $0 < x < l$ we have

$$\frac{x^k}{k!} \le \frac{x^l}{l!} \frac{1}{2^{k-l}}.$$

Letting $k \to \infty$ shows that $x^k/k! \to 0$. Allowing $-l \le x < 0$ only makes the sign alternate. Similar arguments apply to the other series listed previously, and these are left as exercises.

EXAMPLE 5   To find a series expansion for e^{-2x}, there is no need to start from scratch. Simply replace x by $-2x$ everywhere in 2.4(a):

$$e^{-2x} = \sum_{k=0}^{\infty} \frac{1}{k!} (-2)^k x^k.$$

To find an expansion for $\ln(1-x)$ in powers of x, replace x by $-x$ in 2.4(d):

$$\ln(1-x) = \sum_{k=1}^{\infty} \frac{(-1)^{k+1}}{k} (-1)^k x^k, \quad -1 < -x < 1,$$

$$= -\sum_{k=1}^{\infty} \frac{1}{k} x^k, \quad -1 < x < 1.$$

FIGURE 14.2

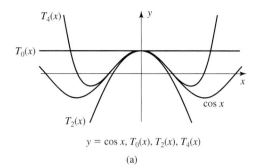

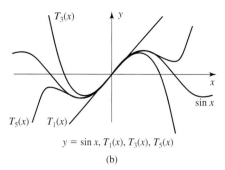

$$y = \cos x, T_0(x), T_2(x), T_4(x)$$

(a)

$$y = \sin x, T_1(x), T_3(x), T_5(x)$$

(b)

EXAMPLE 6   Figure 14.2(a) shows the graph of $\cos x$ along with the Taylor expansion partial sum graphs of $T_0(x) = 1$, $T_2(x) = 1 - \frac{1}{2}x^2$ and $T_4(x) = 1 - \frac{1}{2}x^2 + \frac{1}{24}x^4$.

EXAMPLE 7   Figure 14.2(b) shows the graph of $\sin x$ along with the Taylor expansion partial sum graphs of $T_1(x) = x$, $T_3(x) = x - \frac{1}{6}x^3$ and $T_5(x) = x - \frac{1}{6}x^3 + \frac{1}{120}x^5$.

EXERCISES

In Exercises 1 to 4, write the polynomials in the indicated form; that is, find the coefficients c_k by using Theorem 2.1.

1. $1 + x + x^2 = c_0 + c_1(x - 2) + c_2(x - 2)^2$

2. $2x - x^3 = c_0 + c_1(x - 2) + c_2(x - 2)^2 + c_3(x - 2)^3$

3. $1 + x^2 = c_0 + c_1(x + 1) + c_2(x + 1)^2$

4. $(1 - x) + (1 - x)^2 = c_0 + c_1 x + c_2 x^2$

In Exercises 5 to 8, find the coefficients c_0, c_1, c_2 in the indicated Taylor expansion. Compute the remainder as in Theorem 2.2.

5. $x^{1/3} = c_0 + c_1(x - 1) + c_2(x - 1)^2 + R_2(x)$

6. $x/(1 + x^2) = c_0 + c_1 x + c_2 x^2 + R_2(x)$

7. $1/x = c_0 + c_1(x + 1) + c_2(x + 1)^2 + R_2(x)$

8. $e^{2x} = c_0 + c_1 x + c_2 x^2 + R_2(x)$

In Exercises 9 to 14, find infinite series expansions for each of the following functions about $x = 0$, by modifying one of the series given in Equations 2.4.

9. e^{-x} **10.** $\cos 2x$ **11.** $\sin(x/2)$

12. $\ln(1 - x^2)$ **13.** e^{x^2} **14.** xe^x

Use the definitions of $\cosh x$ and $\sinh x$ along with the Taylor expansion for e^x to establish the Taylor expansions in Exercises 15 and 16.

15. $\cosh x = (e^x + e^{-x})/2 = \displaystyle\sum_{k=0}^{\infty} \frac{x^{2k}}{(2k)!}$

16. $\sinh x = (e^x - e^{-x})/2 = \displaystyle\sum_{k=0}^{\infty} \frac{x^{2k+1}}{(2k + 1)!}$

17. By using a Taylor expansion about $x = 0$, prove the binomial theorem:

$$(x + a)^n = \sum_{k=0}^{n} \binom{n}{k} a^{n-k} x^k$$

where the binomial coefficients are given by

$$\binom{n}{k} = \frac{n!}{k!(n - k)!} = \frac{n(n - 1)\cdots(n - k + 1)}{k(k - 1)\cdots 2 \cdot 1}.$$

In Exercises 18 to 21, sketch using the same coordinate axes the graphs of f and the Taylor approximations T_k. Also estimate the remainder for $-\pi \le x \le \pi$.

18. $f(x) = \cos x$ and $T_2(x) = 1 - \frac{1}{2}x^2$

19. $f(x) = \cos x$ and $T_4(x) = 1 - \frac{1}{2}x^2 + \frac{1}{24}x^4$

20. $f(x) = \sin x$ and $T_3(x) = x - \frac{1}{6}x^3$

21. $f(x) = \sin x$ and $T_5(x) = x - \frac{1}{6}x^3 + \frac{1}{120}x^5$

22. (a) Show that $\displaystyle\sum_{k=0}^{\infty} x^k = \dfrac{1}{1-x}$, $-1 < x < 1$, of Example 1 in Section 1 is the same as the Taylor expansion of $(1-x)^{-1}$ about $x = 0$.

(b) Derive from part (a) the expansion about $x = 1$

$$\frac{1}{x} = \sum_{k=0}^{\infty} (-1)^k (x-1)^k, \quad 0 < x < 2.$$

(c) Use the identity $\dfrac{1}{1+x} = \dfrac{1}{2+(x-1)} = \dfrac{1/2}{1+(x-1)/2}$ to show that

$$\frac{1}{1+x} = \sum_{k=0}^{\infty} (-1)^k 2^{-k-1} (x-1)^k \text{ for } -1 < x < 3.$$

23. (a) If $f(x) = \cos x$, show that the Taylor coefficients of f about $x = 0$ are

$$\frac{f^{(n)}(0)}{n!} = \begin{cases} 0, & n \text{ odd}, \\ (-1)^{n/2}/n!, & n \text{ even}. \end{cases}$$

(b) Using the method of Example 4 of the text, show that

$$\cos x = \sum_{k=0}^{\infty} \frac{(-1)^k}{(2k)!} x^{2k}, \quad -\infty < x < \infty.$$

24. (a) If $f(x) = \sin x$, show that the Taylor coefficients of f about $x = 0$ are

$$\frac{f^{(n)}(0)}{n!} = \begin{cases} (-1)^{(n-1)/2}/n!, & n \text{ odd}, \\ 0, & n \text{ even}. \end{cases}$$

(b) Using the method of Example 4 of the text, show that

$$\sin x = \sum_{k=0}^{\infty} \frac{(-1)^k}{(2k+1)!} x^{2k+1}, \quad -\infty < x < \infty.$$

25. Show that

$$\ln(1+x) = \sum_{k=1}^{\infty} \frac{(-1)^{k+1}}{k} x^k, \quad \text{for } -1 < x < 1.$$

26. Use $x = -\frac{1}{2}$ in Formula 2.4(d) to show that $\ln 2 = \displaystyle\sum_{k=1}^{\infty} \frac{1}{k 2^k}$.

27. Show that $\left(\displaystyle\sum_{k=0}^{\infty} \frac{(-1)^k}{k!} \right) \left(\displaystyle\sum_{k=0}^{\infty} \frac{1}{k!} \right) = 1.$

28. Show that

$$\sum_{k=0}^{\infty} \frac{(-1)^k \pi^{2k}}{(2k)! 2^{2k}} = 0.$$

29. An even function f is a function such that $f(-x) = f(x)$ for all real x, and an odd function is a function such that $f(-x) = -f(x)$ for all real x.

(a) Show that the odd-order derivatives $f^{(2k+1)}(0)$ of an even function are all zero. [For example, $f(x) = \cos x$.]

(b) Show that the even-order derivatives $f^{(2k)}(0)$ of an odd function are all zero. [For example, $f(x) = \sin x$.]

(c) What conclusions can you make about the Taylor expansions about $a = 0$ of even functions and odd functions?

SECTION 3   CONVERGENCE CRITERIA

3A   Convergence of Sequences

For infinite series in general, we need a theory of convergence where we can prove convergence without already knowing the sum of the series. The key property of real numbers that we assume is the following.

3.1 Nondecreasing Sequence Principle.   Let $\{s_n\}$, $n = 1, 2, 3 \ldots$ be a nondecreasing sequence of real numbers, so that $s_n \leq s_{n+1}$ for $n = 1, 2, 3, \ldots$ Then either there is a finite upper bound b such that $s_n \leq b$ for all n, in which case

(i) $\lim_{n \to \infty} s_n = s$,   for some number $s \leq b$,

or else there is no finite upper bound, in which case

(ii) $\lim_{n \to \infty} s_n = +\infty$.

This principle is a fundamental property of real numbers and indeed is built into their very definition. Our attitude here is to accept the principle as a plausible assertion about the real number line. Figure 14.3 illustrates both cases of the principle.

EXAMPLE 1   Every decimal expansion of the form

$$0.b_1 b_2 b_3 \ldots ,$$

where $0 \le b_k \le 9$, is really an infinite series with kth term $a_k = b_k 10^{-k}$. We observe that

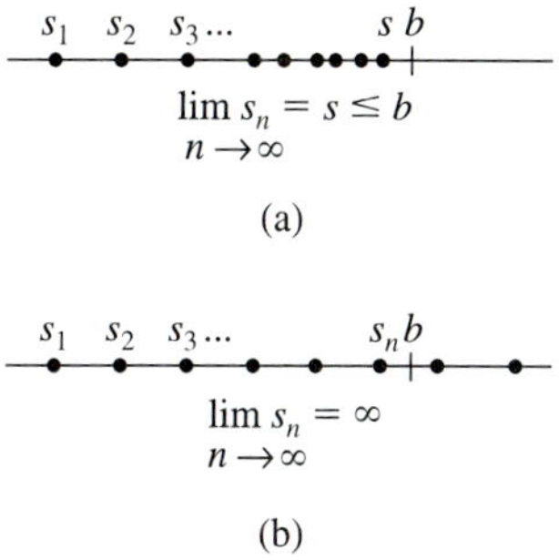

$$s_n = \sum_{k=1}^{n} \frac{b_k}{10^k} \le \sum_{k=1}^{n} \frac{9}{10^k}$$

$$= 9 \sum_{k=1}^{n} \frac{1}{10^k} = 9 \left(-1 + \frac{1 - 10^{-n-1}}{1 - (1/10)} \right) = 1 - \frac{1}{10^n}.$$

Note that $s_{n+1} = s_n + b_{n+1}/10^{n+1} \ge s_n$, and that $s_n \le b = 1$. Hence the nondecreasing sequence principle asserts that the series $\sum_{k=1}^{\infty} b_k 10^{-k}$, and hence the related decimal expansion, converges to some number $s \le 1$. This is what we expect; the largest number we can represent in the decimal form $0.b_1 b_2 b_3 \ldots$ is

$$1 = 0.9999 \ldots .$$

EXAMPLE 2   The infinite series with kth term $a_k = k^{-1} 2^{-k}$ has nth partial sum

$$s_n = \sum_{k=1}^{n} \frac{1}{k 2^k} \le \sum_{k=0}^{n} \frac{1}{2^k}$$

$$= 2(1 - 2^{-n-1}) \le 2.$$

Since $s_{n+1} = s_n + 2^{-n-1}/(n+1) > s_n$, and $s_n \le b = 2$, the nondecreasing sequence principle shows that the series has a sum

$$s = \sum_{k=1}^{\infty} \frac{1}{k 2^k} \le 2.$$

The equations

$$\lim_{n \to \infty} s_n = s \quad \text{and} \quad \lim_{n \to \infty} (s_n - s) = 0$$

are completely equivalent, and very often it's simpler to show that a sequence of numbers $s_n - s$ tends to 0 than to show that s_n tends to s. Here we record two important kinds of sequence that have zero for a limit.

3.2

(a) $\lim_{n \to \infty} \dfrac{1}{n^a} = 0, \quad \text{if } a > 0$

(b) $\lim_{n \to \infty} \dfrac{1}{b^n} = 0, \quad \text{if } |b| > 1$

$s_1 \ s_2 \ s_3 \ldots \qquad \qquad s \ b$

$\lim_{n \to \infty} s_n = s \le b$

(a)

$s_1 \ s_2 \ s_3 \ldots \qquad \qquad s_n b$

$\lim_{n \to \infty} s_n = \infty$

(b)

FIGURE 14.3

The reason in each case is that the denominator, along with its absolute value in (b), tends to infinity as $n \to \infty$. See Exercises 21 and 22 for details.

3B   Sums and Multiples of Series

The general distributive and associative laws for finite sums are

$$c \sum_{k=1}^{n} a_k = \sum_{k=1}^{n} ca_k, \quad \text{and} \quad \sum_{k=1}^{n} a_k + \sum_{k=1}^{n} b_k = \sum_{k=1}^{n} (a_k + b_k).$$

Both laws extend to convergent infinite series as follows.

3.3   Theorem.   Suppose c is a fixed real number and that

$$\sum_{k=1}^{\infty} a_k \quad \text{and} \quad \sum_{k=1}^{\infty} b_k$$

are convergent series of real numbers. Then the series with kth terms ca_k and $a_k + b_k$, respectively, are convergent also, with

$$c \sum_{k=1}^{\infty} a_k = \sum_{k=1}^{\infty} ca_k \quad \text{and} \quad \sum_{k=1}^{\infty} a_k + \sum_{k=1}^{\infty} b_k = \sum_{k=1}^{\infty} (a_k + b_k).$$

Proof.   Use of the distributive and associative laws for finite sums shows that the proof reduces to showing that if

$$s_n = \sum_{k=1}^{n} a_k \quad \text{and} \quad t_n = \sum_{k=1}^{n} b_k,$$

then

$$c \lim_{n \to \infty} s_n = \lim_{n \to \infty} cs_n, \quad \text{and} \quad \lim_{n \to \infty} s_n + \lim_{n \to \infty} t_n = \lim_{n \to \infty} (s_n + t_n).$$

These limit relations are proved in the same way as the more familiar analogues for a continuous variable x, which we assume:

$$c \lim_{x \to \infty} f(x) = \lim_{x \to \infty} cf(x) \quad \text{and}$$

$$\lim_{x \to \infty} f(x) + \lim_{x \to \infty} g(x) = \lim_{x \to \infty} \big(f(x) + g(x)\big). \qquad \blacksquare$$

EXAMPLE 3

We have by Theorem 3.3,

(a) $3 \displaystyle\sum_{k=0}^{\infty} 2^{-k} = \sum_{k=0}^{\infty} 3 \cdot 2^{-k}$

(b) $\displaystyle\sum_{k=0}^{\infty} 2^{-k} + \sum_{k=0}^{\infty} 3^{-k} = \sum_{k=0}^{\infty} (2^{-k} + 3^{-k})$

(c) $2 \displaystyle\sum_{k=0}^{\infty} 2^{-k} + 3 \sum_{k=0}^{\infty} 3^{-k} = \sum_{k=0}^{\infty} (2^{-k+1} + 3^{-k+1})$

It's often convenient to change just finitely many terms in a series, in which case the numerical sum of a convergent series may change, but

3.4   Theorem.   Altering a finite number of terms in an infinite series has no effect on whether the series converges or diverges.

Proof.   Suppose all changes occur among the first M terms, replacing a_k by a_k' for $k = 1, 2, \ldots M$. Then for $n > M$, the new partial sums s_n' differ from s_n by a fixed amount, $s_n' - s_n = d$, independent of n. Hence

$$\lim_{n \to \infty} s_n' = d + \lim_{n \to \infty} s_n,$$

so the two series both converge or both diverge.   ∎

EXAMPLE 4   We saw in Example 2 of Section 1 that

$$\sum_{k=0}^{\infty} \frac{1}{3^k} = 1 + \frac{1}{3} + \frac{1}{3^2} + \frac{1}{3^3} + \cdots = \frac{3}{2}.$$

Leaving out the first two terms, we get

$$\sum_{k=2}^{\infty} \frac{1}{3^k} = \frac{1}{3^2} + \frac{1}{3^3} + \cdots = \frac{3}{2} - 1 - \frac{1}{3} = \frac{1}{6}.$$

Increasing the first term by 2, we get

$$(1+2) + \sum_{k=1}^{\infty} \frac{1}{3^k} = 3 + \frac{1}{3} + \frac{1}{3^2} + \frac{1}{3^3} + \cdots = \frac{3}{2} + 2 = \frac{7}{2}.$$

EXERCISES

Exercises 1 to 8, define s_n for positive integers n. Determine which sequences have limits and which do not. In case the limit value exists, find its value.

1. $s_n = 1 + \dfrac{1}{n^2}$ **2.** $s_n = \dfrac{n}{n+1}$ **3.** $s_n = \dfrac{1+n^2}{n}$

4. $s_n = \dfrac{2^n + 3}{2^n + 2}$ **5.** $s_n = \dfrac{1 + 2^n}{3^n}$ **6.** $s_n = \arctan(n^2)$

7. $s_n = \dfrac{\cos n\pi}{n}$ **8.** $s_n = \dfrac{(2n)!}{2^n n!}$

In Exercise 9 to 12, find a formula for the nth entry in an infinite sequence with the given four values. Then find $\lim_{n \to \infty} s_n$.

9. $\dfrac{1}{2}, \dfrac{2}{3}, \dfrac{3}{4}, \dfrac{4}{5} \cdots$ **10.** $\dfrac{3}{2}, \dfrac{4}{3}, \dfrac{5}{4}, \dfrac{6}{5}, \cdots$

11. $\dfrac{1}{5}, \dfrac{5}{11}, \dfrac{11}{19}, \dfrac{19}{29} \cdots$ **12.** $\dfrac{1}{3}, -\dfrac{3}{5}, \dfrac{5}{7}, -\dfrac{7}{9}, \cdots$

13. Given an infinite sequence of numbers s_n for $n = 1, 2, 3, \ldots$, there is always an infinite *series* $\sum_{k=1}^{\infty} a_k$ that has s_n for its nth partial sum.

(a)   Show that if $a_1 = s_1$ and $a_k = s_k - s_{k-1}$ for $k = 2, 3, 4, \ldots$, then

$$\sum_{k=1}^{n} a_k = s_n \quad \text{for } n = 1, 2, 3, \ldots.$$

(b)   Find a simple formula for the kth term a_k of an infinite series that has $s_n = 1 + (1/n)$ for its nth partial sum. What is the sum of $\sum_{k=1}^{\infty} a_k$?

14. Let $\sum_{n=1}^{\infty} a_k$ be a series of nonnegative terms. Show that the series either converges or else **diverges to infinity**, in the sense that

$$\lim_{n\to\infty} \sum_{k=1}^{n} a_k = +\infty.$$

15. Find $\lim_{n\to\infty}(n+1)^n/n^n$ by letting $n = 1/x$ in $\ln\left((n+1)^n/n^n\right)$ with $x \to 0^+$.

16. Show that $\lim_{k\to\infty} x^k/k! = 0$ for $x > 0$ by choosing $k > l \geq 2x$ and writing

$$\frac{x^k}{k!} = \frac{x}{1}\cdot\frac{x}{2}\cdots\frac{x}{l}\cdot\frac{x}{l+1}\cdots\frac{x}{k}.$$

Then use the assumption $x/l \leq \frac{1}{2}$. What happens if $x < 0$?

In Exercise 17 to 20, write the given expression as a single infinite series.

17. $\displaystyle\sum_{k=1}^{\infty} 2^{-k} + \sum_{k=1}^{\infty} 2^{-k+1}$

18. $\displaystyle 2\sum_{k=0}^{\infty} 2^{-k} - 4\sum_{k=0}^{\infty} 2^{-k-1}$

19. $\displaystyle\sum_{k=0}^{\infty} (2/3)^k - \sum_{k=0}^{\infty} (2/3)^{k+1}$

20. $\displaystyle\sum_{k=1}^{\infty} \frac{k}{k^3+1} - \sum_{k=1}^{\infty} \frac{1}{k^3+1}$

21. Prove that $\lim_{n\to\infty} n^{-\alpha} = 0$ if $\alpha > 0$. [*Hint:* To make $n^{-\alpha} < \epsilon$, make $n > \epsilon^{-1/\alpha}$.]

22. Prove that $\lim_{n\to\infty} b^{-n} = 0$ if $|b| > 1$. [*Hint:* Show $|b|^n = (1+\delta)^n \geq 1 + n\delta$ by the binomial theorem.]

3C   Series with Nonnegative Terms

There is no universal criterion, other than the definition, for deciding about the convergence or divergence of infinite series. However, the tests we explain next have been found to be useful for large classes of series that have practical importance. Here is the simplest test of all to apply, but the only conclusion we can draw from it is *divergence* of a series.

3.5   Term Test for Divergence.   If $\sum_{k=1}^{\infty} a_k$ converges, then $\lim_{k\to\infty} a_k = 0$. In other words, if a_k fails to tend to 0 as $k \to \infty$, then the series diverges.

Proof.   Let $s_n = \sum_{k=1}^{n} a_k$. By assumption $\lim_{n\to\infty} s_n = s$, for some finite number s. Hence $\lim_{n\to\infty} s_{n-1} = s$ also. It follows from $s_n - s_{n-1} = a_n$ that

$$\lim_{n\to\infty} a_n = \lim_{n\to\infty} (s_n - s_{n-1})$$
$$= \lim_{n\to\infty} s_n - \lim_{n\to\infty} s_{n-1}$$
$$= s - s = 0. \qquad \blacksquare$$

EXAMPLE 5   The series

$$\sum_{k=1}^{\infty} \frac{k}{k+1} = \frac{1}{2} + \frac{2}{3} + \frac{3}{4} + \cdots$$

has kth term $a_k = k/(k+1)$. Since

$$\lim_{k\to\infty} a_k = \lim_{k\to\infty} \frac{k}{k+1}$$
$$= \lim_{k\to\infty} \frac{1}{1 + (1/k)} = 1,$$

the series fails to converge because a_n doesn't tend to 0.

Warning.   It is *not* true, just because $\lim_{n\to x} a_n = 0$, that the series $\sum_{k=1}^{\infty} a_k$ converges. The **harmonic series** $\sum_{k=1}^{\infty}(1/k)$, shown to diverge in Example 6, is a counterexample, because $\lim_{k\to 0} 1/k = 0$, but the series diverges.

There are close analogies between integrals over an infinite interval and infinite series. Thus the **improper integral** of $f(x)$ from a to ∞ is **convergent** if it has a finite value determined by

$$\int_a^{\infty} f(x)\,dx = \lim_{b \to \infty} \int_a^b f(x)\,dx;$$

otherwise, the integral is **divergent**. For example,

$$\int_1^{\infty} e^{-px}\,dx = \lim_{b \to \infty} \int_1^b e^{-px}\,dx$$

$$= \lim_{b \to \infty} \frac{e^{-p} - e^{-pb}}{p} = \frac{e^{-p}}{p}, \quad \text{if } p > 0.$$

If $p < 0$, the computation shows that the improper integral diverges to ∞. (What about $p = 0$?) For another example, consider

$$\int_0^{\infty} \frac{dx}{1 + x^2} = \lim_{b \to \infty} \int_0^b \frac{dx}{1 + x^2}$$

$$= \lim_{b \to \infty} [\arctan b - \arctan 0] = \frac{\pi}{2}.$$

The analogy with series is that to compute an improper integral you first compute "proper" integrals over finite intervals and then find their limit over intervals with length tending to ∞. The next theorem shows that there is a very useful connection between the convergence and divergence of particular infinite series and improper integrals.

3.6   Integral Test.   Let $\sum_{k=1}^{\infty} a_k$ be a series of positive terms, and suppose f is a decreasing function such that $f(k) = a_k$ for $k = 1, 2, 3 \ldots$. Then the series and improper integral,

$$\sum_{k=1}^{\infty} a_k \quad \text{and} \quad \int_1^{\infty} f(x)\,dx,$$

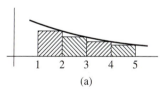

(a)

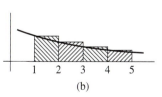

(b)

FIGURE 14.4

either both converge or both diverge.

Proof.   Suppose first that the integral converges. Looking at Figure 14.4(a) shows that, by comparing areas, we have

$$\sum_{k=2}^{N} a_k \leq \int_1^{N-1} f(x)\,dx \leq \int_1^{\infty} f(x)\,dx < \infty.$$

The series converges because the partial sums are increasing (because $a_k > 0$) and bounded above by the value of the integral. Then the nondecreasing sequence principle (3.1) applies.

Figure 14.4(b) shows that

$$\int_1^N f(x)\, dx \leq \sum_{k=1}^N a_k \leq \sum_{k=1}^\infty a_k < \infty,$$

so again Principle 3.1 applies to show that the integral converges to a finite number as $N \to \infty$ if the series converges. ∎

EXAMPLE 6   The **harmonic series** $\sum_{k=1}^\infty (1/k)$ diverges, because with $f(x) = 1/x$, we have $f(k) = 1/k$. But

$$\int_1^\infty \frac{1}{x}\, dx = \lim_{N \to \infty} \int_1^N \frac{1}{x}\, dx$$

$$= \lim_{N \to \infty} \ln N = \infty.$$

The series diverges even though $\lim_{k \to \infty} (1/k) = 0$, as we pointed out in the warning about misapplication of the term test.

EXAMPLE 7   To decide about the **p-series** $\sum_{k=1}^\infty k^{-p}$ for $p > 0$, let $f(x) = 1/x^p$. We have, for $p \neq 1$,

$$\int_1^\infty \frac{1}{x^p}\, dx = \lim_{N \to \infty} \int_1^N \frac{1}{x^p}\, dx$$

$$= \lim_{N \to \infty} \left[\frac{1}{(1-p)x^{p-1}} \right]_1^N$$

$$= \lim_{N \to \infty} \frac{1}{1-p} \left[\frac{1}{N^{p-1}} - 1 \right] = \begin{cases} \dfrac{1}{p-1}, & p > 1, \\ +\infty, & 0 < p < 1. \end{cases}$$

Hence we have convergence for $p > 1$ and divergence for $p < 1$. The case $p = 1$ is the harmonic series, and when $p \leq 0$ the terms of the series fail to tend to zero, so the series diverges by the term test. For $p > 1$, the p-series defines a function $\zeta(p)$ called the **Riemann zeta-function**:

$$\zeta(p) = \sum_{k=1}^\infty \frac{1}{k^p}.$$

For $p = 2$ and $p = 1.1$, the series

$$\zeta(2) = \sum_{k=1}^\infty \frac{1}{k^2} \quad \text{and} \quad \zeta(1.1) = \sum_{k=1}^\infty \frac{1}{k^{1.1}}$$

both converge. But for $p = \frac{1}{2}$ and $p = 0.9$, the series

$$\sum_{k=1}^\infty \frac{1}{\sqrt{k}} \quad \text{and} \quad \sum_{k=1}^\infty \frac{1}{k^{0.9}}$$

both diverge.

Application of the integral test usually depends on being able to compute some indefinite integral, but examples in which the computation is awkward can sometimes be handled by comparison with a related series.

3.7   Comparison Test.   Suppose $0 \le a_k \le b_k$.

    (i) If $\sum_{k=1}^{\infty} b_k$ converges, then $\sum_{k=1}^{\infty} a_k$ converges.
    (ii) If $\sum_{k=1}^{\infty} a_k$ diverges, then $\sum_{k=1}^{\infty} b_k$ diverges.

Proof.   To prove (i), note that

$$s_n = \sum_{k=1}^{n} a_k \le \sum_{k=1}^{n} b_k \le \sum_{k=1}^{\infty} b_k = b.$$

Since $a_k \ge 0$, the partial sums s_n form a nondecreasing sequence, bounded by b. Then $\sum_{k=1}^{\infty} a_k$ converges by Principle 3.1(i). To prove (ii), note that

$$s_n = \sum_{k=1}^{n} a_k \le \sum_{k=1}^{n} b_k,$$

and that by Principle 3.1(ii), $\lim_{n \to \infty} s_n = \infty$. Hence $\sum_{k=1}^{\infty} b_k$ diverges also. ∎

EXAMPLE 8   Consider the series $\displaystyle\sum_{k=2}^{\infty} \frac{1}{k^2 \ln k}$. Since $\ln k \ge 1$ for $k \ge 3$, we have

$$0 \le \frac{1}{k^2 \ln k} \le \frac{1}{k^2}, \quad \text{for } k \ge 3.$$

Since $\sum_{k=3}^{\infty} 1/k^2$ is a convergent p-series by the integral test, we know that $\sum_{k=3}^{\infty} 1/(k^2 \ln k)$ converges by the comparison test. Hence the given series, with the additional term $1/(4 \ln 2)$, converges also.

EXAMPLE 9   Consider the series

$$\sum_{k=1}^{\infty} \frac{1}{k^{1/2}(k+1)^{1/3}}.$$

Since, for $k \ge 1$,

$$\frac{1}{(k+1)^{5/6}} = \frac{1}{(k+1)^{1/2}(k+1)^{1/3}} \le \frac{1}{k^{1/2}(k+1)^{1/3}},$$

the given series diverges by comparison with the divergent p-series for $p = \frac{5}{6}$.

3D   Absolute Convergence
If all terms of a series are nonpositive from some point on, we can simply multiply the series by (-1) and apply the tests of the previous subsection. For a series that

has infinitely many terms a_k of both signs, we try if possible to show that the series with kth term $|a_k|$ converges. If

$$\sum_{k=1}^{\infty} |a_k|$$

converges we say that $\sum_{k=1}^{\infty} a_k$ is **absolutely convergent**. The special terminology is justified because absolute convergence is in general *stronger* than ordinary convergence in the sense that every absolutely convergent series converges, but not conversely. The key theorem is this:

3.8   Theorem.   If $\sum_{k=1}^{\infty} |a_k|$ converges, then so does $\sum_{k=1}^{\infty} a_k$. In other words, absolute convergence implies convergence.

Proof.   Since $0 \le a_k + |a_k| \le 2|a_k|$, the comparison test shows that $\sum_{k=1}^{\infty} (a_k + |a_k|)$ converges, because $2\sum_{k=1}^{\infty} |a_k| = \sum_{k=1}^{\infty} 2|a_k|$ is assumed to converge. Hence

$$\sum_{k=1}^{\infty} a_k = \sum_{k=1}^{\infty} (a_k + |a_k|) - \sum_{k=1}^{\infty} |a_k|$$

converges also, because it is the difference of two convergent series.   ∎

EXAMPLE 10   The series

$$\sum_{k=1}^{\infty} (-1)^k \frac{1}{k^2} = -1 + \frac{1}{2^2} - \frac{1}{3^2} + \frac{1}{4^2} - \cdots$$

converges and even converges absolutely, because

$$\sum_{k=1}^{\infty} \left| (-1)^k \frac{1}{k^2} \right| = \sum_{k=1}^{\infty} \frac{1}{k^2}$$

is a convergent p-series for $p = 2$.

EXAMPLE 11   The series $\sum_{k=1}^{\infty} \dfrac{(-1)^k}{k\sqrt{k+1}}$ converges absolutely by the comparison test for series with positive terms. The reason is that

$$\left| \frac{(-1)^k}{k\sqrt{k+1}} \right| = \frac{1}{k\sqrt{k+1}} \le \frac{1}{k^{3/2}},$$

and $\sum_{k=1}^{\infty} (1/k^{3/2})$ is a convergent p-series with $p = \frac{3}{2}$.

For series with nonnegative terms, convergence is the same as absolute convergence, so a test that proves one proves the other also.

EXAMPLE 12   The geometric series $\sum_{k=0}^{\infty} r^k$ converges to $1/(1-r)$ for $|r| < 1$, by Example 1 of Section 1. We have

$$\sum_{k=0}^{\infty} \frac{1}{2^k} = 2, \quad \text{with } r = \frac{1}{2},$$

and

$$\sum_{k=0}^{\infty} \frac{(-1)^k}{2^k} = \frac{2}{3}, \quad \text{with} \quad r = -1/2.$$

A series $\sum_{k=0}^{\infty}(\pm 1)2^{-k}$ converges absolutely, whatever choice of sign we make in each term because it converges when we always choose the plus sign. However, there is no general way to determine the sum of the series as there is for the two geometric series with all plus signs or with alternating signs.

The following test applies to series with kth term $a_k \neq 0$ from some point on and deals directly with absolute convergence.

3.9   Ratio Test.   Let $\sum_{k=1}^{\infty} a_k$ be a series for which $\lim_{k\to\infty} |a_{k+1}|/|a_k|$ exists.

(i) If $\lim_{k\to\infty} \left| \dfrac{a_{k+1}}{a_k} \right| < 1$, the series converges absolutely.

(ii) If $\lim_{k\to\infty} \left| \dfrac{a_{k+1}}{a_k} \right| > 1$, or is infinite, the series fails to converge.

If the limit of the ratio $|a_{k+1}/a_k|$ of successive terms fails to exist or if the limit is 1, no assertion is being made about convergence of the series. Note also that if a series of positive terms fails to converge absolutely then it fails to converge at all.

Proof.   We'll assume $a_k > 0$ since we are concerned only with terms of the form $|a_k|$ in the proof.

Case (i).   Since the limit of a_{k+1}/a_k is less than 1, there is a number $r < 1$ such that $a_{k+1}/a_k \leq r$ for all sufficiently large values of k, say $k \geq N$. Thus $a_{k+1} \leq ra_k$ for $k = N, N+1, \ldots$. Hence

$$a_{N+k} \leq ra_{N+k-1} \leq r^2 a_{N+k-2} \leq \cdots \leq r^k a_N.$$

Since $0 \leq r < 1$, the series

$$\sum_{k=0}^{\infty} r^k a_N = a_N \sum_{k=0}^{\infty} r^k$$

converges. Hence $\sum_{k=0}^{\infty} a_{N+k}$ converges also by the comparison test. Including the finitely many terms $a_1, \ldots, a_{N-1}$ shows that $\sum_{k=1}^{\infty} a_k$ converges also.

Case (ii).   This time the limit of a_{k+1}/a_k is bigger than 1, so there is a number $r > 1$ for which $a_{k+1}/a_k \geq r$ for k sufficiently large, say $k \geq N$. Thus $a_{k+1} \geq ra_k$ for $k = N, N+1, \ldots$. Hence

$$a_{N+k} \geq ra_{N+k-1} \geq \cdots \geq r^k a_N.$$

Since $r > 1$, r^k tends to ∞ a $k \to \infty$, and the series diverges by the term test.   ∎

EXAMPLE 13

The series $\sum_{k=1}^{\infty} k^2/2^k$ converges because, with $a_k = k^2/2^k$ and $a_{k+1} = (k+1)^2/2^{k+1}$, the ratio test gives

$$\lim_{k \to \infty} \frac{(k+1)^2/2^{k+1}}{k^2/2^k} = \lim_{k \to \infty} \frac{(k+1)^2 2^k}{k^2 2^{k+1}}$$

$$= \lim_{k \to \infty} \left(1 + \frac{1}{k}\right)^2 \cdot \frac{1}{2} = \frac{1}{2} < 1.$$

EXAMPLE 14

Consider the series $\sum_{k=0}^{\infty} x^k/k!$, where $k!$, k **factorial** is $1 \cdot 2 \cdot 3 \cdots k$ if $k \geq 1$, and $0! = 1$. We have $a_k = x^k/k!$ and $a_{k+1} = x^{k+1}/(k+1)!$

$$\lim_{k \to \infty} \left| \frac{a_{k+1}}{a_k} \right| = \lim_{k \to \infty} \left| \frac{x^{k+1}/(k+1)!}{x^k/k!} \right|$$

$$= \lim_{k \to \infty} \frac{k!|x|}{(k+1)!} = \lim_{k \to \infty} \frac{|x|}{k+1} = 0 < 1.$$

Hence the series of terms depending on x converges by case (i) of Test 3.9 for all x. We have already seen that the series converges to e^x.

3E   Alternating Series

An **alternating series** is one in which the terms are alternately positive and negative. The **alternating harmonic series** is an example:

$$\sum_{k=1}^{\infty} (-1)^k \frac{1}{k} = 1 - \frac{1}{2} + \frac{1}{3} - \frac{1}{4} + \cdots .$$

Some of these series converge by the following criterion.

3.10   Leibniz Test.   If $\sum_{k=1}^{\infty} a_k$ is an alternating series such that

(i) $|a_k| \geq |a_{k+1}|$ for $k = 1, 2, 3 \ldots$,

and

(ii) $\lim_{k \to \infty} a_k = 0,$

then the partial sums s_n converge to a sum s, with error $|s - s_n|$ at most $|a_{n+1}|$.

Proof.   Suppose $a_1 = p_1$, $a_2 = -p_2$, $a_3 = p_3$, and so on, with $p_k \geq 0$. Then the partial sum $\sum_{k=1}^{2n} a_k$ is

$$s_{2n} = (p_1 - p_2) + (p_3 - p_4) + \cdots + (p_{2n-1} - p_{2n}),$$

where the terms $p_{2k-1} - p_{2k}$ are all nonnegative, because $p_k = |a_k| \geq |a_{k+1}| = p_{k+1}$. Hence s_{2n} is nondecreasing as n increases. For the same reason, grouping the terms differently shows that

$$p_1 \geq s_{2n} = p_1 - (p_2 - p_3) - \cdots - (p_{2n-2} - p_{2n-1}) - p_{2n}.$$

Hence the partial sums s_{2n} are bounded above and nondecreasing, while the partial sums $s_{2n+1} = p_1 - (p_2 - p_3) - \cdots - (p_{2n-2} - p_{2n-1}) - (p_{2n} - p_{2n+1})$ are nonincreasing. By Principle 3.1 for bounded sequences

$$\lim_{n \to \infty} s_{2n} = \lim_{n \to \infty} \sum_{k=1}^{2n} a_k = s,$$

for some number s. But $s_{2n+1} = s_{2n} + a_{2n+1}$, so since $\lim_{n \to \infty} a_{2n+1} = 0$ by (ii),

$$\lim_{n \to \infty} s_{2n+1} = \lim_{n \to \infty} \sum_{k=1}^{2n+1} a_k = s$$

also. Hence all s_n converge to s. Since $s_{2n} \leq s \leq s_{2n+1}$ it follows that if m is even $s_m \leq s \leq s_{m+1}$ and if m is odd $s_{m+1} \leq s \leq s_m$. Thus $|s - s_m| \leq |a_{m+1}|$.   ■

EXAMPLE 15   The alternating harmonic series converges, because with $a_k = (-1)^{k+1}/k$, we have

$$\text{(i)} \quad \left| \frac{(-1)^{k+1}}{k} \right| > \left| \frac{(-1)^{k+2}}{k+1} \right|,$$

and

$$\text{(ii)} \quad \lim_{k \to \infty} \frac{(-1)^{k+1}}{k} = 0.$$

Note that the alternating harmonic series fails to converge absolutely because $\sum_{k=1}^{\infty} 1/k$ is divergent.

EXAMPLE 16   The series

$$\sum_{k=1}^{\infty} \frac{(-1)^k}{\ln k} = \frac{1}{\ln 2} - \frac{1}{\ln 3} + \frac{1}{\ln 4} - \cdots$$

converges because (i) $1/\ln k \geq 1/\ln(k+1)$ and (ii) $\lim_{k \to \infty} 1/\ln k = 0$. The Leibniz test implies convergence.

EXERCISES

In Exercises 1 to 6, determine the convergence or divergence of the infinite series by using the term test (for divergence) or the integral test (for convergence *or* divergence). Show carefully how the test you use applies in each case.

1. $\displaystyle\sum_{k=1}^{\infty} \frac{k^2}{k^2 + 1}$ 2. $\displaystyle\sum_{k=1}^{\infty} \frac{k}{k^2 + 1}$

3. $\displaystyle\sum_{k=1}^{\infty} k e^{-k}$ 4. $\displaystyle\sum_{k=1}^{\infty} \left(1 + \frac{1}{k}\right)^k$

5. $\displaystyle\sum_{k=1}^{\infty} \frac{1}{k^2 + 1}$ 6. $\displaystyle\sum_{k=2}^{\infty} \frac{1}{k(\ln k)^2}$

In Exercises 7 to 12, determine the convergence or divergence of the infinite series by using the comparison test or the ratio test. Show carefully how the test you use applies in each case.

7. $\displaystyle\sum_{k=1}^{\infty} \frac{2^k}{3^{k+1}}$ 8. $\displaystyle\sum_{k=1}^{\infty} \frac{1}{(k^2 + k)^{3/2}}$

9. $\displaystyle\sum_{n=2}^{\infty} \frac{1}{n^2 \ln n}$ 10. $\displaystyle\sum_{n=1}^{\infty} \frac{1}{n 2^n}$

11. $\displaystyle\sum_{j=1}^{\infty} \frac{j}{j^3+1}$ **12.** $\displaystyle\sum_{j=1}^{\infty} \frac{j+2}{\sqrt{j+1}}$

In Exercises 13 to 18, determine whether the series converges absolutely or not. For those alternating series that fail to converge absolutely, try to apply the Leibniz test for convergence.

13. $\displaystyle\sum_{k=2}^{\infty} \frac{(-1)^k}{k^2 \ln k}$ **14.** $\displaystyle\sum_{k=1}^{\infty} \frac{(-1)^{k+1}}{k^2}$

15. $\displaystyle\sum_{j=1}^{\infty} \frac{(-1)^j}{\sqrt{j}}$ **16.** $\displaystyle\sum_{j=1}^{\infty} \frac{(-1)^j j}{j+1}$

17. $\displaystyle\sum_{m=0}^{\infty} \frac{(-1)^m}{m^2+1}$ **18.** $\displaystyle\sum_{m=0}^{\infty} \frac{(-1)^m m}{\sqrt{m^2+1}}$

In Exercises 19 to 24, determine the convergence, absolute convergence, or divergence of the series.

19. $\displaystyle\sum_{k=2}^{\infty} \frac{(-1)^k}{\ln(1/k)}$ **20.** $\displaystyle\sum_{k=1}^{\infty} \frac{k}{k^3+1}$

21. $\displaystyle\sum_{k=1}^{\infty} \frac{(-2)^k}{k^2+1}$ **22.** $\displaystyle\sum_{k=1}^{\infty} \frac{k!}{(2k)!}$

23. $\displaystyle\sum_{k=1}^{\infty} \left(1+\frac{1}{k}\right) 2^{-k}$ **24.** $\displaystyle\sum_{k=1}^{\infty} (-1)^k \frac{k!}{(k+1)!}$

In Exercises 25 to 30, determine the convergence or divergence of the series.

25. $\displaystyle\sum_{k=1}^{\infty} \frac{k^2}{3^k}$ **26.** $\displaystyle\sum_{k=1}^{\infty} \frac{k^k}{k!}$

27. $\displaystyle\sum_{k=1}^{\infty} \frac{k!}{k^k}$ **28.** $\displaystyle\sum_{k=2}^{\infty} \frac{(-1)^k}{\sqrt{\ln k}}$

29. $\displaystyle\sum_{k=1}^{\infty} \frac{\sin k}{k^2}$ **30.** $\displaystyle\sum_{n=1}^{\infty} \frac{1}{2n^2-n}$

31. (a) Prove that $\sum_{k=1}^{\infty} kx^k$ converges absolutely for $|x| < 1$.
    (b) Prove that $(1-x)\sum_{k=1}^{\infty} kx^k = \sum_{k=1}^{\infty} x^k$, for $|x| < 1$.

32. Prove that $\sum_{k=1}^{\infty} x^k = x/(1-x)$, for $|x| < 1$.

33. (a) Prove that $\sum_{k=1}^{\infty} kx^k = x/(1-x)^2$, for $|x| < 1$.
    (b) Prove that $\sum_{k=1}^{\infty} k2^{-k} = 2$.

In Exercises 34 to 41, determine the real values of x for which the series converges.

34. $\displaystyle\sum_{k=1}^{\infty} \frac{1}{k} x^k$ **35.** $\displaystyle\sum_{k=1}^{\infty} \frac{1}{k^2} x^k$

36. $\displaystyle\sum_{k=1}^{\infty} \frac{\sin kx}{k^3}$ **37.** $\displaystyle\sum_{k=1}^{\infty} \frac{1}{2^k} (x-1)^k$

38. $\displaystyle\sum_{j=1}^{\infty} \frac{1}{j} x^{2j}$ **39.** $\displaystyle\sum_{j=1}^{\infty} \frac{j}{j^2+1} x^j$

40. $\displaystyle\sum_{k=0}^{\infty} (\ln x)^k$ **41.** $\displaystyle\sum_{k=0}^{\infty} \left(\frac{3}{1+x^2}\right)^k$

42. (a) Show that $\zeta(p)$, which is defined by the p-series as $\zeta(p) = \sum_{k=1}^{\infty} k^{-p}$, is decreasing as p increases, for $p > 1$.
    (b) Show that $(1-2^{-p})\zeta(p) = 1+1/3^p+1/5^p+1/7^p+ \cdots$.
    (c) Show that $(1 - 2^{1-p})\zeta(p) = 1 - 1/2^p + 1/3^p - 1/4^p + \cdots$

43. The function $f(x) = x^x$ is defined for $x > 0$ by $f(x) = e^{x \ln x}$.
    (a) Use l'Hôpital's rule to prove that $\lim_{x\to 0+} x^x = 1$, and so conclude that $\lim_{k\to\infty} (1/k)^{(1/k)} = 1$.
    (b) Prove that $\displaystyle\sum_{k=1}^{\infty} \left(\frac{1}{k}\right)^{1/k}$ diverges.
    (c) Prove that $\sum_{k=1}^{\infty} k^{(1/k)}$ diverges.

44. Prove that if a is a real number and $|b| > 1$ then $\lim_{n\to\infty} \frac{n^a}{b^n} = 0$, by applying the ratio test to $\sum_{n=1}^{\infty} n^a b^{-n}$.

45. Prove that $\displaystyle\sum_{k=2}^{\infty} \frac{1}{k(\ln k)^a}$ converges if $a > 1$ and diverges if $a \le 1$.

*46. Prove that if $a_k \ge 0$ and $\sum_{k=1}^{\infty} a_k$ converges then so do $\sum_{k=1}^{\infty} a_{2k}$ and $\sum_{k=0}^{\alpha} a_{(2k+1)}$. Is the conclusion true without the condition $a_k \ge 0$?

SECTION 4   UNIFORM CONVERGENCE

Let $f_k(\mathbf{x})$, $k = 1, 2, 3, \ldots$, be a sequence of real-valued functions defined for all $\mathbf{x}$ in some set S. Then for each $\mathbf{x}$, we consider the series $\sum_{k=1}^{\infty} f_k(\mathbf{x})$. If it converges for each $\mathbf{x}$ in S, we say that the series **converges pointwise** on S. Calling the limit $f(\mathbf{x})$ for each $\mathbf{x}$ in S, we write

$$f(\mathbf{x}) = \sum_{k=1}^{\infty} f_k(\mathbf{x})$$

$$= \lim_{N \to \infty} \sum_{k=1}^{N} f_k(\mathbf{x}).$$

This means that for each $\mathbf{x}$ in S there is a number $f(\mathbf{x})$ such that, given $\epsilon > 0$, there is an integer K sufficiently large that

$$\left| \sum_{k=1}^{N} f_k(\mathbf{x}) - f(\mathbf{x}) \right| < \epsilon,$$

whenever $N \geq K$.

EXAMPLE 1 The series $\sum_{k=0}^{\infty} x^k$ has for its $(N+1)$st partial sum the finite sum

$$\sum_{k=0}^{N} x_k = \begin{cases} \dfrac{1 - x^{N+1}}{1 - x}, & x \neq 1, \\ N+1, & x = 1. \end{cases}$$

Then

$$\sum_{k=0}^{\infty} x^k = \lim_{N \to \infty} \sum_{k=0}^{N} x^k = \frac{1}{1-x}, \quad \text{for } -1 < x < 1.$$

For real values of x outside the interval $(-1, 1)$, the series fails to converge.

The trigonometric series $\sum_{k=1}^{\infty} (\sin kx)/k^2$ converges pointwise for all real x. The reason is that we can compare its terms with those of the convergent series $\sum_{k=1}^{\infty} 1/k^2$, by observing that

$$\left| \frac{\sin kx}{k^2} \right| \leq \frac{1}{k^2}, \quad k = 1, 2, \ldots .$$

The result is that the given series even converges absolutely.

An infinite series $\sum_{k=1}^{\infty} f_k(\mathbf{x})$ that converges for each $\mathbf{x}$ in a set S to a number $f(\mathbf{x})$ defines a function f on S. However, in general we can conclude very little about the properties of f from pointwise convergence alone. For this reason it's sometimes helpful to consider a stronger form of convergence on S. We say that $\sum_{k=1}^{\infty} f_k$ **converges uniformly** to a function f on a set S, if, given $\epsilon > 0$, there is an integer K such that for all $\mathbf{x}$ in S and for all $N \geq K$.

$$\left| \sum_{k=1}^{N} f_k(\mathbf{x}) - f(\mathbf{x}) \right| < \epsilon$$

The definition just given should be compared carefully with that of pointwise convergence. Notice that uniform convergence implies pointwise convergence, but not conversely. Roughly speaking, uniform convergence of a series of functions defined on a set S means that the series converges with at least a certain minimum rate for

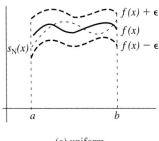

(a) uniform

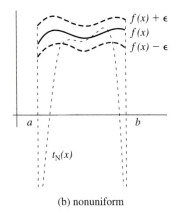

(b) nonuniform

FIGURE 14.5

all points in S. A pointwise convergent series may have points at which the convergence is increasingly slow. Figure 14.5 is a picture of uniform and nonuniform convergence to the same function f; $s_N(x)$ and $t_N(x)$ are Nth partial sums of two series.

To determine that a series converges uniformly, we have the following.

4.1   Weierstrass Test.   Let $\sum_{k=1}^{\infty} f_k$ be a series of real-valued functions defined on a set S. If there is a constant series $\sum_{k=1}^{\infty} p_k$, such that

   1. $|f_k(\mathbf{x})| \leq p_k$ for all $\mathbf{x}$ in S and for $k = 1, 2, \ldots,$

   2. $\displaystyle\sum_{k=1}^{\infty} p_k$ converges,

then $\sum_{k=1}^{\infty} f_k$ converges uniformly to a function f defined on S.

Proof.   The comparison test for series shows that $\sum_{k=1}^{\infty} f_k(\mathbf{x})$ converges (even absolutely) for each $\mathbf{x}$ in S to a number that we'll write $f(\mathbf{x})$. Hence we can write

$$f(\mathbf{x}) - \sum_{k=1}^{N} f_k(\mathbf{x}) = \sum_{k=1}^{\infty} f_k(\mathbf{x}) - \sum_{k=1}^{N} f_k(\mathbf{x}) = \sum_{k=N+1}^{\infty} f_k(\mathbf{x}).$$

It follows that

$$\left| f(\mathbf{x}) - \sum_{k=1}^{N} f_k(\mathbf{x}) \right| \leq \sum_{k=N+1}^{\infty} |f_k(\mathbf{x})| \leq \sum_{k=N+1}^{\infty} p_k.$$

Since $\sum_{k=1}^{\infty} p_k$ converges, we can, given $\epsilon > 0$, find a K such that $\sum_{k=N}^{\infty} p_k < \epsilon$ if $N > K$. This completes the proof, because the number K depends only on ϵ and not on $\mathbf{x}$.   ∎

EXAMPLE 2   The trigonometric series $\sum_{k=1}^{\infty} (\sin kx)/k^2$ converges uniformly for all real x, because

$$\left| \frac{\sin kx}{k^2} \right| \leq \frac{1}{k^2},$$

and $\sum_{k=1}^{\infty} 1/k^2$ converges. However, the power series $\sum_{k=0}^{\infty} x^k$, while it converges pointwise for $-1 < x < 1$, fails to converge uniformly on $(-1, 1)$. See Exercise 6. The Weierstrass test applies on symmetric closed subintervals $[-r, r]$ with $0 < r < 1$ by observing that $|x^k| \leq r^k$ for x on $[-r, r]$ and that $\sum_{k=0}^{\infty} r^k$ converges if $0 \leq r < 1$. Hence the power series converges pointwise on $(-1, 1)$ and uniformly on $[-r, r]$ for any $r < 1$.

The next four theorems are about uniformly convergent series of functions. They all assert that certain limit operations are interchangeable with the summing of a series, provided that certain series converges uniformly. If uniform convergence is replaced by pointwise convergence, then the resulting statements fail to hold in general. See Exercise 9.

4.2   Theorem.   Let $f_1, f_2, f_3, \ldots$ be a sequence of functions defined on a set S in $\mathbb{R}^n$. Suppose $\mathbf{x}_0$ is a limit point of S, and suppose that the limit

$$\lim_{\mathbf{x} \to \mathbf{x}_0} f_k(\mathbf{x})$$

exists for $k = 1, 2, \ldots$. Then

$$\lim_{\mathbf{x} \to \mathbf{x}_0} \sum_{k=1}^{\infty} f_k(\mathbf{x}) = \sum_{k=1}^{\infty} \lim_{\mathbf{x} \to \mathbf{x}_0} f_k(\mathbf{x}),$$

provided the series of numbers on the right converges and the series on the left converges uniformly on S.

Proof.   Let $\lim_{\mathbf{x} \to \mathbf{x}_0} f_k(\mathbf{x}) = a_k$. Then adding and subtracting $\sum_{k=1}^{N} f_k(\mathbf{x})$ and $\sum_{k=1}^{N} a_k$, we get

$$\left| \sum_{k=1}^{\infty} f_k(\mathbf{x}) - \sum_{k=1}^{\infty} a_k \right| \leq \left| \sum_{k=1}^{\infty} f_k(\mathbf{x}) - \sum_{k=1}^{N} f_k(\mathbf{x}) \right|$$

$$+ \left| \sum_{k=1}^{N} f_k(\mathbf{x}) - \sum_{k=1}^{N} a_k \right| + \left| \sum_{k=1}^{N} a_k - \sum_{k=1}^{\infty} a_k \right|. \qquad (1)$$

Now let $\epsilon > 0$. Since $\sum_{k=1}^{\infty} f_k$ converges uniformly, we can choose K such that $N > K$ implies

$$\left| \sum_{k=1}^{\infty} f_k(\mathbf{x}) - \sum_{k=1}^{N} f_k(\mathbf{x}) \right| < \frac{\epsilon}{3}, \quad \text{for all } \mathbf{x} \text{ in } S.$$

Then choose an $N > K$ such that

$$\left| \sum_{k=1}^{N} a_k - \sum_{k=1}^{\infty} a_k \right| < \frac{\epsilon}{3}.$$

Finally, pick $\delta > 0$ so that $|\mathbf{x} - \mathbf{x}_0| < \delta$ implies, via the relation

$$\lim_{\mathbf{x} \to \mathbf{x}_0} \sum_{k=1}^{N} f_k(\mathbf{x}) = \sum_{k=1}^{N} a_k, \quad \text{that} \quad \left| \sum_{k=1}^{N} f_k(\mathbf{x}) - \sum_{k=1}^{N} a_k \right| < \frac{\epsilon}{3}.$$

Then for $\mathbf{x}$ satisfying $|\mathbf{x} - \mathbf{x}_0| < \delta$, the left side of equation (1) is less than ϵ.   ∎

4.3   Corollary.   If $\sum_{k=1}^{\infty} f_k$ is a uniformly convergent series of continuous functions f_k defined on a set S in $\mathbb{R}^n$, then the function f defined by $f(\mathbf{x}) = \sum_{k=1}^{\infty} f_k(\mathbf{x})$ is continuous on S.

In the next two theorems we restrict ourselves to functions of one variable, although by treating one variable at a time, we can apply them to functions of several variables.

4.4   Theorem.   If the series $\sum_{k=1}^{\infty} f_k$ converges uniformly on the interval $[a, b]$, and the functions f_k are continuous on $[a, b]$, then

$$\sum_{k=1}^{\infty} \int_{a}^{b} f_k(x)\, dx = \int_{a}^{b} \left[\sum_{k=1}^{\infty} f_k(x) \right] dx.$$

Proof. By Theorem 4.3 the function $\sum_{k=1}^{\infty} f_k(x)$ is continuous on $[a, b]$ and so is integrable there. We have

$$\int_a^b \left[\sum_{k=1}^{\infty} f_k(x) \right] dx - \sum_{k=1}^{\infty} \int_a^b f_k(x)\, dx = \int_a^b \sum_{k=N+1}^{\infty} f_k(x)\, dx. \qquad (2)$$

Let $\epsilon > 0$, and choose K so large that if $N > K$, then

$$\left| \sum_{k=N+1}^{\infty} f_k(x) \right| < \epsilon(b-a)^{-1}, \quad \text{for all } x \text{ in } [a, b].$$

In general, if $g(x)$ is a continuous function, then

$$\left| \int_a^b g(x)\, dx \right| \leq (b-a) \max_{a \leq x \leq b} |g(x)|,$$

so if $g(x)$ is the sum of the terms $f_k(x)$ from $N+1$ to infinity, then

$$\left| \int_a^b \sum_{k=N+1}^{\infty} f_k(x)\, dx \right| \leq (b-a)\cdot\epsilon\cdot(b-a)^{-1} = \epsilon, \quad \text{for } N > K.$$

Thus the left side of equation (2) is less than ϵ in absolute value for $N > K$. ∎

The interchange of differentiation with the summing of a series requires somewhat more in the way of hypotheses than did the previous theorem on integration.

4.5   Theorem.   Let $f_1, f_2, f_3, \ldots$ be a sequence of continuously differentiable functions defined on an interval $[a, b]$. If $\sum_{k=1}^{\infty} f_k(x) = f(x)$ for all x in $[a, b]$ (pointwise convergence), and if $\sum_{k=1}^{\infty} df_k/dx$ converges uniformly on $[a, b]$, then f is continuously differentiable, and

$$\frac{d}{dx} \sum_{k=1}^{\infty} f_k(x) = \sum_{k=1}^{\infty} \frac{df_k}{dx}(x).$$

Proof.   By the fundamental theorem of calculus

$$\sum_{k=1}^{N} [f_k(x) - f_k(a)] = \sum_{k=1}^{N} \int_a^x f_k'(t)\, dt = \int_a^x \left[\sum_{k=1}^{N} f_k'(t) \right] dt.$$

Using pointwise convergence on the left and uniform convergence on the right to justify letting N tend to infinity in Theorem 4.4, we get $\sum_{k=1}^{\infty} f_k(x) = f(x)$. Hence

$$f(x) - f(a) = \int_a^x \left[\sum_{k=1}^{\infty} f_k'(t) \right] dt.$$

Differentiation of both sides of the last equation gives $f'(x) = \sum\limits_{k=1}^{\infty} f_k'(x)$, which is the conclusion of the theorem.   ∎

EXAMPLE 3   Consider the trigonometric series

$$\sum_{k=1}^{\infty} \frac{\sin kx}{k^4}.$$

The series converges absolutely for all real x, because the terms are dominated by k^{-4}. Furthermore, the series of derivatives of the terms of the given series is

$$\sum_{k=1}^{\infty} \frac{\cos kx}{k^3}.$$

Similarly, this series converges uniformly for all x by the Weierstrass test, because

$$\left| \frac{\cos kx}{k^3} \right| \le \frac{1}{k^3},$$

and because $\sum_{1}^{\infty}(1/k^3)$ converges. Hence by Theorem 4.5,

$$\frac{d}{dx} \sum_{k=1}^{\infty} \frac{\sin kx}{k^4} = \sum_{k=1}^{\infty} \frac{\cos kx}{k^3}.$$

The same kind of argument applies to give

$$\frac{d^2}{dx^2} \sum_{k=1}^{\infty} \frac{\sin kx}{k^4} = -\sum_{k=1}^{\infty} \frac{\sin kx}{k^2}.$$

EXERCISES

1. Show that the series $\sum_{k=0}^{\infty} x^k$ converges uniformly for $-d \le x \le d$ if $0 < d < 1$.

2. (a)   Show that the trigonometric series $\sum_{k=1}^{\infty}(\cos kx/k^2)$ converges uniformly for all real x.
   (b)   Prove that the series of part (a) defines a continuous function for all real x.

3. Show that if a trigonometric series $\dfrac{a_0}{2} + \sum\limits_{k=1}^{\infty}(a_k \cos kx + b_k \sin kx)$ converges uniformly on $[-\pi, \pi]$, then it converges uniformly for *all* real x.

4. (a)   Show that if $|c_k| \le B$ for some fixed number B, then the series

$$u(x, t) = \sum_{k=1}^{\infty} c_k e^{-k^2 t} \sin kx$$

   is a solution of $u_{xx} = u_t$ satisfying $u(0, t) = u(\pi, t) = 0$ when $t > 0$ and x is in $[0, \pi]$. [*Hint:* For arbitrary $\delta > 0$, apply Theorem 4.5 with $t \ge \delta$.]
   (b)   Show that, if $u(x, t)$ in part (a) is defined for $t = 0$ by a series convergent for each x, then $u(x, t)$ is continuous on the set S in $\mathbb{R}^2$ defined by $0 \le t$, $0 < x \le \pi$.

(c) Show that the function $u(x, t)$ is infinitely often differentiable with respect to both x and t, for $t > 0$.

5. Show that if a trigonometric series as displayed in Exercise 3 satisfies the conditions $|a_k| \leq A/k^2$, $|b_k| \leq B/k^2$, for $k = 1, 2, 3, \ldots$ and fixed constants A and B, then the series converges uniformly for all real x.

6. By considering the partial sums of the power series $\sum_{k=0}^{\infty} x^k$ for $-1 < x < 1$, show that the series fails to converge uniformly on $(-1, 1)$. Show uniform convergence for $-\frac{1}{2} \leq x \leq \frac{1}{2}$.

*7. Show that $\sum_{k=1}^{\infty} (-1)^k (1 - x) x^k$ converges uniformly on $[0, 1]$, but that $\sum_{k=1}^{\infty} (1 - x) x^k$ converges only pointwise

on $[0, 1]$. [*Hint:* For the first part use the error estimate in Theorem 3.10.]

8. (a) Assume that the series $\sum_{k=1}^{\infty} k^2 a_k$ and $\sum_{k=1}^{\infty} k^2 b_k$ both converge absolutely. Show that

$$w(x, t) = \sum_{k=1}^{\infty} \sin kx (a_k \cos kat + b_k \sin kat)$$

is a solution of the 1-dimensional wave equation $a^2 w_{xx} = w_{tt}$. [*Hint:* Use the Weierstrass test and Theorem 4.5.]

(b) Show that the solution $w(x, t)$ of part (a) satisfies the boundary conditions $w(0, t) = w(\pi, t) = 0$ for $t \geq 0$ and an initial condition $w(x, 0) = h(x)$, where h is twice continuously differentiable.

SECTION 5   POWER SERIES

A **power series** is an infinite series of the form

$$\sum_{k=0}^{\infty} a_k (x - a)^k = a_0 + a_1(x - a) + a_2(x - a)^2 + \cdots .$$

Such a series defines a function of x for all x for which the series converges. The Taylor series discussed in Section 4 are examples of power series associated with known functions such as e^x, $\cos x$, and $1/(1 - x)$. Our point of view here will be different in that we'll start with the series, rather than with some other representation for the function, and then study the series directly as a function of x. This point of view is essential to the use of power series in solving ordinary differential equations.

5A   Interval of Convergence
We speak of the series $\sum_{k=0}^{\infty} a_k (x - a)^k$ as being a power series "about" $x = a$, because the set of real numbers x for which such a series converges is always either an interval with its midpoint at $x = a$, or else is the whole real line. We won't prove this in general, but the examples will illustrate it. Figure 14.6(a) shows an interval of convergence; one-half its length is called the **radius of convergence**, denoted by R.

FIGURE 14.6

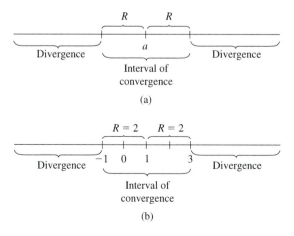

(a)

(b)

EXAMPLE 1   The power series

$$\sum_{k=0}^{\infty} (-1)^k \frac{(x-1)^k}{2^k}$$

is a geometric series of the form $\sum_{k=0}^{\infty} r^k$, with $r = -(x-1)/2$. For a geometric series we saw in Section 1 that it's just when $|r| = |-(x-1)/2| < 1$ that convergence holds, in other words, when $|x - 1| < 2$. Thus the interval of convergence is $-1 < x < 3$, with $x = 1$ as its midpoint. See Figure 14.6(b). The radius of convergence is $R = 2$.

EXAMPLE 2   We can test the power series about $x = 0$ given by

$$\sum_{k=1}^{\infty} \frac{1}{k2^k} x^k$$

for absolute convergence by using the ratio test. The kth term is $a_k = 2^{-k} x^k / k$, so

$$\lim_{k \to \infty} \left| \frac{a_{k+1}}{a_k} \right| = \lim_{k \to \infty} \left| \frac{2^{-k-1} x^{k+1}/(k+1)}{2^{-k} x^k / k} \right|$$

$$= \lim_{k \to \infty} \frac{1}{2} \left(\frac{k}{k+1} \right) |x| = \tfrac{1}{2}|x|.$$

By the ratio test, the series converges absolutely when $\frac{1}{2}|x| < 1$, that is, when $|x| < 2$, and diverges when $\frac{1}{2}|x| > 1$, that is, when $|x| > 2$. Thus the interval of convergence has radius $R = 2$ and is centered at $x = 0$. Because the ratio test gives no information when the limit, in this case $|x|/2$, is equal to 1, we have to check that case separately. The points satisfying $|x|/2 = 1$ are just the points $x = 2$ and $x = -2$. These points are the endpoints of the interval of convergence, and direct substitution into the series shows that at $x = 2$ we have $\sum_{k=1}^{\infty} 1/k$, which diverges; at $x = -2$ we have $\sum_{k=1}^{\infty} (-1)^k/k$, which is a convergent alternating series. Thus the precise interval of convergence is $-2 \le x < 2$.

EXAMPLE 3   The series

$$\sum_{k=0}^{\infty} \frac{1}{k!} x^k$$

is the Taylor expansion of the function e^x, and we showed in Section 2 that the series converges, even absolutely, to e^x for all real values of x. Hence we see that the interval of convergence is $-\infty < x < \infty$, and it's customary to say that the radius of convergence is $R = \infty$.

EXAMPLE 4   The series

$$\sum_{k=0}^{\infty} \frac{1}{3^k(k+2)} x^{2k}$$

has kth term $a_k = x^{2k}/3^k(k+2)$. We apply the ratio test.

$$\lim_{k\to\infty}\left|\frac{a_{k+1}}{a_k}\right| = \lim_{k\to\infty}\left|\frac{x^{2k+2}/3^{k+1}(k+3)}{x^{2k}/3^k(k+2)}\right|$$

$$= \lim_{k\to\infty}|x|^2\frac{k+2}{3(k+3)} = \frac{1}{3}x^2$$

By the ratio test, we have convergence for $|x| < \sqrt{3}$ and divergence for $|x| > \sqrt{3}$. A separate check shows divergence for $x = \pm\sqrt{3}$.

5B   Differentiation and Integration

One reason the power series representation of functions is so useful is that, in the interior of its interval of convergence, we can differentiate and integrate a power series term by term. To see what this means in practice, we first consider some examples. The term "interior" of an interval is meant specifically not to include either endpoint of the interval.

EXAMPLE 5   The Taylor expansion

$$\frac{1}{1-x} = \sum_{k=0}^{\infty} x^k = 1 + x + x^2 + \cdots$$

is valid for $-1 < x < 1$. In the interior of the interval of convergence, that is for $-1 < x < 1$, we integrate both sides and include a constant of integration to get

$$-\ln(1-x) = c + \sum_{k=0}^{\infty}\frac{x^{k+1}}{k+1} = c + x + \frac{x^2}{2} + \frac{x^3}{3} + \cdots .$$

The constant c is determined by setting $x = 0$ in both sides. We get

$$0 = -\ln 1 = 0 + c,$$

so

$$-\ln(1-x) = \sum_{k=1}^{\infty}\frac{x^k}{k} = x + \frac{x^2}{2} + \frac{x^3}{3} + \cdots , \quad -1 < x < 1.$$

Computing the successive derivatives of $-\ln(1-x)$ at $x = 0$ shows that the preceding expansion is just the Taylor expansion of $-\ln(1-x)$ about $x = 0$. Notice that the function $\ln(1-x)$ is defined for all $x < 1$, but that the series fails to converge when $x < -1$. Symmetry of the interval of convergence about $x = 0$ and failure of $\ln(1-x)$ to be bounded near $x = 1$ cause the limitation of the domain of convergence.

EXAMPLE 6   Consider the Taylor expression

$$\sin x = \sum_{k=0}^{\infty}\frac{(-1)^k}{(2k+1)!}x^{2k+1} = x - \frac{x^3}{3!} + \frac{x^5}{5!} - \cdots ,$$

valid for all real x. If we compute derivatives on both sides, we get

$$\cos x = \sum_{k=0}^{\infty} \frac{(-1)^k}{(2k+1)!}(2k+1)x^{2k}$$

$$= \sum_{k=0}^{\infty} \frac{(-1)^k}{(2k)!}x^{2k} = 1 - \frac{x^2}{2!} + \frac{x^4}{4!} - \cdots .$$

This is just the Taylor expansion of $\cos x$. The theorem that justifies the preceding computations is as follows.

5.1   Theorem.   A power series $\sum_{k=0}^{\infty} a_k(x-a)^k$ is arbitrarily often differentiable or integrable term by term in the interior of an interval of convergence of the form $|x-a| < R$, where $R > 0$.

Proof.   The case of an expansion about a point other than $a = 0$ follows from a simple change of variable, as in Exercises 28 and 29. We prove first the part about integration, under the assumption that the series

$$f(x) = \sum_{k=0}^{\infty} a_k x^k$$

represents the function $f(x)$ in the interval $|x| < R$. Assuming $-R < x_1 < R$ we'll prove that

$$\int_0^{x_1} f(x)\, dx = \sum_{k=0}^{\infty} a_k \int_0^{x_1} x^k dx = \sum_{k=0}^{\infty} a_k \frac{x_1^{k+1}}{k+1}.$$

To use Theorem 4.4, we need to verify that the given power series converges uniformly on the interval between 0 and x_1. But if we choose s and r so that $R > s > r > x_1$, then the series converges at $x = s$. Hence its terms tend to zero and so are bounded in absolute value by some number m: $|a_k s^k| \leq m$. Then for $x \leq r < s$, we have

$$|a_k x^k| \leq |a_k| r^k = |a_k| s^k \left(\frac{r}{s}\right)^k$$

$$\leq m \left(\frac{r}{s}\right)^k .$$

The series with kth term $m(r/s)^k$ is a convergent geometric series, because $0 < r/s < 1$. This shows by Theorems 4.1 and 4.3 that the given series converges uniformly on $(-r, r)$ to a continuous function, which is necessarily $f(x)$ because the series is assumed to converge to $f(x)$ at each point x. Because r is a number such that $0 < r < R$, we can include an arbitrary x in $(-R, R)$ in an interval of uniform convergence, so we can integrate term by term on all intervals contained in $[0, x_1]$.

For the differentiation part of the theorem, we start with the same series for $f(x)$ and show that

$$f'(x) = \sum_{k=1}^{\infty} k a_k x^{k-1}.$$

To do this, we apply Theorem 4.5 by showing that the *differentiated* series converges uniformly on every interval $-r \leq x \leq r$, where $0 < r < R$. Choose a number c such that $r < c < R$. Since $\sum_{k=0}^{\infty} a_k c^k$ converges, there is a number b such that $|a_k| c^k \leq b$. Then for x in $[-r, r]$ we have

$$|k a_k x^{k-1}| \leq k |a_k| r^{k-1}$$

$$\leq k \left(\frac{b}{c^k} \right) r^{k-1} = k \left(\frac{b}{c} \right) \left(\frac{r}{c} \right)^{k-1}.$$

The series with kth term $k(b/c)(r/c)^{k-1}$ is geometric and convergent, because $0 < r/c < 1$. Hence the series with kth term $k a_k x^{k-1}$ converges uniformly on $[-r, r]$ by the Weierstrass test. This allows us to differentiate term by term on all intervals $[-r, r]$ with $0 < r < R$. Thus can differentiate at each x such that $-R < x < R$, simply by choosing r so that $|x| < r < R$.   ∎

Theorem 5.1 allows us to differentiate and integrate a power series repeatedly, because the result of performing one such operation on a power series convergent when $|x - a| < R$ is just another power series that is also convergent when $|x - a| < R$.

EXAMPLE 7   Starting with

$$\frac{1}{1-x} = \sum_{k=0}^{\infty} x^k = 1 + x + x^2 + \cdots, \quad |x| < 1,$$

we differentiate once to get

$$\frac{1}{(1-x)^2} = \sum_{k=1}^{\infty} k x^{k-1} = 1 + 2x + 3x^2 + \cdots, \quad |x| < 1.$$

Differentiating again gives

$$\frac{2}{(1-x)^3} = \sum_{k=2}^{\infty} (k-1) k x^{k-2} = 1 \cdot 2 + 2 \cdot 3 x + 3 \cdot 4 x^2 + \cdots, \quad |x| < 1.$$

If we have a power series representation for a function f about a point $x = a$, then the power series is automatically the Taylor series for f about $x = a$. Thus we needn't verify this in practice. Specifically, we have the following theorem.

5.2 Theorem.   If

$$f(x) = \sum_{k=0}^{\infty} a_k (x - a)^k$$

on some interval $|x - a| < R$, then the series is the Taylor series of f about $x = a$. That is,

$$a_n = \frac{1}{n!} f^{(n)}(a), \quad n = 0, 1, 2, \ldots.$$

Proof.   Differentiating n times in the expansion for f knocks out the first n terms, leaving

$$f^{(n)}(x) = \sum_{k=n}^{\infty} k(k-1)\cdots(k-n+1)a_k(x-a)^{k-n}$$

$$= n!a_n + (n+1)!a_{n+1}(x-a) + \cdots.$$

Now set $x = a$ on both sides. All terms on the right become zero except the first, so

$$f^n(a) = n!a_n,$$

which is what we wanted to show.   ∎

5C   Finding Limits by Using Series

A convergent Taylor expansion

$$f(x) = \sum_{k=0}^{\infty} a_k(x-a)^k$$

about some point $x = a$ represents a differentiable, and hence continuous, function f. It follows that we can compute a limit of $f(x)$ as x approaches a by setting $x = a$ to get

$$\lim_{x \to a} f(x) = a_0.$$

This idea applies to calculation of fairly complicated limits, as the following example shows.

EXAMPLE 8   To show that

$$\lim_{x \to 0} \frac{x - \sin x}{x^3} = \frac{1}{6}$$

we just observe that $x - \sin x = x - (x - \frac{1}{6}x^3 + \frac{1}{120}x^5 - \cdots)$. Hence

$$(x - \sin x)/x^3 = \frac{1}{6} - \frac{1}{120}x^5 + \cdots.$$

Taking the limit as $x \to 0$ amounts to setting $x = 0$ in the continuous function represented by the series. The resulting limit is $\frac{1}{6}$.

5D   Products and Quotients

We can multiply and divide power series very much like polynomials. The product of two power series about the same point gives a third series called their **Cauchy product**, which is simply the series formed by collecting equal powers of x:

5.3 $(a_0 + a_1x + a_2x^2 + \cdots)(b_0 + b_1x + b_2x^2 + \cdots)$

$$= a_0b_0 + (a_1b_0 + a_0b_1)x + (a_0b_2 + a_1b_1 + a_2b_0)x^2 + \cdots.$$

The coefficient of x^k is $c_k = a_0b_k + a_1b_{k-1} + \cdots + a_{k-1}b_1 + a_kb_0$, and x may be replaced by $(x - a)$ in all three series. The relevant theorem, which we won't prove here, states that the Cauchy product converges to the correct value in the interior of the common interval of convergence of the two factors.

EXAMPLE 9   Here are several simple examples.

(a) Recall that a polynomial in x is a (finite) power series:

$$(1+x)(1+x+x^2+x^3+\cdots) = 1 + 2x + 2x^2 + 2x^3 + \cdots,$$

which is valid for $-1 < x < 1$.

(b) Multiplying together the powers series about $x = 0$ for $(x-1)^{-1}$ and $\ln(1-x)$, we get, after canceling minus signs,

$$(1 + x + x^2 + x^3 + \cdots)(x + \tfrac{1}{2}x^2 + \tfrac{1}{3}x^3 + \cdots)$$

$$= x + (1 + \tfrac{1}{2})x^2 + (1 + \tfrac{1}{2} + \tfrac{1}{3})x^3 + \cdots = x + \tfrac{3}{2}x + \tfrac{11}{6}x^3 + \cdots,$$

which is valid for $-1 < x < 1$.

(c) Multiplying together the power series for $1/x$ and $\ln x$ about $x = 1$, we get

$$(1 - (x-1) + (x-1)^2 - (x-1)^3 + \cdots)$$

$$\times \left((x-1) - \tfrac{1}{2}(x-1)^2 + \tfrac{1}{3}(x-1)^3 - \cdots \right)$$

$$= (x-1) + (-1 - \tfrac{1}{2})(x-1)^2 + (1 + \tfrac{1}{2} + \tfrac{1}{3})(x-1)^3 - \cdots$$

$$= (x-1) - \tfrac{3}{2}(x-1)^2 + \tfrac{11}{6}(x-1)^3 + \cdots.$$

Examples (b) and (c) show that it may be impossible to find a simple formula for the kth coefficient in a power series.

While division of power series is theoretically the reverse of multiplication, there are some things to watch out for in practice. Suppose we are looking for coefficients a_k that satisfy

$$\frac{c_0 + c_1 x + c_2 x^2 + \cdots}{b_0 + b_1 x + b_2 x^2 + \cdots} = a_0 + a_1 x + a_2 x^2 + \cdots.$$

We do need $b_0 \neq 0$, since $a_0 = c_0/b_0$. (However, if $b_0 = 0$ and $b_1 \neq 0$, we could factor an x from the denominator and then proceed.) Also, if the series in the denominator takes on the value zero for some, possibly complex, value of x, that would limit the convergence interval of the quotient series. With those two observations in mind, we could then proceed to find the a_k by solution of the sequence of equations

$$a_0 b_0 = c_0, \quad a_0 b_1 + a_1 b_0 = c_1, \quad a_0 b_2 + a_1 b_1 + a_2 b_0 = c_2, \ldots$$

for the desired values of the a_k. You can carry this process out as far as you like, because the solution of each equation depends only on the solution of the previous equations in the list. For this reason you can also truncate the series in the numerator and denominator and use polynomial long division to compute a preassigned number of terms.

EXAMPLE 10

To compute the first few terms in the expansion of $(1+e^x)^{-1}$, note that $c_0 = 1, c_1 = c_2 = \cdots = 0$. Also $b_0 = 2, b_k = 1/k!, k = 1, 2, 3, \ldots$. We then solve

$$2a_0 = 1, \quad a_0 + 2a_1 = 0, \quad \tfrac{1}{2}a_0 + a_1 + 2a_2 = 0,$$

$$\tfrac{1}{6}a_0 + \tfrac{1}{2}a_1 + a_2 + 2a_3 = 0, \ldots .$$

The result is $a_0 = \tfrac{1}{2}, a_1 = -\tfrac{1}{4}, a_2 = 0, a_3 = \tfrac{1}{48}, \ldots$. Hence

$$\frac{1}{1+e^x} = \frac{1}{2} - \frac{1}{4}x + \frac{1}{48}x^3 + \cdots .$$

To then find the first three terms in the expansion of $\sin x/(1 + e^x)$, we compute

$$\left(x - \frac{1}{6}x^3 + \cdots\right)\left(\frac{1}{2} - \frac{1}{4}x + \frac{1}{48}x^3 + \cdots\right) = \frac{1}{2}x - \frac{1}{4}x^2 - \frac{1}{12}x^3 \cdots .$$

EXERCISES

Using the ratio test, or by other means, in Exercises 1 to 8 find the interval of convergence of the power series. In case the interval has finite endpoints, determine whether the series converges when x is equal to each of the endpoints, and sketch the interval.

1. $\displaystyle\sum_{k=1}^{\infty} \frac{1}{k}x^k$

2. $\displaystyle\sum_{k=1}^{\infty} \frac{1}{k}(x-2)^k$

3. $\displaystyle\sum_{k=0}^{\infty} \frac{1}{(2k)!}x^{2k}$

4. $\displaystyle\sum_{k=1}^{\infty} k^2(x+1)^k$

5. $\displaystyle\sum_{k=1}^{\infty} \frac{k}{k+1}(x+2)^k$

6. $\displaystyle\sum_{k=0}^{\infty} \frac{1}{\sqrt{k^2+1}}x^k$

7. $\displaystyle\sum_{k=1}^{\infty} \frac{1}{\sqrt{k}}(x+3)^{2k+1}$

8. $\displaystyle\sum_{k=0}^{\infty} 2^k x^{2k}$

In Exercises 9 to 14, use the Taylor expansion $e^x = d\displaystyle\sum_{k=0}^{\infty} \frac{1}{k!}x^k$ to derive the Taylor expansions about the point $a = 0$ for the function.

9. e^{-x}

10. $(e^x + e^{-x})/2 = \cosh x$

11. $(e^x - e^{-x})/2 = \sinh x$

12. xe^x

13. e^{x^2}

14. e^{5x}

In Exercises 15 to 18, use the Taylor expansion $\ln(1 - x) = -\displaystyle\sum_{k=1}^{\infty} \frac{1}{k}x^k$ to derive the Taylor expansion, valid for $|x| < 1$, for the function.

15. $\ln(1 + x)$

16. $\dfrac{1}{2}\ln\left(\dfrac{1+x}{1-x}\right)$

17. $\ln(1 + x^2)$

18. $x^2 \ln(1 - x^3)$

In Exercises 19 to 22, use the Taylor expansion $(1 - x)^{-1} = \sum_{k=0}^{\infty} x^k$ to derive the Taylor expansion for the function about the point a.

19. $\dfrac{1}{1 + x^2}$, about $a = 0$

20. $\dfrac{1}{1 - x^2}$, about $a = 0$

21. $\dfrac{1}{x}$, about $a = 1$

22. $\dfrac{x}{1 - x}$, about $a = 0$

23. Use the relation $d(\arctan x)/dx = 1/(1 + x^2)$ and the result of Exercise 19 to derive the Taylor expansion of $\arctan x$ about $a = 0$.

24. Use the relation $d(1 + x^2)^{-1}/dx = -2x/(1 + x^2)^2$ and the result of Exercise 19 to derive the Taylor expansion of $(1 + x^2)^{-2}$ about $a = 0$.

25. Use the Taylor expansion $(1-x)^{-1} = \sum_{k=0}^{\infty} x^k$ to prove that

$$\frac{x(x+1)}{(1-x)^3} = \sum_{k=1}^{\infty} k^2 x^k, \quad \text{for } |x| < 1.$$

First find the expansion for $(1-x)^{-3}$.

26. Prove that $\sum_{k=0}^{\infty} x^{-k} = x/(x-1)$ for $|x| > 1$.

27. Let α be a real number.

(a) Prove that, if $f(x) = (1+x)^\alpha$, then $f^{(k)}(0) = \alpha(\alpha-1)\cdots(\alpha-k+1)$, so that the Taylor expansion of $(1+x)^\alpha$ about 0 is

$$(1+x)^\alpha = 1 + \sum_{k=1}^{\infty} \frac{\alpha(\alpha-1)\cdots(\alpha-k+1)}{k!} x^k.$$

(b) Write out the first four terms of the expansion in part (a) for $\alpha = 3, \alpha = -3$, and $\alpha = \frac{1}{2}$.

28. Derive the formula

$$\frac{d}{dx}\left(\sum_{k=0}^{\infty} a_k(x-a)^k\right) = \sum_{k=1}^{\infty} ka_k(x-a)^{k-1}, \quad |x-a| < R,$$

from the formula

$$\frac{d}{dx}\left(\sum_{k=0}^{\infty} a_k x^k\right) = \sum_{k=1}^{\infty} ka_k x^{k-1}, \quad |x| < R.$$

29. Derive the formula

$$\int_a^{x_1} \sum_{k=0}^{\infty} a_k(x-a)^k dx = \sum_{k=0}^{\infty} a_k \frac{(x_1-a)^{k+1}}{k+1},$$

$$|x_1 - a| < R,$$

from the formula

$$\int_0^{x_1} \sum_{k=0}^{\infty} a_k x^k dx = \sum_{k=0}^{\infty} a_k \frac{x_1^{k+1}}{k+1}, \quad |x_1| < R.$$

30. (a) Prove that $\lim_{n\to\infty} n \ln\left(1+\frac{a}{n}\right) = a$ by using the Taylor expansion for $\ln(1+x)$ about $x = 0$.

(b) Use part (a) to prove that $\lim_{n\to\infty}\left(1+\frac{a}{n}\right)^n = e^a$.

In Exercises 31 to 34, use Taylor expansions to find the limit.

31. $\displaystyle\lim_{x\to 0} \frac{\ln(1+x^2)}{x^2}$ **32.** $\displaystyle\lim_{x\to 0} \frac{\ln(1-x^5)}{x^5}$

33. $\displaystyle\lim_{x\to 0} \frac{x+\ln(1-x)}{x^2}$ **34.** $\displaystyle\lim_{x\to 0} \frac{\cos x - 1 + x^2/2}{x^4}$

35. Find the Taylor expansion of $f(x) = 1/(x+c)$ about $x = 0$ for $c \neq 0$.

36. Find the Taylor expansion of $g(x) = 1/[(x+c)(x+d)]$ about $x = 0$, for $c \neq 0, d \neq 0$, and $c \neq d$, by expressing g as a sum of two fractions.

37. Prove that, if $f(x) = \sum_{k=0}^{\infty} c_k x^k$ converges in some interval, then

$$f''(x) + f(x) = \sum_{k=0}^{\infty} [c_k + (k+1)(k+2)c_{k+2}]x^k$$

in the interior of the same interval.

38. Use Theorem 5.2 to prove that, if two power series

$$\sum_{k=0}^{\infty} a_k(x-a)^k \quad \text{and} \quad \sum_{k=0}^{\infty} b_k(x-a)^k$$

converge to the same function on an interval containing $x = a$, then $a_k = b_k$ for $k = 0, 1, 2, 3 \ldots$.

SECTION 6   DIFFERENTIAL EQUATIONS

If a differential equation of the form

$$y' = f(x, y)$$

has a solution $y = y(x)$ near $x = x_0$, the equation determines the derivative $y'(x_0)$ from $y'(x_0) = f(x_0, y(x_0))$. If f is sufficiently differentiable even higher derivatives are similarly determined at x_0 as in the following example.

EXAMPLE 1   The equation $y' = y^2$ has $y = (1-x)^{-1}$ for a solution satisfying $y(0) = 1$. The power series expansion

$$y = \frac{1}{1-x} = 1 + x + x^2 + x^3 + \cdots$$

is a Taylor series, so by the Taylor coefficient formulas we also have

$$y = y(0) + \frac{y'(0)}{1!}x + \frac{y''(0)}{2!}x^2 + \frac{y'''(0)}{3!}x^3 + \cdots .$$

Comparison of coefficients of x^k in the two expansions shows that $y^{(k)}(0)/k! = 1$, so $y^{(k)}(0) = k!, k = 0, 1, 2, \ldots$. Suppose, however, that we had no formula for the coefficients to begin with. (Most examples are of this sort.) Starting with the given differential equation, we compute successive derivatives and then simplify by substitution from the earlier equations:

$$y' = y^2,$$

$$y'' = 2yy' = 2y^3,$$

$$y''' = 6y^2y' = 6y^4,$$

$$y^{(4)} = 24y^3y' = 24y^5.$$

The general pattern is evidently $y^{(k)} = k!y^{k+1}$. The formal Taylor expansion for a solution $y = y(x)$ about x_0 with $y(x_0) = y_0$ is then

$$y(x) = y_0 + y_0^2(x - x_0) + y_0^3(x - x_0)^2 + \cdots$$

$$= y_0\left(1 + y_0(x - x_0) + y_0^2(x - x_0)^2 + \cdots\right)$$

$$= \frac{y_0}{1 - y_0(x - x_0)}.$$

We can treat higher-order equations, for example, $y'' = f(x, y, y')$, in a way similar to what we used in the previous example.

EXAMPLE 2   Suppose we want successive derivatives $y^{(k)}(0)$ to a solution of

$$y'' = yy'$$

given that $y(0) = 1$ and $y'(0) = -1$. First compute from the given equation some formulas for higher derivatives. Then simplify by substituting the given values $y(0) = 1, y'(0) = -1$. We find

$$y'' = yy'; \quad y''(0) = -1,$$
$$y''' = yy'' + (y')^2 = y^2y' + (y')^2; \quad y'''(0) = 0,$$
$$y^{(4)} = 2y(y')^2 + y^2y'' + 2y'y'' = 4y(y')^2 + y^3y'; \quad y^{(4)}(0) = 3.$$

The first five terms of the Taylor expansion of $y(x)$ about $x = 0$ then add up to

$$y(0) + \frac{y'(0)}{1!}x + \frac{y''(0)}{2!}x^2 + \frac{y'''(0)}{3!}x^3 + \frac{y^{(4)}(0)}{4!}x^4 = 1 - x - \frac{1}{2}x^2 + \frac{1}{8}x^4.$$

In this example there doesn't seem to be a simple coefficient pattern, but we could compute as many terms as we had time and space for.

EXAMPLE 3   From the differential equation

$$y'' = xy' - y$$

we compute, using substitution after differentiating,

$$
\begin{aligned}
y''' &= xy'' = x(xy' - y) \\
&= x^2 y' - xy, \\
y^{(4)} &= x^2 y'' + xy' - y = x^2(xy' - y) + xy' - y \\
&= (x^3 + x)y' - (x^2 + 1)y.
\end{aligned}
$$

If we denote $y(0)$ by c_0 and $y'(0)$ by c_1, then

$$y''(0) = -c_0, \quad y'''(0) = 0, \quad y^{(4)}(0) = -c_0.$$

Thus the Taylor expansion of $y(x)$ about $x = 0$ has the form

$$y(x) = c_0 + c_1 x - \frac{c_0}{2!}x^2 + 0 - \frac{c_0}{4!}x^4 - \cdots$$

$$= c_0 \left(1 - \frac{1}{2}x^2 - \frac{1}{24}x^4 - \cdots\right) + c_1 x.$$

For comparison, note that if the original differential equation were replaced by

$$y'' = -y$$

then solutions would have the form

$$y(x) = c_0 \cos x + c_1 \sin x,$$

and that $y(0) = c_0$, $y'(0) = c_1$.

EXERCISES

In Exercises 1 to 4, find the first three nonzero terms in the Taylor expansion about $x = 0$ of $y = y(x)$ if y and its derivatives satisfy the given relation.

1. $y' = y^2 + y$, $y(0) = 1$

2. $y' = y^2 + x$, $y(0) = -1$

3. $y' = xy$, $y(0) = 2$

4. $y'' = xy$, $y(0) = 1$, $y'(0) = 0$

5. Find the first four nonzero terms in the Taylor expansion of $y = y(x)$ about $x = 0$ if $y'' = yy'$ and $y(0) = y'(0) = 1$.

6. Find the first four nonzero terms in the Taylor expansion of $y = y(x)$ about $x = 1$ if $y''' = y$ and $y(1) = 2$, $y'(1) = 0$, $y''(1) = 1$.

7. Suppose $y'' = x^2 y$ while $y(0) = c_0$ and $y'(0) = c_1$. Show that $y = y(x)$ has the form

$$y = c_0 \left(1 + \frac{1}{12}x^4 + \cdots\right) + c_1 \left(x + \frac{1}{20}x^5 \cdots\right).$$

8. Show that if $y''' = y^2 y'$ and $y(0) = y'(0) = y''(0) = 1$, then

$$y = 1 + x + \frac{1}{2}x^2 + \frac{1}{6}x^3 + \frac{1}{8}x^4 + \frac{3}{40}x^5 + \cdots.$$

SECTION 7   POWER SERIES SOLUTIONS

The solutions of many of the differential equations we have studied may be represented in terms of their Taylor expansions. For example, polynomials, the elementary transcendental functions $\cos x$, $\sin x$, e^x, and linear combinations of all these have Taylor expansions that are valid for all x. Beyond these examples there is a large and important class of differential equations that has solutions representable by power series, and even if the solution so represented is not a combination of elementary functions at all, the infinite series expansion may serve to define a new function nearly as important as some of the more familiar ones. Furthermore, the partial sums of a series expansion often give useful approximations to the true solution.

Recall that the Taylor expansion of a function f has the form

$$\sum_{k=0}^{\infty} \frac{f^{(k)}(x_0)}{k!}(x - x_0)^k = f(x_0) + \frac{1}{1!}f'(x_0)(x - x_0) + \frac{1}{2!}f''(x_0)(x - x_0)^2 + \cdots .$$

If such a series converges anywhere but at $x = x_0$ then it's absolutely convergent in an interval $x_0 - R < x < x_0 + R$ that is symmetric about x_0, and within such an interval we can treat such series very much like the polynomials in x that arise as special cases. In particular we can add and multiply Taylor expansions about the same point x_0, and a Taylor expansion is differentiable or integrable term by term to produce the Taylor expansion of the derivative or integral of the expanded function within the interval of convergence. A function f is called **analytic** if its Taylor series converges to $f(x)$ for all x in such an interval.

EXAMPLE 1   The Taylor expansions

$$e^x = \sum_{k=0}^{\infty} \frac{x^k}{k!}, \quad -\infty < x < \infty,$$

$$\cos x = \sum_{k=0}^{\infty} \frac{(-1)^k x^{2k}}{(2k)!}, \quad -\infty < x < \infty,$$

$$\sin x = \sum_{k=0}^{\infty} \frac{(-1)^k x^{2k+1}}{(2k+1)!}, \quad -\infty < x < \infty,$$

$$\frac{1}{1-x} = \sum_{k=0}^{\infty} x^k, \quad -1 < x < 1,$$

$$\ln x = \sum_{k=1}^{\infty} \frac{(-1)^{k+1}(x-1)^k}{k}, \quad 0 < x < 2,$$

are computable directly from the general Taylor formula and will converge to the value of the function on the left for each x in the indicated interval. Furthermore, we may compute the derivative of a function such as $\cos 2x$ as follows:

$$\frac{d}{dx}\cos 2x = \frac{d}{dx}\sum_{k=0}^{\infty} \frac{(-1)^k (2x)^{2k}}{(2k)!}$$

$$= \sum_{k=0}^{\infty} \frac{d}{dx} \frac{(-1)^k (2x)^{2k}}{(2k)!}$$

$$= \sum_{k=1}^{\infty} \frac{(-1)^k (2k)(2x)^{2k-1}(2)}{(2k)!}$$

$$= 2 \sum_{k=1}^{\infty} \frac{(-1)^k (2x)^{2k-1}}{(2k-1)!}$$

$$= -2 \sum_{k=0}^{\infty} \frac{(-1)^k (2x)^{2k+1}}{(2k+1)!} = -2 \sin 2x.$$

In the last step we simply replaced k by $k + 1$ throughout to make the expansion look more like the expansion in the preceding examples.

Many of the most important examples of series solutions of differential equations are expressible as power series about the point $x = 0$. Since Taylor expansions about zero are also a little easier to work with, most of our examples will be of that kind. The next example shows how to solve a familiar differential equation using series.

EXAMPLE 2 To solve $y'' + y = 0$, we try to find a solution of the form

$$y(x) = \sum_{k=0}^{\infty} c_k x^k.$$

This form of the expansion is particularly appropriate if we want to solve an initial-value problem with $y(0)$ and $y'(0)$ specified, because then $c_0 = y(0)$ and $c_1 = y'(0)$, if there is a Taylor expansion for the solution about $x = 0$. Proceeding under that assumption for the moment, we compute

$$y'(x) = \sum_{k=1}^{\infty} k c_k x^{k-1},$$

$$y''(x) = \sum_{k=2}^{\infty} (k-1)k c_k x^{k-2}$$

$$= \sum_{k=0}^{\infty} (k+1)(k+2) c_{k+2} x^k.$$

We have shifted the summation index by 2 in the expression for $y''(x)$ to make addition to the series for $y(x)$ more convenient. We find that for $y(x)$ to represent a solution we must have

$$y''(x) + y(x) = \sum_{k=0}^{\infty} c_k x^k + \sum_{k=0}^{\infty} (k+1)(k+2) c_{k+2} x^k$$

$$= \sum_{k=0}^{\infty} [c_k + (k+1)(k+2) c_{k+2}] x^k = 0.$$

Next we use the fundamental property of a Taylor series that for an expansion to be identically zero all the coefficients must be zero. Hence

$$c_{k+2} = -\frac{c_k}{(k+1)(k+2)}, \quad k = 0, 1, 2, \dots .$$

Recalling that we can specify a particular solution by determining the numbers $c_0 = y(0)$, $c_1 = y'(0)$, it's natural to compute recursively

$$c_2 = -\frac{c_0}{1 \cdot 2} = -\frac{c_0}{2!}, \qquad\qquad c_3 = -\frac{c_1}{2 \cdot 3} = -\frac{c!}{3!},$$

$$c_4 = -\frac{c_2}{3 \cdot 4} = \frac{c_0}{4!}, \qquad\qquad c_5 = -\frac{c_3}{4 \cdot 5} = \frac{c_1}{5!},$$

$$c_6 = -\frac{c_4}{5 \cdot 6} = -\frac{c_0}{6!}, \qquad\qquad c_7 = -\frac{c_5}{6 \cdot 7} = -\frac{c_1}{7!},$$

$$\vdots \qquad\qquad\qquad\qquad \vdots$$

$$c_{2k} = -\frac{c_{2k-2}}{(2k-1)(2k)} = (-1)^k \frac{c_0}{(2k)!}, \quad c_{2k+1} = -\frac{c_{2k-1}}{2k(2k+1)}$$

$$= (-1)^k \frac{c_1}{(2k+1)!}.$$

If we take $y(0) = c_0$ and $y'(0) = c_1 = 0$, then only the first column contains nonzero entries, and we get the solution

$$y_0(x) = c_0 - \frac{c_0}{2!}x^2 + \frac{c_0}{4!}x^4 - \cdots + (-1)^k \frac{c_0}{(2k)!}x^{2k} + \cdots$$

$$= c_0 \cos x.$$

On the other hand, the choice $y(0) = c_0 = 0$ and $y'(0) = c_1$ makes the entries in the first column all zero, so we get another solution from the second column:

$$y_1(x) = c_1 - \frac{c_1}{3!}x^3 + \frac{c_1}{5!}x^5 - \cdots + (-1)\frac{c_1}{(2k+1)}x^{2k+1} + \cdots$$

$$= c_1 \sin x.$$

Thus the general solution is $y(x) = c_0 \cos x + c_1 \sin x$ as we expected.

EXAMPLE 3 Solutions of **Airy's equation** $y'' + xy = 0$ are not obtainable in terms of elementary functions, so we try

$$y(x) = \sum_{k=0}^{\infty} c_k x^k,$$

$$y''(x) = \sum_{k=2}^{\infty} (k-1)k c_k x^{k-2}.$$

Substitution into the differential equation gives

$$y''(x) + xy(x) = \sum_{k=2}^{\infty}(k-1)kc_k x^{k-2} + \sum_{k=0}^{\infty}c_k x^{k+1}$$

$$= \sum_{k=0}^{\infty}(k+1)(k+2)c_{k+2}x^k + \sum_{k=1}^{\infty}c_{k-1}x^k = 0,$$

where this time we shifted the index in the first summation up by 2 and in the second summation down by 1 to make the exponents of x agree. Thus we get a single term, the constant $2c_2$, in the first summation that does not correspond to a term in the second summation. We can write then

$$2c_2 + \sum_{k=1}^{\infty}[(k+1)(k+2)c_{k+2} + c_{k+1}]x^k = 0;$$

setting all coefficients equal to zero gives

$$c_2 = 0,$$

$$c_{k+2} = -\frac{c_{k-1}}{(k+1)(k+2)}.$$

We find as a result that the terms are determined in sequences with indexes differing by 3, and that

$$0 = c_2 = c_5 = c_8 = \cdots = c_{3k+2} = \cdots.$$

However, if c_0 or c_1 is not zero, we compute as follows:

$$c_3 = -\frac{c_0}{2\cdot3},$$

$$c_6 = -\frac{c_3}{6\cdot6} = \frac{c_0}{2\cdot3\cdot5\cdot6},$$

$$c_9 = -\frac{c_6}{8\cdot9} = -\frac{c_0}{2\cdot3\cdot5\cdot6\cdot8\cdot9},$$

$$\vdots \qquad\qquad \vdots$$

$$c_{3k} = -\frac{c_{3k-3}}{(3k-1)3k} = \frac{(-1)^k c_0}{2\cdot3\cdot5\cdot6\cdots(3k-1)3k},$$

$$c_4 = -\frac{c_4}{3\cdot4},$$

$$c_7 = -\frac{c_4}{6\cdot7} = \frac{c_1}{3\cdot4\cdot6\cdot7},$$

$$c_{10} = -\frac{c_7}{9\cdot10} = -\frac{c_1}{3\cdot4\cdot6\cdot7\cdot9\cdot10},$$

$$\vdots \qquad\qquad \vdots$$

$$c_{3k+1} = -\frac{c_{3k-2}}{3k(3k+1)} = \frac{(-1)^k c_1}{3\cdot4\cdot6\cdot7\cdots3k(3k+1)}.$$

The solution determined by $y(0) = c_0 = 1$, $y'(0) = c_1 = 0$ is

$$y_0(x) = 1 + \sum_{k=1}^{\infty} \frac{(-1)^k x^{3k}}{2 \cdot 3 \cdot 5 \cdot 6 \cdots (3k-1)3k},$$

and the solution determined by $y(0) = c_0 = 0$, $y'(0) = c_1 = 1$ is

$$y_1(x) = x + \sum_{k=1}^{\infty} \frac{(-1)^k x^{3k+1}}{3 \cdot 4 \cdot 6 \cdot 7 \cdots 3k(3k+1)}.$$

The power series for y_0 and y_1 both converge for all x because the denominators of the kth terms each contain increasing integer factors, $2k$ in number. Hence each series has terms dominated by those of an everywhere convergent series, for example,

$$\left| \frac{(-1)^k x^{3k}}{2 \cdot 3 \cdot 5 \cdot 6 \cdots (3k-1)3k} \right| \leq \frac{|x|^{3k}}{(2k)!}.$$

In practice estimates of this kind are useful for testing the accuracy we get by stopping with a specified number of terms in a Taylor expansion. In this example, to get an estimate when $|x| \leq 1$, we estimate the tail of the factorial series by a geometric series with ratio $1/4n^2$:

$$\begin{aligned}
(2n)! \sum_{k=n}^{\infty} \frac{1}{(2k)!} &= 1 + \frac{1}{(2n+1)(2n+2)} + \frac{1}{(2n+1)\cdots(2n+4)} + \cdots \\
&\leq 1 + \frac{1}{4n^2} + \frac{1}{(4n^2)^2} + \cdots \\
&= \frac{4n^2}{4n^2-1}.
\end{aligned}$$

Thus the error in stopping after $n-1$ terms on the interval $-1 \leq x \leq 1$ is at most

$$\frac{4n^2}{4n^2-1} \cdot \frac{1}{(2n)!}.$$

For $n = 5$, that is, keeping terms of degree 4, the error is at most $3 \cdot 10^{-7}$.

Since the two solutions y_0 and y_1 are linearly independent, the general solution of $y'' + xy = 0$ has the form

$$y(x) = c_0 y_0(x) + c_1 y_1(x).$$

The solutions of $y'' + cy = 0$ have infinitely many zero values if $c > 0$ and at most one such when $c < 0$. Analogously it's true that a solution of the Airy equation has infinitely many positive zeros, but at most one negative zero.

EXAMPLE 4 The Legendre equation of index m is

$$(1 - x^2)y'' - 2xy' + m(m+1)y = 0.$$

To find solutions in the form of power series, we let

$$y(x) = \sum_{k=0}^{\infty} c_k x^k, \quad y'(x) = \sum_{k=1}^{\infty} k c_k x^{k-1}, \quad y''(x) = \sum_{k=2}^{\infty} (k-1) k c_k x^{k-2},$$

so that the c_k are determined by

$$(1-x^2) \sum_{k=2}^{\infty} (k-1) k c_k x^{k-2} - 2x \sum_{k=1}^{\infty} k c_k x^{k-1} + m(m+1) \sum_{k=0}^{\infty} c_k x^k = 0.$$

Shifting the index by 2 in the first sum allows us to write

$$[2c_2 + m(m+1)c_0] + [2 \cdot 3 c_3 - 2c_1 + m(m+1)c_1]x$$

$$+ \sum_{k=2}^{\infty} [(k+1)(k+2)c_{k+2} - ((k-1)k + 2k - m(m+1))c_k]x^k = 0.$$

Setting the coefficient of each power of x equal to zero gives

$$c_2 = -\frac{m(m+1)}{2}c_0, \quad c_3 = -\frac{(m-1)(m+2)}{2 \cdot 3}c_1,$$

$$c_{k+2} = \frac{(k+1+m)(k-m)}{(k+1)(k+2)}c_k, \quad k \geq 2.$$

Since the recurrence relation contains a shift by 2, it's natural to split the coefficients into those of even and those of odd index:

$$c_4 = \frac{(3+m)(2-m)}{3 \cdot 4}c_2 = \frac{(m-2)m(m+3)(m+1)}{4!}c_0,$$

$$c_6 = \frac{(5+m)(4-m)}{5 \cdot 6}c_4 = -\frac{(m-4)(m-2)m(m+5)(m+3)(m+1)}{6!}c_0.$$

$$\vdots$$

and

$$c_5 = \frac{(4+m)(3-m)}{4 \cdot 5}c_3 = \frac{(m-3)(m-1)(m+4)(m+2)}{5!}c_1,$$

$$c_7 = \frac{(6+m)(5-m)}{6 \cdot 7}c_5$$

$$= -\frac{(m-5)(m-3)(m-1)(m+6)(m+4)(m+2)}{7!}c_1,$$

$$\vdots$$

If $m = 2l$ is a positive even integer, then all even coefficients are zero beyond $2l$. Thus the series expansion with even powers reduces to an even polynomial in that case. Similarly, if $m = 2l+1$ is a positive odd integer, the series expansion with odd powers reduces to an odd polynomial. For example, when $m = 4$ and $c_0 = c_1 = 1$, we get the solutions

$$P_4(x) = 1 - \frac{4 \cdot 5}{2!}x^2 + \frac{2 \cdot 4 \cdot 7 \cdot 5}{4!}x^5,$$

$$Q_4(x) = x - \frac{3 \cdot 6}{3!}x^3 + \frac{1 \cdot 3 \cdot 8 \cdot 6}{5!}x^5 - \frac{-1 \cdot 1 \cdot 3 \cdot 10 \cdot 8 \cdot 6}{7!}x^7 + \cdots.$$

Using the ratio test for convergence (see Exercise 7) shows that the infinite series solution converges for $-1 < x < 1$. These two solutions form the basis for the collection of all solutions of the homogeneous Legendre equation of index 4 on the interval $-1 < x < 1$. In general, the Legendre equation of integer index m has two independent solutions of which one is a polynomial and the other is not.

EXERCISES

1. Use the method of power series to derive the general solution

$$y = c_0 \cosh x + c_1 \sinh x$$

to the differential equation $y'' - y = 0$.

2. (a) Show that, if $y = f(x)$ is a solution of $y'' + xy = 0$, then $y = f(-x)$ is a solution of

$$y'' - xy = 0.$$

   (b) Use the result of part (a) together with the result of Example 3 of the text to find a power series expansion for the general solution of $y'' - xy = 0$.

3. (a) Apply the method of power series to solve the first-order differential equation

$$y' + 2xy = 0.$$

   (b) Solve the differential equation in part (a) by finding an exponential multiplier and then integrating.
   (c) Do the results of parts (a) and (b) agree for all x?

4. (a) Apply the power series method to find the general solution of the differential equation

$$y'' - xy' = 0$$

   in the form $y(x) = c_0 y_0(x) + c_1 y_1(x)$.
   (b) Solve the differential equation in part (a) by solving the equivalent system

$$y' = u,$$

$$u' = xu.$$

   (c) Do the results of parts (a) and (b) agree for all x?

5. (a) Apply the power series method to find the general solution of the differential equation

$$y'' + xy' + y = 0$$

   in the form $y(x) = c_0 y_0(x) + c_1 y_1(x)$.
   (b) Show that a special case of the general solution found in part (a) is the solution $y(x) = e^{-(x^2/2)}$.
   (c) Apply the power series method to find a solution of the differential equation

$$y'' + xy' + y = x.$$

   (d) Combine the result of part (a) with that of Exercise 5 to write the general solution of $y'' + xy' + y = x$.

6. Apply the ratio test for series convergence to the series solution found for the Legendre equation in Example 4 of the text. Show that the series converges for $-1 < x < 1$. [*Hint:* Split into even and odd parts; show that each converges separately.)

7. The **Bessel equation** of index n is

$$y'' + \frac{1}{x}y' + \left(1 - \frac{n^2}{x^2}\right)y = 0.$$

   (a) Show that when $n = 0$ the coefficients of a solution of the form $\sum_{k=0}^{\infty} c_k x^k$ satisfy $(k+2)^2 c_{k+2} = -c_k$.
   (b) Show that, if we choose $c_0 = 1, c_1 = 0$ in part (a), we get the solution

$$J_0(x) = \sum_{k=0}^{\infty} (-1)^k 2^{-2k} (k!)^{-2} x^{2k},$$

   called a Bessel function of order 0.
   (c) Show that $J_1(x) = -J_0'(x)$ defines a solution of the Bessel equation of index 1.

8. A Bessel function of integer order n is defined by

$$J_n(x) = \left(\frac{x}{2}\right)^n \sum_{k=0}^{\infty} (-1)^k \frac{x^{2k}}{2^{2k} k! (n+k)!}.$$

(a) Show that J_n satisfies the Bessel equation of index n given in Exercise 7.

(b) Show that, if y_n is a solution of the Bessel equation of index n, and $u_n(x) = \sqrt{x}\, y_n(x)$, then u_n satisfies

$$u'' + \left(1 - \frac{4n^2 - 1}{4x^2}\right) u = 0.$$

SECTION 8   FOURIER SERIES

8A   Introduction

Phenomena that are approximately periodic occur so often in nature that their study has generated a large branch of mathematics known as Fourier analysis, in recognition of one of its originators, Jean-Baptiste Fourier. The prime examples of periodic functions are the sine and cosine functions, and it is these functions that formed the basis of Fourier's own investigations. For the moment disregarding convergence, we define a **trigonometric series** to be one of the form

8.1
$$\frac{a_0}{2} + \sum_{k=1}^{\infty} (a_k \cos kx + b_k \sin kx).$$

In the special circumstances that the coefficients a_k, b_k arise from an integrable function $f(x)$ using the **Euler formulas**

8.2
$$a_k = \frac{1}{\pi} \int_{-\pi}^{\pi} f(x) \cos kx\, dx, \quad b_k = \frac{1}{\pi} \int_{-\pi}^{\pi} f(x) \sin kx\, dx,$$

then the trigonometric series is called the **Fourier series** of f. The coefficients a_k, b_k as given by Equations 8.2 are called the **Fourier coefficients** of f. This choice is justified by Theorem 8.5. The most fundamental question about a Fourier series is the extent to which the series represents the function. The importance of such a representation stems partly from the possibility of incorporating the individual terms of the series into a solution of certain differential equations.

Our first examples illustrate the beautiful way in which the partial sums of a Fourier series attempt to mimic the function f that generates them. A partial sum

8.3
$$S_N(x) = \frac{a_0}{2} + \sum_{k=1}^{N} (a_k \cos kx + b_k \sin kx)$$

is called a **trigonometric polynomial**. Note that each term in S_N is a **periodic function** f of period 2π, that is, $f(x + 2\pi) = f(x)$ for all x. It follows that S_N is also periodic:

$$S_N(x + 2\pi) = S_N(x) \quad \text{for all } x.$$

Note also that the Fourier coefficients a_k, b_k are determined by integral formulas that use the values of $f(x)$ only for $-\pi \leq x \leq \pi$. For these reasons, we'll sometimes restrict attention to values of x in the interval of length 2π between $-\pi$ and π.

8B   Orthogonality

The functions $\cos kx$, $\sin kx$ that occur in a Fourier series are the most important examples of **orthogonal** functions on the interval $-\pi \leq x \leq \pi$. For integers k and l orthogonality of $\cos kx$ and $\sin kx$ means that

8.4
$$\frac{1}{\pi} \int_{-\pi}^{\pi} \cos kx \sin lx \, dx = 0,$$

$$\frac{1}{\pi} \int_{-\pi}^{\pi} \cos kx \cos lx \, dx = \begin{cases} 0, & k \neq l, \\ 1, & k = l \neq 0. \end{cases}$$

$$\frac{1}{\pi} \int_{-\pi}^{\pi} \sin kx \sin lx \, dx = \begin{cases} 0, & k \neq l \text{ or } k = l = 0, \\ 1, & k = l \neq 0. \end{cases}$$

These formulas are usually proved using trigonometric identities, but they can also be proved by first writing the sines and cosines in terms of e^{ikx} and e^{ilx}. (See Exercise 29.) As a sample application of the **orthogonality relations** 8.4, suppose that a trigonometric series satisfies some condition that allows us to integrate it term by term on the interval $-\pi \leq x \leq \pi$. For example, the series might converge uniformly on the interval, or it might be only a finite sum, with all a_k and b_k equal to zero from some point on.

8.5   Theorem.   Suppose the trigonometric series 8.1 converges to a function $f(x)$ and is integrable term-by-term over the interval $[-\pi, \pi]$ to give the integral of $f(x)$. Then the coefficients a_k, b_k of $f(x)$ are given by the Euler formulas 8.2.

Proof.   We denote the sum of the series by

$$f(x) = \frac{a_0}{2} + \sum_{k=1}^{\infty} (a_k \cos kx + b_k \sin kx).$$

Then for a fixed integer $l \geq 0$,

$$\frac{1}{\pi} \int_{-\pi}^{\pi} f(x) \cos lx \, dx = \frac{1}{\pi} \int_{-\pi}^{\pi} \left[\frac{a_0}{2} + \sum_{k=0}^{\infty} (a_k \cos kx + b_k \sin kx) \right] \cos lx \, dx$$

$$= \frac{a_0}{2\pi} \int_{-\pi}^{\pi} \cos lx \, dx + \sum_{k=1}^{\infty} \left[\frac{a_k}{\pi} \int_{-\pi}^{\pi} \cos kx \cos lx \, dx + \frac{b_k}{\pi} \int_{-\pi}^{\pi} \sin kx \cos lx \, dx \right].$$

The first two of Equations 8.4 show all but one of these last terms is zero, the only survivor being the term with the factor a_l. We find for $l \neq 0$,

$$\frac{1}{\pi} \int_{-\pi}^{\pi} f(x) \cos lx \, dx = \frac{a_l}{\pi} \int_{-\pi}^{\pi} \cos lx \cos lx \, dx = a_l.$$

When $l = 0$ the only nonzero term that survives in the sum is the first one. Since the integral of 1 over the interval is 2π, we get a_0. A similar computation in Exercise 30 shows that $\dfrac{1}{\pi} \int_{-\pi}^{\pi} f(x) \sin lx \, dx = b_l$.   ∎

Theorem 8.5 shows that no choices other than Formulas 8.2 for determining the a_k and b_k are possible if we want to represent a reasonably large class of functions $f(x)$ by the trigonometric series of Equation 8.1.

EXAMPLE 1   Let $f(x) = |x|$ for $-\pi \le x \le \pi$. Then

$$a_k = \frac{1}{\pi} \int_{-\pi}^{\pi} |x| \cos kx \, dx, \quad b_k = \frac{1}{\pi} \int_{-\pi}^{\pi} |x| \sin kx \, dx.$$

Now $|x| \sin kx$ has integral zero over $[-\pi, \pi]$, because it's an odd function. Hence $b_k = 0$ for $k = 1, 2, \ldots$. On the other hand, the graph of $|x| \cos kx$ is symmetric about the y-axis, so we can just double the integral over $[0, \pi]$. For $k \neq 0$ we integrate by parts, getting

$$a_k = \frac{2}{\pi} \int_0^{\pi} x \cos kx \, dx$$

$$= \frac{2}{\pi} \left[\frac{x \sin kx}{k} \right]_0^{\pi} - \frac{2}{k\pi} \int_0^{\pi} \sin kx \, dx$$

$$= \left[\frac{2}{k^2 \pi} \cos kx \right]_0^{\pi} = \frac{2}{k^2 \pi} (\cos k\pi - 1)$$

$$= \frac{2}{k^2 \pi} ((-1)^k - 1) = \begin{cases} 0, & k = 2, 4, 6, \ldots, \\ -\dfrac{4}{k^2 \pi}, & k = 1, 3, 5, \ldots. \end{cases}$$

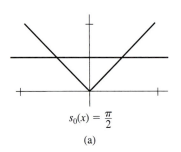

$s_0(x) = \dfrac{\pi}{2}$

(a)

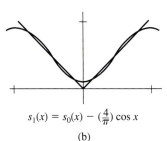

$s_1(x) = s_0(x) - \left(\frac{4}{\pi} \right) \cos x$

(b)

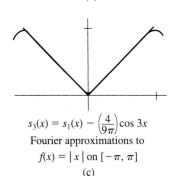

$s_3(x) = s_1(x) - \left(\frac{4}{9\pi} \right) \cos 3x$
Fourier approximations to
$f(x) = |x|$ on $[-\pi, \pi]$

(c)

FIGURE 14.7

When $k = 0$, we have $a_0 = \dfrac{2}{\pi} \int_0^{\pi} x \, dx = \pi$. To summarize,

$$a_0 = \pi, \quad a_k = \begin{cases} 0, & k = 2, 4, 6, \ldots, \\ -\dfrac{4}{k^2 \pi}, & k = 1, 3, 5, \ldots, \end{cases}$$
$$b_k = 0, \quad k = 1, 2, 3, \ldots.$$

Hence the Nth Fourier approximation is given for $N = 1, 3, 5, \ldots$ by the trigonometric polynomial

$$S_N(x) = \frac{\pi}{2} - \frac{4}{\pi} \cos x - \frac{4}{\pi} \frac{\cos 3x}{3^2} - \cdots - \frac{4}{\pi} \frac{\cos Nx}{N^2}.$$

If N is even, we have $S_N(x) = S_{N-1}(x)$. Figure 14.7 shows how the graphs of S_0, S_1, and S_3 approximate that of $|x|$ on $[-\pi, \pi]$; additional terms improve the approximation.

EXAMPLE 2   Let $g(x) = \begin{cases} 1, & 0 \le x \le \pi, \\ -1, & -\pi \le x < 0. \end{cases}$ To compute a_k and b_k we break the interval of integration $[-\pi, \pi]$ at 0:

$$a_k = \frac{-1}{\pi} \int_{-\pi}^{0} \cos kx \, dx + \frac{1}{\pi} \int_0^{\pi} \cos kx \, dx.$$

Since cosine is an *even function*, $\cos(-kx) = \cos kx$, the two integrals are equal, so we get $a_k = 0$. Similarly,

$$b_k = \frac{-1}{\pi} \int_{-\pi}^{0} \sin kx \, dx + \frac{1}{\pi} \int_{0}^{\pi} \sin kx \, dx.$$

Since sine is an *odd function*, $\sin(-x) = -\sin x$, the integrals themselves are negatives of each other, and we get for $k \neq 0$,

$$b_k = \frac{2}{\pi} \int_{0}^{\pi} \sin kx \, dx = \frac{2}{\pi} \left[\frac{-\cos kx}{k} \right]_{0}^{\pi}$$

$$= \frac{2}{k\pi}(-(-1)^k + 1) = \begin{cases} 0, & k \text{ even} \\ \dfrac{4}{k\pi}, & k \text{ odd.} \end{cases}$$

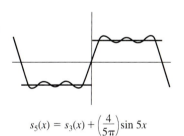

$s_1(x) = \left(\frac{4}{\pi}\right) \sin x$

(a)

In summary,

$$a_k = 0, \qquad k = 0, 1, 2, \ldots,$$

$$b_k = \begin{cases} 0, & k = 2, 4, 6, \ldots, \\ \dfrac{4}{k\pi}, & k = 1, 3, 5, \ldots. \end{cases}$$

Hence for N odd, the Nth Fourier approximation to g is given by

$$S_N(x) = \frac{4}{\pi} \sin x + \frac{4}{\pi} \frac{\sin 3x}{3} + \cdots + \frac{4}{\pi} \frac{\sin Nx}{N}.$$

$s_3(x) = s_1(x) + \left(\frac{4}{3\pi}\right) \sin 3x$

(b)

The graphs of S_1, S_3, and S_5 are shown in Figure 14.8, together with that of $g(x)$.

8C   Convergence of Fourier Series

An important question is whether the Fourier approximations $S_N(x)$ converge as $N \to \infty$ to $f(x)$, where $f(x)$ is the function on $[-\pi, \pi]$ from which the Fourier coefficients are computed. The Fourier series of $f(x)$ is by definition the infinite series

$$\frac{a_0}{2} + \sum_{k=1}^{\infty}(a_k \cos kx + b_k \sin kx),$$

$s_5(x) = s_3(x) + \left(\frac{4}{5\pi}\right) \sin 5x$

(c)

FIGURE 14.8

Step-function approximations $S_N(x)$ for $N = 0, 3, 5$.

where a_k and b_k are given by the Euler Formulas 8.2. Theorem 8.6 gives some conditions on f under which we can use the Fourier series to represent f. Suppose that the graph of f is not only bounded on $[-\pi, \pi]$ but **piecewise monotone**, which means that the interval $[-\pi, \pi]$ breaks into finitely many subintervals, with endpoints $-\pi = x_1 < x_2 < \cdots < x_n = \pi$, such that $f(x)$ is either nondecreasing or nonincreasing on each open subinterval (x_k, x_{k+1}). It's possible to prove that the Fourier series of f will then converge to the 2π-**periodic extension** of $f(x)$ illustrated in Figure 14.9 wherever f is continuous, and at a discontinuity at x_0 will converge to the "average" value

$$\tfrac{1}{2}[f(x_0-) + f(x_0+)].$$

FIGURE 14.9

Typical periodic extension.

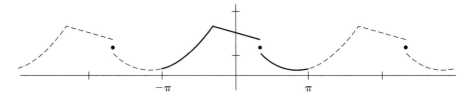

Here $f(x_0-)$ stands for the left-hand limit of f at x_0, and $f(x_0+)$ stands for the right-hand limit. The graph of a typical piecewise monotone function appears in Figure 14.9 with average value at jumps indicated by dots.

8.6   Theorem.   Let f be bounded and piecewise monotone on $[-\pi, \pi]$. Then the Fourier series of f converges at every point x of the interval to $\frac{1}{2}[f(x-)+f(x+)]$. In particular, if f is continuous at x, then the series converges to $f(x)$. At $x = \pm\pi$, the series converges to $(\frac{1}{2})[f(\pi-)+f(-\pi+)]$. (A somewhat stronger version is called **Dirichlet's theorem**; the conclusion is the same, with somewhat weaker assumptions about f.)

Examples 1 and 2 gave an indication of how partial sums of a Fourier series converge. In each of those examples, the function satisfies the condition of piecewise monotonicity; hence the series converges to the value claimed in Theorem 8.6.

EXAMPLE 3

The function g defined in Example 2 is arbitrarily assigned the value 1 at $x = 0$. Theorem 8.6 implies that the Fourier series of g converges as follows:

$$\sum_{k=0}^{\infty} \frac{4}{\pi} \frac{\sin(2k+1)x}{2k+1} = \begin{cases} 1, & 0 < x < \pi, \\ 0, & x = 0 \\ -1 & -\pi < x < 0. \end{cases}$$

To be very specific, we can set $x = \pi/2$ and arrive at the alternating series expansion

$$\sum_{k=0}^{\infty} \frac{(-1)^k}{2k+1} = \frac{\pi}{4}.$$

Theorem 8.6 gives a reason beyond the one in Theorem 8.5 for choosing the coefficients in a trigonometric series according to Euler Formulas 8.2. Assuming piecewise monotonicity the resulting sequence of trigonometric polynomials will converge to the function f and its average value at jumps. Since the partial sums of a Fourier series are themselves periodic functions, the function to which they converge is also periodic, a function we called a periodic extension of f. A periodic extension may differ from the precise definition of $f(x)$ at some points in the interval $-\pi \le x \le \pi$, but changing a value $f(x_0)$ at a point x_0 has no effect on the integral formulas for the Fourier coefficients, so it's customary to make such changes whenever it's convenient in defining a periodic extension of a function from an interval to the entire real number line.

Figure 14.10 shows a function f extended periodically, with period 2π, from the interval $[-\pi, \pi]$ to other values of x. Since the partial sums of the Fourier series are also periodic with period 2π, whatever convergence takes place on $[-\pi, \pi]$ extends periodically to all values of x.

FIGURE 14.10

Periodic extension of
$f(x) = x + \pi$ from $(-\pi, \pi)$
with $S_4(x)$.

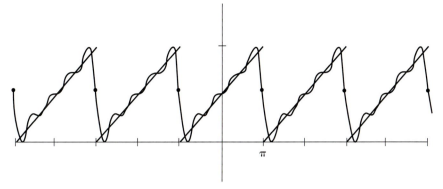

$f(x) = x + \pi$ extended periodically from $(-\pi, \pi)$ and normalized at jumps.
The 4th Fourier partial sum is superimposed.

The coefficient values a_k and b_k that we get from the Euler formulas are independent of the finite values assigned to $f(x)$ at isolated points; this is because the definite integrals in the Euler formulas don't distinguish between two functions that differ at finitely many points. For example, the two functions

$$f(x) = \begin{cases} 0, & -\pi \leq x \leq 0 \\ 1, & 0 < x \leq \pi \end{cases} \quad \text{and} \quad g(x) = \begin{cases} 0, & -\pi \leq x < 0 \\ 1, & 0 \leq x \leq \pi \end{cases}$$

differ only at $x = 0$, where $f(0) = 0$ and $g(0) = 1$. Intuitively speaking the areas under the two graphs should be the same, namely π, and this is an important property of the integral. Since the Fourier series of a piecewise monotone function converges to the average of the right and left limits at each point x, it makes sense simply to redefine such a periodic function to have the average value at each jump discontinuity and refer to this as the **normalized function**.

We restate Theorem 8.6 as follows.

8.7   Theorem.   If a bounded piecewise monotone function is extended periodically and is also normalized to have the value $\frac{1}{2}\big(f(x+) + f(x-)\big)$ at its jump discontinuities, then its Fourier series converges to the function at every point.

Strictly speaking, Theorem 8.7 has implications here only for functions of period 2π, as does Theorem 8.6, but we'll see in the next section that a modified statement is valid for functions with positive period $2p$.

EXAMPLE 4   If the function $f(x) = x + \pi$ is extended periodically from $-\pi < x < \pi$ to other values of x, its graph consists of parallel line segments of slope 1; it remains undefined at odd integer multiples of π since it's initially undefined at $\pm\pi$. To produce a normalized version of the function defined for all x, all we have to do is define the function to have the value π at odd multiples of π. We compute the Fourier series for the functions as originally defined, and the series will converge to the normalized function for all real x. The periodically extended function is shown in Figure 14.10 together with the 4th partial sum of the Fourier expansion. The Fourier coefficients are computed as follows:

$$a_0 = \frac{1}{\pi} \int_{-\pi}^{\pi} (x + \pi) \, dx$$

$$= \frac{1}{\pi} \int_{-\pi}^{\pi} x \, dx + \frac{1}{\pi} \int_{-\pi}^{\pi} \pi \, dx$$

$$= 0 + 2\pi = 2\pi.$$

(Note that the integral of x over an interval symmetric about 0 is always 0.) When $k > 0$,

$$a_k = \frac{1}{\pi} \int_{-\pi}^{\pi} (x + \pi) \cos kx \, dx$$

$$= \frac{1}{\pi} \int_{-\pi}^{\pi} x \cos kx \, dx + \frac{1}{\pi} \int_{-\pi}^{\pi} \pi \cos kx \, dx$$

$$= 0 + 0 = 0.$$

The last integral above is 0 because the indefinite integral is 0 at $\pm\pi$. The previous one is most easily seen to be zero by observing that the integrand $x \cos kx$ is an odd function, so that the integral over $[-\pi, 0]$ is the negative of the integral over $[0, \pi]$. Now for the b_k's,

$$b_k = \frac{1}{\pi} \int_{-\pi}^{\pi} (x + \pi) \sin kx \, dx$$

$$= \frac{1}{\pi} \int_{-\pi}^{\pi} x \sin kx \, dx + \frac{1}{\pi} \int_{-\pi}^{\pi} \pi \sin kx \, dx$$

$$= \frac{2(-1)^{k+1}}{k} + 0 = \frac{2(-1)^{k+1}}{k}.$$

The last integral is zero because the integrand is an odd function. The previous one is computed using integration by parts, with

$$\frac{1}{\pi} \int_{-\pi}^{\pi} x \sin kx \, dx = -\frac{1}{k\pi} x \cos kx \Big|_{-\pi}^{\pi} + \frac{1}{k\pi} \int_{-\pi}^{\pi} \cos kx \, dx$$

$$= -\frac{1}{k} \cos k\pi - \frac{1}{k} \cos(-k\pi) + 0 = \frac{2(-1)^{k+1}}{k}.$$

The full expansion, including the constant $a_0/2$, is then

$$f(x) = \pi + \sum_{k=1}^{\infty} \frac{2(-1)^{k+1}}{k} \sin kx$$

$$= \pi + 2 \sin x - \sin 2x + \tfrac{2}{3} \sin 3x - \tfrac{1}{2} \sin 4x + - \cdots .$$

EXERCISES

The following observations about values of sine and cosine are useful for computing Fourier coefficients; k is always an integer here.

1. $\sin k\pi = 0$

2. $\cos k\pi = (-1)^k$

3. $\sin(k + \frac{1}{2})\pi = (-1)^k$

4. $\cos(k + \frac{1}{2})\pi = 0$

In Exercises 1 to 10, compute the Fourier coefficients of each of the functions and write the corresponding Fourier series in the form of Equation 8.1. Sketch the graph of each function extended to have period 2π on the interval $-2\pi \le x \le 2\pi$. Finally, sketch, relative to the same axes, the graphs of the first three partial sums $S_0(x)$, $S_1(x)$, $S_2(x)$ of the Fourier series.

1. $f(x) = x, -\pi < x \le \pi$

2. $f(x) = \begin{cases} -\pi - x, & -\pi < x < 0, \\ \pi - x, & 0 \le x \le \pi \end{cases}$

3. $f(x) = x^2, -\pi < x \le \pi$

4. $f(x) = |x| + 1, -\pi \le x \le \pi$

5. $f(x) = \begin{cases} 0, & -\pi < x \le 0 \\ 1, & 0 < x \le \pi \end{cases}$

6. $f(x) = x + 1, -\pi < x \le \pi$

7. $f(x) = \begin{cases} -\pi, & -\pi < x < 0, \\ \pi, & 0 \le x \le \pi \end{cases}$

8. $f(x) = 2x + 1, -\pi < x < \pi$

9. $f(x) = -|x|, -\pi \le x \le \pi$

10. $f(x) = \begin{cases} -1, & -\pi < x \le 0 \\ 2, & 0 < x \le \pi \end{cases}$

Exercises 11 to 20. By Theorem 8.6, the Fourier series of each function $f(x)$ in Exercises 1 to 10 converges to some function $F(x)$ whose graph may differ at some points from the graph of $f(x)$. For Exercises 11 to 20, sketch the corresponding $F(x)$ on $-2\pi \le x \le 2\pi$, paying close attention to values at $x = 0, \pm\pi$, and $\pm 2\pi$.

21. Show that if $f(x)$ and $g(x)$ have the Fourier coefficients a_k, b_k and a'_k, b'_k respectively, then $\alpha f(x) + \beta g(x)$, where α and β are constant, has Fourier coefficients $\alpha a_k + \beta a'_k$, $\alpha b_k + \beta b'_k$.

The Nth partial sum of a trigonometric series is called a **trigonometric polynomial** of degree N, and a trigono-

metric polynomial is necessarily the Fourier series of the function it represents. For example, the identity $\cos^2 x = \frac{1}{2} + \frac{1}{2}\cos 2x$ is the Fourier expansion of $\cos^2 x$. In Exercise 22 to 27 find the Fourier series of the function by using appropriate identities, for example the ones in the next exercise.

22. $\sin^2 x$ 23. $\cos^3 x$ 24. $\sin 2x \cos x$

25. $\sin^3 x$ 26. $\cos^4 x - \sin^4 x$ 27. $\sin 5x + \cos 3x$

28. Establish the orthogonality relations in Equations 8.4 of the text, by using the following trigonometric identities to compute the relevant integrals.

$$\cos\alpha\cos\beta = \tfrac{1}{2}\cos(\alpha + \beta) + \tfrac{1}{2}\cos(\alpha - \beta),$$

$$\sin\alpha\cos\beta = \tfrac{1}{2}\sin(\alpha + \beta) + \tfrac{1}{2}\sin(\alpha - \beta),$$

$$\sin\alpha\sin\beta = \tfrac{1}{2}\cos(\alpha - \beta) - \tfrac{1}{2}\cos(\alpha + \beta)$$

*29. Establish the orthogonality relations in Equations 8.4 of the text, by using the identities $\cos nx = \frac{1}{2}(e^{inx} + e^{-inx})$, $\sin nx = \dfrac{1}{2i}(e^{inx} - e^{-inx})$ together with the identity $e^{i(\alpha+\beta)} = e^{i\alpha}e^{i\beta}$ to compute the relevant integrals.

30. Carry out the details of the proof that $\dfrac{1}{\pi}\displaystyle\int_{-\pi}^{\pi} f(x)\sin lx\,dx = b_l$, parallel to the computation in Theorem 8.5.

31. Let $f(x) = \sqrt{|x|}$ for $-\pi \le x \le \pi$. Does f satisfy the hypotheses of Theorem 8.6?

32. Let $f(x)$ be an odd function on $-\pi \le x \le \pi$ (i.e., $f(-x) = -f(x)$) and let $g(x)$ be an even function (i.e., $g(-x) = g(x)$). Let a_k, b_k and a'_k, b'_k be the Fourier coefficients of f and g respectively. Show that

$$a_k = 0. \quad b_k = \frac{2}{\pi}\int_0^\pi f(x)\sin kx\,dx,$$

$$a'_k = \frac{2}{\pi}\int_0^\pi g(x)\cos kx\,dx, \quad b'_k = 0.$$

SECTION 9   APPLIED FOURIER EXPANSIONS

The direct application of Fourier methods to practical problems usually requires adapting the standard formulation presented in the previous section to intervals other than $[-\pi, \pi]$. In the present section we describe some of these adaptations and their application. With these modifications Theorem 8.6 on convergence extends immediately to arbitrary finite intervals.

9A   General Intervals

While the interval $[-\pi, \pi]$ is a natural one for Fourier expansions because it is a period interval for the trigonometric functions, it may be that a function encountered in an application needs to be approximated on some other interval. If the function f to be approximated is defined not on the interval $[-\pi, \pi]$ but on $[-p, p]$, a suitable change in the computation of the approximation is as follows. With f defined on $[-p, p]$, we define

$$f_p(x) = f\left(\frac{px}{\pi}\right), \quad -\pi \le x \le \pi.$$

Then we can compute the Fourier coefficients of f_p by Formula 9.2. The resulting trigonometric polynomials S_N will converge to f_p on $[-\pi, \pi]$ as in Theorems 8.6 and 8.7. To approximate f on $[-p, p]$, we consider

$$S_N\left(\frac{\pi x}{p}\right) = \frac{a_0}{2} + \sum_{k=1}^{N}\left(a_k \cos\frac{k\pi x}{p} + b_k \sin\frac{k\pi x}{p}\right), \quad -p \le x \le p.$$

The coefficients a_k and b_k are computed directly in terms of f by making a change of variable. We have

$$a_k = \frac{1}{\pi}\int_{-\pi}^{\pi} f_p(x)\cos kx\, dx = \frac{1}{\pi}\int_{-\pi}^{\pi} f\left(\frac{px}{\pi}\right)\cos kx\, dx$$

$$= \frac{1}{p}\int_{-p}^{p} f(x)\cos\left(\frac{k\pi x}{p}\right) dx.$$

A similar computation holds for b_k, and we have

9.1 $\qquad a_k = \frac{1}{p}\int_{-p}^{p} f(x)\cos\frac{k\pi x}{p}dx, \quad b_k = \frac{1}{p}\int_{-p}^{p} f(x)\sin\frac{k\pi x}{p}\, dx$

for the coefficients in the Fourier approximation

$$\frac{a_0}{2} + \sum_{k=1}^{N}\left(a_k \cos\frac{k\pi x}{p} + b_k \sin\frac{k\pi x}{p}\right)$$

to the $2p$-periodic extension of the function f defined on $[-p, p]$.

EXAMPLE 1   If

$$h(x) = \begin{cases} 1, & 0 \le x \le p, \\ -1, & -p \le x < 0, \end{cases}$$

then

$$a_k = 0, \qquad\qquad\qquad\qquad k = 0, 1, 2, \ldots,$$

$$b_k = \frac{2}{p}\int_0^{\pi}\sin\frac{k\pi x}{p}dx$$

$$= \frac{2}{\pi}\int_0^{\pi}\sin kx\, dx = \begin{cases} 0, & k = 2, 4, 6, \ldots, \\ \dfrac{4}{k\pi}, & k = 1, 3, 5, \ldots. \end{cases}$$

FIGURE 14.11

Intervals of length $b - a = 2p$.

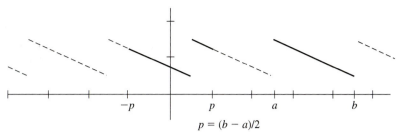

$$p = (b - a)/2$$

Hence the Nth Fourier approximation to h is given, for odd N, by

$$S_N(x) = \frac{4}{\pi} \sin \frac{\pi x}{p} + \frac{4}{3\pi} \sin \frac{3\pi x}{p} + \cdots + \frac{4}{N\pi} \sin \frac{N\pi x}{p}, \quad -p \le x \le p.$$

For a function f defined on an arbitrary interval $a \le x \le b$, it's helpful to think of a periodic extension F of f having period $b - a$ and defined for all real numbers x. Such an extension appears in Figure 14.10. We set $2p = b - a$ so that $p = (b-a)/2$ and $-p = -(b-a)/2$. We then compute the Fourier coefficients of F over the interval $[-p, p]$ according to Formula 9.1. Also, because the integrands in Formula 9.1 have period $2p$, we can use the geometric observation that we can perform the integration over an interval of length $2p = b - a$, in particular, over $[a, b]$ as in Figure 14.11. (See Exercise 7 for a nongeometric proof.) The reinterpretation of Formulas 9.1 is

9.2 $\displaystyle a_k = \frac{2}{b-a} \int_a^b f(x) \cos \frac{2k\pi x}{b-a} dx, \quad b_k = \frac{2}{b-a} \int_a^b f(x) \sin \frac{2k\pi x}{b-a} dx.$

The associated trigonometric polynomials are

$$S_N(x) = \frac{a_0}{2} + \sum_{k=1}^N \left(a_k \cos \frac{2k\pi x}{b-a} + b_k \sin \frac{2k\pi x}{b-a} \right).$$

Equations 9.2 are useful computationally in part because the way $f(x)$ is defined may make it easier to compute its integral over the interval $[a, b]$ rather than $[-p, p]$.

EXAMPLE 2   Let $f(x) = x$, for $0 < x < 1$. We find, integrating by parts for $k \ne 0$,

$$a_k = 2 \int_0^1 x \cos 2k\pi x \, dx$$

$$= 2 \left[x \frac{\sin 2k\pi x}{2k\pi} \right]_0^1 - \frac{2}{2k\pi} \int_0^1 \sin 2k\pi x \, dx = 0,$$

$$b_k = 2 \int_0^1 x \sin 2k\pi x \, dx$$

$$= 2 \left[-x \frac{\cos 2k\pi x}{2k\pi} \right]_0^1 + \frac{2}{2k\pi} \int_0^1 \cos 2k\pi x \, dx$$

$$= -\frac{\cos 2k\pi}{k\pi} = -\frac{1}{k\pi}.$$

Since $a_0 = 2 \displaystyle\int_0^1 x \, dx = 1$, then $a_0/2 = \frac{1}{2}$, and the Fourier series is

$$\frac{1}{2} - \frac{1}{\pi} \left(\sin 2\pi x + \frac{\sin 4\pi x}{2} + \frac{\sin 6\pi x}{3} + \cdots \right).$$

9B   Sine and Cosine Expansions

An expansion in terms only of cosines or only of sines is sometimes more convenient to use than a general Fourier expansion, and is really necessary in Section 10. We start with the observation that the cosine terms in a Fourier expansion are even functions [i.e., $\cos(-k\pi x/p) = \cos(k\pi x/p)$], and that the sine terms are odd [i.e., $\sin(-k\pi x/p) = -\sin(k\pi x/p)$]. It follows that if f is an even periodic function, the product $f(x) \sin(k\pi x/p)$ is odd. (Simple verification.) Therefore for the Fourier sine coefficient b_k, we have by Equation 9.1,

$$b_k = \frac{1}{p} \int_{-p}^{p} f(x) \sin \frac{k\pi x}{p} dx = 0.$$

Hence aside from a possible constant term, **an even function has only cosine terms in its Fourier expansion**. Similarly, if f is an odd periodic function, the product $f(x) \cos(k\pi x/p)$ is also odd; so for the Fourier cosine coefficient we have

$$a_k = \frac{1}{p} \int_{-p}^{p} f(x) \cos \frac{k\pi x}{p} dx = 0.$$

Thus **an odd function has only sine terms in its Fourier expansion.**

Suppose given a function $f(x)$ defined just on the interval $0 \le x \le p$ we want to find a trigonometric series expansion for f consisting only of sine terms, or sometimes only of cosine terms. The trick is to extend the definition of f from the interval $0 \le x \le p$ to all real x in such a way that the extension is periodic of period $2p$ and either is odd or else is even. We then compute the Fourier series of the extension. If f_e is an even periodic extension of f, then f_e will have only cosine terms in its Fourier series in an expansion designed to represent f just on $0 \le x \le p$. Similarly if f_o is an odd periodic extension of f, then f_o has only sine terms in an expansion designed to represent f just for $0 \le x \le p$.

9.3   Sine expansion.   For $f(x)$ on $0 \le x \le p$ we have

$$\sum_{k=1}^{\infty} b_k \sin \frac{k\pi x}{p}, \quad b_k = \frac{2}{p} \int_0^p f(x) \sin \frac{k\pi x}{p} dx.$$

9.4   Cosine expansion.   For $f(x)$ on $0 \le x \le p$ we have

$$\tfrac{1}{2} a_0 + \sum_{k=1}^{\infty} a_k \cos \frac{k\pi x}{p}, \quad a_k = \frac{2}{p} \int_0^p f(x) \cos \frac{k\pi x}{p} dx.$$

We illustrate the method with two examples.

FIGURE 14.12

Even extension of $f(x) = 1 - x$ from $0 \le x \le 2$.

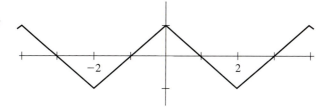

EXAMPLE 3

We'll compute the cosine expansion for the function defined by $f(x) = 1 - x$ for $0 \le x \le 2$. We consider the even periodic extension shown in Figure 14.12. To find the extension we define f_e by $f_e(x) = f(-x)$ for $-2 \le x < 0$, and then extend periodically, with period 4, to the whole x-axis. We use Formula 9.4 to compute the Fourier-cosine expansion of f_e. (Since f_e is even, we know that $b_k = 0$ for all k.) The coefficient formula in 9.4 allows us to write

$$a_k = \frac{1}{2} \int_{-2}^{2} f_e(x) \cos \frac{k\pi x}{2} dx = \int_{0}^{2} f_e(x) \cos \frac{k\pi x}{2} dx.$$

Since on $0 \le x \le 2$, the function f_e is the same as the given function $f(x) = 1 - x$, we integrate by parts for $k > 0$:

$$a_k = \int_{0}^{2} (1 - x) \cos \frac{k\pi x}{2} dx = \left[\frac{2}{k\pi} (1 - x) \sin \frac{k\pi x}{2} \right]_{0}^{2} + \frac{2}{k\pi} \int_{0}^{2} \sin \frac{k\pi x}{2} dx$$

$$= \frac{4}{\pi^2 k^2} [1 - \cos k\pi] = \begin{cases} 0, & k \text{ even,} \\ 8/(\pi^2 k^2), & k \text{ odd.} \end{cases}$$

Finally, $a_0 = \int_{0}^{2} (1 - x) dx = 0$. Thus the cosine expansion of f on $0 \le x \le 2$ has for its general nonzero term

$$\frac{8}{\pi^2 k^2} \cos \left(\frac{k\pi x}{2} \right), \qquad k \text{ odd.}$$

Written out, the expansion of the given function looks like

$$f(x) = \frac{8}{\pi^2} \left(\frac{\cos \pi x/2}{1} + \frac{\cos 3\pi x/2}{9} + \frac{\cos 5\pi x/2}{25} + \cdots \right), \quad 0 \le x \le 2.$$

EXAMPLE 4

Starting with the same function as in Example 3, $f(x) = 1 - x$ for $0 \le x \le 2$, we compute its sine expansion by considering the odd periodic extension shown in Figure 14.13. We first define $f_o(x) = -f(-x)$ for $-2 \le x < 0$, and then extend periodically with period 4. (Since f_o is odd, we know that $a_k = 0$ for all k.) Also, by Equations 9.1 or 9.3,

$$b_k = \frac{1}{2} \int_{-2}^{2} f_o(x) \sin \frac{k\pi x}{2} dx = \int_{0}^{2} f_o(x) \sin \frac{k\pi x}{2} dx.$$

FIGURE 14.13

Odd extension of $f(x) = 1 - x$ from $0 \le x \le 2$.

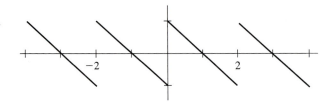

But $f_o(x) = 1 - x$ for $0 \le x \le 2$, so integrating by parts gives

$$b_k = \int_0^2 (1 - x) \sin \frac{k\pi x}{2} dx = \left[-\frac{2}{\pi}(1 - x) \cos k\pi x2 \right]_0^2 - \frac{2}{\pi} \int_0^2 \cos k\pi x2 \, dx$$

$$= \frac{2}{k\pi}(-1)^k + \frac{2}{\pi} - \frac{2}{k\pi} \left[\frac{2}{k\pi} \sin \frac{k\pi x}{2} \right]_0^2 = \begin{cases} 0, & k \text{ odd}, \\ \dfrac{4}{k\pi}, & k \text{ even}. \end{cases}$$

Thus the general nonzero term in the sine expansion is $4/(k\pi) \sin(k\pi x/2)$, for *even* $k > 0$. A careful interpretation of this formula shows that the sine expansion of $f(x)$ is then

$$f(x) = \frac{2}{\pi} \left(\sin \pi x + \frac{\sin 2\pi x}{2} + \frac{\sin 3\pi x}{3} + \cdots \right), \quad 0 < x < 2,$$

with convergence to zero at $x = 0$ and $x = 2$.

Note that Examples 1 and 2 of the previous section are cosine and sine expansions respectively of the given functions restricted to $0 \le x \le \pi$.

The Java Applet FOURIER at Web site http://math.dartmouth.edu/~rewn/ approximates Fourier coefficients using Simpson's rule and then plots graphs of partial sums. An alternative is to use computer algebra software such as Maple, MATLAB or Mathematica to compute Fourier coefficients for elementary functions.

9C Differential Equations

Given a linear differential operator L, for example the harmonic oscillator operator $L = (D^2 + \omega^2)$, we can use Fourier series to solve the nonhomogeneous equation $Ly = f(t)$ if the forcing function f is periodic with period $2p$ and representable by a Fourier series with coefficients a_k, b_k. For example, to solve

$$Ly = \sum_{k=1}^{\infty} b_k \sin \frac{k\pi t}{p},$$

we first find a simple particular solution $y_k(t)$ of the equation

$$Ly = \sin \frac{k\pi t}{p}, \quad k = 1, 2, 3, \ldots .$$

The linearity of L leads us to a formal particular solution of the linear equation as

$$y = \sum_{k=1}^{\infty} b_k y_k(t).$$

The next example is typical of the case in which L is a constant-coefficient operator.

EXAMPLE 5

Suppose we want to solve the forced harmonic oscillator equation $\ddot{y} + \omega^2 y = f(t)$, where $f(t)$ has period $2p = 2$ and is defined on the interval $0 \leq t < 2$ by

$$f(t) = \begin{cases} 1, & 0 \leq t < 1, \\ -1, & 1 \leq t < 2. \end{cases}$$

Figure 14.14(a) shows the graph of f for $0 \leq t < 8$, called a **square wave**.

The square-wave input f has Fourier expansion

$$f(t) = \frac{2}{\pi} \sum_{k=1}^{\infty} \frac{(1 - (-1)^k)}{k} \sin(k\pi t) = \frac{4}{\pi} \sum_{n=0}^{\infty} \frac{1}{2n+1} \sin((2n+1)\pi t),$$

with the understanding that the series converges to 0 at the jump discontinuities. The computation appears in Example 1. The differential equation $\ddot{y} + \omega^2 y = \sin(k\pi t)$ has the particular solution $y_k(t) = (\omega^2 - k^2\pi^2)^{-1} \sin(k\pi t)$. It follows that a solution to $\ddot{y} + \omega^2 y = f(t)$ is formally

$$y(t) = \frac{4}{\pi} \sum_{n=0}^{\infty} \frac{\sin((2n+1)\pi t)}{(2n+1)(\omega^2 - (2n+1)^2\pi^2)}.$$

Note that if ω is an odd multiple of π, one of the terms in the series is undefined and would have to be corrected. Indeed, for ω close to $(2n+1)\pi$ the corresponding term in the series will have a large amplitude. Partial sums to 100 terms are graphed in Figure 14.14 for various values of ω. A formula for the general solution would have to contain additional terms $c_1 \cos \omega t + c_2 \sin \omega t$. The numerical values of the coefficients in the series for $y(t)$ are larger when $n = 0$ than when $n > 0$, particularly for the choices of ω^2 in Figure 14.14 (c), and (d); this explains the dominance of

FIGURE 14.14

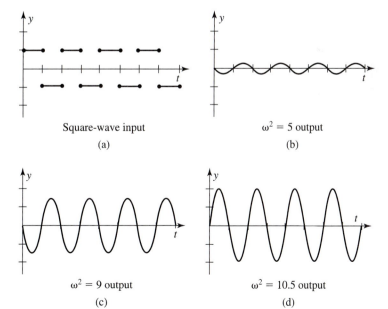

Square-wave input

(a)

$\omega^2 = 5$ output

(b)

$\omega^2 = 9$ output

(c)

$\omega^2 = 10.5$ output

(d)

the first term in the graphs. Note also that Theorem 4.5 applies to the displayed solution, so we can compute $y'(t)$ from term-by-term differentiation, and the output is decidedly smoother than the square-wave input $f(t)$. On intervals $k \leq t < k + 1$ the solution must satisfy $\ddot{y} + \omega^2 y = \pm 1$ so on such intervals $y(t) = c_k \cos \omega t + d_k \sin \omega t \pm \omega^{-2}$, with the pieces fitting together smoothly at $t = k$ to produce a periodic solution.

EXERCISES

1. Find the Fourier series for the function

$$f(x) = -x, \quad -2 < x < 2.$$

To what values will the series converge at $x = 2$ and $x = -2$?

2. Find the Fourier series for the function

$$f(x) = 1 + x, \quad 1 < x < 2.$$

To what values will the series converge at $x = 1$ and $x = 2$?

3. Let f be an **odd function** on $[-p, p]$, that is, $f(-x) = -f(x)$, and let g be an **even function**, that is, $g(-x) = g(x)$. Let a_k, b_k and a'_k, b'_k be the Fourier coefficients of f and g, respectively. Show that

$$a_k = 0, \qquad b_k = \frac{2}{p} \int_0^p f(x) \sin(k\pi x/p)\, dx,$$

$$a'_k = \frac{2}{p} \int_0^p g(x) \cos(k\pi x/p)\, dx, \qquad b'_k = 0.$$

In Exercises 4 to 9, extend the function to an interval to the left of $x = 0$ so that the extended function is odd. In each case sketch the graph of the odd periodic extension of f after normalizing f to have the average value at jump discontinuities. In the same picture sketch also the graph of the sum of the first two nonzero terms of the Fourier expansion of the extended function, which should contain only sine terms.

4. $f(x) = 1, \; 0 < x < \pi$

5. $f(x) = 1 - x, \; 0 < x < 1$

6. $f(x) = x^2, \; 0 < x < \pi$

7. $f(x) = \cos x, \; 0 < x < \pi/2$

8. $f(x) = \begin{cases} 1, & 0 < x < 1, \\ 0, & 1 \leq x < 2 \end{cases}$

9. $f(x) = \begin{cases} 0, & 0 < x < 1, \\ x - 2, & 1 \leq x < 2 \end{cases}$

In Exercises 10 to 15, extend the function to an interval to the left of $x = 0$ so that the extended function is even. In each case sketch the graph of the even periodic extension of f after normalizing f to have the average value at jump discontinuities. In the same picture sketch also the graph of the sum of the first two nonzero terms of the Fourier expansion, which should contain only cosine terms plus perhaps a constant.

10. $f(x) = 1, \; 0 < x < \pi$

11. $f(x) = 1 - x, \; 0 < x < 1$

12. $f(x) = x^2, \; 0 < x < \pi$

13. $f(x) = \sin x, \; 0 < x < \pi/2$

14. $f(x) = \begin{cases} 0, & 0 < x < 1, \\ 1, & 1 \leq x < 2 \end{cases}$

15. $f(x) = \begin{cases} x, & 0 \leq x < 1, \\ 0, & 1 \leq x < 2 \end{cases}$

16. Find
    (a) the Fourier cosine expansion and
    (b) the Fourier sine expansion of the function

$$f(x) = x, \qquad 0 < x < \pi.$$

    (c) Compare the results of (a) and (b) with the complete Fourier expansion of

$$g(x) = x, \qquad -\pi < x < \pi.$$

In Exercises 17 to 24, assume the relevant combinations are defined.

17. Prove that a product of even functions is even.

18. Prove that a product of odd functions is even.

19. Prove that the product of an even function and an odd function is odd.

20. Prove that a linear combination of even functions is even, and a linear combination of odd functions is odd.

21. Show that if f is periodic and differentiable, then f' is periodic.

22. Show by example that if f' is periodic, then f need not be periodic.

23. Show that if f is even and differentiable, then f' is odd. [*Hint:* Consider the limit of $\big(f(-x+h)-f(-x)\big)/h$ as $h \to 0$.]

24. Show that if f is odd and differentiable, then f' is even. See the hint for the previous exercise.

In Exercises 25 and 26, prove that the set of functions $\big\{\sqrt{2/p}\,\sin(k\pi x/p)\big\}_{k=1}^{\infty}$ is an **orthonormal set** on the interval $0 \le x \le p$, that is, show that

$$\frac{2}{p}\int_0^p \sin\frac{k\pi x}{p}\,\sin\frac{l\pi x}{p}\,dx = \begin{cases} 0, & k \neq l, \\ 1, & k = l. \end{cases}$$

25. Use the identity $\sin\alpha\sin\beta = \frac{1}{2}\cos(\alpha-\beta) - \frac{1}{2}\cos(\alpha+\beta)$ to prove the preceding statement directly.

26. Show how the statement above follows from Equations 8.4 of Section 8.

27. Find a formal Fourier expansion for a particular solution to the differential equation $\ddot{y} - a^2 y = f(t)$, where $f(t)$ is the square wave defined in Example 5 of the text.

SECTION 10   HEAT AND WAVE EQUATIONS

Here we show how to use Fourier series to solve problems in 1-dimensional heat conduction and wave motion. As often happens in applications we first find an equation that is satisfied by the physical quantity under study, and then apply some mathematics, in this case Fourier expansions and separation of variables, to solve the equation.

10A   One-Dimensional Heat Equation

Suppose we are given a thin wire of uniform density and length p. Let $u(x,t)$ be the temperature at time t at a point x units from one end. Suppose $0 \le x \le p$ and that $t \ge 0$. We assume that heat transfer takes place only along the direction of the heat conductor and that the temperature at the two ends is held fixed. Thus we can represent the wire as a straight segment along an x-axis and represent temperature as the graph of a function $u = u(x,t)$, as in Figure 14.15.

A basic physical principle of heat conduction is that heat flow is proportional to, and in the direction opposite to, the temperature gradient ∇u. Recall that ∇u is the direction in which the temperature increases most rapidly, so it's reasonable that heat should flow in the opposite direction, from hotter to colder. Since the medium is l-dimensional, represented by a segment of the x-axis, the gradient is just $u_x(x,t)$,

FIGURE 14.15

Temperatures at equally spaced times.

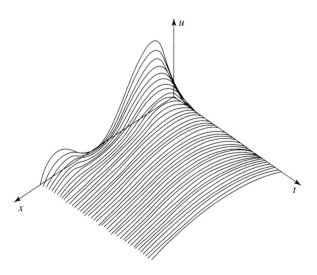

so if $u_x(x, t) > 0$ heat flows to the left at x, while if $u_x(x, t) < 0$ heat flows to the right at x. Thus the rate of change of heat in a segment $[x_1, x_2]$ is

$$k\left[-\frac{\partial}{\partial x}u(x_1, t) + \frac{\partial}{\partial x}u(x_2, t)\right], \tag{1}$$

where the number k is the **heat conductivity** of the wire, assumed to be constant over the length of this wire.

By a version of the Fundamental Theorem of Calculus, the rate of change of heat in the segment in Equation (1) is equal to

$$k\int_{x_1}^{x_2}\frac{\partial^2 u}{\partial x^2}dx. \tag{2}$$

An alternative expression for the rate of change of heat in the segment is

$$\frac{d}{dt}\int_{x_1}^{x_2}c\rho u(x, t)\,dx = c\rho\int_{x_1}^{x_2}\frac{\partial u}{\partial t}(x, t)\,dx, \tag{3}$$

where the constants c and ρ are the **heat capacity** and **density** of the wire per unit of length. Equating the expressions for rate of heat change in the wire in Equations 2 and 3 gives

$$a^2\int_{x_1}^{x_2}\frac{\partial^2 u}{\partial x^2}(x, t)\,dx = \int_{x_1}^{x_2}\frac{\partial u}{\partial t}(x, t)\,dx,$$

where $a^2 = k/c\rho$. Allowing x_2 to vary, we differentiate both sides of this last equation with respect to x_2, to get, after replacing x_2 by x, the

10.1   One-dimensional heat equation $a^2\dfrac{\partial^2 u}{\partial x^2}(x, t) = \dfrac{\partial u}{\partial t}(x, t).$

Equation 10.1 is linear in the sense that if u_1 and u_2 are solutions, then so are linear combinations $c_1u_l + c_2u_2$, the reason being that both $\dfrac{\partial^2}{\partial x^2}$ and $\dfrac{\partial}{\partial t}$ act linearly. To single out particular solutions, we start by imposing two boundary conditions that specify temperature zero at the ends of the wire of length p,

$$u(0, t) = 0 \text{ and } u(p, t) = 0, \ t \geq 0, \tag{4}$$

and one initial condition that specifies the initial temperature at all points x,

$$u(x, 0) = h(x), \ 0 \leq x \leq p. \tag{5}$$

Separation of variables.   The standard way to solve this problem is by an extension to partial differential equations of separation of variables for ordinary

differential equations. The method has many applications, so it's important to understand its principles. In either setting it's sometimes hard to tell in advance whether
the method will work or not. We start by trying to find product solutions of the form

$$u(x, t) = X(x)T(t).$$

The boundary conditions translate into $X(0) = X(p) = 0$. If such product solutions
exist, substitution into $a^2 u_{xx} = u^t$ gives

$$a^2 X''(x)T(t) = X(x)T'(t), \text{ for } 0 \le x \le p, \ 0 < t.$$

Dividing through by $X(x)T(t)$, we get

$$a^2 \frac{X''(x)}{X(x)} = \frac{T'(t)}{T(t)}.$$

Note that if x varies nothing changes on the right. Similarly varying t changes
nothing on the left. Hence both sides must be equal to some constant C. We now
set both sides of the equation equal to a constant, letting $C = -\lambda^2$ for convenience:

$$a^2 X'' + \lambda^2 X = 0, \quad T' + \lambda^2 T = 0.$$

The first of these equations has solutions

$$X(x) = c_1 \cos(\lambda/a)x + +c_2 \sin(\lambda/a)x.$$

But the boundary condition $X(0) = 0$ implies $c_1 = 0$, and $X(p) = 0$ then implies
$c_2 \sin(\lambda/a)p = 0$. Unless we make $c_2 = 0$ also, we can satisfy this condition only
by choosing λ so that $(\lambda/a)p = k\pi$, where k is an integer. That is, we must take
$\lambda = (ka\pi)/p$. The result is that $X(x)$ has the form

$$X(x) = c_2 \sin(k\pi/p)x, \quad k = 1, 2, \ldots .$$

With $\lambda = k\pi/p$, the differential equation for $T(t)$ is now $T' + (ka\pi/p)^2 T = 0$, and
its solutions are

$$T(t) = ce^{-(k^2 a^2 \pi^2 / p^2)t}.$$

Except for a constant factor, the product solutions $u_k(x, t) = X_k(x)T_k(t)$ are

$$u_k(x, t) = e^{-(k^2 a^2 \pi^2 / p^2)t} \sin(k\pi/p)x, \quad k = 1, 2, \ldots .$$

Since the heat equation is linear, linear combinations of the functions $u_k(x, t)$,

$$u_N(x, t) = \sum_{k=1}^{N} b_k e^{-(k^2 a^2 \pi^2 / p^2)t} \sin(k\pi/p)x,$$

are also solutions. But recall that we still have to satisfy an initial condition $u(x, 0) = h(x)$. This amounts to setting $t = 0$ in the previous equation and requiring the coefficients b_k to be chosen so that

$$h(x) = \sum_{k=1}^{N} b_k \sin(k\pi/p)x.$$

It's important to understand the separation technique, because it has many applications, some of which are in Section 10C and in the review problems at the end of the chapter.

If the function $h(x)$ satisfies the conditions of Theorem 8.6 of Section 8, we can let N tend to infinity in $u_N(x, t)$ and get a Fourier series representation that we incorporate using the Fourier sine expansion Formula 9.3 into a

10.2   Solution formula for $a^2 \dfrac{\partial^2 u}{\partial x^2} = \dfrac{\partial u}{\partial t}$, $u(0, t) = u(0, p) = 0$, $u(x, 0) = h(x)$.

$$u(x, t) = \sum_{k=1}^{\infty} b_k e^{-(k^2 a^2 \pi^2 / p^2)t} \sin(k\pi/p)x, \quad \text{where} \quad b_k = \frac{2}{p} \int_0^p h(x) \sin\frac{k\pi x}{p} dx.$$

Note that the decreasing exponential factors make the series converge very rapidly, so we'll be able to differentiate the series term-by-term often enough with respect to x and t to verify that we do have a solution. (See Theorems 4.1 and 4.5.)

EXAMPLE 1

To be more specific about solving the heat equation, we assume for simplicity that $p = \pi$. Recall that to solve $a^2 u_{xx} = u_t$ with boundary condition $u(0, t) = u(\pi, t) = 0$ and initial condition $u(x, 0) = h(x)$, we want in general to be able to represent $h(x)$ by an infinite series of the form

$$h(x) = \sum_{k=1}^{\infty} b_k \sin kx. \tag{6}$$

Suppose, for example, that $h(x)$ is given for x in $[0, \pi]$ by

$$h(x) = \begin{cases} x, & 0 \le x \le \pi/2, \\ \pi - x, & \pi/2 \le x \le \pi. \end{cases}$$

To make Equation 6 represent the Fourier sine expansion of $h(x)$, we extend h to the interval $-\pi \le x \le \pi$ so that the cosine terms in the expansion of h will all be zero, leaving only the sine terms to be computed. We do this by extending the graph of h symmetrically about the origin. According to Formula 9.3,

$$b_k = \frac{2}{\pi} \int_0^\pi h(x) \sin kx\, dx = \frac{2}{\pi} \int_0^{\pi/2} x \sin kx\, dx + \frac{2}{\pi} \int_{\pi/2}^\pi (\pi - x) \sin kx\, dx$$

$$= \frac{2}{\pi} \int_0^{\pi/2} x \sin kx\, dx + \frac{2}{\pi} \int_{-\pi/2}^0 x \sin kx\, dx$$

$$= \frac{4}{\pi} \int_0^{\pi/2} x \sin kx \, dx = \frac{4}{k^2\pi} \sin\left(\frac{k\pi}{2}\right) = \begin{cases} 0, & k = 0, 2, 4, \ldots, \\ \dfrac{4}{k^2\pi}, & k = 1, 5, 9, \ldots, \\ \dfrac{-4}{k^2\pi}, & k = 3, 7, 11, \ldots. \end{cases}$$

Theorem 8.6 then implies that

$$h(x) = \frac{4}{\pi}\left(\frac{\sin x}{1^2} - \frac{\sin 3x}{3^2} + \frac{\sin 5x}{5^2} - \frac{\sin 7x}{7^2} + - \cdots\right), \quad \text{for } 0 \le x \le \pi.$$

From Equation 10.2 we expect that the solution to our problem is

$$u(x, t) = \frac{4}{\pi}\left(e^{-a^2 t}\frac{\sin x}{1^2} - e^{-3^2 a^2 t}\frac{\sin 3x}{3^2} + e^{-5^2 a^2 t}\frac{\sin 5x}{5^2} - + \cdots\right).$$

To verify that $u(x, t)$ satisfies $a^2 u_{xx} = u_t$ for $t > 0$ we use Theorem 4.5, noting that the exponential factors provide the required uniform convergence. Theorem 4.2 shows that $\lim_{t \to 0} u(x, t) = h(x)$ for $0 \le x \le \pi$. The graphs of $h(x)$ and $u(x, t)$ appear in Figure 14.16.

10B   Steady-State Solutions

A solution $u(x, t) = v(x)$ of the heat equation that is independent of t is called a **steady-state solution**, because it doesn't vary with time. The heat equation for such functions becomes simply $v''(x) = 0$, and all solutions are necessarily of the form $u(x, t) = v(x) = \alpha + \beta x$, where α and β are constant. Solutions of this type are useful for solving the time-dependent problem when we have **nonhomogeneous boundary conditions**, which have the form

$$u(0, t) = u_0, \ u(p, t) = u_1, \ t > 0,$$

where at least one of u_0 and u_1 is a nonzero constant. The idea is to choose α, β in the steady-state solution $v(x) = \alpha + \beta x$ so that

$$v(0) = u_0, \qquad v(p) = u_1.$$

Then

$$u(x, t) = w(x, t) + v(x)$$

FIGURE 14.16

Time-varying temperatures for equally spaced x.

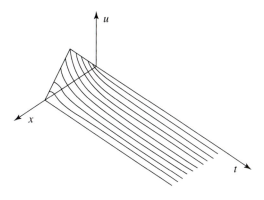

will satisfy $u(0, t) = u_0$, $u(p, t) = u_1$ if $w(x, t)$ is a solution of the heat equation satisfying **homogeneous conditions**

$$w(0, t) = w(p, t) = 0, \qquad t > 0.$$

Note that the function $w + v$ is indeed a solution of the heat equation, because both w and v are solutions and because the heat operator $a^2 D_{xx} - D_t$ acts linearly.

To solve problems of the form

$$a^2 u_{xx} = u_t, \ u(0, t) = u_0, \ u(p, t) = u_1, \ u(x, 0) = h(x),$$

we first find a steady-state solution $v(x) = \alpha + \beta x$. We need $v(0) = u_0 = \alpha$ and $v(p) = u_1 = \alpha + \beta p$. Thus $\alpha = u_0$, and $\beta = (u_1 - u_0)/p$. Then solve the heat equation $a^2 w_{xx} = w_t$ with boundary conditions $w(0, t) = w(p, t) = 0$ and initial condition $w(x, 0) = h(x) - v(x)$. The solution to the original problem has the form $u(x, t) = v(x) + w(x, t)$, or

10.3 $$u(x, t) = v(x) + \sum_{k=1}^{\infty} b_k e^{-k^2(\pi^2 a^2/p^2)t} \sin \frac{k\pi x}{p}, \quad \text{where}$$

$$b_k = \frac{2}{p} \int_0^p \left(h(x) - v(x) \right) \sin \frac{k\pi x}{p} dx.$$

The reason $u(x, 0) = h(x)$ is that when $t = 0$ the series represents $h(x) - v(x)$. All terms of the series are zero when x is 0 or p, so $u(0, t) = v(0) = u_0$ and $u(p, t) = v(p) = u_1$.

EXAMPLE 2   To solve the problem

$$a^2 u_{xx} = u_t, \ u(0, t) = 10, u(5, t) = 30, \ u(x, 0) = h(x) = 10 - 2x, \ 0 < x < 5,$$

we first find the steady-state solution $v(x) = 10 + 4x$. The desired solutions

$$u(x, t) = (10 + 4x) + w(x, t)$$

come from finding solutions $w(x, t)$ that satisfy homogeneous boundary conditions

$$w(0, t) = w(5, t) = 0, \qquad t \geq 0,$$

and an initial condition

$$u(x, 0) = w(x, 0) + v(x) = h(x);$$

this last condition is just

$$w(x, 0) = h(x) - v(x) = -6x.$$

Thus our solution $u(x, t)$ has the form of Equation 10.3, where the b_k are Fourier sine coefficients

$$b_k = \frac{2}{5} \int_0^5 -6x \sin \frac{k\pi x}{5} dx = \frac{60(-1)^k}{k\pi}.$$

The solution to the problem is thus

$$u(x, t) = (10 + 4x) - (60/\pi) \sum_{k=1}^{\infty} \frac{(-1)^{k+1}}{k} e^{-k^2 a^2 \pi^2 t/25} \sin \frac{k\pi x}{5}.$$

EXERCISES

Solve the heat equation $a^2 u_{xx} = u_t$ with boundary and initial conditions 1 through 6.

1. $u(0, t) = u(p, t) = 0$; $u(x, 0) = \sin(\pi x/p)$, $0 < x < p$

2. $u(0, t) = u(1, t) = 0$; $u(x, 0) = x$, $0 < x < 1$

3. $u(0, t) = u(1, t) = 0$; $u(x, 0) = 1 - x$, $0 < x < 1$

4. $u(0, t) = u(\pi, t) = 0$; $u(x, 0) = x(\pi - x)$, $0 < x < \pi$

5. $u(0, t) = u(\pi, t) = 0$; $u(x, 0) = \sin x + \frac{1}{2}\sin 2x$, $0 < x < \pi$

6. $u(0, t) = u(2, t) = 0$; $u(x, 0) = \begin{cases} 1, & 0 < x < 1, \\ 0, & 1 < x < 2 \end{cases}$

Find steady-state solutions $u(x, t) = v(x)$ of the heat equation $a^2 u_{xx} = u_t$ that satisfy each of the conditions 7 through 10.

7. $u(0, t) = -1$, $u(2, t) = 1$

8. $u(0, t) = 0$, $u(100, t) = 100$

9. $u_x(0, t) = 1$, $u(1, t) = 2$

10. $u_x(0, t) = -1$, $u(1, t) = 3$

Solve the heat equation $a^2 u_{xx} = u_t$, given and initial conditions 11 through 14.

11. $u(0, t) = 1$, $u(p, t) = 3$; $u(x, 0) = \sin(\pi x/p)$, $0 < x < p$

12. $u(0, t) = -1$, $u(2, t) = 1$; $u(x, 0) = x$, $0 < x < 2$

13. $u(0, t) = 0$, $u(1, t) = 1$; $u(x, 0) = 1 - x$, $0 < x < 1$

14. $u(0, t) = -1$, $u(1, t) = 2$; $u(x, 0) = 3x - 1$, $0 < x < 1$

Find all solutions of the form $u(x, t) = X(x)T(t)$ for each of the equations 15 through 18:

15. $u_{xx} + u_x = u_t$

16. $u_{xx} - u_x = u_t$

17. $xu_x = 2u_t$

18. $u_{xx} = u_{tt}$

Find the steady-state solution to each of the problems 19 through 22.

19. $u_{xx} = u_t + 2$; $u(0, t) = 1$, $u(1, t) = 2$

20. $u_{xx} = u_t + u$; $u(0, t) = 0$, $u(2, t) = 0$

21. $u_{xx} = u_t + x$; $u(0, t) = 1$, $u(1, t) = 2$

22. $u_{xx} = u_t - 2x + 1$; $u(0, t) = 1$, $u(1, t) = 0$

Insulated endpoints. We can interpret the 1-dimensional heat flow problem

$$u_{xx} = u_t, \ u_x(0, t) = u_x(p, t) = 0,$$

$$u(x, 0) = f(x), \ 0 \le x \le p,$$

as requiring the conducting medium to have insulated ends at $x = 0$ and $x = p$. Because the temperature gradient u_x is always 0 at the endpoints, there is no heat flow past those points. The solution involves Fourier cosine series. The next four exercises are about problems of this kind.

23. (a) Show that product solutions $u(x, t) = T(t)X(x)$ of the insulated endpoint problem have the form
$$u_k(x, t) = a_k e^{-k^2\pi^2 t/p^2} \cos k(\pi/p)x.$$

(b) Use the Fourier cosine expansion for $f(x)$ on $0 \le x \le p$ to solve the boundary value problem for a general initial temperature $f(x)$.

(c) What is the steady-state temperature function $u(x, \infty)$ for $0 \le x \le p$?

24. Solve the heat equation $a^2 u_{xx} = u_t$ with insulated end conditions $u_x(0, t) = u_x(1, t) = 0$ and initial condition $u(x, 0) = x$ for $0 < x < 1$.

25. Solve the heat equation $a^2 u_{xx} = u_t$ with insulated end conditions $u_x(0, t) = u_x(1, t) = 0$ and initial condition $u(x, 0) = 1$ for $0 < x < 1$.

26. Solve the heat equation $a^2 u_{xx} = u_t$ with insulated end conditions $u_x(0, t) = u_x(\pi, t) = 0$ and initial condition $u(x, 0) = \cos x$ for $0 < x < \pi$.

27. The partial differential equation $tu_{xx} = u_t$ has product solutions of the form $u(x, t) = X(x)T(t)$.

(a) Find two ordinary differential equations satisfied by $X(x)$ and $T(t)$ respectively. (The equation for X should not contain t and the equation for T should not contain x. Take care to allow for separation constant $\lambda = 0$.)

(b) Solve the ordinary differential equations found in part (a), and use the results to specify the general form of the solutions $X(x)T(t)$.

28. The partial differential equation $txu_x = u_t$ has product solutions of the form $u(x, t) = X(x)T(t)$ for $x > 0$ and $t > 0$. Find the general form of all such solutions.

29. Suppose that $u(x, t)$ satisfies $a^2 u_{xx} = u_t$ and that $u(x, t_0)$ is concave up as a function of x for $x_0 < x < x_1$. Show that for each x with $x_0 < x < x_1$ there is a time $t(x)$ such that $u(x, t)$ increases as t increases from t_0 to $t(x)$. What if $u(x, t_0)$ is concave down?

30. Verify that the partial differential operator $L = a^2 D_{xx} - D_t$, defined by $L(u) = a^2 u_{xx} - u_t$ is linear in its action on twice differentiable functions u, v, that is, show that $L(cu + dv) = cL(u) + dL(v)$ for constants c and d.

31. The assumption that the separation constant C for the problem $a^2 u_{xx} = u_{tt}$, $u(0, t) = u(p, t) = 0$ has the special form $-\lambda^2$ is convenient but not essential. Derive the product solutions $u_k(x, t) = e^{-k^2(a^2\pi^2/p^2)t} \sin(k\pi/p)x$ using C instead of $-\lambda^2$.

10C   One-Dimensional Wave Equation

Think of a stretched elastic string of length p and uniform density ρ placed along an x-axis in $\mathbb{R}^3$. Suppose that the ends of the string are fixed at $x = 0$ and $x = p$ by opposite forces of magnitude F. If the string is made to vibrate starting at time $t = 0$ our problem is to predict the position at time $t > 0$ of a point $\mathbf{x}(s, t)$ on the string a distance s along the string from the end fixed at $x = 0$. Figure 14.17 shows an exaggerated string shape with a unit tangent $\mathbf{t}(s) = d\mathbf{x}(s)/ds$ at $\mathbf{x}(s, t)$.

We imagine the string partitioned into short pieces of length Δs and then derive two different expressions for the total force vector acting on a typical segment of the subdivision. To find the tension force $\mathbf{T}$ acting on the small piece, note that the opposing forces at $\mathbf{x}(s)$ and $\mathbf{x}(s + \Delta s)$ are

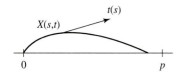

$$F\mathbf{t}(s + \Delta s) \text{ and } -F\mathbf{t}(s), \text{ so } \mathbf{T} = F[\mathbf{t}(s + \Delta s) - \mathbf{t}(s)].$$

FIGURE 14.17
String and tangent.

But by Newton's law, the force $\mathbf{T}$ acting on the small piece equals mass $\rho \Delta s$ times acceleration $\mathbf{a}$, so $\mathbf{T} = (\rho \Delta s)\mathbf{a}$. Hence

$$\rho \mathbf{a} = \frac{\mathbf{T}}{\Delta s} = F\left[\frac{\mathbf{t}(s + \Delta s) - \mathbf{t}(s)}{\Delta s}\right].$$

Letting $\Delta s \to 0$ gives $\rho \mathbf{a} = F\dfrac{\partial \mathbf{t}}{\partial s}(s, t)$. By definition $\mathbf{t} = \dfrac{\partial \mathbf{x}}{\partial s}$ and $\mathbf{a} = \dfrac{\partial^2 \mathbf{x}}{\partial t^2}$, so

$$\rho \mathbf{a} = F\frac{\partial \mathbf{t}(s, t)}{\partial s} = F\frac{\partial^2 \mathbf{x}(s, t)}{\partial s^2}, \quad \text{and also} \quad \rho \mathbf{a} = \rho\frac{\partial^2 \mathbf{x}(s, t)}{\partial t^2}.$$

Equating the two formulas for $\rho \mathbf{a}$ and dividing by ρ gives the

10.4   Vector wave equation $(F/\rho)\dfrac{\partial^2 \mathbf{x}(s, t)}{\partial s^2} = \dfrac{\partial^2 \mathbf{x}(s, t)}{\partial t^2}.$

Before proceeding we'll pause to interpret this equation. From Chapter 8, Section 3 recall that $|\partial^2 \mathbf{x}/\partial s^2| = |\partial \mathbf{t}/\partial s| = \kappa(s)$ is the curvature of the string at $\mathbf{x}(s)$. Thus the magnitude of the acceleration on the right side increases with increasing force F and curvature κ, and decreases with increasing density ρ. Equation 10.4 is linear, but it requires us to use arc length s along the string as an independent variable, something that's very hard to measure in practice.

Setting $F/\rho = a^2$ we write the vector differential equation as a system of three scalar equations

$$a^2\frac{\partial^2 x}{\partial s^s} = \frac{\partial^2 x}{\partial t^2}, \quad a^2\frac{\partial^2 y}{\partial s^2} = \frac{\partial^2 y}{\partial t^2}, \quad a^2\frac{\partial^2 z}{\partial s^2} = \frac{\partial^2 x}{\partial t^2}.$$

The motion's x-axis component is usually slight, indeed zero where we assumed the ends are fixed, so the equation for $x(s, t)$ is usually set aside. Between the other two equations there is little difference in physical significance unless we make some other special assumption. To be specific we assume the string has been set in motion so that movement is entirely in the xy-plane, so $z(s, t) = 0$, and we're left with only the middle equation for y. Equation 3.6 in Chapter 8, Section 3 shows that $y_{ss} = y_{xx}/(1 + y_x^2)^{3/2}$. If the slopes y_x are small enough, we can replace the nonlinear expression y_{ss} by the slightly larger y_{xx}, so with a slight loss of precision we replace the system of three equations by one linear equation for $y(x, t)$:

10.5   One-dimensional linear wave equation $a^2 \dfrac{\partial^2 y}{\partial x^2}(x, t) = \dfrac{\partial y}{\partial t}(x, t).$

This differential equation doesn't specify the displacements $y(x, t)$ of a string of length p unless we impose some boundary and initial conditions:

$$y(0, t) = y(p, t) = 0 \text{ for } t > 0,$$

and

$$y(x, 0) = f(x) \quad \text{and} \quad \frac{\partial y}{\partial t}(x, 0) = g(x).$$

The first pair of equations holds the string on the x-axis at $x = 0$ and $x = p$. The second pair specifies for $0 \le x \le p$ the initial shape of the string, perhaps from plucking as on a harp string, and its initial velocity, perhaps from hammering as on a piano string.

Separation of variables.   As in solving the heat equation we use separation of variables and rely now on the linearity of Equation 10.5 for constructing solutions that satisfy the boundary and initial conditions. Start by setting

$$y(x, t) = X(x)T(t),$$

so Equation 10.5 becomes

$$a^2 X''(x)T(t) = X(x)T''(t), \quad \text{or} \quad a^2 \frac{X''(x)}{X(x)} = \frac{T''(t)}{T(t)}.$$

The right side of the second equation is independent of t because the left side is independent of t, so both sides are constant. For convenience in treating the heat equation we chose a special form for the constant; this wasn't really necessary as we'll show here by calling the separation constant simply λ. We write

$$\frac{X''(x)}{X(x)} = \lambda \quad \text{and} \quad \frac{T''(t)}{T(t)} = a^2\lambda.$$

The first equation is $X'' = \lambda X$ and has solutions

$$X(x) = c_1 e^{\sqrt{\lambda}x} + c_2 e^{-\sqrt{\lambda}x}.$$

The boundary conditions $y(0, t) = y(p, t) = 0$ require $X(0) = X(p) = 0$, so

$$c_1 + c_2 = 0 \quad \text{and} \quad c_1 e^{\sqrt{\lambda}p} + c_2 e^{-\sqrt{\lambda}p} = 0.$$

Solving for c_1 and c_2 shows that to get nonzero solutions we must have

$$c_1 = -c_2 \quad \text{and} \quad e^{\sqrt{\lambda}p} = e^{-\sqrt{\lambda}p}, \quad \text{or} \quad e^{2\sqrt{\lambda}p} = 1.$$

Allowing for complex exponents and complex values for c_1 and c_2, we see that for some integer k we must have $2\sqrt{\lambda}p = 2k\pi i$. Thus $\sqrt{\lambda} = k\pi i/p$, and the corresponding solutions $X(x)$ have the form

$$X_k(x) = c_1 e^{(k\pi i/p)x} - c_2 e^{-(k\pi i/p)x}$$

$$= 2c_1 i \sin \frac{k\pi}{p} x = b_k \sin \frac{k\pi}{p} x.$$

Since we now know that $\lambda = (k\pi i/p)^2 = -(k\pi/p)^2$, the equation for T is

$$T'' + (k\pi a/p)^2 T = 0 \quad \text{with solutions} \quad T_k(t) = C_k \cos \frac{k\pi a}{p} t + D_k \sin \frac{k\pi a}{p} t.$$

Letting $A_k = C_k b_k$ and $B_k = D_k b_k$, the product solutions are

$$y(x, t) = \left[A_k \cos \frac{k\pi a}{p} t + B_k \sin \frac{k\pi a}{p} t \right] \sin \frac{k\pi}{p} x.$$

We now form finite or infinite sums of these terms and try to satisfy the initial conditions by choosing A_k and B_k to be the appropriate Fourier sine coefficients.

10.6   Solution formula for $a^2 \dfrac{\partial^2 y}{\partial x^2} = \dfrac{\partial^2 y}{\partial t^2}$, **satisfying** $y(0, t) = y(p, t) = 0$ **and** $y(x, 0) = f(x)$, $\dfrac{\partial y}{\partial t}(x, 0) = g(x)$.

$$y(x, t) = \sum_{k=1}^{\infty} \left[A_k \cos \frac{k\pi a}{p} t + B_k \sin \frac{k\pi a}{p} t \right] \sin \frac{k\pi}{p} x, \quad \text{where}$$

$$A_k = \frac{2}{p} \int_0^p f(x) \sin \frac{k\pi}{p} x \, dx, \quad \text{and} \quad B_k = \frac{2}{k\pi a} \int_0^p g(x) \sin \frac{k\pi}{p} x \, dx.$$

Figure 14.18 shows equally time-spaced string positions for a plucked string.

EXAMPLE 3   A simple example of Formula 10.6 is a string that's initially stationary, so $g(x) = 0$. If the initial displacement is $f(x) = A \sin(\pi x/p)$, for $0 \leq x \leq p$, the boundary and initial conditions will be

$$y(0, t) = y(p, t) = 0, \quad \text{and} \quad y(x, 0) = A \sin \frac{\pi x}{p}, \quad \frac{\partial y}{\partial t}(x, 0) = 0.$$

FIGURE 14.18
Plucked string.

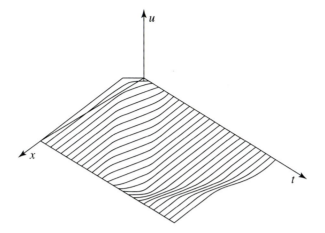

Since $g(x) = 0$, all $B_k = 0$. Since $f(x)$ is a one-term trigonometric polynomial of the same type as the general solution formula, $A_1 = A$ and $A_k = 0$ for $k \neq 1$. The solution according to Formula 10.6, and graphed in Figure 14.18, is

$$y(x, t) = A \cos \frac{\pi a t}{p} \sin \frac{\pi x}{p}.$$

If we want to relax our assumption that the string's motion is confined to a plane, all we have to do is reintroduce the equation $a^2 z_{zz} = x_{tt}$ along with its own initial conditions and solve that problem in the same way. Thus we'd have a vector solution $\big(y(x, t), \ z(x, t)\big)$ with the same independent variables x and t.

EXERCISES

Solve the wave equation $a^2 u_{xx} = u_{tt}$ with boundary and initial conditions 1 to 4.

1. $u(0, t) = u(\pi, t) = 0$, $u(x, 0) = \sin x$, $u_t(x, 0) = 0$

2. $u(0, t) = u(\pi, t) = 0$,
$$u(x, 0) = \begin{cases} x, & 0 < x \le \pi/2, \\ \pi - x, & \pi/2 < x < \pi, \end{cases} \quad u_t(x, 0) = 0$$

3. $u(0, t) = u(\pi, t) = 0$, $u(x, 0) = 0$, $u_t(x, 0) = \begin{cases} 0, & 0 < x \le \pi/2, \\ 1, & \pi/2 < x < \pi \end{cases}$

4. $u(0, t) = u(1, \ t) = 0$, $u(x, 0) = x(1 - x)$, $u_t(x, 0) = \sin \pi x$

5. The **nonhomogeneous wave equation**

$$a^2 u_{xx} = u_{tt} + g$$

incorporates g, the acceleration of gravity, into the vibrating string problem.

(a) Find the equilibrium solutions $u(x, t) = v(x)$ of the nonhomogeneous equation.

(b) Among the solutions found in part (a), select the solution that satisfies the boundary conditions $u(0, t) = u(p, t) = 0$.

(c) Explain how to modify the Fourier solution method for $a^2 u_{xx} = u_{tt}$ to cover the solution of the nonhomogeneous problem.

6. The d'Alembert solution to the wave equation. This method predates the Fourier series method, but is not so popular because it's not so widely applicable. Let $U(x)$ and $V(x)$ be twice differentiable functions for all real x.

(a) Show that $u(x, t) = U(x + at) + V(x - at)$ defines a solution of $a^2 u_{xx} = u_{tt}$ that is valid for all real x and all real t.

(b) Assuming $f(x)$ twice differentiable, show that

$$u(x, t) = \tfrac{1}{2}[f(x + at) + f(x - at)]$$

is a solution to the wave equation of the form described in part (a) that also satisfies the initial conditions $u(x, 0) = f(x)$, $u_t(x, 0) = 0$.

(c) Assuming $g'(x)$ continuous, show that

$$u(x, t) = \frac{1}{2a} \int_{x-at}^{x+at} g(s)\, ds$$

is a solution to the wave equation of the form described in part (a) that also satisfies the initial conditions $u(x, 0) = 0$, $u_t(x, 0) = g(x)$.

(d) Combine the results of parts (b) and (c) to find a solution formula for the wave equation subject to the initial conditions $u(x, 0) = f(x)$, $u_t(x, 0) = g(x)$.

7. (a) Show that the d'Alembert solution $U(x + at) + V(x - at)$ described in the previous exercise represents the sum of two wave motions, the first moving left with speed $a > 0$, the other moving right with the same speed.

(b) Show for the general term in the solution of Equation 10.6 that

$$\left[A \cos(k\pi a/p)t + B \sin(k\pi a/p)t\right] \sin(k\pi/p)x$$
$$= \sqrt{A^2 + B^2}\, \cos(k\pi at/p - \theta) \sin(k\pi x/p),$$

where θ depends on A and B. This term is called a **standing wave**.

(c) Show that the term in part (b) also has the d'Alembert form

$$\tfrac{1}{2}\sqrt{A^2 + B^2}\left[\sin\left((k\pi/p)(x + at) - \theta\right)\right.$$
$$\left.+ \sin\left((k\pi/p)(x - at) + \theta\right)\right].$$

[*Hint:* Use $\sin(a + b) + \sin(a - b) = 2 \sin a \cos b$.]

8. This exercise imposes boundary conditions and initial *velocity* zero on d'Alembert's solution described in Exercise 6.

(a) Let $f(x)$ be twice differentiable for $0 \le x \le p$, with $f(0) = f(p) = 0$. Extend $f(x)$ to the interval $-p < x < 0$ by defining

$$f(x) = -f(-x), \qquad \text{if } -p < x < 0.$$

Then extend $f(x)$ to $-\infty < x < \infty$ to have period $2p$. Show that $f(x)$ so extended is not only periodic, but odd, that is, $f(-x) = -f(x)$ for all real x.

(b) Sketch the odd periodic extension of $f(x) = x(1 - x)$, as described in part (a), from $0 \le x \le 1$ to $-\infty < x < \infty$.

(c) Show that if $f(x)$ is odd and has period $2p$ for $-\infty < x < \infty$, then

$$u(x, t) = \tfrac{1}{2}[f(x + at) + f(x - at)]$$

is a d'Alembert solution to $a^2 u_{xx} = u_{tt}$ that satisfies the boundary conditions $u(0, t) = u(p, t) = 0$ and initial conditions $u(x, 0) = f(x)$ and $u_t(x, 0) = 0$.

9. This exercise imposes boundary conditions and initial *displacement* zero on d'Alembert's solution described in Exercise 6.

(a) Let $G(x)$ be twice differentiable for $0 \le x \le p$. with $G'(0) = G'(p) = 0$. Extend $G(x)$ to the interval $-p < x < 0$ by defining

$$G(x) = G(-x), \qquad \text{if } -p < x < 0.$$

Then extend $G(x)$ to $-\infty < x < \infty$ to have period $2p$. Show that $G(x)$ so extended is not only periodic, but even, that is, $G(-x) = G(x)$ for $-\infty < x < \infty$.

(b) Sketch the even periodic extension of $G(x) = x^2(1 - x)^2$, as described in part (a), from $0 \le x \le 1$ to $-\infty < x < \infty$.

(c) Show that if $G(x)$ is even and has period $2p$ for $-\infty < x < \infty$, then the function

$$u(x, t) = (1/2a)[G(x + at) - G(x - at)]$$

is a d'Alembert solution to $a^2 u_{xx} = u_{tt}$ that satisfies the boundary conditions $u(0, t) = u(p, t) = 0$ and initial conditions $u(x, 0) = 0$ and $u_t(x, 0) = G'(x)$. Why did we assume $G'(0) = G'(p) = 0$?

10. The 1-dimensional wave equation with linear friction constant k is

$$a^2 u_{xx} = u_{tt} + ku_t,$$

where k is a positive constant. Find all *bounded* solutions $u(x, t) = X(x)T(t)$.

11. The partial differential equation $tu_{xx} = u_{tt}$ has product solutions of the form $u(x, t) = X(x)T(t)$. Find ordinary differential equations, each dependent on a parameter λ, satisfied by $X(x)$ and $T(t)$ respectively. The equation for X is to be independent of t and the equation for T is to be independent of x.

Chapter 14 REVIEW

In Exercises 1 to 10, use what you know of specific infinite series to identify a sum in closed form for the given series, determining also its domain of convergence.

1. $\sum_{k=0}^{\infty}(-1)^k(x-5)^k$

2. $\sum_{k=0}^{\infty}(x+1)^{2k}$

3. $\sum_{k=0}^{\infty}(x^2+1)^{-k}$

4. $\sum_{k=0}^{\infty}(-1)^k(x-1)^k/k!$

5. $\sum_{k=0}^{\infty}(x+1)^{2k}/k!$

6. $\sum_{k=0}^{\infty}2^k(x-1)^k/k!$

7. $\sum_{k=0}^{\infty}(-1)^k(x-5)^{2k}/(2k)!$

8. $\sum_{k=0}^{\infty}(-1)^k(x+1)^{2k+1}/(2k+1)!$

9. $\sum_{k=0}^{\infty}(x^2-1)^{-2k}/(2k)!$

10. $\sum_{k=1}^{\infty}(-1)^{k+1}(x^2-1)^k/k$

For the functions in Exercise 11 to 22, find the Taylor expansion about the indicated point a, and state the domain of convergence. In some cases you may be able to do this most easily by some means other than by the Taylor formula $c_k = f^{(k)}(a)/k!$.

11. $(1-x^3)^{-1}; a = 0$ **12.** $(2x-x^2)^{-1}; a = 1$

13. $\ln(1-2x); a = 0$ **14.** $e^{-x^3}; a = 0$

15. $e^{x-1}; a = 1$ **16.** $e^{x-1}; a = 0$

17. $\cos(2x); a = 0$ **18.** $\sin x^2; a = 0$

19. $\sin(x+\pi); a = 0$ **20.** $\sin x + \sin 2x; a = 0$

21. $e^x - e^{2x}; a = 0$ **22.** $(1+x)^{-1}(1-x)^{-1}; a = 0$

State all real values of x for which the series in Exercise 23 to 28 converges.

23. $\sum_{k=1}^{\infty}k^2 x^k$ **24.** $\sum_{k=1}^{\infty}k(x^2-1)^k$

25. $\sum_{k=1}^{\infty}(x/k)^k$ **26.** $\sum_{k=1}^{\infty}\sin kx/k^2$

27. $\sum_{k=1}^{\infty}2^{-k}\cos kx$ **28.** $\sum_{k=0}^{\infty}e^{kx}/k^2$

Test the series in Exercises 29 to 34 for convergence. If the series converges, state whether it also converges absolutely.

29. $\sum_{k=1}^{\infty}(k+1)/k^2$ **30.** $\sum_{k=1}^{\infty}(k/2)^k$

31. $\sum_{k=1}^{\infty}(-1)^k e^k/k^2$ **32.** $\sum_{k=1}^{\infty}k!/k^2$

33. $\sum_{k=1}^{\infty}(k/2^k)^k$ **34.** $\sum_{k=1}^{\infty}(-1)^k e^k/k^k$

35. (a) Let L be defined as an operator by $L(u) = u_{xx} - u_t$. Show that L is a linear operator and conclude that linear combinations of solutions of the heat equation are also solutions.

  (b) Show that boundary conditions of the form $u(a,t) = u(b,t) = 0$ are linear in the sense that if two functions satisfy the conditions, then so does a linear combination of the two functions.

  (c) Show that a boundary condition of the form $u(a,t) = 1$ is not linear in the sense of part (b).

  (d) Show that the initial condition $u(x,t) = f(x)$ is not linear in the sense of part (a) unless $f(x)$ is identically zero.

36. Example 4 in Chapter 9, Section 6 shows that the 2-dimensional Laplace equation $\nabla^2 u = 0$ in polar coordinates is

$$r^2 u_{rr} + r u_r + u_{\theta\theta} = 0.$$

  (a) Letting $u(r,\theta) = R(r)\Theta(\theta)$, show that separation of variables leads to two equations, a **harmonic oscillator equation** and an **Euler equation**,

$$\Theta'' + \lambda^2\Theta = 0, \quad \text{and} \quad r^2 R'' + r R' - \lambda^2 R = 0.$$

  (b) Show that $\Theta'' + \lambda^2\Theta = 0$ has solutions satisfying $\Theta(0) = \Theta(2\pi)$ if $\lambda = k$ for integer k, with solutions $a_k \cos k\theta + b_k \sin k\theta$. Thus Θ takes the same value at polar angles $\theta = 0$ and $\theta = 2\pi$.

  (c) Show that the Euler equation has solutions r^k and r^{-k} for integer k, but that negative exponents are ruled out by the boundary condition that $u(r,\theta)$ should be finite at the origin.

  (d) Show that if $f(\theta)$ has Fourier series representation

$$f(\theta) = \frac{a_0}{2} + \sum_{k=1}^{\infty}(a_k \cos k\theta + b_k \sin k\theta)$$

on $0 \le \theta \le 2\pi$, then for $0 \le r \le 1$ the function

$$u(r,\theta) = \frac{a_0}{2} + \sum_{k=1}^{\infty}r^k(a_k \cos k\theta + b_k \sin k\theta)$$

defines a function $u(r,\theta)$ that solves the Laplace equation in the interior of the unit disk and also satisfies the boundary condition $u(1,\theta) = f(\theta)$.

37. The Laplace equation in spherical coordinates is

$$\nabla^2 \overline{u} = \frac{\partial^2 \overline{u}}{\partial r^2} + \frac{2}{r} \frac{\partial \overline{u}}{\partial r} + \frac{1}{r^2} \frac{\partial^2 \overline{u}}{\partial \phi^2}$$

$$+ \frac{\cos \phi}{r^2 \sin \phi} \frac{\partial \overline{u}}{\partial \phi} + \frac{1}{r^2 \sin^2 \phi} \frac{\partial^2 \overline{u}}{\partial \theta^2} = 0.$$

Assuming there is a function of the form $u(r, \phi, \theta) = R(r)\Phi(\phi)\Theta(\theta)$ that satisfies this partial differential equation, find second-order ordinary differential equations satisfied by R, Φ, and Θ. [*Hint:* First multiply by $r^2 \sin^2 \phi$.]

APPENDIX

FINDING INDEFINITE INTEGRALS

The table at the end of this appendix lists some frequently occurring integrals. As a supplement to the table, you may find it useful to use a symbolic calculator or software that provides some indefinite integrals. If you don't see how to compute an indefinite integral directly and don't find it else where, you may find that one of the following techniques works. Integration constants are omitted, since they're not the main issue here.

I   IDENTITY SUBSTITUTIONS

Rewriting the integrand using an algebraic, trigonometric, exponential, or logarithmic identity will sometimes convert an apparently intractable integrand into an amenable one.

EXAMPLE 1   The integral $\int (e^x + e^{3x})^2 \, dx$ can be rewritten by squaring out the binomial to get

$$\int (e^x + e^{3x})^2 \, dx = \int (e^{2x} + 2e^{4x} + e^{6x}) \, dx$$
$$= \tfrac{1}{2}e^{2x} + \tfrac{1}{2}e^{4x} + \tfrac{1}{6}e^{6x}.$$

EXAMPLE 2   To integrate $\cos^2 x$, recall the trigonometric identity $\cos 2x = 2\cos^2 x - 1$, which is equivalent to $\cos^2 x = \tfrac{1}{2}(1 + \cos 2x)$. Thus Formula 30 in the table follows for $a = 1$ from

$$\int \cos^2 x \, dx = \tfrac{1}{2} \int (1 + \cos 2x) \, dx = \tfrac{1}{2}x + \tfrac{1}{4} \sin 2x.$$

Identities to facilitate the integration of rational functions $P(x)/Q(x)$, where $P(x)$ and $Q(x)$ are polynomials, are derived by **partial fraction decomposition**, described in detail in Chapter 11, Section 5 on Laplace transforms.

EXAMPLE 3   A partial fraction decomposition is used to compute Formula 6 in the table:

$$\int \frac{1}{(x-a)(x-b)} \, dx = \int \left(\frac{1}{(a-b)(x-a)} - \frac{1}{(a-b)(x-b)} \right) dx$$
$$= \frac{\ln|x-a|}{a-b} - \frac{\ln|x-b|}{a-b} = \frac{1}{a-b} \ln \left| \frac{x-a}{x-b} \right|.$$

II   SUBSTITUTION FOR THE INTEGRATION VARIABLE

Awkwardness in an integrand can sometimes be circumvented by a substitution. If the given integration variable is x, a substitution $x = g(u)$, $dx = g'(u)du$ may simplify the integrand enough that the corresponding integral in u can be computed. Then replace u by its equivalent in terms of x using the inverse relation $u = h(x)$, where $g(h(x)) = x$.

EXAMPLE 1   An awkward occurrence of $\sqrt{x}$ can be circumvented by letting $x = u^2$, $dx = 2u\,du$. For example,

$$\int \frac{dx}{\sqrt{x}+1} = \int \frac{2u\,du}{u+1} = 2\int \left(1 - \frac{1}{u+1}\right)du.$$

The last step comes from division of u by $u + 1$. Now integrate with respect to u and reintroduce x using the inverse relation $u = \sqrt{x}$ to get

$$\int \frac{dx}{\sqrt{x}+1} = 2\big(u - \ln(u+1)\big) = 2\big(\sqrt{x} - \ln(\sqrt{x}+1)\big).$$

EXAMPLE 2   To integrate $\sqrt{1-x^2}$, set $x = \sin u$, $dx = \cos u\,du$. Then

$$\int \sqrt{1-x^2}\,dx = \int \sqrt{1-\sin^2 u}\,\cos u\,du$$

$$= \int \cos^2 u\,du = \tfrac{1}{2}u + \tfrac{1}{4}\sin 2u, \text{ by Example 2.}$$

Since $\sin 2u = 2\sin u \cos u$, we find $\int \sqrt{1-x^2}\,dx = \tfrac{1}{2}\arcsin x + \tfrac{1}{2}x\sqrt{1-x^2}$.

III   SUBSTITUTION FOR A PART OF THE INTEGRAND

Here we try to write a given integral in the form $\int f(g(x))g'(x)\,dx$ and set $u = g(x)$, $du = g'(x)\,dx$ in the hope that we can compute the indefinite integral $F(u) = \int f(u)du$. If $F'(u) = f(u)$, the result is

$$\int f(g(x))g'(x)\,dx = \int f(u)\,du = F(g(x)).$$

EXAMPLE 1   In the integral $\int \cos^2 x \sin x\,dx$ we note the square of a function, namely $g(x) = \cos x$, multiplied by a function, $\sin x$, which is easily modified to be the derivative $g'(x) = -\sin x$. By including the constant factor -1 in the integrand and compensating with a "$-$" before the integral, we rewrite the integral as

$$\int \cos^2 x \sin x\,dx = -\int (\cos x)^2(-\sin x)\,dx.$$

It's now natural to think of substituting u for $g(x) = \cos x$ and du for $g'(x)\,dx = (-\sin x)\,dx$ to get

$$\int \cos^2 x \sin x \, dx = -\int u^2 \, du$$

$$= -\tfrac{1}{3}u^2 = -\tfrac{1}{3}\cos^3 x.$$

IV   INTEGRATION BY PARTS

This technique is one of the most important, because of its frequent use in deriving other general formulas; it is embodied in the formula

$$\int f(x)g'(x)\,dx = f(x)g(x) - \int f'(x)g(x)\,dx,$$

which follows from the product rule for differentiation. To apply the method, you need to recognize the integrand of a given integral as a product of two functions; one of them, $f(x)$, you differentiate and the other one, $g'(x)$, you try to identify as a function you can integrate easily. If one choice for $f(x)$ and $g'(x)$ fails to work you may want to try another. Formulas 17, 19, and 23 in the table can be computed by a single application of integration by parts. Formulas 21, 24, and 29 are computed by repeated integration by parts.

EXAMPLE 1   In $\int x \sin x \, dx$, a good choice is $f(x) = x$ and $g'(x) = \sin x$, because $f'(x) = 1$ simplifies the remaining integration, and $g(x) = -\cos x$ is easy to integrate. Thus

$$\int x \sin x \, dx = (x)(-\cos x) - \int (1)(-\cos x)\,dx$$

$$= -x\cos x + \sin x.$$

V   INTEGRAL TABLE

1. $\displaystyle \int (ax+b)^n \, dx = \frac{1}{a(n+1)}(ax+b)^{n+1}, n \neq -1$

2. $\displaystyle \int \frac{dx}{ax+b} = \frac{1}{a}\ln|ax+b|$

3. $\displaystyle \int x(ax+b)^n \, dx = \frac{1}{a^2(n+2)}(ax+b)^{n+2} - \frac{b}{a^2(n+1)}(ax+b)^{n+1},$
   $n \neq -1, -2$

4. $\displaystyle \int \frac{x\,dx}{ax+b} = \frac{x}{a} - \frac{b}{a^2}\ln|ax+b|$

5. $\displaystyle \int \frac{x\,dx}{(ax+b)^2} = \frac{b}{a^2(ax+b)} + \frac{1}{a^2}\ln|ax+b|$

6. $\displaystyle \int \frac{1}{(x-a)(x-b)}\,dx = \frac{1}{a-b}\ln\left|\frac{x-a}{x-b}\right|$

7. $\displaystyle \int \frac{dx}{(ax+b)(cx+d)} = \frac{1}{ad-bc} \ln \left| \frac{ax+b}{cx+d} \right|, \; ad-bc \neq 0$

8. $\displaystyle \int \frac{xdx}{(ax+b)^2} = \frac{b}{2a^2((ax+b)^2)} - \frac{1}{a^2(ax+b)}$

9. $\displaystyle \int x\sqrt{ax+b}\, dx = \frac{2(3ax-2b)}{15a^2}(ax+b)^{3/2}$

10. $\displaystyle \int x^2 \sqrt{ax+b}\, dx = \frac{2(15a^2 x^2 - 12abx + 8b^2)(ax+b)^{3/2}}{105a^3}$

11. $\displaystyle \int \frac{xdx}{\sqrt{ax+b}} = \frac{2(ax-2b)}{3a^2}\sqrt{ax+b}$

12. $\displaystyle \int \frac{dx}{a^2+x^2} = \frac{1}{a} \arctan \frac{x}{a}$

13. $\displaystyle \int \frac{dx}{a^2-x^2} = \frac{1}{2a} \ln \left| \frac{a+x}{a-x} \right|$

14. $\displaystyle \int \frac{dx}{x(ax+b)} = \frac{1}{2b} \ln \left| \frac{x^2}{ax^2+b} \right|$

15. $\displaystyle \int \sqrt{p^2-x^2}\, dx = \tfrac{1}{2}x\sqrt{p^2-x^2} + \tfrac{1}{2}p^2 \arcsin(x/p)$

16. $\displaystyle \int \sqrt{x^2 \pm p^2}\, dx = \tfrac{1}{2}x\sqrt{x^2 \pm p^2} \pm \tfrac{1}{2}p^2 \ln \left(x + \sqrt{x^2 \pm p^2} \right)$

17. $\displaystyle \int \frac{dx}{\sqrt{p^2-x^2}} = \arcsin(x/p)$

18. $\displaystyle \int \frac{dx}{\sqrt{x^2 \pm p^2}} = \ln \left(x + \sqrt{p^2 \pm x^2} \right)$

19. $\displaystyle \int e^{ax}\, dx = \frac{1}{a}e^{ax}$

20. $\displaystyle \int xe^{ax}\, dx = \frac{1}{a^2}(ax-1)e^{ax}$

21. $\displaystyle \int x^2 e^{ax}\, dx = \frac{1}{a^3}(a^2 x^2 - 2ax + 2)e^{ax}$

22. $\displaystyle \int \ln ax\, dx = x \ln ax - x$

23. $\displaystyle \int x \ln ax\, dx = \tfrac{1}{2}x^2 \ln ax - \tfrac{1}{4}x^2$

24. $\displaystyle \int x^2 \ln ax\, dx = \tfrac{1}{3}x^3 \ln ax - \tfrac{1}{9}x^3$

25. $\displaystyle \int \sin ax\, dx = -\frac{1}{a} \cos ax$

26. $\displaystyle\int x \sin axdz = \frac{1}{a^2} \sin ax - \frac{1}{a}x \cos ax$

27. $\displaystyle\int x^2 \sin ax\, dx = \frac{2}{a^2}x \sin ax + \frac{2}{a^3} \cos ax - \frac{1}{a}x^2 \cos ax$

28. $\displaystyle\int \sin^2 ax\, dx = \frac{x}{2} - \frac{\sin 2ax}{4a}$

29. $\displaystyle\int \sin^3 ax\, dx = -\frac{1}{a} \cos ax + \frac{\cos^3 ax}{3a}$

30. $\displaystyle\int \cos ax\, dx = \frac{1}{a} \sin ax$

31. $\displaystyle\int x \cos ax\, dx = \frac{1}{a^2} \cos ax + \frac{1}{a}x \sin ax$

32. $\displaystyle\int x^2 \cos ax\, dx = \frac{2}{a^2}x \cos ax - \frac{2}{a^3} \sin ax + \frac{1}{a}x^2 \sin ax$

33. $\displaystyle\int \cos^2 ax\, dx = \frac{x}{2} + \frac{\sin 2ax}{4a}$

34. $\displaystyle\int \cos^3 ax\, dx = \frac{1}{a} \sin ax - \frac{\sin^3 ax}{3a}$

35. $\displaystyle\int \sin ax \sin bx\, dx = \frac{\sin(a - b)x}{2(a - b)} - \frac{\sin(a + b)x}{2(a + b)}, \; |a| \neq |b|$

36. $\displaystyle\int \cos ax \cos bx\, dx = \frac{\sin(a - b)x}{2(a - b)} + \frac{\sin(a + b)x}{2(a + b)}, \; |a| \neq |b|$

37. $\displaystyle\int \sin ax \cos bx\, dx = -\frac{\cos(a - b)x}{2(a - b)} - \frac{\cos(a + b)x}{2(a + b)}, \; |a| \neq |b|$

38. $\displaystyle\int \tan ax\, dx = -\frac{1}{a} \ln \cos ax$

39. $\displaystyle\int \tan^2 ax\, dx = \frac{1}{a} \tan ax - x$

40. $\displaystyle\int \tan^3 ax\, dx = \frac{1}{2a} \tan^2 ax + \frac{1}{a} \ln \cos ax$

41. $\displaystyle\int \sec ax\, dx = \frac{1}{a} \ln \tan(ax/2 + \pi/4)$

42. $\displaystyle\int \sec^2 ax\, dx = \frac{1}{a} \tan ax$

43. $\displaystyle\int \sec^3 ax\, dx = \frac{1}{2a} \tan ax \sec ax + \frac{1}{2a} \ln \tan(ax/2 + \pi/4)$

44. $\displaystyle\int \tan ax \sec ax\, dx = \frac{1}{a} \sec ax$

45. $\displaystyle\int \csc ax \, dx = \frac{1}{a} \ln |\tan(ax/2)|$

46. $\displaystyle\int \cot ax \, dx = \frac{1}{a} \ln |\sin ax|$

47. $\displaystyle\int e^{ax} \sin bx \, dx = \frac{e^{ax}}{a^2 + b^2}(a \sin bx - b \cos bx)$

48. $\displaystyle\int e^{ax} \cos bx \, dx = \frac{e^{ax}}{a^2 + b^2}(a \cos bx + b \sin bx)$

ANSWERS TO ODD-NUMBERED EXERCISES

CHAPTER 1: VECTORS

Section 1: Coordinate Vectors

Exercise Set 1 (pgs. 7–8)

1. (a) $(-1, 6)$.　(b) $(0, 14)$.　(c) $(4, -6)$.

3. (a) $(9, -3, 0)$.　(b) $(18, 9, -13)$.　(c) $(1, 1, -4)$.

5. $5\mathbf{i} - 8\mathbf{j}$.　　　　　　　　　　　　**7.** $(1 - 2c)\mathbf{i} + (4 - d)\mathbf{j}$.

9. $(a, b) = (3, 2)$ is the only solution.

11. No a and b satisfy $a\mathbf{x} + b\mathbf{y} = (3, 0, 0)$. The only possibility is $c = 5$, $a = b = 1$.

13. (a) $-\mathbf{x} + 2\mathbf{y} - \mathbf{z} = 5\mathbf{i}$.　(b) $6\mathbf{x} - 2\mathbf{y} + \mathbf{z} = 5\mathbf{j}$.　(c) $-4\mathbf{x} + 3\mathbf{y} + \mathbf{z} = 5\mathbf{k}$.

15. Let $\mathbf{x} = (x_1, \dots, x_n)$ be a vector in $\mathbb{R}^n$ and let r and s be real numbers. Then apply the definitions.

17. Apply the definitions.　　　　　　**19.** Apply the definitions.

21. By inspection, $(-2, 3) = -2\mathbf{e}_1 + 3\mathbf{e}_2$.　　**23.** $(2, -7) = -\frac{5}{2}(1, 1) + \frac{9}{2}(1, -1)$.

25. Let $\mathbf{x} = 2\mathbf{i}$, $\mathbf{y} = \mathbf{i} - 3\mathbf{j}$ and $\mathbf{z} = 3\mathbf{i} + 2\mathbf{j} - 2\mathbf{k}$.
　　　(a) $\mathbf{i} = \frac{1}{2}\mathbf{x}$.　(b) $\mathbf{j} = \frac{1}{6}\mathbf{x} - \frac{1}{3}\mathbf{y}$.　(c) $\mathbf{k} = \frac{11}{12}\mathbf{x} - \frac{1}{3}\mathbf{y} - \frac{1}{2}\mathbf{z}$.

27. 700 ink, 90000 paper, 5500 binding.　　**29.** $\frac{1}{4}(\mathbf{x}(2) + \mathbf{x}(8) + \mathbf{x}(14) + \mathbf{x}(20))$.

Section 2: Geometric Vectors

Exercise Set 2A–E (pgs. 16–17)

1. $(-1/2, 3/2)$.　　　　　　　　　　**3.** $(1, 0, 1)$.

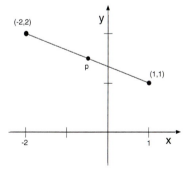

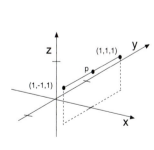

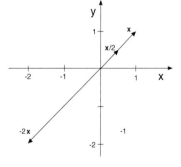

Exercise 1　　　　　　Exercise 3　　　　　　Exercise 5

5. $2\sqrt{2}$.　　　　　　　　　　　**7.** 6.

9. $\mathbf{x} + \mathbf{y}\ \ = (1, 1) + (1, -1) = (2, 0),$
$\mathbf{x} - \mathbf{y}\ \ = (1, 1) - (1, -1) = (0, 2),$
$\mathbf{x} + 2\mathbf{y} = (1, 1) + 2(1, -1) = (3, -1).$

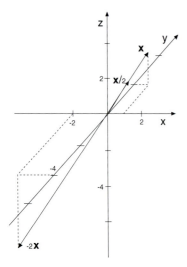

Exercise 7

11. $\mathbf{x} + \mathbf{y}\ \ = (1, 1, 1) + (1, 1, -1) = (2, 2, 0),$
$\mathbf{x} - \mathbf{y}\ \ = (1, 1, 1) - (1, 1, -1) = (0, 0, 2),$
$\mathbf{x} + 2\mathbf{y} = (1, 1, 1) + 2(1, 1, -1) = (3, 3, -1).$

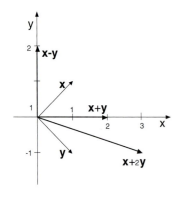

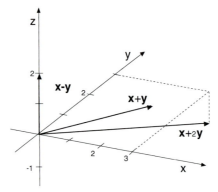

Exercise 9   Exercise 11

13.

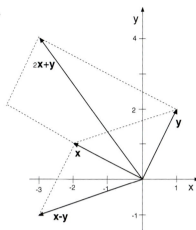

15.

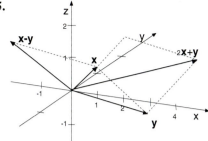

17.

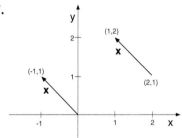

19.

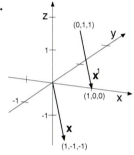

21.

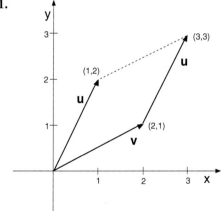

23.

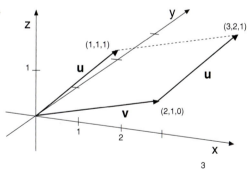

25. *Hint:* Rearrange terms.   **27.** $(0, 0)$.

29. *Hint:* Arrange vector arrows tail-to tip.

31. (a) *Hint:* Consider $t\mathbf{a} + (1 - t)\mathbf{b} - \mathbf{a}$ and $t\mathbf{a} + (1 - t)\mathbf{b} - \mathbf{b}$.

    (b) *Hint:* Use ideas from part (a).

    (c) *Hint:* Use ideas from part (b).

33. About $0.321\mathbf{i} + 0.883\mathbf{j} + 0.342\mathbf{k}$.

35. About $(-1557, -2928, 2237)$.   **37.** About 2420 miles.

Section 3: Lines and Planes

Exercise Set 3AB (pgs. 23–24)

1. $\mathbf{x} = t(2, 1) + (-1, 2)$.

3. $\mathbf{x} = t(1, 0, 1) + (1, 2, 2)$.

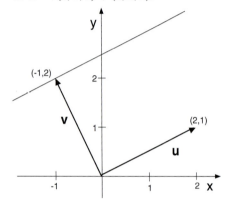

Exercise 1

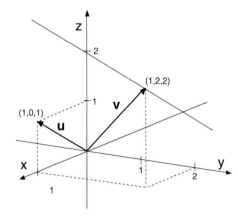

Exercise 3

5. (a)

    (b) $\mathbf{p} = (-5/3, 8/3)$.

    (c) *Hint:* Show a contradiction.

7. Lines are parallel and equal.   **9.** Lines are parallel but not equal.

11. Linearly independent.   **13.** Linearly independent.

15. $\mathbf{x} = t_1(1, 1, 0) + t_2(0, 1, 1)$.   **17.** $\mathbf{x} = t_1(1, -1, 0) + t_2(1, 0, -1) + (1, 0, 0)$.

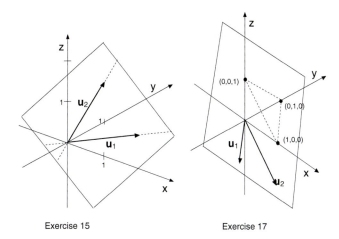

Exercise 15   Exercise 17

19. (a) $(2, -2, 0)$. **(b)** $\mathbf{x} = t_1(1, -1, 2) + t_2(2, -2, 0) + (1, 2, 1)$.

21. *Hint:* Find the midpoints.

23. (a) Use either coordinates or the laws. **(b)** *Hint:* Divide by the scalar. **(c)** *Hint:* Suppose first that the conditions hold.

25. Pick two points in $S + T$.

Section 4: Dot Products

Exercise Set 4A–C (pgs. 31–32)

1. 10   **3.** -5

5. (a) 1. **(b)** 1. **(c)** $\pi/4$.

7. (a) 8. **(b)** 3. **(c)** 0.4759 rad. $\approx 27.3°$.

9. Angle: 0.6147 radians ($\approx 35.2°$. Distance $\approx l = r\theta = 4000(0.6147) \approx 2459$ miles.

11. (a) Positivity *Hint:* Sum of squares is nonegative. **(b) Symmetry** Routine check. **(c) Additivity** Routine check. **(d) Homogeneity** Routine check.

13. (a) $\frac{2}{\sqrt{3}}$. **(b)** $(2/3, 2/3, 2/3)$; $(1/3, -5/3, 4/3)$.

15. (a) $-\frac{9}{\sqrt{14}}$. **(b)** $(-9/14, -27/14, 18/14)$; $(37/14, -15/14, -4/14)$.

17. $|A - B| = \sqrt{101}$, $|A - C| = \sqrt{126}$, $|A - C| = 5$, AB acute, AC right, BC acute.

19. (a) Routine calculation. **(b)** Routine calculation. **(c)** $1/\sqrt{6}$

21. Routine calculation.   **23.** $3120/\sqrt{11} \approx 941$ watts.

25. The total work done is the same either way: $3750\sqrt{3} + 1/\sqrt{2} \approx 7244$ foot-pounds.

27. *Hint:* Consider two cases.   **29.** Follow the outlined steps.

Section 5: Euclidean Geometry

Exercise Set 5AB (pgs. 36–37)

1. $y = 3$.

3. $x + 2y + 4z = -1$. (*Note:* The vector $\mathbf{N}$ in the diagram below is parallel to the given perpendicular but is only half as long. The right triangle with $\mathbf{N}$ as one leg has its other leg in the plane and its hypotenuse parallel with the z-axis.)

5. $(x, y, z) = (2/5 + 1, -2/5, 4(2/5) - 2) = (7/5, -2/5, -2/5)$.

7. Every point on the line is in the plane.

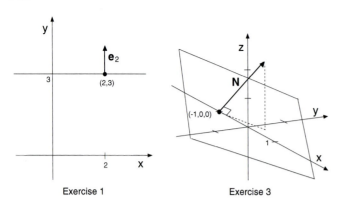

Exercise 1   Exercise 3

9. $(x, y, z) = (-1/3 + 1, 1, -1/3 + 1) = (2/3, 1, 2/3)$.

11. Angle is $90°$ with cosine zero.   **13.** Angle is $90°$ with cosine zero.

15.

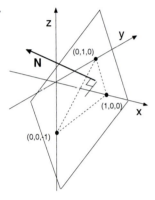

$x + y - z = 1$

17.

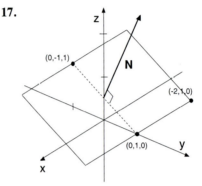

$y + 2z = 1$

19. $3x - 2y + 5z = 9$.

21. (a) $x - 2y + z = 0$.   **(b)** $x = 1$ or $z = 1$ and many others. The three points lie on a line parallel to the y-axis so don't determine a unique plane.

23. $1/\sqrt{5}$.   **25.** $1/\sqrt{3}$. $(1, 0, -1)$ is below P.

27. $3/\sqrt{14}$. $(1, 0, -1)$ is below P.
To locate the origin relative to P, set $x = y = 0$ in the equation for P to find out where the plane crosses the z-axis. This gives $3z = 1$, or $z = 1/3$. Thus, $(0, 0, 1/3)$ is on the plane and $(0, 0, 0)$ is one-third unit below it; i.e., the origin is below P. It follows that $(1, 0, -1)$ and the origin are on the same side of P.

29. $|d|$.

Section 6: The Cross Product

Exercise Set 6 (pgs. 42–44)

1. $-\mathbf{e}_3$.   **3.** $(1, -2, 1)$.

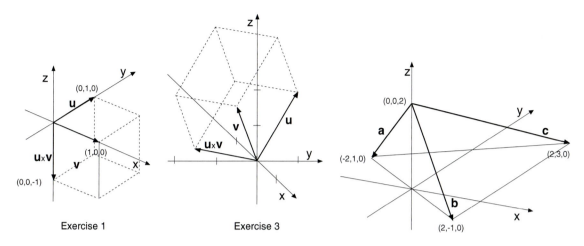

Exercise 1   Exercise 3

5. $\sqrt{2}$.   **7.** $3\sqrt{3}/2$.

9. $2x - 5y + 2z = 5$.   **11.** Routine check.

13. **(a)** $\mathbf{a} = (-2, 1, -2), \mathbf{b} = (2, -1, -2), \mathbf{c} = (2, 3, -2)$ are shown on the right with their tails at the apex.
   (b) $\mathbf{u} = \mathbf{a} \times \mathbf{b} = (-4, -8, 0), \mathbf{v} = \mathbf{b} \times \mathbf{c} = (8, 0, 8), \mathbf{w} = \mathbf{c} \times \mathbf{a} = (-4, 8, 8)$.
   (c) $\cos \alpha = 0, \cos \beta = 1/\sqrt{2}, \cos \gamma = 2/3$.

15. **(a)** Routine computation.   **(b)** *Hint:* Use a little trigonometry.   **(c)** *Hint:* More trigonometry.

17. Routine computation.   **19.** 3.

21. $V(B) = 17$.

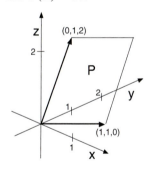

Exercise 19

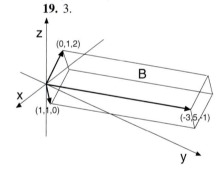

Exercise 21

23. Follow the steps.   **25.** **(a)** Unequal. In (b), (c), (d) pairs are equal.

27. Use the hint in the exercise.

Chapter 1 Review (pgs. 44–45)

1. c.   **3.** $3\mathbf{k}$.

5. $2\mathbf{e}_1 - \mathbf{e}_2 + 3\mathbf{e}_3 + 2\mathbf{e}_4$.   **7.** $(4, 1, -2) = -2(1, 2, 3) + (6, 5, 4)$.

9.

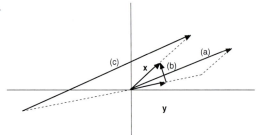

11. $s(-1, 1) + (1, 2)$, $t(5, 7) + (4, 5)$, intersecting at $(3/2, 3/2)$.

13. $f(t) = t\left(6/\sqrt{38}, -1/\sqrt{38}, -1/\sqrt{38}\right) + (-5, 3, 4)$.

15. **(a)** *Hint:* Show $\mathbf{d}(t)$ is always a scalar multiple of a fixed vector.   **(b)** If $\mathbf{p}_1(t) = t\mathbf{v}_1 + \mathbf{u}_1$ and $\mathbf{p}_2(t) = t\mathbf{v}_2 + \mathbf{u}_2$ the collision is at the time when these are equal if that's positive, otherwise no collision.

17. K and M are parallel and are the same line. L and M are parallel but are not the same line.

[**19.** $s(3, -2, 0) + t(0, 2, -5) + (3, 0, 0)$. **21.** $s(1, 0, 0) + t(0, 1, 0) + (1, 2, 3)$.

23. $\mathbf{x} = t(3, 1, 2)$. (The values $t = 0, 1, -2$ give the three points.)

25. $\mathbf{x} = s(-2, 0, 2) + t(-3, 0, 1) + (1, 2, 3)$.

27. **(a)** Angle between $\mathbf{a}$ and $\mathbf{b}$ is less than the angle between $\mathbf{a}$ and $\mathbf{c}$.   **(b)** $(0, -1, 1)$.   **(c)** $\sqrt{2}/2$.   **(d)** $\sqrt{11}/2$.

29. 18 units.

31. **(a)** $6/\sqrt{13}$ units.   **(b)** $\mathbf{x} = t(3, -2) + (3, 0)$.   **(c)** $\pm\left(8/\sqrt{13}, 12/\sqrt{13}\right)$.

33. **(a)** $\mathbf{x} = t(1, -2)$.   **(b)** $\mathbf{n} = \left(1/\sqrt{5}, -2/\sqrt{5}\right)$, $c = -2/\sqrt{5}$.   **(c)** $\sqrt{5}$; the origin and $(3, 5)$ are not on the same side of L.

35. $\mathbf{x} = t(0, 3, -3) + (8/3, 0, 1/3)$; $(8/3, 2/3, -1/3)$.

CHAPTER 2: EQUATIONS AND MATRICES

Section 1: Systems of Linear Equations

Exercise Set 1A (pgs. 51–52)

1. Two lines intersecting in the point $(3/2, -1/2)$.   **3.** Three planes intersecting in $(0, 0, 0)$.

5. No solutions. Each pair of planes intersects in a different line parallel to $(1, -1, 0)$.

7. Intersect at $(4, -2, 7)$.   **9.** Intersect at $(7, 11)$.

11. **(a)** $(a, b, c) = (-5/6, 19/6, 1)$ and $f(x) = -\frac{5}{6}x^2 + \frac{19}{6}x + 1$.
   (b) $y = 11/2$. The path is a straight line, not a parabola.

13. 8 cubic yards or 32 tons of sand, 2 cubic yards or 2 tons of cinders.

15. $\mathbf{b} = 2\mathbf{a}_1 - \frac{3}{2}\mathbf{a}_2$.   **17.** $\frac{2}{3}\mathbf{a}_1 - \frac{1}{3}\mathbf{a}_3$.

Exercise Set 1B (p. 58)

1. Voltages at junctions $J_1 = A$, J_2, J_3, $J_4 = B$, J_5, J_6 in the diagram

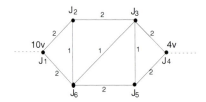

are 10, $\frac{244}{31}$, $\frac{202}{31}$, 4, $\frac{190}{31}$, $\frac{232}{31}$, current of $\frac{72}{31} \approx 2.32$ amps flows in at A and out at B.

3. With junctions labelled $J_1, \ldots, J_8$ as shown,

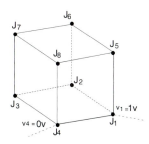

let v_i be the voltage at J_i. With fixed voltages $v_1 = 1$ and $v_4 = 0$, other voltages are $v_2 = \frac{9}{14}$, $v_3 = \frac{5}{14}$, $v_5 = \frac{9}{14}$, $v_6 = \frac{4}{7}$, $v_7 = \frac{3}{7}$, $v_8 = \frac{5}{14}$. Current of $\frac{12}{7} \approx 1.71$ amps flows in at J_1 and out at J_4.

5. Magnitudes proportional to $2\sqrt{5}, \sqrt{2}, \sqrt{10}$.

7. No. Any resultant (a, b, c) of forces acting in the given directions is a linear combination with positive coefficients, and must have $a \geq b \geq c \geq 0$.

9. $p_2 = \frac{5}{7}$.

11. $p_1 = p_2 = p_3 = p_4 = 1/2$.

13. $p_1 = \frac{16}{29}$, $p_2 = \frac{15}{29}$, $p_3 = \frac{17}{29}$, $p_4 = \frac{23}{29}$.

15. $r = (-a + 3, a - 2, -a + 4, a)$, for any a.

17. $A = \frac{1}{3}h$, $B = \frac{4}{3}h$, $C = \frac{1}{3}h$.

Section 2: Matrix Methods

Exercise Set 2A–C (pgs. 69–70)

1. $\begin{pmatrix} 3 & -2 \\ 1 & -3 \end{pmatrix} \begin{pmatrix} x \\ y \end{pmatrix} = \begin{pmatrix} 1 \\ 2 \end{pmatrix}$. $x = -1/7$, $y = -5/7$.

3. $\begin{pmatrix} 1 & 1 & 0 \\ 0 & 1 & -1 \\ 1 & 0 & 1 \end{pmatrix} \begin{pmatrix} x \\ y \\ z \end{pmatrix} = \begin{pmatrix} 1 \\ 1 \\ 0 \end{pmatrix}$.

$(x, y, z) = (-t, t + 1, t) = t(-1, 1, 1) + (0, 1, 0)$.

5. $\begin{matrix} x + 2y = 1 \\ 3x + y = 0 \end{matrix}$, $\begin{pmatrix} x \\ y \end{pmatrix} = \begin{pmatrix} -1/5 \\ 3/5 \end{pmatrix}$.

7. $\begin{matrix} x + z = 0 \\ y = 1 \\ x + y = 0 \end{matrix}$, $\mathbf{x} = \begin{pmatrix} -1 \\ 1 \\ 1 \end{pmatrix}$.

9. $x = 24/15$, $y = 13/15$, $z = -7/15$.

11. (a) $\begin{pmatrix} 1 & 0 & -1 \\ 0 & 1 & -1 \end{pmatrix}$, (b) $\mathbf{x} = t(1, 1, 1)$. (c) $\mathbf{x} = t(1, 1, 1) + (3/5, -1/5, 0)$.

13. (a) $\begin{pmatrix} 0 & 0 & 1 \\ 0 & 1 & 0 \\ 1 & 0 & 0 \end{pmatrix}$, (b) $\mathbf{x} = \mathbf{0}$, (c) $\mathbf{x} = (0, 1, 0)$.

15. $\mathbf{i} = -3(1, 2) + 2(2, 3)$, $\mathbf{j} = 2(1, 2) - (2, 3)$.

17. $(1, 2, 1, 0) + 2(2, -1, 0, 1)$.

19. $(0, 0, 0, -1, 2)$.

21. $\mathbf{v} = -\mathbf{a} + 2\mathbf{b}$.

23. $\mathbf{v} = -\frac{3}{7}\mathbf{a} - \frac{1}{7}\mathbf{b} - \frac{2}{7}\mathbf{c}$.

25. $t(0, 4, 8) + (-4, -2, 0)$.

27. Show that if the lines are non-parallel, equations have a unique solution (whether lines intersect or not). Show that if the lines are parallel, they lie in a plane, and any line that is in the plane and perpendicular to both given lines will do.

29. *Hint:* $A(\mathbf{w} - \mathbf{v}) = A\mathbf{w} - A\mathbf{v}$.

31. *Hint:* $A(t_1\mathbf{x}_1 + t_2\mathbf{x}_2) = t_1 A\mathbf{x}_1 + t_2 A\mathbf{x}_2 = t_1\mathbf{b}_1 + t_2\mathbf{b}_2$.

Exercise Set 2D (p. 73)

1. $r(1, 0, 0, -3) + s(0, 1, 0, 2) + t(0, 0, 1, -1) +$
$(0, 0, 0, 3).$

3. $\begin{pmatrix} 1 & 0 & 0 \\ 0 & 1 & 0 \\ 0 & 0 & 0 \\ 0 & 0 & 1 \end{pmatrix}.$

5. (a) $\{\mathbf{a}, \mathbf{b}, \mathbf{d}\}$ and $\{\mathbf{a}, \mathbf{c}, \mathbf{d}\}$ independent, $\{\mathbf{b}, \mathbf{c}, \mathbf{d}\}$ dependent.

7. *Hint:* When is one vector a scalar multiple of another?

9. $t(7, -3, 1).$ **11.** Solutions are $s(-1, 0, 0, 1) + t(0, 1, -1, 0).$

13. *Hint:* First show that every row and column contains a leading entry.

Section 3: Matrix Algebra

Exercise Set 3A–D (pgs. 80–81)

1. $\begin{pmatrix} -4 & 4 \\ -1 & -1 \end{pmatrix}.$

3. $\begin{pmatrix} 0 & 7 \\ 2 & 8 \end{pmatrix}.$

5. $\begin{pmatrix} 1 & 6 \\ 1 & 4 \end{pmatrix}.$

7. $\begin{pmatrix} 7 & 13 \\ 5 & 9 \end{pmatrix}.$

9. $\begin{pmatrix} 2 & 4 \\ 2 & 4 \end{pmatrix}.$

11. $\begin{pmatrix} -3 & -1 & -4 \\ -5 & 6 & -9 \end{pmatrix}.$

13. BA not defined; B has 3 columns, A has 2 rows.

15. $\begin{pmatrix} 4 & -16 & 2 \\ 0 & 0 & 0 \\ -3 & 3 & 3 \end{pmatrix}.$

17. $\begin{pmatrix} -11 & 19 & -8 \\ -9 & 28 & -23 \end{pmatrix}.$

19. DC is not defined, D has 2 columns, C has 3 rows.

21. X and Y are 2-by-3.

23. X is 2-by-2, Y is 2-by-3.

25. X and Y are 2-by-3.

27. $C\mathbf{i} = \begin{pmatrix} -2 \\ 0 \\ 2 \end{pmatrix}$, $C\mathbf{j} = \begin{pmatrix} 0 \\ 3 \\ 3 \end{pmatrix}$, $C\mathbf{k} = \begin{pmatrix} 1 \\ 0 \\ -1 \end{pmatrix}.$

29. $\begin{pmatrix} 2 \\ 5 \\ 8 \end{pmatrix}.$

31. (38).

33. $\begin{pmatrix} 1 & -1 & 1 \\ -1 & 1 & -1 \\ -1 & 1 & -1 \end{pmatrix}.$

35. AO is defined when O is n-by-p for some p and is then m-by-p. OA is defined when O is p-by-m for some p and is then p-by-n.

37. (a) definition of matrix product.
(b) If $\mathbf{c}$ has entries $c_1, \ldots, c_n$ and $\mathbf{r}$ has entries $r_1, \ldots, r_n$ then $M = \mathbf{cr}$ has entries $m_{ij} = c_i r_j$, for $i, j = 1, \ldots, n.$

39. $A^2 = \begin{pmatrix} -1 & 4 \\ -8 & 7 \end{pmatrix}$, $A^3 = \begin{pmatrix} -9 & 11 \\ -22 & 13 \end{pmatrix}$, $p(A) = \begin{pmatrix} -2 & 5 \\ -10 & 8 \end{pmatrix}.$

41. $A^2 = \begin{pmatrix} 1 & 0 & 0 \\ 0 & 4 & 0 \\ 0 & 0 & 9 \end{pmatrix}$, $A^3 = \begin{pmatrix} -1 & 0 & 0 \\ 0 & 8 & 0 \\ 0 & 0 & 27 \end{pmatrix}$,

$p(A) = \begin{pmatrix} 8 & 0 & 0 \\ 0 & 5 & 0 \\ 0 & 0 & 12 \end{pmatrix}.$

43. $UV = O$, $VU = \begin{pmatrix} 10 & -20 \\ 5 & -10 \end{pmatrix}.$

45. **(a)** Use distributive laws for matrix multiplication.
   (b) Use distributive laws to show $(A + B)^2 = A^2 + AB + BA + B^2$. This equals $A^2 + 2AB + B^2$ if and only if $AB = BA$, so any matrices with $AB \neq BA$, such as those in Exercise 43, are examples.

47. $X^2 = \begin{pmatrix} 4 & 3a \\ 0 & 1 \end{pmatrix}$, $p(X) = O$.

49. **(a)** For the given numbers, $p(x) = x^2 + 2$ and $A = \begin{pmatrix} 1 & -3 \\ 1 & -1 \end{pmatrix}$. $A^2 = \begin{pmatrix} -2 & 0 \\ 0 & -2 \end{pmatrix}$ so $p(A) = O$.

   (b) Start with $A^2 = \begin{pmatrix} a^2 + bc & ab + bd \\ ca + dc & cb + d^2 \end{pmatrix}$ and go on from there.

Section 4: Inverse Matrices

Exercise Set 4A–C (pgs. 86–88)

1. $\begin{pmatrix} 2 & -1 \\ -1 & 1 \end{pmatrix}$.

3. $\begin{pmatrix} 16/3 & -20/3 \\ -20/3 & 40/3 \end{pmatrix}$.

5. $A^{-1} = \begin{pmatrix} 4/11 & 1/11 \\ -3/11 & 2/11 \end{pmatrix}$, $\mathbf{x} = \begin{pmatrix} 5/11 \\ -1/11 \end{pmatrix}$.

7. $\begin{pmatrix} 1 & 0 & 0 \\ -13 & 1 & -5 \\ 2 & 0 & 1 \end{pmatrix}$.

9. No inverse; row reduction gives a row of zeros.

11. $\begin{pmatrix} 1 & 2 \\ 5 & 6 \end{pmatrix}^{-1} = \begin{pmatrix} -3/2 & 1/2 \\ 5/4 & -1/4 \end{pmatrix}$, $X = \begin{pmatrix} 1/2 & 11/2 & -6 \\ -1/4 & -17/4 & 5 \end{pmatrix}$.

13. *Hint:* Simplify $(AB)(B^{-1}A^{-1})$ using associativity.

15. *Hint:* Evaluate $(I + A + A^2)(I - A)$.

17. $\begin{pmatrix} 1 & 0 & 0 \\ 0 & 1/2 & 0 \\ 0 & 0 & 1 \end{pmatrix}$.

19. $\begin{pmatrix} 1 & -1 & 1 & -1/2 \\ 0 & 1/2 & 0 & -1/8 \\ 0 & 0 & 1 & 0 \\ 0 & 0 & 0 & 1/4 \end{pmatrix}$.

21. Apply definition.

23. Apply definition.

25. Use the results of Exercises 21 and 22.

27. Use a trigonometric identity.

29. Note that $\frac{1}{3} + \frac{1}{2} + \frac{1}{6} = \frac{1}{3} + \frac{4}{6} = 1$.

31. *Hint:* The rows of Q^t are the columns of Q and vice versa.

33. *Hints:* **(a)** $(A + A^t)^t = A^t + A$. **(b)** What is $(A + A^t) + (A - A^t)$? **(c)** What is $A^t(A^{-1})^t$?

35. *Hint:* $p(x_k)$ is equal to the dot product of $(a_0, \ldots, a_n)$ and $(1, x_k, x_k^2, \ldots, x_k^n)$.

Section 5: Determinants

Exercise Set 5A–E (pgs. 98–99)

1. $\det A = 24$, $\det(2A) = 192$.

3. $\det(2A) = 2^n \det A$.

5. 32.

7. $\det A = 7$, $\det B = 2$, $\det AB = \det BA = 14$.

9. *Hint:* First use row expansion to show $\det I = 1$.

11. -12.

13. $\begin{pmatrix} 1 & 0 & 0 \\ -13 & 1 & -5 \\ 2 & 0 & 1 \end{pmatrix}$.

15. $\begin{pmatrix} 0 & 1 & 0 \\ 7/4 & -11/2 & 2 \\ -3/4 & 5/2 & -1 \end{pmatrix}$.

17. $\det A = 0$; not invertible.

19. $\begin{pmatrix} 1 & -1 & 1 & -3/4 \\ 0 & 1/2 & 0 & -1/8 \\ 0 & 0 & 1 & -1/4 \\ 0 & 0 & 0 & 1/4 \end{pmatrix}$.

21. The determinant is $(1 - t)(2 - t)(3 - t)$. The matrix fails to have an inverse when $t = 1, 2$, or 3.

23. The determinant is $-4(t - 4)$. The matrix fails to have an inverse when $t = 4$.

25. Expand by the first row and compare with the definition of the cross product.

27. *Hint:* If $A = I$, what is the result of expansion by the first row?

Chapter 2 Review (pgs. 99–101)

1. Dimensions of A and B don't match.

3. $\begin{pmatrix} 6 & -5 & -2 \\ 2 & 1 & 2 \end{pmatrix}$.

5. Dimensions of A and EB don't match.

7. D^{-1} does not exist.

9. $C - A$ must be invertible. If it is, then $X = 2(C - A)^{-1}B$.

11. $X = 3I$ is always a solution, unique if A is invertible.

13. Sometimes true (e.g. if $A = I$), sometimes false, e.g. if $A = \begin{pmatrix} 1 & 0 \\ 0 & 0 \end{pmatrix}$, $B = \begin{pmatrix} 0 & 0 \\ 1 & 0 \end{pmatrix}$.

15. Sometimes true (e.g. if $A = I$), sometimes false, e.g. if $A = \begin{pmatrix} 0 & 0 \\ 1 & 0 \end{pmatrix}$, $B = \begin{pmatrix} 1 & 0 \\ 0 & 0 \end{pmatrix}$.

17. Always true.

19. True if $A = I$, sometimes false, e.g. if $A = \begin{pmatrix} 0 & 1 \\ 1 & 0 \end{pmatrix}$.

21. (a) $t\begin{pmatrix} -1 \\ -1 \\ 1 \end{pmatrix}$, **(b)** $t\begin{pmatrix} -1 \\ -1 \\ 1 \end{pmatrix} + \begin{pmatrix} 1 \\ 2 \\ 0 \end{pmatrix}$, **(c)** no solutions.
There are solutions for $\mathbf{b} = (b_1, b_2, b_3)$ when $b_3 = 0$.

23. No solution unless $b = -2$; if $b = -2$ then $x = 1$, $y = 0$, $z = 2$ is one of many solutions.

25. $\mathbf{v} = -\mathbf{a} + 2\mathbf{b}$.

27. Impossible.

29. $\begin{pmatrix} 1/13 & -5/13 \\ 2/13 & 3/13 \end{pmatrix}$.

31. Not invertible.

33. Not invertible.

35. $\begin{pmatrix} -1/5 & 6/5 & -4/5 & -8/5 & 16/5 \\ 3/5 & -3/5 & 2/5 & 4/5 & -8/5 \\ -1/5 & 1/5 & 1/5 & 2/5 & -4/5 \\ -1/5 & 1/5 & 1/5 & -3/5 & 6/5 \\ 1/5 & -1/5 & -1/5 & 3/5 & -1/5 \end{pmatrix}$.

37. (a) Compute the products.
(b) In DA each row of A is multiplied by the corresponding diagonal entry of D, in AD each column is multiplied.
(c) B is 3-by-3 in all cases. If a, b, c are all different, B is diagonal. If $a = b = c$, B is arbitrary. If $a = b \neq c$ then $b_{31} = b_{32} = b_{13} = b_{23} = 0$.

39. $a = 2$, $b = -3$, $c = 4$.

41. $t(-1, 0, -1, 1)$.

43. 9.

45. -20.

CHAPTER 3: VECTOR SPACES & LINEARITY

Section 1: Linear Functions on $\mathbb{R}^n$

Exercise Set 1A–C (pgs. 110–112)

1. $\begin{pmatrix} 1 & 2 \\ 2 & 4 \end{pmatrix}$, not one-to-one.

3. $\begin{pmatrix} 1 & 3 \\ 2 & 2 \\ 3 & 1 \end{pmatrix}$, one-to-one.

5. $f(\mathbf{e}_1) = \begin{pmatrix} 1/2 \\ 1 \end{pmatrix}$, $f(\mathbf{e}_2) = \begin{pmatrix} 3/2 \\ 0 \end{pmatrix}$.

7. $f(\mathbf{e}_1) = \begin{pmatrix} 2 \\ 3 \end{pmatrix}$, $f(\mathbf{e}_2) = \begin{pmatrix} 0 \\ -2 \end{pmatrix}$, $f(\mathbf{e}_3) = \begin{pmatrix} -1 \\ 1 \end{pmatrix}$.

9. *Hint:* Calculate $R_\theta \mathbf{x} \cdot R_\theta \mathbf{x}$.

11. $\begin{pmatrix} 2 & -1 \\ 7 & -1 \end{pmatrix}$, domain = range = $\mathbb{R}^2$.

13. $\begin{pmatrix} 4 & 0 & -1 \\ 0 & 2 & 1 \\ -2 & 0 & 5 \end{pmatrix}$, domain = range = $\mathbb{R}^3$.

15. **(a)** *Hint:* Image of (x, y) is (y, x).

(b) $\begin{pmatrix} 0 & -1 \\ -1 & 0 \end{pmatrix}$

(c) $\begin{pmatrix} -1 & 0 \\ 0 & -1 \end{pmatrix} = -I$, 180° rotation, (same as reflection through the origin).

17. **(a)** Partial answer: Multiplication by U leaves the x-axis fixed and rotates $\mathbf{e}_2$ and $\mathbf{e}_3$ 90° in the yz-plane.

(b) $UVU^{-1} = \begin{pmatrix} 0 & -1 & 0 \\ 1 & 0 & 0 \\ 0 & 0 & 1 \end{pmatrix}$, $VUV^{-1} \begin{pmatrix} 0 & 1 & 0 \\ -1 & 0 & 0 \\ 0 & 0 & 1 \end{pmatrix}$.

The two products represent rotations of 90° in opposite directions about the z-axis.

19. $\begin{pmatrix} 9/49 & 18/49 & 6/49 \\ 18/49 & 36/49 & 12/49 \\ 6/49 & 12/49 & 4/49 \end{pmatrix}$.

21. Line $t(1, 3)$.

23. Line $t(1, -1)$.

25. Plane $s(1, -2, 0) + t(-2, -1, -5)$.

27. Both make sense.

29. Only $g \circ f$ makes sense.

31. $\mathbf{a} = \left(f(\mathbf{e}_1), \ldots, f(\mathbf{e}_n) \right)$.

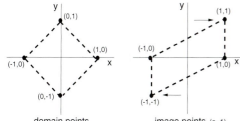

domain points   image points (a=1)

33. **(a)** $\begin{pmatrix} 1 & a \\ 0 & 1 \end{pmatrix}$.

(b) $f_a(1, 0) = (1, 0)$, $f_a(1, 0) = (a, 1)$, $f_a(-1, 0) = (-1, 0)$, $f_a(0, -1) = (-a, -1)$.

(c) *Hint:* If $a > 0$ and $y > 0$ then $x + ay > x$. **(d)** The x-axis. **(e)** Horizontal lines. **(f)** f_{a+b}.

Section 2: Vector Spaces

Exercise Set 2AB (pgs. 118–119)

1. Subspace.

3. Subspace.

5. Not a subspace.

7. Subspace.

9. Not a subspace.

11. $(1, 0, 1)$, $(0, 1, 2)$ (not the only correct answer).

13. *Hint:* Use additivity and homogeneity of the dot product.

15. Sequences that are 0 after the nth term; sequences that are 0 after some finite number of terms; the sequence in which every term is 1.

17. Subspace.

19. Subspace.

21. Subspace.

23. Subspace.

25. *Hint:* $(f_1 + f_2)(a) = f_1(a) + f_2(a)$, $(rf_1)(a) = rf_1(a)$.

27. Partial answer: T is in the span of S since $(3, 5, 4) = (2, 3, 1) + (1, 2, 3)$ and $(1, 1, -2) = (2, 3, 1) - (1, 2, 3)$.

29. *Hint:* Look at Theorems 2.1 and 2.4.

31. *Hint:* If a is in A but not in B, and b is in B but not in A, show that $a + b$ is in neither.

33. **(a)** $f(x) = \sqrt{x - a}$ is one example, $|x - (a + b)/2|$ is another.
  (b) $f_1(x) = \int_a^x f(t)\,dt$, where f is as in part (a).
  (c) Take $f_k(x) = (x - a)^{k+1/2}$, or more generally, $f_k(x) = \int_a^x f_{k-1}(t)\,dt$.

35. *Hint:* $f((x, y, z)) = x + 2y + z$.

37. Always true.

39. True for S_1 and S_2 the x- and y-axes in $\mathbb{R}^3$, false for the x- and y-axes in $\mathbb{R}^2$.

41. True. If $\mathbf{x}$ is in S, then $2\mathbf{x}$ is not.

Section 3: Linear Functions

Exercise Set 3A–C (pgs. 125–126)

1. $\begin{pmatrix} 1 & 2 \\ 2 & 1 \end{pmatrix}$.

3. $\begin{pmatrix} 1 & 0 \\ -1 & 3 \end{pmatrix}$.

5. One-to-one. $f^{-1}(x_1, x_2, x_3, \dots) = \frac{1}{2}(x_1, x_2, x_3, \dots)$. Domain of f^{-1} is all of $\mathbb{R}^\infty$.

7. Not one-to-one.

9. $(g \circ f)(x_1, x_2, \dots) = (f \circ g)(x_1, x_2, \dots) = (2x_1, 4x_2, 6x_3, \dots)$.

11. $(g \circ p)(x_1, x_2, x_3, \dots) = (0, 2x_1, 3x_2, 4x_3, \dots)$, $(p \circ g)(x_1, x_2, x_3, \dots) = (0, x_1, 2x_2, 3x_3, \dots)$.

13. Use the right distributive law and the scalar commutativity laws for matrix multiplication (Theorem 3.2 in Chapter 2).

15. $Du(x) = 6x^2 - 4$, $xu(x) = 2x^4 - 4x^2$, $D(xu(x)) = 8x^3 - 8x$, $xDu(x) = x(6x^2 - 4) = 6x^3 - 4x$.

17. **(a)** $(Dx - xD)u = (xu' + u) - xu' = u$.
  (c) No. $(D^2 - x^2)u = D^2u - x^2u = u'' - x^2u$, but $(D + x)(D - x)u = (D + x)(u' - xu) = D(u' - xu) + x(u' - xu) = (u'' - xu' - u) + xu' - x^2u = u'' - (1 + x^2)u$.

19. *Hint:* Use some basic integration formulas.

21. 11.

23. $(0, 2)$.

25. $2x^2 + 3x$.

27. No. $f'(0)$ is not defined.

29. No. Every function in the image of L has value 0 at 0.

31. No inverse.

33. $f^{-1}(\mathbf{y})$ has matrix $\begin{pmatrix} 2/3 & -7/9 \\ 1/3 & -2/9 \\ 0 & 1/3 \end{pmatrix}$. Domain of f^{-1} is $\mathbb{R}^2$.

35. No inverse.

Section 4: Image and Null-Space

Exercise Set 4A–C (pgs. 130–131)

1. Range: $\mathbb{R}^3$. Image: plane through O spanned by $(1, 0, 1)$ and $(0, 1, 1)$. Linear, null-space $\{0\}$.

3. Range: $\mathbb{R}^3$. Image: plane $u(1, 0, 2) + v(0, 1, 1) + (0, 0, 1)$. Not linear.

5. Range and image: $\mathbb{R}^2$. Linear, null-space $t(1, 2, -3)$.

7. Image: $C(-\infty, \infty)$. Not linear.

9. Range: $C^{(1)}(-\infty, \infty)$. Image: subspace of f with $f(0) = 0$. Linear, null-space $\{0\}$.

11. Image $\mathbb{R}^2$, null-space: $\{0\}$.

13. Image: the plane spanned by $(1, 0, 1)$ and $(4, 1, 1)$. Null-space: the line $t(1, 0, -2)$.

15. **(a)** $t(5, 2)$.   **(c)** $(1, -1)$; $t(5, 2) + (1, -1)$.

17. *Hint:* Show that $f(\mathbf{x}) = (f(\mathbf{e}_1), \dots, f(\mathbf{e}_n)) \cdot \mathbf{x}_0$.

19. **(a)** Use definition of linearity and properties of D.
   (b) $(D - 1)(x + 1) = (x + 1)' - (x + 1) = 1 - (x + 1) = -x$. All solutions: $y(x) = ce^x + x + 1$.

21. **(b)** If Gu is the zero function, taking derivatives gives $tu(t) = 0$ for all t, so $u(t) = 0$ for $t \neq 0$. Since u is continuous, it is the zero function, so G is one-to-one.
   (c) Polynomials of the form $x^2 p(x)$ with $p(x)$ in P_n.
   (d) Polynomials of the form $x^2 p(x)$ with $p(x)$ in P.
   (e) The domain of G^{-1} is the same as the image of G, and consists of continuously differentiable functions g with $g(0) = g'(0) = 0$. For such a g, taking $f(t) = g'(t)/t$ for $t \neq 0$ and $f(0) = 0$ gives a continuous $f = G^{-1}(g)$.
   (f) The constant function 1.

23. **(a)** The reflection of (x, y) is $(-x, y)$ and $(Ru)(x) = u(-x)$.
   (b) $(R^2)(u(x)) = u(-(-x)) = u(x)$.
   (c) *Hint:* If $u(x) = u(-x)$ then $Ru = u$.
   (d) Image the odd functions, null-space the even functions.
   (e) $F_e^2 = F_e$, $F_o^2 = F_o$.

Section 5: Coordinates and Dimension

Exercise Set 5AB (pgs. 137–138)

1. Spanning: $(x, y) = \frac{1}{2}(y - x)(-1, 1) + \frac{1}{2}(x + y)(1, 1)$. Independence: neither vector is a scalar multiple of the other.

3. Partial answer. Spanning: $(x, y, z) = (x - y)(1, 0, 0) + (y - z)(1, 1, 0) + z(1, 1, 1)$.

5. 1.   **7.** 2.

9. If $ae^x + be^{2x} + ce^{3x} = 0$ for all x, we can, for instance, set $x = 0$, $x = \ln 2$ and $x = \ln 3$ to get
$$\begin{pmatrix} 1 & 1 & 1 \\ 2 & 4 & 8 \\ 3 & 9 & 27 \end{pmatrix} \begin{pmatrix} a \\ b \\ c \end{pmatrix} = \mathbf{0}.$$ Row reduction shows that the only solution is $a = b = c = 0$, so the given functions are linearly independent. Another proof is to multiply the equation by e^{-x} and take the limit as x goes to $-\infty$ to show $a = 0$, and then similarly show b and then c are 0.

11. If $a \cos x + b \sin x = 0$ for all x, putting $x = 0$ and $x = \pi/2$ gives $b = 0$ and $a = 0$, so $\sin x$ and $\cos x$ are linearly independent.

13. **(b)** for e^x, $(1, 1)$; for e^{-x}, $(1, -1)$.   **15.** **(b)** $(1, -1, 1)$.

17. For $\cos^2 x$, $(1/2, 0, 0, 1/2, 0)$; for $\sin^2 x$, $(1/2, 0, 0, -1/2, 0)$.

19. Partial answer: A product $f(x)g(x)$ is a linear combination of terms of the form $\cos ax \cos bx$, $\cos ax \sin bx$, and $\sin ax \sin bx$, with $a \leq p$ and $b \leq q$. The trigonometric identity $\cos ax \cos bx = \frac{1}{2}\cos(a - b)x - \frac{1}{2}\cos(a + b)x$ shows directly that if $a > b$ then $\cos ax \cos bx$ is in T_{p+q}. What about other terms? What if $a \leq b$?

21. $\{e^x, e^{-x}\}$.   **23.** $\{\cos x, \sin x, \sin 2x\}$.

25. Image: span of $\{(2, 1), (1, 2)\}$. Null-space: $\{0\}$.

27. Image: span of $\{(2, 0, 1), (4, 1, 3)\}$; null-space $t(1, -1, 1)$.

29. 1.

31. Let $p_1(x), \ldots, p_k(x)$ have different degrees $d_1, \ldots, d_k$, and suppose them ordered so that $d_1 < d_2, \cdots < d_k$. Suppose c_m is the last non-zero coefficient in a linear combination $q(x) = c_1 p_1(x) + \cdots + c_k p_k(x)$. Then the coefficient of x^m is non-zero in $c_m p_m(x)$ and zero in all other terms, so $q(x)$ is not the zero polynomial.

33. One possibility is $p(x) = x$.

35. Yes, $(3, -4, 5, 2) = 2(1, -2, 1, 1) - (2, 1, -2, 1) + (3, 1, 1, 1)$.

37. Yes, $\cos 2x = 2\cos^2 x - 1$.   **39.** $\begin{pmatrix} -2 & -5 \\ 3 & 3 \\ 0 & 2 \end{pmatrix}$.

41. (a) $S(rp(x)) = S(rp)(x) = (rp)(x + 1) = rp(x + 1) = r Sp(x)$, and similarly for additivity, so S is linear.

(b) For $p(x) = a_0 + a_1 x + a_2 x^2$, $p(x + 1) = (a_0 + a_1 + a_2) + (a_1 + 2a_2)x + a_2 x^2$, $(Dp)(x) = a_1 + 2a_2 x$, and $\frac{1}{2}(D^2 p)(x) = a_2$, so $p(x + 1) = p(x) + (Dp)(x) + \frac{1}{2}(D^2 p)(x)$.

(c) By Taylor's formula $p(x + h) = p(x) + p'(a)h + \frac{1}{2!}p''(x)^2 + \cdots + \frac{1}{n!}p^{(n)}(x)h^n$ for all polynomials of degree $\leq n$, since all their derivatives of order $> n$ are zero. Putting $h = 1$ shows that S and $I + D + \frac{1}{2!}D^2 + \cdots + \frac{1}{n!}D^n$ have the same effect on polynomials in $\mathcal{P}_n$.

Exercise Set 5C (p. 142–143)

1. (a) E_{ij} has 1 in row i column j, where all other E_{pq} have 0, so is not a linear combination of them. An m-by-n matrix with entries a_{ij} is the sum of $a_{ij} E_{ij}$, so the E_{ij} are a spanning set. Thus $\{E_{ij}\}$ is a basis for the space of m-by-n matrices, and the dimension is mn.   (b) n.

3. *Hint:* A linear combination $c_1 f(\mathbf{x}_1) + \cdots + c_k f(\mathbf{x}_k) = f(c_1 \mathbf{x}_1 + \cdots + c_k \mathbf{x}_k)$ is $\mathbf{0}$ if and only if $c_1 \mathbf{x}_1 + \cdots + c_k \mathbf{x}_k = \mathbf{0}$. Dimensions of domain and image are equal.

5. *Hint:* Either the intersection contains two linearly independent vectors or it doesn't.

7. A basis for $\mathcal{V}$ is a linearly independent subset of $\mathcal{W}$. Apply Theorem 5.6.

9. (a) Let $\{\mathbf{b}_1, \ldots, \mathbf{b}_n\}$ be a basis for $\mathbb{R}^n$ with $\{\mathbf{b}_1, \ldots, \mathbf{b}_k\}$ a basis for $\mathcal{S}$. One possible f is defined by
$f(c_1 \mathbf{b}_1 + \cdots + c_k \mathbf{b}_k + c_{k+1} \mathbf{b}_{k+1} + \cdots + c_n \mathbf{b}_n) = c_{k+1} \mathbf{b}_{k+1} + \cdots + c_n \mathbf{b}_n$.

(b) *Hint:* Let $f_k(\mathbf{x})$ be the kth coordinate of $f(\mathbf{x})$ for f as in part (a). Consider the null-spaces of $f_{k+1}, \ldots, f_n$.

11. *Hint:* Show that in the notation of Theorem 5.10, k is the dimension of the null-space of f and r is the dimension of the image.

13. (a) By Exercise 7 in this section dim(image of f) $\leq \dim \mathcal{W} < \mathcal{V}$ so by Exercise 11, dim(null-space of f) > 0.

(b) *Hint:* If $A\mathbf{x} = \mathbf{0}$ then $BA\mathbf{x} = \mathbf{0}$. What does part (a) say about the null-spaces of the operators defined by multiplication by A and BA?

Section 6: Eigenvalues & Eigenvectors

Exercise Set 6A (pg. 148)

1. $\begin{pmatrix} 2 \\ 1 \end{pmatrix}$ and $\begin{pmatrix} -4 \\ -2 \end{pmatrix}$ are associated with $\lambda = 7$ and $\begin{pmatrix} -2 \\ 1 \end{pmatrix}$ is associated with $\lambda = -5$. The others are not eigenvectors.

3. $\lambda = 2, \begin{pmatrix} 2 \\ 1 \end{pmatrix}; \lambda = -2, \begin{pmatrix} -2 \\ 1 \end{pmatrix}$.   **5.** $\lambda = 0, \begin{pmatrix} -2 \\ 1 \end{pmatrix}; \lambda = 4, \begin{pmatrix} 2 \\ 1 \end{pmatrix}$.

7. $\lambda = 0, \begin{pmatrix} 1 \\ 0 \\ 0 \end{pmatrix}; \lambda = 1, \begin{pmatrix} 0 \\ 1 \\ 0 \end{pmatrix}; \lambda = 2, \begin{pmatrix} 1 \\ 0 \\ 2 \end{pmatrix}$.

9. If $f(\mathbf{u}) = \lambda \mathbf{u}$ then $\mathbf{u} = f^{-1}(\lambda \mathbf{u}) = \lambda f^{-1}(\mathbf{u})$ so $f^{-1}(\mathbf{u}) = \lambda^{-1} \mathbf{u}$.

11. *Hints:*

(a) $(e^{\pm kx})'' = k^2 e^{\pm kx}$.

(b) $(\sin kx)'' = -k^2 \sin kx$, $(\cos kx)'' = -k^2 \cos kx$.

(c) $1, x$.

(d) For $\lambda = k^2 > 0$ we want $c_1 e^{kx} + c_2 e^{-kx}$ to be 0 when $x = 0$ and $x = \pi$, so $c_1 + c_2 = 0$ and $c_1 e^{k\pi} + c_2 e^{-k\pi} = 0$. Then $c_2 = -c_1$, and $c_1 e^{k\pi} - c_1 e^{-k\pi} = 0$, or $c_1 e^{-k\pi}(e^{2k\pi} - 1) = 0$. $e^{-k\pi} \neq 0$, and $e^{2k\pi} - 1 \neq 0$ because $k \neq 0$ so $c_1 = c_2 = 0$. Thus $c_1 e^{kx} + c_2 e^{-kx}$ is the zero function and therefore not an eigenvector. For $\lambda = 0$, $c_1 + c_2 x = 0$ when $x = 0$ and $x = \pi$ also implies $c_1 = -c_2 = 0$. For $\lambda = -k^2 < 0$, $c_1 \cos kx + c_2 \sin kx = 0$ when $x = 0$ only if $c_1 = 0$, and is 0 when $x = \pi$ only if $c_2 \sin k\pi = 0$. This is possible with $c_2 \neq 0$ only when k is an integer and $\lambda = -k^2$.

13. $\lambda_1 = 1 + \sqrt{2}$, $\mathbf{u_1} = \begin{pmatrix} \sqrt{2} \\ 1 \end{pmatrix}$; $\lambda_2 = 1 - \sqrt{2}$, $\mathbf{u_2} = \begin{pmatrix} -\sqrt{2} \\ 1 \end{pmatrix}$. G stretches by a factor of $1 + \sqrt{2}$ in the direction of $(\sqrt{2}, 1)$ and by $1 - \sqrt{2}$ in the direction of $(-\sqrt{2}, 1)$, with reversal of direction because $1 - \sqrt{2} < 0$.

15. $x(t) = 2c_1 e^{2t} - 2c_2 e^{-2t}$, $y(t) = c_1 e^{2t} + c_2 e^{-2t}$.

17. $x(t) = -2c_1 + 2c_2 e^{4t}$, $y(t) = c_1 + c_2 e^{4t}$.

Exercise Set 6BC (pgs. 154–155)

1. $\lambda_1 = \frac{1}{2}(1 + \sqrt{5})$, $\lambda_2 = \frac{1}{2}(1 - \sqrt{2})$. Theorem 6.7 guarantees a basis of eigenvectors.

3. No real eigenvalues, so no basis of eigenvectors.

5. $\lambda_1 = \sqrt{10}$, $\lambda_1 = -\sqrt{10}$ and $\lambda_3 = -1$. Theorem 6.7 guarantees a basis of eigenvectors.

7. (a) The eigenvalues of R_θ are $\cos \theta \pm i \sin \theta$ and are real only when $\sin \theta = 0$, so $\theta = 0$ or $\theta = \pi$. For $\theta = 0$ every non-zero vector is an eigenvector associated with the eigenvalue 1, and for $\theta = \pi$ every non-zero vector is an eigenvector associated with the eigenvalue -1.

(b) For $R_\theta \mathbf{u}$ to be a real multiple of $\mathbf{u}$, it must have either the same or the opposite direction.

9. *Partial answer:* $U = \begin{pmatrix} 2 & 1 \\ 1 & 1 \end{pmatrix}$.

11. Basis $\begin{pmatrix} 1 \\ 1 \end{pmatrix}, \begin{pmatrix} 3 \\ 2 \end{pmatrix}$; matrix $\begin{pmatrix} 1 & 0 \\ 0 & 2 \end{pmatrix}$.

13. Basis $\begin{pmatrix} 1 \\ 1 \\ 1 \end{pmatrix}, \begin{pmatrix} 0 \\ 1 \\ 1 \end{pmatrix}, \begin{pmatrix} 0 \\ 0 \\ 1 \end{pmatrix}$; matrix $\begin{pmatrix} -1 & 0 & 0 \\ 0 & 0 & 0 \\ 0 & 0 & 1 \end{pmatrix}$.

15. $U = \begin{pmatrix} 1 & 0 \\ 2 & 1 \end{pmatrix}$, $\Lambda = \begin{pmatrix} 0 & 0 \\ 0 & 2 \end{pmatrix}$.

17. $U = \begin{pmatrix} 1 & 1 \\ 1 + i\sqrt{3} & 1 - i\sqrt{3} \end{pmatrix}$, $\Lambda = \begin{pmatrix} 1 + i\sqrt{3} & 0 \\ 0 & 1 - i\sqrt{3} \end{pmatrix}$.

19. $A = \begin{pmatrix} 1 & 0 \\ 0 & -1 \end{pmatrix}$. Eigenvalues $1, -1$, eigenvectors e^x and e^{-x}.

21. $A = \begin{pmatrix} 0 & -1 \\ 1 & 0 \end{pmatrix}$. Eigenvalues $i, -i$; eigenvectors $\cos x + i \sin x$, $\cos x - i \sin x$.

Section 7: Inner Products

Exercise Set 7A (pg. 158)

1. An inner product.

3. An inner product.

5. Ellipse with axes from $(-1/\sqrt{3}, 0)$ to $(1/\sqrt{3}, 0)$ and from $(0, -1/\sqrt{2})$ to $(0, 1/\sqrt{2})$.

7. *Partial Hint:* $\cos kx \cos lx = \frac{1}{2} \cos(k + l)x + \frac{1}{2} \cos(k - l)x$. There are similar formulas for the other products.

9. $\langle 2\mathbf{e_1} + 2\mathbf{e_2} - \mathbf{e_3}, 2\mathbf{e_1} + 2\mathbf{e_2} - \mathbf{e_3} \rangle = 4\langle \mathbf{e_1}, \mathbf{e_1} \rangle + 4\langle \mathbf{e_2}, \mathbf{e_2} \rangle + \langle \mathbf{e_3}, \mathbf{e_3} \rangle + 8\langle \mathbf{e_1}, \mathbf{e_2} \rangle - 4\langle \mathbf{e_1}, \mathbf{e_3} \rangle - 4\langle \mathbf{e_2}, \mathbf{e_3} \rangle = -3 < 0$, so positivity fails.

11. (a) *Hint:* Use linearity properties of the integral, and note that $f(x)^2 \geq 0$.

(b) $\left| \int_{-\pi}^{\pi} f(x) g(x) \, dx \right| \leq \left(\int_{-\pi}^{\pi} f^2(x) \, dx \right)^{1/2} \left(\int_{-\pi}^{\pi} g^2(x) \, dx \right)^{1/2}$.

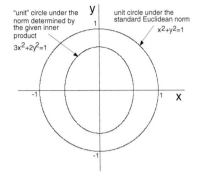

13. *Hint:* $\|\mathbf{x} - \mathbf{y}\|^2 = \langle \mathbf{x} - \mathbf{y}, \mathbf{x} - \mathbf{y} \rangle$. Use additivity of the inner product.

Exercise Set 7B (pg. 167)

1. $\mathbf{x}_1 = (1, 1, 1)$, $\mathbf{x}_2 = \left(-1, \frac{1}{2}, \frac{1}{2}\right)$, $\mathbf{x}_3 = (0, 1, -1)$ form an orthogonal basis. Then $\{(1/\sqrt{3})\mathbf{x}_1, (1/\sqrt{3/2})\mathbf{x}_2, (1/\sqrt{2})\mathbf{x}_3\}$ is an orthonormal basis.

3. $\{(1, 2, 1, 1), (-1, 0, 1, 0), (4, 1, 4, -10), (3, -4, 3, 2)\}$, or any scalar multiples of these.

5. Just apply the process.

7. *Hint:* Use Theorem 7.6 to express the dot product of the matrix columns in terms of the inner product.

9. Use Gram-Schmidt (for the ordinary dot product) to find the orthonormal basis
   $\{\mathbf{u}_1 = (1/\sqrt{3}, 1/\sqrt{3}, 1/\sqrt{3}), \mathbf{u}_2 = (1/\sqrt{6}, -2/\sqrt{6}, 1/\sqrt{6}), \mathbf{u}_3 = (1/\sqrt{2}, 0, -1/\sqrt{2})\}$. Distance $= \sqrt{2}$.

11. *Hint:* Use additivity and homogeneity of the inner product.

13. *Hint:* Use additivity and homogeneity of the dot product, and distributivity of matrix multiplication. Positivity may fail, for example with $A = O$ or $A = -I$.

15. *Hint:* $a = \langle \mathbf{e}_1, \mathbf{e}_1 \rangle > 0$. Write out $\langle (x, y), (x, y) \rangle$ and complete the square.

17. Additional hints: For a complex vector $\mathbf{y}$, $\bar{\mathbf{y}} \cdot \mathbf{y} > 0$ unless $\mathbf{y} = \mathbf{0}$. λ is real if $\lambda = \bar{\lambda}$.

Exercise Set 7C (pgs. 170–171)

1. To verify the axis of rotation, check that $A\mathbf{u}_1 = \mathbf{u}_1$. $M = \begin{pmatrix} 1 & 0 & 0 \\ 0 & -1 & 0 \\ 0 & 0 & -1 \end{pmatrix}$. Angle is π.

3. **(a)** $R = \begin{pmatrix} 1 & 0 & 0 \\ 0 & 0 & -1 \\ 0 & 1 & 0 \end{pmatrix}$, $S = \begin{pmatrix} 0 & 0 & -1 \\ 0 & 1 & 0 \\ 1 & 0 & 0 \end{pmatrix}$. **(b)** $SR = \begin{pmatrix} 0 & -1 & 0 \\ 0 & 0 & -1 \\ 1 & 0 & 0 \end{pmatrix}$. Axis $(1, -1, 1)$.

5. $\begin{pmatrix} 4/5 & 1/5 & -2\sqrt{2}/5 \\ 1/5 & 4/5 & 2\sqrt{2}/5 \\ 2\sqrt{2}/5 & -2\sqrt{2}/5 & 3/5 \end{pmatrix}$.

7. *Hints:*
   (a) $a^2 + b^2 = c^2 + d^2 = 1$, and since $ac + bd = 0$, $a^2 c^2 = b^2 d^2$.
   (b) Since $a^2 + b^2 = 1$ there is a θ such that $\cos\theta = a$ and $\sin\theta = b$.
   (c) $(a + 1, b)$ is an eigenvector for 1, and $(a - 1, b)$ for -1. These vectors are orthogonal (because $a^2 + b^2 = 1$) and can be normalized to be orthonormal.

9. *Hints:*
   (a) The first column is as shown because $\mathbf{u}_1$ is an eigenvector for $\lambda = \pm 1$. The rest of the first row is 0 because columns 2 and 3 are orthogonal to column 1.

(b) Since $\lambda = 1$, f leaves points on the line through $\mathbf{u}_1$ fixed. Apply Exercise 7 to the submatrix $\begin{pmatrix} a & b \\ c & d \end{pmatrix}$. In case (b) of Exercise 7, f is a rotation about the axis $\mathbf{u}_1$; in case (c) it is a reflection in the plane spanned by $\mathbf{u}_1$ and the line of reflection in the $\mathbf{u}_2\mathbf{u}_3$-plane.

(c) In case (b) of Exercise 7, f is the composition of a rotation with axis $\mathbf{u}_1$ with reflection in the $\mathbf{u}_2\mathbf{u}_3$-plane. In case (c), f is a reflection in the line of reflection in the $\mathbf{u}_2\mathbf{u}_3$-plane.

Chapter 3 Review (pg. 171)

1. Dependent.

3. Independent.

5. $\begin{pmatrix} 0 & 0 & 0 \\ 0 & 0 & 0 \\ -1 & 0 & 1 \end{pmatrix}$.

7. $\begin{pmatrix} 1 & 1 & -2 \\ 1 & 2 & -3 \end{pmatrix}$.

9. $\begin{pmatrix} 1/2 & 1/2 & -1/2 \\ -1 & 2 & -1 \\ -1/2 & 1/2 & 1/2 \end{pmatrix}$.

11. $\begin{pmatrix} -1 & 3 & 0 & 0 \\ 0 & 1 & 0 & 0 \\ 2 & 2 & 0 & 0 \\ 0 & 0 & -1 & 3 \\ 0 & 0 & 0 & 1 \\ 0 & 0 & 2 & 2 \end{pmatrix}$.

13. $\begin{pmatrix} -1 & 1 & -1 & 1 \\ 1 & -1 & 1 & -1 \\ -1 & 1 & 2 & -2 \\ 1 & -1 & -2 & 2 \end{pmatrix}$.

15. (a) $R = \begin{pmatrix} 1 & 0 & 0 \\ 0 & 0 & -1 \\ 0 & 1 & 0 \end{pmatrix}$, $S = \begin{pmatrix} \cos\theta & 0 & -\sin\theta \\ 0 & 1 & 0 \\ -\sin\theta & 0 & \cos\theta \end{pmatrix}$. (b) Rotation of angle θ about the z-axis.

17. $\lambda_1 = -5$, $\mathbf{u}_1 = \begin{pmatrix} 1 \\ 2 \end{pmatrix}$; $\lambda_2 = 2$, $\mathbf{u}_2 = \begin{pmatrix} -3 \\ 1 \end{pmatrix}$. Basis.

19. $\lambda_1 = 1$, $\mathbf{u}_1 = \begin{pmatrix} 1 \\ 0 \\ 0 \end{pmatrix}$; $\lambda_2 = 2$, $\mathbf{u}_2 = \begin{pmatrix} 2 \\ 1 \\ 1 \end{pmatrix}$; $\lambda_3 = 3$, $\mathbf{u}_3 = \begin{pmatrix} -2 \\ 0 \\ 1 \end{pmatrix}$. Basis.

21. (a) $0, \pm i\sqrt{a^2 + b^2 + c^2}$.

(b) (c, b, a) is an eigenvector for 0.

(c) Eigenvalues $0, \pm 3i$. Basis of eigenvectors $\begin{pmatrix} -2 \\ 2 \\ 1 \end{pmatrix}$, $\begin{pmatrix} 1+3i \\ -1+3i \\ 4 \end{pmatrix}$, $\begin{pmatrix} 1-3i \\ -1-3i \\ 4 \end{pmatrix}$.

23. *Hint:* Put $\mathbf{x} = M\mathbf{x} + (\mathbf{x} - M\mathbf{x})$, and show that $\mathbf{x} - M\mathbf{x}$ is in the null-space of M.

CHAPTER 4: DERIVATIVES

Exercise Set 1A–D (pgs. 182–184)

1. $f'(t) = (2t, 3t^2)$, $\mathbf{t}(s) = s(4, 12) + (5, 9)$

3. $f'(t) = \begin{pmatrix} e^t \\ 2e^{2t} \end{pmatrix}$, $\mathbf{t}(s) = s\begin{pmatrix} e^{-1} \\ 2e^{-2} \end{pmatrix} + \begin{pmatrix} e^{-1} \\ e^{-2} \end{pmatrix}$

5. $f'(t) = (-\sin t, -2\sin 2t, -3\sin 3t, -4\sin 4t)$; $\mathbf{t}(s) = sf'(\pi/2) + f(\pi/2) = s(-1, 0, 3, 0) + (0, -1, 0, 1)$

7.

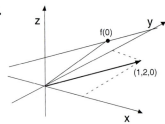

9.

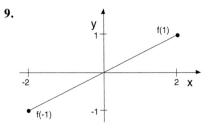

11.

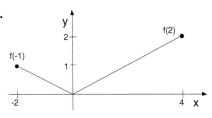

13. $T(1/2) = 21/64$, $T'(1/2) = 27/16$

17.

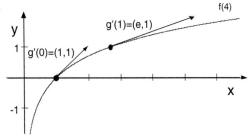

19. $(1/2, 1/4, 1/8)$; No

21. $2\sqrt{1 + 4\sin^2 t}$.

23. $v(t) = 5$, $l(\gamma) = 20$

25. $v(t) = \sqrt{1 + 9t}$, $l(\gamma) = \frac{14}{3}$

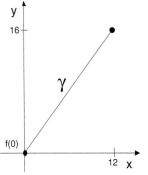

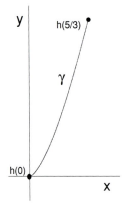

27. $(a\cos\omega t + b\cos\delta t, a\sin\omega t + b\sin\delta t)$

39.

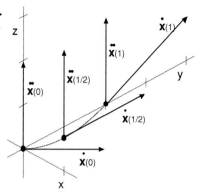

41.

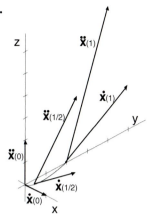

43. $T(t) = 32 - \frac{3}{1000}(-16t^2 + 300t)$, $T_{\min} \approx 27.8°$ F

Exercise Set 1E (pg. 185)

1.

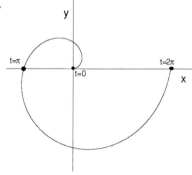

3.

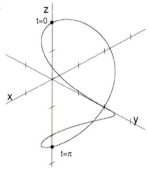

7.

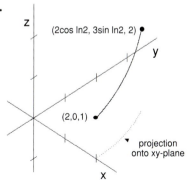

9.

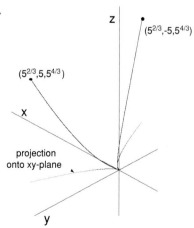

Exercise Set 1F (pgs. 187–188)

1. $F(t) = \left(\frac{1}{3}t^3 + t, \frac{1}{4}t^4 - t\right) + \mathbf{c}$, $\mathbf{c} = (2/3, 11/4)$

3. $F(t) = (t \sin t + \cos t, -t \cos t + \sin t) + \mathbf{c}$, $\mathbf{c} = (0, 1)$

5. $F(t) = \left(t, \frac{1}{3}t^3, -t, \frac{1}{3}t^3\right) + \mathbf{c}$, $\mathbf{c} = (1, 5/3, 3, 5/3)$

7. $\left(\frac{1}{2}t^2 + 2, -\frac{1}{3}t^3 + 1\right)$

9. $\mathbf{x}(t) = \left(\sin t - 2, -\frac{1}{2}\cos 2t + \frac{1}{2}\right)$

11. $\mathbf{x}(t) = \left(\frac{1}{2}t^2 + \frac{1}{2}, \frac{1}{2}t^2 - \frac{3}{2}, \frac{1}{3}t^3 + \frac{2}{3}\right)$

13. $\mathbf{x}(t) = \left(\frac{1}{6}t^3 + t + 2, -\frac{1}{12}t^4 + t + 1\right)$

15. $\tan\theta = h(1/a + 1/b)$, $v_0^2 = abg/(2h\cos^2\theta)$.

17. $\mathbf{v}(0) = (20, -40)$; $v_0 \approx 45$ ft/sec; Clark's speed at time of rescue: 122 ft/sec; victim's speed at time of rescue: 80 ft/sec

21. $\theta \approx 54.4°$; maximum height ≈ 26.2 miles

25. (a) $(1, 1)$ (b) $(1/2, 1/3, 1/4)$

Exercise Set 2AB (pgs. 192–193)

1.

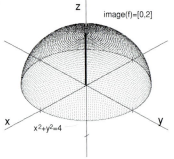

3.

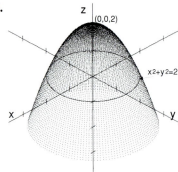

5.

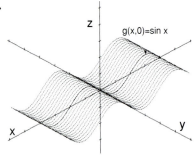

7.

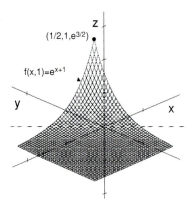

9.

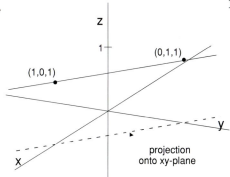

projection
onto xy-plane

11.

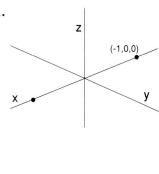

13.

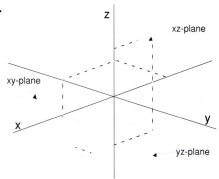

15.

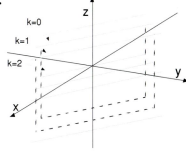

17.

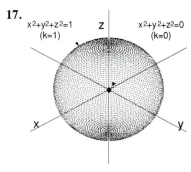

19.

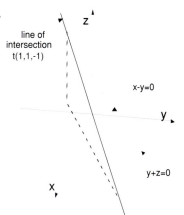

21.

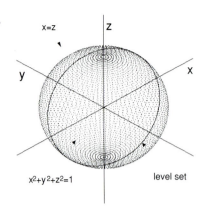

x=z

$x^2+y^2+z^2=1$ level set

23. thin film

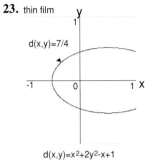

$d(x,y)=7/4$

$d(x,y)=x^2+2y^2-x+1$

25. (a)

D: $x^2+y^2=4,\ 0<z<5$

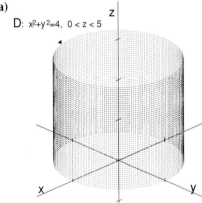

(b)

$T(x,y,z)=x^2+y^2-z=-1$
$1<z\leqslant5$

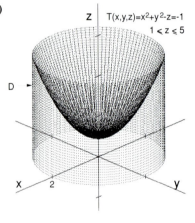

Exercise Set 2C (pgs. 195–196)

1.

$z=-y^2+4,\ x=-2$

$z=-y^2+4,\ x=2$

$(2,-2,0)$ $(-2,2,0)$

$z=x^2-4,\ y=-2$

$z=x^2-4,\ y=2$

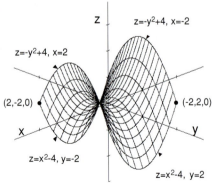

3.

$(0,0,1)$

$(0,2,-1)$

$(2,0,-1)$

$(2,2,-3)$

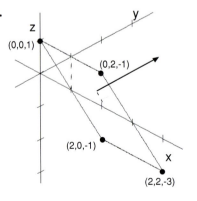

5.

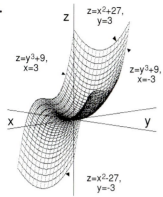

$z=x^2+27$, $y=3$

$z=y^3+9$, $x=3$

$z=y^3+9$, $x=-3$

$z=x^2-27$, $y=-3$

7.

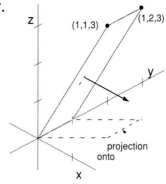

$(1,1,3)$ $(1,2,3)$

projection onto

9.

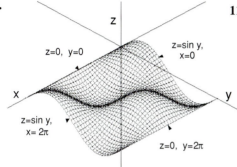

$z=0$, $y=0$

$z=\sin y$, $x=0$

$z=\sin y$, $x=2\pi$

$z=0$, $y=2\pi$

11.

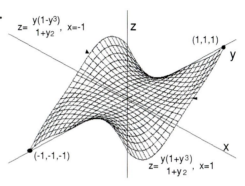

$z=\dfrac{y(1-y^3)}{1+y^2}$, $x=-1$

$(1,1,1)$

$(-1,-1,-1)$

$z=\dfrac{y(1+y^3)}{1+y^2}$, $x=1$

13. $P(x,y) = H(x)H(y)H(1-y)H(y-x)$

15. $P(x,y) = H(1-x^2-y^2)H(x-y)$

17. $P(x,y) = H(x)H(1-x)H(y)H(1-y)$

19.

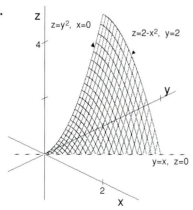

$z=y^2$, $x=0$

$z=2-x^2$, $y=2$

$y=x$, $z=0$

21.

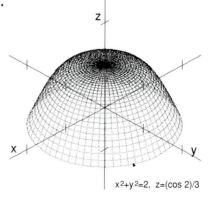

$x^2+y^2=2$, $z=(\cos 2)/3$

23.

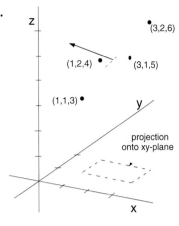

25.

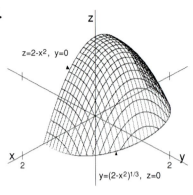

Exercise Set 2D (pg. 198)

1.

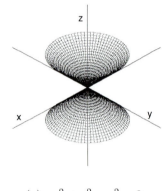

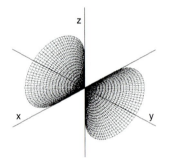

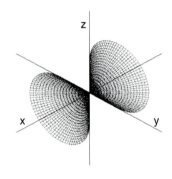

     (a) $x^2 + y^2 - z^2 = 0$ (b) $x^2 + z^2 - y^2 = 0$ (c) $y^2 + z^2 - x^2 = 0$

3. The length of the axes of the ellipsoid of level 2 increases by a factor of $\sqrt{2}$; the vertex (saddle point) of the elliptic paraboloid (hyperbolic paraboloid) of level 1 is $(0, 0, -1/c)$.

7. Q is an elliptic paraboloid.

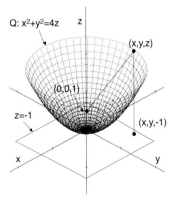

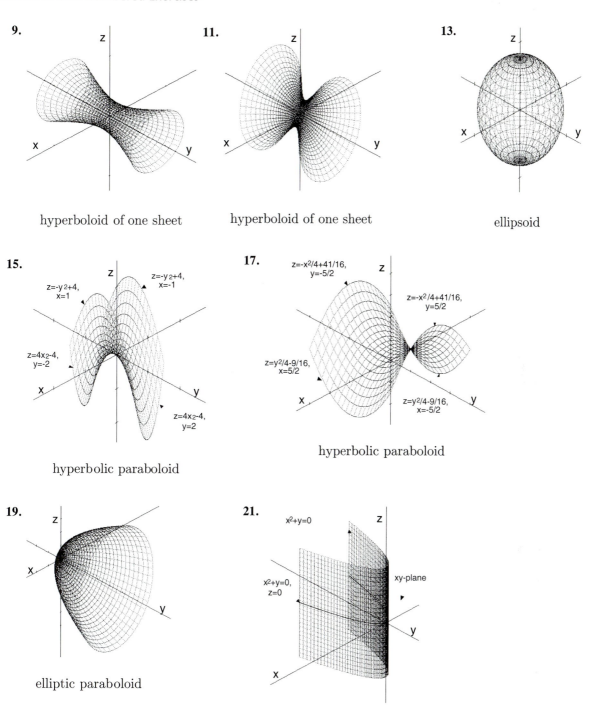

9.

hyperboloid of one sheet

11.

hyperboloid of one sheet

13.

ellipsoid

15.

$z=-y^2+4,$ $x=1$

$z=-y^2+4,$ $x=-1$

$z=4x^2-4,$ $y=-2$

$z=4x^2-4,$ $y=2$

hyperbolic paraboloid

17.

$z=-x^2/4+41/16,$ $y=-5/2$

$z=-x^2/4+41/16,$ $y=5/2$

$z=y^2/4-9/16,$ $x=5/2$

$z=y^2/4-9/16,$ $x=-5/2$

hyperbolic paraboloid

19.

elliptic paraboloid

21.

$x^2+y=0$

$x^2+y=0,$ $z=0$

xy-plane

parabolic cylinder

23.   **25.**

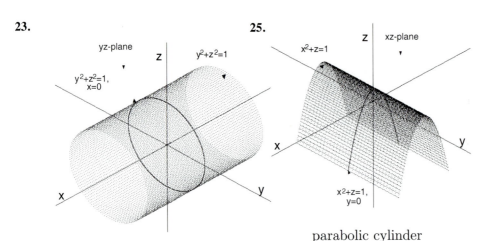

circular cylinder

parabolic cylinder

27.

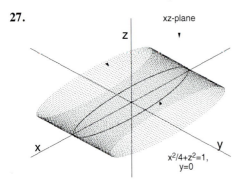

elliptic cylinder

Exercise Set 3A–C (pgs. 203–204)

1. $f_x = 2x + x\cos(x + y) + \sin(x + y)$, $f_y = x\cos(x + y)$

3. $f_x = e^{x+y+1}$, $f_y = e^{x+y+1}$ **5.** $f_x = yx^{y-1}$, $f_y = x^y \ln x$

7. $f_x = 2xy + y^2$, $f_y = x^2 + 2xy$, $f_x(1, -1) = -1$, $f_y(1, -1) = -1$, $x + y + z = 0$.

9. $f_x = -2x/(x^2 + y^2)^2$, $f_y = -2y/(x^2 + y^2)^2$, $f_x(1, 1) = -\frac{1}{2}$, $f_y(1, 1) = -\frac{1}{2}$; $x + y + 2z = 3$.

11. $f_{yx} = f_{xy} = 1 + 6xy^2$ **13.** $f_{yx} = f_{xy} = 8xy/(x^2 + y^2)^3$

15. $f_x = (2x + x^2)e^{x+y+z}\cos y$, $f_y = x^2 e^{x+y+z}(\cos y - \sin y)$, $f_z = x^2 e^{x+y+z}\cos y$

17. $f_x = 2x/(z^2 + w^2)$, $f_y = -2y/(z^2 + w^2)$, $f_z = 2z(y^2 - x^2)/(z^2 + w^2)^2$; $f_w = 2w(y^2 - x^2)/(z^2 + w^2)^2$

19. $f_x = 1$, $f_y = 2z$, $f_z = 2y$ **21.** $f_{yxx} = 2/(x + y)^3$

29. $\Gamma(x, y) = -\frac{\sqrt{2}}{2}x - \frac{\sqrt{2}}{2}y + \sqrt{2}$

31. $\Gamma(x, y) = 1$

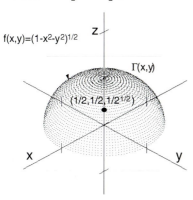

f(x,y)=(1-x²-y²)¹/²

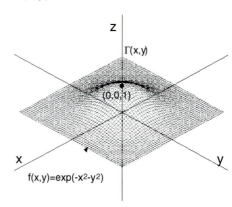

f(x,y)=exp(-x²-y²)

33. $(1, 1, \sqrt{2})t + (1/2, 1/2, \sqrt{2}/2)$

43. D is the xy-plane; u is harmonic on D

47. D is the xy-plane with the y-axis deleted; u is harmonic on D

49. D is the xy-plane; u is not harmonic on D

41. D is the xy-plane; u is not harmonic on D

45. D is the xy-plane; u is not harmonic on D

Exercise Set 4AB (pgs. 210–212)

1. $f_x = \begin{pmatrix} 1 \\ 1 \\ 2x \end{pmatrix}$, $f_y = \begin{pmatrix} 1 \\ -1 \\ 2y \end{pmatrix}$

3. $f_x = \begin{pmatrix} y \\ 1 \end{pmatrix}$, $f_y = \begin{pmatrix} x \\ 1 \end{pmatrix}$

5. $f_x = (e^x, 0, e^{x+y})$, $f_y = (0, e^y, e^{x+y})$.

7. $g_u(1, 1) = (1, 0, 2)$, $g_v(1, 1) = (0, 1, 2)$

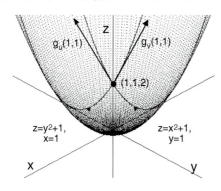

9. $g_u(\pi/4, \pi/4) = (-1/2, 1/2, 0)$;

$g_v(\pi/4, \pi/4) = (1/2, 1/2, -\sqrt{2}/2)$

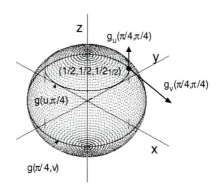

11.

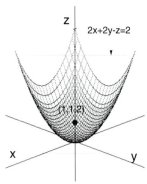

2x+2y-z=2

(1,1,2)

13. $x+y+2^{1/2}z=2$

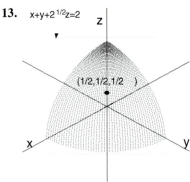

(1/2,1/2,1/2)

15. (a)

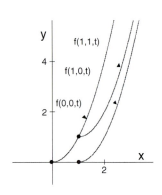

f(1,1,t)

f(1,0,t)

f(0,0,t)

(b) $f_t(-1, 0, t)$ and $f_t(-1, 0, t)$ are the velocity vectors at time t of a point starting at $(-1, 0)$,

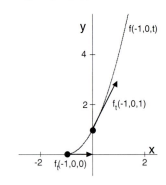

f(-1,0,t)

$f_t(-1,0,1)$

-2 $f_t(-1,0,0)$

17. (a) $f_1(x, y) = x$ $(x \neq 0)$, $f_2(x, y) = (x + y)^2/(4x)$

(b) $f_x(x, y) = \left(1, \left(2(x + y) - (x + y)^2\right)/(4x^2)\right)$, $f_y(x, y) = \left(0, (x + y)/(2x)\right)$

(c) The image in the uv-plane of the given region is bounded by four curves: $v = u$ $(0 < u \leq 4)$, $v = 16/u$ $(4 \leq u \leq 8)$, $v = u - 8 + 16/u$ $(u \leq 4 \leq 8)$, $v = 0$ $(0 < u \leq 4)$

Exercise Set 4C (pg. 213)

1.

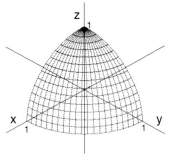

3.

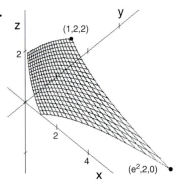

(1,2,2)

$(e^2,2,0)$

5.

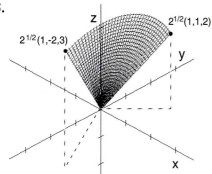

7. u is the angle made by the positive x-axis and the projection of the vector f onto the xy-plane; v is the angle made by the positive z-axis and the vector f.

9.

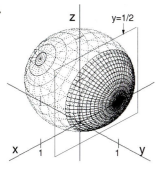

11.

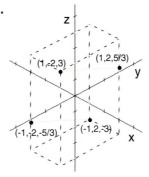

13. $(x, y, z) = \left(\cos u \cosh v, \sin u \cosh .v, (1/\sqrt{2}) \sinh v\right)$

15. $(x, y, z) = (\cos u \sinh v, \sin u \sinh v, -\cosh v)$

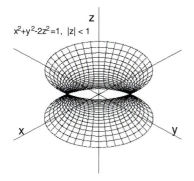

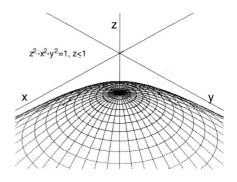

Chapter 4 Review (pgs. 214–215)

1. $\dot{\mathbf{x}}(t) = e^t(\cos t - \sin t)\mathbf{i} + e^t(\cos t + \sin t)\mathbf{j} + e^t\mathbf{k}$,
$\ddot{\mathbf{x}}(t) = (-2e^t \sin t)\mathbf{i} + (2e^t \cos t)\mathbf{j} + e^t\mathbf{k}$,
$\mathbf{t}(t) = \frac{1}{\sqrt{3}}(\cos t - \sin t)\mathbf{i} + \frac{1}{\sqrt{3}}(\cos t + \sin t)\mathbf{j} + \frac{1}{\sqrt{3}}\mathbf{k}$

3. (a) $\dot{\mathbf{x}}(0) = (5/\sqrt{2}, 5/\sqrt{2}, 0)$, $\ddot{\mathbf{x}}(0) = (0, 0, -5)$ **(b)** 10π

(d)

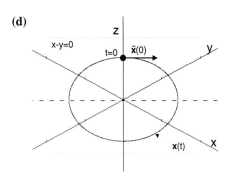

5. max $= \sqrt{a^2 + c^2}$, min $= \sqrt{b^2 + c^2}$

7. $\lambda_1 =$ curve parametrized on $[0, \pi/2]$ and $\lambda_2 =$ curve parametrized on $[-\pi/2, 0]$. $l(\lambda_1) = (e^{\pi/2} - 1)\sqrt{2}$, $l(\lambda_2) = (1 - e^{-\pi/2})\sqrt{2}$

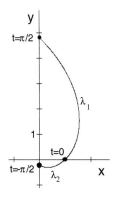

9. $2x + 12y - z = 17$

23. **(a)** ellipses if $k > 0$, single point $(0, 0, 0)$ if $k = 0$ **(b)** parabolas

25. $F(x, y) = y - x^2$

27. $h(x) = (x, x^2)$

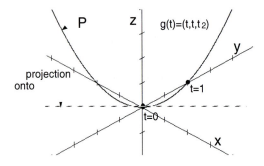

CHAPTER 5: DIFFERENTIABILITY

Section 1: Limits And Continuity

Exercise Set 1A–C (pgs. 224–225)

1.

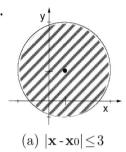

(a) $|\mathbf{x} - \mathbf{x}_0| \leq 3$

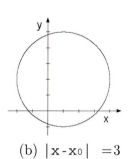

(b) $|\mathbf{x} - \mathbf{x}_0| = 3$

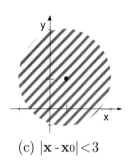

(c) $|\mathbf{x} - \mathbf{x}_0| < 3$

3. The interior of S is S and, therefore, S is open; the boundary of S is the circle of radius 3 centered at $(1, 2)$.

5. Let $S = \{(x, y) | 0 < x < 3, 0 < y < 2\}$. The interior of S is S and, therefore, S is open. The boundary of S consists of the four line segments l_1, l_2, l_3, l_4, where l_1 has endpoints $(0, 0)$ and $(3, 0)$; l_2 has endpoints $(3, 0)$ and $(3, 2)$; l_3 has endpoints $(3, 2)$ and $(0, 2)$; and l_4 has endpoints $(0, 2)$ and $(0, 0)$.

7. The set $S = \{(x, y) | x^2 + 2y^2 < 1\}$ contains all points inside the ellipse $\mathcal{E} : x^2 + 2y^2 = 1$. The interior of S is S and, therefore, S is open. The boundary of S is $\mathcal{E}$.

9. The set $S = \{(x, y) | x^2 + y^2 > 0\}$ is the xy-plane with the origin deleted. The interior of S is S and, therefore, S is open. The boundary of S consists of the single point $(0, 0)$.

11. The given set $S = \{(x, y) | x > y\}$ is the region below the line $y = x$ in the xy-plane. The interior of S is S and, therefore, S is open. The boundary of S consists of all points on the line $y = x$.

13. Lines and planes in $\mathbb{R}^3$ are not open subsets of $\mathbb{R}^3$ because no point on a line or a plane is an interior point. For example, in the case of a line, if $\mathbf{x}_0$ is a point on the line then every neighborhood of $\mathbf{x}_0$ contains points not on the line. No point on a line is an interior point (in fact, they are all boundary points). A similar observation holds for planes.

15. Since $\begin{pmatrix} 1 & 3 \\ 0 & 2 \end{pmatrix} \begin{pmatrix} x \\ y \end{pmatrix} = \begin{pmatrix} x + 3y \\ 2y \end{pmatrix}$, the given function can be written as $f(x, y) = \begin{pmatrix} x + 3y \\ 2y \end{pmatrix}$. Thus, the domain space and the range space are both of dimension 2 (i.e., $n = m = 2$). The real-valued coordinate functions of f are $f_1(x, y) = x + 3y$ and $f_1(x, y) = 2y$.

17. The domain space of $f(t) = (t, t^2, t^3, t^4)$ has dimension 1, and the range space has dimension 4 (i.e., $n = 1$, $m = 4$). The real-valued coordinate functions of f are $f_1(t) = t$, $f_2(t) = t^2$, $f_3(t) = t^3$ and $f_4(t) = t^4$.

19. The domain space and range space of $f(x, y, z) = (2x, 2y, 2z)$ are both of dimension 3 (i.e., $n = m = 3$). The real valued coordinate functions of f are $f_1(x, y, z) = 2x$, $f_2(x, y, z) = 2y$ and $f_3(x, y, z) = 2z$.

21. The coordinate functions of the given function f are

$$f_1(x, y) = \frac{y}{x^2 + 1} \qquad \text{and} \qquad f_2(x, y) = \frac{x}{y^2 - 1}.$$

f_1 is continuous everywhere on the xy-plane. f_2 is continuous on the xy-plane except on the horizontal lines $y = \pm 1$, where $\lim_{\mathbf{x} \to \mathbf{x}_0} f_2(\mathbf{x})$ fails to exist. It follows that $\lim_{\mathbf{x} \to \mathbf{x}_0} f(\mathbf{x})$ fails to exist for $\mathbf{x}_0$ on the horizontal lines $y = \pm 1$. (*Note:* For points of the form $\mathbf{x}_0 = (x_0, \pm 1)$, where $x_0 \neq 0$, $\lim_{\mathbf{x} \to \mathbf{x}_0} f_2(\mathbf{x})$ fails to exist because $f_2(\mathbf{x}_0)$ is infinitely large. But for points of the form $\mathbf{x}_0 = (0, \pm 1)$, $\lim_{\mathbf{x} \to \mathbf{x}_0} f_2(\mathbf{x})$ fails to exist because its value depends on the direction from which we approach $\mathbf{x}_0$.)

23. There is only one coordinate function of the given function f; namely,

$$f(x, y) = \begin{cases} x/\sin x + y, & \text{if } x \neq 0; \\ 2 + y, & \text{if } x = 0. \end{cases}$$

For points $\mathbf{x}_0$ on the vertical lines $l_n : x = n\pi$ (n a nonzero integers), $\lim_{\mathbf{x} \to \mathbf{x}_0} f(\mathbf{x})$ fails to exist, but this limit does exist at all other points in the xy-plane. (*Note:* f is not continuous at points of the form $(0, y_0)$, but $\lim_{\mathbf{x} \to (0, y_0)} f(\mathbf{x})$ *does* exist and equals $1 + y_0$.)

25. The coordinate functions of the given function f are $f_1(u, v) = uv/(1 - u^2 - v^2)$ and $f_2(u, v) = 1/(2 - u^2 - v^2)$. f_1 is continuous everywhere on the uv-plane except on the circle $\mathcal{C}_1 : u^2 + v^2 = 1$. For points $\mathbf{x}_0$ on $\mathcal{C}_1$, $\lim_{\mathbf{x} \to \mathbf{x}_0} f(\mathbf{x})$ fails to exist. f_2 is continuous everywhere on the uv-plane except on the circle $\mathcal{C} : u^2 + v^2 = 2$. For points $\mathbf{x}_0$ on $\mathcal{C}_2$, $\lim_{\mathbf{x} \to \mathbf{x}_0} f(\mathbf{x})$ fails to exist. It follows that f is continuous everywhere on the uv-plane except on the two circles $\mathcal{C}_1$ and $\mathcal{C}_2$, and $\lim_{\mathbf{x} \to \mathbf{x}_0} f(\mathbf{x})$ fails to exist for points $\mathbf{x}_0$ on these two circles.

27. The coordinate functions of the given function f are $f_1(u, v) = 3u - 4v$ and $f_2(u, v) = u + 8$, both of which are continuous on the uv-plane. Thus f is continuous on the uv-plane.

29. The given function f is continuous on the xy-plane *except* possibly at $(0, 0)$. However, it was shown in Example 6 that $\lim_{\mathbf{x} \to (0,0)} f(x, y)$ fails to exist, so that f can't be continuous at $(0, 0)$ (regardless of how $f(0, 0)$ is defined).

31. For $\mathbf{x} \in \mathbb{R}^n$, the function $f(\mathbf{x}) = |\mathbf{x}|/(1 - |\mathbf{x}|^2)$ isn't continuous for points $\mathbf{x} \in \mathbb{R}^n$ that are 1 unit from the origin. That is, f isn't continuous on the n-dimensional unit sphere.

33. (a) The translation $T(x, y) = (x + y) + (1, 1)$ takes each point in the xy-plane and moves it a distance of $|(1, 1)| = \sqrt{2}$ units along a line parallel to the line $y = x$.
(b) *Hint:* Use Theorem 1.4

35. *Hint:* Each time you include another open set you may need a smaller ball inside.

37. *Hint:* Use the definition of length.

39. *Hint:* Use the triangle inequality.

41. *Hint:* Assume the contrary and reach a contradiction.

Section 2: Real-Valued Functions

Exercise Set 2AB (pg. 232)

1. $(f_x, f_y) = (2x, -2y)$.

3. $(f_x, f_y) = (1, 2)$.

5. $(f_x, f_y, f_z) = (1, 1, -2z)$.

7. $(f_{x_1}, f_{x_2}) = (2x_1, 8x_2^3)$.

9. $z = 3x - 3y$.

11. $z = x + 2y$.

13. $w = x + y - 2z + 1$.

15. $w = x + 2y + 2z - 2$.

17. When $x = 0$ and/or $y = 0$.

19. When $x + y = 0$.

21. Routine calculation.

23. *Hint:* Use the definition of differentiability.

25. *Hint:* Use the definition of $f'(0)$.

Section 3: Directional Derivatives

Exercise Set 3AB (pgs. 236–237)

1. $4/\sqrt{3}$.

3. 1.

5. $-2/\sqrt{5}$.

7. $4e^2/\sqrt{10}$.

9. $2/\sqrt{5}$.

11. $4/\sqrt{3}$.

13. $\pm\left(-3/\sqrt{3}+4/\sqrt{3}+1/\sqrt{3}=2/\sqrt{3}\right)$.

15. Routine calculation.

17. Routine calculation.

19. *Hint:* If $|\mathbf{y}-\mathbf{x}| \neq 0$, divide by it.

21. Use the hint.

23. $f(h,k,l) \approx 1+0+\frac{1}{2!}(2h^2+2k^2-2l^2) = 1+h^2+k^2-l^2$.

Section 4: Vector-Valued Functions

Exercise Set 4A–C (pgs. 243–245)

1. $\begin{pmatrix} y & x \\ 1 & 1 \end{pmatrix}$.

3. $\begin{pmatrix} 1 & \cos y & 0 \\ 0 & 1 & -\sin z \\ 1 & 1 & 1 \end{pmatrix}$.

5. $(x^2+2x)e^x$.

7. $\begin{pmatrix} v & u & 0 \\ 0 & w & v \\ w & 0 & u \end{pmatrix}$.

9. $\begin{pmatrix} 2x & 2y \\ 2x & -2y \\ y & x \end{pmatrix}$.

11. $\begin{pmatrix} 2x & -2y \\ 2y & 2x \end{pmatrix}$.

13. $\begin{pmatrix} 2 & 0 \\ 0 & 2 \end{pmatrix}$.

15. $\begin{pmatrix} 2 & 2 \end{pmatrix}$.

17. $\begin{pmatrix} \sqrt{2}/2 \\ -\sqrt{2}/2 \end{pmatrix}$.

19. $\begin{pmatrix} -1 & 0 \\ 0 & -1 \\ 0 & 1 \end{pmatrix}$.

21. $\begin{pmatrix} -1 & 0 \\ 0 & -1 \\ 0 & 1 \end{pmatrix}$.

23. **(a)** P is the projection of the vector (x,y,z) onto the xy-plane. **(b)** $\begin{pmatrix} 1 & 0 & 0 \\ 0 & 1 & 0 \end{pmatrix}$.

25. **(a)** $f(\mathbf{x}_0+\mathbf{y}_1) = \begin{pmatrix} 0 \\ 1.1 \end{pmatrix}$, $f(\mathbf{x}_0+\mathbf{y}_2) = \begin{pmatrix} 0.1 \\ 1.1 \end{pmatrix}$, $f(\mathbf{x}_0+\mathbf{y}_3) = \begin{pmatrix} 0.11 \\ 1.2 \end{pmatrix}$.

(b) $T(x,y) = \begin{pmatrix} y \\ x+y \end{pmatrix}$.

(c) $f(\mathbf{x}_0+\mathbf{y}_1) =\approx \begin{pmatrix} 0 \\ 1.1 \end{pmatrix}$, $f(\mathbf{x}_0+\mathbf{y}_2) =\approx \begin{pmatrix} 0.1 \\ 1.1 \end{pmatrix}$, $f(\mathbf{x}_0+\mathbf{y}_3) \approx \begin{pmatrix} 0.1 \\ 1.2 \end{pmatrix}$.

27. The 3-by-3 matrix in the definition of f.

29. *Hint:* Use the definition of the derivative matrix in both parts.

Section 5: Newton's Method

Exercise Set 5 (pg. 250)

1. (a) The graph of $f(x) = x^{1/3} - x$, $-2 < x < 2$ is shown on the right.
   (b) The tangent lines l_1, l_2 and l_3 to the graph of f for $x_0 = 3/4$, $x_0 = -3/4$ and $x_0 = -1/4$ (resp.) are shown on the graph given in part (a). Their respective points of tangency p_1, p_2 and p_3 are also shown.
   (c) The equation $x^{1/3} - x = 0$ can be factored as $x^{1/3}(1 - x^{1/3})(1 + x^{1/3}) = 0$, from which we obtain the three solutions 1, -1 and 0.
   (c) 1, -1, -1 respectively.
   (d) This choice gives a solution remote from the starting value.

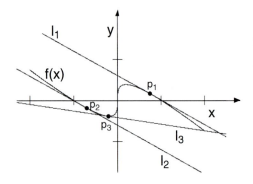

3. (a) Routine calculation.   (b) $x_3 = x_4 = 0.739085133$.

5. $\mathbf{x}_9 = \mathbf{x}_{10} = (0.980222741, 1.993801602, -0.874024343)$.

Chapter 5 Review (pgs. 250–251)

1. (a) Open.   (b) Not closed.   (d) Equals the set.   (e) Nonnegative x and y axes.

3. (c) Neither.   (d) The set with the semicircle deleted.   (e) The semicircle together with the segment $-1 \le y \le 1$ on the y-axis.

5. (a) Open.   (b) Closed.   (d) Equals the set.   (e) Empty.

7. (a) Open.   (b) Not closed.   (d) Equals the set.   (e) Three parts of planes: $z = 3$ where $x \ge 0$ and $y \ge 1$, $y = 1$ where $x \ge 0$ and $z \ge 3$, $x = 0$ where $y \ge 1$ and $z \ge 3$.

9. (a) The interior of S is the solid unit sphere in $\mathbb{R}^3$ without its "skin," and (b) the boundary of S is its "skin". It follows that the smallest closed set containing S is the solid unit sphere in $\mathbb{R}^3$ together with its "skin."

11. $H \circ f$ has no points of discontinuity.

13. The points of discontinuity of $H \circ f$ are the points on the line $y = x$.

15. No points of discontinuity.

17. $\left(-2x(x^2 + y^2)^{-2}, -2y(x^2 + y^2)^{-2}\right)$.

19. $(1, -1, 0)$.   21. $(y + w, x + z, y + w, z + x)$.

23. $f_{xx} = e^x \sin y$, $f_{yy} = -e^x \sin y$, $f_{xy} = f_{yx} = e^x \cos y$.

25. $f_{xx} = yze^x$, $f_{yy} = f_{zz} = 0$, $f_{xy} = f_{yx} = ze^x$, $f_{xz} = f_{zx} = ye^x$, $f_{yz} = f_{zy} = e^x$.

27. $f_{xx} = 12x^2$, $f_{yy} = 6y$, $f_{zz} = 2$, $f_{xy} = f_{yx} = f_{xz} = f_{zx} = f_{yz} = f_{zy} = 0$.

29. $\begin{pmatrix} 1 & 0 \\ 0 & 1 \\ 1 & 1 \\ 1 & 0 \end{pmatrix}$.   31. $\begin{pmatrix} x & -y \\ y & x \end{pmatrix}$.

33. $\begin{pmatrix} v & u & 0 \\ 0 & w & v \\ w & 0 & u \\ vw & uw & uv \end{pmatrix}$.

35. (a) $4\cos\theta - 4\sin\theta$.   (b) $(1/\sqrt{2}, -1/\sqrt{2})$.

CHAPTER 6: VECTOR DIFFERENTIAL CALCULUS

Exercise Set 1A–C (pgs. 257–259)

1. $\nabla f(\mathbf{x}) = (2x, -2y)$

3. $\nabla f(\mathbf{x}) = (1, 2)$

5. $\nabla f(\mathbf{x}) = (1, 1, -2z)$

7. $\nabla f(\mathbf{x}) = (2x, -3y^2)$,
$|\nabla f(\mathbf{x})| = \sqrt{4x^2 + 9y^4}$,
$|\nabla f(1, 1)| = \sqrt{13}$, $\mathbf{u} = \left(2/\sqrt{13}, -3/\sqrt{13}\right)$

9. $\nabla h(\mathbf{x}) = (y \sin z, x \sin z, xy \cos z)$,
$|\nabla h(\mathbf{x})| = \sqrt{(x^2 + y^2) \sin^2 z + x^2 y^2 \cos^2 z}$,
$\nabla h(1, 2, \pi) = (0, 0, -2)$, $\mathbf{u} = (0, 0, -1)$

11.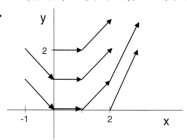

13.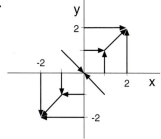

15. $\nabla f(x, y) = (y, x + 2y)$

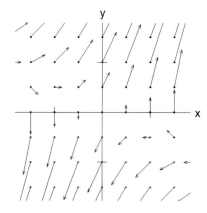

17. $\nabla f(x, y, z) = (2x, 2y, 0)$.

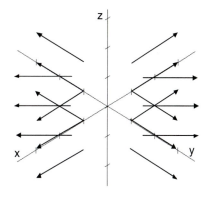

19. normal vector $(2, 2, 0)$; tangent plane $x + y = 2$

21. normal vector $\mathbf{e}_1$; tangent hyperplane $x_1 = 1$ (x_1 is the first-coordinate variable for points in $\mathbb{R}^n$)

23. normal vector $(1, 1, 1)$; tangent plane $x + y + z = 3$

25. (b) normal vector $(-1 - e, -1 - e, 1)$; tangent plane $x + y - z/(1 + e) = 1$.

27. $F'(0) = 3$ **29.** $g'(\pi) = -1$

31. (a) $x^2 - y^2 = 8$ **(b)** $y = (x/3)^{3/2}$, $t \geq 0$ **(c)** $(6, -2)$ **(d)** $(6, 3)$ **(e)** maximum radiation at $t = \sqrt{6}$; radiation decreases for $t > \sqrt{6}$ and is zero for $t \geq 3$.

35. If $f(x, y, z) = \frac{1}{2}x^2 + \frac{1}{2}y^2 + \frac{1}{2}z^2$ **37.** $f(x, y) = \frac{1}{3}x^3 + \frac{1}{3}y^3$

Exercise Set 1D (p. 261)

1.

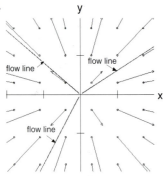

3.

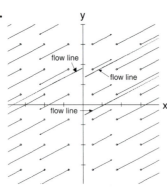

5.

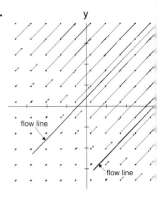

Exercise Set 2A (pgs. 269–271)

1. (a) $f'(x, y) = \begin{pmatrix} 2x + y & x \\ 0 & 2y \end{pmatrix}$, $g'(f(x, y)) = \begin{pmatrix} 1 & 1 \\ 2 & 0 \\ 0 & 2y^2 + 4 \end{pmatrix}$

(b) $(g \circ f)'(1, 1) = \begin{pmatrix} 3 & 3 \\ 6 & 2 \\ 0 & 12 \end{pmatrix}$, $(g \circ f)'(0, 0) = \begin{pmatrix} 0 & 0 \\ 0 & 0 \\ 0 & 0 \end{pmatrix}$

3. $d(g \circ f)/dt(2) = 14$

5. (a) $(F \circ f)'(u, v) = (4u + 4u^3 - 4uv^2, 4v - 4vu^2 + 4v^3)$ **(b)** $\partial w/\partial u = 4u + 4u^3 - 4uv^2$, $\partial w/\partial v = 4v - 4vu^2 + 4v^3$

7. $\partial w/\partial r = \sqrt{2}$; $\partial w/\partial \theta = 0$

9. *Hint:* If $g(\mathbf{u}_0) = \mathbf{x}_0$, use the chain rule to compute $(F \circ g)'(\mathbf{u}_0)$.

13. *Hint:* $f(tx, ty)$ can be computed in two ways.

15. $\partial^2(f \circ g)/\partial v \partial u(1, 1) = 2$

21. $g \circ f$ can possibly be defined; $f \circ g$ cannot be defined.

23. $g \circ f$ cannot be defined; $f \circ g$ can possibly be defined.

25. $g \circ f$ can possibly be defined; $f \circ g$ cannot be defined.

Exercise Set 2B (pgs. 274–275)

9. (a), (b) *Hint:* Use a parametric representation for a line and and a circle in the uv-plane, then apply f.
(c) $\det(f') = -2(u^2 + v^2)$.

11. *Hint:* Choose $\epsilon > 0$ such that $\epsilon < |f'(x_0)|$ and use the continuity of f' to show that $0 < |f'(x)|$ on some open interval containing x_0. Then use the mean-value theorem.

13. *Hint:* $(F^{-1} \circ F)(\mathbf{x}) = \mathbf{x}$ for all $\mathbf{x}$ in some neighborhood of $\mathbf{x}_0$. Now apply the chain rule to this equation.

Exercise Set 3 (pgs. 281–283)

1. (a) $dy/dx = -x/y$, $(y \neq 0)$, $dy/dx(1/\sqrt{2}, 1/\sqrt{2}) = -1$; $dy/dx(0, 1)$ makes sense but $dy/dx(-1, 1)$ does not.
(b) $dx/dy = -y/x$, $(x \neq 0)$, $dx/dy(1/\sqrt{2}, -1/\sqrt{2}) = 1$; $dx/dy(1, 0)$ makes sense but $dx/dy(0, 1)$ does not.

(c) $y = \pm\sqrt{1 - x^2}$

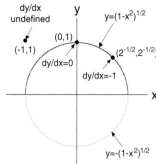

(d) $x = \pm\sqrt{1 - y^2}$

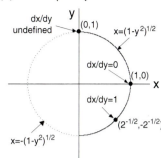

3. $dy/dx(-1, 1) = 1$, $dx/dy(-1, 1) = 1$

5. $dy/dx(-1, 1/4) = -1/4$, $dx/dy(-1, 1/4) = -4$

7. **(a)** $dx/dz(1, 1, -1) = -1/2$, $dy/dz(1, 1, -1) = 3/2$ **(b)** $dy/dx(1, 1, -1) = -3$, $dz/dx(1, 1, -1) = -2$
   (c) $dx/dy(1, 1, -1) = -1/3$, $dz/dy(1, 1, -1) = 2/3$

9. $F(x, y, z) = \begin{pmatrix} x^2y + yz \\ xyz + 1 \end{pmatrix}$ for all three parts.

   For part (a), $\mathbf{x} = z$, $\mathbf{y} = \begin{pmatrix} x \\ y \end{pmatrix}$, $F_{\mathbf{x}} = \begin{pmatrix} y \\ xy \end{pmatrix}$, $F_{\mathbf{y}} = \begin{pmatrix} 2xy & x^2 + z \\ yz & xz \end{pmatrix}$.

   For part (b), $\mathbf{x} = x$, $\mathbf{y} = \begin{pmatrix} y \\ z \end{pmatrix}$, $F_{\mathbf{x}} = \begin{pmatrix} 2xy \\ yz \end{pmatrix}$, $F_{\mathbf{y}} = \begin{pmatrix} x^2 + z & y \\ xz & xy \end{pmatrix}$.

   For part (c), $\mathbf{x} = y$, $\mathbf{y} = \begin{pmatrix} x \\ z \end{pmatrix}$, $F_{\mathbf{x}} = \begin{pmatrix} x^2 + z \\ xz \end{pmatrix}$, $F_{\mathbf{y}} = \begin{pmatrix} 2xy & y \\ yz & xy \end{pmatrix}$.

11. $\partial x/\partial u(1, -1, 1, 1, -1) = 0$, $\partial y/\partial u(1, -1, 1, 1, -1) = 1$

13. $9x + (6\sqrt{11})y + 8z = 36$.

15. **(a)** $f'(1, 1) = \begin{pmatrix} -5/3 & 4 \\ 3 & -5 \\ -1/3 & -1 \end{pmatrix}$ **(b)** $14x + 9y + 11z = 12$

17. **(b)** $x_1(y, z) = \sqrt{a^2 - y^2 - z^2}$, $x_2(y, z) = -\sqrt{a^2 - y^2 - z^2}$, $y_1(x, z) = \sqrt{a^2 - x^2 - z^2}$, $y_2(x, z) = -\sqrt{a^2 - x^2 - z^2}$,
    $z_1(x, y) = \sqrt{a^2 - x^2 - y^2}$, $z_2(x, y) = -\sqrt{a^2 - x^2 - y^2}$

19. **(a)** $(x, 0, z)$, $(x, y, -2x)$ **(b)** $(x, -1/(5x^2), -2x)$ **(c)** $x_1(y, z) = \left(-yz + \sqrt{y(5yz^2 + 4)}\right)/(2y)$,
    $x_2(y, z) = \left(-yz - \sqrt{y(5yz^2 + 4)}\right)/(2y)$

Exercise Set 4A–D (pgs. 292–293)

1. $(2, 1)$

3. $(x, y) = (-1/4, -1/4)$

5. $(x, y, z) = (-2/5, 0, 1/5)$

7. no critical points

9. maximum value at $(1, 1)$; minimum value at $(-1, -1)$

11. maximum value at $(3, -4)$ and $(-3, 4)$; minimum value at $(4, -3)$ and $(-4, 3)$

13. maximum value at $(2/\sqrt{5}, 1/\sqrt{10})$ and $(-2/\sqrt{5}, -1/\sqrt{10})$; minimum value at $(0, 0)$.

15. $(-1, 0, 1)$ and $(-1, 0, -1)$

17. $(0, t, t)$, $|t| \leq \sqrt{2}$

19. $(1/\sqrt{3}, 1/\sqrt{3}, 1/\sqrt{3})$

21. $-1/\sqrt{2}$

23. 6 by 6 by 3

25. $(5V/6)^{1/3}$ by $(5V/6)^{1/3}$ by $(36V/25)^{\frac{1}{3}}$.

27. **(a)** $1 + 1/\sqrt{2}$ **(b)** 1

29. $(27/19, -7/19, 7/19, -3/19)$

31. *Hint:* Minimize the function $f(x) = (a_1 + \cdots + a_N + x)/(N + 1) - (a_1 \cdots a_N x)^{1/(N+1)}$ for $x > 0$ and use induction.

33. **(a)** $|h| < 1$ **(b)** maximum and minimum of f on C_h are $\sqrt{1 - h^2}$ and $-\sqrt{1 - h^2}$, respectively.   **(c)** 0 is both the maximum and minimum value of f on C_1.

35. The given plane is tangent to the given sphere at the point $(1, 1, 1)$. So, $f(1, 1, 1) = 1$ is both the maximum and minimum of f subject to the given conditions.

37. **(a)** length $= 3$, width $=$ height $= 6$ **(b)** length $=$ height $= 3(2)^{1/3}$, width $= 6(2)^{1/3}$

Exercise Set 4E (pgs. 297–298)

3. $(1, -2)$ (minimum)

5. $(1, 2)$ (saddle)

7. $(0, 0)$ (minimum)

9. $(0, 0)$ (saddle)

11. $(-1/2, 4)$ (maximum)

13. $(0, 0)$ (saddle)

17. **(b)** $(0, 0, 0)$ is a saddle point.

Exercise Sct 4F (pgs. 302–303)

3. local extreme point $(1/\sqrt{6}, 1/\sqrt{6})$ gives the local maximum value $\frac{3}{\sqrt{6}}e^{-1/2}$; local extreme point $(-1/\sqrt{6}, -1/\sqrt{6})$ gives the local minimum value $-\frac{3}{\sqrt{6}}e^{-1/2}$

5. no extreme points $((0, 0)$ is a saddle point)

7. The point 0.653271187 gives the extreme value 0.396652961; the point 3.292310007 gives the extreme value -0.000002945.

9. **(a)** 1.49546 by 1.77438 by 0.72437 **(b)** $1/\sqrt{2}$ by $1/\sqrt{2}$ by $2e^{-1}$

Exercise Set 5A–D (pgs. 308–309)

1.

3.

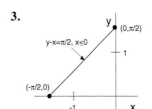

5.

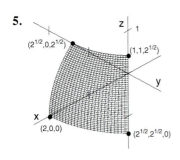

7.

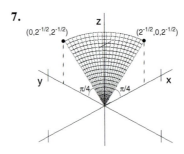

9. $R = \{ (r \cos \theta, r \sin \theta, z) \mid 0 \le r \le 1, -\pi/2 \le \theta \le \pi/2 \}$ **11.**

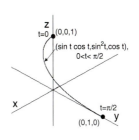

13.

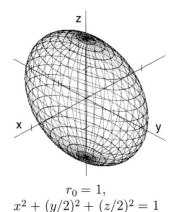

$r_0 = 1,$
$x^2 + (y/2)^2 + (z/2)^2 = 1$

$\phi_0 = \pi/3,$
$4x^2 + y^2 = 3z^2$

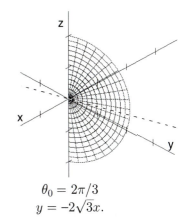

$\theta_0 = 2\pi/3$
$y = -2\sqrt{3}x.$

15. $(-2t \sin t \sin t^2 + \cos t \cos t^2, 2t \sin t \cos t^2 + \cos t \sin t^2, -\sin t)$

Chapter 6 Review (pgs. 309–311)

1. $f'(x, y, z) = (2x, 2y, -2z)$

3. $g'(\pi/3, \pi^2/36) = \begin{pmatrix} -\sqrt{3}/2 & 0 \\ 1/2 & 0 \\ 0 & 1 \end{pmatrix}$

5. $\partial K / \partial \mathbf{u}(1, \sqrt{3}) = 2$

7. in the direction of $(3, 3, 2)$.

9. (a)

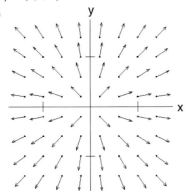

(b) rate of increase is constant at 1.

15. $\partial f / \partial s(2, 1) = 80$

17. (b) $dP/dt(0) = -21/16$ **(c)** $dP/dt(30) = -3/7^3$

19. $z_x(1, 1, 1) = 1$, $z_y(1, 1, 1) = 2$

21. (a), (b) tangent vector: $(5, -4, 1)$

23. (b) the result of part (a) implies Theorem 1.2.

25. $x_u(2, 1, -2, -1) = 3$, $y_u(2, 1, -2, -1) = 1/2$

27. $\partial x/\partial y(-2, 1, -1, 2) = 4/9$, $\partial z/\partial y(-2, 1, -1, 2) = 7/9$

29. maximum at $(\sqrt{14}/4, 1/4)$, $(-\sqrt{14}/4, 1/4)$; minimum at $(0, 1/\sqrt{2})$, $(0, -1/\sqrt{2})$.

31. $3\sqrt{3}$ cubic units   **35.** $(0, 0, 0)$

37. The only critical point $(-1, -1)$ is a saddle point

39. minimum value of $f(\mathbf{x})$ is $-|\mathbf{x}_0|$. The same results are obtained if the restriction $|\mathbf{x}| = 1$ is replace by $|\mathbf{x}| \le 1$.

41. **(a)** $(1, 1)$, $(-1, -1)$ **(b)** maximum value is -1, minimum value is -2.

43. **(a)** $(0, 0)$, $(3/2, -9/4)$ **(b)** maximum value is 5.

CHAPTER 7: MULTIPLE INTEGRATION

Section 1: Iterated Integrals

Exercise Set 1A-D (pgs. 321–322)

1. $-\frac{79}{72}$.

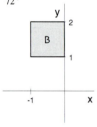

3. $\frac{11}{3}$.

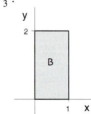

5. $\frac{67}{28}$.

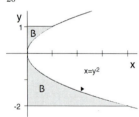

7. $\frac{2}{3}$.

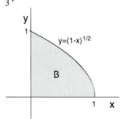

9. $\frac{1}{6}$.

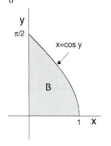

11. $\frac{1}{6}$.

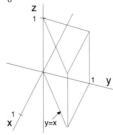

13. 1.

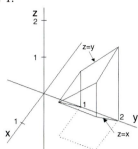

15. $2\pi + 6$.

17. $-\sin 1 - \cos 1 + \frac{3}{2}$.

19. (a) $A(B) = \int_0^2 \left[\int_{\sin \pi x}^{4x-2x^2} 1 \, dy \right] dx.$

(b) $\int_0^2 \left[\int_{\sin \pi x}^{4x-2x^2} f(x, y) \, dy \right] dx.$

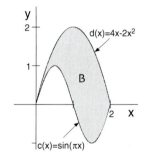

21. $\frac{71}{140}$

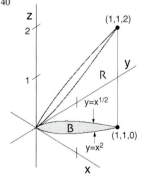

23. $\frac{8}{3}a^3$.

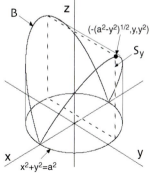

25. (b) πa^3.

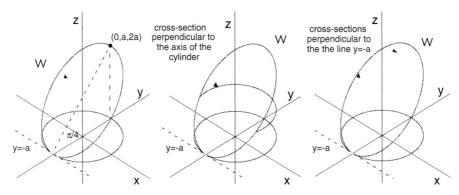

27. Let F_n be the value of the expression for $n \geq 2$; $F_2 = 1/2$ and $F_n = 7/12$ for $n \geq 3$.

29. (a) $(1 - \delta^{1/2})^2$. **(b)** 4.

Section 2: Multiple Integrals

Exercise Set 2A-E (pgs. 332–333)

1. π.

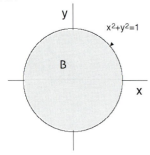

3. π.

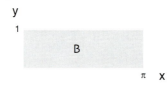

5. 6

7. $\frac{35}{3}$.

9. 2.

11. 2π.

13. $V(B) = \int_{-1}^{1} dx \int_{-2\sqrt{1-x^2}}^{2\sqrt{1-x^2}} dy \int_{0}^{4-4x^2-y^2} dz = 4\pi$.

$$V(B) = \int_{-1}^{1} dx \int_{-2\sqrt{1-x^2}}^{2\sqrt{1-x^2}} (4 - 4x^2 - y^2)\, dy.$$

15. 16π.

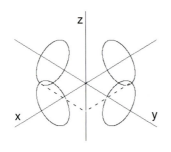

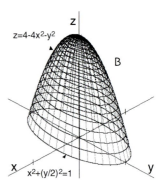

17. $\pi h^2(a - h/3)$.   **19.** $45\pi/2$ lb.

21. About 252.4 lb.

Section 3: Integration Theorems

Exercise Set 3 (pgs. 336–337)

1. $\frac{31}{6}$.   **3.** *Hint:* $\int_{B_r} f(\mathbf{x}_0)\,dV = V(B)f(\mathbf{x}_0)$.

5. Use the definition of the double integral.   **7.** (a) and (b) are routine. For (c) use the hint.

9. $2y$.   **11.** $h'(x) = \int_0^x e^{u^2}\,du;\ h''(x) = e^{x^2}$.

13. Use the hint.

Section 4: Change Of Variable

Exercise Set 4A-D (pgs. 346–348)

1. $\frac{1}{2}(e^4 - 1)$.

3. (a) B is the part of the first quadrant of the xy-plane bounded by the circle of radius 2 centered at the origin.
(b) $\frac{4\pi}{3}$. (c) $\frac{8}{3}$.

5. $M(A) = 20\pi$ grams.   **7.** π.

9. $2a^2$.   **11.** $\frac{\pi^2}{4}$.

13. πa^4.   **15.** $\frac{2\pi}{3}$.

17. Use spherical coordintes.   **19.** Use cylindrical coordinates.

21. (a) R_{xy} is the upper half $(y \geq 0)$ of the annular region of inner radiuus 1 and outer radius 4 centered at the origin.
(b) 3π.

23. (b) $1 + u^2$. (c) $\frac{99}{2}$. (a)   **25.** $\frac{16\pi^2}{3} + 6\pi\sqrt{3}$.

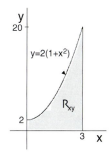

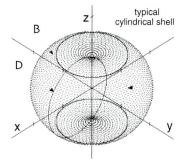

27. $b = a\sqrt{1 - k^{2/3}}$.

Section 5: Centroids and Moments

Exercise Set 5 (pgs. 352–353)

1. $-\frac{1}{6}$.

3. $(0, 1)$.

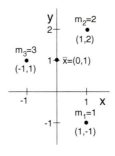

5. $\frac{2}{3}$.

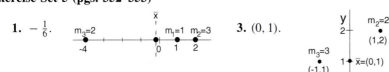

7. $\left(\frac{3(\pi+2)}{32}, \frac{3(\pi+2)}{32}\right) \approx (0.48, 0.48)$.

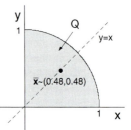

9. $(3/4, 12/5)$.

11. *Hint:* Center the ball at the origin.

13. (a) $\bar{\mathbf{x}} = \left(\frac{4(a^3-b^3)}{3\pi(a^2-b^2)}, \frac{4(a^3-b^3)}{3\pi(a^2-b^2)}\right)$.   (b) $\bar{\mathbf{x}} = \left(\frac{4a}{3\pi}, \frac{4a}{3\pi}\right)$.

15. The centroid of R is on the line of symmetry at distance $4a/3\pi$ units from the flat edge.

17. (a) Use the hint.   (b) *Hint:* Break $M_P(B)$ into two integrals.

19. $2\mu b^4/3$.   21. Use the hint.

23. *Hint:* Use basic properties of integrals.

Section 6: Improper Integrals

Exercise Set 6 (pgs. 358–359)

1. $1/3$.   3. $1/4$.

5. $(1 - e^{-1})^2$.   7. $-\pi$.

9. $\pi/2$.   11. $\begin{cases} +\infty, & \alpha > -1 \\ -1/(\alpha + 1), & \alpha < -1. \end{cases}$

13. (a) $k = 2/\pi$.   (b) $\frac{1}{3} - \frac{3\sqrt{3}}{8\pi} \approx 0.13$.   (c) $(0, 0)$.

15. (a) *Hint:* The iterated integral is the square of an integral.   (b) Routine.   (c) *Hint:* Change Variable to get rid of the m in the exponent.   (d) *Hint:* Same as for (c).

17. *Hint:* Use the result of part (a) of Exercise 15.   (b) *Hint:* Compute $\int_0^\infty v^3 e^{-\alpha v^3} dv$.

Section 7: Numerical Integration

Exercise Set 7 (pgs. 362–363)

1. With $p = 86$, $q = 87$ about 0.5333000. Since $8/15 = 0.5\overline{3}$, the above approximation is accurate to four decimal places.

3. Simpson with $p = q = 2$ gives the exact value -20. The midpoint approximation doesn't give the exact value but rather approaches it from below as p and q increase.

5. The exact value is $\pi/2 \approx 1.570796327$. Because the given region R is not a rectangle, the function that was used for the approximation was $F(x, y) = (1 - x^2 - y^2)H(1 - x^2 - y^2)$, where $H(x)$ is the *Heaviside unit step function* which is 1 fot $x \geq 1$ and otherwise, iintroduced in Chapter 4. Thus, $F(x, y) = f(x, y)$ inside the unit disk and is zero elsewhere. The Simpson approximation and the midpoint approximation were then applied to $F(x, y)$ on the rectangle

$-1 \leq x \leq 1, -1 \leq y \leq 1$ for the values $p = q = 100$ and $p = q = 150$. The two methods produced 1.57082, 1.57080 respectively.

7. The region R given here, is the part of the region defined in Exercise 5 above that lies in the first quadrant of the xy-plane. Since the integrand is the same here as it was there, and since it is symmetric with respect to the origin, it follows that the exact value of the integral is one-fourth the value found in Exercise 5; namely $\pi/8 \approx 0.392699082$. As in Exercise 5 above, the given region R is not a rectangle. Therefore, just as in Exercise 5, the function that was used for the approximations was $F(x, y) = (1 - x^2 - y^2)H(1 - x^2 - y^2)$. However, here we applied the Simpson approximation and the midpoint approximation on the rectangle $0 \leq x \leq 1, 0 \leq y \leq 1$. Six decimal place accuracy was first achieved with the Simpson approximation at $p = q = 136$, while the same accuracy was first achieved with the midpoint approximation at $p = q = 70$.

9. **(a)** π.   **(b)** $G(0, 1.000, 0, 1.000) \approx 2.23098516$, $G(0, 2.000, 0, 2.000) \approx 3.11227036$, $G(0, 2.600, 0, 2.600) \approx 3.140110976$, $G(0, 3.575, 0, 3.575) \approx 3.14158996$, $G(0, 3.600, 0, 3.600) \approx 3.1415904$.
   (c) The list suggests that $G(0, a, 0, a)$ first approximates π to four-decimal places at about $a = 3.58$.

11. The exact value is $\pi/2$. Simpson with $p = q = r = 30$ gives 1.5733481677, accurate to only two plaes.

13. The exact value is $3/2$ and Simpson with $p = q = r = 2$ gives that with 10-place accuracy.

15. **(a)** The exact value is $7/6$.

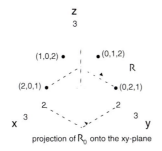

projection of R_0 onto the xy-plane

(b) If $H(x)$ is the Heaviside function, then $H(3 - x - y - z)$ has the value 1 for points below the plane $x + y + z = 3$ and is zero elsewhere. The smallest rectangle, call it R_0, containing R is $0 \leq x \leq 2, 0 \leq y \leq 2, 1 \leq z \leq 2$ (see figure). We integrate the Heaviside unit step function $H(3 - x - y - z)$ over R_0. When Simpson's rule in three dimensions was applied to this function over R_0, the approximation for $p = q = r = 50$ was 1.176576, which is accurate to only one decimal place. When the midpoint approximation in three dimensions was applied with $p = q = r = 50$, the result was 1.166400, which is accurate to three decimal places. Simpson's rule is usually better for smooth functions, but not here H since is not continous.

(c) The apparent superiority of the midpoint approximation suggested by the result of part (b), compelled us to forego the use of the Simpson approximation in favor of the midpoint approximation, which was applied to the function $(x^4 + y^4 + z^4)H(3 - x - y - z)$ over the rectangle R_0 described in part (b). Using $p = q = r = 50$, the result was 6.679043. In order to check this answer, one can directly compute

$$\int_R (x^4 + y^4 + z^4)\, dx\, dy\, dz = \int_1^2 dz \int_0^{3-z} dy \int_0^{3-y-z} (x^4 + y^4 + z^4)\, dx = \frac{1403}{210} \approx 6.680952381.$$

We see that our midpoint approximation is accurate to only one decimal place (although rounding to two places produces two-place accuracy).

(*Note:* In previous exercises, values of p and q were tried for much larger values than were tried here. The reason is that the number of operations required to carry out both the Simpson approximation and the midpoint approximation in dimension 3 is roughly proportional to the cube of the number od subintervals used. Thus, all things being equal, computation time much longer in three dimensions than it is in one or two dimensions.)

Chapter 7 Review (pgs. 363–366)

1. $2/3$.

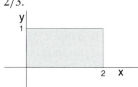

3. $e^2 - 3$.

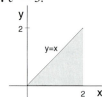

5. 0.

7. $1/2$.

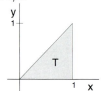

9. $1/3$. The region Q is shown below on the left.

11. $\pi/6$. The unit disk is in the middle figure below.

13. $20/3$. The square S in $\mathbb{R}^2$ of side length 2 centered at $(1, 0)$ is below on the right.

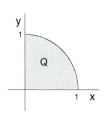

Exercise 9

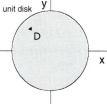

Exercise 11

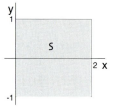

Exercise 13

15. $15/8$.

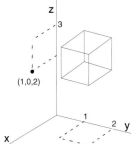

17. π.

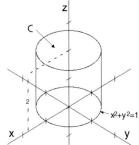

19. $4\pi/5$; B is the solid ball of radius 1 centered at the origin.

Note: The regions of integration in Exercises 21 and 23 below are truncated cones with heights 1 and 2 respectively, so is shown only once.

21. $\pi/10$.

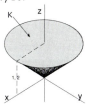

23. $8\pi/3$.

25. $3616/35$.

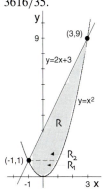

27. $16ab^2/3$.

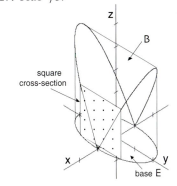

29. $4\pi abc/3$.

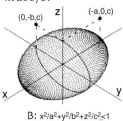

31. $8\pi/3$.

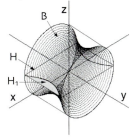

33. $1/(6abc)$.

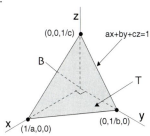

35. (b) $\int_0^{\sqrt{2}} \int_0^{\sqrt{1-y^2/2}} 2x \, dx \, dy$.   (c) $2\sqrt{2}/3$.

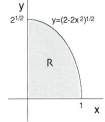

37. 477/20.

39. 4/9.

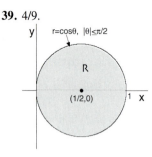

41. (a) $\displaystyle\int_0^1 \left[\int_{-\sqrt{1-x^2}}^{\sqrt{1-x^2}} \left[\int_0^x z\,dz \right] dy \right] dx.$ **(b)** $\displaystyle\int_0^1 \left[\int_{-\arccos z}^{\arccos z} \left[\int_{z\sec\theta}^1 z\,r\,dr \right] d\theta \right] dz.$ **(c)** $\pi/16.$

43. 1/32.

45. $a = -2, b = 2, c = 3 - \sqrt{4 - y^2}, d = 3 - \sqrt{4 - y^2}.$

47. 10.   **49.** $V(C) = 117\pi/8 + 18.$

51. $\pi.$ **53.** $3\pi/2.$

55. $4\pi/3.$ **57.** $\alpha < 1.$

59. $\alpha < 3/2.$

61. (b) $\displaystyle\int_{-1}^1 dx \int_{-\sqrt{1-x^2}}^{\sqrt{1-x^2}} dy \int_0^{\sqrt{1-x^2-y^2}} dz.$ **(c)** $\displaystyle\int_0^{2\pi} d\theta \int_0^{\pi/2} \sin\phi\,d\phi \int_0^2 r^2\,dr.$ **(d)** $2\pi/3.$

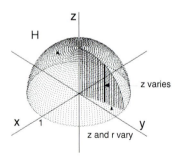

CHAPTER 8: INTEGRALS AND DERIVATIVES ON CURVES

Exercise Set 1AB (pgs. 376–377)

1. 4/3   **3.** on γ_1: π; on γ_2: 2π

5. 0   **7.** 3π

9. on γ_1: $1/2 + \pi/4$; on γ_2: 1   **11.** on γ_1: 0; on γ_2: 0

13. 0   **15.** 0

17. 478/15

19. (a)   **(b)** 0

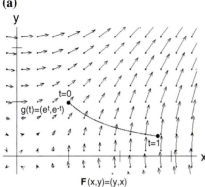

21. *Hint:* Let $\phi(u) = u^2$.

25. Because the flow lines of the field **F** are concentric circles about the origin, the line integral of **F** around an elliptical path γ centered at the origin is not affected by rotating γ about the origin. Also, the angle θ between the velocity vector of a continuously differentiable parametrization of γ and the field always satisfies $0 \le \theta < \pi/2$ so that the integrand of the line integral around γ is always positive and therefore $\int_\gamma \mathbf{F} \cdot d\mathbf{x} \ne 0$.

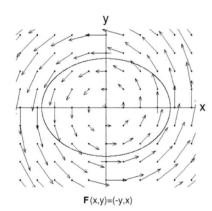

27. (a) See graph for Exer. 25 in this section   **(b)** πa^2 **(c)** ab **(d)** $\frac{1}{2}ab$; the answers are the areas of the enclosed regions.

29. -18 **31.** 24

33. (a)

(b) *Hint:* Try the line segment with endpoints $(0, 1)$, $(1, 0)$, and the half circle joining these points.

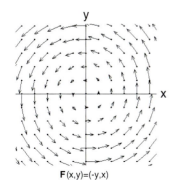

Exercise Set 2 (pgs. 382–383)

1. $\ln(\sec 1 + \tan 1)$ **3.** $335/27$

5. (b) $l(\gamma) = 4$

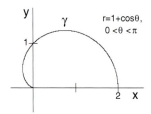

7. $\sqrt{a^2 + b^2}\left(2\pi a^2 + \frac{1}{3}(8b^2\pi^3)\right)$ **9. (a)** 5   **(b)** 126

13. (a) $\int_0^{2\pi} \sqrt{a^2 \cos^2 t + b^2 \sin^2 t}\, dt$ **(c)** ≈ 9.6888

15. $|g'(t)| = \sqrt{2}$, $h(s) = \left(\cos(s/\sqrt{2}), \sin(s/\sqrt{2}), s/\sqrt{2}\right)$, $0 \le s$

17. (a) *Hint:* Use $l(\gamma) = a\theta$ in the formula for $\mathbf{p}_0$ and interpret the result.

Exercise Set 3 (pgs. 385–386)

3. speed   **5.** Yes, by the Cauchy-Schwarz inequality.

9.

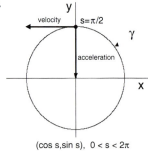

13. (b) $\kappa_{\max} = 1$, $\kappa_{\min} = 0$ **(c)** minimum curvature at $x = 0$, maximum curvature at $x = 56^{1/6}$ and $x = -56^{1/6}$.

15. (b) $6/\left(|t|(4 + 9t^2)^{3/2}\right)$; ∞

Exercise Set 4 (pgs. 394–395)

1. $\operatorname{div} \mathbf{F} = -y\sin(xy) + x\cos(xy)$, $\operatorname{curl} \mathbf{F} = y\cos(xy) + x\sin(xy)$

3. $\operatorname{div} \mathbf{F} = -1$, $\operatorname{curl} \mathbf{F} = 2$

5. $\operatorname{div} \mathbf{F} = 4x^3 - 4xy(x - y) + 4y^3$, $\operatorname{curl} \mathbf{F} = 4x^3 - 4xy(x + y) - 4y^3$

7. the points in $\mathbb{R}^2$ that lie below the line $y = x$

9. the points in $\mathbb{R}^2$ above the x-axis

17. (b) $\mathbf{G} = (y, 0, 0)$ **(c)** $\mathbf{F} = (z, x, y)$

19. (b) neither expansion nor contraction occurs and volume is preserved

21. If $\Delta f > 0$ (< 0) in a region then a given mass is moved to a region of larger (smaller) area and therefore density decreases (increases), and if $\Delta f = 0$ (i.e., f is harmonic) in a region then a given mass is moved to a region of the same area and density is preserved.

Chapter 8 Review (pgs. 395–396)

1. $1/2$ **3.** 63

5. $-\pi$ **7.** $1/3$

9. $I(\gamma_2) = I(\gamma_3)$ because the parametrizations are equivalent. But $I(\gamma_1)$ has a different value because the parametrization is not equivalent to either of the other two.

11. $9/2$

15. (a) $2\pi^2$ **(b)** $2\pi t_0$; If $t_0 = 0$ then the vector field is perpendicular to the path traced by $g(t)$; and if $t_0 = 1$ then the field vectors point in the same direction as $g'(t)$.

17. 2304 kg

19. *Hint:* Use the formula for curvature given in Exercise 13 in Section 3.

21. (a) maximum curvature when $a = b$, minimum curvature when $a = 0$ **(b)** maximum curvature when $b = 0$, no minimum curvature

CHAPTER 9: VECTOR FIELD THEORY

Exercise Set 1ABC (pgs. 408–409)

1. $-\pi$ **3.** 0

5. 1 **7.** 1

9. 0

11. *Hint:* Use Green's Theorem and the field $\mathbf{F} = (F, G) = (-y, x)$.

13. (a) $\nabla f = \left(-y/(x^2 + y^2), x/(x^2 + y^2)\right),\ x > 0$

15. *Hint:* Use Gauss's Theorem in the plane.

17. (a) regions not containing the origin flow into regions of the same area   **(b)** circulation is zero   **(c)** 2π (**F** is not continuous on the interiors of circles centered at the origin so that Stokes's Theorem in the plane does not apply for these regions; i.e., parts (b) and (c) are not contradictory)

Exercise Set 2A–D (pgs. 418–419)

1. (a) $U(x, y, z) = gz$ **(b)** $\mathbf{x}(t) = \left(v_1 t, v_2 t, -\frac{1}{2}gt^2 + v_3 t\right),\ t \geq 0$

3. not a gradient field   **5.** not a gradient field

9. No   **11.** $\sin xy$

13. $\frac{1}{2}\ln(x^2 + y^2)$ **17.** $\ln(x^2 + y^2)$

19. $f(x, y) = xe^y$ **21.** $f(x, y, z) = xy + xz + yz$

23. $f(x, y) = xy^2 + x^2 y$ **27. (c)** $-\frac{1}{2}k\omega^2(b^2 - a^2)$

Exercise Set 3A–D (pgs. 429–431)

1. (a)   **(b)** $\sqrt{14}/2$

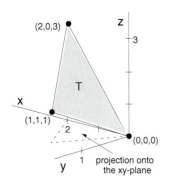

3. (a)   **(b)** $\pi(5\sqrt{5}-1)/12$; yes

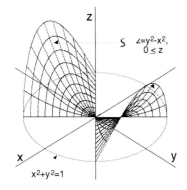

5. (a) $(6^{3/2}-2^{3/2})/12$ **(b)** 1

7. 4π **9. (a)** -2 **(b)** 0

15. (a) *Hint:* S is the graph of a continuously differentiable function $g(x,y)=z$. Now use the result of Exercise 14 in this section with $f=g$.

17. The pair of parametrizations $g_1(u,v)=(\cos u,\sin u,v)$, $0\le u\le 2\pi$, $0\le v\le 1$ (cylindrical side) and $g_2(u,v)=(v\cos u,v\sin u,0)$, $0\le u\le 2\pi$, $0\le v\le 1$ (bottom) are coherent.

19. The pair of parametrizations $g_1(u,v)=(v,u,u)$, $0\le u\le 1$, $0\le v\le 1$ and $g_2(u,v)=(v,u-1,1-u)$, $0\le u\le 1$, $0\le v\le 1$ are coherent and parametrize the two sides of the trough.

21. 2π

23. Change g_1 as given above in the answer to Exercise 17 to $f(u,v)=(\cos v,\sin v,u)$, $0\le u\le 1$, $0\le v\le 2\pi$ (cylinder) and keep the parametrization g_2 as shown there. The resulting integral has value -2π.

27. (a) $g(x,\theta)=\big(x,f(x)\cos\theta,f(x)\sin\theta\big)$, $a\le x\le b$, $0\le\theta\le 2\pi$

29. 0

31. *Hint:* Let $\mathbf{n}(x,y,z)$ be the unit normal to S at the point (x,y,z) and show that $\nabla f\cdot\mathbf{n}$ is identically zero on S.

Exercise Set 4AB (pgs. 437–438)

1. $2x+2y+2z$ **3.** 0

5. **7.** *Hint:* Use the parametrizations $f(u, v) = (u\cos v, u\sin v, 1)$, $0 \le u \le 1$, $0 \le v \le 2\pi$ for the top of R, $g(u, v) = (\cos u, \sin u, v)$, $0 \le u \le 2\pi$, $0 \le v \le 1$ for the cylindrical side of R, and $h(u, v) = (v\cos u, v\sin u, 0)$, $0 \le u \le 2\pi$, $0 \le v \le 1$ for the bottom of R.

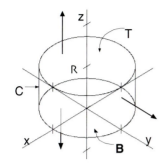

9. 32π

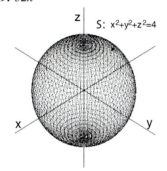

11. 0

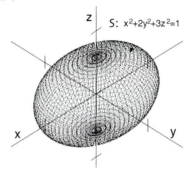

13. *Hint:* Routine calculation

15. **(b)** $f(x, y, z) = x^2$ **(c)** $\Delta f = \partial^2 f/\partial x_1^2 + \cdots + \partial^2 f/\partial x_n^2$

17. 0

19. *Hint:* Use Gauss's Theorem with the field $\mathbf{F}(x, y, z) = (x, y, z)$.

21. **(a)** *Hint:* Show that $\mathbf{F}(\mathbf{x}_0) \cdot \mathbf{n}(\mathbf{x}_0) = |\mathbf{F}(\mathbf{x}_0)|$ and then use Theorem 5.2 in Chapter 1.
   (b) *Hint:* Use part (a) with Gauss's Theorem and the formula $V(S) = 4\pi(abc)/3$.

23. *Hint:* A direct result of Gauss's Theorem and the definition of flux.

25. *Hint:* A direct result of Gauss's Theorem and the definition of flux.

27. **(a)** *Hint:* For each fixed $\mathbf{y}_0$ in R, let $\mathbf{H}_{\mathbf{y}_0}(\mathbf{x}) = (\mathbf{y}_0 - \mathbf{x})|\mathbf{y}_0 - \mathbf{x}|^{-3}$ for $\mathbf{x}$ not in R. Show that div $\mathbf{H}_{\mathbf{y}_0}(\mathbf{x}) = 0$ for all $\mathbf{x}$ not in R. **(b)** *Hint:* Let S_a be a sphere of radius a with a large enough to enclose S, choose the standard spherical coordinate parametrization of S_a and show that $\int_{S_a} \mathbf{H}_{\mathbf{y}_0} \cdot d\mathbf{S} = -4\pi$, then use the Surface Independence Principle on S and S_a.

Section 5: Stokes's Theorem

Exercise Set 5ABC (pgs. 447–449)

1. $(-2y - 1, -2z - 1, -2x - 1)$ **3.** $(0, 0, 0)$

5.

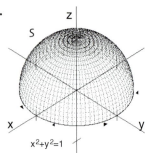

7.

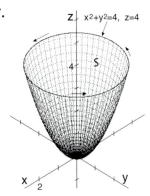

9. Parametrize the border of S by $f(t) = (\cos t, \sin t, 0)$; $\int_S \text{curl } \mathbf{F} \cdot d\mathbf{S} = -\pi$.

11. **(b)** Border of S consists of all points on the either of the two circles $x^2 + y^2 = 1$ and $x^2 + y^2 = 4$.   **(c)** 3π

13. *Hint:* Use the results of Exercise 25 in Section 3 and Exercise 12 in this section.

15. *Hint:* Using the coordinate functions of $\mathbf{F}$, compute $\mathbf{F}'(\mathbf{x}) - [\mathbf{F}'(\mathbf{x})]^t$ and curl $(\mathbf{x}) \times \mathbf{y}$, for $\mathbf{x}, \mathbf{y}$ in $\mathbb{R}^3$. Then use the results of Exercise 33 in Section 4 of Chapter 2.

19. 0   **21.** $\mathbf{F}(x, y) = \left(-y/(x^2 + y^2), x/(x^2 + y^2)\right)$

23. **(a)** *Hint:* Using coordinate functions, find the equations that must hold in order for curl $\mathbf{G} = \mathbf{F}$

25. $\mathbf{G}(x, y, z) = (z^2/2 - xy, -yz, c)$, c constant

27. $\mathbf{G}(x, y, z) = (-yz - 3xy, -xz, c)$, c constant

29. *Hint:* Show that curl $\mathbf{G}(\mathbf{x}) - \text{curl } \mathbf{H}(\mathbf{x}) = \text{curl}(\mathbf{G} - \mathbf{H})(\mathbf{x})$. Then use Theorem 5.4.

31. $\mathbf{G}(\mathbf{x}) = \frac{1}{3}(z^2 - xy, x^2 - yz, y^2 - xz)$

33. $\mathbf{G}(\mathbf{x}) = \frac{1}{3}(-yz - 3xy, 3x^2 - xz, 2xy)$

Exercise Set 6ABC (pgs. 456–457)

1. *Hint:* routine computation   **3.** *Hint:* routine computation

5. *Hint:* routine computation   **7.** *Hint:* routine computation

9. *Hint:* routine computation

11. *Hint:* Use identity (4) in the text with $f = 1/|\mathbf{x}|$ and $\mathbf{F} = \mathbf{v} \times \mathbf{x}$.

13. *Hint:* Use Exercise 12 in this section and identity (10) in the text.

15. *Hint:* Use the hint in the text.

17. *Hint:* Use Theorem 1.3 in Chapter 8 and let the lower limit of integration go to ∞.

19. *Hint:* Use the chain rule to write f_x, f_y and f_z in terms of $\overline{f}_r$. The rest is lengthy but routine computation.

21. *Hint:* routine (but lengthy) computation

Chapter 9 Review (pgs. 457–459)

1. **(a)** $f(x, y) = x^3 y + y^3$ **5.** 48π

7. **(a)** 0   **(b)** -2π

9. **(a)** $2(1 - \alpha)/(x^2 + y^2)^\alpha$ **(b)** positive circulation (counterclockwise flow) if $\alpha < 1$, negative circulation (clockwise flow) if $\alpha > 1$, and no flow if $\alpha = 1$.   **(c)** circulation is positive for all $\alpha > 0$ with maximum value of 2π when $\alpha = 1$.

11. *Hint:* Use Green's Theorem with the four pairs of functions $\left(F(x, y), G(x, y)\right) = (0, x)$, $(-y, 0)$, $(x, 0)$, $(0, y)$

13. **(a)** $g(x, y) = \left(x, y, 2 + \sqrt{1 - x^2 - y^2}\right)$ (hemisphere), $h(x, y) = \left(y, x\sqrt{x^2 + y^2}\right)$ (cone), for $x^2 + y^2 \le 1$ **(b)** surface area: $(2 + \sqrt{5})\pi$; volume: $4\pi/3$ **(c)** 4π

15. (a) *Hint:* Use identities (4) and (9) in Section 6.   **(b)** 0

17. $-1/3$ **19.** $\int_0^1 \int_0^1 2\, du\, dv = 2$

21. (a) *Hint:* Use Gauss's Theorem.   **(b)** *Hint:* Apply Gauss's Theorem to each of the fields $(x, 0, 0)$, $(0, y, 0)$, $(0, 0, z)$.

23. A third surface S in B can always be found that intersects S_1 and S_2 only on their common border, and parametrizations of S_1, S_2 and S can be found such that all normals are pointing in the proper directions to apply Gauss's Theorem to the closed piecewise smooth surfaces $S_1 \cup S$ and $S_2 \cup S$. The Surface Independence Principle can then be used to conclude that the flux across S_1 and the flux across S_2 are both equal to the flux across S.

CHAPTER 10: FIRST-ORDER DIFFERENTIAL EQUATIONS

Section 1: Direction Fields

Exercise Set 1A (pg. 466)

1. $y(x) = 3e^x - 1$, $-\infty < x < \infty$.

3. $y(x) = 6e^5 e^{-x} = 6e^{5-x}$, $-\infty < x < \infty$.

5. $y(x) = \left(\frac{7}{2} - x\right)^{-1}$, $x \le 7/2$.

7. $y' = y/x$, $(x_0, y_0) = (1, 2)$

9. $dy/dx = y + x$, $(x_0, y_0) = (1, -1)$

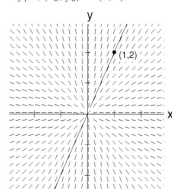

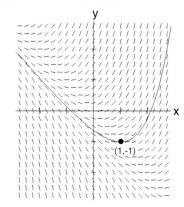

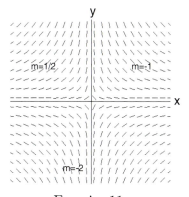

Exercise 11

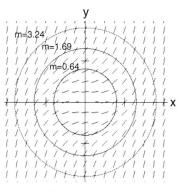

Exercise 13

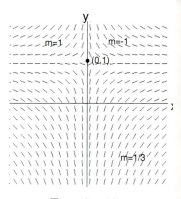

Exercise 15

17. $y' = x^3$

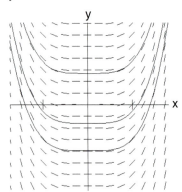

19. $y' = 1/(1 + x^4)$

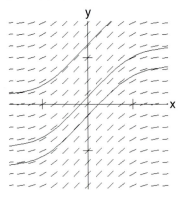

21. $y' = (1 - x^3)^{1/3}$

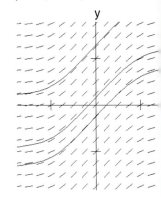

23. (a) *Hint:* Use the Fundamental Theorem of Calculus.   **(b)** Use the hint.

25. *Hint:* If $y(x)$ is a solution so is $y(x + \alpha)$. Theorem 1.1 doesn't apply because the derivative of $\sqrt{1 - y^2}$ isn't bounded near $y = \pm 1$.

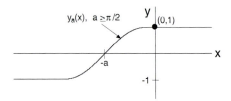
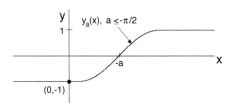

Exercise Set 1B (pg. 469)

1. $y' = \sin(x - y)$

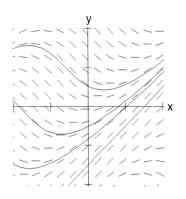

3. $y' = \sqrt{9 - y^3}$

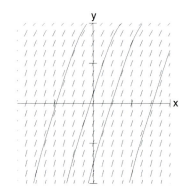

5. $y' = \sin(x^2 + y^2)$

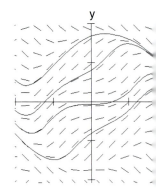

7. $y' = \cos(x^2 + y^2)$

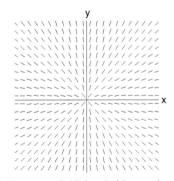

9. $y' = (1 + x^4)^{-1}$

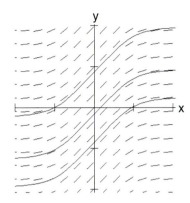

11. $y' = e^{-y^2}$

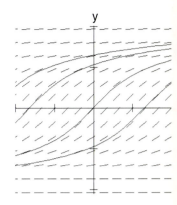

Section 2: Applied Integration

Exercise Set 2AB (pgs. 478–480)

1. $y(x) = \frac{1}{2}x^2 - \frac{1}{3}x^3 + 1, \quad -\infty < x < \infty.$

3. $y(x) = -\frac{1}{2}\ln(1 - x^2) + 1, \quad |x| < 1.$

5. $y(x) = -\sin x + 2x + 1, \quad -\infty < x < \infty.$

7. $z(t) = te^t - e^t + 2, \quad -\infty < t < \infty.$ **9.** $x(t) = e^t, \quad -\infty < t < \infty.$

11. $t_{max} = 5000/32 = 156.25$ seconds. $y_{max} = 390625$ ft ≈ 74 miles.

13. (a) $t = 1$ second at an altitude of $y_1(1) = -16 + 5000 = 4984$ feet.   **(b)** $v_0 = 400$ ft/sec.

15. -340 ft/sec, where the negative sign indicates that the weight is thrown downward.

17. $y(t) = 2e^{2t}, \quad -\infty < t < \infty.$ **19.** $y(x) = \left((3/2)x^2 + (13/2)\right)^{1/3}.$

21. (a) *Hint:* Find dV/dt.   **(b)** $r(t) = \sqrt{1 - \frac{1}{2}t^2}.$ **(c)** 1.414 hours, about 1 hr, 25 min.

23. (a) $S(t) = 150e^{-t/50}, \quad t \geq 0, \ S$ is decreasing, and $\lim_{t \to \infty} S(t) = 0.$
     (b) $S(t) = 100 + 50e^{-t/100}, \quad t \geq 0, \ S(t)$ is decreasing, $\lim_{t \to \infty} S(t) = 100.$

25. $k = 0.$

27. *Hint:* Consider the product of the two slopes.   **(b)** $y(x) = C_1 x, \quad x \neq 0.\ x^2 + y^2 = C_2, \quad y \neq 0,$ where $C_2 = 2K_2 > 0.$

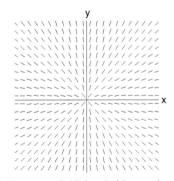

Direction field for $dy/dx = y/x$

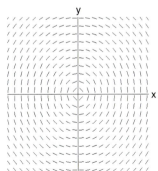

Direction field for $dy/dx = -x/y.$

29. (a) $GM = 96000$ mi^3/sec^2. (b) About $x6.93$ mi/sec ≈ 24900 mi/hr. (c) About 6.20 mi/sec ≈ 22300 mi/hr.

31. (a) Use the hint. (b) Routine calculation. (c) Routine calculation.

33. (a) For $bt + c > 0$, $b = -(a/A)\sqrt{g/2}$. The constants $c = -\sqrt{(h(0))}$, $b = (a/A)\sqrt{g/2}$ also work. (b) Letting $h(0) = h_0$, $t_e = (A/a)\sqrt{2h_0/g}$. If $A = 25\pi$ sq/ft (tank is 10 feet in diameter), $a = \pi/16$ sq ft (hole diameter is 6″), $h_0 = 20$ feet, and $g = 32.2$ ft/sec^2 then $t_e \approx 446$ sec ≈ 7.4 minutes.

Section 3: Linear Equations

Exercise Set 3AB (pgs. 487–488)

1. $M(x) = e^{2x}$.

3. $M(x) = x^2$.

5. $s(t) = 1 - e^{-t^2/2}$.

7. $y(x) = 0$.

9. $S(t) = -40 - \frac{120}{41}e^{-2t} + \frac{1760}{41}e^{t/20}$.

11. Find $M(t)$.

13. $u(t) = \frac{k}{k-2}e^{-2t} + \left(10 - \frac{k}{k-2}\right)e^{-kt}$, $t \geq 0$.

15. (a) $k = \frac{\ln 4}{15} \approx 0.0924$. (b) 22.5 min.

17. Routine calculation.

19. (a) $S(t) = 10100 - 100t - 10090e^{-t/100}$, $0 \leq t \leq 1$; $S(1) \approx 10.397$ pounds. (b) About 73.2 hours.

21. $S(10) \approx 14.35$ pounds.

23. (a) $S(50) = 155$ pounds. (b) About 29.4 min

Chapter 10 Review (pgs. 488–489)

1. $y(x) = \frac{x}{2} + \frac{C}{x}$. Note that $x \neq 0$ if $C \neq 0$, but that x can take any value if $C = 0$.

3. $y(t) = e^{t^2/2} \int e^{t-t^2/2} dt + Ce^{t^2/2}$.

5. $\frac{1}{4}\ln(y^4 + 1) = e^x + C$, $C \geq 0$ or $y = \left(Ke^{4e^x} - 1\right)^{1/4}$, $K \geq 1$.

7. $y(x) = \frac{x^3}{2} + Cx^3 e^{-2x}$, derived assuming $x \neq 0$, but the solution is valid for all x.

9. $x(t) = \frac{1}{5}t^3 - \frac{12}{5t^2}$, $t > 0$.

11. $x(t) = 1/(3t + C)$ and $x(t) \equiv 0$.

13. $x(t) = t + \tan(t + C)$.

15. (a) Increasing over the xy-plane. (b) $x < y$, above the line $y = x$. (c) Yes. (d) Yes.
    (e) $y(x) = \ln(e^x + C) = x + \ln(1 + Ce^{-x})$ gives the same conclusions but with more work.

17. (a) Isoclines are $y = (c - k)/a$.

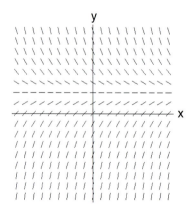

19. If $S_0 = 0$ or $S_0 = 50$, $S(t) = 200 - 200e^{-t/100}$ or $S(t) = 200 - 150e^{-t/100}$.

CHAPTER 11: SECOND-ORDER EQUATIONS

Exercise Set 1AB (pgs. 498–499)

1. $-e^{-2x}$ **3.** $27e^{3x}$

5. $-2\sin x$

7. Char. eq.: $r^2 + r - 6 = 0$; roots: $r_1 = 2$, $r_2 = -3$; general solution: $y(x) = c_1 e^{2x} + c_2 e^{-3x}$; particular solution:
$$y(x) = \frac{8}{5}e^{2x} + \frac{2}{5}e^{-3x}$$

9. Char. eq.: $r^2 + 2r + 1 = 0$; roots: $r_1 = r_2 = -1$; general solution: $y(x) = c_1 e^{-x} + c_2 x e^{-x}$; particular solution:
$y(x) = e^{-x} + 3x e^{-x}$

11. Char. eq.: $r^2 - r = 0$; roots: $r_1 = 0$, $r_2 = 1$; general solution: $y(x) = c_1 + c_2 e^{x}$; particular solution:
$y(x) = (e - e^x)/(e - 1)$

13. Char. eq.: $2r^2 - 3r + 1 = 0$; roots: $r_1 = 1$, $r_2 = 1/2$; general solution: $y(x) = c_1 e^{x} + c_2 e^{x/2}$; particular solution:
$y(x) \equiv 0$

15. $y'' + 2y' + y = 0$,
$y(x) = c_1 e^{-x} + c_2 x e^{-x}$

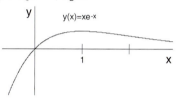

17. $y'' = 0$,
$y(x) = c_1 + c_2 x$

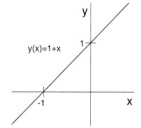

19. $y'' + 2y' + y = 0$,
$y(x) = c_1 e^{-x} + c_2 x e^{-x}$

21. $(D^2 + 2D + 1)y = 0$; $D^2 + 2D + 1 = (D + 1)(D + 1)$

23. $(2D^2 - 1)y = 0$; $2D^2 - 1 = 2(D - 1/\sqrt{2})(D + 1/\sqrt{2})$

25. $D^2 y = 0$; $D^2 = DD$

27. $z = c$ (c constant); $y(x) = c_1 + c_2 e^{3x}$

29. $z = -1 + ce^x$ (c constant); $y(x) = -1 + c_1 e^x + c_2 e^{-x}$

31. **(b)** $y(x) = c_1 x + c_2/x$ **(c)** *Hint:* Apply each operator to the nonzero constant function $y(x) = c$.
(d) $y(x) = c_1 \ln|x| + c_2$

33. **(b)** *Hint:* Write $\cosh \beta x$ and $\sinh \beta x$ in terms of exponential functions.

35. $y(x) = e^x - 2xe^x$ $y(x) = e^x - xe^x$ $y(x) = e^x$

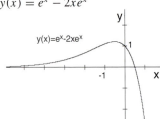

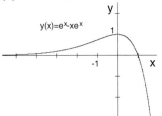

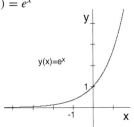

37. (a) *Hint:* Write Newton's equation $F = ma$ in two ways and equate the results.   **(b)** g
   (c) $y(t) = y_0 \cosh \sqrt{g/l}\, t + v_0 \sqrt{l/g} \sinh \sqrt{g/l}\, t$ **(d)** *Hint:* Set $v_0 = 0$ in the solution in part (c) and use an inverse hyperbolic function to solve $y(t) = l$ for t.

39. (b) *Hint:* Find the roots of the characteristic equation.   **(c)** *Hint:* Use l'Hôpital's rule.

Exercise Set 2A (pgs. 506–507)

1. $x = \pi/2$

3. $x = -\pi/4$

5. $c_1 = 1/2,\ c_2 = 1/2$

7. $c_1 = -i\pi/2,\ c_2 = i\pi/2$

9. (a) *Hint:* Use the definition of the complex exponential and the periodicity of the sine and cosine functions.
   (b) $p = 2\pi/\beta$

11. $y(x) = 1 + c_1 \cos x + c_2 \sin x$

13. $y(x) = c_1 \cos \sqrt{2}\, x + c_2 \sin \sqrt{2}\, x$

15. roots: $r_{1,2} = \pm i\sqrt{2}$; general solution: $y(x) = c_1 \cos \sqrt{2}\, x + c_2 \sin \sqrt{2}\, x$; $c_1 = 0,\ c_2 = 1/\sqrt{2}$

17. roots: $r_{1,2} = 1 \pm i$; general solution: $y(x) = e^x(c_1 \cos x + c_2 \sin x)$; $c_1 = c_2 = 0$

19. roots: $r_1 = 1/2,\ r_2 = -1$; general solution: $y(x) = c_1 e^{x/2} + c_2 e^{-x}$; $c_1 = 4/3,\ c_2 = -4/3$

21. roots: $r_{1,2} = -1/4 \pm i\sqrt{7}/4$; general solution $y(x) = e^{-x/4}\left(c_1 \cos \frac{\sqrt{7}}{4} x + c_2 \sin \frac{\sqrt{7}}{4} x \right)$; $c_1 = c_2 = 0$

23. $y'' + 4y = 0$

25. $y'' - 2y' + 5y = 0$

27. $y'' + \dfrac{1}{4}y = 0$

29. $y'' + 9y = 0$

33. *Hint:* Treat the cases of unequal roots and of equal roots separately.

35. (a) *Hint:* Use the identity $\sin(a + b) = \sin a \cos b + \cos a \sin b$
   (b) $A = 2, \theta = \pi/6$ **(c)** *Hint:* Use the cofunction identity $\sin(\pi/2 - a) = \cos a$.

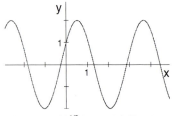

$y(x) = \cos 2x + (3)^{1/2} \sin 2x = 2\sin(2x + \pi/6)$

37. *Hint:* Break the integral into its real and imaginary parts.

39. $y(x) = c_1 e^{(1+i)x/\sqrt{2}} + c_2 e^{-(1+i)x/\sqrt{2}}$, c_1 and c_2 are real or complex constants.

41. *Hint:* routine computation

43. *Hint:* routine computation

45. (a) *Hint:* Use the chain rule to find $d^2 y_c/dx^2$ and dy_c/dx.   **(b)** *Hint:* Use the chain rule to find $d^k y_c/dx^k$ for $k = 1, \ldots, n$.   **(c)** *Hint:* Use the fact that $y(x + c)$ and $y'(x + c)$ are differentiable on $a - c < x < b - c$.

47. *Hint:* For $\beta \neq 0$, show that if the given functions are not linearly independent on some open interval I then $\tan \beta x$ is constant on some open subinterval of I. If $\beta = 0$ then the given functions are not linearly independent.

Exercise Set 2BC (pgs. 512–513)

1. $y(x) = c_1 e^{-x} + c_2 e^{x/2} \cos \dfrac{\sqrt{3}}{2} x + c_3 e^{x/2} \sin \dfrac{\sqrt{3}}{2} x$

3. $y(x) = c_1 + c_2 e^{\sqrt{2}x} + c_3 e^{-\sqrt{2}x}$

5. $y(x) = c_1 + c_2 e^{4x} + c_3 e^{-4x}$

7. $y(x) = c_1 e^x + c_2 e^{-x} + c_3 e^{2x} + c_4 e^{-2x}$

9. $y(x) = c_1 + c_2 x + c_3 e^x$

11. $y' - 5y = 0$

13. $y''' = 0$

15. $y''' - y'' = 0$

17. $y^{(9)} - y^{(6)} = 0$

19. $y'' + 16y = 0$

21. $y(x) = y^{(6)} + 48y^{(4)} + 768y'' + 4096y = 0$

23. $y^{(4)} - 4y''' + 8y'' - 8y' + 4y = 0$

25. $y^{(6)} = 0$

27. $y'' + 2y' + 2y = 0$

29. $y''' - y'' = 0$

31. $y''' - y'' + y' - y = 0$

33. $y''' + y'' + 4y' + 4y = 0$

35. $y''' + 9y = 0$

37. $y''' - 2y'' + 5y' = 0$

39. $c_1 = 0$, $c_2 = c_3 = 1$

41. $c_1 = 5/2$, $c_2 = -5/2$, $c_3 = -1/2$

43. $y(x) = c_1 e^x + c_2 e^{-x} + c_3 \cos x + c_4 \sin x$

45. $y(x) = c_1 + c_2 e^{2^{1/3}x} + e^{-2^{-2/3}x} \left(c_3 \cos(2^{-2/3}\sqrt{3}\, x) + c_4 \sin(2^{-2/3}\sqrt{3}\, x) \right)$

47. $y(x) = c_1 e^{2x} + c_2 e^{-2x} + c_3 e^x + c_4 e^{-x}$

49. Any constant-coefficient linear equation having $y(x) = \cos x + \sin 2x$ as a solution must have a characteristic equation having the four roots $\pm i$ and $\pm 2i$, so the order of the D.E. must be at least four. The given function is a solution of $y^{(4)} + 5y'' + 4y = 0$.

51. (a) $y(x) = c_1 + c_2 x + c_3 x^2 + c_4 x^3 - \dfrac{1}{24} P x^4$, $0 \le x \le L$ **(b)** maximum downward vertical deflection:
$\approx -0.005416122 P L^4$

(c) $P = 0.02$ and $L = 10$ feet.

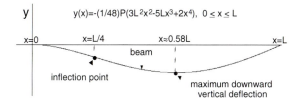

Exercise Set 3AB (pgs. 520–521)

1. gen. sol.: $y(x) = c_1 e^x + c_2 e^{-x} + \dfrac{1}{3} e^{2x}$, part. sol.: $y(x) = -\dfrac{1}{3} e^{-x} + \dfrac{1}{3} e^{2x}$

3. gen. sol.: $y(x) = c_1 e^{-x} + c_2 x e^{-x} + \dfrac{1}{4} e^x$, part. sol.: $y(x) = -\dfrac{1}{4} e^{-x} + \dfrac{1}{2} x e^{-x} + \dfrac{1}{4} e^x$

5. gen. sol.: $y(x) = c_1 e^x + c_2 e^{-x} + \dfrac{1}{2} x e^x - x$, part. sol.: $y(x) = \dfrac{3}{4} e^x - \dfrac{3}{4} e^{-x} + \dfrac{1}{2} x e^x - x$

7. gen. sol.: $y(x) = c_1 \cos x + c_2 \sin x + \dfrac{1}{2} x \sin x$, part. sol.: $y(x) = \sin x + \dfrac{1}{2} x \sin x$

9. gen. sol.: $y(x) = c_1 \cos x + c_2 \sin x + \dfrac{1}{4} x \cos x + \dfrac{1}{4} x^2 \sin x$, part. sol.: $y(x) = \dfrac{3}{4} \sin x + \dfrac{1}{4} x \cos x + \dfrac{1}{4} x^2 \sin x$

11. $y(x) = \dfrac{1}{3} x e^x + c_1 e^x + c_2 e^{-2x}$

13. $y(x) = -\dfrac{1}{2} i e^{ix} + c_1 e^{ix} + c_2 e^{-ix}$

15. $y'' - 3y' + 2y = 0$

17. $y'' = 0$

19. $y^{(4)} + 18y'' + 81y = 0$

21. $y^{(4)} - 4y''' + 8y'' - 8y' + 4y = 0$

23. $y_p(x) = A\cos x + B\sin x$

25. $y_p(x) = Axe^x$

27. $y_p(x) = Ax^2e^x + Bx^3e^x$

29. gen. sol.: $y(x) = c_1 e^{-2x} + c_2 x e^{-2x} - \dfrac{3}{4} + \dfrac{3}{4}x$,   part. sol.: $y(x) = \dfrac{3}{4}e^{-2x} + \dfrac{7}{4}xe^{-2x} - \dfrac{3}{4} + \dfrac{3}{4}x$

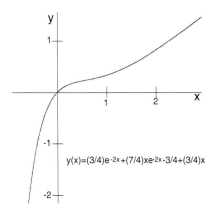

y(x)=(3/4)e-2x+(7/4)xe-2x-3/4+(3/4)x

31. gen. sol.: $y(x) = c_1 e^{-x}\cos x + c_2 e^{-x}\sin x + \dfrac{1}{5}e^x$, part. sol.: $y(x) = -\dfrac{1}{5}e^{-x}\cos x + \dfrac{3}{5}e^{-x}\sin x + \dfrac{1}{5}e^x$

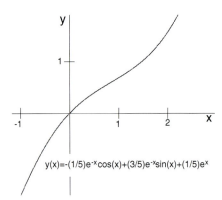

y(x)=-(1/5)e-xcos(x)+(3/5)e-xsin(x)+(1/5)ex

33. $y_p(x) = Ax^2 e^{2x} + Bxe^{2x}$

35. $y_p(x) = Axe^{2x} + Bx^2 e^{2x} + Cxe^{3x}$

37. $y_p(x) = Axe^{2x} + B\cos x + C\sin x$

39. $y_p(x) = Ax + Bx^2 + Cx^3 + Dxe^x$

41. $y_p(x) = Ax^3 + Bx^4 + Cx^5 + Dx^6$

45. (b) $\lim_{k \to 0^+} \dot{y}(t) = v_0 + gt$

Exercise Set 3CD (pgs. 528–530)

1. $y_1(x) = e^{2x}$, $u(x) = e^{-x} + c_1 x + c_2$

3. $y_1(x) = x$, $u(x) = \dfrac{1}{3}x^3 + c_1 x^2 + c_2$

5. $y_1(x) = e^x$, $y_2(x) = e^{-2x}$, $y_p(x) = \dfrac{1}{4}e^{2x}$, $y(x) = c_1 e^x + c_2 e^{-2x} + \dfrac{1}{4}e^{2x}$

7. $y_1 = \cos x$, $y_2 = \sin x$, $y_p(x) = \cos x \ln(\cos x) + x \sin x$, $y(x) = c_1 \cos x + c_2 \sin x + x \sin x + \cos x \ln(\cos x)$

9. $y_1 = 1$, $y_2 = x$, $y_p(x) = x^2 e^x - 4xe^x + 6e^x$, $y(x) = c_1 + c_2 x + x^2 e^x - 4xe^x + 6e^x$

11. $y_p(x) = -\dfrac{1}{2}x^2 e^x$

13. $y_p(x) = e^{-2x}(a + e^x)(-1 + \ln|a + e^x|)$

15. homo. sols.: $y_1(x) = e^{-x}$, $y_2(x) = e^{-2x}$; $G(x,t) = e^{-(x-t)} - e^{-2(x-t)}$; part. sol.:

$$y(x) = 4e^{-x} - 3e^{-2x} + \begin{cases} 0, & x < 1, \\ \frac{1}{2} - e^{1-x} + \frac{1}{2}e^{2(1-x)}, & 1 \le x. \end{cases}$$

17. homo. sols.: $y_1(x) = 1$ and $y_2(x) = \ln x$; $G(x,t) = t \ln x - t \ln t$; part. sol.: $y(x) = \ln x + x - 1$

19. *Hint:* Use the hint in the text.

21. (i) $G(x,t) = \dfrac{1}{r_1 - r_2}(e^{r_1(x-t)} - e^{r_2(x-t)})$, (ii) $G(x,t) = (x-t)e^{r(x-t)}$, (iii) $G(x,t) = \dfrac{1}{\beta}e^{\alpha(x-t)}\sin\beta(x-t)$

23. $y_p(x) = -\dfrac{1}{2}x_0^2 x - x_0 x^2 + \dfrac{1}{2}x^3$, $x/x_0 > 0$

25. $y_p(x) = (x_0 - 1)e^x + (e^{-x_0})e^{2x} - xe^x$

27. *Hint:* Use the hint in the text.

29. $y(x) = c_1 x + c_2 x^{-1}$, $x > 0$

31. $y(x) = c_1 x^{-1} + c_2 x^{-1}\ln x$, $x > 0$

33. (a) *Hint:* Expand the given determinant about the first column.
   (b) $y'' - (2\cot x)y' + (2\cot^2 + 1)y = 0$, $0 < x < \pi$ (c) $(x-1)y'' - xy' + y = 0$

Exercise Set 4A–C (pgs. 531–540)

1. (a) critically damped (b) $x(t) = c_1 e^{-t} + c_2 t e^{-t}$

3. (a) harmonic (b) $x(t) = c_1 \cos 3t + c_2 \sin 3t$

5. (a) underdamped (b) $x(t) = e^{-at/2}\left(c_1 \cos\dfrac{a\sqrt{3}}{2}t + c_2 \sin\dfrac{a\sqrt{3}}{2}t\right)$

7. $c_1 = 0$, $c_2 = 1/2$;

$\ddot{x} + 4x = 0$

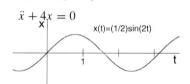

x(t)=(1/2)sin(2t)

9. $c_1 = -1$, $c_2 = 0$;

$\ddot{x} + 4\dot{x} + 4x = 0$

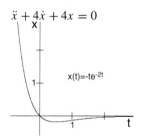

x(t)=-te⁻²ᵗ

11. $c_1 = -5/2$, $c_2 = 3/2$;

$\ddot{x} + 6\dot{x} + 8x = 0$

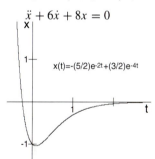

x(t)=-(5/2)e⁻²ᵗ+(3/2)e⁻⁴ᵗ

13. $x_p(t) = \dfrac{1}{10}\cos t + \dfrac{3}{10}\sin t$, $t \approx 5.01$

15. (a) $h = 6$ lbs/ft (b) $h \approx 8.9$ kg/m (c) 80 lbs (d) Hooke's Law is not valid over this range of compression.

17. (a) $k^2 < 8h$ (b) $m > 1/4$ (c) $k = \sqrt{3}$ (d) $0 < k < 2$

19. $A = \sqrt{13}$; frequency $= 1/(2\pi)$; $\phi = \arctan(3/2) \approx 0.9828$ radians

21. $A = \sqrt{5}$; frequency $= 1/2$; $\phi = \arctan(1/2) \approx 0.4636$ radians

23. $\pi/6$ radians   **25.** $\pi/4$ radians

29. (a) $k = 2mV_0/(eE)$, $h = mV_0^2/(eE)^2$ (b) *Hint:* By part (a), k/m and h/m depend only on the parameters V_0 and E.

31. (a) *Hint:* Show that the steady-state response is of the form $x_p(t) = At\cos\omega_0 t + Bt\sin\omega_0 t$ (b) ,(c) *Hint:* Show that the amplitude of the steady-state response is $\dfrac{|a_0|}{\sqrt{(h - \omega^2 m)^2 + \omega^2 k^2}}$

33. (a) *Hint:* Show that the real parts of the roots of the characteristic equation are negative. (b) *Hint:* See Example 5 in the text. (c) *Hint:* Write the general solution as the sum of the transient solution and the steady-state solution and then use the triangle inequality.

35. (a) *Hint:* Straightforward substitution **(b)** $x_p(t) = \dfrac{a_2}{4}t\sin 2t + \displaystyle\sum_{k=0,k\neq 2}^{n} \dfrac{a_k}{4-k^2}\cos 2t$ $(n \geq 2)$, $x_p(t) = \dfrac{a_0}{4}$ $(n = 0)$,

$x_p(t) = \dfrac{a_0}{4} + \dfrac{a_1}{3}\cos t$ $(n = 1)$

37. (a) $(\alpha - \beta)/\omega$ **(b)** $\alpha - \beta - \pi/2$

39. $m_0 = 9/8$ (no oscillation for $m = 9/8$)

41. $m = 1/26$ and $k = 1/13$

43. No such m exists.

Exercise Set 5 (pgs. 547–548)

5. $\dfrac{4s}{(s^2+4)^2}$

7. $\dfrac{2}{s^3} + \dfrac{2}{s^2} - \dfrac{1}{s}$

9. $\dfrac{2}{(s-3)^2} + \dfrac{1}{s-3}$

11. $\dfrac{1}{2}e^t - \dfrac{1}{2}e^{-t}$

13. $\dfrac{1}{3}e^{2t}\sin 3t$

15. $t\sin 2t$

17. $y(t) = -1 - t + 3e^t$

19. $y(t) = \dfrac{3}{13}\cos 2t + \dfrac{2}{13}\sin 2t - \dfrac{3}{13}e^{-3t}$

21. $y(t) = -\dfrac{4}{37}\cos 3t - \dfrac{2}{111}\sin 3t - 4 + \dfrac{152}{37}e^{t/2}$

23. (a), (b) *Hint:* Apply the definition of Laplace transform.

(c)

(d) $y(t) = 1 + \dfrac{1}{2}(t-a)^2 H(t-a)$.

y

1

H(t)-H(t-1)

1   t

25. (b) *Hint:* Use induction and the result of part (a).

Exercise Set 6 (pgs. 552–553)

1. $t - 1 + e^{-t}$

3. t

5. $-1 + t + e^{-t}$

7. $\dfrac{1}{2}(t-2)^2 H(t-2)$

9. $\dfrac{1}{9} - \dfrac{1}{9}e^{-3t} - \dfrac{1}{3}te^{-3t}$

11. $e^{-t}\sin t$

13. $H(t-1) + 1$

15. $y(t) = -1 + \dfrac{7}{10}e^t + \dfrac{13}{10}e^{-t} - \dfrac{1}{5}\sin 2t$

17. $y(t) = 5 - t + \dfrac{1}{2}t^2 - 3e^{-t} - te^{-t}$

19. $y(t) = -1 + 2e^{-t/2}\cosh\dfrac{\sqrt{5}}{2}t$

21. (a) *Hint:* Use induction on $n \geq 0$. **(b)** Use Theorem 6.1

23. (a),(b),(c) *Hint:* Use the hints in the text.

Exercise Set 7AB (pgs. 557–558)

1. $y(t) = \ln t$, $t > 0$

3. $y(t) = \ln(t+1)$, $t > -1$

5. $y(t) = \dfrac{1}{16}t^4 + \dfrac{3}{4}\ln t + \dfrac{15}{16}$, $t > 0$

7. $y(t) = e^t$, $-\infty < t < \infty$

9. $y = 2 - \sqrt{1 + 2t}$, $t < 1/2$ **11.** $y(t) = \ln(\cosh t)$, $-\infty < t < \infty$

13. Both $y_1(t) = -\frac{1}{2}t^2$ and $y(t) = 0$ are solutions. This does not contradict Theorem 7.1 because, in this case, the function f in the theorem is $f(t, y, \dot{y}) = -t^{-2}\dot{y}^2$, which is not continuous at $t = 0$.

15. **(a)** $y_1(t) = \sin t$ (for $\ddot{y} = -y$), $y_2(t) = \sinh t$ (for $\ddot{y} = y$)

   (b)   **(c)** $y_2(it) = iy_1(t)$

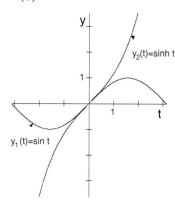

17. **(b)** *Hint:* Show that $\frac{1}{2}\dot{y}^2(t)$ is always positive.   **(c)** *Hint:* Show that oscillatory motion cannot occur if

$$-\cos y_0 + \frac{1}{2}z_0^2 = 1$$

Exercise Set 7C (pg. 561)

1. $y(t) = 1 + \cos t$

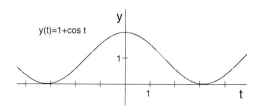

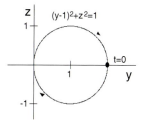

3. $y(t) = \frac{1}{2}t^2$

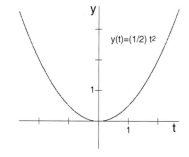

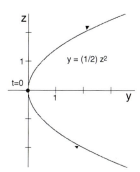

5. $y(t) = 2$

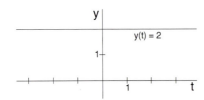

7. $(y - 1)^2 + z^2 = C, \ C \geq 0$

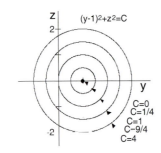

9. $y = \dfrac{1}{2}z^2 + C$

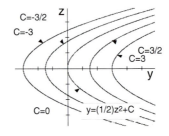

11. $y^2 + z^2 = C, \ C \geq 0$

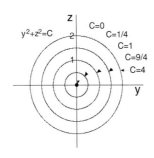

13. $z^2 - y^4 = C$

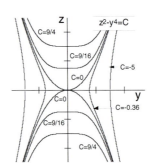

15. **(a)** Equilibrium solutions are the points on the y-axis.
   (b) Since all solutions with non-zero initial velocity move away from the origin, all equilibrium points are unstable.

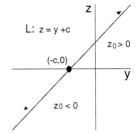

Exercise Set 8AB (pgs. 566–568)

1. **(a)** 0.00, 2.68, 4.35, 5.75, 6.99, 8.14, 9.20, 10.22, 11.17;
   the zeros do not change as long as $a \neq 0$ ($a = 0$ gives the zero solution)
   (b) If $a \neq 0$ then $y(t) \neq 0$ for all $t < 0$.

3. The zeros of $\sin t$ are $t \approx 3.14\,k$, k an integer. The zeros of the solution of $\ddot{y} = -\sin y$, $y(0) = 0$, $\dot{y}(0) = 1$ are $t \approx 3.34\,k$, k an integer.

5. **(a)** $(0.00, 0.00)$, $(0.25, 0.24)$, $(0.50, 0.46)$, $(0.75, 0.63)$, $(1.00, 0.75)$, $(1.25, 0.82)$, $(1.50, 0.86)$, $(1.75, 0.89)$, $(2.00, 0.93)$, $(2.25, 0.98)$, $(2.50, 1.05)$, $(2.75, 1.17)$, $(3.00, 1.34)$, $(3.25, 1.58)$, $(3.50, 1.93)$, $(3.75, 2.44)$, $(4.00, 3.15)$, $(4.25, 4.16)$, $(4.50, 5.54)$, $(4.75, 7.39)$, $(5.00, 9.74)$, $(5.25, 12.52)$, $(5.50, 15.51)$, $(5.75, 18.33)$, $(6.00, 20.50)$, $(6.25, 21.58)$

(b) $(0.00, 0.000)$, $(0.05, 0.049)$, $(0.10, 0.095)$, $(0.15, 0.139)$, $(0.20, 0.181)$, $(0.25, 0.220)$, $(0.30, 0.257)$, $(0.35, 0.292)$, $(0.40, 0.325)$, $(0.45, 0.356)$, $(0.50, 0.385)$, $(0.55, 0.412)$, $(0.60, 0.437)$, $(0.65, 0.461)$, $(0.70, 0.483)$, $(0.75, 0.503)$, $(0.80, 0.521)$, $(0.85, 0.538)$, $(0.90, 0.553)$, $(0.95, 0.567)$, $(1.00, 0.579)$

7. (a)   **(b)** $k \approx 0.6$

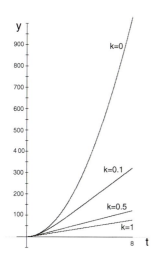

9. $\dot{x}(0) \approx 3.67$

11. (a) $\theta(1.24) = 0.1572$, $\theta(6.2) = 0.154875$, $\theta(11.16) = 0.152584$ **(b)** 0, 2.48, 4.96, 7.44, 9.92, 12.40, 14.88
(c) 0, 3.1, 6.18, 9.25, 12.3

13. (a) $k(t) = 0.2(1 - e^{-0.1t})$, $h = 5$, $m = 1$ **(b)** $k = 0$, $h(t) = 5(1 - e^{-0.2t})$, $m = 1$

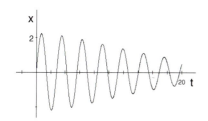

15. $\ddot{y} = -y - y^3 - \frac{1}{2}\dot{y} + \frac{3}{10}\cos t$, $y(0) = 0$, $\dot{y}(0) = 3$, $0 \leq t \leq 40$

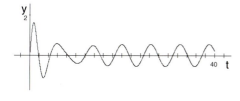
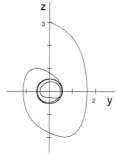

17.

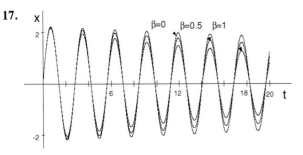

Chapter 11 Review (pgs. 568–570)

1. $y(x) = c_1 e^{-x} + c_2 x e^{-x} + \dfrac{1}{2} x^2 e^{-x} + \dfrac{3}{4} e^x$

3. $y(x) = c_1 e^x + c_2 e^{-x} + \dfrac{1}{2} \sin x$

5. $y(x) = c_1 e^{-x} \cos(\sqrt{2}\,t) + c_2 e^{-x} \sin(\sqrt{2}\,t) + \dfrac{1}{3}$

7. $y(x) = \dfrac{1}{24} x^4 + c_1' x^2 + c_2 x + c_3$

9. $y(x) = \cos 3x - \dfrac{1}{6} x \cos 3x + \dfrac{1}{18} \sin 3x$

11. $y(x) = -\cos x + c_2 \sin x$

13. $\{e^{r_1 x}, e^{\alpha x} \cos \beta x, e^{\alpha x} \sin \beta x\}, \{e^{r_1 x}, e^{r_2 x}, e^{r_3 x}\}, \{e^{r_1 x}, x e^{r_1 x}, e^{r_2 x}\}, \{e^{r_1 x}, x e^{r_1 x}, x^2 e^{r_1 x}\}$, where α, β are real ($\beta \neq 0$) and r_1, r_2, r_3 are different real numbers.

17. *Hint:* If k is a complex constant then $e^{kx} \neq 0$ for all real x and e^{kx} is constant on an open interval of the real line if, and only if $k = 0$.

19. $A = 1/R, \ \alpha = \pi/2$

21. (b) *Hint:* $e^{\pm x} > 0$ for all real x.   **(c)** $x = \dfrac{1}{2} \ln(3/2) + i(2n + 1)\pi/2$, n an integer.

25. (a) *Hint:* Expand $(D - r)\big(D - (r + h)\big)y = 0$.
   (b) *Hint:* Find the roots of the characteristic equation.
   (c) *Hint:* Use the definition of $d(e^{rx})/dr$ as the limit of a quotient.

27. $z = y_1 + 2y_2 - 4y_3$ **29.** $z = 0$

31. yes   **33.** no

35. yes   **37.** no

CHAPTER 12: INTRODUCTION TO SYSTEMS

Section 1: Vector Fields

Exercise Set 1ABC (pgs. 578–580)

1. $\big(x(t), y(t)\big) = (-1 + 2e^t, 2e^t)$.   **3.** $\big(x(t), y(t), z(t)\big) = (0, e^{t/2}, -e^{t/3})$.

5. (a) $F(t, \mathbf{x}) = (t, y)$.   **(b)** $|d\mathbf{x}/dy| = |t|$.

7. The vector field $\mathbf{F}(x, y) = (x + 1, y)$ **9.** The vector field $\mathbf{F}(x, y, z) = (x, y/2, z/3)$.

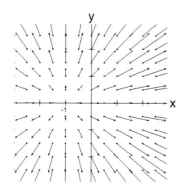

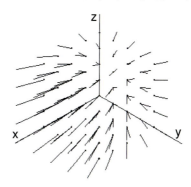

11.

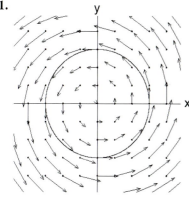

13. Use the hint.

15. $dy/dt = z, dz/dt = -y^2 - z^2 + e^t$.

17. $dy/dt = z, dz/dt = x, dw/dt = w^2 - yz - t$.

19. $dx/dt = (t + y)/2, dy/dt = (t - y)/2$.

21. $dx/dt = x - 3y + t, dy/dt = -3x + y - t$.

23.

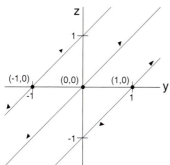

25. (a) Simple susitution.   **(b)** $x = c_1 e^{3t} + c_2 e^{-t}, y = 2x - 2y = 2c_1 e^{3t} - 2c_2 e^{-t}$.

27. $y = \ln(e^x + C)$.

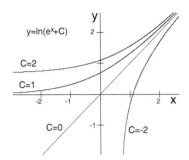

29. $x = \frac{1}{2}y^2 + C$.

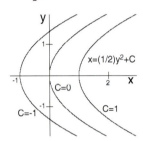

31. (a) $x(t) = (z_0/k)(1 - e^{-kt})$. $y(t) = (kw_o + g)/k^2(1 - e^{-kt}) - g/kt$.
(b) *Hint:* Solve $\dot{y}(t) = 0$. **(c)** Routine calculation. **(d)** Routine calculation.
33. (a) *Hint:* Use the differential equations. **(b)** *Hint:* $p \neq 0$. **(c)** *Hint:* Use part (b). **(d)** $t_0 = \ln 2/(p + q)$.

Exercise Set 1D (pgs. 584–585)

1. Domain is $t < 1/a$.   **3.** No.
5. (a) Use Theorem 1.6.   **(b)** $H(x, y) = -(1/2)(x^2 + y^2)$ **(c)** Show H is constant on flow lines.
7. Use Theorem 1.6.   **9. (a)** Routine.   **(b)** Fairly complicated.
11. (a) Routine.   **(b)** $(1/t)\int_0^t \operatorname{div} \mathbf{F}(T_u(\mathbf{x}))\,du$.

Section 2: Linear Systems

Exercise Set 2AB (pgs. 592–594)

1. Nonlinear.   **3.** Linear.
5. $x(t) = \cosh 2t + 3 \sinh 2t$, $y(t) = -2 \sinh 2t$.
7. *Hint:* Try $x = at + b$, $y = ct + d$.
9. $y(t) = (1/3)e^{-t} - 2e^t + (2/3)e^{2t}$.
11. $A(t) = \begin{pmatrix} 1 & 2t/3 \\ 0 & t/3 \end{pmatrix}$, $\mathbf{b}(t) = \begin{pmatrix} t^2/3 \\ -t^2/3 \end{pmatrix}$.
13. (a) Routine calculation. **(b)** Matrix must be square. **(c)** Suppose $\mathbf{b}(t) \neq \mathbf{0}$.
15. Nonlinear.   **17.** Nonlinear.
19. $(x(t), y(t)) = (-c_1 \sin t + c_2 \cos t - 1 - t, c_1 \cos t + c_2 \sin t + 1 - t)$.
21. $dx/dt = -x + 2y$, $dy/dt = x - y$.

23. $dx/dt = \frac{1}{2}(\sin t + \cos t), dy/dt = (1/2)(\sin t - \cos t)$.

25. $x(t) = 1, y(t) = 0$.

27. $x(t) = \frac{1}{2}e^t - \frac{1}{4}e^{\frac{\sqrt{2}}{2}t}\left((1+\sqrt{2})\cos\frac{\sqrt{2}}{2}t - \sin\frac{\sqrt{2}}{2}t\right) - \frac{1}{4}e^{-\frac{\sqrt{2}}{2}t}\left((1-\sqrt{2})\cos\frac{\sqrt{2}}{2}t + \sin\frac{\sqrt{2}}{2}t\right)$,

$y(t) = -\frac{1}{2}e^t + \frac{1}{4}e^{\frac{\sqrt{2}}{2}t}\left(\cos\frac{\sqrt{2}}{2}t + (1+\sqrt{2})\sin\frac{\sqrt{2}}{2}t\right) + \frac{1}{4}e^{-\frac{\sqrt{2}}{2}t}\left(\cos\frac{\sqrt{2}}{2}t + (-1+\sqrt{2})\sin\frac{\sqrt{2}}{2}t\right)$.

29. $\dot{x} = u, \dot{y} = v, \dot{u} = v, \dot{v} = -u$.

31. $\dot{x} = u, \dot{y} = v, \dot{u} = y + e^t, \dot{v} = -x$.

33. $x(t) = 0, y(t) = c$.

35. $x(t) = ce^{t/2}, y(t) = ce^{t/2}$.

37. (a) $x(t) = c_1 e^{\sqrt{2}t} + c_2 e^{-\sqrt{2}t}, y(t) = c_3 \cos\sqrt{2}t + c_4 \sin\sqrt{2}t$.

$z(t) = \frac{\sqrt{2}}{2}\left(c_1 e^{\sqrt{2}t} - c_2 e^{-\sqrt{2}t} - c_3 \sin\sqrt{2}t + c_4 \cos\sqrt{2}t\right)$.

$w(t) = \frac{\sqrt{2}}{2}\left(c_1 e^{\sqrt{2}t} - c_2 e^{-\sqrt{2}t} + c_3 \sin\sqrt{2}t - c_4 \cos\sqrt{2}t\right)$.

(b) $\ddot{z} = 2w, \ddot{w} = 2z$.

Section 3: Applications

Exercise Set 3 (pgs. 601–606)

1. (a) $dy/dt = (4/100)z - (4/100)y, dz/dt = (1/100)y - (4/100)z$.
(b) $y(t) = 25e^{-t/50} - 15e^{-3t/50}, z(t) = 12.5e^{-t/50} + 7.5e^{-3t/50}$.
(c) $y_{max} \approx y(15) \approx 12$. From $t = 15$ both derease to zero.

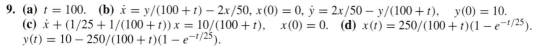

3. (a) Follows from (b). (b) $\begin{pmatrix} 1 & -1 \\ 1 & 1 \end{pmatrix}\begin{pmatrix} c_1 \\ c_3 \end{pmatrix} = \begin{pmatrix} x_0 \\ y_0 \end{pmatrix}$,

$\begin{pmatrix} 1 & -\sqrt{3} \\ 1 & \sqrt{3} \end{pmatrix}\begin{pmatrix} c_2 \\ c_4 \end{pmatrix} = \begin{pmatrix} u_0 \\ v_0 \end{pmatrix}$.

5. (a) *Hint:* $y(t) = 0$ is the unique solution of one equation.
(b) The planet will collide with the star.

7. (a) For $0 \le t \le 50$ $x(t) + y(t) = t, \quad 0 \le t \le 50$.
(b) $\dot{x} = y/50 - x/(50+t), \quad x(0) = 0, \dot{y} = 1 + x/(50+t) - y/50, \quad y(0) = 0$.
(c) $\dot{x} = y/50 - x/50, \dot{y} = 1 + x/100 - y/50, t \ge 50$.

9. (a) $t = 100$. (b) $\dot{x} = y/(100+t) - 2x/50, x(0) = 0, \dot{y} = 2x/50 - y/(100+t), \quad y(0) = 10$.
(c) $\dot{x} + (1/25 + 1/(100+t))x = 10/(100+t), \quad x(0) = 0$. (d) $x(t) = 250/(100+t)(1 - e^{-t/25})$.
$y(t) = 10 - 250/(100+t)(1 - e^{-t/25})$.

11. $\mu_1 = \sqrt{(5-\sqrt{5})/2} \approx 1.1756$ and $\mu_2 = \sqrt{(5+\sqrt{5})/2} \approx 1.9021$.

13. $\mu_1 = \sqrt{(8-\sqrt{2})/2} \approx 1.8146$ and $\mu_2 = \sqrt{(8+\sqrt{2})/2} \approx 2.1696$.

15. (a) $U(x, y) = (1/2)(k_1 + k_2)x^2 - k_2xy + (1/2)(k_2 + k_3)y^2$.
(b) *Hint:* Multiply $m_1\ddot{x} = -U_x$ by $\dot{x}$, and $m_2\dot{y} = -U_y$ by $\dot{y}$ then add.

17. *Hint:* Equate two expressions for the acceleraion at the surface.

19. $v_e \approx 1.1086 \times 10^4$ m/s. **21.** *Hint:* Integrate $\ddot{x} = -g$.

23. (a) Use the hint. (b) *Hint:* $\frac{G(m_1+m_2)}{x_0^2} = \frac{v^2}{x_0}$. (c) $v_e = \sqrt{2}v_1$. (d) About 27.28 days.

25. (a) $\omega^2 = k/a^3$. (b) *Hint:* $a\omega$ is the orbital speed. (c) *Hint:* Let the orbit be $r = f(\theta)$ in polar coordinates.

27. (a) Use the hint. (b) Use what worked in part (a).

29. (a) Routine arithmetic. (b) *Hint:* Differentiate $\dot{x}y - \dot{y}x$. (c) *Hint:* Use the cahin rule. (d) *Hint:* For the last part consider $A(t + \tau) - A(t)$. (e) Apply Green's Theorem to a region swept out by a radius in time t.

31. Follow the steps. **33.** Use the chain rule.

Section 4: Numerical Methods

Exercise Set 4AB (pgs. 612–615)

1. **(a)** & **(b)** Routine calculations.
   (c)

t	IE x	formula x	IE y	formula y
0.0	1.00000	1.00000	2.00000	2.00000
0.1	1.20500	1.20533	2.11000	2.11017
0.2	1.42202	1.42273	2.24105	2.24146
0.3	1.65324	1.65437	2.39446	2.39519
0.4	1.90095	1.90257	2.57175	2.57289
0.5	2.16763	2.16981	2.77471	2.77634

3. **(a)** Routine calculation. **(b)** & **(c)** The table shows the comparison.

t	IE x	IE y	t	formula y	formula x
0.0	2.00000	1.00000	0.0	1.00000	2.00000
0.2	2.70993	1.46714	0.2	1.46715	2.70996
0.4	3.68521	2.10158	0.4	2.10163	3.68528
0.6	5.00851	2.96431	0.6	2.96442	5.00866
0.8	6.78615	4.13513	0.8	4.13532	6.78640
1.0	9.15444	5.71797	1.0	5.71828	9.15484

5. The improved Euler method was used with step size $h = 0.001$. Of the 5,000 values generated, the left-hand table below records every 500th value.

7. The improved Euler method was used with step size $h = 0.001$ to compute a numerical approximation of the solution. Of the 1500 values generated, the right-hand table below records every 150th value.

t	$x(t)$	t	$x(t)$
0.0	0.500000	0.00	0.000000
0.5	0.591083	0.15	0.150008
1.0	0.662880	0.30	0.300243
1.5	0.720503	0.45	0.451852
2.0	0.767312	0.60	0.607861
2.5	0.805669	0.75	0.774373
3.0	0.837303	0.90	0.962467
3.5	0.863521	1.05	1.19204
4.0	0.885335	1.20	1.50098
4.5	0.903540	1.35	1.97104
5.0	0.918770	1.50	2.81982

   Table for Exercise 5   Table for Exercise 7

9.

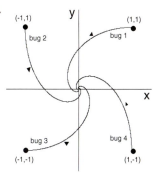

11. Let $f(x, y) = x^2 + \frac{1}{2}y^2$.

(a) The level curves of f are the trajectories of the system

$$\dot{x} = -f_y(x, y) = -y,$$
$$\dot{y} = f_x(x, y) = 2x,$$

and are shown in the top sketch on the right. The improved Euler method was used with step size $h = 0.001$.

(b) Curves perpendicular to the level sets of f are the trajectories of the system

$$\dot{x} = f_x(x, y) = 2x,$$
$$\dot{y} = f_y(x, y) = y,$$

and are shown in the bottom sketch on the right.

(c) The system shown in part (b) is uncoupled. By inspection, the general solutions are seen to be $x(t) = c_1 e^{2t}$ and $y(t) = c_2 e^t$. If initial conditions are given by $x(t_0) = x_0$, $y(t_0) = y_0$ then we are led to the specific solution

$$x(t) = x_0 e^{2(t-t_0)} \qquad \text{and} \qquad y(t) = y_0 e^{t-t_0}.$$

If $x_0 = y_0 = 0$ then the trajectory is the single point $(0, 0)$, which corresponds to the identically zero solution. If $x_0 = 0$ and $y_0 \neq 0$ then the trajectory consists of all positive mulriples of y_0, which is the positive y axis if $y_0 > 0$ and the negative y-axis if $y_0 < 0$. In a similar fashion, If $x_0 \neq 0$ and $y_0 = 0$ then the solution trajectory is the positive x-axis if $x_0 > 0$ and the negative x-axis if $x_0 < 0$.

If $x_0 \neq 0$ and $y_0 \neq 0$ then note first that the corresponding trajectory stays in the same quadrant for all t. Second, the solution equations can be written as $(1/x_0)x = e^{2(t-t_0)}$ and $(1/y_0)y = e^{t-t_0}$, so that $(1/y_0)^2 y^2 = e^{2(t-t_0)} = (1/x_0)x$, or $x = Cy^2$, where $C = x_0/y_0^2$.

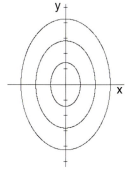

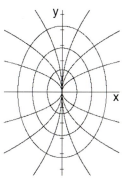

13. (a) The process stops when tank 1 first becomes full at time $t = 50$ minutes.
$\dot{x} = 3y/(100+t) - 7x/(50+t), x(0) = x_0,$
$\dot{y} = 2 + 3x/(50+t) - 4y/(100+t), y(0) = y_0, 0 \le t \le 50.$
(b) The improved Euler method with step size $h = 0.01$ was used to plot the solutions.

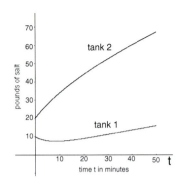

15.

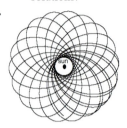 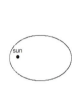 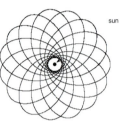

    (a) $p = 1.9$ **(b)** $p = 2$ **(c)** $p = 2.1$

17. (a) $\dot{x} = y,\ \dot{y} = \alpha(1 - x^2)y - x.$
(b)

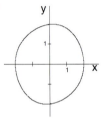 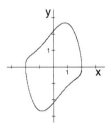 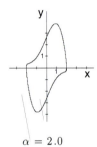

    $\alpha = 0.1$ $\alpha = 1.0$ $\alpha = 2.0$

(c)

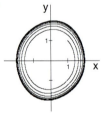 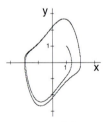 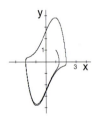

    $\alpha = 0.1$ $\alpha = 1.0$ $\alpha = 2.0$

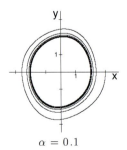 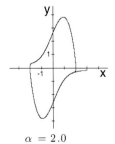

    $\alpha = 0.1$ $\alpha = 1.0$ $\alpha = 2.0$

19. The improved Euler method with step size $h = 0.005$ was used to plot five phase curves of the hard spring oscillator equation $\ddot{y} = -\gamma y^3 + \delta y$ for various pairs (γ, δ). There are three equilibrium solutions; namely, $y = 0$, $y = \sqrt{\gamma/\delta}$ and $y = -\sqrt{\gamma/\delta}$, whose phase curves are the three points $(0, 0)$, $(\sqrt{\gamma/\delta}, 0)$ and $(-\sqrt{\gamma/\delta}, 0)$. The results are shown below.

(a) $\gamma = \delta = 1$ **(b)** $\gamma = 1, \delta = 2$ **(c)** $\gamma = 2, \delta = 1$

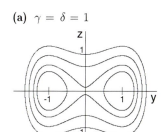   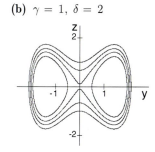   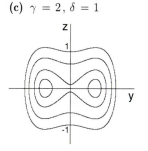

In Exercises 21, 23, and 25, the improved Euler method with step size $h = 0.001$ was used to plot the solution $y = y(t)$ of the oscillator $\ddot{y} + k(t)\dot{y} + h(t)y = \sin t$, $y(0) = 0$, $\dot{y}(0) = 1$, where the damping factor $k(t)$ and the spring stiffness $h(t)$ are time dependent. The results are shown below.

21.   **23.**

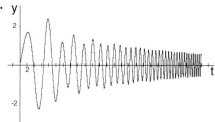   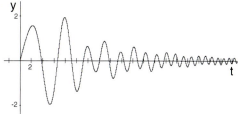

$\ddot{y} + e^{t/2}y = \sin t$, $y(0) = 0$, $\dot{y}(0) = 1$ $\ddot{y} + (1/10)\dot{y} + e^{t/2}y = \sin t$, $y(0) = 0$, $\dot{y}(0) = 0$

25.

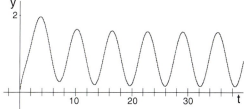

$\ddot{y} + \dot{y} + \left(1/(1 + t^2)\right)y = \sin t$, $y(0) = 0$, $\dot{y}(0) = 1$.

27. Consider the Lotka-Volterra system $\dot{H} = (3 - 2P)H$, $\dot{P} = (\frac{1}{3}H - 1)P$.

(a) Using the improved Euler method with step size $h = 0.001$, various values of t_1 for the interval $0 \le t \le t_1$ were used to sketch a phase curve of the solution correponding to the initial conditions $H(0) = 3$, $P(0) = \frac{3}{2}$. It was found that the corresponding phase curve almost closed with $t_1 = 3.6$ and was closed when $t_1 = 3.7$. The "PLOT H, P" command was then changed to "PRINT T, H, P" on the interval $0 \le t \le 3.7$. The numerical evidence suggested that the orbit time was approximately 3.666 time units (truncated to three places).

(b) The result of part (a) shows that the graphs of $H(t)$ and $P(t)$ are periodic of period ≈ 3.666 time units. Using the same program as was used in part (a), the "PRINT T, H, P" command was suppressed and the commands "PLOT T, H" and "PLOT T, P" were sumultaneously activated to plot one period of the graphs of H and P on the same set of axes. The result is in the figure on the left.

(c) Here, the "PLOT H, P" command was used to plot the trajectory in the HP-plane of the solution found in part (a). The result is shown below in the figure on the right. The arrows indicate the direction of the trajectory.

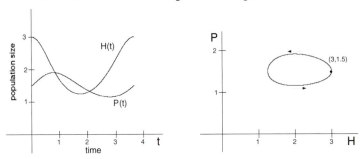

29. Set $a = 3$, $b = d = 2$, $c = 1$, $L = 4$ and $M = 3$ in the given refinement of the Lotka-Volterra equations. The conditions $H(0) = 3$, $P(0) = \frac{1}{2}$ then determine the unique solution of the initial-value problem

$$\dot{H} = (3 - 2P)H(4 - H), \quad H(0) = 3, P(0) = 1/2;$$
$$\dot{P} = (H - 2)P(3 - P).$$

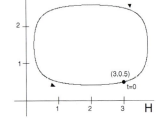

The plot of the trajectory of the solution is shown on the right. The arrows indicate the direction of the trajectory.

31. Set $a = 3$, $b = d = 2$, $c = 1$, $L = 4$ and $M = 3$ in the given refinement of the Lotka-Volterra equations. The conditions $H(0) = 3$, $P(0) = 2$ then determine the unique solution of the initial-value problem

$$\dot{H} = (3 - 2P)H(4 - H), \quad H(0) = 3, P(0) = 2;$$
$$\dot{P} = (H - 2)P(3 - P).$$

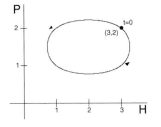

The plot of the trajectory of the solution is shown on the right. The arrows indicate the direction of the trajectory.

33. First observe that $H(0) = 3$ for each of the initial-value problems suggested by Exercises 29, 30, 31, 32 in this section.. Using the improved Euler method with step size $h = 0.001$, various values of t_1 for the interval $0 \leq t \leq t_1$ were used to sketch a phase curve of the four solutions coreponding to the initial conditions $H(0) = 3$, $P(0) = P_0$, where $P_0 = \frac{1}{2}, \frac{3}{2}, 2, \frac{5}{2}$. Visually comparing the curves for various values of t_1 allowed us to hone in on an estimate t_0 of the actual orbit time t_* such that $t_0 - 0.01 < t_* < t_0$. Once, t_0 was determined for a given P_0, the "PLOT H, P" command was suppressed and the "PRINT T, H, P" command was used on the time interval $0 \leq t \leq t_0$. Toward the end of this printout the columns for $H(t)$ and $P(t)$ contained values close to $H(t) = 3$ and $P(t) = P_0$. The corresponding values of t were therefore close to t_*. Using this method, the values of t_* for the four orbits were estimated to be

$$t_* \approx \begin{cases} 1.911, & P_0 = 1/2; \\ 1.562, & P_0 = 3/2; \\ 1.630, & P_0 = 2; \\ 1.911, & P_0 = 5/2. \end{cases}$$

Chapter 12 Review (pgs. 615–616)

1. The system is uncoupled. $\big(x(t), y(t)\big) = \big(\tan(t + \pi/4), 2e^t\big)$, $-3\pi/4 < t < \pi/4$.

3. $\mathbf{x}(t) = \big((3/2)e^t + (1/2)e^{-t}, (3/2)e^t - (1/2)e^{-t}\big)$, $-\infty < t < \infty$.

5. $(x(t), y(t)) = (e^{-5t}, 2e^{-5t})$. **7.** $(x(t), y(t)) = (e^{2t}, 2e^{2t} - t)$.

9. With $\omega_1 = \sqrt{(-5 + \sqrt{17})/2}$ and $\omega_2 = \sqrt{(-5 - \sqrt{17})/2}$,

$$x(t) = \frac{3}{2}\left(1 + \frac{7}{\sqrt{17}}\right)\cos\omega_1 t + \frac{3}{2}\left(1 - \frac{7}{\sqrt{17}}\right)\cos\omega_2 t,$$

$$y(t) = \frac{3}{2}\left(1 + \frac{15}{\sqrt{17}}\right)\cos\omega_1 t + \frac{3}{2}\left(1 - \frac{15}{\sqrt{17}}\right)\cos\omega_2 t.$$

11. $(x(t), y(t), z(t)) = (-\sin t, \cos t, -e^t)$.

13. If $\mathbf{x}_* = \mathbf{x}_*(t)$ solves $\dot{\mathbf{x}} = F(\mathbf{x})$, then $\mathbf{x}(t) = \mathbf{x}_*(-t)$, solves $\dot{\mathbf{x}} = -F(\mathbf{x})$.

15. *Hint:* From Chapter 5, Section 1, $\nabla f(x, y)$ is orthogonal to a level curve through (x, y).

17. *Hint:* Show that $\int_{\mathbf{x}_0}^{\mathbf{x}_1} \nabla f \cdot d\mathbf{x} = \int_{t_0}^{t_1} |\dot{\mathbf{x}}(t)|^2\, dt$, and use the Fundamental Theorem of Calculus for line intrgrals.

19. *Hint:* Solve the specific Hamiltonian system explicitly.

CHAPTER 13: MATRIX METHODS

Section 1: Eigenvalues & Eigenvectors

Exercise Set 1AB (pgs. 624–625)

1. $2, (1, 2); 3, (3, 5)$.

3. $-\frac{5}{2} + i\frac{\sqrt{15}}{2}, (-6, 3 - i\sqrt{15}); -\frac{5}{2} - i\frac{\sqrt{15}}{2}, (-6, 3 + i\sqrt{15})$

5. $1, (1, 0, 0); 2, (1, 0, 1)$.

7. $-e^t \begin{pmatrix} 1 \\ 2 \end{pmatrix} + 2e^{-t} \begin{pmatrix} 1 \\ 1 \end{pmatrix}$.

9. $\begin{pmatrix} e^{5t} \\ e^{5t} \end{pmatrix}$.

11. (a) $\mathbf{x}_h(t) = c_1 e^{-2t} \begin{pmatrix} 1 \\ -1 \\ -1 \end{pmatrix} + c_2 e^t \begin{pmatrix} -1 \\ 4 \\ 1 \end{pmatrix} + c_3 e^{3t} \begin{pmatrix} 1 \\ 2 \\ 1 \end{pmatrix}$. **(b)** $\mathbf{x}_p = \begin{pmatrix} -5/2 \\ 9/2 \\ 3/2 \end{pmatrix}$.

(c) $\mathbf{x}(t) = -\frac{1}{2} e^{-2t} \begin{pmatrix} 1 \\ -1 \\ -1 \end{pmatrix} - 2e^t \begin{pmatrix} -1 \\ 4 \\ 1 \end{pmatrix} + 2e^{3t} \begin{pmatrix} 1 \\ 2 \\ 1 \end{pmatrix} + \begin{pmatrix} -3/2 \\ 5/2 \\ 1/2 \end{pmatrix}$.

13. (a) $x(t) = c_1 t e^{-t} + c_2 e^{-t}, y(t) = c_1 e^{-t}$. **(b)** -1, nonzero multiples of $(1, 0)$.

15. $\det(A - \lambda I) = \lambda^2 + a\lambda + b$, which has the same roots as the characteristic equation.

17. (a) *Hint:* $U\mathbf{e}_k = \mathbf{u}_k$ **(b)** $A = \begin{pmatrix} -3 & 4 \\ -6 & 7 \end{pmatrix}$. **(c)** Use the definition of eigenvector.

(d) $\dot{x} = -3x + 4y, \dot{y} = -6x + 7y$.

Section 2: Matrix Exponentials

Exercise Set 2A–C (pgs. 630–631)

1. $\begin{pmatrix} e^{-t} & 0 \\ 0 & e^t \end{pmatrix}$.

3. $\begin{pmatrix} 0 & e^t \\ 0 & e^t \end{pmatrix}$.

5. $\begin{pmatrix} e^t & 0 & te^t \\ 0 & e^t & 0 \\ 0 & 0 & e^t \end{pmatrix}$.

7. (a) $\begin{pmatrix} e^t & te^t \\ 0 & e^t \end{pmatrix} \begin{pmatrix} e^{-t} & -te^{-t} \\ 0 & e^{-t} \end{pmatrix} = I$. **(b)** $\begin{pmatrix} e^t & te^t \\ 0 & e^t \end{pmatrix} \begin{pmatrix} e^s & se^s \\ 0 & e^s \end{pmatrix} = \begin{pmatrix} e^{t+s} & (t+s)e^{t+s} \\ 0 & e^{t+s} \end{pmatrix}$.

(c) *Hint:* both equal $\begin{pmatrix} e^t & te^t + e^t \\ 0 & e^t \end{pmatrix}$. **(d)** $\begin{pmatrix} -e^t + 2te^t \\ 2e^t \end{pmatrix}$.

9. $e^{2t} I$.

11. $e^{tA} = \begin{pmatrix} -e^t + 2e^{-t} & e^t - e^{-t} \\ -2e^t + 2e^{-t} & 2e^t - e^{-t} \end{pmatrix}$.

13. $e^{tA} = \begin{pmatrix} e^{2t} \cos t & -e^{2t} \sin t \\ e^{2t} \sin t & e^{2t} \cos t \end{pmatrix}$.

15. $e^{tA} = \frac{1}{15} \begin{pmatrix} 15e^{4t} \cosh \sqrt{15}t & 5\sqrt{15}e^{4t} \sinh \sqrt{15}t \\ 3\sqrt{15}e^{4t} \sinh \sqrt{15}t & 15e^{4t} \cosh \sqrt{15}t \end{pmatrix}$.

17. $e^{tA} = \begin{pmatrix} 3e^{2t} - 2e^{3t} & -e^{2t} + e^{3t} \\ 6e^{2t} - 6e^{3t} & -2e^{2t} + 3e^{3t} \end{pmatrix}$.

19. $e^{tA} = \frac{1}{2} \begin{pmatrix} 2e^{-t} & 0 & 0 \\ 0 & e^t + e^{2t} & e^t - e^{2t} \\ 0 & e^t - e^{2t} & e^t + e^{2t} \end{pmatrix}$.

21. (a) *Hint:* In the series for e^{itA}, the even terms are real and the odd terms imaginary.

(b), (c) *Hint:* $\cos tA = \frac{1}{2}(e^{itA} + e^{-itA})$ and $\sin tA = -\frac{i}{2}(e^{itA} - e^{-itA})$.

23. *Hint:* Use part (b) of Exercise 21. To show that this is the most general solution, show that $\mathbf{c}_1$ and $\mathbf{c}_2$ can be chosen to match any given values for $\mathbf{x}(0)$ and $\dot{\mathbf{x}}(0)$.

25. One possible example: $A = \begin{pmatrix} 1 & 0 \\ 1 & 0 \end{pmatrix}$, $B = \begin{pmatrix} 0 & 1 \\ 0 & 1 \end{pmatrix}$.

$e^A e^B = \begin{pmatrix} e & e^2 - e \\ e - 1 & e^2 - e + 1 \end{pmatrix}$, $e^{A+B} = \frac{1}{2} \begin{pmatrix} e^2 + 1 & e^2 - 1 \\ e^2 - 1 & e^2 + 1 \end{pmatrix}$.

Exercise Set 2D (pgs. 636–637)

1. $e^{tA} = \begin{pmatrix} -5e^{2t} + 6e^{3t} & 3e^{2t} - 3e^{3t} \\ -10e^{2t} + 10e^{3t} & 6e^{2t} - 5e^{3t} \end{pmatrix}$; $\mathbf{x}(t) = \begin{pmatrix} -19e^{2t} + 21e^{3t} \\ -38e^{2t} + 35e^{3t} \end{pmatrix}$.

3. $e^{tA} = \begin{pmatrix} e^t & te^t & -e^t + e^{2t} + te^t \\ 0 & e^t & e^t - e^{2t} - 2te^t \\ 0 & 0 & e^{2t} \end{pmatrix}$; $\mathbf{x}(t) = \begin{pmatrix} te^t \\ e^t \\ 0 \end{pmatrix}$.

5. $e^{tA} = \begin{pmatrix} \frac{1}{2}(e^{2t} + 1) & \frac{1}{2}(e^{2t} - 1) & \frac{1}{2}(e^{2t} - 1) & 0 \\ e^t - \frac{1}{2}(e^{2t} + 1) & e^t - \frac{1}{2}(e^{2t} + 1) & -\frac{1}{2}(e^{2t} - 1) & 0 \\ -e^t + e^{2t} & -e^t + e^{2t} & e^{2t} & 0 \\ e^t + te^t - \frac{1}{2}(e^{2t} + 1) & te^t - \frac{1}{2}(e^{2t} - 1) & -\frac{1}{2}(e^{2t} - 1) & e^t + te^t - \frac{1}{2}(e^{2t} - 1) \end{pmatrix}$,

which can also be expressed as $e^{tA} = e^t \begin{pmatrix} \cosh t & \sinh t & \sinh t & 0 \\ 1 - \cosh t & 1 - \sinh t & -\sinh t & 0 \\ -1 + e^t & -1 + e^t & e^t & 0 \\ 1 + t - \cosh t & t - \sinh t & -\sinh t & 1 + t - \sinh t \end{pmatrix}$.

7. Exponential matrix is $e^{3t} \begin{pmatrix} 1 - t & 0 & t \\ -t & 1 & t \\ -t & 0 & 1 + t \end{pmatrix}$.

9. (a) Eigenvectors $(1, 0)$ and $(0, \alpha - \beta)$ are linearly independent if $\alpha \neq \beta$)

(b) *Hint:* The only solutions of $\begin{pmatrix} 0 & 1 \\ 0 & 0 \end{pmatrix}\begin{pmatrix} u \\ v \end{pmatrix} = \begin{pmatrix} 0 \\ 0 \end{pmatrix}$ are multiples of $(1, 0)$.

(c) $\begin{pmatrix} e^{\alpha t} & \frac{e^{\beta t} - e^{\alpha t}}{\beta - \alpha} \\ 0 & e^{\beta t} \end{pmatrix}$; $\begin{pmatrix} e^{\alpha t} & te^{\alpha t} \\ 0 & e^{\alpha t} \end{pmatrix}$.

11. *Hint:* $I^k = I$ for all integers k.

13. $P(\lambda) = \lambda^2 - 5\lambda - 2$ so $A^2 - 5A - 2I = 0$. Then $A^2 = 5A + 2I = \begin{pmatrix} 7 & 10 \\ 15 & 22 \end{pmatrix}$, $A^3 = 5A^2 + 2A = \begin{pmatrix} 37 & 54 \\ 81 & 118 \end{pmatrix}$.

15. $A^{-1} = A^2 - 3A + 3I = \begin{pmatrix} 1 & 0 & 0 \\ -13 & 1 & -5 \\ 2 & 0 & 1 \end{pmatrix}$.

17. $A^{-1} = \frac{1}{4}(A^2 + 5A - 26I) = \begin{pmatrix} 0 & 1 & 0 \\ \frac{7}{4} & -\frac{11}{2} & 2 \\ -\frac{3}{4} & \frac{5}{2} & -1 \end{pmatrix}$.

19. $\det A = 0$, no inverse.

21. $A^{-1} = \frac{1}{8}(-A^3 + 8A^2 - 21A + 22I) = \begin{pmatrix} 1 & -1 & 1 & -\frac{3}{4} \\ 0 & \frac{1}{2} & 0 & -\frac{1}{8} \\ 0 & 0 & 1 & -\frac{1}{4} \\ 0 & 0 & 0 & \frac{1}{4} \end{pmatrix}$.

23. $A^{-1} = e^{-2t}\left(A^2 - (1 + 2e^t)A + (2e^t + e^{2t})I\right) = \begin{pmatrix} 1 & 0 & 0 \\ 0 & e^{-t} & -te^{-t} \\ 0 & 0 & e^{-t} \end{pmatrix}$.

Exercise Set 2E (pgs. 639–640)

1. $A = \begin{pmatrix} 0 & 0 \\ 0 & 1 \end{pmatrix}$.

3. $A = \begin{pmatrix} 1 & 0 & 0 \\ 0 & 2 & 0 \\ 0 & 0 & 3 \end{pmatrix}$.

5. *Hint:* Set $t = 0$.

7. *Hint:* Compare the square of the matrix with its value when t is replaced by $2t$.

9. *Hint:* Multiply the linear combination $c_1\mathbf{x}_1 + \cdots + c_n\mathbf{x}_n = 0$, where $\mathbf{x}_k$ is the kth column of A, by A^{-1}.

11. *Hint:* $\mathbf{x}_1(t) = c_2\mathbf{x}_2(t) + \cdots + c_m\mathbf{x}_m(t)$ if and only if $(-1)\mathbf{x}_1(t) + c_2\mathbf{x}_2 + \cdots + c_m\mathbf{x}_m(t) = 0$.

Section 3: Nonhomogeneous Systems

Exercise Set 3A–C (pgs. 644–645)

1. $\begin{pmatrix} -\frac{5}{6}e^{3t} - \frac{1}{2}e^t + \frac{1}{3} \\ -\frac{2}{3}e^{2t} - \frac{1}{3}e^{-t} \end{pmatrix}$.

3. $\begin{pmatrix} -1 - \frac{1}{4}e^t - te^t + \frac{5}{4}e^{5t} \\ \frac{5}{4}e^{5t} - \frac{1}{4}e^t \end{pmatrix}$.

5. (b) *Hint:* $\mathbf{x}_0 = X(t_0)\mathbf{c}$.

7. $x(t) = c_1e^{3t} + \frac{1}{3} - \frac{1}{2}e^t$, $y(t) = c_2e^{2t} - \frac{1}{3}e^{-t}$.

9. $x(t) = (c_1 - c_2)e^t + c_2e^{5t} - 1 - te^t$, $y(t) = c_2e^{5t} - \frac{1}{4}e^t$.

11. $\begin{pmatrix} t^4/4 \\ 3t^3/4 \end{pmatrix}$.

13. (a) *Hint:* $X^{-1}(t)\mathbf{x}_k(t) = \mathbf{e}_k$. **(b)** $\dot{x} = -x + e^t y$, $\dot{y} = -2e^t x + 3y$.

15. Use the definition of fundamental matrix.

17. *Hint:* Apply the result of Exercise 16(b) to the power series expansion of $e^{A(t)}$.

19. Fill in the details of the given line of reasoning.

Section 4: Equilibrium & Stability

Exercise Set 4A (pgs. 652–653)

1. $\lambda_1 = -1, \lambda_2 = 1$; Saddle, type II.

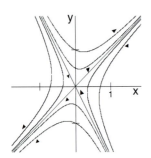

3. $\lambda_{1,2} = 2 \pm i$; Unstable spiral, type IV.

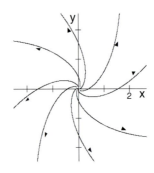

5. $\lambda_{1,2} = \pm\sqrt{10}$; Saddle, type II.

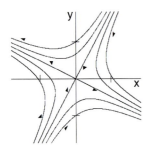

7. $\lambda_1 = -1, \lambda_2 = 2$; Saddle, type II.

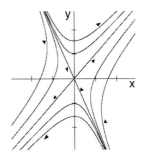

9. Roots $1, -1$; Saddle, type II.

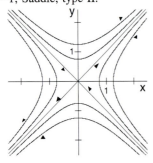

11. Roots $-\frac{1}{2} \pm i\frac{\sqrt{3}}{2}$; Asymptotically stable spiral, type VI.

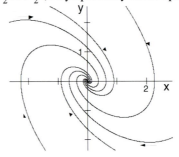

13. $k = 0$, stable center, type V; $0 < k < 2$, asyptotically stable spiral, type VI; $k = 2$, asymptotically stable star, type VIII; $k > 2$, asymptotically stable node, type III.

15. $k = 0$, stable center, type V; $0 < k < \sqrt{8}$, asyptotically stable spiral, type VI; $k = \sqrt{8}$, asymptotically stable star, type VIII; $k > \sqrt{8}$, asymptotically stable node, type III.

17. At a constant equilibrium solution, $\dot{x} = \dot{y} = 0$, so there are infinitely many of them when $ax + by = cx + dy = 0$ has infinitely many solutions, which happens if and only if $\det\begin{pmatrix} a & b \\ c & d \end{pmatrix} = ad - bd = 0$.

19. The trajectories are the same, but traversed in the opposite direction (time reversal).

21. (a) $x(t) = e^{-t/2}\left[c_1 \cos \frac{\sqrt{3}}{2}t + c_2 \sin \frac{\sqrt{3}}{2}t\right]$,

$y(t) = \frac{1}{2}e^{-t/2}\left[(\sqrt{3}c_2 - c_1)\cos \frac{\sqrt{3}}{2}t - (\sqrt{3}c_1 + c_2)\sin \frac{\sqrt{3}}{2}t\right]$,

$z(t) = c_3e^{-t} + \frac{1}{2}e^{-t/2}\left[(c_1 - \sqrt{3}c_2)\cos \frac{\sqrt{3}}{2}t + (\sqrt{3}c_1 + c_2)\sin \frac{\sqrt{3}}{2}t\right]$.

(b) All terms in the solution have exponential factors that go to zero as t goes to $+\infty$.

(c) The solution for $z(t)$ contains a term c_3e^t which is unbounded as t goes to $+\infty$. The equilibrium point $(0, 0, 0)$ is unstable.

Exercise Set 4B (pgs. 657–658)

1. The solutions of the system of linear equations $A\mathbf{x} = \mathbf{0}$ form a line of equilibrium solutions of the autonomous system.

3. *Hint:* At an equilibrium point, $\sigma(y - x) = 0$, $\rho x - y - xz = 0$, and $-\beta z + xy = 0$. Then $x = y$, so $x(\rho - 1 - z) = 0$.

5. (a) *Hint:* At an equilibrium point, $y + \alpha x(x^2 + y^2) = 0$ and $-x + \alpha y(x^2 + y^2) = 0$. For $\alpha \neq 0$, multiply the first equation by x, the second by y, and add.

(b) The linearized system at $(0, 0)$ is $\dot{x} = y$, $\dot{y} = -x$, with eigenvalues $\pm i$ so it has a stable center at $(0, 0)$.

(c) *Hint:* With $x = r\cos\theta$ and $y = r\sin\theta$, the chain rule gives $\dot{x} = \dot{r}\cos\theta - \dot{\theta}r\sin\theta$, $\dot{y} = \dot{r}\sin\theta + \dot{\theta}r\cos\theta$. Substitute in the given differential equations, multiply one by $\cos\theta$, the other by $\sin\theta$, and add.

(d) $\theta(t) = -t + c_1$, $r(t) = (c_2 - 2\alpha t)^{1/2}$.

7. (a) $(A, B/A)$. **(b)** The figure shows the case $A = 1$, $B = 3$ with an equilibrium point at $(1, 3)$.

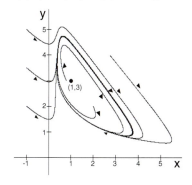

9. Equilibrium points are at $(0, 0)$ and on the circle $1 - x^2 - y^2 = 0$. *Hint:* $\frac{d}{dt}(x^2 + y^2) = 2(x\dot{x} + y\dot{y}) = 0$.

11. Equilibrium points: $(-1, 0)$, $(0, 0)$, $(1, 0)$. $(-1, 0)$ and $(1, 0)$ are stable, with derivative matrix $\begin{pmatrix} -2 & 0 \\ 0 & -2 \end{pmatrix}$, and

eigenvalues $-2, -2$. $(0, 0)$ is unstable, with derivative matrix $\begin{pmatrix} 1 & 0 \\ 0 & -1 \end{pmatrix}$, and eigenvalues $1, -1$.

13. (a) $x(t) = \cos t$, $y(t) = \sin t$ is a solution. **(b)** Evaluate $(d/dt)(y/x)$ in terms of $\dot{x}$ and $\dot{y}$.

(c) *Hint:* $\tan\theta = y/x$ **(e)** *Hint:* $(d/dt)r^2 = (d/dt)(x^2 + y^2) = 2(x\dot{x} + y\dot{y})$

15. (a) $\dot{x} = y$, $\dot{y} = -x + \alpha y$.

(b) The type of equilibrium at $(0, 0)$ for the linearized system is: for $0 < \alpha < 2$, unstable spiral; for $\alpha = 2$, unstable star; for $\alpha > 2$, unstable node; for $\alpha = 0$ stable center; for $-2 < \alpha < 0$, stable spiral; for $\alpha = -2$, stable star; for $\alpha < -2$, stable node.

Chapter 13 Review (pgs. 658–659)

1. $x(t) = \frac{3}{4}e^t + \frac{9}{4}e^{-t} - \frac{3}{2}te^t - 2$, $y(t) = \frac{3}{4}e^t + \frac{9}{4}e^{-t} - \frac{1}{2}te^t - 1$.

3. $x(t) = \frac{3}{2}e^t + te^t - \frac{1}{6}e^{-t} - \frac{1}{3}e^{2t}$, $y(t) = -\frac{1}{3}e^{-t} + \frac{1}{3}e^{2t}$.

5. $x(t) = \frac{5+\sqrt{5}}{10}e^{t(7-\sqrt{5})/2} + \frac{5-\sqrt{5}}{10}e^{t(7+\sqrt{5})/2}$; $y(t) = -\frac{5-\sqrt{5}}{10}e^{t(7-\sqrt{5})/2} - \frac{5+\sqrt{5}}{10}e^{t(7+\sqrt{5})/2}$.

7. $x(t) = e^t - 2te^t$, $y(t) = 2e^t - 2te^t$, $z(t) = -2te^t$.

9. $x(t) = e^t$, $y(t) = e^t + e^{2t}$, $z(t) = e^t - e^{2t}$.

11. $x(t) = 2e^{2t} - 1$, $y(t) = 2e^{2t}$, $z(t) = 0$, $w(t) = te^t + e^t$.
$2e^{-\lambda t} + e^{-i\lambda t} + e^{i\lambda t} + e^{\lambda t}, -e^{-\lambda t} - ie^{i\lambda t} + ie^{i\lambda t} - e^{\lambda t}w^3, -- e^{-\lambda t} + e^{-i\lambda t} + e^{i\lambda t} - e^{\lambda t}w^2, -e^{-\lambda t} + ie^{i\lambda t} - ie^{i\lambda t} - e^{\lambda t}w$

13. $\dot{\mathbf{x}} = A\mathbf{x}$, where $\mathbf{x} = (x, y, \dot{x}, \dot{y})$ and $A = \begin{pmatrix} 0 & 0 & 1 & 0 \\ 0 & 0 & 0 & 1 \\ 1 & 1 & 0 & 0 \\ 1 & -1 & 0 & 0 \end{pmatrix}$. Eigenvalues are $\lambda = 2^{1/4}, i\lambda, -\lambda$, and $-i\lambda$. By

Theorem 2.4,

$$e^{tA} = b_0 + b_1 A + b_2 A^2 + b_3 A^3 = \begin{pmatrix} b_0 + b_2 & b_2 & b_1 + b_3 & b_3 \\ b_2 & b_0 - b_2 & b_3 & b_1 - b_3 \\ b_1 + 2b_3 & b_1 & b_0 + b_2 & b_2 \\ b_1 & -b_1 + 2b_3 & b_2 & b_0 - b_2 \end{pmatrix}, \text{ where } b_0, b_1, b_2, b_3 \text{ are functions}$$

satisfying $e^{\lambda t} = b_0 + \lambda b_1 + \lambda^2 b_2 + \lambda^3 b_3$, $e^{i\lambda t} = b_0 + i\lambda b_1 - \lambda^2 b_2 - i\lambda^3 b_3$, $e^{-\lambda t} = b_0 - \lambda b_1 + \lambda^2 b_2 - \lambda^3 b_3$, and $e^{-i\lambda t} = b_0 - i\lambda b_1 - \lambda^2 b_2 + i\lambda^3 b_3$. This gives b_1, b_2, b_3, b_4 as linear combinations of $e^{\lambda t}, e^{i\lambda t}, e^{-\lambda t}, e^{-i\lambda t}$ that can be written more compactly as $b_0 = \frac{1}{2}(\cos \lambda t + \cosh \lambda t)$, $b_1 = \frac{1}{4}\lambda^3(\sin \lambda t + \sinh \lambda t)$, $b_2 = -\frac{1}{4}\lambda^2(\cos \lambda t - \cosh \lambda t)$, $b_3 = -\frac{1}{4}\lambda(\sin \lambda t - \sinh \lambda t)$. In terms of initial conditions at $t = 0$,
$x(t) = \left(\frac{1}{2}(\cos \lambda t + \cosh \lambda t) - \frac{1}{4}\sqrt{2}(\cos \lambda t - \cosh \lambda t)\right)x(0) - \frac{1}{4}\sqrt{2}(\cos \lambda t - \cosh \lambda t)y(0) + \left(\frac{1}{4}2^{3/4}(\sin \lambda t + \sinh \lambda t) - \frac{1}{4}2^{1/4}(\sin \lambda t - \sinh \lambda t)\right)\dot{x}(0) - \frac{1}{4}2^{1/4}(\sin \lambda t + \sinh \lambda t)\dot{y}(0)$,
$y(t) = -\frac{1}{4}\sqrt{2}(\cos \lambda t + \cosh \lambda t)x(0) + \left(\frac{1}{2}(\cos \lambda t + \cosh \lambda t) + \frac{1}{4}\sqrt{2}(\cos \lambda t - \cosh \lambda t)\right)y(0) - \frac{1}{4}2^{1/4}(\sin \lambda t - \sinh \lambda t)\dot{x}(0) + \left(\frac{1}{4}2^{3/4}(\sin \lambda t + \sinh \lambda t) + \frac{1}{4}2^{1/4}(\sin \lambda t - \sinh \lambda t)\right)\dot{y}(0)$, where λ stands for $2^{1/4}$.

15. **(a)** Letting $\mathbf{x}_0 = (a_1, \ldots, a_n)$ and $\mathbf{z}_0 = (b_1, \ldots, b_n)$, the system is equivalent to the sequence of systems $\dot{x}_k = y_k, \dot{y}_k = -x_k, x_k(0) = a_k, y_k(0) = b_k$ for $k = 1, \ldots, n$.
(b) $x_k(t) = a_k \cos t + b_k \sin t$ for $k = 1, \ldots, n$.
(c) The exponential matrix has n 2-by-2 blocks $\begin{pmatrix} \cos t & \sin t \\ -\sin t & \cos t \end{pmatrix}$ on its diagonal and is zero elsewhere.

CHAPTER 14: INFINITE SERIES

Exercise Set 1 (pgs. 662–664)

1. $5/6$

3. $31/16$

5. $a_k = 1/k, k \geq 1$

7. $a_k = \dfrac{1}{k(k+2)}, k \geq 1$

9. $a_k = \dfrac{k+3}{(k+1)3^{k+1}}$,

11. *Hint:* Follow Example 3 in the text. Sum of the infinite series is $1/4$.

13. *Hint:* Use induction. The infinite series diverges to ∞.

15. $6/5$

17. $1/\pi$

19. $e^2/(e^2 - 1)$

21. $8\sum_{k=1}^{\infty}(1/10)^k = 8/9$

23. $123\sum_{k=1}^{\infty}(1/1000)^k = 41/333$

25. *Hint:* First show that the assertion holds for numbers ω satisfying $0 < \omega < 1$.

27. $\sum_{k=1}^{\infty}\dfrac{1}{2^k k!} = \dfrac{1}{2} + \dfrac{1}{8} + \dfrac{1}{48} + \cdots$

29. $\sum_{k=1}^{\infty}\dfrac{(k-1)!}{(2k)!} = \dfrac{1}{2} + \dfrac{1}{24} + \dfrac{1}{360} + \cdots$

31. $\displaystyle\sum_{k=0}^{\infty}(-1)^k\frac{2^k k!}{(2k+1)!}=1-\frac{1}{3}+\frac{1}{15}-\cdots$

33. $s_1=1,\ s_2=3/2,\ s_3=5/3,\ \lim_{n\to\infty}s_n=2$

35. $s_1=2/9,\ s_2=4/27,\ s_3=8/81,\ \lim_{n\to\infty}s_n=0$

37. $s_1=1,\ s_2=1025/1026,\ s_3=29525/29526,\ \lim_{n\to\infty}s_n=1$

39. *Hint:* The given series is a convergent geometric series that begins with $k=2$.

41. *Hint:* Write the given series as a telescoping series and use Example 2 of the text.

43. $\displaystyle\sum_{k=0}^{\infty}\frac{3}{r}\left(1-\frac{1}{r}\right)^k=3$, for $1/2<r<1$.

45. **(a)** *Hint:* Show that all subintervals remaining after the nth step are of the same length and that the number of such subintervals is twice the number remaining after the $(n-1)$st step.

(b) *Hint:* For a given $\epsilon>0$ and any positive integer n satisfying $n>\ln\epsilon/\ln(2/3)$, show that $0<(2/3)^n<\epsilon$.

Exercise Set 2AB (pgs. 669–670)

1. $7+5(x-2)+(x-2)^2$

3. $2-2(x+1)+(x+1)^2$

5. $c_0=1,\ c_1=1/3,\ c_2=-1/9,\ R_2(x)=\dfrac{5}{81}c^{-8/3}(x-1)^3$, for some c between 1 and x.

7. $c_0=-1,\ c_1=-1,\ c_2=1,\ R_2(x)=-c^{-4}(x+1)^3$ for some c between -1 and x.

9. $\displaystyle\sum_{k=0}^{\infty}\frac{(-1)^k}{k!}x^k$

11. $\displaystyle\sum_{k=0}^{\infty}\frac{(-1)^k}{2^{2k+1}(2k+1)!}x^{2k+1}$

13. $\displaystyle\sum_{k=0}^{\infty}\frac{1}{k!}x^{2k}$

15. *Hint:* Combine the series for e^x and e^{-x} and simplify.

17. *Hint:* Use Theorem 2.1 with $f(x)=(x+a)^n$.

19. $|R_4(x)|\le\pi^5/120$

21. $|R_5(x)|=\le\pi^6/720$

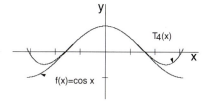

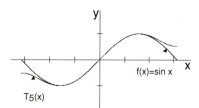

23. **(b)** *Hint:* Follow the given directions.

25. *Hint:* Compute the nth degree Taylor polynomial with remainder $R_n(x)$ of $\ln(1+x)$ about $x=0$ and show that $R_n(x)$ tends to 0 as $n\to\infty$.

27. *Hint:* Let $x=1$ and $x=-1$ in Formula 2.4(a) in the text.

29. **(a)** *Hint:* Differentiate both sides of the equation $f(x)=f(-x)$ n times and let $x=0$.

(b) Differentiate both sides of the equation $f(-x)=-f(x)$ n times and let $x=0$.

(c) The terms in the Taylor expansion about $x=0$ of an even function contain only even powers of x; and the terms in the Taylor expansions about $x=0$ of an odd function contain only odd powers of x.

Exercise Set 3AB (pgs. 673–674)

1. $\lim_{n\to\infty}s_n=1$

3. $\lim_{n\to\infty}s_n=\infty$

5. $\lim_{n\to\infty}s_n=0$

7. $\lim_{n\to\infty}s_n=0$

9. $s_n=n/(n+1),\ n=1,2,\ldots,\ \lim_{n\to\infty}s_n=1$

11. $s_n = \dfrac{n^2 + n - 1}{n^2 + 3n + 1}$, $n = 1, 2, \ldots$, $\lim_{n \to \infty} s_n = 1$

13. (a) *Hint:* $\sum_{k=1}^{n} a_k$ is a telescoping sum. **(b)** $a_1 = 2$, $a_k = -\dfrac{1}{k(k-1)}$, $k \geq 2$, $\sum_{k=1}^{\infty} a_k = 1$

15. e **17.** $\sum_{k=1}^{\infty} 3/2^k$

19. $\sum_{k=0}^{\infty} 2^k/3^{k+1}$ **21.** *Hint:* Use the given hint.

Exercise Set 3CDE (pgs. 681–682)

1. diverges (term test)   **3.** converges (integral test)

5. converges (integral test)   **7.** converges (comparison test, ratio test)

9. converges (comparison test)   **11.** converges (comparison test)

13. converges absolutely   **15.** converges

17. converges absolutely   **19.** converges

21. diverges   **23.** converges absolutely

25. converges   **27.** converges

29. converges

31. (a) *Hint:* Use the ratio test.
  (b) *Hint:* Multiply the partial sum $\sum_{k=1}^{n} kx^k$ by $1 - x$, combine the resulting terms and then let $n \to \infty$.

33. (a) *Hint:* Multiply the partial sum $\sum_{k=1}^{n} kx^k$ by $(1 - x)^2$, combine the resulting terms and then let $n \to \infty$.
  (b) Let $x = 1/2$ in part (a).

35. $|x| \leq 1$ **37.** $-1 < x < 3$

39. $-1 \leq x < 1$ **41.** $|x| > \sqrt{2}$

43. (a) *Hint:* Follow the given steps. **(b)** *Hint:* Use the term test and part (a). **(c)** *Hint:* Use part (b) and the fact that $k^{1/k} \geq (1/k)^{1/k}$.

45. *Hint:* Use the integral test.

Exercise Set 4 (pgs. 687–688)

1. *Hint:* Show that $\sum_{k=0}^{\infty} d^k$ converges and use the Weierstrass test.

3. *Hint:* Show that the given series has period 2π on $\mathbb{R}$.

5. *Hint:* Show that the absolute value of the kth term is dominated by $(A + B)/k^2$ and apply the Weierstrass test.

7. *Hint:* For the first series, use the first derivative test to find the maximum value of $(1 - x)x^{n+1}$. For the second series, use Exercise 32 in the last section to show pointwise convergence on $[0, 1]$ to a function $f(x)$, and then show that if N is any given positive integer then $|s_N - f(x)|$ can be made arbitrarily close to 1.

Exercise Set 5A–D (pgs. 695–696)

1. $-1 \leq x < 1$ **3.** $-\infty < x < \infty$

5. $-3 < x < -1$ **7.** $-4 < x < -2$

9. $\sum_{k=0}^{\infty} \dfrac{(-1)^k}{k!} x^k$ **11.** $\sum_{k=0}^{\infty} \dfrac{1}{(2k+1)!} x^{2k+1}$

13. $\displaystyle\sum_{k=1}^{\infty} \frac{1}{k!} x^{2k}$

15. $\displaystyle\sum_{k=1}^{\infty} \frac{(-1)^{k+1}}{k} x^k$

17. $\displaystyle\sum_{k=1}^{\infty} \frac{(-1)^{k+1}}{k} x^{2k}$

19. $\displaystyle\sum_{k=0}^{\infty} (-1)^k x^{2k}$

21. $\displaystyle\sum_{k=0}^{\infty} (-1)^k (x-1)^k$

23. $\displaystyle\sum_{k=0}^{\infty} \frac{(-1)^k}{2k+1} x^{2k+1}$

25. *Hint:* Find the Taylor expansions for $1/(1-x)^2$ and $1/(1-x)^3$, and then use the partial fraction decomposition of $x(x+1)/(1-x)^3$.

27. **(a)** *Hint:* Follow the given steps. **(b)** $1 + 3x + 3x^2 + x^3$ ($\alpha = 3$), $1 - 3x + 6x^2 - 10x^3$ ($\alpha = -3$),
$1 + \frac{1}{2}x - \frac{1}{8}x^2 + \frac{1}{16}x^3$ ($\alpha = 1/2$)

29. *Hint:* Use the change of variable $t = x + a$.

31. 1   **33.** 1/2

35. $\displaystyle\sum_{k=0}^{\infty} \frac{(-1)^k}{c^{k+1}} x^k$

37. *Hint:* Shift the index of summation in the series for $f''(x)$ and then add the series for $f''(x)$ and $f(x)$ by combining like powers of x.

Exercise Set 6 (pg. 698)

1. $1 + 2x + 3x^2$

3. $2 + x^2 + \frac{1}{4}x^4$

5. $1 + x + \frac{1}{2}x^2 + \frac{1}{3}x^3 + \cdots$

7. *Hint:* Find the first four nonzero terms of $y(x)$ and separate the result into two series.

Exercise Set 7 (pgs. 705–706)

1. *Hint:* Follow the steps in Example 2 of the text.

3. **(a)** $y(x) = c_0 \displaystyle\sum_{k=0}^{\infty} \frac{(-1)^k}{k!} x^{2k}$ **(b)** $y(x) = ce^{-x^2}$ **(c)** Yes

5. **(a)** $y_h(x) = c_0 \displaystyle\sum_{k=0}^{\infty} \frac{(-1)^k}{2^k k!} x^{2k} + c_1 \sum_{k=0}^{\infty} \frac{(-1)^k 2^k k!}{(2k+1)!} x^{2k+1}$ **(b)** *Hint:* Set $c_0 = 1$, $c_1 = 0$ in part (a). **(c)** $y_p(x) = \frac{1}{2}x$

(d) $y(x) = \frac{x}{2} + c_0 e^{-x^2/2} + c_1 \displaystyle\sum_{k=0}^{\infty} \frac{(-1)^k 2^k k!}{(2k+1)!} x^{2k+1}$

7. **(a), (b)** *Hint:* Apply the power series method to the Bessel equation of index 0.

(c) *Hint:* Show that $x J_0 = -x J_0'' - J_0'$ (there is no need to use the series representation of $J_0(x)$ found in part (b))

Exercise Set 8ABC (pgs. 712–713)

1. $\displaystyle\sum_{k=1}^{\infty} \frac{2(-1)^{k+1}}{k}\sin kx$

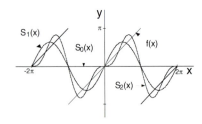

3. $\displaystyle\frac{\pi^2}{3} + \sum_{k=1}^{\infty} \frac{4(-1)^k}{k^2}\cos kx$

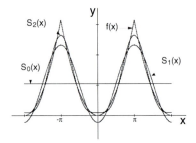

5. $\displaystyle\frac{1}{2} + \sum_{k=0}^{\infty} \frac{2}{(2k+1)\pi}\sin(2k+1)x$

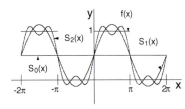

7. $\displaystyle\sum_{k=0}^{\infty} \frac{4}{2k+1}\sin(2k+1)x$

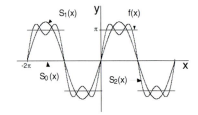

9. $\displaystyle-\frac{\pi}{2} + \sum_{k=0}^{\infty} \frac{4}{(2k+1)^2\pi}\cos(2k+1)x$

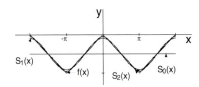

11.

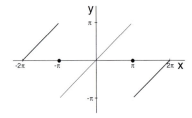

13.

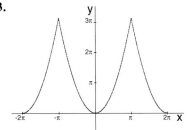

15.

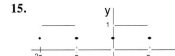

17.

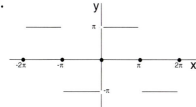

19.

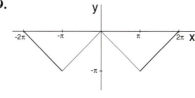

21. *Hint:* Use Equations 8.2 in the text and the linearity of the integral.

23. $\dfrac{3}{4}\cos x + \dfrac{1}{4}\cos 3x$ **25.** $\dfrac{3}{4}\sin x - \dfrac{1}{4}\sin 3x$

27. $\cos 3x + \sin 5x$

29. *Hint:* Straightforward but tedious integration   **31.** Yes.

Exercise Set 9ABC (pgs. 720–721)

1. $\displaystyle\sum_{k=1}^{\infty} \frac{4(-1)^k}{k\pi}\sin\frac{k\pi x}{2}$ **3.** *Hint:* routine computation

5. $f_o(x) = \begin{cases} 2n+1-x, & 2n < x < 2n+2; \\ 0, & x = 2n, \end{cases} = \displaystyle\sum_{k=1}^{\infty}\frac{2}{k\pi}\sin k\pi x$

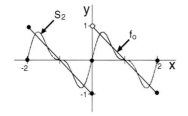

7. $f_o(x) = \begin{cases} \cos x, & 2n\pi < x < (2n+1)\pi; \\ -\cos x, & (2n-1)\pi < x < 2n\pi; \\ 0, & x = n\pi, \end{cases} = \displaystyle\sum_{k=1}^{\infty}\frac{8k}{(4k^2-1)\pi}\sin 2kx$

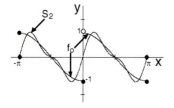

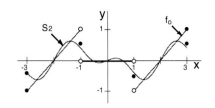

9. $f_o(x) = \begin{cases} 0, & 4n-1 < x < 4n+1; \\ x - 2 - 4n, & 4n+1 < x < 4n+3; \\ \dfrac{1}{2}, & x = 4n-1; \\ -\dfrac{1}{2}, & x = 4n+1, \end{cases}$

$$= \sum_{k=1}^{\infty} \left(-\frac{2}{k\pi} \cos \frac{k\pi}{2} - \frac{4}{k^2\pi^2} \sin \frac{k\pi}{2} \right) \sin \frac{k\pi x}{2}$$

11. $f_e(x) = \begin{cases} 1 - 2n + x, & 2n-1 \le x \le 2n; \\ 1 + 2n - x, & 2n < x < 2n+1, \end{cases} = \frac{1}{2} + \sum_{k=0}^{\infty} \frac{4}{(2k+1)^2\pi^2} \cos(2k+1)\pi x$

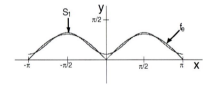

13. $f_e(x) = \begin{cases} \sin x, & 2n\pi < x < (2n+1)\pi; \\ -\sin x, & (2n-1)\pi \le x \le 2n\pi, \end{cases} = \frac{2}{\pi} + \sum_{k=1}^{\infty} \frac{-4}{(4k^2-1)\pi} \cos 2kx$

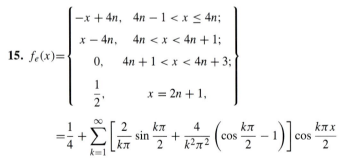

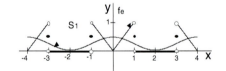

15. $f_e(x) = \begin{cases} -x + 4n, & 4n-1 < x \le 4n; \\ x - 4n, & 4n < x < 4n+1; \\ 0, & 4n+1 < x < 4n+3; \\ \dfrac{1}{2}, & x = 2n+1, \end{cases}$

$$= \frac{1}{4} + \sum_{k=1}^{\infty} \left[\frac{2}{k\pi} \sin \frac{k\pi}{2} + \frac{4}{k^2\pi^2} \left(\cos \frac{k\pi}{2} - 1 \right) \right] \cos \frac{k\pi x}{2}$$

17. *Hint:* routine computation   **19.** *Hint:* routine computation

21. *Hint:* Consider the limit of $\big(f(x + p + h) - f(x + p)\big)/h$ as $h \to 0$.

23. *Hint:* Use the hint in the text. **25.** *Hint:* routine computation

27. $y_p(x) = \frac{4}{\pi} \sum_{n=0}^{\infty} \frac{1}{(2n+1)\big[(2n+1)^2\pi^2 + a^2\big]} \sin(2n+1)\pi t$

Exercise Set 10AB (pgs. 727–728)

1. $u(x, t) = e^{-(\pi^2 a^2/p^2)t} \sin \frac{\pi x}{p}, \ 0 \leq x \leq p, t \geq 0$

3. $u(x, t) = \sum_{k=1}^{\infty} \frac{2}{k\pi} e^{-k^2 a^2 \pi^2 t} \sin k\pi x, \ 0 \leq x \leq 1, t \geq 0$

5. $u(x, t) = e^{-a^2 t} \sin x + \frac{1}{2} e^{-4a^2 t} \sin 2x, \ 0 \leq x \leq \pi, t \geq 0$

7. $v(x) = -1 + x$ **9.** $v(x) = 1 + x$

11. $u(x, t) = 1 + \frac{2x}{p} + \left(1 - \frac{8}{\pi}\right) e^{-(a^2\pi^2/p^2)t} \sin \frac{\pi x}{p} + \sum_{k=2}^{\infty} \frac{2}{k\pi}[3(-1)^k - 1]e^{-k^2(a^2\pi^2/p^2)t} \sin \frac{k\pi x}{p}, \quad 0 \leq x \leq p, t > 0$

13. $u(x, t) = x + \frac{2}{\pi} \sum_{k=1}^{\infty} \frac{1}{k} e^{-4k^2 a^2 \pi^2 t} \sin 2k\pi x, \ 0 \leq x \leq 1, t \geq 0$

15. $u(x, t) = \begin{cases} \big(c_1 e^{r_1 x} + c_2 e^{r_2 x}\big)e^{\lambda t}, & \lambda \neq -1/4; \\ \big(c_1 e^{-x/2} + c_2 x e^{-x/2}\big)e^{-t/4}, & \lambda = -1/4. \end{cases}$

17. $u(x, t) = cx^\lambda e^{\lambda t/2}, x > 0$ **19.** $v(x) = x^2 + 1$

21. $v(x) = \frac{1}{6} x^3 + \frac{5}{6} x + 1$

23. (b) $u(x, t) = \frac{1}{p} \int_0^p f(x)\, dx + \sum_{k=1}^{\infty} \left(\frac{2}{p} \int_0^p f(x) \cos \frac{k\pi x}{p}\, dx\right) e^{-k^2\pi^2 t/p^2} \cos \frac{k\pi x}{p}$ **(c)** $u(x, \infty) = \frac{1}{p} \int_0^p f(x)\, dx$

25. $u(x, t) = 1, 0 \leq x \leq 1, t \geq 0$

27. (a) $X'' + \lambda^2 X = 0, T' + \lambda^2 t T = 0$ **(b)** If $\lambda = 0$ then $X(x) = c_1 + c_2 x, T(t) = c_0, u(x, t) = C_1 + C_2 x$.

If $\lambda \neq 0$ then $X(x) = A \cos \lambda(x - \alpha), T(t) = c_0 e^{-\lambda^2 t^2/2}, u(x, t) = Ce^{-\lambda^2 t^2/2} \cos \lambda(x - \alpha)$.

29. *Hint:* $u(x, t)$ concave up implies $u_{xx}(x, t_0) > 0$, which implies $u_t(x, t_0) > 0$, which implies u is increasing. Use a similar argument if $u(x, t)$ is concave down.

31. *Hint:* Except for the obvious minor changes, use the derivation of the product solutions that was used in the text.

Exercise Set 10C (pgs. 731–732)

1. $u(x, t) = \cos at \sin x$

3. $u(x, t) = \sum_{k=1}^{\infty} \frac{2}{k^2 \pi a} \left[(-1)^{k+1} + \cos \frac{k\pi}{2}\right] \sin kat \sin kx$

5. (a) $v(x) = \frac{g}{2a^2} x^2 + c_1 x + c_2$ **(b)** $v(x) = \frac{g}{2a^2} x(x - p)$ **(c)** First solve the homogeneous equation $a^2 u_{xx} = u_{tt}$, $u(0, t) = u(p, t) = 0$, and let $w(x, t)$ denote the solution. The solution of the nonhomogeneous problem is then $u(x, t) = w(x, t) + v(x)$, where $v(x)$ is given in part (b).

7. (a) *Hint:* Let t_0 be a fixed time and for $t > t_0$ let $b = a(t - t_0)$. Show that $U(x + at_0) = U\big((x - b) + at\big)$ and $V(x - at_0) = V\big((x + b) - at\big)$. **(b)** *Hint:* Use the identity $\cos(\alpha - \beta) = \cos \alpha \cos \beta + \sin \alpha \sin \beta$. **(c)** *Hint:* Use the identity $2 \sin \alpha \cos \beta = \sin(\alpha + \beta) + \sin(\alpha - \beta)$.

9. (a) *Hint:* routine computation

(b)

(c) *Hint:* Show that $G''(0)$ and $G''(p)$ exist as two-sided derivatives. The condition $G'(0) = G'(p) = 0$ insures that G is continuously differentiable on all of $\mathbb{R}$.

11. $X'' - \lambda X = 0$, $T'' - \lambda t T = 0$

Chapter 14 Review (pgs. 733–734)

1. $1/(x - 4)$, $4 < x < 6$

3. $1 + 1/x^2$, $x \neq 0$

5. $e^{(x+1)^2}$, $-\infty < x < \infty$

7. $\cos(x - 5)$, $-\infty < x < \infty$

9. $\cosh\left(\dfrac{1}{x^2 - 1}\right)$, $|x| \neq 1$

11. $\displaystyle\sum_{k=0}^{\infty}(x^3)^k$, $-1 < x < 1$

13. $\displaystyle\sum_{k=1}^{\infty}\dfrac{-2^k}{k}x^k$, $-\dfrac{1}{2} \leq x < \dfrac{1}{2}$

15. $\displaystyle\sum_{k=0}^{\infty}\dfrac{1}{k!}(x - 1)^k$, $-\infty < x < \infty$

17. $\displaystyle\sum_{k=0}^{\infty}\dfrac{(-1)^k 4^k}{(2k)!}x^{2k}$, $-\infty < x < \infty$

19. $\displaystyle\sum_{k=0}^{\infty}\dfrac{(-1)^{k+1}}{(2k + 1)!}x^{2k+1}$, $-\infty < x < \infty$

21. $\sum_{k=0}^{\infty}\left(\dfrac{1-2^k}{k!}\right)x^k$, $-\infty < x < \infty$

23. $|x| < 1$

25. $-\infty < x < \infty$

27. $-\infty < x < \infty$

29. diverges

31. diverges

33. absolutely convergent

35. (a), (b), (c), (d) Hints: routine computations.

37. $r^2 R'' + 2r R' - \gamma R = 0$, $(\sin\phi)\Phi'' + (\sin\phi\cos\phi)\Phi' + (\gamma\sin^2\phi - \lambda)\Phi = 0$, $\Theta'' + \lambda\Theta = 0$

INDEX